$$\int e^x \sin x \, dx = e^x \sin x - e^x \cos x - \int e^x \sin x \, dx$$

D	AD
$\sin x$	e^x
$\cos x$	e^x
$-\sin x$	e^x

Advanced Engineering Mathematics

Third Edition

Advanced Engineering Mathematics

Erwin Kreyszig

Professor of Mathematics
Ohio State University
Columbus, Ohio

John Wiley and Sons, Inc.
New York · London · Sydney · Toronto

Library of Congress Catalogue Card Number: 71-172951

ISBN 0-471-50728-8

Printed in the United States of America.

10 9 8 7 6 5 4 3 2 1

Preface

Purpose of the book. This book is intended to introduce students of engineering and physics to those areas of mathematics which, from a modern point of view, seem to be the most important in connection with practical problems. Topics are chosen according to the frequency of occurrence in applications. New ideas of modern mathematical training, as expressed in recent symposia on engineering education, were taken into account. The book should suit those institutions that have offered extended mathematical training for a long time as well as those that intend to follow the general trend of broadening the program of instruction in mathematics.

A course in elementary calculus is the sole prerequisite.

The material included in the book has formed the basis of various lecture courses given to undergraduate and graduate students of engineering, physics, and mathematics in this country, in Canada, and in Europe.

Changes in the third edition. The present edition differs essentially from the first and second editions as follows.

The problems have been changed.

A new chapter on numerical methods for engineers (Chap. 18) has been added. It is suitable for a course meeting three hours a week.

The material on linear algebra and analysis (vectors and matrices in Chaps. 5 to 8) has been revised and amplified. The new presentation emphasizes algebraic as well as geometric aspects. Chapter 5 also contains an elementary introduction to vector spaces, inner product spaces, and some other concepts that are important in functional analysis and its applications.

These major changes are suggested and motivated by the development of engineering mathematics during the past decade, especially the widespread application of automatic computers to the solution of engineering problems and the increasing importance of methods of linear algebra.

Other improvements include a more compact consideration of Bessel functions and Legendre polynomials. In particular, the orthogonality of these functions is now discussed much earlier (in Chap. 3) in connection with the Sturm-Liouville theory. Appendix 1, containing references to the literature, has been updated and extended. A collection of frequently used formulas for elementary and higher functions has been added as Appendix 3; this replaces and extends Chap. 0 in the previous editions. The numerical tables have been placed together in Appendix 4 in order to facilitate the use of the book.

Content and arrangement. The arrangement of the subject matter in major parts can be seen from the diagram.

Much space is devoted to ordinary differential equations, linear algebra and vector analysis, and complex analysis, probably the three most important fields for engineers. But also the length of the other chapters on Fourier series, partial differential equations, numerical methods, etc., is such that the presentation may serve as a text for courses of the usual type.

To facilitate the use of parts of the book, the chapters are kept as independent of one another as possible.

The chapters are subdivided into relatively short sections. Each section includes typical examples and problems illustrating the concepts, methods, and results and their engineering applications.

Historical notes, references to original literature, and about 400 figures are included in the text.

References. A list of some books for reference and further study can be found at the end of the book, on page 789. Some formulas for special functions are included in Appendix 3, on page 826.

Problems and answers. The book contains more than 3500 carefully selected problems, which range from simple routine exercises to practical applications of considerable complexity. Answers to odd-numbered problems are included at the end of the book, starting on page 794.

Tables of functions are included in Appendix 4, on page 833.

Suggestions for a sequence of courses. The material may be taken in sequence and is then suitable for four consecutive semester courses, meeting 3–5 hours a week as follows:

First semester.   Ordinary differential equations (Chaps. 1–4).
Second semester.   Linear algebra and vector analysis (Chaps. 5–8).
Third semester.   Fourier series and partial differential equations (Chaps. 9, 10), Numerical methods (Chap. 18).
Fourth semester.   Complex analysis (Chaps. 11–17).

A course on **engineering statistics** (3–5 hours a week; Chap. 19) may be taught during any of those four semesters or afterwards.

Independent one-semester courses. The book is also suitable for various independent one-semester courses meeting 3 hours a week; for example:

Introduction to ordinary differential equations (Chaps. 1, 2).
Laplace transformation (Chap. 4).
Vector algebra and calculus (Chaps. 5, 7).
Matrices and systems of linear equations (Chap. 8).
Fourier series and partial differential equations (Chaps. 9, 10).
Complex analysis (Chaps. 11–16).
Numerical analysis (Chap. 18).

Shorter courses. Sections that may be omitted in a shorter course are indicated at the beginning of each chapter.

Principles for selection of topics. Which topics should be contained in a book of the present type, and how should these topics be arranged and presented?

To find some answer to these basic questions we may take a look at the historical development of engineering mathematics. This development shows the following two interesting features:

1. Mathematics has become more and more important to engineering science, and it is easy to conjecture that this trend will also continue in the future. In fact, problems in modern engineering are so complex that most of them cannot be solved solely by using physical intuition and past experience. The empirical approach has been successful in the solution of many problems in the past, but fails as soon as high speeds, large forces, high temperatures or other abnormal conditions are involved, and the situation becomes still more critical by the fact that various modern materials (plastics, alloys, etc.) have unusual physical properties. Experimental work has become complicated, time-consuming, and expensive. Here mathematics offers aid in planning constructions and experiments, in evaluating experimental data, and in reducing the work and cost of finding solutions.

2. Mathematical methods which were developed for purely theoretical reasons suddenly became of great importance in engineering mathematics. Examples are the theory of matrices, conformal mapping, and the theory of differential equations having periodic solutions.

What are the reflections of this situation on the teaching of engineering mathematics? Since the engineer will need more and more mathematics, should we try to include more and more topics in our courses, devoting less and less time to each topic? Or should we concentrate on a few carefully selected basic ideas of general practical importance which are especially suitable for teaching the student mathematical thinking and developing his own creative ability?

Sixty or eighty years ago no one was able to predict that conformal mapping or matrices would ever be of importance in the mathematical part of engineering work. Similarly, it is difficult to conjecture which new mathematical theories will have applications to engineering twenty or thirty years from now. But no matter what happens in that respect, if a student has a good training in the fundamentals of mathematics he will meet the future needs because he will be able to get acquainted with new methods by his own further study.

It follows that the most important objective and purpose in engineering mathematics seems to be that the student becomes familiar with mathematical thinking. He should learn to recognize the guiding principles and ideas "behind the scenes," which are more important than formal manipulations. He should get the impression that mathematics is not a collection of tricks and recipes but a systematic science of practical importance, resting on a relatively small number of basic concepts and involving powerful unifying methods. He should soon convince himself of the necessity for applying mathematical procedures to engineering problems, and he will find that the theory and its applications are related to each other like a tree and its fruits.

The student will see that the application of mathematics to an engineering problem consists essentially of three phases:

1. Translation of the given physical information into a mathematical form. In this way we obtain a mathematical model of the physical situation. This model may be a differential equation, a system of linear equations, or some other mathematical expression.

2. Treatment of the model by mathematical methods. This will lead to the solution of the given problem in mathematical form.

3. Interpretation of the mathematical result in physical terms.

All three steps seem to be of equal importance, and the presentation in this book is such that it will help the student to develop skill in carrying out all these steps. In this connection, preference has been given to applications which are of a general nature.

In some considerations it will be unavoidable to rely on results whose proofs are beyond the level of a book of the present type. In any case these points are marked distinctly, because hiding difficulties and oversimplifying matters is no real help to the student in preparing him for his professional work.

These are some of the guiding principles I used in selecting and presenting the material in this book. I made the choice with greatest care, using past and present teaching and research experience and resisting the temptation to include "everything which is important" in engineering mathematics.

Particular efforts have been made in presenting the topics as simply, clearly, and accurately as possible; this refers also to the choice of the notations. In each chapter the level increases gradually, avoiding jumps and accumulations of difficult theoretical considerations.

Acknowledgement. I am indebted to many of my former teachers, colleagues, and students for advice and help in preparing this book. Various parts of the manuscript were distributed to my classes in mimeographed form and returned to me, together with suggestions for improvement. Discussions with various engineers and mathematicians were of great help to me; I want to mention in particular Professors S. Bergman, P. L. Chambré, A. Cronheim, J. W. Dettman, R. G. Helsel, E. C. Klipple, H. Kuhn, H. B. Mann, I. Marx, W. D. Munroe, H.-W. Pu, T. Rado(†), H. A. Smith, J. P. Spencer, J. Todd, H. J. Weiss, A. Wilansky, all in this country, Professor H. S. M. Coxeter of Toronto, and Professors B. Baule, H. Behnke, H. Florian, H. Graf, F. Hohenberg, K. Klotter, M. Pinl, F. Reutter, C. Schmieden, H. Unger, A. Walther(†), H. Wielandt, all in Europe. I can offer here only an inadequate acknowledgment of my appreciation.

Finally, I want to thank John Wiley and Sons for their effective cooperation and their great care in preparing this edition of the book.

Suggestions of many readers were evaluated in preparing the present edition. Any further comment and suggestion for improvement of the book will be gratefully received.

                                                        Erwin Kreyszig

Contents

Contents

Advanced Engineering Mathematics

CHAPTER 1

Ordinary Differential Equations of the First Order

Differential equations are of fundamental importance in engineering mathematics because many physical laws and relations appear mathematically in the form of such equations. We shall consider various physical and geometrical problems which lead to differential equations and the most important standard methods for solving such equations. These methods will in general involve integration.

We shall pay particular attention to the derivation of differential equations from given physical situations. This transition from the physical problem to a corresponding "mathematical model" is of great practical importance and will be illustrated by typical examples.

The first four chapters of the book are devoted to ordinary differential equations, and in the present chapter we shall start with the simplest of these equations, the so-called equations of the first order. Corresponding **numerical methods** for obtaining approximate solutions are included in Sec. 18.6, which is independent of the other sections in Chap. 18.

> *Prerequisite for the present chapter:* integral calculus.
> *Sections which may be omitted in a shorter course:* 1.8–1.12.
> *References:* Appendix 1, Part B.
> *Answers to Problems:* Appendix 2.

1.1 Basic Concepts and Ideas

In this section we shall define and explain the basic concepts that matter in connection with differential equations and illustrate these concepts by examples. Then we shall consider two simple practical problems taken from physics and geometry. This will give us a first idea of the nature and purpose of differential equations and their application.

By an **ordinary differential equation** we mean a relation which involves one or several derivatives of an unspecified function y of x with respect to x; the relation may also involve y itself, given functions of x, and constants.

1

For example,

(1) $$y' = \cos x,$$

(2) $$y'' + 4y = 0,$$

(3) $$x^2 y''' y' + 2e^x y'' = (x^2 + 2)y^2$$

are ordinary differential equations.

The term *ordinary* distinguishes it from a *partial differential equation*, which involves partial derivatives of an unspecified function of two or more independent variables. For example,

$$\frac{\partial^2 u}{\partial x^2} + \frac{\partial^2 u}{\partial y^2} = 0$$

is a partial differential equation. In the present chapter we shall consider only ordinary differential equations.

An ordinary differential equation is said to be of **order** n, if the nth derivative of y with respect to x is the highest derivative of y in that equation.

The notion of the order of a differential equation leads to a useful classification of the equations into equations of first order, second order, etc. Thus (1) is a first-order equation, (2) is of the second order, and (3) is of the third order.

In the present chapter we shall consider first-order equations. Equations of second and higher order will be discussed in Chaps. 2–4.

A function

(4) $$y = g(x)$$

is called a **solution** of a given first-order differential equation on some interval, say, $a < x < b$ (perhaps infinite) if it is defined and differentiable throughout the interval and is such that the equation becomes an identity when y and y' are replaced by g and g', respectively.

For example, the function

$$y = g(x) = e^{2x}$$

is a solution of the first-order differential equation

$$y' = 2y$$

for all x, because by differentiation we obtain

$$g' = 2e^{2x},$$

and by inserting g and g' we see that the equation reduces to the identity

$$2e^{2x} = 2e^{2x}.$$

Sometimes a solution of a differential equation will appear as an implicit function, that is, implicitly given in the form

$$G(x, y) = 0,$$

and is then called an *implicit solution*, in contrast to the *explicit solution* (4).

Fig. 1. Solutions of $y' = \cos x$

For example,

$$x^2 + y^2 = 1 \qquad\qquad (y \neq 0)$$

is an implicit solution of the differential equation

$$yy' = -x$$

on the interval $-1 < x < 1$, as the student may verify.

The principal task of the theory of differential equations is to find all the solutions of a given differential equation and investigate their properties. We shall discuss various standard methods which were developed for that purpose.

A differential equation may have many solutions. Let us illustrate this fact by the following examples.

Example 1. Each of the functions

$$y = \sin x, \qquad y = \sin x + 3, \qquad y = \sin x - \tfrac{4}{5}$$

is a solution of the differential equation (1),

$$y' = \cos x,$$

and we know from calculus that any solution of the equation is of the form

(5) $$y = \sin x + c$$

where c is a constant. If we regard c as arbitrary, then (5) represents the totality of all solutions of the equation (cf. Fig. 1).

Example 2. The student may verify that each of the functions

(6) $$y = e^x, \qquad y = 2e^x, \qquad y = -\tfrac{6}{5}e^x$$

is a solution of the differential equation

$$y' = y$$

for all x. We shall see later that any solution of this equation is of the form

(7) $$y = ce^x$$

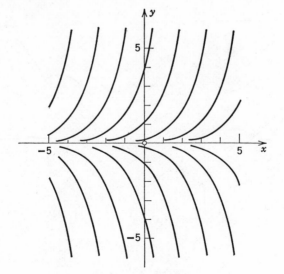

Fig. 2. Solutions of $y' = y$

where c is a constant. Formula (7), with arbitrary c, thus represents the totality of all solutions of the equation (cf. Fig. 2).

Our examples illustrate that a differential equation may (and, in general, will) have more than one solution, even infinitely many solutions, which can be represented by a single formula involving an arbitrary constant c. It is customary to call such a function which contains an arbitrary[1] constant a **general solution** of the corresponding differential equation of the first order. If we assign a definite value to that constant, then the solution so obtained is called a **particular solution**.

Thus (7) is a general solution of the equation $y' = y$, and (6) are particular solutions.

In the following sections we shall develop various methods for obtaining general solutions of first-order equations. We shall see that each such method will yield a general solution of an equation which is unique, except for notation, and will then be called *the* general solution of that equation.

We mention that in some cases there may be further solutions of a given equation which cannot be obtained by assigning a definite value to the arbitrary constant in the general solution; such a solution is then called a **singular solution** of the equation.

For example, the equation

(8) $$y'^2 - xy' + y = 0$$

has the general solution

$$y = cx - c^2$$

---

[1] The range of the constant may have to be restricted in some cases to avoid imaginary expressions or other degeneracies.

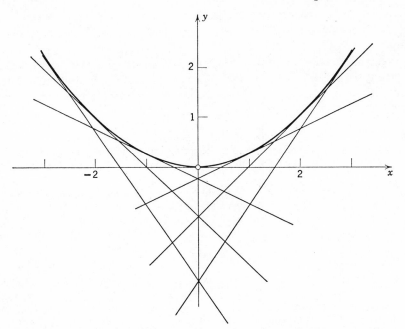

Fig. 3. Singular solution and particular solutions of the equation (8)

representing a family of straight lines where each line corresponds to a definite value of c. A further solution is

$$y = \frac{x^2}{4}$$

and since this solution cannot be obtained by assigning a definite value to c in the general solution, it is a singular solution. Obviously, each particular solution represents a tangent to the parabola represented by the singular solution (Fig. 3).

Singular solutions will rarely occur in engineering problems.

It should be noted that in some books the notion *general solution* means a formula which includes *all* solutions of an equation, that is, both the particular and the singular ones. We do not adopt this definition for two reasons. First of all, it is frequently quite difficult to prove that a particular formula includes all solutions, and that definition of a general solution is therefore rather useless from the practical point of view. Furthermore, we shall see that a large and very important class of equations (the so-called linear differential equations) does not have singular solutions, and our definition of a general solution can be easily generalized to equations of higher order so that the resulting notion includes all solutions of a differential equation which is linear.

We shall see that the conditions under which a given differential equation has solutions are fairly general. But we should note that there are simple equations which do not have solutions at all, and others which do not have a general solution. For example, the equation $y'^2 = -1$ does not have a solution for real y, as is obvious; the equation $|y'| + |y| = 0$ does not have a general solution, because its only solution is $y \equiv 0$.

Differential equations are of great importance in engineering, because many physical laws and relations appear mathematically in the form of differential equations.

Let us consider a simple physical example which will illustrate the typical steps leading from the physical situation to its mathematical formulation and solution. This may be the easiest way to obtain a first idea of the nature and purpose of differential equations and their applications.

Example 3 (Radioactivity, exponential decay). Experiments show that a radioactive substance decomposes at a rate proportional to the amount present. Starting with a given amount of substance, say, 2 grams, at a certain time, say, $t = 0$, what can be said about the amount available at a later time?

1st Step (Mathematical description of the physical process by a differential equation). We denote the amount of substance still present at time t by $y(t)$. The rate of change is dy/dt. According to the law governing the process of radiation, dy/dt is proportional to y. Thus,

$$(9) \qquad \frac{dy}{dt} = ky$$

where k is a definite physical constant whose numerical value is known for various radioactive substances. (For example, in the case of radium, $k \approx -1.4 \cdot 10^{-11} \sec^{-1}$.) Clearly, since the amount of substance is positive and decreases with time, dy/dt is negative, and so is k. We see that the physical process under consideration is described mathematically by an ordinary linear differential equation of the first order. Whenever a physical law involves a rate of change of a function, such as velocity, acceleration, etc., it will lead to a differential equation. For this reason differential equations occur frequently in physical and engineering problems.

2nd Step (Solving the differential equation). At this early stage of our discussion no systematic method for solving (9) is at our disposal. However, we see that if (9) has a solution $y(t)$, its derivative must be proportional to y. From calculus we remember that exponential functions have this property. In fact, the function e^{kt} or more generally

$$(10) \qquad y(t) = ce^{kt}$$

where c is any constant, is a solution of (9) for all t, as can readily be verified by substituting (10) into (9). Since (10) involves an arbitrary constant, it is a general solution of the first order equation (9). [We shall see that (10) includes all solutions of (9), that is, (9) does not have singular solutions.]

3rd Step (Determination of a particular solution). It is clear that our physical process has a unique behavior, and we may therefore expect that by using further given information we shall be able to select a definite numerical value of c in (10) so that the resulting particular solution will describe the process properly. The amount of substance $y(t)$ still present at time t will depend on the initial amount of substance given. This amount is 2 grams at $t = 0$. Hence we have to specify the value of c so that $y = 2$ when $t = 0$. By inserting this *initial condition*

$$(11) \qquad y(0) = 2$$

in (10) we obtain

$$y(0) = ce^0 = 2 \qquad \text{or} \qquad c = 2.$$

If we use this value of c, then (10) becomes

$$(12) \qquad y(t) = 2e^{kt}.$$

This particular solution of (9) characterizes the amount of substance still present at any time $t \geqq 0$. The physical constant k is negative, and $y(t)$ decreases, as is shown in Fig. 4.

Fig. 4. Radioactivity (exponential decay)

4th Step (Checking). From (12) we have

$$\frac{dy}{dt} = 2ke^{kt} = ky \qquad \text{and} \qquad y(0) = 2e^0 = 2.$$

The function (12) satisfies the equation (9) as well as the initial condition (11).

The student should never forget to carry out the important final step which shows him whether the function is (or is not) the solution of the problem.

Let us illustrate that geometrical problems may also lead to differential equations.

Example 4. Find the curve through the point $(1, 1)$ in the xy-plane having at each of its points the slope $-y/x$. Clearly the function representing the desired curve must be a solution of the differential equation

$$(13) \qquad\qquad y' = -\frac{y}{x}.$$

We shall soon learn how to solve such an equation. For the time being the student may verify that

$$y = \frac{c}{x}$$

is a solution of (13) for any value of the constant c. Some of the corresponding curves are shown in Fig. 5. Now we must have $y = 1$ when $x = 1$, and this yields $c = 1$. Hence the solution of our problem is the hyperbola

$$y = \frac{1}{x}.$$

Fig. 5. Solutions of $y' = -y/x$

Problems

In each case verify that the given function is a solution of the differential equation.
1. $xy' - 3y = 0$, $y = cx^3$
2. $y'' + 4y = 0$, $y = A \cos 2x + B \sin 2x$
3. $y'' - y = 0$, $y = c_1 e^x + c_2 e^{-x}$
4. $y''' = 6$, $y = x^3 + ax^2 + bx + c$
5. $y'' + 2y' + 2 = 0$, $y = e^{-x}(A \cos x + B \sin x)$

Solve the following differential equations.
6. $y' = \sin x$ 7. $y' = e^x$ 8. $y'' = 3$ 9. $y'' = \cos 2x$

In each case verify that the given function is a solution of the differential equation. Graph the corresponding curves for some values of the constant c.
10. $y' + y = 0$, $y = ce^{-x}$ 11. $y' + y = 2$, $y = ce^{-x} + 2$
12. $xy' - 4y = 0$, $y = cx^4$ 13. $yy' = -x$, $x^2 + y^2 = c$
14. $y' + 2xy = 0$, $y = ce^{-x^2}$ (*"bell-shaped curves"*)

In each case verify that the given function is a solution of the corresponding differential equation and determine c such that the resulting particular solution satisfies the given condition. Graph this solution.
15. $xy' = y$, $y = cx$, $y = \pi$ when $x = 2$
16. $y' = 1$, $y = x + c$, $y = 0$ when $x = 7$
17. $y' + 2xy = 0$, $y = ce^{-x^2}$, $y = 0.5$ when $x = 0$
18. $y' + y = 2$, $y = ce^{-x} + 2$, $y = 3.2$ when $x = 0$
19. $y' = 3x^2$, $y = x^3 + c$, $y = -1$ when $x = 1$

Find a first-order differential equation involving both y' and y for which the given function is a solution.
20. $y = -e^{-2x}$ 21. $y = x^3 - 4$ 22. $y = \sin 2x$

23. Find and graph the curve in the xy-plane passing through the point $(2, 2)$ and having at each of its points the slope $-1/x^2$.
24. (**Falling body**) If a body falls in vacuum due to the action of gravity, starting at time $t = 0$ with initial velocity 0, experiments show that its velocity is proportional to the time. State this physical law in terms of a first-order differential equation, and show that by solving it one obtains the familiar law

$$s(t) = \frac{g}{2} t^2$$

where s is the distance of the body from its initial position and g is the constant of proportionality between velocity and time.
25. (**Falling body**)   What is $s(t)$ in Prob. 24 if the initial velocity is v_0?

1.2 Geometrical Considerations. Isoclines

We shall now start with a systematic treatment of first-order differential equations. Any such equation may be written in the *implicit form*

(1) $$F(x, y, y') = 0.$$

Not always, but in many cases we shall be able to write a first-order differential equation in the **explicit form**

(2) $$y' = f(x, y).$$

Before we discuss various standard methods for solving equations of the form (2) we shall demonstrate that (2) has a very simple geometric interpretation. This will immediately lead to a useful graphical method for obtaining a rough picture of the particular solutions of (2), without actually solving the equation.

We assume that the function f is defined in some region of the xy-plane, so that to each point in that region it assigns one (and only one) value. The solutions of (2) can be plotted as curves in the xy-plane. We do not know the solutions, but we see from (2) that a solution passing through a point (x_0, y_0) must have the slope $f(x_0, y_0)$ at this point. This suggests the following method. We first graph some of the curves in the xy-plane along which $f(x, y)$ is constant. These curves are called *curves of constant slope*, or **isoclines** of (2). Along any such isocline, given by

$$f(x, y) = k = const$$

we draw a number of parallel short line segments **(lineal elements)** with slope k, which is the slope of solution curves of (2) at any point of that isocline. This we do for all isoclines which we graphed before. In this way we obtain a field of lineal elements, called the **direction field** of (2). With the help of the lineal elements we can now easily graph approximation curves to the (unknown) solution curves of the given equation (2) and thus obtain a qualitatively correct picture of these solution curves.

Let us consider a simple illustrative example.

Example 1.   Graph the direction field of the first-order differential equation

(3) $y' = xy$

and an approximation to the solution curve through the point $(1, 1)$. The isoclines are the equilateral hyperbolas $xy = k$ together with the two coordinate axes. The result is shown in Fig. 6. In the next section we shall see that (3) can be easily solved, so that our example merely

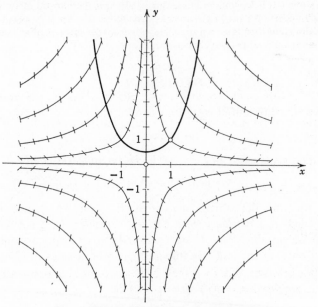

Fig. 6. Direction field of the equation (3)

illustrates the technical details of the method of direction field. This method is particularly helpful in engineering problems involving a first-order differential equation whose solution cannot be expressed in terms of known functions or is a very complicated expression.

Problems

1. Verify that $y = ce^{x^2/2}$ is a solution of (3) for any c. For what value of c do we obtain the solution shown in Fig. 6?

Draw good direction fields for the following differential equations and plot several approximate solution curves.

2. $y' = 2x$ **3.** $y' = y$ **4.** $yy' + x = 0$
5. $y' = x^2$ **6.** $y' = \cos y$ **7.** $y' = 2y/x$
8. $y' = 2y + x$ **9.** $y' = y^2$ **10.** $y' = x^2 + y^2$

Using direction fields, plot an approximate solution curve of the given differential equation satisfying the given condition.

11. $y' = x$, $y = 6$ when $x = -2$
12. $4y' + y = 0$, $y = -1$ when $x = 0$
13. $y' = x + y$, $y = 0$ when $x = 0$
14. $y' + y^2 = 0$, $y = 0.25$ when $x = 5$
15. $9yy' + 4x = 0$, $y = -4$ when $x = 3$

16. Give two examples of differential equations (2) whose isoclines are concentric circles with center at the origin.

17. Using direction fields, graph the curve which passes through the point $(0, 1)$ and has the slope $-2xy$.

18. A body B moves on a straight line L. Let $s(t)$ be its distance from a fixed point O on L. Suppose that at each instant the velocity of B equals $1/s(t)$, and $s = 1$ when $t = 0$. Find the corresponding differential equation. Graph the direction field and the (approximate) solution curve $s(t)$ of the problem.

19. Using a direction field, find an approximate value of the integral of $\sin(x^2)$ from 0 to 1 (which cannot be evaluated by elementary methods; cf. (38) in Appendix 3).

20. Using a direction field, find approximate values of the integral of x^2 from 0 to $b = 0.2$, 0.4, 0.6, 0.8, 1.0, and compare them with the exact values.

1.3   Separable Equations

Many first-order differential equations can be reduced to the form

$$(1) \qquad\qquad g(y)y' = f(x)$$

by algebraic manipulations. Since $y' = dy/dx$ we find it convenient to write

$$(2) \qquad\qquad g(y)\, dy = f(x)\, dx,$$

but we keep in mind that this is merely another way of writing (1). Such an equation is called an *equation with separable variables*, or a **separable equation,** because in (2) the variables x and y are *separated* so that x appears only on the right and y appears only on the left. By integrating on both sides of (2) we obtain

$$(3) \qquad\qquad \int g(y)\, dy = \int f(x)\, dx + c.$$

If we assume that f and g are continuous functions, the integrals in (3) will exist, and by evaluating these integrals we obtain the general solution of (1).

Example 1. Solve the differential equation

$$9yy' + 4x = 0.$$

By separating variables we have

$$9y\, dy = -4x\, dx.$$

By integrating on both sides we obtain the general solution

$$\tfrac{9}{2}y^2 = -2x^2 + \tilde{c} \qquad \text{or} \qquad \frac{x^2}{9} + \frac{y^2}{4} = c \qquad \left(c = \frac{\tilde{c}}{18} \right).$$

The solution represents a family of ellipses.

Example 2. Solve the differential equation

$$y' = -2xy.$$

By separating variables we have

$$\frac{dy}{y} = -2x\, dx \qquad\qquad (y \neq 0).$$

Integration yields

(4) $$\ln |y| = -x^2 + \tilde{c}.$$

In fact, the result on the left may be verified by differentiation as follows. When $y > 0$, then $(\ln y)' = y'/y$. When $y < 0$, then $-y > 0$ and $(\ln(-y))' = -y'/(-y) = y'/y$. Now $y = |y|$ when $y > 0$ and $-y = |y|$ when $y < 0$. Hence we may combine both formulas, finding $(\ln |y|)' = y'/y$.

   It is of great importance that the constant of integration is introduced immediately when the integration is carried out. From (4) we obtain

$$|y| = e^{-x^2 + \tilde{c}}$$

or by noting that $e^{a+b} = e^a e^b$ and setting $e^{\tilde{c}} = c$ when $y > 0$ and $e^{\tilde{c}} = -c$ when $y < 0$,

$$y = ce^{-x^2}.$$

This solution represents a family of so-called **bell-shaped curves** (Fig. 7). We may also admit $c = 0$, which gives the solution $y \equiv 0$.

Fig. 7. Solutions of $y' = -2xy$ ("bell-shaped curves") in the upper half plane ($y > 0$)

Example 3. Solve the differential equation

$$y' = 1 + y^2.$$

By separating variables and integrating we obtain

$$\frac{dy}{1 + y^2} = dx, \quad \text{arc tan } y = x + c, \quad y = \tan(x + c).$$

In many engineering applications we are not interested in the general solution of a given differential equation but only in the particular solution $y(x)$ satisfying a given **initial condition,** say, the condition that at some point x_0 the solution $y(x)$ has a prescribed value y_0, briefly

(5) $$y(x_0) = y_0.$$

A first-order differential equation together with an initial condition is called an **initial value problem.** To solve such a problem we must find the particular solution of the equation which satisfies the given initial condition.

Example 4. Solve the initial value problem

$$(x^2 + 1)y' + y^2 + 1 = 0, \quad y(0) = 1.$$

1st Step. Separating variables, we find

$$\frac{dy}{1 + y^2} = -\frac{dx}{1 + x^2}.$$

By integration we obtain

$$\text{arc tan } y = -\text{arc tan } x + c$$

or

$$\text{arc tan } y + \text{arc tan } x = c.$$

Taking the tangent on both sides, we have

(6) $$\tan(\text{arc tan } y + \text{arc tan } x) = \tan c.$$

Now the addition formula for the tangent is

$$\tan(a + b) = \frac{\tan a + \tan b}{1 - \tan a \tan b}.$$

For $a = \text{arc tan } y$ and $b = \text{arc tan } x$ this becomes

$$\tan(\text{arc tan } y + \text{arc tan } x) = \frac{y + x}{1 - xy}.$$

Consequently, (6) may be written

(7) $$\frac{y + x}{1 - xy} = \tan c.$$

2nd Step. We determine c from the initial condition. Setting $x = 0$ and $y = 1$ in (7), we have $1 = \tan c$, so that

$$\frac{y + x}{1 - xy} = 1 \quad \text{or} \quad y = \frac{1 - x}{1 + x}.$$

3rd Step. Check the result.

Example 5 (Newton's law of cooling[2]). A copper ball is heated to a temperature of 100°C. Then at time $t = 0$ it is placed in water which is maintained at a temperature of 30°C. At the end of 3 minutes the temperature of the ball is reduced to 70°C. Find the time at which the temperature of the ball is reduced to 31°C.

Physical information. Experiments show that the rate of change of the temperature T of the ball is proportional to the difference between T and the temperature of the surrounding medium (*Newton's law of cooling*).

Heat flows so rapidly in copper that at any time the temperature is practically the same at all points of the ball.

1st Step. The mathematical formulation of Newton's law of cooling in our case is

$$(8) \qquad \frac{dT}{dt} = -k(T - 30)$$

where we denoted the constant of proportionality by $-k$ in order that $k > 0$.

2nd Step. The general solution of (8) is obtained by separating variables; we find

$$T(t) = ce^{-kt} + 30.$$

3rd Step. The given initial condition is $T(0) = 100$. The particular solution satisfying this condition is

$$T(t) = 70e^{-kt} + 30.$$

4th Step. k can be determined from the given information $T(3) = 70$. We obtain

$$T(3) = 70e^{-3k} + 30 = 70 \qquad \text{or} \qquad k = \tfrac{1}{3}\ln\tfrac{7}{4} = 0.1865.$$

Using this value of k we see that the temperature $T(t)$ of the ball is

$$T(t) = 70e^{-0.1865t} + 30.$$

It follows that the value $T = 31°C$ is reached when

$$70e^{-0.1865t} = 1 \qquad \text{or} \qquad 0.1865t = \ln 70, \qquad t = \frac{\ln 70}{0.1865} = 22.78,$$

that is, after approximately 23 minutes.

5th Step. Check the result.

Example 6 (Velocity of escape from the earth). Find the minimum initial velocity of a body which is fired in radial direction from the earth and is supposed to escape from the earth. Neglect the air resistance and the gravitational pull of other celestial bodies.

1st Step. According to **Newton's law of gravitation,** the gravitational force, and therefore also the acceleration a of the body, is proportional to $1/r^2$ where r is the distance between the body and the center of the earth. Thus

$$a(r) = \frac{dv}{dt} = \frac{k}{r^2}$$

where v is the velocity and t is the time. Since v is decreasing, $a < 0$ and thus $k < 0$. Let R be the radius of the earth. When $r = R$ then $a = -g$, the acceleration of gravity at the surface of the earth. Note that the minus sign occurs because g is positive and the attraction of gravity acts

[2] Sir ISAAC NEWTON (1642–1727), English physicist and mathematician. Newton and the German mathematician and philosopher, GOTTFRIED WILHELM LEIBNIZ (1646–1716), invented (independently) the differential and integral calculus. Newton discovered many basic physical laws and introduced the method of investigating physical problems by means of calculus. His work is of greatest importance to both mathematics and physics.

in the negative direction (the direction towards the center of the earth). Thus

$$-g = a(R) = \frac{k}{R^2} \quad \text{and} \quad a(r) = -\frac{gR^2}{r^2}.$$

Now $v = dr/dt$, and differentiation yields

$$a = \frac{dv}{dt} = \frac{dv}{dr}\frac{dr}{dt} = \frac{dv}{dr}v.$$

Hence the differential equation for the velocity is

$$\frac{dv}{dr}v = -\frac{gR^2}{r^2}.$$

2nd Step. Separating variables and integrating, we obtain

(9)
$$v\, dv = -gR^2\frac{dr}{r^2} \quad \text{and} \quad \frac{v^2}{2} = \frac{gR^2}{r} + c.$$

3rd Step. On the earth's surface, $r = R$ and $v = v_0$, the initial velocity. For these values of r and v, formula (9) becomes

$$\frac{v_0^2}{2} = \frac{gR^2}{R} + c, \quad \text{and} \quad c = \frac{v_0^2}{2} - gR.$$

By inserting this expression for c into (9) we have

(10)
$$v^2 = \frac{2gR^2}{r} + v_0^2 - 2gR.$$

4th Step. If $v^2 = 0$, then $v = 0$, the body stops, and it is clear from the physical situation, that then the velocity will change from positive to negative and the body will return to the earth. Consequently, we have to choose v_0 so large that this cannot happen. If we choose

(11)
$$v_0 = \sqrt{2gR},$$

then in (10) the expression $v_0^2 - 2gR$ is zero and v^2 is never zero for any r. However, if we choose a smaller value for the initial velocity v_0, then $v = 0$ for a certain r. The expression in (11) is called the *velocity of escape* from the earth. Since $R = 3960$ statute miles and $g = 32.17$ ft/sec^2 = 0.006 09 miles/sec^2,

$$v_0 = \sqrt{2gR} \approx 7 \text{ miles/sec}.$$

5th Step. Check the result.

Example 7 (Torricelli's law[3]). A funnel (Fig. 8), whose angle at the outlet is $60°$ and whose outlet has a cross-sectional area of 0.5 cm^2, contains water. At time $t = 0$ the outlet is opened and the water flows out. Determine the time when the funnel will be empty, assuming that the initial height of water is $h(0) = 10$ cm.

1st Step. The volume of water which flows out during a short interval of time Δt is

$$\Delta V = 0.5\, v\, \Delta t$$

where v is the velocity of the outflowing water. From *Torricelli's law*, the velocity with which a liquid issues from an orifice is

$$v = 0.6\sqrt{2gh}$$

[3] EVANGELISTA TORRICELLI (1608–1647), Italian physicist. The "contraction factor" 0.6 was introduced by BORDA in 1766 because of the fact that the cross section of the outflowing stream of liquid is somewhat smaller than that of the orifice.

Fig. 8. Funnel

where $g = 980$ (cm/sec²) is the acceleration of gravity at the surface of the earth and h is the instantaneous height of the liquid above the orifice. Hence

$$\Delta V = 0.3 \sqrt{2gh} \, \Delta t.$$

ΔV must equal the change ΔV^* of the volume of water in the funnel. Now

$$\Delta V^* = -\pi r^2 \Delta h$$

where Δh is the decrease of the height $h(t)$ of the water and $r = h \tan 30° = h/\sqrt{3}$ is the radius of the funnel at height $h(t)$, so that

$$\Delta V^* = -\pi \frac{h^2}{3} \Delta h.$$

The minus sign appears because the volume of water in the funnel decreases.

By equating ΔV and ΔV^* we have

$$0.3\sqrt{2gh} \, \Delta t = -\frac{\pi}{3} h^2 \, \Delta h \quad \text{or} \quad \frac{\Delta h}{\Delta t} = -\frac{0.9\sqrt{2g}}{\pi} h^{-3/2}.$$

If we now let Δt approach zero, we obtain the differential equation

$$\frac{dh}{dt} = -kh^{-3/2}$$

where

$$k = \frac{0.9\sqrt{2g}}{\pi} = 12.7.$$

2nd Step. By separating variables and integrating we have

(12) $h^{3/2} \, dh = -k \, dt \quad \text{and} \quad \tfrac{2}{5} h^{5/2} = -kt + c.$

3rd Step. The initial condition is $h(0) = 10$. Setting $t = 0$ and $h = 10$ in the general solution given by the last formula in (12), we obtain

$$\tfrac{2}{5} 10^{5/2} = c.$$

If we use this value of c, the last formula in (12) becomes

(13) $$\tfrac{2}{5}h^{5/2} = -kt + \tfrac{2}{5}10^{5/2}.$$

4th Step. By solving (13) for t and substituting $k = 12.7$ we find

$$t = \frac{2}{5k}\,(10^{5/2} - h^{5/2}) \approx 10 - 0.0315\,h^{5/2}.$$

Hence the funnel will be empty ($h = 0$) at $t \approx 10$ sec.

Problems

1. Why is it important to introduce the constant of integration immediately when the integration is performed?

2. We have $yy' = (y^2)'/2$. How can this be used for solving the differential equation in Ex. 1?

Find the general solution of the following equations (where a, b, k, θ, and ω are constants).

3. $y' = ky$
4. $y' = y^2$
5. $y' - 2y + a = 0$
6. $xy' + by = 0$
7. $(x - 1)y' = 2x^3 y$
8. $(x + 2)y' - xy = 0$
9. $y' = 2x^{-1}\sqrt{y - 1}$
10. $y' + e^{2x}y^2 = 0$
11. $y' = y \cot 2x$
12. $y' + \csc y = 0$
13. $y' = (1 + x)(1 + y^2)$
14. $yy' = 0.5 \sin^2 \omega x$
15. $y' = y \tanh x$
16. $y' + 3y \sin \omega x = 0$
17. $\sin 2x\, dy = y \cos 2x\, dx$
18. $y^2\, dy - \cos^2 x\, dx = 0$
19. $x \ln x\, dy - y\, dx = 0$
20. $(1 - \cos\theta)\, dr = r \sin\theta\, d\theta$

21. Using (3), show that the solution of (1) satisfying (5) is obtained from

(14) $$\int_{y_0}^{y} g(y^*)\, dy^* = \int_{x_0}^{x} f(x^*)\, dx^*.$$

22. Using (14), find the solution of the differential equation in Ex. 4 satisfying $y(0) = 0$.

23. Use (14) to obtain the solution y given in Ex. 4.

Solve the following initial value problems.

24. $(x + 1)y' = 2y$, $y(0) = 1$
25. $y' = y \tan 2x$, $y(0) = 2$
26. $y'x \ln x = y$, $y(2) = \ln 4$
27. $2xy' = 3y$, $y(1) = 4$
28. $(x^2 + 1)yy' = 1$, $y(0) = -3$
29. $y' = 2e^x y^3$, $y(0) = 0.5$
30. $y' = 3x^2 e^{-y}$, $y(-1) = 0$
31. $dr/dt = -rt$, $r(0) = r_0$
32. $y' = y^2 \sin x$, $y(\pi) = 0.2$
33. $y' = \sec y$, $y(0) = 0$
34. $yy' = xe^{y^2}$, $y(1) = 0$
35. $xyy' = y + 2$, $y(2) = 0$

Geometrical applications. In each case find all curves in the xy-plane having the given property

36. The normals pass through the origin.
37. The tangents pass through the origin.
38. The normal at each point (x, y) intersects the x-axis at $(x + 1, 0)$.
39. The tangent at each point (x, y) intersects the x-axis at $(x - 1, 0)$.
40. The slope at each point P equals twice the slope of the straight line through P and the origin.
41. The slope at each point P equals the reciprocal of the slope of the straight line through P and the origin.

42. The slope of the normal at each point (x, y) equals the ratio x/y.
43. The segment of each tangent between the coordinate axis is bisected by the point of tangency.

Physical applications

44. **(Velocity of escape)** Consider the following modification of Ex. 6. The body is carried by a rocket and is separated from the rocket at a distance of 200 miles from the earth's surface. Determine the minimum velocity at this point sufficient for escape from the earth.
45. **(Torricelli's law)** Consider a spherical tank of radius $R = 50$ cm which contains water and has an outlet of radius $r_0 = 5$ cm at the bottom. At time $t = 0$ the outlet is opened and the water flows out. Determine the time when the tank will be empty, assuming that the initial height of water is $h(0) = R = 50$ cm.
46. In Prob. 45, determine the time necessary to let 50% of the water flow out. Why is this time less than half of the time necessary to drain the tank completely?
47. **(Half-life)** Experiments show that radium disintegrates at a rate proportional to the amount of radium instantaneously present. Its *half-life*, that is, the time in which 50% of a given amount will disappear, is 1590 years. What per cent will disappear in 1 year?
48. **(Exponential growth)** Suppose that in a culture of yeast at each instant the time rate of change of active ferment $y(t)$ is proportional to the amount present. If $y(t)$ doubles in 2 hours, how much can be expected at the end of 8 hours at the same rate of growth?
49. **(Exponential decay)** Experiments show that the rate of inversion of cane sugar in dilute solution is proportional to the concentration $y(t)$ of the unaltered sugar. Suppose that the concentration is 1/100 at $t = 0$ and is 1/300 at $t = 10$ [hours]. Find $y(t)$.
50. The **law of mass action** states that, if the temperature is kept constant, the velocity of a chemical reaction is proportional to the product of the concentrations of the substances which are reacting. In the bimolecular reaction

$$A + B \to M,$$

a moles per liter of a substance A and b moles of a substance B are combined. If y is the number of moles per liter which have reacted after time t, the rate of reaction is given by

$$y' = k(a - y)(b - y).$$

Solve this equation, assuming that $a \neq b$.
51. A tank contains 100 gal of brine in which 10 lb of salt is dissolved. Fresh water runs into the tank at the rate of 5 gal/min, and the mixture, kept practically uniform by stirring, runs out at the same rate. How much salt will there be in the tank at the end of 2 hours?
52. Two liquids are boiling in a vessel. It is found that the ratio of the quantities of each passing off as vapor at any instant is proportional to the ratio of the quantities x and y still in the liquid state. Show that $\dfrac{dy}{dx} = k\dfrac{y}{x}$ and solve this equation.
53. **(Newton's law of cooling)** Consider Ex. 5. Find the time at which the temperature of the ball is reduced to $31°C$, assuming that at the end of 5 minutes the temperature of the ball is reduced to $60°C$ while all the other conditions remain as in Ex. 5.
54. **(Evaporation)** A wet porous substance in the open air loses its moisture at a rate proportional to the moisture content. If a sheet hung in the wind loses half its moisture during the first hour, when will it have lost 99%, weather conditions remaining the same?
55. **Lambert's**[4] **law of absorption** states that the absorption of light in a very thin transparent layer is proportional to the thickness of the layer and to the amount incident on that layer. Formulate this in terms of a differential equation and solve it.

---

[4] JOHANN HEINRICH LAMBERT (1728–1777), German physicist and mathematician.

1.4   *Equations Reducible to Separable Form*

Certain first-order differential equations are not separable but can be made separable by a simple change of variables. This holds for equations of the form[5]

$$(1) \qquad\qquad y' = g\left(\frac{y}{x}\right)$$

where g is any given function of y/x, for example $(y/x)^3$, $\sin (y/x)$, etc. The form of the equation suggests that we set

$$\frac{y}{x} = u,$$

remembering that y and u are functions of x. Then $y = ux$, and by differentiation,

$$(2) \qquad\qquad y' = u + u'x.$$

By inserting this into (1) and noting that $g(y/x) = g(u)$ we have

$$u + u'x = g(u).$$

Now we may separate the variables u and x, finding

$$\frac{du}{g(u) - u} = \frac{dx}{x}.$$

If we integrate and replace u by y/x, we obtain the general solution of (1).

Example 1. Solve

$$2xyy' - y^2 + x^2 = 0.$$

Dividing by x^2, we have

$$2\frac{y}{x}y' - \left(\frac{y}{x}\right)^2 + 1 = 0.$$

If we set $u = y/x$ and use (2), the equation becomes

$$2u(u + u'x) - u^2 + 1 = 0 \qquad \text{or} \qquad 2xuu' + u^2 + 1 = 0.$$

Separating variables, we find

$$\frac{2u\,du}{1 + u^2} = -\frac{dx}{x}$$

and by integration

$$\ln (1 + u^2) = -\ln |x| + c^* \qquad \text{or} \qquad 1 + u^2 = \frac{c}{x}.$$

Replacing u by y/x, we finally obtain

$$x^2 + y^2 = cx \qquad \text{or} \qquad \left(x - \frac{c}{2}\right)^2 + y^2 = \frac{c^2}{4}.$$

---

[5] These equations are sometimes called **homogeneous equations**. We shall not use this terminology but reserve the term "homogeneous" for a much more important purpose (cf. Sec. 1.7).

Sometimes the form of a given differential equation suggests other simple substitutions, as the following example illustrates.

Example 2.   Solve the first-order differential equation

$$(2x - 4y + 5)y' + x - 2y + 3 = 0.$$

We set $x - 2y = v$. Then $y' = \frac{1}{2}(1 - v')$ and the equation takes the form

$$(2v + 5)v' = 4v + 11.$$

Separating variables and integrating, we find

$$\left(1 - \frac{1}{4v + 11}\right) dv = 2\, dx \qquad \text{and} \qquad v - \tfrac{1}{4}\ln|4v + 11| = 2x + c^*.$$

Since $v = x - 2y$, this may be written

$$4x + 8y + \ln|4x - 8y + 11| = c.$$

Further simple substitutions are illustrated by the equations in Probs. 11–16.

Problems

Find the general solution of the following equations.

1. $xy' = 3y - 4x$

2. $2xyy' = 3y^2 + x^2$

3. $xy' = 2x + 2y$

4. $x^2y' = y^2 + 5xy + 4x^2$

5. $xy' - y = x^2 \sec(y/x)$

6. $x^2y' = y^2 + xy + x^2$

7. $xy' = (y - x)^3 + y$

8. $(xy' - y)\cos(2y/x) = -3x^4$

9. $y' = \dfrac{y - x}{y + x}$

10. $y' = \dfrac{y + x}{y - x}$

Using the indicated transformations, find the general solution of the following equations.

11. $y' = \dfrac{y - x + 1}{y - x + 5}$ $(y - x = v)$

12. $y' = \dfrac{1 - 2y - 4x}{1 + y + 2x}$ $(y + 2x = v)$

13. $y' = \tan(x + y) - 1$ $(x + y = v)$

14. $y' = (x + e^y - 1)e^{-y}$ $(x + e^y = w)$

15. $2x^2yy' = \tan(x^2y^2) - 2xy^2$ $(x^2y^2 = z)$

16. $2xyy' = \exp(x^2 + y^2) - 2x^2$ $(x^2 + y^2 = z)$

17. Four flies sit at the corners of a card table, facing inward. They start walking simultaneously at the same rate, each directing its motion steadily toward the fly on its right. Find the path of each.

18. Show that a differential equation of the form $y' = f(ax + by + c)$ can be reduced to a separable equation by setting $v = ax + by + c$.

1.5   *Exact Differential Equations*

A first-order differential equation of the form

(1) $$M(x, y)\, dx + N(x, y)\, dy = 0$$

is said to be **exact** if the left-hand side is the total or exact differential

(2) $$du = \frac{\partial u}{\partial x}\, dx + \frac{\partial u}{\partial y}\, dy$$

of some function $u(x, y)$. Then the differential equation (1) can be written

$$du = 0.$$

By integration we immediately obtain the general solution of (1) in the form

(3) $$u(x, y) = c.$$

Comparing (1) and (2), we see that (1) is exact, if there is some function $u(x, y)$ such that

(4)   (a) $\dfrac{\partial u}{\partial x} = M$,   (b) $\dfrac{\partial u}{\partial y} = N$.

Suppose that M and N are defined and have continuous first partial derivatives in a region in the xy-plane whose boundary is a closed curve having no self-intersections. Then from (4),

$$\frac{\partial M}{\partial y} = \frac{\partial^2 u}{\partial x\, \partial y}, \qquad \frac{\partial N}{\partial x} = \frac{\partial^2 u}{\partial y\, \partial x}$$

and by assumption of continuity the second derivatives are equal. Thus

(5) $$\frac{\partial M}{\partial y} = \frac{\partial N}{\partial x}.$$

This condition is not only necessary but also sufficient[6] for $M\, dx + N\, dy$ to be a total differential.

If (1) is exact, the function $u(x, y)$ can be found by guessing or in the following systematic way. From (4a) we have[7]

(6) $$u = \int M\, dx + k(y);$$

in this integration, y is to be regarded as a constant, and $k(y)$ plays the role of a "constant" of integration. To determine $k(y)$, we derive $\partial u/\partial y$ from (6), use (4b) to get dk/dy, and integrate.

Example 1. Solve

$$xy' + y + 4 = 0.$$

If we write the equation in the form (1), that is,

$$(y + 4)\, dx + x\, dy = 0,$$

we see that $M = y + 4$, $N = x$, and (5) holds, so that the equation is exact. Now from (4b) we have $N = \partial u/\partial y = x$, and by integration

$$u = xy + k(x).$$

---

[6] We shall prove this fact at another occasion (in Sec. 8.12); the proof can also be found in some books on elementary calculus; cf. Ref. [A14] in Appendix 1.

[7] Or from (4b), $u = \int N\, dy + k(x)$, etc; see Ex. 1.

To determine $k(x)$ we differentiate u and apply (4a), finding

$$\frac{\partial u}{\partial x} = y + \frac{dk}{dx} = M = y + 4.$$

Thus $dk/dx = 4$ and $k = 4x + c^*$. By inserting this into the equation $u = xy + k = const$ and solving for y, we obtain the general solution of our equation,

$$y = \frac{c}{x} - 4.$$

Intelligent students, who know that thinking is important in mathematics, write the given equation in the form

$$y\,dx + x\,dy = -4\,dx,$$

recognize the left side as the total differential of xy, and integrate, finding $xy = -4x + c$, which is equivalent to our previous result.

Example 2. Solve

$$2x \sin 3y\,dx + 3x^2 \cos 3y\,dy = 0.$$

The equation is exact, and from (6) we obtain

$$u = \int 2x \sin 3y\,dx + k(y) = x^2 \sin 3y + k(y).$$

Therefore $\partial u/\partial y = 3x^2 \cos 3y + dk/dy$, and $dk/dy = 0$ or $k = c^* = const$. The general solution is $u = const$ or $x^2 \sin 3y = c$.

Problems

Find the exact differential equations that have the following general solutions.
1. $x^2 + y^2 = c$
2. $e^x \cos y = c$
3. $\sinh(x^3 y) = c$
4. $\ln x \sin \omega y = c$
5. $(x - a)(b - y) = c$
6. $\ln(x^2 + y^2) + 2 \arctan(y/x) = c$

Show that the following equations are exact and solve them.
7. $y\,dx + x\,dy = 0$
8. $(x\,dy - y\,dx)/x^2 = 0$
9. $\sinh x \cos y\,dx = \cosh x \sin y\,dy$
10. $(y - x)\,dx + (y + x)\,dy = 0$
11. $(2x + e^y)\,dx + xe^y\,dy = 0$
12. $(\tan y - \sin x)\,dx + x \sec^2 y\,dy = 0$

In each case, find whether the equation is exact and solve it.
13. $(1 + x^2)\,dy + 2xy\,dx = 0$
14. $x\,dy - 4y\,dx = 0$
15. $y\,dx + x(1 + y)\,dy = 0$
16. $(2y\,dx + dy)e^{2x} = 0$
17. $(3y \cos 3x\,dx - \sin 3x\,dy)/y^2 = 0$
18. $\sin \beta y\,dx + \beta \cos \beta y\,dy = 0$

Solve the following equations (*a*) by the present method, (*b*) by separating variables, (*c*) by inspection.
19. $x\,dx + y\,dy = 0$
20. $y \sinh x\,dx + \cosh x\,dy = 0$

Under what conditions are the following equations exact? Solve the exact equations.
21. $(ax + by)\,dx + (kx + ly)\,dy = 0$
22. $(x^3 + xy^2)\,dx + (ax^2 y + h(x, y))\,dy = 0$

Under what conditions are the following equations exact?
23. $f(x)g(y)\,dx + h(y)\,dy = 0$
24. $[f(x) + g(y)]\,dx + [h(x) + l(y)]\,dy = 0$

Solve the following initial value problems.

25. $(y \, dx + dy)e^x = 0$, $y(0) = 2$

26. $(x - 3) \, dx + (y - 3) \, dy = 0$, $y(3) = 0$

27. $[(x + x^2 y) \, dx + x^3 \, dy] \, e^{2xy} = 0$, $y(1) = \ln 2$

28. $2(y \sin 2x + \cos 2x) \, dx = \cos 2x \, dy$, $y(\pi) = 0$

29. $\beta \cos \alpha x \cos \beta y \, dy - \alpha \sin \alpha x \sin \beta y \, dx = 0$, $y(0) = \pi/2\beta$

30. $(e^{-x}/y) \, dy - e^{-x} \ln |y| \, dx = 0$, $y(0) = e$

31. $(xe^{xy} + 2y)y' + ye^{xy} = 0$, $y(0) = 1$

32. $\cosh x \sec y \, dx + \sinh x \sin y \sec^2 y \, dy = 0$, $y(0) = 3\pi/2$

33. $x \exp (x^2 - y^2) \, dx = y(\exp (x^2 - y^2) + 1) \, dy$, $y(0) = 0$

Find and graph the curve having the given slope y' and passing through the given point.

34. $y' = \dfrac{1 - x}{y - 1}$, $(0, 1)$ **35.** $y' = \dfrac{3 - x}{9y - 9}$, $(3, 0)$

1.6   *Integrating Factors*

Sometimes a given differential equation

(1) $$P(x, y) \, dx + Q(x, y) \, dy = 0$$

is not exact but can be made exact by multiplying it by a suitable function $F(x, y)$ ($\not\equiv 0$). This function is then called an **integrating factor** of (1). With some experience, integrating factors can be found by inspection. For this purpose the student should keep in mind the differentials listed in Ex. 2 below. In some important special cases, integrating factors can be determined in a systematic way which is illustrated in the subsequent section.

Example 1.   Solve

$$x \, dy - y \, dx = 0.$$

The differential equation is not exact. An integrating factor is $F = 1/x^2$, and we obtain

$$F(x)(x \, dy - y \, dx) = \frac{x \, dy - y \, dx}{x^2} = d\left(\frac{y}{x}\right) = 0, \qquad y = cx.$$

Theorem 1.   *If* (1) *is not exact and has a general solution $u(x, y) = c$, then there exists an integrating factor (even infinitely many such factors).*

Proof.   By differentiating $u(x, y) = c$ we have

$$du = \frac{\partial u}{\partial x} \, dx + \frac{\partial u}{\partial y} \, dy = 0.$$

Comparing this and (1), we see that

$$\frac{\partial u}{\partial x} : \frac{\partial u}{\partial y} = P : Q$$

must hold identically. Hence there exists a function $F(x, y)$ such that

$$\frac{\partial u}{\partial x} = FP, \qquad \frac{\partial u}{\partial y} = FQ.$$

Using these expressions for $\partial u/\partial x$ and $\partial u/\partial y$, we have

(2) $$du = FP\,dx + FQ\,dy = F(P\,dx + Q\,dy).$$

This shows that F is an integrating factor of (1). Furthermore, multiplication of (2) by a function $H(u)$ yields the expression

$$H(u)F(P\,dx + Q\,dy) = H(u)\,du$$

which is exact. Hence $H(u)F(x, y)$ is another integrating factor, and since H is arbitrary, there are infinitely many integrating factors of (1).

Example 2. Find integrating factors of $x\,dy - y\,dx = 0$; cf. Ex. 1. Since

$$d\left(\frac{x}{y}\right) = \frac{y\,dx - x\,dy}{y^2}, \qquad d\left(\ln\frac{y}{x}\right) = \frac{x\,dy - y\,dx}{xy}, \qquad d\left(\arctan\frac{y}{x}\right) = \frac{x\,dy - y\,dx}{x^2 + y^2}$$

the functions $1/y^2$, $1/xy$, and $1/(x^2 + y^2)$ are such factors. The corresponding solutions

$$\frac{x}{y} = c, \qquad \ln\frac{y}{x} = c, \qquad \arctan\frac{y}{x} = c$$

are essentially the same, because each represents a family of straight lines through the origin.

Problems

Show that the given function is an integrating factor and solve:
1. $2y\,dx + x\,dy = 0$, x
2. $4y\,dx + 3x\,dy = 0$, x^3y^2
3. $x\,dy - y\,dx = 0$, $1/x^2$
4. $(k + 1)y\,dx + (m + 1)x\,dy = 0$, $x^k y^m$
5. $\sin y\,dx + \cos y\,dy = 0$, e^x
6. $y^2\,dx + (1 + xy)\,dy = 0$, e^{xy}

7. Show that $F = xy^2$ is an integrating factor of $2y\,dx + 3x\,dy = 0$ and find further integrating factors of this equation.

Find integrating factors F and solve:
8. $dx + (y/x)^2\,dy = 0$
9. $2y\,dx + x\,dy = 0$
10. $(y - 1)\,dx - x\,dy = 0$
11. $(3y + 4xy)\,dx - 2x\,dy = 0$
12. $\sinh 2y\,dx + x\cosh 2y\,dy = 0$
13. $y\cos x\,dx + 2\sin x\,dy = 0$

Find integrating factors F and solve the following initial value problems.
14. $y\,dx - x\,dy = 0$, $y(1) = 2$
15. $2\sin y\,dx + \cos y\,dy = 0$, $y(0) = \pi/2$
16. $2x^2\,dx - 3xy^2\,dy = 0$, $y(1) = 0$
17. $2\,dx + \sec x\cos y\,dy = 0$, $y(0) = 0$

Under what conditions are the following functions F integrating factors of (1)? *Hint.* Assume $F(P\,dx + Q\,dy) = 0$ to be exact and apply (5) in Sec. 1.5.
18. $F = x^k$
19. $F = y^m$
20. $F = x^k y^m$
21. $F = e^x$
22. $F = y^m e^x$
23. $F = e^{xy}$

24. Using the result of Prob. 20, solve Probs. 8–11.

25. Show that if $\dfrac{1}{Q}\left(\dfrac{\partial P}{\partial y} - \dfrac{\partial Q}{\partial x}\right)$ is a function of x only, say, $f(x)$, then (1) has an integrating

factor depending only on x, namely, $F(x) = \exp \int f(x)\,dx$ (where $\exp u = e^u$).

1.7 **Linear First-Order Differential Equations**

A first-order differential equation is said to be **linear** if it can be written in the form

(1) $$y' + f(x)y = r(x).$$

The characteristic feature of this equation is that it is linear in y and y', whereas f and r may be *any* given functions of x.

If $r(x) \equiv 0$, the equation is said to be **homogeneous;** otherwise, it is said to be **nonhomogeneous.** (Here, $r \equiv 0$ means $r = 0$ for all x in the domain of r.)

Let us find a formula for the general solution of (1) in some interval I, assuming that f and r are continuous in I. For the homogeneous equation

(2) $$y' + f(x)y = 0$$

this is very simple. By separating variables we have

$$\frac{dy}{y} = -f(x)\,dx \quad \text{and thus} \quad \ln|y| = -\int f(x)\,dx + c^*$$

or

(3) $$y(x) = ce^{-\int f(x)\,dx} \qquad (c = \pm e^{c^*} \quad \text{when} \quad y \gtrless 0);$$

here we may also take $c = 0$ and obtain the *trivial solution* $y \equiv 0$.

To solve the nonhomogeneous equation (1), let us write it in the form

$$(fy - r)\,dx + dy = 0$$

and show that we can find an integrating factor $F(x)$ *depending only on x*. If such a factor exists,

$$F(x)(fy - r)\,dx + F(x)\,dy = 0$$

must be exact. Now for this equation the condition (5) of Sec. 1.5 assumes the form

$$\frac{\partial}{\partial y}\,[F(fy - r)] = \frac{dF}{dx} \quad \text{or} \quad Ff = \frac{dF}{dx}.$$

By separating variables and integrating we obtain

$$\ln|F| = \int f(x)\,dx$$

and may take

$$F(x) = e^{h(x)} \quad \text{where} \quad h(x) = \int f(x)\,dx.$$

This shows that $F(x)$ is an integrating factor of (1). Let us now multiply (1) by this factor, finding

$$e^h(y' + fy) = e^h r.$$

Since $h' = f$, this may be written

$$\frac{d}{dx}(ye^h) = e^h r.$$

Now we can integrate on both sides and obtain

$$ye^h = \int e^h r \, dx + c.$$

By dividing both sides by e^h we have the desired formula

(4) $$y(x) = e^{-h}\left[\int e^h r \, dx + c\right], \qquad h = \int f(x) \, dx,$$

which represents the general solution of (1) in the form of an integral. (The choice of the values of the constants of integration is immaterial; cf. Prob. 2 below.)

Example 1. Solve the linear differential equation

$$y' - y = e^{2x}.$$

Here

$$f = -1, \qquad r = e^{2x}, \qquad h = \int f \, dx = -x$$

and from (4) we obtain the general solution

$$y(x) = e^x\left[\int e^{-x}e^{2x} \, dx + c\right] = e^x[e^x + c] = ce^x + e^{2x}.$$

Example 2. Solve

$$xy' + y + 4 = 0.$$

We write this equation in the form (1), that is,

$$y' + \frac{1}{x}y = -\frac{4}{x}.$$

Hence, $f = \dfrac{1}{x}$, $r = -\dfrac{4}{x}$ and, therefore,

$$h = \int f \, dx = \ln|x|, \qquad e^h = x, \qquad e^{-h} = \frac{1}{x}.$$

From this and (4) we obtain the general solution

$$y(x) = \frac{1}{x}\left[\int x\left(-\frac{4}{x}\right) dx + c\right] = \frac{c}{x} - 4,$$

in agreement with Ex. 1 of Sec. 1.5.

Of course, in simple cases such as Exs. 2, 3, and 5 (below) we may solve the equation without using (4).

Example 3. Solve the linear equation

$$xy' + y = \sin x.$$

We can write the equation in the form

$$d(xy) = \sin x \, dx$$

and integrate on both sides, finding

$$xy = -\cos x + c \qquad \text{or} \qquad y = \frac{1}{x}(c - \cos x).$$

Let us now illustrate the procedure of solving initial value problems.

Example 4. Solve the initial value problem

$$y' + y \tan x = \sin 2x, \qquad y(0) = 1.$$

Here $f = \tan x$, $r = \sin 2x = 2 \sin x \cos x$, and

$$\int f \, dx = \int \tan x \, dx = \ln |\sec x|.$$

From this we see that in (4),

$$e^h = \sec x, \qquad e^{-h} = \cos x, \qquad e^h r = 2 \sin x$$

and the general solution of our equation is

$$y(x) = \cos x \left[2 \int \sin x \, dx + c \right] = c \cos x - 2 \cos^2 x.$$

According to the initial condition, $y = 1$ when $x = 0$, that is

$$1 = c - 2 \qquad \text{or} \qquad c = 3,$$

and the solution of our initial value problem is

$$y = 3 \cos x - 2 \cos^2 x.$$

Example 5. Solve the initial value problem

$$x^2 y' + 2xy - x + 1 = 0, \qquad y(1) = 0.$$

The equation may be written as $d(x^2 y) = (x - 1) \, dx$. By integration we obtain

$$x^2 y = \tfrac{1}{2} x^2 - x + c \qquad \text{or} \qquad y(x) = \frac{1}{2} - \frac{1}{x} + \frac{c}{x^2}.$$

From this and the initial condition we have $y(1) = \tfrac{1}{2} - 1 + c = 0$ or $c = \tfrac{1}{2}$. Thus

$$y(x) = \frac{1}{2} - \frac{1}{x} + \frac{1}{2x^2}.$$

Very often the solutions of differential equations in engineering applications are not elementary functions. For instance, the integral in (4) may be not elementary. If it is not a tabulated function,[8] we may develop the integrand in a power

---

[8] Whether tables of a certain function were computed can be found in the extremely useful book Ref. [A6] listed in Appendix 1.

series and integrate term by term, or we may apply a method of numerical integration (cf. Sec. 18.5) or a numerical method for solving first-order differential equations (cf. Sec. 18.6).

In applications the independent variable x will often be the time; the function $r(x)$ on the right-hand side of (1) may represent a force, and the solution $y(x)$ a displacement, a current, or some other variable physical quantity. In engineering mathematics $r(x)$ is frequently called the **input,** and $y(x)$ is called the **output** or *response to the input* (and the initial conditions). For instance, in electrical engineering the differential equation may govern the behavior of an electric circuit and the output $y(x)$ is obtained as the solution of that equation corresponding to the input $r(x)$. In the field of analogue computers the physical system may be an integrating unit; then the corresponding differential equation is $y' = r(x)$, and the output $y(x)$ corresponding to an input $r(x)$ is the integral of $r(x)$,

$$y(x) = \int_{x_0}^{x} r(x^*) \, dx^*.$$

The system may be represented by a "block diagram" as shown in Fig. 9. Various examples will occur in our further consideration.

Fig. 9. Simple block diagram

Problems

1. Show that $e^{-\ln x} = 1/x$ (but not $-x$) and $e^{-\ln(\sec x)} = \cos x$.

2. Show that the choice of the value of the constant of integration in $\int f \, dx$ [cf. (4)] is immaterial.

3. Perform the steps that lead to (4) for a special linear differential equation, say, $y' - (2/x)y = x^3$.

Find the general solutions of the following differential equations.

4. $y' - y = 3$ 5. $y' + 2xy = 0$ 6. $y' + 2xy = x$

7. $y' + 2y = 6e^x$ 8. $xy' = y + (x+1)^2$ 9. $xy' + y = x + x^3$

10. $y' + ky = e^{-kx}$ 11. $y' - (2/x)y = x^2 e^x$ 12. $y' = (y-1)\cot x$

13. $y' + y = \sin x$ 14. $(1 - x^2)y' + xy = x$ 15. $y' - 2y = \cos 3x$

Find the general solution of $y' + (2/x)y = r(x)$ where

16. $r(x) = 2 - 3x + x^3$ 17. $r(x) = e^{x^2}/x$ 18. $r(x) = x^{-2} \sinh 3x$

Solve the following initial value problems.

19. $y' + y = (x+1)^2$, $y(0) = 0$

20. $y' - 2y = 2\cosh 2x + 4$, $y(0) = -1.25$

21. $xy' - 3y = x^4(e^x + \cos x) - 2x^2$, $y(\pi) = e^\pi + 2/\pi$

22. In Prob. 11, $r(x) = x^2 e^x$ is a solution of the equation. For what equations of the form (1) does it happen that $r(x)$ is a solution?

Prove that the homogeneous linear differential equation (2) has the following important properties. [In the next chapter (Secs. 2.3 and 2.9) we shall see that homogeneous linear differential equations of higher order enjoy similar properties.]

23. Equation (2) has the "*trivial solution*" $y \equiv 0$.

24. If y_1 is a solution of (2), then $y = cy_1$ (c any constant) is a solution of (2).

25. If y_1 and y_2 are solutions of (2), then their sum $y = y_1 + y_2$ is a solution of (2).

Prove and illustrate with examples that the nonhomogeneous linear differential equation (1) has the following important properties.

26. If y_1 is a solution of (1) and y_2 is a solution of (2), then $y = y_1 + y_2$ is a solution of (1).

27. The difference $y = y_1 - y_2$ of two solutions y_1 and y_2 of (1) is a solution of (2).

28. If y_1 is a solution of (1), then $y = cy_1$ is a solution of $y' + fy = cr$.

29. If y_1 is a solution of $y_1' + fy_1 = r_1$ and y_2 is a solution of $y_2' + fy_2 = r_2$ (with the same f), then $y = y_1 + y_2$ is a solution of $y' + fy = r_1 + r_2$.

30. Show that in the direction field of (1), the lineal elements on any vertical line $x = x_0$ all pass through a single point with coordinates $\xi = x_0 + 1/f(x_0)$, $\eta = r(x_0)/f(x_0)$. Using this interesting property, find and graph an approximate solution curve of the initial value problem $y' = 2xy + 1$, $y(1) = 1$.

1.8   *Variation of Parameters*

There is another interesting way of obtaining the general solution of the linear differential equation

$$(1) \qquad\qquad y' + f(x)y = r(x).$$

We have seen that a solution of the corresponding homogeneous equation is

$$(2) \qquad\qquad v(x) = e^{-\int f(x)\,dx}.$$

Using this function $v(x)$, let us try to determine a function $u(x)$ such that

$$(3) \qquad\qquad y(x) = u(x)v(x)$$

is the general solution of (1). This attempt is suggested by the form of the general solution $cv(x)$ of the homogeneous equation and consists in replacing the parameter c by a variable $u(x)$. Therefore, this approach, credited to Lagrange,[9] is called the **method of variation of parameters.** It can be generalized to equations of higher order where it is of great importance. Our present simple case is a good occasion for getting acquainted with this basic method.

By substituting (3) into (1) we obtain

$$u'v + u(v' + fv) = r.$$

Since v is a solution of the homogeneous equation, this reduces to

$$u'v = r \qquad \text{or} \qquad u' = \frac{r}{v}.$$

---

[9] JOSEPH LOUIS LAGRANGE (1736–1813), a great French mathematician, spent 20 years of his life in Prussia and then returned to Paris. His important major work was in the calculus of variations, celestial and general mechanics, differential equations, and algebra.

Integration yields

$$u = \int \frac{r}{v} \, dx + c.$$

We thus obtain the result

(4) $$y = uv = v\left(\int \frac{r}{v} \, dx + c\right),$$

and from (2) we see that this is identical with (4) of the previous section.

Problems

Apply the method of variation of parameters to the following equations.

    1. $y' - y = x$ **2.** $y' + 2y = e^{-2x}$ **3.** $xy' - 2y = x^4$

    4. Solve $y' - 2xy = -2x$, (*a*) by (4) in Sec. 1.7, (*b*) by variation of parameters, (*c*) by separating variables.

    5. Show that the transformation $x = \phi(t)$ (where ϕ has a continuous derivative) transforms (1) into a linear differential equation involving y and dy/dt.

    6. Show that the transformation $y = a(x)u + b(x)$ with differentiable $a \not\equiv 0$ and b transforms (1) into a linear differential equation involving u and du/dx.

Taking x as the dependent variable, solve:

    7. $y' = 1/(e^y + x)$ **8.** $(\cos y - 2x)y' = 1$

    9. $(2x + y^4)y' - y = 0$ **10.** $y' = y/[x + (y + 1)^2]$

Reduction of nonlinear equations to linear form. The following problems illustrate that *certain nonlinear first-order differential equations may be reduced to linear form by a suitable change of the dependent variable.*

    11. Solve $y' + y = y^2$, (*a*) by setting $y = 1/u$ and solving the resulting linear differential equation for u, (*b*) by separation of variables.

    12. (Bernoulli equation) The differential equation

$$y' + f(x)y = g(x)y^a \qquad\qquad (a \text{ any real number})$$

is called the **Bernoulli**[10] **equation.** For $a = 0$ and $a = 1$ the equation is linear, and otherwise it is nonlinear. Set $[y(x)]^{1-a} = u(x)$ and show that the equation assumes the linear form

$$u' + (1 - a)f(x)u = (1 - a)g(x).$$

Solve the following Bernoulli equations.

    13. $y' + xy = xy^{-1}$ **14.** $y' + y = xy^{-1}$

    15. $2xy' = 10x^3y^5 + y$ **16.** $y' + x^{-1}y = x^{-1}y^2$

    17. Solve Probs. 13 and 16 by separation of variables.

Applying suitable substitutions, reduce to linear form and solve:

    18. $e^y y' - e^y = 2x - x^2$ **19.** $y' \cos y + x \sin y = 2x$

    20. $2(y + 1)y' - \dfrac{2}{x}(y + 1)^2 = x^4$, $y(1) = \sqrt{2/3} - 1$

---

[10] JACOB BERNOULLI (1654–1705), Swiss mathematician, who contributed to the theory of elasticity and to the theory of mathematical probability. The method for solving Bernoulli's equation was discovered by Leibniz in 1696.

1.9 *Electric Circuits*

Linear first-order differential equations have various applications in physics and engineering practice. To illustrate this, let us consider some standard examples in connection with electric circuits. These and similar considerations are important, because they help all students to learn how to express physical situations in terms of mathematical relations. The transition from the physical situation to a corresponding mathematical model is always the first step in engineering mathematics, and this important step requires experience and training which can only be gained by considering typical examples from various fields.

The simplest electric circuit is a series circuit in which we have a source of electric energy (*electromotive force*) such as a generator or a battery, and a resistor, which uses energy, for example an electric light bulb (Fig. 10). If we close the switch, a current I will flow through the resistor, and this will cause a *voltage drop*, that is, the electric potential at the two ends of the resistor will be different; this potential difference or voltage drop can be measured by a voltmeter. Experiments show that the following law holds.

The voltage drop E_R across a resistor is proportional to the instantaneous current I, say

(1) $$E_R = RI$$ *(Ohm's law)*

where the constant of proportionality R is called the **resistance** of the resistor. The current I is measured in *amperes*, the resistance R in *ohms*, and the voltage E_R in *volts*.[11]

Source — Resistor

Switch

Fig. 10. Circuit

The other two important elements in more complicated circuits are *inductors* and *capacitors*. An inductor opposes a change in current, having an inertia effect in electricity similar to that of mass in mechanics; we shall consider this analogy later (Sec. 2.14). Experiments yield the following law.

The voltage drop E_L across an inductor is proportional to the instantaneous time rate of change of the current I, say,

(2) $$E_L = L\frac{dI}{dt}$$

[11] These and the subsequent units are named after ANDRÉ MARIE AMPÈRE (1775–1836), French physicist, GEORG SIMON OHM (1789–1854), German physicist, ALLESSANDRO VOLTA (1745–1827), Italian physicist, JOSEPH HENRY (1797–1878), American physicist, MICHAEL FARADAY (1791–1867), English physicist, and CHARLES AUGUSTIN DE COULOMB (1736–1806), French physicist.

Fig. 11. *RL*-circuit   **Fig. 12.** Current in an *RL*-circuit due to a constant
electromotive force

where the constant of proportionality L is called the **inductance** of the inductor
and is measured in *henrys;* the time t is measured in seconds.

A capacitor is an element which stores energy. Experiments yield the following
law.

*The voltage drop E_C across a capacitor is proportional to the instantaneous electric
charge Q on the capacitor,* say,

(3*) $$E_C = \frac{1}{C} Q$$

where C is called the **capacitance** and is measured in *farads;* the charge Q is measured in *coulombs.* Since

(3′) $$I(t) = \frac{dQ}{dt}$$

this may be written

(3) $$E_C = \frac{1}{C} \int_{t_0}^{t} I(t^*) \, dt^*.$$

The current $I(t)$ in a circuit may be determined by solving the equation (or
equations) resulting from the application of **Kirchhoff's**[12] **second law:**

*The algebraic sum of all the instantaneous voltage drops around any closed loop
is zero, or the voltage impressed on a closed loop is equal to the sum of the voltage
drops in the rest of the loop.*

Example 1 (*RL*-circuit). For the "*RL*-circuit" in Fig. 11 we obtain from Kirchhoff's second law
and (1), (2)

(4) $$L \frac{dI}{dt} + RI = E(t).$$

Case A (*Constant electromotive force*). If $E = E_0 = $ *const*, then (4) in Sec. 1.7 yields the
general solution

$$I(t) = e^{-\alpha t} \left[\frac{E_0}{L} \int e^{\alpha t} \, dt + c \right] \qquad (\alpha = R/L)$$

(5)

$$= \frac{E_0}{R} + ce^{-(R/L)t}.$$

[12] GUSTAV ROBERT KIRCHHOFF (1824–1887), German physicist. Later we shall also
need **Kirchhoff's first law:** At any point of a circuit the sum of the inflowing currents is equal to
the sum of the outflowing currents.

Fig. 13. Phase angle δ in (6) as a function of $\omega L/R$

The last term approaches zero as t tends to infinity, so that $I(t)$ tends to the limit value E_0/R; after a sufficiently long time, I will practically be constant. The particular solution corresponding to the initial condition $I(0) = 0$ is (cf. Fig. 12)

$$(5^*) \qquad I(t) = \frac{E_0}{R}(1 - e^{-(R/L)t}) = \frac{E_0}{R}(1 - e^{-t/\tau_L})$$

where $\tau_L = L/R$ is called the **inductive time constant** of the circuit (cf. Probs. 4–6).

Case B (Periodic electromotive force). If $E(t) = E_0 \sin \omega t$, then, by (4) in Sec. 1.7, the general solution of (4) in this current section is

$$I(t) = e^{-\alpha t}\left[\frac{E_0}{L}\int e^{\alpha t}\sin \omega t\, dt + c\right] \qquad (\alpha = R/L).$$

Integration by parts yields

$$I(t) = ce^{-(R/L)t} + \frac{E_0}{R^2 + \omega^2 L^2}(R \sin \omega t - \omega L \cos \omega t).$$

This may be written (cf. (14) in Appendix 3)

$$(6) \qquad I(t) = ce^{-(R/L)t} + \frac{E_0}{\sqrt{R^2 + \omega^2 L^2}}\sin(\omega t - \delta), \qquad \delta = \arctan\frac{\omega L}{R}.$$

The exponential term will approach zero as t approaches infinity. This means that after a sufficiently long time the current $I(t)$ executes practically harmonic oscillations. [Cf. Fig. 14.] Figure 13 shows the phase angle δ as a function of $\omega L/R$. If $L = 0$, then $\delta = 0$, and the oscillations of $I(t)$ are in phase with those of $E(t)$.

An electrical (or dynamical) system is said to be in the **steady state** when the variables describing its behavior are periodic functions of the time or constant, and it is said to be in the **transient state** (or *unsteady state*) when it is not in the steady

Fig. 14. Current in an RL-circuit due to a sinusoidal electromotive force, as given by (6)
(with $\delta = \pi/4$)

state. The corresponding variables are called *steady state functions* and *transient functions*, respectively.

In Ex. 1, Case A, the function E_0/R is the steady state function or **steady state solution** of (4), and in Case B the steady state solution is represented by the last term in (6). Before the circuit (practically) reaches the steady state it is in the transient state. It is clear that such an interim or transient period occurs because inductors and capacitors store energy, and the corresponding inductor currents and capacitor voltages cannot be changed instantly. Similar situations arise in various physical systems. For example, if a radio receiver having heater-type vacuum tubes is turned on, a transient interval of time must elapse during which the tubes change from "cold" to "hot."

Example 2 (*RC-circuit*). By applying Kirchhoff's second law and (1), (3) to the *RC*-circuit in Fig. 15 we obtain the equation

$$(7) \qquad RI + \frac{1}{C}\int I\,dt = E(t).$$

To get rid of the integral we differentiate the equation with respect to t, finding

$$(8) \qquad R\frac{dI}{dt} + \frac{1}{C}I = \frac{dE}{dt}.$$

According to (4) in Sec. 1.7 this differential equation has the general solution

$$(9) \qquad I(t) = e^{-t/RC}\left(\frac{1}{R}\int e^{t/RC}\frac{dE}{dt}\,dt + c\right).$$

Case A (*Constant electromotive force*). If E is constant, then $\dfrac{dE}{dt} = 0$, and (9) assumes the simple form (Fig. 16)

$$(10) \qquad I(t) = ce^{-t/RC} = ce^{-t/\tau_C}$$

where $\tau_C = RC$ is called the **capacitive time constant** of the circuit.

Case B (*Sinusoidal electromotive force*). If $E(t) = E_0 \sin \omega t$, then

$$\frac{dE}{dt} = \omega E_0 \cos \omega t,$$

Fig. 15. *RC*-circuit

Fig. 16. Current in an *RC*-circuit due to a constant electromotive force

and by inserting this in (9) and integrating by parts we find

$$I(t) = ce^{-t/RC} + \frac{\omega E_0 C}{1 + (\omega RC)^2}(\cos \omega t + \omega RC \sin \omega t)$$

(11)

$$= ce^{-t/RC} + \frac{\omega E_0 C}{\sqrt{1 + (\omega RC)^2}} \sin(\omega t - \delta),$$

where $\tan \delta = -1/\omega RC$. The first term decreases steadily as t increases, and the last term represents the steady-state current, which is sinusoidal. The graph of $I(t)$ is similar to that in Fig. 14.

More complicated circuits and the analogy between electrical and mechanical vibrations will be considered in connection with second-order differential equations in Sec. 2.14.

Problems

1. Derive (5) from (4).
2. Derive (6) from (4).
3. Obtain from (5) the solutions satisfying $I(0) = 0.5E_0/R$, E_0/R, $2E_0/R$. Sketch the corresponding curves.
4. Show that the inductive time constant $\tau_L = L/R$ has the (physical) dimensions of time and is the time t at which the current (5*) reaches about 63% of its final value.
5. Show that the current (5*) builds up to half its theoretical maximum value in $\tau_L \ln 2 \approx 0.7 \tau_L$ sec.
6. In Ex. 1, Case A, let $E_0 = 100$ volts, $R = 1000$ ohms, and $L = 4$ henrys. Find τ_L. Graph (5*) and the voltage drops E_R and E_L [cf. (1) and (2)].
7. In Ex. 1, Case A, let $R = 20$ ohms, $L = 0.04$ millihenry ($4 \cdot 10^{-5}$ henry), and $I(0) = 0$. Find the time when the current reaches 99% of its final value.
8. What L should we choose in an RL-circuit with $R = 100$ ohms for (5*) to reach 99.9% of its final value at $t = 0.01$ sec?
9. Obtain from (6) the particular solution satisfying the initial condition $I(0) = 0$.
10. For what initial condition does (6) yield the steady-state solution?

Find the current $I(t)$ in the RL-circuit in Fig. 11 under the following assumptions.

11. $R = 1$ ohm, $L = 10$ henrys, $E = 6$ volts when $0 \leq t \leq 10$ sec, $E = 0$ when $t > 10$ sec, $I(0) = 6$ amperes
12. $E = 110$ volts, $R = 22$ ohms, $L = (10 - t)$ henrys when $0 \leq t \leq 10$ sec, $L = 0$ when $t > 10$ sec, $I(0) = 5$ amperes
13. $E = 4$ volts, $R = 2$ ohms, $L = (100 - t)$ henrys when $0 \leq t \leq 100$ sec, $L = 0$ when $t > 100$ sec, $I(0) = 0$
14. $R = 1$ ohm, $L = 100$ henrys, E periodic as shown in Fig. 17, $I(0) = 0$

15. Verify that (11) is a solution of (8) with $E = E_0 \sin \omega t$.
16. Obtain from (11) the particular solution satisfying the initial condition $I(0) = 0$.

Fig. 17. Problem 14

17. For what initial condition does (11) yield the steady state solution?

18. In Ex. 2, Case A, let $E = 200$ volts, $R = 10$ ohms, $C = 0.1$ farad, and $I(0) = 2$ amperes. Find the capacitive time constant τ_C. Find and graph $I(t)$ and the voltage drops $E_R(t)$ and $E_C(t)$.

19. Find the current in the *RC*-circuit in Fig. 15 assuming that $R = 100$ ohms, $C = 0.01$ farad, $E(t) = 110 \sin 314t$ volts, and $I(0) = 0$.

20. Using (3'), introduce $Q(t)$ as the dependent variable in (7) and solve the resulting equation for the cases of a constant and a sinusoidal electromotive force.

1.10   *Families of Curves.*
Orthogonal Trajectories

If for each fixed real value of c the equation

$$(1) \qquad\qquad F(x, y, c) = 0$$

represents a curve in the xy-plane and if for variable c it represents infinitely many curves, then the totality of these curves is called a **one-parameter family of curves,** and c is called the *parameter* of the family.

Example 1. The equation

$$(2) \qquad\qquad F(x, y, c) = x + y + c = 0$$

represents a family of parallel straight lines; each line corresponds to precisely one value of c. The equation

$$(3) \qquad\qquad F(x, y, c) = x^2 + y^2 - c^2 = 0$$

represents a family of concentric circles of radius c with center at the origin.

The general solution of a first-order differential equation involves a parameter c and thus represents a family of curves. This yields a possibility for representing many one-parameter families of curves by such differential equations. The practical use of such representations will become obvious from our further consideration.

Example 2. By differentiating (2) we see that

$$y' + 1 = 0$$

is the differential equation of that family of straight lines. Similarly, the differential equation of the family (3) is

$$y' = -x/y.$$

If the equation obtained by differentiating (1) still contains the parameter c, then we have to eliminate c from this equation by using (1).

Example 3. The differential equation of the family of parabolas

$$(4) \qquad\qquad y = cx^2$$

is obtained by differentiating (4),

$$(5) \qquad\qquad y' = 2cx,$$

and by eliminating c from (5). From (4) we have $c = y/x^2$, and by substituting this into (5) we find the desired result

(6) $$y' = 2y/x.$$

In many engineering applications, a family of curves is given, and it is required to find another family whose curves intersect each of the given curves at right angles.[13] Then the curves of the two families are said to be *mutually orthogonal*, they form an *orthogonal net*, and the curves of the family to be obtained are called the **orthogonal trajectories** of the given curves (and conversely); cf. Fig. 18.

Let us mention some familiar examples. The meridians on the earth's surface are the orthogonal trajectories of the parallels. On a map the curves of steepest descent are the orthogonal trajectories of the contour lines. In electrostatics the equipotential lines and the lines of electric force are orthogonal trajectories of each other. An illustrative example is shown in Fig. 19. We shall see later that orthogonal trajectories play an important role in various fields of physics, for example, in hydrodynamics and heat conduction.

Given a family $F(x, y, c) = 0$ which can be represented by a differential equation

(7) $$y' = f(x, y)$$

we may find the corresponding orthogonal trajectories as follows. From (7) we see that a curve of the given family which passes through a point (x_0, y_0) has the slope $f(x_0, y_0)$ at this point. The slope of the orthogonal trajectory through (x_0, y_0) at this point should be the negative reciprocal of $f(x_0, y_0)$, that is, $-1/f(x_0, y_0)$, because this is the condition for the tangents of the two curves at (x_0, y_0) to be perpendicular. Consequently, the differential equation of the orthogonal trajectories is

(8) $$y' = - \frac{1}{f(x, y)},$$

and the trajectories are obtained by solving this equation.

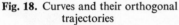

Fig. 18. Curves and their orthogonal
trajectories

Fig. 19. Equipotential lines and lines
of electric force (dashed) between two
concentric cylinders

---

[13] Remember that the angle of intersection of two curves is defined to be the angle between the tangents of the curves at the point of intersection.

Fig. 20. Parabolas and their orthogonal trajectories in Example 4

It can be shown that, under relatively general conditions, a family of curves has orthogonal trajectories, but we shall not discuss this question here.

Example 4.   Find the orthogonal trajectories of the parabolas in Ex. 3. From (6) we see that the differential equation of the orthogonal trajectories is

$$y' = -\frac{1}{2y/x} = -\frac{x}{2y}.$$

By separating variables and integrating we find that the orthogonal trajectories are the ellipses (Fig. 20)

$$\frac{x^2}{2} + y^2 = c^*.$$

Example 5.   Find the orthogonal trajectories of the circles

(9) $$x^2 + (y - c)^2 = c^2.$$

We first determine the differential equation of the given family. By differentiating (9) with respect to x we find

(10) $$2x + 2(y - c)y' = 0.$$

We have to eliminate c. Solving (9) for c, we have

$$c = \frac{x^2 + y^2}{2y}.$$

By inserting this into (10) and simplifying we get

$$x + \frac{y^2 - x^2}{2y}y' = 0 \qquad \text{or} \qquad y' = \frac{2xy}{x^2 - y^2}.$$

From this and (8) we see that the differential equation of the orthogonal trajectories is

$$y' = -\frac{x^2 - y^2}{2xy} \qquad \text{or} \qquad 2xyy' - y^2 + x^2 = 0.$$

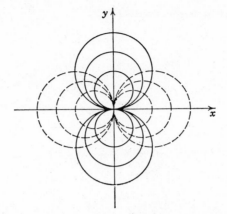

Fig. 21. Circles and their orthogonal trajectories (dashed) in Example 5

The orthogonal trajectories obtained by solving this equation (cf. Ex. 1 in Sec. 1.4) are the circles (Fig. 21)

$$(x - \tilde{c})^2 + y^2 = \tilde{c}^2.$$

Problems

Represent the following families of curves in the form (1).
  1. All nonvertical straight lines through the point $(-2, 3)$.
  2. All circles through the origin and tangent to the y-axis.
  3. All ellipses with foci -1 and 1 on the x-axis.
  4. The hyperbolas obtained by translating $y = 1/x$ in x-direction.

What families of curves are represented by the following equations?
  5. $3y + 2x + c = 0$ **6.** $(x - c)^2 + y^2 = 9$ **7.** $y - (x - c)^2 = 0$
  8. $x^2 + c^2 y^2 = 1$ **9.** $(x - c)^2 + y^2 = c^2/2$ **10.** $e^{x^2} y = c$

What families of curves are represented by the following differential equations?
  11. $(1 + x^2)y' + 2xy = 0$ **12.** $xy' = 3y + 2x$ **13.** $y' = 2y + \cos 3x$

Represent the following families of curves by differential equations.
  14. $y = ce^{-x}$ **15.** $y = e^{-cx}$ **16.** $x^2 + c^2 y^2 = c^2$
  17. $y = \sin(x - c)$ **18.** $y = c \cos x$ **19.** $y = \tan(x + c)$
  20. $y = \sqrt{x - c}$ **21.** $x^2 + 4y^2 = c$ **22.** $y = cxe^x$

Find the orthogonal trajectories of the following curves and graph some of the curves and their trajectories.
  23. $y = x + c$ **24.** $xy = c$ **25.** $x^2 + y^2 = c^2$
  26. $x^2 - y^2 = c^2$ **27.** $y = cx^2$ **28.** $y = ce^{-x^2}$
  29. $y = \sqrt{x + c}$ **30.** $y = \sqrt{c + \ln x}$ **31.** $y = ce^{-2x}$

Find the orthogonal trajectories of the following curves.
  32. $e^x \cos y = c$ **33.** $\cos x \sinh y = c$ **34.** $y^2 + 2xy - x^2 = c$

  35. The lines of electric force of two oppositely charged sources of the same strength at $(-1, 0)$ and $(1, 0)$ are the circles through $(-1, 0)$ and $(1, 0)$. Show that they have the representation $x^2 + (y - c)^2 = 1 + c^2$ and find the equipotential lines (orthogonal trajectories).

36. Show that (8) may be written in the form

$$\frac{dx}{dy} = -f(x, y).$$

Using this result, find the orthogonal trajectories of the curves $y = x + ce^{-x}$.

37. Show that the orthogonal trajectories of the family $g(x, y) = c$ can be obtained from the differential equation

$$\frac{dy}{dx} = \frac{\partial g/\partial y}{\partial g/\partial x}.$$

38. Using the result of Prob. 37, solve Prob. 24.

39. Show that if $u(x, y) = c = const$ represents a family of curves, a representation of its orthogonal trajectories of the form $v(x, y) = c = const$ can be obtained from

$$\frac{\partial u}{\partial x} = \frac{\partial v}{\partial y}, \qquad \frac{\partial u}{\partial y} = -\frac{\partial v}{\partial x}.$$

(These so-called *Cauchy-Riemann differential equations* are basic in complex analysis and will be considered in Chap. 11.)

40. Using the result in Prob. 39, solve Prob. 32.

1.11 *Picard's Iteration Method*

There are various differential equations of the first order which cannot be solved by one of the standard methods discussed before or by another elementary method for obtaining exact solutions.[14] Such equations occur quite frequently in engineering problems. Then one may apply an approximation method for obtaining approximate solutions, for instance, a numerical method such as those to be discussed in Sec. 18.6. In practical problems, approximate solutions will often be sufficient.

In this section we shall consider an approximation method, which is called **Picard's iteration method**[15] and gives approximate solutions of an initial value problem which is of the form

$$(1) \qquad\qquad y' = f(x, y), \qquad y(x_0) = y_0$$

and is assumed to have a unique solution on some interval containing x_0. Picard's method is of great theoretical value, as we shall see in the next section. Its practical value is limited because it involves integrations which may sometimes be complicated.

The basic idea of Picard's method is very simple. By integration we see that (1) may be written in the form

$$(2) \qquad\qquad y(x) = y_0 + \int_{x_0}^{x} f[t, y(t)]\, dt,$$

---

[14] Reference [B9] in Appendix 1 includes more than 1500 important differential equations and their solutions, arranged in systematic order and accompanied by numerous references to original literature.

[15] EMILE PICARD (1856–1941), French mathematician who made important contributions to the theory of complex analytic functions and differential equations.

where t denotes the variable of integration. In fact, when $x = x_0$ the integral is zero and $y = y_0$, so that (2) satisfies the initial condition in (1); furthermore, by differentiating (2) we obtain the differential equation in (1).

To find approximations to the solution $y(x)$ of (2) we proceed as follows. We substitute the crude approximation $y = y_0 = const$ on the right; this yields the presumably better approximation

$$y_1(x) = y_0 + \int_{x_0}^{x} f(t, y_0)\, dt.$$

Then we substitute $y_1(x)$ in the same way to get

$$y_2(x) = y_0 + \int_{x_0}^{x} f[t, y_1(t)]\, dt,$$

etc. The nth step of this iteration yields

(3)
$$y_n(x) = y_0 + \int_{x_0}^{x} f[t, y_{n-1}(t)]\, dt.$$

In this way we obtain a sequence of approximations

$$y_1(x), \qquad y_2(x), \cdots, \qquad y_n(x), \cdots,$$

and we shall see in the next section that the conditions under which this sequence converges to the solution $y(x)$ of (1) are relatively general.

An **iteration method** is a method that yields a sequence of approximations to an (unknown) function, say $y_1, y_2, \cdots$, where the nth approximation, y_n, is obtained in the nth step by using one (or several) of the previous approximations, and the operation performed in each step is the same. In the simplest case, y_n is obtained from y_{n-1}; denoting the operation by T, we may write

$$y_n = T(y_{n-1}).$$

Picard's method is of this type, because (3) may be written

$$y_n(x) = T(y_{n-1}(x)) = y_0 + \int_{x_0}^{x} f(t, y_{n-1}(t))\, dt.$$

To illustrate the method, let us apply it to an equation which we can readily solve exactly, so that we may compare the approximations with the exact solution.

Example 1. Find approximate solutions to the initial value problem

$$y' = 1 + y^2, \qquad y(0) = 0.$$

In this case, $x_0 = 0$, $y_0 = 0$, $f(x, y) = 1 + y^2$, and (3) becomes

$$y_n(x) = \int_{0}^{x} [1 + y_{n-1}^2(t)]\, dt = x + \int_{0}^{x} y_{n-1}^2(t)\, dt.$$

Fig. 22. Approximate solutions in Example 1

Starting from $y_0 = 0$, we thus obtain (cf. Fig. 22)

$$y_1(x) = x + \int_0^x 0 \cdot dt = x$$

$$y_2(x) = x + \int_0^x t^2 \, dt = x + \tfrac{1}{3}x^3$$

$$y_3(x) = x + \int_0^x \left(t + \frac{t^3}{3}\right)^2 dt = x + \tfrac{1}{3}x^3 + \tfrac{2}{15}x^5 + \tfrac{1}{63}x^7$$

etc. Of course, we can obtain the exact solution of our present problem by separating variables (Ex. 3 in Sec. 1.3), finding

$$(4) \qquad\qquad y(x) = \tan x = x + \tfrac{1}{3}x^3 + \tfrac{2}{15}x^5 + \tfrac{17}{315}x^7 + \cdots. \qquad \left(-\frac{\pi}{2} < x < \frac{\pi}{2}\right).$$

The first three terms of $y_3(x)$ and the series in (4) are the same. The series in (4) converges for $|x| < \pi/2$, and all we may expect is that our sequence $y_1, y_2, \cdots$ converges to a function which is the solution of our problem for $|x| < \pi/2$. This illustrates that the problem of convergence is of practical importance.

Problems

1. Show that in the case of the initial value problem $y' = y$, $y(0) = 1$, Picard's iteration method gives approximations which approach $y = e^x$ (the exact solution) as $n \to \infty$.

2. Apply Picard's method to $y' = xy$, $y(0) = 2$. Graph y_1, y_2, y_3 and the exact solution for $-2 \leqq x \leqq 2$. Compute $y_1(1)$, $y_2(1)$, $y_3(1)$ and compare the values with the exact value $y(1) = 2\sqrt{e} \approx 3.297$.

Apply Picard's iteration method to the following initial value problems.

3. $y' = x + y$, $y(0) = 0$

4. $y' = 2xy$, $y(0) = \pi$

5. $y' - xy = 1$, $y(0) = 1$

6. $y' - xy = 1$, $y(0) = 0$

7. $y' = 2xy - 1$, $y(0) = 0$

8. $y' = x - y^2$, $y(0) = 0.5$

9. $y' = x(y - x^2 + 2)$, $y(0) = 0$

10. Show that if f in (1) does not depend on y, the approximations obtained by Picard's method are identical with the exact solution.

1.12   Existence and Uniqueness of Solutions

The initial value problem

$$|y'| + |y| = 0, \qquad y(0) = 1$$

has no solution because $y \equiv 0$ is the only solution of the differential equation. The initial value problem

$$y' = x, \qquad y(0) = 1$$

has precisely one solution, namely, $y = \frac{1}{2}x^2 + 1$. The initial value problem

$$xy' = y - 1, \qquad y(0) = 1$$

has infinitely many solutions, namely, $y = 1 + cx$ where c is arbitrary. From these examples we see that an initial value problem

(1)
$$y' = f(x, y), \qquad y(x_0) = y_0$$

may have none, precisely one, or more than one solution. This leads to the following two fundamental questions.

Problem of existence.   *Under what conditions does an initial value problem of the form (1) have at least one solution?*

Problem of uniqueness.   *Under what conditions does that problem have a unique solution, that is, only one solution?*

Theorems which state such conditions are called **existence theorems** and **uniqueness theorems,** respectively.

Of course, our three examples are so simple that we can find the answer to those two questions by inspection, without using any theorems. However, it is clear that in more complicated cases—for example, when the equation cannot be solved by elementary methods—existence and uniqueness theorems will be of great importance. As a matter of fact, a more advanced course in differential equations consists mainly of considerations concerning the existence, uniqueness, and general behavior of solutions of various types of differential equations, and many of these considerations have far-reaching practical consequences. For example, results about the existence and uniqueness of periodic solutions of certain differential equations, which were obtained by Poincaré for entirely theoretical reasons about 90 years ago, are nowadays the base of many practical investigations in nonlinear mechanics.

We want to mention that in many branches of mathematics existence and uniqueness theorems are of similar importance. Let us illustrate this by a familiar example in linear algebra. The three systems of linear equations

$$(a) \quad \begin{matrix} x + y = 1 \\ x + y = 0 \end{matrix} \qquad (b) \quad \begin{matrix} x + y = 1 \\ x - y = 0 \end{matrix} \qquad (c) \quad \begin{matrix} x + y = 1 \\ 2x + 2y = 2 \end{matrix}$$

have, respectively, none, precisely one, and infinitely many solutions. This is immediately clear, and we need no theorems. However, if we have a system of many equations in many unknowns, there is no immediate answer to the question of whether the system has a solution at all and, if so, how many solutions there are. Then the need for existence and uniqueness theorems becomes obvious.

In connection with differential equations, a student who is exclusively interested in applications and does not like theoretical considerations (an attitude which will prevent him from having good success in the applied area) may be inclined to reason in the following manner. The physical problem corresponding to a certain differential equation has a unique solution, and the same must therefore be true for the differential equation. In simpler cases he may be right. However, he should keep in mind that the differential equation is merely an abstraction of the reality obtained by disregarding certain physical facts which seem to be of minor influence, and in complicated physical situations there may be no way of judging a priori the importance of various factors. In such a case there will then be no a priori guarantee that the differential equation leads to a faithful picture of the reality.

In the case of an initial value problem of the form

(1) $$y' = f(x, y), \qquad y(x_0) = y_0$$

the conditions for the existence and uniqueness of the solution are quite simple. If f is continuous in some region of the xy-plane containing the point (x_0, y_0), then the problem (1) has at least one solution. If moreover the partial derivative $\partial f / \partial y$ exists and is continuous in that region, then the problem (1) has precisely one solution. This solution can then be obtained by Picard's iteration method. Let us formulate these three statements in a precise manner.

Existence theorem. *If $f(x, y)$ is continuous at all points (x, y) in some rectangle* (cf. Fig. 23 on the next page)

$$R: \quad |x - x_0| < a, \qquad |y - y_0| < b$$

and bounded[16] in R, say,

(2) $$|f(x, y)| \leq K \qquad\qquad \text{for all } (x, y) \text{ in } R,$$

---

[16] A function $f(x, y)$ is said to be **bounded** when (x, y) varies in a region in the xy-plane, if there is a number K such that $|f| \leq K$ when (x, y) is in that region. For example, $f = x^2 + y^2$ is bounded, with $K = 2$ if $|x| < 1$ and $|y| < 1$. The function $f = \tan(x + y)$ is not bounded for $|x + y| < \pi/2$.

Fig. 23. Rectangle R in the existence and uniqueness theorems

then the initial value problem (1) *has at least one solution* $y(x)$, *which is defined at least for all* x *in the interval* $|x - x_0| < \alpha$ *where* α *is the smaller of the two numbers* a *and* b/K.

Uniqueness theorem. *If* $f(x, y)$ *and* $\partial f/\partial y$ *are continuous for all* (x, y) *in that rectangle* R *and bounded, say,*

$$(3) \qquad (a) \quad |f| \leqq K, \qquad (b) \quad \left| \frac{\partial f}{\partial y} \right| \leqq M \qquad \text{for all } (x, y) \text{ in } R,$$

then the initial value problem (1) *has only one solution* $y(x)$, *which is defined at least for all* x *in that interval* $|x - x_0| < \alpha$. *This solution can then be obtained by Picard's iteration method, that is, the sequence* $y_0, y_1, \cdots, y_n, \cdots$ *where*

$$y_n(x) = y_0 + \int_{x_0}^x f[t, y_{n-1}(t)]\, dt, \qquad n = 1, 2, \cdots,$$

converges to that solution $y(x)$.

Since the proofs of these theorems require familiarity with uniformly convergent series and other concepts to be considered later in this book, we shall not present these proofs at this time but refer the student to Ref. [B8] in Appendix 1. However, we want to include some remarks and examples which may be helpful for a good understanding of the two theorems.

Since $y' = f(x, y)$, the condition (2) implies that $|y'| \leqq K$, that is, the slope of any solution curve $y(x)$ in R is at least $-K$ and at most K. Hence a solution curve which passes through the point (x_0, y_0) must lie in the shaded region in Fig. 24 bounded by the lines l_1 and l_2 whose slopes are $-K$ and K, respectively. Depending on the form of R, two different cases may arise. In the first case, shown in Fig. 24*a*, we have $b/K \geqq a$ and therefore $\alpha = a$ in the existence theorem, which then asserts that the solution exists for all x between $x_0 - a$ and $x_0 + a$. In the second case, shown in Fig. 24*b*, we have $b/K < a$. Therefore $\alpha = b/K$, and all we can conclude from the theorems is that the solution exists for all x between $x_0 - b/K$ and $x_0 + b/K$; for larger or smaller x's the solution curve may leave the rectangle R, and since nothing is assumed about f outside R, nothing can be concluded about the solution for those corresponding values of x.

Example 1. Consider the problem

$$y' = 1 + y^2, \qquad y(0) = 0 \qquad\qquad \text{(cf. Ex. 1 in Sec. 1.11)}$$

and take $R: |x| < 5, |y| < 3$. Then

$$a = 5, b = 3, |f| = |1 + y^2| \leqq K = 10, |\partial f/\partial y| = 2|y| \leqq M = 6, \alpha = b/K = 0.3 < a.$$

In fact, the solution $y = \tan x$ of the problem is discontinuous at $x = \pm\pi/2$, and there is no continuous solution valid in the entire interval $|x| < 5$ from which we started.

The conditions in the two theorems are sufficient conditions rather than necessary ones, and can be lessened. For example, by the mean value theorem of differential calculus we have

$$f(x, y_2) - f(x, y_1) = (y_2 - y_1) \frac{\partial f}{\partial y}\bigg|_{y=\tilde{y}}$$

where (x, y_1) and (x, y_2) are assumed to be in R, and $\tilde{y}$ is a suitable value between y_1 and y_2. From this and (3b) it follows that

$$(4) \qquad\qquad |f(x, y_2) - f(x, y_1)| \leqq M|y_2 - y_1|,$$

and it can be shown that (3b) may be replaced by the weaker condition (4) which is known as a **Lipschitz condition.**[17] However, continuity of $f(x, y)$ is not enough to guarantee the *uniqueness* of the solution. This may be illustrated by the following example.

Example 2. The initial value problem

$$y' = \sqrt{|y|}, \qquad y(0) = 0$$

has the two solutions

$$y \equiv 0 \qquad \text{and} \qquad y^* = \begin{cases} x^2/4 & \text{when } x \geqq 0 \\ -x^2/4 & \text{when } x \leqq 0 \end{cases}$$

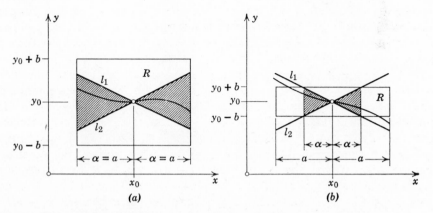

Fig. 24. The condition (2) of the existence theorem. (*a*) First case. (*b*) Second case

[17] RUDOLF LIPSCHITZ (1831–1903), German mathematician, who contributed also to algebra and differential geometry.

although $f(x, y) = \sqrt{|y|}$ is continuous for all y. The Lipschitz condition (4) is violated in any region which includes the line $y = 0$, because for $y_1 = 0$ and positive y_2 we have

$$(5) \qquad \frac{|f(x, y_2) - f(x, y_1)|}{|y_2 - y_1|} = \frac{\sqrt{y_2}}{y_2} = \frac{1}{\sqrt{y_2}}, \qquad (\sqrt{y_2} > 0)$$

and this can be made as large as we please by choosing y_2 sufficiently small, whereas (4) requires that the quotient on the left-hand side of (5) does not exceed a fixed constant M.

Numerical methods for first-order differential equations are presented in Sec. 18.6, which is independent of the other sections in Chap. 18 and may be studied now.

Problems

1. Find all solutions of the initial value problem $xy' = 2y$, $y(0) = 0$. Graph some of them.
2. How many solutions does the initial value problem $xy' = 3y$, $y(0) = 1$, have? Explain.
3. Consider the differential equation $xy' = 2y$ and find all initial conditions $y(x_0) = y_0$ such that the resulting initial value problem has (*a*) no solutions, (*b*) more than one solution, (*c*) precisely one solution.
4. Perform the same task as in Prob. 3, for the differential equation $(x^2 - x)y' = (2x - 1)y$.
5. Find further examples similar to those in Probs. 3 and 4.
6. Show that both $y \equiv 0$ and $y = -0.25x^2$ are solutions of the problem
$$y' = -0.5x + \sqrt{0.25x^2 + y}, \qquad y(0) = 0,$$
in the range $x \geqq 0$. Which of these solutions do we obtain by Picard's method if we start from $y_0 = 0$?
7. Find all solutions of the initial value problem $y' = 2\sqrt{y}$, $y(1) = 0$. Which of them do we obtain by Picard's method if we start from $y_0 = 0$?
8. Show that $f(x, y) = |\sin y| + x$ satisfies a Lipschitz condition (4), with $M = 1$, over the whole xy-plane, but $\partial f / \partial y$ does not exist when $y = 0$.
9. Does $f(x, y) = |x| + |y|$ satisfy a Lipschitz condition in the xy-plane? Does $\partial f / \partial y$ exist?
10. Does $2\sqrt{y}$ in Prob. 7 satisfy a Lipschitz condition?

CHAPTER 2

Ordinary Linear Differential Equations

The ordinary differential equations may be divided into two large classes, the so-called linear equations and the nonlinear equations. While nonlinear equations (of the second and higher orders) are rather difficult, linear equations are much simpler in many respects, because various properties of their solutions can be characterized in a general way, and standard methods are available for solving many of these equations. In the present chapter we shall consider linear differential equations and their applications. These equations play an important role in engineering mathematics, for example, in connection with mechanical vibrations and electric circuits and networks.

Much space will be devoted to linear equations of the second order. This seems to be justified by the following facts. First of all, second-order equations are the most important ones from the practical point of view. Furthermore, the necessary theoretical considerations will become easier if we first concentrate upon second-order equations. Once the student understands what is needed to handle second-order equations, he will easily become familiar with the generalizations of the concepts, methods, and results to differential equations of the third and higher orders.

Important engineering applications in connection with mechanical and electrical oscillations will be considered in Secs. 2.6 and 2.13–2.15.

Numerical methods for solving second-order differential equations are included in Sec. 18.7, which is independent of the other sections in Chap. 18.

(Legendre's, Bessel's, and the hypergeometric equations will be considered in Chap. 3.)

Prerequisite for this chapter: Chap. 1, in particular Secs. 1.7 and 1.8.
Sections which may be omitted in a shorter course: 2.8–2.10, 2.15, 2.16.
References: Appendix 1, Part B.
Answers to problems: Appendix 2.

2.1 Homogeneous Linear Equations of the Second Order

The student has already met with linear differential equations of the first order (Secs. 1.7–1.9), and we shall now define and consider linear differential equations of the second order.

A second-order differential equation is said to be **linear** if it can be written in the form

(1) $$y'' + f(x)y' + g(x)y = r(x).$$

The characteristic feature of this equation is that it is linear in *y and its derivatives*, while *f*, *g*, and *r* may be any given functions of *x*. Any differential equation of the second order which cannot be written in the form (1) is said to be **nonlinear.** For example

(2) $$y'' + 4y = e^{-x} \sin x$$

and

(3) $$x^2 y'' + xy' + (x^2 - 1)y = 0$$

are linear differential equations, while

$$y'' y + y' = 0$$

and

$$y'' + \sqrt{y} = 0$$

are nonlinear differential equations.

Linear differential equations of the second order play a basic role in many different engineering problems. We shall see that some of these equations are very simple because their solutions are elementary functions. Others are more complicated, their solutions being important higher functions, such as Bessel and hypergeometric functions.

Our further discussion will show that equation (1) has somewhat simpler properties if the function $r(x)$ on the right-hand side is identically zero. Then equation (1) becomes

(4) $$y'' + f(x)y' + g(x)y = 0$$

and is said to be **homogeneous.** If $r(x) \not\equiv 0$, equation (1) is said to be **nonhomogeneous.** These notions are quite similar to those in the case of linear equations of the first order in Sec. 1.7.

For example, (2) is nonhomogeneous and (3) is homogeneous.

The functions *f* and *g* in (1) and (4) are called the **coefficients** of the equations.

In all our considerations x will be assumed to vary in some arbitrary fixed range, for example, in some finite interval or on the entire x-axis. All the assumptions and statements will refer to such a fixed range, which need not be specified in each case.

A function

$$y = \phi(x)$$

is called a **solution** of a (linear or nonlinear) differential equation of the second order *on some interval* (perhaps infinite), if $\phi(x)$ is defined and twice differentiable throughout that interval and is such that the equation becomes an identity when we replace the unspecified function y and its derivatives in the equation by ϕ and its derivatives.

This definition is analogous to that in the case of first-order differential equations in Sec. 1.1.

Example 1. The functions $y = \cos x$ and $y = \sin x$ are solutions of the homogeneous linear differential equation

$$y'' + y = 0$$

for all x, since

$$(\cos x)'' + \cos x = -\cos x + \cos x = 0$$

and similarly for $y = \sin x$. We can even go an important step further. If we multiply the first solution by a constant, for example, by 3, the resulting function $y = 3 \cos x$ is also a solution because

$$(3 \cos x)'' + 3 \cos x = 3[(\cos x)'' + \cos x] = 0.$$

It is clear that instead of 3 we may choose any other constant, for example, -5 or $\frac{2}{9}$. We may even multiply $\cos x$ and $\sin x$ by different constants, say, by 2 and -8, respectively, and add the resulting functions, having

$$y = 2 \cos x - 8 \sin x,$$

and this function is also a solution of our homogeneous equation for all x, because

$$(2 \cos x - 8 \sin x)'' + 2 \cos x - 8 \sin x = 2[(\cos x)'' + \cos x] - 8[(\sin x)'' + \sin x] = 0.$$

This example illustrates the very important fact that we may obtain new solutions of the homogeneous linear equation (4) from known solutions by multiplication by constants and by addition. Of course, this is a great practical and theoretical advantage, because this property will enable us to generate more complicated solutions from simple solutions. We may characterize this basic property as follows.

Fundamental Theorem 1. *If a solution of the* **homogeneous linear** *differential equation* (4) *on some interval J is multiplied by any constant, the resulting function is also a solution of* (4) *on J. The sum of two solutions of* (4) *on J is also a solution of* (4) *on that interval.*[1]

Proof. We assume that $\phi(x)$ is a solution of (4) on some interval J and show that then $y = c\phi(x)$ is also a solution of (4) on J. If we substitute $y = c\phi(x)$ into (4), the left-hand side of (4) becomes

$$(c\phi)'' + f(c\phi)' + gc\phi = c[\phi'' + f\phi' + g\phi].$$

Since ϕ satisfies (4), the expression in brackets is zero, and the first statement of the theorem is proved. The proof of the last statement is quite simple, too, and is left to the reader.

---

[1] This theorem is sometimes called the **superposition principle** or *linearity principle*.

The student should know this highly important theorem as well as he knows his multiplication table, but he should not forget that *the theorem* **does not hold** *for nonhomogeneous linear equations or nonlinear equations*, as may be illustrated by the following two examples.

Example 2. Substitution shows that the functions

$$y = 1 + \cos x \quad \text{and} \quad y = 1 + \sin x$$

are solutions of the nonhomogeneous linear equation

$$y'' + y = 1,$$

but the functions

$$2(1 + \cos x) \quad \text{and} \quad (1 + \cos x) + (1 + \sin x)$$

are not solutions of that equation.

Example 3. Substitution shows that the functions

$$y = x^2 \quad \text{and} \quad y = 1$$

are solutions of the nonlinear differential equation

$$y''y - xy' = 0,$$

but the functions

$$-x^2 \quad \text{and} \quad x^2 + 1$$

are not solutions of that equation.

Problems

Important general properties of linear differential equations. Prove and illustrate with examples:

1. The sum $y = y_1 + y_2$ of a solution y_1 of (1) and a solution y_2 of (4) is a solution of (1).
2. The difference of two solutions of (1) is a solution of (4).
3. Equation (4) [but not (1)] has the *trivial solution* $y \equiv 0$.
4. The sum of two solutions of (1) is **not** a solution of (1) but satisfies $y'' + fy' + gy = 2r$.
5. A multiple $y = cy_1$ $(c \neq 1)$ of a solution y_1 of (1) is **not** a solution of (1) but satisfies $y'' + fy' + gy = cr$.

Second-order differential equations reducible to the first order. Problems 6–19 illustrate that certain second-order differential equations can be reduced to the first order.

6. If in a second-order equation the dependent variable y does not appear explicitly, the equation is of the form $F(x, y', y'') = 0$. Show that by setting $y' = z$ we obtain a first-order differential equation in z and from its solution the solution of the original equation by integration.

Reduce to the first order and solve:

7. $y'' = y'$	8. $xy'' = 2y'$	9. $y'' + 9y' = 0$
10. $y'' = 2y' \coth 2x$	11. $xy'' + y' = 0$	12. $xy'' + y' = y'^2$

13. If in a second-order differential equation the independent variable x does not appear explicitly, the equation is of the form $F(y, y', y'') = 0$. Show that this equation can be reduced to a first-order equation in $z = y'$ in which y is the independent variable. *Hint.* Using the chain rule of differentiation, show that $y'' = (dz/dy) z$.

Reduce to the first order and solve:

14. $yy'' + y'^2 = 0$	15. $y'' + y'^2 = 0$	16. $yy'' - y'^2 = 0$
17. $yy'' + y'^2 = y'$	18. $y'' = 2y'^2/y$	19. $y'' + (1 + y^{-1})y'^2 = 0$

20. A particle moves on a straight line so that the product of its velocity and acceleration is constant, say, 2 ft²/sec³. At the time $t = 0$ its displacement from the origin is 5 ft and its velocity is zero. Find its position and velocity when $t = 9$ sec.

21. A particle moves on a straight line so that its acceleration is equal to three times its velocity. At $t = 0$ its displacement from the origin is 1 ft and its velocity is 1.5 ft/sec. Find the time when the displacement is 10 ft.

22. (Catenary) Find the solution of $y'' = k\sqrt{1 + y'^2}$ satisfying the *"boundary conditions"* $y(-2) = 0$, $y(2) = 0$. (The curve to be obtained is called a *catenary*, because it is the curve in which a chain—or an inextensible flexible homogeneous cable—hangs under its own weight, where k is a constant depending on the weight; in fact, the above differential equation can be obtained by considering the forces acting on a small portion of the chain. *"Boundary conditions"* in general will be discussed in Sec. 3.8 and *"initial conditions"* in Secs. 2.8 and 2.9.)

23. Find the curve in the xy-plane which passes through the point $(1, 1)$, intersects the line $y = x$ at a right angle, and satisfies $y''x + 2y' = 0$.

24. Find the curve $y(x)$ which satisfies $y'' + y' = 0$, intersects the y-axis at $y = 1$, and approaches the x-axis as $x \to \infty$.

25. Find the curve $y(x)$ through $(0, 0)$ for which $y'' = 12\sqrt{y}$ and the tangent at $(0, 0)$ is the x-axis.

2.2 *Homogeneous Second-Order Equations with Constant Coefficients*

We shall now consider linear differential equations of the second order whose coefficients are constant. These equations have important engineering applications, especially in connection with mechanical and electrical vibrations, as we shall see in Secs. 2.6, 2.13, and 2.14.

We start with homogeneous equations, that is, equations of the form

(1) $$y'' + ay' + by = 0$$

where a and b are constants. *We assume that a and b are real* and the range of x considered is the entire x-axis.

How to solve equation (1)? We remember that the solution of the *first-order* homogeneous linear equation with constant coefficients

$$y' + ky = 0$$

is an exponential function, namely,

$$y = ce^{-kx}.$$

It seems natural to conjecture that

(2) $$y = e^{\lambda x}$$

may be a solution of (1) if λ is properly chosen. Substituting (2) and its derivatives

$$y' = \lambda e^{\lambda x} \quad \text{and} \quad y'' = \lambda^2 e^{\lambda x}$$

into our equation (1), we obtain

$$(\lambda^2 + a\lambda + b)e^{\lambda x} = 0.$$

So (2) is a solution of (1), if λ is a solution of the quadratic equation

(3) $$\lambda^2 + a\lambda + b = 0.$$

This equation is called the **characteristic equation** (or *auxiliary equation*) of (1). Its roots are

(4) $$\lambda_1 = \tfrac{1}{2}(-a + \sqrt{a^2 - 4b}), \qquad \lambda_2 = \tfrac{1}{2}(-a - \sqrt{a^2 - 4b}).$$

From our derivation it follows that the functions

(5) $$y_1 = e^{\lambda_1 x} \quad \text{and} \quad y_2 = e^{\lambda_2 x}$$

are solutions of (1). The student may check this result by substituting (5) into (1).

From elementary algebra we know that, since a and b are real, the characteristic equation may have

> **(Case I)** *two distinct real roots,*
>
> **(Case II)** *two complex conjugate roots, or*
>
> **(Case III)** *a real double root.*

These cases will be discussed separately. For the time being let us illustrate each case by a simple example.

Example 1. Find solutions of the equation

$$y'' + y' - 2y = 0.$$

The characteristic equation is

$$\lambda^2 + \lambda - 2 = 0.$$

The roots are 1 and -2, and we obtain the two solutions

$$y_1 = e^x \quad \text{and} \quad y_2 = e^{-2x}.$$

Example 2. Find solutions of the equation

$$y'' + y = 0.$$

The characteristic equation is

$$\lambda^2 + 1 = 0.$$

The roots are $i\,(= \sqrt{-1})$ and $-i$, and we obtain the two solutions

$$y_1 = e^{ix} \quad \text{and} \quad y_2 = e^{-ix}.$$

In Sec. 2.4 we shall see that it is quite simple to obtain real solutions from such complex solutions.

Example 3. Find solutions of the equation

$$y'' - 2y' + y = 0.$$

The characteristic equation

$$\lambda^2 - 2\lambda + 1 = 0$$

has the real double root 1, and we obtain at first only one solution

$$y_1 = e^x.$$

The case of a real double root will be discussed in Sec. 2.5.

Problems

Find solutions of the following differential equations.

1. $y'' - y' - 2y = 0$ 2. $y'' + 4y = 0$ 3. $y'' + 5y' = 0$

4. $y'' + 5y' + 6y = 0$ 5. $y'' + 0.25y = 0$ 6. $y'' - 2y' - 8y = 0$

7. $y'' + 16y = 0$ 8. $y'' - 4y' = 0$ 9. $y'' + 4y' + 5y = 0$

10. $y'' - 2y' + 2y = 0$ 11. $y'' + \omega^2 y = 0$ 12. $y'' - 2y' + 5y = 0$

13. Find solutions of $y'' + 3y' = 0$ (*a*) by the present method, (*b*) by reduction to the first order.

14. Show that Cases I, II, and III occur if and only if $a^2 > 4b$, $a^2 = 4b$, and $a^2 < 4b$, respectively.

Find a differential equation of the form (1) for which the following functions are solutions.

15. e^{3x}, e^{-2x} 16. e^{4ix}, e^{-4ix} 17. $e^{(2+i)x}, e^{(2-i)x}$

18. $e^{\lambda_1 x}, e^{\lambda_2 x}$ 19. $e^{-2x}, 1$ 20. $e^{-(\alpha+i\omega)x}, e^{-(\alpha-i\omega)x}$

2.3   *General Solution.*   *Fundamental System*

In our further discussion of linear equations with constant coefficients, it will be necessary to know about some important general properties of solutions of linear differential equations. For this purpose we first introduce the following two concepts.

A solution of a differential equation of the second order (linear or not) is called a **general solution** if it contains two arbitrary[2] independent constants. Here *independence* means that the solution cannot be reduced to a form containing only one arbitrary constant or none. If we assign definite values to those two constants, then the solution so obtained is called a **particular solution.**

These notions are obvious generalizations of those in the case of first-order differential equations (Sec. 1.1).

It will be shown later (in Secs. 2.8 and 2.11) that if a differential equation of the second order is linear (homogeneous or not), then a general solution includes *all* solutions of the equation; any solution containing no arbitrary constants can be obtained by assigning definite values to the arbitrary constants in that general solution (cf. also the general remark in Sec. 1.1, pp. 4–5).

In the present section we consider the homogeneous linear equation

(1) $$y'' + f(x)y' + g(x)y = 0,$$

and we want to show that a general solution of such an equation can be readily obtained if two suitable solutions y_1 and y_2 are known.

Clearly, if $y_1(x)$ and $y_2(x)$ are solutions of (1) on some interval I, then, by Fundamental Theorem 1 in Sec. 2.1,

(2) $$y(x) = c_1 y_1(x) + c_2 y_2(x)$$

---

[2] The range of the constants may have to be restricted in some cases to avoid imaginary expressions or other degeneracies.

where c_1 and c_2 are arbitrary constants, is a solution of (1) on I. Since it contains two arbitrary constants, it is a general solution of (1), provided it cannot be reduced to an expression containing less than two arbitrary constants.

Under what conditions will such a reduction be possible? To answer this question, we first introduce the following basic concepts, which are important in other considerations, too.

Two functions $y_1(x)$ and $y_2(x)$ are said to be **linearly dependent** *on an open interval I* where both functions are defined, if they are proportional on I, that is, if[3]

(3)   (a) $y_1 = ky_2$ or   (b) $y_2 = ly_1$

holds for all x on I; here k and l are numbers, zero or not. If the functions are not proportional on I, they are said to be **linearly independent** *on I*.

Note that these concepts of linear dependence and independence of functions refer to an infinite set (the points of I), but not merely to a single point.

If at least one of the functions y_1 and y_2 is identically zero on I, then the functions are linearly dependent on I. In any other case the functions are linearly dependent on I if and only if the quotient y_1/y_2 is constant on I. Hence if y_1/y_2 depends on x on I, then y_1 and y_2 are linearly independent on I.

In fact, if $y_1 \equiv 0$, then (3a) holds with $k = 0$; and if $y_2 \equiv 0$, then (3b) holds with $l = 0$. This proves the first statement, and the other statements are immediate consequences of (3), too. Of course, in the last statement, we admit that y_1/y_2 may become infinite, namely, for those x for which $y_2 = 0$.

Example 1. The functions

$$y_1 = 3x \quad \text{and} \quad y_2 = x$$

are linearly dependent on any interval, because $y_1/y_2 = 3 = const$. The functions

$$y_1 = x^2 \quad \text{and} \quad y_2 = x$$

are linearly independent on any interval because $y_1/y_2 = x \neq const$, and so are the functions

$$y_1 = x \quad \text{and} \quad y_2 = x + 1.$$

If the solutions y_1 and y_2 in (2) are linearly dependent on I, then (3a) or (3b) holds and we see that we may then reduce (2) to one of the forms

$$y = Ay_2 \qquad (A = c_1k + c_2)$$

and

$$y = By_1 \qquad (B = c_1 + lc_2)$$

containing only one arbitrary constant. Hence in this case (2) is certainly not a general solution of (1) on I.

---

[3] Note that if (3a) holds for a $k \neq 0$, we may divide by k, finding $y_2 = y_1/k$, so that in this case (3a) implies (3b) (with $l = 1/k$). However, if $y_1 \equiv 0$, then (3a) holds for $k = 0$, but does not imply (3b); in fact, in this case (3b) will not hold unless $y_2 \equiv 0$, too. (An interval I is said to be **open** if its two end points are not regarded as points belonging to I. It is said to be **closed** if the end points are regarded as points of I.)

On the other hand, if y_1 and y_2 are linearly independent on I, then they are not proportional, and no such reduction can be performed. Hence, in this case, (2) is a general solution of (1) on I.

Two linearly independent solutions of (1) on I are called a **fundamental system** or a **basis** of solutions on I. Using this notion, we may formulate our result as follows.

Theorem 1. *The solution*

$$y(x) = c_1 y_1(x) + c_2 y_2(x) \qquad (c_1, c_2 \text{ arbitrary})$$

is a general solution of the differential equation (1) *on an interval I of the x-axis if and only if the functions y_1 and y_2 constitute a fundamental system of solutions of* (1) *on I.*

y_1 *and* y_2 *constitute such a fundamental system if and only if their quotient*[4] y_1/y_2 *is not constant on I but depends on x.*

Example 2. The functions

$$y_1 = e^x \qquad \text{and} \qquad y_2 = e^{-2x}$$

considered in Ex. 1, Sec. 2.2, are solutions of

$$y'' + y' - 2y = 0.$$

Since y_1/y_2 is not constant, these solutions constitute a fundamental system, and the corresponding general solution for all x is

$$y = c_1 y_1 + c_2 y_2 = c_1 e^x + c_2 e^{-2x}.$$

Example 3. The functions

$$y_1 = e^x \qquad \text{and} \qquad y_2 = 3e^x$$

are solutions of the equation in Ex. 2, but since their quotient is constant, they do not form a fundamental system.

We are now prepared for continuing our discussion of differential equations with constant and coefficients, and shall, therefore, postpone further theoretical considerations until Sec. 2.8.

Problems

Find a general solution of the following differential equations.

1. $y'' - 4y' + 3y = 0$ 2. $y'' - 16y = 0$ 3. $y'' + 16y = 0$
4. $y'' - y' - 6y = 0$ 5. $y'' + 2y' = 0$ 6. $y'' - 2y' + 2y = 0$
7. $y'' + 4y' + 5y = 0$ 8. $y'' - y' - 12y = 0$ 9. $y'' + \omega^2 y = 0$

10. Verify that both $y = c_1 e^{2x} + c_2 e^{-2x}$ and $\tilde{y} = b_1 \cosh 2x + b_2 \sinh 2x$ are general solutions of $y'' - 4y = 0$. For what b_1, b_2 is $\tilde{y} = y$?

---

[4] Of course, here we exclude the trivial case $y_2 \equiv 0$ in which y_1 and y_2 do not constitute a fundamental system, and we also admit that y_1/y_2 may become infinite (namely, for those values of x for which $y_2 = 0$).

Find a differential equation of the form $y'' + ay' + by = 0$ for which the given functions constitute a fundamental system of solutions. Test linear independence by means of Theorem 1.

11. e^x, e^{4x} 12. e^{-3x}, e^{3x} 13. e^{-x}, e^{-2x}

14. $e^{\pi x}, 1$ 15. $e^{(1+2i)x}, e^{(1-2i)x}$ 16. $e^{(-2+i)x}, e^{(-2-i)x}$

Are the following functions linearly dependent or independent on the given interval?

17. $\cos x, \sin x$ $(-\infty < x < \infty)$ 18. e^{2x}, e^{-2x} $(-\infty < x < \infty)$

19. $e^{\lambda x}, xe^{\lambda x}$ $(-\infty < x < \infty)$ 20. $1, e^{4x}$ $(-\infty < x < \infty)$

21. x^2, x^3 $(0 < x < 1)$ 22. $x + 1, x - 1$ $(0 < x < 1)$

23. $x^2 - 4, -3x^2 + 12$ $(x > 0)$ 24. $\ln x, \ln x^2$ $(x > 0)$

25. $\ln x, (\ln x)^2$ $(x > 0)$ 26. $x^3, x^2 |x|$ $(0 < x < 1)$

27. $x^3, x^2 |x|$ $(-1 < x < 1)$ 28. $\sin 2x, \sin x \cos x$ $(x > 0)$

29. Show that linear dependence of two functions on an interval I implies linear dependence of the functions on any subinterval of I.

30. Illustrate with an example that the statement in Prob. 29 with "dependence" replaced by "independence" is not generally true.

2.4   Complex Roots of the Characteristic Equation. Initial Value Problem

In Sec. 2.2 we saw that solutions of the homogeneous linear equation with constant coefficients

$$(1) \qquad\qquad y'' + ay' + by = 0 \qquad\qquad (a, b \text{ real})$$

can be obtained by a purely algebraic procedure, namely, by determining the roots λ_1, λ_2 of the corresponding characteristic equation

$$(2) \qquad\qquad \lambda^2 + a\lambda + b = 0,$$

and these solutions are of the form

$$(3) \qquad\qquad y_1 = e^{\lambda_1 x}, \qquad y_2 = e^{\lambda_2 x}.$$

If $\lambda_1 \neq \lambda_2$, the quotient y_1/y_2 is not constant, and from Theorem 1 in Sec. 2.3 it follows that these solutions constitute a fundamental system for all x. The corresponding general solution is

$$(4) \qquad\qquad y = c_1 e^{\lambda_1 x} + c_2 e^{\lambda_2 x}.$$

If these distinct roots are real (Case I in Sec. 2.2), then the solutions (3) are real. However, if λ_1 and λ_2 are complex conjugate (Case II in Sec. 2.2), say,

$$\lambda_1 = p + iq, \qquad \lambda_2 = p - iq,$$

where p and q are real and $q \neq 0$, then the solutions (3) are complex,

$$y_1 = e^{(p+iq)x}, \qquad y_2 = e^{(p-iq)x},$$

and it is of great practical interest to obtain real solutions from these complex solutions.

This can be done by applying the *Euler formulas*[5]

$$e^{i\theta} = \cos\theta + i\sin\theta, \qquad e^{-i\theta} = \cos\theta - i\sin\theta,$$

taking $\theta = qx$. In fact, from the first Euler formula we obtain

$$y_1 = e^{(p+iq)x} = e^{px}e^{iqx} = e^{px}(\cos qx + i\sin qx),$$

and from the second Euler formula,

$$y_2 = e^{(p-iq)x} = e^{px}e^{-iqx} = e^{px}(\cos qx - i\sin qx).$$

From these two expressions for y_1 and y_2 we now derive

$$\tfrac{1}{2}(y_1 + y_2) = e^{px}\cos qx,$$

$$\frac{1}{2i}(y_1 - y_2) = e^{px}\sin qx.$$

We see that these two functions on the right-hand sides are real, and from Theorem 1 in Sec. 2.1 it follows that they are solutions of the differential equation (1). Furthermore, since their quotient is not a constant, they are linearly independent on any interval and thus constitute a fundamental system of solutions of (1) on the entire x-axis. The corresponding general solution is

(5) $$y(x) = e^{px}(A\cos qx + B\sin qx)$$

where A and B are arbitrary constants.

Example 1. Find a general solution of the equation

$$y'' - 2y' + 10y = 0.$$

The characteristic equation

$$\lambda^2 - 2\lambda + 10 = 0$$

has the roots

$$\lambda_1 = p + iq = 1 + 3i \qquad \text{and} \qquad \lambda_2 = p - iq = 1 - 3i.$$

Thus $p = 1, q = 3$, and the answer is

$$y = e^x(A\cos 3x + B\sin 3x).$$

Example 2. A general solution of the equation

$$y'' + \omega^2 y = 0 \qquad\qquad (\omega \text{ constant, not zero})$$

is

$$y = A\cos\omega x + B\sin\omega x.$$

In many practical cases it is required to find a particular solution $y(x)$ of (1) which has a given value K at a point $x = x_0$ and whose derivative has a given value L at $x = x_0$; that is,

$$y = K \qquad \text{and} \qquad y' = L \qquad \text{when} \quad x = x_0,$$

---

[5] Students not familiar with these important formulas from elementary calculus will find a proof in Sec. 11.5. Of course, once we obtained (5) we may verify directly that it satisfies (1) in the present case.

or more briefly,

(6) $$y(x_0) = K, \qquad y'(x_0) = L.$$

The equation (1), together with the two conditions (6), constitutes what is called an **initial value problem.** To solve such a problem we must find a particular solution of (1) satisfying (6). We shall see later (in Sec. 2.8) that such a problem has a unique solution; practically speaking, this means that by the use of the conditions (6) we can assign definite values to the arbitrary constants in a general solution of (1), the result being a unique particular solution satisfying (1) and (6). Let us illustrate the simple practical procedure, which is quite similar to that in the case of a differential equation of the first order.

Example 3. Solve the initial value problem

$$y'' - 2y' + 10y = 0, \qquad y(0) = 4, \qquad y'(0) = 1.$$

A fundamental system of solutions is

$$e^x \cos 3x \qquad \text{and} \qquad e^x \sin 3x$$

(cf. Ex. 1). The corresponding general solution is

$$y(x) = e^x(A \cos 3x + B \sin 3x),$$

and $y(0) = A$. By differentiation,

$$y'(x) = e^x[(A + 3B) \cos 3x + (B - 3A) \sin 3x],$$

and $y'(0) = A + 3B$. From this and the given initial conditions,

$$y(0) = A = 4,$$

$$y'(0) = A + 3B = 1.$$

Hence $A = 4$, $B = -1$, and the answer is

(7) $$y = e^x(4 \cos 3x - \sin 3x).$$

The reader may derive the same answer from another general solution (cf. Prob. 35).

Example 4. Solve the initial value problem

$$y'' + y' - 2y = 0, \qquad y(0) = 3, \qquad y'(0) = 0.$$

A general solution is

$$y(x) = c_1 e^x + c_2 e^{-2x},$$

and

$$y'(x) = c_1 e^x - 2c_2 e^{-2x}.$$

From this and the initial conditions,

$$y(0) = c_1 + c_2 = 3, \qquad y'(0) = c_1 - 2c_2 = 0.$$

Hence $c_1 = 2$, $c_2 = 1$, and the answer is

$$y = 2e^x + e^{-2x}.$$

Problems

Verify that the following functions are solutions of the given differential equations and represent them in the form (5).

1. $y = c_1 e^{ix} + c_2 e^{-ix}$, $y'' + y = 0$
2. $y = c_1 e^{3ix} + c_2 e^{-3ix}$, $y'' + 9y = 0$
3. $y = c_1 e^{(1+i)x} + c_2 e^{(1-i)x}$, $y'' - 2y' + 2y = 0$
4. $y = c_1 e^{(-1+3i)x} + c_2 e^{(-1-3i)x}$, $y'' + 2y' + 10y = 0$
5. $y = c_1 e^{(-5+2i)x} + c_2 e^{(-5-2i)x}$, $y'' + 10y' + 29y = 0$
6. $y = c_1 e^{(0.5+i\sqrt{2})x} + c_2 e^{(0.5-i\sqrt{2})x}$, $y'' - y' + 2.25y = 0$
7. $y = c_1 e^{i\omega x} + c_2 e^{-i\omega x}$, $y'' + \omega^2 y = 0$
8. $y = c_1 e^{-(\alpha - i\omega)x} + c_2 e^{-(\alpha + i\omega)x}$, $y'' + 2\alpha y' + (\alpha^2 + \omega^2)y = 0$

In each case find a real-valued general solution.

9. $y'' + 25y = 0$
10. $y'' - 25y = 0$
11. $y'' + 3y = 0$
12. $y'' - 3y = 0$
13. $y'' + y' - 2y = 0$
14. $y'' - 6y' + 10y = 0$
15. $y'' + 2y' + 10y = 0$
16. $y'' + y' + 0.5y = 0$
17. $y'' + 5y' + 12.5y = 0$
18. $y'' + \pi^2 y = 0$
19. $y'' + 2\alpha y' + (\alpha^2 + 1)y = 0$
20. $y'' + 2y' + (\omega^2 + 1)y = 0$

21. Assuming that (5) is a solution of (1), express a and b in terms of p and q.

Find differential equations of the form (1) having the following solutions. (*Hint.* Use the result of Prob. 21.)

22. $y = A \cos \pi x + B \sin \pi x$
23. $y = e^x(A \cos 2x + B \sin 2x)$
24. $y = e^{-2x}(A \cos \omega x + B \sin \omega x)$
25. $y = e^{-\alpha x}(A \cos 3x + B \sin 3x)$

In each case find a differential equation (1) for which the given function is a solution and determine A and B such that the given initial conditions are satisfied.

26. $y = A \cos 3x + B \sin 3x$, $y(0) = 4$, $y'(0) = -3$
27. $y = e^x(A \cos x + B \sin x)$, $y(0) = 1$, $y'(0) = 3$
28. $y = e^{-2x}(A \cos 0.5x + B \sin 0.5x)$, $y(0) = 0$, $y'(0) = 0.5$

Solve the following initial value problems.

29. $y'' + 2y' + 2y = 0$, $y(0) = 1$, $y'(0) = -1$
30. $y'' - 4y' + 3y = 0$, $y(0) = -1$, $y'(0) = -5$
31. $y'' + 3y = 0$, $y(0) = 2$, $y'(0) = 3\sqrt{3}$
32. $y'' + 4y' + (4 + \omega^2)y = 0$, $y(0) = 1$, $y'(0) = \omega - 2$
33. $y'' - 4y = 0$, $y(0) = 2$, $y'(0) = 0$
34. $y'' + 2\alpha y' + (\alpha^2 + \pi^2)y = 0$, $y(0) = 3$, $y'(0) = -3\alpha$

35. Consider Ex. 3 and start from another general solution, say

$$y(x) = c_1 e^{(1+3i)x} + c_2 e^{(1-3i)x}.$$

Show that from the initial conditions we obtain $c_1 = 2 + i/2$, $c_2 = 2 - i/2$ and the resulting particular solution may be simplified so that it becomes identical with (7), which illustrates that the choice of the general solution is immaterial.

2.5 Double Root
of the Characteristic Equation

To complete our discussion of the solutions of homogeneous linear differential equations with constant coefficients

(1) $$y'' + ay' + by = 0$$

we shall now consider the case when the characteristic equation

(2) $$\lambda^2 + a\lambda + b = 0$$

has a double root. This case is sometimes called the **critical case.** It arises if and only if the discriminant of (2) is zero:

$$a^2 - 4b = 0, \quad \text{and then} \quad b = \tfrac{1}{4}a^2.$$

The root is $\lambda = -a/2$, and we obtain at first only one solution

(3) $$y_1 = e^{\lambda x} \qquad\qquad \left(\lambda = -\frac{a}{2}\right).$$

To find another solution $y_2(x)$ we may apply the method of variation of parameters (cf. Sec. 1.8), starting from

$$y_2(x) = u(x)y_1(x) \qquad \text{where} \qquad y_1(x) = e^{-ax/2}.$$

In the case of a double root, the differential equation (1) has the form

$$y'' + ay' + \tfrac{1}{4}a^2 y = 0.$$

Substituting y_2 in this equation and collecting terms, we obtain

$$u(y_1'' + ay_1' + \tfrac{1}{4}a^2 y_1) + u'(2y_1' + ay_1) + u''y_1 = 0.$$

Since y_1 is a solution, the expression in the first parentheses is zero. Since

$$2y_1' = 2\left(-\frac{a}{2}\right)e^{-ax/2} = -ay_1,$$

the expression in the last parentheses is zero, and our equation reduces to $u''y_1 = 0$. Hence $u'' = 0$. A solution is $u = x$. Consequently,

(4) $$y_2(x) = xe^{\lambda x} \qquad\qquad \left(\lambda = -\frac{a}{2}\right).$$

We see that y_1 and y_2 are linearly independent, and our result may be formulated as follows.

Theorem 1 (Double root).   *In the case of a double root of (2) the functions (3) and (4) are solutions of (1). They constitute a fundamental system. The corresponding general solution is*

(5) $$y = (c_1 + c_2 x)e^{\lambda x} \qquad\qquad \left(\lambda = -\frac{a}{2}\right).$$

Note that if λ is a *simple* root of (2), then (4) is *not* a solution of (1).

Example 1. Solve

$$y'' + 8y' + 16y = 0.$$

The characteristic equation has the double root $\lambda = -4$. Hence a fundamental system of solutions is

$$e^{-4x} \quad \text{and} \quad xe^{-4x}$$

and the corresponding general solution is

$$y = (c_1 + c_2 x)e^{-4x}.$$

Example 2. Solve the initial value problem

$$y'' - 4y' + 4y = 0, \qquad y(0) = 3, \qquad y'(0) = 1.$$

A general solution of the differential equation is

$$y(x) = (c_1 + c_2 x)e^{2x}.$$

By differentiation we obtain

$$y'(x) = c_2 e^{2x} + 2(c_1 + c_2 x)e^{2x}.$$

From this and the initial conditions it follows that

$$y(0) = c_1 = 3, \qquad y'(0) = c_2 + 2c_1 = 1.$$

Hence $c_1 = 3$, $c_2 = -5$, and the answer is

$$y = (3 - 5x)e^{2x}.$$

This completes the discussion of all three cases, and we may sum up our result in tabular form:

Case	Roots of (2)	Fundamental system of (1)	General solution of (1)
I	Distinct real λ_1, λ_2	$e^{\lambda_1 x}, e^{\lambda_2 x}$	$y = c_1 e^{\lambda_1 x} + c_2 e^{\lambda_2 x}$
II	Complex conjugate $\lambda_1 = p + iq, \lambda_2 = p - iq$	$e^{px} \cos qx$ $e^{px} \sin qx$	$y = e^{px}(A \cos qx + B \sin qx)$
III	Real double root $\lambda \, (= -a/2)$	$e^{\lambda x}, xe^{\lambda x}$	$y = (c_1 + c_2 x)e^{\lambda x}$

Problems

1. Verify that $y = e^{-x}$ is a solution of $y'' + 2y' + y = 0$. Using the method of variation of parameters, find another solution.
2. Verify that $y = e^{-x}$ is a solution of $y'' - y = 0$. Using the method of variation of parameters, find another solution.
3. Verify by direct calculation that if (2) has a double root, then (4) is a solution of (1).
4. Verify by direct calculation that if (2) does not have a double root, then (4) is **not** a solution of (1).

Find a general solution of the following differential equations.

5. $y'' - 8y' + 16y = 0$

6. $y'' + 6y' + 9y = 0$

7. $y'' + 9y = 0$

8. $y'' + 4y' + 4y = 0$

9. $y'' + y' + 0.25y = 0$

10. $y'' + 2\sqrt{\pi}\, y' + \pi y = 0$

11. $y'' - 5y' + 6.25y = 0$

12. $y'' - 2\sqrt{3}\, y' + 3y = 0$

13. $y'' + 2y' + 4y = 0$

14. $y'' + 2\alpha y' + \alpha^2 y = 0$

Find a differential equation of the form (1) for which the given functions constitute a fundamental system of solutions.

15. e^{-4x}, xe^{-4x}

16. $1, e^{-4x}$

17. $e^{-\alpha x}, xe^{-\alpha x}$

18. $e^{-x}\cos \sqrt{5}\, x, e^{-x}\sin \sqrt{5}\, x$

19. $\cos 0.5x, \sin 0.5x$

20. $\cosh 2x, \sinh 2x$

Solve the following initial value problems.

21. $y'' - 4y' + 4y = 0$, $y(0) = 0$, $y'(0) = -3$

22. $y'' + 2y' + y = 0$, $y(0) = 1$, $y'(0) = 0$

23. $y'' - 2ky' + k^2y = 0$, $y(0) = \sqrt{2}$, $y'(0) = k\sqrt{2}$

24. $y'' + 6y' + 9y = 0$, $y(0) = -4$, $y'(0) = 14$

25. $y'' - 2y' + 5y = 0$, $y(0) = 1$, $y'(0) = 1$

2.6   Free Oscillations

We shall now consider an important engineering application of linear homo-geneous differential equations with constant coefficients. The problem will be taken from mechanics, but we shall see later that it has a complete analogue in electric circuits. In our discussion we shall gain some more experience in setting up mathematical models of physical situations. We shall also see that the three cases (I–III) in Sec. 2.2 correspond to three different types of motions.

We take an ordinary spring which resists compression as well as extension and suspend it vertically from a fixed support (Fig. 25). At the lower end of the spring we attach a body of mass m. We assume m to be so large that we may disregard the mass of the spring. If we pull the body down a certain distance and then release it, it undergoes a motion. We assume that the body moves strictly vertically.

Fig. 25. Mechanical system under consideration

We want to determine the motion of our mechanical system. For this purpose we consider the forces[6] acting on the body during the motion. This will lead to a differential equation, and by solving this equation we shall then obtain the displacement as a function of the time.

We may choose the downward direction as the positive direction and regard forces which act downward as positive and upward forces as negative.

The most obvious force acting on the body is the *attraction of gravity*

(1) $$F_1 = mg$$

where m is the mass of the body and g ($= 980$ cm/sec^2) is the acceleration of gravity.

The next force to be considered is the *spring force* exerted by the spring if the spring is stretched. Experiments show that this force is proportional to the stretch, say,

$$F = ks$$ (**Hooke's[7] law**)

where s is the stretch; the constant of proportionality k is called the *spring modulus*. If $s = 1$, then $F = k$. The stiffer the spring is, the larger is k.

When the body is at rest we describe its position as the *static equilibrium position*. Clearly in this position the spring is stretched by an amount s_0 such that the resultant of the corresponding spring force and the gravitational force (1) is zero. Hence this spring force acts upward and its magnitude ks_0 is equal to that of F_1; that is,

(2) $$ks_0 = mg.$$

Let $y = y(t)$ denote the displacement of the body from the static equilibrium position, with the positive direction downward (Fig. 25).

From Hooke's law it follows that the spring force corresponding to a displacement y is

(3) $$F_2 = -ks_0 - ky,$$

the resultant of the spring force $-ks_0$ when the body is in static equilibrium position and the additional spring force $-ky$ caused by the displacement. Note that

[6] The most important systems of units are as follows.

System of Units	Length	Mass	Time	Force
Cgs system	centimeter (cm)	gram (gm)	second (sec)	dyne
Mks system	meter	kilogram (kg)	second (sec)	newton (nt)
Engineering system	foot (ft)	slug	second (sec)	pound (lb)

1 ft = 30.4800 cm = 0.304800 meter,   1 slug = 14,594 g = 14.594 kg,
1 lb = 444,822 dynes = 4.44822 nt
For further details see, e.g., D. Halliday and R. Resnick, *Physics*. New York: Wiley, 1966.

[7] ROBERT HOOKE (1635–1703), English physicist.

the sign of the last term in (3) is properly chosen, because when y is positive, $-ky$ is negative and, according to our agreement, represents an upward force, while for negative y the force $-ky$ represents a downward force.

The resultant of the forces F_1 and F_2 given by (1) and (3) is

$$F_1 + F_2 = mg - ks_0 - ky,$$

and because of (2) this becomes

(4) $$F_1 + F_2 = -ky.$$

Undamped system. If the damping of the system is so small that it can be disregarded, then (4) is the resultant of all the forces acting on the body. The differential equation will now be obtained by the use of **Newton's second law**

$$\text{Mass} \times \text{Acceleration} = \text{Force}$$

where *force* means the resultant of the forces acting on the body at any instant. In our case, the acceleration is $\ddot{y} = d^2y/dt^2$ and that resultant is given by (4). Thus

$$m\ddot{y} = -ky.$$

Hence the motion of our system is governed by the linear differential equation with constant coefficients

(5) $$m\ddot{y} + ky = 0.$$

The general solution is

(6) $$y(t) = A \cos \omega_0 t + B \sin \omega_0 t \qquad \text{where } \omega_0 = \sqrt{k/m}.$$

The corresponding motion is called a **harmonic oscillation.** Figure 26 shows typical forms of (6) corresponding to various initial conditions.

By applying the addition formula for the cosine, the student may verify that (6) can be written [cf. also (13) in Appendix 3]

(6*) $$y(t) = C \cos (\omega_0 t - \delta) \qquad \left(C = \sqrt{A^2 + B^2}, \quad \tan \delta = \frac{B}{A} \right).$$

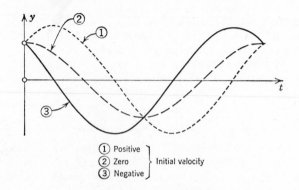

① Positive ⎫
② Zero   ⎬ Initial velocity
③ Negative ⎭

Fig. 26. Harmonic oscillations

k Spring

m Body

c Dashpot

Fig. 27. Damped system

Since the period of the trigonometric functions in (6) is $2\pi/\omega_0$, the body executes $\omega_0/2\pi$ cycles per second. The quantity $\omega_0/2\pi$ is called the **frequency** of the oscillations. The larger k is, the higher is the frequency, in agreement with our intuition that the stiffer the spring is, the more rapidly the body oscillates.

Damped system. If we connect the mass to a dashpot (Fig. 27), then we have to take the corresponding viscous damping into account. The corresponding damping force has the direction opposite to the instantaneous motion, and we shall assume that it is proportional to the velocity $\dot{y} = dy/dt$ of the body. This is generally a good approximation, at least for small velocities. Thus the damping force is of the form

$$F_3 = -c\dot{y}.$$

Let us show that the *damping constant c is positive.* If $\dot{y}$ is positive, the body moves downward (in the positive y-direction) and $-c\dot{y}$ must be an upward force, that is, by agreement, $-c\dot{y} < 0$, which implies $c > 0$. For negative $\dot{y}$ the body moves upward and $-c\dot{y}$ must represent a downward force, that is, $-c\dot{y} > 0$ which implies $c > 0$.

The resultant of the forces acting on the body is now [cf. (4)]

$$F_1 + F_2 + F_3 = -ky - c\dot{y}.$$

Hence, by Newton's second law,

$$m\ddot{y} = -ky - c\dot{y},$$

and we see that the motion of the damped mechanical system is governed by the linear differential equation with constant coefficients

(7) $$m\ddot{y} + c\dot{y} + ky = 0.$$

The corresponding characteristic equation is

$$\lambda^2 + \frac{c}{m}\lambda + \frac{k}{m} = 0.$$

The roots are

$$\lambda_{1,2} = -\frac{c}{2m} \pm \frac{1}{2m}\sqrt{c^2 - 4mk}.$$

(a)

① Positive ⎤
② Zero   ⎬ Initial velocity
③ Negative ⎦

(b)

Fig. 28. Typical motions in the overdamped case. (*a*) Positive initial displacement. (*b*) Negative initial displacement

Using the abbreviated notations

(8) $$\alpha = \frac{c}{2m} \quad \text{and} \quad \beta = \frac{1}{2m}\sqrt{c^2 - 4mk},$$

we can write

$$\lambda_1 = -\alpha + \beta \quad \text{and} \quad \lambda_2 = -\alpha - \beta.$$

The form of the solution of (7) will depend on the damping, and, as in Sec. 2.2, we may now distinguish between the following three cases:

> *Case I.* $c^2 > 4mk$.   *Distinct real roots* λ_1, λ_2. *(Overdamping.)*
> *Case II.* $c^2 < 4mk$.   *Complex conjugate roots.* *(Underdamping.)*
> *Case III.* $c^2 = 4mk$.   *A real double root.* *(Critical damping.)*

Let us discuss these three cases separately.

Case I. Overdamping.   When the damping constant c is so large that $c^2 > 4mk$, then λ_1 and λ_2 are distinct real roots, the general solution of (7) is

(9) $$y(t) = c_1 e^{-(\alpha-\beta)t} + c_2 e^{-(\alpha+\beta)t},$$

and we see that in this case the body does not oscillate. For $t > 0$ both exponents in (9) are negative because $\alpha > 0$, $\beta > 0$, and $\beta^2 = \alpha^2 - k/m < \alpha^2$. Hence both terms in (9) approach zero as t approaches infinity. Practically speaking, after a sufficiently long time the mass will be at rest in the static equilibrium position ($y = 0$). Figure 28 shows (9) for some typical initial conditions.

Case II. Underdamping. This is the most interesting case. If the damping constant c is so small that $c^2 < 4mk$, then β in (8) is pure imaginary, say

$$(10) \qquad \beta = i\omega^* \qquad \text{where} \qquad \omega^* = \frac{1}{2m}\sqrt{4mk - c^2} \qquad (>0).$$

(We write ω^* to reserve ω for Sec. 2.13.) The roots of the characteristic equation are complex conjugate,

$$\lambda_1 = -\alpha + i\omega^*, \qquad \lambda_2 = -\alpha - i\omega^*,$$

and the general solution is

$$(11) \qquad y(t) = e^{-\alpha t}(A\cos\omega^* t + B\sin\omega^* t) = Ce^{-\alpha t}\cos(\omega^* t - \delta)$$

where $C^2 = A^2 + B^2$ and $\tan\delta = B/A$ [as in (6*)].

This solution represents damped oscillations. Since $\cos(\omega^* t - \delta)$ varies between -1 and 1, the curve of the solution lies between the curves $y = Ce^{-\alpha t}$ and $y = -Ce^{-\alpha t}$ in Fig. 29, touching these curves when $\omega^* t - \delta$ is an integral multiple of π.

The frequency is $\omega^*/2\pi$ cycles per second. From (10) we see that the smaller c (>0) is, the larger is ω^* and the more rapid the oscillations become. As c

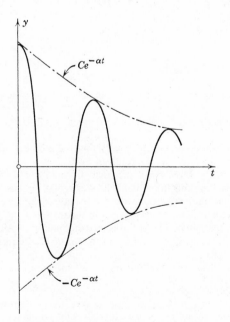

Fig. 29. Damped oscillation in Case II

(1) Positive ⎫
(2) Zero   ⎬ Initial velocity
(3) Negative ⎭

Fig. 30. Critical damping

approaches zero, ω^* approaches the value $\omega_0 = \sqrt{k/m}$ corresponding to the harmonic oscillation (6).

Case III. Critical damping. If $c^2 = 4mk$, then $\beta = 0$, $\lambda_1 = \lambda_2 = -\alpha$, and the general solution is

(12) $$y(t) = (c_1 t + c_2)e^{-\alpha t}.$$

Since the exponential function is never zero and $c_1 t + c_2$ can have at most one positive zero, it follows that the motion can have at most one passage through the equilibrium position ($y = 0$). If the initial conditions are such that c_1 and c_2 have the same sign, there is no such passage at all. This is similar to Case I. Figure 30 shows typical forms of (12).

Problems

1. How does the frequency of the harmonic oscillation depend on the mass? On the initial conditions?
2. Determine the harmonic oscillation (6) that starts with initial displacement y_0 and initial velocity v_0.
3. Find the harmonic oscillation (6*) satisfying the initial conditions $y(0) = y_0$, $\dot{y}(0) = v_0$. (a) Proceed directly; (b) use the result of Prob. 2.
4. Find and graph the three harmonic oscillations (6*) with $\omega_0 = 1$ for which the initial displacement is 1 in. and the initial velocities are -1 in./sec, 0, and 1 in./sec. What are the displacements and velocities when $t = \pi$ and $t = 2\pi$?
5. Represent the three functions in Prob. 4 in the form (6).
6. Graph the harmonic oscillations (6) corresponding to initial displacement 1, initial velocity 0, mass 1, and various values of the spring modulus, say $k = 0.25, 1, 4, 9$. What are the frequencies?
7. Determine and graph the motions $y(t)$ of the mechanical system described by (7) corresponding to initial displacement 1, initial velocity 0, mass 1, spring modulus 1, and various values of the damping constant, say, $c = 0, 1, 2, 10$.

Show that the **overdamped motion** (9) has the following properties.

8. The mass cannot pass through $y = 0$ more than once.

9. If the initial conditions are such that c_1 and c_2 have the same sign, the mass does not pass through $y = 0$.
10. If the mass starts from $y = 0$, it cannot pass through $y = 0$ again.
11. To the initial displacement y_0 and initial velocity v_0 there correspond the values $c_1 = [(1 + \alpha/\beta)y_0 + v_0/\beta]/2$ and $c_2 = [(1 - \alpha/\beta)y_0 - v_0/\beta]/2$ in (9).

Show that the **underdamped motion** (11) has the following properties.

12. To the initial displacement y_0 and initial velocity v_0 there corresponds $A = y_0$ and $B = (\alpha y_0 + v_0)/\omega^*$ in (11).
13. The frequency is independent of the initial conditions but depends on the damping and decreases as c increases.
14. The maxima and minima occur at equidistant values of t, the distance between two consecutive maxima being $2\pi/\omega^*$.
15. The ratio of two consecutive maximum amplitudes is constant, the natural logarithm of the ratio being $\Delta = 2\pi\alpha/\omega^*$. The number Δ is called the **logarithmic decrement** of the oscillation.
16. Find the logarithmic decrement of the oscillation $y = e^{-t} \cos t$. Find the values of t corresponding to the maxima and minima. To check your result, graph $y(t)$.
17. How does the position of the maxima and minima of the oscillation $y = e^{-\alpha t} \cos \omega^* t$ depend on the damping?
18. Determine and graph the three damped oscillations of the form

$$y = e^{-t}(A \cos t + B \sin t) = Ce^{-t} \cos (t - \delta)$$

corresponding to the initial displacement 1 and initial velocities $-1, 0, 1$. Compare the motions with those in Prob. 4.

Show that the **critically damped motion** (12) has the following properties.

19. To the initial displacement y_0 and initial velocity v_0 there corresponds the solution $y = [(v_0 + \alpha y_0)t + y_0]e^{-\alpha t}$.
20. A maximum or minimum at some instant $t > 0$ occurs if and only if y_0 and v_0 have the same sign.

2.7 *Cauchy Equation*

The so-called **Cauchy equation** or **Euler equation**[8]

(1) $$x^2 y'' + axy' + by = 0 \qquad (a, b \text{ constant})$$

can also be solved by purely algebraic manipulations. By substituting

(2) $$y = x^m$$

and its derivatives into the equation (1) we find

$$x^2 m(m - 1)x^{m-2} + axmx^{m-1} + bx^m = 0.$$

By omitting the common power x^m, which is not zero when $x \neq 0$, we obtain the

---

[8] LEONHARD EULER (1707–1783) was an enormously creative Swiss mathematician. He contributed to almost all branches of mathematics and its application to physical problems. His important books on algebra and calculus include numerous results of his own research work. The great French mathematician, AUGUSTIN LOUIS CAUCHY (1789–1857), is the father of modern analysis. He exercised a great influence on the theory of infinite series, complex analysis, and differential equations.

auxiliary equation

(3) $$m^2 + (a - 1)m + b = 0.$$

If the roots m_1 and m_2 of this equation are different, then the two functions

$$y_1(x) = x^{m_1} \quad \text{and} \quad y_2(x) = x^{m_2}$$

constitute a fundamental system of solutions of (1) for all x for which these functions are defined. The corresponding general solution is

(4) $$y = c_1 x^{m_1} + c_2 x^{m_2} \qquad (c_1, c_2 \text{ arbitrary}).$$

Example 1.   Solve

$$x^2 y'' - 1.5xy' - 1.5y = 0.$$

The auxiliary equation is

$$m^2 - 2.5m - 1.5 = 0.$$

The roots are $m_1 = -0.5$ and $m_2 = 3$. Hence a fundamental system of real solutions for all positive x is

$$y_1 = \frac{1}{\sqrt{x}}, \qquad y_2 = x^3$$

and the corresponding general solution for all those x is

$$y = \frac{c_1}{\sqrt{x}} + c_2 x^3.$$

The auxiliary equation (3) has a *double root* $m_1 = m_2$ if, and only if,

$$b = \tfrac{1}{4}(1 - a)^2, \quad \text{and then} \quad m_1 = m_2 = \frac{1 - a}{2}.$$

In this *critical case* we may obtain a second solution by applying the method of variation of parameters. The procedure is similar to that in Sec. 2.5 and yields (cf. Prob. 14)

$$y_2 = uy_1 = (\ln x)y_1.$$

Thus, writing m in place of m_1, we see that

(5) $$y_1 = x^m \quad \text{and} \quad y_2 = x^m \ln x \qquad \left(m = \frac{1 - a}{2} \right)$$

are solutions of (1) in the case of a double root m of (3). Since these solutions are linearly independent, they constitute a fundamental system of real solutions of (1) for all positive x, and the corresponding general solution is

(6) $$y = (c_1 + c_2 \ln x) x^m, \qquad (c_1, c_2 \text{ arbitrary}).$$

Example 2.   Solve

$$x^2 y'' - 3xy' + 4y = 0.$$

The auxiliary equation has the double root $m = 2$. Hence a fundamental system of real solutions for all positive x is x^2, $x^2 \ln x$, and the corresponding general solution is

$$y = (c_1 + c_2 \ln x)x^2.$$

Problems

1. Verify directly by substitution that y_2 in (5) is a solution of (1) if (3) has a double root, but $x^{m_1} \ln x$ and $x^{m_2} \ln x$ are **not** solutions of (5) if the roots m_1 and m_2 of (3) are different.

Find a general solution of the following equations.

2. $x^2 y'' - 2xy' + 2y = 0$ **3.** $x^2 y'' + xy' - 4y = 0$
4. $x^2 y'' - 2xy' + 1.25y = 0$ **5.** $x^2 y'' + 3xy' + y = 0$
6. $xy'' + 4y' = 0$ **7.** $x^2 y'' - 6y = 0$
8. $x^2 y'' + 0.25y = 0$ **9.** $x^2 y'' + (1 - k - k^{-1})xy' + y = 0$
10. $x^2 y'' - 5xy' + 9y = 0$ **11.** $xy'' + y' = 0$

12. (**Equations with constant coefficients**) Setting $x = e^t$ ($x > 0$), transform the Cauchy equation (1) into the equation $\ddot{y} + (a - 1)\dot{y} + by = 0$, whose coefficients are constant.

13. Show that $x = e^t$ in Prob. 12 transforms (2) into (2), Sec. 2.2, and (6) into (5), Sec. 2.5, except for notation.

14. Carry out the derivation of (5) by the method of variation of parameters.

Solve the following initial value problems.

15. $x^2 y'' - 3xy' + 3y = 0$, $y(1) = 1$, $y'(1) = -1$
16. $x^2 y'' + xy' - 0.25y = 0$, $y(4) = 2$, $y'(4) = -0.25$
17. $x^2 y'' - 4xy' + 4y = 0$, $y(1) = 4$; $y'(1) = 13$
18. $x^2 y'' - 5xy' + 8y = 0$, $y(1) = 5$, $y'(1) = 18$
19. $x^2 y'' - 3xy' + 4y = 0$, $y(1) = 1$, $y'(1) = 3$
20. $x^2 y'' - 6y = 0$, $y(1) = 4$, $y'(1) = -8$

2.8 *Existence and Uniqueness of Solutions*

An **initial value problem** *for a second-order differential equation* consists of the equation and two **initial conditions,** one for the solution $y(x)$ and one for its derivative $y'(x)$. We shall now consider initial value problems of the form

(1) $y'' + f(x)y' + g(x)y = 0$, $y(x_0) = K_1$, $y'(x_0) = K_2$

involving a second-order homogeneous linear differential equation; K_1 and K_2 are given constants. The important result will be that continuity of the coefficients f and g is sufficient for the existence and uniqueness of the solution of the problem (1).

Existence and uniqueness theorem. *If $f(x)$ and $g(x)$ are continuous functions on an open interval I and x_0 is in I, then the initial value problem (1) has a unique solution $y(x)$ on I.*

The *proof of existence* uses the same prerequisites as those of the existence theorem in Sec. 1.12 and will not be presented here; it can be found in Ref. [B8] in Appendix 1.

Proof of uniqueness.[9]   Assuming that (1) has two solutions $y_1(x)$ and $y_2(x)$ on I we show that their difference

$$y(x) = y_1(x) - y_2(x)$$

is identically zero on I; then $y_1 \equiv y_2$ on I, which implies uniqueness.

[9] This proof was suggested by my colleague, Prof. A. D. Ziebur.

Since the equation in (1) is homogeneous and linear, y is a solution of that equation on I, and since y_1 and y_2 satisfy the same initial conditions, y satisfies the conditions

(2) $$y(x_0) = 0, \qquad y'(x_0) = 0.$$

We consider the function

$$z(x) = y(x)^2 + y'(x)^2$$

and its derivative

$$z' = 2yy' + 2y'y''.$$

From the differential equation we have

$$y'' = -fy' - gy.$$

By substituting this in the expression for z' we obtain

(3) $$z' = 2yy' - 2fy'^2 - 2gyy'.$$

Now, since y and y' are real,

$$(y \pm y')^2 = y^2 \pm 2yy' + y'^2 \geqq 0,$$

and from this we immediately obtain the two inequalities

(4)   (a) $2yy' \leqq y^2 + y'^2 = z$,   (b) $-2yy' \leqq y^2 + y'^2 = z$.

For the last term in (3) we now have

$$-2gyy' \leqq |-2gyy'| = |g|\,|2yy'| \leqq |g|\,z.$$

Using this result and applying (4a) to the term $2yy'$ in (3), we find

$$z' \leqq z + 2\,|f|\,y'^2 + |g|\,z.$$

Since $y'^2 \leqq y^2 + y'^2 = z$, we obtain from this

$$z' \leqq (1 + 2\,|f| + |g|)\,z$$

or, denoting the function in parentheses by h,

(5) $$z' \leqq hz \qquad\qquad \text{for all } x \text{ on } I.$$

Similarly, from (3) and (4) it follows that

(6) $$\begin{aligned} -z' &= -2yy' + 2fy'^2 + 2gyy' \\ &\leqq z + 2\,|f|\,z + |g|\,z = hz. \end{aligned}$$

The inequalities (5) and (6) are equivalent to the inequalities

(7) $$z' - hz \leqq 0, \qquad z' + hz \geqq 0.$$

Integrating factors for the two expressions on the left are

$$F_1 = e^{-\int h(x)\,dx} \qquad \text{and} \qquad F_2 = e^{\int h(x)\,dx}.$$

The integrals in the exponents exist since h is continuous. Since F_1 and F_2 are positive, we thus obtain from (7)

$$F_1(z' - hz) = (F_1z)' \leq 0 \qquad \text{and} \qquad F_2(z' + hz) = (F_2z)' \geq 0,$$

which means that F_1z is nonincreasing and F_2z is nondecreasing on I. Since by (2), $z(x_0) = 0$ we thus obtain when $x \leq x_0$

$$F_1z \geq (F_1z)_{x_0} = 0, \qquad F_2z \leq (F_2z)_{x_0} = 0$$

and similarly, when $x \geq x_0$,

$$F_1z \leq 0, \qquad F_2z \geq 0.$$

Dividing by F_1 and F_2 and noting that these functions are positive, we altogether have

$$z \leq 0, \qquad z \geq 0 \qquad\qquad \text{for all } x \text{ on } I.$$

This implies $z = y^2 + y'^2 \equiv 0$ on I. Hence $y \equiv 0$ or $y_1 \equiv y_2$ on I, and the proof is complete.

The remaining part of the present section will be devoted to some important consequences of this theorem.

To prepare for this consideration let us return to the concepts of linear dependence and independence of functions introduced in Sec. 2.3. Let $y_1(x)$ and $y_2(x)$ be given functions defined on an open interval I and consider the relation

$$k_1 y_1(x) + k_2 y_2(x) = 0$$

on I, where k_1 and k_2 are constants to be determined. This relation certainly holds for $k_1 = 0, k_2 = 0$. It may happen that we can find a $k_1 \neq 0$ such that the relation holds on I. Then we may divide by k_1 and solve for y_1:

$$y_1 = -\frac{k_2}{k_1} y_2.$$

Similarly, if that relation holds for a $k_2 \neq 0$ on I we may solve for y_2:

$$y_2 = -\frac{k_1}{k_2} y_1.$$

We see that in both cases y_1 and y_2 are linearly dependent on I.

However, if $k_1 = 0, k_2 = 0$ is the only pair of numbers for which that relation holds on I, we cannot solve for y_1 or y_2, and the functions are linearly independent on I.

Consequently, *two functions $y_1(x)$ and $y_2(x)$ are linearly dependent on an interval I where both functions are defined if and only if we can find constants k_1, k_2, not both zero such that*

$$k_1 y_1(x) + k_2 y_2(x) = 0$$

for all x on I.

This result will be used in the subsequent proof.

In our further consideration we shall need another criterion for testing linear dependence and independence of two solutions $y_1(x)$ and $y_2(x)$ of a differential equation of the form

(8) $$y'' + f(x)y' + g(x)y = 0.$$

This criterion uses the determinant

$$W(y_1, y_2) = \begin{vmatrix} y_1 & y_2 \\ y_1' & y_2' \end{vmatrix} = y_1 y_2' - y_2 y_1',$$

which is called the *Wronski determinant*[10] or briefly the **Wronskian** of y_1 and y_2. It can be stated as follows.

Theorem 2. *Suppose that the coefficients $f(x)$ and $g(x)$ of the differential equation (8) are continuous on an open interval I. Then any two solutions of (8) on I are linearly dependent on I if and only if their Wronskian W is zero for some $x = x_0$ in I. (If $W = 0$ for $x = x_0$, then $W \equiv 0$ on I.)*

Proof. (a) Let $y_1(x)$ and $y_2(x)$ be linearly dependent solutions of (8) on *I*. Then there are constants k_1 and k_2, not both zero, such that

$$k_1 y_1 + k_2 y_2 = 0$$

holds on *I*. By differentiation we obtain

$$k_1 y_1' + k_2 y_2' = 0.$$

Suppose that $k_1 \neq 0$. Then by multiplying the first equation by y_2'/k_1 and the last by $-y_2/k_1$ and adding the resulting equations, we obtain

$$y_1 y_2' - y_2 y_1' = 0, \qquad \text{that is,} \qquad W(y_1, y_2) = 0.$$

If $k_1 = 0$, then $k_2 \neq 0$, and the argument is similar.

(b) Conversely, suppose that $W(y_1, y_2) = 0$ for some $x = x_0$ in *I*. Consider the system of linear equations

$$\begin{aligned} k_1 y_1(x_0) + k_2 y_2(x_0) &= 0 \\ k_1 y_1'(x_0) + k_2 y_2'(x_0) &= 0 \end{aligned}$$

(9)

in the unknowns k_1, k_2. Since the system is homogeneous and the determinant of the system is just the Wronskian $W[y_1(x_0), y_2(x_0)]$ and $W = 0$, the system has a solution k_1, k_2 where k_1 and k_2 are not both zero (cf. Sec. 6.6). Using these numbers k_1, k_2, we introduce the function

$$y(x) = k_1 y_1(x) + k_2 y_2(x).$$

By Fundamental Theorem 1 in Sec. 2.1 the function $y(x)$ is a solution of (8) on *I*. From (9) we see that it satisfies the initial conditions $y(x_0) = 0$, $y'(x_0) = 0$. Now another solution of (8) satisfying the same initial conditions is $y^* \equiv 0$. Since f

---

[10] Introduced by HÖNE (1778–1853), who changed his name to Wronski. Second-order determinants should be familiar to the reader from elementary calculus; otherwise he may consult Sec. 6.6, which is independent of the other sections in Chap. 6.

and g are continuous, the uniqueness theorem is applicable and we conclude that $y^* \equiv y$, that is,

$$k_1 y_1 + k_2 y_2 \equiv 0$$

on I. Since k_1 and k_2 are not both zero, this means that y_1 and y_2 are linearly dependent on I. Furthermore, from this and the first part of the proof it follows that $W(y_1, y_2) \equiv 0$ on I.

Example 1. The functions $y_1 = \cos \omega x$ and $y_2 = \sin \omega x$ are solutions of the equation

$$y'' + \omega^2 y = 0 \qquad\qquad (\omega \neq 0).$$

A simple calculation shows that their Wronskian has the value ω. Hence these functions constitute a fundamental system of solutions for all x.

Further examples are included in Secs. 2.3–2.5. The following example illustrates the fact that *the assumption of continuity of the coefficients in Theorem 2 cannot be omitted.*

Example 2. The functions

$$y_1 = x^3 \quad \text{and} \quad y_2 = \begin{cases} x^3 & \text{when } x \geq 0 \\ -x^3 & \text{when } x \leq 0 \end{cases}$$

are solutions of the differential equation

(10) $$y'' - \frac{3}{x} y' + \frac{3}{x^2} y = 0$$

for all x, as can be readily verified by substituting each of the functions and its derivatives into (10). The reader may show that the functions are linearly independent on the x-axis, although their Wronskian is identically zero. This does not contradict Theorem 2, because the coefficients of (10) are not continuous at $x = 0$.

Of course, Ex. 2 also illustrates the fact that *if u_1 and u_2 are any differentiable functions of x, the condition $W(u_1, u_2) = 0$ is only necessary for linear dependence of u_1 and u_2* (as follows from part **(a)** of the proof of Theorem 2), *but not sufficient.*

Let us now show that *an equation of the form* (8) *whose coefficients are continuous on an open interval I has a fundamental system of solutions on I, and, in fact, infinitely many such systems.*

Proof. From the existence theorem it follows that (8) has a solution $y_1(x)$ satisfying the initial conditions

$$y_1(x_0) = 1, \qquad y_1'(x_0) = 0,$$

and a solution $y_2(x)$ satisfying the initial conditions

$$y_2(x_0) = 0, \qquad y_2'(x_0) = 1,$$

where x_0 is any fixed point in I. At x_0 we have $W(y_1, y_2) = 1$. Hence, by Theorem 2, the two solutions are linearly independent on I and form a fundamental system of (8) on I. Another fundamental system is ay_1, by_2 where $a \neq 0$, $b \neq 0$ are any constants, so that there are infinitely many such systems. This completes the proof.

We shall now prove the important fact that a general solution of an equation of the form (8) with continuous coefficients includes the totality of all solutions of (8).

Theorem 3. *Let*

$$(11) \qquad\qquad y(x) = c_1 y_1(x) + c_2 y_2(x)$$

be a general solution of the differential equation (8) *on an open interval I where the coefficients f and g are continuous. Let Y(x) be any solution of* (8) *on I containing no arbitrary constants. Then Y(x) is obtained from* (11) *by assigning suitable values to the constants c_1 and c_2.*

Proof. Let $Y(x)$ be any solution of (8) on I containing no arbitrary constants, and let $x = x_0$ be any fixed point in I. We first show that we can find values of c_1 and c_2 in (11) such that

$$y(x_0) = Y(x_0), \qquad y'(x_0) = Y'(x_0)$$

written out

$$(12) \qquad \begin{aligned} c_1 y_1(x_0) + c_2 y_2(x_0) &= Y(x_0), \\ c_1 y_1{}'(x_0) + c_2 y_2{}'(x_0) &= Y'(x_0). \end{aligned}$$

In fact, this is a system of linear equations in the unknowns c_1, c_2. Its determinant is the Wronskian of y_1 and y_2 at $x = x_0$. Since (11) is a general solution, y_1 and y_2 are linearly independent on I, and from Theorem 2 it follows that their Wronskian is not zero. Hence the system has a unique solution $c_1 = c_1{}^*, c_2 = c_2{}^*$ which can be obtained by Cramer's rule (Sec. 6.6). By using these constants we obtain from (11) the particular solution

$$y^*(x) = c_1{}^* y_1(x) + c_2{}^* y_2(x).$$

From (12) we see that

$$(13) \qquad\qquad y^*(x_0) = Y(x_0), \qquad y^{*\prime}(x_0) = Y'(x_0).$$

From this and the uniqueness theorem we conclude that y^* and Y must be identical on I, and the proof is complete.

Problems

In each case find a linear homogeneous differential equation of the second order satisfied by the pair of given functions. Find the Wronskian of the functions. Using Theorem 2, show that the functions are linearly independent on any interval. (In Probs. 4 and 5, assume $x > 0$.)

1. e^x, e^{-x} **2.** $1, e^{4x}$ **3.** $\cos 2x, \sin 2x$ **4.** $x, x \ln x$

5. $\sqrt{x}, 1/\sqrt{x}$ **6.** $1, x$ **7.** $e^x \cos x, e^x \sin x$ **8.** e^{2x}, xe^{2x}

9. Show that y_1 and y_2 in Ex. 2 constitute a fundamental system of solutions of (10), but $W(y_1, y_2) \equiv 0$.

Find the Wronskian of the following important fundamental systems which appeared in the previous sections.

    10. x^{m_1}, x^{m_2} **11.** $x^m, x^m \ln x$ **12.** $e^{\lambda_1 x}, e^{\lambda_2 x}$

    13. $e^{\lambda x}, x e^{\lambda x}$ **14.** e^{iqx}, e^{-iqx} **15.** $\cos qx, \sin qx$

    16. $e^{(p+iq)x}, e^{(p-iq)x}$ **17.** $e^{px}\cos qx, e^{px}\sin qx$

    18. Let y_1, y_2 be a fundamental system of solutions of a differential equation satisfying the assumptions of Theorem 2. Show that $z_1 = a_{11}y_1 + a_{12}y_2$ and $z_2 = a_{21}y_1 + a_{22}y_2$ constitute a fundamental system of solutions of that equation if and only if the determinant of the coefficients a_{jk} is not zero.

Illustrate Prob. 18 with the following examples:

    19. $y_1 = e^x, y_2 = e^{-x}, z_1 = \cosh x, z_2 = \sinh x$

    20. $y_1 = e^{(p+iq)x}, y_2 = e^{(p-iq)x}, z_1 = e^{px}\cos qx, z_2 = e^{px}\sin qx$

2.9   *Homogeneous Linear Equations of Arbitrary Order*

A differential equation of nth order is said to be **linear** if it can be written in the form

(1) $$y^{(n)} + f_{n-1}(x)y^{(n-1)} + \cdots + f_1(x)y' + f_0(x)y = r(x)$$

where the function r on the right-hand side and the *coefficients* $f_0, f_1, \cdots, f_{n-1}$ are any given functions of x, and $y^{(n)}$ is the nth derivative of y with respect to x, etc. Any differential equation of order n which cannot be written in the form (1) is said to be **nonlinear.**

If $(x) \equiv 0$, equation (1) becomes

(2) $$y^{(n)} + f_{n-1}(x)y^{(n-1)} + \cdots + f_1(x)y' + f_0(x)y = 0$$

and is said to be **homogeneous.** If $r(x)$ is not identically zero, the equation is said to be **nonhomogeneous.**

A function $y = \phi(x)$ is called a **solution** of a (linear or nonlinear) differential equation of nth order *on an interval I*, if $\phi(x)$ is defined and n times differentiable on I and is such that the equation becomes an identity when we replace the unspecified function y and its derivatives in the equation by ϕ and its corresponding derivatives.

We shall now see that our considerations in Secs. 2.3 and 2.8 can be readily extended to linear homogeneous equations of arbitrary order n.

Existence and uniqueness theorem. *If $f_0(x), \cdots, f_{n-1}(x)$ in (2) are continuous functions on an open interval I, then the initial value problem consisting of the equation (2) and the n **initial conditions***

$$y(x_0) = K_1, \qquad y'(x_0) = K_2, \qquad \cdots, \qquad y^{(n-1)}(x_0) = K_n$$

has a unique solution $y(x)$ on I; here x_0 is any fixed point in I, and $K_1, \cdots, K_n$ are given numbers.

The proof of existence can be found in Ref. [B8] in Appendix 1, and uniqueness can be proved by a slight generalization of the proof in Sec. 2.8.

To extend the considerations of Sec. 2.3 we need the concepts of linear dependence and independence of n functions, say, $y_1(x), \cdots, y_n(x)$. These functions are said to be **linearly dependent** *on some interval I* where they are defined, if (at least) one of them can be represented on I as a "**linear combination**" of the other $n - 1$ functions, that is, as a sum of those functions, each multiplied by a constant (zero or not). If none of the functions can be represented in this way, they are said to be **linearly independent** *on I*.

This definition includes that in Sec. 2.3 for two functions as a particular case. n linearly dependent functions on I are also called a **linearly dependent set** *of functions on I*. Similarly for linearly independent functions.

For example, the functions $y_1 = x$, $y_2 = x^2$, $y_3 = x^3$ are linearly independent on any interval. The functions $y_1 = x$, $y_2 = 3x$, $y_3 = x^2$ are linearly dependent on any interval, because $y_2 = 3y_1 + 0y_3$.

n functions $y_1(x), \cdots, y_n(x)$ are linearly dependent on an interval I where they are defined if and only if we can find constants $k_1, \cdots, k_n$, not all zero, such that the relation

(3) $$k_1 y_1(x) + \cdots + k_n y_n(x) = 0$$

holds for all x on I.

In fact, if (3) holds for some $k_1 \neq 0$, we may divide by k_1 and express y_1 as a linear combination of $y_2, \cdots, y_n$:

$$y_1 = -\frac{1}{k_1}(k_2 y_2 + \cdots + k_n y_n),$$

which proves linear dependence. Similarly, if (3) holds for some $k_i \neq 0$, we may express y_i as a linear combination of the other functions. On the other hand, if $k_1 = 0, \cdots, k_n = 0$ is the only set of constants for which (3) holds on I, then we cannot solve for any of the functions in terms of the others, and the functions are linearly independent. This proves the statement.

A solution of a differential equation of the order n (linear or not) is called a **general solution** if it contains n arbitrary[11] independent constants. Here *independence* means that the solution cannot be reduced to a form containing less than n arbitrary constants. If we assign definite values to the n constants, then the solution so obtained is called a **particular solution** of that equation.

A set of n linearly independent solutions $y_1(x), \cdots, y_n(x)$ of the linear homogeneous equation (2) on I is called a **fundamental system** or a **basis** of solutions of (2) on I.

If $y_1, \cdots, y_n$ is such a fundamental system, then

(4) $$y(x) = c_1 y_1(x) + \cdots + c_n y_n(x) \qquad (c_1, \cdots, c_n \text{ arbitrary})$$

is a general solution of (2) on I. In fact, since (2) is linear and homogeneous, (4)

[11] Cf. footnote 2 in Sec. 2.3.

is a solution of (2), and since $y_1, \cdots, y_n$ are linearly independent functions on I, none of them can be expressed as a linear combination of the others, that is, (4) cannot be reduced to a form containing less than n arbitrary constants.

The test for linear dependence and independence of solutions (Theorem 2 in Sec. 2.8) can be generalized to nth order equations as follows.

Theorem 2.   *Suppose that the coefficients* $f_0(x), \cdots, f_{n-1}(x)$ *of* (2) *are continuous on an open interval* I. *Then* n *solutions* $y_1, \cdots, y_n$ *of* (2) *on* I *are linearly dependent on* I *if and only if their* **Wronskian**

$$
(5) \qquad W(y_1, \cdots, y_n) = \begin{vmatrix} y_1 & y_2 & \cdots & y_n \\ y_1' & y_2' & \cdots & y_n' \\ \cdot & \cdot & \cdots & \cdot \\ y_1^{(n-1)} & y_2^{(n-1)} & \cdots & y_n^{(n-1)} \end{vmatrix}
$$

is zero for some $x = x_0$ *in* I. *(If* $W = 0$ *at* $x = x_0$, *then* $W \equiv 0$ *on* I.)

The idea of the proof is similar to that of Theorem 2 in Sec. 2.8.

A general solution of an equation of the form (2) with continuous coefficients includes the totality of all solutions of (2). In fact, the following generalization of Theorem 3 in Sec. 2.8 holds true.

Theorem 3.   *Let* (4) *be a general solution of* (2) *on an open interval* I *where* $f_0(x)$, $\cdots, f_{n-1}(x)$ *are continuous, and let* $Y(x)$ *be any solution of* (2) *on* I *involving no arbitrary constants. Then* $Y(x)$ *is obtained from* (4) *by assigning suitable values to the arbitrary constants* $c_1, \cdots, c_n$.

The idea of the proof is similar to that of Theorem 3 in Sec. 2.8.

Example 1.   The functions $y_1 = e^{-x}$, $y_2 = e^x$, and $y_3 = e^{2x}$ are solutions of the equation

$$
(6) \qquad\qquad y''' - 2y'' - y' + 2y = 0.
$$

The Wronskian is

$$
W(e^{-x}, e^x, e^{2x}) = \begin{vmatrix} e^{-x} & e^x & e^{2x} \\ -e^{-x} & e^x & 2e^{2x} \\ e^{-x} & e^x & 4e^{2x} \end{vmatrix} = 6e^{2x} \neq 0,
$$

which shows that the functions constitute a fundamental system of solutions of (6) for all x, and the corresponding general solution is

$$
y = c_1 e^{-x} + c_2 e^x + c_3 e^{2x}.
$$

Problems

Are the following functions linearly dependent or independent on the x-axis?

1. x, x^2, x^4
2. $\cos^2 x, \sin^2 x, -1$
3. e^x, e^{-x}, e^{-2x}
4. $x, x + 1, x + 2$
5. $e^x, e^{-x}, 0$
6. $\cos^2 x, \sin^2 x, \cos 2x$
7. $e^x, xe^x, x^2 e^x$
8. $\sin x, \sin 2x, \sin 3x$
9. $x, 2x, x^2$

10. Find an interval I on which the three functions x^3, $|x|^3$, and 1 are linearly independent and a subinterval of I on which the functions are linearly dependent. What fact does this illustrate?

11. If n functions are linearly dependent on an interval I, prove that they are also linearly dependent on any subinterval of I.

12. If one of the functions of a set of functions on an interval I is a constant multiple of another function of the set on I, prove that the set is linearly dependent on I.

13. If a set of p functions is linearly dependent on an interval I, prove that a set of n ($\geq p$) functions on I containing that set is linearly dependent on I.

14. If $y \equiv 0$ is a function of a set of functions on an interval I, prove that the set is linearly dependent on I.

Show that the given functions constitute a fundamental system of solutions of the corresponding differential equation on the x-axis. (*Hint.* In Probs. 15 and 17–20, use Theorem 2.)

15. $1, x, x^2, x^3, \qquad y^{\mathrm{IV}} = 0$

16. $x, x^2, x^3, \qquad x^3 y''' - 3x^2 y'' + 6xy' - 6y = 0$

17. $e^{-x}, xe^{-x}, x^2 e^{-x}, x^3 e^{-x}, \qquad y^{\mathrm{IV}} + 4y''' + 6y'' + 4y' + y = 0$

18. $\cosh x, \sinh x, \cos x, \sin x, \qquad y^{\mathrm{IV}} - y = 0$

19. $\cos x, \sin x, \cos 2x, \sin 2x, \qquad y^{\mathrm{IV}} + 5y'' + 4y = 0$

20. $e^x, xe^x, e^{-2x}, \qquad y''' - 3y' + 2y = 0$

2.10 *Homogeneous Linear Equations of Arbitrary Order with Constant Coefficients*

The method in Sec. 2.2 may easily be extended to a linear homogeneous equation of any order n with constant coefficients

(1) $$y^{(n)} + a_{n-1}y^{(n-1)} + \cdots + a_1 y' + a_0 y = 0.$$

Substituting $y = e^{\lambda x}$ and its derivatives into (1), we obtain the *characteristic equation*

(2) $$\lambda^n + a_{n-1}\lambda^{n-1} + \cdots + a_1\lambda + a_0 = 0.$$

If this equation has n distinct roots $\lambda_1, \cdots, \lambda_n$, then the n solutions

(3) $$y_1 = e^{\lambda_1 x}, \qquad \cdots, \qquad y_n = e^{\lambda_n x}$$

constitute a fundamental system for all x, and the corresponding general solution of (1) is

(4) $$y = c_1 e^{\lambda_1 x} + \cdots + c_n e^{\lambda_n x}.$$

If a *double root* occurs, say, $\lambda_1 = \lambda_2$, then $y_1 = y_2$ in (3) and we take y_1 and $y_2 = xy_1$ as two linearly independent solutions corresponding to that root; this is as in Sec. 2.5.

If a *triple root* occurs, say, $\lambda_1 = \lambda_2 = \lambda_3$, then $y_1 = y_2 = y_3$ in (3) and three linearly independent solutions corresponding to that root are

(5) $$y_1, \qquad xy_1, \qquad x^2 y_1;$$

these solutions are obtained by the method of variation of parameters. More generally, if λ is a *root of order* m, then

(6) $$e^{\lambda x}, \qquad xe^{\lambda x}, \qquad \cdots, \qquad x^{m-1}e^{\lambda x}$$

are m linearly independent solutions of (1) corresponding to that root.

Let us illustrate these simple facts by two examples.

Example 1. Solve the differential equation

$$y''' - 2y'' - y' + 2y = 0.$$

The roots of the characteristic equation

$$\lambda^3 - 2\lambda^2 - \lambda + 2 = 0$$

are -1, 1, and 2, and the corresponding general solution (4) is

$$y = c_1 e^{-x} + c_2 e^x + c_3 e^{2x}.$$

Example 2. Solve the differential equation

$$y^{(5)} - 3y^{(4)} + 3y''' - y'' = 0.$$

The characteristic equation

$$\lambda^5 - 3\lambda^4 + 3\lambda^3 - \lambda^2 = 0$$

has the roots $\lambda_1 = \lambda_2 = 0$ and $\lambda_3 = \lambda_4 = \lambda_5 = 1$, and the answer is

(7) $$y = c_1 + c_2 x + (c_3 + c_4 x + c_5 x^2)e^x.$$

Problems

1. Show that e^{-x}, xe^{-x}, $x^2 e^{-x}$ are solutions of $y''' + 3y'' + 3y' + y = 0$. Starting from $y_1 = e^{-x}$, apply the method of variation of parameters to obtain a general solution.

2. Show that e^x is a solution of $y^{IV} - 4y''' + 6y'' - 4y' + y = 0$ and obtain a general solution by the method of variation of parameters.

Find equations of the form (1) having the following fundamental systems.

3. e^x, e^{2x}, e^{3x} **4.** $e^x, e^{-x}, \cos x, \sin x$ **5.** $1, x, x^2$

6. The functions in formula (6) with $\lambda = -1$.

Find a general solution of the following differential equations.

7. $y''' - y' = 0$ **8.** $y^{IV} - 13y'' + 36y = 0$ **9.** $y''' - y'' - y' + y = 0$

10. $y^{IV} + 4y = 0$ **11.** $y^{IV} - 16y = 0$ **12.** $y''' + 6y'' + 11y' + 6y = 0$

13. (Cauchy equation) The *Cauchy equation* or *Euler equation* of the third order is

$$x^3 y''' + ax^2 y'' + bxy' + cy = 0.$$

Generalizing the method of Sec. 2.7, show that $y = x^m$ is a solution of the equation if and only if m is a root of the auxiliary equation

$$m^3 + (a - 3)m^2 + (b - a + 2)m + c = 0.$$

14. Solve $x^3 y''' - 3x^2 y'' + 6xy' - 6y = 0$.

15. (Reduction of order) It may sometimes be relatively easy to find one solution of a differential equation. Then the order of the equation can be reduced, and this may facilitate the task of finding further solutions. To understand this, show that if $y_1(x)$ is a solution of

$$y'' + f(x)y' + g(x)y = 0,$$

another solution $y_2(x) = u(x)y_1(x)$ is obtained by solving the following first-order equation in $z = u'$:

$$y_1 u'' + (2y_1' + fy_1)u' = 0.$$

16. Show that $y_1 = e^{x^2}$ is a solution of $y'' - 4xy' + (4x^2 - 2)y = 0$ and find a second independent solution, using the method in Prob. 15.

17. (Reduction of order) Show that if $y_1(x)$ is a solution of

$$y''' + f(x)y'' + g(x)y' + h(x)y = 0,$$

then other solutions $y_2(x) = u(x)y_1(x)$ can be obtained by solving the following second-order equation in $z = u'$:

$$y_1 u''' + (3y_1' + fy_1)u'' + (3y_1'' + 2fy_1' + gy_1)u' = 0.$$

Using the method of reduction of order, solve:

18. $y''' - y'' + 4y' - 4y = 0, \qquad y_1 = e^x$

19. $y''' + 6y'' + 12y' + 8y = 0, \qquad y_1 = e^{-2x}$

20. $x^3 y''' + x^2 y'' - 2xy' + 2y = 0, \qquad y_1 = x^2$

2.11 *Nonhomogeneous Linear Equations*

So far we have considered *homogeneous* linear equations. We shall now discuss linear equations which are *nonhomogeneous*. For the sake of simplicity we concentrate on second-order equations

(1) $$y'' + f(x)y' + g(x)y = r(x),$$

but it will be obvious that the results in this section can easily be generalized to linear equations of any order.

We shall first show that a general solution $y(x)$ of (1) can be obtained in a simple way from a general solution $y_h(x)$ of the corresponding homogeneous equation

(2) $$y'' + f(x)y' + g(x)y = 0,$$

namely, by adding to y_h any solution $\tilde{y}$ of (1) involving no arbitrary constant:

(3) $$y(x) = y_h(x) + \tilde{y}(x).$$

In fact, since y_h involves two arbitrary constants, so does y and will therefore be a general solution of (1) as defined in Sec. 2.3, provided y is a solution of (1) at all. To show this, we substitute (3) into (1). Then the left-hand side of (1) becomes

$$(y_h + \tilde{y})'' + f(y_h + \tilde{y})' + g(y_h + \tilde{y}).$$

This may be written

$$(y_h'' + fy_h' + gy_h) + \tilde{y}'' + f\tilde{y}' + g\tilde{y}.$$

Since y_h is a solution of (2), the expression in the parentheses is zero. Since $\tilde{y}$ satisfies (1), the sum of the other terms equals $r(x)$, and the statement is proved.

The remaining practical question is how to obtain a solution $\tilde{y}$ of (1). Two corresponding methods will be discussed in Secs. 2.12 and 2.16.

We shall now prove that a general solution of an equation (1) with continuous functions f, g, and r represents the totality of all solutions of (1) as follows.

Theorem 1. *Suppose that $f(x)$, $g(x)$, and $r(x)$ in (1) are continuous[12] functions on an open interval I. Let $Y(x)$ be any solution of (1) on I containing no arbitrary constants. Then $Y(x)$ is obtained from (3) by assigning suitable values to the two arbitrary constants contained in the general solution $y_h(x)$ of (2). In (3), the function $\tilde{y}(x)$ is any solution of (1) on I containing no arbitrary constants.*

Proof. We set $Y - \tilde{y} = y^*$. Then

$$y^{*''} + f y^{*'} + g y^* = (Y'' + f Y' + g Y) - (\tilde{y}'' + f \tilde{y}' + g \tilde{y}) = r - r = 0,$$

that is, y^* is a solution of (2). Now y^* does not contain arbitrary constants. Hence, by Theorem 3, Sec. 2.8 (with Y replaced by y^*), it can be obtained from y_h by assigning suitable values to the arbitrary constants in y_h. From this, since $Y = y^* + \tilde{y}$, the statement follows.

This theorem shows that any solution of (1) containing no arbitrary constants is a particular solution of (1) in the sense of our definition in Sec. 2.3, and we may now formulate our results as follows.

Theorem 2. *A general solution $y(x)$ of the linear nonhomogeneous differential equation (1) is the sum of a general solution $y_h(x)$ of the corresponding homogeneous equation (2) and an arbitrary particular solution $y_p(x)$ of (1):*

(4)
$$y(x) = y_h(x) + y_p(x).$$

Problems

In each case verify that $y_p(x)$ is a solution of the given differential equation and find a general solution.

1. $y'' - y = 2 \cos x$, $y_p = -\cos x$
2. $y'' - y' - 2y = -4x$, $y_p = 2x - 1$
3. $y'' + 9y = 17e^{-5x}$, $y_p = 0.5e^{-5x}$
4. $x^2 y'' - 3xy' + 3y = 3 \ln x - 4$, $y_p = \ln x$
5. $y'' + 3y' - 4y = -6.8 \sin x$, $y_p = \sin x + 0.6 \cos x$
6. $y'' + 3y' - 4y = 10e^x$, $y_p = 2xe^x$
7. $4x^2 y'' + y = (1 - x^2) \cos 0.5x$, $y_p = \cos 0.5x$
8. $y'' - 4y' + 3y = 4e^{3x}$, $y_p = 2xe^{3x}$
9. $y'' - 6y' + 9y = 4e^{3x}$, $y_p = 2x^2 e^{3x}$
10. $y'' + 6y' + 9y = e^{-3x}/(x^2 + 1)$, $y_p = (x \arctan x - 0.5 \ln (x^2 + 1))e^{-3x}$

11. Extend Theorems 1 and 2 to a linear differential equation of any order n.
12. To illustrate that the choice of y_p in (4) is immaterial, show that $y_{p1} = -\cos x$ and $y_{p2} = e^x - \cos x$ are particular solutions of $y'' - y = 2 \cos x$, find the corresponding general solutions (4), and show that one of them can be expressed in terms of the other.
13. Show that two general solutions (4) of (1) which correspond to different y_p can be transformed into each other.

---

[12] This suffices for the existence of y_h on I, as we know, and for the existence of $\tilde{y}$ on I, as will become obvious in Sec. 2.16.

14. Show that if y_1 is a solution of (1) with $r = r_1$ and y_2 is a solution of (1) with $r = r_2$, then $y = y_1 + y_2$ is a solution of (1) with $r = r_1 + r_2$.

15. To illustrate Prob. 14, find a general solution of the following differential equation. (*Hint.* Use Probs. 5 and 6.)

$$y'' + 3y' - 4y = 5e^x - 6.8 \sin x$$

2.12   *A Method for Solving Nonhomogeneous Linear Equations*

In the last section we have shown that for obtaining a general solution of a linear nonhomogeneous equation we need a particular solution $y_p(x)$ of the equation. How can we find such a function y_p?

One method is the so-called **method of undetermined coefficients.** The advantage of this method is that it is much simpler and more economical than another general method to be discussed later (in Sec. 2.16). The disadvantage is that it is applicable only to certain linear equations. But since some of these equations are of great practical importance—for example, in connection with vibrational problems— the method is frequently used in engineering mathematics.

The method is suitable for equations with constant coefficients

(1) $$y'' + ay' + by = r(x)$$

when $r(x)$ is such that the form of a particular solution $y_p(x)$ of (1) may be guessed;[13] for example, r may be a single power of x, a polynomial, an exponential function, a sine or cosine, or a sum of such functions. The method consists in assuming for y_p an expression similar to that of $r(x)$ containing unknown coefficients which are to be determined by inserting y_p and its derivatives into (1).

Let us start with some illustrative examples.

Example 1. Solve the nonhomogeneous equation

(2) $$y'' + 4y = 12.$$

We try $y_p = K$, a constant. Then $y_p'' = 0$. By substitution, $4K = 12$. Thus $y_p = K = 3$. A general solution of the corresponding homogeneous equation is

$$y_h = A \cos 2x + B \sin 2x.$$

Hence, by Theorem 2 in the preceding section, a general solution of (2) is

$$y = y_h + y_p = A \cos 2x + B \sin 2x + 3.$$

Example 2. Solve the nonhomogeneous equation

(3) $$y'' + 4y = 8x^2.$$

We try $y_p = Kx^2$, but this attempt will fail. We have $y_p'' = 2K$. By substitution in (3) we obtain

$$2K + 4Kx^2 = 8x^2.$$

[13] The student will see in Sec. 2.16 that our present method actually results from the general method.

For this to hold for all x, the coefficients of each power of x on both sides (x^2 and x^0) must be the same:

$$4K = 8 \qquad \text{and} \qquad 2K = 0,$$

which has no solution. Let us try

$$y_p = Kx^2 + Lx + M. \qquad \text{Then} \qquad y_p{}'' = 2K.$$

Substitution in (3) yields

$$2K + 4(Kx^2 + Lx + M) = 8x^2.$$

Equating the coefficients of x^2, x, and x^0 on both sides, we have $4K = 8$, $4L = 0$, $2K + 4M = 0$. Thus $K = 2$, $L = 0$, $M = -1$. Hence $y_p = 2x^2 - 1$, and a general solution of (3) is

$$y = A \cos 2x + B \sin 2x + 2x^2 - 1.$$

Example 3. Solve the nonhomogeneous equation

(4) $$\qquad\qquad\qquad y'' - y' - 2y = 10 \cos x.$$

The student may try $y_p = K \cos x$ and see that this attempt fails. We try

$$y_p = K \cos x + M \sin x.$$

Then

$$y_p{}' = -K \sin x + M \cos x,$$
$$y_p{}'' = -K \cos x - M \sin x.$$

By inserting this into (4) we obtain

$$(-3K - M) \cos x + (K - 3M) \sin x = 10 \cos x.$$

By equating the coefficients of $\cos x$ and $\sin x$ on both sides we have

$$-3K - M = 10, \qquad K - 3M = 0.$$

Hence $K = -3$, $M = -1$, and $y_p = -3 \cos x - \sin x$. A general solution of the corresponding homogeneous equation is

$$y_h = c_1 e^{-x} + c_2 e^{2x}.$$

Altogether,

$$y = y_h + y_p = c_1 e^{-x} + c_2 e^{2x} - 3 \cos x - \sin x.$$

In the first column of Table 1 are listed some of the possible terms of $r(x)$. The second column gives the form of y_p, provided the number listed in the last column is not one of the roots of the characteristic equation of the homogeneous equation corresponding to (1).

If $r(x)$ is a sum of functions in the first column, choose for y_p the sum of the functions in the corresponding lines.

Example 4. Find a particular solution of the equation

$$y'' - 3y' + 2y = 4x + e^{3x}.$$

Table 1. Method of Undetermined Coefficients

Term in $r(x)$	Choice for y_p	
$kx^n \ (n = 0, 1, \cdots)$	$K_n x^n + K_{n-1} x^{n-1} + \cdots + K_1 x + K_0$	0
ke^{px}	Ce^{px}	p
$k \cos qx$	$\Big\}\ \ K \cos qx + M \sin qx$	$\Big\{\ \ iq$
$k \sin qx$		iq

We start from

$$y_p = K_1 x + K_0 + Ce^{3x}.$$

By substitution we find $K_1 = 2$, $K_0 = 3$, $C = \frac{1}{2}$, and therefore

$$y_p = 2x + 3 + \frac{1}{2}e^{3x}.$$

Modification rule. *If the number listed in the last column of Table 1 is a root of the characteristic equation of the homogeneous equation corresponding to* (1), *the function in the second column of the table must be multiplied by x^m where m is the multiplicity of that root in that equation.* (Hence for a second-order equation, m may equal 1 or 2.)

This rule, most likely first discovered by experimenting, can be justified by the general method to be discussed in Sec. 2.16.

Example 5.   Solve the nonhomogeneous equation

(5) $$y'' - 2y' + y = e^x + x.$$

The characteristic equation of the corresponding homogeneous equation is

(6) $$\lambda^2 - 2\lambda + 1 = 0.$$

Since 0 is not a root of (6), the function x on the right-hand side of (5) indicates the particular solution choice

$$K_1 x + K_0.$$

Since 1 is a double root of (6), that is, a root of multiplicity 2, according to the modification rule the function e^x calls for the particular solution

$$Cx^2 e^x \qquad\qquad \text{(instead of } Ce^x\text{)}.$$

Altogether,

$$y_p = K_1 x + K_0 + Cx^2 e^x.$$

Substituting this into (5), we obtain

$$y_p'' - 2y_p' + y_p = 2Ce^x + K_1 x - 2K_1 + K_0 = e^x + x.$$

Hence $C = \frac{1}{2}$, $K_1 = 1$, $K_0 = 2$, and a general solution of (5) is

$$y = y_h + y_p = (c_1 x + c_2)e^x + \tfrac{1}{2}x^2 e^x + x + 2.$$

Problems

Find a particular solution of the following differential equations.

1. $y'' + y = x^2$
2. $y'' - 4y = 6e^x$
3. $y'' + 2y' + 3y = x$
4. $y'' + y = 3\cos 2x$
5. $y'' + y = -15\sin 4x$
6. $y'' + 4y' + 3y = \sin x + 2\cos x$
7. $y'' - y' + y = e^x + x - 1$
8. $y'' - 9y = e^{2x} + \sin x$
9. $y'' + 4y' + 4y = 18\cosh x$
10. $y'' - 6y' + 9y = 3x - 2\sinh x$

Find a general solution of the following differential equations.

11. $y'' + y = e^x$
12. $y'' - y = e^x$
13. $y'' - 2y' + y = e^x$
14. $y'' + y' + y = x^4 + 4x^3 + 12x^2$
15. $y'' - 4y' + 9y = 10e^{2x} - 12\cos 3x$
16. $y'' - 2y' + 2y = 2e^x \cos x$

Solve the following initial value problems.

17. $y'' + 25y = 5x,$ $y(0) = 5,$ $y'(0) = -4.8$

18. $y'' + y' - 2y = 14 + 2x - 2x^2,$ $y(0) = 0,$ $y'(0) = 0$

19. $y'' - 4y' + 3y = 4e^{3x},$ $y(0) = -1,$ $y'(0) = 3$

20. $y'' + 4y' + 4y = 4\cos x + 3\sin x,$ $y(0) = 1,$ $y'(0) = 0$

21. $y'' + y' - 2y = 3e^x,$ $y(0) = 2,$ $y'(0) = 3$

22. $y'' - y' - 2y = 10\sin x,$ $y(0) = 1,$ $y'(0) = -3$

23. $y'' + 2y' - y = 2e^x + 2x^2 - 8x - 4,$ $y(0) = 1,$ $y'(0) = 1$

24. $y'' + y' - 2y = -6\sin 2x - 18\cos 2x,$ $y(0) = 2,$ $y'(0) = 2$

25. The method of undetermined coefficients can be applied to certain linear first-order differential equations, too, and may sometimes be simpler than the usual method (Sec. 1.7). Using both methods, solve $y' - y = x^5$ and $y' + 2y = \cos 2x$.

2.13 Forced Oscillations. Resonance

In Sec. 2.6 we considered free oscillations of a body on a spring as shown in Fig. 31. These motions are governed by the homogeneous equation

$$(1) \qquad\qquad m\ddot{y} + c\dot{y} + ky = 0$$

where m is the mass of the body, c is the damping constant, and k is the spring modulus. We shall now extend our consideration, assuming that a variable force $r(t)$ acts on the system. In this way we shall become familiar with further interesting facts which are fundamental in engineering mathematics, in particular with resonance.

We remember that the equation (1) was obtained by considering the forces acting on the body and using Newton's second law. From this it is immediately clear that the differential equation corresponding to the present situation is obtained from (1) by adding the force $r(t)$; this yields

$$m\ddot{y} + c\dot{y} + ky = r(t).$$

$r(t)$ is called the **input** or **driving force,** and a corresponding solution is called an **output** or a **response** *of the system to the driving force.* (Cf. also Sec. 1.7.) The resulting motion is called a **forced motion,** in contrast to the **free motion** corresponding to (1), which is a motion in the absence of an external force $r(t)$.

Fig. 31. Mass on a spring

Of particular interest are periodic inputs. We shall consider a sinusoidal input, say,

$$r(t) = F_0 \cos \omega t \qquad (F_0 > 0, \, \omega > 0).$$

More complicated periodic inputs will be considered later in Sec. 9.7.

The differential equation now under consideration is

(2) $$m\ddot{y} + c\dot{y} + ky = F_0 \cos \omega t.$$

We want to determine and discuss its general solution, which represents the general output of our vibrating system. Since a general solution of the corresponding homogeneous equation (1) is known from Sec. 2.6, we must now determine a particular solution $y_p(t)$ of (2).

This can be done by the method of undetermined coefficients (Sec. 2.12), starting from

(3) $$y_p(t) = a \cos \omega t + b \sin \omega t.$$

By differentiating this function we have

$$\dot{y}_p = -\omega a \sin \omega t + \omega b \cos \omega t, \qquad \ddot{y}_p = -\omega^2 a \cos \omega t - \omega^2 b \sin \omega t.$$

Substituting these expressions into (2) and collecting the cosine and the sine terms, we obtain

$$[(k - m\omega^2)a + \omega cb] \cos \omega t + [-\omega ca + (k - m\omega^2)b] \sin \omega t = F_0 \cos \omega t.$$

By equating the coefficients of the cosine and sine terms on both sides we have

(4)
$$(k - m\omega^2)a + \qquad \omega cb \qquad = F_0$$
$$-\omega ca \quad + (k - m\omega^2)b = 0.$$

The solution of this system of equations in the unknowns a and b is (cf. Sec. 6.6)

$$a = F_0 \frac{k - m\omega^2}{(k - m\omega^2)^2 + \omega^2 c^2}, \qquad b = F_0 \frac{\omega c}{(k - m\omega^2)^2 + \omega^2 c^2}$$

provided the denominator is not zero. If we set $\sqrt{k/m} = \omega_0 \, (>0)$ as in Sec. 2.6, this becomes

(5) $$a = F_0 \frac{m(\omega_0^2 - \omega^2)}{m^2(\omega_0^2 - \omega^2)^2 + \omega^2 c^2}, \qquad b = F_0 \frac{\omega c}{m^2(\omega_0^2 - \omega^2)^2 + \omega^2 c^2}.$$

We thus obtain the general solution

(6) $$y(t) = y_h(t) + y_p(t),$$

where y_h is a general solution of (1) and y_p is given by (3) with coefficients (5).

We shall now discuss the behavior of the mechanical system, distinguishing between the two cases $c = 0$ (no damping) and $c > 0$, which will correspond to two different types of the output.

   Case 1. Undamped forced oscillations.   If there is no damping, then $c = 0$ and, assuming that $\omega \neq \omega_0$, we obtain from (3) and (5)

(7) $$y_p(t) = \frac{F_0}{m(\omega_0{}^2 - \omega^2)} \cos \omega t = \frac{F_0}{k[1 - (\omega/\omega_0)^2]} \cos \omega t.$$

From this and (6*) in Sec. 2.6 we have the general solution

(8) $$y(t) = C \cos (\omega_0 t - \delta) + \frac{F_0}{m(\omega_0{}^2 - \omega^2)} \cos \omega t.$$

This output represents a superposition of two harmonic oscillations; the frequencies are the "natural frequency" $\omega_0/2\pi$ [cycles/sec] of the system (that is, the frequency of the free undamped motion) and the frequency $\omega/2\pi$ of the input.

   From (7) we see that the maximum amplitude of y_p is

(9) $$a_0 = \frac{F_0}{k} \rho \qquad \text{where} \qquad \rho = \frac{1}{1 - (\omega/\omega_0)^2}.$$

It depends on ω and ω_0. As $\omega \to \omega_0$, the quantities ρ and a_0 tend to infinity. This phenomenon of excitation of large oscillations by matching input and natural frequencies ($\omega = \omega_0$) is known as **resonance,** and is of basic importance in the study of vibrating systems (cf. below).   The quantity ρ is called the *resonance factor* (Fig. 32).   From (9) we see that ρ/k is the ratio of the amplitudes of the function y_p and the input.

   In the case of resonance equation (2) becomes

(10) $$\ddot{y} + \omega_0{}^2 y = \frac{F_0}{m} \cos \omega_0 t.$$

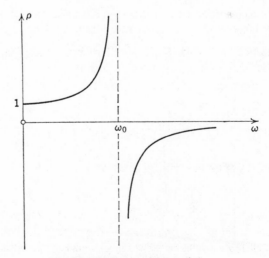

Fig. 32. Resonance factor $\rho(\omega)$

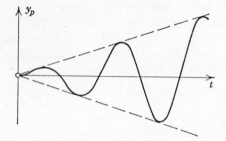

Fig. 33. Particular solution in the case of resonance

From the modification rule in the preceding section we conclude that a particular solution of (10) has the form

$$y_p(t) = t(a \cos \omega_0 t + b \sin \omega_0 t).$$

By substituting this into (10) we find $a = 0$, $b = F_0/2m\omega_0$, and (Fig. 33)

(11) $$y_p(t) = \frac{F_0}{2m\omega_0} \, t \sin \omega_0 t.$$

Another interesting and highly important type of oscillation is obtained when ω is close to ω_0. Take, for example, the particular solution [cf. (8)]

(12) $$y(t) = \frac{F_0}{m(\omega_0{}^2 - \omega^2)} (\cos \omega t - \cos \omega_0 t) \qquad (\omega \neq \omega_0)$$

corresponding to the initial conditions $y(0) = 0$, $\dot{y}(0) = 0$. This may be written [cf. (12) in Appendix 3]

$$y(t) = \frac{2F_0}{m(\omega_0{}^2 - \omega^2)} \sin \frac{\omega_0 + \omega}{2} \, t \sin \frac{\omega_0 - \omega}{2} \, t.$$

Since ω is close to ω_0 the difference $\omega_0 - \omega$ is small, so that the period of the last sine function is large, and we obtain an oscillation of the type shown in Fig. 34.

Case 2. Damped forced oscillations. If there is damping, then $c > 0$, and we know from Sec. 2.6 that the general solution y_h of (1) approaches zero as t

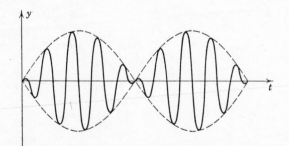

Fig. 34. Forced undamped oscillation when the difference of the input and natural frequencies is small (**"beats"**)

approaches infinity (practically, after a sufficiently long time); that is, the general solution (6) of (2) now represents the **transient solution** and tends to the **steady-state solution** y_p. *Hence, after a sufficiently long time, the ouput corresponding to a purely sinusoidal input will practically be a harmonic oscillation whose frequency is that of the input. This is what happens in practice, because no physical system is completely undamped.*

While in the undamped case the amplitude y_p approaches infinity as ω approaches ω_0, this will not happen in the present case. *The amplitude will always be finite,* but may have a maximum for some ω, depending on c. This may be called *practical resonance.* It is of great importance because it shows that some input may excite oscillations with such a large amplitude that the system may be destroyed. Such cases happened in practice, in particular in earlier times when less was known about resonance. Cars, ships, airplanes, and bridges are vibrating mechanical systems, and it is sometimes rather difficult to find constructions which are completely free of undesired resonance effects.

To investigate the amplitude of y_p as a function of ω, we write (3) in the form

$$(13) \qquad y_p(t) = C^* \cos(\omega t - \eta)$$

where, according to (5),

$$(14) \qquad C^*(\omega) = \sqrt{a^2 + b^2} = \frac{F_0}{\sqrt{m^2(\omega_0{}^2 - \omega^2)^2 + \omega^2 c^2}},$$

$$\tan \eta = \frac{b}{a} = \frac{\omega c}{m(\omega_0{}^2 - \omega^2)}.$$

Let us determine the maximum of $C^*(\omega)$. By equating $dC^*/d\omega$ to zero we find

$$[-2m^2(\omega_0{}^2 - \omega^2) + c^2]\omega = 0.$$

The expression in brackets is zero for

$$(15) \qquad c^2 = 2m^2(\omega_0{}^2 - \omega^2).$$

For sufficiently large damping ($c^2 > 2m^2\omega_0{}^2 = 2mk$) equation (15) has no real solution, and C^* decreases in a monotone way as ω increases (Fig. 35). If $c^2 \leqq 2mk$, equation (15) has a real solution $\omega = \omega_{\max}$, which increases as c decreases and approaches ω_0 as c approaches zero. The amplitude $C^*(\omega)$ has a maximum at $\omega = \omega_{\max}$, and by inserting $\omega = \omega_{\max}$ into (14) we find that

$$(16) \qquad C^*(\omega_{\max}) = \frac{2mF_0}{c\sqrt{4m^2\omega_0{}^2 - c^2}}.$$

We see that $C^*(\omega_{\max})$ is finite when $c > 0$. Since $dC^*(\omega_{\max})/dc < 0$ when $c^2 < 2mk$, the value of $C^*(\omega_{\max})$ increases as c ($\leqq \sqrt{2mk}$) decreases and approaches

Fig. 35. Amplification C^*/F_0 as a function of ω for $m = 1$, $k = 1$, and various values of the damping constant c

infinity as c approaches zero, in agreement with our result in Case 1. Figure 35 shows the **amplification** C^*/F_0 (ratio of the amplitudes of output and input) as a function of ω for $m = 1$, $k = 1$, and various values of the damping constant c.

The angle η in (14) is called the **phase angle** or **phase lag** (Fig. 36) because it measures the lag of the output with respect to the input. If $\omega < \omega_0$, then $\eta < \pi/2$; if $\omega = \omega_0$, then $\eta = \pi/2$, and if $\omega > \omega_0$, then $\eta > \pi/2$.

Fig. 36. Phase lag η as a function of ω for $m = 1$, $k = 1$, and various values of the damping constant c

Problems

In each case find the steady-state oscillations.

1. $\ddot{y} + 4\dot{y} + 3y = 32.5 \cos 2t$
2. $\ddot{y} + 3\dot{y} + 2y = 20 \sin t$
3. $2\ddot{y} + 4\dot{y} + y = 3 \sin t + \sin 3t$
4. $3\ddot{y} + \dot{y} + 2y = 2 \cos t - 52 \sin 2t$
5. $2\ddot{y} + 2\dot{y} + 3y = \cos 3t - 5 \sin t$
6. $\ddot{y} + \dot{y} + y = \cos t + 13 \cos 2t$

In each case find the transient oscillations.

7. $\ddot{y} + 2\dot{y} + 2y = \cos t$
8. $\ddot{y} + 2\dot{y} + y = 50 \sin 3t$
9. $\ddot{y} + 2\dot{y} + 5y = -\sin t$
10. $\ddot{y} + 0.5\dot{y} + 2y = 5 \cos t$

Solve the following initial value problems.

11. $\ddot{y} + y = -9 \sin 2t$, $\qquad y(0) = 1$, $\qquad \dot{y}(0) = 0$
12. $\ddot{y} + y = \cos \omega t$, $\qquad y(0) = y_0$, $\qquad \dot{y}(0) = v_0$, $\qquad \omega^2 \neq 1$
13. $\ddot{y} + 2\dot{y} + 2y = \sin 2t - 2 \cos 2t$, $\qquad y(0) = 0$, $\qquad \dot{y}(0) = 0$
14. $\ddot{y} + 4y = \sin t + \frac{1}{3} \sin 3t + \frac{1}{5} \sin 5t$, $\qquad y(0) = 1$, $\qquad \dot{y}(0) = \frac{3}{35}$
15. $\ddot{y} + 4\dot{y} + 20y = 23 \sin t - 15 \cos t$, $\qquad y(0) = 0$, $\qquad \dot{y}(0) = -1$
16. $\ddot{y} + 9y = 8 \sin t$, $\qquad y(0) = 1$, $\qquad \dot{y}(0) = 1$

17. Find a particular solution of the differential equation

$$m\ddot{y} + c\dot{y} + ky = K \sin \omega t \qquad\qquad (c \neq 0).$$

18. In (12) let ω approach ω_0. Show that this leads to a solution of the form (11).
19. Solve the initial value problem $\ddot{y} + y = \cos \omega t$, $\omega^2 \neq 1$, $y(0) = 0$, $\dot{y}(0) = 0$, and graph the maximum amplitude as a function of ω.
20. Show that the solution in Prob. 19 may be written

$$y(t) = 2(1 - \omega^2)^{-1} \sin [0.5(1 + \omega)t] \sin [0.5(1 - \omega)t],$$

and plot good graphs of $y(t)$ with $\omega = 0.5, 0.9, 1.1, 2$.

2.14 Electric Circuits

The last section was devoted to the study of a mechanical system which is of great practical interest. We shall now consider a similarly important electrical system, which may be regarded as a basic building block in electrical networks. This consideration will also provide a striking example of the important fact that entirely different physical systems may correspond to the same differential equation, and thus illustrate the role which mathematics plays in unifying various phenomena of entirely different physical nature. We shall obtain a correspondence between mechanical and electrical systems which is not merely qualitative but strictly quantitative in the sense that to a given mechanical system we can construct an electric circuit whose current will give the exact values of the displacement in the mechanical system when suitable scale factors are introduced. The practical importance of such an analogy between mechanical and electrical systems is almost obvious. The analogy may be used for constructing an "electrical model" of a given mechanical system; in many cases this will be an essential simplification, because electric circuits are easy to assemble and currents and voltages are easy to measure, whereas the construction of a mechanical model may be complicated and expensive, and the measurement of displacements will be time-consuming and inaccurate.

Fig. 37. *RLC*-circuit

We shall consider the "*RLC*-circuit" in Fig. 37. Simpler circuits were considered in Sec. 1.9, and it will be assumed that the student is familiar with those simpler cases, because they will immediately provide the starting point for obtaining the differential equation of our present circuit. We remember that the differential equations of those circuits were obtained by equating the sum of the voltage drops across the elements of the circuits to the impressed electromotive force. This was done by using Kirchhoff's second law, which therefore plays a role in circuits similar to that of Newton's second law (Sec. 2.6) in mechanical systems.

To obtain the differential equation of the circuit in Fig. 37, we may use the equation of an *RC*-circuit [cf. (7), Sec. 1.9] and add to it the voltage drop $L\dot{I}$ across the inductor in our present circuit. This yields

$$(1') \qquad L\dot{I} + RI + \frac{1}{C}\int I \, dt = E(t) = E_0 \sin \omega t.$$

Differentiating this equation with respect to t, we obtain the differential equation of our *RLC*-circuit in the form

$$\textbf{(1)} \qquad L\ddot{I} + R\dot{I} + \frac{1}{C}I = E_0\omega \cos \omega t.$$

This equation and the equation (2) in Sec. 2.13 are essentially the same. This shows that our RLC-circuit is the electrical analogue of the mechanical system in Sec. 2.13, and we note the following correspondence between the electrical and mechanical quantities:

$$\text{Inductance } L \quad \leftrightarrow \quad \text{mass } m$$

$$\text{Resistance } R \quad \leftrightarrow \quad \text{damping constant } c$$

$$\text{Reciprocal of capacitance } 1/C \quad \leftrightarrow \quad \text{spring modulus } k$$

$$\left.\begin{array}{c}\text{Derivative of}\\ \text{electromotive force } E_0\omega \cos \omega t\end{array}\right\} \quad \leftrightarrow \quad \text{driving force } F_0 \cos \omega t \ (input)$$

$$\text{Current } I(t) \quad \leftrightarrow \quad \text{displacement } y(t) \ (output)$$

To obtain a particular solution of (1) we may proceed as in Sec. 2.13. By substituting

$$(2) \qquad I_p(t) = a \cos \omega t + b \sin \omega t$$

into (1) we obtain

(3) $$a = \frac{-E_0 S}{R^2 + S^2}, \qquad b = \frac{E_0 R}{R^2 + S^2}$$

where S is the so-called **reactance**, given by the expression

(4) $$S = \omega L - \frac{1}{\omega C}.$$

In any practical case, $R \neq 0$ so that the denominator in (3) is not zero. The result is that (2), with a and b given by (3), is a particular solution of (1).

Using (3), we may write I_p in the form

(5) $$I_p(t) = I_0 \sin (\omega t - \theta)$$

where [cf. (14) in Appendix 3]

$$I_0 = \sqrt{a^2 + b^2} = \frac{E_0}{\sqrt{R^2 + S^2}}, \qquad \tan \theta = -\frac{a}{b} = \frac{S}{R}.$$

The quantity $\sqrt{R^2 + S^2}$ is called the **impedance**.

The characteristic equation of the homogeneous differential equation corresponding to (1) is

$$\lambda^2 + \frac{R}{L} \lambda + \frac{1}{LC} = 0$$

and has the roots $\lambda_1 = -\alpha + \beta$ and $\lambda_2 = -\alpha - \beta$ where

$$\alpha = \frac{R}{2L}, \qquad \beta = \frac{1}{2L}\sqrt{R^2 - \frac{4L}{C}}.$$

As in Sec. 2.13 we conclude that if $R > 0$ (which, of course, is true in any practical case) the general solution $I_h(t)$ of the homogeneous equation approaches zero as t approaches infinity (practically: after a sufficiently long time). Hence, the transient current $I = I_h + I_p$ tends to the steady state current I_p, and *after a sufficiently long time the output will practically be a harmonic oscillation, which is given by (5) and whose frequency is that of the input.*

Problems

Find the current $I(t)$ in the LC-circuit in Fig. 38, where
    1. $L = 0.4$ henry, $C = 0.1$ farad, $E = 110 \sin \omega t$ volts, $\omega^2 \neq 25$
    2. $L = 0.2$ henry, $C = 0.05$ farad, $E = 100$ volts
    3. $L = 2.5$ henrys, $C = 10^{-3}$ farad, $E = 10t^2$ volts

    4. Let $I(t)$ be the current in the LC-circuit in Fig. 38 and $q(0)$ the charge in the capacitor at $t = 0$. Show that

(6) $$\dot{I}(0) = \frac{E(0)}{L} - \frac{q(0)}{LC}.$$

Fig. 38. *LC*-circuit   **Fig. 39.** *RLC*-circuit   **Fig. 40.** *RC*-circuit

Using (6), find the current $I(t)$ in the *LC*-circuit in Fig. 38, assuming zero initial current and charge and

   5. $L = 1$ henry, $C = 0.25$ farad, $E = 30 \sin t$ volts
   6. $L = 10$ henrys, $C = \frac{1}{90}$ farad, $E = 10 \cos 2t$ volts
   7. $L = 10$ henrys, $C = 0.1$ farad, $E = 10t$ volts

Find the steady-state current in the *RLC*-circuit in Fig. 39, where

   8. $R = 4$ ohms, $L = 1$ henry, $C = 2 \cdot 10^{-4}$ farad, $E = 220$ volts
   9. $R = 20$ ohms, $L = 10$ henrys, $C = 0.05$ farad, $E = 50 \sin t$ volts
   10. $R = 10$ ohms, $L = 2$ henrys, $C = 0.5$ farad, $E = 10.9 \cos 2t$ volts

Find the transient current in the *RLC*-circuit in Fig. 39, where

   11. $R = 20$ ohms, $L = 5$ henrys, $C = 10^{-2}$ farad, $E = 85 \sin 4t$ volts
   12. $R = 200$ ohms, $L = 100$ henrys, $C = 0.005$ farad, $E = 500 \sin t$ volts
   13. $R = 16$ ohms, $L = 8$ henrys, $C = \frac{1}{8}$ farad, $E = 100 \cos 2t$ volts

   14. Show that if at $t = 0$ the charge in the capacitor in Fig. 39 is $Q(0)$, then

$$(7) \qquad\qquad \dot{I}(0) = \frac{E(0)}{L} - \frac{R}{L} I(0) - \frac{Q(0)}{LC} \qquad\qquad (LC \neq 0).$$

Using (7), find the current in the *RLC*-circuit in Fig. 39, assuming zero initial current and charge, and

   15. $R = 6$ ohms, $L = 1$ henry, $C = 0.04$ farad, $E = 24 \cos 5t$ volts
   16. $R = 160$ ohms, $L = 20$ henrys, $C = 0.002$ farad, $E = 481 \sin 10t$ volts
   17. $R = 80$ ohms, $L = 20$ henrys, $C = 0.01$ farad, $E = 100$ volts

   18. Show that if $E(t)$ in Fig. 38 has a jump of magnitude J at $t = a$, then $\dot{I}(t)$ has a jump of magnitude J/L at $t = a$, whereas $I(t)$ is continuous at $t = a$.

Using the result of Prob. 18, find the current $I(t)$ in the *LC*-circuit in Fig. 38, assuming $L = 1$ henry, $C = 1$ farad, zero initial current and charge, and

   19. $E = \begin{cases} t & \text{when} \ \ 0 < t < a \\ 0 & \text{when} \ \ t > a \end{cases}$

   20. $E = \begin{cases} 1 & \text{when} \ \ 0 < t < a \\ 0 & \text{when} \ \ t > a \end{cases}$

   21. $E = \begin{cases} 1 - e^{-t} & \text{when} \ \ 0 < t < \pi \\ 0 & \text{when} \ \ t > \pi \end{cases}$

   22. $E = \begin{cases} t & \text{when} \ \ 0 < t < a \\ a & \text{when} \ \ t > a \end{cases}$

   23. Show that if $E(t)$ in Fig. 40 has a jump of magnitude J at $t = a$, then the current $I(t)$ in the circuit has a jump of magnitude J/R at $t = a$.

Using the result of Prob. 23, find the current $I(t)$ in the RC-circuit in Fig. 40, assuming $R = 1$ ohm, $C = 1$ farad, zero initial charge on the capacitor, and

24. $E = \begin{cases} t & \text{when} \quad 0 < t < a \\ 0 & \text{when} \quad t > a \end{cases}$

25. $E = \begin{cases} t & \text{when} \quad 0 < t < a \\ a & \text{when} \quad t > a \end{cases}$

2.15 Complex Method for Obtaining Particular Solutions

Given an equation of the form (1) in Sec. 2.14, for example,

(1) $$\ddot{I} + \dot{I} + 2I = 6 \cos t,$$

we know that we can obtain a particular solution $I_p(t)$ by the method of undetermined coefficients, that is, by substituting

$$I_p(t) = a \cos t + b \sin t$$

into (1) and determining a and b. The result will be

$$I_p(t) = 3 \cos t + 3 \sin t.$$

Engineers often prefer a simple and elegant complex method for obtaining $I_p(t)$. In this method we start from the differential equation

(2) $$\ddot{I} + \dot{I} + 2I = 6e^{it} \qquad\qquad (i = \sqrt{-1}).$$

and determine a complex particular solution of the form

(3) $$I_p{}^*(t) = Ke^{it}.$$

By substituting this function and its derivatives

$$\dot{I}_p{}^* = iKe^{it}, \qquad \ddot{I}_p{}^* = -Ke^{it}$$

into the equation (2) we have

$$(-1 + i + 2)Ke^{it} = 6e^{it}.$$

Solving for K, we obtain

$$K = \frac{6}{1 + i} = \frac{6(1 - i)}{(1 + i)(1 - i)} = 3 - 3i.$$

Hence

$$I_p{}^*(t) = (3 - 3i)e^{it} = (3 - 3i)(\cos t + i \sin t)$$

is a solution of (2); here the last expression is obtained by means of the Euler formula (Sec. 2.4). The real part of $I_p{}^*$ is

$$I_p(t) = 3 \cos t + 3 \sin t$$

and this function is a solution of the real part of the differential equation (2), that is, of the given differential equation (1). In fact, the function is identical with that obtained above. This illustrates the practical procedure of the complex method.

Our equation (1) is a particular case of the equation [cf. (1), Sec. 2.14]

$$(4) \qquad L\ddot{I} + R\dot{I} + \frac{1}{C}I = E_0\omega \cos \omega t$$

which we shall now consider, assuming that $R \neq 0$. The corresponding complex equation is

$$(5) \qquad L\ddot{I} + R\dot{I} + \frac{1}{C}I = E_0\omega e^{i\omega t}.$$

The function on the right suggests a particular solution of the form

$$(6) \qquad I_p{}^*(t) = Ke^{i\omega t} \qquad\qquad (i = \sqrt{-1}).$$

By substituting this function and its derivatives

$$\dot{I}_p{}^* = i\omega K e^{i\omega t} \qquad \ddot{I}_p{}^* = -\omega^2 K e^{i\omega t}$$

into (5) we immediately have

$$\left(-\omega^2 L + i\omega R + \frac{1}{C}\right)Ke^{i\omega t} = E_0\omega e^{i\omega t}.$$

Solving this equation for K, we obtain

$$(7) \qquad K = \frac{E_0\omega}{\dfrac{1}{C} - \omega^2 L + i\omega R} = \frac{E_0}{iZ}$$

where Z is the so-called **complex impedance,** given by

$$(8) \qquad Z = R + i\left(\omega L - \frac{1}{\omega C}\right).$$

It follows that (6) with K given by (7) is a solution of (5).

We see that the imaginary part of Z is the reactance S defined by (4), Sec. 2.14, and $|Z|$ is the impedance defined in connection with (5), Sec. 2.14. Thus (cf. Fig. 41)

$$Z = |Z|e^{i\theta} \quad \text{where} \quad \tan \theta = \frac{S}{R}.$$

Fig. 41. Complex impedance Z

Consequently, formula (6) with K according to (7) can be written in the form

$$I_p^*(t) = \frac{E_0}{iZ} e^{i\omega t} = -i\frac{E_0}{|Z|} e^{i(\omega t - \theta)}.$$

The real part is

(9) $$I_p(t) = \frac{E_0}{|Z|} \sin(\omega t - \theta) = \frac{E_0}{\sqrt{R^2 + S^2}} \sin(\omega t - \theta),$$

and this is a solution of the real part of the differential equation (5), that is, $I_p(t)$ is a solution of (4). We see that this function is identical with (5) in the previous section. Since

$$\sin(\omega t - \theta) = \sin \omega t \cos \theta - \cos \omega t \sin \theta$$

and furthermore

$$\cos \theta = \frac{\text{Re } Z}{|Z|} = \frac{R}{\sqrt{R^2 + S^2}}, \qquad \sin \theta = \frac{\text{Im } Z}{|Z|} = \frac{S}{\sqrt{R^2 + S^2}},$$

the solution can be written in the form

(10) $$I_p(t) = \frac{E_0}{R^2 + S^2} [R \sin \omega t - S \cos \omega t].$$

This is identical with (2) in the previous section, where a and b are given by (3) in that section.

Problems

Determine the steady-state output of the following differential equations by means of the complex method.

1. $\ddot{y} + 2\dot{y} + 4y = 6 \cos 3t$
2. $\ddot{y} + 25y = \cos t$
3. $\ddot{y} + 3\dot{y} + 16y = 2 \cos 4t$
4. $\ddot{y} + \dot{y} + 9y = -3 \sin 3t$
5. $\ddot{y} + 2\dot{y} + 2y = \cos 2t$
6. $\ddot{y} + \dot{y} + 4y = 8 \sin 2t$

Using the complex method, determine the steady-state current $I_p(t)$ in the RLC-circuit governed by (4), where

7. $L = 2, R = 4, C = \frac{1}{8}, E_0 = 10, \omega = 5$
8. $L = 4, R = 20, C = 0.5, E_0 = 10, \omega = 10$
9. $L = 15, R = 25, C = 0.05, E_0 = 100, \omega = 4$

10. Find the complex impedance Z and the reactance in Prob. 9.

2.16 General Method for Solving Nonhomogeneous Equations

The method of undetermined coefficients in Sec. 2.12 is a simple procedure for obtaining particular solutions of nonhomogeneous linear differential equations, and we have seen in Secs. 2.13–2.15 that it has important applications in connection with vibrations. However, it is limited to certain simpler types of differential equations. The method to be discussed in this section will be more general but more complicated than that method.

We consider linear differential equations of the form

$$(1) \qquad y'' + f(x)y' + g(x)y = r(x)$$

assuming that f, g, and r are continuous on an open interval I. We shall obtain a particular solution of (1) by using the method of **variation of parameters**[14] as follows.

We know that the corresponding homogeneous differential equation

$$(2) \qquad y'' + f(x)y' + g(x)y = 0$$

has a general solution $y_h(x)$ on I, which is of the form

$$y_h(x) = c_1 y_1(x) + c_2 y_2(x).$$

The method consists in replacing c_1 and c_2 by functions $u(x)$ and $v(x)$ to be determined so that the resulting function

$$(3) \qquad y_p(x) = u(x)y_1(x) + v(x)y_2(x)$$

is a particular solution of (1) on I. By differentiating (3) we obtain

$$y_p' = u'y_1 + uy_1' + v'y_2 + vy_2'.$$

We shall see that we can determine u and v such that

$$(4) \qquad u'y_1 + v'y_2 = 0.$$

This reduces the expression for y_p' to the form

$$(5) \qquad y_p' = uy_1' + vy_2'.$$

By differentiating this function we obtain

$$(6) \qquad y_p'' = u'y_1' + uy_1'' + v'y_2' + vy_2''.$$

By substituting (3), (5), and (6) into (1) and collecting terms containing u and terms containing v we readily obtain

$$u(y_1'' + fy_1' + gy_1) + v(y_2'' + fy_2' + gy_2) + u'y_1' + v'y_2' = r.$$

Since y_1 and y_2 are solutions of the homogeneous equation (2), this reduces to

$$u'y_1' + v'y_2' = r.$$

Equation (4) is
$$u'y_1 + v'y_2 = 0.$$

This is a system of two linear algebraic equations for the unknown functions u' and v'. The solution is obtained by Cramer's rule (cf. Sec. 6.6). We find

$$(7) \qquad u' = -\frac{y_2 r}{W}, \qquad v' = \frac{y_1 r}{W},$$

---

[14] The student may wish to review the simpler application of the method in Sec. 1.8.

where

$$W = y_1 y_2' - y_1' y_2$$

is the Wronskian of y_1 and y_2. Clearly, $W \neq 0$, because the functions y_1 and y_2 constitute a fundamental system. By integrating (7) we have

$$u = -\int \frac{y_2 r}{W} \, dx, \qquad v = \int \frac{y_1 r}{W} \, dx.$$

These integrals exist because $r(x)$ is continuous. Substituting these expressions for u and v into (3), we obtain the desired solution of (1),

(8) $$y_p(x) = -y_1 \int \frac{y_2 r}{W} \, dx + y_2 \int \frac{y_1 r}{W} \, dx.$$

Note that if the constants of integration in (8) are left arbitrary, then (8) represents a general solution of (1).

Example. Solve the differential equation

(9) $$y'' + y = \sec x.$$

The method of undetermined coefficients (Sec. 2.12) cannot be applied. The functions

$$y_1 = \cos x, \qquad y_2 = \sin x$$

constitute a fundamental system of the homogeneous equation, and their Wronskian is

$$W(y_1, y_2) = \cos x \cos x - (-\sin x) \sin x = 1.$$

From (8) we thus obtain the following particular solution of (9):

$$y_p = -\cos x \int \sin x \sec x \, dx + \sin x \int \cos x \sec x \, dx$$

$$= \cos x \ln (\cos x) + x \sin x.$$

The corresponding general solution of the differential equation (9) is

$$y = y_h + y_p = [c_1 + \ln (\cos x)] \cos x + (c_2 + x) \sin x.$$

Numerical methods for second-order differential equations are presented in Sec. 18.7, which is similar to Sec. 18.6 and is independent of the other sections of Chap. 18, so that it can be studied now.

Problems

Find a general solution of the following differential equations.

1. $y'' + 4y = \sec 2x$
2. $y'' + y = \csc x$
3. $y'' + 2y' + y = 4e^{-x} \ln x$
4. $y'' - 2y' + y = -e^x \sin x$
5. $y'' - 2y' + y = 2e^x/x^3$
6. $y'' + 4y' + 4y = e^{-2x}/x^2$
7. $y'' + 2y' + 2y = e^{-x}/\cos^3 x$
8. $y'' - 4y' + 5y = e^{2x}/\sin x$
9. $y'' - 2y' + y = e^x x^{3/2}$
10. $y'' - 4y' + 4y = (3x^2 + 2)e^x$
11. $y'' - 4y' + 4y = e^{2x}/x$
12. $y'' + 6y' + 9y = e^{-3x}/(x^2 + 1)$
13. $x^2 y'' + xy' - y = 10$
14. $x^2 y'' + xy' - y = 8x^3$
15. $x^2 y'' - 2xy' + 2y = 6/x$
16. $x^2 y'' - 2y = 3x^2$
17. $x^2 y'' - 4xy' + 6y = 42/x^4$
18. $x^2 y'' - 4xy' + 6y = -x^4 \sin x$
19. $x^2 y'' - 2xy' + 2y = x^3 \cos x$
20. $xy'' - y' = (3 + x)x^2 e^x$

CHAPTER 3

Power Series Solutions of Differential Equations. Orthogonal Functions

In the preceding chapter we have seen that linear homogeneous differential equations with *constant coefficients* can be solved by algebraic methods, and the solutions are elementary functions known from calculus. In the case of equations with *variable coefficients*, the situation is more complicated and the solutions may be nonelementary functions. Bessel's equation, Legendre's equation, and the hypergeometric equation are of this type. Since these and other equations and their solutions play an important role in engineering mathematics, we shall now consider a method for solving such equations. The solutions will appear in the form of power series, and the method is, therefore, known as the *power series method*. We shall also consider some basic properties of the solutions, so that the student may get acquainted with these "higher transcendental functions" and with some standard procedures used in connection with special functions.

The last three sections are devoted to the orthogonality of Legendre polynomials, Bessel functions, and other sets of solutions of certain boundary value problems, which are known as *Sturm-Liouville problems*.

> *Prerequisite for this chapter:* Chap. 2.
> *Sections which may be omitted in a shorter course:* 3.2, 3.6, 3.9.
> *References:* Appendix 1, Part B.
> *Answers to problems:* Appendix 2.

3.1 The Power Series Method

We shall now consider solving differential equations by the so-called *power series method* which yields solutions in the form of power series. It is a very effective standard procedure in connection with linear differential equations whose coefficients are variable.

We first remember that a **power series**[1] (in powers of $x - a$) is an infinite series of the form

(1)
$$\sum_{m=0}^{\infty} c_m(x - a)^m = c_0 + c_1(x - a) + c_2(x - a)^2 + \cdots,$$

where $c_0, c_1, \cdots$ are constants, called the **coefficients** of the series, a is a constant, called the **center,** and x is a variable.

If in particular $a = 0$, we obtain a *power series in powers of* x

(2)
$$\sum_{m=0}^{\infty} c_m x^m = c_0 + c_1 x + c_2 x^2 + c_3 x^3 + \cdots.$$

We shall assume in this section that all variables and constants are real.

Familiar examples of power series are the Maclaurin series

$$\frac{1}{1 - x} = \sum_{m=0}^{\infty} x^m = 1 + x + x^2 + \cdots \qquad (|x| < 1, \textit{geometric series}),$$

$$e^x = \sum_{m=0}^{\infty} \frac{x^m}{m!} = 1 + x + \frac{x^2}{2!} + \frac{x^3}{3!} + \cdots,$$

$$\cos x = \sum_{m=0}^{\infty} \frac{(-1)^m x^{2m}}{(2m)!} = 1 - \frac{x^2}{2!} + \frac{x^4}{4!} - + \cdots,$$

$$\sin x = \sum_{m=0}^{\infty} \frac{(-1)^m x^{2m+1}}{(2m + 1)!} = x - \frac{x^3}{3!} + \frac{x^5}{5!} - + \cdots.$$

The basic idea of the power series method for solving differential equations is very simple and natural. We shall describe the practical procedure and illustrate it by simple examples, postponing the mathematical justification of the method to the next section.

A differential equation being given, we first represent all given functions in the equation by power series in powers of x (or in powers of $x - a$, if solutions in the form of power series in powers of $x - a$ are wanted). Then we assume a solution in the form of a power series, say,

(3)
$$y = c_0 + c_1 x + c_2 x^2 + c_3 x^3 + \cdots = \sum_{m=0}^{\infty} c_m x^m$$

and insert this series and the series obtained by termwise differentiation,

(4)
(a) $y' = c_1 + 2c_2 x + 3c_3 x^2 + \cdots = \sum_{m=1}^{\infty} m c_m x^{m-1}$

(b) $y'' = 2c_2 + 3 \cdot 2c_3 x + 4 \cdot 3c_4 x^2 + \cdots = \sum_{m=2}^{\infty} m(m - 1)c_m x^{m-2}$

etc., into the equation. Collecting like powers of x, we may write the resulting

---

[1] The term "power series" alone usually refers to a series of the form (1), but does not include series of negative powers of x such as $c_0 + c_1 x^{-1} + c_2 x^{-2} + \cdots$ or series involving fractional powers of x. Note that in (1) we write, for convenience, $(x - a)^0 = 1$, even when $x = a$.

equation in the form

(5) $$k_0 + k_1 x + k_2 x^2 + \cdots = 0$$

where the constants k_0, k_1, $\cdots$ are expressions containing the unknown co-efficients c_0, c_1, $\cdots$ in (3). In order that (5) hold for all x in some interval, we must have

$$k_0 = 0, \qquad k_1 = 0, \qquad k_2 = 0, \cdots.$$

From these equations we may then determine the coefficients c_0, c_1, $\cdots$ successively.

Let us now illustrate the practical procedure for some simple equations which can also be solved by elementary methods.

Example 1. Solve

$$y' - y = 0.$$

In the first step, we insert (3) and (4a) into the equation:

$$(c_1 + 2c_2 x + 3c_3 x^2 + \cdots) - (c_0 + c_1 x + c_2 x^2 + \cdots) = 0.$$

Then we collect like powers of x, finding

$$(c_1 - c_0) + (2c_2 - c_1)x + (3c_3 - c_2)x^2 + \cdots = 0.$$

Equating the coefficient of each power of x to zero, we have

$$c_1 - c_0 = 0, \qquad 2c_2 - c_1 = 0, \qquad 3c_3 - c_2 = 0, \cdots.$$

Solving these equations, we may express c_1, c_2, $\cdots$ in terms of c_0 which remains arbitrary:

$$c_1 = c_0, \qquad c_2 = \frac{c_1}{2} = \frac{c_0}{2!}, \qquad c_3 = \frac{c_2}{3} = \frac{c_0}{3!}, \cdots.$$

With these values (3) becomes

$$y = c_9 + c_0 x + \frac{c_0}{2!} x^2 + \frac{c_0}{3!} x^3 + \cdots,$$

and we see that we have obtained the familiar general solution

$$y = c_0 \left(1 + x + \frac{x^2}{2!} + \frac{x^3}{3!} + \cdots \right) = c_0 e^x.$$

Example 2. Solve

$$y'' + y = 0.$$

By inserting (3) and (4b) into the equation we obtain

$$(2c_2 + 3 \cdot 2c_3 x + 4 \cdot 3c_4 x^2 + \cdots) + (c_0 + c_1 x + c_2 x^2 + \cdots) = 0.$$

Collecting like powers of x, we find

$$(2c_2 + c_0) + (3 \cdot 2c_3 + c_1)x + (4 \cdot 3c_4 + c_2)x^2 + \cdots = 0.$$

Equating the coefficient of each power of x to zero, we have

$$2c_2 + c_0 = 0, \qquad 3 \cdot 2c_3 + c_1 = 0, \qquad 4 \cdot 3c_4 + c_2 = 0, \cdots.$$

Solving these equations, we see that c_2, c_4, $\cdots$ may be expressed in terms of c_0, and c_3, c_5, $\cdots$ may be expressed in terms of c_1:

$$c_2 = -\frac{c_0}{2!}, \qquad c_3 = -\frac{c_1}{3!}, \qquad c_4 = -\frac{c_2}{4 \cdot 3} = \frac{c_0}{4!}, \cdots;$$

c_0 and c_1 are arbitrary. With these values (3) becomes

$$y = c_0 + c_1 x - \frac{c_0}{2!} x^2 - \frac{c_1}{3!} x^3 + \frac{c_0}{4!} x^4 + \frac{c_1}{5!} x^5 + \cdots.$$

This can be written

$$y = c_0 \left(1 - \frac{x^2}{2!} + \frac{x^4}{4!} - + \cdots \right) + c_1 \left(x - \frac{x^3}{3!} + \frac{x^5}{5!} - + \cdots \right)$$

and we recognize the familiar general solution

$$y = c_0 \cos x + c_1 \sin x.$$

Example 3. Solve

$$(x + 1)y' - (x + 2)y = 0.$$

By inserting (3) and (4a) into the equation we obtain

$$(x + 1)(c_1 + 2c_2 x + 3c_3 x^2 + \cdots) - (x + 2)(c_0 + c_1 x + c_2 x^2 + \cdots) = 0.$$

By performing the indicated multiplications we get

$$
\begin{aligned}
&c_1 x + 2c_2 x^2 + 3c_3 x^3 + 4c_4 x^4 + 5c_5 x^5 + \cdots + && s c_s x^s + \cdots \\
+\ &c_1 + 2c_2 x + 3c_3 x^2 + 4c_4 x^3 + 5c_5 x^4 + 6c_6 x^5 + \cdots + (s+1)c_{s+1} x^s + \cdots \\
-\ &c_0 x - c_1 x^2 - c_2 x^3 - c_3 x^4 - c_4 x^5 - \cdots - && c_{s-1} x^s - \cdots \\
-\ &2c_0 - 2c_1 x - 2c_2 x^2 - 2c_3 x^3 - 2c_4 x^4 - 2c_5 x^5 - \cdots - && 2c_s x^s - \cdots = 0.
\end{aligned}
$$

Since this must be an identity in x, the sum of the coefficients of each power of x must be zero; we thus obtain

(6)   (a) $c_1 - 2c_0 = 0$ (b) $2c_2 - c_1 - c_0 = 0$,   etc.,

and in general

(7) $s c_s + (s + 1)c_{s+1} - c_{s-1} - 2c_s = 0.$

By solving (6a) for c_1 and (7) for c_{s+1} we have

(8)   (a) $c_1 = 2c_0$,   (b) $c_{s+1} = \dfrac{1}{s+1}[c_{s-1} + (2-s)c_s]$, $s = 1, 2, \cdots.$

Formula (8b) is a **recursion formula,** from which we may now determine $c_2, c_3, \cdots$, successively; if we wish, we may arrange the calculation in tabular form:

s	c_{s-1}	$(2-s)c_s$	Sum	$s+1$	$c_{s+1} = \dfrac{\text{Sum}}{s+1}$	c_{s+1} in terms of c_0
						$c_1 = 2c_0$
1	c_0	c_1	$c_0 + c_1$	2	$\dfrac{c_0}{2} + \dfrac{c_1}{2}$	$c_2 = \frac{3}{2}c_0$
2	c_1	0	c_1	3	$\dfrac{c_1}{3}$	$c_3 = \frac{2}{3}c_0$
3	c_2	$-c_3$	$c_2 - c_3$	4	$\dfrac{c_2}{4} - \dfrac{c_3}{4}$	$c_4 = \frac{5}{24}c_0$
.	..	..	...	.	...	...

With these values (3) becomes

$$y = c_0(1 + 2x + \tfrac{3}{2}x^2 + \tfrac{2}{3}x^3 + \tfrac{5}{24}x^4 + \cdots),$$

and the reader may verify that the explicitly written terms are the first few terms of the Maclaurin series of

$$y = c_0(1 + x)e^x,$$

the general solution of the equation obtained by separating variables.

Problems

Apply the power series method to the following differential equations.

1. $y' = 2y$	**2.** $y' + y = 0$	**3.** $y' - ky = 0$	**4.** $y' - 2xy = 0$
5. $y' = 3x^2y$	**6.** $(x + 1)y' = 2y$	**7.** $(x - 2)y' = 3y$	**8.** $(1 - x^2)y' = y$
9. $y'' - y = 0$	**10.** $y'' + 4y = 0$	**11.** $y'' - y' = 0$	**12.** $y'' - 9y = 0$

(More problems of this type are included at the end of Sec. 3.2.)

3.2 Theoretical Basis of the Power Series Method

We have seen that the power series method yields solutions of differential equations in the form of power series. The solution y of a given equation is assumed in the form of a power series with undetermined coefficients, and the coefficients are determined successively by inserting that series and the series for the derivatives of y into the given equation.

The practical usefulness of the method becomes obvious if we remember that power series may be used for computing values of the solutions. Furthermore, many general properties of the solutions can be derived from their power series; this will be seen in our further discussion.

The power series method involves various operations on power series, for example, differentiation, addition, and multiplication of power series. Therefore, to justify the method we have to consider the theoretical basis of these operations. This will be done in the present section. The discussion will involve some concepts and facts which may be already known to the student from elementary calculus, as well as other concepts which are not considered in elementary classes. [Corresponding proofs and further details (which will not be used in this chapter) can be found in Chap. 14.]

A **power series** is an infinite series of the form

$$\textbf{(1)} \qquad \sum_{m=0}^{\infty} c_m(x - a)^m = c_0 + c_1(x - a) + c_2(x - a)^2 + \cdots,$$

and we assume that the variable x, the *center a*, and the *coefficients* $c_0, c_1, \cdots$ are real, as in the previous section.

The expression

$$\textbf{(2)} \qquad s_n(x) = c_0 + c_1(x - a) + c_2(x - a)^2 + \cdots + c_n(x - a)^n$$

(n a positive integer) is called the nth **partial sum** of the series (1). Clearly if we omit the terms of s_n from (1), the remaining expression is

$$(3) \qquad R_n(x) = c_{n+1}(x - a)^{n+1} + c_{n+2}(x - a)^{n+2} + \cdots,$$

and this expression is called the **remainder** *of* (1) *after the term* $c_n(x - a)^n$.

For example, in the case of the geometric series

$$1 + x + x^2 + \cdots + x^n + \cdots$$

we have

$$s_1 = 1 + x, \qquad\qquad R_1 = x^2 + x^3 + x^4 + \cdots,$$
$$s_2 = 1 + x + x^2, \qquad R_2 = x^3 + x^4 + x^5 + \cdots, \text{ etc.}$$

In this way we have now associated with (1) the sequence of the partial sums $s_1(x), s_2(x), \cdots$. It may happen that for some $x = x_0$ this sequence converges, say,

$$\lim_{n \to \infty} s_n(x_0) = s(x_0).$$

Then we say that the series (1) **converges,** or *is convergent, at* $x = x_0$; the number $s(x_0)$ is called the **value** or *sum* of (1) at x_0, and we write

$$s(x_0) = \sum_{m=0}^{\infty} c_m(x_0 - a)^m.$$

If that sequence is divergent at $x = x_0$, then the series (1) is said to **diverge,** or to *be divergent*, at $x = x_0$.

We remember that a *sequence* $s_1, s_2, \cdots$ is said to *converge* to a number s, or to *be convergent* with the *limit s*, if for each given positive number ϵ (no matter how small, but not zero) we can find a number N such that

$$(4) \qquad\qquad |s_n - s| < \epsilon \qquad \text{for each } n > N.$$

Geometrically speaking, (4) means that s_n with $n > N$ lies between $s - \epsilon$ and $s + \epsilon$ (Fig. 42, p. 108). Of course, N will in general depend on the choice of ϵ.

Now in our case, $s = s_n + R_n$ or $R_n = s - s_n$. Hence

$$|s_n - s| = |R_n|$$

in (4), and convergence at $x = x_0$ means that we can make $|R_n(x_0)|$ as small as we please, by taking n large enough. In other words, in the case of convergence, $s_n(x_0)$ is an approximation of $s(x_0)$, and the error $|R_n(x_0)|$ of the approximation can be made smaller than any preassigned positive number ϵ by taking n sufficiently large.

If we choose $x = x_0 = a$ in (1), the series reduces to the single term c_0 because the other terms are zero. This shows that the series (1) converges at $x = a$. In some cases this may be the only value of x for which (1) converges. If there are other values of x for which the series converges, these values form an interval, the **convergence interval,** having the midpoint $x = a$. This interval may be finite, as in Fig. 43. Then the series converges for all x in the interior of the interval, that is, for all x for which

$$(5) \qquad\qquad\qquad |x - a| < R$$

Fig. 42. Convergence of a sequence   **Fig. 43.** Convergence interval (5) of a
power series with center a

and diverges when $|x - a| > R$. The interval may also be infinite, that is, the series may converge for all x.

The quantity R in Fig. 43 is called the **radius of convergence** of (1); it is the distance of each end point of the convergence interval from the center a. If the series converges for all x, then we set $R = \infty$ (and $1/R = 0$).

The radius of convergence can be determined from the coefficients of the series by means of each of the formulas

(6) $(a) \quad \dfrac{1}{R} = \lim_{m \to \infty} \sqrt[m]{|c_m|}$ $(b) \quad \dfrac{1}{R} = \lim_{m \to \infty} \left| \dfrac{c_{m+1}}{c_m} \right|$

provided the limits in (6) exist. (Proofs of these facts can be found in Sec. 14.5.)

For each x for which (1) converges, it has a certain value $s(x)$ depending on x; if the radius of convergence R of (1) is not zero, we write

$$s(x) = \sum_{m=0}^{\infty} c_m (x - a)^m \qquad\qquad (|x - a| < R)$$

and we say that the series (1) *represents* the function $s(x)$ in the interval of convergence.

Example 1. In the case of the **geometric series**

$$\frac{1}{1 - x} = \sum_{m=0}^{\infty} x^m = 1 + x + x^2 + \cdots \qquad\qquad (|x| < 1).$$

In fact, $c_m = 1$ for all m, and from (6) we obtain $R = 1$, that is, the geometric series converges and represents $1/(1 - x)$ when $|x| < 1$.

Example 2. In the case of the series

$$e^x = \sum_{m=0}^{\infty} \frac{x^m}{m!} = 1 + x + \frac{x^2}{2!} + \cdots$$

we have $c_m = 1/m!$. Hence in (6b),

$$\frac{c_{m+1}}{c_m} = \frac{1/(m + 1)!}{1/m!} = \frac{1}{m + 1} \to 0 \qquad\qquad \text{as } m \to \infty,$$

which means $R = \infty$; the series converges for all x.

Example 3. In the case of the series

$$\sum_{m=0}^{\infty} m! \, x^m = 1 + x + 2x^2 + 6x^3 + \cdots$$

we have $c_m = m!$, and in (6b),

$$\frac{c_{m+1}}{c_m} = \frac{(m+1)!}{m!} = m + 1 \to \infty \qquad \text{as } m \to \infty.$$

Thus $R = 0$, and the series converges only at the center $x = 0$.

We shall now consider the operations on power series which are used in connection with the power series method.

A power series may be differentiated term by term. More precisely: if

$$y(x) = \sum_{m=0}^{\infty} c_m(x - a)^m$$

converges for $|x - a| < R$, where $R > 0$, then the series obtained by differentiating term by term also converges for those x and represents the derivative y' of y, that is

$$y'(x) = \sum_{m=1}^{\infty} mc_m(x - a)^{m-1}.$$

(Proof in Sec. 14.6, Theorems 3 and 5.)

Two power series may be added term by term. More precisely: if the series

(7) $$\sum_{m=0}^{\infty} b_m(x - a)^m \quad \text{and} \quad \sum_{m=0}^{\infty} c_m(x - a)^m$$

have positive radii of convergence and their sums are $f(x)$ and $g(x)$, then the series

$$\sum_{m=0}^{\infty} (b_m + c_m)(x - a)^m$$

converges and represents $f(x) + g(x)$ for each x which lies in the interior of the convergence interval of each of the given series (Proof in Sec. 14.5).

Two power series may be multiplied term by term. More precisely: Suppose that the series (7) have positive radii of convergence and let $f(x)$ and $g(x)$ be their sums. Then the series obtained by multiplying each term of the first series by each term of the second series and collecting like powers of $x - a$, that is,

$$b_0c_0 + (b_0c_1 + b_1c_0)(x - a) + \cdots = \sum_{n=0}^{\infty} (b_0c_n + b_1c_{n-1} + \cdots + b_nc_0)(x - a)^n$$

converges and represents $f(x)g(x)$ for each x in the interior of the convergence interval of each of the given series. (Proof in Sec. 14.5, Theorem 3.)

If a power series has a positive radius of convergence and a sum which is identically zero throughout its interval of convergence, then each coefficient of the series is zero. (Proof in Sec. 14.6, Theorem 2.)

These properties of power series form the theoretical basis of the power series method. The remaining question is whether a given differential equation has solutions representable by power series at all. We may answer this question by using the following concept.

A function $f(x)$ is said to be **analytic** *at a point* $x = a$ if it can be represented by a power series in powers of $x - a$ with radius of convergence $R > 0$.

Using this notion, we may formulate a basic criterion of the desired type as follows.

Theorem 1. *If the functions f, g, and r in the differential equation*

$$(8) \qquad\qquad y'' + f(x)y' + g(x)y = r(x)$$

are analytic at $x = a$, *then every solution* $y(x)$ *of* (8) *is analytic at* $x = a$ *and can thus be represented by a power series in powers of* $x - a$ *with radius of convergence*[2] $R > 0$.

The proof requires advanced methods of complex analysis and can be found in Ref. [B8] in Appendix 1. In applying this theorem it is important to write the linear equation in the form (8), with 1 as the coefficient of y''.

Problems

Apply the power series method to the following differential equations.

1. $y' = 3(y + 1)/x$
2. $y' = (1 + 1/x)y$
3. $(x + 1)y' - (2x + 3)y = 0$
4. $(x - 3)y' - xy = 0$
5. $xy' - (x + 2)y = -2x^2 - 2x$
6. $y'' = y + x$
7. $y'' - 3y' + 2y = 0$
8. $y'' + y = 2x^2 + x$
9. $(1 - x^2)y'' - 2xy' + 2y = 0$
10. $y'' - 4xy' + (4x^2 - 2)y = 0$

11. Show that $y' = (y/x) + 1$ cannot be solved for y as a power series in x. Solve this equation for y as a power series in powers of $x - 1$. (*Hint.* Introduce $t = x - 1$ as a new independent variable and solve the resulting equation for y as a power series in t.) Compare the result with that obtained by the appropriate elementary method.

Solve for y as a power series in powers of $x - 1$:

12. $y' = ky$
13. $y'' + y = 0$
14. $y' + y/x = 0$

15. Show that $xy'' + y' + xy = 0$ does not satisfy the assumptions in Theorem 1, but has a power series in powers of x as a solution. Does this contradict Theorem 1?

Find the radius of convergence of the following power series.

16. $\displaystyle\sum_{m=0}^{\infty} (-1)^m x^{2m}$
17. $\displaystyle\sum_{m=0}^{\infty} \frac{x^m}{4^m}$
18. $\displaystyle\sum_{m=0}^{\infty} \frac{x^{2m}}{(2m)!}$

19. $\displaystyle\sum_{m=0}^{\infty} \frac{(-1)^m x^{2m+1}}{(2m+1)!}$
20. $\displaystyle\sum_{m=2}^{\infty} \frac{m(m-1)}{3^m} x^m$
21. $\displaystyle\sum_{m=0}^{\infty} \frac{(-1)^m}{k^m} x^{2m}$

22. $\displaystyle\sum_{m=0}^{\infty} \frac{(-1)^m}{5^m} (x - 5)^m$
23. $\displaystyle\sum_{m=0}^{\infty} 3^m (x - 2)^m$
24. $\displaystyle\sum_{m=0}^{\infty} m^{2m} x^m$

25. $\displaystyle\sum_{m=0}^{\infty} \frac{(-1)^m}{m^m} (x - 1)^m$
26. $\displaystyle\sum_{m=0}^{\infty} \frac{(3m)!}{(m!)^3} x^m$
27. $\displaystyle\sum_{m=0}^{\infty} \frac{x^{2m}}{2^{2m}(m!)^2}$

28. $\displaystyle\sum_{m=0}^{\infty} \frac{x^{m^2}}{2^m}$
29. $\displaystyle\sum_{m=1}^{\infty} \frac{1}{3^m m^2} (x + 1)^m$
30. $\displaystyle\sum_{m=0}^{\infty} \frac{(x - 2)^{2m}}{m!}$

---

[2] R is at least equal to the distance between the point $x = a$ and that point (or those points) closest to $x = a$ at which one of the functions f, g, r, *as functions of a complex variable*, is not analytic. (Note that that point may not lie on the x-axis, but somewhere in the complex plane.)

3.3   Legendre's Equation. Legendre Polynomials

Legendre's differential equation[3]

(1) $$(1 - x^2)y'' - 2xy' + n(n + 1)y = 0$$

arises in numerous physical problems, particularly in boundary value problems for spheres. The parameter n in (1) is a given real number. Any solution of (1) is called a **Legendre function.**

Dividing (1) by $1 - x^2$, we obtain the standard form (8), Sec. 3.2, and we see that the coefficients of the resulting equation are analytic at $x = 0$, so that we may apply the power series method. Substituting

(2) $$y = \sum_{m=0}^{\infty} c_m x^m$$

and its derivatives into (1) and denoting the constant $n(n + 1)$ by k we obtain

$$(1 - x^2) \sum_{m=2}^{\infty} m(m - 1)c_m x^{m-2} - 2x \sum_{m=1}^{\infty} mc_m x^{m-1} + k \sum_{m=0}^{\infty} c_m x^m = 0.$$

By writing the first expression as two separate series we have

$$\sum_{m=2}^{\infty} m(m - 1)c_m x^{m-2} - \sum_{m=2}^{\infty} m(m - 1)c_m x^m - 2 \sum_{m=1}^{\infty} mc_m x^m + k \sum_{m=0}^{\infty} c_m x^m = 0,$$

written out

$$2 \cdot 1c_2 + 3 \cdot 2c_3 x + 4 \cdot 3c_4 x^2 + \cdots + (s + 2)(s + 1)c_{s+2} x^s + \cdots$$
$$- 2 \cdot 1c_2 x^2 - \cdots \qquad - s(s - 1)c_s x^s - \cdots$$
$$- 2 \cdot 1c_1 x - 2 \cdot 2c_2 x^2 - \cdots \qquad - 2sc_s x^s - \cdots$$
$$+ kc_0 \quad + kc_1 x \quad + kc_2 x^2 + \cdots \qquad + kc_s x^s + \cdots = 0.$$

Since this must be an identity in x if (2) is to be a solution of (1), the sum of the coefficients of each power of x must be zero; remembering that $k = n(n + 1)$ we thus have

(3a)
$$2c_2 + n(n + 1)c_0 = 0,$$
$$6c_3 + [-2 + n(n + 1)]c_1 = 0,$$

and in general, when $s = 2, 3, \cdots$,

(3b) $$(s + 2)(s + 1)c_{s+2} + [-s(s - 1) - 2s + n(n + 1)]c_s = 0.$$

The expression in brackets $[\cdots]$ can be written

$$(n - s)(n + s + 1).$$

---

[3] ADRIEN MARIE LEGENDRE (1752–1833), French mathematician, who made important contributions to the theory of numbers and elliptic functions.

We thus obtain from (3)

(4) $$c_{s+2} = -\frac{(n-s)(n+s+1)}{(s+2)(s+1)} c_s \qquad (s = 0, 1, \cdots).$$

This is a recursion formula, giving each coefficient in terms of the second one preceding it, except for c_0 and c_1, which are left as arbitrary constants. We find successively

$$c_2 = -\frac{n(n+1)}{2!} c_0 \qquad\qquad c_3 = -\frac{(n-1)(n+2)}{3!} c_1$$

$$c_4 = -\frac{(n-2)(n+3)}{4\cdot 3} c_2 \qquad\qquad c_5 = -\frac{(n-3)(n+4)}{5\cdot 4} c_3$$

$$= \frac{(n-2)n(n+1)(n+3)}{4!} c_0 \qquad\qquad = \frac{(n-3)(n-1)(n+2)(n+4)}{5!} c_1$$

etc. By inserting these values for the coefficients into (2) we obtain

(5) $$y(x) = c_0 y_1(x) + c_1 y_2(x)$$

where

(6) $$y_1(x) = 1 - \frac{n(n+1)}{2!} x^2 + \frac{(n-2)n(n+1)(n+3)}{4!} x^4 - + \cdots$$

and

(7) $$y_2(x) = x - \frac{(n-1)(n+2)}{3!} x^3 + \frac{(n-3)(n-1)(n+2)(n+4)}{5!} x^5 - + \cdots.$$

These series converge for $|x| < 1$. Since (6) contains even powers of x only, while (7) contains odd powers of x only, the ratio y_1/y_2 is not a constant, and y_1 and y_2 are linearly independent solutions. Hence (5) is a general solution of (1) on the interval $-1 < x < 1$.

In many applications the parameter n in Legendre's equation will be a non-negative integer. Then the right hand side of (4) is zero when $s = n$, and, therefore, $c_{n+2} = 0, c_{n+4} = 0, c_{n+6} = 0, \cdots$. Hence, if n is even, $y_1(x)$ reduces to a polynomial of degree n. If n is odd, the same is true with respect to $y_2(x)$. These polynomials, multiplied by some constants, are called **Legendre polynomials.** Since they are of great practical importance, let us consider them in more detail. For this purpose we write (4) in the form

(8) $$c_s = -\frac{(s+2)(s+1)}{(n-s)(n+s+1)} c_{s+2} \qquad (s \leq n-2)$$

and may then express all the nonvanishing coefficients in terms of the coefficient c_n of the highest power of x of the polynomial. The coefficient c_n is then arbitrary. It is customary to choose $c_n = 1$ when $n = 0$ and

(9) $$c_n = \frac{(2n)!}{2^n (n!)^2} = \frac{1 \cdot 3 \cdot 5 \cdots (2n-1)}{n!}, \qquad n = 1, 2, \cdots,$$

the reason being that for this choice of c_n all those polynomials will have the value 1 when $x = 1$; this follows from (14) in Prob. 9. We then obtain from (8) and (9)

(9*)
$$c_{n-2} = -\frac{n(n-1)}{2(2n-1)} c_n = -\frac{n(n-1)(2n)!}{2(2n-1)2^n(n!)^2}$$

$$= -\frac{n(n-1)2n(2n-1)(2n-2)!}{2(2n-1)2^n n(n-1)! \, n(n-1)(n-2)!},$$

that is,

$$c_{n-2} = -\frac{(2n-2)!}{2^n(n-1)!(n-2)!}.$$

Similarly,

$$c_{n-4} = -\frac{(n-2)(n-3)}{4(2n-3)} c_{n-2} = \frac{(2n-4)!}{2^n 2! \, (n-2)!(n-4)!}$$

etc., and in general, when $n - 2m \geqq 0$,

(10)
$$c_{n-2m} = (-1)^m \frac{(2n-2m)!}{2^n m! \, (n-m)!(n-2m)!}.$$

The resulting solution of Legendre's equation is called the **Legendre polynomial** *of degree n* and is denoted by $P_n(x)$; from (10) we obtain

(11)
$$P_n(x) = \sum_{m=0}^{M} (-1)^m \frac{(2n-2m)!}{2^n m! \, (n-m)!(n-2m)!} x^{n-2m}$$

$$= \frac{(2n)!}{2^n(n!)^2} x^n - \frac{(2n-2)!}{2^n 1! \, (n-1)!(n-2)!} x^{n-2} + - \cdots,$$

where $M = n/2$ or $(n-1)/2$, whichever is an integer. In particular (Fig. 44),

(11′)
$$P_0(x) = 1, \qquad\qquad\qquad P_1(x) = x,$$
$$P_2(x) = \tfrac{1}{2}(3x^2 - 1), \qquad\qquad P_3(x) = \tfrac{1}{2}(5x^3 - 3x),$$
$$P_4(x) = \tfrac{1}{8}(35x^4 - 30x^2 + 3), \qquad P_5(x) = \tfrac{1}{8}(63x^5 - 70x^3 + 15x),$$

etc.

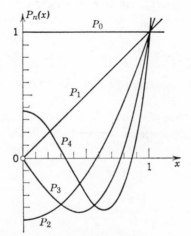

Fig. 44. Legendre polynomials

Problems

1. Using Fig. 44 and formula (11'), graph $P_0, P_1, \cdots, P_5$ in the interval $-1 \leq x \leq 1$. Count the number of zeros.
2. Find and graph $P_6(x)$.
3. Using (11'), verify by substitution that $P_0, \cdots, P_5$ satisfy Legendre's equation.
4. Derive (11') from (11).
5. Show that the series (6) and (7) have radius of convergence 1.
6. Using (11), show that $P_n(-x) = (-1)^n P_n(x)$ and $P_n{}'(-x) = (-1)^{n+1} P_n{}'(x)$.
7. Applying the binomial theorem to $(x^2 - 1)^n$, differentiating n times term by term, and comparing with (11), show that

(12) $$P_n(x) = \frac{1}{2^n n!} \frac{d^n}{dx^n} [(x^2 - 1)^n] \qquad \text{(Rodrigues' formula).}$$

8. Using (12) and integrating n times by parts, show that

(13) $$\int_{-1}^{1} P_n{}^2(x) \, dx = \frac{2}{2n + 1} \qquad (n = 0, 1, \cdots).$$

9. Show that

(14) $$\frac{1}{\sqrt{1 - 2xu + u^2}} = \sum_{n=0}^{\infty} P_n(x) u^n.$$

Hint. Start from the binomial expansion of $1/\sqrt{1 - v}$, set $v = 2xu - u^2$, multiply the powers of $2xu - u^2$ out, collect all the terms involving u^n, and verify that the sum of these terms is $P_n(x) u^n$.

10. Let A_1 and A_2 be two points in space (Fig. 45, $r_2 > 0$). Using (14), show that

$$\frac{1}{r} = \frac{1}{\sqrt{r_1{}^2 + r_2{}^2 - 2r_1 r_2 \cos \theta}} = \frac{1}{r_2} \left[P_0 + P_1(\cos \theta) \frac{r_1}{r_2} + P_2(\cos \theta) \left(\frac{r_1}{r_2}\right)^2 + \cdots \right].$$

Fig. 45. Problem 10

3.4   Extended Power Series Method. Indicial Equation

Several second-order differential equations, which are highly important in many applications, have coefficients which are not analytic at $x = 0$ (definition in Sec. 3.2), but are such that the following theorem can be applied.

Theorem 1.   *Any differential equation of the form*

(1) $$y'' + \frac{a(x)}{x} y' + \frac{b(x)}{x^2} y = 0,$$

where the functions $a(x)$ and $b(x)$ are analytic at $x = 0$, has at least one solution *which can be represented in the form*

(2)
$$y(x) = x^r \sum_{m=0}^{\infty} c_m x^m = x^r(c_0 + c_1 x + c_2 x^2 + \cdots) \qquad (c_0 \neq 0)$$

where the exponent r may be any (real or complex) number (and is chosen so that $c_0 \neq 0$).[4]

The *proof* requires advanced methods of complex analysis and can be found in Ref. [B8], Appendix 1. The following method for solving (1) is called the **Frobenius' method.**[5]

To solve (1), we write it in the somewhat more convenient form

(1')
$$x^2 y'' + x a(x) y' + b(x) y = 0.$$

We first expand $a(x)$ and $b(x)$ in power series,

$$a(x) = a_0 + a_1 x + a_2 x^2 + \cdots, \qquad b(x) = b_0 + b_1 x + b_2 x^2 + \cdots.$$

Then we differentiate (2) term by term, finding

$$y'(x) = \sum_{m=0}^{\infty} (m + r) c_m x^{m+r-1} = x^{r-1}[r c_0 + (r + 1) c_1 x + \cdots],$$

$$y''(x) = \sum_{m=0}^{\infty} (m + r)(m + r - 1) c_m x^{m+r-2}$$

$$= x^{r-2}[r(r - 1) c_0 + (r + 1) r c_1 x + \cdots].$$

By inserting all these series into (1') we readily obtain

$$x^r[r(r - 1) c_0 + \cdots] + (a_0 + a_1 x + \cdots) x^r (r c_0 + \cdots)$$
$$+ (b_0 + b_1 x + \cdots) x^r (c_0 + c_1 x + \cdots) = 0.$$

We now equate the sum of the coefficients of each power of x to zero, as before. This yields a system of equations involving the unknown coefficients c_m. The smallest power is x^r, and the corresponding equation is

$$[r(r - 1) + a_0 r + b_0] c_0 = 0.$$

Since by assumption $c_0 \neq 0$, we obtain

(3)
$$r^2 + (a_0 - 1) r + b_0 = 0.$$

This important quadratic equation is called the **indicial equation** of the differential equation (1). We shall see that our method will yield a fundamental system of solutions; one of the solutions will always be of the form (2), but for the form of the other solution there will be three different possibilities corresponding to the following cases.

---

[4] In this theorem, the variable x may be replaced by $x - a$ where a is any number.
[5] GEORG FROBENIUS (1849–1917), German mathematician, who also made important contributions to the theory of matrices and groups.

Case 1. *The roots of the indicial equation are distinct and do not differ by an integer*
$(1, 2, 3, \cdots)$.[6]
Case 2. *The indicial equation has a double root.*
Case 3. *The roots of the indicial equation differ by an integer.*

Let us discuss these cases separately.

Case 1. Distinct roots of the indicial equation not differing by an integer. This
is the simplest case. Let r_1 and r_2 be the roots of (3). If we insert $r = r_1$ into the
aforementioned system of equations and determine the coefficients c_1, c_2, $\cdots$
successively, as before, then we obtain a solution

$$(4) \qquad y_1(x) = x^{r_1}(c_0 + c_1 x + c_2 x^2 + \cdots).$$

It can be shown (cf. Ref. [B8]) that in the present case the differential equation
has another independent solution $y_2(x)$ of a similar form, namely

$$(5) \qquad y_2(x) = x^{r_2}(c_0{}^* + c_1{}^* x + c_2{}^* x^2 + \cdots),$$

which can be obtained by inserting $r = r_2$ into that system of equations and
determining the coefficients $c_0{}^*$, $c_1{}^*$, $\cdots$ successively. Linear independence of
y_1 and y_2 follows from the fact that y_1/y_2 is not constant because $r_1 - r_2$ is not an
integer. Hence y_1 and y_2 form a fundamental system of (1) on the interval of
convergence of both series.

Example 1 (Case 1). We consider the differential equation

$$x^2 y'' + (x^2 + \tfrac{5}{36})y = 0.$$

By substituting (2) into this equation we obtain

$$(6) \qquad \sum_{m=0}^{\infty} (m + r)(m + r - 1)c_m x^{m+r} + \sum_{m=0}^{\infty} c_m x^{m+r+2} + \frac{5}{36} \sum_{m=0}^{\infty} c_m x^{m+r} = 0.$$

By equating the sum of the coefficients of x^r to zero we obtain the indicial equation

$$r(r - 1) + \tfrac{5}{36} = 0.$$

The roots are $r_1 = \tfrac{5}{6}$ and $r_2 = \tfrac{1}{6}$. By equating the sum of the coefficients of x^{r+s} in (6) to zero, we
find

$$(7) \qquad \begin{array}{lll} (a) & [(r + 1)r + \tfrac{5}{36}]c_1 = 0 & (s = 1), \\[2mm] (b) & (s + r)(s + r - 1)c_s + c_{s-2} + \tfrac{5}{36}c_s = 0 & (s = 2, 3, \cdots). \end{array}$$

Let us first determine a solution $y_1(x)$ corresponding to $r_1 = \tfrac{5}{6}$. For this value of r, equation
(7a) yields $c_1 = 0$. Equation (7b) may be written

$$(8) \qquad s(s + \tfrac{2}{3})c_s + c_{s-2} = 0.$$

From $c_1 = 0$ it follows that $c_3 = 0$, $c_5 = 0$, $\cdots$, successively. If we set $s = 2p$ in (8), we obtain

$$c_{2p} = -\frac{3}{4} \cdot \frac{c_{2p-2}}{p(3p + 1)} \qquad (p = 1, 2, \cdots).$$

---

[6] Note that this case includes complex conjugate roots r_1 and $r_2 = \bar{r}_1$, because $r_1 - r_2 = r_1 - \bar{r}_1 = 2i \operatorname{Im} r_1 \neq \pm 1, \pm 2, \cdots$.

From this formula we see that the coefficients are

$$c_2 = -\frac{3}{4} \cdot \frac{c_0}{4}$$

$$c_4 = -\frac{3}{4} \cdot \frac{c_2}{2 \cdot 7} = \left(\frac{3}{4}\right)^2 \frac{c_0}{2! \, 4 \cdot 7}$$

$$c_6 = -\frac{3}{4} \cdot \frac{c_4}{3 \cdot 10} = -\left(\frac{3}{4}\right)^3 \frac{c_0}{3! \, 4 \cdot 7 \cdot 10} \qquad \text{etc.}$$

This yields the solution

$$(9) \quad y_1(x) = c_0 \sum_{p=0}^{\infty} (-1)^p \left(\frac{3}{4}\right)^p \frac{x^{2p+5/6}}{p! \, 1 \cdot 4 \cdots (3p+1)} = c_0 x^{5/6} \left(1 - \frac{3}{16} x^2 + \frac{9}{896} x^4 - + \cdots\right).$$

Another solution $y_2(x)$ corresponding to the root $r_2 = \frac{1}{6}$ may be obtained in a similar fashion. Denoting the coefficients of this solution by $c_0{}^*, c_1{}^*, \cdots$, instead of (7) we now have

$$(10) \qquad \begin{array}{ll} (a) & [(r_2+1)r_2 + \frac{5}{36}]c_1{}^* = 0 \\[2mm] (b) & (s+r_2)(s+r_2-1)c_s{}^* + c_{s-2}^* + \frac{5}{36}c_s{}^* = 0 \qquad (s = 2, 3, \cdots). \end{array}$$

From (10a) we see that $c_1{}^* = 0$, and (10b) may be written

$$(11) \qquad s(s - \tfrac{2}{3})c_s{}^* + c_{s-2}^* = 0 \qquad (s = 2, 3, \cdots).$$

Since $c_1{}^* = 0$, it follows that $c_3{}^* = 0, c_5{}^* = 0, \cdots$, successively. If we set $s = 2p$ in (11), we obtain

$$c_{2p}{}^* = -\frac{3}{4} \cdot \frac{c_{2p-2}^*}{p(3p-1)} \qquad (p = 1, 2, \cdots).$$

From this formula we find

$$c_2{}^* = -\frac{3}{4} \cdot \frac{c_0{}^*}{2}$$

$$c_4{}^* = -\frac{3}{4} \cdot \frac{c_2{}^*}{2 \cdot 5} = \left(\frac{3}{4}\right)^2 \frac{c_0{}^*}{2! \, 2 \cdot 5}$$

etc., and the corresponding solution is

$$(12) \quad y_2(x) = c_0{}^* + c_0{}^* \sum_{p=1}^{\infty} (-1)^p \left(\frac{3}{4}\right)^p \frac{x^{2p+1/6}}{p! \, 2 \cdot 5 \cdots (3p-1)} = c_0{}^* x^{1/6} \left(1 - \frac{3}{8} x^2 + - \cdots\right).$$

The series (9) and (12) converge for all x, and y_1 and y_2 are linearly independent. Hence these functions constitute a fundamental system for all values of x. (In the next section we shall see that our equation is closely related to Bessel's equation with the parameter $\nu = \frac{1}{3}$.)

*Case 2. **Double root of the indicial equation.*** The indicial equation (3) has a double root r if, and only if, $(a_0 - 1)^2 - 4b = 0$, and then $r = (1 - a_0)/2$. We may determine a first solution

$$(13) \qquad y_1(x) = x^r(c_0 + c_1 x + c_2 x^2 + \cdots) \qquad \left(r = \frac{1 - a_0}{2}\right)$$

as before. To find another solution we may apply the method of variation of parameters, that is, we replace the constant c in the solution cy_1 by a function

$u(x)$ to be determined such that

(14) $$y_2(x) = u(x)y_1(x)$$

is a solution of (1). By inserting (14) and the derivatives

$$y_2' = u'y_1 + uy_1', \qquad y_2'' = u''y_1 + 2u'y_1' + uy_1''$$

into the differential equation (1') we obtain

$$x^2(u''y_1 + 2u'y_1' + uy_1'') + xa(u'y_1 + uy_1') + buy_1 = 0.$$

Since y_1 is a solution of (1'), the sum of the terms involving u is zero, and the last equation reduces to

$$x^2 y_1 u'' + 2x^2 y_1' u' + xay_1 u' = 0.$$

By dividing by $x^2 y_1$ and inserting the power series for a we obtain

$$u'' + \left(2\frac{y_1'}{y_1} + \frac{a_0}{x} + \cdots\right)u' = 0.$$

Here and in the following the dots designate terms which are constant or involve positive powers of x. Now from (13) it follows that

$$\frac{y_1'}{y_1} = \frac{x^{r-1}[rc_0 + (r+1)c_1 x + \cdots]}{x^r[c_0 + c_1 x + \cdots]}$$

$$= \frac{1}{x}\left(\frac{rc_0 + (r+1)c_1 x + \cdots}{c_0 + c_1 x + \cdots}\right) = \frac{r}{x} + \cdots.$$

Hence the last equation can be written

(15) $$u'' + \left(\frac{2r + a_0}{x} + \cdots\right)u' = 0.$$

Since $r = (1 - a_0)/2$ the term $(2r + a_0)/x$ equals $1/x$, and by dividing by u' we thus have

$$\frac{u''}{u'} = -\frac{1}{x} + \cdots.$$

By integration we obtain

$$\ln u' = -\ln x + \cdots \qquad \text{or} \qquad u' = \frac{1}{x} e^{(\cdots)}.$$

Expanding the exponential function in powers of x and integrating once more, we see that the expression for u will be of the form

$$u = \ln x + k_1 x + k_2 x^2 + \cdots.$$

By inserting this into (14) we have the following result.

Form of the solutions in the case of a double root of the indicial equation. *If the indicial equation (3) has a double root r, then linearly independent solutions of (1)* *are of the form* (13) *and*

(16) $$y_2(x) = y_1(x) \ln x + x^r \sum_{m=1}^{\infty} A_m x^m.$$

The simplest illustration of this case is given by the Cauchy equation (Sec. 2.7) in the critical case (cf. also Ex. 5, below).

*Example 2 (**Case 2**).* Solve the differential equation

(17) $$x(x-1)y'' + (3x-1)y' + y = 0.$$

Writing this equation in the standard form (1), we see that it satisfies the assumptions in Theorem 1. By inserting (2) and its derivatives into (17) we obtain

(18) $$\sum_{m=0}^{\infty} (m+r)(m+r-1)c_m x^{m+r} - \sum_{m=0}^{\infty} (m+r)(m+r-1)c_m x^{m+r-1}$$
$$+ 3\sum_{m=0}^{\infty} (m+r)c_m x^{m+r} - \sum_{m=0}^{\infty} (m+r)c_m x^{m+r-1} + \sum_{m=0}^{\infty} c_m x^{m+r} = 0.$$

The smallest power is x^{r-1}; by equating the sum of its coefficients to zero we have

$$[-r(r-1) - r]c_0 = 0 \qquad \text{or} \qquad r^2 = 0.$$

Hence this indicial equation has the double root $r = 0$. We insert this value into (18) and equate the sum of the coefficients of the power x^s to zero, finding

$$s(s-1)c_s - (s+1)sc_{s+1} + 3sc_s - (s+1)c_{s+1} + c_s = 0$$

or $c_{s+1} = c_s$. Hence $c_0 = c_1 = c_2 = \cdots$, and by choosing $c_0 = 1$ we obtain the solution

$$y_1(x) = \sum_{m=0}^{\infty} x^m = \frac{1}{1-x}.$$

To derive a second solution we substitute $y_2 = uy_1$ and its derivatives into (17), finding

$$x(x-1)(u''y_1 + 2u'y_1' + uy_1'') + (3x-1)(u'y_1 + uy_1') + uy_1 = 0.$$

Since y_1 is a solution of (17), this reduces to

$$x(x-1)(u''y_1 + 2u'y_1') + (3x-1)u'y_1 = 0.$$

Inserting the expressions for y_1 and y_1' and simplifying, we obtain the differential equation

$$xu'' + u' = 0 \qquad \text{or} \qquad \frac{u''}{u'} = -\frac{1}{x}.$$

By integrating twice we finally obtain $u = \ln x$, and a second solution for all positive x is

$$y_2 = uy_1 = \frac{\ln x}{1-x}.$$

Case 3. Roots of the indicial equation differing by an integer. If the roots r_1 and r_2 of the indicial equation (3) differ by an integer, say, $r_1 = r$ and $r_2 = r - p$, where p is a positive integer, then we may always determine one solution as before, namely, the solution

$$y_1(x) = x^{r_1}(c_0 + c_1 x + c_2 x^2 + \cdots)$$

corresponding to the root r_1. However, for the root r_2 there may arise difficulties, and it may not be possible to determine a solution y_2 by the method used in Case 1 (cf. Ex. 3, below). To determine y_2, we may proceed as in Case 2. The first steps are literally the same and yield equation (15). We determine $2r + a_0$ in (15). From elementary algebra we know that the coefficient $a_0 - 1$ in (3) equals $-(r_1 + r_2)$. In our case, $r_1 = r$, $r_2 = r - p$, and, therefore, $a_0 - 1 = p - 2r$. Hence in (15) we have $2r + a_0 = p + 1$, and we thus obtain

$$\frac{u''}{u'} = -\left(\frac{p+1}{x} + \cdots\right).$$

The further steps are as in Case 2. Integrating, we find

$$\ln u' = -(p+1)\ln x + \cdots \qquad \text{or} \qquad u' = x^{-(p+1)}e^{(\cdots)}$$

where the dots stand for some series of nonnegative powers of x. By expanding the exponential function as before we obtain a series of the form

$$u' = \frac{1}{x^{p+1}} + \frac{k_1}{x^p} + \cdots + \frac{k_p}{x} + k_{p+1} + k_{p+2}x + \cdots.$$

By integrating once more we have

$$u = -\frac{1}{px^p} - \cdots + k_p \ln x + k_{p+1}x + \cdots.$$

Multiplying this expression by the series

$$y_1 = x^{r_1}(c_0 + c_1 x + \cdots)$$

and remembering that $r_1 - p = r_2$ we see that $y_2 = uy_1$ is of the form

(19) $$y_2(x) = k_p y_1(x) \ln x + x^{r_2} \sum_{m=0}^{\infty} C_m x^m \qquad (x > 0).$$

We may sum up our result as follows.

Form of the solutions when the roots of the indicial equation differ by an integer. *If the roots of the indicial equation* (3) *differ by an integer, then to the root with the larger real part* (say, r_1) *there corresponds a solution of* (1) *of the form* (4), *and to the other root* (r_2) *there corresponds a solution of the form* (19). *The solutions form a fundamental system of* (1).

While for a double root of (3) the second solution always contains a logarithmic term, the coefficient k_p in (19) may be zero and in the present case the logarithmic term may be missing (cf. Ex. 4).[7]

Example 3 (Case 3). We consider

$$(x^2 - 1)x^2 y'' - (x^2 + 1)xy' + (x^2 + 1)y = 0.$$

[7] A general theory of convergence of those series will not be presented here, but in each individual case convergence may be tested in the usual way.

By substituting (2) and its derivatives into the differential equation we first have

$$(x^2-1)\sum_{m=0}^{\infty}(m+r)(m+r-1)c_m x^{m+r}$$

$$-(x^2+1)\sum_{m=0}^{\infty}(m+r)c_m x^{m+r}+(x^2+1)\sum_{m=0}^{\infty}c_m x^{m+r}=0.$$

Performing the indicated multiplications and simplifying, we obtain

(20) $$\sum_{m=0}^{\infty}(m+r-1)^2 c_m x^{m+r+2}-\sum_{m=0}^{\infty}(m+r+1)(m+r-1)c_m x^{m+r}=0.$$

By equating the coefficient of x^r to zero we see that the indicial equation is

$$(r+1)(r-1)=0.$$

The roots $r_1=1$ and $r_2=-1$ differ by an integer. By equating the coefficient of x^{r+1} in (20) to zero we have

$$-(r+2)rc_1=0.$$

For $r=r_1$ as well as for $r=r_2$ this implies $c_1=0$. By equating the sum of the coefficients of x^{s+r+2} in (20) to zero we find

(21) $$(s+r-1)^2 c_s=(s+r+3)(s+r+1)c_{s+2}$$ $$(s=0,1,\cdots).$$

Inserting $r=r_1=1$ and solving for c_{s+2}, we see that

$$c_{s+2}=\frac{s^2}{(s+4)(s+2)}c_s$$ $$(s=0,1,\cdots).$$

Since $c_1=0$, it follows that $c_3=0$, $c_5=0$, etc. For $s=0$ we obtain $c_2=0$, and from this by taking $s=2,4,\cdots$, we have $c_4=0$, $c_6=0$, etc. Hence to the larger root $r_1=1$ there corresponds the solution

$$y_1=c_0 x.$$

Let us determine another solution. If we insert the smaller root $r=r_2=-1$ into (21), we get

$$(s-2)^2 c_s=(s+2)sc_{s+2}$$ $$(s=0,1,\cdots).$$

With $s=0$ this becomes $4c_0=0$, which implies $c_0=0$. This contradicts our initial assumption $c_0\neq 0$. In fact, we shall now see that our equation does not have a second independent solution $y_2(x)$ of the form (2) corresponding to the smaller root r_2. From (19) we see that $y_2(x)$ must be of the form

$$y_2(x)=kx\ln x+\frac{1}{x}\sum_{m=0}^{\infty}C_m x^m.$$

Substituting this and the derivatives into the differential equation, we obtain

$$(x^2-1)x^2\left(\frac{k}{x}+\sum_{m=0}^{\infty}(m-1)(m-2)C_m x^{m-3}\right)$$

$$-(x^2+1)x\left(k\ln x+k+\sum_{m=0}^{\infty}(m-1)C_m x^{m-2}\right)+(x^2+1)\left(kx\ln x+\sum_{m=0}^{\infty}C_m x^{m-1}\right)=0.$$

The logarithmic terms drop out, and simplification yields

$$-2kx+\sum_{m=0}^{\infty}(m-2)^2 C_m x^{m+1}-\sum_{m=0}^{\infty}m(m-2)C_m x^{m-1}=0.$$

By equating the sum of the coefficients of x^s to zero we obtain $C_1 = 0$ ($s = 0$), $-2k + 4C_0 = 0$ ($s = 1$), and

$$(s - 3)^2 C_{s-1} - (s^2 - 1)C_{s+1} = 0 \qquad\qquad (s = 2, 3, \cdots).$$

Solving for C_{s+1}, we have

$$C_{s+1} = \frac{(s - 3)^2}{s^2 - 1}\, C_{s-1} \qquad\qquad (s = 2, 3, \cdots).$$

Since $C_1 = 0$, we see that $C_3 = 0$, $C_5 = 0$, etc. $s = 3$ yields $C_4 = 0$, hence $C_6 = 0$, $C_8 = 0$, etc. $k = 2C_0$ and C_2 remain arbitrary. Consequently,

$$y_2(x) = 2C_0 x \ln x + \frac{1}{x}\,(C_0 + C_2 x^2).$$

Since the last term is $C_2 y_1/c_0$, we may take $C_2 = 0$. With $C_0 = \frac{1}{2}$,

$$y_2(x) = x \ln x + \frac{1}{2x}.$$

In Cases 2 and 3 it may sometimes be simpler to obtain a second independent solution by the method of variation of parameters.

Let us illustrate this for the equation under consideration. Since x is a solution, we may set

$$y_2(x) = xu(x).$$

Substituting this into the equation and simplifying, we find

$$(x^3 - x)u'' + (x^2 - 3)u' = 0.$$

Using partial fractions, we obtain

$$\frac{u''}{u'} = \frac{3 - x^2}{x^3 - x} = -\frac{3}{x} + \frac{1}{x + 1} + \frac{1}{x - 1}.$$

Integrating on both sides, we have

$$\ln u' = -3 \ln x + \ln (x + 1) + \ln (x - 1) = \ln \frac{x^2 - 1}{x^3}.$$

Taking exponentials and integrating again, we find

$$u = \ln x + 1/(2x^2).$$

This yields the same expression as before, namely,

$$y_2(x) = xu(x) = x \ln x + \frac{1}{2x}.$$

Example 4 (Case 3, second solution without logarithmic term). We consider the differential equation

$$x^2 y'' + xy' + (x^2 - \tfrac{1}{4})y = 0.$$

Substituting (2) and the derivatives into this equation, we obtain

$$\sum_{m=0}^{\infty} [(m + r)(m + r - 1) + (m + r) - \tfrac{1}{4}]c_m x^{m+r} + \sum_{m=0}^{\infty} c_m x^{m+r+2} = 0.$$

By equating the coefficient of x^r to zero we get the indicial equation

$$r(r - 1) + r - \tfrac{1}{4} = 0 \qquad \text{or} \qquad r^2 = \tfrac{1}{4}.$$

The roots $r_1 = \frac{1}{2}$ and $r_2 = -\frac{1}{2}$ differ by an integer. By equating the sum of the coefficients of

x^{s+r} to zero we find

(22)
$$\text{(a)} \qquad [(r+1)r + (r+1) - \tfrac{1}{4}]c_1 = 0 \qquad\qquad (s = 1)$$
$$\text{(b)} \qquad [(s+r)(s+r-1) + s + r - \tfrac{1}{4}]c_s + c_{s-2} = 0 \qquad (s = 2, 3, \cdots).$$

For $r = r_1 = \frac{1}{2}$, equation (22a) yields $c_1 = 0$, and (22b) becomes

$$(23) \qquad\qquad\qquad (s+1)sc_s + c_{s-2} = 0.$$

From this and $c_1 = 0$ we obtain $c_3 = 0$, $c_5 = 0$, etc. Solving (23) for c_s and setting $s = 2p$, we get

$$c_{2p} = -\frac{c_{2p-2}}{2p(2p+1)} \qquad\qquad (p = 1, 2, \cdots).$$

Hence the nonzero coefficients are

$$c_2 = -\frac{c_0}{3!}, \qquad c_4 = -\frac{c_2}{4\cdot 5} = \frac{c_0}{5!}, \qquad c_6 = -\frac{c_0}{7!}, \qquad \text{etc.,}$$

and we obtain the solution

$$(24) \qquad y_1(x) = c_0 x^{1/2} \sum_{m=0}^{\infty} \frac{(-1)^m x^{2m}}{(2m+1)!} = c_0 x^{-1/2} \sum_{m=0}^{\infty} \frac{(-1)^m x^{2m+1}}{(2m+1)!} = c_0 \frac{\sin x}{\sqrt{x}}.$$

From (19) we see that a second independent solution is of the form

$$y_2(x) = ky_1(x) \ln x + x^{-1/2} \sum_{m=0}^{\infty} C_m x^m.$$

If we substitute this and the derivatives into the differential equation, we see that the three expressions involving $\ln x$ and the expressions ky_1 and $-ky_1$ drop out. Simplifying the remaining equation, we thus obtain

$$2kxy_1{}' + \sum_{m=0}^{\infty} m(m-1)C_m x^{m-1/2} + \sum_{m=0}^{\infty} C_m x^{m+3/2} = 0.$$

From (24) we find $2kxy' = -kc_0 x^{1/2} + \cdots$. Since there is no further term involving $x^{1/2}$ and $c_0 \neq 0$, we must have $k = 0$. The sum of the coefficients of the power $x^{s-1/2}$ is

$$s(s-1)C_s + C_{s-2} \qquad\qquad (s = 2, 3, \cdots).$$

Equating this to zero and solving for C_s, we have

$$C_s = -\frac{C_{s-2}}{s(s-1)} \qquad\qquad (s = 2, 3, \cdots).$$

We thus obtain

$$C_2 = -\frac{C_0}{2!}, \qquad C_4 = -\frac{C_2}{4\cdot 3} = \frac{C_0}{4!}, \qquad C_6 = -\frac{C_0}{6!}, \qquad \text{etc.,}$$

$$C_3 = -\frac{C_1}{3!}, \qquad C_5 = -\frac{C_3}{5\cdot 4} = \frac{C_1}{5!}, \qquad C_7 = -\frac{C_1}{7!}, \qquad \text{etc.}$$

We may take $C_1 = 0$, because the odd powers would yield $C_1 y_1/c_0$. Then

$$(25) \qquad y_2(x) = C_0 x^{-1/2} \sum_{m=0}^{\infty} \frac{(-1)^m x^{2m}}{(2m)!} = C_0 \frac{\cos x}{\sqrt{x}}.$$

$y_1(x)$ and $y_2(x)$ constitute a fundamental system of solutions for all $x \neq 0$.

Example 5 (*Cauchy equation*). In Sec. 2.7 we have seen that the Cauchy equation

$$(26) \qquad\qquad\qquad x^2 y'' + a_0 xy' + b_0 y = 0 \qquad\qquad (a_0, b_0 \text{ constant})$$

can be solved by substituting $y = x^r$ into (26). This yields the auxiliary equation

$$r^2 + (a_0 - 1)r + b_0 = 0.$$

Obviously, this is the indicial equation. If $r_1 \neq r_2$, then $y_1 = x^{r_1}$, $y_2 = x^{r_2}$ form a fundamental system; this corresponds to our present Cases 1 and 3. If $r_1 = r_2 = r$, then the solutions

$$y_1 = x^r, \qquad y_2 = y_1 \ln x = x^r \ln x$$

form a fundamental system for all positive x, as follows by substituting (16) into (26), in agreement with the result in Sec. 2.7.

Problems

Using the Frobenius method, find a fundamental system of solutions of the following equations. In each case, try to identify the series solutions as expansions of known functions.

1. $x(1 - x)y'' + 2(1 - 2x)y' - 2y = 0$ **2.** $16x^2y'' + 3y = 0$

3. $xy'' + y' - xy = 0$ **4.** $xy'' + 2y' + xy = 0$

5. $x^2y'' + 6xy' + (6 - x^2)y = 0$ **6.** $x^2y'' + 6xy' + (x^2 + 6)y = 0$

7. $x^2y'' + 0.5xy' - 1.5y = 0$ **8.** $xy'' - y = 0$

9. $xy'' + (1 - 2x)y' + (x - 1)y = 0$ **10.** $x^2y'' + x^3y' + (x^2 - 2)y = 0$

11. $x^2y'' + 4xy' + (x^2 + 2)y = 0$ **12.** $(x + 1)^2y'' + (x + 1)y' - y = 0$

13. $xy'' + 3y' + 4x^3y = 0$ **14.** $2x(x - 1)y'' - (x + 1)y' + y = 0$

15. $x(x + 1)^2y'' + (1 - x^2)y' + (x - 1)y = 0$

16. $(1 + x)x^2y'' - (1 + 2x)xy' + (1 + 2x)y = 0$

17. $2x(x - 1)y'' - (4x^2 - 3x + 1)y' + (2x^2 - x + 2)y = 0$

18. $x^2(x + 1)^2y'' - (5x^2 + 8x + 3)xy' + (9x^2 + 11x + 4)y = 0$

19. $2x(1 - 2x)y'' + (12x^2 - 4x + 1)y' - 2(4x^2 + 1)y = 0$

20. $2x^2(1 - 2x)y'' + (4x^2 - 8x + 5)xy' + (2x^2 - x + 1)y = 0$

21. (Gauss's[8] hypergeometric equation)   The differential equation

(27) $$x(1 - x)y'' + [c - (a + b + 1)x]y' - aby = 0$$

$(a, b, c$ constant) is known as **Gauss's hypergeometric equation.** Show that the roots of the corresponding indicial equation are $r_1 = 0$ and $r_2 = 1 - c$.

22. Show that for $r_1 = 0$ in Prob. 21 the Frobenius method yields the solution

$$y_1(x) = 1 + \frac{ab}{1!\,c}\,x + \frac{a(a + 1)b(b + 1)}{2!\,c(c + 1)}\,x^2$$

(28) $$+ \frac{a(a + 1)(a + 2)b(b + 1)(b + 2)}{3!\,c(c+1)(c+2)}\,x^3 + \cdots$$

$$(c \neq 0, -1, -2, \cdots).$$

This series is called the **hypergeometric series;** its sum $y_1(x)$ is usually denoted by $F(a, b, c; x)$ and is called the **hypergeometric function.** It satisfies numerous relations, and many elementary functions are special cases of $F(a, b, c; x)$; cf. the following problems.

[8] CARL FRIEDRICH GAUSS (1777–1855), great German mathematician, whose work was of basic importance in algebra, number theory, differential equations, differential geometry, non-Euclidean geometry, complex analysis, numerical analysis, and theoretical mechanics. He also paved the way for a general and systematic use of complex numbers.

Using (28), prove the following statements and relations.

23. The series (28) converges for $|x| < 1$.

24. $F(1, b, b; x) = 1 + x + x^2 + \cdots$, the geometric series.

25. If a or b is a negative integer, (28) reduces to a polynomial.

26. $\dfrac{dF(a, b, c; x)}{dx} = \dfrac{ab}{c} F(a + 1, b + 1, c + 1; x)$,

$$\frac{d^2 F(a, b, c; x)}{dx^2} = \frac{a(a + 1)b(b + 1)}{c(c + 1)} F(a + 2, b + 2, c + 2; x), \text{ etc.}$$

27. $(1 + x)^n = F(-n, b, b; -x)$, $\ln(1 + x) = xF(1, 1, 2; -x)$

28. $\arcsin x = xF(\tfrac{1}{2}, \tfrac{1}{2}, \tfrac{3}{2}; x^2)$, $\ln \dfrac{1 + x}{1 - x} = 2xF(\tfrac{1}{2}, 1, \tfrac{3}{2}; x^2)$

29. Show that for $r_2 = 1 - c$ in Prob. 21 the Frobenius method yields the solution

(29)
$$y_2(x) = x^{1-c} \left[1 + \frac{(a - c + 1)(b - c + 1)}{1! \, (-c + 2)} x \right.$$
$$\left. + \frac{(a - c + 1)(a - c + 2)(b - c + 1)(b - c + 2)}{2! \, (-c + 2)(-c + 3)} x^2 + \cdots \right]$$
$$(c \neq 2, 3, 4, \cdots).$$

30. Show that in Prob. 29,

$$y_2(x) = x^{1-c} F(a - c + 1, b - c + 1, 2 - c; x).$$

31. Show that if $c \neq 0, \pm 1, \pm 2, \cdots$, the functions y_1 and y_2 (Probs. 22, 29) constitute a fundamental system of solutions of the hypergeometric equation (27).

Solve, in terms of hypergeometric functions, the following differential equations.

32. $3x(1 - x)y'' + (1 - 5x)y' + y = 0$

33. $4x(x - 1)y'' + 2(4x - 1)y' + y = 0$

34. $2x(x - 1)y'' + (6x + 1)y' + 2y = 0$

35. Consider the differential equation

(30)
$$(t^2 + At + B)\ddot{y} + (Ct + D)\dot{y} + Ky = 0$$

where A, B, C, D, K are constants and $t^2 + At + B$ has distinct zeros t_1 and t_2. Show that by introducing the new independent variable $x = (t - t_1)/(t_2 - t_1)$ the equation (30) becomes the hypergeometric equation where $Ct_1 + D = -c(t_2 - t_1)$, $C = a + b + 1$, $K = ab$.

3.5 Bessel's Equation.
Bessel Functions of the First Kind

One of the most important differential equations in applied mathematics is **Bessel's differential equation**[9]

(1)
$$x^2 y'' + xy' + (x^2 - \nu^2)y = 0$$

where the parameter ν is a given number. We assume that ν is real and non-negative. This equation is of the type characterized in Theorem 1, Sec. 3.4, and

[9] FRIEDRICH WILHELM BESSEL (1784–1846), German astronomer and mathematician.

therefore has a solution of the form

$$(2) \qquad\qquad y(x) = \sum_{m=0}^{\infty} c_m x^{m+r} \qquad\qquad (c_0 \neq 0).$$

Substituting this expression and its derivatives into Bessel's equation, we have

$$\sum_{m=0}^{\infty} (m + r)(m + r - 1)c_m x^{m+r} + \sum_{m=0}^{\infty} (m + r)c_m x^{m+r}$$

$$+ \sum_{m=0}^{\infty} c_m x^{m+r+2} - \nu^2 \sum_{m=0}^{\infty} c_m x^{m+r} = 0.$$

By equating the sum of the coefficients of x^{s+r} to zero we find

$$(a) \qquad\qquad r(r - 1)c_0 + rc_0 - \nu^2 c_0 = 0 \qquad\qquad (s = 0)$$

$$(3)\ (b) \qquad\qquad (r + 1)rc_1 + (r + 1)c_1 - \nu^2 c_1 = 0 \qquad\qquad (s = 1)$$

$$(c) \quad (s + r)(s + r - 1)c_s + (s + r)c_s + c_{s-2} - \nu^2 c_s = 0 \quad (s = 2, 3, \cdots).$$

From (3*a*) we obtain the indicial equation

$$(4) \qquad\qquad\qquad (r + \nu)(r - \nu) = 0.$$

The roots are $r_1 = \nu\ (\geq 0)$ and $r_2 = -\nu$.

Let us first determine a solution corresponding to the root r_1. For this value of r, equation (3*b*) yields $c_1 = 0$. Equation (3*c*) may be written

$$(s + r + \nu)(s + r - \nu)c_s + c_{s-2} = 0,$$

and for $r = \nu$ this takes the form

$$(5) \qquad\qquad\qquad (s + 2\nu)sc_s + c_{s-2} = 0.$$

Sinc $c_1 = 0$ and $\nu \geq 0$, it follows that $c_3 = 0$, $c_5 = 0, \cdots$, successively. If we set $s = 2m$ in (5), we obtain

$$(6) \qquad\qquad c_{2m} = -\frac{1}{2^2 m(\nu + m)} c_{2m-2}, \qquad m = 1, 2, \cdots$$

and may determine the coefficients $c_2, c_4, \cdots$ successively.

c_0 is arbitrary. It is customary to put

$$(7) \qquad\qquad\qquad c_0 = \frac{1}{2^\nu \Gamma(\nu + 1)}$$

where $\Gamma(\nu + 1)$ is the **gamma function.** Some basic formulas for this important function are included in Appendix 3. For the present purpose it suffices to know that $\Gamma(\alpha)$ is defined by the integral

$$(8) \qquad\qquad\qquad \Gamma(\alpha) = \int_0^{\infty} e^{-t} t^{\alpha-1}\, dt \qquad\qquad (\alpha > 0).$$

By integration by parts we obtain

$$\Gamma(\alpha + 1) = \int_0^{\infty} e^{-t} t^{\alpha}\, dt = -e^{-t} t^{\alpha} \Big|_0^{\infty} + \alpha \int_0^{\infty} e^{-t} t^{\alpha-1}\, dt.$$

The first expression on the right is zero, and the integral on the right is $\Gamma(\alpha)$. This yields the basic relation

(9)
$$\Gamma(\alpha + 1) = \alpha\Gamma(\alpha).$$

Since

$$\Gamma(1) = \int_0^\infty e^{-t} \, dt = 1,$$

we conclude from (9) that

$$\Gamma(2) = \Gamma(1) = 1!, \qquad \Gamma(3) = 2\Gamma(2) = 2!, \cdots$$

and in general

(10)
$$\Gamma(k + 1) = k! \qquad (k = 0, 1, \cdots).$$

This shows that the gamma function may be regarded as a generalization of the factorial function known from elementary calculus.

Returning to our problem, we obtain from (6), (7), and (9)

$$c_2 = -\frac{c_0}{2^2(\nu + 1)} = -\frac{1}{2^{2+\nu}1! \, \Gamma(\nu + 2)}$$

$$c_4 = -\frac{c_2}{2^2 2(\nu + 2)} = \frac{1}{2^{4+\nu}2! \, \Gamma(\nu + 3)}$$

and so on; in general,

(11)
$$c_{2m} = \frac{(-1)^m}{2^{2m+\nu}m! \, \Gamma(\nu + m + 1)}.$$

By inserting these coefficients in (2) and remembering that $c_1 = 0$, $c_3 = 0, \cdots$ we obtain a particular solution of Bessel's equation which is denoted by $J_\nu(x)$:

(12)
$$J_\nu(x) = x^\nu \sum_{m=0}^\infty \frac{(-1)^m x^{2m}}{2^{2m+\nu}m! \, \Gamma(\nu + m + 1)}.$$

This solution of (1) is known as the **Bessel function of the first kind** *of order* ν. The ratio test shows that the series converges for all x.

Integer values of ν are frequently denoted by n. Thus, for $n \geqq 0$,

(13)
$$J_n(x) = x^n \sum_{m=0}^\infty \frac{(-1)^m x^{2m}}{2^{2m+n}m! \, (n + m)!}.$$

Replacing ν by $-\nu$ in (12), we have

(14)
$$J_{-\nu}(x) = x^{-\nu} \sum_{m=0}^\infty \frac{(-1)^m x^{2m}}{2^{2m-\nu}m! \, \Gamma(m - \nu + 1)}.$$

Since Bessel's equation involves ν^2, the functions J_ν and $J_{-\nu}$ are solutions of the equation for the same ν. If ν is not an integer, they are linearly independent, because the first term in (12) and the first term in (14) are finite nonzero multiples of x^ν and $x^{-\nu}$, respectively. This yields the following result.

Theorem 1 (General solution). *If v is not an integer, a general solution of Bessel's equation for all $x \neq 0$ is*

(15) $$y(x) = a_1 J_v(x) + a_2 J_{-v}(x).$$

However, if v is an integer, then (15) is not a general solution. This follows from

Theorem 2 (Linear dependence). *For integer $v = n$ the Bessel functions $J_n(x)$ and $J_{-n}(x)$ are linearly dependent, because*

(16) $$J_{-n}(x) = (-1)^n J_n(x)$$ $(n = 1, 2, \cdots).$

Proof. We use (14) and let v approach a positive integer n. Then the gamma functions in the coefficients of the first n terms become infinite (cf. Fig. 389 in Appendix 3), the coefficients become zero, and the summation starts with $m = n$. Since in this case $\Gamma(m - n + 1) = (m - n)!$ [cf. (10)], we obtain

$$J_{-n}(x) = \sum_{m=n}^{\infty} \frac{(-1)^m x^{2m-n}}{2^{2m-n} m! \, (m-n)!} = \sum_{s=0}^{\infty} \frac{(-1)^{n+s} x^{2s+n}}{2^{2s+n} (n+s)! \, s!}$$

where $m = n + s$ and $s = m - n$. From (13) we see that the last series represents $(-1)^n J_n(x)$. This completes the proof.

A general solution of the Bessel equation with integer $v = n$ will be obtained in the next section. Numerical values of J_0 and J_1 are included in Appendix 4.

Problems

1. Show that (cf. Fig. 46)

(17) $$J_0(x) = 1 - \frac{x^2}{2^2(1!)^2} + \frac{x^4}{2^4(2!)^2} - \frac{x^6}{2^6(3!)^2} + - \cdots$$

and

(18) $$J_1(x) = \frac{x}{2} - \frac{x^3}{2^3 1! \, 2!} + \frac{x^5}{2^5 2! \, 3!} - \frac{x^7}{2^7 3! \, 4!} + - \cdots.$$

Fig. 46. Bessel functions of the first kind

2. Show that for small $|x|$ we have $J_0(x) \approx 1 - 0.25x^2$. Using this formula, compute $J_0(x)$ for $x = 0.1, 0.2, \cdots, 1.0$ and determine the relative error by comparing with Table A2 in Appendix 4.

3. Show that the series in (12) converges for all x.

4. How many terms of (17) are needed to compute $J_0(1)$ with an error less than 1 unit of the fifth decimal place? How many terms would be needed to compute $\ln 2$ from the Maclaurin series of $\ln (1 + x)$ with the same degree of accuracy?

5. It can be shown that for large x,

(19)
$$J_{2n}(x) \approx (-1)^n (\pi x)^{-\frac{1}{2}} (\cos x + \sin x)$$
$$J_{2n+1}(x) \approx (-1)^{n+1} (\pi x)^{-\frac{1}{2}} (\cos x - \sin x).$$

Using (19), graph $J_0(x)$ for large x, compute approximate values of the first five positive zeros of $J_0(x)$, and compare them with the more accurate values 2.405, 5.520, 8.654, 11.792, 14.931.

Differential equations reducible to Bessel's equation. There are various differential equations which can be reduced to Bessel's equation. To illustrate this, use the indicated substitutions and solve, in terms of Bessel functions, the following equations.

6. $x^2 y'' + xy' + (\lambda^2 x^2 - \nu^2)y = 0$ $(\lambda x = z)$
7. $4x^2 y'' + 4xy' + (x - \nu^2)y = 0$ $(\sqrt{x} = z)$
8. $x^2 y'' + xy' + 4(x^4 - \nu^2)y = 0$ $(x^2 = z)$
9. $xy'' - y' + xy = 0$ $(y = xu)$
10. $y'' + xy = 0$ $(y = u\sqrt{x}, \frac{2}{3}x^{3/2} = z)$
11. $xy'' + (1 + 2k)y' + xy = 0$ $(y = u/x^k)$
12. $x^2 y'' + (1 - 2\nu)xy' + \nu^2(x^{2\nu} + 1 - \nu^2)y = 0$ $(y = x^\nu u, x^\nu = z)$

Show that the given functions are solutions of the corresponding equations.

13. $y'' + k^2 xy = 0$, $y = \sqrt{x} \, J_{1/3}(2kx^{3/2}/3)$
14. $y'' + k^2 x^2 y = 0$, $y = \sqrt{x} \, J_{1/4}(kx^2/2)$
15. $y'' + k^2 x^4 y = 0$, $y = \sqrt{x} \, J_{1/6}(kx^3/3)$

16. (Elimination of first derivative)   Consider the differential equation

$$y'' + f(x)y' + g(x)y = 0.$$

Substitute $y(x) = u(x)v(x)$ and determine v such that the resulting second-order differential equation does not involve u'.

17. Show that for the Bessel equation the substitution in Prob. 16 is $y = ux^{-\frac{1}{2}}$ and gives

(20)
$$x^2 u'' + (x^2 + \tfrac{1}{4} - \nu^2)u = 0.$$

18. Using (20), find a general solution of (1) with $\nu = \frac{1}{2}$.

19. Using (12) and the relation $\Gamma(\frac{1}{2}) = \sqrt{\pi}$, show that

(21)
$$J_{1/2}(x) = \sqrt{\frac{2}{\pi x}} \sin x, \qquad J_{-1/2}(x) = \sqrt{\frac{2}{\pi x}} \cos x.$$

Relations between Bessel functions. Bessel functions are important in engineering applications. (An illustration will be given in Sec. 10.10.) In using them, it is necessary to know that they satisfy various relations. (Ref. [B14], the standard book on Bessel functions, and Ref. [A5] in Appendix 1 contain an incredibly large number of formulas.) Some important relations are included in the following problems. Formulas (24) and (25) are useful for computing tables.

20. Using (12) and (9), show that

(22)
$$[x^\nu J_\nu(x)]' = x^\nu J_{\nu-1}(x).$$

21. Using (12), show that

(23)
$$[x^{-\nu} J_\nu(x)]' = -x^{-\nu} J_{\nu+1}(x).$$

22. Derive from (22) and (23) the **recurrence relation**

(24)
$$J_{\nu-1}(x) + J_{\nu+1}(x) = \frac{2\nu}{x} J_\nu(x).$$

23. Derive from (22) and (23) the **recurrence relation**

(25)
$$J_{\nu-1}(x) - J_{\nu+1}(x) = 2J_\nu'(x).$$

24. Derive the Bessel equation (1) from (22) and (23).

25. Using (22) and (23) and Rolle's theorem, show that between two consecutive zeros of $J_0(x)$ there is precisely one zero of $J_1(x)$.

26. Show that between any two consecutive positive zeros of $J_n(x)$ there is precisely one zero of $J_{n+1}(x)$.

27. Using (21) and (24), show that

$$J_{3/2}(x) = \sqrt{\frac{2}{\pi x}}\left(\frac{\sin x}{x} - \cos x\right), \qquad J_{-3/2}(x) = -\sqrt{\frac{2}{\pi x}}\left(\frac{\cos x}{x} + \sin x\right).$$

28. Conclude from (21) and (24) that the *Bessel functions* $J_\nu(x)$, $\nu = \pm\frac{1}{2}, \pm\frac{3}{2}, \pm\frac{5}{2}, \cdots$ *are elementary functions.*

29. Using Table A2 in Appendix 4 and (24), compute $J_2(x)$ for $x = 0, 0.1, 0.2, \cdots, 1.0$.

Using formulas (22)–(25), show that

30. $J_0'(x) = -J_1(x)$

31. $J_1'(x) = J_0(x) - x^{-1}J_1(x)$

32. $J_2'(x) = \frac{1}{2}[J_1(x) - J_3(x)]$

33. $J_2'(x) = (1 - 4x^{-2})J_1(x) + 2x^{-1}J_0(x)$

34. $\displaystyle\int x^\nu J_{\nu-1}(x)\,dx = x^\nu J_\nu(x) + c$

35. $\displaystyle\int x^{-\nu} J_{\nu+1}(x)\,dx = -x^{-\nu}J_\nu(x) + c$

36. $\displaystyle\int J_{\nu+1}(x)\,dx = \int J_{\nu-1}(x)\,dx - 2J_\nu(x)$

Using the formulas in Probs. 34–36 and, if necessary, integration by parts, evaluate

37. $\displaystyle\int J_3(x)\,dx$

38. $\displaystyle\int J_5(x)\,dx$

39. $\displaystyle\int x^{-3}J_4(x)\,dx$

40. $\displaystyle\int x^3 J_0(x)\,dx$

(In other cases, one may be left with the integral of $J_0(x)$, which cannot be evaluated in finite form, but has been tabulated; cf. Ref. [B14] in Appendix 1.)

3.6 *Bessel Functions of the Second Kind*

For integer $\nu = n$, the Bessel functions $J_n(x)$ and $J_{-n}(x)$ are linearly dependent (Sec. 3.5) and do not form a fundamental system. We shall now obtain a second independent solution, starting with the case $n = 0$. In this case Bessel's equation may be written

(1) $$xy'' + y' + xy = 0,$$

the indicial equation has the double root $r = 0$, and we see from (16) in Sec. 3.4, that the desired solution must have the form

(2) $$y_2(x) = J_0(x) \ln x + \sum_{m=1}^{\infty} A_m x^m.$$

We substitute y_2 and its derivatives

$$y_2' = J_0' \ln x + \frac{J_0}{x} + \sum_{m=1}^{\infty} m A_m x^{m-1}$$

$$y_2'' = J_0'' \ln x + \frac{2J_0'}{x} - \frac{J_0}{x^2} + \sum_{m=1}^{\infty} m(m-1) A_m x^{m-2}$$

into (1). Then the logarithmic terms disappear because J_0 is a solution of (1), the other two terms containing J_0 cancel, and we find

$$2J_0' + \sum_{m=1}^{\infty} m(m-1) A_m x^{m-1} + \sum_{m=1}^{\infty} m A_m x^{m-1} + \sum_{m=1}^{\infty} A_m x^{m+1} = 0.$$

From (12) in Sec. 3.5 we obtain the power series of J_0' in the form

$$J_0'(x) = \sum_{m=1}^{\infty} \frac{(-1)^m 2m x^{2m-1}}{2^{2m}(m!)^2} = \sum_{m=1}^{\infty} \frac{(-1)^m x^{2m-1}}{2^{2m-1} m! \, (m-1)!}.$$

By inserting this series we have

$$\sum_{m=1}^{\infty} \frac{(-1)^m x^{2m-1}}{2^{2m-2} m! \, (m-1)!} + \sum_{m=1}^{\infty} m^2 A_m x^{m-1} + \sum_{m=1}^{\infty} A_m x^{m+1} = 0.$$

We first show that the A_m with odd subscripts are all zero. The coefficient of the power x^0 is A_1, and so, $A_1 = 0$. By equating the sum of the coefficients of the power x^{2s} to zero we obtain

$$(2s+1)^2 A_{2s+1} + A_{2s-1} = 0, \qquad s = 1, 2, \cdots.$$

Since $A_1 = 0$, we thus obtain $A_3 = 0$, $A_5 = 0$, $\cdots$, successively. We now equate the sum of the coefficients of x^{2s+1} to zero. For $s = 0$ this gives

$$-1 + 4A_2 = 0 \qquad \text{or} \qquad A_2 = \tfrac{1}{4}.$$

For the other values of s we obtain

$$\frac{(-1)^{s+1}}{2^{2s}(s+1)! \, s!} + (2s+2)^2 A_{2s+2} + A_{2s} = 0 \qquad (s = 1, 2, \cdots).$$

For $s = 1$ this yields

$$\tfrac{1}{8} + 16A_4 + A_2 = 0 \qquad \text{or} \qquad A_4 = -\tfrac{3}{128}$$

and in general

(3) $$A_{2m} = \frac{(-1)^{m-1}}{2^{2m}(m!)^2}\left(1 + \frac{1}{2} + \frac{1}{3} + \cdots + \frac{1}{m}\right), \qquad m = 1, 2, \cdots.$$

Using the short notation

(4)
$$h_m = 1 + \frac{1}{2} + \cdots + \frac{1}{m}$$

and inserting (3) and $A_1 = A_3 = \cdots = 0$ into (2), we obtain the result

(5)
$$y_2(x) = J_0(x) \ln x + \sum_{m=1}^{\infty} \frac{(-1)^{m-1} h_m}{2^{2m}(m!)^2} x^{2m}$$

$$= J_0(x) \ln x + \tfrac{1}{4}x^2 - \tfrac{3}{128}x^4 + - \cdots.$$

Since J_0 and y_2 are linearly independent functions, they form a fundamental system of (1). Of course, another fundamental system is obtained by replacing y_2 by an independent particular solution of the form $a(y_2 + bJ_0)$ where $a \ (\neq 0)$ and b are constants. It is customary to choose $a = 2/\pi$ and $b = \gamma - \ln 2$, where $\gamma = 0.577\ 215\ 664\ 90 \cdots$ is the so-called **Euler constant**, which is defined as the limit of

$$1 + \frac{1}{2} + \cdots + \frac{1}{s} - \ln s$$

as s approaches infinity. The standard particular solution thus obtained is known as the **Bessel function of the second kind** *of order zero* (Fig. 47) or **Neumann's function** *of order zero* and is denoted by $Y_0(x)$. Thus

(6)
$$Y_0(x) = \frac{2}{\pi} \left[J_0(x) \left(\ln \frac{x}{2} + \gamma \right) + \sum_{m=1}^{\infty} \frac{(-1)^{m-1} h_m}{2^{2m}(m!)^2} x^{2m} \right],$$

where h_m is defined by (4).

If $\nu = n = 1, 2, \cdots$, a second solution can be obtained by similar manipulations, starting from (19), Sec. 3.4. It turns out that also in these cases the solution contains a logarithmic term.

The situation is not yet completely satisfactory, because the second solution is defined differently, depending on whether the order ν is integral or not. To provide uniformity of formalism and numerical tabulation, it is desirable to adopt a form of the second solution which is valid for all values of the order. This is the

Fig. 47. Bessel functions of the second kind. (A small corresponding table is included in Appendix 4.)

reason for introducing a standard second solution $Y_\nu(x)$ defined for all ν by the formula

(7)

$$(a) \quad Y_\nu(x) = \frac{1}{\sin \nu\pi}[J_\nu(x) \cos \nu\pi - J_{-\nu}(x)]$$

$$(b) \quad Y_n(x) = \lim_{\nu \to n} Y_\nu(x).$$

This function is known as the **Bessel function of the second kind** *of order* ν or **Neumann's function**[10] *of order* ν.

For nonintegral order ν, the function $Y_\nu(x)$ is evidently a solution of Bessel's equation because $J_\nu(x)$ and $J_{-\nu}(x)$ are solutions of that equation. Since for those ν the solutions J_ν and $J_{-\nu}$ are linearly independent and Y_ν involves $J_{-\nu}$, the functions J_ν and Y_ν are linearly independent. Furthermore, it can be shown that the limit in (7b) exists and Y_n is a solution of Bessel's equation for integral order; cf. Ref. [B14] in Appendix 1. We shall see that the series development of $Y_n(x)$ contains a logarithmic term. Hence $J_n(x)$ and $Y_n(x)$ are linearly independent solutions of Bessel's equation. The series development of $Y_n(x)$ can be obtained by inserting the series (12) and (14), Sec. 3.5, for $J_\nu(x)$ and $J_{-\nu}(x)$ into (7a) and then letting ν approach n; for details see Ref. [B14]; the result is

(8)

$$Y_n(x) = \frac{2}{\pi}J_n(x)\left(\ln\frac{x}{2} + \gamma\right) + \frac{x^n}{\pi}\sum_{m=0}^{\infty}\frac{(-1)^{m-1}(h_m + h_{m+n})}{2^{2m+n}m!(m+n)!}x^{2m}$$
$$-\frac{x^{-n}}{\pi}\sum_{m=0}^{n-1}\frac{(n-m-1)!}{2^{2m-n}m!}x^{2m}$$

where $x > 0$, $n = 0, 1, \cdots$, and

$$h_0 = 0, \qquad h_s = 1 + \frac{1}{2} + \frac{1}{3} + \cdots + \frac{1}{s} \qquad (s = 1, 2, \cdots),$$

and when $n = 0$ the last sum in (8) is to be replaced by 0. For $n = 0$ the representation (8) takes the form (6). Furthermore, it can be shown that

$$Y_{-n}(x) = (-1)^n Y_n(x).$$

We may formulate our result, as follows.

Theorem 1 (General solution). *A general solution of Bessel's equation for all values of ν is*

(9)

$$y(x) = C_1 J_\nu(x) + C_2 Y_\nu(x).$$

We finally mention that there is a practical need for solutions of Bessel's equation which are complex for real values of x. For this reason the solutions

(10)

$$H_\nu^{(1)}(x) = J_\nu(x) + iY_\nu(x)$$
$$H_\nu^{(2)}(x) = J_\nu(x) - iY_\nu(x)$$

---

[10] CARL NEUMANN (1832–1925), German mathematician and physicist. The solutions $Y_\nu(x)$ are sometimes denoted by $N_\nu(x)$; in Ref. [B14] they are called Weber's functions. Euler's constant in (6) is often denoted by C or $\ln \gamma$ (instead of γ).

are frequently used and have been tabulated (cf. Ref. [A6] in Appendix 1). These linearly independent functions are called **Bessel functions of the third kind** *of order* *v* or *first and second* **Hankel functions**[11] *of order v.*

Problems

1. Show that for small $x > 0$ we have $Y_0(x) \approx 2\pi^{-1}(\ln \frac{1}{2}x + \gamma)$. Using this formula, compute an approximate value of the smallest positive zero of $Y_0(x)$ and compare it with the more accurate value 0.9.

2. It can be shown that, for large x,

$$(11) \qquad Y_n(x) \approx \sqrt{2/\pi x} \sin (x - \tfrac{1}{2}n\pi - \tfrac{1}{4}\pi).$$

Using (11) and Table A2 in Appendix 4, graph $Y_0(x)$ and $Y_1(x)$ for $0 < x \leq 15$. Using (11), compute approximate values of the first three positive zeros of $Y_0(x)$ and compare these values with the more accurate values 0.89, 3.96, and 7.09.

Using the indicated substitutions, find a general solution.

3. $x^2 y'' + xy' + (x^2 - 4)y = 0$ **4.** $xy'' + y' + xy = 0$

5. $xy'' + y' + \frac{1}{4}y = 0$ $(\sqrt{x} = z)$ **6.** $x^2 y'' + xy' + 4(x^4 - 1)y = 0$ $(x^2 = z)$

7. $x^2 y'' + xy' + 4(x^2 - 1)y = 0$ **8.** $x^2 y'' + (x^2 + \frac{1}{4})y = 0$ $(y = u\sqrt{x})$

9. $x^2 y'' + xy' + (k^2 x^2 - n^2)y = 0$ $(x = z/k)$

10. $xy'' + (1 + 2n)y' + xy = 0$ $(y = x^{-n}u)$

11. $x^2 y'' + (1 - 2n)xy' + n^2(x^{2n} + 1 - n^2)y = 0$ $(y = x^n u, \ x^n = z)$

12. Show that the Hankel functions (10) constitute a fundamental system of solutions of Bessel's equation for any v.

13. **(Modified Bessel functions)** The function $I_v(x) = i^{-v}J_v(ix)$, $i = \sqrt{-1}$, is called the *modified Bessel function of the first kind of order v.* Show that $I_v(x)$ is a solution of the differential equation

$$(12) \qquad x^2 y'' + xy' - (x^2 + v^2)y = 0$$

and gas the representation

$$(13) \qquad I_v(x) = \sum_{m=0}^{\infty} \frac{x^{2m+v}}{2^{2m+v} m! \, \Gamma(m + v + 1)}.$$

14. Show that $I_v(x)$ is real for all real x (and real v), $I_v(x) \neq 0$ for all real $x \neq 0$, and $I_{-n}(x) = I_n(x)$, where n is any integer.

15. **(Modified Bessel functions)** Show that another solution of the differential equation (12) is the so-called *modified Bessel function of the third kind* (sometimes: *of the second kind*)

$$(14) \qquad K_v(x) = \frac{\pi}{2 \sin v\pi} \, [I_{-v}(x) - I_v(x)].$$

3.7   Orthogonal Sets of Functions

Legendre polynomials (Sec. 3.3) and Bessel functions enjoy a property which is called **orthogonality** and is of general importance in engineering mathematics. In the present section we shall introduce the corresponding concepts and notations.

---

[11] HERMANN HANKEL (1839–1873), German mathematician.

The next section will be devoted to certain boundary value problems ("Sturm-Liouville problems") whose solutions form orthogonal sets. Finally, in Sec. 3.9 we shall apply the results of Sec .3.8 to Legendre polynomials and Bessel functions.

Let $g_m(x)$ and $g_n(x)$ be two real-valued functions which are defined on an interval $a \leqq x \leqq b$ and are such that the integral of the product $g_m(x)g_n(x)$ over that interval exists. We shall denote this integral by (g_m, g_n). This is a simple standard notation which is widely used. Thus

(1)
$$(g_m, g_n) = \int_a^b g_m(x)g_n(x)\, dx.$$

The functions are said to be **orthogonal** *on the interval* $a \leqq x \leqq b$ if the integral (1) is zero, that is,

(2)
$$(g_m, g_n) = \int_a^b g_m(x)g_n(x)\, dx = 0 \qquad\qquad (m \neq n).$$

A set of real-valued functions $g_1(x), g_2(x), g_3(x), \cdots$ is called an **orthogonal set** *of functions on an interval* $a \leqq x \leqq b$ if these functions are defined on that interval and if all the integrals (g_m, g_n) exist and are zero for all pairs of distinct functions in the set.

The nonnegative square root of (g_m, g_m) is called the **norm** of $g_m(x)$ and is generally denoted by $\|g_m\|$; thus

(3)
$$\|g_m\| = \sqrt{(g_m, g_m)} = \sqrt{\int_a^b g_m{}^2(x)\, dx}.$$

Throughout our discussion we shall make the following

General assumption. *All occurring functions are bounded and are such that the occurring integrals exist, and the norms are not zero.*

Clearly, an orthogonal set $g_1, g_2, \cdots$ on an interval $a \leqq x \leqq b$ whose functions have norm 1 satisfies the relations

(4) $$(g_m, g_n) = \int_a^b g_m(x)g_n(x)\, dx = \begin{cases} 0 & \text{when} \quad m \neq n \\ 1 & \text{when} \quad m = n \end{cases} \qquad \begin{pmatrix} m = 1, 2, \cdots \\ n = 1, 2, \cdots \end{pmatrix}.$$

Such a set is called an **orthonormal set** *of functions on the interval* $a \leqq x \leqq b$.

Obviously, from an *orthogonal* set we may obtain an *orthonormal* set by dividing each function by its norm on the interval under consideration.

Example 1. The functions $g_m(x) = \sin mx$, $m = 1, 2, \cdots$ form an orthogonal set on the interval $-\pi \leqq x \leqq \pi$, because [cf. (11) in Appendix 3]

$$(g_m, g_n) = \int_{-\pi}^{\pi} \sin mx \sin nx\, dx$$

(5) $(m \neq n)$.

$$= \frac{1}{2}\int_{-\pi}^{\pi} \cos(m - n)x\, dx - \frac{1}{2}\int_{-\pi}^{\pi} \cos(m + n)x\, dx = 0$$

The norm $\|g_m\|$ equals $\sqrt{\pi}$, because

$$\|g_m\|^2 = \int_{-\pi}^{\pi} \sin^2 mx \, dx = \pi \qquad\qquad (m = 1, 2, \cdots).$$

Hence the corresponding orthonormal set consists of the functions

$$\frac{\sin x}{\sqrt{\pi}}, \quad \frac{\sin 2x}{\sqrt{\pi}}, \quad \frac{\sin 3x}{\sqrt{\pi}}, \quad \cdots.$$

Example 2. The functions

$$1, \quad \cos x, \quad \sin x, \quad \cos 2x, \quad \sin 2x, \quad \cdots$$

form an orthogonal set on the interval $-\pi \leqq x \leqq \pi$. This follows from (5), the analogous relation for the cosine functions, and

$$\int_{-\pi}^{\pi} \cos mx \sin nx \, dx = \frac{1}{2} \int_{-\pi}^{\pi} \sin (m + n)x \, dx - \frac{1}{2} \int_{-\pi}^{\pi} \sin (m - n)x \, dx = 0$$

for all $m, n = 0, 1, \cdots$. The corresponding orthonormal set is

$$\frac{1}{\sqrt{2\pi}}, \quad \frac{\cos x}{\sqrt{\pi}}, \quad \frac{\sin x}{\sqrt{\pi}}, \quad \frac{\cos 2x}{\sqrt{\pi}}, \quad \frac{\sin 2x}{\sqrt{\pi}}, \quad \cdots.$$

Orthogonal sets yield important types of series developments in a relatively simple fashion. In fact, let $g_1(x), g_2(x), \cdots$ be any orthogonal set of functions on an interval $a \leqq x \leqq b$ and let $f(x)$ be a given function which can be represented in terms of the g_j's by a convergent series

$$\textbf{(6)} \qquad\qquad f(x) = \sum_{n=1}^{\infty} c_n g_n(x) = c_1 g_1(x) + c_2 g_2(x) + \cdots$$

This series is called a **generalized Fourier series** of $f(x)$ and its coefficients $c_1, c_2, \cdots$ are called the **Fourier constants** *of $f(x)$ with respect to that orthogonal set of functions.*

Because of the orthogonality, these constants can be determined in a very simple fashion. In fact, multiplying both sides of (6) by $g_m(x)$ (m fixed), integrating over $a \leqq x \leqq b$, and assuming that term-by-term integration is permissible,[12] we have

$$(f, g_m) = \int_a^b fg_m \, dx = \sum_{n=1}^{\infty} c_n(g_n, g_m) = \sum_{n=1}^{\infty} c_n \int_a^b g_n g_m \, dx.$$

The integral for which $n = m$ equals $(g_m, g_m) = \|g_m\|^2$, whereas all the other integrals on the right are zero, because of the orthogonality. Hence

$$\textbf{(7')} \qquad\qquad (f, g_m) = c_m \|g_m\|^2,$$

and the desired formula for the Fourier constants is

$$\textbf{(7)} \qquad\qquad c_m = \frac{(f, g_m)}{\|g_m\|^2} = \frac{1}{\|g_m\|^2} \int_a^b f(x) g_m(x) \, dx \qquad (m = 1, 2, \cdots).$$

---

[12] This is justified, for instance, in the case of uniform convergence (cf. Theorem 3 in Sec. 15.4).

Example 3 (Fourier series). In the case of the orthogonal set in Ex. 2, representation (6) may be written

$$(8) \qquad (x)f = a_0 + \sum_{n=1}^{\infty} (a_n \cos nx + b_n \sin nx)$$

and (7) becomes

$$a_0 = \frac{1}{2\pi} \int_{-\pi}^{\pi} f(x)\, dx$$

$$(9) \qquad a_n = \frac{1}{\pi} \int_{-\pi}^{\pi} f(x) \cos nx\, dx$$

$$(n = 1, 2, \cdots)$$

$$b_n = \frac{1}{\pi} \int_{-\pi}^{\pi} f(x) \sin nx\, dx$$

If the series in (8) converges and represents $f(x)$, it is called the **Fourier series** of $f(x)$. Its coefficients are called the **Fourier coefficients** of $f(x)$ and formulas (9) are called the **Euler formulas** for these coefficients. Since Fourier series are particularly important in engineering mathematics, we shall devote a whole chapter to them (Chap. 9) and present some basic applications in Chap. 10 in connection with partial differential equations.

Some important sets of real functions $g_1, g_2, \cdots$ occurring in applications are not orthogonal but have the property that for some function $p(x)$,

$$(10) \qquad \int_a^b p(x)g_m(x)g_n(x)\, dx = 0 \qquad \text{when} \quad m \neq n.$$

Such a set is then said to be *orthogonal with respect to the* **weight function** $p(x)$ *on the interval $a \leq x \leq b$.* The *norm* of g_m is now defined as

$$(11) \qquad \|g_m\| = \sqrt{\int_a^b p(x)g_m{}^2(x)\, dx}$$

and if the norm of each function g_m is 1, the set is said to be *orthonormal* on that interval with respect to $p(x)$.

If we set $h_m = \sqrt{p}\, g_m$, then (10) becomes

$$\int_a^b h_m(x)h_n(x)\, dx = 0 \qquad (m \neq n),$$

that is, the functions h_m form an orthogonal set in the usual sense. Clearly, if these functions are to be real, the weight function must be nonnegative. Important examples will be considered in the following sections.

If $g_1(x), g_2(x), \cdots$ is an orthogonal set on an interval $a \leq x \leq b$ with respect to a weight function p, and if a given function $f(x)$ can be represented by a **generalized Fourier series** [cf. (6)]

$$(12) \qquad f(x) = c_1 g_1(x) + c_2 g_2(x) + \cdots,$$

then the Fourier constants $c_1, c_2, \cdots$ of $f(x)$ with respect to that set can be determined as before; the only difference is that we now have to multiply that series by pg_m (instead of g_m) before integrating. The other steps are as before and give

$$(13) \qquad c_m = \frac{1}{\|g_m\|^2} \int_a^b p(x)f(x)g_m(x)\, dx \qquad (m = 1, 2, \cdots),$$

where the norm is now defined by (11).

Problems

In each case show that the given set is orthogonal on the given interval I and determine the corresponding orthonormal set.

1. $1, \cos x , \cos 2x, \cos 3x, \cdots,$ \qquad $I: 0 \leqq x \leqq 2\pi$
2. $\sin x, \sin 2x, \sin 3x, \cdots,$ \qquad $I: 0 \leqq x \leqq \pi$
3. $\sin \pi x, \sin 2\pi x, \sin 3\pi x, \cdots,$ \qquad $I: -1 \leqq x \leqq 1$
4. $1, \cos 2x, \cos 4x, \cos 6x, \cdots,$ \qquad $I: 0 \leqq x \leqq \pi$
5. $1, \cos \dfrac{2n\pi}{T} x \;(n = 1, 2, \cdots),$ \qquad $I: 0 \leqq x \leqq T$
6. $\sin \dfrac{2n\pi}{T} x \;(n = 1, 2, \cdots),$ \qquad $I: -T/2 \leqq x \leqq T/2$
7. $P_0(x), P_1(x), P_2(x), P_3(x),$ \qquad $I: -1 \leqq x \leqq 1$ (cf. Sec. 3.3)

8. Determine constants $a_0, b_0, \cdots, c_2$ so that the functions $g_1 = a_0,\; g_2 = b_0 + b_1 x,$ $g_3 = c_0 + c_1 x + c_2 x^2$ form an orthonormal set on the interval $-1 \leqq x \leqq 1$. Compare the result with that of Prob. 7.
9. Show that if the functions $g_1(x), g_2(x), \cdots$ form an orthogonal set on an interval $a \leqq x \leqq b$, then the functions $g_1(ct + k), g_2(ct + k), \cdots, c \neq 0,$ form an orthogonal set on the interval $(a-k)/c \leqq t \leqq (b - k)/c.$
10. Using the result of Prob. 9, derive the orthogonality property of the set in Prob. 5 from that in Prob. 1.

3.8 Sturm-Liouville Problem

Various important orthogonal sets of functions arise as solutions of second-order differential equations of the form

$$(1) \qquad [r(x)y']' + [q(x) + \lambda p(x)]y = 0$$

on some interval $a \leqq x \leqq b$, satisfying conditions of the form

$$(2) \qquad \text{(a) } k_1 y + k_2 y' = 0 \quad \text{at } x = a \qquad \text{(b) } l_1 y + l_2 y' = 0 \quad \text{at } x = b;$$

here λ is a parameter, and k_1, k_2, l_1, l_2 are given real constants, at least one in each condition (2) being different from zero.

The equation (1) is known as the **Sturm-Liouville**[13] **equation,** and we shall see that Legendre's equation, Bessel's equation, and other important equations can be written in the form (1). The conditions (2) refer to the boundary points $x = a$ and $x = b$ of that interval and, therefore, are called **boundary conditions.** A differential

---

[13] JACQUES CHARLES FRANCOIS STURM (1803–1855), Swiss mathematician, JOSEPH LIOUVILLE (1809–1882), French mathematician, known by his important research work in complex analysis.

equation together with boundary conditions constitutes what is known as a **boundary value problem.** Our boundary value problem (1), (2) is called a **Sturm-Liouville problem.**

Of course, this problem has the trivial solution $y \equiv 0$ for any value of the parameter λ. Solutions $y \not\equiv 0$ are called *characteristic functions* or **eigenfunctions** of the problem, and the values of λ for which such solutions exist, are called *characteristic values* or **eigenvalues** of the problem.

Example 1. Find the eigenvalues and eigenfunctions of the Sturm-Liouville problem

$$(3) \qquad (a) \;\; y'' + \lambda y = 0 \qquad (b) \;\; y(0) = 0, \qquad y(\pi) = 0.$$

For negative $\lambda = -\nu^2$ the general solution of the equation is

$$y(x) = c_1 e^{\nu x} + c_2 e^{-\nu x}.$$

From (3b) we obtain $c_1 = c_2 = 0$ and $y \equiv 0$, which is not an eigenfunction. For $\lambda = 0$ the situation is similar. For positive $\lambda = \nu^2$ the general solution is

$$y(x) = A \cos \nu x + B \sin \nu x.$$

From the first boundary condition we obtain $y(0) = A = 0$. The second boundary condition then yields

$$y(\pi) = B \sin \nu\pi = 0 \qquad \text{or} \qquad \nu = 0, \pm 1, \pm 2, \cdots.$$

For $\nu = 0$ we have $y \equiv 0$. For $\lambda = \nu^2 = 1, 4, 9, 16, \cdots$, taking $B = 1$, we obtain

$$y(x) = \sin \nu x \qquad\qquad \nu = 1, 2, \cdots.$$

These are the eigenfunctions of the problem, and the eigenvalues are $\lambda = \nu^2$ where $\nu = 1, 2, \cdots$.

It can be shown that under rather general conditions on the functions p, q, and r in (1) the Sturm-Liouville problem (1), (2) has infinitely many eigenvalues; the corresponding rather complicated theory can be found in Ref. [B8]; cf. Appendix 1.

Furthermore, the eigenfunctions are orthogonal, as follows.

Theorem 1 (Orthogonality). *Let the functions p, q, and r in the Sturm-Liouville equation* (1) *be real-valued and continuous on the interval $a \leq x \leq b$. Let $y_m(x)$ and $y_n(x)$ be eigenfunctions of the Sturm-Liouville problem* (1), (2) *corresponding to distinct eigenvalues λ_m and λ_n, respectively. Then y_m and y_n are orthogonal on that interval with respect to the weight function p.*

If $r(a) = 0$, then (2a) *can be dropped from the problem. If $r(b) = 0$, then* (2b) *can be dropped. If $r(a) = r(b)$, then* (2) *can be replaced by*

$$(4) \qquad\qquad y(a) = y(b), \qquad y'(a) = y'(b).$$

Proof. y_m satisfies

$$(ry_m')' + (q + \lambda_m p)y_m = 0,$$

and y_n satisfies

$$(ry_n')' + (q + \lambda_n p)y_n = 0.$$

Multiplying the first equation by y_n and the second by $-y_m$ and adding, we get

$$(\lambda_m - \lambda_n)py_m y_n = y_m(ry_n')' - y_n(ry_m')'$$
$$= [(ry_n')y_m - (ry_m')y_n]'$$

where the last equality can be readily verified by carrying out the indicated differentiation in the last expression. Integrating over x from a to b, we obtain

$$(5) \qquad (\lambda_m - \lambda_n) \int_a^b p y_m y_n \, dx = \left[r(y_n' y_m - y_m' y_n) \right]_a^b,$$

because the function in the brackets is continuous on $a \leq x \leq b$. The expression on the right equals

$$(6) \qquad \begin{aligned} &r(b)[y_n'(b)y_m(b) - y_m'(b)y_n(b)] \\ &-r(a)[y_n'(a)y_m(a) - y_m'(a)y_n(a)]. \end{aligned}$$

Case 1. If $r(a) = 0$ and $r(b) = 0$, then the expression in (6) is zero. Hence the expression on the left-hand side of (5) must be zero, and since λ_m and λ_n are distinct, we obtain the desired orthogonality

$$(7) \qquad \int_a^b p(x) y_m(x) y_n(x) \, dx = 0 \qquad\qquad (m \neq n)$$

without the use of the boundary conditions (2).

Case 2. Let $r(b) = 0$, but $r(a) \neq 0$. Then the first line in (6) is zero. We consider the remaining expression in (6). From (2a) we have

$$k_1 y_n(a) + k_2 y_n'(a) = 0,$$
$$k_1 y_m(a) + k_2 y_m'(a) = 0.$$

Let $k_2 \neq 0$. Then by multiplying the first equation by $y_m(a)$ and the last by $-y_n(a)$ and adding, we have

$$k_2 [y_n'(a) y_m(a) - y_m'(a) y_n(a)] = 0.$$

Since $k_2 \neq 0$, the expression in brackets must be zero. This expression is identical with that in the last line of (6). Hence (6) is zero, and from (5) we obtain (7). If $k_2 = 0$, then by assumption $k_1 \neq 0$, and the argument of proof is similar.

Case 3. If $r(a) = 0$, but $r(b) \neq 0$, the proof is similar to that in Case 2, but instead of (2a) we now have to use (2b).

Case 4. If $r(a) \neq 0$ and $r(b) \neq 0$, we have to use both boundary conditions (2) and proceed as in Cases 2 and 3.

Case 5. Let $r(a) = r(b)$. Then (6) takes the form

$$r(b)[y_n'(b)y_m(b) - y_m'(b)y_n(b) - y_n'(a)y_m(a) + y_m'(a)y_n(a)].$$

We may use (2) as before and conclude that the expression in brackets is zero. However, we immediately see that this would also follow from (4), so that we may replace (2) by (4). Hence (5) yields (7), as before. This completes the proof of Theorem 1.

Example 2. The differential equation in Ex. 1 is of the form (1) where $r = 1$, $q = 0$, and $p = 1$. From Theorem 1 it follows that the eigenfunctions are orthogonal on the interval $0 \leq x \leq \pi$.

Example 3. The reader may show that the functions

$$1, \quad \cos x, \quad \sin x, \quad \cos 2x, \quad \sin 2x, \quad \cdots$$

occurring in the Fourier series in Ex. 3, Sec. 3.7, are the eigenfunctions of the Sturm-Liouville problem

$$y'' + \lambda y = 0, \qquad y(\pi) = y(-\pi), \qquad y'(\pi) = y'(-\pi).$$

Hence from Theorem 1 it follows that these functions form an orthogonal set on the interval $-\pi \leqq x \leqq \pi$. Note that the boundary conditions of our present problem are of the form (4).

Further examples will be included in the next section.

A generalized Fourier series (cf. Sec. 3.7) in which the orthogonal set is a set of eigenfunctions is called an **eigenfunction expansion.**

The eigenvalues of Sturm-Liouville problems have the following interesting property.

Theorem 2 (Real eigenvalues). *If the Sturm-Liouville problem* (1), (2) *satisfies the condition stated in Theorem* 1 *and p is positive in the whole interval* $a \leqq x \leqq b$ *(or negative everywhere in that interval), then all the eigenvalues of the problem are real.*

Proof. Let $\lambda = \alpha + i\beta$ be an eigenvalue of the problem and let

$$y(x) = u(x) + iv(x)$$

be a corresponding eigenfunction; here α, β, u, and v are real. Substituting this into (1) we have

$$(ru' + irv')' + (q + \alpha p + i\beta p)(u + iv) = 0.$$

This complex equation is equivalent to the following pair of equations for the real and imaginary parts:

$$(ru')' + (q + \alpha p)u - \beta pv = 0,$$
$$(rv')' + (q + \alpha p)v + \beta pu = 0.$$

Multiplying the first equation by v and the second by $-u$ and adding, we obtain

$$-\beta(u^2 + v^2)p = u(rv')' - v(ru')'$$
$$= [(rv')u - (ru')v]'.$$

Integrating over x from a to b, we find

$$-\beta \int_a^b (u^2 + v^2)p \, dx = \left[r(uv' - u'v) \right]_a^b.$$

From the boundary conditions it follows that the expression on the right is zero; this can be shown in a fashion similar to that in the proof of Theorem 1. Since y is an eigenfunction, $u^2 + v^2 \not\equiv 0$. Since y and p are continuous and $p > 0$ (or $p < 0$) for all x between a and b, the integral on the left is not zero. Hence, $\beta = 0$, which means that $\lambda = \alpha$ is real. This completes the proof.

Theorem 2 is illustrated by Exs. 2 and 3, and further examples are included in the next section and the corresponding problem set.

Problems

1. Carry out the details of the proof of Theorem 1 in Cases 3 and 4.

2. Carry out the steps of the proof of Theorem 1 for the Sturm-Liouville problem in Ex. 1.

Find the eigenfunctions of the following Sturm-Liouville problems.

3. $y'' + \lambda y = 0$, $y(0) = 0$, $y'(l) = 0$ **4.** $y'' + \lambda y = 0$, $y'(0) = 0$, $y'(l) = 0$

Find a Sturm-Liouville problem whose eigenfunctions are:

5. $1, \cos x, \cos 2x, \cos 3x, \cdots$ **6.** $1, \cos \dfrac{2n\pi}{T} x, \sin \dfrac{2n\pi}{T} x$, $(n = 1, 2, \cdots)$

3.9 *Orthogonality of Legendre Polynomials and Bessel Functions*

Example 1 (***Legendre polynomials***).   Legendre's equation (1) in Sec. 3.3 can be written

$$[(1 - x^2)y']' + \lambda y = 0, \qquad\qquad \lambda = n(n + 1)$$

and is, therefore, a Sturm-Liouville equation (1), Sec. 3.8, with $r = 1 - x^2$, $q = 0$, and $p = 1$. Since $r = 0$ when $x = \pm 1$, no boundary conditions are needed to form a Sturm-Liouville problem on the interval $-1 \leq x \leq 1$. We know that, for $n = 0, 1, \cdots$, the Legendre polynomials $P_n(x)$ are solutions of the problem. Hence these are the eigenfunctions, and from Theorem 1 in Sec. 3.8 it follows that they are orthogonal on that interval, that is,

(1) $$\int_{-1}^{1} P_m(x)P_n(x)\, dx = 0 \qquad\qquad (m \neq n).$$

The norm is (cf. Prob. 8 in Sec. 3.3)

(2) $$\|P_m\| = \sqrt{\int_{-1}^{1} P_m{}^2(x)\, dx} = \sqrt{\frac{2}{2m + 1}} \qquad\qquad m = 0, 1, \cdots.$$

Example 2 (***Bessel functions***).   The orthogonality of the Bessel functions J_n is quite important in engineering applications, for example in connection with vibrations of circular membranes (to be considered in Sec. 10.10). $J_n(s)$ satisfies the Bessel equation (Sec. 3.5)

$$s^2 \ddot{J}_n + s\dot{J}_n + (s^2 - n^2)J_n = 0$$

where the dots denote derivatives with respect to s. We assume that n is a nonnegative integer. Setting $s = \lambda x$ where λ is a constant, not zero, we have $dx/ds = 1/\lambda$, and, by the chain rule,

$$\dot{J}_n = J_n{}'/\lambda, \qquad \ddot{J}_n = J_n{}''/\lambda^2,$$

where primes denote derivatives with respect to x. By substitution,

$$x^2 J_n{}''(\lambda x) + x J_n{}'(\lambda x) + (\lambda^2 x^2 - n^2)J_n(\lambda x) = 0.$$

Dividing by x, we may write this equation in the form

(3) $$[x J_n{}'(\lambda x)]' + \left(-\frac{n^2}{x} + \lambda^2 x\right) J_n(\lambda x) = 0.$$

We see that for each fixed n this is a Sturm-Liouville equation (1), Sec. 3.8, with the parameter written as λ^2 instead of λ, and

$$p(x) = x, \qquad q(x) = -n^2/x, \qquad r(x) = x.$$

Since $r(x) = 0$ at $x = 0$, it follows from Theorem 1 in the last section that those solutions of (3) on a given interval $0 \leq x \leq R$ which satisfy the boundary condition

(4) $$J_n(\lambda R) = 0$$

form an orthogonal set on that interval with respect to the weight function $p(x) = x$. (Note that for $n \neq 0$ the function q is discontinuous at $x = 0$, but this does not affect the proof of that theorem.) It can be shown (Ref. [B14]) that $J_n(s)$ has infinitely many real zeros; let $\alpha_{1n} < \alpha_{2n} < \alpha_{3n} \cdots$ denote the positive zeros of $J_n(s)$. Then (4) holds for

$$
(5) \qquad \lambda R = \alpha_{mn} \qquad \text{or} \qquad \lambda = \lambda_{mn} = \frac{\alpha_{mn}}{R} \qquad (m = 1, 2, \cdots),
$$

and we obtain the following result.

Theorem 1. *For each fixed $n = 0, 1, \cdots$ the Bessel functions $J_n(\lambda_{1n}x)$, $J_n(\lambda_{2n}x)$, $J_n(\lambda_{3n}x)$, $\cdots$, with λ_{mn} given by (5), form an orthogonal set on the interval $0 \leq x \leq R$ with respect to the weight function $p(x) = x$, that is,*

$$
(6) \qquad \int_0^R x J_n(\lambda_{mn}x) J_n(\lambda_{kn}x) \, dx = 0, \qquad (k \neq m).
$$

Hence we have obtained infinitely many orthogonal sets, each corresponding to one of the fixed values of n.

Since $p(x) = x$, we see from (11)–(13) in Sec. 3.7 that if a given function $f(x)$ has an eigenfunction expansion in terms of one of those orthogonal sets, the expansion is

$$
(7) \qquad f(x) = c_1 J_n(\lambda_{1n}x) + c_2 J_n(\lambda_{2n}x) + \cdots.
$$

This is called a **Fourier-Bessel series.** We shall prove that

$$
(8) \qquad \|J_n(\lambda_{mn}x)\|^2 = \int_0^R x J_n^2(\lambda_{mn}x) \, dx = \frac{R^2}{2} J_{n+1}^2(\lambda_{mn}R),
$$

where $\lambda_{mn} = \alpha_{mn}/R$. By (13) in Sec. 3.7 this implies that the Fourier constants c_m of f in (7) are

$$
(9) \qquad c_m = \frac{2}{R^2 J_{n+1}^2(\alpha_{mn})} \int_0^R x f(x) J_n(\lambda_{mn}x) \, dx, \qquad m = 1, 2, \cdots.
$$

We prove (8). Multiplying (3) by $2x J_n'(\lambda x)$, we obtain an equation which can be written

$$
\{[x J_n'(\lambda x)]^2\}' + (\lambda^2 x^2 - n^2)\{J_n^2(\lambda x)\}' = 0.
$$

Integrating over x from 0 to R we find

$$
(10) \qquad [x J_n'(\lambda x)]^2 \Big|_0^R = - \int_0^R (\lambda^2 x^2 - n^2)\{J_n^2(\lambda x)\}' \, dx.
$$

From (23) in Prob. 21, Sec. 3.5, writing s and n instead of x and ν, and denoting the derivative with respect to s by a dot, we have

$$
-n s^{-n-1} J_n(s) + s^{-n} \dot{J}_n(s) = -s^{-n} J_{n+1}(s).
$$

Multiplying by s^{n+1} and setting $s = \lambda x$, we obtain

$$
\lambda x J_n'(\lambda x) \frac{1}{\lambda} = n J_n(\lambda x) - \lambda x J_{n+1}(\lambda x),
$$

where the prime denotes the derivative with respect to x. Hence the left-hand side of (10) equals

$$
\Big[[n J_n(\lambda x) - \lambda x J_{n+1}(\lambda x)]^2 \Big]_{x=0}^R .
$$

If $\lambda = \lambda_{mn}$ then $J_n(\lambda R) = 0$, and since $J_n(0) = 0$ $(n = 1, 2, \cdots)$, that left-hand side becomes

$$
(11) \qquad \lambda_{mn}^2 R^2 J_{n+1}^2(\lambda_{mn}R).
$$

If we integrate by parts, the right-hand side of (10) becomes

$$-\left[(\lambda^2 x^2 - n^2)J_n{}^2(\lambda x)\right]_0^R + 2\lambda^2 \int_0^R x J_n{}^2(\lambda x)\,dx.$$

When $\lambda = \lambda_{mn}$, the first expression is zero at $x = R$. It is also zero at $x = 0$ because $\lambda^2 x^2 - n^2 = 0$ when $n = x = 0$, and $J_n(\lambda x) = 0$ when $x = 0$ and $n = 1, 2, \cdots$. From this and (11) we readily obtain (8). This completes the proof.

Problems

Legendre polynomials. Represent the following polynomials in terms of Legendre polynomials.

1. $5x^3 + x$ 2. $10x^3 - 3x^2 - 5x - 1$

3. $1, x, x^2, x^3$ 4. $35x^4 + 15x^3 - 30x^2 - 15x + 3$

In each case, obtain the first few terms of the expansion of $f(x)$ in terms of Legendre polynomials and graph the first three partial sums.

5. $f(x) = \begin{cases} 0 & \text{if } -1 < x < 0 \\ x & \text{if } 0 < x < 1 \end{cases}$ 6. $f(x) = \begin{cases} 0 & \text{if } -1 < x < 0 \\ 1 & \text{if } 0 < x < 1 \end{cases}$

7. $f(x) = |x|$ if $-1 < x < 1$

8. Show that the functions $P_n(\cos\theta)$, $n = 0, 1, \cdots$, form an orthogonal set on the interval $0 \leqq \theta \leqq \pi$ with respect to the weight function $\sin\theta$.

9. **(Tchebichef[14] polynomials)** The functions

$$T_n(x) = \cos(n \arccos x), \qquad U_n(x) = \sin(n \arccos x), \qquad n = 0, 1, \cdots$$

are called *Tchebichef polynomials of the first and second kind*, respectively. Show that

$$T_0 = 1, \qquad T_1(x) = x, \qquad T_2(x) = 2x^2 - 1, \qquad T_3(x) = 4x^3 - 3x,$$
$$U_0 = 0, \qquad U_1(x) = \sqrt{1 - x^2}, \qquad U_2(x) = 2x\sqrt{1 - x^2}, \qquad U_3(x) = (4x^2 - 1)\sqrt{1 - x^2}.$$

10. Show that the Tchebichef polynomials $T_n(x)$ are orthogonal on the interval $-1 \leqq x \leqq 1$ with respect to the weight function $p(x) = 1/\sqrt{1 - x^2}$. *Hint.* To evaluate the integral, set $\arccos x = \theta$.

11. **(Laguerre[15] polynomials)** The functions

$$L_0 = 1, \qquad L_n(x) = \frac{e^x}{n!}\frac{d^n(x^n e^{-x})}{dx^n}, \qquad n = 1, 2, \cdots$$

are called *Laguerre polynomials*. Show that

$$L_1(x) = 1 - x, \qquad L_2(x) = 1 - 2x + x^2/2, \qquad L_3(x) = 1 - 3x + 3x^2/2 - x^3/6.$$

12. Show that

$$L_n(x) = \sum_{m=0}^{n} \frac{(-1)^m}{m!}\binom{n}{m}x^m = 1 - nx + \frac{n(n-1)}{4}x^2 - + \cdots + \frac{(-1)^n}{n!}x^n.$$

[14] PAFNUTI TCHEBICHEF (1821–1894), Russian mathematician, particularly known for his important contributions to the theory of numbers.
[15] EDMOND LAGUERRE (1834–1886), French mathematician who did research work in geometry and the theory of infinite series.

13. $L_n(x)$ satisfies Laguerre's differential equation

$$xy'' + (1 - x)y' + ny = 0.$$

Verify this fact for $n = 0, 1, 2, 3$.

14. Verify by direct integration that L_0, $L_1(x)$, $L_2(x)$ are orthogonal on the positive axis $0 \leqq x < \infty$ with respect to the weight function $p(x) = e^{-x}$.

15. Prove that the set of all Laguerre polynomials is orthogonal on $0 \leqq x < \infty$ with respect to the weight function $p(x) = e^{-x}$. *Hint.* Consider the integral of $e^{-x}L_m(x)L_n(x)$ from 0 to ∞ with $m < n$. Since the highest power in L_m is x^m, conclude that it suffices to show that the integral of $e^{-x}x^kL_n(x)$ ($k < n$) from 0 to ∞ is zero. Prove this by repeated integration by parts.

16. (Hermite[16] polynomials) The functions

$$H_0 = 1, \qquad H_n(x) = (-1)^n e^{x^2/2} \frac{d^n e^{-x^2/2}}{dx^n}, \qquad n = 1, 2, \cdots$$

are called *Hermite polynomials.* Show that

$$H_1(x) = x, \qquad H_2(x) = x^2 - 1, \qquad H_3(x) = x^3 - 3x, \qquad H_4(x) = x^4 - 6x^2 + 3.$$

(As is true for many special functions, the literature contains more than one notation, and one often defines as Hermite polynomials the expressions

$$H_0{}^* = 1, \qquad H_n{}^*(x) = (-1)^n e^{x^2} \frac{d^n e^{-x^2}}{dx^n}.$$

This differs from our definition, which is preferably used in applications.)

17. Show that the Hermite polynomials are related to the coefficients of the Maclaurin series

$$e^{tx - t^2/2} = \sum_{n=0}^{\infty} a_n(x)t^n$$

by the formula $H_n(x) = n! \, a_n(x)$. *Hint.* Note that $tx - t^2/2 = x^2/2 - (x - t)^2/2$. (The exponential function on the left is called the *generating function* of the H_n.)

18. Show that the Hermite polynomials satisfy the relation

$$H_{n+1}(x) = xH_n(x) - H_n'(x).$$

19. Differentiating the generating function in Prob. 17 with respect to x, show that

$$H_n'(x) = nH_{n-1}(x).$$

Using this and the formula in Prob. 18 (with n replaced by $n - 1$), prove that $H_n(x)$ satisfies the differential equation

$$y'' - xy' + ny = 0.$$

20. Solve the differential equation in Prob. 19 by the power series method.

21. Using the equation in Prob. 19, show that $w = e^{-x^2/4}H_n(x)$ is a solution of **Weber's[17] equation**

$$w'' + (n + \tfrac{1}{2} - \tfrac{1}{4}x^2)w = 0 \qquad\qquad (n = 0, 1, \cdots).$$

22. Show that the Hermite polynomials are orthogonal on the x-axis $-\infty < x < \infty$ with respect to the weight function $p(x) = e^{-x^2/2}$. *Hint.* Use the formulas in Probs. 16 and 19 and integrate by parts.

---

[16] CHARLES HERMITE (1822–1901), French mathematician, is known by his work in algebra and number theory.

[17] HEINRICH WEBER (1842–1913), German mathematician.

23. **(Bessel functions)** Graph $J_0(\lambda_{10}x)$, $J_0(\lambda_{20}x)$, $J_0(\lambda_{30}x)$, and $J_0(\lambda_{40}x)$ for $R = 1$ in the interval $0 \leq x \leq 1$. (Use Table A2 in Appendix 4.)

24. Graph $J_1(\alpha_{11}x)$, $J_1(\alpha_{21}x)$, $J_1(\alpha_{31}x)$, and $J_1(\alpha_{41}x)$ in the interval $0 \leq x \leq 1$. (Use Table A2 in Appendix 4.)

Develop the following functions $f(x)$ $(0 < x < R)$ in a Fourier-Bessel series of the form

$$f(x) = c_1 J_0(\lambda_{10}x) + c_2 J_0(\lambda_{20}x) + c_3 J_0(\lambda_{30}x) + \cdots$$

and graph the first few partial sums.

25. $f(x) = 1$. *Hint.* Use (22), Sec. 3.5.   26. $f(x) = \begin{cases} 1 & \text{if} \quad 0 < x < R/2 \\ 0 & \text{if} \quad R/2 < x < R \end{cases}$

27. $f(x) = \begin{cases} k & \text{if} \quad 0 < x < a \\ 0 & \text{if} \quad a < x < R \end{cases}$ 28. $f(x) = \begin{cases} 0 & \text{if} \quad 0 < x < R/2 \\ k & \text{if} \quad R/2 < x < R \end{cases}$

29. $f(x) = 1 - x^2$ $(R = 1)$. *Hint.* Use (22), Sec. 3.5, and integration by parts.

30. $f(x) = R^2 - x^2$ 31. $f(x) = x^2$ 32. $f(x) = x^4$

33. Show that $f(x) = x^n$ $(0 < x < 1, n = 0, 1, \cdots)$ can be represented by the Fourier-Bessel series

$$x^n = \frac{2J_n(\alpha_{1n}x)}{\alpha_{1n}J_{n+1}(\alpha_{1n})} + \frac{2J_n(\alpha_{2n}x)}{\alpha_{2n}J_{n+1}(\alpha_{2n})} + \cdots.$$

34. Find a representation of $f(x) = x^n$ $(0 < x < R, n = 0, 1, \cdots)$ similar to that in Prob. 33.

35. Represent $f(x) = x^3$ $(0 < x < 2)$ by a Fourier-Bessel series involving J_3.

CHAPTER 4

Laplace Transformation

The Laplace transformation is a powerful method for solving linear differential equations arising in engineering mathematics. It consists essentially of three steps. In the first step, the given differential equation is transformed into an algebraic equation (*subsidiary equation*). Then the latter is solved by purely algebraic manipulations. Finally, the solution of the subsidiary equation is transformed back so that it becomes the required solution of the original differential equation.

In this way the Laplace transformation reduces the problem of solving a differential equation to an algebraic problem. Another advantage is that it takes care of initial conditions so that in initial value problems the determination of a general solution is avoided. Similarly, if we apply the Laplace transformation to a nonhomogeneous equation, we obtain the solution directly, that is, without first solving the corresponding homogeneous equation.

In the present chapter we shall consider the Laplace transformation from a practical point of view and illustrate its use by important engineering applications.

> *Prerequisite for this chapter:* Chap. 2.
> *References:* Appendix 1, Part B.
> *Answers to problems:* Appendix 2.

4.1 Laplace Transform. Inverse Transform. Linearity

Let $f(t)$ be a given function which is defined for all positive values of t. We multiply $f(t)$ by e^{-st} and integrate with respect to t from zero to infinity. Then, if the resulting integral exists, it is a function of s, say, $F(s)$:

$$F(s) = \int_0^\infty e^{-st}f(t)\, dt.$$

The function $F(s)$ is called the **Laplace transform**[1] of the original function $f(t)$,

---

[1] PIERRE SIMON DE LAPLACE (1749–1827), great French mathematician, who developed the foundation of potential theory and made important contributions to celestial mechanics and probability theory.

and will be denoted by $\mathcal{L}(f)$. Thus

(1)
$$F(s) = \mathcal{L}(f) = \int_0^\infty e^{-st}f(t)\,dt.$$

The described operation on $f(t)$ is called the *Laplace transformation*.

Furthermore the original function $f(t)$ in (1) is called the *inverse transform* or **inverse** of $F(s)$ and will be denoted by $\mathcal{L}^{-1}(F)$; that is, we shall write

$$f(t) = \mathcal{L}^{-1}(F).$$

We shall denote the original function by a lower case letter and its transform by the same letter in capital.

Example 1. Let $f(t) = 1$ when $t > 0$. Then

$$\mathcal{L}(f) = \mathcal{L}(1) = \int_0^\infty e^{-st}\,dt = -\frac{1}{s}e^{-st}\Big|_0^\infty;$$

hence, when $s > 0$,

$$\mathcal{L}(1) = \frac{1}{s}.$$

Example 2. Let $f(t) = e^{at}$ when $t > 0$, where a is a constant. Then

$$\mathcal{L}(e^{at}) = \int_0^\infty e^{-st}e^{at}\,dt = \frac{1}{a-s}e^{-(s-a)t}\Big|_0^\infty;$$

hence, when $s - a > 0$,

$$\mathcal{L}(e^{at}) = \frac{1}{s-a}.$$

We shall see that the Laplace transformation possesses various general properties, in consequence of which transforms of many functions can be obtained in very simple ways. The most important property of the Laplace transformation is stated in the following theorem.

Theorem 1 (Linearity property).   *The Laplace transformation is a linear operation, that is, for any functions $f(t)$ and $g(t)$ whose Laplace transforms exist and any constants a and b we have*

$$\mathcal{L}\{af(t) + bg(t)\} = a\mathcal{L}(f) + b\mathcal{L}(g).$$

Proof.   By definition,

$$\mathcal{L}\{af(t) + bg(t)\} = \int_0^\infty e^{-st}[af(t) + bg(t)]\,dt$$

$$= a\int_0^\infty e^{-st}f(t)\,dt + b\int_0^\infty e^{-st}g(t)\,dt = a\mathcal{L}(f) + b\mathcal{L}(g).$$

Example 3.   Let $f(t) = \cosh at = (e^{at} + e^{-at})/2$. Then, using Theorem 1 and the result of Ex. 2, we find

$$\mathcal{L}(\cosh at) = \frac{1}{2}\mathcal{L}(e^{at}) + \frac{1}{2}\mathcal{L}(e^{-at}) = \frac{1}{2}\left(\frac{1}{s-a} + \frac{1}{s+a}\right);$$

that is, when $s > a$,

$$\mathcal{L}(\cosh at) = \frac{s}{s^2 - a^2}.$$

A short list of some important elementary functions and their Laplace transforms is given in Table 2, and a more extensive list in Sec. 4.10. Cf. also the references in Appendix 1, Part B.

Table 2. Some Elementary Functions $f(t)$ and Their Laplace Transforms $\mathscr{L}(f)$

	$f(t)$	$\mathscr{L}(f)$		$f(t)$	$\mathscr{L}(f)$
1	1	$1/s$	6	e^{at}	$\dfrac{1}{s-a}$
2	t	$1/s^2$	7	$\cos \omega t$	$\dfrac{s}{s^2+\omega^2}$
3	t^2	$2!/s^3$	8	$\sin \omega t$	$\dfrac{\omega}{s^2+\omega^2}$
4	t^n $(n=1,2,\cdots)$	$\dfrac{n!}{s^{n+1}}$	9	$\cosh at$	$\dfrac{s}{s^2-a^2}$
5	t^a (a positive)	$\dfrac{\Gamma(a+1)}{s^{a+1}}$	10	$\sinh at$	$\dfrac{a}{s^2-a^2}$

Once we know the transforms in Table 2, nearly all the transforms we shall need can be obtained through the use of some simple general theorems which we shall consider in the subsequent sections.

Formulas 1, 2, and 3 in Table 2 are special cases of formula 4. Formula 4 follows from formula 5 and $\Gamma(n+1)=n!$, where n is a nonnegative integer [cf. (10) in Sec. 3.5]. Formula 5 can be proved by starting from the definition

$$\mathscr{L}(t^a) = \int_0^\infty e^{-st} t^a \, dt,$$

setting $st = x$, and using (8) in Sec. 3.5; then

$$\mathscr{L}(t^a) = \int_0^\infty e^{-x}\left(\frac{x}{s}\right)^a \frac{dx}{s} = \frac{1}{s^{a+1}}\int_0^\infty e^{-x} x^a \, dx = \frac{\Gamma(a+1)}{s^{a+1}} \qquad (s>0).$$

Formula 6 was proved in Ex. 2. To prove the formulas 7 and 8 we set $a = i\omega$ in formula 6. Then

$$\mathscr{L}(e^{i\omega t}) = \frac{1}{s-i\omega} = \frac{s+i\omega}{(s-i\omega)(s+i\omega)} = \frac{s+i\omega}{s^2+\omega^2} = \frac{s}{s^2+\omega^2} + i\frac{\omega}{s^2+\omega^2}.$$

On the other hand, by Theorem 1,

$$\mathscr{L}(e^{i\omega t}) = \mathscr{L}(\cos \omega t + i \sin \omega t) = \mathscr{L}(\cos \omega t) + i\mathscr{L}(\sin \omega t).$$

Equating the real and imaginary parts of these two equations, we obtain the formulas 7 and 8. Formula 9 was proved in Ex. 3, and formula 10 can be proved in a similar manner.

Further transforms may be obtained by the following important property of the Laplace transformation.

Theorem 2 (First shifting theorem).   *If* $\mathscr{L}(f) = F(s)$ *when* $s > \alpha$, *then*

$$\mathscr{L}\{e^{at}f(t)\} = F(s - a) \qquad\qquad (s > \alpha + a);$$

that is, the substitution of $s - a$ *for* s *in the transform corresponds to the multiplication of the original function by* e^{at}.

Proof.   By definition,

$$F(s) = \int_0^\infty e^{-st}f(t)\, dt$$

and, therefore,

$$F(s - a) = \int_0^\infty e^{-(s-a)t}f(t)\, dt = \int_0^\infty e^{-st}[e^{at}f(t)]\, dt = \mathscr{L}\{e^{at}f(t)\}.$$

Example 4.   By applying Theorem 2 to the formulas 4, 7, and 8 in Table 2 we obtain the following results.

$f(t)$	$\mathscr{L}(f)$
$e^{at}t^n$	$\dfrac{n!}{(s - a)^{n+1}}$
$e^{at} \cos \omega t$	$\dfrac{s - a}{(s - a)^2 + \omega^2}$
$e^{at} \sin \omega t$	$\dfrac{\omega}{(s - a)^2 + \omega^2}$

We shall now state simple sufficient conditions for the existence of the Laplace transform of a function. These conditions are that f is piecewise continuous and $|f|$ does not increase too rapidly as t approaches infinity. Let us first define piecewise continuity.

A function $f(t)$ is said to be **piecewise continuous** on a finite interval $a \leq t \leq b$, if it is defined on that interval and is such that the interval can be subdivided into finitely many intervals, in each of which $f(t)$ is continuous and has finite limits as t approaches either end point of the interval of subdivision from the interior.

It follows from the definition that finite jumps are the only discontinuities which a piecewise continuous function may have; these are known as *ordinary discontinuities*. Furthermore, it is clear that the class of piecewise continuous functions includes every continuous function.

Theorem 3 (Existence theorem).   *Let* $f(t)$ *be a function which is piecewise continuous on every finite interval in the range* $t \geq 0$ *and satisfies*

(2) $|f(t)| \leq Me^{\alpha t}$ *for all* $t \geq 0$

and for some constants α *and* M. *Then the Laplace transform of* $f(t)$ *exists for all* $s > \alpha$.

Proof. Since $f(t)$ is piecewise continuous, $e^{-st}f(t)$ is integrable over any finite interval on the t-axis, and from (2),

$$|\mathscr{L}(f)| = \left| \int_0^\infty e^{-st}f(t)\,dt \right| \leq \int_0^\infty e^{-st}|f(t)|\,dt \leq \int_0^\infty e^{-st}Me^{\alpha t}\,dt$$

$$= M\int_0^\infty e^{-(s-\alpha)t}\,dt = \frac{M}{s-\alpha} \qquad (s > \alpha).$$

This completes the proof.

The conditions in Theorem 3 are practical for most applications, and it is simple to find whether a given function satisfies an inequality of the form (2). For example,

$$(3) \qquad\qquad \cosh t < e^t, \qquad t^n < n!\,e^t \quad (n = 0, 1, \cdots) \qquad \text{for all } t > 0,$$

and any function which is bounded in absolute value for all $t \geq 0$, such as the sine and cosine function of a real variable, satisfies that condition. An example of a function which does not satisfy a relation of the form (2), is the exponential function e^{t^2}, because, no matter how large we choose the numbers M and α in (2),

$$e^{t^2} > Me^{\alpha t} \qquad\qquad \text{for all } t > t_0$$

where t_0 is a sufficiently large number, depending on M and α.

It should be noted that the conditions in Theorem 3 are sufficient rather than necessary. For example, the function $1/\sqrt{t}$ is infinite at $t = 0$, but its transform exists; in fact, from the definition and $\Gamma(\tfrac{1}{2}) = \sqrt{\pi}$ [cf. (30) in Appendix 3] we obtain

$$\mathscr{L}(t^{-\frac{1}{2}}) = \int_0^\infty e^{-st}t^{-\frac{1}{2}}\,dt = \frac{1}{\sqrt{s}}\int_0^\infty e^{-x}x^{-\frac{1}{2}}\,dx = \frac{1}{\sqrt{s}}\,\Gamma(\tfrac{1}{2}) = \sqrt{\frac{\pi}{s}}.$$

If the Laplace transform of a given function exists, it is uniquely determined. Conversely, it can be shown that two functions having the same transform cannot differ over an interval of positive length, although they may differ at various isolated points (cf. Ref. [B15] in Appendix 1). Since this is of no importance in applications, we may say that the inverse of a given transform is essentially unique. In particular, if two *continuous* functions have the same transform, they are completely identical.

Problems

Find the Laplace transforms of the following functions, where a, b, T, etc. are constants.

1. $3t + 4$ 2. $at + b$ 3. $t^2 + at + b$ 4. $(a + bt)^2$
5. $2te^t$ 6. t^2e^{-2t} 7. e^{at+b} 8. $e^{-t}\cos 2t$
9. $\sin(2n\pi t/T)$ 10. $\sin(\omega t + \theta)$ 11. $\cos(\omega t + \theta)$ 12. $\sin^2 t$
13. $\cos^2 t$ 14. $e^{-2t}\sin n\pi t$ 15. $e^{-t}\sin(\omega t + \theta)$ 16. $\cosh^2 3t$

17.   18.   19.   20.

Representing the hyperbolic functions in terms of exponential functions and applying Theorem 2, show that

21. $\mathscr{L}(\cosh at \cos at) = \dfrac{s^3}{s^4 + 4a^4}$

22. $\mathscr{L}(\cosh at \sin at) = \dfrac{a(s^2 + 2a^2)}{s^4 + 4a^4}$

23. $\mathscr{L}(\sinh at \cos at) = \dfrac{a(s^2 - 2a^2)}{s^4 + 4a^4}$

24. $\mathscr{L}(\sinh at \sin at) = \dfrac{2a^2 s}{s^4 + 4a^4}$

25. Derive the three results given in Ex. 4.

26. Obtain the answer to Prob. 19 from those to Probs. 17 and 18.

27. Obtain formula 6 in Table 2 from formulas 9 and 10.

28. Obtain formula 6 in Table 2 from formula 1 and Theorem 2.

29. Obtain formula 10 in Table 2 from formula 6.

30. Derive formulas 7 and 8 in Table 2 by integrating by parts.

31. Using $\cosh x = \cos ix$ and $\sinh x = -i \sin ix$, $i = \sqrt{-1}$, obtain formulas 9 and 10 in Table 2 from formulas 7 and 8.

32. Using $\cos x = \cosh ix$ and $\sin x = -i \sinh ix$, derive formulas 7 and 8 in Table 2 from formulas 9 and 10.

33. Prove formula 4 in Table 2 by induction.

34. Prove (3).

35. Extend Theorem 1 to linear combinations of more than two functions.

Find $f(t)$ if $\mathscr{L}(f)$ equals

36. $\dfrac{5}{s + 3}$

37. $\dfrac{2}{s^2 + 16}$

38. $\dfrac{s + 1}{s^2 + 1}$

39. $\dfrac{s - 4}{s^2 - 4}$

40. $\dfrac{1}{(s + 9)^2}$

41. $\dfrac{s + 2}{(s + 2)^2 + 1}$

42. $\dfrac{1}{s^4}$

43. $\dfrac{s}{(s - 1)^2 - 4}$

44. $\dfrac{a_1}{s} + \dfrac{a_2}{s^2} + \cdots + \dfrac{a_n}{s^n}$

45. $\dfrac{1}{(s - a)(s - b)}$ $(a \neq b)$

4.2 Laplace Transforms of Derivatives and Integrals

We shall now see that, roughly speaking, differentiation and integration of $f(t)$ corresponds to multiplication and division of the Laplace transform $F(s) = \mathscr{L}(f)$ by s. The great importance of this property of the Laplace transformation is obvious, because in this way operations of calculus may be replaced by simple algebraic operations on the transforms. We start with the following theorem.

Theorem 1 [Differentiation of $f(t)$]. *Suppose that $f(t)$ is continuous for all $t \geq 0$, satisfies (2), p. 150, for some α and M, and has a derivative $f'(t)$ which is piecewise continuous on every finite interval in the range $t \geq 0$. Then the Laplace transform of the derivative $f'(t)$ exists when $s > \alpha$, and*

(1) $$\mathscr{L}(f') = s\mathscr{L}(f) - f(0) \qquad (s > \alpha).$$

Proof. We first consider the case when $f'(t)$ is continuous for all $t \geq 0$. Then, by definition and by integrating by parts,

$$\mathscr{L}(f') = \int_0^\infty e^{-st} f'(t)\, dt = [e^{-st} f(t)]_0^\infty + s \int_0^\infty e^{-st} f(t)\, dt.$$

Since f satisfies (2), Sec. 4.1, the integrated portion on the right is zero at the upper limit when $s > \alpha$, and at the lower limit it is $-f(0)$. The last integral is $\mathscr{L}(f)$, the existence for $s > \alpha$ being a consequence of Theorem 3 in Sec. 4.1. This proves that the expression on the right exists when $s > \alpha$, and is equal to $-f(0) + s\mathscr{L}(f)$. Consequently, $\mathscr{L}(f')$ exists when $s > \alpha$, and (1) holds.

If $f'(t)$ is merely piecewise continuous, the proof is quite similar; in this case, the range of integration in the original integral must be broken up into parts such that f' is continuous in each such part.

This theorem may be extended to piecewise continuous functions $f(t)$, but in place of (1) we then obtain the formula (1*) in Prob. 14 at the end of the current section.

By applying (1) to the second-order derivative $f''(t)$ we obtain

$$\mathscr{L}(f'') = s\mathscr{L}(f') - f'(0)$$
$$= s[s\mathscr{L}(f) - f(0)] - f'(0);$$

that is,

(2) $$\mathscr{L}(f'') = s^2\mathscr{L}(f) - sf(0) - f'(0).$$

Similarly,

(3) $$\mathscr{L}(f''') = s^3\mathscr{L}(f) - s^2f(0) - sf'(0) - f''(0),$$

etc. By induction we thus obtain the following extension of Theorem 1.

Theorem 2 (Derivative of any order n). *Let $f(t)$ and its derivatives $f'(t), f''(t), \cdots,$ $f^{(n-1)}(t)$ be continuous functions for all $t \geq 0$, satisfying (2), p. 150, for some α and M, and let the derivative $f^{(n)}(t)$ be piecewise continuous on every finite interval in the range $t \geq 0$. Then the Laplace transform of $f^{(n)}(t)$ exists when $s > \alpha$, and is given by the formula*

(4) $$\mathscr{L}(f^{(n)}) = s^n\mathscr{L}(f) - s^{n-1}f(0) - s^{n-2}f'(0) - \cdots - f^{(n-1)}(0).$$

An obvious application of Theorems 1 and 2 will be the transformation of derivatives in differential equations; this problem will be considered in Sec. 4.3. However, these theorems may also be used for determining transforms; cf. the following examples.

Example 1. Let us give another proof of formula 3 in Table 2 (Sec. 4.1). Let $f(t) = t^2$. Then $f(0) = 0, f'(0) = 0, f''(t) = 2$. Since $\mathscr{L}(2) = 2/s$, we obtain from (2)

$$\mathscr{L}(f'') = \mathscr{L}(2) = \frac{2}{s} = s^2\mathscr{L}(f) \quad \text{or} \quad \mathscr{L}(t^2) = \frac{2}{s^3}.$$

Example 2. Let $f(t) = \sin^2 t$. Find $\mathscr{L}(f)$. We have $f(0) = 0$,

$$f'(t) = 2\sin t \cos t = \sin 2t$$

and (1) gives

$$\mathscr{L}(\sin 2t) = \frac{2}{s^2 + 4} = s\mathscr{L}(f) \quad \text{or} \quad \mathscr{L}(\sin^2 t) = \frac{2}{s(s^2 + 4)}.$$

Example 3. Let $f(t) = t \sin t$. Find $\mathscr{L}(f)$. We have $f(0) = 0$,

$$f'(t) = \sin t + t \cos t, \qquad f'(0) = 0,$$
$$f''(t) = 2 \cos t - t \sin t.$$

From this and (2) we obtain

$$\mathscr{L}(f'') = 2\mathscr{L}(\cos t) - \mathscr{L}(f) = s^2 \mathscr{L}(f).$$

Using $\mathscr{L}(\cos t) = s/(s^2 + 1)$ and solving for $\mathscr{L}(f)$, we have

$$\mathscr{L}(f) = \mathscr{L}(t \sin t) = \frac{2\mathscr{L}(\cos t)}{s^2 + 1} = \frac{2s}{(s^2 + 1)^2}.$$

We have seen that differentiation of $f(t)$ corresponds to multiplication of $\mathscr{L}(f)$ by s (and addition of a constant). We shall now prove that *integration of* $f(t)$ corresponds to *division of $\mathscr{L}(f)$ by s*.

Theorem 3 [Integration of $f(t)$]. *If $f(t)$ is piecewise continuous and satisfies an inequality of the form* (2), *Sec.* 4.1, *then*

$$(5) \qquad\qquad \mathscr{L}\left\{\int_0^t f(\tau)\,d\tau\right\} = \frac{1}{s}\,\mathscr{L}\{f(t)\} \qquad\qquad (s > 0, s > \alpha).$$

Proof. Suppose that $f(t)$ is piecewise continuous and satisfies (2), Sec. 4.1, for some α and M. Clearly, if (2) holds for some negative α, it also holds for positive α, and we may assume that α is positive. Then the integral

$$g(t) = \int_0^t f(\tau)\,d\tau$$

is continuous, and by using (2) we obtain

$$|g(t)| \leqq \int_0^t |f(\tau)|\,d\tau \leqq M \int_0^t e^{\alpha \tau}\,d\tau = \frac{M}{\alpha}\,(e^{\alpha t} - 1) \qquad\qquad (\alpha > 0).$$

Furthermore, $g'(t) = f(t)$, except for points at which $f(t)$ is discontinuous. Hence $g'(t)$ is piecewise continuous on each finite interval, and, according to Theorem 1,

$$\mathscr{L}\{f(t)\} = \mathscr{L}\{g'(t)\} = s\mathscr{L}\{g(t)\} - g(0) \qquad\qquad (s > \alpha).$$

Clearly, $g(0) = 0$, and, therefore,

$$\mathscr{L}(g) = \frac{1}{s}\,\mathscr{L}(f).$$

This completes the proof.

Example 4. Let $\mathscr{L}(f) = \dfrac{1}{s^2(s^2 + \omega^2)}$. Find $f(t)$. From Table 2 in Sec. 4.1 we have

$$\mathscr{L}^{-1}\left(\frac{1}{s^2 + \omega^2}\right) = \frac{1}{\omega}\sin \omega t.$$

From this and Theorem 3 it follows that

$$\mathscr{L}^{-1}\left\{\frac{1}{s}\left(\frac{1}{s^2 + \omega^2}\right)\right\} = \frac{1}{\omega}\int_0^t \sin \omega\tau\,d\tau = \frac{1}{\omega^2}(1 - \cos \omega t).$$

Applying Theorem 3 once more, we obtain the desired answer

$$\mathcal{L}^{-1}\left\{\frac{1}{s^2}\left(\frac{1}{s^2+\omega^2}\right)\right\} = \frac{1}{\omega^2}\int_0^t (1-\cos\omega\tau)\,d\tau = \frac{1}{\omega^2}\left(t - \frac{\sin\omega t}{\omega}\right).$$

Problems

1. Check the result in Ex. 2, using $\sin^2 t = (1-\cos 2t)/2$.
2. Using (1), find $\mathcal{L}(\cos^2 \omega t)$.
3. Using (1), find $\mathcal{L}(\sin^2 \omega t)$.
4. Using (2), derive formula 7 in Table 2 (Sec. 4.1).
5. Using (2), derive formula 8 in Table 2 (Sec. 4.1).
6. Using (1), obtain $\mathcal{L}(\sinh 3t)$ from $\mathcal{L}(\cosh 3t)$.

Using Theorems 1 and 2, show that

7. $\mathcal{L}(t\cos\omega t) = \dfrac{s^2-\omega^2}{(s^2+\omega^2)^2}$　　　　8. $\mathcal{L}(t\sin\omega t) = \dfrac{2\omega s}{(s^2+\omega^2)^2}$

9. $\mathcal{L}(t\cosh at) = \dfrac{s^2+a^2}{(s^2-a^2)^2}$　　　10. $\mathcal{L}(t\sinh at) = \dfrac{2as}{(s^2-a^2)^2}$

Using the formulas in Probs. 7 and 8, show that

11. $\mathcal{L}^{-1}\left(\dfrac{1}{(s^2+\omega^2)^2}\right) = \dfrac{1}{2\omega^3}(\sin\omega t - \omega t\cos\omega t)$

12. $\mathcal{L}^{-1}\left(\dfrac{s^2}{(s^2+\omega^2)^2}\right) = \dfrac{1}{2\omega}(\sin\omega t + \omega t\cos\omega t)$

13. Carry out the details of the proof of Theorem 1, assuming that $f'(t)$ has finite jumps at $t_1, t_2, \cdots, t_m$, where $0 < t_1 < t_2 < \cdots < t_m$.
14. Show that if $f(t)$ is continuous, except for an ordinary discontinuity at $t = a \ (>0)$, the other conditions remaining the same as in Theorem 1, then (cf. Fig. 48)

(1*)　　　　　$\mathcal{L}(f') = s\mathcal{L}(f) - f(0) - [f(a+0) - f(a-0)]e^{-as}$.

Fig. 48. Problem 14

Graph the following functions. Using (1*), find their Laplace transform.
15. $f(t) = t$ when $0 < t < 1$, $f(t) = 0$ otherwise
16. $f(t) = 1$ when $1 < t < 2$, $f(t) = 0$ otherwise
17. $f(t) = t - 1$ when $1 < t < 2$, $f(t) = 0$ otherwise
18. $f(t) = t$ when $0 < t < 1$, $f(t) = 1$ when $1 < t < 2$, $f(t) = 0$ otherwise

Using Theorem 3, find $f(t)$ if $\mathcal{L}(f)$ equals

19. $\dfrac{1}{s^2+s}$　　　20. $\dfrac{4}{s^3-4s}$　　　21. $\dfrac{1}{s^3+4s}$　　　22. $\dfrac{1}{s^2+as}$

23. $\dfrac{1}{s}\left(\dfrac{s-a}{s+a}\right)$ **24.** $\dfrac{1}{s^2}\left(\dfrac{s-a}{s+a}\right)$ **25.** $\dfrac{8}{s^4-4s^2}$ **26.** $\dfrac{1}{s^2}\left(\dfrac{s-2}{s^2+4}\right)$

27. $\dfrac{1}{s^4-2s^3}$ **28.** $\dfrac{2s-\pi}{s^3(s-\pi)}$ **29.** $\dfrac{1}{s^2}\left(\dfrac{s+1}{s^2+1}\right)$ **30.** $\dfrac{1}{s^4(s^2+\pi^2)}$

4.3 Transformation of Ordinary Differential Equations

We shall now see that ordinary linear differential equations with constant coefficients can be reduced to algebraic equations of the transform. For example, consider the equation

(1) $$y''(t) + \omega^2 y(t) = r(t)$$

where $r(t)$ and ω are given. Applying the Laplace transformation and using (2), Sec. 4.2, we obtain

(2) $$s^2 Y(s) - sy(0) - y'(0) + \omega^2 Y(s) = R(s),$$

where $Y(s)$ is the Laplace transform of the (unknown) function $y(t)$, and $R(s)$ is the Laplace transform of $r(t)$. This algebraic equation is called the **subsidiary equation** of the given differential equation. Its solution is clearly

(3) $$Y(s) = \frac{sy(0) + y'(0)}{s^2 + \omega^2} + \frac{R(s)}{s^2 + \omega^2}.$$

Note that the first term on the right is completely determined by means of given initial conditions, say, $y(0) = k_1$, $y'(0) = k_2$.

The last step of the procedure is to determine the inverse $\mathscr{L}^{-1}(Y) = y(t)$ which is then the desired solution of (1).

Note that (2) was obtained from (1) by means of Theorem 2, Sec. 4.2. Instead of justifying this step by showing that the final result $y(t)$ satisfies the conditions of that theorem, it is simpler and more natural to see by substitution whether $y(t)$ satisfies the given equation and the initial conditions.

Systems of differential equations will be considered in Sec. 4.5.

Example 1. Find the solution of the differential equation

$$y'' + 9y = 0$$

satisfying the initial conditions $y(0) = 0$, $y'(0) = 2$. From (2) we obtain the subsidiary equation

$$s^2 Y(s) - 2 + 9 Y(s) = 0.$$

Solving for $Y(s)$, we have

$$Y(s) = \frac{2}{s^2 + 9} = \frac{2}{3}\left(\frac{3}{s^2 + 9}\right).$$

From this and Table 2 in Sec. 4.1 we find

$$y(t) = \mathscr{L}^{-1}(Y) = \tfrac{2}{3}\sin 3t.$$

By substitution we see that this function satisfies the given equation and the initial conditions.

Example 2. A small body of mass $m = 2$ is attached at the lower end of an elastic spring whose upper end is fixed, the spring modulus being $k = 10$. Let $y(t)$ be the displacement of the body from the position of static equilibrium. Determine the free vibrations of the body, starting from the initial position $y(0) = 2$ with the initial velocity $y'(0) = -4$, asuming that there is damping proportional to the velocity, the damping constant being $c = 4$.

The motion is described by the solution $y(t)$ of the initial value problem

$$y'' + 2y' + 5y = 0, \qquad y(0) = 2, \qquad y'(0) = -4,$$

cf. (7), Sec. 2.6. Using (1) and (2) in Sec. 4.2, we obtain the subsidiary equation

$$s^2 Y - 2s + 4 + 2(sY - 2) + 5Y = 0.$$

The solution is

$$Y(s) = \frac{2s}{(s+1)^2 + 2^2} = 2\frac{s+1}{(s+1)^2 + 2^2} - \frac{2}{(s+1)^2 + 2^2}.$$

Now

$$\mathscr{L}^{-1}\left(\frac{s}{s^2 + 2^2}\right) = \cos 2t, \qquad \mathscr{L}^{-1}\left(\frac{2}{s^2 + 2^2}\right) = \sin 2t.$$

From this and Theorem 2 in Sec. 4.1 we obtain the expected type of solution

$$y(t) = \mathscr{L}^{-1}(Y) = e^{-t}(2\cos 2t - \sin 2t).$$

Of course, the present examples are merely for explaining the method. Equations for which the present method is better than the usual procedure will be considered in Secs. 4.8 and 4.9.

Problems

Solve the following initial value problems by means of the Laplace transformation.

1. $y'' + y = 0$, $\qquad y(0) = 1$, $\qquad y'(0) = -1$
2. $4y'' + \pi^2 y = 0$, $\qquad y(0) = 2$, $\qquad y'(0) = 0$
3. $y'' - 4y = 0$, $\qquad y(0) = 0$, $\qquad y'(0) = -6$
4. $16y'' - 9y = 0$, $\qquad y(0) = 3$, $\qquad y'(0) = 3.75$
5. $y'' + \omega^2 y = 0$, $\qquad y(0) = A$, $\qquad y'(0) = B$ $\quad$ (ω real, not zero)
6. $y'' - a^2 y = 0$, $\qquad y(0) = k_1$, $\qquad y'(0) = k_2$ $\quad$ (a real, not zero)
7. $y'' + 2y' + 2y = 0$, $\qquad y(0) = 0$, $\qquad y'(0) = 1$
8. $y'' + 4y' + 4y = 0$, $\qquad y(0) = 2$, $\qquad y'(0) = -3$
9. $y'' + 2y' + 17y = 0$, $\qquad y(0) = 0$, $\qquad y'(0) = 12$
10. $y'' - 4y' + 5y = 0$, $\qquad y(0) = 1$, $\qquad y'(0) = 2$
11. $9y'' - 6y' + y = 0$, $\qquad y(0) = 3$, $\qquad y'(0) = 1$
12. $y'' + y' = 0$, $\qquad y(0) = 0$, $\qquad y'(0) = 4$
13. $y'' - ky' = 0$, $\qquad y(0) = 2$, $\qquad y'(0) = k$
14. $y'' - 2y' + 10y = 0$, $\qquad y(0) = 3$, $\qquad y'(0) = 3$
15. $4y'' - 4y' + 37y = 0$, $\qquad y(0) = 3$, $\qquad y'(0) = 1.5$
16. $4y'' - 8y' + 5y = 0$, $\qquad y(0) = 0$, $\qquad y'(0) = 1$
17. $y'' + y' + 1.25y = 0$, $\qquad y(0) = 1$, $\qquad y'(0) = -0.5$
18. $2y'' - y' + 2.125y = 0$, $\qquad y(0) = 0$, $\qquad y'(0) = 1$
19. $y'' + ky' - 2k^2 y = 0$, $\qquad y(0) = 2$, $\qquad y'(0) = 2k$

20. Show that the subsidiary equation of the differential equation

$$y'' + ay' + by = r(t) \qquad\qquad (a, b \text{ constant})$$

has the solution

$$Y(s) = \frac{(s+a)y(0) + y'(0)}{s^2 + as + b} + \frac{R(s)}{s^2 + as + b}$$

where $R(s)$ is the Laplace transform of $r(t)$.

4.4 *Partial Fractions*

Our discussion in the previous section shows that in many cases of practical importance the solution $Y(s)$ of the subsidiary equation of a given differential equation will be of the form

$$Y(s) = \frac{G(s)}{H(s)}$$

where $G(s)$ and $H(s)$ are polynomials in s. We shall now see that in such a case we can determine $y(t) = \mathscr{L}^{-1}(Y)$ by expressing $Y(s)$ in terms of partial fractions. Throughout this section we shall make the following

Assumption. *$G(s)$ and $H(s)$ have no common factors and have real coefficients. The degree of $G(s)$ is lower than that of $H(s)$.*

Let $s = a$ be a root of $H(s) = 0$. We shall state formulas for the inverse transform of the partial fraction or fractions corresponding to a in the representation of $Y(s)$ in terms of partial fractions. The proofs will be given below and examples in the next section.

Notation. (1a), (2a), $\cdots$, (5a) (see below) are representations of $Y(s) = G(s)/H(s)$, and in each case $W(s)$ denotes the sum of the partial fractions corresponding to all the (unrepeated or repeated) linear factors of $H(s)$ which are not under consideration.

Case 1. Unrepeated factor $s - a$. In this case,

(1a)
$$Y(s) = \frac{G(s)}{H(s)} = \frac{A}{s - a} + W(s).$$

The inverse transform is

(1b)
$$\mathscr{L}^{-1}(Y) = Ae^{at} + \mathscr{L}^{-1}(W)$$

where A is given by either of the two expressions

(1c)
$$A = Q_a(a) \quad \text{or} \quad A = \frac{G(a)}{H'(a)}$$

and $Q_a(s)$ is the function which is left after removing the factor $s - a$ from $H(s)$ in $Y(s)$, that is

(1d)
$$Q_a(s) = \frac{(s - a)G(s)}{H(s)}.$$

The subscript a of Q_a emphasizes the important fact that Q_a depends on a and changes when we proceed from one linear factor to another.

Case 2. Repeated factor $(s - a)^m$. In this case,

(2a)
$$Y(s) = \frac{G(s)}{H(s)} = \frac{A_m}{(s - a)^m} + \frac{A_{m-1}}{(s - a)^{m-1}} + \cdots + \frac{A_1}{s - a} + W(s).$$

The inverse transform of this expression is

(2b) $\quad \mathcal{L}^{-1}(Y) = e^{at}\left(A_m \dfrac{t^{m-1}}{(m-1)!} + A_{m-1}\dfrac{t^{m-2}}{(m-2)!} + \cdots + A_2\dfrac{t}{1!} + A_1\right)$

$$+ \mathcal{L}^{-1}(W)$$

where the constants $A_1, \cdots, A_m$ *are given by the formula*

(2c) $\quad A_m = Q_a(a), \qquad A_k = \dfrac{1}{(m-k)!}\dfrac{d^{m-k}Q_a(s)}{ds^{m-k}}\bigg|_{s=a} \qquad (k = 1, \cdots, m-1);$

here,

(2d) $$Q_a(s) = \dfrac{(s-a)^m G(s)}{H(s)}$$

Complex factors. The previous cases include real as well as complex numbers a, but if a is complex, other formulas are preferable from the practical point of view. Let

$$a = \alpha + i\beta \qquad \text{and} \qquad \bar{a} = \alpha - i\beta \qquad\qquad (\alpha, \beta \text{ real}, \beta \neq 0).$$

Suppose that $s = a$ is a root of $H(s) = 0$. Then since $H(s)$ has real coefficients, $s = \bar{a}$ is also a root of $H(s) = 0$.

   Case 3. Unrepeated complex factor $s - a$. *Writing the partial fraction corresponding to* $s - a$ *and* $s - \bar{a}$ *explicitly and noting that*

$$(s-a)(s-\bar{a}) = (s-\alpha)^2 + \beta^2,$$

we have

(3a) $$Y(s) = \dfrac{G(s)}{H(s)} = \dfrac{As + B}{(s-\alpha)^2 + \beta^2} + W(s) \qquad\qquad (A, B \text{ real}).$$

The inverse transform is

(3b) $$\mathcal{L}^{-1}(Y) = \dfrac{1}{\beta}e^{\alpha t}(T_a \cos \beta t + S_a \sin \beta t) + \mathcal{L}^{-1}(W)$$

where S_a *and* T_a *are the real and imaginary parts of*

(3c) $$R_a(a) = S_a + iT_a;$$

here

(3d) $$R_a(s) = [(s-\alpha)^2 + \beta^2]\dfrac{G(s)}{H(s)}.$$

   Case 4. Repeated complex factor $(s - a)^2$. *Writing the partial fractions corresponding to* $(s-a)^2$ *and* $(s-\bar{a})^2$ *explicitly, we have*

(4a) $$Y(s) = \dfrac{G(s)}{H(s)} = \dfrac{As + B}{[(s-\alpha)^2 + \beta^2]^2} + \dfrac{Cs + D}{(s-\alpha)^2 + \beta^2} + W(s)$$

where A, B, C, D *are real. The inverse transform is*

(4b) $\quad \mathcal{L}^{-1}(Y) = \dfrac{1}{2\beta^3}e^{\alpha t}[(T_a - \beta S_a{}^* - \beta S_a t)\cos \beta t$

$$+ (S_a + \beta T_a{}^* + \beta T_a t)\sin \beta t] + \mathcal{L}^{-1}(W)$$

where

(4c) $R_a(a) = S_a + iT_a, \qquad R_a'(a) = S_a{}^* + iT_a{}^*$

and

(4d) $R_a(s) = [(s - \alpha)^2 + \beta^2]^2 \dfrac{G(s)}{H(s)}.$

Case 5. Repeated complex factor $(s - a)^m$. *Writing the partial fractions corresponding to $(s - a)^m$ and $(s - \bar{a})^m$ explicitly, we have*

(5a) $Y(s) = \dfrac{G(s)}{H(s)} = \sum\limits_{k=1}^{m} \left\{ \dfrac{A_k}{(s - a)^k} + \dfrac{B_k}{(s - \bar{a})^k} \right\} + W(s).$

The transform is

(5b) $\mathscr{L}^{-1}(Y) = 2e^{\alpha t} \sum\limits_{k=1}^{m} \dfrac{t^{k-1}}{(k - 1)!\,(m - k)!}\, (Q_{k1} \cos \beta t - Q_{k2} \sin \beta t) + \mathscr{L}^{-1}(W)$

where

(5c) $Q_a(a)^{(m-k)} = Q_{k1} + iQ_{k2} = \dfrac{d^{m-k}Q_a(s)}{ds^{m-k}}\bigg|_{s=a}, \qquad Q_a(a)^{(0)} = Q_a(a),$

and

(5d) $Q_a(s) = \dfrac{(s - a)^m G(s)}{H(s)}.$

Remark. When $m = 2$, the previous formulas are preferable. Case 5 with $m > 2$ is of minor practical importance, and we present those formulas merely for the sake of completeness.

Proofs of the formulas

Proof in Case 1. Multiplying (1a) by $s - a$ and using (1d), we have

$$Q_a(s) = A + (s - a)W(s).$$

Letting $s \to a$ and noting that $W(s)$ does not contain the factor $s - a$, we obtain $A = Q_a(a)$, the first formula in (1c). If we write (1d) in the form

$$Q_a(s) = \dfrac{G(s)}{H(s)/(s - a)}$$

and let $s \to a$, then $G(s) \to G(a)$ while the denominator appears as an indeterminate of the form $0/0$; evaluating it in the usual way, we obtain

$$\lim_{s \to a} \dfrac{H(s)}{s - a} = \lim_{s \to a} \dfrac{H'(s)}{(s - a)'} = H'(a).$$

This proves the second formula in (1c). The inverse transform of $A/(s - a)$ is Ae^{at}, and the proof is complete.

Proof in Case 2. Multiplying (2a) by $(s - a)^m$ and using (2d), we have

$$Q_a(s) = A_m + (s - a)A_{m-1} + (s - a)^2 A_{m-2} + \cdots + (s - a)^m W(s).$$

Hence $Q_a(a) = A_m$. Differentiation yields

$$Q_a'(s) = A_{m-1} + \text{further terms containing factors } s - a.$$

Hence $Q_a'(a) = A_{m-1}$ because those terms are zero when $s = a$. Differentiating again we find $Q_a''(a) = 2! A_{m-2}$, etc. This proves (2c). The inverse transform of $A_k/(s-a)^k$ is $A_k e^{at} t^{k-1}/(k-1)!$. This yields (2b), and the proof is complete.

Proof in Case 3. Multiplying (3a) by $(s-\alpha)^2 + \beta^2$, using (3d), and letting $s \to a$, we obtain

$$R_a(a) = aA + B = (\alpha + i\beta)A + B.$$

The real and imaginary parts on both sides must be equal. Using (3c), we thus have

$$S_a = \alpha A + B, \qquad T_a = \beta A.$$

Hence in (3a),

$$\frac{As + B}{(s-\alpha)^2 + \beta^2} = \frac{A(s-\alpha) + \alpha A + B}{(s-\alpha)^2 + \beta^2} = \frac{1}{\beta}\frac{T_a(s-\alpha) + \beta S_a}{(s-\alpha)^2 + \beta^2}.$$

From this, Table 2, and Theorem 2 in Sec. 4.1 we obtain (3b). This completes the proof.

Proof in Case 4. Introducing the notation

$$p(s) = (s-a)(s-\bar{a}) = (s-\alpha)^2 + \beta^2,$$

multiplying (4a) by p^2, and using (4d), we have

$$R_a(s) = As + B + (Cs + D)p(s) + p^2(s)W(s).$$

We now let $s \to a$. Since $p(a) = 0$, we see that then we simply obtain

(6) $$R_a(a) = aA + B = (\alpha + i\beta)A + B.$$

Differentiating $R_a(s)$ with respect to s, we get

$$R_a'(s) = A + (Cs + D)p'(s) + \text{terms containing a factor } p(s).$$

Since $p(a) = 0$ and $p'(a) = 2(a-\alpha) = 2i\beta$, it follows that

(7) $$R_a'(a) = A + 2i\beta[C(\alpha + i\beta) + D].$$

In (6) and (7) the real and imaginary parts on both sides must be equal. Using (4c) we thus obtain

(8) $$S_a = \alpha A + B, \qquad T_a = \beta A$$

and furthermore

$$S_a{}^* = A - 2\beta^2 C, \qquad T_a{}^* = 2\alpha\beta C + 2\beta D,$$

which yields

(9) $$C = (T_a - \beta S_a{}^*)/2\beta^3, \qquad \alpha C + D = T_a{}^*/2\beta.$$

From Table 2 and Theorem 2 in Sec. 4.1 we see that the partial fraction

$$\frac{Cs + D}{(s - \alpha)^2 + \beta^2} = \frac{C(s - \alpha) + \alpha C + D}{(s - \alpha)^2 + \beta^2}$$

in (4a) has the inverse transform

$$e^{\alpha t}[\beta C \cos \beta t + (D + \alpha C) \sin \beta t]/\beta.$$

The other partial fraction

$$\frac{As + B}{[(s - \alpha)^2 + \beta^2]^2} = \frac{A(s - \alpha) + \alpha A + B}{[(s - \alpha)^2 + \beta^2]^2}$$

has the inverse transform

$$e^{\alpha t}[(A\beta^2 t + \alpha A + B) \sin \beta t - (\alpha A + B)\beta t \cos \beta t]/2\beta^3,$$

as follows from Example 3 and Problem 11 in Sec. 4.2 and Theorem 2 in Sec. 4.1. By inserting (8) and (9) and simplifying we obtain (4b), and the proof is complete.

   Proof in Case 5.   From (2b) we see that the inverse transform of $A_k(s - a)^{-k} + B_k(s - \bar{a})^{-k}$ in (5a) is

$$\text{(10)} \quad \frac{t^{k-1}}{(k - 1)!} (A_k e^{at} + B_k e^{\bar{a}t})$$

$$= \frac{t^{k-1}}{(k - 1)!} e^{\alpha t}[(A_k + B_k) \cos \beta t + i(A_k - B_k) \sin \beta t].$$

Using the notation in (5c), we have from (2c)

$$A_k = \frac{1}{(m - k)!} Q_a(a)^{(m-k)} = \frac{Q_{k1} + iQ_{k2}}{(m - k)!}.$$

Similarly, since $G(s)$ and $H(s)$ have real coefficients,

$$B_k = \frac{1}{(m - k)!} Q_{\bar{a}}(\bar{a})^{(m-k)} = \frac{Q_{k1} - iQ_{k2}}{(m - k)!}.$$

By addition and subtraction we find

$$A_k + B_k = 2Q_{k1}/(m - k)!, \qquad A_k - B_k = 2iQ_{k2}/(m - k)!.$$

If we insert this into (10), we obtain (5b), and the proof is complete.

4.5   *Examples and Applications*

Example 1.   Find the inverse transformation of

$$Y(s) = \frac{G(s)}{H(s)} = \frac{s + 1}{s^3 + s^2 - 6s}.$$

The denominator has the three distinct linear factors s, $s - 2$, and $s + 3$, corresponding to the roots $a_1 = 0$, $a_2 = 2$, and $a_3 = -3$. Hence

$$Y(s) = \frac{s+1}{s(s-2)(s+3)} = \frac{A_1}{s} + \frac{A_2}{s-2} + \frac{A_3}{s+3},$$

and $G(s) = s + 1$, $H'(s) = 3s^2 + 2s - 6$. Using (1c) in Sec. 4.4 we thus obtain

$$A_1 = \frac{G(0)}{H'(0)} = -\frac{1}{6}, \quad A_2 = \frac{G(2)}{H'(2)} = \frac{3}{10}, \quad A_3 = \frac{G(-3)}{H'(-3)} = -\frac{2}{15}.$$

From this and (1b) in Sec. 4.4 we see that the inverse transform is

$$\mathscr{L}^{-1}(Y) = -\tfrac{1}{6} + \tfrac{3}{10}e^{2t} - \tfrac{2}{15}e^{-3t}.$$

Example 2. Solve the initial value problem

$$y'' - 3y' + 2y = 4t + e^{3t}, \quad y(0) = 1, \quad y'(0) = -1.$$

From Table 2, Sec. 4.1, and (1) and (2) in Sec. 4.2 we see that the subsidiary equation is

$$s^2 Y - s + 1 - 3(sY - 1) + 2Y = \frac{4}{s^2} + \frac{1}{s-3}.$$

Solving for Y and representing Y in terms of partial fractions, we have

$$Y(s) = \frac{G(s)}{H(s)} = \frac{s^4 - 7s^3 + 13s^2 + 4s - 12}{s^2(s-3)(s^2 - 3s + 2)}$$

$$= \frac{A_2}{s^2} + \frac{A_1}{s} + \frac{B}{s-3} + \frac{C}{s-2} + \frac{D}{s-1}.$$

We first determine the constants A_1 and A_2 in the partial fractions corresponding to the root $a = 0$. In (2d) and (2c), Sec. 4.4,

$$Q_0(s) = \frac{G(s)}{\cdot(s-3)(s^2 - 3s + 2)}, \quad A_2 = Q_0(0) = 2, \quad A_1 = Q_0'(0) = 3.$$

Using (1c) in Sec. 4.4 we obtain the remaining coefficients

$$B = \frac{G(s)}{s^2(s^2 - 3s + 2)}\bigg|_{s=3} = \tfrac{1}{2}, \quad C = \frac{G(s)}{s^2(s-3)(s-1)}\bigg|_{s=2} = -2,$$

$$D = \frac{G(s)}{s^2(s-3)(s-2)}\bigg|_{s=1} = -\tfrac{1}{2}.$$

From (1b) and (2b) in Sec. 4.4 we see that the solution is

$$y(t) = \mathscr{L}^{-1}(Y) = 2t + 3 + \tfrac{1}{2}e^{3t} - 2e^{2t} - \tfrac{1}{2}e^t.$$

The student may solve the problem by the classical method and convince himself that the present approach is much simpler and much more rapid.

Example 3 (Free oscillations). Determine the free damped oscillation corresponding to the initial value problem

$$y'' + 2y' + 5y = 0, \quad y(0) = 2, \quad y'(0) = -4.$$

The solution of the subsidiary equation is (cf. Ex. 2 in Sec. 4.3)

$$Y(s) = \frac{2s}{s^2 + 2s + 5} = \frac{2s}{(s-a)(s-\bar{a})}$$

where $a = \alpha + i\beta = -1 + 2i$, so that $\alpha = -1$ and $\beta = 2$. Furthermore, in (3d) and (3c), Sec. 4.4,

$$R_a(s) = 2s, \qquad R_a(a) = S_a + iT_a = 2a = 2(\alpha + i\beta) = -2 + 4i,$$

that is, $S_a = -2$ and $T_a = 4$. From this and (3b) in Sec. 4.4,

$$y(t) = \mathscr{L}^{-1}(Y) = e^{-t}(2 \cos 2t - \sin 2t).$$

Example 4 (Forced oscillations, resonance). Solve the initial value problem

$$my'' + ky = F_0 \sin pt, \qquad y(0) = 0, \quad y'(0) = 0.$$

From Sec. 2.13 we know that this equation governs the forced oscillations of a body of mass m attached at the lower end of an elastic spring whose upper end is fixed (Fig. 49). k is the spring modulus, and $F_0 \sin pt$ is the driving force or input. Setting $\omega_0 = \sqrt{k/m}$, we may write the equation in the form

$$y'' + \omega_0^2 y = K \sin pt \qquad\qquad \left(K = \frac{F_0}{m}\right).$$

The subsidiary equation is

(1)
$$s^2 Y + \omega_0^2 Y = K \frac{p}{s^2 + p^2}.$$

No resonance. If $p^2 \neq \omega_0^2$, the solution of (1) and its representation in terms of partial fractions is

(2)
$$Y(s) = \frac{Kp}{(s^2 + \omega_0^2)(s^2 + p^2)} = \frac{As + B}{s^2 + \omega_0^2} + \frac{Ms + N}{s^2 + p^2}.$$

We first determine A and B in the fraction corresponding to $a_1 = \alpha_1 + i\beta_1 = i\omega_0$. In (3c), Sec. 4.4,

$$R_{a_1}(a_1) = S_{a_1} + iT_{a_1} = \left.\frac{Kp}{s^2 + p^2}\right|_{s=a_1} = \frac{Kp}{p^2 - \omega_0^2}$$

so that $T_{a_1} = 0$. From this and (3b) in Sec. 4.4 we see that the inverse transform of the first fraction on the right-hand side of (2) is

$$\frac{Kp}{\omega_0(p^2 - \omega_0^2)} \sin \omega_0 t.$$

We determine M and N in the last fraction, which corresponds to $a_2 = \alpha_2 + i\beta_2 = ip$. In (3c), Sec. 4.4, we now have

$$R_{a_2}(a_2) = S_{a_2} + iT_{a_2} = \left.\frac{Kp}{s^2 + \omega_0^2}\right|_{s=a_2} = \frac{Kp}{\omega_0^2 - p^2}.$$

Fig. 49. Vibrating system.
Forced oscillations

$y = 0$ (Equilibrium position)

$F_0 \sin pt$

Fig. 50. Example 5

k

$0 \cdots m_1 = 1$

y_1

k

$0 \cdots m_2 = 1$

y_2

k

Hence $T_{a_2} = 0$. From (3b) in Sec. 4.4 we see that the inverse transform of that fraction is

$$-\frac{K}{p^2 - \omega_0{}^2}\sin pt.$$

By adding these two transforms we obtain the desired solution

$$y(t) = \mathscr{L}^{-1}(Y) = \frac{K}{p^2 - \omega_0{}^2}\left(\frac{p}{\omega_0}\sin \omega_0 t - \sin pt\right).$$

The vibration is a superposition of two harmonic motions, one of which has the frequency of the driving force and the other has the frequency of the freely vibrating system.

Resonance. If $p = \omega_0$, the solution of (1) is

$$Y(s) = \frac{K\omega_0}{(s^2 + \omega_0{}^2)^2}.$$

The denominator has the double roots $a = \alpha + i\beta = i\omega_0$ and $\bar{a} = -i\omega_0$, that is, $\alpha = 0$ and $\beta = \omega_0$. In (4c) and (4d), Sec. 4.4,

$$R_a(s) = K\omega_0, \qquad S_a = K\omega_0, \qquad T_a = 0, \qquad R_a{}'(s) = 0,$$

and (4b) in Sec. 4.4 yields the solution

(3) $$\qquad y(t) = \mathscr{L}^{-1}(Y) = \frac{K}{2\omega_0{}^2}(\sin \omega_0 t - \omega_0 t \cos \omega_0 t).$$

Systems of differential equations may also be solved by means of the Laplace transformation; cf. the following example.

Example 5 (*Vibration of two masses*). The vibrations of the mechanical system of two masses shown in Fig. 50 are governed by the differential equations

(4)
$$y_1{}'' = -ky_1 + k(y_2 - y_1)$$
$$y_2{}'' = -k(y_2 - y_1) - ky_2$$

where k is the spring modulus of each of the three springs, y_1 and y_2 are the displacements of the masses from their position of static equilibrium; the masses of the springs and the damping are neglected. The derivation of (4) is similar to that of the differential equation in Sec. 2.6.

We shall determine the solution corresponding to the initial conditions $y_1(0) = 1$, $y_2(0) = 1$, $y_1{}'(0) = \sqrt{3k}$, $y_2{}'(0) = -\sqrt{3k}$. Let $Y_1 = \mathscr{L}(y_1)$ and $Y_2 = \mathscr{L}(y_2)$. Then from (2) in Sec. 4.2 and the initial conditions we obtain the subsidiary equations

$$s^2 Y_1 - s - \sqrt{3k} = -kY_1 + k(Y_2 - Y_1)$$
$$s^2 Y_2 - s + \sqrt{3k} = -k(Y_2 - Y_1) - kY_2.$$

This system of linear algebraic equations in the unknowns Y_1 and Y_2 may be written

$$(s^2 + 2k)Y_1 - \qquad kY_2 \qquad = s + \sqrt{3k}$$
$$-kY_1 \quad + (s^2 + 2k)Y_2 = s - \sqrt{3k}.$$

Cramer's rule (Sec. 6.6) yields the solution

$$Y_1 = \frac{(s + \sqrt{3k})(s^2 + 2k) + k(s - \sqrt{3k})}{(s^2 + 2k)^2 - k^2}$$

$$Y_2 = \frac{(s^2 + 2k)(s - \sqrt{3k}) + k(s + \sqrt{3k})}{(s^2 + 2k)^2 - k^2}.$$

The representations in terms of partial fractions are

(5) $$Y_1 = \frac{s}{s^2 + k} + \frac{\sqrt{3k}}{s^2 + 3k}, \qquad Y_2 = \frac{s}{s^2 + k} - \frac{\sqrt{3k}}{s^2 + 3k}.$$

Hence the solution of our initial value problem is

$$y_1(t) = \mathscr{L}^{-1}(Y_1) = \cos \sqrt{k}\, t + \sin \sqrt{3k}\, t$$

$$y_2(t) = \mathscr{L}^{-1}(Y_2) = \cos \sqrt{k}\, t - \sin \sqrt{3k}\, t.$$

Problems

Find $f(t)$ if $\mathscr{L}(f)$ equals

1. $\dfrac{s + 12}{s^2 + 4s}$

2. $\dfrac{s - 3}{s^2 - 1}$

3. $\dfrac{3s}{s^2 + 2s - 8}$

4. $\dfrac{2s^2 + 5s - 1}{s^3 - s}$

5. $\dfrac{s^2 - 10s - 25}{s^3 - 25s}$

6. $\dfrac{s^2 - 6s + 4}{s^3 - 3s^2 + 2s}$

7. $\dfrac{3s - 1}{s^2 + 9}$

8. $\dfrac{s - 1}{s^2 + 1}$

9. $\dfrac{s}{s^2 - 2s + 2}$

10. $\dfrac{s}{s^2 - 2\pi s + 2\pi^2}$

11. $\dfrac{3s^2 - 2s - 1}{(s - 3)(s^2 + 1)}$

12. $\dfrac{s + 1}{s^2 + 4s + 13}$

13. $\dfrac{10 - 4s}{(s - 2)^2}$

14. $\dfrac{s^2 + s - 2}{(s + 1)^3}$

15. $\dfrac{s^3 + 3s^2 - s - 3}{(s^2 + 2s + 5)^2}$

16. $\dfrac{s^3 - 7s^2 + 14s - 9}{(s - 1)^2(s - 2)^3}$

17. $\dfrac{s^2 - 6s + 7}{(s^2 - 4s + 5)^2}$

18. $\dfrac{s^4 + 3(s + 1)^3}{s^4(s + 1)^3}$

19. Check the result in Ex. 1 by starting from $(s + 1)/(s - 2)(s + 3)$ and applying Theorem 3 in Sec. 4.2.
20. Derive (4) in Example 5.
21. Verify the representation (5) in Example 5.

Show that

22. $\mathscr{L}^{-1}\left\{\dfrac{s^3}{s^4 + 4a^4}\right\} = \cosh at \cos at$

23. $\mathscr{L}^{-1}\left\{\dfrac{s}{s^4 + 4a^4}\right\} = \dfrac{1}{2a^2} \sinh at \sin at$

24. $\mathscr{L}^{-1}\left\{\dfrac{s^2}{s^4 + 4a^4}\right\} = \dfrac{1}{2a}(\cosh at \sin at + \sinh at \cos at)$

25. $\mathscr{L}^{-1}\left\{\dfrac{1}{s^4 + 4a^4}\right\} = \dfrac{1}{4a^3}(\cosh at \sin at - \sinh at \cos at)$

Solve the following initial value problems by means of the Laplace transformation.

26. $y'' + y = 2$, $y(0) = 0$, $y'(0) = 3$
27. $y'' + y = 3 \cos 2t$, $y(0) = 0$, $y'(0) = 0$
28. $y'' - 4y = 8t^2 - 4$, $y(0) = 5$, $y'(0) = 10$
29. $y'' - 9y = -8e^t$, $y(0) = 0$, $y'(0) = 10$
30. $y'' + y = 2 \cos t$, $y(0) = 2$, $y'(0) = 0$
31. $y'' - 4y' + 3y = 0$, $y(0) = 3$, $y'(0) = 7$
32. $y'' - 2y' + 2y = 0$, $y(0) = 0$, $y'(0) = 1$

33. $y'' - 2y' + 5y = 0,$ $y(0) = 0,$ $y'(0) = 2$
34. $y'' - y' - 12y = 0,$ $y(0) = 4,$ $y'(0) = -5$
35. $y'' + 2y' + 2y = 0,$ $y(0) = 1,$ $y'(0) = -3$
36. $y'' - 2y' + 2y = 2\cos 2t - 4\sin 2t,$ $y(0) = 0,$ $y'(0) = 0$
37. $y'' + y' - 2y = 1 - 2t,$ $y(0) = 0,$ $y'(0) = 4$
38. $y'' + 2y' + y = e^{-2t},$ $y(0) = 0,$ $y'(0) = 0$
39. $y'' + 25y = 10(\cos 5t - 2\sin 5t),$ $y(0) = 1,$ $y'(0) = 2$
40. $y'' + 4y = 4(\cos 2t - \sin 2t),$ $y(0) = 1,$ $y'(0) = 3$
41. $y'' + y = 2(\cos t + \sin t),$ $y(0) = 0,$ $y'(0) = -1$
42. $y'' - 2y' + 5y = 8\sin t - 4\cos t,$ $y(0) = 1,$ $y'(0) = 3$

Solve the following initial value problems involving systems of differential equations.

43. $y_1' = -y_2,$ $y_2' = y_1,$ $y_1(0) = 1,$ $y_2(0) = 0$
44. $y_1' + y_2 = 2\cos t,$ $y_1 + y_2' = 0, y_1(0) = 0, y_2(0) = 1$
45. $y_1' + y_1 = y_2'' + 4y_2, y_1' + y_2' = \cos t + 2\cos 2t, y_1(0) = 0, y_2(0) = 0, y_1'(0) = 1,$
    $y_2'(0) = 2$
46. $y_1' = y_1 + 3y_2,$ $y_2'' = 4y_1 - 4e^t,$ $y_1(0) = 2,$ $y_1'(0) = 3,$ $y_2(0) = 1,$ $y_2'(0) = 2$
47. $y_1' + y_1 = y_2' + y_2, y_1'' + y_2'' = e^t, y_1(0) = 0, y_1'(0) = 1, y_2(0) = 1, y_2'(0) = 0$
48. $y_1'' + y_2 = -5\cos 2t,$ $y_2'' + y_1 = 5\cos 2t,$ $y_1(0) = 1,$ $y_1'(0) = 1,$ $y_2(0) = -1,$
    $y_2'(0) = 1$
49. $y_1' + y_2' = 2\sinh t, y_2' + y_3' = e^t, y_3' + y_1' = 2e^t + e^{-t}, y_1(0) = 1, y_2(0) = 1,$
    $y_3(0) = 0$
50. $-2y_1' + y_2' + y_3' = 0, y_1' + y_2' = 4t + 2, y_2' + y_3 = t^2 + 2,$
    $y_1(0) = y_2(0) = y_3(0) = 0$

4.6 Differentiation and Integration of Transforms

It can be shown that if $f(t)$ satisfies the conditions of the existence theorem in Sec. 4.1, then the derivative of the corresponding transform

$$F(s) = \mathscr{L}(f) = \int_0^\infty e^{-st} f(t)\, dt$$

with respect to s can be obtained by differentiating under the integral sign with respect to s (proof in Ref. [A3]); thus

$$F'(s) = -\int_0^\infty e^{-st}[tf(t)]\, dt.$$

Consequently, if $\mathscr{L}(f) = F(s)$, then

(1) $$\mathscr{L}\{tf(t)\} = -F'(s);$$

differentiation of the transform of a function corresponds to the multiplication of the function by $-t$.

This property of the Laplace transformation enables us to obtain new transforms from given ones.

Example 1. We shall derive the following three formulas:

	$\mathscr{L}(f)$	$f(t)$
(2)	$\dfrac{1}{(s^2 + \beta^2)^2}$	$\dfrac{1}{2\beta^3}(\sin \beta t - \beta t \cos \beta t)$
(3)	$\dfrac{s}{(s^2 + \beta^2)^2}$	$\dfrac{t}{2\beta}\sin \beta t$
(4)	$\dfrac{s^2}{(s^2 + \beta^2)^2}$	$\dfrac{1}{2\beta}(\sin \beta t + \beta t \cos \beta t)$

From (1) and formula 8 (with $\omega = \beta$) in Table 2, Sec. 4.1, we obtain

$$\mathscr{L}(t \sin \beta t) = \frac{2\beta s}{(s^2 + \beta^2)^2}\,.$$

By dividing by 2β we obtain (3). From (1) and formula 9 (with $\omega = \beta$) in Table 2, Sec. 4.1, we find

$$(5) \qquad \mathscr{L}(t \cos \beta t) = -\frac{(s^2 + \beta^2) - 2s^2}{(s^2 + \beta^2)^2} = \frac{s^2 - \beta^2}{(s^2 + \beta^2)^2} = \frac{1}{s^2 + \beta^2} - \frac{2\beta^2}{(s^2 + \beta^2)^2}$$

where the last equality may be readily verified by direct calculation. Hence

$$t \cos \beta t = \mathscr{L}^{-1}\left\{\frac{1}{s^2 + \beta^2} - \frac{2\beta^2}{(s^2 + \beta^2)^2}\right\} = \frac{1}{\beta}\sin \beta t - 2\beta^2 \mathscr{L}^{-1}\left\{\frac{1}{(s^2 + \beta^2)^2}\right\},$$

and we see that the last term on the right equals $f(t)$ in (2). Finally, by subtracting and adding β^2 in the numerator we have

$$\frac{s^2}{(s^2 + \beta^2)^2} = \frac{s^2 - \beta^2}{(s^2 + \beta^2)^2} + \frac{\beta^2}{(s^2 + \beta^2)^2}\,.$$

From (5) and (2) we see that the two terms on the right are the inverse transforms of $t \cos \beta t$ and $(\sin \beta t - t \cos \beta t)/2\beta$, respectively. This yields (4).

Similarly, if $f(t)$ satisfies the conditions of the existence theorem in Sec. 4.1 and the limit of $f(t)/t$, as t approaches 0 from the right, exists, then

$$(6) \qquad \mathscr{L}\left\{\frac{f(t)}{t}\right\} = \int_s^\infty F(\tilde{s})\, d\tilde{s} \qquad\qquad (s > \alpha);$$

in this manner, *integration of the transform of a function $f(t)$ corresponds to the division of $f(t)$ by t.*

In fact, from the definition it follows that

$$\int_s^\infty F(\tilde{s})\, d\tilde{s} = \int_s^\infty \left[\int_0^\infty e^{-\tilde{s}t} f(t)\, dt\right] d\tilde{s},$$

and it can be shown (cf. Ref. [A3]) that under the above assumptions the order of integration can be reversed, that is,

$$\int_s^\infty F(\tilde{s})\, d\tilde{s} = \int_0^\infty \left[\int_s^\infty e^{-\tilde{s}t} f(t)\, d\tilde{s}\right] dt = \int_0^\infty f(t)\left[\int_s^\infty e^{-\tilde{s}t}\, d\tilde{s}\right] dt.$$

The integral over $\tilde{s}$ on the right equals e^{-st}/t when $s > \alpha$, and, therefore,

$$\int_s^\infty F(\tilde{s})\, d\tilde{s} = \int_0^\infty e^{-st} \frac{f(t)}{t}\, dt = \mathscr{L}\left\{\frac{f(t)}{t}\right\} \qquad (s > \alpha).$$

Example 2.   Find the inverse of $\ln\left(1 + \dfrac{\omega^2}{s^2}\right)$.

By differentiation we first have

$$-\frac{d}{ds} \ln\left(1 + \frac{\omega^2}{s^2}\right) = \frac{2\omega^2}{s(s^2 + \omega^2)} \equiv F(s).$$

Writing $F(s)$ in terms of partial fractions, we obtain

$$f(t) = \mathscr{L}^{-1}(F) = \mathscr{L}^{-1}\left\{\frac{2}{s} - 2\frac{s}{s^2 + \omega^2}\right\} = 2 - 2\cos \omega t.$$

This function satisfies the conditions under which (6) holds. Therefore,

$$\mathscr{L}^{-1}\left\{\ln\left(1 + \frac{\omega^2}{s^2}\right)\right\} = \int_s^\infty F(\tilde{s})\, d\tilde{s} = \frac{f(t)}{t}.$$

Our result is

$$\mathscr{L}^{-1}\left\{\ln\left(1 + \frac{\omega^2}{s^2}\right)\right\} = \frac{2}{t}(1 - \cos \omega t).$$

Problems

1. Find $\mathscr{L}(t^n e^{at})$ by the use of Theorem 2, Sec. 4.1, and by repeated application of (1).

Find the Laplace transform of the following functions by means of (1).

2. $t \cos 2t$	**3.** te^{2t}	**4.** $t \cosh t$
5. $t^2 e^t$	**6.** $t \sinh 2t$	**7.** $t^2 \sinh 2t$
8. $t^2 \sin t$	**9.** $t^2 \cos \omega t$	**10.** $te^{-2t} \sin \omega t$

11. Use (3) to solve the boundary value problem $y'' + y = \cos t$, $y(0) = 0$, $y'(0) = 0$.

Using (6), find $f(t)$ if $\mathscr{L}(f)$ equals

12. $\dfrac{1}{(s+1)^2}$	**13.** $\dfrac{2}{(s-a)^3}$	**14.** $\dfrac{2s}{(s^2-4)^2}$
15. $\ln \dfrac{s}{s-1}$	**16.** $\ln \dfrac{s+a}{s-a}$	**17.** $\ln \dfrac{s+a}{s+b}$
18. $\ln \dfrac{s^2+1}{(s-1)^2}$	**19.** arc cot $(s+1)$	**20.** arc cot $\dfrac{s}{\omega}$

4.7   Unit Step Function

The unit step function $u_a(t)$ is defined as follows (Fig. 51 on p. 170):

(1)
$$u_a(t) = \begin{cases} 0 & \text{when } t < a \\ 1 & \text{when } t > a \end{cases} \qquad (a \geqq 0).$$

Fig. 51. Unit step function $u_a(t)$

Fig. 52. Unit step function $u_0(t)$

In particular when $a = 0$ (Fig. 52),

$$(2) \qquad u_0(t) = \begin{cases} 0 & \text{when } t < 0 \\ 1 & \text{when } t > 0. \end{cases}$$

The transform of $u_a(t)$ is readily obtained from the definition of the Laplace transformation. In fact,

$$\mathscr{L}\{u_a(t)\} = \int_0^\infty e^{-st} u_a(t)\, dt = \int_0^a e^{-st} 0\, dt + \int_a^\infty e^{-st} 1\, dt = -\frac{1}{s} e^{-st}\Big|_a^\infty,$$

that is, assuming $s > 0$,

$$(3) \qquad \mathscr{L}\{u_a(t)\} = \frac{e^{-as}}{s}.$$

The unit step function is very important. It can be considered the basic building block of certain functions whose knowledge greatly increases the use of the Laplace transformation.

Example 1. Find the transform of the function $f(t)$ shown in Fig. 53. Obviously

$$f(t) = k[u_a(t) - u_b(t)],$$

and from (3) we obtain the result

$$\mathscr{L}(f) = \frac{k}{s}(e^{-as} - e^{-bs}).$$

Example 2 (Periodic square wave). Represent the square wave function $f(t)$, shown in Fig. 54, in terms of unit step functions and find its transform. Obviously,

$$f(t) = k[u_0(t) - 2u_a(t) + 2u_{2a}(t) - 2u_{3a}(t) + - \cdots].$$

Fig. 53. $f(t)$ in Example 1   **Fig. 54.** Example 2

Fig. 55. Example 3

Assuming that termwise integration is permissible and using (3), we readily obtain

$$\mathscr{L}(f) = k\left[\frac{1}{s} - 2\frac{e^{-as}}{s} + 2\frac{e^{-2as}}{s} - + \cdots\right]$$

$$= \frac{k}{s}[1 - 2e^{-as}(1 - e^{-as} + e^{-2as} - + \cdots)].$$

The expression in the parentheses $(\cdots)$ is a geometric series; using the well-known formula

$$\sum_{n=0}^{\infty} x^n = 1 + x + x^2 + \cdots = \frac{1}{1-x} \qquad\qquad |x| < 1$$

and setting $x = -e^{-as}$, we find

$$\mathscr{L}(f) = \frac{k}{s}\left(1 - \frac{2e^{-as}}{1 + e^{-as}}\right) = \frac{k}{s}\left(\frac{1 - e^{-as}}{1 + e^{-as}}\right) = \frac{k}{s}\frac{e^{-as/2}(e^{as/2} - e^{-as/2})}{e^{-as/2}(e^{as/2} + e^{-as/2})} = \frac{k}{s}\tanh\frac{as}{2}$$

$$(a > 0, s > 0).$$

Example 3. Find the transform of the function $g(t)$ shown in Fig. 55. We see that $g(t)$ is the integral of the function $f(t)$ considered in Ex. 2, where $k = 1$. Hence, by Theorem 3 in Sec. 4.2,

$$\mathscr{L}(g) = \frac{1}{s}\mathscr{L}(f) = \frac{1}{s^2}\tanh\frac{as}{2}.$$

Problems

Represent the following functions in terms of unit step functions and find their Laplace transforms.

7.

(Periodic, period 3)

8.

(Periodic, period 3)

9.

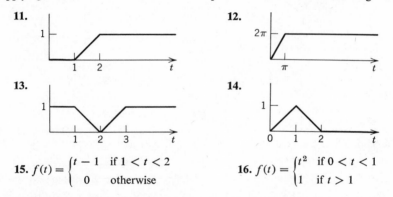

(Staircase function)

10.

(Staircase function)

Applying Theorem 3 in Sec. 4.2, find the Laplace transforms of the following functions.

11.

12.

13.

14.

15. $f(t) = \begin{cases} t - 1 & \text{if } 1 < t < 2 \\ 0 & \text{otherwise} \end{cases}$

16. $f(t) = \begin{cases} t^2 & \text{if } 0 < t < 1 \\ 1 & \text{if } t > 1 \end{cases}$

17. Solve Prob. 15 by using the definition of the Laplace transform and integration.

Find and graph $f(t)$ if $\mathscr{L}(f)$ equals

18. e^{-s}/s

19. $2(e^{-2s} - e^{-4s})/s$

20. $4(1 - e^{-3s})/s$

21. e^{-as}/s^2

22. $(e^{-s} - e^{-2s} - e^{-3s} + e^{-4s})/s^2$

23. $(e^{-s} + e^{-2s} - 3e^{-3s} + e^{-6s})/s^2$

24. A capacitor of capacitance C is charged so that its potential is V_0. At $t = 0$ the switch in Fig. 56 is closed and the capacitor starts to discharge through the resistor of resistance R. Using the Laplace transform find the charge $q(t)$ on the capacitor.

25. Find the current $i(t)$ in the circuit in Fig. 57, assuming that no current flows when $t \leqq 0$ and the switch is closed at $t = 0$.

Fig. 56. Problem 24

Fig. 57. Problem 25

4.8 Shifting on the t-Axis

In Sec. 4.1 it was shown that

$$\text{if } F(s) = \mathcal{L}\{f(t)\}, \quad \text{then} \quad F(s - a) = \mathcal{L}\{e^{at}f(t)\}.$$

This theorem answers the question how the function changes when the argument *s* of the *transform* is replaced by $s - a$ (*shifting on the s-axis*). Conversely, we may ask how the *transform* of a function $f(t)$ changes when the argument *t* of the *function* is replaced by $t - a$ (*shifting on the t-axis*).

The latter question is of importance in connection with systems which are acted upon by a disturbance beginning not at $t = 0$ but at some later time $t = a$.

Suppose that the transform

$$F(s) = \mathcal{L}\{f(\tau)\} = \int_0^\infty e^{-s\tau}f(\tau)\,d\tau$$

of a given function $f(\tau)$ exists. Multiplying by e^{-as}, where *a* is some positive constant, we have

$$e^{-as}F(s) = \int_0^\infty e^{-s(\tau+a)}f(\tau)\,d\tau.$$

Substituting $\tau + a = t$ in the integral, we obtain

$$e^{-as}F(s) = \int_a^\infty e^{-st}f(t - a)\,dt.$$

We want to write this integral as an integral from 0 to ∞. For this purpose we may replace $f(t - a)$ by the function which is zero on the interval $0 \le t < a$ and is equal to $f(t - a)$ when $t > a$. From the definition of the unit step function it follows that the function

$$f(t - a)u_a(t) = \begin{cases} 0 & (t < a) \\ f(t - a) & (t > a) \end{cases}$$

has the desired property. (Figure 58 shows an illustrative example.) Therefore,

$$e^{-as}F(s) = \int_0^\infty e^{-st}f(t - a)u_a(t)\,dt = \mathcal{L}\{f(t - a)u_a(t)\}.$$

We may sum up our result as follows.

Theorem 1 (Second shifting theorem). *If*

$$F(s) = \mathcal{L}\{f(t)\},$$

then for any positive constant a,

(1) $$e^{-as}F(s) = \mathcal{L}\{f(t - a)u_a(t)\}$$

where the unit step function $u_a(t)$ is defined by (1), Sec. 4.7.

Fig. 58. $f(t - a)u_a(t)$ where $f(t) = \cos t$

Example 1. Find the inverse of e^{-3s}/s^3.

Since $\mathscr{L}^{-1}(1/s^3) = t^2/2$ (cf. Table 2 in Sec. 4.1), Theorem 1 gives (cf. Fig. 59)

$$\mathscr{L}^{-1}(e^{-3s}/s^3) = \tfrac{1}{2}(t - 3)^2 u_3(t).$$

Example 2 (*Response of an RC-circuit to a single square wave*). Find the current $i(t)$ flowing in the circuit in Fig. 60, if a single square wave with voltage of height V_0 is applied. The circuit is assumed to be quiescent before the square wave is applied.

The equation of the circuit is (cf. Sec. 1.9)

$$Ri(t) + \frac{q(t)}{C} = Ri(t) + \frac{1}{C}\int_0^t i(\tau)\,d\tau = v(t)$$

where $v(t)$ can be represented in terms of two unit step functions:

$$v(t) = V_0[u_a(t) - u_b(t)].$$

Using Theorem 3 in Sec. 4.2 and formula (3) in Sec. 4.7, we obtain the subsidiary equation

$$RI(s) + \frac{I(s)}{sC} = \frac{V_0}{s}[e^{-as} - e^{-bs}].$$

Fig. 59. Example 1   **Fig. 60.** Example 2

Fig. 61. Voltage and current in Example 2

The solution of this equation may be written

$$I(s) = F(s)(e^{-as} - e^{-bs}) \qquad \text{where} \qquad F(s) = \frac{V_0/R}{s + 1/RC}.$$

From Table 2 in Sec. 4.1 we have

$$\mathscr{L}^{-1}(F) = \frac{V_0}{R} e^{-t/RC}.$$

Hence Theorem 1 yields the solution (Fig. 61)

$$i(t) = \mathscr{L}^{-1}(I) = \mathscr{L}^{-1}\{e^{-as}F(s)\} - \mathscr{L}^{-1}\{e^{-bs}F(s)\}$$

$$= \frac{V_0}{R} [e^{-(t-a)/RC}u_a(t) - e^{-(t-b)/RC}u_b(t)];$$

that is, $i = 0$ when $t < a$, and

$$i(t) = \begin{cases} K_1 e^{-t/RC} & \text{when } a < t < b \\ (K_1 - K_2)e^{-t/RC} & \text{when } t > b \end{cases}$$

where $K_1 = V_0 e^{a/RC}/R$ and $K_2 = V_0 e^{b/RC}/R$.

Example 3 (Response of a damped vibrating system to a single square wave). Determine the response of the damped vibrating system corresponding to the equation

$$y'' + 3y' + 2y = r(t)$$

where $r(t) = 1$ when $0 < t < 1$ and 0 otherwise (Fig. 62); assume that $y(0) = 0$ and $y'(0) = 0$.
We have

$$r(t) = u_0(t) - u_1(t).$$

The subsidiary equation is

$$s^2 Y + 3s Y + 2Y = \frac{1}{s} (1 - e^{-s}).$$

Solving for Y, we obtain

$$Y(s) = F(s)(1 - e^{-s}) \qquad \text{where} \qquad F(s) = \frac{1}{s(s + 1)(s + 2)}.$$

Fig. 62. Input in Example 3   **Fig. 63.** Output in Example 3

Using the method of partial fractions [cf. Sec. 4.4], we find

$$f(t) = \mathscr{L}^{-1}(F) = \tfrac{1}{2} - e^{-t} + \tfrac{1}{2}e^{-2t}.$$

Therefore, by Theorem 1, we have

$$\mathscr{L}^{-1}\{e^{-s}F(s)\} = f(t-1)u_1(t) = \begin{cases} 0 & (t < 1) \\ \tfrac{1}{2} - e^{-(t-1)} + \tfrac{1}{2}e^{-2(t-1)} & (t > 1) \end{cases}$$

This yields the solution (cf. Fig. 63)

$$y(t) = \mathscr{L}^{-1}(Y) = f(t) - f(t-1)u_1(t) = \begin{cases} \tfrac{1}{2} - e^{-t} + \tfrac{1}{2}e^{-2t} & (0 \leq t < 1) \\ K_1 e^{-t} - K_2 e^{-2t} & (t \geq 1) \end{cases}$$

where $K_1 = e - 1$ and $K_2 = (e^2 - 1)/2$.

Problems

Graph the following functions and find their Laplace transforms.

1. $(t - \pi)u_\pi(t)$ **2.** $(t - 1)^2 u_1(t)$ **3.** $tu_2(t)$

4. $u_1(t)\sin \pi(t - 1)$ **5.** $u_\pi(t)\sin t$ **6.** $u_{2\pi}(t)\cos t$

7. $e^{-2t}u_1(t)$ **8.** $e^t u_1(t)$ **9.** $t^2 u_1(t)$

In each case graph the given function, which is assumed to be zero outside the given interval, and find its Laplace transform.

10. t $(0 < t < 2)$ **11.** t^2 $(0 < t < 1)$

12. t^2 $(0 < t < 3)$ **13.** e^t $(0 < t < 1)$

14. $K \sin \omega t$ $(0 < t < \pi/\omega)$ **15.** $K \cos \omega t$ $(0 < t < \pi/\omega)$

16. $K \cos \omega t$ $(0 < t < 2\pi/\omega)$ **17.** $1 - e^{-t}$ $(0 < t < \pi)$

Find and graph the inverse Laplace transforms of the following functions.

18. $2e^{-0.5s}/s$ **19.** $se^{-\pi s}/(s^2 + 4)$ **20.** $e^{-s}/(s^2 + \omega^2)$

21. $e^{-2s}/(s - 2)$ **22.** e^{-2s}/s^2 **23.** e^{-s}/s^4

24. $s(1 + e^{-\pi s})/(s^2 + 1)$ **25.** $e^{-\pi s}/(s^2 + 2s + 2)$ **26.** e^{-s} arc cot s

27. If $\mathscr{L}\{f(t)\} = F(s)$, show that $\mathscr{L}\{f(at)\} = F(s/a)/a$. Using this formula, derive the transform of cos ωt from that of cos t.

28. If $F(s) = \mathscr{L}\{f(t)\}$, show that $F(as - b) = \mathscr{L}\left\{\dfrac{1}{a}e^{bt/a}f\left(\dfrac{t}{a}\right)\right\}$, $(a > 0)$.

Find the inverse of the following transforms.

29. $\dfrac{1}{(2s - 4)^4}$ **30.** $\dfrac{12}{(4s + 3)^2 - 9}$ **31.** $\dfrac{\pi^2 s + 3\pi}{(\pi s + 3)^2 + \pi^2}$

Find the current $i(t)$ in the *LC*-circuit in Fig. 64, assuming $L = 1$ henry, $C = 1$ farad, zero initial current and charge on the capacitor, and $v(t)$ as follows.

32. $v = 1$ when $0 < t < a$ and 0 otherwise

33. $v = t$ when $0 < t < 1$ and 0 otherwise

34. $v = t$ when $0 < t < 1$ and $v = 1$ when $t > 1$

35. $v = 1 - e^{-t}$ when $0 < t < \pi$ and 0 otherwise

Fig. 64. *LC*-circuit

4.9   *Periodic Functions*

The Laplace transforms of some special periodic functions were already considered in Sec. 4.7, and we shall now approach the problem of determining such transforms in a systematic way.

Let $f(t)$ be a function which is defined for all positive t and has the period p (>0), that is,

$$f(t + p) = f(t) \qquad\qquad \text{for all } t > 0.$$

If $f(t)$ is piecewise continuous over an interval of length p, then its Laplace transform exists, and we can write the integral from zero to infinity as the series of integrals over successive periods:

$$\mathscr{L}(f) = \int_0^\infty e^{-st} f(t)\, dt = \int_0^p e^{-st} f\, dt + \int_p^{2p} e^{-st} f\, dt + \int_{2p}^{3p} e^{-st} f\, dt + \cdots.$$

If we substitute $t = \tau + p$ in the second integral, $t = \tau + 2p$ in the third integral, $\cdots$, $t = \tau + (n-1)p$ in the nth integral, $\cdots$, then the new limits are 0 and p. Since

$$f(\tau + p) = f(\tau), \qquad f(\tau + 2p) = f(\tau),$$

etc., we thus obtain

$$\mathscr{L}(f) = \int_0^p e^{-s\tau} f(\tau)\, d\tau + \int_0^p e^{-s(\tau+p)} f(\tau)\, d\tau + \int_0^p e^{-s(\tau+2p)} f(\tau)\, d\tau + \cdots.$$

The factors which do not depend on τ can be taken out from under the integral signs; this gives

$$\mathscr{L}(f) = [1 + e^{-sp} + e^{-2sp} + \cdots] \int_0^p e^{-s\tau} f(\tau)\, d\tau.$$

The series in brackets $[\cdots]$ is a geometric series whose sum is $1/(1 - e^{-ps})$ (cf. also Ex. 2 in Sec. 4.7). The following result is therefore established.

Theorem 1 (Transform of periodic functions). *The Laplace transform of a piecewise continuous periodic function $f(t)$ with period p is*

(1)
$$\mathscr{L}(f) = \frac{1}{1 - e^{-ps}} \int_0^p e^{-st} f(t)\, dt \qquad\qquad (s > 0).$$

Example 1 (Half-wave rectifier). Find the Laplace transform of the following function $f(t)$ with period $p = 2\pi/\omega$:

$$f(t) = \begin{cases} \sin \omega t & \text{when} \quad 0 < t < \pi/\omega, \\ 0 & \text{when} \quad \pi/\omega < t < 2\pi/\omega. \end{cases}$$

Note that this function is the half-wave rectification of $\sin \omega t$ (Fig. 65 on p. 178). From (1) we obtain

$$\mathscr{L}(f) = \frac{1}{1 - e^{-2\pi s/\omega}} \int_0^{\pi/\omega} e^{-st} \sin \omega t\, dt.$$

Fig. 65. Half-wave rectification of sin ωt

Integrating by parts or noting that the integral is the imaginary part of the integral

$$\int_0^{\pi/\omega} e^{(-s+i\omega)t}\, dt = \frac{1}{-s+i\omega} e^{(-s+i\omega)t} \Big|_0^{\pi/\omega} = \frac{-s-i\omega}{s^2+\omega^2}(-e^{-s\pi/\omega}-1)$$

we obtain the result

$$\mathscr{L}(f) = \frac{\omega(1+e^{-\pi s/\omega})}{(s^2+\omega^2)(1-e^{-2\pi s/\omega})} = \frac{\omega}{(s^2+\omega^2)(1-e^{-\pi s/\omega})}.$$

Example 2 (Saw-tooth wave). Find the Laplace transform of the function (Fig. 66)

$$f(t) = \frac{k}{p}t \qquad \text{when } 0 < t < p, \qquad f(t+p) = f(t).$$

Integration by parts yields

$$\int_0^p e^{-st}t\, dt = -\frac{t}{s}e^{-st}\Big|_0^p + \frac{1}{s}\int_0^p e^{-st}\, dt$$

$$= -\frac{p}{s}e^{-sp} - \frac{1}{s^2}(e^{-sp}-1),$$

and from (1) we thus obtain the result

$$\mathscr{L}(f) = \frac{k}{ps^2} - \frac{ke^{-ps}}{s(1-e^{-ps})} \qquad\qquad (s>0).$$

Fig. 66. Saw-tooth wave   **Fig. 67.** Staircase function

Example 3 (Staircase function). Find the Laplace transform of the staircase function (Fig. 67)

$$g(t) = kn \qquad\qquad [np < t < (n+1)p, \qquad n = 0, 1, 2, \cdots].$$

Since $g(t)$ is the difference of the functions $h(t) = kt/p$ (whose transform is k/ps^2) and $f(t)$ in Ex. 2, we obtain

$$\mathscr{L}(g) = \mathscr{L}(h) - \mathscr{L}(f) = \frac{ke^{-ps}}{s(1-e^{-ps})} \qquad\qquad (s>0).$$

The next example illustrates the application of the Laplace transformation to a linear nonhomogeneous differential equation which has a periodic function on the right-hand side.

Fig. 68. Input $r(t)$ in Example 4

Example 4 (*Forced oscillations*). Solve the differential equation

(2) $$y'' + 2y' + 10y = r(t)$$

where (Fig. 68)

$$r(t) = \begin{cases} 1 & (0 < t < \pi) \\ -1 & (\pi < t < 2\pi) \end{cases} \qquad r(t + 2\pi) = r(t).$$

We have (cf. Ex. 2 in Sec. 4.7)

$$R(s) = \mathscr{L}(r) = \frac{1}{s}\,[1 - 2e^{-\pi s} + 2e^{-2\pi s} - 2e^{-3\pi s} + - \cdots].$$

The subsidiary equation of (2) is

$$s^2 Y - sy(0) - y'(0) + 2[sY - y(0)] + 10Y = R(s).$$

By solving for Y we obtain

(3) $$Y(s) = \frac{(s + 2)y(0) + y'(0)}{s^2 + 2s + 10} + \frac{R(s)}{s^2 + 2s + 10}.$$

The roots of the denominator are

$$a = -1 + 3i \qquad \text{and} \qquad \bar{a} = -1 - 3i.$$

The inverse transform of the first term in (3) is

$$\mathscr{L}^{-1}\!\left\{\frac{(s + 2)y(0) + y'(0)}{s^2 + 2s + 10}\right\} = \mathscr{L}^{-1}\!\left\{\frac{(s + 1)y(0)}{(s + 1)^2 + 9} + \frac{y(0) + y'(0)}{(s + 1)^2 + 9}\right\}$$

(4)
$$= e^{-t}\!\left(y(0)\cos 3t + \frac{y(0) + y'(0)}{3}\sin 3t\right).$$

This is the solution of the homogeneous equation corresponding to (2). The last term in (3) is

$$\frac{R(s)}{s^2 + 2s + 10} = \frac{1}{s[(s + 1)^2 + 9]}\,\{1 - 2e^{-\pi s} + - \cdots\}.$$

Now we know that

$$\mathscr{L}^{-1}\!\left\{\frac{1}{(s + 1)^2 + 9}\right\} = \frac{1}{3}\,e^{-t}\sin 3t.$$

Therefore, according to Theorem 3 in Sec. 4.2,

(5) $$\mathscr{L}^{-1}\!\left\{\frac{1}{s}\!\left(\frac{1}{(s + 1)^2 + 9}\right)\right\} = \frac{1}{3}\int_0^t e^{-\tau}\sin 3\tau\,d\tau = \frac{1}{10}\,[1 - h(t)]$$

where

(6) $$h(t) = e^{-t}(\cos 3t + \tfrac{1}{3}\sin 3t).$$

Denoting the function on the right-hand side of (5) by $f(t)$, that is,

$$f(t) = \frac{1}{10} - \frac{h(t)}{10},$$

and applying Theorem 1, Sec. 4.8, we thus obtain

(7) $\mathcal{L}^{-1}\left\{\dfrac{R(s)}{s^2 + 2s + 10}\right\} = f(t) - 2f(t - \pi)u_\pi(t) + 2f(t - 2\pi)u_{2\pi}(t) - + \cdots.$

Now in the terms on the right,

$$h(t - \pi) = e^{-(t-\pi)}[\cos 3(t - \pi) + \tfrac{1}{3}\sin 3(t - \pi)] = -h(t)e^\pi$$

and similarly,

$$h(t - 2\pi) = h(t)e^{2\pi}, \qquad h(t - 3\pi) = -h(t)e^{3\pi},$$

etc. It follows that in (7),

$$f(t - \pi) = \frac{1}{10} - \frac{h(t - \pi)}{10} = \frac{1}{10} + \frac{h(t)}{10}e^\pi$$

$$f(t - 2\pi) = \frac{1}{10} - \frac{h(t - 2\pi)}{10} = \frac{1}{10} - \frac{h(t)}{10}e^{2\pi}$$

etc., and the right-hand side of (7) equals

$$\frac{1}{10} - \frac{h(t)}{10} \qquad\qquad\qquad\text{when } 0 < t < \pi,$$

$$-\frac{1}{10} - \frac{h(t)}{10} - \frac{h(t)}{5}e^\pi \qquad\qquad\text{when } \pi < t < 2\pi,$$

$$\frac{1}{10} - \frac{h(t)}{10} - \frac{h(t)}{5}(e^\pi + e^{2\pi}) \qquad\text{when } 2\pi < t < 3\pi,$$

$$\frac{(-1)^n}{10} + \frac{h(t)}{10} - \frac{h(t)}{5}(1 + e^\pi + \cdots + e^{n\pi}) \qquad\text{when } n\pi < t < (n + 1)\pi.$$

Summing the finite geometric progression in the parentheses, we see that the expression in the last line may be written

$$\frac{(-1)^n}{10} + \frac{h(t)}{10} - \frac{h(t)}{5}\left(\frac{e^{(n+1)\pi} - 1}{e^\pi - 1}\right).$$

If we use formula (6), this becomes

(8)
$$\left[\frac{1}{10} + \frac{1}{5(e^\pi - 1)}\right]e^{-t}\left(\cos 3t + \frac{1}{3}\sin 3t\right)$$

$$+ \left[\frac{(-1)^n}{10} - \frac{1}{5(e^\pi - 1)}\right]e^{-[t-(n+1)\pi]}\left(\cos 3t + \frac{1}{3}\sin 3t\right),$$

and the solution of (2) on the interval $n\pi < t < (n + 1)\pi$ is the sum of (8) and the right-hand side of (4).

Clearly the function in (4) and the function in the first line of (8) approach zero as t approaches infinity. The function in the last line of (8) is a periodic function of t with period 2π and thus represents the *steady-state solution* of our problem. To see this more distinctly we denote that function by $y_2(t)$ and set $t - n\pi = \tau$. Then $t = \tau + n\pi$, and τ ranges from 0 to π as t ranges from $n\pi$ to $(n + 1)\pi$. Also

$$\cos 3t = \cos (3\tau + 3n\pi) = \cos 3\tau \cos 3n\pi = (-1)^n \cos 3\tau,$$

$$\sin 3t = \sin (3\tau + 3n\pi) = \sin 3\tau \cos 3n\pi = (-1)^n \sin 3\tau,$$

Fig. 69. Steady-state output (9) in Example 4

and the function y_2 becomes

$$y_2 = (-1)^n \left[\frac{1}{10} - \frac{e^{-\tau+\pi}}{5(e^\pi - 1)} \left(\cos 3\tau + \frac{1}{3} \sin 3\tau \right) \right]$$

(9)

$$\approx (-1)^n [0.1 - 0.22 e^{-\tau} \cos (3\tau - 0.32)]$$

where $0 < \tau < \pi$ and $t = \tau + n\pi$, $n = 0, 1, \cdots$. Figure 69 shows this function; the graph for $0 < t < \pi$ is obtained from (9) with $n = 0$ and $t = \tau$, and for $\pi < t < 2\pi$ it is obtained from (9) with $n = 1$ and $t = \tau + \pi$.

We have the surprising result that the output y_2 tends to oscillate more rapidly than the input $r(t)$. The reason for this unexpected behavior of our system will become obvious from our later consideration of Fourier series in Sec. 9.7.

Problems

Graph the following functions $f(t)$ which are assumed to have the period 2π and find their Laplace transforms.

1. $f(t) = \pi - t$ $(0 < t < 2\pi)$ **2.** $f(t) = t$ $(0 < t < 2\pi)$

3. $f(t) = t^2$ $(0 < t < 2\pi)$ **4.** $f(t) = 4\pi^2 - t^2$ $(0 < t < 2\pi)$

5. $f(t) = e^t$ $(0 < t < 2\pi)$ **6.** $f(t) = \sin 0.5t$ $(0 < t < 2\pi)$

7. $f(t) = \begin{cases} t & \text{if } 0 < t < \pi \\ 0 & \text{if } \pi < t < 2\pi \end{cases}$ **8.** $f(t) = \begin{cases} 1 & \text{if } 0 < t < \pi \\ -1 & \text{if } \pi < t < 2\pi \end{cases}$

9. $f(t) = \begin{cases} t & \text{if } 0 < t < \pi \\ \pi - t & \text{if } \pi < t < 2\pi \end{cases}$ **10.** $f(t) = \begin{cases} 0 & \text{if } 0 < t < \pi \\ t - \pi & \text{if } \pi < t < 2\pi \end{cases}$

11. How can the answer to Prob. 9 be obtained from the answers to Probs. 7 and 10?

12. Solve Prob. 10 by applying Theorem 1, Sec. 4.8, to Prob. 7.

Find the Laplace transforms of the following functions.

13. Half-wave rectification of $-\sin \omega t$ (Fig. 70). (*a*) Use direct calculation. (*b*) Apply Theorem 1, Sec. 4.8, to the result of Ex. 1.

Fig. 70. Problem 13

14. Full-wave rectification $|\sin \omega t|$ of $\sin \omega t$ (Fig. 71)

Fig. 71. Problem 14

15. $|\cos \omega t|$

16. Show that the solution y_2 in Ex. 4 is continuous at 0 and π.

17. Apply Theorem 1 to the function $f(t) = 1$, which is periodic with any period p.

18. Find the steady-state current in the circuit in Fig. 72.

Fig. 72. Problem 18

19. Solve Prob. 18 without the use of the Laplace transformation.

20. Find the current $i(t)$ in the circuit in Fig. 73, assuming that $i(0) = 0$.

(a)   (b)

Fig. 73. Problem 20

21. Find the steady-state current in the circuit in Fig. 73b, if $v(t) = t$ when $0 < t < 1$ and $v(t + 1) = v(t)$ as shown in Fig. 74.

Fig. 74. Problem 21

22. Find the ramp-wave response of the *RC*-circuit in Fig. 75, assuming that the circuit is quiescent at $t = 0$.

23. Solve Prob. 22 without the use of the Laplace transformation. Explain the reason for the jump of $i(t)$ at $t = 1$.

24. Solve Prob. 22 by means of the Laplace transformation, starting from the equation of the form $Ri' + (1/C)i = v'$.

25. Find the current $i(t)$ in the *RC*-circuit in Fig. 75 when $v(t) = \sin \omega t \ (0 < t < \pi/\omega)$, $v(t) = 0 \ (t > \pi/\omega)$, and $i(0) = 0$.

Fig. 75. Problem 22

26. Steady current is flowing in the circuit in Fig. 76 with the switch closed. At $t = 0$ the switch is opened. Find the current $i(t)$.

Fig. 76. Problem 26

27. Figure 77 shows a system for automatic control of pressure. $y(t)$ is the displacement, where $y = 0$ corresponds to the equilibrium position due to a given constant pressure.

Fig. 77. Problem 27   **Fig. 78.** Displacement $y(t)$ in Problem 27

We make the following assumptions. The damping of the system is proportional to the velocity y'. For $t < 0$, the system is at rest. At $t = 0$ the pressure is suddenly increased in the form of a unit step function. Show that the corresponding differential equation is

$$my'' + cy' + ky = Pu_0(t)$$

(m = effective mass of the moving parts, c = damping constant, k = spring modulus, P = force due to the increase of pressure at $t = 0$). Using the Laplace transformation, show that if $c^2 < 4mk$, then (Fig. 78)

$$y(t) = \frac{P}{k}[1 - e^{-\alpha t}\sqrt{1 + (\alpha/\omega^*)^2}\cos(\omega^* t + \theta)]$$

where $\alpha = c/2m$, $\omega^* = \sqrt{(k/m) - \alpha^2}$ (>0), $\tan\theta = -\alpha/\omega^*$.

28. Solve Prob. 27 when $c^2 > 4mk$.
29. Solve Prob. 27 when $c^2 = 4mk$.
30. Two flywheels (moments of inertia M_1 and M_2) are connected by an elastic shaft (moment of inertia negligible), and are rotating with constant angular velocity ω. At $t = 0$ a constant retarding couple P is applied to the first wheel. Find the subsequent angular velocity $v(t)$ of the other wheel.
31. Same conditions as in Prob. 30, but the retarding couple is applied during the interval $0 < t < 1$ only. Find $v(t)$.
32. The circuits in Fig. 79 are coupled by mutual inductance M, and at $t = 0$ the currents $i_1(t)$ and $i_2(t)$ are zero. Applying Kirchhoff's second law (Sec. 1.9), show that the

Fig. 79. Problem 32

differential equations for the currents are

$$L_1 i_1' + R_1 i_1 = M i_2' + V_0 u_0(t)$$

$$L_2 i_2' + R_2 i_2 = M i_1'$$

where $u_0(t)$ is the unit step function. Setting $\mathscr{L}(i_1) = I_1(s)$ and $\mathscr{L}(i_2) = I_2(s)$, show that the subsidiary equations are

$$L_1 s I_1 + R_1 I_1 = M s I_2 + \frac{V_0}{s}$$

$$L_2 s I_2 + R_2 I_2 = M s I_1.$$

Assuming that $A \equiv L_1 L_2 - M^2 > 0$, show that the expression for I_2, obtained by solving these algebraic equations, may be written

$$I_2 = \frac{K}{(s + \alpha)^2 + \omega^{*2}}$$

where $K = A^{-1}V_0 M$, $\alpha = (2A)^{-1}(R_1 L_2 + R_2 L_1)$, $\omega^{*2} = A^{-1}R_1 R_2 - \alpha^2$. Show that

$$i_2(t) = \mathscr{L}^{-1}(I_2) = \begin{cases} (K/\omega^*)e^{-\alpha t}\sin\omega^* t & \text{if } \omega^{*2} > 0 \\ Kte^{-\alpha t} & \text{if } \omega^{*2} = 0 \\ (K/\beta)e^{-\alpha t}\sinh\beta t & \text{if } \omega^{*2} = -\beta^2 < 0. \end{cases}$$

33. Find $i_1(t)$ in Prob. 32, assuming that $\omega^{*2} > 0$.
34. Show that when $A = 0$ in Prob. 32, then $i_2(t) = K_0 e^{-at}$ where $K_0 = BV_0M$, $a = BR_1R_2$, $B = (R_1L_2 + R_2L_1)^{-1}$.
35. Suppose that in the first circuit in Fig. 79 the switch is closed and a steady current V_0/R_1 is flowing in it. At $t = 0$ the switch is opened. Find the secondary current $i_2(t)$. *Hint.* Note that $i_1(0) = V_0/R_1$ and $i_1 = 0$ when $t > 0$.

4.10 *Table of Some Laplace Transforms*

Remark. For a more extensive list of Laplace transforms and their inverse transforms, see Ref. [B5] in Appendix 1.

	$F(s) = \mathscr{L}\{f(t)\}$	$f(t)$
1	$1/s$	1
2	$1/s^2$	t
3	$1/s^n,\quad (n = 1, 2, \cdots)$	$t^{n-1}/(n-1)!$
4	$1/\sqrt{s}$	$1/\sqrt{\pi t}$
5	$1/s^{3/2}$	$2\sqrt{t/\pi}$
6	$1/s^a \quad (a > 0)$	$t^{a-1}/\Gamma(a)$
7	$\dfrac{1}{s-a}$	e^{at}
8	$\dfrac{1}{(s-a)^2}$	te^{at}
9	$\dfrac{1}{(s-a)^n} \quad (n = 1, 2, \cdots)$	$\dfrac{1}{(n-1)!}t^{n-1}e^{at}$
10	$\dfrac{1}{(s-a)^k} \quad (k > 0)$	$\dfrac{1}{\Gamma(k)}t^{k-1}e^{at}$
11	$\dfrac{1}{(s-a)(s-b)} \quad (a \neq b)$	$\dfrac{1}{(a-b)}(e^{at} - e^{bt})$
12	$\dfrac{s}{(s-a)(s-b)} \quad (a \neq b)$	$\dfrac{1}{(a-b)}(ae^{at} - be^{bt})$
13	$\dfrac{1}{s^2 + \omega^2}$	$\dfrac{1}{\omega}\sin \omega t$
14	$\dfrac{s}{s^2 + \omega^2}$	$\cos \omega t$
15	$\dfrac{1}{s^2 - a^2}$	$\dfrac{1}{a}\sinh at$
16	$\dfrac{s}{s^2 - a^2}$	$\cosh at$
17	$\dfrac{1}{(s-a)^2 + \omega^2}$	$\dfrac{1}{\omega}e^{at}\sin \omega t$
18	$\dfrac{s-a}{(s-a)^2 + \omega^2}$	$e^{at}\cos \omega t$

Table of Some Laplace Transforms (*continued*)

	$F(s) = \mathscr{L}\{f(t)\}$	$f(t)$
19	$\dfrac{1}{s(s^2 + \omega^2)}$	$\dfrac{1}{\omega^2}(1 - \cos \omega t)$
20	$\dfrac{1}{s^2(s^2 + \omega^2)}$	$\dfrac{1}{\omega^3}(\omega t - \sin \omega t)$
21	$\dfrac{1}{(s^2 + \omega^2)^2}$	$\dfrac{1}{2\omega^3}(\sin \omega t - \omega t \cos \omega t)$
22	$\dfrac{s}{(s^2 + \omega^2)^2}$	$\dfrac{t}{2\omega}\sin \omega t$
23	$\dfrac{s^2}{(s^2 + \omega^2)^2}$	$\dfrac{1}{2\omega}(\sin \omega t + \omega t \cos \omega t)$
24	$\dfrac{s}{(s^2 + a^2)(s^2 + b^2)}$ $(a^2 \neq b^2)$	$\dfrac{1}{b^2 - a^2}(\cos at - \cos bt)$
25	$\dfrac{1}{s^4 + 4a^4}$	$\dfrac{1}{4a^3}(\sin at \cosh at - \cos at \sinh at)$
26	$\dfrac{s}{s^4 + 4a^4}$	$\dfrac{1}{2a^2}\sin at \sinh at$
27	$\dfrac{1}{s^4 - a^4}$	$\dfrac{1}{2a^3}(\sinh at - \sin at)$
28	$\dfrac{s}{s^4 - a^4}$	$\dfrac{1}{2a^2}(\cosh at - \cos at)$
29	$\sqrt{s - a} - \sqrt{s - b}$	$\dfrac{1}{2\sqrt{\pi t^3}}(e^{bt} - e^{at})$
30	$\dfrac{1}{\sqrt{s + a}\,\sqrt{s + b}}$	$e^{-(a+b)\,t/2}I_0\left(\dfrac{a - b}{2}\,t\right)$ (cf. Sec. 3.6)
31	$\dfrac{1}{\sqrt{s^2 + a^2}}$	$J_0(at)$ (cf. Sec. 3.5)
32	$\dfrac{s}{(s - a)^{3/2}}$	$\dfrac{1}{\sqrt{\pi t}}e^{at}(1 + 2at)$
33	$\dfrac{1}{(s^2 - a^2)^k}$ $(k > 0)$	$\dfrac{\sqrt{\pi}}{\Gamma(k)}\left(\dfrac{t}{2a}\right)^{k-1/2}I_{k-1/2}(at)$ (cf. Sec. 3.6).
34	$\dfrac{1}{s}e^{-k/s}$	$J_0(2\sqrt{kt})$ (cf. Sec. 3.5)
35	$\dfrac{1}{\sqrt{s}}e^{-k/s}$	$\dfrac{1}{\sqrt{\pi t}}\cos 2\sqrt{kt}$
36	$\dfrac{1}{s^{3/2}}e^{k/s}$	$\dfrac{1}{\sqrt{\pi k}}\sinh 2\sqrt{kt}$

Table of Some Laplace Transforms (*continued*)

	$F(s) = \mathcal{L}\{f(t)\}$	$f(t)$
37	$\dfrac{1}{s} \ln s$	$-\ln t - \gamma$ ($\gamma \approx 0.5772$; cf. Sec. 3.6)
38	$\ln \dfrac{s-a}{s-b}$	$\dfrac{1}{t}(e^{bt} - e^{at})$
39	$\ln \dfrac{s^2 + \omega^2}{s^2}$	$\dfrac{2}{t}(1 - \cos \omega t)$
40	$\ln \dfrac{s^2 - a^2}{s^2}$	$\dfrac{2}{t}(1 - \cosh at)$
41	$\arctan \dfrac{\omega}{s}$	$\dfrac{1}{t} \sin \omega t$
42	$\dfrac{1}{s} \operatorname{arc cot} s$	$\mathrm{Si}(t)$ (cf. Appendix 3)

CHAPTER **5**

Linear Algebra Part I: Vectors

Two main factors have affected the development of engineering mathematics during the past decade. These factors are the widespread application of automatic computers to engineering problems and the increasing use of linear algebra and linear analysis, for instance, in handling large problems in systems analysis.

The next four chapters (5–8) are devoted to linear algebra (Chaps. 5, 6) and linear analysis (Chaps. 7, 8). The first two of these chapters include the theory and application of vectors and vector spaces (Secs. 5.1–5.4), dot products and inner product spaces (Sec. 5.5, 5.6), vector products and scalar triple products (Secs. 5.7–5.9), matrices and linear equations (Secs. 6.1–6.5, 6.8–6.12), determinants and Cramer's rule (Secs. 6.6, 6.7, 6.11), quadratic and Hermitian forms (Sec. 6.13), and eigenvalues and eigenvectors (Secs. 6.14, 6.15). Chapter 7 contains vector differential calculus (scalar and vector fields, curves, velocity, directional derivative, gradient, divergence, curl), and Chap. 8 is devoted to vector integral calculus (line, surface, and triple integrals and their transformation by means of the integral theorems of Green, Gauss, and Stokes).

In the present chapter we shall consider basic concepts and methods of vector algebra and its applications to physical and geometrical problems. The usefulness of vectors in engineering mathematics results from the fact that many physical quantities—for example, forces and velocities—may be represented by vectors, and in several respects the rules of vector calculation are as simple as the rules governing the system of real numbers.

It is true that any problem that can be solved by the use of vectors can also be treated by nonvectorial methods, but vector analysis is a shorthand which simplifies many calculations considerably. Furthermore, it is a way of visualizing physical and geometrical quantities and relations between them. For all those reasons extensive use is made of vector notation in modern engineering literature.

Prerequisite for this chapter: determinants of third order (cf. Sec. 6.6).
Section which may be omitted in a shorter course: Sec. 5.6.
References: Appendix 1, Part C.
Answers to problems: Appendix 2.

5.1 *Scalars and Vectors*

In physics and geometry there are quantities each of which is completely specified when its magnitude—that is, its size or number of units according to some scale—is given. Examples are the mass of a body, the charge of an electron, the specific heat of water, the resistance of a resistor, the diameter of a circle, the area of a triangle, and the volume of a cube. Each of these quantities is described by a single number (after a suitable choice of units of measure). Such a quantity is called a **scalar**.[1]

However, there are other physical and geometrical quantities which cannot be described by a single number, because they require for their complete characterization the specification of a direction as well as a magnitude.

For example, forces in mechanics are quantities of this type. We know that we may represent a force graphically by an arrow, or *directed line segment*, which indicates the direction of the force and whose length is equal to the magnitude of the force according to some suitable scale. Figure 80 shows the force of attraction for the earth's motion around the sun. The instantaneous velocity of the earth may also be represented by an arrow of suitable length and direction, and this illustrates that a velocity is also a quantity which is characterized by a magnitude and a direction.

Figure 81 shows a translation (displacement without rotation) of a triangle in the plane. This motion can be characterized by its magnitude (distance travelled by each point of the triangle) and its direction. We may represent the translation graphically by a directed line segment whose *initial point* is the original position P of a point of the triangle and whose *terminal point* is the new position Q of that point after the translation. If we do this for each point of the triangle, we obtain a family of directed line segments which have the *same length* and the *same direction* (that is, are parallel and are directed in the same sense). We may say that each of these directed line segments "carries" a point of the triangle from its original position to its new position.

This situation suggests the following definition of a vector.

Fig. 80. Force and velocity   **Fig. 81.** Translation

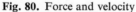

[1] Clearly, if a number characterizes a scalar which has a physical or geometrical meaning, this number should be independent of the choice of coordinates; since this point will not matter in the present chapter, we may discuss it later (in Sec. 7.1).

Definition. *A directed line segment is called a* **vector**. *Its length is called the* **length** *of the vector and its direction is called the* **direction** *of the vector. Two vectors are equal if and only if they have the same length and the same direction.*

The length of a vector is also called the **Euclidean norm** *or* **magnitude** *of the vector.*

Vectors will be denoted by lowercase boldfaced letters,[2] such as **a**, **b**, **v**, and the length of a vector **a** by |**a**|.

Of course, if we want, we may combine the two parts of this definition and define a vector to be the collection of all directed line segments having a given length and a given direction.

From the definition we see that a vector may be arbitrarily translated (that is, displaced without rotation) or, what amounts to the same thing, its initial point may be chosen in an arbitrary fashion. Clearly, if we choose a certain point as the initial point of a given vector, the terminal point of the vector is uniquely determined.

If two vectors **a** and **b** are equal, we write

$$\mathbf{a} = \mathbf{b},$$

and if they are different, we may write

$$\mathbf{a} \neq \mathbf{b}.$$

Any vector may be represented graphically as an arrow of suitable length and direction, as is illustrated in Fig. 82.

A vector of length 1 is called a **unit vector**.

For the sake of completeness, we mention that in physics and geometry there are situations where we want to impose restrictions on the position of the initial point of a vector. For example, as is known from mechanics, a force acting on a rigid body may be equally well applied to any point of the body on its line of action. This suggests the concept of a **sliding vector**, that is, a vector whose initial point can be any point on a straight line which is parallel to the vector. A force acting on an elastic body is a vector whose initial point cannot be changed at all. In fact, if we choose another point of application of the force, the effect of the force will in general be different. This suggests the notion of a **bound vector**, that is, a vector having a certain fixed initial point (*point of application*). Since these concepts occur in the literature, the student should know about them, but they will not be of particular importance in our further considerations.

Equal vectors,	Vectors having	Vectors having	Vectors having
a = **b**	the same length but different direction	the same direction but different length	different length and different direction

Fig. 82. Vectors

[2] This is customary in printed work; in handwritten work one may characterize vectors by arrows, for example $\vec{a}$ (in place of **a**), $\vec{b}$, etc.

5.2   *Components of a Vector*

A point in three-dimensional space is a geometric object, but if we introduce a coordinate system, we may describe it by (or even regard it as) an ordered triple of numbers (called its coordinates). Similarly, in the last section we defined a vector in a geometrical fashion (using the notion of a directed segment), but if we use a coordinate system, we may describe vectors in algebraic terms.

In fact, let us introduce a coordinate system in space whose axes are three mutually perpendicular straight lines. On all three axes we choose the same scale. Then the three *unit points* on the axes, whose coordinates are $(1, 0, 0)$, $(0, 1, 0)$, and $(0, 0, 1)$, have the same distance from the *origin*, the point of intersection of the axes. The rectangular coordinate system thus obtained is called a **Cartesian coordinate system** in space (cf. Fig. 83).

We consider now a vector **a** obtained by directing a line segment PQ such that P is the initial point and Q is the terminal point (Fig. 84). Let (x_1, y_1, z_1) and (x_2, y_2, z_2) be the coordinates of P and Q, respectively. Then the numbers

$$(1) \qquad a_1 = x_2 - x_1, \qquad a_2 = y_2 - y_1, \qquad a_3 = z_2 - z_1$$

are called the **components** of the vector **a** with respect to that Cartesian coordinate system.

By definition, the length $|\mathbf{a}|$ of the vector **a** is the distance $\overline{PQ}$, and from (1) and the theorem of Pythagoras it follows that

$$(2) \qquad |\mathbf{a}| = \sqrt{a_1{}^2 + a_2{}^2 + a_3{}^2}.$$

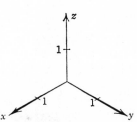

Fig. 83. Cartesian coordinate system

Example 1. The vector **a** with initial point P: $(3, 1, 4)$ and terminal point Q: $(1, -2, 4)$ has the components

$$a_1 = 1 - 3 = -2, \quad a_2 = -2 - 1 = -3, \quad a_3 = 4 - 4 = 0$$

and the length

$$|\mathbf{a}| = \sqrt{(-2)^2 + (-3)^2 + 0^2} = \sqrt{13}.$$

If we choose $(-1, 5, 8)$ as the initial point of **a**, then the corresponding terminal point is $(-3, 2, 8)$.

Fig. 84. Components of a vector

Fig. 85. Position vector **r** of a point A having the coordinates x, y, z

Observe from (1) that if we choose the initial point of a vector to be the origin, then its components are equal to the coordinates of the terminal point and the vector is then called the **position vector** of the terminal point (with respect to our coordinate system) and is usually denoted by **r**. Cf. Fig. 85.

From (1) we also see immediately that *the components a_1, a_2, a_3 of the vector* **a** *are independent of the choice of the initial point of* **a**, because if we translate **a**, then corresponding coordinates of P and Q are altered by the same amount. *Hence, a fixed Cartesian coordinate system being given, each vector is uniquely determined by the ordered triple of its components with respect to that coordinate system.*

If we introduce the *null vector or* **zero vector 0**, defined as the vector with components $0, 0, 0$, then any ordered triple of real numbers, including the triple $0, 0, 0$, can be chosen as the components of a vector.

The coordinate system being fixed, the correspondence between the ordered triples of real numbers and the vectors in space is one-to-one, that is, to each such triple there corresponds a vector in space that has those three numbers as its components with respect to that coordinate system, and conversely.

It follows that two vectors **a** *and* **b** *are equal if and only if corresponding components of these vectors are equal. Consequently, a vector equation*

$$\mathbf{a} = \mathbf{b}$$

is equivalent to the three equations

$$a_1 = b_1, \qquad a_2 = b_2, \qquad a_3 = b_3$$

for the components a_1, a_2, a_3 and b_1, b_2, b_3 of the vectors with respect to a given Cartesian coordinate system.

There is yet another consequence of that one-to-one correspondence between all ordered triples of real numbers and all vectors as defined in Sec. 5.1 in a geometrical fashion. That correspondence implies that a vector could equally well be *defined* as an ordered triple of real numbers (called the *components* of the vector). Starting from this definition, we would then obtain the geometrical interpretation of a vector which we used as a definition in Sec. 5.1.

Clearly, if a vector is a physical or geometric object, its length and direction should be independent of the particular choice of a coordinate system, and this requirement implies a certain transformation behavior of the components under the transition to another coordinate system. We shall discuss this problem later (in Sec. 7.9).

In the next chapter we shall also generalize the concept of a vector to include ordered n-tuples of numbers, where n is any fixed positive integer.

Problems

In each case, find the components and the length of the vector **v** with given initial point P: (x_1, y_1, z_1) and terminal point Q: (x_2, y_2, z_2), and sketch **v**.

1. P: $(1, 0, 0)$, Q: $(4, 2, 0)$	**2.** P: $(6, -1, 0)$, Q: $(3, 3, 0)$
3. P: $(4, 0, -1)$, Q: $(1, 0, 2)$	**4.** P: $(0, 0, 0)$, Q: (a, b, c)
5. P: $(0, a, 0)$, Q: $(0, b, 0)$	**6.** P: $(3, -2, 1)$, Q: $(1, 2, -4)$
7. P: $(8, 6, 1)$, Q: $(-8, 6, 1)$	**8.** P: $(-1, -1, -1)$, Q: $(3, 0, 0)$
9. P: $(-1, 7, 5)$, Q: $(-1, 7, 5)$	**10.** P: $(-3, -2, 1)$, Q: $(3, 2, -1)$

In each case, the components v_1, v_2, v_3 of a vector **v** and a particular initial point P are given. Find the corresponding terminal point and the length of **v**. Sketch **v**.

11. $1, -1, 0$, P: $(2, 1, 0)$	**12.** $0, 0, 1$, P: $(-3, 2, 0)$
13. $6, 2, 1$, P: $(-6, -2, -1)$	**14.** $2, 2, 2$, P: $(4, 4, 0)$
15. $3, 1, 2$, P: $(3, 1, 2)$	**16.** $5, 0, -3$, P: $(1, 0, 3)$
17. $1, 2, 3$, P: $(0, 0, 0)$	**18.** $0, 0, 0$, P: $(1, -1, -2)$
19. $\frac{1}{2}, 1, \frac{3}{2}$, P: $(-\frac{1}{2}, 1, \frac{1}{2})$	**20.** $2, -4, 6$, P: $(4, -2, 6)$

5.3 Addition of Vectors, Multiplication by Scalars

Since we want to do calculations with vectors, we should now introduce algebraic operations for vectors. We shall define two such operations, which are called addition of vectors and multiplication of vectors by scalars.

Experiments show that the resultant of two forces can be determined by the familiar parallelogram law (Fig. 86). This suggests the following definition of vector addition.

Given two vectors **a** *and* **b**, *put the initial point of* **b** *at the terminal point of* **a**; *then the* **sum** *of* **a** *and* **b** *is defined as the vector* **c** *drawn from the initial point of* **a** *to the terminal point of* **b** (Fig. 87), *and we write*

$$c = a + b.$$

If in some fixed coordinate system, **a** has the components a_1, a_2, a_3 and **b** has the components b_1, b_2, b_3, then *the components* c_1, c_2, c_3 *of the sum vector* $\mathbf{c} = \mathbf{a} + \mathbf{b}$

Fig. 86. Resultant of two forces (parallelogram law)

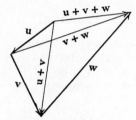

Fig. 87. Vector addition   **Fig. 88.** Commutativity   **Fig. 89.** Associativity of
of vector addition   vector addition

are obtained by adding corresponding components of **a** *and* **b**; thus

(1) $$c_1 = a_1 + b_1, \qquad c_2 = a_2 + b_2, \qquad c_3 = a_3 + b_3.$$

From the definition or from (1) we see that vector addition has the following properties. For any vectors (cf. Figs. 88, 89)

(2)

    (*a*) $\mathbf{a} + \mathbf{b} = \mathbf{b} + \mathbf{a}$ (*commutativity*)

    (*b*) $(\mathbf{u} + \mathbf{v}) + \mathbf{w} = \mathbf{u} + (\mathbf{v} + \mathbf{w})$ (*associativity*)

    (*c*) $\mathbf{a} + \mathbf{0} = \mathbf{0} + \mathbf{a} = \mathbf{a}$

    (*d*) $\mathbf{a} + (-\mathbf{a}) = \mathbf{0}$

where $-\mathbf{a}$ denotes the vector having the length $|\mathbf{a}|$ and the direction opposite to that of **a**.

In (2*b*) we may simply write $\mathbf{u} + \mathbf{v} + \mathbf{w}$, and similarly for sums of more than three vectors. Instead of $\mathbf{a} + \mathbf{a}$ we also write $2\mathbf{a}$, and so on. This (and the notation $-\mathbf{a}$ used before) suggests that we define the second algebraic operation (multiplication of vectors by scalars) in the following way.

Let **a** *be any vector and q any real number. Then the vector q***a** *is defined as follows.*
*The length of q***a** *is* $|q|\ |\mathbf{a}|$.
If $\mathbf{a} \neq \mathbf{0}$ *and* $q > 0$, *then q***a** *has the direction of* **a**.
If $\mathbf{a} \neq \mathbf{0}$ *and* $q < 0$, *then q***a** *has the direction opposite to* **a**.
If $\mathbf{a} = \mathbf{0}$ *or* $q = 0$ *(or both), then q***a** $= \mathbf{0}$.

Simple examples are shown in Fig. 90.

Clearly, if **a** has the components a_1, a_2, a_3, then $q\mathbf{a}$ has the components qa_1, qa_2, qa_3 (with respect to the same coordinate system). Furthermore, from the definitions

Fig. 90. Multiplication of vectors by numbers

Fig. 91. Difference of vectors

it follows that for any vectors and real numbers,

(3)

    (*a*) $q(\mathbf{a} + \mathbf{b}) = q\mathbf{a} + q\mathbf{b}$

    (*b*) $(c + k)\mathbf{a} = c\mathbf{a} + k\mathbf{a}$

    (*c*) $c(k\mathbf{a}) = (ck)\mathbf{a}$ (written $ck\mathbf{a}$)

    (*d*) $1\mathbf{a} = \mathbf{a}.$

The reader may prove that (2) and (3) imply that for any **a**,

(4)

    (*a*) $0\mathbf{a} = \mathbf{0}$

    (*b*) $(-1)\mathbf{a} = -\mathbf{a}.$

Instead of $\mathbf{b} + (-\mathbf{a})$ we simply write $\mathbf{b} - \mathbf{a}$ (Fig. 91).

A Cartesian coordinate system being given, we may now represent a vector **a** with components a_1, a_2, a_3 as the sum of three vectors parallel to the coordinate axes. For this purpose we associate with that coordinate system three unit vectors **i**, **j**, **k** which have the positive directions of the three coordinate axes. Then (Fig. 92)

(5) $$\mathbf{a} = a_1\mathbf{i} + a_2\mathbf{j} + a_3\mathbf{k}.$$

Figure 92 shows the vectors **i**, **j**, **k** when the origin is chosen as their common initial point. These vectors are mutually perpendicular. Instead of *perpendicular* the term **orthogonal** is also used, and we say that **i**, **j**, **k** form a *triple of orthogonal unit vectors*. This triple is also known as the *fundamental orthogonal triad* associated with that coordinate system.

Example. With respect to a given coordinate system, let

$$\mathbf{a} = 4\mathbf{i} + \mathbf{k} \quad \text{and} \quad \mathbf{b} = 2\mathbf{i} - 4\mathbf{j} + 3\mathbf{k}.$$

Then

$$2\mathbf{b} = 4\mathbf{i} - 8\mathbf{j} + 6\mathbf{k}, \quad -\mathbf{b} = -2\mathbf{i} + 4\mathbf{j} - 3\mathbf{k}, \quad 1.5\mathbf{a} - \mathbf{b} = 4\mathbf{i} + 4\mathbf{j} - 1.5\mathbf{k}.$$

Fig. 92. The unit vectors **i**, **j**, **k** and the representation (5)

Problems

Let $a = 2i - j + 3k$, $b = i + j - k$, $c = 4k$. Find

1. $a + b, b + a$
2. $-a, 2a, 0.5a$
3. $3a - 2b + 4c$
4. $2a + 2b, 2(a + b)$
5. $3b - 6c, 3(b - 2c)$
6. $|a + b|, |a| + |b|$
7. $a/|a|, -a/|a|$
8. $a/|a|, b/|b|, c/|c|$
9. $|a + b| + |c|, |a| + |b + c|$
10. $|a - b|, |a| - |b|$
11. $(a + b) + c, a + (b + c)$

12. What laws do Probs. 1, 4, and 11 illustrate?

In each case, find the resultant of the given forces.

13. $p = 3i + 2j + k$, $q = 4i - j - k$, $u = -7i - j$
14. $p = i + j + k$, $q = i - 2j + 3k$, $u = -2i + j + k$

15. Determine a force p such that p, $q = 3j - 4k$, and $u = i - j$ are in equilibrium.
16. Determine three forces p, q, u in the direction of the coordinate axes such that p, q, u, $v = 3i - 2j + k$, and $w = -2k$ are in equilibrium. Are p, q, u uniquely determined?
17. If $|p| = 1$ and $|q| = 2$, what can be said about the length and direction of the resultant?
18. Prove (4a).
19. Let a, b, c be the edge vectors of a parallelepiped. What are $a + b, a + b + c, a - b$, and $a - b - c$ in relation to the parallelepiped?

Using vectors, prove the following propositions.

20. The four diagonals of a cube have the same length.
21. The diagonals of a parallelogram bisect each other.
22. The four diagonals of a parallelepiped meet and bisect each other.
23. The line which passes through one vertex of a parallelogram and the midpoint of an opposite side divides one of the diagonals in the ratio $1:2$.
24. The sum of the vectors drawn from the center of a regular polygon to its vertices is the null vector.
25. The medians of a triangle meet at a point P which divides each median in the ratio $1:2$.

5.4   Vector Spaces. Linear Dependence and Independence

The set V of all vectors with the two algebraic operations of vector addition and multiplication of vectors by scalars defined on V suggests and motivates an algebraic structure which is called a *vector space* or *linear space* and is important because it includes many sets (sets of vectors, matrices, functions, operators, etc.) which occur quite often in mathematics and its applications. In fact, the concept of a vector space is basic in *functional analysis* (abstract modern analysis), which has applications to differential equations, numerical analysis, and other fields of practical interest to the engineer. (Cf. Ref. [A13] in Appendix 1.) By definition, a *real vector space* is a set of (unspecified) objects on which there are defined two algebraic operations satisfying (2) and (3) in Sec. 5.3; in detail:

A nonempty set V of elements $a, b, \cdots$ is called a **real vector space** or **real linear space** if in V there are defined two algebraic operations (called *vector addition* and *multiplication by a scalar*) as follows.

I. Vector addition associates with every pair of elements **a** and **b** of V a unique element of V, called the *sum* of **a** and **b** and denoted by $\mathbf{a} + \mathbf{b}$, such that the following axioms are satisfied.

I.1 For any two elements **a** and **b** of V,

$$\mathbf{a} + \mathbf{b} = \mathbf{b} + \mathbf{a}.$$

I.2 For any three elements **u**, **v**, **w** of V,

$$(\mathbf{u} + \mathbf{v}) + \mathbf{w} = \mathbf{u} + (\mathbf{v} + \mathbf{w}) \qquad (\text{written } \mathbf{u} + \mathbf{v} + \mathbf{w}).$$

I.3 There is a unique element in V, called the *zero element* and denoted by **0**, such that for every **a** in V,

$$\mathbf{a} + \mathbf{0} = \mathbf{a}.$$

I.4 For every **a** in V there is a unique element in V which is denoted by $-\mathbf{a}$ and is such that

$$\mathbf{a} + (-\mathbf{a}) = \mathbf{0}.$$

II. Multiplication by a scalar. The real numbers are called *scalars*. Multiplication of an element of V by a scalar associates with every **a** in V and every scalar q a unique element of V, called the *product* of q and **a** and denoted by $q\mathbf{a}$ (or $\mathbf{a}q$) such that the following axioms are satisfied.

II.1 For any scalar q and any **a** and **b** in V,

$$q(\mathbf{a} + \mathbf{b}) = q\mathbf{a} + q\mathbf{b}.$$

II.2 For any scalars c and k and any **a** in V,

$$(c + k)\mathbf{a} = c\mathbf{a} + k\mathbf{a}.$$

II.3 For any scalars c and k and any **a** in V,

$$c(k\mathbf{a}) = (ck)\mathbf{a} \qquad\qquad (\text{written } ck\mathbf{a}).$$

II.4 For any **a** in V,

$$1\mathbf{a} = \mathbf{a}.$$

The elements of V are called **vectors**.[3]

A *complex vector space* is obtained if, instead of real numbers, we take complex numbers as scalars.

We do not want to sidetrack our attention from our actual present goal, which is the study of vector algebra, so we shall confine ourselves to illustrating the generality of the notion of a vector space with some typical examples included in the problem set below, and we shall discuss further examples in connection with matrices in the next chapter. Note that we obtained the **abstract concept** of a vector space by the principle of defining it in terms of some of the most important properties of our concrete model, the set of all vectors in three-dimensional space.

---

[3] Regardless of what they actually are, but this convention causes no confusion because in any specific case the nature of those elements is clear from the context.

This is the principle by which many abstract mathematical concepts are derived from concrete models. Of course, in each case the selection of properties to be stated in the form of axioms needs experience.

If $\mathbf{a}_{(1)}, \cdots, \mathbf{a}_{(m)}$ are any vectors in V, then any element of V of the form

$$q_1\mathbf{a}_{(1)} + \cdots + q_m\mathbf{a}_{(m)} \qquad (q_1, \cdots, q_m \text{ scalars})$$

is called a **linear combination** of those vectors. The set S of all these linear combinations is called the **span** of $\mathbf{a}_{(1)}, \cdots, \mathbf{a}_{(m)}$ and is denoted by span $(\mathbf{a}_{(1)}, \cdots, \mathbf{a}_{(m)})$, and we say that S is **spanned** or *generated* by those m vectors.

Using the notion of linear combination, we can now introduce the basic concepts of linear dependence and independence of vectors.

A set of m vectors $\mathbf{a}_{(1)}, \cdots, \mathbf{a}_{(m)}$ is called a **linearly dependent set** if at least one of the vectors can be represented as a linear combination of the others (with scalars which may be zero or not). The set is called a **linearly independent set** if none of the vectors can be represented in that fashion.

Example 1. The vectors $\mathbf{a} = \mathbf{i} + 2\mathbf{j} + \mathbf{k}$, $\mathbf{b} = 3\mathbf{k}$, $\mathbf{c} = 2\mathbf{i} + 4\mathbf{j}$ constitute a linearly dependent set because $6\mathbf{a} - 2\mathbf{b} - 3\mathbf{c} = 0$; we have $\mathbf{a} = \frac{1}{3}\mathbf{b} + \frac{1}{2}\mathbf{c}$. The vectors $\mathbf{i}, \mathbf{j}, \mathbf{k}$ defined in the last section constitute a linearly independent set.

If a vector space V is such that it contains a linearly independent set B of n vectors, whereas any set of $n + 1$ or more vectors in V is linearly dependent, then V is said to have **dimension** n (or to be *n-dimensional*), and B is called a **basis** of V. Then any vector $\mathbf{v}$ of V can be represented as a linear combination of the n vectors of the basis in a unique fashion.

For example, the set of all vectors in space as defined in Sec. 5.1 forms a vector space. It is three-dimensional because a basis is $\mathbf{i}, \mathbf{j}, \mathbf{k}$ (cf. Sec. 5.3).

There is an important criterion for linear dependence and independence. To obtain it, we consider the vector equation

$$(1) \qquad k_1\mathbf{a}_{(1)} + k_2\mathbf{a}_{(2)} + \cdots + k_m\mathbf{a}_{(m)} = \mathbf{0},$$

where $\mathbf{a}_{(1)}, \cdots, \mathbf{a}_{(m)}$ are given vectors and $k_1, \cdots, k_m$ are scalars. This equation certainly holds for $k_1 = k_2 = \cdots = k_m = 0$. If we can find scalars $k_1, \cdots, k_m$, not all zero, such that (1) holds, then we can divide by a scalar $k_i \neq 0$ and represent the corresponding vector $\mathbf{a}_{(i)}$ as a linear combination of the others. (For example, if $k_1 \neq 0$, we obtain from (1)

$$\mathbf{a}_{(1)} = l_2\mathbf{a}_{(2)} + \cdots + l_m\mathbf{a}_{(m)} \qquad \text{where} \qquad l_j = -k_j/k_1.)$$

Hence, in this case the vectors constitute a linearly dependent set. However, if

Fig. 93. Collinear vectors

Fig. 94. Coplanar vectors

$k_1 = k_2 = \cdots = k_m = 0$ is the only set of scalars for which (1) holds, we cannot solve (1) for any of the vectors, and the vectors constitute a linearly independent set. This proves the following criterion.

Theorem 1 (Linear dependence). *A set of vectors* $\mathbf{a}_{(1)}, \cdots, \mathbf{a}_{(m)}$ *is linearly dependent if and only if* (1) *holds for a set of scalars* $k_1, \cdots, k_m$, *not all zero.*

This necessary and sufficient condition for linear dependence is often taken as a *definition* of linear dependence.

If one of those vectors is the zero vector, that set is linearly dependent, because if, say $\mathbf{a}_{(1)} = \mathbf{0}$, then (1) holds for any k_1 and $k_2 = \cdots = k_m = 0$.

Two vectors in three-dimensional space which form a linearly dependent set are **collinear**; that is, if we let their initial points coincide, they lie in the same line (Fig. 93). If three such vectors **a**, **b**, **c** form a linearly dependent set, they are collinear or **coplanar**; that is, if we let their initial points coincide, they lie in the same plane (Fig. 94). In fact, linear dependence implies that one of the vectors is a linear combination of the other two vectors, for instance, $\mathbf{a} = k\mathbf{b} + l\mathbf{c}$, and thus lies in the plane determined by **b** and **c** (or in a straight line if **b** and **c** are collinear). A set of four or more vectors in three-dimensional space is linearly dependent, because all those vectors form a vector space of dimension three.

Problems

Show that each of the following sets constitutes a vector space. Determine its dimension and find a basis.

1. The set of all vectors whose second component is zero.
2. The set of all ordered pairs of numbers, with the usual addition $(a_1, a_2) + (b_1, b_2) = (a_1 + b_1, a_2 + b_2)$ and scalar multiplication $c(a_1, a_2) = (ca_1, ca_2)$.
3. The set of all vectors of the form $c(\mathbf{i} + 2\mathbf{j} - \mathbf{k})$, where c is a scalar.
4. The set of all vectors of the form $b(\mathbf{i} - \mathbf{k}) + c\mathbf{j}$, where b and c are scalars.
5. The set of all ordered n-tuples of numbers $(a_1, \cdots, a_n)$ with addition defined by $(a_1, \cdots, a_n) + (b_1, \cdots, b_n) = (a_1 + b_1, \cdots, a_n + b_n)$ and multiplication by a scalar defined by $c(a_1, \cdots, a_n) = (ca_1, \cdots, ca_n)$.
6. The set of all real numbers.

7. span $(\mathbf{u}, \mathbf{v})$, where $\mathbf{u} = \mathbf{i} + \mathbf{j}$ and $\mathbf{v} = \mathbf{i} - \mathbf{j}$.

8. span $(\mathbf{i}, 3\mathbf{i} - \mathbf{k}, 2\mathbf{k})$.

9. All functions of the form $y(x) = a \cos x + b \sin x$ (where a and b are numbers) with the usual addition and multiplication by numbers.

10. The set of all quadratic, linear, and constant polynomials in x, with the usual addition and multiplication.

11. Show that the representation of a vector $\mathbf{a}$ in a vector space V as a linear combination of basis vectors $\mathbf{e}_{(1)}, \cdots, \mathbf{e}_{(n)}$ is unique.

12. Give examples of bases of the space of all vectors in three-dimensional space.

13. If a set $\mathbf{a}_{(1)}, \cdots, \mathbf{a}_{(m)}$ is linearly independent, show that every (nonempty) subset of that set is linearly independent.

14. (**Subspace**) A nonempty subset W of a vector space V is called a *subspace* of V if W is itself a vector space with respect to the algebraic operations defined in V. Give examples of one- and two-dimensional subspaces of the space of all vectors in three-dimensional space.

15. Show that a subset W of a vector space V is a subspace of V if and only if for any $\mathbf{a}$ and $\mathbf{b}$ in W and any scalars c and q the vector $c\mathbf{a} + q\mathbf{b}$ is in W.

5.5 *Inner Product (Dot Product)*

The **inner product**, **dot product**, or **scalar product** of two vectors $\mathbf{a}$ and $\mathbf{b}$ in three-dimensional space is written $\mathbf{a} \cdot \mathbf{b}$ and is defined as

(1)
$$\mathbf{a} \cdot \mathbf{b} = |\mathbf{a}|\,|\mathbf{b}| \cos \gamma \qquad (\text{when } \mathbf{a} \neq 0, \mathbf{b} \neq 0),$$
$$\mathbf{a} \cdot \mathbf{b} = 0 \qquad (\text{when } \mathbf{a} = 0 \text{ or } \mathbf{b} = 0)$$

here γ $(0 \leqq \gamma \leqq \pi)$ *is the angle between* $\mathbf{a}$ *and* $\mathbf{b}$ *(computed when the vectors have their initial point coinciding).* (Cf. Fig. 95.)

The value of the inner product is a scalar (a real number), and this motivates the term "scalar product." Since the cosine in (1) may be positive, zero or negative, the same is true for the inner product (cf. Fig. 95). Since that cosine is zero if and only if $\gamma = \pi/2$, we have the following important result.

Theorem 1 (Orthogonality). *Two nonzero vectors are orthogonal if and only if their inner product (their scalar product) is zero.*

Setting $\mathbf{b} = \mathbf{a}$ in (1), we have $\mathbf{a} \cdot \mathbf{a} = |\mathbf{a}|^2$, and this shows that the length (Euclidean norm) of a vector can be written in terms of the inner product,

(2)
$$|\mathbf{a}| = \sqrt{\mathbf{a} \cdot \mathbf{a}} \qquad (\geqq 0).$$

Fig. 95. Angle between vectors

From this and (1) we obtain the useful formula

(3) $$\cos \gamma = \frac{\mathbf{a} \cdot \mathbf{b}}{|\mathbf{a}|\,|\mathbf{b}|} = \frac{\mathbf{a} \cdot \mathbf{b}}{\sqrt{\mathbf{a} \cdot \mathbf{a}}\,\sqrt{\mathbf{b} \cdot \mathbf{b}}}.$$

From the definition we also see that the inner product has the properties

(4)

    (a) $[q_1\mathbf{a} + q_2\mathbf{b}] \cdot \mathbf{c} = q_1\mathbf{a} \cdot \mathbf{c} + q_2\mathbf{b} \cdot \mathbf{c}$ *(Linearity)*

    (b) $\mathbf{a} \cdot \mathbf{b} = \mathbf{b} \cdot \mathbf{a}$ *(Symmetry)*

    (c) $\mathbf{a} \cdot \mathbf{a} \geqq 0$

          $\mathbf{a} \cdot \mathbf{a} = 0$ if and only if $\mathbf{a} = \mathbf{0}$ *(Positive-definiteness).*

Hence *dot multiplication is commutative* [cf. (4b)] *and is distributive with respect to vector addition;* in fact, from (4a) with $q_1 = 1$ and $q_2 = 1$ we have

(4a*) $$(\mathbf{a} + \mathbf{b}) \cdot \mathbf{c} = \mathbf{a} \cdot \mathbf{c} + \mathbf{b} \cdot \mathbf{c} \qquad (Distributivity).$$

Furthermore, from (1) and $|\cos \gamma| \leqq 1$ we see that

(5) $$|\mathbf{a} \cdot \mathbf{b}| \leqq |\mathbf{a}|\,|\mathbf{b}| \qquad (Schwarz[4]\ inequality).$$

Using this and (2), the reader may prove

(6) $$|\mathbf{a} + \mathbf{b}| \leqq |\mathbf{a}| + |\mathbf{b}| \qquad \textbf{(Triangle inequality).}$$

A simple direct calculation with inner products shows that

(7) $$|\mathbf{a} + \mathbf{b}|^2 + |\mathbf{a} - \mathbf{b}|^2 = 2(|\mathbf{a}|^2 + |\mathbf{b}|^2) \qquad (Parallelogram\ equality).$$

If vectors **a** and **b** are represented in terms of components, say,

$$\mathbf{a} = a_1\mathbf{i} + a_2\mathbf{j} + a_3\mathbf{k} \qquad \text{and} \qquad \mathbf{b} = b_1\mathbf{i} + b_2\mathbf{j} + b_3\mathbf{k},$$

their inner product is given by the simple basic formula

(8) $$\mathbf{a} \cdot \mathbf{b} = a_1 b_1 + a_2 b_2 + a_3 b_3.$$

We prove (8). Since **i**, **j**, and **k** are unit vectors, we have from (1)

$$\mathbf{i} \cdot \mathbf{i} = 1, \qquad \mathbf{j} \cdot \mathbf{j} = 1, \qquad \mathbf{k} \cdot \mathbf{k} = 1,$$

and since they are orthogonal, it follows from Theorem 1 that

$$\mathbf{i} \cdot \mathbf{j} = 0, \qquad \mathbf{j} \cdot \mathbf{k} = 0, \qquad \mathbf{k} \cdot \mathbf{i} = 0.$$

Hence if we substitute those representations of **a** and **b** into $\mathbf{a} \cdot \mathbf{b}$ and use (4a*) and (4b), we first have a sum of nine inner products,

$$\mathbf{a} \cdot \mathbf{b} = a_1 b_1 \mathbf{i} \cdot \mathbf{i} + a_1 b_2 \mathbf{i} \cdot \mathbf{j} + \cdots + a_3 b_3 \mathbf{k} \cdot \mathbf{k}.$$

Since six of these products are zero, we obtain (8).

Before we consider some applications of inner products, let us introduce one more concept. Suppose that **a** and **b** $(\neq \mathbf{0})$ are any given vectors and let γ denote

[4] HERMANN AMANDUS SCHWARZ (1843–1921), German mathematician, who contributed to complex analysis and differential geometry.

Fig. 96. Component of a vector **a** in the direction of a vector **b**

the angle between them. Then the real number

$$p = |\mathbf{a}| \cos \gamma$$

is called the **component** *of* **a** *in the direction of* **b** or the **projection** *of* **a** *in the direction of* **b**. If **a** $= \mathbf{0}$, then γ is undefined, and we set $p = 0$.

It follows that $|p|$ is the length of the orthogonal projection of **a** on a straight line l in the direction of **b**. p may be positive, zero, or negative (Fig. 96).

From this definition we see that in particular the components of a vector **a** in the directions of the unit vectors **i**, **j**, **k** of the fundamental triad associated with a Cartesian coordinate system are the components a_1, a_2, a_3 of **a** as defined in Sec. 5.2. This shows that our present use of the term "component" is merely a slight generalization of the previous one.

From (3) we obtain

(9) $$p = |\mathbf{a}| \cos \gamma = \frac{\mathbf{a} \cdot \mathbf{b}}{|\mathbf{b}|} \qquad\qquad (\mathbf{b} \neq \mathbf{0})$$

and if in particular **b** is a unit vector, then we simply have

(10) $$p = \mathbf{a} \cdot \mathbf{b}.$$

The following examples may illustrate the usefulness of inner products. Various other applications will be considered later.

Example 1 (*Work done by a force*). Consider a particle on which a constant force **a** acts. Let the particle be given a displacement **d**. Then the work W done by **a** in the displacement is defined as the product of $|\mathbf{d}|$ and the component of **a** in the direction of **d**, that is,

(11) $$W = |\mathbf{a}|\,|\mathbf{d}| \cos \alpha = \mathbf{a} \cdot \mathbf{d}$$

where α is the angle between **d** and **a** (Fig. 97).

Fig. 97. Work done by a force

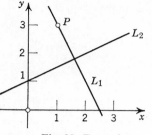

Fig. 98. Example 2

Example 2 (Orthogonal straight lines in a plane). Find a representation of the straight line L_1 through the point P: $(1, 3)$ in the xy-plane and perpendicular to the line L_2 represented by $x - 2y + 2 = 0$.

Any straight line L_1 in the xy-plane can be represented in the form $a_1x + a_2y = c$. If $c = 0$, then L_1 passes through the origin. If $c \neq 0$, then $a_1x + a_2y = 0$ represents a line L_1^* through the origin and parallel to L_1. The position vector of a point on L_1^* is $\mathbf{r} = x\mathbf{i} + y\mathbf{j}$. Introducing the vector $\mathbf{a} = a_1\mathbf{i} + a_2\mathbf{j}$, we see from (8) that the representation of L_1^* may be written

$$\mathbf{a} \cdot \mathbf{r} = 0.$$

Certainly $\mathbf{a} \neq \mathbf{0}$ and, by Theorem 1, the vector $\mathbf{a}$ is perpendicular to $\mathbf{r}$ and, therefore, perpendicular to the line L_1^*. It is called a **normal vector** to L_1^*. Since L_1 and L_1^* are parallel, $\mathbf{a}$ is also a normal vector to L_1. Hence two lines L_1 and L_2: $b_1x + b_2y = d$ are perpendicular or orthogonal if and only if their normal vectors $\mathbf{a}$ and $\mathbf{b} = b_1\mathbf{i} + b_2\mathbf{j}$ are orthogonal, that is, $\mathbf{a} \cdot \mathbf{b} = 0$. Note that this implies that the slopes of the lines are negative reciprocals.

In our case $\mathbf{b} = \mathbf{i} - 2\mathbf{j}$, and a vector perpendicular to $\mathbf{b}$ is $\mathbf{a} = 2\mathbf{i} + \mathbf{j}$. Hence the representation of L_1 must be of the form $2x + y = c$, and by substituting the coordinates of P we obtain $c = 5$. This yields the solution (cf. Fig. 98)

$$y = -2x + 5.$$

Example 3 (Normal vector to a plane). Find a unit vector perpendicular to the plane $4x + 2y + 4z = -7$.

Any plane in space can be represented in the form

$$(12) \qquad\qquad a_1x + a_2y + a_3z = c.$$

The position vector of a point in the plane is $\mathbf{r} = x\mathbf{i} + y\mathbf{j} + z\mathbf{k}$. Introducing the vector $\mathbf{a} = a_1\mathbf{i} + a_2\mathbf{j} + a_3\mathbf{k}$ and using (8), we may write (12) in the form

$$(13) \qquad\qquad \mathbf{a} \cdot \mathbf{r} = c.$$

Fig. 99. Normal vector to a plane

Certainly $\mathbf{a} \neq \mathbf{0}$, and the unit vector in the direction of $\mathbf{a}$ is

$$\mathbf{n} = \frac{\mathbf{a}}{|\mathbf{a}|}.$$

Dividing by $|\mathbf{a}|$, we obtain from (13)

(14) $\mathbf{n} \cdot \mathbf{r} = p$ where $p = \dfrac{c}{|\mathbf{a}|}.$

From (10) we see that p is the projection of $\mathbf{r}$ in the direction of $\mathbf{n}$, and this projection has the same constant value $c/|\mathbf{a}|$ for the position vector $\mathbf{r}$ of any point in the plane. Clearly this holds if and only if $\mathbf{n}$ is perpendicular to the plane. $\mathbf{n}$ is called a **unit normal vector** to the plane (the other being $-\mathbf{n}$). Furthermore from this and the definition of projection it follows that $|p|$ is the distance of the plane from the origin. Representation (14) is called **Hesse's**[5] **normal form** of a plane. An illustration is given in Fig. 99. In our case,

$$\mathbf{a} = 4\mathbf{i} + 2\mathbf{j} + 4\mathbf{k}, \quad c = -7, \quad |\mathbf{a}| = 6, \quad \mathbf{n} = \tfrac{1}{6}\mathbf{a} = \tfrac{2}{3}\mathbf{i} + \tfrac{1}{3}\mathbf{j} + \tfrac{2}{3}\mathbf{k},$$

and the plane has the distance 7/6 from the origin.

Problems

Let $\mathbf{a} = 2\mathbf{i} + \mathbf{j} + 3\mathbf{k}$, $\mathbf{b} = \mathbf{i} - 4\mathbf{k}$, and $\mathbf{c} = 3\mathbf{i} - \mathbf{j} + 2\mathbf{k}$. Find

1. $\mathbf{a} \cdot \mathbf{b}, \mathbf{b} \cdot \mathbf{a}$ 2. $|\mathbf{a}|, |\mathbf{b}|, |\mathbf{c}|$ 3. $(\mathbf{a} + \mathbf{b}) \cdot \mathbf{c}, \mathbf{a} \cdot \mathbf{c} + \mathbf{b} \cdot \mathbf{c}$
4. $3\mathbf{a} \cdot 2\mathbf{c}, 6(\mathbf{a} \cdot \mathbf{c})$ 5. $(\mathbf{a} - \mathbf{b}) \cdot \mathbf{c}$ 6. $(\mathbf{a} - \mathbf{c}) \cdot \mathbf{b}, \mathbf{a} \cdot \mathbf{b} - \mathbf{c} \cdot \mathbf{b}$
7. $|\mathbf{a} + \mathbf{b}|, |\mathbf{a} + \mathbf{c}|$ 8. $|\mathbf{b} + \mathbf{c}|, |\mathbf{b}| + |\mathbf{c}|$ 9. $|\mathbf{a} + \mathbf{b} + \mathbf{c}|$

10. What laws do Probs. 1, 3, and 6 illustrate?
11. Find $\mathbf{a} \cdot \mathbf{b}$ where $\mathbf{a} = 2\mathbf{i}$ and $\mathbf{b} = \mathbf{i} + \mathbf{j}$, $\mathbf{b} = \mathbf{j}$, $\mathbf{b} = -\mathbf{i} + \mathbf{j}$, and sketch a figure similar to Fig. 95.
12. If $\mathbf{a}$ is given and $\mathbf{a} \cdot \mathbf{b} = \mathbf{a} \cdot \mathbf{c}$, can we conclude that $\mathbf{b} = \mathbf{c}$?
13. Find all vectors $\mathbf{v}$ such that $\mathbf{a} \cdot \mathbf{v} = 0$, where $\mathbf{a} = \mathbf{i} + 2\mathbf{j}$.
14. For what values of b_2 are $\mathbf{a} = \mathbf{i} - \mathbf{j} + 2\mathbf{k}$ and $\mathbf{b} = 2\mathbf{i} + b_2\mathbf{j} + \mathbf{k}$ orthogonal vectors?
15. Show that $\mathbf{a} = \mathbf{i} - \mathbf{j} + 2\mathbf{k}$, $\mathbf{b} = 4\mathbf{j} + 2\mathbf{k}$, and $\mathbf{c} = -10\mathbf{i} - 2\mathbf{j} + 4\mathbf{k}$ are orthogonal.
16. Using vectors, show that if the diagonals of a rectangle are orthogonal, the rectangle must be a square.
17. Under what conditions will the diagonals of a parallelogram be orthogonal?
18. Are the four diagonals of a cube orthogonal?
19. Find and sketch a plane through the origin and orthogonal to the plane $x + y + z = 2$.
20. Show that the planes $x + 2y - 3z = 4$ and $5x - y + z = -1$ are orthogonal.
21. Find all vectors $\mathbf{v} = v_1\mathbf{i} + v_2\mathbf{j}$ such that $\mathbf{a} \cdot \mathbf{v} = 2$, where $\mathbf{a} = 2\mathbf{i} + \mathbf{j}$. Sketch some of the vectors $\mathbf{v}$.

Find the work done by a force $\mathbf{p}$ acting on a particle, if the particle is displaced from a point A to a point B along a straight segment AB, where

22. $\mathbf{p} = 2\mathbf{i} + \mathbf{j}$, A: $(0, 0, 0)$, B: $(0, 1, 0)$
23. $\mathbf{p} = \mathbf{i} + 2\mathbf{j}$, A: $(4, -7, 3)$, B: $(4, -7, 8)$
24. $\mathbf{p} = \mathbf{i} + \mathbf{j} + \mathbf{k}$, A: $(1, -1, 2)$, B: $(2, 1, 3)$

25. Suppose that a particle is displaced from a point A to a point B along a straight segment AB. Show that the work done by two forces acting on the particle is equal to the work done by the resultant of the forces.

In each case find the component of $\mathbf{a}$ in the direction of $\mathbf{b}$.

26. $\mathbf{a} = \mathbf{i} + \mathbf{j} - \mathbf{k}, \mathbf{b} = 3\mathbf{i} + 7\mathbf{k}$ 27. $\mathbf{a} = 4\mathbf{j} + 3\mathbf{k}, \mathbf{b} = -4\mathbf{j} - 3\mathbf{k}$
28. $\mathbf{a} = 4\mathbf{i} + \mathbf{j}, \mathbf{b} = -\mathbf{i} + 4\mathbf{j} + 3\mathbf{k}$ 29. $\mathbf{a} = \mathbf{i} + \mathbf{j} + 2\mathbf{k}, \mathbf{b} = 6\mathbf{k}$

[5] LUDWIG OTTO HESSE (1811–1874), German mathematician, who contributed to the theory of curves and surfaces.

Let $\mathbf{a} = \mathbf{i} + \mathbf{j} + \mathbf{k}$, $\mathbf{b} = -\mathbf{i} + \mathbf{j}$, and $\mathbf{c} = 3\mathbf{i} + \mathbf{k}$. Find the cosine of the angle between the following vectors.

   30. $\mathbf{a}, \mathbf{b}$ **31.** $\mathbf{a}, \mathbf{b} + \mathbf{c}$ **32.** $\mathbf{a} + \mathbf{b}, \mathbf{c}$ **33.** $\mathbf{a} + \mathbf{c}, \mathbf{a} - \mathbf{c}$

   34. Find all unit vectors $\mathbf{u} = u_1\mathbf{i} + u_2\mathbf{j} + u_3\mathbf{k}$ such that the angle between $\mathbf{u}$ and $\mathbf{v} = 2\mathbf{i} + \mathbf{j}$ equals the angle between $\mathbf{u}$ and $\mathbf{w} = 2\mathbf{i} + 4\mathbf{j}$.

   35. Find the angle between the planes $x + y + z = 1$ and $3x - 2y + z = 4$.

   36. Find the angles of the triangle with vertices A: $(0, 0, 0)$, B: $(1, 2, 3)$, C: $(4, -1, 3)$.

   37. Deduce the law of cosines by using vectors $\mathbf{a}$, $\mathbf{b}$, and $\mathbf{a} - \mathbf{b}$.

   38. Let $\mathbf{a} = \cos \alpha \, \mathbf{i} + \sin \alpha \, \mathbf{j}$, and $\mathbf{b} = \cos \beta \, \mathbf{i} + \sin \beta \, \mathbf{j}$ where $0 \leqq \alpha \leqq \beta \leqq 2\pi$. Show that $\mathbf{a}$ and $\mathbf{b}$ are unit vectors. Use (8) to obtain the trigonometric identity for $\cos (\beta - \alpha)$.

   39. Prove the triangle inequality (6).

   40. Prove the parallelogram equality (7).

5.6   *Inner Product Spaces*

In Sec. 5.4 we obtained the definition of a vector space by taking some of the basic properties of vector addition and multiplication by scalars in three-dimensional space. It is quite interesting that the dot product (Sec. 5.5) may be used in a similar fashion and motivates the concept of a *real inner product space*. By definition, this is a real vector space on which there is defined an inner product satisfying (4), Sec. 5.5; in detail:

A real vector space V is called a **real inner product space** (or *real pre-Hilbert space*) if it has the following property. With every pair of vectors $\mathbf{a}$ and $\mathbf{b}$ in V there is associated a real number, which is denoted by $(\mathbf{a}, \mathbf{b})$ and is called the **inner product** (or *scalar product*) of $\mathbf{a}$ and $\mathbf{b}$, such that the following axioms are satisfied.

   I. For any scalars q_1 and q_2 and any vectors $\mathbf{a}$, $\mathbf{b}$, $\mathbf{c}$ in V,

$$(q_1\mathbf{a} + q_2\mathbf{b}, \mathbf{c}) = q_1(\mathbf{a}, \mathbf{c}) + q_2(\mathbf{b}, \mathbf{c}) \qquad (Linearity).$$

   II. For any vectors $\mathbf{a}$ and $\mathbf{b}$ in V,

$$(\mathbf{a}, \mathbf{b}) = (\mathbf{b}, \mathbf{a}) \qquad (Symmetry).$$

   III. For any $\mathbf{a}$ in V,

$$(\mathbf{a}, \mathbf{a}) \geqq 0, \text{ and}$$
$$(\mathbf{a}, \mathbf{a}) = 0 \quad \text{if and only if} \quad \mathbf{a} = \mathbf{0} \qquad (Positive\text{-}definiteness).$$

Theorem 1 in Sec. 5.5 suggests that any vectors $\mathbf{a}$ and $\mathbf{b}$ in V such that

$$(\mathbf{a}, \mathbf{b}) = 0$$

are said to be **orthogonal**.

Using the inner product, with any element $\mathbf{a}$ in V we may associate a number which is denoted by $\|\mathbf{a}\|$, is defined by

$$\|\mathbf{a}\| = \sqrt{(\mathbf{a}, \mathbf{a})} \qquad (\geqq 0),$$

and is called the **norm** of $\mathbf{a}$. From (2), Sec. 5.5, we see that this generalizes the concept of a length. In fact, the dot product is an inner product in the sense of our new definition,

$$(\mathbf{a}, \mathbf{b}) = \mathbf{a} \cdot \mathbf{b},$$

and in our new notation, formula (2), Sec. 5.5, can be written

$$\|\mathbf{a}\| = |\mathbf{a}| = \sqrt{(\mathbf{a}, \mathbf{a})} = \sqrt{\mathbf{a} \cdot \mathbf{a}}.$$

From the axioms of an inner product and that definition of a norm, one can derive (cf. (5)–(7) in Sec. 5.5)

$$|(\mathbf{a}, \mathbf{b})| \leqq \|\mathbf{a}\| \, \|\mathbf{b}\| \qquad (\textit{Schwarz inequality}),$$

from this

$$\|\mathbf{a} + \mathbf{b}\| \leqq \|\mathbf{a}\| + \|\mathbf{b}\| \qquad (\textit{Triangle inequality})$$

and by a simple direct calculation

$$\|\mathbf{a} + \mathbf{b}\|^2 + \|\mathbf{a} - \mathbf{b}\|^2 = 2(\|\mathbf{a}\|^2 + \|\mathbf{b}\|^2) \qquad (\textit{Parallelogram equality}).$$

We shall not go into details, because these belong to more advanced courses (in functional analysis, in particular Hilbert spaces, approximation theory etc.), but we want to illustrate the generality of the concept of an inner product space by two important examples. The first is *n-dimensional Euclidean space*, which is the inner product space of all real ordered *n*-tuples $\mathbf{a} = (a_1, \cdots, a_n)$, $\mathbf{b} = (b_1, \cdots, b_n)$, etc., with the inner product (suggested by (8) in Sec. 5.5)

$$(\mathbf{a}, \mathbf{b}) = a_1 b_1 + a_2 b_2 + \cdots + a_n b_n.$$

The second example is the inner product space of all continuous function $f(x)$, $g(x)$, $\cdots$ defined on an interval $\alpha \leqq x \leqq \beta$ with the inner product defined by

$$(f, g) = \int_\alpha^\beta f(x) g(x) \, dx.$$

Note that in this case orthogonality is identical with orthogonality as defined in Sec. 3.7.

5.7   *Vector Product (Cross Product)*

Various applications suggest the introduction of another kind of vector multiplication in which the product of two vectors is again a vector. This so-called **vector product** or **cross product** of two vectors **a** and **b** is written

$$\mathbf{a} \times \mathbf{b}$$

and is a vector **v** which is defined as follows.

If **a** *and* **b** *have the same or opposite direction or one of these vectors is the zero vector, then* $\mathbf{v} = \mathbf{0}$.

In any other case, **v** *is the vector whose length is equal to the area of the parallelogram with* **a** *and* **b** *as adjacent sides and whose direction is perpendicular to both* **a** *and* **b** *and is such that* **a**, **b**, **v**, *in this order, form a right-handed triple or right-handed triad, as shown in Fig.* 100.

Fig. 100. Vector product

The term **right-handed** comes from the fact that the vectors **a**, **b**, **v**, in this order, assume the same sort of orientation as the thumb, index finger, and middle finger of the right hand when these are held as shown in Fig. 101. We may also say that if **a** is rotated into the direction of **b** through the angle α $(<\pi)$, then **v** advances in the same direction as a right-handed screw would if turned in the same way (cf. Fig. 102).

The parallelogram with **a** and **b** as adjacent sides has the area $|\mathbf{a}|\,|\mathbf{b}|\sin\gamma$ (cf. Fig. 100). We thus obtain

(1) $$|\mathbf{v}| = |\mathbf{a}|\,|\mathbf{b}|\sin\gamma.$$

Let $\mathbf{a} \times \mathbf{b} = \mathbf{v}$ and let $\mathbf{b} \times \mathbf{a} = \mathbf{w}$. Then, by definition, $|\mathbf{v}| = |\mathbf{w}|$, and in order that **b**, **a**, **w** form a right-handed triple we must have $\mathbf{w} = -\mathbf{v}$ (Fig. 103). Hence,

(2) $$\mathbf{b} \times \mathbf{a} = -(\mathbf{a} \times \mathbf{b});$$

that is, *cross-multiplication of vectors is* **not commutative** *but anticommutative.* The order of the factors in a vector product is, therefore, of great importance and must be carefully observed.

From the definition it follows that for any constant k,

(3) $$(k\mathbf{a}) \times \mathbf{b} = k(\mathbf{a} \times \mathbf{b}) = \mathbf{a} \times (k\mathbf{b}).$$

Furthermore, cross-multiplication is distributive with respect to vector addition,

Fig. 101. Right-handed triple of vectors **a**, **b**, **v**

Fig. 102. Right-handed screw

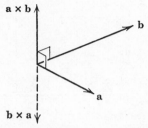

Fig. 103. Anticommutativity of cross multiplication

that is,

(4)
$$\mathbf{a} \times (\mathbf{b} + \mathbf{c}) = (\mathbf{a} \times \mathbf{b}) + (\mathbf{a} \times \mathbf{c}),$$

$$(\mathbf{a} + \mathbf{b}) \times \mathbf{c} = (\mathbf{a} \times \mathbf{c}) + (\mathbf{b} \times \mathbf{c}).$$

The proof will be given in the next section. More complicated cross products will be considered later, but we want to mention now that cross-multiplication is **not associative**, that is, in general

$$\mathbf{a} \times (\mathbf{b} \times \mathbf{c}) \neq (\mathbf{a} \times \mathbf{b}) \times \mathbf{c}.$$

From (1) in this section and (1) in Sec. 5.5 it follows that

$$|\mathbf{v}|^2 = |\mathbf{a}|^2 |\mathbf{b}|^2 \sin^2 \gamma = |\mathbf{a}|^2 |\mathbf{b}|^2 (1 - \cos^2 \gamma) = (\mathbf{a} \cdot \mathbf{a})(\mathbf{b} \cdot \mathbf{b}) - (\mathbf{a} \cdot \mathbf{b})^2.$$

Hence we obtain the formula

(5)
$$|\mathbf{a} \times \mathbf{b}| = \sqrt{(\mathbf{a} \cdot \mathbf{a})(\mathbf{b} \cdot \mathbf{b}) - (\mathbf{a} \cdot \mathbf{b})^2},$$

which is convenient for computing the length of a vector product.

5.8 Vector Products in Terms of Components

We shall now represent a vector product in terms of the components of its factors with respect to a Cartesian coordinate system. In this connection it is important to note that there are two types of such systems, depending on the orientation of the axes, namely, right-handed and left-handed. The definitions are as follows.

A Cartesian coordinate system is called **right-handed**, if the corresponding unit vectors **i, j, k** in the positive directions of the axes form a right-handed triple (Fig. 104*a*); it is called **left-handed**, if these vectors form a *left-handed triple*, that is, assume the same sort of orientation as the thumb, index finger, and middle finger of the left hand (Fig. 104*b*).

In applications we preferably use right-handed systems.

With respect to a given right-handed or left-handed Cartesian coordinate system. let a_1, a_2, a_3 and b_1, b_2, b_3 be the components of two vectors **a** and **b**, respectively, *(a)* Right-handed.   *(b)* Left-handed.

Fig. 104. The two types of Cartesian coordinate systems

We want to express the components v_1, v_2, v_3 of the product vector

$$\mathbf{v} = \mathbf{a} \times \mathbf{b}$$

in terms of the components of $\mathbf{a}$ and $\mathbf{b}$. We need only consider the case when $\mathbf{v} \neq \mathbf{0}$. Since $\mathbf{v}$ is perpendicular to both $\mathbf{a}$ and $\mathbf{b}$, it follows from Theorem 1, Sec. 5.5, that $\mathbf{a} \cdot \mathbf{v} = 0$ and $\mathbf{b} \cdot \mathbf{v} = 0$ or, by (8) in Sec. 5.5,

(1)
$$a_1 v_1 + a_2 v_2 + a_3 v_3 = 0,$$
$$b_1 v_1 + b_2 v_2 + b_3 v_3 = 0.$$

Multiplying the first equation by b_3, the last by a_3, and subtracting, we obtain

$$(a_3 b_1 - a_1 b_3) v_1 = (a_2 b_3 - a_3 b_2) v_2.$$

Multiplying the first equation of (1) by b_1, the last by a_1, and subtracting, we obtain

$$(a_1 b_2 - a_2 b_1) v_2 = (a_3 b_1 - a_1 b_3) v_3.$$

We readily see that these two equations are satisfied by

(2) $v_1 = c(a_2 b_3 - a_3 b_2),$ $v_2 = c(a_3 b_1 - a_1 b_3),$ $v_3 = c(a_1 b_2 - a_2 b_1),$

where c is a constant. The reader may verify by inserting that (2) also satisfies (1). Now each of the equations in (1) represents a plane through the origin in $v_1 v_2 v_3$-space. The vectors $\mathbf{a}$ and $\mathbf{b}$ are normal vectors of these planes (cf. Ex. 3 in Sec. 5.5). Since $\mathbf{v} \neq \mathbf{0}$, these vectors are not parallel and the two planes do not coincide. Hence their intersection is a straight line L through the origin. Since (2) is a solution of (1) and, for varying c, represents a straight line, we conclude that (2) represents L, and every solution of (1) must be of the form (2). In particular, the components of $\mathbf{v}$ must be of this form, where c is to be determined. From (2) we obtain

$$|\mathbf{v}|^2 = v_1^2 + v_2^2 + v_3^2 = c^2 [(a_2 b_3 - a_3 b_2)^2 + (a_3 b_1 - a_1 b_3)^2 + (a_1 b_2 - a_2 b_1)^2].$$

This can be written

$$|\mathbf{v}|^2 = c^2[(a_1{}^2 + a_2{}^2 + a_3{}^2)(b_1{}^2 + b_2{}^2 + b_3{}^2) - (a_1 b_1 + a_2 b_2 + a_3 b_3)^2],$$

as the reader may readily verify. Using (8) in Sec. 5.5, we thus have

$$|\mathbf{v}|^2 = c^2[(\mathbf{a} \cdot \mathbf{a})(\mathbf{b} \cdot \mathbf{b}) - (\mathbf{a} \cdot \mathbf{b})^2].$$

By comparing this with (5) in the last section we conclude that $c = \pm 1$.

From now on it will matter whether the coordinate system is right-handed or left-handed. We first consider the case of a right-handed system, and we want to show that then $c = +1$. This can be done as follows.

If we change the lengths and directions of $\mathbf{a}$ and $\mathbf{b}$ continuously and such that at the end $\mathbf{a} = \mathbf{i}$ and $\mathbf{b} = \mathbf{j}$ (Fig. 104a), then $\mathbf{v}$ will change its length and direction continuously, and at the end, $\mathbf{v} = \mathbf{i} \times \mathbf{j} = \mathbf{k}$. Obviously we may effect the change so that both $\mathbf{a}$ and $\mathbf{b}$ remain different from the zero vector and are not parallel at any instant. Then $\mathbf{v}$ is never equal to the zero vector, and since the change is continuous and c can only assume the values $+1$ or -1, it follows that at the end c must have the same value as before. Now at the end $\mathbf{a} = \mathbf{i}$, $\mathbf{b} = \mathbf{j}$, $\mathbf{v} = \mathbf{k}$ and, therefore, $a_1 = 1$, $b_2 = 1$, $v_3 = 1$, and the other components in (2) are zero. Hence, from (2), $v_3 = c = +1$. Noting that the expressions in parentheses in (2) may be written in the form of second-order determinants, we may sum up our result as follows.

With respect to a right-handed Cartesian coordinate system,

$$\mathbf{a} \times \mathbf{b} = (a_2 b_3 - a_3 b_2)\mathbf{i} + (a_3 b_1 - a_1 b_3)\mathbf{j} + (a_1 b_2 - a_2 b_1)\mathbf{k}$$

or, written in terms of second-order determinants,[6]

$$(3) \qquad \mathbf{a} \times \mathbf{b} = \mathbf{i}\begin{vmatrix} a_2 & a_3 \\ b_2 & b_3 \end{vmatrix} + \mathbf{j}\begin{vmatrix} a_3 & a_1 \\ b_3 & b_1 \end{vmatrix} + \mathbf{k}\begin{vmatrix} a_1 & a_2 \\ b_1 & b_2 \end{vmatrix},$$

where a_1, a_2, a_3 and b_1, b_2, b_3 are the components of $\mathbf{a}$ and $\mathbf{b}$, respectively.

For memorizing, it is useful to note that (3) can be interpreted as the expansion of the determinant

$$(4) \qquad \mathbf{a} \times \mathbf{b} = \begin{vmatrix} \mathbf{i} & \mathbf{j} & \mathbf{k} \\ a_1 & a_2 & a_3 \\ b_1 & b_2 & b_3 \end{vmatrix}$$

by the first row, but we should keep in mind that this is not an ordinary determinant because the elements of the first row are vectors.

---

[6] Second- and third-order determinants are usually considered in elementary calculus. Readers unfamiliar with them may consult Sec. 6.6.

In a *left-handed* Cartesian coordinate system, $\mathbf{i} \times \mathbf{j} = -\mathbf{k}$ (Fig. 104*b*), and the above reasoning leads to $c = -1$ in (2). Hence *with respect to a left-handed Cartesian coordinate system,*

(5)
$$\mathbf{a} \times \mathbf{b} = - \begin{vmatrix} \mathbf{i} & \mathbf{j} & \mathbf{k} \\ a_1 & a_2 & a_3 \\ b_1 & b_2 & b_3 \end{vmatrix}.$$

Example 1. With respect to a right-handed Cartesian coordinate system, let $\mathbf{a} = 4\mathbf{i} - \mathbf{k}$ and $\mathbf{b} = -2\mathbf{i} + \mathbf{j} + 3\mathbf{k}$. Then

$$\mathbf{a} \times \mathbf{b} = \begin{vmatrix} \mathbf{i} & \mathbf{j} & \mathbf{k} \\ 4 & 0 & -1 \\ -2 & 1 & 3 \end{vmatrix} = \mathbf{i} - 10\mathbf{j} + 4\mathbf{k}.$$

We may now prove the **distributive law** (4) in the previous section. From (3) it follows that the first component of $\mathbf{a} \times (\mathbf{b} + \mathbf{c})$ is

$$\begin{vmatrix} a_2 & a_3 \\ b_2 + c_2 & b_3 + c_3 \end{vmatrix} = a_2(b_3 + c_3) - a_3(b_2 + c_2)$$

$$= (a_2 b_3 - a_3 b_2) + (a_2 c_3 - a_3 c_2)$$

$$= \begin{vmatrix} a_2 & a_3 \\ b_2 & b_3 \end{vmatrix} + \begin{vmatrix} a_2 & a_3 \\ c_2 & c_3 \end{vmatrix}.$$

The expression on the right is the first component of $\mathbf{a} \times \mathbf{b} + \mathbf{a} \times \mathbf{c}$. For the other two components of that vector the consideration is similar. This proves the first of the relations (4) in the previous section, and the second relation in (4) can be proved by the same argument.

The reader may prove the following criterion for linear dependence and independence of two vectors.

Theorem 1. *Two vectors form a linearly dependent set if and only if their vector product is the zero vector.*

The definition of cross-multiplication is suggested by many applications, two of which may be illustrated by the following examples. Further physical and geometrical applications will be considered later.

Example 2 (*Moment of a force*). In mechanics the moment m of a force $\mathbf{p}$ about a point Q is defined as the product $m = |\mathbf{p}|\, d$ where d is the (perpendicular) distance between Q and the line of action L of $\mathbf{p}$ (Fig. 105). If $\mathbf{r}$ is the vector from Q to any point A on L, then $d = |\mathbf{r}| \sin \gamma$

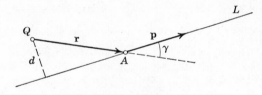

Fig. 105. Moment of a force

Fig. 106. Rotation of a rigid body

(Fig. 105) and

$$m = |\mathbf{r}| \, |\mathbf{p}| \sin \gamma.$$

Since γ is the angle between $\mathbf{r}$ and $\mathbf{p}$,

$$m = |\mathbf{r} \times \mathbf{p}|,$$

as follows from (1) in the last section. The vector

(6) $\mathbf{m} = \mathbf{r} \times \mathbf{p}$

is called the **moment vector** or **vector moment** of $\mathbf{p}$ about Q. Its magnitude is m, and its direction is that of the axis of the rotation about Q which $\mathbf{p}$ has the tendency to produce.

Example 3 (Velocity of a rotating body).   A rotation of a rigid body B in space can be simply and uniquely described by a vector $\mathbf{w}$ as follows. The direction of $\mathbf{w}$ is that of the axis of rotation and such that the rotation appears clockwise, if one looks from the initial point of $\mathbf{w}$ to its terminal point. The length of $\mathbf{w}$ is equal to the **angular speed** ω (>0) of the rotation, that is, the linear (or tangential) speed of a point of B divided by its distance from the axis of rotation.

Let P be any point of B and d its distance from the axis. Then P has the speed ωd. Let $\mathbf{r}$ be the position vector of P referred to a coordinate system with origin 0 on the axis of rotation. Then $d = |\mathbf{r}| \sin \gamma$ where γ is the angle between $\mathbf{w}$ and $\mathbf{r}$. Therefore,

$$\omega d = |\mathbf{w}| \, |\mathbf{r}| \sin \gamma = |\mathbf{w} \times \mathbf{r}|.$$

From this and the definition of vector product we see that the velocity vector $\mathbf{v}$ of P can be represented in the form (Fig. 106)

(7) $\mathbf{v} = \mathbf{w} \times \mathbf{r}.$

This simple formula is useful for determining $\mathbf{v}$ at any point of B.

Problems

With respect to a right-handed Cartesian coordinate system, let $\mathbf{a} = \mathbf{i} + \mathbf{j}$, $\mathbf{b} = -\mathbf{i} + 2\mathbf{j}$, and $\mathbf{c} = 2\mathbf{i} + 3\mathbf{j} + \mathbf{k}$. Find

1. $\mathbf{a} \times \mathbf{b}, \mathbf{b} \times \mathbf{a}$
2. $\mathbf{b} \times \mathbf{c}, |\mathbf{b} \times \mathbf{c}|, |\mathbf{c} \times \mathbf{b}|$
3. $\mathbf{a} \times \mathbf{c}, |\mathbf{a} \times \mathbf{c}|, \mathbf{a} \cdot \mathbf{c}$
4. $2\mathbf{a} \times 3\mathbf{b}, 3\mathbf{a} \times 2\mathbf{b}, 6\mathbf{a} \times \mathbf{b}$
5. $(\mathbf{a} + \mathbf{b}) \times \mathbf{c}, \mathbf{a} \times \mathbf{c} + \mathbf{b} \times \mathbf{c}$
6. $(\mathbf{a} + \mathbf{b}) \times \mathbf{c}, \mathbf{a} \times (\mathbf{b} + \mathbf{c})$
7. $(\mathbf{a} + \mathbf{c}) \times \mathbf{c}, \mathbf{a} \times \mathbf{c}$
8. $(\mathbf{a} - \mathbf{b}) \times (\mathbf{a} - \mathbf{c}), (\mathbf{a} - \mathbf{b}) \times (\mathbf{c} - \mathbf{a})$
9. $(\mathbf{a} \cdot \mathbf{b})\mathbf{c}, (\mathbf{a} \times \mathbf{b}) \cdot \mathbf{c}$
10. $(\mathbf{a} \times \mathbf{b}) \times \mathbf{c}, \mathbf{a} \times (\mathbf{b} \times \mathbf{c})$

11. What properties of cross multiplication do Probs. 1, 5, and 10 illustrate?
12. Show that $\mathbf{i} \times \mathbf{j} = \mathbf{k}$, $\mathbf{k} \times \mathbf{i} = \mathbf{j}$, $\mathbf{j} \times \mathbf{k} = \mathbf{i}$, assuming right-handed coordinates. Find $(\mathbf{i} \times \mathbf{j}) \times \mathbf{k}, \mathbf{i} \times (\mathbf{j} \times \mathbf{k})$.

Find two unit vectors perpendicular to both vectors.

13. $\mathbf{i}, \mathbf{j}$
14. $3\mathbf{i} + 2\mathbf{j}, -\mathbf{i} - \mathbf{j}$
15. $3\mathbf{k}, 2\mathbf{j} + \mathbf{k}$
16. $\mathbf{i} - \mathbf{j} + 2\mathbf{k}, 2\mathbf{i} + 3\mathbf{k}$
17. $2\mathbf{j} - 3\mathbf{k}, 2\mathbf{i}$
18. $3\mathbf{i} + 2\mathbf{j} + 5\mathbf{k}, 3\mathbf{j} - \mathbf{k}$

Find a unit normal vector to the plane through the following points.
19. $(2, 0, 0), (0, 2, 0), (0, 0, 2)$ 20. $(0, 0, 0), (1, 1, 1), (4, -1, 0)$
21. $(1, 3, 5), (3, 3, 0), (-2, 0, 6)$ 22. $(1, 0, -5), (-3, 0, 8), (2, 0, -7)$

23. Find a vector **v** parallel to the intersection of the planes $x - y + z = 5$ and $2x + y - 3z = 4$.

24. Find a vector **v** parallel to the plane $x + y + z = 3$ and perpendicular to the line $y = x$, $z = 0$.

Are the following vectors parallel? Orthogonal?
25. $4\mathbf{i} - 7\mathbf{j}, -2\mathbf{k}$ 26. $-4\mathbf{i} + 2\mathbf{k}, 2\mathbf{i} - \mathbf{k}$
27. $\mathbf{i} + \mathbf{j}, \mathbf{j} + 2\mathbf{k}$ 28. $\mathbf{i} \times \mathbf{j}, \mathbf{j} \times 2\mathbf{k}$
29. $(\mathbf{i} \times \mathbf{j}) \times \mathbf{j}, \mathbf{j} \times (\mathbf{j} \times \mathbf{k})$ 30. $\mathbf{i} + 2\mathbf{j} + 3\mathbf{k}, 3\mathbf{i} + 2\mathbf{j} + \mathbf{k}$

Using (5) in Sec. 5.7, find $|\mathbf{a} \times \mathbf{b}|$, where
31. $\mathbf{a} = -\mathbf{j} + 2\mathbf{k}, \mathbf{b} = 3\mathbf{i} + 4\mathbf{j}$ 32. $\mathbf{a} = 2\mathbf{i} + \mathbf{k}, \mathbf{b} = \mathbf{i} + \mathbf{j} - \mathbf{k}$
33. $\mathbf{a} = \mathbf{i} + \mathbf{j} + 2\mathbf{k}, \mathbf{b} = -\mathbf{i} + 3\mathbf{k}$ 34. $\mathbf{a} = 3\mathbf{i} + 2\mathbf{j} + \mathbf{k}, \mathbf{b} = -\mathbf{i} + 4\mathbf{k}$

Find the area of the parallelogram of which the given vectors are adjacent sides.
35. $\mathbf{i}, \mathbf{i} + \mathbf{j}$ 36. $3\mathbf{i}, 4\mathbf{k}$ 37. $\mathbf{j} - \mathbf{k}, \mathbf{i} - \mathbf{k}$
38. $\mathbf{i} + \mathbf{j}, \mathbf{i} + \mathbf{j} + \mathbf{k}$ 39. $\mathbf{i} + \mathbf{j} - \mathbf{k}, \mathbf{i} - \mathbf{j} + \mathbf{k}$ 40. $3\mathbf{i} + 2\mathbf{j} + \mathbf{k}, \mathbf{i} + 2\mathbf{j} + 3\mathbf{k}$

Find the area of the parallelogram that has the following vertices in the xy-plane.
41. $(0, 0), (4, 1), (2, 3), (6, 4)$ 42. $(-1, 2), (2, 0), (4, 3), (7, 1)$
43. $(2, -3), (1, 1), (5, -6), (4, -2)$ 44. $(4, 4), (-1, 9), (6, 6), (1, 11)$

Find the area of the triangle that has the following vertices.
45. $(6, -1, 3), (6, 1, 1), (3, 3, 3)$ 46. $(1, 2, 3), (3, 5, 4), (4, 3, 5)$
47. $(1, 3, 0), (0, 2, 5), (-1, 0, 2)$ 48. $(2, 2, 2), (5, 2, 4), (-2, 4, -1)$

49. Do $\mathbf{a} \times \mathbf{b} = 0$ and $\mathbf{a} \cdot \mathbf{b} = 0$ together imply that $\mathbf{a} = 0$ or $\mathbf{b} = 0$?
50. Determine all vectors **v** such that $\mathbf{i} \times \mathbf{v} = \mathbf{k}$.
51. If $\mathbf{c} = \mathbf{a} \times \mathbf{b}$ and $\mathbf{b} = \mathbf{a} \times \mathbf{c}$, show that $\mathbf{b} = 0$ and $\mathbf{c} = 0$.
52. Prove Theorem 1.

A force **p** acts on a line through a point A. Find the moment vector **m** and the moment m of **p** about a point Q, where
53. $\mathbf{p} = \mathbf{k}$, A: $(0, 0, 0)$, Q: $(0, 0, 5)$
54. $\mathbf{p} = \mathbf{i} - 2\mathbf{j}$, A: $(1, 1, 1)$, Q: $(2, -1, 3)$
55. $\mathbf{p} = 2\mathbf{i} + 4\mathbf{j} + \mathbf{k}$, A: $(4, 2, -1)$, Q: $(0, 1, 2)$

5.9 Scalar Triple Product. Other Repeated Products

Repeated products of vectors that have three or more factors occur frequently in applications. The most important of these products is the **scalar triple product** or *mixed triple product* $\mathbf{a} \cdot (\mathbf{b} \times \mathbf{c})$ of three vectors. With respect to any right-handed Cartesian coordinate system, let

$$\mathbf{a} = a_1\mathbf{i} + a_2\mathbf{j} + a_3\mathbf{k}, \qquad \mathbf{b} = b_1\mathbf{i} + b_2\mathbf{j} + b_3\mathbf{k}, \qquad \mathbf{c} = c_1\mathbf{i} + c_2\mathbf{j} + c_3\mathbf{k}.$$

Then from (4), Sec. 5.8, it follows that

$$\mathbf{a} \cdot (\mathbf{b} \times \mathbf{c}) = (a_1 \mathbf{i} + a_2 \mathbf{j} + a_3 \mathbf{k}) \cdot \begin{vmatrix} \mathbf{i} & \mathbf{j} & \mathbf{k} \\ b_1 & b_2 & b_3 \\ c_1 & c_2 & c_3 \end{vmatrix}.$$

From this and (8), Sec. 5.5, we see that the scalar triple product takes the form

$$(1) \qquad \mathbf{a} \cdot (\mathbf{b} \times \mathbf{c}) = \begin{vmatrix} a_1 & a_2 & a_3 \\ b_1 & b_2 & b_3 \\ c_1 & c_2 & c_3 \end{vmatrix}.$$

The scalar triple product $\mathbf{a} \cdot (\mathbf{b} \times \mathbf{c})$ will be denoted by $(\mathbf{a}\,\mathbf{b}\,\mathbf{c})$.

Since interchanging of two rows reverses the sign of the determinant, we have

$$(2) \qquad (\mathbf{a}\,\mathbf{b}\,\mathbf{c}) = -(\mathbf{b}\,\mathbf{a}\,\mathbf{c}), \qquad \text{etc.}$$

Interchanging twice, we obtain

$$(3) \qquad (\mathbf{a}\,\mathbf{b}\,\mathbf{c}) = (\mathbf{b}\,\mathbf{c}\,\mathbf{a}) = (\mathbf{c}\,\mathbf{a}\,\mathbf{b}).$$

Now, by definition

$$(\mathbf{a}\,\mathbf{b}\,\mathbf{c}) = \mathbf{a} \cdot (\mathbf{b} \times \mathbf{c}), \qquad (\mathbf{c}\,\mathbf{a}\,\mathbf{b}) = \mathbf{c} \cdot (\mathbf{a} \times \mathbf{b}),$$

and since dot multiplication is commutative, the last expression is equal to $(\mathbf{a} \times \mathbf{b}) \cdot \mathbf{c}$. Therefore,

$$(4) \qquad \mathbf{a} \cdot (\mathbf{b} \times \mathbf{c}) = (\mathbf{a} \times \mathbf{b}) \cdot \mathbf{c}.$$

Furthermore, for any constant k,

$$(5) \qquad (k\mathbf{a}\,\mathbf{b}\,\mathbf{c}) = k(\mathbf{a}\,\mathbf{b}\,\mathbf{c}).$$

The absolute value of the scalar triple product $(\mathbf{a}\,\mathbf{b}\,\mathbf{c})$ *has a simple geometrical interpretation. It is equal to the volume of the parallelepiped P with* $\mathbf{a}$, $\mathbf{b}$, $\mathbf{c}$ *as adjacent edges.*

Indeed, from (1) in Sec. 5.5 we obtain

$$(\mathbf{a}\,\mathbf{b}\,\mathbf{c}) = \mathbf{a} \cdot (\mathbf{b} \times \mathbf{c}) = |\mathbf{a}|\,|\mathbf{b} \times \mathbf{c}| \cos \beta$$

where β is the angle between $\mathbf{a}$ and the product vector $\mathbf{b} \times \mathbf{c}$. Now $|\mathbf{b} \times \mathbf{c}|$ is the area of the base of P, and the altitude h of P is equal to the absolute value of $|\mathbf{a}| \cos \beta$ (Fig. 107). This proves our statement.

Fig. 107. Geometrical interpretation of a scalar triple product

From this geometrical consideration it follows that the value of the scalar triple product is a real number which is independent of the choice of right-handed Cartesian coordinates in space. But we should keep in mind that for left-handed Cartesian coordinate systems we have to use (5) instead of (4), Sec. 5.8, and this leads to a minus sign in front of the determinant in (1). We may also say that the value of the determinant is invariant under transformations of right-handed into right-handed, or left-handed into left-handed, Cartesian coordinate systems, but is multiplied by -1 under a transition from a right-handed to a left-handed system (or conversely).

*Example 1 (**Tetrahedron**).* Find the volume of the tetrahedron with **a**, **b**, **c** as adjacent edges where, with respect to right-handed Cartesian coordinates,

$$\mathbf{a} = \mathbf{i} + 2\mathbf{k}, \qquad \mathbf{b} = 4\mathbf{i} + 6\mathbf{j} + 2\mathbf{k}, \qquad \mathbf{c} = 3\mathbf{i} + 3\mathbf{j} - 6\mathbf{k}.$$

The volume V of the parallelepiped having **a**, **b**, **c** as adjacent edges is obtained from the scalar triple product

$$(\mathbf{a}\ \mathbf{b}\ \mathbf{c}) = \begin{vmatrix} 1 & 0 & 2 \\ 4 & 6 & 2 \\ 3 & 3 & -6 \end{vmatrix} = -54$$

and $V = 54$; the minus sign indicates that the vectors **a**, **b**, **c**, in this order, form a left-handed triple. The volume of the tetrahedron is $\frac{1}{6}$ of that of the parallelepiped, namely, 9.

Fig. 108. Tetrahedron

From the geometrical interpretation of a scalar triple product we also obtain a useful criterion for linear dependence and independence of three vectors. These vectors form a linearly dependent set if and only if they are coplanar (including the special case that they are collinear; cf. Sec. 5.4). This implies the following theorem.

Theorem 1 (Linear dependence). *Three vectors form a linearly dependent set if and only if their scalar triple product is zero.*

Other repeated products that occur in applications may be expressed in terms of dot, vector, and scalar triple products. An important corresponding formula is

(6) $$\mathbf{b} \times (\mathbf{c} \times \mathbf{d}) = (\mathbf{b} \cdot \mathbf{d})\mathbf{c} - (\mathbf{b} \cdot \mathbf{c})\mathbf{d}$$

(proof below). It implies the **identity of Lagrange**

(7) $$(\mathbf{a} \times \mathbf{b}) \cdot (\mathbf{c} \times \mathbf{d}) = (\mathbf{a} \cdot \mathbf{c})(\mathbf{b} \cdot \mathbf{d}) - (\mathbf{a} \cdot \mathbf{d})(\mathbf{b} \cdot \mathbf{c})$$

(whose proof is left to the reader; cf. Prob. 31, below), as well as (cf. Prob. 32)

(8) $$(\mathbf{a} \times \mathbf{b}) \times (\mathbf{c} \times \mathbf{d}) = (\mathbf{a}\ \mathbf{b}\ \mathbf{d})\mathbf{c} - (\mathbf{a}\ \mathbf{b}\ \mathbf{c})\mathbf{d}.$$

Before we prove (6), we note that (6) also implies

$$(\mathbf{b} \times \mathbf{c}) \times \mathbf{d} = -\mathbf{d} \times (\mathbf{b} \times \mathbf{c}) = (\mathbf{d} \cdot \mathbf{b})\mathbf{c} - (\mathbf{d} \cdot \mathbf{c})\mathbf{b}.$$

This shows that in general $\mathbf{b} \times (\mathbf{c} \times \mathbf{d})$ and $(\mathbf{b} \times \mathbf{c}) \times \mathbf{d}$ are different vectors, that is, *cross-multiplication is* **not associative** and the parentheses in (6) are important and cannot be omitted. For instance, with respect to a right-handed Cartesian coordinate system,

$$(\mathbf{i} \times \mathbf{j}) \times \mathbf{j} = \mathbf{k} \times \mathbf{j} = -\mathbf{i} \qquad \text{but} \qquad \mathbf{i} \times (\mathbf{j} \times \mathbf{j}) = \mathbf{0}.$$

***Proof of* (6).** We choose a right-handed Cartesian coordinate system such that the x-axis has the direction of $\mathbf{d}$ and the xy-plane contains $\mathbf{c}$. Then the vectors in (6) are of the form

$$\mathbf{b} = b_1\mathbf{i} + b_2\mathbf{j} + b_3\mathbf{k}, \qquad \mathbf{c} = c_1\mathbf{i} + c_2\mathbf{j}, \qquad \mathbf{d} = d_1\mathbf{i}.$$

Hence, $\mathbf{c} \times \mathbf{d} = -c_2 d_1 \mathbf{k}$ and, furthermore,

$$\mathbf{b} \times (\mathbf{c} \times \mathbf{d}) = \begin{vmatrix} \mathbf{i} & \mathbf{j} & \mathbf{k} \\ b_1 & b_2 & b_3 \\ 0 & 0 & -c_2 d_1 \end{vmatrix} = -b_2 c_2 d_1 \mathbf{i} + b_1 c_2 d_1 \mathbf{j}.$$

On the other hand, we obtain

$$(\mathbf{b} \cdot \mathbf{d})\mathbf{c} - (\mathbf{b} \cdot \mathbf{c})\mathbf{d} = b_1 d_1 (c_1\mathbf{i} + c_2\mathbf{j}) - (b_1 c_1 + b_2 c_2)d_1\mathbf{i} = b_1 c_2 d_1\mathbf{j} - b_2 c_2 d_1\mathbf{i}.$$

This proves (6) for our special coordinate system. Now the length and direction of a vector and a vector product, and the value of an inner product are independent of the choice of the coordinates. Furthermore, the representation of $\mathbf{b} \times (\mathbf{c} \times \mathbf{d})$ in terms of $\mathbf{i}, \mathbf{j}, \mathbf{k}$ will be the same for right-handed and left-handed systems, because of the double cross-multiplication. Hence, (6) holds in any Cartesian coordinate system, and the proof is complete.

Problems

Find the value of the scalar triple product of the given vectors (taken in the given order), whose components are referred to right-handed Cartesian coordinates.

1. $\mathbf{i}, \mathbf{j}, \mathbf{k}$
2. $\mathbf{j}, \mathbf{i}, \mathbf{k}$
3. $3\mathbf{i} - 4\mathbf{j}, 2\mathbf{j} + \mathbf{k}, -6\mathbf{i} + 8\mathbf{j}$
4. $\mathbf{k}, 3\mathbf{i} - \mathbf{j} + 2\mathbf{k}, 5\mathbf{i} + 3\mathbf{j} - \mathbf{k}$
5. $\mathbf{i} + \mathbf{j}, \mathbf{j} + \mathbf{k}, \mathbf{i} + \mathbf{k}$
6. $4\mathbf{i} + 3\mathbf{j} - \mathbf{k}, 0, 2\mathbf{k}$
7. $-\mathbf{i}, 3\mathbf{i} - \mathbf{j} + 5\mathbf{k}, 2\mathbf{i} - 3\mathbf{k}$
8. $\mathbf{i} - \mathbf{j} - \mathbf{k}, 4\mathbf{i} + 2\mathbf{j} + \mathbf{k}, 3\mathbf{i} + 2\mathbf{j} + 4\mathbf{k}$

Find the volume of the parallelepiped that has the given vectors as adjacent edges.

9. $\mathbf{i}, 2\mathbf{j}, -3\mathbf{k}$
10. $\mathbf{i} + \mathbf{j}, \mathbf{i} - \mathbf{j}, \mathbf{i} + 2\mathbf{j} + 4\mathbf{k}$
11. $8\mathbf{j}, 3\mathbf{i} + 2\mathbf{k}, \mathbf{i} + \mathbf{j} - \mathbf{k}$
12. $3\mathbf{i} - 2\mathbf{j}, \mathbf{i} + \mathbf{j} - \mathbf{k}, 3\mathbf{i} - \mathbf{j} + 2\mathbf{k}$
13. $\mathbf{i} + 4\mathbf{j}, 6\mathbf{k}, 2\mathbf{j} + 3\mathbf{k}$
14. $2\mathbf{i} - 6\mathbf{k}, \mathbf{j} + \mathbf{k}, \mathbf{i} + \mathbf{j}$

Find the volume of the tetrahedron that has the following vertices.

15. $(2, 1, 1), (3, 3, 4), (0, 1, 5), (1, 2, 2)$
16. $(0, 1, 2), (5, 5, 6), (1, 2, 1), (3, 3, 1)$
17. $(1, 0, 7), (2, 1, 8), (1, 0, 3), (2, 2, 9)$
18. $(2, 2, 2), (3, 2, 2), (2, 3, 2), (2, 2, 3)$

Are the following vectors linearly dependent or independent?

19. $i + j, i + 2j$
20. $2i - 4j, -3i + 6j + k$
21. $i - 3j + 5k, -2i + 6j - 10k$
22. $i, j, k, -i + 4j - 3k$
23. $3i - 2j, 4i + j, 2i - 5j$
24. $i + j, j + k, k + i$
25. $2j + k, -j + 4k, i - j$
26. $i - 6j + 2k, 2j + 7k, -2i + 12j - 4k$

27. Determine λ such that $a = i + j + k$, $b = 2i - 4k$, and $c = i + \lambda j + 3k$ are coplanar.

28. A system of three homogeneous linear equations in three unknowns has a non-trivial solution if and only if the determinant of the coefficients of the system is zero. Using this familiar theorem, prove Theorem 1.

29. Do the points $(4, -2, 1)$, $(5, 1, 6)$, $(2, 2, -5)$, and $(3, 5, 0)$ lie in a plane?

30. Determine λ and μ such that the points $(-1, 3, 2)$, $(-4, 2, -2)$, and $(5, \lambda, \mu)$ lie on a straight line.

31. Prove (7). *Hint.* Take the dot product of a and (6).

32. Derive (8) from (6).

33. Find $(i \times j) \times k, i \times (j \times k), (i \times j) \cdot k, i \cdot (j \times k)$.

With respect to a right-handed Cartesian coordinate system, let $a = 3i + 2k$, $b = 2i + 2j - k$, $c = i - 2j + 3k$, and $d = 2i - j + 5k$. Find

34. $a \times (b \times c), (a \times b) \times c$
35. $4c \times (3a \times b), 12c \times (a \times b)$
36. $(2b + a) \times (c \times d)$
37. $(a \times b) \cdot (c \times d), (b \times a) \cdot (d \times c)$

38. Show that $b \times (c \times d) + c \times (d \times b) + d \times (b \times c) = 0$.

39. Let the pairs a, b, and c, d each determine a plane. Show that these planes are parallel if and only if $(a \times b) \times (c \times d) = 0$.

40. What does it mean geometrically if in Prob. 39, $(a \times b) \cdot (c \times d) = 0$?

CHAPTER 6

Linear Algebra Part II: Matrices and Determinants

A matrix is a set of numbers arranged in a rectangular array. Such sets and arrangements occur in various branches of applied mathematics. In many cases they are sets of coefficients of linear transformations (examples in Sec. 6.1) or systems of linear equations (simultaneous linear equations) arising, for instance, from electrical networks, frameworks in mechanics, curve fitting in statistics, and transportation problems. Matrices are useful because they enable us to consider an array of many numbers as a single object, denote it by a single symbol, and perform calculations with these symbols in a very compact form. The "mathematical shorthand" thus obtained is very elegant and powerful and is suitable for various practical problems. It entered engineering mathematics about forty years ago and is of increasing importance in various branches.

In this chapter we first introduce matrices and related concepts (Sec. 6.1), define the algebraic operations for matrices (Secs. 6.2–6.4), and consider systems of linear equations (solution by elimination in Sec. 6.5, existence in Sec. 6.9, solution by Cramer's rule in Sec. 6.11), determinants (Secs. 6.6 and 6.7), bilinear and other forms (Sec. 6.13), and matrix eigenvalue problems (Secs. 6.14 and 6.15). Other important concepts in this chapter are the rank of a matrix (Secs. 6.8 and 6.10) and the inverse (Sec. 6.12).

Numerical methods are presented in Secs. 18.8–18.13 which are independent of the other sections in Chap. 18, so that they can be studied immediately after the corresponding material in the present chapter.

> *Prerequisite for this chapter:* Chap. 5.
> *Sections which may be omitted in a shorter course:* 6.13–6.15.
> *References:* Appendix 1, Part C.
> *Answers to problems:* Appendix 2.

6.1 Basic Concepts

Matrices are sets of numbers arranged as rectangular arrays and enclosed in parentheses; for instance,

$$\begin{pmatrix} 2 & 0.4 & 8 \\ 5 & -3 & 1 \end{pmatrix}, \quad \begin{pmatrix} 0 \\ 4 \end{pmatrix}, \quad (a_1 \quad a_2 \quad a_3)$$

(general definition below). Such sets and arrangements occur as sets of coefficients of linear transformations or systems of linear equations. For example, consider

(1)
$$20x_1 - 3x_2 = y_1$$
$$0.5x_1 + 7x_2 = y_2$$

If y_1 and y_2 are given numbers, this is a system of two linear equations in two unknowns x_1, x_2. The coefficients are 20, -3, 0.5, 7 and, to recognize their position we may arrange them in the way they occur in the equations, that is, in the form of a matrix,

(2)
$$\begin{pmatrix} 20 & -3 \\ 0.5 & 7 \end{pmatrix}.$$

If x_1, x_2 and y_1, y_2 are variable quantities (for instance, coordinates corresponding to two different coordinate systems in the plane), then our pair of formulas (1) represents a linear transformation. This transformation is completely determined by its four coefficients, that is, by the matrix (2). We may denote the matrix by a single letter, for instance **A**, and write it in the form

$$\mathbf{A} = \begin{pmatrix} a_{11} & a_{12} \\ a_{21} & a_{22} \end{pmatrix}$$

where $a_{11} = 20$, $a_{12} = -3$, $a_{21} = 0.5$, $a_{22} = 7$, and the notation has the advantage that we can immediately see the position of a coefficient a_{jk} because the first subscript is the number of the row (horizontal line) and the second is the number of the column (vertical line) of a_{jk}. Let us now give the general definition.

A rectangular array of (real or complex) numbers of the form

$$\begin{pmatrix} a_{11} & a_{12} & \cdots & a_{1n} \\ a_{21} & a_{22} & \cdots & a_{2n} \\ \cdot & \cdot & \cdots & \cdot \\ a_{m1} & a_{m2} & \cdots & a_{mn} \end{pmatrix}$$

is called a **matrix**. The numbers $a_{11}, \ldots, a_{mn}$ are called the **elements** of the matrix. The horizontal lines are called **rows** or *row vectors*, and the vertical lines are called **columns** or *column vectors* of the matrix. A matrix with m rows and n columns is called an $(m \times n)$ matrix (read "m by n matrix").

Matrices will be denoted by capital (upper case) bold-faced letters **A**, **B**, etc., or by (a_{jk}), (b_{ik}), etc., that is, by writing the general element of the matrix, enclosed in parentheses.

In the **double-subscript notation** *for the elements, the first subscript always denotes the row and the second subscript the column containing the given element.*

A matrix

$$(a_1 \quad a_2 \quad \cdots \quad a_n)$$

having only one row is called a **row matrix** or *row vector*. A matrix

having only one column is called a **column matrix** or *column vector*. Row and column matrices will be denoted by small (lower case) bold-faced letters. This agrees with our notation for vectors (Chap. 5), whose components may be written in the form of a row matrix or a column matrix.

The elements of a matrix may be real or complex numbers. If all the elements of a matrix are real, the matrix is called a **real matrix**.

A matrix having the same number of rows and columns is called a **square matrix**, and the number of rows is called its **order**. The diagonal containing the elements $a_{11}, a_{22}, \cdots, a_{nn}$ of a square matrix of order n is called the **principal diagonal**. Square matrices are of particular importance.

A definition of **equality of matrices** is suggested by that for vectors in Chap. 5 (cf. Sec. 5.2, at the end). Two matrices $\mathbf{A} = (a_{jk})$ and $\mathbf{B} = (b_{jk})$ are *equal* if and only if $\mathbf{A}$ and $\mathbf{B}$ have the same number of rows and the same number of columns and corresponding elements are equal, that is

$$a_{jk} = b_{jk} \qquad \text{for all occurring } j \text{ and } k.$$

Then we write

$$\mathbf{A} = \mathbf{B}.$$

Any matrix obtained by omitting some rows and columns from a given $(m \times n)$ matrix $\mathbf{A}$ is called a **submatrix** of $\mathbf{A}$.

As a matter of convention and convenience the notion "submatrix" includes $\mathbf{A}$ itself (as the matrix obtained from $\mathbf{A}$ by omitting no rows or columns).

Example 1. The matrix

$$\begin{pmatrix} a_{11} & a_{12} & a_{13} \\ a_{21} & a_{22} & a_{23} \end{pmatrix}$$

contains three (2×2) submatrices, namely,

$$\begin{pmatrix} a_{11} & a_{12} \\ a_{21} & a_{22} \end{pmatrix}, \qquad \begin{pmatrix} a_{11} & a_{13} \\ a_{21} & a_{23} \end{pmatrix}, \qquad \begin{pmatrix} a_{12} & a_{13} \\ a_{22} & a_{23} \end{pmatrix},$$

two (1×3) submatrices (the two row vectors), three (2×1) submatrices (the column vectors), six (1×2) submatrices, namely,

$$(a_{11} \quad a_{12}), \qquad (a_{11} \quad a_{13}), \qquad (a_{12} \quad a_{13}),$$
$$(a_{21} \quad a_{22}), \qquad (a_{21} \quad a_{23}), \qquad (a_{22} \quad a_{23}),$$

and six (1×1) submatrices, $(a_{11}), (a_{12}), \cdots, (a_{23})$.

6.2 *Addition of Matrices, Multiplication by Numbers*

Since we want to do calculations with matrices, we shall now introduce algebraic operations for matrices. We shall define two such operations, which are called *addition of matrices* and *multiplication of matrices by numbers*. These definitions are motivated and suggested by various applications (to be considered later in this chapter) as well as by the corresponding definitions for vectors in Sec. 5.3, which they are supposed to include as special cases.

Addition of matrices is defined only for matrices having the same number of rows and the same number of columns, and is then defined as follows. The **sum** of two $(m \times n)$ matrices $\mathbf{A} = (a_{jk})$ and $\mathbf{B} = (b_{jk})$ is the $(m \times n)$ matrix $\mathbf{C} = (c_{jk})$ with elements

(1) $$c_{jk} = a_{jk} + b_{jk} \qquad \begin{aligned} j &= 1, \cdots, m \\ k &= 1, \cdots, n \end{aligned}$$

and is written

$$\mathbf{C} = \mathbf{A} + \mathbf{B}.$$

We see that $\mathbf{A} + \mathbf{B}$ is obtained by adding corresponding elements of $\mathbf{A}$ and $\mathbf{B}$.

Example 1. If

$$\mathbf{A} = \begin{pmatrix} -4 & 6 & 3 \\ 0 & 1 & 2 \end{pmatrix} \quad \text{and} \quad \mathbf{B} = \begin{pmatrix} 5 & -1 & 0 \\ 3 & 1 & 0 \end{pmatrix}, \quad \text{then} \quad \mathbf{A} + \mathbf{B} = \begin{pmatrix} 1 & 5 & 3 \\ 3 & 2 & 2 \end{pmatrix}.$$

We define the $(m \times n)$ **zero matrix 0** to be the $(m \times n)$ matrix with all elements zeros. Then we see from the definition that matrix addition enjoys properties which are quite similar to those of addition of real numbers,

(2)

(*a*) $\mathbf{A} + \mathbf{B} = \mathbf{B} + \mathbf{A}$

(*b*) $(\mathbf{U} + \mathbf{V}) + \mathbf{W} = \mathbf{U} + (\mathbf{V} + \mathbf{W})$ (written $\mathbf{U} + \mathbf{V} + \mathbf{W}$)

(*c*) $\mathbf{A} + \mathbf{0} = \mathbf{A}$

(*d*) $\mathbf{A} + (-\mathbf{A}) = \mathbf{0}$

where $-\mathbf{A} = (-a_{jk})$ is the $(m \times n)$ matrix obtained by multiplying every element of $\mathbf{A}$ by -1 and is called the **negative** of $\mathbf{A}$.

Instead of $\mathbf{A} + (-\mathbf{B})$ we simply write $\mathbf{A} - \mathbf{B}$ and call this matrix the **difference** of $\mathbf{A}$ and $\mathbf{B}$. Obviously, its elements are obtained by subtracting corresponding elements of $\mathbf{A}$ and $\mathbf{B}$.

Multiplication of matrices by numbers is defined as follows. The **product** of an $(m \times n)$ matrix $\mathbf{A}$ by a number q is denoted by $q\mathbf{A}$ or $\mathbf{A}q$ and is the $(m \times n)$ matrix obtained by multiplying every element of $\mathbf{A}$ by q, that is

$$(3) \qquad q\mathbf{A} = \mathbf{A}q = \begin{pmatrix} qa_{11} & qa_{12} & \cdots & qa_{1n} \\ qa_{21} & qa_{22} & \cdots & qa_{2n} \\ \cdot & \cdot & \cdots & \cdot \\ qa_{m1} & qa_{m2} & \cdots & qa_{mn} \end{pmatrix}.$$

Example 2. If

$$\mathbf{A} = \begin{pmatrix} 2.7 & -1.8 \\ 0.9 & 3.6 \end{pmatrix}, \quad \text{then} \quad \mathbf{A} + \mathbf{A} = 2\mathbf{A} = \begin{pmatrix} 5.4 & -3.6 \\ 1.8 & 7.2 \end{pmatrix} \quad \text{and} \quad \frac{10}{9}\mathbf{A} = \begin{pmatrix} 3 & -2 \\ 1 & 4 \end{pmatrix}.$$

From the definitions we see that for any $(m \times n)$ matrices (with fixed m and n) and any numbers,

$$(4) \qquad \begin{aligned} &(a) & q(\mathbf{A} + \mathbf{B}) &= q\mathbf{A} + q\mathbf{B} \\ &(b) & (c + k)\mathbf{A} &= c\mathbf{A} + k\mathbf{A} \\ &(c) & c(k\mathbf{A}) &= (ck)\mathbf{A} \qquad \text{(written } ck\mathbf{A}) \\ &(d) & 1\mathbf{A} &= \mathbf{A}. \end{aligned}$$

Note that $(-1)\mathbf{A} = -\mathbf{A}$, the negative of $\mathbf{A}$.

Formulas (2) and (4) express the properties which are characteristic for a vector space (cf. Sec. 5.4). This gives

Theorem 1 (Vector spaces of matrices). *All real (or complex) $(m \times n)$ matrices (with m and n fixed) form a real (or complex) vector space of dimension mn. A basis consists of the mn matrices $\mathbf{E}_{rs}$ $(r = 1, \cdots, m;\ s = 1, \cdots, n)$, where $\mathbf{E}_{rs}$ is the $(m \times n)$ matrix with $e_{rs} = 1$ and all the other elements zero.*

Problems on addition and multiplication by numbers are included at the end of the next section.

6.3 *Transpose of a Matrix.*
Special Matrices

In this short section we want to define another simple operation on a matrix and introduce certain special types of matrices which are of practical importance.

Transposition of a matrix is defined as follows. The **transpose** $\mathbf{A}^\mathsf{T}$ of an $(m \times n)$ matrix $\mathbf{A} \doteq (a_{jk})$ is the $(n \times m)$ matrix obtained by interchanging the rows and columns in $\mathbf{A}$, that is

$$(1) \qquad \mathbf{A}^\mathsf{T} = (a_{kj}) = \begin{pmatrix} a_{11} & a_{21} & \cdots & a_{m1} \\ a_{12} & a_{22} & \cdots & a_{m2} \\ \cdot & \cdot & \cdots & \cdot \\ a_{1n} & a_{2n} & \cdots & a_{mn} \end{pmatrix}.$$

The reader may prove that

$$(2) \qquad (\mathbf{A} + \mathbf{B})^\mathsf{T} = \mathbf{A}^\mathsf{T} + \mathbf{B}^\mathsf{T}.$$

Example 1. If

$$\mathbf{A} = \begin{pmatrix} 5 & -8 & 1 \\ 4 & 0 & 0 \end{pmatrix}, \qquad \text{then} \qquad \mathbf{A}^\mathsf{T} = \begin{pmatrix} 5 & 4 \\ -8 & 0 \\ 1 & 0 \end{pmatrix}.$$

Example 2. If

$$\mathbf{b} = (7 \quad 5 \quad -2), \qquad \text{then} \qquad \mathbf{b}^\mathsf{T} = \begin{pmatrix} 7 \\ 5 \\ -2 \end{pmatrix}.$$

A **real** square matrix $\mathbf{A} = (a_{jk})$ is said to be **symmetric** if it is equal to its transpose,

(3) $\qquad\qquad \mathbf{A}^\mathsf{T} = \mathbf{A}, \qquad\quad \text{that is,} \qquad a_{kj} = a_{jk} \qquad\qquad (j, k = 1, \cdots, n).$

A **real** square matrix $\mathbf{A} = (a_{jk})$ is said to be **skew-symmetric** if

(4) $\qquad\qquad \mathbf{A}^\mathsf{T} = -\mathbf{A}, \qquad\quad \text{that is,} \qquad a_{kj} = -a_{jk} \qquad\qquad (j, k = 1, \cdots, n).$

(Similar notions for complex matrices will be defined later, in Sec. 6.13). Note that for $k = j$ in (4) we have $a_{jj} = -a_{jj}$, which implies that the elements in the principal diagonal of a skew-symmetric matrix are all zero.

Any real square matrix $\mathbf{A}$ may be written as the sum of a symmetric matrix $\mathbf{R}$ and a skew-symmetric matrix $\mathbf{S}$, where

$$\mathbf{R} = \tfrac{1}{2}(\mathbf{A} + \mathbf{A}^\mathsf{T}) \qquad \text{and} \qquad \mathbf{S} = \tfrac{1}{2}(\mathbf{A} - \mathbf{A}^\mathsf{T}).$$

Example 3. The matrices

$$\mathbf{A} = \begin{pmatrix} -3 & 1 & 5 \\ 1 & 0 & -2 \\ 5 & -2 & 4 \end{pmatrix} \qquad \text{and} \qquad \mathbf{B} = \begin{pmatrix} 0 & -4 & 1 \\ 4 & 0 & -5 \\ -1 & 5 & 0 \end{pmatrix}$$

are symmetric and skew-symmetric, respectively. The matrix

$$\mathbf{A} = \begin{pmatrix} 2 & 3 \\ 5 & -1 \end{pmatrix}$$

may be written in the form $\mathbf{A} = \mathbf{R} + \mathbf{S}$ where

$$\mathbf{R} = \tfrac{1}{2}(\mathbf{A} + \mathbf{A}^\mathsf{T}) = \begin{pmatrix} 2 & 4 \\ 4 & -1 \end{pmatrix} \qquad \text{and} \qquad \mathbf{S} = \tfrac{1}{2}(\mathbf{A} - \mathbf{A}^\mathsf{T}) = \begin{pmatrix} 0 & -1 \\ 1 & 0 \end{pmatrix}$$

are symmetric and skew-symmetric, respectively.

A square matrix $\mathbf{A} = (a_{jk})$ whose elements above the principal diagonal (or below the principal diagonal) are all zero is called a **triangular matrix**.
For example,

$$\mathbf{T}_1 = \begin{pmatrix} 1 & 0 & 0 \\ -2 & 3 & 0 \\ 5 & 0 & 2 \end{pmatrix} \qquad \text{and} \qquad \mathbf{T}_2 = \begin{pmatrix} 1 & 6 & -1 \\ 0 & 2 & 3 \\ 0 & 0 & 4 \end{pmatrix}$$

are triangular matrices.

A square matrix $\mathbf{A} = (a_{jk})$ whose elements above and below the principal diagonal are all zero, that is, $a_{jk} = 0$ for all $j \neq k$, is called a **diagonal matrix**.

For example,

$$\begin{pmatrix} 2 & 0 & 0 \\ 0 & 1 & 0 \\ 0 & 0 & -4 \end{pmatrix} \quad \text{and} \quad \begin{pmatrix} 2 & 0 & 0 \\ 0 & 2 & 0 \\ 0 & 0 & 2 \end{pmatrix}$$

are diagonal matrices.

A diagonal matrix whose elements in the principal diagonal are all 1 is called a **unit matrix** (for reasons to be explained in Sec. 6.4) and is denoted by $\mathbf{I}$. For example, the three-rowed unit matrix is

$$\mathbf{I} = \begin{pmatrix} 1 & 0 & 0 \\ 0 & 1 & 0 \\ 0 & 0 & 1 \end{pmatrix}.$$

Problems

Let $\mathbf{A} = \begin{pmatrix} 2 & -3 \\ 0 & 4 \end{pmatrix}$, $\mathbf{B} = \begin{pmatrix} -5 & 2 \\ 2 & 1 \end{pmatrix}$, $\mathbf{C} = \begin{pmatrix} 0 & 1 & -2 \\ 3 & 0 & 4 \end{pmatrix}$, $\mathbf{D} = \begin{pmatrix} 1 & 2 & 4 \\ -1 & 5 & -3 \end{pmatrix}$. Find

1. $\mathbf{A} + \mathbf{B}, \mathbf{B} + \mathbf{A}$
2. $\mathbf{C} + \mathbf{D}, \mathbf{D} + \mathbf{C}$
3. $\mathbf{A} - \mathbf{B}, \mathbf{B} - \mathbf{A}$
4. $\mathbf{C} - \mathbf{D}, \mathbf{D} - \mathbf{C}$
5. $2\mathbf{A} - 4\mathbf{B}, 2(\mathbf{A} - 2\mathbf{B})$
6. $\mathbf{A}^\mathsf{T}, \mathbf{B}^\mathsf{T}, (\mathbf{B}^\mathsf{T})^\mathsf{T}$
7. $\mathbf{C}^\mathsf{T}, \mathbf{D}^\mathsf{T}, (\mathbf{D}^\mathsf{T})^\mathsf{T}$
8. $(\mathbf{A} + \mathbf{B})^\mathsf{T}, \mathbf{A}^\mathsf{T} + \mathbf{B}^\mathsf{T}$
9. $(3\mathbf{A})^\mathsf{T}, 3\mathbf{A}^\mathsf{T}$
10. $\mathbf{A} + \mathbf{A}^\mathsf{T}, \mathbf{A} - \mathbf{A}^\mathsf{T}$
11. $(\mathbf{C} + \mathbf{D})^\mathsf{T}, \mathbf{C}^\mathsf{T} + \mathbf{D}^\mathsf{T}$
12. $(\mathbf{C} - \mathbf{D})^\mathsf{T}, \mathbf{C}^\mathsf{T} - \mathbf{D}^\mathsf{T}$

13. Is $\mathbf{C} + \mathbf{C}^\mathsf{T}$ defined?
14. Is $\mathbf{A} + \mathbf{C}$ defined?
15. Are $\mathbf{A}$ and $\mathbf{B}$ symmetric?
16. Is $\mathbf{A} + \mathbf{A}^\mathsf{T}$ symmetric?
17. Is $\mathbf{A} - \mathbf{A}^\mathsf{T}$ skew-symmetric?
18. Represent $\mathbf{A}$ as the sum of a symmetric and a skew-symmetric matrix.
19. Represent $\mathbf{A} - \mathbf{B}$ as the sum of a symmetric and a skew-symmetric matrix.
20. Find all real n-rowed square matrices which are both symmetric and skew-symmetric.

Let $\mathbf{K} = \begin{pmatrix} 1 & 3 & 5 \\ 0 & 4 & 2 \\ 0 & 0 & 6 \end{pmatrix}$, $\mathbf{M} = \begin{pmatrix} 2 & 0 & 0 \\ -1 & 1 & 0 \\ 4 & -3 & 0 \end{pmatrix}$, $\mathbf{N} = \begin{pmatrix} 6 & 7 \\ 0 & -2 \\ 3 & 8 \end{pmatrix}$. Find

21. $\mathbf{K} + \mathbf{M}, \mathbf{M} + \mathbf{K}$
22. $\mathbf{K} - \mathbf{M}, \mathbf{M} - \mathbf{K}$
23. $4\mathbf{K} - 3\mathbf{M}$
24. $\mathbf{K} + \mathbf{K}^\mathsf{T}, \mathbf{K} - \mathbf{K}^\mathsf{T}$
25. $(\mathbf{K} + \mathbf{M})^\mathsf{T}, \mathbf{K}^\mathsf{T} + \mathbf{M}^\mathsf{T}$
26. $(3\mathbf{N})^\mathsf{T}, 3\mathbf{N}^\mathsf{T}$

27. Are $\mathbf{K} + \mathbf{M} + \mathbf{N}, \mathbf{M} + \mathbf{N}, \mathbf{K} + \mathbf{M}^\mathsf{T}, \mathbf{N} + \mathbf{N}^\mathsf{T}, \mathbf{N} + \mathbf{C}^\mathsf{T}$ (see before) defined?
28. Represent $\mathbf{K}$ as the sum of a symmetric and a skew-symmetric matrix.
29. Prove (2) and (4) in Sec. 6.2.
30. Prove that $(\mathbf{A} + \mathbf{B})^\mathsf{T} = \mathbf{A}^\mathsf{T} + \mathbf{B}^\mathsf{T}, (a\mathbf{A})^\mathsf{T} = a\mathbf{A}^\mathsf{T}, (\mathbf{A}^\mathsf{T})^\mathsf{T} = \mathbf{A}$.

6.4 *Matrix Multiplication*

We shall now define the multiplication of a matrix by a matrix. The definition will be suggested by the use of matrices in connection with linear transformations. To illustrate this, let us start with a very simple case.

We consider three coordinate systems in the plane which we denote as the x_1x_2-system, the y_1y_2-system, and the z_1z_2-system, and we assume that these systems are related by transformations

(1)
$$x_1 = a_{11}y_1 + a_{12}y_2$$
$$x_2 = a_{21}y_1 + a_{22}y_2$$

and

(2)
$$y_1 = b_{11}z_1 + b_{12}z_2$$
$$y_2 = b_{21}z_1 + b_{22}z_2.$$

which are called *linear transformations*. Clearly, the x_1x_2-coordinates can be obtained directly from the z_1z_2-coordinates by a single linear transformation of the form

(3)
$$x_1 = c_{11}z_1 + c_{12}z_2$$
$$x_2 = c_{21}z_1 + c_{22}z_2$$

whose coefficients can be found by inserting (2) into (1),

$$x_1 = a_{11}(b_{11}z_1 + b_{12}z_2) + a_{12}(b_{21}z_1 + b_{22}z_2)$$
$$x_2 = a_{21}(b_{11}z_1 + b_{12}z_2) + a_{22}(b_{21}z_1 + b_{22}z_2)$$

and comparing this and (3); the result is

$$c_{11} = a_{11}b_{11} + a_{12}b_{21} \qquad c_{12} = a_{11}b_{12} + a_{12}b_{22}$$
$$c_{21} = a_{21}b_{11} + a_{22}b_{21} \qquad c_{22} = a_{21}b_{12} + a_{22}b_{22}$$

or briefly

(4)
$$c_{jk} = a_{j1}b_{1k} + a_{j2}b_{2k} = \sum_{i=1}^{2} a_{ji}b_{ik} \qquad\qquad j, k = 1, 2.$$

Now the coefficient matrices of the transformations (1) and (2) are

$$\mathbf{A} = \begin{pmatrix} a_{11} & a_{12} \\ a_{21} & a_{22} \end{pmatrix} \quad \text{and} \quad \mathbf{B} = \begin{pmatrix} b_{11} & b_{12} \\ b_{21} & b_{22} \end{pmatrix},$$

and we define the product $\mathbf{AB}$ (in this order) of $\mathbf{A}$ and $\mathbf{B}$ to be the coefficient matrix

$$\mathbf{C} = \begin{pmatrix} c_{11} & c_{12} \\ c_{21} & c_{22} \end{pmatrix}$$

of the "composed transformation" (3), that is,

$$\mathbf{C} = \mathbf{AB}$$

where the elements of $\mathbf{C}$ are given by (4).

If we write $\mathbf{A}$ in terms of row vectors, say,

$$\mathbf{A} = \begin{pmatrix} \mathbf{a}_1 \\ \mathbf{a}_2 \end{pmatrix} \quad \text{where} \quad \begin{matrix} \mathbf{a}_1 = (a_{11} \quad a_{12}) \\ \mathbf{a}_2 = (a_{21} \quad a_{22}) \end{matrix}$$

and **B** in terms of column vectors, say,

$$\mathbf{B} = (\mathbf{b}_1 \quad \mathbf{b}_2) \qquad \text{where} \qquad \mathbf{b}_1 = \begin{pmatrix} b_{11} \\ b_{21} \end{pmatrix}, \qquad \mathbf{b}_2 = \begin{pmatrix} b_{12} \\ b_{22} \end{pmatrix},$$

then the elements of **C** may simply be written as dot products:

$$c_{11} = \mathbf{a}_1 \cdot \mathbf{b}_1, \qquad c_{12} = \mathbf{a}_1 \cdot \mathbf{b}_2,$$
$$c_{21} = \mathbf{a}_2 \cdot \mathbf{b}_1, \qquad c_{22} = \mathbf{a}_2 \cdot \mathbf{b}_2.$$

Example 1. Let

$$\mathbf{A} = \begin{pmatrix} 2 & 1 \\ 3 & 4 \end{pmatrix} \qquad \text{and} \qquad \mathbf{B} = \begin{pmatrix} 1 & -2 \\ 5 & 3 \end{pmatrix}.$$

Then, by (4),

$$\mathbf{AB} = \begin{pmatrix} 2 & 1 \\ 3 & 4 \end{pmatrix} \begin{pmatrix} 1 & -2 \\ 5 & 3 \end{pmatrix} = \begin{pmatrix} 2 \cdot 1 + 1 \cdot 5 & 2 \cdot (-2) + 1 \cdot 3 \\ 3 \cdot 1 + 4 \cdot 5 & 3 \cdot (-2) + 4 \cdot 3 \end{pmatrix} = \begin{pmatrix} 7 & -1 \\ 23 & 6 \end{pmatrix}.$$

To arrive at the definition of the multiplication of two matrices in the general case, we now extend the preceding consideration to two more general linear transformations, say, the transformation

$$
\begin{aligned}
x_1 &= a_{11}y_1 + \cdots + a_{1n}y_n \\
x_2 &= a_{21}y_1 + \cdots + a_{2n}y_n \\
&\,\cdots\cdots\cdots\cdots\cdots\cdots\cdots \\
x_m &= a_{m1}y_1 + \cdots + a_{mn}y_n
\end{aligned}
$$

(5)

and the transformation

$$
\begin{aligned}
y_1 &= b_{11}z_1 + \cdots + b_{1p}z_p \\
y_2 &= b_{21}z_1 + \cdots + b_{2p}z_p \\
&\,\cdots\cdots\cdots\cdots\cdots\cdots\cdots \\
y_n &= b_{n1}z_1 + \cdots + b_{np}z_p.
\end{aligned}
$$

(6)

By inserting (6) into (5) we find that $x_1, \cdots, x_m$ can be expressed in terms of $z_1, \cdots, z_p$ in the form

$$
\begin{aligned}
x_1 &= c_{11}z_1 + \cdots + c_{1p}z_p \\
x_2 &= c_{21}z_1 + \cdots + c_{2p}z_p \\
&\,\cdots\cdots\cdots\cdots\cdots\cdots\cdots \\
x_m &= c_{m1}z_1 + \cdots + c_{mp}z_p
\end{aligned}
$$

(7)

where the coefficients are given by the formula

$$c_{jk} = a_{j1}b_{1k} + a_{j2}b_{2k} + \cdots + a_{jn}b_{nk}.$$

The requirement that the product **AB** (in this order) of the coefficient matrices $\mathbf{A} = (a_{jk})$ and $\mathbf{B} = (b_{jk})$ of the transformations (5) and (6) be the coefficient matrix $\mathbf{C} = (c_{jk})$ of the "composed transformation" (7) leads to the following definition.

Definition. *Let* $\mathbf{A} = (a_{jk})$ *be an* $(m \times n)$ *matrix and* $\mathbf{B} = (b_{jk})$ *an* $(r \times p)$ *matrix, then the product* **AB** *(in this order) is defined only when* $r = n$ *and is the* $(m \times p)$

matrix $\mathbf{C} = (c_{jk})$ *whose elements are*

$$(8) \qquad c_{jk} = a_{j1}b_{1k} + a_{j2}b_{2k} + \cdots + a_{jn}b_{nk} = \sum_{i=1}^{n} a_{ji}b_{ik}.$$

Instead of $\mathbf{AA}$ we simply write $\mathbf{A}^2$, etc.

We see that c_{jk} is the scalar product (dot product) of the j-th row vector of the first matrix, $\mathbf{A}$, and the k-th column vector of the second matrix, $\mathbf{B}$; that is, if we write

$$\mathbf{A} = \begin{pmatrix} \mathbf{a}_1 \\ \mathbf{a}_2 \\ \cdot \\ \cdot \\ \cdot \\ \mathbf{a}_m \end{pmatrix} \qquad and \qquad \mathbf{B} = (\mathbf{b}_1 \quad \mathbf{b}_2 \quad \cdots \quad \mathbf{b}_p)$$

where $\mathbf{a}_1, \cdots, \mathbf{a}_m$ are the row vectors of $\mathbf{A}$ and $\mathbf{b}_1, \cdots, \mathbf{b}_p$ are the column vectors of $\mathbf{B}$, then

$$(9) \qquad \mathbf{C} = \begin{pmatrix} \mathbf{a}_1 \cdot \mathbf{b}_1 & \mathbf{a}_1 \cdot \mathbf{b}_2 & \cdots & \mathbf{a}_1 \cdot \mathbf{b}_p \\ \mathbf{a}_2 \cdot \mathbf{b}_1 & \mathbf{a}_2 \cdot \mathbf{b}_2 & \cdots & \mathbf{a}_2 \cdot \mathbf{b}_p \\ \cdot & \cdot & \cdots & \\ \mathbf{a}_m \cdot \mathbf{b}_1 & \mathbf{a}_m \cdot \mathbf{b}_2 & \cdots & \mathbf{a}_m \cdot \mathbf{b}_p \end{pmatrix}.$$

The process of matrix multiplication is, therefore, conveniently referred to as **the multiplication of rows into columns.** Cf. Fig. 109.

Example 2. Let

$$\mathbf{A} = \begin{pmatrix} 3 & 2 & -1 \\ 0 & 4 & 6 \end{pmatrix} \qquad and \qquad \mathbf{B} = \begin{pmatrix} 1 & 0 & 2 \\ 5 & 3 & 1 \\ 6 & 4 & 2 \end{pmatrix}.$$

Then, by (8), the product $\mathbf{AB}$ is

$$\mathbf{AB} = \begin{pmatrix} 3 & 2 & -1 \\ 0 & 4 & 6 \end{pmatrix} \begin{pmatrix} 1 & 0 & 2 \\ 5 & 3 & 1 \\ 6 & 4 & 2 \end{pmatrix} = \begin{pmatrix} 7 & 2 & 6 \\ 56 & 36 & 16 \end{pmatrix},$$

whereas the product $\mathbf{BA}$ is not defined.

Fig. 109. Matrix multiplication. $\mathbf{AB} = \mathbf{C}$

The multiplication of matrices by vectors and the matrix multiplication of vectors may be illustrated by the following examples.

Example 3.

$$\begin{pmatrix} 3 & 4 & 2 \\ 6 & 0 & -1 \\ -5 & -2 & 1 \end{pmatrix} \begin{pmatrix} 1 \\ 3 \\ 2 \end{pmatrix} = \begin{pmatrix} 19 \\ 4 \\ -9 \end{pmatrix}$$

Example 4.

$$(3 \quad 6 \quad 1) \begin{pmatrix} 1 \\ 2 \\ 4 \end{pmatrix} = (19), \qquad \begin{pmatrix} 1 \\ 2 \\ 4 \end{pmatrix} (3 \quad 6 \quad 1) = \begin{pmatrix} 3 & 6 & 1 \\ 6 & 12 & 2 \\ 12 & 24 & 4 \end{pmatrix}$$

Note that the first of these products corresponds to the dot product as defined in Sec. 5.5.

Desk calculations can be checked by the use of the sums of the elements of the columns. In Ex. 2,

$$\begin{pmatrix} 3 & 2 & -1 \\ 0 & 4 & 6 \end{pmatrix} \begin{pmatrix} 1 & 0 & 2 \\ 5 & 3 & 1 \\ 6 & 4 & 2 \end{pmatrix} = \begin{pmatrix} 7 & 2 & 6 \\ 56 & 36 & 16 \end{pmatrix}$$

Sum   3   6   5   63   38   22

Check:

$$3 \cdot 1 + 6 \cdot 5 + 5 \cdot 6 = 63$$
$$3 \cdot 0 + 6 \cdot 3 + 5 \cdot 4 = 38$$
$$3 \cdot 2 + 6 \cdot 1 + 5 \cdot 2 = 22$$

Figure 110 shows an arrangement of the matrices **A** and **B** in Ex. 2 and their product **AB**, which is very convenient for desk calculations. The point is that each element c_{jk} of the product matrix occupies the intersection of that row of the first matrix and that column of the last matrix which are used for computing c_{jk}. The extension of the arrangement to computations of products of three and more matrices is obvious (Fig. 111).

Theorem 1 (Properties of matrix multiplication). (a) *Matrix multiplication is associative and is distributive with respect to addition of matrices, that is,*

(10)
$$\begin{aligned} (k\mathbf{A})\mathbf{B} = k(\mathbf{AB}) = \mathbf{A}(k\mathbf{B}) & \quad (written\ k\mathbf{AB}\ or\ \mathbf{A}k\mathbf{B}) \\ \mathbf{A}(\mathbf{BC}) = (\mathbf{AB})\mathbf{C} & \quad (written\ \mathbf{ABC}) \\ (\mathbf{A} + \mathbf{B})\mathbf{C} = \mathbf{AC} + \mathbf{BC} & \\ \mathbf{C}(\mathbf{A} + \mathbf{B}) = \mathbf{CA} + \mathbf{CB} & \end{aligned}$$

provided **A**, **B**, *and* **C** *are such that the expressions on the left are defined.* (*k is any number.*)

(b) *Matrix multiplication is* **not commutative**; *that is, if* **A** *and* **B** *are matrices such that both* **AB** *and* **BA** *are defined, then*

$$\mathbf{AB} \neq \mathbf{BA} \qquad in\ general.$$

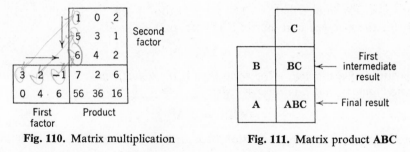

Fig. 110. Matrix multiplication   **Fig. 111.** Matrix product **ABC**

(**c**) *The cancellation law is not true, in general, that is*

$$\mathbf{AB} = 0 \quad \textit{does not necessarily imply} \quad \mathbf{A} = 0 \textit{ or } \mathbf{B} = 0.$$

Proof. The proof of (*a*) is left as an exercise. (*b*) can be seen from Example 4 (above) or

$$\begin{pmatrix} 1 & 0 \\ 0 & 0 \end{pmatrix} \begin{pmatrix} 0 & 1 \\ 1 & 0 \end{pmatrix} = \begin{pmatrix} 0 & 1 \\ 0 & 0 \end{pmatrix} \quad \text{but} \quad \begin{pmatrix} 0 & 1 \\ 1 & 0 \end{pmatrix} \begin{pmatrix} 1 & 0 \\ 0 & 0 \end{pmatrix} = \begin{pmatrix} 0 & 0 \\ 1 & 0 \end{pmatrix}.$$

(*c*) can be seen from

$$\begin{pmatrix} 1 & 1 \\ 2 & 2 \end{pmatrix} \begin{pmatrix} -1 & 1 \\ 1 & -1 \end{pmatrix} = \begin{pmatrix} 0 & 0 \\ 0 & 0 \end{pmatrix}.$$

Properties (*b*) and (*c*) are quite unusual because they have no counterparts in the usual multiplication of numbers and should, therefore, be carefully observed. Property (*c*) will be considered in more detail in Sec. 6.10. The reader may use the above example to illustrate that $\mathbf{AB} = 0$ does *not* imply $\mathbf{BA} = 0$.

Property (*b*) implies that *the order of the factors in a matrix product must be carefully observed*, and only in certain cases the equation $\mathbf{AB} = \mathbf{BA}$ will hold. To be precise we say that, in the product $\mathbf{AB}$, the matrix $\mathbf{B}$ is **premultiplied** by the matrix $\mathbf{A}$, or, alternatively, that $\mathbf{A}$ is **postmultiplied** by $\mathbf{B}$.

A diagonal matrix whose diagonal elements are all equal is called a **scalar matrix**. If

$$\mathbf{S} = \begin{pmatrix} k & 0 & \cdots & 0 \\ 0 & k & \cdots & \cdot \\ \cdot & \cdot & \cdots & \cdot \\ 0 & 0 & \cdots & k \end{pmatrix}$$

is an n-rowed scalar matrix and $\mathbf{A}$ is any n-rowed square matrix, then

(11) $$\mathbf{AS} = \mathbf{SA} = k\mathbf{A};$$

that is, the multiplication of an n-rowed square matrix $\mathbf{A}$ by an n-rowed scalar matrix has the same effect as the multiplication of $\mathbf{A}$ by a number, and $\mathbf{S}$ commutes with any n-rowed square matrix. The simple proof is left to the reader. In particular, for the n-rowed unit matrix $\mathbf{I}$ (cf. Sec. 6.3) we have

$$\mathbf{AI} = \mathbf{IA} = \mathbf{A}.$$

Transposition of a product. *The transpose of a product equals the product of the transposed factors, taken in reverse order,*

$$(12) \qquad\qquad (\mathbf{AB})^\mathsf{T} = \mathbf{B}^\mathsf{T}\mathbf{A}^\mathsf{T}.$$

The proof follows from the definition of matrix multiplication and is left to the reader.

Column vectors of a product. If $\mathbf{b}_1, \cdots, \mathbf{b}_p$ are the column vectors of a matrix $\mathbf{B}$ and the product $\mathbf{AB}$ is defined, then from the definition of matrix multiplication it follows immediately that the product matrix $\mathbf{AB}$ has the column vectors $\mathbf{Ab}_1, \cdots,$ $\mathbf{Ab}_p$,

$$(13) \qquad\qquad \mathbf{AB} = (\mathbf{Ab}_1 \quad \mathbf{Ab}_2 \quad \cdots \quad \mathbf{Ab}_p).$$

Linear transformations. By using the definition of matrix multiplication we may now write the linear transformation (5) in the simple form

$$(5') \qquad\qquad\qquad \mathbf{x} = \mathbf{Ay}$$

where $\mathbf{A} = (a_{jk})$ and

$$\mathbf{x} = \begin{pmatrix} x_1 \\ x_2 \\ \cdot \\ \cdot \\ \cdot \\ x_m \end{pmatrix}, \qquad \mathbf{y} = \begin{pmatrix} y_1 \\ y_2 \\ \cdot \\ \cdot \\ \cdot \\ y_n \end{pmatrix}.$$

Similarly, (6) may be written

$$(6') \qquad\qquad\qquad \mathbf{y} = \mathbf{Bz},$$

where $\mathbf{z}$ is the column matrix with elements $z_1, \cdots, z_p$. By inserting (6') into (5') and using the associative law of matrix multiplication (cf. Theorem 1) we obtain

$$(7') \qquad\qquad \mathbf{x} = \mathbf{A}(\mathbf{Bz}) = \mathbf{ABz} = \mathbf{Cz},$$

in agreement with (7).

We mention that this consideration can be generalized and use the opportunity to introduce a few concepts of general interest. Let X and Y be any vector spaces (cf. Sec. 5.4). To each vector $\mathbf{x}$ in X we assign a single, fully determined vector $\mathbf{y}$ in Y. Then we say that a **mapping** (or **transformation** or **operator**) of X into Y is given. Such a mapping is denoted by a capital letter, say F. The vector $\mathbf{y}$ in Y assigned to a vector $\mathbf{x}$ in X is called the **image** of $\mathbf{x}$ and is denoted by $F(\mathbf{x})$ [or $F\mathbf{x}$, without parentheses]. More generally, if S is any subset of X, the set of the images of all the elements of S is called the **image** of S and is denoted by $F(S)$.

F is said to be **injective** or **one-to-one** if different vectors in *X* have different images in *Y*. The mapping *F* is said to be **surjective** or a mapping of *X* **onto** *Y* if every vector **y** in *Y* is the image of at least one vector in *X*. The mapping *F* is said to be **bijective** if *F* is both injective and surjective.

F is called a **linear mapping** or **linear transformation** if for any vectors **x** and **z** in *X* and any scalar α,

(14)
$$F(\mathbf{x} + \mathbf{z}) = F(\mathbf{x}) + F(\mathbf{z})$$

$$F(\alpha\mathbf{x}) = \alpha F(\mathbf{x}).$$

For instance, (5) and (6) are linear transformations in the sense of this more general definition.

Let *X* be *n*-dimensional and $\mathbf{e}_{(1)}, \cdots, \mathbf{e}_{(n)}$ a basis of *X*. Then any vector **x** in *X* has a unique representation

$$\mathbf{x} = x_1\mathbf{e}_{(1)} + \cdots + x_n\mathbf{e}_{(n)},$$

which we may write $\mathbf{x} = (x_1 \cdots x_n)^\mathsf{T}$. If *F* is any linear transformation of *X* into *Y*, then

$$F(\mathbf{x}) = F(x_1\mathbf{e}_{(1)} + \cdots + x_n\mathbf{e}_{(n)}) = x_1 F(\mathbf{e}_{(1)}) + \cdots + x_n F(\mathbf{e}_{(n)}).$$

Hence a linear transformation of a vector space X into a vector space Y is uniquely determined by the images of the elements of a basis of X.

(5) and (5′) show that any matrix determines a linear transformation. Conversely, if we choose bases

$$\mathbf{e}_{(1)} = \begin{pmatrix} 1 \\ 0 \\ 0 \\ \cdot \\ \cdot \\ \cdot \\ 0 \end{pmatrix}, \mathbf{e}_{(2)} = \begin{pmatrix} 0 \\ 1 \\ 0 \\ \cdot \\ \cdot \\ \cdot \\ 0 \end{pmatrix}, \cdots, \mathbf{e}_{(n)} = \begin{pmatrix} 0 \\ 0 \\ \cdot \\ \cdot \\ \cdot \\ 0 \\ 1 \end{pmatrix}$$

and

$$\mathbf{g}_{(1)} = \begin{pmatrix} 1 \\ 0 \\ 0 \\ \cdot \\ \cdot \\ \cdot \\ 0 \end{pmatrix}, \cdots, \mathbf{g}_{(m)} = \begin{pmatrix} 0 \\ 0 \\ \cdot \\ \cdot \\ \cdot \\ 0 \\ 1 \end{pmatrix}$$

in *X* and *Y*, respectively, and *F* is any linear transformation of *X* into *Y*, there is an $m \times n$ matrix **A** such that for any **x** in *X* and its image $\mathbf{y} = F(\mathbf{x})$,

$$\mathbf{y} = \mathbf{A}\mathbf{x}.$$

In fact, *F* is determined by the images $\mathbf{y}_{(1)} = F(\mathbf{e}_{(1)}), \cdots, \mathbf{y}_{(n)} = F(\mathbf{e}_{(n)})$; writing

$\mathbf{y}_{(j)} = (\eta_1^{(j)}, \cdots, \eta_m^{(j)})$ and setting $\mathbf{y}_{(1)} = \mathbf{A}\mathbf{e}_{(1)}$, that is,

$$
\begin{pmatrix} \eta_1^{(1)} \\ \cdot \\ \cdot \\ \cdot \\ \eta_m^{(1)} \end{pmatrix} = \begin{pmatrix} a_{11} & \cdots & a_{1n} \\ a_{21} & \cdots & a_{2n} \\ \cdot & \cdots & \cdot \\ a_{m1} & \cdots & a_{mn} \end{pmatrix} \begin{pmatrix} 1 \\ 0 \\ \cdot \\ \cdot \\ 0 \end{pmatrix}
$$

we can determine the elements of the first column of $\mathbf{A}$, namely, $a_{11} = \eta_1^{(1)}$, $a_{21} = \eta_2^{(1)}$, etc. From $\mathbf{y}_{(2)} = \mathbf{A}\mathbf{e}_{(2)}$ we determine the elements of the second column of $\mathbf{A}$, and so on. We say that the matrix $\mathbf{A}$ *represents* the linear transformation F with respect to those bases. The purpose of such a representation is to replace one object of study by another object whose properties are more readily apparent.

Problems

Let $\mathbf{A} = \begin{pmatrix} 4 & 6 & -1 \\ 3 & 0 & 2 \\ 1 & -2 & 5 \end{pmatrix}$, $\mathbf{B} = \begin{pmatrix} 2 & 4 \\ 0 & 1 \\ -1 & 2 \end{pmatrix}$, $\mathbf{C} = \begin{pmatrix} 3 \\ 1 \\ 2 \end{pmatrix}$.

Find those of the following expressions which are defined.

1. $\mathbf{AB}$, $\mathbf{BA}$, $\mathbf{A^TB}$, $\mathbf{B^TA}$ 2. $\mathbf{BC}$, $\mathbf{CB}$, $\mathbf{B^TC}$, $\mathbf{C^TB}$ 3. $\mathbf{A}^2$, $\mathbf{B}^2$, $\mathbf{C}^2$

4. $\mathbf{AA^T}$, $\mathbf{A^TA}$ 5. $\mathbf{BB^T}$, $\mathbf{B^TB}$ 6. $\mathbf{CC^T}$, $\mathbf{C^TC}$

7. $(\mathbf{AB})^T$, $\mathbf{B^TA^T}$ 8. $\mathbf{C^TAB}$, $(\mathbf{AB})^T\mathbf{C}$ 9. $\mathbf{A}^2 + 3\mathbf{A} - 2\mathbf{I}$

10. Prove (10).   11. Prove (11).   12. Prove (12).

13. Let $\mathbf{A} = \begin{pmatrix} 2 & -1 & 0 \\ 0 & -2 & 1 \\ 1 & 0 & 1 \end{pmatrix}$ and $\mathbf{B} = \begin{pmatrix} -2 & 1 & -1 \\ 1 & 2 & -2 \\ 2 & -1 & -4 \end{pmatrix}$. Find $\mathbf{AB}$ and $\mathbf{BA}$.

14. The so-called Pauli spin matrices are

$$
\mathbf{S}_x = \begin{pmatrix} 0 & 1 \\ 1 & 0 \end{pmatrix}, \quad \mathbf{S}_y = \begin{pmatrix} 0 & -i \\ i & 0 \end{pmatrix}, \quad \mathbf{S}_z = \begin{pmatrix} 1 & 0 \\ 0 & -1 \end{pmatrix} \quad (i = \sqrt{-1}).
$$

Show that $\mathbf{S}_x\mathbf{S}_y = i\mathbf{S}_z$, $\mathbf{S}_y\mathbf{S}_x = -i\mathbf{S}_z$, $\mathbf{S}_x^2 = \mathbf{S}_y^2 = \mathbf{S}_z^2 = \mathbf{I}$.

15. (Associative algebra) A *real* (or complex) *associative algebra* is defined to be a set A of elements $\mathbf{A}, \mathbf{B}, \mathbf{C}, \cdots$ such that: (*a*) A is a real (or complex) vector space (cf. Sec. 5.4). (*b*) A multiplication is defined on the elements of A such that with any ordered pair of elements $\mathbf{A}, \mathbf{B}$ in A, there is associated a unique element of A, written $\mathbf{AB}$ and called the *product* of $\mathbf{A}$ and $\mathbf{B}$, and this multiplication satisfies (10). Show that the set of all $(n \times n)$ matrices (n fixed) is an associative algebra.

16. Represent each of the transformations

$$
\begin{aligned} x_1 &= 3y_1 + 2y_2 \\ x_2 &= -y_1 + 4y_2 \end{aligned} \quad \text{and} \quad \begin{aligned} y_1 &= z_1 + 2z_2 \\ y_2 &= 3z_1 \end{aligned}
$$

by the use of matrices and find the composite transformation which expresses x_1, x_2 in terms of z_1, z_2.

17. Let $\mathbf{x} = \begin{pmatrix} x_1 \\ x_2 \end{pmatrix}$, $\mathbf{y} = \begin{pmatrix} y_1 \\ y_2 \end{pmatrix}$, $\mathbf{A} = \begin{pmatrix} \cos\theta & -\sin\theta \\ \sin\theta & \cos\theta \end{pmatrix}$. Show that the linear transformation $\mathbf{y} = \mathbf{Ax}$ represents a rotation of the Cartesian $x_1 x_2$-coordinate system in the plane about the origin.

18. Interpret the transformation $\mathbf{x} = \mathbf{Bz}$ geometrically, where

$$\mathbf{x} = \begin{pmatrix} x_1 \\ x_2 \end{pmatrix}, \qquad \mathbf{z} = \begin{pmatrix} z_1 \\ z_2 \end{pmatrix}, \qquad \mathbf{B} = \begin{pmatrix} 1 & 0 \\ 0 & 2 \end{pmatrix}.$$

19. Consider the transformations $\mathbf{y} = \mathbf{Ax}$ and $\mathbf{x} = \mathbf{Bz}$ in Probs. 17 and 18. Find the composite transformation which expresses $\mathbf{y}$ in terms of $\mathbf{z}$. What does it mean geometrically that $\mathbf{AB} \neq \mathbf{BA}$?

20. Show that in Prob. 17,

$$\mathbf{A}^n = \begin{pmatrix} \cos n\theta & -\sin n\theta \\ \sin n\theta & \cos n\theta \end{pmatrix}.$$

What does this result mean geometrically?

6.5 Systems of Linear Equations. Gauss Elimination

A *system of m linear equations* (or *set of m simultaneous linear equations*) *in n unknowns* $x_1, \cdots, x_n$ is a set of equations of the form

(1)

$$\begin{aligned}
a_{11}x_1 + \cdots + a_{1n}x_n &= b_1 \\
a_{21}x_1 + \cdots + a_{2n}x_n &= b_2 \\
&\cdots\cdots\cdots\cdots\cdots \\
a_{m1}x_1 + \cdots + a_{mn}x_n &= b_m
\end{aligned}$$

The a_{ik} are given numbers, which are called the **coefficients** of the system. The b_i are also given numbers. If the b_i are all zero, then (1) is called a **homogeneous system**. If at least one b_i is not zero, then (1) is called a **nonhomogeneous system**.

A **solution** of (1) is a set of numbers $x_1, \cdots, x_n$ which satisfy all the m equations. A **solution vector** of (1) is a vector $\mathbf{x}$ whose components constitute a solution of (1). If (1) is homogeneous, it has at least the **trivial solution** $x_1 = 0, \cdots, x_n = 0$.

From the definition of matrix multiplication we see that the m equations of (1) may be written as a single vector equation

(2) $$\mathbf{Ax} = \mathbf{b},$$

where the **coefficient matrix** $\mathbf{A} = (a_{ik})$ is the $(m \times n)$ matrix

$$\mathbf{A} = \begin{pmatrix} a_{11} & a_{12} & \cdots & a_{1n} \\ a_{21} & a_{22} & \cdots & a_{2n} \\ \cdot & \cdot & \cdots & \cdot \\ a_{m1} & a_{m2} & \cdots & a_{mn} \end{pmatrix}, \quad \text{and} \quad \mathbf{x} = \begin{pmatrix} x_1 \\ \cdot \\ \cdot \\ \cdot \\ x_n \end{pmatrix} \quad \text{and} \quad \mathbf{b} = \begin{pmatrix} b_1 \\ \cdot \\ \cdot \\ \cdot \\ b_m \end{pmatrix}$$

are column vectors. We assume that the coefficients a_{ik} are not all zero, so that **A** is not a zero matrix. Note that **x** has n components, whereas **b** has m components. The matrix

$$\mathbf{B} = \begin{pmatrix} a_{11} & \cdots & a_{1n} & b_1 \\ \cdot & \cdots & \cdot & \cdot \\ \cdot & \cdots & \cdot & \cdot \\ \cdot & \cdots & \cdot & \cdot \\ a_{m1} & \cdots & a_{mn} & b_m \end{pmatrix}$$

is called the **augmented matrix** of the system (1). We see that **B** is obtained by augmenting **A** by the column **b**. The matrix **B** determines the system (1) completely, because it contains all the given numbers appearing in (1).

Example 1. If $m = n = 2$, we have two equations in two unknowns x_1, x_2

$$a_{11}x_1 + a_{12}x_2 = b_1$$
$$a_{21}x_1 + a_{22}x_2 = b_2$$

If we interpret x_1, x_2 as coordinates in the x_1x_2-plane, then each of the two equations represents a straight line, and (x_1, x_2) is a solution if and only if the point P with coordinates x_1, x_2 lies on both lines. Hence there are three possible cases:

(a) No solution if the lines are parallel.
(b) Precisely one solution if they intersect.
(c) Infinitely many solutions if they coincide.

For instance,

$x + y = 1$	$x + y = 1$	$x + y = 1$
$x + y = 0$	$x - y = 0$	$2x + 2y = 2$
Case (*a*)	Case (*b*)	Case (*c*)

If the system is homogeneous, Case (*a*) cannot happen, because then those two straight lines pass through the origin, whose coordinates 0, 0 constitute the trivial solution. The reader may consider three equations in three unknowns as representations of three planes in space and discuss the various possible cases in a similar fashion.

Our simple example illustrates that a system (1) may not always have a solution, and relevant problems are as follows. Does a given system (1) have a solution? Under what conditions does it have precisely one solution? If it has more than one solution, how can the set of all solutions be characterized? How can the solutions be obtained?

In the present section we shall discuss a method for obtaining solutions by a systematic process of elimination, which is called the **Gauss elimination** (or **Gauss algorithm**). We first explain the process by an example. Since a system of linear equations is completely determined by its augmented matrix, the process of elimination can be performed by merely considering the matrices. To see this correspondence, we shall write systems of equations and augmented matrices side by side.

Example 2. Solve the system of three linear equations in four unknowns

$$(3) \quad \begin{aligned} 3.0x_1 + 2.0x_2 + 2.0x_3 - 5.0x_4 &= 8.0 \\ 0.6x_1 + 1.5x_2 + 1.5x_3 - 5.4x_4 &= 2.7 \\ 1.2x_1 - 0.3x_2 - 0.3x_3 + 2.4x_4 &= 2.1 \end{aligned} \qquad \begin{pmatrix} 3.0 & 2.0 & 2.0 & -5.0 & 8.0 \\ 0.6 & 1.5 & 1.5 & -5.4 & 2.7 \\ 1.2 & -0.3 & -0.3 & 2.4 & 2.1 \end{pmatrix}$$

First step. *Elimination of x_1 from the second and third equation* by subtracting

$$0.6/3.0 = 0.2 \text{ times the first equation from the second equation,}$$
$$1.2/3.0 = 0.4 \text{ times the first equation from the third equation.}$$

This gives a new system of equations

$$(4) \quad \begin{aligned} 3.0x_1 + 2.0x_2 + 2.0x_3 - 5.0x_4 &= 8.0 \\ 1.1x_2 + 1.1x_3 - 4.4x_4 &= 1.1 \\ -1.1x_2 - 1.1x_3 + 4.4x_4 &= -1.1 \end{aligned} \qquad \begin{pmatrix} 3.0 & 2.0 & 2.0 & -5.0 & 8.0 \\ 0 & 1.1 & 1.1 & -4.4 & 1.1 \\ 0 & -1.1 & -1.1 & 4.4 & -1.1 \end{pmatrix}$$

Second step. *Elimination of x_2 from the third equation of* (4) by subtracting

$$-1.1/1.1 = -1 \text{ times the second equation from the third equation.}$$

This gives

$$\begin{aligned} 3.0x_1 + 2.0x_2 + 2.0x_3 - 5.0x_4 &= 8.0 \\ 1.1x_2 + 1.1x_3 - 4.4x_4 &= 1.1 \\ 0 &= 0 \end{aligned} \qquad \begin{pmatrix} 3.0 & 2.0 & 2.0 & -5.0 & 8.0 \\ 0 & 1.1 & 1.1 & -4.4 & 1.1 \\ 0 & 0 & 0 & 0 & 0 \end{pmatrix}$$

From the second equation, $x_2 = 1 - x_3 + 4x_4$. From this and the first equation, $x_1 = 2 - x_4$. Since x_3 and x_4 remain arbitrary, we have infinitely many solutions; if we choose a value of x_3 and a value of x_4, then the corresponding values of x_1 and x_2 are uniquely determined.

The method illustrated in this example is called the **Gauss elimination** or **Gauss algorithm**.[1] In terms of general formulas for the general system (1) it can be formulated as follows.

First step. *Elimination of x_1 from the second, third, $\cdots$ m-th equation.* We may assume that the order of the equations and the order of the unknowns in each equation are such that $a_{11} \neq 0$. The variable x_1 can then be eliminated from the second, $\cdots$, *m*th equation by subtracting

$$a_{21}/a_{11} \text{ times the first equation from the second equation,}$$
$$a_{31}/a_{11} \text{ times the first equation from the third equation,}$$

---

[1] We present the Gauss elimination in its classical form; further reduction of the amount of computational work was obtained by Banachiewicz and Cholesky. These and other methods are discussed in Ref. [C1]. Cf. also Sec. 18.8.

etc. This gives a new system of equations of the form

$$a_{11}x_1 + a_{12}x_2 + \cdots + a_{1n}x_n = b_1$$
$$c_{22}x_2 + \cdots + c_{2n}x_n = b_2{}^*$$

(5)

$$c_{m2}x_2 + \cdots + c_{mn}x_n = b_m{}^*.$$

Any solution of (1) is a solution of (5) and conversely, because each equation of (5) was derived from two equations of (1) and, by reversing the process, (1) may be obtained from (5).

Second step. *Elimination of x_2 from the third, $\cdots$, m-th equation in* (5). If the coefficients $c_{22}, \cdots, c_{mn}$ in (5) are not all zero, we may assume that the order of the equations and the unknowns is such that $c_{22} \neq 0$. Then we may eliminate x_2 from the third, fourth, $\cdots$, mth equation of (5) by subtracting

c_{32}/c_{22} times the second equation from the third equation,

c_{42}/c_{22} times the second equation from the fourth equation, etc.

The further steps are now obvious. In the third step we eliminate x_3, in the fourth step we eliminate x_4, etc.

This process will terminate only when no equations are left or when the coefficients of all the unknowns in the remaining equations are zero. We then have a system of the form

$$a_{11}x_1 + a_{12}x_2 + \cdots\cdots + a_{1n}x_n = b_1$$
$$c_{22}x_2 + \cdots\cdots + c_{2n}x_n = b_2{}^*$$

(6)
$$k_{rr}x_r + \cdots + k_{rn}x_n = \tilde{b}_r$$
$$0 = \tilde{b}_{r+1}$$

$$0 = \tilde{b}_m$$

where $r \leqq m$. We see that there are three possible cases:

   (a) No solution if $r < m$ and one of the numbers $\tilde{b}_{r+1}, \cdots, \tilde{b}_m$ is not zero.

   (b) Precisely one solution if $r = n$ and $\tilde{b}_{r+1}, \cdots, \tilde{b}_m$, if present, are zero. This solution is obtained by solving the nth equation of (6) for x_n, then the $(n-1)$th equation for x_{n-1} and so on up to the line.

   (c) Infinitely many solutions if $r < n$ and $\tilde{b}_{r+1}, \cdots, \tilde{b}_m$, if present, are zero. Then any of these solutions is obtained by choosing values at pleasure for the unknowns $x_{r+1}, \cdots, x_n$, solving the rth equation for x_r, then the $(r-1)$th equation for x_{r-1} and so on up to the line. Ex. 2 illustrates this case.

The operations in the Gauss elimination are called **elementary operations**. *Elementary operations for equations* are:

Interchange of two equations.
Multiplication of an equation by a nonzero constant.
Addition of a constant multiple of one equation to another equation.

Instead of equations we may consider augmented matrices (cf. Ex. 2). Then we have the following corresponding *elementary row operations for matrices:*

Interchange of two rows.
Multiplication of a row by a nonzero constant.
Addition of a constant multiple of one row to another row.

A system S_1 of linear equations is said to be **equivalent** to a system S_2 of linear equations if S_1 can be obtained from S_2 by (finitely many) elementary operations. From our discussion we have the following result.

Theorem 1. *Equivalent systems of linear equations have the same sets of solutions.*

A matrix A_1 is said to be **row-equivalent** to a matrix A_2 if A_1 can be obtained from A_2 by finitely many row operations. Thus the augmented matrices of (1) and (6) are row-equivalent.

These discussions and concepts will be helpful in Sec. 6.9, where we shall give a more detailed characterization of sets of solutions of simultaneous linear equations.

Problems

Solve the following systems of linear equations.

1. $5x - 2y = 1$
$6x + 8y = 22$

2. $3x + y = -5$
$2x + 3y = 6$

3. $x - 2y = -8$
$5x + 3y = -1$

4. $-x + 2y = 4$
$3x + 4y = 38$

5. $2x + 3y = 4$
$3x + 2y = -4$

6. $3x + 2y = -17$
$10x + y = 0$

7. $x + 2y - 8z = 0$
$2x - 3y + 5z = 0$
$3x + 2y - 12z = 0$

8. $7x - y - 2z = 0$
$9x - y - 3z = 0$
$2x + 4y - 7z = 0$

9. $3x - y + z = -2$
$x + 5y + 2z = 6$
$2x + 3y + z = 0$

10. $2y - z = -1$
$x + 3z = 11$
$2x - 4y + 2z = 6$

11. $5x + 3y - 3z = -1$
$3x + 2y - 2z = -1$
$2x - y + 2z = 8$

12. $9x + 4y + 3z = -1$
$5x + y + 2z = 1$
$7x + 3y + 4z = 1$

13. $3x - 3y - 7z = -4$
$x - y + 2z = 3$

14. $7x - 4y - 2z = -6$
$16x + 2y + z = 3$

15. $x - 3y + 2z = 2$
$5x - 15y + 7z = 10$

16. $3w - 15x - 3y - 11z = -27$
$5w - 25x - 5y - 7z = -45$

17. $7w + 2x - 2y - 6z = 6$
$w + 3x - 3y - 9z = -10$

18. $3w - 6x - y - z = 0$
$w - 2x + 5y - 3z = 0$
$2w - 4x + 3y - z = 3$

19. $4w + 3x - 9y + z = 1$
$-w + 2x - 13y + 3z = 3$
$3w - x + 8y - 2z = -2$

6.6 *Determinants of Second and Third Order*

This section is independent of the other sections in this chapter. It is for reference in connection with other chapters, and it also serves to motivate the discussions in the next section.

Determinants of second order arise in connection with systems of two linear equations

(1)
$$a_{11}x_1 + a_{12}x_2 = b_1$$
$$a_{21}x_1 + a_{22}x_2 = b_2$$

in two unknowns x_1, x_2. To solve this system, we may multiply the first equation by a_{22}, the second by $-a_{12}$, and add, finding

$$(a_{11}a_{22} - a_{21}a_{12})x_1 = b_1a_{22} - b_2a_{12}.$$

Then we multiply the first equation of (1) by $-a_{21}$, the second by a_{11}, and add again, finding

$$(a_{11}a_{22} - a_{21}a_{12})x_2 = a_{11}b_2 - a_{21}b_1.$$

If $a_{11}a_{22} - a_{21}a_{12}$ is not zero, we may divide and obtain the desired result

(2)
$$x_1 = \frac{b_1a_{22} - b_2a_{12}}{a_{11}a_{22} - a_{21}a_{12}}, \qquad x_2 = \frac{a_{11}b_2 - a_{21}b_1}{a_{11}a_{22} - a_{21}a_{12}}.$$

The expression in the denominators is written in the form

$$\begin{vmatrix} a_{11} & a_{12} \\ a_{21} & a_{22} \end{vmatrix}$$

and is called a **determinant of second order**; thus

(3)
$$\begin{vmatrix} a_{11} & a_{12} \\ a_{21} & a_{22} \end{vmatrix} = a_{11}a_{22} - a_{21}a_{12}.$$

The four numbers a_{11}, a_{12}, a_{21}, a_{22} are called the **elements** of the determinant. The elements in a horizontal line are said to form a **row** and the elements in a vertical line are said to form a **column** of the determinant.

We may now write the solution (2) of the system (1) in the form

(4)
$$x_1 = \frac{D_1}{D}, \qquad x_2 = \frac{D_2}{D} \qquad\qquad (D \neq 0)$$

where

$$D = \begin{vmatrix} a_{11} & a_{12} \\ a_{21} & a_{22} \end{vmatrix}, \qquad D_1 = \begin{vmatrix} b_1 & a_{12} \\ b_2 & a_{22} \end{vmatrix}, \qquad D_2 = \begin{vmatrix} a_{11} & b_1 \\ a_{21} & b_2 \end{vmatrix}.$$

This formula is called **Cramer's rule**.[2] Note that D_1 is obtained by replacing the

[2] GABRIEL CRAMER (1704–1752), Italian mathematician, also known by his contributions to the theory of curves.

first column of D by the column with elements b_1, b_2, and D_2 is obtained by re-placing the second column of D by that column.

If both b_1 and b_2 are zero, the system is said to be **homogeneous.** *In this case it has at least the "trivial solution" $x_1 = 0$, $x_2 = 0$. It has further solutions if and only if $D = 0$.*

If at least b_1 or b_2 is not zero, the system is said to be **nonhomogeneous.** *Then if $D \neq 0$, it has precisely one solution, which is obtained from* (4).

Determinants of third order occur in connection with systems of three linear equations

$$(5) \quad \begin{array}{l} a_{11}x_1 + a_{12}x_2 + a_{13}x_3 = b_1 \\ a_{21}x_1 + a_{22}x_2 + a_{23}x_3 = b_2 \\ a_{31}x_1 + a_{32}x_2 + a_{33}x_3 = b_3 \end{array}$$

in three unknowns x_1, x_2, x_3. To obtain an equation involving only x_1, we multiply the equations by

$$a_{22}a_{33} - a_{32}a_{23}, \qquad -(a_{12}a_{33} - a_{32}a_{13}), \qquad a_{12}a_{23} - a_{22}a_{13},$$

respectively. We see that these expressions may be written as second-order determinants:

$$M_{11} = \begin{vmatrix} a_{22} & a_{23} \\ a_{32} & a_{33} \end{vmatrix}, \qquad -M_{21} = -\begin{vmatrix} a_{12} & a_{13} \\ a_{32} & a_{33} \end{vmatrix}, \qquad M_{31} = \begin{vmatrix} a_{12} & a_{13} \\ a_{22} & a_{23} \end{vmatrix}.$$

Adding the resulting equations, we obtain

$$(6) \quad (a_{11}M_{11} - a_{21}M_{21} + a_{31}M_{31})x_1 = b_1 M_{11} - b_2 M_{21} + b_3 M_{31}.$$

Two further equations containing only x_2 and x_3, respectively, may be obtained in a similar manner.

To simplify our notation, we now define a **determinant of third order** by the equation

$$(7) \quad D = \begin{vmatrix} a_{11} & a_{12} & a_{13} \\ a_{21} & a_{22} & a_{23} \\ a_{31} & a_{32} & a_{33} \end{vmatrix} = a_{11}\begin{vmatrix} a_{22} & a_{23} \\ a_{32} & a_{33} \end{vmatrix} - a_{21}\begin{vmatrix} a_{12} & a_{13} \\ a_{32} & a_{33} \end{vmatrix} + a_{31}\begin{vmatrix} a_{12} & a_{13} \\ a_{22} & a_{23} \end{vmatrix}.$$

We see that

$$D = a_{11}M_{11} - a_{21}M_{21} + a_{31}M_{31},$$

the coefficient of x_1 in (6), and if we write out the second-order determinants in (7), we obtain

$$(8) \quad D = a_{11}a_{22}a_{33} - a_{11}a_{32}a_{23} + a_{21}a_{32}a_{13} - a_{21}a_{12}a_{33} + a_{31}a_{12}a_{23} - a_{31}a_{22}a_{13}.$$

Obviously the determinant on the right-hand side of (7) which is multiplied by a_{i1}, $i = 1, 2$, or 3, is obtained from D by omitting the first column and the ith row of D.

We see that (6) may now be written

$$Dx_1 = D_1 \qquad \text{where} \qquad D_1 = \begin{vmatrix} b_1 & a_{12} & a_{13} \\ b_2 & a_{22} & a_{23} \\ b_3 & a_{32} & a_{33} \end{vmatrix}.$$

Similarly, the aforementioned equation containing only x_2 may be written

$$Dx_2 = D_2 \qquad \text{where} \qquad D_2 = \begin{vmatrix} a_{11} & b_1 & a_{13} \\ a_{21} & b_2 & a_{23} \\ a_{31} & b_3 & a_{33} \end{vmatrix},$$

and the equation containing only x_3 may be written

$$Dx_3 = D_3 \qquad \text{where} \qquad D_3 = \begin{vmatrix} a_{11} & a_{12} & b_1 \\ a_{21} & a_{22} & b_2 \\ a_{31} & a_{32} & b_3 \end{vmatrix}.$$

Note that the elements of D are arranged in the same order as they occur as coefficients in the equations of (5), and $D_j, j = 1, 2,$ or 3, is obtained from D by replacing the jth column by the column with elements b_1, b_2, b_3, the expressions on the right sides of the equations of (5).

It follows that if $D \neq 0$, then system (5) has the unique solution

(9) $$x_1 = \frac{D_1}{D}, \qquad x_2 = \frac{D_2}{D}, \qquad x_3 = \frac{D_3}{D} \qquad \textbf{(Cramer's rule).}$$

If (5) is homogeneous, that is, $b_1 = b_2 = b_3 = 0$, it has at least the trivial solution $x_1 = x_2 = x_3 = 0$, and nontrivial solutions exist if and only if $D = 0$.

If (5) is nonhomogeneous and $D \neq 0$, it has precisely one solution, which is obtained from (9).

Example 1. Apply Cramer's rule to the system

$$\begin{aligned} 2x_1 - \quad x_2 + 2x_3 &= \quad 2 \\ x_1 + 10x_2 - 3x_3 &= \quad 5 \\ -x_1 + \quad x_2 + \quad x_3 &= -3. \end{aligned}$$

The determinant of the system is

$$D = \begin{vmatrix} 2 & -1 & 2 \\ 1 & 10 & -3 \\ -1 & 1 & 1 \end{vmatrix} = 46.$$

The determinants in the numerators in (9) are

$$D_1 = \begin{vmatrix} 2 & -1 & 2 \\ 5 & 10 & -3 \\ -3 & 1 & 1 \end{vmatrix} = 92, \quad D_2 = \begin{vmatrix} 2 & 2 & 2 \\ 1 & 5 & -3 \\ -1 & -3 & 1 \end{vmatrix} = 0, \quad D_3 = \begin{vmatrix} 2 & -1 & 2 \\ 1 & 10 & 5 \\ -1 & 1 & -3 \end{vmatrix} = -46.$$

Therefore, $x_1 = 2$, $x_2 = 0$, and $x_3 = -1$.

We shall now list the most important properties of our determinants. The proofs follow from (7) by direct calculation. (In the next section we shall see that determinants of any order n have quite similar properties.)

(A) *The value of a determinant is not altered if its rows are written as columns in the same order.* Example:

(10)
$$\begin{vmatrix} 1 & 3 & 0 \\ 2 & 6 & 4 \\ -1 & 0 & 2 \end{vmatrix} = \begin{vmatrix} 1 & 2 & -1 \\ 3 & 6 & 0 \\ 0 & 4 & 2 \end{vmatrix} = -12.$$

(B) *If any two rows (or two columns) of a determinant are interchanged, the value of the determinant is multiplied by* -1. Example:

(11)
$$\begin{vmatrix} 2 & 6 & 4 \\ 1 & 3 & 0 \\ -1 & 0 & 2 \end{vmatrix} = - \begin{vmatrix} 1 & 3 & 0 \\ 2 & 6 & 4 \\ -1 & 0 & 2 \end{vmatrix} = 12.$$

The second-order determinant obtained from D [cf. (7)] by deleting one row and one column is called the **minor** of the element which belongs to the deleted row and column. Example: The minors of a_{21} and a_{22} in D are

$$\begin{vmatrix} a_{12} & a_{13} \\ a_{32} & a_{33} \end{vmatrix} \quad \text{and} \quad \begin{vmatrix} a_{11} & a_{13} \\ a_{31} & a_{33} \end{vmatrix},$$

respectively. The **cofactor** of the element of D in the ith row and the kth column is defined as $(-1)^{i+k}$ times the minor of that element. Example: The cofactors of a_{21} and a_{22} are

$$- \begin{vmatrix} a_{12} & a_{13} \\ a_{32} & a_{33} \end{vmatrix} \quad \text{and} \quad \begin{vmatrix} a_{11} & a_{13} \\ a_{31} & a_{33} \end{vmatrix},$$

respectively. The signs $(-1)^{i+k}$ form a checkerboard pattern:

$$\begin{matrix} + & - & + \\ - & + & - \\ + & - & + \end{matrix}$$

Furthermore, we see that we may now write (7) in the form

$$D = a_{11}C_{11} + a_{21}C_{21} + a_{31}C_{31},$$

where C_{i1} is the cofactor of a_{i1} in D. From this and the properties **(A)** and **(B)** we obtain the following property.

(C) *The determinant D may be developed by any row or column, that is, it may be written as the sum of the three elements of any row (or column) each multiplied by its cofactor.* For example, the development of D by its second row is

$$D = -a_{21}\begin{vmatrix} a_{12} & a_{13} \\ a_{32} & a_{33} \end{vmatrix} + a_{22}\begin{vmatrix} a_{11} & a_{13} \\ a_{31} & a_{33} \end{vmatrix} - a_{23}\begin{vmatrix} a_{11} & a_{12} \\ a_{31} & a_{32} \end{vmatrix}.$$

From **(C)** we obtain

(D) *A factor of the elements of any row (or column) can be placed before the determinant.* Example:

$$\begin{vmatrix} 4 & 6 & 1 \\ 3 & -9 & 2 \\ -1 & 12 & 5 \end{vmatrix} = \begin{vmatrix} 4 & 2 \cdot 3 & 1 \\ 3 & -3 \cdot 3 & 2 \\ -1 & 4 \cdot 3 & 5 \end{vmatrix} = 3\begin{vmatrix} 4 & 2 & 1 \\ 3 & -3 & 2 \\ -1 & 4 & 5 \end{vmatrix}.$$

From the properties **(B)** and **(D)** we may draw the following conclusion.

(E) *If corresponding elements of two rows (or columns) of a determinant are proportional, the value of the determinant is zero.*

The following property is basic for simplifying determinants to be evaluated.

(F) *The value of a determinant remains unaltered if the elements of one row (or column) are altered by adding to them any constant multiple of the corresponding elements in any other row (or column).* Example:

$$\begin{vmatrix} -6 & 21 & -30 \\ 1 & -3 & 5 \\ 2 & 7 & -4 \end{vmatrix} = \begin{vmatrix} -6 + 1 \cdot 7 & 21 - 3 \cdot 7 & -30 + 5 \cdot 7 \\ 1 & -3 & 5 \\ 2 & 7 & -4 \end{vmatrix}$$

$$= \begin{vmatrix} 1 & 0 & 5 \\ 1 & -3 & 5 \\ 2 & 7 & -4 \end{vmatrix} = \begin{vmatrix} 1 & 0 & 5 \\ 0 & -3 & 0 \\ 2 & 7 & -4 \end{vmatrix} = -3\begin{vmatrix} 1 & 5 \\ 2 & -4 \end{vmatrix} = 42.$$

(G) *If each element of a row (or column) of a determinant is expressed as a binomial, the determinant can be written as the sum of two determinants.* Example:

$$(12) \qquad \begin{vmatrix} 4x + 3 & 2 & 1 \\ x & 3 & 4 \\ 2x - 1 & 1 & -1 \end{vmatrix} = \begin{vmatrix} 4x & 2 & 1 \\ x & 3 & 4 \\ 2x & 1 & -1 \end{vmatrix} + \begin{vmatrix} 3 & 2 & 1 \\ 0 & 3 & 4 \\ -1 & 1 & -1 \end{vmatrix}$$

By applying the product rule of differentiation we obtain the following property.

(H) *If the elements of a determinant are differentiable functions of a variable, the derivative of the determinant may be written as a sum of three determinants,*

$$\frac{d}{dx}\begin{vmatrix} f & g & h \\ p & q & r \\ u & v & w \end{vmatrix} = \begin{vmatrix} f' & g' & h' \\ p & q & r \\ u & v & w \end{vmatrix} + \begin{vmatrix} f & g & h \\ p' & q' & r' \\ u & v & w \end{vmatrix} + \begin{vmatrix} f & g & h \\ p & q & r \\ u' & v' & w' \end{vmatrix},$$

where primes denote derivatives with respect to x.

Problems

Evaluate the following determinants.

1. $\begin{vmatrix} 1 & 2 \\ 4 & 3 \end{vmatrix}$

2. $\begin{vmatrix} -4 & 2 \\ -2 & 5 \end{vmatrix}$

3. $\begin{vmatrix} \cos\theta & \sin\theta \\ -\sin\theta & \cos\theta \end{vmatrix}$

4. $\begin{vmatrix} \cos n\theta & \sin n\theta \\ -\sin n\theta & \cos n\theta \end{vmatrix}$

5. $\begin{vmatrix} 16 & 22 & 4 \\ 4 & -3 & 2 \\ 12 & 25 & 2 \end{vmatrix}$

6. $\begin{vmatrix} 4 & 6 & 5 \\ 0 & 1 & -7 \\ 0 & 0 & 6 \end{vmatrix}$

7. $\begin{vmatrix} 5 & 1 & 8 \\ 15 & 3 & 6 \\ 10 & 4 & 2 \end{vmatrix}$

8. $\begin{vmatrix} 1 & a & a^2 \\ 1 & b & b^2 \\ 1 & c & c^2 \end{vmatrix}$

9. $\begin{vmatrix} a-b & m-n & x-y \\ b-c & n-p & y-z \\ c-a & p-m & z-x \end{vmatrix}$

10. $\begin{vmatrix} b^2+c^2 & ab & ca \\ ab & c^2+a^2 & bc \\ ca & bc & a^2+b^2 \end{vmatrix}$

6.7   Determinants of Arbitrary Order

From the last section we know that determinants of second and third order appear in connection with systems of two and three linear equations in two and three unknowns, and we shall see that determinants of arbitrary order n are suggested by systems of n linear equations in n unknowns and are scalars associated with n-rowed square matrices. They have various applications in engineering mathematics, although their importance has decreased because they are not practical in numerical work. In fact, Cramer's rule (cf. Sec. 6.6 and below) is certainly not the *practical* answer to the problem of solving those equations numerically, and there are better methods for that purpose (cf. Secs. 6.5, 18.8, 18.9).

Determinants can be defined in several equivalent ways. We shall use a definition which is suggested by the process of solving linear equations by means of determinants and generalizes that in the last section. Important properties of determinants will be considered in this section and the relation to systems of linear equations in Sec. 6.11.

A determinant of order n is a square array of n^2 quantities enclosed between vertical bars,

(1)
$$D = \begin{vmatrix} a_{11} & a_{12} & \cdots & a_{1n} \\ a_{21} & a_{22} & \cdots & a_{2n} \\ \cdot & \cdot & \cdots & \cdot \\ \cdot & \cdot & \cdots & \cdot \\ a_{n1} & a_{n2} & \cdots & a_{nn} \end{vmatrix}$$

which has a certain value defined below. The quantities $a_{11}, \cdots, a_{nn}$ which are numbers (or sometimes functions), are called the **elements** of the determinant. The horizontal lines of elements are called **rows**; the vertical lines are called **columns**. The sloping line of elements extending from a_{11} to a_{nn} is called the **principal diagonal** of the determinant.

By deleting the ith row and the kth column from the determinant D we obtain an $(n - 1)$th order determinant (a square array of $n - 1$ rows and $n - 1$ columns between vertical bars), which is called the **minor** of the element a_{ik} (which belongs to the deleted row and column) and is denoted by M_{ik}.

The minor M_{ik} multiplied by $(-1)^{i+k}$ is called the **cofactor** of a_{ik} in D and will be denoted by C_{ik}; thus

(2)
$$C_{ik} = (-1)^{i+k} M_{ik}.$$

For example, in the third-order determinant

$$\begin{vmatrix} a_{11} & a_{12} & a_{13} \\ a_{21} & a_{22} & a_{23} \\ a_{31} & a_{32} & a_{33} \end{vmatrix}$$

we have

$$C_{11} = M_{11} = \begin{vmatrix} a_{22} & a_{23} \\ a_{32} & a_{33} \end{vmatrix}, \qquad C_{32} = -M_{32} = -\begin{vmatrix} a_{11} & a_{13} \\ a_{21} & a_{23} \end{vmatrix}, \text{ etc.}$$

We are now in a position to define an nth-order determinant in a way which is most convenient for practical purposes.

Definition. *The symbol*

$$D = \begin{vmatrix} a_{11} & a_{12} & \cdots & a_{1n} \\ a_{21} & a_{22} & \cdots & a_{2n} \\ \cdot & \cdot & \cdots & \cdot \\ \cdot & \cdot & \cdots & \cdot \\ a_{n1} & a_{n2} & \cdots & a_{nn} \end{vmatrix}$$

is called a **determinant of order** *n and means the sum of the products of the elements of any row or column and their respective cofactors; that is,*

(3a) $\qquad D = a_{i1}C_{i1} + a_{i2}C_{i2} + \cdots + a_{in}C_{in} \qquad (i = 1, 2, \cdots, \text{or } n)$

or

(3b) $\qquad D = a_{1k}C_{1k} + a_{2k}C_{2k} + \cdots + a_{nk}C_{nk} \qquad (k = 1, 2, \cdots, \text{or } n).$

In this way, D is defined in terms of n determinants of order $n - 1$, each of which is, in turn, defined in terms of $n - 1$ determinants of order $n - 2$, and so on; we finally arrive at second-order determinants, in which the cofactors of the elements are single elements of D. We see that *if the elements of D are numbers, the value of D will be a number.*

Furthermore, it follows from the definition that we may **develop** *D by any row or column*, that is, choose in (3) the elements of any row or column, similarly when developing the cofactors in (3), and so on. Consequently, before we accept this definition we have to show that *it is unambiguous, that is, yields the same value of D no matter which columns or rows we choose.*

For a second-order determinant

$$D = \begin{vmatrix} a_{11} & a_{12} \\ a_{21} & a_{22} \end{vmatrix}$$

this is immediately clear, because then we have only four possible forms of (3), namely, the development by

$$\text{the first row: } D = a_{11}a_{22} + a_{12}(-a_{21}),$$
$$\text{the last row: } D = a_{21}(-a_{12}) + a_{22}a_{11},$$
$$\text{the first column: } D = a_{11}a_{22} + a_{21}(-a_{12}),$$
$$\text{and the last column: } D = a_{12}(-a_{21}) + a_{22}a_{11}.$$

We see that we always obtain the same value of D, which value agrees with (3) in Sec. 6.6.

We shall now prove that *our definition of a determinant D of arbitrary order n yields the same value of D no matter which row or column is chosen.*

We shall prove first that *the same result is obtained no matter which row is chosen.*

The proof proceeds by induction. The statement is true for a second order determinant (see before). Assuming that it is true for a determinant of order $n - 1$, we prove that it is true for a determinant D of order n.

For this purpose we expand D, given in the definition, in terms of each of two arbitrary rows, say the ith and the jth, and compare the results. Without loss of generality let us assume $i < j$.

First expansion. We expand D by the ith row. A typical term in this expansion is

(4) $\qquad a_{ik}C_{ik} = a_{ik} \cdot (-1)^{i+k}M_{ik}.$

Fig. 112. Cases I and II of the two expansions of D

The minor M_{ik} of a_{ik} in D is an $(n-1)$th order determinant. By the induction hypothesis we may expand it by any row. We expand it by the row corresponding to the jth row of D. This row contains the elements a_{jl} $(l \neq k)$. It is the $(j-1)$th row of M_{ik}, because M_{ik} does not contain elements of the ith row of D, and $i < j$. We have to distinguish between two cases as follows.

Case I. If $l < k$, then the element a_{jl} belongs to the lth column of M_{ik} (cf. Fig. 112). Hence the term involving a_{jl} in this expansion is

$$(5) \qquad a_{jl} \cdot (\text{cofactor of } a_{jl} \text{ in } M_{ik}) = a_{jl} \cdot (-1)^{(j-1)+l} M_{ikjl}$$

where M_{ikjl} is the minor of a_{jl} in M_{ik}. Since this minor is obtained from M_{ik} by deleting the row and column of a_{jl}, it is obtained from D by deleting the ith and jth rows and the kth and lth columns of D. We insert the expansions of the M_{ik} into that of D. Then it follows from (4) and (5) that the terms of the resulting representation of D are of the form

$$(6\text{I}) \qquad\qquad a_{ik}a_{jl} \cdot (-1)^b M_{ikjl} \qquad\qquad (l < k)$$

where

$$b = i + k + j + l - 1.$$

Case II. If $l > k$, the only difference is that then a_{jl} belongs to the $(l-1)$th column of M_{ik}, because M_{ik} does not contain elements of the kth column of D, and $k < l$. This causes an additional minus sign in (5), and, instead of (6I), we therefore obtain

$$(6\text{II}) \qquad\qquad -a_{ik}a_{jl} \cdot (-1)^b M_{ikjl} \qquad\qquad (l > k)$$

where b is the same as before.

Second expansion. We now expand D at first by the jth row. A typical term in this expansion is

$$(7) \qquad\qquad a_{jl}C_{jl} = a_{jl} \cdot (-1)^{j+l} M_{jl}.$$

By the induction hypothesis we may expand the minor M_{jl} of a_{jl} in D by its ith row, which corresponds to the ith row of D, since $j > i$.

Case I. If $k > l$, the element a_{ik} in that row belongs to the $(k-1)$th column of M_{jl}, because M_{jl} does not contain elements of the lth column of D, and $l < k$

(cf. Fig. 112). Hence the term involving a_{ik} in this expansion is

$$(8) \qquad a_{ik} \cdot (\text{cofactor of } a_{ik} \text{ in } M_{jl}) = a_{ik} \cdot (-1)^{i+(k-1)} M_{ikjl},$$

where the minor M_{ikjl} of a_{ik} in M_{jl} is obtained by deleting the ith and jth rows and the kth and lth columns of D [and is, therefore, identical with M_{ikjl} in (5), so that our notation is consistent]. We insert the expansions of the M_{jl} into that of D. It follows from (7) and (8) that this yields a representation whose terms are identical with those given by (6I) when $l < k$.

Case II. If $k < l$, then a_{ik} belongs to the kth column of M_{jl}, we obtain an additional minus sign, and the result agrees with that characterized by (6II).

We have shown that the two expansions of D consist of the same terms, and this proves our statement concerning rows.

The proof of the statement concerning *columns* is quite similar; if we expand D in terms of two arbitrary columns, say, the kth and the lth, we find that the general term involving $a_{jl}a_{ik}$ is exactly the same as before. This proves that not only all column expansions of D yield the same value, but also that their common value is equal to the common value of the row expansions of D.

This completes the proof and shows that *our definition of an nth order determinant is unambiguous.*

Example 1. Let

$$D = \begin{vmatrix} 1 & 3 & 0 \\ 2 & 6 & 4 \\ -1 & 0 & 2 \end{vmatrix}.$$

The development by the first row is

$$D = 1 \begin{vmatrix} 6 & 4 \\ 0 & 2 \end{vmatrix} - 3 \begin{vmatrix} 2 & 4 \\ -1 & 2 \end{vmatrix} = 1(12 - 0) - 3(4 + 4) = -12.$$

The development by the first column is

$$D = 1 \begin{vmatrix} 6 & 4 \\ 0 & 2 \end{vmatrix} - 2 \begin{vmatrix} 3 & 0 \\ 0 & 2 \end{vmatrix} - 1 \begin{vmatrix} 3 & 0 \\ 6 & 4 \end{vmatrix} = 12 - 12 - 12 = -12,$$

etc.

From our definition we may now readily obtain the most important properties of determinants, as follows.

Since the same value is obtained whether we expand a determinant by any row or any column we have

Theorem 1. *The value of a determinant is not altered if its rows are written as columns, in the same order.*

An example is given in formula (10) of the preceding section.

Theorem 2. *If all the elements of one row (or one column) of a determinant are multiplied by the same factor k, the value of the new determinant is k times the value of the given determinant.*

Proof. Expand the determinant by that row (or column) whose elements are multiplied by k.

Note that kD equals a determinant obtained by multiplying the elements of *just one row* (or column) of D by k, whereas in the case of a matrix, $k\mathbf{A}$ equals the matrix obtained by multiplying *all the elements of* $\mathbf{A}$ by k.

Theorem 2 can be used for simplifying a given determinant; cf. the following example.

Example 2.

$$\begin{vmatrix} 1 & 3 & 0 \\ 2 & 6 & 4 \\ -1 & 0 & 2 \end{vmatrix} = 2 \begin{vmatrix} 1 & 3 & 0 \\ 1 & 3 & 2 \\ -1 & 0 & 2 \end{vmatrix} = 6 \begin{vmatrix} 1 & 1 & 0 \\ 1 & 1 & 2 \\ -1 & 0 & 2 \end{vmatrix} = 12 \begin{vmatrix} 1 & 1 & 0 \\ 1 & 1 & 1 \\ -1 & 0 & 1 \end{vmatrix} = -12$$

From Theorem 2, with $k = 0$, or directly, by expanding, we obtain

Theorem 3. *If all the elements of a row (or a column) of a determinant are zero, the value of the determinant is zero.*

Theorem 4. *If each element of a row (or a column) of a determinant is expressed as a binomial, the determinant can be written as the sum of two determinants,* for example,

$$\begin{vmatrix} a_1 + d_1 & b_1 & c_1 \\ a_2 + d_2 & b_2 & c_2 \\ a_3 + d_3 & b_3 & c_3 \end{vmatrix} = \begin{vmatrix} a_1 & b_1 & c_1 \\ a_2 & b_2 & c_2 \\ a_3 & b_3 & c_3 \end{vmatrix} + \begin{vmatrix} d_1 & b_1 & c_1 \\ d_2 & b_2 & c_2 \\ d_3 & b_3 & c_3 \end{vmatrix}.$$

Proof.   Expand the determinant by the row (or column) whose terms are binomials.

Another example is given in (12), Sec. 6.6. The generalization of Theorem 4 to the case of elements which are sums of more than two terms is obvious.

Theorem 5. *If any two rows (or two columns) of a determinant are interchanged, the value of the determinant is multiplied by* -1.

Proof.   Consider the determinant D in the definition. Carry out the development of D whose terms are of the form (6I) and (6II) ("First expansion," p. 245). Let D^* denote the determinant obtained from D by interchanging the ith and jth rows of D. Then the ith row of D^* consists of the elements a_{jl}, $l = 1, \cdots, n$, and the jth row consists of the elements a_{ik}, $k = 1, \cdots, n$. Develop D^* by its ith row and then each occurring $(n - 1)$th order determinant by its row corresponding to the jth row of D^*. The terms of the resulting representation of D^* will be quite similar to those of D, given by (6I) and (6II). However, there is one essential difference. The element a_{jl} now appears already in the first step of the development. Hence, if $k > l$, the element a_{ik} appearing when developing the minor of a_{jl} in D^*, will belong to the $(k - 1)$th column of that minor, because the minor does not contain elements of the lth column of D^* and $k > l$. This causes an additional minus sign in the formula for D^* corresponding to the formula (5) for D, and this minus sign will appear in the formula corresponding to (6I). Otherwise there will be no difference between these two formulas, because they will both contain $a_{ik}a_{jl}$, and M_{ikjl} will be as before, since it does not contain elements from the ith and jth rows of D and D^*, the only two rows by which these determinants differ from each other. Similarly, if $l > k$, an additional minus sign will appear in the formula corresponding to (6II). This completes the proof.

An example is given in formula (11) of the preceding section.

Theorem 6. *If corresponding elements of two rows (or two columns) of a determinant are proportional, the value of the determinant is zero.*

Proof. Let the elements of the ith and jth rows of D be proportional, say, $a_{ik} = ca_{jk}$, $k = 1, \cdots, n$. If $c = 0$, then $D = 0$. Let $c \neq 0$. By Theorem 2,

$$D = cB$$

where the ith and jth rows of B are identical. Interchange these rows. Then, by Theorem 5, B goes over into $-B$. On the other hand, since the rows are identical, the new determinant is still B. Thus $B = -B$, $B = 0$, and $D = 0$.

Example 3.

$$\begin{vmatrix} 3 & 6 & -4 \\ 1 & -1 & 3 \\ -6 & -12 & 8 \end{vmatrix} = 0$$

Before evaluating a determinant it is advisable to simplify it. This may be done by Theorem 2 and

Theorem 7. *The value of a determinant is left unchanged, if the elements of a row (or column) are altered by adding to them any constant multiple of the corresponding elements in any other row (or column).*

Proof. Apply Theorem 4 to the determinant that results from the given addition. This yields a sum of two determinants; one is the original determinant and the other contains two proportional rows. According to Theorem 6, the second determinant is zero, and the proof is complete.

Example 4. Evaluate

$$D = \begin{vmatrix} 1 & 24 & 21 & 93 \\ 2 & -37 & -1 & 194 \\ -2 & 35 & 0 & -171 \\ -3 & 177 & 63 & 234 \end{vmatrix}.$$

Add the second row to the third, add three times the first row to the last, subtract two times the first row from the second, and develop the resulting determinant by the first column:

$$D = \begin{vmatrix} 1 & 24 & 21 & 93 \\ 0 & -85 & -43 & 8 \\ 0 & -2 & -1 & 23 \\ 0 & 249 & 126 & 513 \end{vmatrix} = \begin{vmatrix} -85 & -43 & 8 \\ -2 & -1 & 23 \\ 249 & 126 & 513 \end{vmatrix}.$$

Add three times the first row to the last row:

$$D = \begin{vmatrix} -85 & -43 & 8 \\ -2 & -1 & 23 \\ -6 & -3 & 537 \end{vmatrix}.$$

Subtract two times the second column from the first, and then develop the resulting determinant by the first column:

$$D = \begin{vmatrix} 1 & -43 & 8 \\ 0 & -1 & 23 \\ 0 & -3 & 537 \end{vmatrix} = \begin{vmatrix} -1 & 23 \\ -3 & 537 \end{vmatrix} = -537 + 69 = -468.$$

The determinant of the elements of an *n*-rowed square matrix $\mathbf{A} = (a_{jk})$ is called the **determinant of the matrix A** and will be denoted by det A; thus

$$\det \mathbf{A} = \begin{vmatrix} a_{11} & \cdots & a_{1n} \\ \cdot & \cdots & \cdot \\ \cdot & \cdots & \cdot \\ \cdot & \cdots & \cdot \\ a_{n1} & \cdots & a_{nn} \end{vmatrix}.$$

A square matrix **A** is said to be **singular** if $\det \mathbf{A} = 0$, and **nonsingular** if $\det \mathbf{A} \neq 0$.

We shall now prove the important theorem that the determinant of the product of two *n*-rowed square matrices equals the product of the determinants of the matrices. As a theorem on determinants this means that the product of two *n*th-order determinants can be written as an *n*th-order determinant whose elements are obtained in the same fashion as the elements of a product matrix.

Theorem 8 (Determinant of a product of matrices). *Let* $\mathbf{A} = (a_{ik})$ *and* $\mathbf{B} = (b_{ik})$ *be n-rowed square matrices. Then*

(9) $$\det (\mathbf{AB}) = \det \mathbf{A} \det \mathbf{B},$$

which is equivalent to

(9*) $$\begin{vmatrix} a_{11} & \cdots & a_{1n} \\ \cdot & \cdots & \cdot \\ \cdot & \cdots & \cdot \\ a_{n1} & \cdots & a_{nn} \end{vmatrix} \begin{vmatrix} b_{11} & \cdots & b_{1n} \\ \cdot & \cdots & \cdot \\ \cdot & \cdots & \cdot \\ b_{n1} & \cdots & b_{nn} \end{vmatrix} = \begin{vmatrix} c_{11} & \cdots & c_{1n} \\ \cdot & \cdots & \cdot \\ \cdot & \cdots & \cdot \\ c_{n1} & \cdots & c_{nn} \end{vmatrix},$$

where the element c_{ik} in the ith row and kth column is

(9') $$c_{ik} = a_{i1}b_{1k} + a_{i2}b_{2k} + \cdots + a_{in}b_{nk},$$

the scalar product of the ith row vector of the first determinant and the kth column vector of the second determinant.

Proof of Theorem 8 for $n = 2$.   Let

$$D_1 = \begin{vmatrix} a_{11} & a_{12} \\ a_{21} & a_{22} \end{vmatrix}, \qquad D_2 = \begin{vmatrix} b_{11} & b_{12} \\ b_{21} & b_{22} \end{vmatrix}, \qquad \text{and } D = D_1 D_2.$$

We first show that

(10)
$$D = D_1 D_2 = \begin{vmatrix} a_{11} & a_{12} & 0 & 0 \\ a_{21} & a_{22} & 0 & 0 \\ -1 & 0 & b_{11} & b_{12} \\ 0 & -1 & b_{21} & b_{22} \end{vmatrix}.$$

To prove (10), develop the determinant on the right by the first row and then the resulting two determinants again by their first rows:

$$D = a_{11} \begin{vmatrix} a_{22} & 0 & 0 \\ 0 & b_{11} & b_{12} \\ -1 & b_{21} & b_{22} \end{vmatrix} - a_{12} \begin{vmatrix} a_{21} & 0 & 0 \\ -1 & b_{11} & b_{12} \\ 0 & b_{21} & b_{22} \end{vmatrix}$$

$$= a_{11}a_{22} \begin{vmatrix} b_{11} & b_{12} \\ b_{21} & b_{22} \end{vmatrix} - a_{12}a_{21} \begin{vmatrix} b_{11} & b_{12} \\ b_{21} & b_{22} \end{vmatrix} = (a_{11}a_{22} - a_{12}a_{21})D_2 = D_1 D_2.$$

We now transform the determinant in (10) as follows. We add the third row, multiplied by a_{11}, and the fourth row, multiplied by a_{12}, to the first row; then the first row takes the form

$$0 \qquad 0 \qquad a_{11}b_{11} + a_{12}b_{21} \qquad a_{11}b_{12} + a_{12}b_{22}.$$

We now add the third row, multiplied by a_{21}, and the fourth row, multiplied by a_{22}, to the second row; then the second row becomes

$$0 \qquad 0 \qquad a_{21}b_{11} + a_{22}b_{21} \qquad a_{11}b_{12} + a_{22}b_{22}.$$

Altogether,

$$D = \begin{vmatrix} 0 & 0 & a_{11}b_{11} + a_{12}b_{21} & a_{11}b_{12} + a_{12}b_{22} \\ 0 & 0 & a_{21}b_{11} + a_{22}b_{21} & a_{21}b_{12} + a_{22}b_{22} \\ -1 & 0 & b_{11} & b_{12} \\ 0 & -1 & b_{21} & b_{22} \end{vmatrix}.$$

Developing this determinant by the first column, and then the resulting third-order determinant again by its first column, we obtain

$$D = D_1 D_2 = \begin{vmatrix} a_{11}b_{11} + a_{12}b_{21} & a_{11}b_{12} + a_{12}b_{22} \\ a_{21}b_{11} + a_{22}b_{21} & a_{21}b_{12} + a_{22}b_{22} \end{vmatrix},$$

which is (9*) with $n = 2$.

Proof of Theorem 8 for arbitrary n. The steps are quite similar to those in the case $n = 2$. Let $D_1 = \det \mathbf{A}$, $D_2 = \det \mathbf{B}$, and $D = D_1 D_2$, as before. We first show that

$$
D = \begin{vmatrix}
a_{11} & \cdot & \cdots & a_{1n} & 0 & \cdots & 0 \\
\cdot & \cdot & \cdots & \cdot & \cdot & \cdots & \cdot \\
\cdot & \cdot & \cdots & \cdot & \cdot & \cdots & \cdot \\
a_{n1} & \cdot & \cdots & a_{nn} & 0 & \cdots & 0 \\
-1 & 0 & \cdots & 0 & b_{11} & \cdots & b_{1n} \\
0 & -1 & \cdots & 0 & \cdot & \cdots & \cdot \\
\cdot & \cdot & \cdots & \cdot & \cdot & \cdots & \cdot \\
0 & 0 & \cdots & -1 & b_{n1} & \cdots & b_{nn}
\end{vmatrix}.
$$

To prove this, develop the determinant by the first row, the resulting determinants of order $2n - 1$ again by their first rows, etc. After n steps, the result will be of the form

$$
(\cdots) \begin{vmatrix}
b_{11} & \cdots & b_{1n} \\
\cdot & \cdots & \cdot \\
b_{n1} & \cdots & b_{nn}
\end{vmatrix}
$$

where $(\cdots)$ is the representation of the first of the given determinants in terms of products of its elements. Then D is transformed as follows. Add to the first row

the $(n + 1)$th row multiplied by a_{11},

the $(n + 2)$th row multiplied by a_{12},

· ·

the $2n$th row multiplied by a_{1n}.

The first row will then become

$$
\underbrace{0 \cdots 0}_{\substack{\text{first } n \\ \text{elements}}} \quad \underbrace{a_{11}b_{11} + a_{12}b_{21} + \cdots + a_{1n}b_{n1}}_{(n + 1)\text{th element} = c_{11}} \quad \cdots \quad \underbrace{a_{11}b_{1n} + \cdots + a_{1n}b_{nn}}_{\text{last element} = c_{1n}}
$$

Transforming the second, third, · · ·, nth row in a similar fashion, we altogether obtain

$$
D = \begin{vmatrix}
0 & \cdots & 0 & c_{11} & \cdots & c_{1n} \\
\cdot & \cdots & \cdot & \cdot & \cdots & \cdot \\
0 & \cdots & 0 & c_{n1} & \cdots & c_{nn} \\
-1 & \cdots & 0 & b_{11} & \cdots & b_{1n} \\
\cdot & \cdots & \cdot & \cdot & \cdots & \cdot \\
0 & \cdots & -1 & b_{n1} & \cdots & b_{nn}
\end{vmatrix}
$$

where c_{ik} is given by (9'). Developing this determinant by the first column, the resulting $(2n - 1)$th order determinant again by its first column, etc., the result will be the determinant on the right side of (9*) multiplied by $(-1)^n$ and, if n is odd, by another -1 because then the cofactor of the element -1 in the first column is equal to minus one times the minor of this element. Hence, if n is odd, that determinant is multiplied by the factor $(-1)^{n+1} = +1$, and, if n is even, by $(-1)^n = +1$. This completes the proof.

Example 5.

$$\begin{vmatrix} 2 & 4 & 3 \\ 6 & 10 & 14 \\ 4 & 7 & 9 \end{vmatrix} \begin{vmatrix} 4 & 0 & 5 \\ -2 & 1 & -1 \\ 3 & 0 & 4 \end{vmatrix} = \begin{vmatrix} 9 & 4 & 18 \\ 46 & 10 & 76 \\ 29 & 7 & 49 \end{vmatrix}$$

Problems

Evaluate the following determinants.

1. $\begin{vmatrix} b^2 + c^2 & a^2 & a^2 \\ b^2 & c^2 + a^2 & b^2 \\ c^2 & c^2 & a^2 + b^2 \end{vmatrix}$

2. $\begin{vmatrix} a^2 + 1 & ab & ac \\ ab & b^2 + 1 & bc \\ ac & bc & c^2 + 1 \end{vmatrix}$

3. $\begin{vmatrix} 1 & 2 & -1 & 2 \\ 3 & 0 & 1 & 5 \\ 1 & -2 & 0 & 3 \\ -2 & -4 & 1 & 6 \end{vmatrix}$

4. $\begin{vmatrix} -2 & 8 & -3 & 7 \\ -4 & 1 & 4 & -2 \\ -5 & -8 & 4 & -2 \\ 8 & 16 & -6 & 15 \end{vmatrix}$

5. Prove that the definition of a determinant D of order n yields the same value of D no matter which column is chosen [cf. (3b)]. *Hint.* Model your work somewhat after that for rows given in the text.

6. Carry out the proof of Theorem 2 in detail.

7. Carry out the proof of Theorem 4 in detail.

8. Write the product of the determinants in Probs. 6 and 7, Sec. 6.6, as a determinant.

9. Verify (9) for the matrices in Prob. 13, Sec. 6.4.

10. Show that det $(k\mathbf{A}) = k^n$ det $\mathbf{A}$, where $\mathbf{A}$ is an n-rowed square matrix.

6.8 Rank of a Matrix. Row Equivalence

The concept of the rank of a matrix will become important in connection with the existence and uniqueness of solutions of simultaneous equations. It can be defined in several equivalent ways. We use the following definition.

A matrix $\mathbf{A}$ is said to have **rank** r if it contains at least one r-rowed square submatrix (cf. Sec. 6.1) with a nonzero determinant, whereas the determinant of any square submatrix having $r + 1$ or more rows, possibly contained in $\mathbf{A}$, is zero. $\mathbf{A}$ is said to have *rank* 0 if all its elements are zero.

It is clear that the rank r of an $(m \times n)$ matrix can at most be equal to the smaller of the numbers m and n, but it may be less.

An n-rowed square matrix $\mathbf{A}$ has rank $r < n$ if and only if det $\mathbf{A} = 0$. In this case, $\mathbf{A}$ is called a **singular matrix**. The matrix has rank $r = n$ if and only if det $\mathbf{A} \neq 0$ and is then called a **nonsingular matrix**. Cf. also Sec. 6.7, p. 250.

Example 1. The matrix

$$\mathbf{A} = \begin{pmatrix} 1 & -2 & -1 \\ -3 & 3 & 0 \\ 2 & 2 & 4 \end{pmatrix}$$

is singular because det $\mathbf{A} = 0$. Its rank is 2 because, for example, a two-rowed square submatrix with a nonzero determinant is

$$\begin{pmatrix} 1 & -2 \\ -3 & 3 \end{pmatrix}.$$

Example 2. The matrix

$$\mathbf{A} = \begin{pmatrix} 4 & 2 & 1 & 3 \\ 6 & 3 & 4 & 7 \\ 2 & 1 & 0 & 1 \end{pmatrix}$$

contains the four (3×3) submatrices

$$\begin{pmatrix} 4 & 2 & 1 \\ 6 & 3 & 4 \\ 2 & 1 & 0 \end{pmatrix}, \quad \begin{pmatrix} 4 & 2 & 3 \\ 6 & 3 & 7 \\ 2 & 1 & 1 \end{pmatrix}, \quad \begin{pmatrix} 4 & 1 & 3 \\ 6 & 4 & 7 \\ 2 & 0 & 1 \end{pmatrix}, \quad \begin{pmatrix} 2 & 1 & 3 \\ 3 & 4 & 7 \\ 1 & 0 & 1 \end{pmatrix},$$

whose determinants are zero. Hence rank $\mathbf{A} < 3$. In fact, rank $\mathbf{A} = 2$ because a (2×2) submatrix with nonzero determinant is

$$\begin{pmatrix} 4 & 1 \\ 6 & 4 \end{pmatrix}.$$

Theorem 1 (Rank of transpose). *The rank of the transpose of a matrix is the same as that of the original matrix.*

Proof. Let rank $\mathbf{A} = r$, and let $\mathbf{R}$ be an r-rowed square submatrix of $\mathbf{A}$ with det $\mathbf{R} \neq 0$. Clearly $\mathbf{R}^{\mathsf{T}}$ is a submatrix of $\mathbf{A}^{\mathsf{T}}$. By Theorem 1 in the last section, det $\mathbf{R}^{\mathsf{T}} = $ det $\mathbf{R}$. Thus rank $\mathbf{A}^{\mathsf{T}} \geqq r$. On the other hand, if $\mathbf{A}$ contains an $(r + 1)$-rowed square submatrix $\mathbf{S}$, then, by the definition of rank, det $\mathbf{S} = 0$. Since $\mathbf{S}$ corresponds to $\mathbf{S}^{\mathsf{T}}$ in $\mathbf{A}^{\mathsf{T}}$, and det $\mathbf{S}^{\mathsf{T}} = 0$, it follows that $\mathbf{A}^{\mathsf{T}}$ cannot contain an $(r + 1)$-rowed square submatrix with a nonzero determinant. Consequently, rank $\mathbf{A}^{\mathsf{T}} \leqq r$. Altogether rank $\mathbf{A}^{\mathsf{T}} = r$, and the proof is complete.

We shall now prove that the rank of a matrix is an invariant with respect to elementary operations (cf. Sec. 6.5).

Theorem 2 (Row-equivalence). *Row-equivalent matrices have the same rank.*

Proof. We prove that an elementary operation cannot *raise* the rank r of a matrix $\mathbf{A}$. If $\mathbf{A}$ has maximum possible rank, this is clear. Let r be smaller and consider any $(r + 1)$-rowed square submatrix $\mathbf{B}$ of $\mathbf{A}$. By the definition of rank, det $\mathbf{B} = 0$. Let $\tilde{\mathbf{B}}$ be the matrix obtained from $\mathbf{B}$ by applying an elementary operation to $\mathbf{A}$. If $\tilde{\mathbf{B}} = \mathbf{B}$ then det $\tilde{\mathbf{B}} = 0$. Let $\tilde{\mathbf{B}} \neq \mathbf{B}$. (*a*) Consider the operation of the interchange of two rows of **A**. If both of them have elements in **B**, then det $\tilde{\mathbf{B}} = -\det \mathbf{B} = 0$, by Theorem 5 in Sec. 6.7. Otherwise $\tilde{\mathbf{B}}$ is equal to another $(r + 1)$-rowed square submatrix of **A** (with one row possibly not in the natural order). Hence det $\tilde{\mathbf{B}} = 0$. (*b*) Under the elementary operation of the multiplication of a row of **A** by a nonzero number k we have det $\tilde{\mathbf{B}} = k \det \mathbf{B} = 0$. (*c*) Under the addition of a constant multiple of a row of **A** to another row the matrix $\tilde{\mathbf{B}}$ differs from **B** by a row, say, $\tilde{\mathbf{b}}$, of the form $\tilde{\mathbf{b}} = \mathbf{b} + k\mathbf{c}$, where **b** is the corresponding row of **B**. From Theorems 2 and 4 in Sec. 6.7 we thus have det $\tilde{\mathbf{B}} = \det \mathbf{B} + k \det \mathbf{C} = k \det \mathbf{C}$, where **C** either has two identical rows (if **c** is in **B**) or is an $(r + 1)$-rowed square submatrix of the matrix **A** (with one row possibly not in the natural order). In either case, det $\mathbf{C} = 0$, and det $\tilde{\mathbf{B}} = 0$. This completes the proof that an elementary operation cannot *raise* the rank of a matrix **A**. Let us now show that an elementary operation cannot *lower* that rank. By the *inverse* of an elementary operation we mean the operation which undoes the effect of the given operation. Such an inverse exists and is an elementary operation. Hence, if an elementary operation could lower the rank, its inverse would raise the rank, but this is impossible by what has just been proved. This completes the proof of Theorem 2.

Obviously, Theorem 2 implies that if we want to determine the rank of a given matrix, we may first simplify the matrix by elementary operations. More important, the theorem will be the basis of our discussion of existence, uniqueness, and structure of solutions of simultaneous linear equations in the next section.

Theorem 3.   *The rank of a matrix equals the maximum number of linearly independent row vectors or column vectors.*

This will be proved in Sec. 6.10. (For linear dependence of vectors see Sec. 5.4.)

Problems

Find all submatrices of the following matrices.

1. $\begin{pmatrix} 3 & -1 \\ 2 & 0 \\ 4 & 5 \end{pmatrix}$
   2. $\begin{pmatrix} 1 & 3 & 5 \\ 0 & -2 & 6 \\ 4 & 0 & 7 \end{pmatrix}$
   3. $\begin{pmatrix} 4 & 2 & 1 & 3 \\ 0 & 5 & 7 & 0 \\ 1 & 9 & 7 & 1 \end{pmatrix}$

Determine the rank of the following matrices.

4. $\begin{pmatrix} -4 & 1 \\ 3 & 2 \end{pmatrix}$
   5. $\begin{pmatrix} 1 & 7 & -1 \\ 3 & 0 & 4 \end{pmatrix}$
   6. $\begin{pmatrix} 4 & 2 & 0 & 3 \\ 6 & 1 & 0 & 0 \end{pmatrix}$

7. $\begin{pmatrix} 4 & 2 \\ 1 & 0 \\ -3 & 10 \end{pmatrix}$
   8. $\begin{pmatrix} 3 & 1 & 4 \\ 0 & 5 & 8 \\ -3 & 4 & 4 \end{pmatrix}$
   9. $\begin{pmatrix} 8 & 1 & 3 & 6 \\ 0 & 3 & 2 & 2 \\ -8 & -1 & -3 & 4 \end{pmatrix}$

10. (*a*) Show that the product of two $(n \times n)$ matrices is singular if and only if at least one of the two matrices is singular. (*b*) Show that the sum of two nonsingular $(n \times n)$ matrices may be singular, and the sum of two singular matrices may be nonsingular.

6.9 *Systems of Linear Equations: Existence and General Properties of Solutions*

A practical method for solving systems of linear equations is given in Sec. 6.5. We know that such a system may have no solutions, precisely one solution, or infinitely many solutions, and we shall now formulate necessary and sufficient conditions for the existence of solutions. We shall also characterize basic general properties of sets of solutions.

Fundamental Theorem 1. *A system of m linear equations*

(1)
$$a_{11}x_1 + a_{12}x_2 + \cdots + a_{1n}x_n = b_1$$
$$a_{21}x_1 + a_{22}x_2 + \cdots + a_{2n}x_n = b_2$$
$$\cdots\cdots\cdots\cdots\cdots\cdots\cdots\cdots\cdots\cdots\cdots\cdots$$
$$a_{m1}x_1 + a_{m2}x_2 + \cdots + a_{mn}x_n = b_m$$

in n unknowns $x_1, \cdots, x_n$ has solutions if and only if the coefficient matrix $\mathbf{A}$ and the augmented matrix $\mathbf{B}$, that is (cf. *also Sec.* 6.5)

$$\mathbf{A} = \begin{pmatrix} a_{11} & \cdots & a_{1n} \\ \cdot & \cdots & \cdot \\ \cdot & \cdots & \cdot \\ a_{m1} & \cdots & a_{mn} \end{pmatrix} \quad and \quad \mathbf{B} = \begin{pmatrix} a_{11} & \cdots & a_{1n} & b_1 \\ \cdot & \cdots & \cdot & \cdot \\ \cdot & \cdots & \cdot & \cdot \\ a_{m1} & \cdots & a_{mn} & b_m \end{pmatrix},$$

have the same rank. If this rank r equals n, there is precisely one solution. If $r < n$, there are infinitely many solutions which are obtained by determining r suitable unknowns (*whose submatrix of coefficients must have rank r*) *in terms of the remaining $n - r$ unknowns, to which arbitrary values can be assigned. If solutions exist, they can all be obtained by the Gauss elimination* (cf. *Sec.* 6.5). (This elimination may be started without first looking at the ranks of $\mathbf{A}$ and $\mathbf{B}$.)

Proof.   To the given system we apply the Gauss elimination (Sec. 6.5), but modify it as follows. We may assume that the order of the equations and the order of the unknowns in each equation are such that $a_{11} \neq 0$. [If necessary, we first rearrange those orders; this does not affect the ranks of the matrices (by Theorems 1 and 2 in Sec. 6.8) or the set of solutions (by Theorem 1 in Sec. 6.5).] We divide the first equation by a_{11} and then eliminate x_1 from equations 2 through m. For reasons similar to those given before, we may assume that x_2 in the resulting second equation has a nonzero coefficient. We divide the equation by this coefficient and then eliminate x_2 from the new equations 3 through m *as well as from the old first equation.* (This is the aforementioned modification of the procedure in Sec. 6.5.) In the next step, we eliminate x_3 from all but the third equation. Continuing in this fashion, we come to an end after finitely many steps when we have used all m equations or when the remaining equations contain no more variables. The resulting system is similar to (6) in Sec. 6.5, but even simpler. It is of the form

$$x_1 + \qquad \tilde{a}_{1,r+1}x_{r+1} + \tilde{a}_{1,r+2}x_{r+2} + \cdots + \tilde{a}_{1n}x_n = \tilde{b}_1$$
$$x_2 + \qquad \tilde{a}_{2,r+1}x_{r+1} + \tilde{a}_{2,r+2}x_{r+2} + \cdots + \tilde{a}_{2n}x_n = \tilde{b}_2$$

(2)
$$x_r + \tilde{a}_{r,r+1}x_{r+1} + \tilde{a}_{r,r+2}x_{r+2} + \cdots + \tilde{a}_{rn}x_n = \tilde{b}_r$$
$$0 = \tilde{b}_{r+1}$$

$$0 = \tilde{b}_m$$

where $r \leqq m$, and if $r < m$, then $\tilde{b}_{r+1}, \cdots, \tilde{b}_m$ may or may not be all zero. This is called the **echelon form** of a system of linear equations. Let $\tilde{\mathbf{A}}$ be the coefficient matrix of (2) and $\tilde{\mathbf{B}}$ the augmented matrix of (2). Theorem 2 in Sec. 6.8 implies that rank $\tilde{\mathbf{A}}$ = rank $\mathbf{A}$ and rank $\tilde{\mathbf{B}}$ = rank $\mathbf{B}$. We see that if at least one of $\tilde{b}_{r+1}$, $\cdots, \tilde{b}_m$ is not zero, then rank $\tilde{\mathbf{B}} >$ rank $\tilde{\mathbf{A}}$, and rank $\mathbf{B} \neq$ rank $\mathbf{A}$. We also see that then the system is inconsistent and has no solution because at least one of the equations $0 = \tilde{b}_{r+1}, \cdots, 0 = \tilde{b}_m$ is false. On the other hand, if $\tilde{b}_{r+1}, \cdots, \tilde{b}_m$ are all zero, then rank $\tilde{\mathbf{B}}$ = rank $\tilde{\mathbf{A}} = r$, hence rank $\mathbf{B}$ = rank $\mathbf{A} = r$. In this case, if $r = n$, we have a unique solution, and if $r < n$, we obtain infinitely many solutions by assigning any values to $x_{r+1}, \cdots, x_n$ and determining $x_1, \cdots, x_r$ from equations 1 through r in (2). This completes the proof.

If (1) has solutions, they can be obtained by the Gauss elimination (Sec. 6.5) and, to avoid misunderstandings, let us say again that the elimination can be started without first looking at the rank of the matrices $\mathbf{A}$ and $\mathbf{B}$. If we obtain a solution, then $\mathbf{A}$ and $\mathbf{B}$ must have the same rank, and if we arrive at a contradiction, these ranks are different and no solution exists.

The system in (1) is said to be **homogeneous** if all the b_j's on the right-hand side are zero. Otherwise it is said to be **nonhomogeneous**. (Cf. also Sec. 6.5.) From Fundamental Theorem 1 we readily obtain the following results.

Theorem 2 (Homogeneous system). *A homogeneous system of linear equations*

(3)
$$a_{11}x_1 + a_{12}x_2 + \cdots + a_{1n}x_n = 0$$
$$a_{21}x_1 + a_{22}x_2 + \cdots + a_{2n}x_n = 0$$
$$a_{m1}x_1 + a_{m2}x_2 + \cdots + a_{mn}x_n = 0$$

always has the **trivial solution** $x_1 = 0, \cdots, x_n = 0$. *Nontrivial solutions exist if and only if* rank $\mathbf{A}$ *is less than* n. *If rank* $\mathbf{A} = r < n$, *the solutions form a vector space of dimension* $n - r$ *(cf. Sec. 5.4). In particular, if* $\mathbf{x}_{(1)}$ *and* $\mathbf{x}_{(2)}$ *are solution vectors of (3), then* $\mathbf{x} = c_1\mathbf{x}_{(1)} + c_2\mathbf{x}_{(2)}$, *where* c_1 *and* c_2 *are any constants, is a solution vector of (3). (This* **does not hold** *for nonhomogeneous systems.)*

Proof. The first proposition is obvious and is in agreement with the fact that for a homogeneous system the matrix of the coefficients and the augmented matrix have the same rank. The solution vectors form a vector space because if $\mathbf{x}_{(1)}$ and $\mathbf{x}_{(2)}$ are any of them, then $\mathbf{Ax}_{(1)} = \mathbf{0}$, $\mathbf{Ax}_{(2)} = \mathbf{0}$, and this implies $\mathbf{A}(\mathbf{x}_{(1)} + \mathbf{x}_{(2)}) = \mathbf{Ax}_{(1)} + \mathbf{Ax}_{(2)} = \mathbf{0}$ as well as $\mathbf{A}(c\mathbf{x}_{(1)}) = c\mathbf{Ax}_{(1)} = \mathbf{0}$, where c is arbitrary. If rank $\mathbf{A} = r < n$, then Fundamental Theorem 1 implies that we can choose $n - r$ suitable unknowns, call them $x_{r+1}, \cdots, x_n$, in an arbitrary fashion, and any solution is obtained in this way. It follows that a basis of solutions is $\mathbf{y}_{(1)}, \cdots,$ $\mathbf{y}_{(n-r)}$, where the solution vector $\mathbf{y}_{(j)}$, $j = 1, \cdots, n - r$, is obtained by choosing $x_{r+j} = 1$ and the other $x_{r+1}, \cdots, x_n$ zero; the corresponding $x_1, \cdots, x_r$ are then determined. This proves that the vector space of all solutions has dimension $n - r$ and completes the proof.

The next two theorems are special cases of Theorem 2, which are particularly important in applications.

Theorem 3. *A homogeneous system of n linear equations in the same number of unknowns has nontrivial solutions if and only if the matrix of the coefficients is singular* (*that is, its determinant is zero; cf. p.* 250).

Theorem 4. *A homogeneous system of linear equations with fewer equations than unknowns always has nontrivial solutions.*

If a nonhomogeneous system of linear equations has solutions, their totality can be characterized as follows.

Theorem 5 (Nonhomogeneous system). *If a nonhomogeneous system of linear equations of the form* (1) *has solutions, then all these solutions are of the form*

$$\mathbf{x} = \mathbf{x}_0 + \mathbf{x}_h$$

where $\mathbf{x}_0$ *is any fixed solution of* (1) *and* $\mathbf{x}_h$ *runs through all the solutions of the corresponding homogeneous system* (3).

Proof. Let $\mathbf{x}$ be any given solution of (1) and $\mathbf{x}_0$ an arbitrarily chosen solution of (1). Then $\mathbf{Ax} = \mathbf{b}$, $\mathbf{Ax}_0 = \mathbf{b}$ and, therefore,

$$\mathbf{A}(\mathbf{x} - \mathbf{x}_0) = \mathbf{Ax} - \mathbf{Ax}_0 = \mathbf{0}.$$

This shows that the difference $\mathbf{x} - \mathbf{x}_0$ between any solution $\mathbf{x}$ of (1) and any fixed solution $\mathbf{x}_0$ of (1) is a solution of (3), say, $\mathbf{x}_h$. Hence all solutions of (1) are obtained by letting $\mathbf{x}_h$ run through all the solutions of the homogeneous system (3), and the proof is complete.

6.10   Linear Dependence and Rank. Singular Matrices

Linear dependence and independence of vectors has been defined in Sec. 5.4, and we can now prove the following useful criterion.

Theorem 1 (Linear dependence and independence).   *p vectors* $\mathbf{x}_{(1)}, \cdots, \mathbf{x}_{(p)}$ *are linearly independent if the matrix with column vectors* $\mathbf{x}_{(1)}, \cdots, \mathbf{x}_{(p)}$ *has rank p; they are linearly dependent if that rank is less than p.*

Proof.   Those vectors are linearly dependent if there are numbers $k_1, \cdots, k_p$, not all zero, such that

$$k_1 \mathbf{x}_{(1)} + \cdots + k_p \mathbf{x}_{(p)} = \mathbf{0}.$$

Denoting the components of the vector $\mathbf{x}_{(j)}$ by $x_{1j}, \cdots, x_{nj}$, we may write this vector equation in the form

$$
\begin{aligned}
x_{11}k_1 + \cdots + x_{1p}k_p &= 0 \\
x_{21}k_1 + \cdots + x_{2p}k_p &= 0 \\
&\cdots\cdots\cdots\cdots\cdots\cdots \\
x_{n1}k_1 + \cdots + x_{np}k_p &= 0
\end{aligned}
$$

In this system we may regard $k_1, \cdots, k_p$ as unknowns. From Theorem 2 in Sec. 6.9 it follows that the system has a nontrivial solution $k_1, \cdots, k_p$ if, and only if, the coefficient matrix has rank less than p. From this the statements follow.

An immediate consequence of Theorem 1 is:

Theorem 2.   *p vectors with* $n < p$ *components are always linearly dependent.*

In fact, then the coefficient matrix of the system considered just before has rank $r \leqq n < p$, and this proves Theorem 2.

By combining Theorem 1 (above) and Theorem 1 in Sec. 6.8, we may now characterize the rank of a matrix in terms of linearly independent column vectors or row vectors; this characterization is often used as a definition of the rank.

Theorem 3 (Rank of a matrix).   *The rank of a matrix equals the maximum number of linearly independent column vectors or row vectors of the matrix.*

As another important application of our results obtained in the preceding section, let us consider matrix multiplication and use Theorem 3, Sec. 6.9, to obtain some more information about the strange fact that the cancellation law is not true, in general; that is

$$\mathbf{AB} = \mathbf{0} \qquad \text{does not necessarily imply} \qquad \mathbf{A} = \mathbf{0} \text{ or } \mathbf{B} = \mathbf{0}.$$

This was stated in Theorem 1 of Sec. 6.4 and illustrated by the example

$$\begin{pmatrix} 1 & 1 \\ 2 & 2 \end{pmatrix}\begin{pmatrix} -1 & 1 \\ 1 & -1 \end{pmatrix} = \begin{pmatrix} 0 & 0 \\ 0 & 0 \end{pmatrix}.$$

From Theorem 8 in Sec. 6.7, it follows that if $\mathbf{A}$ and $\mathbf{B}$ are n-rowed square matrices such that $\mathbf{AB} = \mathbf{0}$, then $\mathbf{A}$ or $\mathbf{B}$ must be singular (why?). In our example, even both matrices are singular, and we shall now prove that this is typical.

Theorem 4.   *If* $\mathbf{A}$ *and* $\mathbf{B}$ *are n-rowed square matrices, not zero matrices, such that* $\mathbf{AB} = \mathbf{0}$, *then both* $\mathbf{A}$ *and* $\mathbf{B}$ *are singular.*

Proof. Let $\mathbf{B} = (\mathbf{b}_1 \mathbf{b}_2 \cdots \mathbf{b}_n)$ be the representation of $\mathbf{B}$ in terms of column vectors. Then from (13) in Sec. 6.4 and $\mathbf{AB} = \mathbf{0}$ we have

$$\mathbf{AB} = (\mathbf{Ab}_1 \cdots \mathbf{Ab}_n) = (\mathbf{0} \cdots \mathbf{0}),$$

that is

(1) $$\mathbf{Ab}_k = \mathbf{0}, \qquad\qquad\qquad k = 1, \cdots, n,$$

where $\mathbf{0}$ designates a column vector of the $(n \times n)$ zero matrix. Since $\mathbf{B}$ is not a zero matrix, $\mathbf{b}_k \neq \mathbf{0}$ for at least one k. Regarding the corresponding vector equation (1) as a system of n linear equations in the "unknowns" $b_{1k}, \cdots, b_{nk}$, the components of $\mathbf{b}_k$, we conclude that $\mathbf{A}$ must be singular, because this is a necessary and sufficient condition for the existence of a nontrivial solution (Theorem 3 in Sec. 6.9). The proof that $\mathbf{B}$ must be singular follows by applying the same argument to the transposed equation

$$\mathbf{B}^{\mathsf{T}}\mathbf{A}^{\mathsf{T}} = \mathbf{0},$$

obtained from $\mathbf{AB} = \mathbf{0}$ by the use of (12) in Sec. 6.4.

By interchanging the factors in the preceding example we obtain

$$\begin{pmatrix} -1 & 1 \\ 1 & -1 \end{pmatrix}\begin{pmatrix} 1 & 1 \\ 2 & 2 \end{pmatrix} = \begin{pmatrix} 1 & 1 \\ -1 & -1 \end{pmatrix}.$$

Hence, if both $\mathbf{A}$ and $\mathbf{B}$ are singular and $\mathbf{AB} = \mathbf{0}$, it does *not* follow that $\mathbf{BA} = \mathbf{0}$. From the proof of Theorem 4 we immediately have

Theorem 5. *If* $\mathbf{AB} = \mathbf{0}$ *and* $\mathbf{A}$ *is nonsingular, then* $\mathbf{B} = \mathbf{0}$.

Furthermore, if $\mathbf{AB} = \mathbf{0}$ and we set $\mathbf{B} = \mathbf{S} - \mathbf{T}$, then

$$\mathbf{AB} = \mathbf{A}(\mathbf{S} - \mathbf{T}) = \mathbf{0} \qquad \text{or} \qquad \mathbf{AS} = \mathbf{AT}.$$

From this and Theorem 5 we obtain

Theorem 6. *If* $\mathbf{A}$ *is nonsingular, the matrix equation*

$$\mathbf{AS} = \mathbf{AT} \qquad implies \qquad \mathbf{S} = \mathbf{T}.$$

Problems

In each case, find whether the given vectors are linearly dependent or independent.

1. $(2, 1), (1, 2)$
2. $(3, 4), (-9, -12)$
3. $(0, 0), (2, -2)$
4. $(-3, 7, 8), (0, 2, 1)$
5. $(2, 0, 4), (3, 0, 2), (8, 0, -1)$
6. $(0, 1, 0), (1, 0, 0), (1, 1, 0)$
7. $(3, 2, 7), (2, 4, 1), (1, -2, 6)$
8. $(1, 2, 3), (4, -1, 6), (5, 1, 9)$
9. $(1, 2, 3), (2, 3, 1), (3, 1, 2)$
10. $(2, -1, 3), (3, -2, 1), (5, 3, 4)$

11. Find an $\mathbf{x}$ such that $\mathbf{x}$, $(1, 0, 0, 0), (0, 1, 0, 0), (0, 0, 1, 0)$ are linearly independent.
12. For what values of k are $(1, 2, 3), (4, 5, 6), (7, 8, k)$ linearly dependent?
13. For what values of k are $(-4, -3, -2), (-1, k, 1), (2, 3, 4)$ linearly dependent?
14. Show that in Theorem 1, "column vectors" may be replaced by "row vectors."
15. Let $\mathbf{A}$ and $\mathbf{B}$ be n-rowed square matrices such that $\mathbf{AB} = \mathbf{0}$. Using Theorem 8 in Sec. 6.7, show that then $\mathbf{A}$ or $\mathbf{B}$ is singular.

6.11   *Systems of Linear Equations:* *Solution by Determinants*

The definition of a determinant of order n (Sec. 6.7) is suggested by systems of n linear equations in n unknowns, and in the present section we shall derive *Cramer's rule*, which represents solutions as quotients of determinants. For numerical work, other formulas are preferable (cf. Secs. 6.5, 18.8, and 18.9), but Cramer's rule is of practical interest, for instance, if the equations contain parameters.

Cramer's theorem.   *If the determinant $D = \det \mathbf{A}$ of a system of n linear equations*

(1)
$$
\begin{aligned}
a_{11}x_1 + a_{12}x_2 + \cdots + a_{1n}x_n &= b_1 \\
a_{21}x_1 + a_{22}x_2 + \cdots + a_{2n}x_n &= b_2 \\
&\cdots\cdots\cdots\cdots\cdots\cdots\cdots\cdots \\
a_{n1}x_1 + a_{n2}x_2 + \cdots + a_{nn}x_n &= b_n
\end{aligned}
$$

in the same number of unknowns $x_1, \cdots, x_n$ is not zero, the system has precisely one solution. This solution is given by the formulas

(2)
$$
x_1 = \frac{D_1}{D}, \qquad x_2 = \frac{D_2}{D}, \cdots, \qquad x_n = \frac{D_n}{D}
$$
(Cramer's rule)

where D_k is the determinant obtained from D by replacing in D the kth column by the column with the elements $b_1, \cdots, b_n$.

   Hence if (1) *is homogeneous and $D \neq 0$, it has only the trivial solution $x_1 = 0$, $x_2 = 0, \cdots, x_n = 0$.*

Proof.   From Fundamental Theorem 1 in Sec. 6.9 it follows that (1) has a unique solution, because rank $\mathbf{A} \leqq$ rank $\mathbf{B} \leqq n$, and

$$
D = \det \mathbf{A} =
\begin{vmatrix}
a_{11} & \cdots & a_{1n} \\
\cdot & \cdots & \cdot \\
\cdot & \cdots & \cdot \\
a_{n1} & \cdots & a_{nn}
\end{vmatrix} \neq 0
$$

implies rank $\mathbf{A} = n$ (cf. Sec. 6.8), so that rank $\mathbf{A} =$ rank $\mathbf{B} = n$. We prove (2). If we develop D by the kth column, we obtain

(3)
$$
D = a_{1k}C_{1k} + a_{2k}C_{2k} + \cdots + a_{nk}C_{nk}
$$

where C_{ik} is the cofactor of the element a_{ik} in D. If we replace the elements of the kth column of D by any other numbers, we obtain a new determinant, say, $\tilde{D}$. Clearly, its development by the kth column will be of the form (3), with $a_{1k}, \cdots,$ a_{nk} replaced by those new elements and the cofactors C_{ik} as before. In particular, if we choose as new elements the elements $a_{1l}, \cdots, a_{nl}$ of the lth column of D

(where $l \neq k$), then the development of the resulting determinant $\tilde{D}$ becomes

(4) $$a_{1l}C_{1k} + a_{2l}C_{2k} + \cdots + a_{nl}C_{nk} = 0 \qquad (l \neq k),$$

because $\tilde{D}$ has two identical columns and is zero (cf. Theorem 6 in Sec. 6.7). If we multiply the first equation in (1) by C_{1k}, the second by $C_{2k}, \cdots$, the last by C_{nk} and add the resulting equations, we first obtain

$$C_{1k}(a_{11}x_1 + \cdots + a_{1n}x_n) + \cdots + C_{nk}(a_{n1}x_1 + \cdots + a_{nn}x_n)$$
$$= b_1 C_{1k} + \cdots + b_n C_{nk}.$$

The expression on the left may be written

$$x_1(a_{11}C_{1k} + \cdots + a_{n1}C_{nk}) + \cdots + x_n(a_{1n}C_{1k} + \cdots + a_{nn}C_{nk}).$$

From (3) we conclude that the factor of x_k in this representation is equal to D, and from (4) it follows that the factor of x_l ($l \neq k$) is zero. Thus

$$x_k D = b_1 C_{1k} + b_2 C_{2k} + \cdots + b_n C_{nk}.$$

Since $D \neq 0$, we may divide, finding

(5) $$x_k = \frac{1}{D}(b_1 C_{1k} + b_2 C_{2k} + \cdots + b_n C_{nk}) = \frac{D_k}{D}$$

where $k = 1, \cdots, n$. This is identical with (2), and the proof is complete.

Remark. If in this proof we do not want to use Fundamental Theorem 1, we may reason as follows. If (1) has a solution, it must be given by (5). It remains to show that (5) does satisfy (1). Developing D by the ith row, we have

(6) $$D = a_{i1}C_{i1} + a_{i2}C_{i2} + \cdots + a_{in}C_{in}.$$

If we replace the ith row of D by the jth row, without changing the other rows (including the jth), the resulting determinant is zero, and by developing it by the row which has been changed we obtain

(7) $$a_{j1}C_{i1} + a_{j2}C_{i2} + \cdots + a_{jn}C_{in} = 0 \qquad (j \neq i).$$

By substituting (5) into the ith equation we have

$$a_{i1}x_1 + \cdots + a_{in}x_n$$
$$= \frac{1}{D}[a_{i1}(b_1 C_{11} + \cdots + b_n C_{n1}) + \cdots + a_{in}(b_1 C_{1n} + \cdots + b_n C_{nn})]$$
$$= \frac{1}{D}[b_1(a_{i1}C_{11} + \cdots + a_{in}C_{1n}) + \cdots + b_n(a_{i1}C_{n1} + \cdots + a_{in}C_{nn})].$$

From (6) we see that in this representation, b_i is multiplied by D, and from (7) it follows that b_j ($j \neq i$) is multiplied by zero. Hence the expression on the right reduces to $(1/D)b_i D = b_i$, which means that (5) satisfies (1). This completes the extended proof.

An example is included in Sec. 6.6.

We want to mention that it is possible (although not very practical) to apply Cramer's rule to systems of m linear equations

$$
\begin{aligned}
a_{11}x_1 + a_{12}x_2 + \cdots + a_{1n}x_n &= b_1 \\
a_{21}x_1 + a_{22}x_2 + \cdots + a_{2n}x_n &= b_2 \\
&\cdots\cdots\cdots\cdots\cdots \\
a_{m1}x_1 + a_{m2}x_2 + \cdots + a_{mn}x_n &= b_m
\end{aligned}
$$

in n unknowns. If the matrix of the coefficients $\mathbf{A}$ and the augmented matrix $\mathbf{B}$ both have rank r, we know from Fundamental Theorem 1 that we may assign arbitrary values to $n - r$ suitable unknowns, call them $x_{r+1}, \cdots, x_n$, such that the submatrix of the coefficients of the other unknowns $x_1, \cdots, x_r$ has rank r. Theorem 3 in Sec. 6.10 implies that $\mathbf{A}$ and $\mathbf{B}$ have r linearly independent row vectors, say the first r row vectors (after a rearrangement if necessary), and if $r < m$, then any of the other row vectors is a linear combination of those. It follows that the corresponding $m - r$ equations can be reduced to the form $0 = 0$ by elementary operations. From this and Theorem 1 in Sec. 6.5 we see that we may omit those $m - r$ equations from our system. We can now write the reduced system in the form

$$
\begin{aligned}
a_{11}x_1 + \cdots + a_{1r}x_r &= b_1 - (a_{1,r+1}x_{r+1} - \cdots + a_{1n}x_n) \\
a_{21}x_1 + \cdots + a_{2r}x_r &= b_2 - (a_{2,r+1}x_{r+1} + \cdots + a_{2n}x_n) \\
&\cdots\cdots\cdots\cdots\cdots\cdots\cdots\cdots\cdots \\
a_{r1}x_1 + \cdots + a_{rr}x_r &= b_r - (a_{r,r+1}x_{r+1} + \cdots + a_{rn}x_n)
\end{aligned}
$$

(where, if $r = n$, the expressions on the right are $b_1, \cdots, b_r$) and solve it for $x_1, \cdots, x_r$ by Cramer's rule.

Example 1.   In the case of the system

$$
\begin{aligned}
3x_1 + 2x_2 + 2x_3 - 5x_4 &= 8 \\
2x_1 + 5x_2 + 5x_3 - 18x_4 &= 9 \\
4x_1 - x_2 - x_3 + 8x_4 &= 7
\end{aligned}
$$

the matrix of the coefficients and the augmented matrix have rank 2. We may omit the last equation (why?) and write the reduced system in the form

$$
\begin{aligned}
3x_1 + 2x_2 &= 8 - 2x_3 + 5x_4 \\
2x_1 + 5x_2 &= 9 - 5x_3 + 18x_4.
\end{aligned}
$$

Then Cramer's rule gives $x_1 = 2 - x_4$, $x_2 = 1 - x_3 + 4x_4$, where x_3 and x_4 are arbitrary.

Problems

Solve the following systems of linear equations, using (a) Cramer's rule, (b) the Gauss elimination

1. $\quad 4x - y = 7$
$\quad -2x + 5y = 1$

2. $\quad 3x - 2y = 9$
$\quad -x + 6y = -3$

3. $\quad x + 2y = 6$
$\quad 3x + 2y = 10$

4. $\quad 2x + 3y = -3$
$\quad 3x - 2y = 15$

5. $\quad x + y + z = 0$
$\quad 2x + 5y + 3z = 1$
$\quad -x + 2y + z = 2$

6. $\quad 2x + y + 2z = -1$
$\quad x \quad\;\; + z = -1$
$\quad -x + 3y - 2z = 7$

7. $\quad x - y + 2z = 2$
$\quad 3x + y - z = 3$
$\quad 2x + 2y - 3z = 1$

8. $3x + y + 2z = 6$
$-2x - 2y + 2z = 0$
$4x + 5y + 6z = 22$

9. $x \quad\quad + 2z = 2$
$3x + y - 4z = -4$
$5x + 2y \quad\quad = 0$

10. $2x + y + 2z = 2$
$x + y + z = 0$
$3x + 2y + 4z = 1$

11. Using Kirchhoff's laws, show that the equations for the unknown currents I_1, I_2, I_3 in Fig. 113 are

$$I_1 - I_2 - I_3 = 0, \quad R_2I_2 - R_3I_3 = 0, \quad (R_1 + R_4)I_1 + R_2I_2 = E_0$$

and determine these currents in terms of R_1, R_2, R_3, R_4, and E_0.

Fig. 113. Network in Problem 11

Using Kirchhoff's laws, find the currents in the following networks.

12.

13.

14.

15. (Wheatstone bridge) Show that if $R_x/R_3 = R_1/R_2$ in Fig. 114, then $I = 0$. (R_0 is the resistance of the instrument by which I is measured.)

16. Show that the straight line through two given points P_1: (x_1, y_1) and P_2: (x_2, y_2) in the xy-plane is

$$\begin{vmatrix} x & y & 1 \\ x_1 & y_1 & 1 \\ x_2 & y_2 & 1 \end{vmatrix} = 0.$$

Fig. 114. Wheatstone bridge

17. Find the straight line through the points $(3, 5)$ and $(-4, 8)$.

18. Derive from the formula in Prob. 16 the familiar formula

$$\frac{x - x_1}{x_1 - x_2} = \frac{y - y_1}{y_1 - y_2}.$$

19. Show that three points (x_1, y_1), (x_2, y_2), and (x_3, y_3) lie on a straight line if and only if

$$\begin{vmatrix} x_1 & y_1 & 1 \\ x_2 & y_2 & 1 \\ x_3 & y_3 & 1 \end{vmatrix} = 0.$$

20. Show that the equation of the circle through three given points (x_1, y_1), (x_2, y_2), and (x_3, y_3) in the xy-plane is

$$\begin{vmatrix} x^2 + y^2 & x & y & 1 \\ x_1^2 + y_1^2 & x_1 & y_1 & 1 \\ x_2^2 + y_2^2 & x_2 & y_2 & 1 \\ x_3^2 + y_3^2 & x_3 & y_3 & 1 \end{vmatrix} = 0.$$

6.12 *The Inverse of a Matrix*

The **inverse** of an n-rowed square matrix $\mathbf{A} = (a_{jk})$ is denoted by $\mathbf{A}^{-1}$ and is an n-rowed square matrix such that

(1) $$\mathbf{A}\mathbf{A}^{-1} = \mathbf{A}^{-1}\mathbf{A} = \mathbf{I},$$

where $\mathbf{I}$ is the n-rowed unit matrix. $\mathbf{A}^{-1}$ exists (and is then unique) if and only if $\mathbf{A}$ is nonsingular (cf. p. 250), that is, if and only if det $\mathbf{A} \neq 0$. It is then given by the formula

$$
(2) \qquad \mathbf{A}^{-1} = \frac{1}{\det \mathbf{A}} \begin{pmatrix} A_{11} & A_{21} & \cdots & A_{n1} \\ A_{12} & A_{22} & \cdots & A_{n2} \\ \cdot & \cdot & \cdots & \cdot \\ A_{1n} & A_{2n} & \cdots & A_{nn} \end{pmatrix},
$$

where A_{jk} is the cofactor (Sec. 6.7) of a_{jk} in $\det \mathbf{A}$. Note that in $\mathbf{A}^{-1}$ the cofactor A_{jk} occupies the same place as a_{kj} (not a_{jk}) does in $\mathbf{A}$.

Let us motivate the concept of an inverse and prove the above proposition about the existence as well as the formula (2). For this purpose we start from $\mathbf{A}$ and consider a corresponding **linear transformation**

$$
(3) \qquad \mathbf{y} = \mathbf{Ax},
$$

where $\mathbf{x}$ and $\mathbf{y}$ are column vectors with n components $x_1, \cdots, x_n$ and $y_1, \cdots, y_n$, respectively. If the inverse matrix $\mathbf{A}^{-1}$ exists, we may premultiply (3) by $\mathbf{A}^{-1}$ and have $\mathbf{A}^{-1}\mathbf{y} = \mathbf{A}^{-1}\mathbf{Ax} = \mathbf{Ix} = \mathbf{x}$, the **inverse transformation**

$$
(4) \qquad \mathbf{x} = \mathbf{A}^{-1}\mathbf{y},
$$

which expresses $\mathbf{x}$ in terms of $\mathbf{y}$, and involves the inverse $\mathbf{A}^{-1}$.

For a given $\mathbf{y}$, the equation (3) may be regarded as a system of n linear equations in n unknowns $x_1, \cdots, x_n$, and we know that it has a unique solution if and only if $\mathbf{A}$ has rank n, that is, $\mathbf{A}$ is nonsingular (cf. Fundamental Theorem 1 in Sec. 6.9). This solution (4) ($\mathbf{x}$ expressed in terms of $\mathbf{y}$) is then given by Cramer's rule (Sec. 6.11) in the form

$$
x_1 = D_1/D, \cdots, x_n = D_n/D, \qquad \text{where} \qquad D = \det \mathbf{A}
$$

and D_k is obtained from D by replacing the kth column of D by the column $y_1, \cdots, y_n$. By developing D_1 by the first column, D_2 by the second column, etc., we see that the solution may be written

$$
x_1 = \frac{1}{\det \mathbf{A}}(y_1 A_{11} + \cdots + y_n A_{n1})
$$
$$
\cdot
$$
$$
\cdot
$$
$$
\cdot
$$
$$
x_n = \frac{1}{\det \mathbf{A}}(y_1 A_{1n} + \cdots + y_n A_{nn})
$$

where A_{jk} is the cofactor of a_{jk} in $\det \mathbf{A}$. Substituting these expressions for the components of $\mathbf{x}$ into (4), we see that $\mathbf{A}^{-1}$ must have the form (2). This completes the proof of all our propositions at the beginning of this section.

For a two-rowed square matrix we obtain from (2)

$$
(5) \qquad \mathbf{A} = \begin{pmatrix} a_{11} & a_{12} \\ a_{21} & a_{22} \end{pmatrix}, \qquad \mathbf{A}^{-1} = \frac{1}{\det \mathbf{A}}\begin{pmatrix} a_{22} & -a_{12} \\ -a_{21} & a_{11} \end{pmatrix}.
$$

In the case of a diagonal matrix we simply have

$$(6) \qquad A = \begin{pmatrix} a_{11} & \cdots & 0 \\ & \cdots & \\ & \cdots & \\ 0 & \cdots & a_{nn} \end{pmatrix}, \qquad A^{-1} = \begin{pmatrix} 1/a_{11} & \cdots & 0 \\ & \cdots & \\ & \cdots & \\ 0 & \cdots & 1/a_{nn} \end{pmatrix};$$

the elements of A^{-1} in the principal diagonal are the reciprocals of those of A.

Example 1.

$$A = \begin{pmatrix} 3 & 1 \\ 2 & 4 \end{pmatrix}, \qquad A^{-1} = \frac{1}{10}\begin{pmatrix} 4 & -1 \\ -2 & 3 \end{pmatrix} = \begin{pmatrix} 0.4 & -0.1 \\ -0.2 & 0.3 \end{pmatrix}.$$

Example 2.

$$A = \begin{pmatrix} -3 & 6 & -11 \\ 3 & -4 & 6 \\ 4 & -8 & 13 \end{pmatrix}, \qquad \det A = 10, \qquad A^{-1} = \begin{pmatrix} -0.4 & 1.0 & -0.8 \\ -1.5 & 0.5 & -1.5 \\ -0.8 & 0 & -0.6 \end{pmatrix}.$$

The inverse of the inverse is the given matrix A, that is

$$(7) \qquad\qquad\qquad (A^{-1})^{-1} = A.$$

The simple proof is left to the reader (Prob. 11).

The inverse of a product AC can be obtained by inverting each factor and multiplying the results *in reverse order:*

$$\textbf{(8)} \qquad\qquad\qquad (AC)^{-1} = C^{-1}A^{-1}.$$

To prove (8), we start from (1), with A replaced by AC, that is,

$$AC(AC)^{-1} = I.$$

By premultiplying this by A^{-1} and using $A^{-1}A = I$ we obtain

$$C(AC)^{-1} = A^{-1}.$$

If we premultiply this by C^{-1}, the result follows.

Of course, (8) may be generalized to products of more than two matrices; by induction we obtain

$$(9) \qquad\qquad (AC \cdots PQ)^{-1} = Q^{-1}P^{-1} \cdots C^{-1}A^{-1}.$$

Problems

1. Check the result in Examples 1 and 2 by means of (1).

In each case find the inverse of the given matrix and check the result.

2. $\begin{pmatrix} -1 & 5 \\ 2 & 3 \end{pmatrix}$
    3. $\begin{pmatrix} \cos\theta & \sin\theta \\ -\sin\theta & \cos\theta \end{pmatrix}$
    4. $\begin{pmatrix} 1 & 2 \\ 3 & 4 \end{pmatrix}$

5. $\begin{pmatrix} 0.5 & -0.1 & 0.5 \\ 0 & 0.2 & 0 \\ -0.5 & 0.3 & -1.5 \end{pmatrix}$
    6. $\begin{pmatrix} 0 & 0 & c \\ 0 & b & 0 \\ a & 0 & 0 \end{pmatrix}$
    7. $\begin{pmatrix} 2 & 0 & -1 \\ 5 & 1 & 0 \\ 0 & 1 & 3 \end{pmatrix}$

In each case find the inverse of the given linear transformation.

8. $x^* = x + 2y + 5z$
$\quad\; y^* = \quad\; -y + 2z$
$\quad\; z^* = 2x + 4y + 11z$

9. $x^* = 2x + 4y + z$
$\quad\; y^* = x + 2y + z$
$\quad\; z^* = 3x + 4y + 2z$

10. Show that the inverse of $\mathbf{B} = k\mathbf{A}$ is $\mathbf{B}^{-1} = \mathbf{A}^{-1}/k$.

11. Prove (7).

12. Show that the inverse of $\mathbf{A}^2$ is $(\mathbf{A}^{-1})^2$.

13. Find the inverse of the square of the matrices in Probs. 3 and 7.

14. Find the general formula for the inverse of a three-rowed nonsingular square matrix $\mathbf{A} = (a_{jk})$.

15. Show that $(\mathbf{A}^{-1})^{\mathsf{T}} = (\mathbf{A}^{\mathsf{T}})^{-1}$.

Four-terminal network. Consider a *four-terminal network* (as shown in the figure) in which an input signal is applied to one pair of terminals and the output signal is taken from the other

Four-terminal network

pair. Assume that the network is linear, that is, the currents i_1 and i_2 are linear functions of u_1 and u_2, say,

$$(10) \qquad \begin{aligned} i_1 &= a_{11}u_1 + a_{12}u_2 \\ i_2 &= a_{21}u_1 + a_{22}u_2. \end{aligned}$$

16. Show that (10) may be written $\mathbf{i} = \mathbf{Au}$, where

$$\mathbf{i} = \begin{pmatrix} i_1 \\ i_2 \end{pmatrix}, \qquad \mathbf{A} = \begin{pmatrix} a_{11} & a_{12} \\ a_{21} & a_{22} \end{pmatrix}, \qquad \mathbf{u} = \begin{pmatrix} u_1 \\ u_2 \end{pmatrix}$$

and $\mathbf{u} = \mathbf{A}^{-1}\mathbf{i}$.

17. Show that the input potential and current may be expressed as linear functions of the output potential and current, say,

$$(11) \quad \begin{aligned} u_1 &= t_{11}u_2 + t_{12}i_2, \\ i_1 &= t_{21}u_2 + t_{22}i_2, \end{aligned} \quad \text{or} \quad \mathbf{v}_1 = \mathbf{Tv}_2 \quad \text{where} \quad \mathbf{v}_1 = \begin{pmatrix} u_1 \\ i_1 \end{pmatrix}, \mathbf{v}_2 = \begin{pmatrix} u_2 \\ i_2 \end{pmatrix},$$

and the "transmission matrix" $\mathbf{T}$ is

$$\mathbf{T} = \begin{pmatrix} t_{11} & t_{12} \\ t_{21} & t_{22} \end{pmatrix} = \frac{1}{a_{21}} \begin{pmatrix} -a_{22} & 1 \\ -\det \mathbf{A} & a_{11} \end{pmatrix}.$$

Hint. To verify the last expression, start from (10); express u_1 in terms of u_2 and i_2; then express i_1 in terms of u_2 and i_2; finally, compare the resulting representations with (11).

18. Express $\mathbf{A}$ in terms of the elements of $\mathbf{T}$.

In each case show that the given transmission matrix $\mathbf{T}$ corresponds to the indicated four-terminal network.

24. If two four-terminal networks are connected in cascade as shown in the figure, prove that the resulting network may be regarded as a four-terminal network for which $\mathbf{v}_1 = \mathbf{T}\mathbf{v}_2$ where

$$\mathbf{v}_1 = \begin{pmatrix} u_1 \\ i_1 \end{pmatrix}, \qquad \mathbf{v}_2 = \begin{pmatrix} u_2 \\ i_2 \end{pmatrix}, \qquad \mathbf{T} = \mathbf{T}_1\mathbf{T}_2.$$

Four-terminal networks in cascade

25. Applying the result of Prob. 24, obtain the matrix in Prob. 22 from those in Prob. 20 (with $Z = Z_1$) and Prob. 21 (with $Z = Z_2$). Obtain the matrix in Prob. 23 from those in Probs. 20 and 21.

6.13 *Bilinear, Quadratic, Hermitian and Skew-Hermitian Forms*

An expression of the form

(1) $$B = \sum_{j=1}^{n} \sum_{k=1}^{n} a_{jk} x_j y_k$$

(where the a_{jk} are numbers) is called a **bilinear form** in the $2n$ variables $x_1, \cdots, x_n$ and $y_1, \cdots, y_n$. Writing B out, we have

(1')
$$\begin{aligned}
B = \quad & a_{11}x_1y_1 + a_{12}x_1y_2 + \cdots + a_{1n}x_1y_n \\
+ & a_{21}x_2y_1 + a_{22}x_2y_2 + \cdots + a_{2n}x_2y_n \\
& \cdots\cdots\cdots\cdots\cdots\cdots\cdots\cdots\cdots\cdots \\
+ & a_{n1}x_ny_1 + a_{n2}x_ny_2 + \cdots + a_{nn}x_ny_n.
\end{aligned}$$

The n-rowed square matrix $\mathbf{A} = (a_{jk})$ is called the *coefficient matrix* of the form. Introducing the vectors

$$\mathbf{x} = \begin{pmatrix} x_1 \\ \cdot \\ \cdot \\ \cdot \\ x_n \end{pmatrix} \quad \text{and} \quad \mathbf{y} = \begin{pmatrix} y_1 \\ \cdot \\ \cdot \\ \cdot \\ y_n \end{pmatrix},$$

we see from the definition of matrix multiplication that we may write

$$(2) \qquad\qquad B = \mathbf{x}^\mathsf{T} \mathbf{A} \mathbf{y}$$

where $\mathbf{x}^\mathsf{T} = (x_1 \cdots x_n)$ is the transpose of $\mathbf{x}$.

Example 1.   If $\mathbf{A}$ is the unit matrix $\mathbf{I}$, then

$$B = \mathbf{x}^\mathsf{T} \mathbf{I} \mathbf{y} = \mathbf{x}^\mathsf{T} \mathbf{y} = x_1 y_1 + x_2 y_2 + \cdots + x_n y_n;$$

that is, in this case, B is the inner product (scalar product) of the vectors $\mathbf{x}$ and $\mathbf{y}$.

If $\mathbf{y} = \mathbf{x}$, then (1) is called a **quadratic form** in the n variables $x_1, \cdots, x_n$. Denoting this form by Q, we have

$$(3) \qquad\qquad Q = \sum_{j=1}^{n} \sum_{k=1}^{n} a_{jk} x_j x_k.$$

Writing Q out and taking corresponding terms $a_{jk} x_j x_k$ and $a_{kj} x_k x_j$ together, we obtain

$$\begin{aligned} Q = a_{11} x_1{}^2 &+ (a_{12} + a_{21}) x_1 x_2 + \cdots + (a_{1n} + a_{n1}) x_1 x_n \\ &+ a_{22} x_2{}^2 \qquad + \cdots + (a_{2n} + a_{n2}) x_2 x_n \\ &+ \cdots\cdots\cdots\cdots\cdots\cdots \\ &\qquad\qquad\qquad\qquad + a_{nn} x_n{}^2. \end{aligned}$$

We may now set

$$\tfrac{1}{2}(a_{jk} + a_{kj}) = c_{jk}.$$

Then $c_{kj} = c_{jk}$ and $c_{jk} + c_{kj} = a_{jk} + a_{kj}$ so that we may write

$$Q = \sum_{j=1}^{n} \sum_{k=1}^{n} c_{jk} x_j x_k.$$

If $\mathbf{A}$ is real, then the coefficient matrix $\mathbf{C} = (c_{jk})$ in this new representation is a real symmetric matrix (cf. Sec. 6.3). This shows that *any real quadratic form Q in n variables $x_1, \cdots, x_n$ may be written*

$$Q = \mathbf{x}^\mathsf{T} \mathbf{C} \mathbf{x}$$

where $\mathbf{C}$ is a real symmetric matrix. The reader may prove that the correspondence between these forms Q and symmetric matrices $\mathbf{C}$ is then one-to-one, that is, to each such form Q there corresponds exactly one symmetric matrix $\mathbf{C}$, and conversely.

Example 2. The quadratic form

$$Q = \mathbf{x}^\mathsf{T}\mathbf{A}\mathbf{x} = 2x_1^2 + x_1x_2 - 3x_2^2$$

has the coefficient matrix

$$\mathbf{A} = (a_{jk}) = \begin{pmatrix} 2 & 1 \\ 0 & -3 \end{pmatrix}.$$

Hence the corresponding symmetric coefficient matrix is

$$\mathbf{C} = \tfrac{1}{2}[\mathbf{A} + \mathbf{A}^\mathsf{T}] = \begin{pmatrix} 2 & \tfrac{1}{2} \\ \tfrac{1}{2} & -3 \end{pmatrix}.$$

We introduce the following notation. *Let* $\mathbf{A} = (a_{jk})$ *be any matrix. Then* $\bar{\mathbf{A}}$ *denotes the matrix* $(\bar{a}_{jk})$ *which is obtained from* $\mathbf{A}$ *by replacing each element* a_{jk} *by its complex conjugate* $\bar{a}_{jk}$.

A square matrix $\mathbf{A} = (a_{jk})$ for which the transpose equals the complex conjugate, that is,

(4) $$\mathbf{A}^\mathsf{T} = \bar{\mathbf{A}} \qquad \text{(that is, } a_{kj} = \bar{a}_{jk})$$

is called a **Hermitian**[3] **matrix**.

From (4) we see that the elements in the principal diagonal of a Hermitian matrix are always real. Furthermore, if all the elements of a Hermitian matrix $\mathbf{A}$ are real, then (4) assumes the form $\mathbf{A}^\mathsf{T} = \mathbf{A}$, which means that *a real Hermitian matrix is a symmetric matrix, and so Hermitian matrices are a natural generalization of real symmetric matrices.*

A form

$$H = \bar{\mathbf{x}}^\mathsf{T}\mathbf{A}\mathbf{x} \qquad\qquad \text{(A Hermitian)}$$

is called a **Hermitian form**; here the n components of the vector $\mathbf{x}$ may be real or complex variables. Obviously, this is a generalization of a real quadratic form.

From the definition of matrix multiplication it follows that

$$H = \sum_{j=1}^{n}\sum_{k=1}^{n} a_{jk}\bar{x}_j x_k.$$

Example 3. The matrix

$$\mathbf{A} = \begin{pmatrix} 2 & 3+i \\ 3-i & 1 \end{pmatrix}$$

is Hermitian, and the corresponding Hermitian form is

$$H = \bar{\mathbf{x}}^\mathsf{T}\mathbf{A}\mathbf{x} = (\bar{x}_1 \quad \bar{x}_2)\begin{pmatrix} 2 & 3+i \\ 3-i & 1 \end{pmatrix}\begin{pmatrix} x_1 \\ x_2 \end{pmatrix}$$

$$= 2\bar{x}_1x_1 + (3+i)\bar{x}_1x_2 + (3-i)\bar{x}_2x_1 + \bar{x}_2x_2$$

$$= 2\,|x_1|^2 + |x_2|^2 + 2\,\mathrm{Re}\,[(3+i)\bar{x}_1x_2]$$

where Re denotes the real part. The last expression shows that for every choice of $\mathbf{x}$ the value of H is a real number. We shall now prove that any Hermitian form has this remarkable property.

[3] CHARLES HERMITE (1822–1901), French mathematician.

Theorem 1. *For every choice of the vector* **x** *the value of a Hermitian form* $H = \bar{\mathbf{x}}^{\mathsf{T}}\mathbf{A}\mathbf{x}$ (**A** *Hermitian*) *is a real number.*

Proof. Using (4), we obtain

$$\bar{H} = \overline{(\bar{\mathbf{x}}^{\mathsf{T}}\mathbf{A}\mathbf{x})} = \mathbf{x}^{\mathsf{T}}\bar{\mathbf{A}}\bar{\mathbf{x}} = \mathbf{x}^{\mathsf{T}}\mathbf{A}^{\mathsf{T}}\bar{\mathbf{x}}.$$

The expression on the right is a scalar. Hence transposition does not change its value. Using (12) in Sec. 6.4, we thus have

$$\mathbf{x}^{\mathsf{T}}\mathbf{A}^{\mathsf{T}}\bar{\mathbf{x}} = (\mathbf{x}^{\mathsf{T}}\mathbf{A}^{\mathsf{T}}\bar{\mathbf{x}})^{\mathsf{T}} = \bar{\mathbf{x}}^{\mathsf{T}}\mathbf{A}\mathbf{x} = H.$$

Hence, $\bar{H} = H$, which means that H is real. This completes the proof.

A square matrix $\mathbf{A} = (a_{jk})$ for which

$$(5) \qquad\qquad \mathbf{A}^{\mathsf{T}} = -\bar{\mathbf{A}} \qquad \text{(that is, } a_{kj} = -\bar{a}_{jk})$$

is called a **skew-Hermitian matrix**. Obviously this is a generalization of a real skew-symmetric matrix, because if all the elements of a skew-Hermitian matrix **A** are real, then (5) assumes the form $\mathbf{A}^{\mathsf{T}} = -\mathbf{A}$; that is, **A** is then a real skew-symmetric matrix (cf. Sec. 6.3).

A form

$$(6) \qquad\qquad S = \bar{\mathbf{x}}^{\mathsf{T}}\mathbf{A}\mathbf{x} \qquad\qquad (\mathbf{A} \text{ skew-Hermitian})$$

is called a **skew-Hermitian form**. The reader may prove the following theorem.

Theorem 2. *For every choice of* **x** *the value of a skew-Hermitian form is a pure imaginary number or zero.*

Problems

Quadratic forms. Find a real symmetric matrix **C** such that $Q = \mathbf{x}^{\mathsf{T}}\mathbf{C}\mathbf{x}$, where Q equals

1. $6x_1^2 - 4x_1x_2 + 2x_2^2$

2. $(x_1 - x_2)^2$

3. $14x_1x_2 - x_2^2$

4. $x_1^2 + 2x_1x_2 + 3x_2^2 + 6x_2x_3 + 2x_3^2$

5. $(x_1 + x_2 + x_3)^2$

6. $-3x_1^2 + 4x_1x_2 - x_2^2 + 2x_1x_3 - 5x_3^2$

Hermitian forms. Find $H = \bar{\mathbf{x}}^{\mathsf{T}}\mathbf{A}\mathbf{x}$, where

7. $\mathbf{A} = \begin{pmatrix} 1 & 0 \\ 0 & 1 \end{pmatrix}, \mathbf{x} = \begin{pmatrix} 2+i \\ 1-i \end{pmatrix}$

8. $\mathbf{A} = \begin{pmatrix} 1 & i \\ -i & 1 \end{pmatrix}, \mathbf{x} = \begin{pmatrix} 1 \\ i \end{pmatrix}$

9. $\mathbf{A} = \begin{pmatrix} a & b+ic \\ b-ic & k \end{pmatrix}, \mathbf{x} = \begin{pmatrix} x_1 \\ x_2 \end{pmatrix}$

10. $\mathbf{A} = \begin{pmatrix} 2 & 1+i \\ 1-i & 1 \end{pmatrix}, \mathbf{x} = \begin{pmatrix} 1 \\ 2 \end{pmatrix}$

11. $\mathbf{A} = \begin{pmatrix} 0 & i & 0 \\ -i & 1 & -2i \\ 0 & 2i & 2 \end{pmatrix}, \mathbf{x} = \begin{pmatrix} i \\ 1 \\ -i \end{pmatrix}$

12. $\mathbf{A} = \begin{pmatrix} 3 & -i & 0 \\ i & 0 & 2i \\ 0 & -2i & 4 \end{pmatrix}, \mathbf{x} = \begin{pmatrix} x_1 \\ x_2 \\ x_3 \end{pmatrix}$

Skew-Hermitian forms. Find $S = \bar{\mathbf{x}}^T \mathbf{A} \mathbf{x}$, where

13. $\mathbf{A} = \begin{pmatrix} i & 0 \\ 0 & i \end{pmatrix}, \mathbf{x} = \begin{pmatrix} x_1 \\ x_2 \end{pmatrix}$

14. $\mathbf{A} = \begin{pmatrix} i & 1 \\ -1 & 2i \end{pmatrix}, \mathbf{x} = \begin{pmatrix} 1 \\ i \end{pmatrix}$

15. $\mathbf{A} = \begin{pmatrix} 2i & 3i \\ 3i & 0 \end{pmatrix}, \mathbf{x} = \begin{pmatrix} 4i \\ -5 \end{pmatrix}$

16. $\mathbf{A} = \begin{pmatrix} i\alpha & b \\ -\bar{b} & i\delta \end{pmatrix}, \mathbf{x} = \begin{pmatrix} x_1 \\ x_2 \end{pmatrix}$

17. Show that any square matrix may be written as the sum of a Hermitian matrix and a skew-Hermitian matrix.

18. If $\mathbf{A}$ and $\mathbf{B}$ are n-rowed Hermitian matrices and a and b are any *real* numbers, show that $\mathbf{C} = a\mathbf{A} + b\mathbf{B}$ is a Hermitian matrix.

19. Let $\mathbf{A}$ and $\mathbf{B}$ be skew-Hermitian matrices. Under what conditions is $\mathbf{C} = a\mathbf{A} + b\mathbf{B}$ a skew-Hermitian matrix? (a and b are numbers.)

20. Show that the elements of the principal diagonal of a skew-Hermitian matrix are pure imaginary or zero.

21. Prove Theorem 2.

22. Show that if the variables of a quadratic form $Q = \mathbf{x}^T \mathbf{C} \mathbf{x}$ undergo a linear transformation, say, $\mathbf{x} = \mathbf{P}\mathbf{y}$, then $Q = \mathbf{y}^T \mathbf{A} \mathbf{y}$ where $\mathbf{A} = \mathbf{P}^T \mathbf{C} \mathbf{P}$.

23. Consider the quadratic form $Q = \mathbf{x}^T \mathbf{C} \mathbf{x}$ ($\mathbf{C}$ symmetric) in three variables x_1, x_2, x_3. Find $\partial Q / \partial x_1$, $\partial Q / \partial x_2$, $\partial Q / \partial x_3$ and show that these are the components of the vector $2\mathbf{C}\mathbf{x}$.

24. **(Definiteness)** A real quadratic form $Q = \mathbf{x}^T \mathbf{C} \mathbf{x}$ and its symmetric matrix $\mathbf{C} = (c_{jk})$ are said to be **positive definite** if $Q > 0$ for all $(x_1, \cdots, x_n) \neq (0, \cdots, 0)$. A necessary and sufficient condition for positive definiteness is that all the determinants

$$C_1 = c_{11}, \quad C_2 = \begin{vmatrix} c_{11} & c_{12} \\ c_{21} & c_{22} \end{vmatrix}, \quad C_3 = \begin{vmatrix} c_{11} & c_{12} & c_{13} \\ c_{21} & c_{22} & c_{23} \\ c_{31} & c_{32} & c_{33} \end{vmatrix}, \cdots, \quad C_n = \det \mathbf{C}$$

are positive (cf. Ref. [C3]). Show that the form in Prob. 1 is positive definite.

25. Test the forms in Probs. 2 and 4 for positive definiteness.

6.14 *Eigenvalues. Eigenvectors*

Let $\mathbf{A} = (a_{jk})$ be a given square n-rowed matrix and consider the vector equation

(1) $$\mathbf{A}\mathbf{x} = \lambda \mathbf{x}$$

where λ is a number.

It is clear that the zero vector $\mathbf{x} = \mathbf{0}$ is a solution of (1) for any value of λ. A value of λ for which (1) has a solution $\mathbf{x} \neq \mathbf{0}$ is called an **eigenvalue**[4] or **characteristic value** (or *latent root*) of the matrix $\mathbf{A}$. The corresponding solutions $\mathbf{x} \neq \mathbf{0}$ of (1) are called **eigenvectors** or **characteristic vectors** of $\mathbf{A}$ corresponding to that eigenvalue λ. The set of the eigenvalues is called the **spectrum** of $\mathbf{A}$. The largest of the absolute values of the eigenvalues of $\mathbf{A}$ is called the **spectral radius** of $\mathbf{A}$.

The problem of determining the eigenvalues and eigenvectors of a matrix is called an *eigenvalue problem*.[5] Problems of this type occur in connection with

---

[4] German: Eigenwert

[5] More precisely: an algebraic eigenvalue problem, because there are other eigenvalue problems involving a differential equation (see Secs. 3.8 and 10.3) or an integral equation.

physical and technical applications. Therefore, the student should know the fundamental ideas and concepts which are important in this field of mathematics. During the last two decades various new methods for the approximate determination of eigenvalues have been developed and other methods which have been known for still a longer time have been put into a form which is suitable for electronic computers. Cf. Ref. [C1] in Appendix 1.

Let us mention two simple problems which lead to an equation of the form (1).

Example 1. The vertical motion of the mechanical system in Fig. 115 (no damping, masses of springs neglected) is governed by the simultaneous differential equations

(2)
$$\ddot{y}_1 = -3y_1 + 2(y_2 - y_1)$$
$$\ddot{y}_2 = -2(y_2 - y_1)$$

where $y_1(t)$ and $y_2(t)$ are the displacements of the two masses such that $y_1 = 0$, $y_2 = 0$ correspond to the positions of static equilibrium. The derivation of these equations is similar to that in Sec. 2.6. We may write the system in the form

$$\ddot{y}_1 = -5y_1 + 2y_2$$
$$\ddot{y}_2 = 2y_1 - 2y_2$$

or as a single vector equation

$$\ddot{\mathbf{y}} = \mathbf{A}\mathbf{y},$$

where
$$\mathbf{y} = \begin{pmatrix} y_1 \\ y_2 \end{pmatrix} \quad \text{and} \quad \mathbf{A} = \begin{pmatrix} -5 & 2 \\ 2 & -2 \end{pmatrix}.$$

To solve this equation we substitute

(3)
$$\mathbf{y} = \mathbf{x}e^{\omega t}.$$

This yields

$$\omega^2 \mathbf{x}e^{\omega t} = \mathbf{A}\mathbf{x}e^{\omega t}$$

or

$$\mathbf{A}\mathbf{x} = \lambda\mathbf{x} \qquad\qquad \text{where } \lambda = \omega^2.$$

Hence, for (3) to be a solution of (2), not identically zero, $\omega^2 = \lambda$ must be an eigenvalue of $\mathbf{A}$ and the vector $\mathbf{x}$ in (3) must be a corresponding eigenvector.

Fig. 115. Example 1

Example 2. Given a linear transformation

$$\mathbf{y} = \mathbf{Ax} \qquad \text{(A real)},$$

does there exist a real vector $\mathbf{x} \neq \mathbf{0}$ for which the corresponding vector $\mathbf{y}$ is real and has the same direction and sense as $\mathbf{x}$? This means we want to find a vector $\mathbf{x}$ such that

$$\mathbf{y} = \mathbf{Ax} = \lambda\mathbf{x} \qquad (\lambda \text{ real and positive}).$$

Clearly, such a vector exists if and only if $\mathbf{A}$ has a real positive eigenvalue. $\mathbf{x}$ is then an eigenvector corresponding to that eigenvalue.

Let us consider (1). If $\mathbf{x}$ is any vector, then the vectors $\mathbf{x}$ and $\mathbf{Ax}$ will, in general, be linearly independent. If $\mathbf{x}$ is an eigenvector, then $\mathbf{x}$ and $\mathbf{Ax}$ are linearly dependent; corresponding components of $\mathbf{x}$ and $\mathbf{Ax}$ are then proportional, the factor of proportionality being the eigenvalue λ.

We shall now demonstrate that *any n-rowed square matrix has at least* 1 *and at most n distinct (real or complex) eigenvalues.*

For this purpose we write (1) out:

$$a_{11}x_1 + \cdots + a_{1n}x_n = \lambda x_1$$
$$a_{21}x_1 + \cdots + a_{2n}x_n = \lambda x_2$$
$$\cdots\cdots\cdots\cdots\cdots\cdots\cdots$$
$$a_{n1}x_1 + \cdots + a_{nn}x_n = \lambda x_n.$$

By transferring the terms on the right-hand side to the left-hand side we obtain

(4)
$$(a_{11} - \lambda)x_1 + a_{12}x_2 + \cdots + a_{1n}x_n = 0$$
$$a_{21}x_1 + (a_{22} - \lambda)x_2 + \cdots + a_{2n}x_n = 0$$
$$\cdots\cdots\cdots\cdots\cdots\cdots\cdots\cdots\cdots\cdots\cdots\cdots$$
$$a_{n1}x_1 + a_{n2}x_2 + \cdots + (a_{nn} - \lambda)x_n = 0.$$

According to Theorem 3 in Sec. 6.9 this homogeneous system of linear equations has a nontrivial solution if and only if the corresponding determinant of the coefficients is zero:

(5)
$$D(\lambda) = \det(\mathbf{A} - \lambda\mathbf{I}) = \begin{vmatrix} a_{11} - \lambda & a_{12} & \cdots & a_{1n} \\ a_{21} & a_{22} - \lambda & \cdots & a_{2n} \\ \cdot & \cdot & \cdots & \cdot \\ a_{n1} & a_{n2} & \cdots & a_{nn} - \lambda \end{vmatrix} = 0.$$

$D(\lambda)$ is called the **characteristic determinant**, and (5) is called the **characteristic equation** corresponding to the matrix $\mathbf{A}$. By developing $D(\lambda)$ we obtain a polynomial of nth degree in λ. This is called the **characteristic polynomial** corresponding to $\mathbf{A}$.

We have thus obtained the following important result.

Theorem 1. *The eigenvalues of a square matrix* $\mathbf{A}$ *are the roots of the corresponding characteristic equation* (5).

An eigenvalue which is a root of mth order of the characteristic polynomial is called an *eigenvalue of mth* **order** of the corresponding matrix.

Once the eigenvalues have been determined, corresponding eigenvectors can be determined from the system (4). Since the system is homogeneous, it is clear that *if* **x** *is an eigenvector of* **A**, *then* k**x**, *where* k *is any constant, not zero, is also an eigenvector of* **A** *corresponding to the same eigenvalue.*

Example 3. Determine the eigenvalues and eigenvectors of the matrix

$$\mathbf{A} = \begin{pmatrix} 5 & 4 \\ 1 & 2 \end{pmatrix}.$$

The characteristic equation

$$D(\lambda) = \begin{vmatrix} 5 - \lambda & 4 \\ 1 & 2 - \lambda \end{vmatrix} = \lambda^2 - 7\lambda + 6 = 0$$

has the roots $\lambda_1 = 6$ and $\lambda_2 = 1$. For $\lambda = \lambda_1$ the system (4) assumes the form

$$-x_1 + 4x_2 = 0$$
$$x_1 - 4x_2 = 0.$$

Thus $x_1 = 4x_2$, and

$$\mathbf{x}_1 = \begin{pmatrix} 4 \\ 1 \end{pmatrix}$$

is an eigenvector of **A** corresponding to the eigenvalue λ_1. In the same way we find that an eigenvector of **A** corresponding to λ_2 is

$$\mathbf{x}_2 = \begin{pmatrix} 1 \\ -1 \end{pmatrix}.$$

$\mathbf{x}_1$ and $\mathbf{x}_2$ are linearly independent vectors.

Example 4. The matrix

$$\mathbf{A} = \begin{pmatrix} a & b \\ -b & a \end{pmatrix} \qquad (a, b \text{ real}, b \neq 0)$$

has the conjugate complex eigenvalues $\lambda_1 = a + ib$ and $\lambda_2 = a - ib$. Corresponding eigenvectors are

$$\mathbf{x}_1 = \begin{pmatrix} 1 \\ i \end{pmatrix} \qquad \text{and} \qquad \mathbf{x}_2 = \begin{pmatrix} 1 \\ -i \end{pmatrix}.$$

If $b = 0$, then $\lambda_1 = \lambda_2 = a$; the matrix **A** has just one eigenvalue (which is then of the second order), and every vector $\mathbf{x} \neq \mathbf{0}$ with two components is an eigenvector of **A**.

Example 5. The matrix

$$\mathbf{A} = \begin{pmatrix} -2 & 2 & -3 \\ 2 & 1 & -6 \\ -1 & -2 & 0 \end{pmatrix}$$

has the eigenvalues $\lambda_1 = 5$ and $\lambda_2 = \lambda_3 = -3$. The vector

$$\mathbf{x}_1 = \begin{pmatrix} 1 \\ 2 \\ -1 \end{pmatrix}$$

is an eigenvector of **A** corresponding to the eigenvalue 5, and the vectors

$$\mathbf{x}_2 = \begin{pmatrix} -2 \\ 1 \\ 0 \end{pmatrix} \quad \text{and} \quad \mathbf{x}_3 = \begin{pmatrix} 3 \\ 0 \\ 1 \end{pmatrix}$$

are two linearly independent eigenvectors of **A** corresponding to the eigenvalue -3. This agrees with the fact that, for $\lambda = -3$, the matrix $\mathbf{A} - \lambda\mathbf{I}$ has rank 1 and so, by Theorem 2 in Sec. 6.9, a fundamental system of solutions of the corresponding system (4), viz.,

$$x_1 + 2x_2 - 3x_3 = 0$$
$$2x_1 + 4x_2 - 6x_3 = 0$$
$$-x_1 - 2x_2 + 3x_3 = 0$$

consists of two linearly independent vectors.

Example 5 shows that *to an eigenvalue there may correspond several linearly independent eigenvectors*. The maximum number of linearly independent eigenvectors corresponding to an eigenvalue λ is called the **multiplicity** of λ.

Problems

Find the eigenvalues and eigenvectors of the following matrices.

1. $\begin{pmatrix} 1 & 0 \\ 0 & -3 \end{pmatrix}$ **2.** $\begin{pmatrix} -2 & 2 \\ -8 & 8 \end{pmatrix}$ **3.** $\begin{pmatrix} 0 & 0 \\ 0 & 0 \end{pmatrix}$ **4.** $\begin{pmatrix} 1 & 2 \\ -8 & 11 \end{pmatrix}$

5. $\begin{pmatrix} 3 & 4 \\ 4 & -3 \end{pmatrix}$ **6.** $\begin{pmatrix} 1 & 0 \\ 2 & -1 \end{pmatrix}$ **7.** $\begin{pmatrix} 4 & -2 \\ 1 & 1 \end{pmatrix}$ **8.** $\begin{pmatrix} 14 & -10 \\ 5 & -1 \end{pmatrix}$

9. $\begin{pmatrix} 1 & 0 & 0 \\ 0 & 1 & 0 \\ 0 & 0 & 1 \end{pmatrix}$ **10.** $\begin{pmatrix} a & 0 & 0 \\ 0 & b & 0 \\ 0 & 0 & c \end{pmatrix}$ **11.** $\begin{pmatrix} 3 & 0 & 0 \\ 5 & 4 & 0 \\ 3 & 6 & 1 \end{pmatrix}$ **12.** $\begin{pmatrix} 0 & 1 & 0 \\ 1 & 0 & 0 \\ 0 & 0 & 1 \end{pmatrix}$

Find the eigenvalues of the following matrices.

13. $\begin{pmatrix} 1.5 & -1.0 & -2.0 \\ -0.4 & 1.2 & 0.4 \\ -0.3 & -0.6 & -0.2 \end{pmatrix}$ **14.** $\begin{pmatrix} 13 & 0 & -15 \\ -3 & 4 & 9 \\ 5 & 0 & -7 \end{pmatrix}$ **15.** $\begin{pmatrix} 26 & 2 & 4 \\ -2 & 21 & 2 \\ 2 & 4 & 28 \end{pmatrix}$

Let $\lambda_1, \cdots, \lambda_n$ be the eigenvalues of a given matrix **A**. Prove the following propositions.

16. The transpose $\mathbf{A}^T$ has the eigenvalues $\lambda_1, \cdots, \lambda_n$.

17. If **A** is triangular, the elements of the principal diagonal are the eigenvalues of **A**.

18. The inverse $\mathbf{A}^{-1}$ exists if and only if $\lambda_j \neq 0$ ($j = 1, \cdots, n$).

19. The inverse $\mathbf{A}^{-1}$ has the eigenvalues $1/\lambda_1, \cdots, 1/\lambda_n$.

20. The matrix $\mathbf{A} - k\mathbf{I}$ has the eigenvalues $\lambda_1 - k, \cdots, \lambda_n - k$.

21. The matrix $k\mathbf{A}$ has the eigenvalues $k\lambda_1, \cdots, k\lambda_n$.

22. The matrix $\mathbf{A}^2$ has the eigenvalues $\lambda_1^2, \cdots, \lambda_n^2$.

23. The matrix $\mathbf{A}^m$ (m a nonnegative integer) has the eigenvalues $\lambda_1^m, \cdots, \lambda_n^m$.

24. The matrix $k_m\mathbf{A}^m + k_{m-1}\mathbf{A}^{m-1} + \cdots + k_1\mathbf{A} + k_0\mathbf{I}$ has the eigenvalues

$$k_m\lambda_j^m + k_{m-1}\lambda_j^{m-1} + \cdots + k_1\lambda_j + k_0 \qquad (j = 1, \cdots, n).$$

25. Verify the statement in Prob. 19 for the matrices in Probs. 5 and 15.

26. Using Prob. 20, obtain the eigenvalues in Prob. 4 from those in Prob. 2.

27. Show that the vertical motion of the mechanical system in Fig. 116 (no damping, masses of springs neglected) is governed by the simultaneous differential equations

$$m_1 \ddot{y}_1 = -k_1 y_1 + k_2(y_2 - y_1)$$
$$m_2 \ddot{y}_2 = -k_2(y_2 - y_1)$$

where dots denote derivatives with respect to the time t.

Fig. 116. Mechanical system in Problem 27

28. Show that the system of equations in Prob. 27 may be written as a single vector equation $\ddot{\mathbf{y}} = \mathbf{A}\mathbf{y}$, where

$$\mathbf{y} = \begin{pmatrix} y_1 \\ y_2 \end{pmatrix} \quad \text{and} \quad \mathbf{A} = \begin{pmatrix} -\dfrac{k_1 + k_2}{m_1} & \dfrac{k_2}{m_1} \\[2mm] \dfrac{k_2}{m_2} & -\dfrac{k_2}{m_2} \end{pmatrix}$$

29. To solve the equation in Prob. 28, substitute

$$\mathbf{y} = \mathbf{x}e^{\omega t}$$

and show that this leads to the eigenvalue problem

$$\mathbf{A}\mathbf{x} = \lambda\mathbf{x} \quad \text{where} \quad \lambda = \omega^2.$$

30. In Prob. 29, let $m_1 = m_2 = 1$, $k_1 = 3$, $k_2 = 2$. Find the eigenvalues and show that corresponding eigenvectors are

$$\mathbf{x}_1 = \begin{pmatrix} 1 \\ 2 \end{pmatrix}, \quad \mathbf{x}_2 = \begin{pmatrix} -2 \\ 1 \end{pmatrix}.$$

Conclude that this leads to the following solution of the equation in Prob. 28 (with m_1, m_2, k_1, k_2 as indicated):

$$\mathbf{y}(t) = a_1 \mathbf{x}_1 \cos t + b_1 \mathbf{x}_1 \sin t + a_2 \mathbf{x}_2 \cos \sqrt{6}\, t + b_2 \mathbf{x}_2 \sin \sqrt{6}\, t,$$

where a_1, b_1, a_2, b_2 are arbitrary constants.

31. Show that by imposing the initial conditions $y_1(0) = 1$, $y_2(0) = 2$, $\dot{y}_1(0) = -2\sqrt{6}$, $\dot{y}_2(0) = \sqrt{6}$ or, in vector form,

$$\mathbf{y}(0) = \begin{pmatrix} 1 \\ 2 \end{pmatrix}, \qquad \dot{\mathbf{y}}(0) = \begin{pmatrix} -2\sqrt{6} \\ \sqrt{6} \end{pmatrix},$$

we obtain from the solution in Prob. 30 the particular solution

$$\mathbf{y}(t) = \mathbf{x}_1 \cos t + \mathbf{x}_2 \sin \sqrt{6}t,$$

that is,

$$y_1 = \cos t - 2 \sin \sqrt{6}t, \qquad y_2 = 2 \cos t + \sin \sqrt{6}t.$$

Verify by substitution that these functions are solutions of the differential equations in Prob. 27 with $m_1 = m_2 = 1$, $k_1 = 3$, $k_2 = 2$.

32. In Prob. 30, find and graph the solutions satisfying the initial conditions $y_1(0) = 1$, $y_2(0) = 2$, $\dot{y}_1(0) = 0$, $\dot{y}_2(0) = 0$.

33. Show that, for any positive k_1, k_2, m_1, m_2, the eigenvalues of the matrix in Prob. 28 are negative real.

Find the solutions of the differential equations in Prob. 27, assuming that
34. $m_1 = m_2 = 1$, $k_1 = 9$, $k_2 = 6$, $y_1(0) = -2$, $y_2(0) = 1$, $\dot{y}_1(0) = 0$, $\dot{y}_2(0) = 0$.
35. $m_1 = m_2 = 1$, $k_1 = 6$, $k_2 = 4$, $y_1(0) = 0$, $y_2(0) = 0$, $\dot{y}_1(0) = \sqrt{2}$, $\dot{y}_2(0) = 2\sqrt{2}$.

In each case find the solution satisfying the given initial conditions.
36. $\dot{y}_1 = 4y_1 - 2y_2$ $y_1(0) = 3, y_2(0) = 2$
  $\dot{y}_2 = y_1 + y_2$

37. $\dot{y}_1 = 3y_1 + 4y_2$ $y_1(0) = 1, y_2(0) = 3$
  $\dot{y}_2 = 4y_1 - 3y_2$

38. $\dot{y}_1 = y_1 + 2y_2$ $y_1(0) = 1, y_2(0) = 1$
  $\dot{y}_2 = -8y_1 + 11y_2$

39. $\dot{y}_1 = y_2$
  $\dot{y}_2 = y_1 + 3y_3$ $y_1(0) = 2, y_2(0) = 0, y_3(0) = 2$
  $\dot{y}_3 = y_2$

40. $\dot{y}_1 = 3y_1$
  $\dot{y}_2 = 5y_1 + 4y_2$ $y_1(0) = 2, y_2(0) = -10, y_3(0) = -27$
  $\dot{y}_3 = 3y_1 + 6y_2 + y_3$

6.15   *Eigenvalues of Hermitian, Skew-Hermitian, and Unitary Matrices*

A square matrix $\mathbf{A} = (a_{jk})$ for which

(1) $$\mathbf{A}^\mathsf{T} = \bar{\mathbf{A}}^{-1}$$

is called a **unitary matrix**.

A real unitary matrix **A** is called an **orthogonal matrix**. For such a matrix, (1) assumes the form

(2) $$\mathbf{A}^{\mathsf{T}} = \mathbf{A}^{-1},$$

that is, *an orthogonal matrix is a real matrix for which the inverse equals its transpose.*

A system of vectors $\mathbf{x}_1, \cdots, \mathbf{x}_n$ for which

(3) $$\bar{\mathbf{x}}_j{}^{\mathsf{T}}\mathbf{x}_k = \delta_{jk} = \begin{cases} 0 & \text{when } j \neq k \\ 1 & \text{when } j = k \end{cases} \qquad (j, k = 1, \cdots, n)$$

is called a **unitary system**.

Clearly, if these vectors are real, then $\bar{\mathbf{x}}_j{}^{\mathsf{T}}\mathbf{x}_k = \mathbf{x}_j{}^{\mathsf{T}}\mathbf{x}_k$ is the inner product (scalar product) of $\mathbf{x}_j$ and $\mathbf{x}_k$ in the elementary sense, and the condition (3) means that $\mathbf{x}_1, \cdots, \mathbf{x}_n$ are orthogonal unit vectors. *Hence a unitary system of real vectors is a system of orthogonal (mutually perpendicular) unit vectors.*

Theorem 1. *The column vectors (and also the row vectors) of a unitary matrix form a unitary system.*

Proof. Let **A** be a unitary matrix with column vectors $\mathbf{a}_1, \cdots, \mathbf{a}_n$. Then from (1) it follows that

$$\mathbf{A}^{-1}\mathbf{A} = \bar{\mathbf{A}}^{\mathsf{T}}\mathbf{A} = \begin{pmatrix} \bar{\mathbf{a}}_1{}^{\mathsf{T}} \\ \cdot \\ \cdot \\ \cdot \\ \bar{\mathbf{a}}_n{}^{\mathsf{T}} \end{pmatrix} (\mathbf{a}_1 \cdots \mathbf{a}_n) = \begin{pmatrix} \bar{\mathbf{a}}_1{}^{\mathsf{T}}\mathbf{a}_1 & \bar{\mathbf{a}}_1{}^{\mathsf{T}}\mathbf{a}_2 & \cdots & \bar{\mathbf{a}}_1{}^{\mathsf{T}}\mathbf{a}_n \\ \bar{\mathbf{a}}_2{}^{\mathsf{T}}\mathbf{a}_1 & \bar{\mathbf{a}}_2{}^{\mathsf{T}}\mathbf{a}_2 & \cdots & \bar{\mathbf{a}}_2{}^{\mathsf{T}}\mathbf{a}_n \\ \cdot & \cdot & \cdots & \cdot \\ \bar{\mathbf{a}}_n{}^{\mathsf{T}}\mathbf{a}_1 & \bar{\mathbf{a}}_n{}^{\mathsf{T}}\mathbf{a}_2 & \cdots & \bar{\mathbf{a}}_n{}^{\mathsf{T}}\mathbf{a}_n \end{pmatrix} = \mathbf{I}$$

so that

$$\bar{\mathbf{a}}_j{}^{\mathsf{T}}\mathbf{a}_k = \begin{cases} 0 & \text{when } j \neq k, \\ 1 & \text{when } j = k. \end{cases}$$

This shows that the column vectors of a unitary matrix form a unitary system. From (1) it follows that the inverse $\mathbf{A}^{-1}$ of a unitary matrix **A** is unitary (cf. Prob. 1) and, furthermore, the column vectors of $\mathbf{A}^{-1}$ are the conjugates of the row vectors of **A**. This shows that the row vectors of **A** form a unitary system. Theorem 1 is proved.

With respect to orthogonal matrices, Theorem 1 means that *the row vectors (and also the column vectors) of an orthogonal matrix* $\mathbf{A} = (a_{jk})$ *form a system of orthogonal unit vectors, that is*

(4) $$\sum_{m=1}^{n} a_{jm}a_{km} = \delta_{jk}.$$

Orthogonal matrices are of great practical importance because **orthogonal transformations**, that is, linear transformations $\mathbf{y} = \mathbf{A}\mathbf{x}$ with an orthogonal matrix as the coefficient matrix, have the following property.

Theorem 2. *An orthogonal transformation leaves the inner product*

$$(\mathbf{x}, \mathbf{y}) = \mathbf{x}^\mathsf{T}\mathbf{y} \quad \text{and thus the norm} \quad \|\mathbf{x}\| = \sqrt{\mathbf{x}^\mathsf{T}\mathbf{x}}$$

invariant. In particular, in three-dimensional Euclidean space an orthogonal transformation preserves the length of any vector and the angle between any two vectors; hence it is a rotation (possibly combined with a reflection), and the coefficient matrix of a rotation of a Cartesian coordinate system is an orthogonal matrix.

Proof. Let **A** be orthogonal and $\mathbf{u} = \mathbf{A}\mathbf{x}$ and $\mathbf{v} = \mathbf{A}\mathbf{y}$. Using (2) in this section and (12) in Sec. 6.4, we obtain

$$(\mathbf{u}, \mathbf{v}) = \mathbf{u}^\mathsf{T}\mathbf{v} = (\mathbf{A}\mathbf{x})^\mathsf{T}\mathbf{A}\mathbf{y} = \mathbf{x}^\mathsf{T}\mathbf{A}^\mathsf{T}\mathbf{A}\mathbf{y} = \mathbf{x}^\mathsf{T}\mathbf{A}^{-1}\mathbf{A}\mathbf{y} = \mathbf{x}^\mathsf{T}\mathbf{I}\mathbf{y} = \mathbf{x}^\mathsf{T}\mathbf{y} = (\mathbf{x}, \mathbf{y}).$$

We shall now consider the eigenvalues of Hermitian, skew-Hermitian, and unitary matrices, and prove the general characterization illustrated in Fig. 117.

Theorem 3. **(a)** *The eigenvalues of a Hermitian matrix are real.* **(b)** *The eigenvalues of a skew-Hermitian matrix are pure imaginary or zero.* **(c)** *The eigenvalues of a unitary matrix have the absolute value* 1.

Proof. **(a)** and **(b)**. Let λ be an eigenvalue of **A**. Then by definition there is a vector $\mathbf{x} \neq \mathbf{0}$ such that $\mathbf{A}\mathbf{x} = \lambda\mathbf{x}$. From this we obtain

$$\bar{\mathbf{x}}^\mathsf{T}\mathbf{A}\mathbf{x} = \bar{\mathbf{x}}^\mathsf{T}\lambda\mathbf{x} = \lambda\bar{\mathbf{x}}^\mathsf{T}\mathbf{x}.$$

Since $\mathbf{x} \neq \mathbf{0}$, it follows that $\bar{\mathbf{x}}^\mathsf{T}\mathbf{x} \neq 0$ and we may divide, finding

$$\lambda = \frac{\bar{\mathbf{x}}^\mathsf{T}\mathbf{A}\mathbf{x}}{\bar{\mathbf{x}}^\mathsf{T}\mathbf{x}}.$$

The denominator is real. If **A** is Hermitian, the numerator is real (Theorem 1 in Sec. 6.13) and so is λ. If **A** is skew-Hermitian, the numerator is pure imaginary or zero (Theorem 2 in Sec. 6.13) and so is λ.

(c) Let **U** be unitary, λ an eigenvalue of **U**, and **x** a corresponding eigenvector.

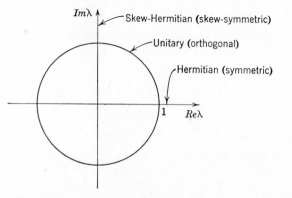

Fig. 117. Location of the eigenvalues of Hermitian, skew-Hermitian, and unitary matrices in the complex λ-plane

Then

(5) $$\mathbf{U}\mathbf{x} = \lambda\mathbf{x}.$$

Taking the conjugate transpose and using (12) in Sec. 6.4, we obtain

$$(\bar{\mathbf{U}}\bar{\mathbf{x}})^\mathsf{T} = \bar{\mathbf{x}}^\mathsf{T}\bar{\mathbf{U}}^\mathsf{T} = \bar{\lambda}\bar{\mathbf{x}}^\mathsf{T}.$$

Since $\mathbf{U}$ is unitary, $\bar{\mathbf{U}}^\mathsf{T}=\mathbf{U}^{-1}$, and our equation can be written

(6) $$\bar{\mathbf{x}}^\mathsf{T}\mathbf{U}^{-1} = \bar{\lambda}\bar{\mathbf{x}}^\mathsf{T}.$$

By postmultiplying the left-hand side of (6) by the left-hand side of (5) and the right-hand side of (6) by the right-hand side of (5) we obtain

$$\bar{\mathbf{x}}^\mathsf{T}\mathbf{U}^{-1}\mathbf{U}\mathbf{x} = \bar{\lambda}\bar{\mathbf{x}}^\mathsf{T}\lambda\mathbf{x} = \lambda\bar{\lambda}\bar{\mathbf{x}}^\mathsf{T}\mathbf{x}.$$

In this equation we have $\mathbf{U}^{-1}\mathbf{U} = \mathbf{I}$ and $\lambda\bar{\lambda} = |\lambda|^2$. Hence the equation becomes

$$\bar{\mathbf{x}}^\mathsf{T}\mathbf{x} = |\lambda|^2\,\bar{\mathbf{x}}^\mathsf{T}\mathbf{x}.$$

Since $\mathbf{x} \neq \mathbf{0}$, it follows that $\bar{\mathbf{x}}^\mathsf{T}\mathbf{x} \neq 0$. Hence we must have $|\lambda|^2 = 1$. Thus $|\lambda| = 1$, and the proof is complete.

An important particular case of Theorem 3 is

Theorem 4.   *The eigenvalues of a symmetric matrix are real. The eigenvalues of a skew-symmetric matrix are pure imaginary or zero. The eigenvalues of an orthogonal matrix have the absolute value* 1 *and are real or complex conjugates in pairs.*

Numerical methods for matrix eigenvalue problems are presented in Secs. 18.12 and 18.13, which are independent of the other sections in Chap. 18, so that they can be studied now.

Problems

1. Show that the inverse of a unitary matrix is a unitary matrix.
2. Show that the product of two n-rowed unitary matrices is a unitary matrix.
3. Show that if $\mathbf{A}$ is orthogonal, then $\det \mathbf{A} = 1$ or -1.
4. Find a real matrix which has real eigenvalues but is not symmetric. Does this contradict Theorem 4?

Show that the following matrices are unitary and determine their eigenvalues and eigenvectors.

5. $\begin{pmatrix} i/2 & \sqrt{3}/2 \\ \sqrt{3}/2 & i/2 \end{pmatrix}$
   6. $\begin{pmatrix} 1/\sqrt{3} & i\sqrt{2/3} \\ -i\sqrt{2/3} & -1/\sqrt{3} \end{pmatrix}$
   7. $\begin{pmatrix} 1/\sqrt{2} & i/\sqrt{2} \\ -i/\sqrt{2} & -1/\sqrt{2} \end{pmatrix}$

8. Verify the statement in Prob. 2 for the matrices in Probs. 5 and 6.

Show that the following matrices are orthogonal and determine their eigenvalues.

9. $\begin{pmatrix} 0 & 1 & 0 \\ 1 & 0 & 0 \\ 0 & 0 & 1 \end{pmatrix}$
   10. $\begin{pmatrix} \cos\theta & -\sin\theta & 0 \\ \sin\theta & \cos\theta & 0 \\ 0 & 0 & 1 \end{pmatrix}$
   11. $\begin{pmatrix} \cos\theta & -\sin\theta \\ \sin\theta & \cos\theta \end{pmatrix}$

12. Interpret the transformation $\mathbf{y} = \mathbf{Ax}$ geometrically, where $\mathbf{A}$ is the matrix in Prob. 11 and the components of $\mathbf{x}$ and $\mathbf{y}$ are Cartesian coordinates.

13. **(Orthogonality)** Prove that the eigenvectors of a real symmetric matrix corresponding to different eigenvalues are orthogonal. Give an example.

14. **(Modal matrix and spectral matrix)** Let $\mathbf{A}$ be a given square matrix, and let $\mathbf{X}$ be the matrix whose column vectors $\mathbf{x}_1, \cdots, \mathbf{x}_n$ are eigenvectors of $\mathbf{A}$ corresponding to the eigenvalues $\lambda_1, \cdots, \lambda_n$ of $\mathbf{A}$. Show that

$$\mathbf{AX} = (\lambda_1 \mathbf{x}_1 \cdots \lambda_n \mathbf{x}_n) = \mathbf{XD},$$

where $\mathbf{D}$ is the diagonal matrix with elements $\lambda_1, \cdots, \lambda_n$ in the principal diagonal. ($\mathbf{X}$ is called the *modal matrix* and $\mathbf{D}$ the *spectral matrix* of $\mathbf{A}$.)

15. If the $\mathbf{x}_i$ in Prob. 14 are linearly independent, show that $\mathbf{X}^{-1}\mathbf{AX} = \mathbf{D}$.

16. **(Vibrations)** Show that the vertical vibrations of the mechanical system in the figure (masses of springs and damping neglected) are governed by the system of differential equations

$$\begin{aligned} \ddot{y}_1 &= -ky_1 + k(y_2 - y_1) \\ \ddot{y}_2 &= -k(y_2 - y_1) - ky_2 \end{aligned} \quad \text{thus} \quad \ddot{\mathbf{y}} = \mathbf{Ay} \quad \text{where} \quad \mathbf{A} = \begin{pmatrix} -2k & k \\ k & -2k \end{pmatrix}.$$

Problem 16 (System in static equilibrium) Problem 18 (System in static equilibrium)

17. To solve the vector equation in Prob. 16, substitute $\mathbf{y} = \mathbf{x}e^{\omega t}$. Show that this leads to the eigenvalue problem $\mathbf{Ax} = \lambda\mathbf{x}$ where $\lambda = \omega^2$. Find the eigenvalues and eigenvectors. Find the solution of the vector equation that satisfies the initial conditions $y_1(0) = 1$, $y_2(0) = 1$, $\dot{y}_1(0) = \sqrt{3k}$, $\dot{y}_2(0) = -\sqrt{3k}$.

18. Find the system of differential equations governing the motions of the (undamped) mechanical system in the figure, write it in vector form $\ddot{\mathbf{y}} = \mathbf{Ay}$, substitute $\mathbf{y} = \mathbf{x}e^{\omega t}$ to obtain $\mathbf{Ax} = \lambda\mathbf{x}$ ($\lambda = \omega^2$), and determine the eigenvalues of $\mathbf{A}$. What does it mean mechanically that the eigenvalues are negative?

19. **(Trace)** The sum of the elements in the principal diagonal of a square matrix $\mathbf{C} = (c_{ik})$ is called the *trace* of $\mathbf{C}$. Thus

$$\text{trace } \mathbf{C} = c_{11} + c_{22} + \cdots + c_{nn}.$$

Let $\mathbf{A} = (a_{ik})$ and $\mathbf{B} = (b_{ik})$ be two n-rowed square matrices. Show that

$$\text{trace } \mathbf{AB} = \sum_{i=1}^{n}\sum_{l=1}^{n} a_{il}b_{li}.$$

20. (**Collineatory transformation**) Consider the linear transformation $\mathbf{y} = \mathbf{Ax}$ where $\mathbf{A}$ is square and set $\mathbf{x} = \mathbf{T\tilde{x}}$, $\mathbf{y} = \mathbf{T\tilde{y}}$ where $\mathbf{T}$ is nonsingular. Show that then

$$\mathbf{\tilde{y}} = \mathbf{\tilde{A}\tilde{x}} \qquad \text{where} \qquad \mathbf{\tilde{A}} = \mathbf{T}^{-1}\mathbf{AT}.$$

(This transformation which transforms $\mathbf{A}$ into $\mathbf{\tilde{A}}$ is called a *collineatory transformation*.)

21. Show that in Prob. 20, $\det \mathbf{\tilde{A}} = \det \mathbf{A}$. *Hint.* Use (9) in Sec. 6.7.
22. Show that in Prob. 20, trace $\mathbf{\tilde{A}} =$ trace $\mathbf{A}$.
23. Show that $\mathbf{A}$ and $\mathbf{\tilde{A}}$ in Prob. 20 have the same eigenvalues.
24. Verify the statement of Prob. 23 for the matrices

$$\mathbf{A} = \begin{pmatrix} 4 & -2 \\ 1 & 1 \end{pmatrix} \qquad \text{and} \qquad \mathbf{\tilde{A}} = \mathbf{T}^{-1}\mathbf{AT} \qquad \text{where} \qquad \mathbf{T} = \begin{pmatrix} 2 & 1 \\ 5 & 3 \end{pmatrix}.$$

25. If $\mathbf{x}$ is an eigenvector of $\mathbf{A}$ corresponding to an eigenvalue λ, show that $\mathbf{y} = \mathbf{T}^{-1}\mathbf{x}$ is an eigenvector of $\mathbf{\tilde{A}} = \mathbf{T}^{-1}\mathbf{AT}$ corresponding to the same eigenvalue λ.

CHAPTER 7

Vector Differential Calculus. Vector Fields

Vector fields can be given in terms of vector functions (Sec. 7.1) and have various physical and geometrical applications. The basic concepts of differential calculus can be extended to vector functions in a simple and natural fashion (Sec. 7.2). Vector functions are useful for representing and investigating curves (Secs. 7.3–7.5) and their application in mechanics (Sec. 7.6). Physically and geometrically important concepts in connection with scalar and vector fields are the gradient (Sec. 7.8), divergence (Sec. 7.10), and curl (Sec. 7.11). (Corresponding integral theorems will be considered in Chap. 8.)

Prerequisite for this chapter: Chap. 5.
Sections which may be omitted in a shorter course: 7.4–7.6, 7.9.
References: Appendix 1, Part C.
Answers to problems: Appendix 2.

7.1 Scalar Fields and Vector Fields

A **scalar function** is a function which is defined at a certain set of points in space and whose values are real numbers depending only on the points in space but not on the particular choice of the coordinate system. In most applications the domain of definition D of a scalar function f will be a curve, a surface, or a three-dimensional region in space. The function f associates with each point in D a scalar, a real number, and we say that a **scalar field** is given in D.

If we introduce coordinates x, y, z, then f may be represented in terms of the coordinates, and we write $f(x, y, z)$, but we should keep in mind that the value of f at any point P is independent of the particular choice of coordinates. In order to indicate this fact it is also customary to write $f(P)$ instead of $f(x, y, z)$. The function f may also depend on parameters such as time.

Example 1. The distance $f(P)$ of any point P from a fixed point P_0 in space is a scalar function whose domain of definition D is the whole space. $f(P)$ defines a scalar field in space. If we introduce a Cartesian coordinate system and P_0 has the coordinates x_0, y_0, z_0, then f is given by the

Fig. 118. Field of tangent vectors of a curve

well-known formula

$$f(P) = f(x, y, z) = \sqrt{(x - x_0)^2 + (y - y_0)^2 + (z - z_0)^2}$$

where x, y, z are the coordinates of P. If we replace the given Cartesian coordinate system by another such system, then the values of the coordinates of P and P_0 will in general change, but $f(P)$ will have the same value as before. Hence, $f(P)$ is a scalar function. The direction cosines of the line through P and P_0 are not scalars because their values will depend on the choice of the coordinate system.

Example 2. The temperature T within a body B is a scalar function. It defines a scalar field, namely, the temperature field in B. The function T may depend on the time or other parameters. Other examples of scalar fields are the pressure within a region through which a compressible fluid is flowing, and the density of the air of the earth's atmosphere.

If to each point P of a certain set of points in space (for example, the points of a curve, a surface, or a three-dimensional region) a vector $\mathbf{v}(P)$ is assigned, then a **vector field** is said to be given at those points, and $\mathbf{v}(P)$ is called a **vector function**. Some illustrative examples are shown in Figs. 118 to 120.

If we introduce Cartesian coordinates x, y, z, then we may write

$$\mathbf{v}(x, y, z) = v_1(x, y, z)\mathbf{i} + v_2(x, y, z)\mathbf{j} + v_3(x, y, z)\mathbf{k},$$

but we should keep in mind that $\mathbf{v}$ depends only on the points of its domain of definition, and at any such point defines the same vector for every choice of the coordinate system.

Example 3 (Velocity field). At any instant the velocity vectors $\mathbf{v}(P)$ of a rotating body B constitute a vector field, the so-called *velocity field* of the rotation. If we introduce a Cartesian coordinate system having the origin on the axis of rotation, then (cf. Ex. 3 in Sec. 5.8)

(1) $$\mathbf{v}(x, y, z) = \mathbf{w} \times (x\mathbf{i} + y\mathbf{j} + z\mathbf{k})$$

Fig. 119. Field of normal vectors of a surface

Fig. 120. Velocity field of a rotating body

Fig. 121. Gravitational field

where x, y, z are the coordinates of any point P of B at the instant under consideration. If the coordinates are such that the z-axis is the axis of rotation and **w** points in the positive z-direction, then $\mathbf{w} = \omega\mathbf{k}$ and

$$\mathbf{v} = \begin{vmatrix} \mathbf{i} & \mathbf{j} & \mathbf{k} \\ 0 & 0 & \omega \\ x & y & z \end{vmatrix} = \omega(-y\mathbf{i} + x\mathbf{j}).$$

Example 4 (Field of force). Let a particle A of mass M be fixed at a point P_0 and let a particle B of mass m to be free to take up various positions P in space. Then A attracts B. According to **Newton's law of gravitation** the corresponding gravitational force **p** is directed from P to P_0, and its magnitude is proportional to $1/r^2$ where r is the distance between P and P_0, say,

$$(2) \qquad\qquad |\mathbf{p}| = \frac{c}{r^2}, \qquad\qquad c = GMm,$$

where G ($= 6.67 \cdot 10^{-8}$ cm^3/gm $\cdot$ sec^2) is the gravitational constant. Hence **p** defines a vector field in space. If we introduce Cartesian coordinates such that P_0 has the coordinates x_0, y_0, z_0 and P has the coordinates x, y, z, then

$$r = \sqrt{(x - x_0)^2 + (y - y_0)^2 + (z - z_0)^2} \qquad (\geq 0).$$

Assuming that $r > 0$ and introducing the vector

$$\mathbf{r} = (x - x_0)\mathbf{i} + (y - y_0)\mathbf{j} + (z - z_0)\mathbf{k},$$

we have $|\mathbf{r}| = r$, and $-\mathbf{r}/r$ is a unit vector in the direction of **p**; the minus sign indicates that **p** is directed from P to P_0 (Fig. 121). From this and (2) we obtain

$$(3) \qquad \mathbf{p} = |\mathbf{p}|\left(-\frac{\mathbf{r}}{r}\right) = -c\frac{\mathbf{r}}{r^3} = -c\frac{x - x_0}{r^3}\mathbf{i} - c\frac{y - y_0}{r^3}\mathbf{j} - c\frac{z - z_0}{r^3}\mathbf{k}.$$

This vector function describes the gravitational force acting on B.

Problems

Determine the **level curves** $T(x, y) = const$ (curves of constant temperature or *isotherms*) of the temperature field in the xy-plane given by the following functions. Plot some of the isotherms.

1. $T = x$
2. $T = xy$
3. $T = x^2 - y^2$
4. $T = \ln(x^2 + y^2)$
5. $T = x^3 - 3xy^2$
6. $T = e^{-x}\cos y$
7. $T = \dfrac{x}{x^2 + y^2}$
8. $T = \arctan\dfrac{y}{x}$
9. $T = \sin x \cosh y$

Consider the scalar field determined by $f(x, y) = x^2 + 4y^2$. Find:

10. The level curves. (Graph some of them.)

11. The value of f at the points $(-1, 2)$, $(3, 8)$, and $(0, 5)$.

12. f on the parabola $y = 2x^2$.

13. f on the unit circle $x^2 + y^2 = 1$.

14. The region in which $1 \leq f(x, y) \leq 4$.

Determine the **level surfaces** (surfaces $f(x, y, z) = const$ in xyz-space) of the following functions.

15. $x + y - z$ **16.** $x^2 - y$

17. $x^2 + 2y^2 + 3z^2$ **18.** $(x - 1)^2 + (y + 1)^2 + z^2$

19. $x^2 - y + z^2$ **20.** $z - \sqrt{x^2 + y^2}$

Consider the pressure field in space given by $p(x, y, z) = 3x^2 + y^2 + z^2$. Find:

21. The level surfaces. (Graph some of them.)

22. A formula for the pressure on the surface $xyz = 1$.

23. The level curves in the plane $z = 1$. (Graph some of them.)

24. The region in which $3 \leq p(x, y, z) \leq 12$.

Draw figures (similar to Fig. 121) of the vector fields in the xy-plane given by the following vector functions.

25. $\mathbf{i}$ **26.** $2\mathbf{i} + 3x\mathbf{j}$

27. $2x\mathbf{i} - 2y\mathbf{j}$ **28.** $x^2\mathbf{i} + y^2\mathbf{j}$

29. $(x^2 - y^2)\mathbf{i} - 2xy\mathbf{j}$ **30.** $(x^2 + y^2)\mathbf{i} + \mathbf{j}$

Let $\mathbf{v}(x, y) = (x + y)\mathbf{i} + (x - y)\mathbf{j}$. Find

31. $\mathbf{v}$ at the points of the hyperbola $xy = 1$. (Graph some of these vectors.)

32. The curves on which $\mathbf{v}$ has constant length.

33. The curves on which $\mathbf{v}$ has constant direction.

34. Find the surfaces on which $\mathbf{v}(x, y, z) = y\mathbf{i} + z\mathbf{j} + x\mathbf{k}$ has constant length.

35. Find the surfaces on which $\mathbf{v}(x, y, z) = 2x\mathbf{i} - z\mathbf{j} + 3y\mathbf{k}$ has constant length.

7.2   *Vector Calculus*

The basic concepts of calculus, such as convergence, continuity, and differentiability, can be introduced to vector analysis in a simple and natural way as follows.

An infinite sequence of vectors $\mathbf{a}_{(n)}$, $n = 1, 2, \cdots$, is said to **converge** to a limit vector $\mathbf{a}$ if

(1) $$\lim_{n \to \infty} |\mathbf{a}_{(n)} - \mathbf{a}| = 0,$$

and then we write

(2) $$\lim_{n \to \infty} \mathbf{a}_{(n)} = \mathbf{a}.$$

Clearly, if a Cartesian coordinate system has been introduced, then that sequence of vectors converges to $\mathbf{a}$ if and only if the three sequences of the components of the vectors converge to the corresponding components of $\mathbf{a}$. The simple proof is left to the reader.

Similarly, a vector function $\mathbf{u}(t)$ of a real variable t is said to have the **limit l** as t approaches t_0, if $\mathbf{u}(t)$ is defined in some *neighborhood*[1] of t_0 (possibly except at t_0)

[1] That is, in some interval on the t-axis containing t_0 as an interior point.

and

(3)
$$\lim_{t \to t_0} |\mathbf{u}(t) - \mathbf{l}| = 0.$$

Then we write

(4)
$$\lim_{t \to t_0} \mathbf{u}(t) = \mathbf{l}.$$

A vector function $\mathbf{u}(t)$ is said to be **continuous** at $t = t_0$ if it is defined in some neighborhood of t_0 and

(5)
$$\lim_{t \to t_0} \mathbf{u}(t) = \mathbf{u}(t_0).$$

If we introduce a Cartesian coordinate system, we may represent $\mathbf{u}(t)$ in the form

$$\mathbf{u}(t) = u_1(t)\mathbf{i} + u_2(t)\mathbf{j} + u_3(t)\mathbf{k}.$$

Then $\mathbf{u}(t)$ is continuous at t_0 if and only if its three components are continuous at t_0.

A vector function $\mathbf{u}(t)$ is said to be **differentiable** at a point t if the limit

(6)
$$\mathbf{u}'(t) = \lim_{\Delta t \to 0} \frac{\mathbf{u}(t + \Delta t) - \mathbf{u}(t)}{\Delta t}$$

exists. The vector $\mathbf{u}'(t)$ is called the **derivative** of $\mathbf{u}(t)$. Cf. Fig. 122.

Suppose that a Cartesian coordinate system has been introduced. Then $\mathbf{u}(t)$ is differentiable at a point t if and only if its three components are differentiable at that t, that is, if and only if the derivatives

$$u_m'(t) = \lim_{\Delta t \to 0} \frac{u_m(t + \Delta t) - u_m(t)}{\Delta t} \qquad (m = 1, 2, 3)$$

exist. Then

(7)
$$\mathbf{u}'(t) = u_1'(t)\mathbf{i} + u_2'(t)\mathbf{j} + u_3'(t)\mathbf{k};$$

that is, *to differentiate a vector function one differentiates each component separately.*

The familiar rules of differentiation yield corresponding rules for differentiating vector functions, for example,

$$(c\mathbf{u})' = c\mathbf{u}' \ (c \text{ constant}), \qquad (\mathbf{u} + \mathbf{v})' = \mathbf{u}' + \mathbf{v}'$$

and in particular

(8)
$$(\mathbf{u} \cdot \mathbf{v})' = \mathbf{u}' \cdot \mathbf{v} + \mathbf{u} \cdot \mathbf{v}'$$

(9)
$$(\mathbf{u} \times \mathbf{v})' = \mathbf{u}' \times \mathbf{v} + \mathbf{u} \times \mathbf{v}'$$

(10)
$$(\mathbf{u} \ \mathbf{v} \ \mathbf{w})' = (\mathbf{u}' \ \mathbf{v} \ \mathbf{w}) + (\mathbf{u} \ \mathbf{v}' \ \mathbf{w}) + (\mathbf{u} \ \mathbf{v} \ \mathbf{w}').$$

Fig. 122. Derivative of a vector function

The simple proofs are left to the reader. In (9), the order of the vectors must be carefully observed because cross-multiplication is not commutative.

Example 1. Let $\mathbf{u}(t)$ be a vector function whose magnitude is constant, say, $|\mathbf{u}(t)| = c$. Then $|\mathbf{u}|^2 = \mathbf{u} \cdot \mathbf{u} = c^2$, and $(\mathbf{u} \cdot \mathbf{u})' = 2\mathbf{u} \cdot \mathbf{u}' = 0$, by differentiation [cf. (8)]. This yields the following result. *The derivative of a vector function* $\mathbf{u}(t)$ *of constant magnitude is either the zero vector or is perpendicular to* $\mathbf{u}(t)$.

Important applications of derivatives will be considered in the following sections.

From our considerations the way of introducing partial differentiation to vector analysis is obvious. Let the components of a vector function

$$\mathbf{u} = u_1\mathbf{i} + u_2\mathbf{j} + u_3\mathbf{k}$$

be differentiable functions of n variables $t_1, \cdots, t_n$. Then the partial derivative of $\mathbf{u}$ with respect to t_l is denoted by $\partial \mathbf{u}/\partial t_l$ and is defined as the vector function

$$\frac{\partial \mathbf{u}}{\partial t_l} = \frac{\partial u_1}{\partial t_l}\mathbf{i} + \frac{\partial u_2}{\partial t_l}\mathbf{j} + \frac{\partial u_3}{\partial t_l}\mathbf{k}.$$

Similarly,

$$\frac{\partial^2 \mathbf{u}}{\partial t_l\, \partial t_m} = \frac{\partial^2 u_1}{\partial t_l\, \partial t_m}\mathbf{i} + \frac{\partial^2 u_2}{\partial t_l\, \partial t_m}\mathbf{j} + \frac{\partial^2 u_3}{\partial t_l\, \partial t_m}\mathbf{k},$$

and so on.

Example 2. Let $\mathbf{r}(t_1, t_2) = a \cos t_1\mathbf{i} + a \sin t_1\mathbf{j} + t_2\mathbf{k}$. Then

$$\frac{\partial \mathbf{r}}{\partial t_1} = -a \sin t_1\mathbf{i} + a \cos t_1\mathbf{j}, \qquad \frac{\partial \mathbf{r}}{\partial t_2} = \mathbf{k}.$$

Note that if $\mathbf{r}(t_1, t_2)$ is interpreted as a position vector, it represents a cylinder of revolution of radius a, having the z-axis as axis of rotation. (Representations of surfaces will be considered in Sec. 8.5.)

Problems

Find the first and second derivatives of the following vector functions and the lengths of these derivatives.

1. $\mathbf{a} + \mathbf{b}t^2$ 2. $t\mathbf{i} + t^2\mathbf{j}$ 3. $\cos t\,\mathbf{i} + \sin t\,\mathbf{j}$
4. $3 \cos t\,\mathbf{i} + 2 \sin t\,\mathbf{j}$ 5. $t\mathbf{i} + t^2\mathbf{j} + t^3\mathbf{k}$ 6. $\cos t\,\mathbf{i} + \sin 2t\,\mathbf{j}$
7. $e^t\mathbf{i} + e^{-t}\mathbf{j}$ 8. $e^{-t}(\cos t\,\mathbf{i} + \sin t\,\mathbf{j})$ 9. $\sin 3t\,(\mathbf{i} + \mathbf{j})$

In each case find the first partial derivatives with respect to x, y, z.

10. $x^2\mathbf{i} - z^2\mathbf{j} + y^2\mathbf{k}$ 11. $xy\mathbf{i} + yz\mathbf{j}$ 12. $y\mathbf{i} - z\mathbf{j} + x\mathbf{k}$
13. $e^y\mathbf{i} - e^{-z}\mathbf{j}$ 14. $(x + y)\mathbf{j} + (x - y)\mathbf{k}$ 15. $x^2y\mathbf{i} + y^2z\mathbf{j} + z^2x\mathbf{k}$

16. Using (8) in Sec. 5.5, prove (8). Prove (9).
17. Derive (10) from (8) and (9).
18. Give a direct proof of (10).
19. Find formulas similar to (8) and (9) for $(\mathbf{u} \cdot \mathbf{v})''$ and $(\mathbf{u} \times \mathbf{v})''$.
20. Show that if $\mathbf{u}(t)$ has length 1 and $\mathbf{u}'(t) \neq 0$, then $\mathbf{u}$ and $\mathbf{u}'$ are orthogonal.
21. Show that $(\mathbf{u}\,\mathbf{u}'\,\mathbf{u}'')' = (\mathbf{u}\,\mathbf{u}'\,\mathbf{u}''')$.

Let $\mathbf{u} = t^2\mathbf{i} - 3t\mathbf{k}$, $\mathbf{v} = -4t^3\mathbf{j} + (t + 1)\mathbf{k}$, $\mathbf{w} = 3t^2\mathbf{i} + t\mathbf{j} - t^3\mathbf{k}$. Find

  22. $(\mathbf{u} \cdot \mathbf{v})'$ **23.** $(\mathbf{u} \times \mathbf{v})'$ **24.** $(\mathbf{u}\,\mathbf{v}\,\mathbf{w})'$

  25. $(\mathbf{u} \times \mathbf{u}')'$ **26.** $[\mathbf{u} \times (\mathbf{v} \times \mathbf{w})]'$ **27.** $[\mathbf{u} \times (\mathbf{v} + \mathbf{w})]'$

28. Show that $\mathbf{u}(t) = \mathbf{b}e^{\lambda t} + \mathbf{c}e^{-\lambda t}$ satisfies the equation $\mathbf{u}'' - \lambda^2\mathbf{u} = 0$. (**b** and **c** are constant vectors.)

29. Show that $\left(\dfrac{\mathbf{u}}{|\mathbf{u}|}\right)' = \dfrac{\mathbf{u}'(\mathbf{u} \cdot \mathbf{u}) - \mathbf{u}(\mathbf{u} \cdot \mathbf{u}')}{(\mathbf{u} \cdot \mathbf{u})^{3/2}}$.

30. Differentiate $(t\mathbf{i} + t^2\mathbf{j})/|t\mathbf{i} + t^2\mathbf{j}|$.

7.3   *Curves*

As an important application of vector calculus, let us now consider some basic facts about curves in space. The student will know that curves occur in many considerations in calculus as well as in physics, for example, as paths of moving particles. The consideration will be a part of an important branch of mathematics, which is called **differential geometry** and which may be defined as the study of curves and surfaces by means of calculus. Cf. Ref. [C8] in Appendix 1.

A Cartesian coordinate system being given, we may represent a curve C by a vector function (Fig. 123)

(1) $$\mathbf{r}(t) = x(t)\mathbf{i} + y(t)\mathbf{j} + z(t)\mathbf{k};$$

to each value t_0 of the real variable t there corresponds a point of C having the position vector $\mathbf{r}(t_0)$, that is, the coordinates $x(t_0)$, $y(t_0)$, $z(t_0)$.

A representation of the form (1) is called a **parametric representation** of the curve C, and t is called the *parameter* of this representation. This type of representation is useful in many applications, for example, in mechanics where the variable t may be the time.

Other types of representations of curves in space are

(2) $$y = f(x), \qquad z = g(x)$$

and

(3) $$F(x, y, z) = 0, \qquad G(x, y, z) = 0.$$

Fig. 123. Parametric representation of a curve

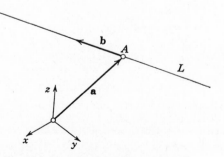

Fig. 124. Parametric representation of a straight line

By setting $x = t$ we may write (2) in the form (1), namely

$$\mathbf{r}(t) = t\mathbf{i} + f(t)\mathbf{j} + g(t)\mathbf{k}.$$

In (3), each equation represents a surface, and the curve is the intersection of the two surfaces.

A **plane curve** is a curve which lies in a plane in space. A curve which is not a plane curve is called a **twisted curve**.

Example 1 (*Straight line*). Any straight line L can be represented in the form

(4) $\mathbf{r}(t) = \mathbf{a} + t\mathbf{b} = (a_1 + tb_1)\mathbf{i} + (a_2 + tb_2)\mathbf{j} + (a_3 + tb_3)\mathbf{k}$

where $\mathbf{a}$ and $\mathbf{b}$ are constant vectors. L passes through the point A with position vector $\mathbf{r} = \mathbf{a}$ and has the direction of $\mathbf{b}$ (Fig. 124). If $\mathbf{b}$ is a unit vector, its components are the *direction cosines* of L, and in this case, $|t|$ measures the distance of the points of L from A.

Example 2 (*Ellipse, circle*). The vector function

(5) $\mathbf{r}(t) = a \cos t\, \mathbf{i} + b \sin t\, \mathbf{j}$

represents an ellipse in the xy-plane with center at the origin and principal axes in the direction of the x and y axes. In fact, since $\cos^2 t + \sin^2 t = 1$, we obtain from (5)

$$\frac{x^2}{a^2} + \frac{y^2}{b^2} = 1, \qquad z = 0.$$

If $b = a$, then (5) represents a *circle* of radius a.

Example 3 (*Circular helix*). The twisted curve C represented by the vector function

(6) $\mathbf{r}(t) = a \cos t\, \mathbf{i} + a \sin t\, \mathbf{j} + ct\mathbf{k}$ $(c \neq 0)$

is called a *circular helix*. It lies on the cylinder $x^2 + y^2 = a^2$. If $c > 0$, the helix is shaped like a right-handed screw (Fig. 125). If $c < 0$, it looks like a left-handed screw (Fig. 126).

The portion between any two points of a curve is often called an **arc** *of a curve*. For the sake of simplicity we shall use the single term "curve" to denote an entire curve as well as an arc of a curve.

A curve may have self-intersections; the points of intersection are called *multiple points* of the curve. Two examples are shown in Fig. 127. A curve having no multiple points is called a **simple curve**.

Fig. 125. Right-handed
circular helix

Fig. 126. Left-handed circular helix

Fig. 127. Curves having double points

Example 4. Ellipses and helices are simple curves. The curve represented by

$$\mathbf{r}(t) = (t^2 - 1)\mathbf{i} + (t^3 - t)\mathbf{j}$$

is not simple since it has a double point at the origin; this point corresponds to the two values $t = 1$ and $t = -1$.

We finally mention that a given curve C may be represented by various vector functions. For example, if C is represented by (1) and we set $t = h(t^*)$, then we obtain a new vector function $\tilde{\mathbf{r}}(t^*) = \mathbf{r}(h(t^*))$ representing C, provided $h(t^*)$ takes on all the values of t occurring in (1).

Example 5. The parabola $y = x^2$ in the xy-plane may be represented by the vector function

$$\mathbf{r}(t) = t\mathbf{i} + t^2\mathbf{j} \qquad\qquad (-\infty < t < \infty).$$

If we set $t = -2t^*$, we obtain another representation of the parabola:

$$\tilde{\mathbf{r}}(t^*) = \mathbf{r}(-2t^*) = -2t^*\mathbf{i} + 4t^{*2}\mathbf{j}.$$

If we set $t = t^{*2}$, we obtain

$$\tilde{\mathbf{r}}(t^*) = t^{*2}\mathbf{i} + t^{*4}\mathbf{j},$$

but this function represents only the portion of the parabola in the first quadrant, because $t^{*2} \geqq 0$ for all t^*.

Problems

Find a parametric representation of the straight line through a point A in the direction of a vector **b**, where

  1. A: $(1, 1, 0)$, $\mathbf{b} = \mathbf{k}$ **2.** A: $(2, 3, 1)$, $\mathbf{b} = \mathbf{i} - \mathbf{j} + 2\mathbf{k}$
  3. A: $(-3, 1, -2)$, $\mathbf{b} = 3\mathbf{i} - \mathbf{k}$ **4.** A: $(3, 4, 1)$, $\mathbf{b} = 3\mathbf{i} + 5\mathbf{j} - \mathbf{k}$

Find a parametric representation of the straight line through the points A and B, where
  5. A: $(2, 0, 4)$, B: $(1, 0, 6)$ **6.** A: $(0, 2, 0)$, B: $(1, 5, -2)$
  7. A: $(3, -1, 5)$, B: $(5, -5, 5)$ **8.** A: $(0, 0, 4)$, B: $(-2, 3, 2)$

Find a parametric representation of the straight line represented by
  9. $x = y$, $y = z$ **10.** $x + y - z = 2$, $2x - 5y + z = 3$
  11. $7x - 3y + z = 14$, $4x - 3y - 2z = -1$ **12.** $2x + 3y + 4z = 0$, $x - y + 3z = 0$

Sketch the curves represented by the following functions $\mathbf{r}(t)$.
  13. $(1 + \cos t)\mathbf{i} + (1 + \sin t)\mathbf{j}$ **14.** $t\mathbf{i} + (3 + t^3)\mathbf{j}$
  15. $t\mathbf{i} + t^2\mathbf{j} + t^3\mathbf{k}$ **16.** $2\cos t\,\mathbf{i} + 2\sin t\,\mathbf{j} + t\mathbf{k}$
  17. $\cos t\,\mathbf{j} + (2 + 2\sin t)\mathbf{k}$ **18.** $3\cos t\,\mathbf{i} + \sin t\,\mathbf{j} + 2t\mathbf{k}$
  19. $(3t\mathbf{i} + 3t^2\mathbf{j})/(1 + t^3)$ (*"Folium of Descartes"*)

Represent the following curves in parametric form and sketch these curves.
  20. $y = 2(x - 1)^4$, $z = 0$ **21.** $(x + 2)^2 + (y - 2)^2 = 4$, $z = 0$
  22. $x^2 + y^2 = 1$, $z = e^x$ **23.** $y = x^2$, $z = x^3$
  24. $y = x^3$, $z = x$ **25.** $x^2 + y^2 = 4$, $z = x^2$

7.4 Arc Length

To define the length of a curve C we may proceed as follows. We inscribe in C a broken line of n chords joining the two end points of C as shown in Fig. 128. This we do for each positive integer n in an arbitrary way but such that the maximum chord-length approaches zero as n approaches infinity. The lengths of these lines of chords can be obtained from the theorem of Pythagoras. If the sequence of these lengths l_1, l_2, $\cdots$ is convergent with the limit l, then C is said to be **rectifiable,** and l is called the **length** of C.

If C can be represented by a continuously differentiable[2] vector function

$$\mathbf{r} = \mathbf{r}(t) \qquad\qquad (a \leqq t \leqq b),$$

then it can be shown that C is rectifiable, and its length l is given by the integral

(1) $$l = \int_a^b \sqrt{\dot{\mathbf{r}} \cdot \dot{\mathbf{r}}}\, dt \qquad\qquad \left(\dot{\mathbf{r}} = \frac{d\mathbf{r}}{dt}\right),$$

whose value is independent of the choice of the parametric representation. The proof is quite similar to that for plane curves usually considered in elementary integral calculus (cf. Ref. [A14]) and can be found in Ref. [C8] in Appendix 1.

If we replace the fixed upper limit b in (1) with a variable upper limit t, the integral becomes a function of t, say, $s(t)$; denoting the variable of integration by t^*, we have

(2) $$s(t) = \int_a^t \sqrt{\dot{\mathbf{r}} \cdot \dot{\mathbf{r}}}\, dt^* \qquad\qquad \left(\dot{\mathbf{r}} = \frac{d\mathbf{r}}{dt^*}\right).$$

This function $s(t)$ is called the **arc length** of C.

The arc length s may serve as a parameter in parametric representations of curves, and we shall see that this will lead to a simplification of various formulas.

The constant a in (2) may be replaced by another constant; that is, the point of the curve corresponding to $s = 0$ may be chosen in an arbitrary manner. The sense corresponding to increasing values of s is called the **positive sense** on C; in this fashion any representation $\mathbf{r}(s)$ or $\mathbf{r}(t)$ of C defines a certain **orientation** of C.

Fig. 128. Length of a curve

---

[2] *"Continuously differentiable"* means that the derivative exists and is continuous; *"twice continuously differentiable"* means that the first and second derivatives exist and are continuous, and so on.

Obviously, there are two ways of *orienting C*, and it is not difficult to see that the transition from one orientation to the opposite orientation can be effected by a transformation of the parameter whose derivative is negative.

From (2) we obtain

(3)
$$\left(\frac{ds}{dt}\right)^2 = \frac{d\mathbf{r}}{dt} \cdot \frac{d\mathbf{r}}{dt} = \left(\frac{dx}{dt}\right)^2 + \left(\frac{dy}{dt}\right)^2 + \left(\frac{dz}{dt}\right)^2.$$

It is customary to write
$$d\mathbf{r} = dx\,\mathbf{i} + dy\,\mathbf{j} + dz\,\mathbf{k}$$
and

(4)
$$ds^2 = d\mathbf{r} \cdot d\mathbf{r} = dx^2 + dy^2 + dz^2.$$

ds is called the **linear element** of C.

Example 1. In the case of the circle
$$\mathbf{r}(t) = a \cos t\,\mathbf{i} + a \sin t\,\mathbf{j}$$
we have $\dot{\mathbf{r}} = -a \sin t\,\mathbf{i} + a \cos t\,\mathbf{j}$, $\dot{\mathbf{r}} \cdot \dot{\mathbf{r}} = a^2$, and therefore,

$$s(t) = \int_0^t a\,dt^* = at.$$

Hence a representation of the circle with the arc length as parameter is

$$\mathbf{r}(s) = a \cos \frac{s}{a}\,\mathbf{i} + a \sin \frac{s}{a}\,\mathbf{j}.$$

The circle is oriented in the counterclockwise sense, which corresponds to increasing values of s. Setting $s = -\bar{s}$, we obtain

$$\bar{\mathbf{r}}(\bar{s}) = a \cos \frac{\bar{s}}{a}\,\mathbf{i} - a \sin \frac{\bar{s}}{a}\,\mathbf{j};$$

we have $ds/d\bar{s} = -1 < 0$, and the circle is now oriented in the clockwise sense. Note that we wrote $\bar{\mathbf{r}}$ because this is a new function; in fact, $\mathbf{r}(-\bar{s}) = \bar{\mathbf{r}}(\bar{s})$.

Problems

1. If a plane curve is represented in the form $y = f(x)$, $z = 0$, show that its length between $x = a$ and $x = b$ is

$$l = \int_a^b \sqrt{1 + y'^2}\,dx.$$

2. Using the formula in Prob. 1, find the length of a circle of radius a.

3. Show that if a plane curve is represented in polar coordinates $\rho = \sqrt{x^2 + y^2}$ and $\theta = \arctan(y/x)$, then $ds^2 = \rho^2\,d\theta^2 + d\rho^2$.

4. Using the formula in Prob. 3, find the length of a circle of radius a.

5. Graph the *cardioid* $\rho = a(1 - \cos \theta)$ and find its length.

6. If a curve is represented by a parametric representation, show that a transformation of the parameter whose derivative is negative reverses the orientation.

Graph the following curves and find their lengths.

  7. *Catenary* $y = \cosh x$, $z = 0$, from $x = 0$ to $x = 1$
  8. *Circular helix* $\mathbf{r}(t) = a \cos t\,\mathbf{i} + a \sin t\,\mathbf{j} + ct\mathbf{k}$ from $(a, 0, 0)$ to $(a, 0, 2\pi c)$
  9. *Semicubical parabola* $y = x^{3/2}$, $z = 0$, from $(0, 0, 0)$ to $(4, 8, 0)$
  10. *Four-cusped hypocycloid* $\mathbf{r}(t) = a \cos^3 t\,\mathbf{i} + a \sin^3 t\,\mathbf{j}$, total length

7.5 Tangent. Curvature and Torsion

The *tangent* to a curve C at a point P of C is defined as the limiting position of the straight line L through P and another point Q of C as Q approaches P along the curve (Fig. 129).

Suppose that C is represented by a continuously differentiable vector function $\mathbf{r}(t)$ where t is any parameter. Let P and Q correspond to t and $t + \Delta t$, respectively. Then L has the direction of the vector

$$[\mathbf{r}(t + \Delta t) - \mathbf{r}(t)]/\Delta t.$$

Hence, if the vector

$$(1) \qquad\qquad \dot{\mathbf{r}} = \lim_{\Delta t \to 0} \frac{\mathbf{r}(t + \Delta t) - \mathbf{r}(t)}{\Delta t}$$

is not the zero vector, it has the direction of the tangent to C at P. It points in the direction of increasing values of t, and its sense, therefore, depends on the orientation of the curve. $\dot{\mathbf{r}}$ is called a *tangent vector* of C at P, and the corresponding unit vector

$$(2) \qquad\qquad \mathbf{u} = \frac{\dot{\mathbf{r}}}{|\dot{\mathbf{r}}|}$$

is called the **unit tangent vector** to C at P.

If in particular C is represented by $\mathbf{r}(s)$, where s is the arc length, it follows from (3), Sec. 7.4, that the derivative $d\mathbf{r}/ds$ is a unit vector, and (2) becomes

$$(3) \qquad\qquad \mathbf{u} = \mathbf{r}' = \frac{d\mathbf{r}}{ds}.$$

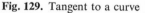

Fig. 129. Tangent to a curve

Fig. 130. Representation of the tangent to a curve

Clearly the position vector of a point T on the tangent is the sum of the position vector $\mathbf{r}$ of P and a vector in the direction of the tangent. Hence a parametric representation of the tangent is (Fig. 130)

$$\text{(4)} \qquad\qquad \mathbf{q}(w) = \mathbf{r} + w\dot{\mathbf{r}}$$

where both $\mathbf{r}$ and $\dot{\mathbf{r}}$ depend on P and the parameter w is a real variable.

Consider a given curve C, represented by a three times continuously differentiable vector function $\mathbf{r}(s)$ (cf. footnote 2 in Sec. 7.4) where s is the arc length. Then

$$\text{(5)} \qquad\qquad \kappa(s) = |\mathbf{u}'(s)| = |\mathbf{r}''(s)| \qquad\qquad (\kappa \geqq 0)$$

is called the **curvature** of C. If $\kappa \neq 0$, the unit vector $\mathbf{p}$ in the direction of $\mathbf{u}'(s)$ is

$$\text{(6)} \qquad\qquad \mathbf{p} = \frac{\mathbf{u}'}{\kappa} \qquad\qquad (\kappa > 0)$$

and is called the **unit principal normal vector** of C. From Ex. 1 in Sec. 7.2 we see that $\mathbf{p}$ is perpendicular to $\mathbf{u}$. The vector

$$\text{(7)} \qquad\qquad \mathbf{b} = \mathbf{u} \times \mathbf{p} \qquad\qquad (\kappa > 0)$$

is called the **unit binormal vector** of C. From the definition of a vector product it follows that $\mathbf{u}, \mathbf{p}, \mathbf{b}$ constitute a right-handed triple of orthogonal unit vectors (Secs. 5.3 and 5.7). This triple is called the **trihedron** of C at the point under consideration (Fig. 131). The three straight lines through that point in the directions of $\mathbf{u}, \mathbf{p}, \mathbf{b}$ are called the **tangent**, the **principal normal**, and the **binormal** of C. Figure 131 also shows the names of the three planes spanned by each pair of those vectors.

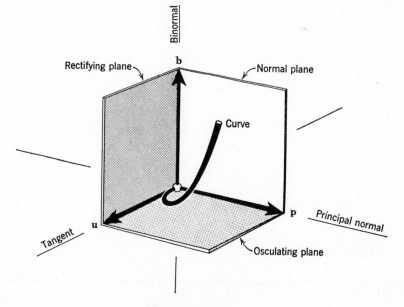

Fig. 131. Trihedron

If the derivative $\mathbf{b}'$ is not the zero vector, it is perpendicular to $\mathbf{b}$ (cf. Ex. 1 in Sec. 7.2). It is also perpendicular to $\mathbf{u}$. In fact, by differentiating $\mathbf{b} \cdot \mathbf{u} = 0$ we have $\mathbf{b}' \cdot \mathbf{u} + \mathbf{b} \cdot \mathbf{u}' = 0$; hence $\mathbf{b}' \cdot \mathbf{u} = 0$ because $\mathbf{b} \cdot \mathbf{u}' = 0$. Consequently, $\mathbf{b}'$ is of the form $\mathbf{b}' = \alpha\mathbf{p}$ where α is a scalar. It is customary to set $\alpha = -\tau$. Then

$$(8) \qquad\qquad \mathbf{b}' = -\tau\mathbf{p} \qquad\qquad (\kappa > 0).$$

The scalar function τ is called the **torsion** of C. Scalar multiplication of both sides of (8) by $\mathbf{p}$ yields

$$(9) \qquad\qquad \tau(s) = -\mathbf{p}(s) \cdot \mathbf{b}'(s).$$

Example 1 (Circular helix). In the case of the circular helix (6) in Sec. 7.3 we obtain the arc length $s = t\sqrt{a^2 + c^2}$. Hence we may represent the helix in the form

$$\mathbf{r}(s) = a \cos \frac{s}{w}\mathbf{i} + a \sin \frac{s}{w}\mathbf{j} + c\frac{s}{w}\mathbf{k} \quad \text{where} \quad w = \sqrt{a^2 + c^2}.$$

It follows that

$$\mathbf{u}(s) = \mathbf{r}'(s) = -\frac{a}{w} \sin \frac{s}{w}\mathbf{i} + \frac{a}{w}\cos\frac{s}{w}\mathbf{j} + \frac{c}{w}\mathbf{k}$$

$$\mathbf{r}''(s) = -\frac{a}{w^2}\cos\frac{s}{w}\mathbf{i} - \frac{a}{w^2}\sin\frac{s}{w}\mathbf{j}$$

$$\kappa = |\mathbf{r}''| = \sqrt{\mathbf{r}'' \cdot \mathbf{r}''} = \frac{a}{w^2} = \frac{a}{s^2 + c^2}$$

$$\mathbf{p}(s) = \frac{\mathbf{r}''(s)}{\kappa(s)} = -\cos\frac{s}{w}\mathbf{i} - \sin\frac{s}{w}\mathbf{j}$$

$$\mathbf{b}(s) = \mathbf{u}(s) \times \mathbf{p}(s) = \frac{c}{w}\sin\frac{s}{w}\mathbf{i} - \frac{c}{w}\cos\frac{s}{w}\mathbf{j} + \frac{a}{w}\mathbf{k}$$

$$\mathbf{b}'(s) = \frac{c}{w^2}\cos\frac{s}{w}\mathbf{i} + \frac{c}{w^2}\sin\frac{s}{w}\mathbf{j}$$

$$\tau(s) = -\mathbf{p}(s) \cdot \mathbf{b}'(s) = \frac{c}{w^2} = \frac{c}{a^2 + c^2}.$$

Hence the circular helix has constant curvature and torsion. If $c > 0$ (right-handed helix, cf. Fig. 125), then $\tau > 0$, and if $c < 0$ (left-handed helix, cf. Fig. 126), then $\tau < 0$.

Since $\mathbf{u}$, $\mathbf{p}$, and $\mathbf{b}$ are linearly independent vectors, we may represent any vector in space as a linear combination of these vectors. Hence if the derivatives $\mathbf{u}'$, $\mathbf{p}'$, and $\mathbf{b}'$ exist, they may be represented in that fashion. The corresponding formulas are

$$\begin{aligned}
(a) \qquad \mathbf{u}' &= \qquad\quad \kappa\mathbf{p} \\
(10) \qquad (b) \qquad \mathbf{p}' &= -\kappa\mathbf{u} \qquad + \tau\mathbf{b} \\
(c) \qquad \mathbf{b}' &= \qquad -\tau\mathbf{p}
\end{aligned}$$

They are called the **formulas of Frenet**. (10a) follows from (6), and (10c) is identical with (8). The derivation of (10b) is left to the reader (cf. Prob. 15).

Problems

Find a parametric representation of the tangent to the following curves at the given point P.

1. $\mathbf{r}(t) = \cos t\,\mathbf{i} + \sin t\,\mathbf{j}$, P: $(-1/\sqrt{2}, 1/\sqrt{2})$

2. $\mathbf{r}(t) = \cos t\,\mathbf{i} + \sin t\,\mathbf{j} + t\mathbf{k},$ P: $(1, 0, 0)$

3. $\mathbf{r}(t) = 2\cos t\,\mathbf{i} + \sin t\,\mathbf{j},$ P: $(\sqrt{2}, 1/\sqrt{2}, 0)$

4. $\mathbf{r}(t) = t^2\mathbf{i} + t^3\mathbf{j},$ P: $(1, 1, 0)$ and P: $(0, 0, 0)$

5. $\mathbf{r}(t) = t\mathbf{i} + 2t^4\mathbf{k},$ P: $(1, 0, 2)$

6. Show that straight lines are the only curves whose unit tangent vector is constant.

7. Show that the tangent in Ex. 1 makes a constant angle with the z-axis.

8. Show that if a curve C is represented by $\mathbf{r}(t)$ where t is any parameter, then its curvature is

$$(5') \qquad\qquad \kappa = \frac{\sqrt{(\dot{\mathbf{r}} \cdot \dot{\mathbf{r}})(\ddot{\mathbf{r}} \cdot \ddot{\mathbf{r}}) - (\dot{\mathbf{r}} \cdot \ddot{\mathbf{r}})^2}}{(\dot{\mathbf{r}} \cdot \dot{\mathbf{r}})^{3/2}}.$$

9. Show that the curvature of a circle of radius a equals $1/a$.

10. Find the curvature of the ellipse $\mathbf{r}(t) = a\cos t\,\mathbf{i} + b\sin t\,\mathbf{j}$.

11. Show that the curvature of a straight line is identically zero.

12. Using $(5')$, show that for a curve $y = y(x)$ in the xy-plane,

$$\kappa = |y''|/(1 + y'^2)^{3/2} \qquad\qquad (y' = dy/dx, \text{ etc.}).$$

13. Show that the torsion of a plane curve (with $\kappa > 0$) is identically zero.

14. Using (7) and (9), show that

$$(9') \qquad\qquad \tau = (\mathbf{u}\ \mathbf{p}\ \mathbf{p}') \qquad\qquad (\kappa > 0),$$

and, because of (6), this may be written

$$(9'') \qquad\qquad \tau = (\mathbf{r}'\ \mathbf{r}''\ \mathbf{r}''')/\kappa^2 \qquad\qquad (\kappa > 0),$$

and, if the curve C is represented by $\mathbf{r}(t)$ where t is any parameter, then

$$(9''') \qquad\qquad \tau = \frac{(\dot{\mathbf{r}}\ \ddot{\mathbf{r}}\ \dddot{\mathbf{r}})}{(\dot{\mathbf{r}} \cdot \dot{\mathbf{r}})(\ddot{\mathbf{r}} \cdot \ddot{\mathbf{r}}) - (\dot{\mathbf{r}} \cdot \ddot{\mathbf{r}})^2} \qquad\qquad (\kappa > 0).$$

15. Prove the Frenet formula $(10b)$.

7.6   *Velocity and Acceleration*

Let $\mathbf{r}(t)$ be the position vector of a moving particle P in space, where t is the time. Then $\mathbf{r}(t)$ represents the path C of P. From the previous section we know that the vector

$$(1) \qquad\qquad \mathbf{v} = \dot{\mathbf{r}} = \frac{d\mathbf{r}}{dt}$$

is tangent to C and, therefore, points in the instantaneous direction of motion of P. From (3) in Sec. 7.4 we see that

$$|\mathbf{v}| = \sqrt{\dot{\mathbf{r}} \cdot \dot{\mathbf{r}}} = \frac{ds}{dt}$$

where s is the arc length, which measures the distance of P from a fixed point $(s = 0)$ on C along the curve. Hence ds/dt is the **speed** of P. The vector $\mathbf{v}$ is, therefore, called the **velocity vector**[3] of the motion.

---

[3] When no confusion is likely to arise, the word "*velocity*" is often used to denote the speed, the magnitude of $\mathbf{v}$.

The derivative of the velocity vector is called the **acceleration vector** and will be denoted by **a**; thus

(2) $$\mathbf{a}(t) = \dot{\mathbf{v}}(t) = \ddot{\mathbf{r}}(t).$$

Example 1 (*Centripetal acceleration*). The vector function

$$\mathbf{r}(t) = R \cos \omega t \, \mathbf{i} + R \sin \omega t \, \mathbf{j} \qquad (\omega > 0)$$

represents a circle C of radius R with center at the origin of the xy-plane and describes a motion of a particle P in the counterclockwise sense. The velocity vector

$$\mathbf{v} = \dot{\mathbf{r}} = -R\omega \sin \omega t \, \mathbf{i} + R\omega \cos \omega t \, \mathbf{j}$$

(cf. Fig. 132) is tangent to C, and its magnitude, the speed

$$|\mathbf{v}| = \sqrt{\dot{\mathbf{r}} \cdot \dot{\mathbf{r}}} = R\omega$$

is constant. The **angular speed** (speed divided by the distance R from the center) is equal to ω. The acceleration vector is

(3) $$\mathbf{a} = \dot{\mathbf{v}} = -R\omega^2 \cos \omega t \, \mathbf{i} - R\omega^2 \sin \omega t \, \mathbf{j} = -\omega^2 \mathbf{r}.$$

We see that there is an acceleration of constant magnitude $|\mathbf{a}| = \omega^2 R$ toward the origin, which is called the **centripetal acceleration**. It results from the fact that the velocity vector is changing direction at a constant rate. The **centripetal force** is $m\mathbf{a}$, where m is the mass of P. The opposite vector $-m\mathbf{a}$ is called the **centrifugal force**, and the two forces are in equilibrium at each instant of the motion.

It is clear that **a** is the rate of change of **v**. In Ex. 1, $|\mathbf{v}| = const$, but $|\mathbf{a}| \neq 0$, which illustrates that the magnitude of **a** is not in general the rate of change of $|\mathbf{v}|$. The reason is that in general **a** is not tangent to the path C. In fact, by applying the chain rule of differentiation to (1), and denoting derivatives with respect to s by primes, we have

$$\mathbf{v} = \frac{d\mathbf{r}}{dt} = \frac{d\mathbf{r}}{ds}\frac{ds}{dt} = \mathbf{r}'\frac{ds}{dt}$$

and by differentiating this again,

(4) $$\mathbf{a} = \frac{d\mathbf{v}}{dt} = \frac{d}{dt}\left(\mathbf{r}'\frac{ds}{dt}\right) = \mathbf{r}''\left(\frac{ds}{dt}\right)^2 + \mathbf{r}'\frac{d^2s}{dt^2}.$$

Since $\mathbf{r}'$ is the unit tangent vector **u** to C (Sec. 7.5) and its derivative $\mathbf{u}' = \mathbf{r}''$ is perpendicular to **u** (Sec. 7.5), the formula (4) is a decomposition of the acceleration

Fig. 132. Centripetal acceleration

Fig. 133. Motion in Example 2

vector into its normal component $\mathbf{r}''\dot{s}^2$ and its tangential component $\mathbf{r}'\ddot{s}$. From this we see that if and only if the normal component is zero, $|\mathbf{a}|$ equals the time rate of change of $|\mathbf{v}| = \dot{s}$ (except for the sign), because then $|\mathbf{a}| = |\mathbf{r}'|\,|\ddot{s}| = |\ddot{s}|$.

Example 2 (Coriolis[4] acceleration). A particle P moves on a disk towards the edge, the position vector being

$$(5) \qquad\qquad \mathbf{r}(t) = t\mathbf{b}$$

where $\mathbf{b}$ is a unit vector, rotating together with the disk with constant angular speed ω in the counterclockwise sense (Fig. 133). Find the acceleration $\mathbf{a}$ of P.

Because of the rotation, $\mathbf{b}$ is of the form

$$(6) \qquad\qquad \mathbf{b}(t) = \cos \omega t\,\mathbf{i} + \sin \omega t\,\mathbf{j}.$$

Differentiating (5), we obtain the velocity

$$(7) \qquad\qquad \mathbf{v} = \dot{\mathbf{r}} = \mathbf{b} + t\dot{\mathbf{b}}.$$

Obviously $\mathbf{b}$ is the velocity of P relative to the disk, and $t\dot{\mathbf{b}}$ is the additional velocity due to the rotation. Differentiating once more, we obtain the acceleration

$$(8) \qquad\qquad \mathbf{a} = \dot{\mathbf{v}} = 2\dot{\mathbf{b}} + t\ddot{\mathbf{b}}.$$

In the last term of (8) we have $\ddot{\mathbf{b}} = -\omega^2\mathbf{b}$, as follows by differentiating (6). Hence this acceleration $t\ddot{\mathbf{b}}$ is directed toward the center of the disk, and from Ex. 1 we see that this is the centripetal acceleration due to the rotation. In fact, the distance of P from the center is equal to t which, therefore, plays the role of R in Ex. 1.

The most interesting and probably unexpected term in (8) is $2\dot{\mathbf{b}}$, the so-called **Coriolis acceleration**, which results from the interaction of the rotation of the disk and the motion of P on the disk. It has the direction of $\dot{\mathbf{b}}$, that is, it is tangential to the edge of the disk and, referred to the fixed xy-coordinate system, it points in the direction of the rotation. If P is a person of mass m_0 walking on the disk according to (5), he will feel a force $-2m_0\dot{\mathbf{b}}$ in the opposite direction, that is, against the sense of the rotation.

Example 3 (Superposition of two rotations). Find the acceleration of a particle P moving on a "meridian" M of a rotating sphere with constant speed relative to the sphere.

The motion of P on M can be described analytically in the form

$$(9) \qquad\qquad \mathbf{r}(t) = R \cos \gamma t\,\mathbf{b} + R \sin \gamma t\,\mathbf{k}$$

where R is the radius of the sphere, γ (>0) the angular speed of P on M, $\mathbf{b}$ a horizontal unit vector in the plane of M (Fig. 134), and $\mathbf{k}$ the unit vector in the positive z-direction. Since $\mathbf{b}$ rotates together with the sphere, it is of the form

$$(10) \qquad\qquad \mathbf{b} = \cos \omega t\,\mathbf{i} + \sin \omega t\,\mathbf{j}$$

where ω (>0) is the angular speed of the sphere and $\mathbf{i}$ and $\mathbf{j}$ are the unit vectors in the positive x and y directions, which are fixed in space. By differentiating (9) we obtain the velocity

$$(11) \qquad\qquad \mathbf{v} = \dot{\mathbf{r}} = R \cos \gamma t\,\dot{\mathbf{b}} - \gamma R \sin \gamma t\,\mathbf{b} + \gamma R \cos \gamma t\,\mathbf{k}.$$

By differentiating this again we obtain the acceleration

$$(12) \qquad\qquad \mathbf{a} = \dot{\mathbf{v}} = R \cos \gamma t\,\ddot{\mathbf{b}} - 2\gamma R \sin \gamma t\,\dot{\mathbf{b}} - \gamma^2 R \cos \gamma t\,\mathbf{b} - \gamma^2 R \sin \gamma t\,\mathbf{k},$$

[4] GUSTAVE GASPARD CORIOLIS (1792–1843), French physicist.

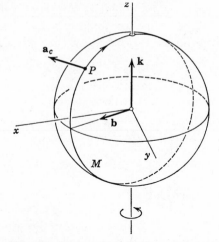

Fig. 134. Superposition of two rotations

where, by (10),

$$\dot{\mathbf{b}} = -\omega \sin \omega t \, \mathbf{i} + \omega \cos \omega t \, \mathbf{j},$$

$$\ddot{\mathbf{b}} = -\omega^2 \cos \omega t \, \mathbf{i} - \omega^2 \sin \omega t \, \mathbf{j} = -\omega^2 \mathbf{b}.$$

From (9) we see that the sum of the last two terms in (12) is equal to $-\gamma^2 \mathbf{r}$, and (12) becomes

(13) $$\mathbf{a} = -\omega^2 R \cos \gamma t \, \mathbf{b} - 2\gamma R \sin \gamma t \, \dot{\mathbf{b}} - \gamma^2 \mathbf{r}.$$

The first term on the right is the centripetal acceleration caused by the rotation of the sphere, and the last term is the centripetal acceleration resulting from the rotation of P on M. The second term is the **Coriolis acceleration**

(14) $$\mathbf{a}_c = -2\gamma R \sin \gamma t \, \dot{\mathbf{b}}.$$

On the "Northern hemisphere," $\sin \gamma t > 0$ [cf. (9)] and because of the minus sign, $\mathbf{a}_c$ is directed opposite to $\dot{\mathbf{b}}$, that is, tangential to the surface of the sphere, perpendicular to M, and opposite to the rotation of the sphere. Its magnitude $2\gamma R \, |\sin \gamma t| \, \omega$ is maximum at the "North pole" and zero at the equator. If P is a fly of mass m_0 walking according to (9), it will feel a force $-m_0 \mathbf{a}_c$, opposite to $m_0 \mathbf{a}_c$; this is quite similar to the force felt in Ex. 2. This force tends to let the fly deviate from the path M to the right. On the "Southern hemisphere," $\sin \gamma t < 0$ and that force acts in the opposite direction, tending to let the fly deviate from M to the left. This effect can be observed in connection with missiles and other projectiles. The flow of the air towards an area of low pressure also shows these deviations.

Problems

Let $\mathbf{r}(t)$ be the position vector of a moving particle, where $t \, (\geqq 0)$ is the time. Describe the geometric shape of the path and find the velocity vector, the speed, and the acceleration vector.

1. $\mathbf{r}(t) = 3t\mathbf{i}$
2. $\mathbf{r}(t) = 4t\mathbf{i} - 4t\mathbf{j} + 2t\mathbf{k}$
3. $\mathbf{r}(t) = 4t^2\mathbf{k}$
4. $\mathbf{r}(t) = (1 + t^3)\mathbf{i} + 2t^3\mathbf{j} + (2 - t^3)\mathbf{k}$
5. $\mathbf{r}(t) = \sin t \, \mathbf{j}$
6. $\mathbf{r}(t) = \cos t \, \mathbf{i} + \sin t \, \mathbf{j} + t\mathbf{k}$
7. $\mathbf{r}(t) = 10e^{-t}\mathbf{i} + 5e^{-t}\mathbf{j}$
8. $\mathbf{r}(t) = 3 \cos 2t \, \mathbf{i} + 3 \sin 2t \, \mathbf{j}$
9. $\mathbf{r}(t) = \cos t^2\mathbf{i} + \sin t^2\mathbf{j}$
10. $\mathbf{r}(t) = 3 \cos 2t \, \mathbf{i} + 2 \sin 3t \, \mathbf{j}$

11. Obtain (3) by differentiating (7), Sec. 5.8.
12. If $\mathbf{r}(t)$ represents the path of a moving particle and t is the time, what does a parametric transformation $t = \varphi(\tilde{t})$ mean in terms of mechanics?
13. Find the centripetal acceleration of the moon towards the earth, assuming that the orbit of the moon is a circle of radius 239,000 miles $= 3.85 \cdot 10^8$ meters and the time for one complete revolution is 27.3 days $= 2.36 \cdot 10^6$ sec.
14. Find the Coriolis acceleration in Ex. 2 with (5) replaced by $\mathbf{r} = t^2\mathbf{b}$.
15. Find the motion for which the acceleration vector is constant.

7.7   *Chain Rule and Mean Value Theorem for Functions of Several Variables*

We shall now consider some facts about functions of several variables which will be needed in the following sections. For the sake of simplicity we shall formulate everything for functions of two variables; the generalization to functions of three or more variables will be obvious. Familiarity of the student with the concept of a partial derivative will be assumed.

A function $f(x, y)$ is said to be **continuous** *at a point* (x_0, y_0) if f is defined in a *neighborhood*[5] of that point and if for any positive number ϵ (no matter how small, but not zero) we can find a positive number δ such that

$$|f(x, y) - f(x_0, y_0)| < \epsilon$$

for all (x, y) in that neighborhood for which

$$(x - x_0)^2 + (y - y_0)^2 < \delta^2.$$

Geometrically speaking, continuity of $f(x, y)$ at (x_0, y_0) means that for each interval of length 2ϵ with midpoint $f(x_0, y_0)$ we can find a circular disk with nonzero radius δ and center (x_0, y_0) in that neighborhood such that for every point (x, y) in the disk the corresponding function value $f(x, y)$ lies in that interval (Fig. 135).

From elementary calculus we know that if w is a differentiable function of x, and x is a differentiable function of t, then

$$\frac{dw}{dt} = \frac{dw}{dx}\frac{dx}{dt}.$$

This so-called chain rule of differentiation can be generalized as follows.

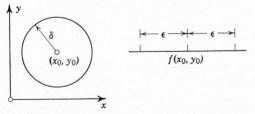

Fig. 135. Continuity of a function of two variables

[5] That is, in some circular disk $(x - x_0)^2 + (y - y_0)^2 < r^2$, $(r > 0)$, in the xy-plane.

Theorem 1 (Chain rule).   *Let $w = f(x, y)$ be continuous and have continuous first partial derivatives at every point of a domain[6] D in the xy-plane. Let $x = x(t)$ and $y = y(t)$ be differentiable functions of a variable in some interval T such that, for each t in T, the point $[x(t), y(t)]$ lies in D. Then $w = f[x(t), y(t)]$ is a differentiable function for all t in T, and*

$$\frac{dw}{dt} = \frac{\partial w}{\partial x}\frac{dx}{dt} + \frac{\partial w}{\partial y}\frac{dy}{dt}.$$ (1)

Proof.   We choose at t in T and Δt so small that $t + \Delta t$ is also in T, and set

(2) $\Delta x = x(t + \Delta t) - x(t), \qquad \Delta y = y(t + \Delta t) - y(t),$

and furthermore

$$\Delta w = f(x + \Delta x, y + \Delta y) - f(x, y).$$

By adding and subtracting a term this may be written

$$\Delta w = [f(x + \Delta x, y + \Delta y) - f(x, y + \Delta y)] + [f(x, y + \Delta y) - f(x, y)].$$

If we apply the mean value theorem for a function of a single variable (cf. Ref. [A14]) to each of the two expressions in the brackets, we obtain

$$\Delta w = \Delta x \frac{\partial f}{\partial x}\bigg|_{x_1, y+\Delta y} + \Delta y \frac{\partial f}{\partial y}\bigg|_{x, y_1}$$ (3)

where x_1 lies between x and $x + \Delta x$, and y_1 lies between y and $y + \Delta y$. Dividing (3) by Δt on both sides, letting Δt approach 0, and noting that $\partial f / \partial x$ and $\partial f / \partial y$ are assumed to be continuous, we obtain (1).

This theorem may now immediately be extended as follows.

Theorem 2.   *Let $w = f(x, y)$ be continuous and have continuous first partial derivatives in a domain D of the xy-plane. Let $x = x(u, v)$ and $y = y(u, v)$ be functions which have first partial derivatives in a domain B of the uv-plane, which is such that, for any point (u, v) in B, the corresponding point $[x(u, v), y(u, v)]$ lies in D. Then the function $w = f(x(u, v), y(u, v))$ is defined in B and has first partial derivatives with respect to u and v in B, and*

(4)

$$\frac{\partial w}{\partial u} = \frac{\partial w}{\partial x}\frac{\partial x}{\partial u} + \frac{\partial w}{\partial y}\frac{\partial y}{\partial u},$$

$$\frac{\partial w}{\partial v} = \frac{\partial w}{\partial x}\frac{\partial x}{\partial v} + \frac{\partial w}{\partial y}\frac{\partial y}{\partial v}.$$

The proof follows immediately from Theorem 1 by keeping one of the two variables u and v constant.

---

[6] A domain D is an open connected point set, where "connected" means that any two points of D can be joined by a broken line of finitely many linear segments all of whose points belong to D, and "open" means that every point of D has a neighborhood all of whose points belong to D. For example, the interior of a rectangle or a circle is a domain.

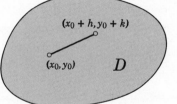

Fig. 136. Mean value theorem

From elementary calculus we know that if a function $f(x)$ is differentiable, then

$$f(x_0 + h) - f(x_0) = h\frac{df}{dx},$$

the derivative being evaluated at a suitable point between x_0 and $x_0 + h$ (cf. Ref. [A14]). This so-called mean-value theorem of differential calculus can be extended to functions of two variables as follows.

Theorem 3 (Mean-value theorem). *Let $f(x, y)$ be continuous and have continuous first partial derivatives in a domain D. Furthermore, let (x_0, y_0) and $(x_0 + h, y_0 + k)$ be points in D such that the line segment joining these points lies in D (Fig. 136). Then*

(5) $$f(x_0 + h, y_0 + k) - f(x_0, y_0) = h\frac{\partial f}{\partial x} + k\frac{\partial f}{\partial y},$$

the partial derivatives being evaluated at a suitable point of that segment.

Proof. Let

$$x = x_0 + th, \qquad y = y_0 + tk \qquad (0 \leq t \leq 1)$$

and

$$F(t) = f(x_0 + th, y_0 + tk).$$

Then

$$f(x_0 + h, y_0 + k) = F(1), \qquad f(x_0, y_0) = F(0).$$

By the mean value theorem for a function of a single variable there is a value t_1 between 0 and 1 such that

(6) $$f(x_0 + h, y_0 + k) - f(x_0, y_0) = F(1) - F(0) = F'(t_1).$$

We now apply Theorem 1. Since $dx/dt = h$ and $dy/dt = k$, we obtain

(7) $$F'(t_1) = \frac{\partial f}{\partial x} h + \frac{\partial f}{\partial y} k,$$

the derivatives on the right being evaluated at the point $(x_0 + t_1 h, y_0 + t_1 k)$, which lies on the line segment with end points (x_0, y_0) and $(x_0 + h, y_0 + k)$. By inserting (7) into (6) we obtain (5), and the proof is complete.

For a function $f(x, y, z)$ of three variables, satisfying conditions analogous to those in Theorem 3, the consideration is quite similar and leads to the formula

(8) $$f(x_0 + h, y_0 + k, z_0 + l) - f(x_0, y_0, z_0) = h\frac{\partial f}{\partial x} + k\frac{\partial f}{\partial y} + l\frac{\partial f}{\partial z},$$

the partial derivatives being evaluated at a suitable point of the segment with end points (x_0, y_0, z_0) and $(x_0 + h, y_0 + k, z_0 + l)$.

Problems

1. Prove Theorem 2.

2. Let $w = f(x, y, z)$, where x, y, z are functions of t. Show that under conditions similar to those in Theorem 1,

$$(9) \qquad \frac{dw}{dt} = \frac{\partial w}{\partial x} \frac{dx}{dt} + \frac{\partial w}{\partial y} \frac{dy}{dt} + \frac{\partial w}{\partial z} \frac{dz}{dt}.$$

3. Let $w = (x + y)/(x - y)$, where $x = \sin t^2$, $y = \cos t^2$. Find dw/dt (*a*) by using (1), (*b*) by substituting x and y in terms of t in the expression for w.

Using (1), find dw/dt, where

4. $w = x + y$, $x = 4(t^2 - 1)$, $y = \ln t$ **5.** $w = \sqrt{x^2 + y^2}$, $x = e^t$, $y = \sin t$

6. $w = x/y$, $x = g(t)$, $y = h(t)$ **7.** $w = x^y$, $x = \cos t$, $y = \sin t$

Using (9) in Prob. 2, find dw/dt, where

8. $w = x^2 + y^2 + z^2$, $x = e^t \cos t$, $y = e^t \sin t$, $z = e^t$

9. $w = xy + yz + zx$, $x = t$, $y = \cos t$, $z = \sin t$

Find $\partial w/\partial u$ and $\partial w/\partial v$, where

10. $w = x^2 + y^2$, $x = u + v$, $y = u - v$

11. $w = e^{x^2 y^2}$, $x = e^u \cos v$, $y = e^u \sin v$

12. $w = x^2 - y^2$, $x = u^2 - v^2$, $y = 2uv$

13. Let $w = f(x, y)$ and $x = r \cos \theta$, $y = r \sin \theta$. Show that, granted sufficient differentiability,

$$w_{xx} + w_{yy} = w_{rr} + \frac{1}{r} w_r + \frac{1}{r^2} w_{\theta\theta}.$$

14. Let $w = f(x, y)$ and $x = r \cos \theta$, $y = r \sin \theta$. Show that, granted sufficient differentiability,

$$\left(\frac{\partial w}{\partial r}\right)^2 + \frac{1}{r^2}\left(\frac{\partial w}{\partial \theta}\right)^2 = \left(\frac{\partial w}{\partial x}\right)^2 + \left(\frac{\partial w}{\partial y}\right)^2.$$

15. Let $w = f(v, z)$ and $v = x + ct$, $z = x - ct$, where c is a constant. Show that, granted sufficient differentiability,

$$c^2 w_{xx} - w_{tt} = 4c^2 w_{vz}.$$

7.8 Directional Derivative. Gradient of a Scalar Field

We consider a scalar field in space given by a scalar function $f(P) = f(x, y, z)$ (cf. Sec. 7.1). We know that the first partial derivatives of f are the rates of change of f in the directions of the coordinate axes. It seems unnatural to restrict attention to these three directions, and we may ask for the rate of change of f in any direction. This simple idea leads to the notion of a directional derivative.

Fig. 137. Directional derivative

To define that derivative we choose a point P in space and a direction at P, given by a unit vector $\mathbf{b}$. Let C be the ray from P in the direction of $\mathbf{b}$, and let Q be a point on C, whose distance from P is s (Fig. 137). Then if the limit

(1) $$\frac{\partial f}{\partial s} = \lim_{s \to 0} \frac{f(Q) - f(P)}{s}$$ 　　　　(s = distance between P and Q)

exists, it is called the **directional derivative** of f at P *in the direction of* $\mathbf{b}$. Obviously, $\partial f/\partial s$ is the rate of change of f at P in the direction of $\mathbf{b}$.

In this way there are now infinitely many directional derivatives of f at P, each corresponding to a certain direction. But, a Cartesian coordinate system being given, we may represent any such derivative in terms of the first partial derivatives of f at P as follows. If P has the position vector $\mathbf{a}$, then the ray C can be represented in the form

(2) 　　　　$$\mathbf{r}(s) = x(s)\mathbf{i} + y(s)\mathbf{j} + z(s)\mathbf{k} = \mathbf{a} + s\mathbf{b}$$　　　　($s \geqq 0$),

and $\partial f/\partial s$ is the derivative of the function $f[x(s), y(s), z(s)]$ with respect to the arc length s of C. Hence, assuming that f has continuous first partial derivatives and applying the chain rule (Theorem 1 in the last section), we obtain

(3) 　　　　$$\frac{\partial f}{\partial s} = \frac{\partial f}{\partial x} x' + \frac{\partial f}{\partial y} y' + \frac{\partial f}{\partial z} z'$$

where primes denote derivatives with respect to s. Now from (2),

$$\mathbf{r}' = x'\mathbf{i} + y'\mathbf{j} + z'\mathbf{k} = \mathbf{b}.$$

This suggests that we introduce the vector

(4) 　　　　$$\operatorname{grad} f = \frac{\partial f}{\partial x}\mathbf{i} + \frac{\partial f}{\partial y}\mathbf{j} + \frac{\partial f}{\partial z}\mathbf{k}$$

and write (3) in the form of a scalar product:

(5) 　　　　$$\frac{\partial f}{\partial s} = \mathbf{b} \cdot \operatorname{grad} f$$　　　　($|\mathbf{b}| = 1$).

The vector grad f is called the **gradient** of the scalar function f.

By introducing the *differential operator*

$$\nabla = \frac{\partial}{\partial x}\mathbf{i} + \frac{\partial}{\partial y}\mathbf{j} + \frac{\partial}{\partial z}\mathbf{k}$$

(read **nabla** or "del") we may write

$$\operatorname{grad} f = \nabla f = \frac{\partial f}{\partial x}\,\mathbf{i} + \frac{\partial f}{\partial y}\,\mathbf{j} + \frac{\partial f}{\partial z}\,\mathbf{k}.$$

The notation ∇f for the gradient is frequently used in the engineering literature. If in particular **b** has the direction of the positive x-axis, then $\mathbf{b} = \mathbf{i}$, and

$$\frac{\partial f}{\partial s} = \mathbf{b} \cdot \operatorname{grad} f = \frac{\partial f}{\partial x}\,\mathbf{i} \cdot \mathbf{i} = \frac{\partial f}{\partial x}.$$

Similarly, the directional derivative in the positive y-direction is $\partial f/\partial y$, etc.

Example 1.   Find the directional derivative $\partial f/\partial s$ of $f(x, y, z) = 2x^2 + 3y^2 + z^2$ at the point P: $(2, 1, 3)$ in the direction of the vector $\mathbf{a} = \mathbf{i} - 2\mathbf{k}$. We obtain

$$\operatorname{grad} f = 4x\mathbf{i} + 6y\mathbf{j} + 2z\mathbf{k}, \qquad \text{and at } P, \qquad \operatorname{grad} f = 8\mathbf{i} + 6\mathbf{j} + 6\mathbf{k}.$$

Since $|\mathbf{a}| = \sqrt{5}$, the unit vector in the direction of **a** is

$$\mathbf{b} = \frac{\mathbf{a}}{|\mathbf{a}|} = \frac{1}{\sqrt{5}}\,\mathbf{i} - \frac{2}{\sqrt{5}}\,\mathbf{k}.$$

Therefore,

$$\frac{\partial f}{\partial s} = (8\mathbf{i} + 6\mathbf{j} + 6\mathbf{k}) \cdot \left(\frac{1}{\sqrt{5}}\,\mathbf{i} - \frac{2}{\sqrt{5}}\,\mathbf{k} \right) = -\frac{4}{\sqrt{5}}.$$

The minus sign indicates that f decreases in the direction under consideration.

We want to show that *the length and direction of* grad f *are independent of the particular choice of Cartesian coordinates.*

This is, of course, not obvious, because (4) involves partial derivatives, which depend on the choice of the coordinates, so that we do not know yet whether the corresponding expression

$$\frac{\partial f}{\partial x^*}\,\mathbf{i}^* + \frac{\partial f}{\partial y^*}\,\mathbf{j}^* + \frac{\partial f}{\partial z^*}\,\mathbf{k}^*$$

with respect to other Cartesian coordinates x^*, y^*, z^* (and corresponding unit vectors $\mathbf{i}^*$, $\mathbf{j}^*$, $\mathbf{k}^*$) will have the same length and direction as (4).

To prove our proposition, we may reason as follows. By the definition of a scalar function, the value of f at a point P depends on P but is independent of the coordinates, and s, the arc length of that ray C, is also independent of the choice of coordinates. Hence $\partial f/\partial s$ is independent of the particular choice of coordinates. From (5) we obtain

$$\frac{\partial f}{\partial s} = |\mathbf{b}|\,|\operatorname{grad} f|\cos \gamma = |\operatorname{grad} f|\cos \gamma$$

where γ is the angle between **b** and grad f. We see that $\partial f/\partial s$ is maximum when $\cos \gamma = 1$, $\gamma = 0$, and then $\partial f/\partial s = |\operatorname{grad} f|$. This shows that the length and direction of grad f are independent of the coordinates, and we have the following result.

Fig. 138. Level surface and gradient

Theorem 1. *Let $f(P) = f(x, y, z)$ be a scalar function having continuous first partial derivatives. Then grad f exists and its length and direction are independent of the particular choice of Cartesian coordinates in space. If at a point P the gradient of f is not the zero vector, it has the direction of maximum increase of f at P.*

Another important geometrical characterization of the gradient can be obtained as follows. Consider a differentiable scalar function $f(x, y, z)$ in space. Suppose that for each constant c the equation

(6) $$f(x, y, z) = c = const$$

represents a surface S in space. Then, by letting c assume all values, we obtain a family of surfaces, which are called the **level surfaces** of the function f. Since, by the definition of a function, our function f has a unique value at each point in space, it follows that through each point in space there passes one, and only one, level surface of f. We remember that a curve C in space may be represented in the form (cf. Sec. 7.3)

(7) $$\mathbf{r}(t) = x(t)\mathbf{i} + y(t)\mathbf{j} + z(t)\mathbf{k}.$$

If we now require that C lies on S, then the functions $x(t)$, $y(t)$, and $z(t)$ in (7) must be such that

$$f[x(t), y(t), z(t)] = c;$$

cf. (6). By differentiating this with respect to t and using the chain rule (Sec. 7.7) we obtain

(8) $$\frac{\partial f}{\partial x}\dot{x} + \frac{\partial f}{\partial y}\dot{y} + \frac{\partial f}{\partial z}\dot{z} = (\text{grad} f) \cdot \dot{\mathbf{r}} = 0,$$

where the vector

$$\dot{\mathbf{r}} = \dot{x}\mathbf{i} + \dot{y}\mathbf{j} + \dot{z}\mathbf{k}$$

is tangent to C (cf. Sec. 7.5). If we consider curves on S passing through a point P of S in various directions, their tangents at P will, in general, lie in the same plane which touches S at P. This plane is called the **tangent plane** of S at P. The straight line through P and perpendicular to the tangent plane is called the **normal** of S at P (Fig. 138). From (8) and Theorem 1, Sec. 5.5, we thus obtain the following result.

Theorem 2 (Gradient and surface normal). *Let f be a scalar function which is defined and differentiable in a domain D in space, and let P be any point in D on a* *level surface S of f. Then if the gradient of f at P is not the zero vector, it is perpendicular to S at P, that is, it has the direction of the normal to S at P.*

Example 2 (*Normal to a plane curve*). The level curves $f = const$ of $f(x, y) = \ln (x^2 + y^2)$ are concentric circles about the origin. The gradient

$$\text{grad } f = \frac{\partial f}{\partial x}\mathbf{i} + \frac{\partial f}{\partial y}\mathbf{j} = \frac{2x}{x^2 + y^2}\mathbf{i} + \frac{2y}{x^2 + y^2}\mathbf{j}$$

has the direction of the normals to the circles, and its direction corresponds to that of the maximum increase of f. For example, at the point P: $(2, 1)$, we have (cf. Fig. 139)

$$\text{grad } f = 0.8\mathbf{i} + 0.4\mathbf{j}.$$

Example 3 (*Normal to a surface*). Find a unit normal vector $\mathbf{n}$ of the cone of revolution $z^2 = 4(x^2 + y^2)$ at the point P: $(1, 0, 2)$. We may regard the cone as the level surface $f = 0$ of the function $f(x, y, z) = 4(x^2 + y^2) - z^2$. Then

$$\text{grad } f = 8x\mathbf{i} + 8y\mathbf{j} - 2z\mathbf{k} \quad \text{and at } P, \quad \text{grad } f = 8\mathbf{i} - 4\mathbf{k}.$$

Hence, by Theorem 2,

$$\mathbf{n} = \frac{\text{grad } f}{|\text{grad } f|} = \frac{2}{\sqrt{5}}\mathbf{i} - \frac{1}{\sqrt{5}}\mathbf{k}$$

is a unit normal vector of the cone at P, and $-\mathbf{n}$ is the other one (cf. Fig. 140).

Some of the vector fields occurring in physics are given by vector functions which can be obtained as the gradients of suitable scalar functions. Such a scalar function is then called a *potential function* or **potential** of the corresponding vector field. The use of potentials simplifies the investigation of those vector fields considerably. To obtain a first impression of this approach to vector fields, let us consider an important example.

Example 4 (*Gravitational field. Laplace's equation*). In Ex. 4, Sec. 7.1, we have seen that, according to Newton's law of gravitation, the force of attraction between two particles is

(9)
$$\mathbf{p} = -c\frac{\mathbf{r}}{r^3} = -c\left(\frac{x - x_0}{r^3}\mathbf{i} + \frac{y - y_0}{r^3}\mathbf{j} + \frac{z - z_0}{r^3}\mathbf{k}\right)$$

where

$$r = \sqrt{(x - x_0)^2 + (y - y_0)^2 + (z - z_0)^2}$$

Fig. 139. Normal to a circle

Fig. 140. Intersection of the xz-plane and the cone in Example 3

is the distance between the two particles and c is a constant. Observing that

(10a)
$$\frac{\partial}{\partial x}\left(\frac{1}{r}\right) = -\frac{2(x - x_0)}{2[(x - x_0)^2 + (y - y_0)^2 + (z - z_0)^2]^{3/2}} = -\frac{x - x_0}{r^3}$$

and similarly

(10b)
$$\frac{\partial}{\partial y}\left(\frac{1}{r}\right) = -\frac{y - y_0}{r^3}, \qquad \frac{\partial}{\partial z}\left(\frac{1}{r}\right) = -\frac{z - z_0}{r^3},$$

we see that $\mathbf{p}$ is the gradient of the scalar function

$$f(x, y, z) = \frac{c}{r} \qquad\qquad (r > 0);$$

that is, f is a potential of that gravitational field.

By differentiating (10) we find

$$\frac{\partial^2}{\partial x^2}\left(\frac{1}{r}\right) = -\frac{1}{r^3} + \frac{3(x - x_0)^2}{r^5}, \qquad \frac{\partial^2}{\partial y^2}\left(\frac{1}{r}\right) = -\frac{1}{r^3} + \frac{3(y - y_0)^2}{r^5},$$

$$\frac{\partial^2}{\partial z^2}\left(\frac{1}{r}\right) = -\frac{1}{r^3} + \frac{3(z - z_0)^2}{r^5}.$$

Since the sum of the three expressions on the right is zero, we see that the potential $f = c/r$ satisfies the equation

(11)
$$\frac{\partial^2 f}{\partial x^2} + \frac{\partial^2 f}{\partial y^2} + \frac{\partial^2 f}{\partial z^2} = 0.$$

This important partial differential equation is called **Laplace's equation**; it will be considered in detail in Chaps. 10 and 17. The expression on the left is called the **Laplacian** of f and is denoted by $\nabla^2 f$ or Δf. The differential operator

$$\nabla^2 = \Delta = \frac{\partial^2}{\partial x^2} + \frac{\partial^2}{\partial y^2} + \frac{\partial^2}{\partial z^2}$$

(read "nabla squared" or "delta") is called the **Laplace operator**. Using this operator, we may write (11) in the form

$$\nabla^2 f = 0.$$

It can be shown that the field of force produced by any distribution of masses is given by a vector function which is the gradient of a scalar function f, and f satisfies (11) in any region of space which is free of matter.

There are other laws in physics which are of the same form as Newton's law of gravitation. For example, in electrostatics the force of attraction (or repulsion) between two particles of opposite (or like) charges Q_1 and Q_2 is

$$\mathbf{p} = k\frac{\mathbf{r}}{r^3} \qquad\qquad \textbf{(Coulomb's law)}$$

where $k = Q_1 Q_2 / 4\pi\epsilon$, and ϵ is the dielectric constant. Hence $\mathbf{p}$ is the gradient of the potential $f = -k/r$, and f satisfies (11).

If the vector function defining a vector field is the gradient of a scalar function, the field is said to be **conservative**, because, as we shall see in Sec. 8.12, in such a field the work done in displacing a particle from a point P_1 to a point P_2 in the field depends only on P_1 and P_2 but not on the path along which the particle is displaced from P_1 to P_2. We shall also see that not every field is conservative.

Problems

Find the gradient ∇f. Graph some level curves $f = const$ and indicate ∇f by arrows at some points of these curves.

1. $f = 2xy$
4. $f = x/y$
2. $f = x^2 - y^2$
5. $f = x/(x^2 + y^2)$
3. $f = x^2 + y^2$
6. $f = $ arc tan (y/x)

Find the gradient ∇f, where f equals

7. $e^x \sin y$
10. $\sin x \sinh y$
8. $x^2 + 3y^2 - 2z^2$
11. $\sin(x^2 + y^2 + z^2)$
9. $(x^2 + y^2 + z^2)^2$
12. $\frac{1}{2} \ln(x^2 + y^2)$

Find and graph a normal vector to the given plane curve at a point P: (x, y).

13. $y = 2x^2$,　　P: $(2, 8)$
15. $y^2 = x^3$,　　P: $(4, -8)$
17. $x^2 + y^2 = 100$,　　P: $(6, 8)$
14. $y = 3x - 4$,　　P: $(1, -1)$
16. $y = 1/x$,　　P: $(4, 1/4)$
18. $x^2 - y^2 = 1$,　　P: $(2, \sqrt{3})$

Find a unit normal vector to the given surface at the given point P: (x, y, z).

19. $x + y + z = 1$,　　P: $(4, 2, -5)$
21. $z = xy$,　　P: $(2, -1, -2)$
23. $x^2 + 3y^2 = 7$,　　P: $(-2, 1, 7)$
20. $x^2 + y^2 + 2z^2 = 26$,　　P: $(2, 2, 3)$
22. $z = \sqrt{x^2 + y^2}$,　　P: $(3, 4, 5)$
24. $z = y^4$,　　P: $(2, 1, 1)$

Find a scalar function f such that $\mathbf{v} = \nabla f$.

25. $\mathbf{v} = x\mathbf{i} + y\mathbf{j}$
27. $\mathbf{v} = 6x\mathbf{i} - 4y\mathbf{j} + 2z\mathbf{k}$
29. $\mathbf{v} = (y + z)\mathbf{i} + (z + x)\mathbf{j} + (x + y)\mathbf{k}$
26. $\mathbf{v} = (x\mathbf{i} + y\mathbf{j})/\sqrt{x^2 + y^2}$
28. $\mathbf{v} = -(x\mathbf{i} + y\mathbf{j} + z\mathbf{k})(x^2 + y^2 + z^2)^{-3/2}$
30. $\mathbf{v} = e^{xyz}(yz\mathbf{i} + xz\mathbf{j} + xy\mathbf{k})$

31. Show that the functions in Probs. 1, 2, 5, and 6 are solutions of Laplace's equation.
32. Find the angle between the gradients in Probs. 1 and 2.
33. Find the angle between the gradients in Probs. 6 and 12.

Granted sufficient differentiability, show that

34. $\nabla(fg) = f\nabla g + g\nabla f$
36. $\nabla(f/g) = (g\nabla f - f\nabla g)/g^2$
38. $\nabla f(r) = f'(r)(x\mathbf{i} + y\mathbf{j} + z\mathbf{k})/r$, where $r = \sqrt{x^2 + y^2 + z^2}$
35. $\nabla(f^n) = nf^{n-1}\nabla f$
37. $\nabla^2(fg) = g\nabla^2 f + 2\nabla f \cdot \nabla g + f\nabla^2 g$

39. Let r be the distance from a fixed point P to a point Q: (x, y, z). What are the level surfaces of r? Show that ∇r is a unit vector in the direction from P to Q.

40. What are the directional derivatives of a function $f(x, y, z)$ in the x, y, and z directions? In the negative x, y, and z directions?

41. Find the directional derivative of $f = x^2 + y^2$ at the point P: $(2, 2)$ in the directions of the vectors $\mathbf{i}$, $\mathbf{i} + \mathbf{j}$, $\mathbf{j}$, $-\mathbf{i} + \mathbf{j}$, $-\mathbf{i}$, $-\mathbf{i} - \mathbf{j}$, $-\mathbf{j}$, and $\mathbf{i} - \mathbf{j}$.

Find the directional derivative of f at P in the direction of $\mathbf{a}$, where

42. $f = x^2 + y^2 + z^2$,　　P: $(1, 2, 3)$,　　$\mathbf{a} = \mathbf{i} + \mathbf{j}$
43. $f = x^2 + y^2 + z^2$,　　P: $(2, 2, 2)$,　　$\mathbf{a} = \mathbf{i} + 2\mathbf{j} - 3\mathbf{k}$
44. $f = 4x^2 - 9y^2 + z^2$,　　P: $(0, -1, 2)$,　　$\mathbf{a} = \mathbf{i} + \mathbf{j} + \mathbf{k}$
45. $f = x/z$,　　P: $(1, -1, 3)$,　　$\mathbf{a} = -\mathbf{i} - \mathbf{k}$
46. $f = x^3 - 3xy^2$,　　P: $(0, 2, 0)$,　　$\mathbf{a} = \mathbf{i} + \mathbf{j} + 2\mathbf{k}$
47. $f = e^{x+y+z}$,　　P: $(0, 0, 0)$,　　$\mathbf{a} = \mathbf{i} + 2\mathbf{j} - 2\mathbf{k}$
48. $f = e^x \cos y$,　　P: $(2, \pi, 0)$,　　$\mathbf{a} = 2\mathbf{i} + 3\mathbf{k}$
49. $f = xyz$,　　P: $(-1, 1, 3)$,　　$\mathbf{a} = \mathbf{i} - 2\mathbf{j} + 2\mathbf{k}$
50. $f = 1/\sqrt{x^2 + y^2 + z^2}$,　　P: $(3, 0, 4)$,　　$\mathbf{a} = \mathbf{i} + \mathbf{j} + \mathbf{k}$

7.9   *Transformation of Coordinate Systems and Vector Components*

We shall now characterize the transformations which carry Cartesian co-ordinate systems into Cartesian coordinate systems and investigate the change of vector components under such a transformation. This problem is basic for theoretical as well as practical reasons.[7]

Let x, y, z and x^*, y^*, z^* be any two systems of Cartesian coordinates. Let

$$(1) \qquad (a) \ \mathbf{v} = v_1\mathbf{i} + v_2\mathbf{j} + v_3\mathbf{k} \qquad \text{and} \qquad (b) \ \mathbf{v} = v_1{}^*\mathbf{i}^* + v_2{}^*\mathbf{j}^* + v_3{}^*\mathbf{k}^*$$

be the representations of a given vector $\mathbf{v}$ in these two coordinate systems; here $\mathbf{i}, \mathbf{j}, \mathbf{k}$ and $\mathbf{i}^*, \mathbf{j}^*, \mathbf{k}^*$ are unit vectors in the positive x, y, z and x^*, y^*, z^* directions, respectively. We want to express the components $v_1{}^*$, $v_2{}^*$, $v_3{}^*$ in terms of the components v_1, v_2, v_3, and conversely.

From (1*a*) we obtain

$$(2) \qquad\qquad \mathbf{i}^* \cdot \mathbf{v} = v_1 \mathbf{i}^* \cdot \mathbf{i} + v_2 \mathbf{i}^* \cdot \mathbf{j} + v_3 \mathbf{i}^* \cdot \mathbf{k}.$$

Similarly, by taking the scalar product of (1*b*) and $\mathbf{i}^*$, we have

$$\mathbf{i}^* \cdot \mathbf{v} = v_1{}^* \mathbf{i}^* \cdot \mathbf{i}^* + v_2{}^* \mathbf{i}^* \cdot \mathbf{j}^* + v_3{}^* \mathbf{i}^* \cdot \mathbf{k}^*.$$

Since the first scalar product on the right is 1 and the others are zero,

$$\mathbf{i}^* \cdot \mathbf{v} = v_1{}^*.$$

From this and (2) it follows that

$$v_1{}^* = \mathbf{i}^* \cdot \mathbf{i} \, v_1 + \mathbf{i}^* \cdot \mathbf{j} \, v_2 + \mathbf{i}^* \cdot \mathbf{k} \, v_3.$$

Similarly, $$v_2{}^* = \mathbf{j}^* \cdot \mathbf{i} \, v_1 + \mathbf{j}^* \cdot \mathbf{j} \, v_2 + \mathbf{j}^* \cdot \mathbf{k} \, v_3,$$

$$v_3{}^* = \mathbf{k}^* \cdot \mathbf{i} \, v_1 + \mathbf{k}^* \cdot \mathbf{j} \, v_2 + \mathbf{k}^* \cdot \mathbf{k} \, v_3.$$

Hence the components of a vector $\mathbf{v}$ in a Cartesian coordinate system can be expressed as linear functions of the components of $\mathbf{v}$ with respect to another Cartesian coordinate system.

To write our transformation formulas in a simpler form we adopt the notation

$$(3) \qquad \begin{array}{ccc} \mathbf{i}^* \cdot \mathbf{i} = a_{11} & \mathbf{i}^* \cdot \mathbf{j} = a_{12} & \mathbf{i}^* \cdot \mathbf{k} = a_{13} \\ \mathbf{j}^* \cdot \mathbf{i} = a_{21} & \mathbf{j}^* \cdot \mathbf{j} = a_{22} & \mathbf{j}^* \cdot \mathbf{k} = a_{23} \\ \mathbf{k}^* \cdot \mathbf{i} = a_{31} & \mathbf{k}^* \cdot \mathbf{j} = a_{32} & \mathbf{k}^* \cdot \mathbf{k} = a_{33}. \end{array}$$

Then we have

$$(4) \qquad \begin{aligned} v_1{}^* &= a_{11}v_1 + a_{12}v_2 + a_{13}v_3 \\ v_2{}^* &= a_{21}v_1 + a_{22}v_2 + a_{23}v_3 \\ v_3{}^* &= a_{31}v_1 + a_{32}v_2 + a_{33}v_3. \end{aligned}$$

---

[7] We shall keep the present section completely independent of Chap. 6, but readers familiar with matrices should recognize that we are dealing with orthogonal transformations and matrices and Theorems 1 and 2 follow from Theorem 2 in Sec. 6.15.

Using summation signs, we may write this more briefly

$$(4') \qquad v_k{}^* = \sum_{l=1}^{3} a_{kl} v_l, \qquad\qquad k = 1, 2, 3.$$

A similar consideration leads to the inverse formulas

$$(5) \qquad \begin{aligned} v_1 &= a_{11} v_1{}^* + a_{21} v_2{}^* + a_{31} v_3{}^* \\ v_2 &= a_{12} v_1{}^* + a_{22} v_2{}^* + a_{32} v_3{}^* \\ v_3 &= a_{13} v_1{}^* + a_{23} v_2{}^* + a_{33} v_3{}^* \end{aligned}$$

or, more briefly,

$$(5') \qquad v_l = \sum_{m=1}^{3} a_{ml} v_m{}^*, \qquad\qquad l = 1, 2, 3.$$

Note that (4) and (5) contain the same coefficients a_{kl}, but these coefficients (except for a_{11}, a_{22}, a_{33}) occupy different positions in (4) and (5).

The geometrical interpretation of those nine coefficients a_{kl} is very simple. Since **i** and **i*** are unit vectors, it follows from (1) in Sec. 5.5 that $a_{11} = $ **i*** $\cdot$ **i** is the cosine of the angle between the positive x^* and x axes. Similarly, $a_{12} = $ **i*** $\cdot$ **j** is the cosine of the angle between the positive x^* and y axes, and so on.

The coefficients a_{kl} satisfy certain important relations which we shall now derive. By inserting (5′) into (4′) we find

$$(6) \qquad v_k{}^* = \sum_{l=1}^{3} a_{kl} v_l = \sum_{l=1}^{3} a_{kl} \sum_{m=1}^{3} a_{ml} v_m{}^* = \sum_{m=1}^{3} v_m{}^* \left(\sum_{l=1}^{3} a_{kl} a_{ml} \right),$$

where $k = 1, 2, 3$. For example, if $k = 1$, this becomes

$$v_1{}^* = v_1{}^* \left(\sum_{l=1}^{3} a_{1l} a_{1l} \right) + v_2{}^* \left(\sum_{l=1}^{3} a_{1l} a_{2l} \right) + v_3{}^* \left(\sum_{l=1}^{3} a_{1l} a_{3l} \right).$$

In order that this relation hold for any vector $\mathbf{v} = v_1{}^* \mathbf{i}^* + v_2{}^* \mathbf{j}^* + v_3{}^* \mathbf{k}^*$, the first sum must be 1 and the other two sums must be zero. For $k = 2$ and $k = 3$ the situation is similar. Consequently, (6) holds for any vector if and only if

$$(7) \qquad \sum_{l=1}^{3} a_{kl} a_{ml} = \begin{cases} 0 & (k \neq m) \\ 1 & (k = m). \end{cases}$$

Using the so-called *Kronecker*[8] *symbol* or **Kronecker delta**

$$\delta_{km} = \begin{cases} 0 & (k \neq m) \\ 1 & (k = m) \end{cases}$$

we may write (7) in the form

$$(7') \qquad \sum_{l=1}^{3} a_{kl} a_{ml} = \delta_{km} \qquad\qquad (k, m = 1, 2, 3).$$

[8] LEOPOLD KRONECKER (1823–1891), German mathematician, who made important contributions to algebra and the theory of numbers.

Forming three vectors with components

$$a_{11}, a_{12}, a_{13} \qquad a_{21}, a_{22}, a_{23} \qquad a_{31}, a_{32}, a_{33}$$

we see that the left side of (7′) is the scalar product of two of these vectors, and (7′) implies that these vectors are orthogonal unit vectors. Hence their scalar triple product has the value $+1$ or -1; that is,

(8)
$$\begin{vmatrix} a_{11} & a_{12} & a_{13} \\ a_{21} & a_{22} & a_{23} \\ a_{31} & a_{32} & a_{33} \end{vmatrix} = \pm 1.$$

We mention, without proof, that if both coordinate systems under consideration are right-handed (or both left-handed), then the determinant has the value $+1$, whereas if one system is right-handed and the other left-handed, the determinant has the value -1. We may now sum up our result as follows.

Theorem 1. *The components v_1, v_2, v_3 and $v_1{}^*$, $v_2{}^*$, $v_3{}^*$ of any vector **v** with respect to any two Cartesian coordinate systems can be obtained from each other by means of (4) and (5), where the coefficients a_{kl} are given by (3) and satisfy (7) and (8).*

From our consideration of vector components we may now immediately obtain the formulas for the transformation of a Cartesian coordinate system into any other such system as follows.

If the xyz and $x^*y^*z^*$ coordinate systems have the same origin, then **v** may be bound at the origin and regarded as the position vector of its terminal point Q. If (x, y, z) and (x^*, y^*, z^*) are the coordinates of Q with respect to the two coordinate systems, then in (4) and (5)

$$v_1 = x, \quad v_2 = y, \quad v_3 = z \qquad \text{and} \qquad v_1{}^* = x^*, \quad v_2{}^* = y^*, \quad v_3{}^* = z^*.$$

Consequently, (4) and (5), written in terms of x, y, z and x^*, y^*, z^* instead of v_1, v_2, v_3 and $v_1{}^*$, $v_2{}^*$, $v_3{}^*$, represent the transformations between those two coordinate systems with common origin.

The most general transformation of a Cartesian coordinate system into another such system may be decomposed into a transformation of the type just considered and a translation. Under a translation, corresponding coordinates differ merely by a constant. We thus obtain

Theorem 2. *The transformation of any Cartesian xyz-coordinate system into any other Cartesian $x^*y^*z^*$-coordinate system is of the form*

(9)
$$\begin{aligned} x^* &= a_{11}x + a_{12}y + a_{13}z + b_1 \\ y^* &= a_{21}x + a_{22}y + a_{23}z + b_2 \\ z^* &= a_{31}x + a_{32}y + a_{33}z + b_3 \end{aligned}$$

and conversely

(10)
$$\begin{aligned} x &= a_{11}x^* + a_{21}y^* + a_{31}z^* + c_1 \\ y &= a_{12}x^* + a_{22}y^* + a_{32}z^* + c_2 \\ z &= a_{13}x^* + a_{23}y^* + a_{33}z^* + c_3 \end{aligned}$$

where the coefficients a_{kl} are given by (3) and satisfy (7) and (8) and b_1, b_2, b_3, c_1, c_2, c_3 are constants.

Important applications of our results will be considered in the following two sections.

Problems

1. Derive (5).
2. Interpret the coefficients in (4′) geometrically.
3. What does it mean geometrically if $b_1 = 0$, $b_2 = 0$, $b_3 = 0$ in (9)? If $a_{jk} = 0$ when $j \neq k$?
4. From (1), Sec. 5.5, it follows that the value of a scalar product is independent of the particular choice of Cartesian coordinates. Using (4) or (5) in this section, show that this follows also from (8), Sec. 5.5.

Determine constants a_{kl} and b_l such that (9) represents:

5. A translation which carries the origin into the point $(1, 3, -4)$.
6. A translation which carries the point $(1, -1, 2)$ into the point $(0, -3, 5)$.
7. A reflection in the plane $y = x$.
8. A reflection in the yz-plane.
9. A rotation about the x-axis through an angle ϕ.
10. A rotation about the z-axis through an angle γ.

7.10   *Divergence of a Vector Field*

Let $\mathbf{v}(x, y, z)$ be a differentiable vector function, where x, y, z are Cartesian coordinates in space, and let v_1, v_2, v_3 be the components of $\mathbf{v}$. Then the function

$$(1) \qquad \operatorname{div} \mathbf{v} = \frac{\partial v_1}{\partial x} + \frac{\partial v_2}{\partial y} + \frac{\partial v_3}{\partial z}$$

is called the **divergence** *of* $\mathbf{v}$ or the *divergence of the vector field defined by* $\mathbf{v}$.

Another common notation for the divergence of $\mathbf{v}$ is $\nabla \cdot \mathbf{v}$,

$$\operatorname{div} \mathbf{v} = \nabla \cdot \mathbf{v} = \left(\frac{\partial}{\partial x} \mathbf{i} + \frac{\partial}{\partial y} \mathbf{j} + \frac{\partial}{\partial z} \mathbf{k} \right) \cdot (v_1 \mathbf{i} + v_2 \mathbf{j} + v_3 \mathbf{k})$$

$$= \frac{\partial v_1}{\partial x} + \frac{\partial v_2}{\partial y} + \frac{\partial v_3}{\partial z},$$

with the understanding that the "product" $(\partial/\partial x)v_1$ in the dot product means the partial derivative $\partial v_1/\partial x$, etc. This is a convenient notation, but nothing more. Note that $\nabla \cdot \mathbf{v}$ means the scalar div $\mathbf{v}$ whereas ∇f means the vector grad f defined in Sec. 7.8.

We shall see later that the divergence has an important physical meaning. Clearly the values of a function which characterizes a physical or geometrical property must be independent of the particular choice of coordinates; that is, those values must be invariant with respect to coordinate transformations.

Theorem 1. *The values of* div **v** *depend only on the points in space (and, of course, on* **v**) *but not on the particular choice of the coordinates in* (1), *so that with respect to other Cartesian coordinates* x^*, y^*, z^* *and corresponding components* v_1^*, v_2^*, v_3^* *of* **v** *the function div* **v** *is given by*

(2)
$$\text{div } \mathbf{v} = \frac{\partial v_1^*}{\partial x^*} + \frac{\partial v_2^*}{\partial y^*} + \frac{\partial v_3^*}{\partial z^*}.$$

Proof. We shall derive (2) from (1). To simplify our formulas in this proof, we adopt the notation

$$x_1 = x, \quad x_2 = y, \quad x_3 = z \quad \text{and} \quad x_1^* = x^*, \quad x_2^* = y^*, \quad x_3^* = z^*.$$

Then we may write formula (9) in Sec. 7.9 in the short form

(3)
$$x_k^* = \sum_{l=1}^{3} a_{kl} x_l + b_k \qquad (k = 1, 2, 3).$$

The chain rule for functions of several variables (cf. Sec. 7.7) yields

(4)
$$\frac{\partial v_l}{\partial x_l} = \sum_{k=1}^{3} \frac{\partial v_l}{\partial x_k^*} \frac{\partial x_k^*}{\partial x_l}.$$

In this sum, $\partial x_k^*/\partial x_l = a_{kl}$, as follows from (3). By (5') in Sec. 7.9,

$$v_l = \sum_{m=1}^{3} a_{ml} v_m^*.$$

By differentiating this we have

$$\frac{\partial v_l}{\partial x_k^*} = \sum_{m=1}^{3} a_{ml} \frac{\partial v_m^*}{\partial x_k^*}.$$

If we substitute these expressions into (4), we obtain

$$\frac{\partial v_l}{\partial x_l} = \sum_{k=1}^{3} \sum_{m=1}^{3} a_{ml} \frac{\partial v_m^*}{\partial x_k^*} a_{kl} \qquad (l = 1, 2, 3).$$

By adding these three formulas we find

$$\text{div } \mathbf{v} = \sum_{l=1}^{3} \frac{\partial v_l}{\partial x_l} = \sum_{k=1}^{3} \sum_{m=1}^{3} \sum_{l=1}^{3} a_{kl} a_{ml} \frac{\partial v_m^*}{\partial x_k^*}.$$

Because of (7') in Sec. 7.9 this reduces to

$$\text{div } \mathbf{v} = \sum_{k=1}^{3} \sum_{m=1}^{3} \delta_{km} \frac{\partial v_m^*}{\partial x_k^*} = \frac{\partial v_1^*}{\partial x_1^*} + \frac{\partial v_2^*}{\partial x_2^*} + \frac{\partial v_3^*}{\partial x_3^*}.$$

We see that the expression on the right is identical with that in (2), and the proof is complete.

If $f(x, y, z)$ is a twice differentiable scalar function, then

$$\text{grad } f = \frac{\partial f}{\partial x} \mathbf{i} + \frac{\partial f}{\partial y} \mathbf{j} + \frac{\partial f}{\partial z} \mathbf{k}$$

and by (1),

$$\operatorname{div}(\operatorname{grad} f) = \frac{\partial^2 f}{\partial x^2} + \frac{\partial^2 f}{\partial y^2} + \frac{\partial^2 f}{\partial z^2}.$$

The expression on the right is the Laplacian of f (cf. Sec. 7.8). Thus

(5) $\operatorname{div}(\operatorname{grad} f) = \nabla^2 f.$

Example 1. The gravitational force **p** in Ex. 4, Sec. 7.8, is the gradient of the scalar function $f(x, y, z) = c/r$, which satisfies Laplace's equation $\nabla^2 f = 0$. According to (5), this means that $\operatorname{div} \mathbf{p} = 0$ $(r > 0)$.

The following example, taken from hydrodynamics, may serve as an introductory illustration of the physical significance of the divergence of a vector field. A more detailed physical interpretation of the divergence will be given later (in Sec. 8.9).

Example 2 (*Motion of a compressible fluid*). We consider the motion of a fluid in a region R having no *sources* or *sinks* in R, that is, no points at which fluid is produced or disappears. The concept of *fluid state* is meant to cover also gases and vapors. Fluids in the restricted sense, or liquids, have a small compressibility which can be neglected in many problems. Gases and vapors have a large compressibility, that is, their density ρ (= mass per unit volume) depends on the coordinates x, y, z in space (and may depend on the time t). We assume that our fluid is compressible.

We consider the flow through a small rectangular parallelepiped W of dimensions[9] Δx, Δy, Δz with edges parallel to the coordinate axes (Fig. 141). W has the volume $\Delta V = \Delta x \, \Delta y \, \Delta z$. Let

$$\mathbf{v} = v_1 \mathbf{i} + v_2 \mathbf{j} + v_3 \mathbf{k}$$

be the velocity vector of the motion. We set

(6) $\mathbf{u} = \rho \mathbf{v} = u_1 \mathbf{i} + u_2 \mathbf{j} + u_3 \mathbf{k}$

and assume that **u** and **v** are continuously differentiable vector functions of x, y, z, and t. Let us calculate the change in the mass included in W by considering the **flux** across the boundary, that is, the total loss of mass leaving W per unit time. Consider the flow through the left-hand face of W, whose area is $\Delta x \Delta z$. The components v_1 and v_3 of **v** are parallel to that face and contribute nothing to that flow. Hence the mass of fluid entering through that face during a short time interval Δt is given approximately by

$$(\rho v_2)_y \Delta x \, \Delta z \, \Delta t = u_2(y) \, \Delta x \, \Delta z \, \Delta t$$

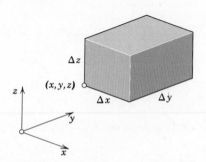

Fig. 141. Physical interpretation of the divergence

[9] It is a standard usage to indicate small quantities by Δ; this has, of course, nothing to do with the Laplacian.

where the subscript y indicates that this expression refers to the left-hand face. The mass of fluid leaving the parallelepiped W through the opposite face during the same time interval is approximately $u_2(y + \Delta y)\, \Delta x\, \Delta z\, \Delta t$. The difference

$$\Delta u_2 \, \Delta x \, \Delta z \, \Delta t = \frac{\Delta u_2}{\Delta y} \, \Delta V \, \Delta t \qquad\qquad [\Delta u_2 = u_2(y + \Delta y) - u_2(y)]$$

is the approximate loss of mass. Two similar expressions are obtained by considering the other two pairs of parallel faces of W. If we add these three expressions, we find that the total loss of mass in W during the time interval Δt is approximately

$$\left(\frac{\Delta u_1}{\Delta x} + \frac{\Delta u_2}{\Delta y} + \frac{\Delta u_3}{\Delta z} \right) \Delta V \, \Delta t$$

where

$$\Delta u_1 = u_1(x + \Delta x) - u_1(x) \qquad \text{and} \qquad \Delta u_3 = u_3(z + \Delta z) - u_3(z).$$

This loss of mass in W is caused by the time rate of change of the density and is thus equal to

$$-\frac{\partial \rho}{\partial t} \, \Delta V \, \Delta t.$$

If we equate both expressions, divide the resulting equation by $\Delta V \, \Delta t$, and let Δx, Δy, Δz, and Δt approach zero, then we obtain

$$\text{div } \mathbf{u} = \text{div } (\rho \mathbf{v}) = -\frac{\partial \rho}{\partial t}$$

or

(7) $$\frac{\partial \rho}{\partial t} + \text{div } (\rho \mathbf{v}) = 0.$$

This important relation is called the *condition for the conservation of mass* or the **continuity equation** *of a compressible fluid flow*.

If the flow is *steady*, that is, independent of time t, then $\partial \rho / \partial t = 0$, and the continuity equation is

(8) $$\text{div } (\rho \mathbf{v}) = 0.$$

If the fluid is incompressible, then the density ρ is constant, and (8) becomes

(9) $$\text{div } \mathbf{v} = 0.$$

This relation is known as the **condition of incompressibility**. It expresses the fact that the balance of outflow and inflow for a given volume element is zero at any time. Clearly the assumption that the flow has no sources or sinks in R is essential to our argument.

Problems

1. Verify the result in Ex. 1 by direct calculation.
2. Verify (5) when $f(x, y) = \ln (x^2 + y^2)$ and show that $\nabla^2 f = 0$.

Find the divergence of the following vector functions.

3. $ax\mathbf{i} + by\mathbf{j} + cz\mathbf{k}$
4. $x^2\mathbf{i} + y^2\mathbf{j} + z^2\mathbf{k}$
5. $y^2\mathbf{i} + z^2\mathbf{j} + x^2\mathbf{k}$
6. $e^{xyz}(\mathbf{i} + \mathbf{j} + \mathbf{k})$
7. $xyz(x\mathbf{i} + y\mathbf{j} + z\mathbf{k})$
8. $\cos x \cosh y \, \mathbf{i} + \sin x \sinh y \, \mathbf{j}$
9. $(x\mathbf{i} + y\mathbf{j} + z\mathbf{k})/(x^2 + y^2 + z^2)^{3/2}$
10. $(x\mathbf{i} + y\mathbf{j})/(x^2 + y^2)$

Show that

11. $\text{div } (k\mathbf{v}) = k \text{ div } \mathbf{v} \qquad (k \text{ constant})$
12. $\text{div } (f\mathbf{v}) = f \text{ div } \mathbf{v} + \mathbf{v} \cdot \nabla f$
13. $\text{div } (f \nabla g) = f\nabla^2 g + \nabla f \cdot \nabla g$
14. $\text{div } (f \nabla g) - \text{div } (g \nabla f) = f\nabla^2 g - g\nabla^2 f$

15. Verify the formula in Prob. 12 when $f = e^{xyz}$ and $\mathbf{v} = ax\mathbf{i} + by\mathbf{j} + cz\mathbf{k}$.
16. Obtain the answer to Prob. 9 from the formula in Prob. 12.
17. Verify the formula in Prob. 13 when $f = x + y + z$ and $g = xyz$.
18. Find the directional derivative of div $(x^5\mathbf{i} + y^5\mathbf{j} + z^5\mathbf{k})$ at the point P: $(2, 2, 1)$ in the direction of the outer normal to the sphere $x^2 + y^2 + z^2 = 9$.
19. Consider a steady flow whose velocity vector is $\mathbf{v} = y\mathbf{i}$. Show that it has the following properties. The flow is incompressible. The particles which at time $t = 0$ are in the cube bounded by the planes $x = 0$, $x = 1$, $y = 0$, $y = 1$, $z = 0$, $z = 1$ occupy at $t = 1$ the volume 1.
20. Consider a steady flow having the velocity $\mathbf{v} = x\mathbf{i}$. Show that the individual particles have the position vectors $\mathbf{r}(t) = c_1 e^t\mathbf{i} + c_2\mathbf{j} + c_3\mathbf{k}$ where c_1, c_2, c_3 are constants, the flow is compressible, and e is the volume occupied by the particles which at $t = 0$ fill the cube in Prob. 19.

7.11   *Curl of a Vector Field*

Let x, y, z be right-handed Cartesian coordinates in space, and let

$$\mathbf{v}(x, y, z) = v_1\mathbf{i} + v_2\mathbf{j} + v_3\mathbf{k}$$

be a differentiable vector function. Then the function

(1)
$$\operatorname{curl} \mathbf{v} = \nabla \times \mathbf{v} = \begin{vmatrix} \mathbf{i} & \mathbf{j} & \mathbf{k} \\ \dfrac{\partial}{\partial x} & \dfrac{\partial}{\partial y} & \dfrac{\partial}{\partial z} \\ v_1 & v_2 & v_3 \end{vmatrix}$$

$$= \left(\frac{\partial v_3}{\partial y} - \frac{\partial v_2}{\partial z}\right)\mathbf{i} + \left(\frac{\partial v_1}{\partial z} - \frac{\partial v_3}{\partial x}\right)\mathbf{j} + \left(\frac{\partial v_2}{\partial x} - \frac{\partial v_1}{\partial y}\right)\mathbf{k}$$

is called the **curl** *of the vector function* $\mathbf{v}$ or the *curl of the vector field defined by* $\mathbf{v}$. In the case of a left-handed Cartesian coordinate system, the symbolic determinant in (1) is preceded by a minus sign, in agreement with (5) in Sec. 5.8.

Instead of curl $\mathbf{v}$ the notation rot $\mathbf{v}$ is also used in the literature.

Theorem 1.   *The length and direction of* curl $\mathbf{v}$ *are independent of the particular choice of Cartesian coordinate systems in space.*

The proof of this theorem is included in the answer to Prob. 17.

The curl plays an important role in many applications. Its significance will be explained in more detail in Sec. 8.11. Now we confine ourselves to some simple examples and remarks.

Example 1 (Rotation of a rigid body).   We have seen in Ex. 3, Sec. 5.8, that a rotation of a rigid body B about a fixed axis in space can be described by a vector $\mathbf{w}$ of magnitude ω in the direction of the axis of rotation, where ω (> 0) is the angular speed of the rotation, and $\mathbf{w}$ is directed so that the rotation appears clockwise if we look in the direction of $\mathbf{w}$. According to (7), Sec. 5.8, the velocity field of the rotation can be represented in the form

$$\mathbf{v} = \mathbf{w} \times \mathbf{r}$$

where $\mathbf{r}$ is the position vector of a moving point with respect to a Cartesian coordinate system having the origin on the axis of rotation. Let us choose right-handed Cartesian coordinates such that $\mathbf{w} = \omega\mathbf{k}$; that is, the axis of rotation is the z-axis. Then (cf. Ex. 3 in Sec. 7.1)

$$\mathbf{v} = \mathbf{w} \times \mathbf{r} = -\omega y\mathbf{i} + \omega x\mathbf{j}$$

and, therefore,

$$\operatorname{curl} \mathbf{v} = \begin{vmatrix} \mathbf{i} & \mathbf{j} & \mathbf{k} \\ \dfrac{\partial}{\partial x} & \dfrac{\partial}{\partial y} & \dfrac{\partial}{\partial z} \\ -\omega y & \omega x & 0 \end{vmatrix} = 2\omega\mathbf{k},$$

that is,

(2) $$\operatorname{curl} \mathbf{v} = 2\mathbf{w}.$$

Hence, in the case of a rotation of a rigid body, the curl of the velocity field has the direction of the axis of rotation, and its magnitude equals twice the angular speed ω of the rotation.

Note that our result does not depend on the particular choice of the Cartesian coordinate system.

For any twice continuously differentiable scalar function f,

(3) $$\operatorname{curl} (\operatorname{grad} f) = \mathbf{0},$$

as can easily be verified by direct calculation. *Hence if a vector function is the gradient of a scalar function, its curl is the zero vector.* Since the curl characterizes the rotation in a field, we also say more briefly that *gradient fields describing a motion are irrotational.* (If such a field occurs in some other connection, not as a velocity field, it is usually called *conservative*; cf. at the end of Sec. 7.8.)

Example 2. The gravitational field in Ex. 4, Sec. 7.8, has curl $\mathbf{p} = \mathbf{0}$. The field in Ex. 1, this section, is not irrotational. A similar velocity field is obtained by stirring coffee in a cup.

Problems

Find curl $\mathbf{v}$ where, with respect to right-handed Cartesian coordinates,

1. $\mathbf{v} = x\mathbf{i} + 3xy\mathbf{j}$ 2. $\mathbf{v} = z\mathbf{i} + x\mathbf{j} + y\mathbf{k}$
3. $\mathbf{v} = y^2\mathbf{i} + z^2\mathbf{j} + x^2\mathbf{k}$ 4. $\mathbf{v} = yz\mathbf{i} + zx\mathbf{j} + xy\mathbf{k}$
5. $\mathbf{v} = x\mathbf{i} + yz\mathbf{j} - (x^2 + z^2)\mathbf{k}$ 6. $\mathbf{v} = (x\mathbf{i} + y\mathbf{j} + z\mathbf{k})/(x^2 + y^2 + z^2)^{3/2}$

7. Verify by direct calculation that in Ex. 2, curl $\mathbf{p} = \mathbf{0}$.
8. With respect to right-handed Cartesian coordinates, let $\mathbf{u} = z\mathbf{i} + x\mathbf{j} + y\mathbf{k}$, $\mathbf{v} = (x - y)^2\mathbf{i} + (y - z)^2\mathbf{j} + (z - x)^2\mathbf{k}$, and $\mathbf{w} = 2xy\mathbf{i} + 4yz\mathbf{j} - xz\mathbf{k}$. Find curl $\mathbf{u}$, curl $\mathbf{v}$, curl $\mathbf{w}$, curl $(\mathbf{u} - \mathbf{v})$, curl $(\mathbf{u} \times \mathbf{w})$, grad $(\mathbf{u} \cdot \mathbf{v})$, and div $(\mathbf{u} \times \mathbf{v})$.

Show that, granted sufficient differentiability,

9. curl $(\mathbf{u} + \mathbf{v}) = $ curl $\mathbf{u} + $ curl $\mathbf{v}$ 10. curl $(\operatorname{grad} f) = \mathbf{0}$
11. div $(\mathbf{u} \times \mathbf{v}) = \mathbf{v} \cdot $ curl $\mathbf{u} - \mathbf{u} \cdot $ curl $\mathbf{v}$ 12. div (curl $\mathbf{v}) = 0$
13. curl $(f\mathbf{v}) = \operatorname{grad} f \times \mathbf{v} + f$ curl $\mathbf{v}$ 14. div $(g\nabla f \times f\nabla g) = 0$

15. Find curl $(\mathbf{u} \times \mathbf{v})$ and div $(\mathbf{u} \times \mathbf{v})$, where $\mathbf{u} = y\mathbf{i} + z\mathbf{j} + x\mathbf{k}$ and $\mathbf{v} = xy\mathbf{i} + yz\mathbf{j} + zx\mathbf{k}$.
16. Using Prob. 13, show that for a vector $\mathbf{w}$ of constant direction, curl $\mathbf{w}$ is orthogonal to $\mathbf{w}$.
17. Prove Theorem 1.

In each case the velocity $\mathbf{v}$ of a steady fluid motion is given. Find curl $\mathbf{v}$. Is the motion incompressible? Find the paths of the particles.

18. $\mathbf{v} = 2x\mathbf{i} + 2y\mathbf{j}$ 19. $\mathbf{v} = -2y\mathbf{i} + 2x\mathbf{j}$ 20. $\mathbf{v} = y^2\mathbf{i}$

CHAPTER 8

Line and Surface Integrals.
Integral Theorems

In this chapter we shall define line integrals and surface integrals and consider some important applications of such integrals, which occur frequently in connection with physical and engineering problems. We shall see that a line integral is a natural generalization of a definite integral, and a surface integral is a generalization of a double integral.

Line integrals can be transformed into double integrals (Sec. 8.4) or into surface integrals (Sec. 8.10), and conversely. Triple integrals can be transformed into surface integrals (Sec. 8.8). These transformations are of great practical importance. The corresponding formulas of Gauss, Green, and Stokes serve as powerful tools in many applications as well as in theoretical problems. We shall see that they also lead to a better understanding of the physical meaning of the divergence and the curl of a vector function.

> *Prerequisites for this chapter:* elementary integral calculus and Chap. 7.
> *Sections which may be omitted in a shorter course:* 8.6, 8.9, 8.11.
> *References:* Appendix 1, Part C.
> *Answers to problems:* Appendix 2.

8.1 *Line Integral*

The concept of a line integral is a simple and natural generalization of the concept of à definite integral

$$(1) \qquad \qquad \int_a^b f(x)\, dx.$$

In (1) we integrate along the x-axis, and the integrand $f(x)$ is a function defined at each point between a and b. We shall see that, in the case of a line integral, we integrate along a curve in space (or in the plane), and the integrand is a function defined at the points of that curve.

We recall that, by definition, a function is a **single-valued** relation, that is, to each point in its domain of definition it assigns a *single* value in its range.

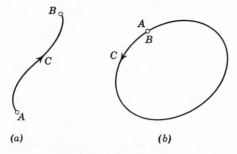

(a)   *(b)*

Fig. 142. Oriented curve

The way of defining a line integral is quite similar to the familiar way of defining a definite integral known from calculus. We may proceed as follows.

We consider a curve C in space and orient C by choosing one of the two directions along C as the *positive direction*; the opposite direction along C is then called the *negative direction*. Let A be the initial point and B the terminal point of C under the chosen orientation. (These points may coincide as in Fig. 142b; then C is closed.)

We may now represent C by a parametric representation

(2) $$\mathbf{r}(s) = x(s)\mathbf{i} + y(s)\mathbf{j} + z(s)\mathbf{k} \qquad (a \leqq s \leqq b),$$

where s is the arc length of C (cf. Sec. 7.4) and A and B correspond to $s = a$ and $s = b$, respectively.

We assume that $\mathbf{r}(s)$ is continuous and has a continuous first derivative which is different from the zero vector for all s under consideration. Then C has a unique tangent at each of its points. A curve satisfying these assumptions will be called a **smooth curve**.

Let $f(x, y, z)$ be a given function which is defined (at least) at each point of C, and is a continuous function of s. We subdivide C into n portions in an arbitrary manner (Fig. 143); let $P_0 (= A)$, $P_1, \cdots, P_{n-1}$, $P_n (= B)$ be the end points of these portions, and let

$$s_0 (= a) < s_1 < s_2 < \cdots < s_n (= b)$$

be the corresponding values of s. Then we choose an arbitrary point on each

Fig. 143. Subdivision of C

portion, say, a point Q_1 between P_0 and P_1, a point Q_2 between P_1 and P_2, etc. Taking the values of f at these points $Q_1, Q_2, \cdots, Q_n$, we form the sum

$$J_n = \sum_{m=1}^{n} f(x_m, y_m, z_m) \, \Delta s_m$$

where x_m, y_m, z_m are the coordinates of Q_m and

$$\Delta s_m = s_m - s_{m-1}.$$

This we do for each $n = 2, 3, \cdots$ in a completely independent manner, but so that the greatest Δs_m approaches zero as n approaches infinity. In this way we obtain a sequence of real numbers $J_2, J_3, \cdots$. The limit of this sequence is called the **line integral** *of f along C from A to B*, and is denoted by

$$\int_C f(x, y, z) \, ds.$$

The curve C is called the *path of integration.*

Since, by assumption, f is continuous and C is smooth, that limit exists and is independent of the choice of those subdivisions and points Q_m. In fact, the position of a point P on C is determined by the corresponding value of the arc length s; since A and B correspond to $s = a$ and $s = b$, respectively, we thus have

$$\int_C f(x, y, z) \, ds = \int_a^b f[x(s), y(s), z(s)] \, ds.$$

Hence the line integral is equal to the definite integral on the right, for which the statement holds, as is known from calculus.

Throughout this book all paths of integration for line integrals will be assumed to be **piecewise smooth**, *that is, will consist of finitely many smooth curves.*

From the definition it follows that familiar properties of ordinary definite integrals are equally valid for line integrals:

$$(a) \quad \int_C kf \, ds = k \int_C f \, ds \qquad (k \text{ constant})$$

(3)
$$(b) \quad \int_C (f + g) \, ds = \int_C f \, ds + \int_C g \, ds$$

$$(c) \quad \int_C f \, ds = \int_{C_1} f \, ds + \int_{C_2} f \, ds$$

where in (3c) the path C is subdivided into two arcs C_1 and C_2 which have the same orientation as C (Fig. 144). In (3b) the orientation of C is the same in all three integrals. If the sense of integration along C is reversed, the value of the integral is multiplied by -1.

Examples will be considered in the following sections.

Fig. 144. Formula (3c)

We finally mention that for a line integral over a *closed* path C the symbol

$$\oint_C \qquad \left(\text{instead of } \int_C\right)$$

is sometimes used in the literature.

8.2   *Evaluation of Line Integrals*

To evaluate a line integral we first write it as an ordinary definite integral. This is quite simple and can be done by using the given representation of the path of integration C.

If C is represented in the form

$$\mathbf{r}(s) = x(s)\mathbf{i} + y(s)\mathbf{j} + z(s)\mathbf{k} \qquad\qquad (a \leqq s \leqq b),$$

[cf. (2) in Sec. 8.1], then we immediately have

$$\int_C f(x, y, z)\, ds = \int_a^b f[x(s), y(s), z(s)]\, ds,$$

the integral on the right being a definite integral.

In most cases the representation of C will be of the form

(1) $$\mathbf{r}(t) = x(t)\mathbf{i} + y(t)\mathbf{j} + z(t)\mathbf{k} \qquad\qquad (t_0 \leqq t \leqq t_1)$$

or may easily be converted to this form; here, t denotes any parameter, and $\mathbf{r}(t)$ is continuous and has a continuous derivative. Then, using the familiar substitution formula for definite integrals, we obtain

$$\int_a^b f[x(s), y(s), z(s)]\, ds = \int_{t_0}^{t_1} f[x(t), y(t), z(t)] \frac{ds}{dt}\, dt.$$

Therefore, if C is represented in the form (1), then

(2) $$\int_C f(x, y, z)\, ds = \int_{t_0}^{t_1} f[x(t), y(t), z(t)] \frac{ds}{dt}\, dt$$

where, according to (3) in Sec. 7.4,

$$\frac{ds}{dt} = \sqrt{\dot{\mathbf{r}} \cdot \dot{\mathbf{r}}} = \sqrt{\dot{x}^2 + \dot{y}^2 + \dot{z}^2}.$$

Example 1. Evaluate $\displaystyle\int_C xy^3\, ds$, where C is the segment of the line $y = 2x$ in the xy-plane from A: $(-1, -2, 0)$ to B: $(1, 2, 0)$.

Representing C in the form[1]

$$\mathbf{r}(t) = t\mathbf{i} + 2t\mathbf{j} \qquad\qquad (-1 \leqq t \leqq 1)$$

we have

$$\mathbf{r} = \mathbf{i} + 2\mathbf{j} \qquad \text{and} \qquad \frac{ds}{dt} = \sqrt{\mathbf{r}\cdot\mathbf{r}} = \sqrt{5}.$$

On C, $xy^3 = t(2t)^3 = 8t^4$, and therefore

$$\int_C xy^3\,ds = 8\sqrt{5}\int_{-1}^{1} t^4\,dt = \frac{16}{\sqrt{5}} \approx 7.16.$$

Example 2. Evaluate $\displaystyle\int_C (x^2 + y^2 + z^2)^2\,ds$ where C is the arc of the circular helix (Sec. 7.3)

$$\mathbf{r}(t) = \cos t\,\mathbf{i} + \sin t\,\mathbf{j} + 3t\mathbf{k}$$

from $A\colon (1, 0, 0)$ to $B\colon (1, 0, 6\pi)$. We find $ds/dt = \sqrt{10}$. On C,

$$(x^2 + y^2 + z^2)^2 = [\cos^2 t + \sin^2 t + (3t)^2]^2 = (1 + 9t^2)^2.$$

Since C corresponds to $0 \leqq t \leqq 2\pi$, we thus have

$$\int_C (x^2 + y^2 + z^2)^2\,ds = \sqrt{10}\int_0^{2\pi} (1 + 9t^2)^2\,dt$$

$$= \sqrt{10}\,[2\pi + 6(2\pi)^3 + \tfrac{81}{5}(2\pi)^5] \approx 506\,400.$$

Line integrals which involve empirically given functions or lead to complicated definite integrals may be evaluated by using numerical or graphical methods of integration.

In many applications the integrands of line integrals are of the form

$$(3) \qquad g(x, y, z)\frac{dx}{ds}, \qquad g(x, y, z)\frac{dy}{ds}, \qquad \text{or} \qquad g(x, y, z)\frac{dz}{ds},$$

where dx/ds, dy/ds, and dz/ds are the derivatives of the functions occurring in the parametric representation of the path of integration. Then we simply write

$$(4) \qquad\qquad \int_C g(x, y, z)\frac{dx}{ds}\,ds = \int_C g(x, y, z)\,dx,$$

and similarly in the other two cases. For sums of these types of integrals along the same path C we adopt the simplified notation

$$(5) \qquad\qquad \int_C f\,dx + \int_C g\,dy + \int_C h\,dz = \int_C (f\,dx + g\,dy + h\,dz).$$

Using the representation of C, we may eliminate two of the three independent variables in the integrand and then evaluate the resulting definite integral in which the remaining independent variable is the variable of integration.

Example 3. Evaluate the line integral $I = \displaystyle\int_C [x^2y\,dx + (x - z)\,dy + xyz\,dz]$ where C is the arc of the parabola $y = x^2$ in the plane $z = 2$ from $A\colon (0, 0, 2)$ to $B\colon (1, 1, 2)$; cf. Fig. 145.

---

[1] Of course, since $x = t$ we may write x in place of t.

Fig. 145. Paths in Examples 3 and 4

Since $y = x^2$, we have $dy/dx = 2x$ or $dy = 2x\,dx$. Since $z = 2$ is constant, it follows that the integral of the last term in the integrand is zero. Thus

$$I = \int_0^1 [x^2x^2\,dx + (x - 2)2x\,dx] = \int_0^1 (x^4 + 2x^2 - 4x)\,dx = -\tfrac{17}{15}.$$

Example 4. Evaluate the line integral I in Ex. 3 where C is the segment of the straight line $y = x,\ z = 2$ from A: $(0, 0, 2)$ to B: $(1, 1, 2)$. Now $dy = dx$, and

$$I = \int_0^1 (x^3 + x - 2)\,dx = -\tfrac{5}{4}.$$

In Ex. 3 and Ex. 4 the integrands and the end points of the paths are the same, but the values of I are different. This illustrates the important fact that, *in general, the value of a line integral of a given function depends not only on the end points but also on the geometric shape of the path of integration.* We shall consider this basic fact in Sec. 8.12.

In many cases, the functions f, g, h in (5) are the components v_1, v_2, v_3 of a vector function

$$\mathbf{v} = v_1\mathbf{i} + v_2\mathbf{j} + v_3\mathbf{k}.$$

Then

$$v_1\,dx + v_2\,dy + v_3\,dz = \left(v_1\frac{dx}{ds} + v_2\frac{dy}{ds} + v_3\frac{dz}{ds}\right)ds,$$

the expression in parentheses being the scalar product of the vector $\mathbf{v}$ and the unit tangent vector

$$\frac{d\mathbf{r}}{ds} = \frac{dx}{ds}\mathbf{i} + \frac{dy}{ds}\mathbf{j} + \frac{dz}{ds}\mathbf{k} \qquad \text{(cf. Sec. 7.5)},$$

where $\mathbf{r}(s)$ represents the path of integration C. Therefore,

(6) $$\int_C (v_1\,dx + v_2\,dy + v_3\,dz) = \int_C \mathbf{v}\cdot\frac{d\mathbf{r}}{ds}\,ds.$$

This is sometimes written

$$\int_C \mathbf{v}\cdot\frac{d\mathbf{r}}{ds}\,ds = \int_C \mathbf{v}\cdot d\mathbf{r}$$

where

$$d\mathbf{r} = dx\,\mathbf{i} + dy\,\mathbf{j} + dz\,\mathbf{k},$$

but we should note that this is merely a convenient notation.

Example 5 (*Work done by a force*).   Consider a particle on which a variable force **p** acts. Let the particle be displaced along a given path C in space. Then the work W done by **p** in this displacement is given by the line integral

$$(7) \qquad\qquad W = \int_C \mathbf{p} \cdot d\mathbf{r},$$

the integration being taken in the sense of the displacement. This definition of work is suggested by that in Ex. 1, Sec. 5.5, together with the definition of a line integral as the limit of a sum.
    We may introduce the time t as the variable of integration. Then

$$d\mathbf{r} = \frac{d\mathbf{r}}{dt}\, dt = \mathbf{v}\, dt$$

where **v** is the velocity vector (cf. Sec. 7.6). Hence the line integral (7) becomes

$$(7') \qquad\qquad W = \int_{t_0}^{t_1} \mathbf{p} \cdot \mathbf{v}\, dt$$

where t_0 and t_1 are the initial and final values of t. Furthermore, by Newton's second law (Sec. 2.6) we have

$$\mathbf{p} = m\ddot{\mathbf{r}} = m\dot{\mathbf{v}},$$

where m is the mass of the particle. If we insert this into (7'), we obtain

$$W = \int_{t_0}^{t_1} m\dot{\mathbf{v}} \cdot \mathbf{v}\, dt = \int_{t_0}^{t_1} \frac{d}{dt}\left(\frac{m}{2}\mathbf{v} \cdot \mathbf{v}\right) dt = \int_{t_0}^{t_1} \frac{d}{dt}\left(\frac{m}{2}|\mathbf{v}|^2\right) dt = \frac{m}{2}|\mathbf{v}|^2 \Big|_{t_0}^{t_1};$$

that is, *the work done equals the gain in kinetic energy*. This is a basic law in mechanics.

Problems

Orienting C so that the sense of integration becomes the positive sense on C, evaluate $\int_C (3x^2 + 3y^2)\, ds$:

1. Over the path $y = x$ from $(0, 0)$ to $(1, 1)$.
2. Over the path $y = -x$ from $(-1, 1)$ to $(1, -1)$.
3. Over the path $y = 3x$ from $(0, 0)$ to $(2, 6)$.
4. Over the y-axis from $(0, 0)$ to $(0, 1)$ and then parallel to the x-axis from $(0, 1)$ to $(1, 1)$.
5. Over the x-axis from $(0, 0)$ to $(1, 0)$ and then parallel to the y-axis from $(1, 0)$ to $(1, 1)$.
6. Clockwise along the circle $x^2 + y^2 = 1$ from $(0, 1)$ to $(1, 0)$.
7. Counterclockwise along the circle $x^2 + y^2 = 1$ from $(1, 0)$ to $(0, 1)$.

Evaluate $\int_C (y^2\, dx - x^2\, dy)$:

8. Counterclockwise along the circle $x^2 + y^2 = 1$ from $(0, 1)$ to $(1, 0)$.
9. Counterclockwise along the circle $x^2 + y^2 = 1$ from $(1, 0)$ to $(0, 1)$.
10. Counterclockwise around the circle $x^2 + y^2 = 1$ from $(0, 1)$ to $(0, 1)$.
11. Along the straight line from $(0, 1)$ to $(1, 0)$.

Find the work done by the force $\mathbf{p} = 4xy\mathbf{i} - 8y\mathbf{j} + 2\mathbf{k}$ in the displacement:

12. Along the curve $y = 2x$, $z = 2$, from $(0, 0, 2)$ to $(3, 6, 2)$.
13. Along the straight line $y = 2x$, $z = 0$, from $(3, 6, 0)$ to $(0, 0, 0)$.
14. Along the straight line $y = 2x$, $z = 0$, from $(0, 0, 0)$ to $(3, 6, 0)$.
15. Along the straight line $y = 2x$, $z = 2x$ from $(0, 0, 0)$ to $(3, 6, 6)$.
16. Along the parabola $y = 2x^2/3$, $z = 0$ from $(0, 0, 0)$ to $(3, 6, 0)$.
17. Around the circle $x^2 + y^2 = 4$, $z = 0$ from $(2, 0, 0)$ to $(2, 0, 0)$.

18. Using the trapezoidal rule (Sec. 18.5), evaluate $\int_C f(x, y)\, ds$ along the straight line $y = x$ in the xy-plane from $(0, 0)$ to $(1, 1)$, where $f(x, y)$ is given by the values $f(0, 0) = 1.0$, $f(\frac{1}{4}, \frac{1}{4}) = 1.5, f(\frac{1}{2}, \frac{1}{2}) = 1.7, f(\frac{3}{4}, \frac{3}{4}) = 1.5, f(1, 1) = 1.0$.

19. Let **p** be a vector function defined at all points of a curve C, and suppose that $|\mathbf{p}|$ is bounded, say, $|\mathbf{p}| < M$ on C, where M is some positive number. Show that

(8)
$$\left| \int_C \mathbf{p} \cdot d\mathbf{r} \right| < Ml$$

where l is the length of C.

20. Using (8), find an upper bound for the absolute value of the work W done by the force $\mathbf{p} = x\mathbf{i} + y^2\mathbf{j}$ in the displacement along the straight line from $(0, 0, 0)$ to $(1, 1, 0)$. Find W by integration and compare the results.

8.3 *Double Integrals*

In our further consideration we shall need the concept of a double integral. Although the reader will be familiar with double integrals from calculus, we shall now present a brief review.

In the case of a definite integral

$$\int_a^b f(x)\, dx$$

the integrand is a function $f(x)$ which exists for all x in an interval $a \leq x \leq b$ of the x-axis. In the case of a double integral the integrand is a function $f(x, y)$ which is given for all (x, y) in a closed bounded[2] region R of the xy-plane.

The definition of the double integral is quite similar to that of the definite integral. We subdivide the region R by drawing parallels to the x and y axes (Fig. 146). We number those rectangles which are within R from 1 to n. In each such rectangle we choose a point, say, (x_k, y_k) in the kth rectangle, and then we form the sum

$$J_n = \sum_{k=1}^{n} f(x_k, y_k)\, \Delta A_k$$

where ΔA_k is the area of the kth rectangle. This we do for larger and larger positive integers n in a completely independent manner but so that the length of the maximum diagonal of the rectangles approaches zero as n approaches infinity. In this fashion we obtain a sequence of real numbers $J_{n_1}, J_{n_2}, \cdots$. Assuming that

Fig. 146. Subdivision of R

---

[2] "Closed" means that the boundary is part of the region, and "bounded" means that the region can be enclosed in a circle of sufficiently large radius.

Fig. 147. Formula (1)

$f(x, y)$ is continuous in R and R is bounded by finitely many smooth curves (cf. Sec. 8.1) it can be shown[3] that the sequence converges and its limit is independent of the choice of subdivisions and corresponding points (x_k, y_k). This limit is called the **double integral** *of* $f(x, y)$ *over the region* R, and is denoted by the symbol

$$\iint\limits_{R} f(x, y)\, dx\, dy.$$

From the definition it follows that double integrals enjoy properties which are quite similar to those of definite integrals. Let f and g be functions of x and y, defined and continuous in a region R. Then

$$\iint\limits_{R} kf\, dx\, dy = k \iint\limits_{R} f\, dx\, dy \qquad\qquad (k \text{ constant})$$

(1)
$$\iint\limits_{R} (f + g)\, dx\, dy = \iint\limits_{R} f\, dx\, dy + \iint\limits_{R} g\, dx\, dy$$

$$\iint\limits_{R} f\, dx\, dy = \iint\limits_{R_1} f\, dx\, dy + \iint\limits_{R_2} f\, dx\, dy \qquad (\text{cf. Fig. 147}).$$

Furthermore, there exists at least one point (x_0, y_0) in R such that

(2)
$$\iint\limits_{R} f(x, y)\, dx\, dy = f(x_0, y_0)\, A,$$

where A is the area of R; this is called the **mean value theorem** *for double integrals.*

Double integrals over a region R *may be evaluated by two successive integrations* as follows.

Suppose that R can be described by inequalities of the form

$$a \leqq x \leqq b, \qquad g(x) \leqq y \leqq h(x)$$

(Fig. 148), so that $y = g(x)$ and $y = h(x)$ represent the boundary of R. Then

(3)
$$\iint\limits_{R} f(x, y)\, dx\, dy = \int_a^b \left[\int_{g(x)}^{h(x)} f(x, y)\, dy \right] dx.$$

We first integrate the inner integral

$$\int_{g(x)}^{h(x)} f(x, y)\, dy.$$

---

[3] Cf. Ref. [A3] in Appendix 1.

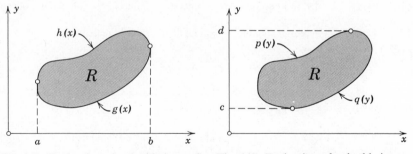

Fig. 148. Evaluation of a double integral   **Fig. 149.** Evaluation of a double integral

In this definite integral, x plays the role of a parameter, and the result of the integration will be a function of x, say, $F(x)$. By integrating $F(x)$ over x from a to b we then obtain the value of the double integral in (3).

Similarly, if R can be described by inequalities of the form

$$c \leqq y \leqq d, \qquad p(y) \leqq x \leqq q(y)$$

(Fig. 149), then we obtain

(4)
$$\iint\limits_{R} f(x, y)\, dx\, dy = \int_c^d \left[\int_{p(y)}^{q(y)} f(x, y)\, dx \right] dy;$$

we now integrate first over x and then the resulting function of y over y from c to d.

If R cannot be represented by those inequalities, but can be subdivided into finitely many portions which have that property, we may integrate $f(x, y)$ over each portion separately and add the results; this will give us the value of the double integral of $f(x, y)$ over that region R.

Double integrals have various geometrical and physical applications. For example, the **area** A of R is

$$A = \iint\limits_{R} dx\, dy.$$

The **volume** V beneath the surface $z = f(x, y)$ (>0) and above the region R in the xy-plane is (Fig. 150)

$$V = \iint\limits_{R} f(x, y)\, dx\, dy,$$

Fig. 150. Double integral as volume

because the term $f(x_k, y_k) \, \Delta A_k$ in J_n on p. 329 represents the volume of a rectangular parallelepiped with base ΔA_k and altitude $f(x_k, y_k)$.

Let $f(x, y)$ be the density (= mass per unit area) of a distribution of mass in the xy-plane. Then the total mass M in R is

$$M = \iint\limits_R f(x, y) \, dx \, dy,$$

the **center of gravity** of the mass in R has the coordinates

$$\bar{x} = \frac{1}{M} \iint\limits_R x f(x, y) \, dx \, dy \qquad \text{and} \qquad \bar{y} = \frac{1}{M} \iint\limits_R y f(x, y) \, dx \, dy,$$

the **moments of inertia** I_x and I_y of the mass in R about the x and y axes, respectively, are

$$I_x = \iint\limits_R y^2 f(x, y) \, dx \, dy, \qquad I_y = \iint\limits_R x^2 f(x, y) \, dx \, dy,$$

and the *polar moment of inertia* about the origin is

$$I_0 = I_x + I_y = \iint\limits_R (x^2 + y^2) f(x, y) \, dx \, dy.$$

Example 1. Let $f(x, y) = 1$ be the density of mass in the region R: $0 \le y \le \sqrt{1 - x^2}$, $0 \le x \le 1$ (Fig. 151). Find the center of gravity and the moments of inertia I_x, I_y, and I_0. The total mass in R is

$$M = \iint\limits_R dx \, dy = \int_0^1 \left[\int_0^{\sqrt{1-x^2}} dy \right] dx = \int_0^1 \sqrt{1 - x^2} \, dx = \int_0^{\pi/2} \cos^2 \theta \, d\theta = \frac{\pi}{4}$$

($x = \sin \theta$), which is the area of R. The coordinates of the center of gravity are

$$\bar{x} = \frac{4}{\pi} \iint\limits_R x \, dx \, dy = \frac{4}{\pi} \int_0^1 \left[\int_0^{\sqrt{1-x^2}} x \, dy \right] dx = \frac{4}{\pi} \int_0^1 x\sqrt{1 - x^2} \, dx = -\frac{4}{\pi} \int_1^0 z^2 \, dz = \frac{4}{3\pi}$$

($\sqrt{1 - x^2} = z$), and $\bar{y} = \bar{x}$, for reasons of symmetry. Furthermore,

$$I_x = \iint\limits_R y^2 \, dx \, dy = \int_0^1 \left[\int_0^{\sqrt{1-x^2}} y^2 \, dy \right] dx = \frac{1}{3} \int_0^1 \left(\sqrt{1 - x^2} \right)^3 dx$$

$$= \frac{1}{3} \int_0^{\pi/2} \cos^4 \theta \, d\theta = \frac{\pi}{16}, \qquad I_y = \frac{\pi}{16}, \qquad I_0 = I_x + I_y = \frac{\pi}{8} \approx 0.3927.$$

A simpler way of calculating I_x will be shown in Ex. 2.

Fig. 151. Example 1

It will often be necessary to change the variables of integration in double integrals. We know from calculus that in the case of a definite integral

$$\int_a^b f(x)\,dx$$

a new variable of integration u can be introduced by setting

$$x = x(u),$$

where the function $x(u)$ is continuous and has a continuous derivative in some interval $\alpha \leq u \leq \beta$ such that $x(\alpha) = a$, $x(\beta) = b$ [or $x(\alpha) = b$, $x(\beta) = a$] and $x(u)$ varies between a and b when u varies between α and β. Then

(5)
$$\int_a^b f(x)\,dx = \int_\alpha^\beta f(x(u))\frac{dx}{du}\,du.$$

For example, let $f(x) = \sqrt{1 - x^2}$, $a = 0$, $b = 1$, and set $x = \sin u$. Then

$$f[x(u)] = \sqrt{1 - \sin^2 u} = \cos u, \qquad \frac{dx}{du} = \cos u, \qquad \alpha = 0, \qquad \beta = \frac{\pi}{2},$$

and

$$\int_0^1 \sqrt{1 - x^2}\,dx = \int_0^{\pi/2} \cos^2 u\,du = \frac{\pi}{4}.$$

In the case of a double integral

$$\iint_R f(x, y)\,dx\,dy$$

new variables of integration u, v can be introduced by setting

$$x = x(u, v), \qquad y = y(u, v)$$

where the functions $x(u, v)$ and $y(u, v)$ are continuous and have continuous first partial derivatives in some region R^* in the uv-plane so that each point (u_0, v_0) in R^* corresponds to a point $[x(u_0, v_0), y(u_0, v_0)]$ in R and conversely, and furthermore the **Jacobian**[4]

$$J = \frac{\partial(x, y)}{\partial(u, v)} = \begin{vmatrix} \dfrac{\partial x}{\partial u} & \dfrac{\partial x}{\partial v} \\[2mm] \dfrac{\partial y}{\partial u} & \dfrac{\partial y}{\partial v} \end{vmatrix}$$

is either positive throughout R^* or negative throughout R^*. Then

(6)
$$\iint_R f(x, y)\,dx\,dy = \iint_{R^*} f[x(u, v), y(u, v)]\left|\frac{\partial(x, y)}{\partial(u, v)}\right|du\,dv;$$

[4] Named after the German mathematician CARL GUSTAV JACOB JACOBI (1804–1851), who made important contributions to elliptic functions, partial differential equations, mechanics, and the calculus of variations.

Fig. 152. Region in Example 3

that is, the integrand is expressed in terms of u and v, and $dx\,dy$ is replaced by $du\,dv$ times the absolute value of the Jacobian J. For the proof see Ref. [A2] in Appendix 1.

For example, polar coordinates r and θ can be introduced by setting

$$x = r\cos\theta, \qquad y = r\sin\theta.$$

Then

$$J = \frac{\partial(x, y)}{\partial(r, \theta)} = \begin{vmatrix} \cos\theta & -r\sin\theta \\ \sin\theta & r\cos\theta \end{vmatrix} = r,$$

and

(7) $$\iint_R f(x, y)\,dx\,dy = \iint_{R*} f(r\cos\theta, r\sin\theta)r\,dr\,d\theta,$$

where $R*$ is the region in the $r\theta$-plane corresponding to R in the xy-plane.

Example 2. Using (7), we obtain for I_x in Ex. 1

$$I_x = \iint_R y^2\,dx\,dy = \int_0^{\pi/2}\int_0^1 r^2\sin^2\theta\,r\,dr\,d\theta = \int_0^{\pi/2}\sin^2\theta\,d\theta\int_0^1 r^3\,dr = \frac{\pi}{4}\cdot\frac{1}{4} = \frac{\pi}{16}.$$

Example 3. Evaluate the double integral

$$\iint_R (x^2 + y^2)\,dx\,dy$$

where R is the square in Fig. 152. The shape of R suggests the transformation $x + y = u$, $x - y = v$. Then $x = \frac{1}{2}(u + v)$, $y = \frac{1}{2}(u - v)$, the Jacobian is

$$J = \frac{\partial(x, y)}{\partial(u, v)} = \begin{vmatrix} \frac{1}{2} & \frac{1}{2} \\ \frac{1}{2} & -\frac{1}{2} \end{vmatrix} = -\frac{1}{2},$$

R corresponds to the square $0 \leq u \leq 2$, $0 \leq v \leq 2$, and, therefore,

$$\iint_R (x^2 + y^2)\,dx\,dy = \int_0^2\int_0^2 \frac{1}{2}(u^2 + v^2)\frac{1}{2}\,du\,dv = \frac{8}{3}.$$

Problems

Describe the region of integration and evaluate:

1. $\displaystyle\int_0^1\int_x^{2x} (1 + x^2 + y^2)\,dy\,dx$ 2. $\displaystyle\int_0^{\pi/2}\int_{-1}^1 x^2y^2\,dx\,dy$

3. $\int_0^1 \int_{x^2}^x (1 - xy) \, dy \, dx$ **4.** $\int_0^2 \int_0^x e^{x+y} \, dy \, dx$

5. $\int_0^\pi \int_0^{1-\cos\theta} r \, dr \, d\theta$ **6.** $\int_0^1 \int_0^{\sqrt{1-x^2}} (x^2 + y^2) \, dy \, dx$

7. $\int_0^2 \int_{\sinh x^2}^{\cosh x^2} x \, dy \, dx$ **8.** $\int_{-1}^2 \int_{x^2}^{x+2} (x + y) \, dy \, dx$

Find the volume of the following regions in space.

9. The first octant section cut from the region inside the cylinder $x^2 + z^2 = 1$ by the planes $y = 0$, $z = 0$, $x = y$.

10. The tetrahedron cut from the first octant by the plane $3x + 4y + 2z = 12$.

11. The region between the paraboloid $z = 1 - x^2 - y^2$ and the xy-plane.

12. The first octant region bounded by the coordinate planes and the surfaces $y = 1 - x^2$, $z = 1 - x^2$.

13. The region bounded by the cylinders $x^2 + y^2 = 1$ and $y^2 + z^2 = 1$.

14. Find the Jacobians corresponding to a translation $x = u + a$, $y = v + b$ and a dilatation $x = au$, $y = bv$. Give a geometrical reason for the results.

15. Find the Jacobian corresponding to a rotation

$$x = u \cos \phi - v \sin \phi, \qquad y = u \sin \phi + v \cos \phi.$$

Using polar coordinates, evaluate $\iint_R f(x, y) \, dx \, dy$, where

16. $f = e^{-x^2-y^2}$, $R: x^2 + y^2 \leq 1$.

17. $f = 2(x + y)$, $R: x^2 + y^2 \leq 9, x \geq 0$.

18. $f = \cos(x^2 + y^2)$, $R: x^2 + y^2 \leq \pi/2, x \geq 0$.

Find the coordinates $\bar{x}$, $\bar{y}$ of the center of gravity of a mass of density $f(x, y)$ in a region R, where:

19. $f(x, y) = xy$, R the rectangle $0 \leq x \leq 2, 0 \leq y \leq 4$.

20. $f(x, y) = 1$, R the region $x^2 + y^2 \leq a^2$ in the first quadrant.

21. $f(x, y) = x^2 + y^2$, R as in Prob. 20.

Find the moments of inertia I_x, I_y, I_0 of a mass of density $f(x, y) = 1$ in a region R shown in the following figures.

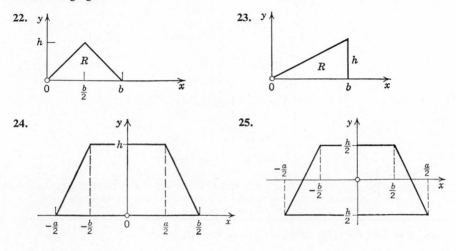

22.   **23.**

24.   **25.**

8.4   *Transformation of Double Integrals into Line Integrals*

Double integrals over a plane region may be transformed into line integrals over the boundary of the region and conversely. This transformation is of practical as well as theoretical interest and can be done by means of the following basic theorem.

Green's[5] theorem in the plane. *Let R be a closed bounded region in the xy-plane whose boundary C consists of finitely many smooth curves. Let $f(x, y)$ and $g(x, y)$ be functions which are continuous and have continuous partial derivatives $\partial f / \partial y$ and $\partial g / \partial x$ everywhere in some domain containing R. Then*

(1)
$$\iint\limits_{R} \left(\frac{\partial g}{\partial x} - \frac{\partial f}{\partial y} \right) dx\, dy = \int_{C} (f\, dx + g\, dy),$$

the integration being taken along the entire boundary C of R such that R is on the left as one advances in the direction of integration (cf. Fig. 153).

Proof. We first prove the theorem for a *special region R* which can be represented in both of the forms

$$a \leqq x \leqq b, \qquad u(x) \leqq y \leqq v(x),$$

and

$$c \leqq y \leqq d, \qquad p(y) \leqq x \leqq q(y)$$

(Fig. 154). Using (3) in Sec. 8.3, we obtain

(2)
$$\iint\limits_{R} \frac{\partial f}{\partial y}\, dx\, dy = \int_{a}^{b} \left[\int_{u(x)}^{v(x)} \frac{\partial f}{\partial y}\, dy \right] dx.$$

We integrate the inner integral:

$$\int_{u(x)}^{v(x)} \frac{\partial f}{\partial y}\, dy = f(x, y) \Big|_{y=u(x)}^{y=v(x)} = f[x, v(x)] - f[x, u(x)].$$

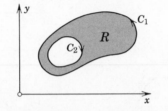

Fig. 153. Region R whose boundary C consists of two parts; C_1 is traversed in the counterclockwise sense, while C_2 is traversed in the clockwise sense.

[5] GEORGE GREEN (1793–1841), English mathematician.

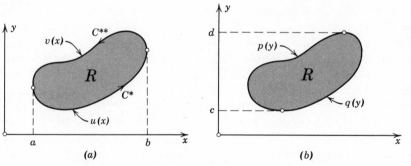

Fig. 154. Example of a special region for Green's theorem

By inserting this into (2) we find

$$\iint\limits_{R} \frac{\partial f}{\partial y}\, dx\, dy = \int_{a}^{b} f[x, v(x)]\, dx - \int_{a}^{b} f[x, u(x)]\, dx$$

$$= -\int_{a}^{b} f[x, u(x)]\, dx - \int_{b}^{a} f[x, v(x)]\, dx.$$

Since $y = u(x)$ represents the oriented curve C^* (Fig. 154a) and $y = v(x)$ represents C^{**}, the integrals on the right may be written as line integrals over C^* and C^{**}, and, therefore,

$$(3) \qquad \iint\limits_{R} \frac{\partial f}{\partial y}\, dx\, dy = -\int_{C^*} f(x, y)\, dx - \int_{C^{**}} f(x, y)\, dx = -\int_{C} f(x, y)\, dx.$$

If portions of C are segments parallel to the y-axis (such as $\tilde{C}$ and $\tilde{\tilde{C}}$ in Fig. 155), then the result is the same as before, because the integrals over these portions are zero and may be added to the integrals over C^* and C^{**} to obtain the integral over the whole boundary C in (3). Similarly, using (4), Sec. 8.3, we obtain (cf. Fig. 154b)

$$\iint\limits_{R} \frac{\partial g}{\partial x}\, dx\, dy = \int_{c}^{d} \left[\int_{p(y)}^{q(y)} \frac{\partial g}{\partial x}\, dx \right] dy = \int_{C} g(x, y)\, dy.$$

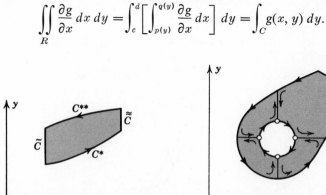

Fig. 155. Proof of Green's theorem   **Fig. 156.** Proof of Green's theorem

From this and (3), the formula (1) follows, and the theorem is proved for special regions.

We now prove the theorem for a region R which itself is not a special region but can be subdivided into finitely many special regions (Fig. 156). In this case we apply the theorem to each subregion and then add the results; the left-hand members add up to the integral over R while the right-hand members add up to the line integral over C plus integrals over the curves introduced for subdividing R. Each of the latter integrals occurs twice, taken once in each direction. Hence these two integrals cancel each other, and we are left with the line integral over C.

The proof thus far covers all regions which are of interest in engineering problems. To prove the theorem for the most general region R satisfying the conditions in the theorem, we must approximate R by a region of the type just considered and then use a limiting process. For details see Ref. [A2] in Appendix 1.

Green's theorem will be of basic importance in our further consideration. For the time being we consider a few simple illustrative examples.

Example 1 (*Area of a plane region as a line integral over the boundary*). In (1), let $f = 0$ and $g = x$. Then

$$\iint\limits_R dx\, dy = \int_C x\, dy.$$

The integral on the left is the area A of R. Similarly, let $f = -y$ and $g = 0$; then from (1),

$$A = \iint\limits_R dx\, dy = -\int_C y\, dx.$$

By adding both formulas we obtain

(4) $$A = \frac{1}{2}\int_C (x\, dy - y\, dx),$$

the integration being taken as indicated in Green's theorem. This interesting formula expresses the area of R in terms of a line integral over the boundary. It has various applications; for example, the theory of certain planimeters is based upon this formula; cf. Ref. [G15] in Appendix 1.

Example 2 (*Area of a plane region in polar coordinates*). Let r and θ be polar coordinates defined by $x = r \cos \theta$, $y = r \sin \theta$. Then

$$dx = \cos \theta\, dr - r \sin \theta\, d\theta, \qquad dy = \sin \theta\, dr + r \cos \theta\, d\theta,$$

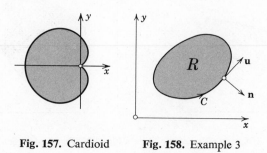

Fig. 157. Cardioid   **Fig. 158.** Example 3

and (4) assumes the form

(5)
$$A = \frac{1}{2} \int_C r^2 \, d\theta.$$

This formula is well known from calculus.

As an application of (5), consider the cardioid $r = a(1 - \cos \theta)$ where $0 \leq \theta \leq 2\pi$ (Fig. 157). We find

$$A = \frac{a^2}{2} \int_0^{2\pi} (1 - \cos \theta)^2 \, d\theta = \frac{3\pi}{2} a^2.$$

Example 3 (*Transformation of a double integral of the Laplacian of a function into a line integral of its normal derivative*). Let $w(x, y)$ be a function which is continuous and has continuous first and second partial derivatives in a domain of the xy-plane containing a region R of the type indicated in Green's theorem. We set $f = -\partial w/\partial y$ and $g = \partial w/\partial x$. Then $\partial f/\partial y$ and $\partial g/\partial x$ are continuous in R, and

(6)
$$\frac{\partial g}{\partial x} - \frac{\partial f}{\partial y} = \frac{\partial^2 w}{\partial x^2} + \frac{\partial^2 w}{\partial y^2} = \nabla^2 w,$$

the Laplacian of w (cf. Sec. 7.8). Furthermore, using those expressions for f and g, we obtain

(7)
$$\int_C (f \, dx + g \, dy) = \int_C \left(f \frac{dx}{ds} + g \frac{dy}{ds} \right) ds = \int_C \left(-\frac{\partial w}{\partial y} \frac{dx}{ds} + \frac{\partial w}{\partial x} \frac{dy}{ds} \right) ds,$$

where s is the arc length of C, and C is oriented as shown in Fig. 158. The integrand of the last integral may be written as the dot product (scalar product) of the vectors

$$\text{grad } w = \frac{\partial w}{\partial x} \mathbf{i} + \frac{\partial w}{\partial y} \mathbf{j} \quad \text{and} \quad \mathbf{n} = \frac{dy}{ds} \mathbf{i} - \frac{dx}{ds} \mathbf{j};$$

that is,

(8)
$$-\frac{\partial w}{\partial y} \frac{dx}{ds} + \frac{\partial w}{\partial x} \frac{dy}{ds} = (\text{grad } w) \cdot \mathbf{n}.$$

The vector $\mathbf{n}$ is a unit normal to C, because the vector

$$\mathbf{u} = \frac{d\mathbf{r}}{ds} = \frac{dx}{ds} \mathbf{i} + \frac{dy}{ds} \mathbf{j}$$

(cf. Sec. 7.5) is the unit tangent vector to C, and $\mathbf{u} \cdot \mathbf{n} = 0$. Furthermore, it is not difficult to see that $\mathbf{n}$ is directed to the exterior of C. From this and (5) in Sec. 7.8 it follows that the expression on the right-hand side of (8) is the derivative of w in the direction of the outward normal to C. Denoting this directional derivative by $\partial w/\partial n$ and taking (6), (7), and (8) into account, we obtain from Green's theorem the useful integral formula

(9)
$$\iint_R \nabla^2 w \, dx \, dy = \int_C \frac{\partial w}{\partial n} \, ds.$$

Further important applications and consequences of Green's theorem in the plane will be considered in the following sections.

Problems

Using Green's theorem, evaluate the following integrals. Check the result by direct calculation.

1. $\int_C (y \, dx + 2x \, dy)$ C: the boundary of the square $0 \leq x \leq 1, 0 \leq y \leq 1$ (counterclockwise)

2. $\int_C [y^3\, dx + (x^3 + 3y^2 x)\, dy]$ C: the boundary of the region between $y = x^2$ and $y = x$, where $0 \leq x \leq 1$ (counterclockwise)

3. $\int_C [2xy\, dx + (e^x + x^2)\, dy]$ C: the boundary of the triangle with vertices $(0, 0)$, $(1, 0)$, $(1, 1)$ (clockwise)

4. $\int_C (-xy^2\, dx + x^2 y\, dy)$ C: the boundary of the region in the first quadrant bounded by $y = 1 - x^2$ (counterclockwise)

Using one of the formulas in Ex. 1, find the area of the following plane regions.

5. The region in the first quadrant bounded by $y = x$ and $y = x^3$.
6. The interior of the ellipse $x^2/a^2 + y^2/b^2 = 1$.
7. The region in the upper half-plane bounded by $y = 1 - x^4$.

Using Green's theorem, evaluate the line integral $\int_C (f\, dx + g\, dy)$ in the counterclockwise sense around the given contour C, where $f\, dx + g\, dy$ equals

8. $y\, dx - x\, dy$, C: the boundary of the square $0 \leq x \leq 1$, $0 \leq y \leq 1$.
9. $(3x^2 + y)\, dx + 4y^2\, dy$, C: the boundary of the triangle with vertices $(0, 0)$, $(1, 0)$, $(0, 2)$.
10. $(x^2 + y^2)\, dy$, C: the boundary of the square $2 \leq x \leq 4$, $2 \leq y \leq 4$.
11. $2xy^3\, dx + 3x^2 y^2\, dy$, C: $x^2 + y^2 = 1$.
12. $(2x - y)\, dx + (x + 3y)\, dy$, C: $x^2 + 4y^2 = 4$.
13. $2 \arctan (y/x)\, dx + \ln (x^2 + y^2)\, dy$, C: $(x - 2)^2 + y^2 = 1$.
14. $(e^x - 3y)\, dx + (e^y + 6x)\, dy$, C: $x^2 + 4y^2 = 4$.
15. $(x^3 - 3y)\, dx + (x + \sin y)\, dy$, C as in Prob. 9.

Using (9), evaluate $\int_C \dfrac{\partial w}{\partial n}\, ds$, where

16. $w = e^x + e^y$, C: the boundary of the square $0 \leq x \leq 2$, $0 \leq y \leq 2$.
17. $w = x^2 + 0.5y^2$, C: the boundary of the region bounded by $y = x^2$ and $y = x + 2$.
18. $w = x^2 y - y^3/3$, C as in Prob. 1.
19. $w = e^x \cos y + x^3$, C: the boundary of the region $x^2 + y^2 \leq 1$ in the first quadrant.

20. If $w(x, y)$ satisfies Laplace's equation $\nabla^2 w = 0$ in a region R, show that

(10)
$$\iint_R \left[\left(\frac{\partial w}{\partial x} \right)^2 + \left(\frac{\partial w}{\partial y} \right)^2 \right] dx\, dy = \int_C w\, \frac{\partial w}{\partial n}\, ds$$

where $\partial w/\partial n$ is defined as in Ex. 3. *Hint.* Model your work somewhat after that of Ex. 3.

21. Using (10), evaluate $\int_C w\, \dfrac{\partial w}{\partial n}\, ds$, where $w = e^x \cos y$ and C is as in Prob. 1.

22. Introducing $\mathbf{v} = g\mathbf{i} - f\mathbf{j}$, show that the formula in Green's theorem may be written

(11)
$$\iint_R \operatorname{div} \mathbf{v}\, dx\, dy = \int_C \mathbf{v} \cdot \mathbf{n}\, ds$$

where $\mathbf{n}$ is the outward unit normal vector to C (Fig. 158) and s is the arc length of C.

23. Verify (11) when $\mathbf{v} = x\mathbf{i} + v\mathbf{j}$ and C is the circle $x^2 + y^2 = 1$.
24. Show that the formula in Green's theorem may be written

(12)
$$\iint_R (\operatorname{curl} \mathbf{v}) \cdot \mathbf{k}\, dx\, dy = \int_C \mathbf{v} \cdot \mathbf{u}\, ds$$

where **k** is a unit vector perpendicular to the xy-plane, **u** is the unit tangent vector to C, and s is the arc length of C.

25. Verify (12) when $\mathbf{v} = -y\mathbf{i} + x\mathbf{j}$ and C is the boundary of the triangle with vertices at $(0, 0)$, $(1, 0)$, $(1, 1)$.

8.5 *Surfaces*

In Sec. 8.7 we shall consider surface integrals. This consideration will require knowledge of some basic facts about surfaces which we shall now explain and illustrate by simple examples.

A surface S may be represented in the form

$$(1) \qquad\qquad f(x, y, z) = 0,$$

where x, y, z are Cartesian coordinates in space. Then the gradient of f is normal to S (cf. Theorem 2 in Sec. 7.8), provided $\operatorname{grad} f \neq \mathbf{0}$. Consequently, for S to have a unique normal at each point whose direction depends continuously on the points of S, we must require that *f has continuous first partial derivatives, and at each point at least one of these three derivatives is not zero.* Then the vector

$$(2) \qquad\qquad \mathbf{n} = \frac{\operatorname{grad} f}{|\operatorname{grad} f|}$$

is a unit normal vector of S (and $-\mathbf{n}$ is the other).

Example 1. The sphere $x^2 + y^2 + z^2 - a^2 = 0$ has the unit normal vector

$$\mathbf{n}(x, y, z) = \frac{x}{a}\mathbf{i} + \frac{y}{a}\mathbf{j} + \frac{z}{a}\mathbf{k}.$$

Sometimes it is convenient to use an *explicit representation*

$$(3) \qquad\qquad z = g(x, y)$$

of a given surface. Clearly, by writing $z - g(x, y) = 0$ we immediately obtain an *implicit representation* of the form (1).

From Sec. 7.3 we know that parametric representations of curves are quite useful, and we shall now see that surfaces may also be represented in parametric form. Suppose that

$$(4) \qquad\qquad \mathbf{r}(u, v) = x(u, v)\mathbf{i} + y(u, v)\mathbf{j} + z(u, v)\mathbf{k}$$

is a vector function of two independent real variables u and v which is continuous and at least once continuously differentiable in a domain of the uv-plane containing a simply connected bounded region R. Then to any point (u_0, v_0) in R there corresponds a point with position vector $\mathbf{r}(u_0, v_0)$ in space. Furthermore, we assume that

$$(5) \qquad\qquad \mathbf{r}_u \times \mathbf{r}_v \neq \mathbf{0}$$

everywhere in R; *here, and in the following, subscripts u and v denote partial derivatives.* We shall see that then $\mathbf{r}(u, v)$ [with (u, v) varying in R]

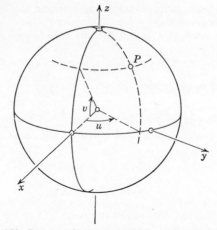

Fig. 159. Parametric representation of the sphere

represents a surface S in space which has a unique normal whose direction depends continuously on the points of S. The representation (4) is called a **parametric representation** of S, and the variables u and v are called the *parameters* of this representation or *coordinates on S*. The curves $u = const$ and $v = const$ form a net on S; they are called *coordinate curves on S*.

If a surface S has a unique normal whose direction depends continuously on the points of S, then S is called a **smooth surface**.

If S is not smooth but can be subdivided into finitely many smooth portions, then it is called a **piecewise smooth surface**.

For example, the surface of a sphere is smooth, while the surface of a cube is piecewise smooth.

Example 2. The sphere in Ex. 1 may be represented in the form

(6) $$\mathbf{r}(u, v) = a \cos v \cos u \, \mathbf{i} + a \cos v \sin u \, \mathbf{j} + a \sin v \, \mathbf{k}$$

where $0 \leqq u \leqq 2\pi$, $-\pi/2 \leqq v \leqq \pi/2$; that is,

$$x = a \cos v \cos u, \qquad y = a \cos v \sin u, \qquad z = a \sin v.$$

The curves $u = const$ and $v = const$ are the "meridians" and "parallels" on S (cf. Fig. 159). The relation (5) is satisfied everywhere except at the "poles" $v = -\pi/2$ and $v = \pi/2$. *The representation (6) is used·in geography for measuring the latitude and longitude of points on the globe.*

Problems

What surfaces are represented by the following parametric representations? What are the corresponding coordinate curves?

1. $\mathbf{r} = u\mathbf{i} + v\mathbf{j}$
2. $\mathbf{r} = (u + v)\mathbf{i} + (u - v)\mathbf{j}$
3. $\mathbf{r} = u \cos v \, \mathbf{i} + u \sin v \, \mathbf{j}$
4. $\mathbf{r} = u\mathbf{i} + v\mathbf{j} + uv\mathbf{k}$
5. $\mathbf{r} = u \cos v \, \mathbf{i} + u \sin v \, \mathbf{j} + u\mathbf{k}$
6. $\mathbf{r} = 3 \cos u \, \mathbf{i} + 3 \sin u \, \mathbf{j} + v\mathbf{k}$
7. $\mathbf{r} = u \cos v \, \mathbf{i} + u \sin v \, \mathbf{j} + u^4\mathbf{k}$
8. $\mathbf{r} = u \cos v \, \mathbf{i} + u \sin v \, \mathbf{j} + v\mathbf{k}$
9. $\mathbf{r} = (4 + 2 \cos v) \cos u \, \mathbf{i} + (4 + 2 \cos v) \sin u \, \mathbf{j} + 2 \sin v \, \mathbf{k}$

Sketch figures of the following surfaces and represent them in the form (1).

10. The elliptic paraboloid $\mathbf{r} = au \cos v \, \mathbf{i} + bu \sin v \, \mathbf{j} + u^2 \mathbf{k}$
11. The ellipsoid $\mathbf{r} = a \cos v \cos u \, \mathbf{i} + b \cos v \sin u \, \mathbf{j} + c \sin v \, \mathbf{k}$
12. The hyperboloid $\mathbf{r} = a \sinh u \cos v \, \mathbf{i} + b \sinh u \sin v \, \mathbf{j} + c \cosh u \, \mathbf{k}$
13. The hyperbolic paraboloid $\mathbf{r} = au \cosh v \, \mathbf{i} + bu \sinh v \, \mathbf{j} + u^2 \mathbf{k}$

Find a parametric representation of the following surfaces.

14. The plane $y = x$ **15.** The yz-plane
16. The parabolic cylinder $z = y^2$ **17.** The plane $x + y + z = 3$
18. The cylinder $x^2 + z^2 = a^2$ **19.** The cylinder $(x - 1)^2 + (y - 2)^2 = 9$
20. The elliptic cylinder $x^2 + 4y^2 = 4$ **21.** $z = e^{-(x^2 + y^2)}$

22. Find the points in Ex. 2, Prob. 5, and Prob. 7 at which (5) is not satisfied.
23. Find a unit normal vector of the surface in Prob. 4.
24. Find a unit normal vector of the surface in Prob. 8.
25. Find a representation of a unit normal vector $\mathbf{n}$ of a surface $z = g(x, y)$.

8.6 Tangent Plane.
First Fundamental Form. Area

If a surface S is represented in the form $\mathbf{r} = \mathbf{r}(u, v)$, then a curve on S can be represented by a pair of continuous functions

(1) $u = g(t), \qquad v = h(t)$

of a real parameter t.

Example 1 (Cylinder of revolution). $\mathbf{r}(u, v) = a \cos u \, \mathbf{i} + a \sin u \, \mathbf{j} + v\mathbf{k}$ represents a cylinder of revolution S of radius a. The equations $u = t, v = ct$ represent a circular helix on S. In fact, by substituting these equations into the representation of S we obtain (cf. Ex. 3 in Sec. 7.3)

$$\mathbf{r}[u(t), v(t)] = a \cos t \, \mathbf{i} + a \sin t \, \mathbf{j} + ct\mathbf{k}.$$

Let S be a smooth surface represented by a vector function $\mathbf{r}(u, v)$ and let C be a curve on S represented in the form (1). Then C, considered as a curve in space, is represented by the vector function

(2) $\mathbf{r}(t) = \mathbf{r}[u(t), v(t)].$

Suppose that both functions in (1) have continuous first derivatives such that for each t at least one of these derivatives is not zero. Then C has a tangent at each of its points whose direction depends continuously on the points, and a tangent vector to C is

$$\dot{\mathbf{r}}(t) \equiv \frac{d\mathbf{r}}{dt} = \mathbf{r}_u \dot{u} + \mathbf{r}_v \dot{v}.$$

From (5) in the last section it follows that the vectors $\mathbf{r}_u \equiv \partial \mathbf{r}/\partial u$ and $\mathbf{r}_v \equiv \partial \mathbf{r}/\partial v$ are linearly independent. Hence they determine a plane. This plane is called the **tangent plane** of S at the corresponding point P of S and will be denoted by $T(P)$. It touches S at P. From (2) it follows that $T(P)$ *contains the tangent to any curve on S through P, at that point P* (see also Sec. 7.8).

Fig. 160. Tangent plane and normal vector

The straight line through P and perpendicular to $T(P)$ is called the **normal** to S at P. Since the vectors $\mathbf{r}_u$ and $\mathbf{r}_v$ lie in $T(P)$, the unit vector

$$(3) \qquad\qquad \mathbf{n} = \frac{\mathbf{r}_u \times \mathbf{r}_v}{|\mathbf{r}_u \times \mathbf{r}_v|}$$

(Fig. 160) is perpendicular to $T(P)$. This vector is called the **unit normal vector** of S at P. Its sense depends on the choice of the coordinates u and v; the transformation $u = -\bar{u}$, $v = \bar{v}$ or any other transformation whose Jacobian (cf. Sec. 8.3) has a negative value, reverses the sense of $\mathbf{n}$.

We shall now determine the *linear element* of a curve C, represented in the form (1), on a surface S, represented by $\mathbf{r}(u, v)$. We have

$$d\mathbf{r} = \mathbf{r}_u \, du + \mathbf{r}_v \, dv,$$

and, by (4) in Sec. 7.4,

$$
\begin{aligned}
ds^2 = d\mathbf{r} \cdot d\mathbf{r} &= (\mathbf{r}_u \, du + \mathbf{r}_v \, dv) \cdot (\mathbf{r}_u \, du + \mathbf{r}_v \, dv) \\
&= \mathbf{r}_u \cdot \mathbf{r}_u \, du^2 + 2\mathbf{r}_u \cdot \mathbf{r}_v \, du \, dv + \mathbf{r}_v \cdot \mathbf{r}_v \, dv^2.
\end{aligned}
$$

If we use the standard notation

$$(4) \qquad\qquad E = \mathbf{r}_u \cdot \mathbf{r}_u \qquad F = \mathbf{r}_u \cdot \mathbf{r}_v \qquad G = \mathbf{r}_v \cdot \mathbf{r}_v$$

the last expression assumes the form

$$(5) \qquad\qquad ds^2 = E \, du^2 + 2F \, du \, dv + G \, dv^2.$$

This quadratic differential form is called the **first fundamental form** of S.

Example 2 (Polar coordinates). The vector function

$$\mathbf{r}(u, v) = u \cos v \, \mathbf{i} + u \sin v \, \mathbf{j}$$

represents the xy-plane, and u and v are polar coordinates. By differentiation we obtain

$$\mathbf{r}_u = \cos v \, \mathbf{i} + \sin v \, \mathbf{j}, \qquad \mathbf{r}_v = -u \sin v \, \mathbf{i} + u \cos v \, \mathbf{j}.$$

Hence $E = 1$, $F = 0$, $G = u^2$, and the first fundamental form corresponding to polar coordinates $u = \rho$ and $v = \theta$ is

$$(6) \qquad\qquad ds^2 = d\rho^2 + \rho^2\, d\theta^2.$$

We shall now see that *the first fundamental form is of basic importance because it enables us to measure lengths, angles between curves, and areas on the corresponding surface S.*

In fact, from (1) and (4) in Sec. 7.4 and (5) in the present section it follows that a curve

$$C\colon u(t),\, v(t), \qquad a \leqq t \leqq b,$$

on a surface $S\colon \mathbf{r}(u, v)$ has the length

$$(7) \qquad l = \int_a^b \sqrt{\dot{\mathbf{r}} \cdot \dot{\mathbf{r}}}\; dt = \int_a^b \frac{ds}{dt}\, dt = \int_a^b \sqrt{E\dot{u}^2 + 2F\dot{u}\dot{v} + G\dot{v}^2}\; dt.$$

This proves the proposition concerning length, and we may now consider the measurement of the angle between two curves

$$C_1\colon u = g(t),\, v = h(t) \qquad \text{and} \qquad C_2\colon u = p(t),\, v = q(t)$$

on a surface $S\colon \mathbf{r}(u, v)$ which intersect at a point P of S. The vectors

$$\mathbf{a} = \frac{d}{dt}\, \mathbf{r}[g(t),\, h(t)] = \mathbf{r}_u \dot{g} + \mathbf{r}_v \dot{h}$$

and

$$\mathbf{b} = \frac{d}{dt}\, \mathbf{r}[p(t),\, q(t)] = \mathbf{r}_u \dot{p} + \mathbf{r}_v \dot{q}$$

at P are tangent to C_1 and C_2, respectively. The angle of intersection between C_1 and C_2 at P is defined as the angle γ between $\mathbf{a}$ and $\mathbf{b}$, and from (3) in Sec. 5.5 we have

$$(8) \qquad\qquad \cos \gamma = \frac{\mathbf{a} \cdot \mathbf{b}}{|\mathbf{a}|\,|\mathbf{b}|}$$

where

$$\mathbf{a} \cdot \mathbf{b} = (\mathbf{r}_u \dot{g} + \mathbf{r}_v \dot{h}) \cdot (\mathbf{r}_u \dot{p} + \mathbf{r}_v \dot{q}) = E\dot{g}\dot{p} + F(\dot{g}\dot{q} + \dot{h}\dot{p}) + G\dot{h}\dot{q}$$

and similarly

$$|\mathbf{a}| = \sqrt{\mathbf{a} \cdot \mathbf{a}} = \sqrt{E\dot{g}^2 + 2F\dot{g}\dot{h} + G\dot{h}^2},$$

$$|\mathbf{b}| = \sqrt{\mathbf{b} \cdot \mathbf{b}} = \sqrt{E\dot{p}^2 + 2F\dot{p}\dot{q} + G\dot{q}^2}.$$

This important result shows that the angle between two intersecting curves on a surface can be expressed in terms of E, F, G and the derivatives of the functions representing the curves, evaluated at the point of intersection.

We have shown that the coefficients E, F, and G of the first fundamental form permit the measurement of lengths and angles on a surface. We shall now demonstrate that they also permit the measurement of areas on a surface.

Fig. 161. Area

The **area** A of a surface S: $\mathbf{r}(u, v)$ is defined by the double integral

(9)
$$A = \iint_R \sqrt{EG - F^2}\, du\, dv$$

over the region R in the uv-plane corresponding to that surface; here the positive sign of the square root is taken. The expression

(10)
$$dA = \sqrt{EG - F^2}\, du\, dv$$

is called the *element of area* of S.

Formula (10) can be made plausible by noting that

$$\Delta A = \sqrt{EG - F^2}\, \Delta u\, \Delta v$$

is the area of a small parallelogram whose sides are the vectors

$$\mathbf{r}_u\, \Delta u \qquad \text{and} \qquad \mathbf{r}_v\, \Delta v$$

(Fig. 161). Indeed, from the definition of a vector product it follows that

$$\Delta A = |\mathbf{r}_u\, \Delta u \times \mathbf{r}_v\, \Delta v| = |\mathbf{r}_u \times \mathbf{r}_v|\, \Delta u\, \Delta v,$$

and from (5) in Sec. 5.7 and (4) in the current section we readily obtain

(11) $$|\mathbf{r}_u \times \mathbf{r}_v|^2 = (\mathbf{r}_u \cdot \mathbf{r}_u)(\mathbf{r}_v \cdot \mathbf{r}_v) - (\mathbf{r}_u \cdot \mathbf{r}_v)^2 = EG - F^2.$$

The integral (9) is obtained by subdividing S into parts $S_1, \cdots, S_n$, approximating each part S_k by a portion of the tangent plane of S at a point in S_k, and forming the sum of all the approximating areas. This is done for each $n = 1, 2, \cdots$ so that the dimensions of the largest S_k approach zero as n approaches infinity. The limit of those sums is the integral (9). For details see Ref. [C8] in Appendix 1.

Example 3 (*Torus*). The vector function

$$\mathbf{r}(u, v) = (a + b \cos v) \cos u\, \mathbf{i} + (a + b \cos v) \sin u\, \mathbf{j} + b \sin v\, \mathbf{k} \qquad (a > b > 0)$$

represents a torus (Fig. 162) which is obtained by rotating a circle C about a fixed straight line A so that the plane of C always passes through A, and C does not intersect A. Using (4), we find

$$E = (a + b \cos v)^2, \qquad F = 0, \qquad G = b^2,$$

$$EG - F^2 = b^2(a + b \cos v)^2,$$

and the total area is

$$A = \int_0^{2\pi} \int_0^{2\pi} b(a + b \cos v)\, du\, dv = 4\pi^2 ab.$$

Fig. 162. Torus

If S is represented in the form

$$z = g(x, y),$$

we may set $x = u$ and $y = v$ and write the representation in parametric form:

$$\mathbf{r}(u, v) = u\mathbf{i} + v\mathbf{j} + g(u, v)\mathbf{k}.$$

Then the partial derivatives of $\mathbf{r}$ with respect to u and v are

(12) $$\mathbf{r}_u = \mathbf{i} + g_u\mathbf{k}, \qquad \mathbf{r}_v = \mathbf{j} + g_v\mathbf{k}.$$

It follows that the coefficients of the first fundamental form are

$$E = 1 + g_u{}^2, \qquad F = g_u g_v, \qquad G = 1 + g_v{}^2.$$

Hence

$$EG - F^2 = 1 + g_u{}^2 + g_v{}^2.$$

Since $u = x$ and $v = y$, the integral (9) now assumes the form

(9′) $$A = \iint\limits_{\bar{S}} \sqrt{1 + \left(\frac{\partial g}{\partial x}\right)^2 + \left(\frac{\partial g}{\partial y}\right)^2}\, dx\, dy$$

where $\bar{S}$ is the orthogonal projection of S into the xy-plane. Clearly,

(10′) $$dA = \sqrt{1 + \left(\frac{\partial g}{\partial x}\right)^2 + \left(\frac{\partial g}{\partial y}\right)^2}\, dx\, dy.$$

We want to show that this may be written

$$(10'') \qquad\qquad\qquad dA = \sec \gamma \, dx \, dy$$

where γ is the *acute* angle between the z-axis and the undirected normal to S.

In fact, the vector product of the vectors in (12) is perpendicular to S; denoting it by **a** and observing that $u = x$, $v = y$, we have

$$\mathbf{a} = \frac{\partial \mathbf{r}}{\partial x} \times \frac{\partial \mathbf{r}}{\partial y} = -\frac{\partial g}{\partial x}\,\mathbf{i} - \frac{\partial g}{\partial y}\,\mathbf{j} + \mathbf{k}, \qquad |\mathbf{a}| = \sqrt{1 + \left(\frac{\partial g}{\partial x}\right)^2 + \left(\frac{\partial g}{\partial y}\right)^2}.$$

Hence $\mathbf{a} \cdot \mathbf{k} = 1$. On the other hand, $\mathbf{a} \cdot \mathbf{k} = |\mathbf{a}| \cos \gamma^*$, where γ^* is the angle between **a** and the positive z-axis. Together $|\mathbf{a}| \cos \gamma^* = 1$. Clearly $\cos \gamma^* > 0$ and therefore $\gamma^* < \pi/2$, that is, γ^* is acute and thus equal to γ. Hence

$$|\mathbf{a}| \cos \gamma = 1 \qquad \text{or} \qquad \sec \gamma = |\mathbf{a}| \qquad\qquad \left(\gamma < \frac{\pi}{2}\right)$$

and (10″) is established.

Problems

1. Show that the tangent plane of a surface S: $\mathbf{r}(u, v)$ at a point P can be represented in the form

$$\mathbf{r}^*(p, q) = \mathbf{r} + p\mathbf{r}_u + q\mathbf{r}_v \qquad\qquad (\mathbf{r}, \mathbf{r}_u, \mathbf{r}_v \text{ refer to } P)$$

or

$$(\mathbf{r}^* - \mathbf{r} \qquad \mathbf{r}_u \qquad \mathbf{r}_v) = 0.$$

Find a representation of the tangent plane and a unit normal vector of a surface S at a point P if S is represented in the form

    2. $z = g(x, y)$ **3.** $f(x, y, z) = 0$ **4.** $x = h(y, z)$

    5. Let $\mathbf{n}(u, v)$ be the unit normal vector (3) of a surface S: $\mathbf{r}(u, v)$. Set $u = -\tilde{u}$, $v = \tilde{v}$ and show that the unit normal vector corresponding to the resulting representation is $-\mathbf{n}(u, v)$.

Find a representation of the tangent plane of the following surfaces at P_0: (x_0, y_0, z_0).

    6. $z = x^2$, P_0: $(1, 0, 1)$ **7.** $y^2 + z^2 = 4$, P_0: $(3, -\sqrt{2}, \sqrt{2})$
    8. $z^2 = x^2 + y^2$, P_0: $(1, 0, 1)$ **9.** $x^2 + y^2 + z^2 = 25$, P_0: $(3, 4, 0)$
10. $3x^2 + 2y^2 + z^2 = 20$, P_0: $(1, 2, 3)$ **11.** $z = x^2 + y^2$, P_0: $(2, 1, 5)$

Find the first fundamental form of the following surfaces.

  12. $\mathbf{r} = u\mathbf{i} + v\mathbf{j}$ **13.** $r = a \cos u \, \mathbf{i} + a \sin u \, \mathbf{j} + v\mathbf{k}$
  14. The sphere (6) in Sec. 8.5   **15.** The torus in Ex. 3
  16. $\mathbf{r} = u\mathbf{i} + v\mathbf{j} + (u^2 + v^2)\mathbf{k}$ **17.** $\mathbf{r} = u \cos v \, \mathbf{i} + u \sin v \, \mathbf{j} + u\mathbf{k}$

  18. **(Orthogonal coordinates)** Show that the coordinate curves $u = const$ and $v = const$ on a surface $\mathbf{r} = \mathbf{r}(u, v)$ intersect at right angles if and only if $F = \mathbf{r}_u \cdot \mathbf{r}_v = 0$.
  19. **(Theorem of Pappus)** Show that the area A in Ex. 3 can be obtained by the theorem of Pappus, which states that the area of a surface of revolution equals the product of the length of a meridian C and the length of the path of the center of gravity of C when C is rotated through the angle 2π.
  20. Using the theorem of Pappus, find the center of gravity $(0, \bar{y})$ of a semicircle $x^2 + y^2 = a^2$, $y \geqq 0$.
  21. Using the theorem of Pappus, find the area of the cone $z^2 = x^2 + y^2$ $(0 \leqq z \leqq 1)$.

Using (9), find the area of the following surfaces.

  22. $x^2 + y^2 = a^2$, $0 \leq z \leq b$ **23.** The cone in Prob. 21
  24. The sphere (6) in Sec. 8.5   **25.** $z = x^2 + y^2$, $0 \leq z \leq 1$

8.7 Surface Integrals

The concept of a surface integral is a natural generalization of the concept of a double integral considered in Sec. 8.3. Surface integrals occur in many applications, for example, in connection with the center of gravity of a curved lamina, the potential due to charges distributed on surfaces, etc.

The definition of a surface integral parallels that of a double integral. Let S be a portion of a surface of finite area, and let $f(x, y, z)$ be a function which is defined and continuous on S. We subdivide S into n parts $S_1, \cdots, S_n$ of areas $\Delta A_1, \cdots, \Delta A_n$. In each part S_k we choose an arbitrary point P_k with coordinates x_k, y_k, z_k and form the sum

$$(1) \qquad J_n = \sum_{k=1}^{n} f(x_k, y_k, z_k) \, \Delta A_k.$$

This we do for each $n = 1, 2, \cdots$ in an arbitrary manner, but so that the largest part S_k shrinks to a point as n approaches infinity. The infinite sequence $J_1, J_2, \cdots$ has a limit which is independent of the choice of subdivisions and points P_k; the proof is similar to that in the case of a double integral. This limit is called the **surface integral** *of* $f(x, y, z)$ *over* S and is denoted by

$$(2) \qquad \iint\limits_{S} f(x, y, z) \, dA.$$

To evaluate the surface integral (2), *we may reduce it to a double integral as follows.* If S is represented in parametric form by a vector function $\mathbf{r}(u, v)$, then $dA = \sqrt{EG - F^2} \, du \, dv$, cf. (10) in the previous section, and

$$(3) \qquad \iint\limits_{S} f(x, y, z) \, dA = \iint\limits_{R} f[x(u, v), y(u, v), z(u, v)] \sqrt{EG - F^2} \, du \, dv$$

where R is the region corresponding to S in the uv-plane.

Similarly, if S is represented in the form $z = g(x, y)$, then from (10′) and (10″) in the last section it follows that

$$\iint\limits_{\bar{S}} f(x, y, z) \, dA = \iint\limits_{\bar{S}} f[x, y, g(x, y)] \sqrt{1 + \left(\frac{\partial g}{\partial x}\right)^2 + \left(\frac{\partial g}{\partial y}\right)^2} \, dx \, dy$$

$$(4)$$

$$\qquad\qquad = \iint\limits_{\bar{S}} f[x, y, g(x, y)] \sec \gamma \, dx \, dy \qquad\qquad (\gamma < \pi/2),$$

where γ is the acute angle between the normal to S and the z-axis, and $\bar{S}$ is the orthogonal projection of S in the xy-plane. Since $g(x, y)$ is a function (hence single-valued, by definition), it follows that different points of S are projected onto different points of $\bar{S}$. The condition $\gamma < \pi/2$ means that the normal of S is nowhere perpendicular to the z-axis.

Example 1 (Moment of inertia). Find the moment of inertia I of a homogeneous spherical lamina S: $x^2 + y^2 + z^2 = a^2$ of mass M about the z-axis.

If a mass is distributed over a surface S and $\mu(x, y, z)$ is the density of the mass ($=$ mass per unit area), then the moment of inertia I of the mass with respect to a given axis L is defined by the surface integral

$$(5) \qquad I = \iint\limits_{S} \mu D^2 \, dA$$

where $D(x, y, z)$ is the distance of the point (x, y, z) from L.

Since, in the present example, μ is constant and S has the area $A = 4\pi a^2$, we have

$$\mu = \frac{M}{A} = \frac{M}{4\pi a^2}.$$

Representing S by (6), Sec. 8.5, we obtain from (4) in the previous section

$$E = a^2 \cos^2 v, \qquad F = 0, \qquad G = a^2,$$

$$dA = \sqrt{EG - F^2} \, du \, dv = a^2 \cos v \, du \, dv.$$

Furthermore, the square of the distance of a point (x, y, z) from the z-axis becomes

$$D^2 = x^2 + y^2 = a^2 \cos^2 v.$$

Hence we obtain the result

$$I = \iint\limits_{S} \mu D^2 \, dA = \frac{M}{4\pi a^2} \int_{-\pi/2}^{\pi/2} \int_{0}^{2\pi} a^4 \cos^3 v \, du \, dv = \frac{Ma^2}{2} \int_{-\pi/2}^{\pi/2} \cos^3 v \, dv = \frac{2Ma^2}{3}.$$

In various applications there occur surface integrals for which the concept of orientation of a surface is essential. Therefore we shall now consider this concept, starting with the case of a smooth surface (cf. Sec. 8.5).

Let S be a smooth surface, and let P be any point of S. Then we may choose a unit normal vector **n** of S at P. The direction of **n** is then called the *positive normal direction* of S at P. Obviously there are two possibilities in choosing **n**.

A smooth surface S is said to be **orientable** if the positive normal direction, when given at an arbitrary point P_0 of S, can be continued in a unique and continuous way to the entire surface.

Hence the surface S is orientable provided there does not exist a closed curve C on S through P_0 such that the positive normal direction reverses when it is displaced continuously from P_0 along C and back to P_0.

A sufficiently small portion of a smooth surface is always orientable. However, this may not hold in the large. There are nonorientable surfaces. A well-known example of such a surface is the **Möbius**[6] **strip** shown in Fig. 163. When a normal vector, which is given at P_0, is displaced continuously along the curve C in Fig. 163, the resulting normal vector upon returning to P_0 is opposite to the original vector

[6] AUGUST FERDINAND MÖBIUS (1790–1868), German mathematician, who made important contributions to the theory of surfaces and projective geometry.

Fig. 163. Möbius strip

at P_0. A model of a Möbius strip can be made by taking a long rectangular piece of paper and sticking the shorter sides together so that the two points A and the two points B in Fig. 163 coincide.

If a smooth surface S is orientable, then we may *orient* S by choosing one of the two possible directions of the normal vector **n**.

If the boundary of S is a simple closed curve C, then we may associate with each of the two possible orientations of S an orientation of C, as shown in Fig. 164a. Using this simple idea, we may now readily extend the concept of orientation to piecewise smooth surfaces as follows.

A piecewise smooth surface S is said to be **orientable** if we can orient each smooth piece of S in such a manner that along each curve C^* which is a common boundary of two pieces S_1 and S_2 the positive direction of C^* relative to S_1 is opposite to the positive direction of C^* relative to S_2.

Figure 164b illustrates the situation for a surface consisting of two smooth pieces.

Let S be a given orientable surface. We orient S by choosing a unit normal vector **n**. Denoting the angles between **n** and the positive x, y, and z axes by α, β,

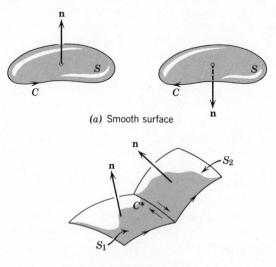

(a) Smooth surface

(b) Piecewise smooth surface

Fig. 164. Orientation of a surface

and γ, respectively, we have

(6) $$\mathbf{n} = \cos \alpha \, \mathbf{i} + \cos \beta \, \mathbf{j} + \cos \gamma \, \mathbf{k}.$$

Let $u_1(x, y, z)$, $u_2(x, y, z)$, and $u_3(x, y, z)$ be given functions which are defined and continuous at every point of S. Then the integrals to be considered are usually written in the form

$$\iint_S u_1 \, dy \, dz, \qquad \iint_S u_2 \, dz \, dx, \qquad \iint_S u_3 \, dx \, dy,$$

and *by definition* this means

(7)
$$\iint_S u_1 \, dy \, dz = \iint_S u_1 \cos \alpha \, dA,$$

$$\iint_S u_2 \, dz \, dx = \iint_S u_2 \cos \beta \, dA,$$

$$\iint_S u_3 \, dx \, dy = \iint_S u_3 \cos \gamma \, dA.$$

It is clear that the value of such an integral depends on the choice of $\mathbf{n}$, that is, on the orientation of S. The transition to the opposite orientation corresponds to the multiplication of the integral by -1, because then the components $\cos \alpha$, $\cos \beta$, and $\cos \gamma$ of $\mathbf{n}$ are multiplied by -1.

Sums of three such integrals may be written in a simple form by using vector notation. In fact, if we introduce the vector

$$\mathbf{u} = u_1 \mathbf{i} + u_2 \mathbf{j} + u_3 \mathbf{k},$$

then we obtain from the definition (7) the formula

(8)
$$\iint_S (u_1 \, dy \, dz + u_2 \, dz \, dx + u_3 \, dx \, dy)$$

$$= \iint_S (u_1 \cos \alpha + u_2 \cos \beta + u_3 \cos \gamma) \, dA = \iint_S \mathbf{u} \cdot \mathbf{n} \, dA.$$

To evaluate the integrals (7) we may reduce them to double integrals over a plane region, as follows.

If S can be represented in the form $z = h(x, y)$ and is oriented such that $\mathbf{n}$ points upward, then γ is acute. Hence from (10″) in the last section and the above definition (7) we immediately have

(9a)
$$\iint_S u_3(x, y, z) \, dx \, dy = + \iint_{\overline{R}} u_3[x, y, h(x, y)] \, dx \, dy$$

where $\bar{R}$ is the orthogonal projection of S in the xy-plane. If $\mathbf{n}$ points downward, then γ is obtuse and we obtain

$$(9b) \qquad \iint_S u_3(x, y, z)\, dx\, dy = -\iint_{\bar{R}} u_3[x, y, h(x, y)]\, dx\, dy.$$

For the other two integrals in (7) the situation is quite similar.

If S is represented in parametric form

$$\mathbf{r}(u, v) = x(u, v)\mathbf{i} + y(u, v)\mathbf{j} + z(u, v)\mathbf{k},$$

then the normal vector is [cf. (3) in the last section]

$$(10) \qquad (a) \quad \mathbf{n} = +\frac{\mathbf{r}_u \times \mathbf{r}_v}{|\mathbf{r}_u \times \mathbf{r}_v|} \quad \text{or} \quad (b) \quad \mathbf{n} = -\frac{\mathbf{r}_u \times \mathbf{r}_v}{|\mathbf{r}_u \times \mathbf{r}_v|},$$

depending on the orientation. Now from (10) and (11) in the last section it follows that

$$dA = \sqrt{EG - F^2}\, du\, dv = |\mathbf{r}_u \times \mathbf{r}_v|\, du\, dv.$$

Since $\cos \gamma = \mathbf{k} \cdot \mathbf{n}$, we thus obtain

$$\cos \gamma\, dA = \mathbf{k} \cdot \mathbf{n}\, dA = \pm \mathbf{k} \cdot (\mathbf{r}_u \times \mathbf{r}_v)\, du\, dv = \pm \begin{vmatrix} 0 & 0 & 1 \\ x_u & y_u & z_u \\ x_v & y_v & z_v \end{vmatrix} du\, dv$$

$$= \pm \frac{\partial(x, y)}{\partial(u, v)}\, du\, dv,$$

where the last expression involves the Jacobian (cf. Sec. 8.3). Hence in (7),

$$(11) \qquad \iint_S u_3(x, y, z)\, dx\, dy = \pm \iint_R u_3[x(u, v), y(u, v), z(u, v)]\, \frac{\partial(x, y)}{\partial(u, v)}\, du\, dv$$

with the plus sign when S is oriented such that (10a) holds and the minus sign in the case of the opposite orientation. Here R is the region corresponding to S in the uv-plane.

Problems

1. Construct a paper model of a Möbius strip. What happens if you cut it along the curve C in Fig. 163?

Representing S in parametric form and using (3), evaluate $\iint_S f(x, y, z)\, dA$, where

2. $f = x + 2,$ $S: x^2 + y^2 = 1, 0 \leq z \leq 1$
3. $f = xy,$ $S: z = xy, 0 \leq x \leq 1, 0 \leq y \leq 1$
4. $f = xy,$ $S: x^2 + y^2 = 4, |z| \leq 1$
5. $f = \arctan (y/x),$ $S: z = x^2 + y^2, 1 \leq z \leq 4$
6. $f = \cos x + \sin y,$ $S:$ the portion of $x + y + z = 1$ in the first octant

7. $f = x(12y - y^4 + z^2)$, $S: z = y^2, 0 \leq x \leq 1, 0 \leq y \leq 1$

8. $f = 6x + z - y^2$, $S: z = y^2, 0 \leq x \leq 1, -1 \leq y \leq 1$

9. Solve Prob. 8 by using (4).

10. The electrostatic potential at $(0, 0, -a)$ of a charge of constant density σ on the hemisphere $S: x^2 + y^2 + z^2 = a^2, z \geq 0$ is

$$U = \iint_S \frac{\sigma}{\sqrt{x^2 + y^2 + (z + a)^2}} \, dA.$$

Show that $U = 2\pi\sigma a(2 - \sqrt{2})$. *Hint.* Represent S in the form

$$\mathbf{r} = a \sin u \cos v \, \mathbf{i} + a \sin u \sin v \, \mathbf{j} + a \cos u \, \mathbf{k}.$$

11. Justify the following formulas for the mass M and the center of gravity $(\bar{x}, \bar{y}, \bar{z})$ of a lamina S of density (mass per unit area) $\sigma(x, y, z)$ in space:

$$M = \iint_S \sigma \, dA, \quad \bar{x} = \frac{1}{M} \iint_S x\sigma \, dA, \quad \bar{y} = \frac{1}{M} \iint_S y\sigma \, dA, \quad \bar{z} = \frac{1}{M} \iint_S z\sigma \, dA.$$

12. Justify the following formulas for the moments of inertia of the lamina in Prob. 11 about the x-, y-, and z-axes, respectively:

$$I_x = \iint_S (y^2 + z^2)\sigma \, dA, \quad I_y = \iint_S (x^2 + z^2)\sigma \, dA, \quad I_z = \iint_S (x^2 + y^2)\sigma \, dA.$$

13. Find a formula for the moment of inertia of the lamina in Prob. 11 about the line $y = x, z = 0$.

Find the moment of inertia of a lamina S of density 1 about an axis A, where

14. $S: x^2 + y^2 = a^2$, $0 \leq z \leq h$, A: the z-axis

15. S as in Prob. 14, A: the line $z = h/2$ in the xz-plane

16. $S: x^2 + y^2 = z^2$, $0 \leq z \leq h$, A: the z-axis

17. S: the torus in Ex. 3, Sec. 8.6, A: the z-axis

18. (**Steiner's**[7] **theorem**) If I_A is the moment of inertia of a mass distribution of total mass M with respect to an axis A through the center of gravity, show that its moment of inertia I_B with respect to an axis B, which is parallel to A and has the distance k from it, is

$$I_B = I_A + k^2 M.$$

19. Using Steiner's theorem, find the moment of inertia of S in Prob. 15 about the x-axis.

20. Using Steiner's theorem, find the moment of inertia of S in Prob. 17 about the line $x = a + b$ in the xz-plane.

8.8 Triple Integrals.
Divergence Theorem of Gauss

The triple integral is a generalization of the double integral introduced in Sec. 8.3. For defining this integral we consider a function $f(x, y, z)$ defined in a bounded

[7] JACOB STEINER (1796–1863), Swiss geometer.

closed[8] region T of space. We subdivide T by planes parallel to the three coordinate planes. Then we number the parallelepipeds inside T from 1 to n. In each such parallelepiped we choose an arbitrary point, say, (x_k, y_k, z_k) in the kth parallelepiped, and form the sum

$$J_n = \sum_{k=1}^{n} f(x_k, y_k, z_k)\, \Delta V_k$$

where ΔV_k is the volume of the kth parallelepiped. This we do for larger and larger positive integers n in an arbitrary manner, but so that the lengths of the edges of the largest parallelepiped of subdivision approach zero as n approaches infinity. In this way we obtain a sequence of real numbers $J_{n_1}, J_{n_2}, \cdots$. We assume that $f(x, y, z)$ is continuous in a domain containing T, and T is bounded by finitely many *smooth surfaces*.[9] Then it can be shown (cf. Ref. [A3]) that the sequence converges to a limit which is independent of the choice of subdivisions and corresponding points (x_k, y_k, z_k). This limit is called the **triple integral** *of* $f(x, y, z)$ *over the region* T, and is denoted by

$$\iiint_T f(x, y, z)\, dx\, dy\, dz \qquad \text{or} \qquad \iiint_T f(x, y, z)\, dV.$$

We shall now show that the triple integral of the divergence of a continuously differentiable vector function $\mathbf{u}$ over a region T in space can be transformed into a surface integral of the normal component of $\mathbf{u}$ over the boundary surface S of T. This can be done by means of *the divergence theorem which is the three-dimensional analogue of Green's theorem in the plane* considered in Sec. 8.4. The divergence theorem is of basic importance in various theoretical and practical considerations.

Divergence theorem of Gauss. *Let T be a closed bounded region in space whose boundary is a piecewise smooth[10] orientable surface S. Let $\mathbf{u}(x, y, z)$ be a vector function which is continuous and has continuous first partial derivatives in some domain containing T. Then*

(1) $$\iiint_T \operatorname{div} \mathbf{u}\, dV = \iint_S u_n\, dA$$

where

(2) $$u_n = \mathbf{u} \cdot \mathbf{n}$$

is the component of $\mathbf{u}$ in the direction of the outer normal of S with respect to T, and $\mathbf{n}$ is the outer unit normal vector of S.

Remark. If we write $\mathbf{u}$ and $\mathbf{n}$ in terms of components, say

$$\mathbf{u} = u_1\mathbf{i} + u_2\mathbf{j} + u_3\mathbf{k} \qquad \text{and} \qquad \mathbf{n} = \cos\alpha\, \mathbf{i} + \cos\beta\, \mathbf{j} + \cos\gamma\, \mathbf{k},$$

so that α, β, and γ are the angles between $\mathbf{n}$ and the positive x, y, and z axes,

---

[8] Explained in footnote 2, p. 329 (with "sphere" instead of "circle").
[9] Cf. Sec. 8.5.
[10] Cf. Sec. 8.5.

respectively, formula (1) takes the form

$$(3^*) \quad \iiint\limits_{T} \left(\frac{\partial u_1}{\partial x} + \frac{\partial u_2}{\partial y} + \frac{\partial u_3}{\partial z}\right) dx\, dy\, dz = \iint\limits_{S} (u_1 \cos \alpha + u_2 \cos \beta + u_3 \cos \gamma)\, dA.$$

Because of (8) in the last section this may be written

$$(3) \quad \iiint\limits_{T} \left(\frac{\partial u_1}{\partial x} + \frac{\partial u_2}{\partial y} + \frac{\partial u_3}{\partial z}\right) dx\, dy\, dz = \iint\limits_{S} (u_1\, dy\, dz + u_2\, dz\, dx + u_3\, dx\, dy).$$

Proof of the divergence theorem. Clearly, (3^*) is true, if the following three relations hold simultaneously:

$$(4) \qquad\qquad\qquad \iiint\limits_{T} \frac{\partial u_1}{\partial x} dx\, dy\, dz = \iint\limits_{S} u_1 \cos \alpha\, dA,$$

$$(5) \qquad\qquad\qquad \iiint\limits_{T} \frac{\partial u_2}{\partial y} dx\, dy\, dz = \iint\limits_{S} u_2 \cos \beta\, dA,$$

$$(6) \qquad\qquad\qquad \iiint\limits_{T} \frac{\partial u_3}{\partial z} dx\, dy\, dz = \iint\limits_{S} u_3 \cos \gamma\, dA.$$

We first prove (6) for a *special region* T which is bounded by a piecewise smooth orientable surface S and has the property that any straight line parallel to any one of the coordinate axes and intersecting T has at most *one* segment (or a single point) in common with T. This implies that T can be represented in the form

$$(7) \qquad\qquad\qquad\qquad g(x, y) \leqq z \leqq h(x, y)$$

where (x, y) varies in the orthogonal projection $\bar{R}$ of T in the xy-plane. Clearly, $z = g(x, y)$ represents the "bottom" S_2 of S (Fig. 165) whereas $z = h(x, y)$ represents the "top" S_1 of S, and there may be a remaining vertical portion S_3 of S. (The portion S_3 may degenerate into a curve, as for a sphere.)

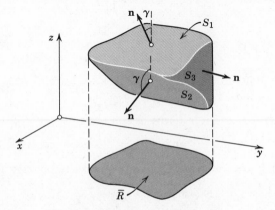

Fig. 165. Example of a special region

To prove (6), we use (7). Since **u** is continuously differentiable in some domain containing T, we have

(8)
$$\iiint\limits_{T} \frac{\partial u_3}{\partial z}\, dx\, dy\, dz = \iint\limits_{R}\left[\int_{g(x,y)}^{h(x,y)} \frac{\partial u_3}{\partial z}\, dz\right] dx\, dy.$$

We integrate the inner integral:

$$\int_{g}^{h} \frac{\partial u_3}{\partial z}\, dz = u_3(x, y, h) - u_3(x, y, g).$$

Hence the left-hand side of (8) equals

(9)
$$\iint\limits_{R} u_3[x, y, h(x, y)]\, dx\, dy - \iint\limits_{R} u_3[x, y, g(x, y)]\, dx\, dy.$$

Let us show that this is equal to the right-hand side of (6). On the lateral portion S_3 of S (Fig. 165) we have $\gamma = \pi/2$ and $\cos \gamma = 0$. Hence this portion does not contribute to the surface integral in (6), and

$$\iint\limits_{S} u_3 \cos \gamma\, dA = \iint\limits_{S_1} u_3 \cos \gamma\, dA + \iint\limits_{S_2} u_3 \cos \gamma\, dA$$

(cf. Fig. 165). On S_1 in Fig. 165 the angle γ is acute. Hence by (10″) at the end of Sec. 8.6 we have $dA = \sec \gamma\, dx\, dy$. Since $\cos \gamma \sec \gamma = 1$, we thus obtain

$$\iint\limits_{S_1} u_3 \cos \gamma\, dA = \iint\limits_{R} u_3[x, y, h(x, y)]\, dx\, dy,$$

which is identical with the first double integral in (9). On S_2, the angle γ is obtuse (Fig. 165) so that $\pi - \gamma$ corresponds to the acute angle in (10″), Sec. 8.6, and, therefore,

$$dA = \sec (\pi - \gamma)\, dx\, dy = -\sec \gamma\, dx\, dy.$$

Hence

$$\iint\limits_{S_2} u_3 \cos \gamma\, dA = -\iint\limits_{R} u_3[x, y, g(x, y)]\, dx\, dy,$$

which is identical with the last term in (9). This proves (6).

The relations (4) and (5) then follow by merely relabelling the variables and using the fact that, by assumption, T has representations similar to (7), namely,

$$\tilde{g}(y, z) \leqq x \leqq \tilde{h}(y, z) \qquad \text{and} \qquad g^*(z, x) \leqq y \leqq h^*(z, x).$$

This establishes the divergence theorem for special regions.

For any region T which can be subdivided into *finitely many* special regions by means of auxiliary surfaces, the theorem follows by adding the result for each part separately; this procedure is analogous to that in the proof of Green's theorem in Sec. 8.4. The surface integrals over the auxiliary surfaces cancel in pairs, and the sum of the remaining surface integrals is the surface integral over the whole boundary S of T; the volume integrals over the parts of T add up to the volume integral over T.

The divergence theorem is now proved for any bounded region which is of interest in practical problems. The extension to the most general region T of the type characterized in the theorem would require a certain limit process; this is similar to the situation in the case of Green's theorem in Sec. 8.4.

Green's theorem in the plane (Sec. 8.4) can be used for evaluating line integrals. Similarly, the divergence theorem is suitable for evaluating surface integrals, as may be illustrated by the following example.

Example 1.   By transforming to a triple integral evaluate

$$I = \iint_S (x^3 \, dy \, dz + x^2 y \, dz \, dx + x^2 z \, dx \, dy)$$

where S is the closed surface consisting of the cylinder $x^2 + y^2 = a^2$ $(0 \leq z \leq b)$ and the circular disks $z = 0$ and $z = b$ $(x^2 + y^2 \leq a^2)$.

In (3) we now have

$$u_1 = x^3, \qquad u_2 = x^2 y, \qquad u_3 = x^2 z.$$

Hence, if we make use of the symmetry of the region T bounded by S, the triple integral in (3) assumes the form

$$\iiint_T (3x^2 + x^2 + x^2) \, dx \, dy \, dz = 4 \cdot 5 \int_0^b \int_0^a \int_0^{\sqrt{a^2 - y^2}} x^2 \, dx \, dy \, dz.$$

The integral over x equals $\frac{1}{3}(a^2 - y^2)^{3/2}$. Setting $y = a \cos t$, we have

$$dy = -a \sin t \, dt, \qquad (a^2 - y^2)^{3/2} = a^3 \sin^3 t,$$

and the integral over y becomes

$$\frac{1}{3} \int_0^a (a^2 - y^2)^{3/2} \, dy = -\frac{1}{3} a^4 \int_{\pi/2}^0 \sin^4 t \, dt = \frac{\pi}{16} a^4.$$

The integral over z contributes the factor b, and therefore

$$I = 4 \cdot 5 \frac{\pi}{16} b a^4 = \frac{5\pi}{4} a^4 b.$$

8.9 Consequences and Applications of the Divergence Theorem

The divergence theorem has various applications and important consequences, some of which may be illustrated by the subsequent examples. In these examples, the regions and functions are assumed to satisfy the conditions under which the divergence theorem holds, and in each case, $\mathbf{n}$ is the *outward* unit normal vector of the boundary surface of the region, as before.

Example 1 (Representation of the divergence independent of the coordinates). Dividing both sides of (1) in Sec. 8.8 by the volume $V(T)$ of the region T, we obtain

$$(1) \qquad \frac{1}{V(T)} \iiint\limits_{T} \operatorname{div} \mathbf{u} \, dV = \frac{1}{V(T)} \iint\limits_{S(T)} u_n \, dA$$

where $S(T)$ is the boundary surface of T. The basic properties of the triple integral are essentially the same as those of the double integral considered in Sec. 8.3. In particular the **mean value theorem** *for triple integrals* asserts that for any continuous function $f(x, y, z)$ in the region T under consideration there is a point Q: (x_0, y_0, z_0) in T such that

$$\iiint\limits_{T} f(x, y, z) \, dV = f(x_0, y_0, z_0) V(T).$$

Setting $f = \operatorname{div} \mathbf{u}$ and using (1), we have

$$(2) \qquad \frac{1}{V(T)} \iiint\limits_{T} \operatorname{div} \mathbf{u} \, dV = \operatorname{div} \mathbf{u}(x_0, y_0, z_0).$$

Let P: (x_1, y_1, z_1) be any fixed point in T, and let T shrink down onto P, so that the maximum distance $d(T)$ of the points of T from P approaches zero. Then Q must approach P, and from (1) and (2) it follows that the divergence of $\mathbf{u}$ at P is

$$(3) \qquad \operatorname{div} \mathbf{u}(x_1, y_1, z_1) = \lim_{d(T) \to 0} \frac{1}{V(T)} \iint\limits_{S(T)} u_n \, dA.$$

This formula is sometimes used as a definition of the divergence. While the definition of the divergence in Sec. 7.10 involves coordinates, the formula (3) is independent of the coordinate system. Hence *from (3) it follows immediately that the divergence is independent of the particular choice of Cartesian coordinates.*

Example 2 (Physical interpretation of the divergence). By means of the divergence theorem we may obtain an intuitive interpretation of the divergence of a vector. For this purpose it is convenient to consider the flow of an incompressible fluid (cf. p. 319) of constant density $\rho = 1$ which is *stationary* or *steady*, that is, does not vary with time. Such a flow is determined by the field of its velocity vector $\mathbf{v}(P)$ at any point P.

Let S be the boundary surface of a region T in space, and let $\mathbf{n}$ be the outward unit normal vector of S. The mass of fluid which flows through a small portion ΔS of S of area ΔA per unit time from the interior of S to the exterior is equal to $v_n \Delta A$ where[11] $v_n = \mathbf{v} \cdot \mathbf{n}$ is the normal component of $\mathbf{v}$ in the direction of $\mathbf{n}$, taken at a suitable point of ΔS. Consequently, the total mass of fluid which flows across S from T to the outside per unit of time is given by the surface integral

$$\iint\limits_{S} v_n \, dA.$$

Hence this integral represents the total flow out of T, and the integral

$$(4) \qquad \frac{1}{V} \iint\limits_{S} v_n \, dA,$$

---

[11] Note that v_n may be negative at a certain point, which means that fluid enters the interior of S at such a point.

where V is the volume of T, represents the average flow out of T. Since the flow is steady and the fluid is incompressible, the amount of fluid flowing outward must be continuously supplied. Hence, if the value of the integral (4) is different from zero, there must be **sources** (*positive sources and negative sources, called* **sinks**) in T, that is, points where fluid is produced or disappears.

If we let T shrink down to a fixed point P in T, we obtain from (4) the *source intensity* at P represented by the right-hand side of (3) (with u_n replaced by v_n). From this and (3) it follows that *the divergence of the velocity vector* **v** *of a steady incompressible flow is the source intensity of the flow at the corresponding point.* There are no sources in T if and only if div **v** $\equiv 0$; in this case,

$$\iint_{S^*} v_n \, dA = 0$$

for any closed surface S^* in T.

Example 3 (Heat flow).   We know that in a body heat will flow in the direction of decreasing temperature. Physical experiments show that the rate of flow is proportional to the gradient of the temperature. This means that the velocity **v** of the heat flow in a body is of the form

$$(5) \qquad\qquad \mathbf{v} = -K \operatorname{grad} U,$$

where $U(x, y, z, t)$ is the temperature, t is the time, and K is called the *thermal conductivity* of the body; in ordinary physical circumstances K is a constant.

Let R be a region in the body and let S be its boundary surface. Then the amount of heat leaving R per unit of time is

$$\iint_S v_n \, dA,$$

where $v_n = \mathbf{v} \cdot \mathbf{n}$ is the component of **v** in the direction of the outer unit normal vector **n** of S. This expression is obtained in a fashion similar to that in the preceding example. From (5) and the divergence theorem we obtain

$$(6) \qquad \iint_S v_n \, dA = -K \iiint_R \operatorname{div}(\operatorname{grad} U) \, dx \, dy \, dz = -K \iiint_R \nabla^2 U \, dx \, dy \, dz;$$

cf. (5) in Sec. 7.10.

On the other hand, the total amount of heat H in R is

$$H = \iiint_R \sigma \rho U \, dx \, dy \, dz,$$

where the constant σ is the specific heat of the material of the body and ρ is the density ($=$ mass per unit volume) of the material. Hence the time rate of decrease of H is

$$-\frac{\partial H}{\partial t} = -\iiint_R \sigma \rho \frac{\partial U}{\partial t} \, dx \, dy \, dz,$$

and this must be equal to the above amount of heat leaving R; from (6) we thus have

$$-\iiint_R \sigma \rho \frac{\partial U}{\partial t} \, dx \, dy \, dz = -K \iiint_R \nabla^2 U \, dx \, dy \, dz$$

or

$$\iiint_R \left(\sigma \rho \frac{\partial U}{\partial t} - K \nabla^2 U \right) dx \, dy \, dz = 0.$$

Since this holds for any region R in the body, the integrand (if continuous) must be zero everywhere; that is,

$$(7) \qquad \frac{\partial U}{\partial t} = c^2 \nabla^2 U \qquad\qquad c^2 = \frac{K}{\sigma \rho}.$$

This partial differential equation is called the **heat equation**; it is fundamental for heat conduction. Methods for solving problems in heat conduction will be considered in Chap. 10.

Example 4 (A basic property of solutions of Laplace's equation). Consider the formula in the divergence theorem:

$$(8^*) \qquad \iiint\limits_{T} \text{div } \mathbf{u} \, dV = \iint\limits_{S} u_n \, dA.$$

Assume that $\mathbf{u}$ is the gradient of a scalar function, say, $\mathbf{u} = \text{grad } f$. Then [cf. (5) in Sec. 7.10]

$$\text{div } \mathbf{u} = \text{div (grad } f) = \nabla^2 f.$$

Furthermore,

$$u_n = \mathbf{u} \cdot \mathbf{n} = \mathbf{n} \cdot \text{grad } f,$$

and from (5) in Sec. 7.8 we see that the right side is the directional derivative of f in the outward normal direction of S. If we denote this derivative by $\partial f / \partial n$, formula (8^*) becomes

$$(8) \qquad \iiint\limits_{T} \nabla^2 f \, dV = \iint\limits_{S} \frac{\partial f}{\partial n} \, dA.$$

Obviously this is the three-dimensional analogue of the formula (9) in Sec. 8.4.

Taking into account the assumptions under which the divergence theorem holds, we immediately obtain from (8) the following result.

Theorem 1. *Let $f(x, y, z)$ be a solution of Laplace's equation*

$$\nabla^2 f = \frac{\partial^2 f}{\partial x^2} + \frac{\partial^2 f}{\partial y^2} + \frac{\partial^2 f}{\partial z^2} = 0$$

in some domain D, and suppose that the second partial derivatives of f are continuous in D. Then the integral of the normal derivative of f over any piecewise smooth[12] closed orientable surface S in D is zero.

Example 5 (Green's theorem). Let f and g be scalar functions such that $\mathbf{u} = f \, \text{grad } g$ satisfies the assumptions of the divergence theorem in some region T. Then

$$\text{div } \mathbf{u} = \text{div } (f \, \text{grad } g) = f \nabla^2 g + \text{grad } f \cdot \text{grad } g$$

(cf. Prob. 13 at the end of Sec. 7.10). Furthermore

$$\mathbf{u} \cdot \mathbf{n} = \mathbf{n} \cdot (f \, \text{grad } g) = f(\mathbf{n} \cdot \text{grad } g).$$

The expression $\mathbf{n} \cdot \text{grad } g$ is the directional derivative of g in the direction of the outward normal vector $\mathbf{n}$ of the surface S in the divergence theorem. If we denote this derivative by $\partial g / \partial n$, the formula in the divergence theorem becomes

$$(9) \qquad \iiint\limits_{T} (f \nabla^2 g + \text{grad } f \cdot \text{grad } g) \, dV = \iint\limits_{S} f \frac{\partial g}{\partial n} \, dA.$$

[12] Cf. Sec. 8.5.

This formula is called **Green's first formula** or (together with the assumptions) the *first form of Green's theorem.*

By interchanging f and g we obtain a similar formula. On subtracting this formula from (9), we find

$$(10) \qquad \iiint\limits_{T} (f \nabla^2 g - g \nabla^2 f) \, dV = \iint\limits_{S} \left(f \frac{\partial g}{\partial n} - g \frac{\partial f}{\partial n} \right) dA.$$

This formula is called **Green's second formula** or (together with the assumptions) the *second form of Green's theorem.*

Example 6 (*Uniqueness of solutions of Laplace's equation*). Suppose that f satisfies the assumptions in Theorem 1 and is zero everywhere on a piecewise smooth closed orientable surface S in D. Then, setting $g = f$ in (9) and denoting the interior of S by T, we obtain

$$\iiint\limits_{T} \text{grad } f \cdot \text{grad } f \, dV = \iiint\limits_{T} |\text{grad } f|^2 \, dV = 0.$$

Since by assumption $|\text{grad } f|$ is continuous in T and on S and is non-negative, it must be zero everywhere in T. Hence $f_x = f_y = f_z = 0$, and f is constant in T and, because of continuity, equal to its value 0 on S. This proves

Theorem 2. *If a function $f(x, y, z)$ satisfies the assumptions of Theorem* 1 *and is zero at all points of a piecewise smooth closed orientable surface S in D, then it vanishes identically in the region T bounded by S.*

This theorem has an important consequence. Let f_1 and f_2 be functions which satisfy the assumptions of Theorem 1 and take on the same values on S. Then their difference $f_1 - f_2$ satisfies those assumptions and has the value 0 everywhere on S. Consequently, from Theorem 2 it follows that $f_1 - f_2 = 0$ throughout T, and we have the following result.

Theorem 3 (Uniqueness theorem). *Let f be a solution of Laplace's equation which has continuous second partial derivatives in a domain D, and let T be a region in D which satisfies the assumptions of the divergence theorem. Then f is uniquely determined in T by its values on the boundary surface S of T.*

Problems

Using triple integrals, find the total mass of a mass distribution of density σ in a region T where
1. $\sigma = x^2 y^2 z^2$, T: $|x| \leq 1, |y| \leq 1, |z| \leq 1$
2. $\sigma = x + y + z$, T: $0 \leq x \leq 1, 0 \leq y \leq 2, 0 \leq z \leq 3$
3. $\sigma = xy$, · T: the tetrahedron with vertices $(0, 0, 0)$, $(1, 0, 0)$, $(0, 1, 0)$, $(0, 0, 1)$
4. $\sigma = z$, T: the region in the first octant bounded by $y = 1 - x^2$ and $z = x$

Consider a mass of density 1 in a region T and find the moment of inertia about the x-axis

$$I_x = \iiint\limits_{T} (y^2 + z^2) \, dx \, dy \, dz,$$

assuming that T is:
5. The rectangular parallelepiped $0 \leq x \leq a$, $-b/2 \leq y \leq b/2$, $-c/2 \leq z \leq c/2$
6. The cube $0 \leq x \leq 1, 0 \leq y \leq 1, 0 \leq z \leq 1$
7. The cylinder $y^2 + z^2 \leq a^2, 0 \leq x \leq h$

8. The cone $y^2 + z^2 \leq x^2$, $0 \leq x \leq h$

9. The interior of the sphere $x^2 + y^2 + z^2 = a^2$

Evaluate the following surface integrals by the divergence theorem, assuming that S is oriented as in that theorem.

10. $\displaystyle\iint_S (x \, dy \, dz + y \, dz \, dx + z \, dx \, dy)$, S: $x^2 + y^2 + z^2 = a^2$

11. $\displaystyle\iint_S (x^2 \, dy \, dz + y^2 \, dz \, dx + z^2 \, dx \, dy)$, S: the surface of the cube in Prob. 6

12. $\displaystyle\iint_S xyz \, dy \, dz$, S as in Prob. 11

13. $\displaystyle\iint_S (yz \, dy \, dz + zx \, dz \, dx + xy \, dx \, dy)$, S: $x^2 + y^2 + z^2 = 1$

14. $\displaystyle\iint_S [(y \cos^2 x + y^3) \, dz \, dx + z(\sin^2 x - 3y^2) \, dx \, dy]$, S: $x^2 + y^2 + z^2 = 4$

15. $\displaystyle\iint_S e^y \, dz \, dx$, S: the surface of the parallelepiped $0 \leq x \leq 3, 0 \leq y \leq 2, 0 \leq z \leq 1$

16. $\displaystyle\iint_S [\sin x \, dy \, dz + (2 - \cos x)y \, dz \, dx]$, S as in Prob. 15

Let T be a closed bounded region in space and S its boundary surface. Using the divergence theorem, prove the following statements.

17. If g is **harmonic**[13] in a domain containing T, then $\displaystyle\iint_S \frac{\partial g}{\partial n} \, dA = 0$.

18. If g is harmonic in a domain containing T, then $\displaystyle\iint_S g \frac{\partial g}{\partial n} \, dA = \iiint_T |\text{grad } g|^2 \, dV$.

19. If g is harmonic in a domain containing T and $\partial g/\partial n = 0$ on S, then g is constant in T.

20. If f and g are harmonic in a domain containing T and $\partial f/\partial n = \partial g/\partial n$ on S, then $f = g + c$ in T, where c is a constant.

21. If f and g are harmonic in a domain containing T, then $\displaystyle\iint_S \left(f \frac{\partial g}{\partial n} - g \frac{\partial f}{\partial n} \right) dA = 0$.

22. Show that the Laplacian can be represented independently of all coordinate systems in the form

$$\nabla^2 f = \lim_{d(T) \to 0} \frac{1}{V(T)} \iint_{S(T)} \frac{\partial f}{\partial n} \, dA$$

where $d(T)$ is the maximum distance of the points of a region T bounded by $S(T)$ from the point at which the Laplacian is evaluated and $V(T)$ is the volume of T. *Hint.* Put $\mathbf{u} = \text{grad } f$ in (3) and use (5), Sec. 7.8, with $\mathbf{b} = \mathbf{n}$, the outer unit normal vector to S.

---

[13] This means that g is a solution of Laplace's equation, and its second partial derivatives are continuous in T.

23. Using the divergence theorem, show that the volume V of a region T bounded by a surface S is

$$V = \iint_S x \, dy \, dz = \iint_S y \, dz \, dx = \iint_S z \, dx \, dy = \frac{1}{3} \iint_S (x \, dy \, dz + y \, dz \, dx + z \, dx \, dy).$$

Verify these formulas for the cylinder $x^2 + y^2 \leq 1$, $0 \leq z \leq h$.

24. Using $\mathbf{u} = x\mathbf{i} + y\mathbf{j} + z\mathbf{k}$ in (1), Sec. 8.8, show that a region T with boundary surface S has the volume

$$V = \frac{1}{3} \iint_S r \cos \theta \, dA$$

where r is the distance of a variable point P: (x, y, z) on S from the origin O and θ is the angle between the directed line OP and the outer normal to S at P.

25. Find the volume of the interior of a sphere of radius a by means of the formula in Prob. 24.

8.10 Stokes's Theorem

In Sec. 8.4 it was shown that double integrals over a plane region can be transformed into line integrals over the boundary curve of the region. Generalizing this result, we shall now consider the corresponding problem in the case of a surface integral.

Stokes's theorem.[14] *Let S be a piecewise smooth*[15] *oriented surface in space and let the boundary of S be a piecewise smooth simple closed curve C. Let* $\mathbf{v}(x, y, z)$ *be a continuous vector function which has continuous first partial derivatives in a domain in space which contains S. Then*

(1)
$$\iint_S (\text{curl } \mathbf{v})_n \, dA = \int_C v_t \, ds;$$

here $(\text{curl } \mathbf{v})_n = (\text{curl } \mathbf{v}) \cdot \mathbf{n}$ *is the component of* curl $\mathbf{v}$ *in the direction of a unit normal vector* $\mathbf{n}$ *of S; the integration around C is taken in the sense shown in Fig. 166, and* v_t *is the component of* $\mathbf{v}$ *in the direction of the tangent vector of C in Fig. 166.*

Proof. We first prove Stokes's theorem for a surface S which can be represented simultaneously in the forms

(2)   (*a*) $z = f(x, y)$,   (*b*) $y = g(x, z)$,   (*c*) $x = h(y, z)$,

where f, g, and h are continuous functions of the respective variables and have continuous first partial derivatives. Let

(3*) $\mathbf{n} = \cos \alpha \, \mathbf{i} + \cos \beta \, \mathbf{j} + \cos \gamma \, \mathbf{k}$

be the "upper" unit normal vector of S (Fig. 167), and let $\mathbf{v} = v_1\mathbf{i} + v_2\mathbf{j} + v_3\mathbf{k}$.

---

[14] GEORGE GABRIEL STOKES (1819–1903), Irish mathematician and physicist, who made important contributions to the theory of infinite series and several branches of theoretical physics.
[15] Cf. Secs. 8.1 and 8.5.

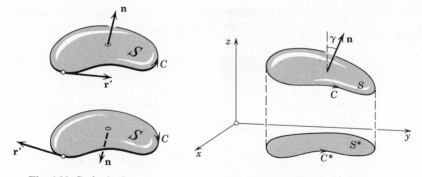

Fig. 166. Stokes's theorem   **Fig. 167.** Proof of Stokes's theorem

If we represent C in the form $\mathbf{r} = \mathbf{r}(s)$ where the arc length s increases in the direction of integration, the unit tangent vector is

$$\frac{d\mathbf{r}}{ds} = \frac{dx}{ds}\mathbf{i} + \frac{dy}{ds}\mathbf{j} + \frac{dz}{ds}\mathbf{k},$$

and, therefore,

$$v_t = \mathbf{v} \cdot \frac{d\mathbf{r}}{ds} = v_1\frac{dx}{ds} + v_2\frac{dy}{ds} + v_3\frac{dz}{ds}.$$

From this we obtain

$$v_t\,ds = \mathbf{v} \cdot \frac{d\mathbf{r}}{ds}\,ds = v_1\,dx + v_2\,dy + v_3\,dz.$$

Consequently, if we use the representation of the curl in terms of right-handed Cartesian coordinates [cf. (1), Sec. 7.11], the formula in Stokes's theorem may be written

$$\iint_S \left[\left(\frac{\partial v_3}{\partial y} - \frac{\partial v_2}{\partial z}\right)\cos\alpha + \left(\frac{\partial v_1}{\partial z} - \frac{\partial v_3}{\partial x}\right)\cos\beta + \left(\frac{\partial v_2}{\partial x} - \frac{\partial v_1}{\partial y}\right)\cos\gamma \right] dA$$

(3)

$$= \int_C (v_1\,dx + v_2\,dy + v_3\,dz),$$

where α, β, and γ are defined in connection with (3*).

We prove that in (3) the integrals over the terms involving v_1 are equal,

(4)
$$\iint_S \left(\frac{\partial v_1}{\partial z}\cos\beta - \frac{\partial v_1}{\partial y}\cos\gamma\right) dA = \int_C v_1\,dx.$$

Let S^* be the orthogonal projection of S in the xy-plane and let C^* be its boundary which is oriented as shown in Fig. 167. Using the representation (2a) of S, we may write the line integral over C as a line integral over C^*:

$$\int_C v_1(x, y, z)\,dx = \int_{C^*} v_1[x, y, f(x, y)]\,dx.$$

We now apply Green's theorem in the plane (Sec. 8.4) to the functions $v_1[x, y, f(x, y)]$ and 0 [instead of f and g in Sec. 8.4]. Then

$$\int_{C^*} v_1[x, y, f(x, y)]\, dx = -\iint_{S^*} \frac{\partial v_1}{\partial y}\, dx\, dy.$$

In the integral on the right,

$$\frac{\partial v_1[x, y, f(x, y)]}{\partial y} = \frac{\partial v_1(x, y, z)}{\partial y} + \frac{\partial v_1(x, y, z)}{\partial z}\frac{\partial f}{\partial y} \qquad [z = f(x, y)]$$

and, thus,

(5) $$\int_C v_1(x, y, z)\, dx = -\iint_{S^*} \left(\frac{\partial v_1}{\partial y} + \frac{\partial v_1}{\partial z}\frac{\partial f}{\partial y} \right) dx\, dy.$$

We prove that the integral on the left-hand side of (4) equals the integral on the right-hand side of (5). In the former integral we introduce x and y as variables of integration. Writing (2a) in the form

$$F(x, y, z) = z - f(x, y) = 0,$$

we obtain

$$\operatorname{grad} F = -\frac{\partial f}{\partial x}\mathbf{i} - \frac{\partial f}{\partial y}\mathbf{j} + \mathbf{k}.$$

Denoting the length of grad F by a, we have

$$a = |\operatorname{grad} F| = \sqrt{1 + \left(\frac{\partial f}{\partial x}\right)^2 + \left(\frac{\partial f}{\partial y}\right)^2}.$$

Since grad F is normal to S, we obtain

$$\mathbf{n} = \pm \frac{\operatorname{grad} F}{a}$$

where $\mathbf{n}$ is defined as before. But the components of both $\mathbf{n}$ and grad F in the positive z-direction are positive, and, therefore,

$$\mathbf{n} = +\frac{\operatorname{grad} F}{a}.$$

Hence, from the preceding representations of $\mathbf{n}$ and grad F in terms of components with respect to the xyz-coordinate system, we see that

$$\cos \alpha = -\frac{1}{a}\frac{\partial f}{\partial x}, \qquad \cos \beta = -\frac{1}{a}\frac{\partial f}{\partial y}, \qquad \cos \gamma = \frac{1}{a}.$$

Furthermore, from (10″) at the end of Sec. 8.6 it follows that in (4), $dA = a\, dx\, dy$.

Hence

$$\iint\limits_{S} \left(\frac{\partial v_1}{\partial z} \cos \beta - \frac{\partial v_1}{\partial y} \cos \gamma \right) dA = \iint\limits_{S^*} \left(\frac{\partial v_1}{\partial z} \left(-\frac{1}{a} \frac{\partial f}{\partial y} \right) - \frac{\partial v_1}{\partial y} \frac{1}{a} \right) a \, dx \, dy.$$

The integral on the right equals the right-hand side of (5), and (4) is proved.

If $-\mathbf{n}$ were chosen for the positive normal direction, then by assumption the sense of integration along C would be reversed and the result would be the same as before. This shows that (4) holds for both choices of the positive normal of S.

Using the representations (2b) and (2c) of S and reasoning exactly as before, we obtain

(6)
$$\iint\limits_{S} \left(\frac{\partial v_2}{\partial x} \cos \gamma - \frac{\partial v_2}{\partial z} \cos \alpha \right) dA = \int_{C} v_2 \, dy$$

and

(7)
$$\iint\limits_{S} \left(\frac{\partial v_3}{\partial y} \cos \alpha - \frac{\partial v_3}{\partial x} \cos \beta \right) dA = \int_{C} v_3 \, dz.$$

By adding (4), (6), and (7) we obtain (1). This proves Stokes's theorem for a surface S which can be represented simultaneously in the forms (2a), (2b), and (2c).

As in the proof of the divergence theorem, our result may be immediately extended to a surface S which can be decomposed into finitely many pieces, each of which is of the type considered before. This covers most of the cases which are of practical interest. The proof in the case of a most general surface S satisfying the assumptions of the theorem would require a limit process; this is similar to the situation in the case of Green's theorem in Sec. 8.4.

8.11 Consequences and Applications of Stokes's Theorem

Example 1 (Green's theorem in the plane as a special case of Stokes's theorem). Let $\mathbf{v} = v_1 \mathbf{i} + v_2 \mathbf{j}$ be a vector function which is continuously differentiable in a domain in the xy-plane containing a simply-connected bounded closed region S whose boundary C is a piecewise smooth simple closed curve. Then, according to (1) in Sec. 7.11,

$$(\operatorname{curl} \mathbf{v})_n = \frac{\partial v_2}{\partial x} - \frac{\partial v_1}{\partial y}.$$

Furthermore, $v_t \, ds = v_1 \, dx + v_2 \, dy$, and (1) in the last section takes the form

$$\iint\limits_{S} \left(\frac{\partial v_2}{\partial x} - \frac{\partial v_1}{\partial y} \right) dA = \int_{C} (v_1 \, dx + v_2 \, dy).$$

This shows that Green's theorem in the plane (Sec. 8.4) is a special case of Stokes's theorem.

Example 2 (Physical interpretation of the curl). Let S_r be a circular disk of radius r and center P bounded by the circle C_r (Fig. 168), and let $\mathbf{v}(Q) \equiv \mathbf{v}(x, y, z)$ be a continuously differentiable

Fig. 168. Example 2

vector function in a domain containing S_r. Then by Stokes's theorem and the mean value theorem for surface integrals,

$$\int_{C_r} v_t \, ds = \iint_{S_r} (\operatorname{curl} \mathbf{v})_n \, dA = [\operatorname{curl} \mathbf{v}(P^*)]_n A_r$$

where A_r is the area of S_r and P^* is a suitable point of S_r. This may be written

$$[\operatorname{curl} \mathbf{v}(P^*)]_n = \frac{1}{A_r} \int_{C_r} v_t \, ds.$$

In the case of a fluid motion with velocity $\mathbf{v}$, the integral

$$\int_{C_r} v_t \, ds$$

is called the **circulation** of the flow around C_r; it measures the extent to which the corresponding fluid motion is a rotation around the circle C_r. If we now let r approach zero, we find

(1) $$[\operatorname{curl} \mathbf{v}(P)]_n = \lim_{r \to 0} \frac{1}{A_r} \int_{C_r} v_t \, ds;$$

that is, the component of the curl in the positive normal direction can be regarded as the *specific circulation* (circulation per unit area) of the flow in the surface at the corresponding point.

Problems

Evaluate $\iint_S (\operatorname{curl} \mathbf{v})_n \, dA$ by direct integration, where

  1. $\mathbf{v} = z\mathbf{i} + x\mathbf{j} + y\mathbf{k}$, S: the square with vertices $(0, 0, 0)$, $(1, 0, 0)$, $(1, 1, 0)$, $(0, 1, 0)$
  2. $\mathbf{v} = z^2\mathbf{i} + x^2\mathbf{j} + y^2\mathbf{k}$, S: the square $0 \leq x \leq 1$, $0 \leq y \leq 1$, $z = 1$
  3. $\mathbf{v} = -y^3\mathbf{i} + x^3\mathbf{j}$, S: the circular disk $x^2 + y^2 \leq 1$, $z = 0$

  4. Verify Stokes's theorem for $\mathbf{v}$ and S in Prob. 3. *Hint.* Use a parametric representation of C.

Using Stokes's theorem, evaluate $\int_C v_t \, ds$ where, with respect to right-handed Cartesian coordinates,

  5. $\mathbf{v} = 2y\mathbf{i} + z\mathbf{j} + 3y\mathbf{k}$, C: the intersection of $x^2 + y^2 + z^2 = 6z$ and $z = x + 3$, oriented in the clockwise sense as viewed from the origin.
  6. $\mathbf{v} = 4z\mathbf{i} - 2x\mathbf{j} + 2x\mathbf{k}$, C: the intersection of $x^2 + y^2 = 1$ and $z = y + 1$ oriented in the clockwise sense as viewed from the origin.
  7. $\mathbf{v} = x^2\mathbf{i} + y^2\mathbf{j} + z^2\mathbf{k}$, C: the intersection of $x^2 + y^2 + z^2 = a^2$ and $z = y^2$.
  8. $\mathbf{v} = y\mathbf{i} + xz^3\mathbf{j} - zy^3\mathbf{k}$, C: the circle $x^2 + y^2 = a^2$, $z = b$ (< 0), oriented in the counterclockwise sense as viewed from the origin.

  9. Show that if $\mathbf{v}$ and S satisfy the assumptions of Stokes's theorem and $\mathbf{v} = \operatorname{grad} f$, then $\int_C v_t \, ds = 0$, where C is the boundary of S.
  10. Evaluate $\int_C v_t \, ds$, $\mathbf{v} = (-y\mathbf{i} + x\mathbf{j})/(x^2 + y^2)$, C: $x^2 + y^2 = 1$ $(z = 0)$, oriented in the clockwise sense. Note that Stokes's theorem cannot be applied. Why?

Using (3) in the preceding section, evaluate the following line integrals which are taken around the given contour C in the clockwise sense as viewed from the origin.

11. $\int_C [(x+y)\,dx + (2x - z)\,dy + (y+z)\,dz]$, C: the boundary of the triangle with vertices $(2, 0, 0)$, $(0, 3, 0)$, $(0, 0, 6)$

12. $\int_C (z\,dx + x\,dy + y\,dz)$, C: the boundary of the triangle with vertices $(1, 0, 0)$, $(0, 1, 0)$, $(0, 0, 1)$

13. $\int_C (yz\,dx + xz\,dy + xy\,dz)$, C: the intersection of $x^2 + y^2 = 1$ and $z = y^2$

14. $\int_C (\sin z\,dx - \cos x\,dy + \sin y\,dz)$, C: the boundary of the rectangle $0 \leq x \leq \pi$, $0 \leq y \leq 1$, $z = 3$

15. $\int_C (e^x\,dx + 2y\,dy - dz)$, C: the intersection of $x^2 + y^2 = 1$ and $z = 1$

8.12 Line Integrals Independent of Path

In Sec. 8.2 we saw that the value of a line integral

$$(1) \qquad \int_C (f\,dx + g\,dy + h\,dz)$$

will in general depend not only on the end points P and Q of the path C but also on C; that is, if we integrate from P to Q along different paths, we shall, in general, obtain different values of the integral. We shall now see under what conditions that value depends only on P and Q but does not depend on the path C from P to Q. This problem is of great importance. We first state the following definition.

Let $f(x, y, z)$, $g(x, y, z)$, and $h(x, y, z)$ be functions which are defined and continuous in a domain D of space. Then a line integral of the form (1) is said to be **independent of path** *in* D, if for every pair of end points P and Q in D the value of the integral is the same for all paths C in D starting from P and ending at Q. This value will then, in general, depend on the choice of the points P and Q but not on the choice of the path joining them. (Remember from the beginning of Sec. 8.1 that, by definition, a function is a single-valued relation; this will matter in our discussion.)

For formulating our results it will be convenient to use the following concepts. An expression of the form

$$(2) \qquad f\,dx + g\,dy + h\,dz,$$

where f, g, h are functions defined in a domain D in space, is called a *first-order* **differential form** in three variables. This form is said to be **exact** or *an exact differential in* D, if it is the differential

$$(3) \qquad du = \frac{\partial u}{\partial x}\,dx + \frac{\partial u}{\partial y}\,dy + \frac{\partial u}{\partial z}\,dz$$

of a differentiable function $u(x, y, z)$ everywhere in D; that is,

$$(4) \qquad f\,dx + g\,dy + h\,dz = du.$$

By comparing (3) and (4) we see that *the form* (2) *is exact in D if and only if there is a differentiable function* $u(x, y, z)$ *such that*

(5)
$$f = \frac{\partial u}{\partial x}, \qquad g = \frac{\partial u}{\partial y}, \qquad h = \frac{\partial u}{\partial z}$$

everywhere in D.

In vector language this means that *the form* (2) *is exact in D if and only if the vector function*

$$\mathbf{v} = f\mathbf{i} + g\mathbf{j} + h\mathbf{k}$$

is the gradient of a function $u(x, y, z)$ *in D:*

(5')
$$\mathbf{v} = \text{grad } u = \frac{\partial u}{\partial x}\mathbf{i} + \frac{\partial u}{\partial y}\mathbf{j} + \frac{\partial u}{\partial z}\mathbf{k}.$$

Theorem 1. *Let* $f(x, y, z)$, $g(x, y, z)$, *and* $h(x, y, z)$ *be continuous in a domain D in space. Then the line integral*

(6)
$$\int_C (f\,dx + g\,dy + h\,dz)$$

is independent of path in D if and only if the differential form under the integral sign is exact in D.

Proof. (a) Suppose that the integral under consideration is independent of path in D. We choose any fixed point P: (x_0, y_0, z_0) and a point Q: (x, y, z) in D. Then we define a function $u(x, y, z)$ by the formula

(7)
$$u(x, y, z) = u_0 + \int_P^Q (f\,dx + g\,dy + h\,dz),$$

where u_0 is a constant and we integrate along an arbitrary path in D from P to Q. Since the integral is independent of path and P is fixed, it does indeed depend only on the coordinates x, y, z of the end point Q and defines a function $u(x, y, z)$ in D. It remains to show that from (7) the relations (5) follow, which means that $f\,dx + g\,dy + h\,dz$ is exact in D. We prove the first of these three relations. Since the integral is independent of path, we may integrate from P to a point Q_1: (x_1, y, z) and then parallel to the x-axis along the line segment from Q_1 to Q (Fig. 169); here Q_1 is chosen so that the whole segment Q_1Q lies in D. Then

$$u(x, y, z) = u_0 + \int_P^{Q_1} (f\,dx + g\,dy + h\,dz) + \int_{Q_1}^Q (f\,dx + g\,dy + h\,dz).$$

We take the partial derivative with respect to x. Since P and Q_1 do not depend on x, that derivative of the first integral is zero. Since on the segment Q_1Q, y, and z are constant, the last integral may be written as the definite integral

$$\int_{x_1}^x f(x, y, z)\,dx.$$

Hence the partial derivative of that integral with respect to x is $f(x, y, z)$, and the

Fig. 169. Proof of Theorem 1

first of the relations (5) is proved. The other two relations can be proved by the same argument.

   (b) Conversely, suppose that $f\,dx + g\,dy + h\,dz$ is exact in D. Then (5) holds in D for some function u. Let C be any path from P to Q in D, and let

$$\mathbf{r}(t) = x(t)\mathbf{i} + y(t)\mathbf{j} + z(t)\mathbf{k} \qquad (t_0 \leqq t \leqq t_1)$$

be a parametric representation of C such that P corresponds to $t = t_0$ and Q corresponds to $t = t_1$. Then

$$\int_P^Q (f\,dx + g\,dy + h\,dz) = \int_P^Q \left(\frac{\partial u}{\partial x}\,dx + \frac{\partial u}{\partial y}\,dy + \frac{\partial u}{\partial z}\,dz\right)$$

$$= \int_{t_0}^{t_1} \frac{du}{dt}\,dt = u[x(t), y(t), z(t)]\Big|_{t=t_0}^{t=t_1} = u(Q) - u(P).$$

This shows that the value of the integral is simply the difference of the values of u at the two end points of C and is, therefore, independent of the path C. Theorem 1 is proved.

   The last formula in the proof,

(8) $$\int_P^Q (f\,dx + g\,dy + h\,dz) = u(Q) - u(P),$$

is the analogue of the usual formula

$$\int_a^b f(x)\,dx = F(x)\Big|_a^b = F(b) - F(a) \qquad [\text{where } F'(x) = f(x)]$$

for evaluating definite integrals which is known from elementary calculus. Formula (8) should be used whenever a line integral is independent of path. The practical use of this formula will be explained at the end of this section.

   We remember that the work W done by a (variable) force $\mathbf{p} = f\mathbf{i} + g\mathbf{j} + h\mathbf{k}$ in the displacement of a particle along a path C in space is given by the line integral (7) in Sec. 8.2,

$$W = \int_C \mathbf{p} \cdot d\mathbf{r} = \int_C (f\,dx + g\,dy + h\,dz),$$

the integration being taken in the sense of the displacement. From Theorem 1 we conclude that W depends only on the end points P and Q of the path C if and only if the differential form under the integral sign is exact; and this form is exact if and only if $\mathbf{p}$ is the gradient of a scalar function u. In this case the field of force given by $\mathbf{p}$ is said to be **conservative** (cf. also at the end of Sec. 7.8).

Example 1. Find the work W done by the force $\mathbf{p} = yz\mathbf{i} + xz\mathbf{j} + xy\mathbf{k}$ in the displacement of a particle along the straight segment C from P: $(1, 1, 1)$ to Q: $(3, 3, 2)$. The work is given by the integral

$$W = \int_C (yz\,dx + xz\,dy + xy\,dz).$$

This integral is of the form (1) where

$$f = yz, \qquad g = xz, \qquad h = xy,$$

and (5) holds for $u = xyz$. Hence W is independent of path and we may use (8), finding

$$W = u(Q) - u(P) = u(3, 3, 2) - u(1, 1, 1) = 18 - 1 = 17.$$

From Theorem 1 we may immediately derive the following important result.

Theorem 2. *Let f, g, h be continuous in a domain D of space. Then the line integral*

$$\int_C (f\,dx + g\,dy + h\,dz)$$

is independent of path in D if and only if it is zero on every simple closed path in D.

Proof. **(a)** Let C be a simple closed path in D and suppose that the integral is independent of path in D. We subdivide C into two arcs C_1 and C_2 (Fig. 170). Then

$$\oint_C (f\,dx + g\,dy + h\,dz)$$

(9)

$$= \int_{C_1} (f\,dx + g\,dy + h\,dz) + \int_{C_2} (f\,dx + g\,dy + h\,dz).$$

Because of independence of path,

$$\int_{C_2} (f\,dx + g\,dy + h\,dz) = \int_{C_1{}^*} (f\,dx + g\,dy + h\,dz)$$

$$= -\int_{C_1} (f\,dx + g\,dy + h\,dz),$$

where $C_1{}^*$ denotes C_1 traced in the reverse direction. From this it follows that the integral on the left-hand side of (9) is zero.

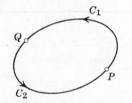

Fig. 170. Proof of Theorem 2

(b) Conversely, suppose that the integral under consideration is zero on every simple closed path in D. Let P and Q be any two points of D, and let C_1 and C_2 be two paths in D which join P and Q and do not cross (Fig. 170). C_1 and C_2 together form a simple closed path C. Therefore,

$$\int_{C_1} (f\,dx + g\,dy + h\,dz) + \int_{C_2} (f\,dx + g\,dy + h\,dz)$$

$$= \oint_C (f\,dx + g\,dy + h\,dz) = 0.$$

From this we obtain

$$\int_{C_2} (f\,dx + g\,dy + h\,dz) = -\int_{C_1} (f\,dx + g\,dy + h\,dz)$$

$$= \int_{C_1^*} (f\,dx + g\,dy + h\,dz).$$

This completes the proof of Theorem 2.

The extension of this theorem to the case of closed paths crossing a finite number of times is immediate.

To make effective practical use of the simple Theorem 1 we need a criterion by which we can decide whether the differential form under the integral sign is exact. Such a criterion will be given in Theorem 3, below. For formulating this theorem we shall need the following concept.

A domain D is called **simply connected** if every closed curve in D can be continuously shrunk to any point in D without leaving D.

For example, the interior of a sphere or a cube, the interior of a sphere with finitely many points removed, and the domain between two concentric spheres are simply connected, while the interior of a torus (cf. Sec. 8.6) and the interior of a cube with one space diagonal removed are not simply connected.

Theorem 3. *Let* $f(x, y, z)$, $g(x, y, z)$, *and* $h(x, y, z)$ *be continuous functions having continuous first partial derivatives in a domain D in space. If the line integral*

(10)
$$\int_C (f\,dx + g\,dy + h\,dz)$$

is independent of path in D (and, consequently, $f\,dx + g\,dy + h\,dz$ is exact in D), then

(11)
$$\frac{\partial h}{\partial y} = \frac{\partial g}{\partial z}, \qquad \frac{\partial f}{\partial z} = \frac{\partial h}{\partial x}, \qquad \frac{\partial g}{\partial x} = \frac{\partial f}{\partial y}$$

or, in vector language,

(11')
$$\text{curl } \mathbf{v} = \mathbf{0} \qquad\qquad (\mathbf{v} = f\mathbf{i} + g\mathbf{j} + h\mathbf{k})$$

everywhere in D. Conversely, if D is simply connected and (11) holds everywhere in D, then the integral (10) is independent of path in D (and, consequently, the differential form $f\,dx + g\,dy + h\,dz$ is exact in D).

Proof.   (a) Suppose that (10) is independent of path in D. Then, by Theorem 1, the form $f\,dx + g\,dy + h\,dz$ is exact in D and there is a function u such that, by (5'),

$$\mathbf{v} = f\mathbf{i} + g\mathbf{j} + h\mathbf{k} = \text{grad } u.$$

in D. From this and (3) in Sec. 7.11 we obtain

$$\text{curl } \mathbf{v} = \text{curl (grad } u) = \mathbf{0}.$$

(b) Conversely, suppose that D is simply connected and (11') holds everywhere in D. Let C be any simple closed path in D. Since D is simply connected, we can find a surface S in D bounded by C, Stokes's theorem (Sec. 8.10) is applicable, and we obtain

$$\int_C (f\,dx + g\,dy + h\,dz) = \int_C v_t\,ds = \iint_S (\text{curl } \mathbf{v})_n\,dA = 0$$

for proper direction on C and normal $\mathbf{n}$ on S. From this and Theorem 2 it follows that the integral (10) is independent of path in D. This completes the proof.

We see that in the case of a line integral

$$\int_C (f\,dx + g\,dy)$$

in the xy-plane, (11) reduces to the single relation

(11*) $$\frac{\partial g}{\partial x} = \frac{\partial f}{\partial y}.$$

The assumption that D be simply connected is essential and cannot be omitted. This can be seen from the following example.

Example 2.   Let

$$f = -\frac{y}{x^2 + y^2}, \qquad g = \frac{x}{x^2 + y^2}, \qquad h = 0.$$

Differentiation shows that (11*) is satisfied in any domain of the xy-plane not containing the origin, for example, in the domain $D: \frac{1}{2} < \sqrt{x^2 + y^2} < \frac{3}{2}$ shown in Fig. 171. Clearly, D is not simply connected. If the integral

$$I = \int_C (f\,dx + g\,dy) = \int_C \frac{-y\,dx + x\,dy}{x^2 + y^2}$$

Fig. 171. Example 2

were independent of path in D, then $I = 0$ on any closed curve in D, for example, on the circle $x^2 + y^2 = 1$. But setting $x = r \cos \theta$, $y = r \sin \theta$, and noting that the circle is represented by $r = 1$, we easily obtain

$$I = \int_0^{2\pi} d\theta = 2\pi,$$

the integration being taken once around the circle in the counterclockwise sense. In fact, since D is not simply connected, we cannot apply Theorem 3 and conclude that I is independent of path in D. Furthermore, we see that

$$f \, dx + g \, dy = du \qquad \text{where} \qquad u = \arctan \frac{y}{x} = \theta,$$

but u is not single-valued in D. On the other hand, if we take the so-called "principal value" of u, defined by $-\pi < u \leq \pi$, then u is not differentiable (not even continuous) at the points of the negative x-axis in D; this was required in connection with the exactness of a differential form.

If a line integral is independent of path, it can be evaluated by means of the formula (8), and the corresponding function $u(x, y, z)$ can be found by a single integration as follows.

Let P: (x_0, y_0, z_0) and Q: (x, y, z), as before. Then we obtain from (8)

$$u(Q) = u(x, y, z) = u(x_0, y_0, z_0) + \int_{(x_0, y_0, z_0)}^{(x, y, z)} (f \, dx + g \, dy + h \, dz);$$

integrating from P to Q along the line segment

$$(12) \qquad x = x_0 + (x - x_0)t, \qquad y = y_0 + (y - y_0)t, \qquad z = z_0 + (z - z_0)t$$

(where $0 \leq t \leq 1$), we obtain $dx = (x - x_0) \, dt$, etc. Therefore,

$$(13) \qquad u(x, y, z) = u(x_0, y_0, z_0) + \int_0^1 [(x - x_0)f + (y - y_0)g + (z - z_0)h] \, dt$$

where $f = f(x, y, z)$, etc., and x, y, z are given by (12).

Another way of determining $u(x, y, z)$ is to integrate one of the relations (5), for example, $\partial u / \partial x = f$, finding

$$u = \int f \, dx + a(y, z),$$

and then to determine the function $a(y, z)$ by differentiating this result with respect to y and z and using the other two relations (5); cf. the following example.

Example 3. Evaluate

$$I = \int_C [2xyz^2 \, dx + (x^2z^2 + z \cos yz) \, dy + (2x^2yz + y \cos yz) \, dz]$$

on any path C from P: $(0, 0, 1)$ to Q: $(1, \pi/4, 2)$.

From Theorem 3 it follows that the integral is independent of path in space. Since $f = 2xyz^2$, we obtain from the first relation (5)

$$(14) \qquad u = \int f \, dx = x^2yz^2 + a(y, z).$$

From this and the second relation (5) it follows that

$$\frac{\partial u}{\partial y} = x^2 z^2 + \frac{\partial a}{\partial y} = g = x^2 z^2 + z \cos yz.$$

Hence we must have

$$\frac{\partial a}{\partial y} = z \cos yz, \qquad a = \sin yz + c(z).$$

From this, the third relation (5), and (14) we obtain

$$\frac{\partial u}{\partial z} = 2x^2 yz + y \cos yz + \frac{dc}{dz} = h = 2x^2 yz + y \cos yz.$$

Hence $dc/dz = 0$, $c = const$ and, therefore,

$$u(x, y, z) = x^2 y z^2 + a = x^2 y z^2 + \sin yz + c.$$

This yields the result

$$I = [x^2 y z^2 + \sin yz + c]\Big|_P^Q = \pi + \sin \frac{\pi}{2} = \pi + 1.$$

Problems

Are the following differential forms exact?

1. $x^2 \, dx + y^2 \, dy + z \, dz$
2. $yz \, dx + xz \, dy + xy \, dz$
3. $y \cos xy \, dx + x \cos xy \, dy + \sin z \, dz$
4. $y \, dx + z \, dy + x \, dz$
5. $x^2 \, dx + z^2 \, dy + y^2 \, dz$
6. $6xy^3 \, dx + 9x^2 y^2 \, dy$

In each case show that the given differential form is exact and find a function u such that the form equals du.

7. $dx + dy + dz$
8. $3x^2 y^2 z \, dx + 2x^3 yz \, dy + x^3 y^2 \, dz$
9. $\cos x \, dx + 2 \sin y \, dy$
10. $2xy \, dx + (x^2 + z^2) \, dy + 2yz \, dz$
11. $2(x \, dx + y \, dy)/(x^2 + y^2)^2$
12. $e^x \cos y \, dx - e^x \sin y \, dy + 2 \, dz$
13. $2(y \, dy + z \, dz) \cosh (y^2 + z^2)$
14. $(y/z) \, dx + (x/z) \, dy - (xy/z^2) \, dz$

Show that the form under the integral sign is exact and evaluate:

15. $\displaystyle\int_{(2,0,0)}^{(1,2,3)} (x \, dx + y \, dy + z \, dz)$

16. $\displaystyle\int_{(1,1,2)}^{(3,-2,-1)} (yz \, dx + xz \, dy + xy \, dz)$

17. $\displaystyle\int_{(0,1,2)}^{(2,\pi,0)} [(y \, dx + x \, dy) \cos xy + dz]$

18. $\displaystyle\int_{(0,1,\pi/2)}^{(\pi,0,-\pi/2)} (\cos x \, dx + \sin z \, dz)$

19. $\displaystyle\int_{(0,0,0)}^{(-1,2,\pi)} (ye^{xy} \cos z \, dx + xe^{xy} \cos z \, dy - e^{xy} \sin z \, dz)$

20. Prove Theorem 2 for closed paths crossing a finite number of times.

CHAPTER **9**

Fourier Series and Integrals

Periodic functions occur frequently in engineering problems. Their representation in terms of simple periodic functions, such as sine and cosine, is a matter of great practical importance, which leads to Fourier series. These series, named after the French physicist JOSEPH FOURIER (1768–1830), are a very powerful tool in connection with various problems involving ordinary and partial differential equations.

In the present chapter we shall discuss basic concepts, facts, and techniques in connection with Fourier series. Illustrative examples and some important engineering applications will be included. Further applications will be presented in the next chapter on partial differential equations and boundary value problems.

The *theory* of Fourier series is rather complicated, but the *application* of these series is simple. Fourier series are, in a certain sense, more universal than Taylor series, because many *discontinuous* periodic functions of practical interest can be developed in Fourier series, but, of course, do not have Taylor series representations.

The last section of the present chapter is devoted to Fourier integrals. Applications to partial differential equations will be considered in the next chapter (in Sec. 10.6).

Prerequisite for this chapter: elementary integral calculus.
Sections which may be omitted in a shorter course: 9.6–9.8.
References: Appendix 1, Part D.
Answers to problems: Appendix 2.

9.1 Periodic Functions.
Trigonometric Series

A function $f(x)$ is said to be **periodic** if it is defined for all real x and if there is some positive number T such that

$$(1) \qquad\qquad f(x + T) = f(x) \qquad\qquad \text{for all } x.$$

Fig. 172. Periodic function

The number T is then called a **period**[1] of $f(x)$. The graph of such a function is obtained by periodic repetition of its graph in any interval of length T (Fig. 172).

From (1) it follows that, if n is any integer,

$$f(x + nT) = f(x) \qquad \text{for all } x,$$

so that any integral multiple nT ($n \neq 0$) of T is also a period. Furthermore, if $f(x)$ and $g(x)$ have the period T, then the function

$$h(x) = af(x) + bg(x) \qquad (a, b \text{ constant})$$

has the period T.

Familiar examples of periodic functions are the sine and cosine functions, and we note that the function $f = c = const$ is also a periodic function in the sense of the definition, because it satisfies (1) for any positive T.

Our problem in the first few sections of this chapter will be the representation of various functions of period 2π in terms of the simple functions

$$1, \qquad \cos x, \sin x, \qquad \cos 2x, \sin 2x, \cdots, \qquad \cos nx, \sin nx, \cdots$$

which have the period 2π (Fig. 173). The series which will arise in this connection will be of the form

$$(2) \qquad a_0 + a_1 \cos x + b_1 \sin x + a_2 \cos 2x + b_2 \sin 2x + \cdots,$$

where $a_0, a_1, a_2, \cdots, b_1, b_2, \cdots$ are real constants. Such a series is called a

Fig. 173. Cosine and sine functions having the period 2π

[1] If a periodic function $f(x)$ has a smallest period T (>0), this is often called the *primitive period* of $f(x)$. For example, the primitive periods of sin x and sin $2x$ are 2π and π, respectively. Examples of periodic functions without primitive period are $f = const$ and $f(x) = 0$ (x rational), $f(x) = 1$ otherwise.

trigonometric series, and the a_n and b_n are called the **coefficients** of the series. We see that each term of the series has the period 2π. Hence *if the series converges, its sum will be a function of period* 2π.

Periodic functions that occur in engineering problems are often rather complicated, and it is desirable to represent these functions in terms of simple periodic functions. We shall see that almost any periodic function $f(x)$ of period 2π that appears in applications, for example, in connection with vibrations, can be represented by a trigonometric series, and we shall derive formulas for the coefficients in (2) in terms of $f(x)$ such that (2) converges and has the sum $f(x)$. Later we shall extend our results to functions of arbitrary period; this extension will turn out to be quite simple.

Problems

Find the smallest positive period T of the following functions.

1. $\cos x$, $\sin x$, $\cos 2x$, $\sin 2x$, $\cos \pi x$, $\sin \pi x$, $\cos 2\pi x$, $\sin 2\pi x$

2. $\cos nx$, $\sin nx$, $\cos \dfrac{2\pi x}{k}$, $\sin \dfrac{2\pi x}{k}$, $\cos \dfrac{2\pi nx}{k}$, $\sin \dfrac{2\pi nx}{k}$

3. If T is a period of f, show that nT, $n = \pm 2, \pm 3, \cdots$ is a period of f.

4. Show that if f and g are periodic of period T, so is $h = af + bg$ (a, b constant). Thus all functions of period T form a vector space.

5. If $f(x)$ is a periodic function of x of period T, show that $f(ax)$, $a \neq 0$, is a periodic function of x of period T/a, and $f(x/b)$, $b \neq 0$, is a periodic function of x of period bT. Verify these results for $f(x) = \cos x$, $a = b = 2$.

6. Show that $f(x) = const$ is periodic of any period T.

Plot accurate graphs of the following functions.

7. $\sin x$, $\sin x + \dfrac{1}{3} \sin 3x$, $\sin x + \dfrac{1}{3} \sin 3x + \dfrac{1}{5} \sin 5x$,

$$f(x) = \begin{cases} -\pi/4 & \text{when } -\pi < x < 0 \\ \pi/4 & \text{when } \quad 0 < x < \pi \end{cases} \quad \text{and} \quad f(x + 2\pi) = f(x)$$

8. $\sin 2\pi x$, $\sin 2\pi x + \dfrac{1}{3} \sin 6\pi x$, $\sin 2\pi x + \dfrac{1}{3} \sin 6\pi x + \dfrac{1}{5} \sin 10\pi x$

9. $\sin x$, $\sin x - \dfrac{1}{2} \sin 2x$, $\sin x - \dfrac{1}{2} \sin 2x + \dfrac{1}{3} \sin 3x$,

$f(x) = x/2$ when $-\pi < x < \pi$ and $f(x + 2\pi) = f(x)$

10. $-\cos x$, $-\cos x + \dfrac{1}{4} \cos 2x$, $-\cos x + \dfrac{1}{4} \cos 2x - \dfrac{1}{9} \cos 3x$,

$f(x) = x^2/4 - \pi^2/12$ when $-\pi < x < \pi$ and $f(x + 2\pi) = f(x)$

11. $f(x) = x^3$ when $-\pi < x < \pi$ and $f(x + 2\pi) = f(x)$

12. $f(x) = e^{|x|}$ when $-\pi < x < \pi$ and $f(x + 2\pi) = f(x)$

Graph the following functions $f(x)$ which are periodic of period 2π and, for $-\pi < x < \pi$, are given by the formulas

13. $f(x) = \begin{cases} x^2 & \text{if } -\pi < x < 0 \\ 0 & \text{if } \quad 0 < x < \pi \end{cases}$ **14.** $f(x) = \begin{cases} 0 & \text{if } -\pi < x < 0 \\ \sin x & \text{if } \quad 0 < x < \pi \end{cases}$

15. $f(x) = \begin{cases} \pi + x & \text{if } -\pi < x < 0 \\ \pi - x & \text{if } \quad 0 < x < \pi \end{cases}$ **16.** $f(x) = \begin{cases} 1 & \text{if } -\pi < x < 0 \\ \cos x/2 & \text{if } \quad 0 < x < \pi \end{cases}$

Evaluate the following integrals where $n = 0, 1, 2, \cdots$. (These are typical examples of integrals which will be needed in our further work.)

17. $\displaystyle\int_0^\pi \sin nx \, dx$ **18.** $\displaystyle\int_{-\pi/2}^0 \cos nx \, dx$ **19.** $\displaystyle\int_{-\pi}^\pi x \sin nx \, dx$

20. $\displaystyle\int_{-\pi/2}^{\pi/2} x \sin nx \, dx$ **21.** $\displaystyle\int_{-\pi/2}^{\pi/2} x \cos nx \, dx$ **22.** $\displaystyle\int_0^\pi x \sin nx \, dx$

23. $\displaystyle\int_{-\pi}^0 e^x \sin nx \, dx$ **24.** $\displaystyle\int_0^\pi e^x \cos nx \, dx$ **25.** $\displaystyle\int_{-\pi}^\pi x^2 \cos nx \, dx$

9.2 Fourier Series.
Euler Formulas

Let us suppose that $f(x)$ is a periodic function with period 2π which can be represented by a trigonometric series

$$(1) \qquad f(x) = a_0 + \sum_{n=1}^\infty (a_n \cos nx + b_n \sin nx).$$

Given such a function $f(x)$, we want to determine the coefficients a_n and b_n in the corresponding series (1).

We first determine a_0. Integrating on both sides of (1) from $-\pi$ to π, we have

$$\int_{-\pi}^\pi f(x) \, dx = \int_{-\pi}^\pi \left[a_0 + \sum_{n=1}^\infty (a_n \cos nx + b_n \sin nx) \right] dx.$$

If term-by-term integration of the series is allowed,[2] then we obtain

$$\int_{-\pi}^\pi f(x) \, dx = a_0 \int_{-\pi}^\pi dx + \sum_{n=1}^\infty \left(a_n \int_{-\pi}^\pi \cos nx \, dx + b_n \int_{-\pi}^\pi \sin nx \, dx \right).$$

The first term on the right equals $2\pi a_0$. All the other integrals on the right are zero, as can be readily seen by performing the integrations. Hence our first result is

$$(2) \qquad a_0 = \frac{1}{2\pi} \int_{-\pi}^\pi f(x) \, dx.$$

We now determine $a_1, a_2, \cdots$ by a similar procedure. We multiply (1) by $\cos mx$, where m is any fixed positive integer, and then integrate from $-\pi$ to π, finding

$$(3) \qquad \int_{-\pi}^\pi f(x) \cos mx \, dx = \int_{-\pi}^\pi \left[a_0 + \sum_{n=1}^\infty (a_n \cos nx + b_n \sin nx) \right] \cos mx \, dx.$$

---

[2] This is justified, for instance, in the case of uniform convergence (cf. Theorem 3 in Sec. 15.4).

By term-by-term integration the right-hand side becomes

$$a_0 \int_{-\pi}^{\pi} \cos mx \, dx + \sum_{n=1}^{\infty} \left[a_n \int_{-\pi}^{\pi} \cos nx \cos mx \, dx + b_n \int_{-\pi}^{\pi} \sin nx \cos mx \, dx \right].$$

The first integral is zero. By applying (11) in Appendix 3 we obtain

$$\int_{-\pi}^{\pi} \cos nx \cos mx \, dx = \frac{1}{2} \int_{-\pi}^{\pi} \cos (n + m)x \, dx + \frac{1}{2} \int_{-\pi}^{\pi} \cos (n - m)x \, dx,$$

$$\int_{-\pi}^{\pi} \sin nx \cos mx \, dx = \frac{1}{2} \int_{-\pi}^{\pi} \sin (n + m)x \, dx + \frac{1}{2} \int_{-\pi}^{\pi} \sin (n - m)x \, dx.$$

Integration shows that the four terms on the right are zero, except for the last term in the first line which equals π when $n = m$. Since in (3) this term is multiplied by a_m, the right-hand side in (3) is equal to $a_m \pi$, and our second result is

$$(4) \qquad\qquad a_m = \frac{1}{\pi} \int_{-\pi}^{\pi} f(x) \cos mx \, dx, \qquad\qquad m = 1, 2, \cdots.$$

We finally determine $b_1, b_2, \cdots$ in (1). If we multiply (1) by $\sin mx$, where m is any fixed positive integer, and then integrate from $-\pi$ to π, we have

$$(5) \quad \int_{-\pi}^{\pi} f(x) \sin mx \, dx = \int_{-\pi}^{\pi} \left[a_0 + \sum_{n=1}^{\infty} (a_n \cos nx + b_n \sin nx) \right] \sin mx \, dx.$$

Integrating term-by-term, we see that the right-hand side becomes

$$a_0 \int_{-\pi}^{\pi} \sin mx \, dx + \sum_{n=1}^{\infty} \left[a_n \int_{-\pi}^{\pi} \cos nx \sin mx \, dx + b_n \int_{-\pi}^{\pi} \sin nx \sin mx \, dx \right].$$

The first integral is zero. The next integral is of the type considered before, and we know that it is zero for all $n = 1, 2, \cdots$. For the last integral we obtain

$$\int_{-\pi}^{\pi} \sin nx \sin mx \, dx = \frac{1}{2} \int_{-\pi}^{\pi} \cos (n - m)x \, dx - \frac{1}{2} \int_{-\pi}^{\pi} \cos (n + m)x \, dx.$$

The last term is zero. The first term on the right is zero when $n \neq m$ and is π when $n = m$. Since in (5) this term is multiplied by b_m, the right-hand side in (5) is equal to $b_m \pi$, and our last result is

$$b_m = \frac{1}{\pi} \int_{-\pi}^{\pi} f(x) \sin mx \, dx, \qquad\qquad m = 1, 2, \cdots.$$

Writing n in place of m we altogether have the so-called **Euler formulas**:

$$\textbf{(a)} \qquad a_0 = \frac{1}{2\pi} \int_{-\pi}^{\pi} f(x) \, dx$$

(6) $\qquad\qquad \textbf{(b)} \qquad a_n = \frac{1}{\pi} \int_{-\pi}^{\pi} f(x) \cos nx \, dx$

$$\qquad\qquad\qquad\qquad\qquad\qquad\qquad\qquad\qquad\qquad\qquad n = 1, 2, \cdots,$$

$$\textbf{(c)} \qquad b_n = \frac{1}{\pi} \int_{-\pi}^{\pi} f(x) \sin nx \, dx$$

Note that because of the periodicity of the integrands the interval of integration in (6) may be replaced by any other interval of length 2π, for instance, by the interval $0 \leq x \leq 2\pi$.

A periodic function $f(x)$ with period 2π being given, we may compute the a_n and b_n by (6) and form the trigonometric series

(7) $$a_0 + a_1 \cos x + b_1 \sin x + \cdots + a_n \cos nx + b_n \sin nx + \cdots.$$

This series is then called the **Fourier series** corresponding to $f(x)$, and its coefficients obtained from (6) are called the **Fourier coefficients** of $f(x)$.

From the definition of a definite integral it follows that, if $f(x)$ is continuous or merely piecewise continuous (continuous except for finitely many finite jumps in the interval of integration), the integrals in (6) exist and we may compute the Fourier coefficients of $f(x)$ by (6). The remaining question, whether the Fourier series thus obtained converges and has the sum $f(x)$, will be considered later in this section.

Let us illustrate the practical use of (6) by a simple example. Numerous other examples will occur in the following sections.

Example 1.   Find the Fourier coefficients of the periodic function $f(x)$ in Fig. 174a. The analytic representation is

$$f(x) = \begin{cases} -k & \text{when} & -\pi < x < 0 \\ k & \text{when} & 0 < x < \pi \end{cases} \quad \text{and} \quad f(x + 2\pi) = f(x).$$

Functions of this type may occur as external forces acting on mechanical systems, electromotive forces in electric circuits, etc.

From (6a) we obtain $a_0 = 0$. From (6b),

$$a_n = \frac{1}{\pi} \int_{-\pi}^{\pi} f(x) \cos nx \, dx = \frac{1}{\pi} \left[\int_{-\pi}^{0} (-k) \cos nx \, dx + \int_{0}^{\pi} k \cos nx \, dx \right]$$

$$= \frac{1}{\pi} \left[-k \frac{\sin nx}{n} \Big|_{-\pi}^{0} + k \frac{\sin nx}{n} \Big|_{0}^{\pi} \right] = 0$$

because $\sin nx = 0$ at $-\pi, 0$, and π for all $n = 1, 2, \cdots$. Similarly, from (6c) we obtain

$$b_n = \frac{1}{\pi} \int_{-\pi}^{\pi} f(x) \sin nx \, dx = \frac{1}{\pi} \left[\int_{-\pi}^{0} (-k) \sin nx \, dx + \int_{0}^{\pi} k \sin nx \, dx \right]$$

$$= \frac{1}{\pi} \left[k \frac{\cos nx}{n} \Big|_{-\pi}^{0} - k \frac{\cos nx}{n} \Big|_{0}^{\pi} \right].$$

Since $\cos(-\alpha) = \cos \alpha$ and $\cos 0 = 1$, this yields

$$b_n = \frac{k}{n\pi} [\cos 0 - \cos(-n\pi) - \cos n\pi + \cos 0] = \frac{2k}{n\pi}(1 - \cos n\pi).$$

Now, $\cos \pi = -1$, $\cos 2\pi = 1$, $\cos 3\pi = -1$ etc., in general,

$$\cos n\pi = \begin{cases} -1 & \text{for odd } n, \\ 1 & \text{for even } n, \end{cases} \quad \text{and thus} \quad 1 - \cos n\pi = \begin{cases} 2 & \text{for odd } n, \\ 0 & \text{for even } n. \end{cases}$$

Hence the Fourier coefficients b_n of our function are

$$b_1 = \frac{4k}{\pi}, \quad b_2 = 0, \quad b_3 = \frac{4k}{3\pi}, \quad b_4 = 0, \quad b_5 = \frac{4k}{5\pi}, \cdots,$$

and since the a_n are zero, the corresponding Fourier series is

$$\frac{4k}{\pi}\left(\sin x + \frac{1}{3}\sin 3x + \frac{1}{5}\sin 5x + \cdots\right).$$

The partial sums are

$$S_1 = \frac{4k}{\pi}\sin x, \qquad S_2 = \frac{4k}{\pi}\left(\sin x + \frac{1}{3}\sin 3x\right), \qquad \text{etc.,}$$

and their graphs in Fig. 174 seem to indicate that the series is convergent and has the sum $f(x)$, the given function. We notice that at $x = 0$ and $x = \pi$, the points of discontinuity of $f(x)$, all partial sums have the value zero, the arithmetic mean of the values $-k$ and k of our function.

(a) The given function $f(x)$

(b) The first three partial sums of the corresponding Fourier series.

Fig. 174. Example 1

Furthermore, assuming that $f(x)$ is the sum of the series and setting $x = \pi/2$, we have

$$f\left(\frac{\pi}{2}\right) = k = \frac{4k}{\pi}\left(1 - \frac{1}{3} + \frac{1}{5} - + \cdots\right)$$

or

$$1 - \frac{1}{3} + \frac{1}{5} - \frac{1}{7} + - \cdots = \frac{\pi}{4}.$$

This is a famous result by Leibniz (obtained about 1673 from geometrical considerations). It illustrates that the values of various series with constant terms can be obtained by evaluating Fourier series at specific points.

The class of functions which can be represented by Fourier series is surprisingly large and general. Corresponding sufficient conditions covering almost any conceivable engineering application are as follows.

Theorem 1.   *If a periodic function $f(x)$ with period 2π is piecewise continuous[3] in the interval $-\pi \leqq x \leqq \pi$ and has a left- and right-hand derivative[4] at each point of that interval, then the corresponding Fourier series (7) [with coefficients (6)] is convergent. Its sum is $f(x)$, except at a point x_0 at which $f(x)$ is discontinuous and the sum of the series is the average of the left- and right-hand limits[4] of $f(x)$ at x_0.*

Remark.   If the Fourier series *corresponding to* a function $f(x)$ converges with the sum $f(x)$ as characterized in Theorem 1, the series will be called the Fourier series *of $f(x)$*, we write

$$f(x) = a_0 + a_1 \cos x + b_1 \sin x + \cdots + a_n \cos nx + b_n \sin nx + \cdots,$$

and we say that $f(x)$ is *represented* by this Fourier series. Since insertion of parentheses in a convergent series yields a new convergent series having the same

[3] Definition in Sec. 4.1.
[4] The **left-hand limit** of $f(x)$ at x_0 is defined as the limit of $f(x)$ as x approaches x_0 from the left and is frequently denoted by $f(x_0 - 0)$. Thus

$$f(x_0 - 0) = \lim_{h \to 0} f(x_0 - h) \text{ as } h \to 0 \text{ through positive values.}$$

The **right-hand limit** is denoted by $f(x_0 + 0)$ and

$$f(x_0 + 0) = \lim_{h \to 0} f(x_0 + h) \text{ as } h \to 0 \text{ through positive values.}$$

The **left-** and **right-hand derivatives** of $f(x)$ at x_0 are defined as the limits of

$$\frac{f(x_0 - h) - f(x_0 - 0)}{-h} \quad \text{and} \quad \frac{f(x_0 + h) - f(x_0 + 0)}{h},$$

respectively, as $h \to 0$ through positive values. Of course if $f(x)$ is continuous at x_0 the last term in both numerators is simply $f(x_0)$.

Fig. 175. Left- and right-hand limits $f(1 - 0) = 1$, $f(1 + 0) = \frac{1}{2}$ of the function

$$f(x) = \begin{cases} x^2 & \text{when } x < 1 \\ x/2 & \text{when } x > 1 \end{cases}$$

sum as the original series (proof in Sec. 14.4), we may write more briefly

$$f(x) = a_0 + \sum_{n=1}^{\infty}(a_n \cos nx + b_n \sin nx).$$

Proof of convergence in Theorem 1 for a continuous function $f(x)$ having continuous first and second derivatives. Integrating $(6b)$ by parts, we obtain

$$a_n = \frac{1}{\pi}\int_{-\pi}^{\pi} f(x)\cos nx\,dx = \frac{f(x)\sin nx}{n\pi}\bigg|_{-\pi}^{\pi} - \frac{1}{n\pi}\int_{-\pi}^{\pi} f'(x)\sin nx\,dx.$$

The first term on the right is zero. A second integration by parts gives

$$a_n = \frac{f'(x)\cos nx}{n^2\pi}\bigg|_{-\pi}^{\pi} - \frac{1}{n^2\pi}\int_{-\pi}^{\pi} f''(x)\cos nx\,dx.$$

The first term on the right is zero because of the periodicity and continuity of $f'(x)$. Since f'' is continuous in the interval of integration, we have

$$|f''(x)| < M$$

for an appropriate constant M. Furthermore, $|\cos nx| \leq 1$. It follows that

$$|a_n| = \frac{1}{n^2\pi}\left|\int_{-\pi}^{\pi} f''(x)\cos nx\,dx\right| < \frac{1}{n^2\pi}\int_{-\pi}^{\pi} M\,dx = \frac{2M}{n^2}.$$

Similarly, $|b_n| < 2M/n^2$ for all n. Hence the absolute value of each term of the Fourier series corresponding to $f(x)$ is at most equal to the corresponding term of the series

$$|a_0| + 2M\left(1 + 1 + \frac{1}{2^2} + \frac{1}{2^2} + \frac{1}{3^2} + \frac{1}{3^2} + \cdots\right)$$

which is convergent. Hence that Fourier series converges and the proof is complete. (Readers already familiar with uniform convergence will see that, by the Weierstrass test in Sec. 15.4, the Fourier series converges uniformly, and our derivation of (6) by integrating term-by-term is then justified by Theorem 3 of that section.)

The proofs of convergence in the case of a piecewise continuous function $f(x)$ and of the last statement of Theorem 1 can be found in more advanced texts, for example in Ref. [D4].

Problems

Find the Fourier series of the function $f(x)$ which is assumed to have the period 2π and plot accurate graphs of $f(x)$ and the first three partial sums.[5]

3. $f(x) = \begin{cases} 1 & \text{if } -\pi/2 < x < \pi/2 \\ -1 & \text{if } \pi/2 < x < 3\pi/2 \end{cases}$

4. $f(x) = \begin{cases} 1 & \text{if } 0 < x < \pi/2 \\ 0 & \text{if } \pi/2 < x < 2\pi \end{cases}$

[5] That is, $a_0 + \sum_{n=1}^{N}(a_n \cos nx + b_n \sin nx)$ for $N = 1, 2, 3$.

5. $f(x) = \begin{cases} -1 & \text{if } -\pi/2 < x < 0 \\ 1 & \text{if } \quad 0 < x < \pi/2 \\ 0 & \text{if } \quad \pi/2 < x < 3\pi/2 \end{cases}$ **6.** $f(x) = \begin{cases} 0 & \text{if } -\pi < x < 0 \\ -1 & \text{if } \quad 0 < x < \pi/2 \\ 1 & \text{if } \pi/2 < x < \pi \end{cases}$

7. $f(x) = x \qquad (-\pi < x < \pi)$ **8.** $f(x) = x^3 \qquad (-\pi < x < \pi)$

9. $f(x) = x^2 \qquad (-\pi < x < \pi)$ **10.** $f(x) = |x| \qquad (-\pi < x < \pi)$

11. $f(x) = \begin{cases} \pi + x & \text{if } -\pi < x < 0 \\ \pi - x & \text{if } \quad 0 < x < \pi \end{cases}$ **12.** $f(x) = \begin{cases} 0 & \text{if } -\pi < x < 0 \\ x & \text{if } \quad 0 < x < \pi \end{cases}$

13. $f(x) = \begin{cases} x & \text{if } -\pi/2 < x < \pi/2 \\ \pi - x & \text{if } \quad \pi/2 < x < 3\pi/2 \end{cases}$ **14.** $f(x) = \begin{cases} x^2 & \text{if } -\pi/2 < x < \pi/2 \\ \pi^2/4 & \text{if } \quad \pi/2 < x < 3\pi/2 \end{cases}$

15. Verify the last statement in Theorem 1 concerning the discontinuities for the function in Prob. 1.

16. Show that if $f(x)$ has the Fourier coefficients a_n, b_n and $g(x)$ has the Fourier coefficients $a_n{}^*$, $b_n{}^*$, then $f(x) + g(x)$ has the Fourier coefficients $a_n + a_n{}^*$, $b_n + b_n{}^*$.

17. If $f(x)$ has the Fourier coefficients a_n, b_n, show that $kf(x)$, where k is a constant, has the Fourier coefficients ka_n, kb_n.

18. Obtain the Fourier series in Prob. 2 from that in Prob. 5.

19. Obtain the Fourier series in Prob. 10 from that in Prob. 11.

20. Obtain the Fourier series in Prob. 6 from that in Prob. 5.

9.3   Functions Having Arbitrary Period

The transition from functions having period 2π to functions having any period T is quite simple, because it can be effected by a change of scale.

In fact, suppose that $f(t)$ has period T. Then we can introduce a new variable x such that $f(t)$, as a function of x, has period 2π. If we set

$$(1) \qquad (a) \quad t = \frac{T}{2\pi} x \qquad \text{so that} \qquad (b) \quad x = \frac{2\pi}{T} t,$$

then $x = \pm\pi$ corresponds to $t = \pm T/2$. This means that f, as a function of x, has period 2π. Hence if f has a Fourier series, this series must be of the form

$$(2) \qquad f(t) = f\left(\frac{T}{2\pi} x\right) = a_0 + \sum_{n=1}^{\infty}(a_n \cos nx + b_n \sin nx),$$

with coefficients obtained from the Euler formulas (6) in the last section:

$$a_0 = \frac{1}{2\pi}\int_{-\pi}^{\pi} f\left(\frac{T}{2\pi} x\right) dx, \qquad a_n = \frac{1}{\pi}\int_{-\pi}^{\pi} f\left(\frac{T}{2\pi} x\right) \cos nx \, dx,$$

$$b_n = \frac{1}{\pi}\int_{-\pi}^{\pi} f\left(\frac{T}{2\pi} x\right) \sin nx \, dx.$$

We could use these formulas directly, but the change to t simplifies calculation. Since

$$x = \frac{2\pi}{T} t, \qquad \text{we have} \qquad dx = \frac{2\pi}{T} dt,$$

and the interval of integration on the x-axis corresponds to the interval

$$-\frac{T}{2} \leq t \leq \frac{T}{2}.$$

Consequently, we see that the Fourier coefficients of the periodic function $f(t)$ of period T are given by the **Euler formulas**

(3)

    (a) $a_0 = \dfrac{1}{T} \displaystyle\int_{-T/2}^{T/2} f(t)\, dt$

    (b) $a_n = \dfrac{2}{T} \displaystyle\int_{-T/2}^{T/2} f(t) \cos \dfrac{2n\pi t}{T}\, dt$

    (c) $b_n = \dfrac{2}{T} \displaystyle\int_{-T/2}^{T/2} f(t) \sin \dfrac{2n\pi t}{T}\, dt$

$n = 1, 2, \cdots.$

The Fourier series (2) with x expressed in terms of t becomes

(4)
$$f(t) = a_0 + \sum_{n=1}^{\infty} \left(a_n \cos \frac{2n\pi}{T} t + b_n \sin \frac{2n\pi}{T} t \right).$$

The interval of integration in (3) may be replaced by any interval of length T, for example, by the interval $0 \leq t \leq T$.

Example 1. Find the Fourier series of the function (cf. Fig. 176)

$$f(t) = \begin{cases} 0 & \text{when} \quad -2 < t < -1 \\ k & \text{when} \quad -1 < t < 1 \qquad T = 4. \\ 0 & \text{when} \quad 1 < t < 2 \end{cases}$$

From (3a) and (3b) we obtain

$$a_0 = \frac{1}{2} \int_0^2 f(t)\, dt = \frac{1}{2} \int_0^1 k\, dt = \frac{k}{2},$$

$$a_n = \int_0^2 f(t) \cos \frac{n\pi}{2} t\, dt = \int_0^1 k \cos \frac{n\pi}{2} t\, dt = \frac{2k}{n\pi} \sin \frac{n\pi}{2}.$$

Thus $a_n = 0$ when n is even, $a_n = 2k/n\pi$ when $n = 1, 5, 9, \cdots$, and $a_n = -2k/n\pi$ when $n = 3, 7, 11, \cdots$. From (3c) we find that $b_n = 0$ for $n = 1, 2, \cdots$. Hence the result is

$$f(t) = \frac{k}{2} + \frac{2k}{\pi} \left(\cos \frac{\pi}{2} t - \frac{1}{3} \cos \frac{3\pi}{2} t + \frac{1}{5} \cos \frac{5\pi}{2} t - + \cdots \right).$$

Fig. 176. Example 1

Fig. 177. Half-wave rectifier

Example 2 (Half-wave rectifier). A sinusoidal voltage $E \sin \omega t$ is passed through a half-wave rectifier which clips the negative portion of the wave (Fig. 177). Find the Fourier series of the resulting periodic function

$$u(t) = \begin{cases} 0 & \text{when} \quad -T/2 < t < 0, \\ E \sin \omega t & \text{when} \qquad 0 < t < T/2, \end{cases} \qquad T = \frac{2\pi}{\omega}.$$

Since $u = 0$ when $-T/2 < t < 0$, we obtain from (3*a*)

$$a_0 = \frac{\omega}{2\pi} \int_0^{\pi/\omega} E \sin \omega t \, dt = \frac{E}{\pi}$$

and from (3*b*), by using (11) in Appendix 3, with $x = \omega t$ and $y = n\omega t$,

$$a_n = \frac{\omega}{\pi} \int_0^{\pi/\omega} E \sin \omega t \cos n\omega t \, dt = \frac{\omega E}{2\pi} \int_0^{\pi/\omega} [\sin(1+n)\omega t + \sin(1-n)\omega t] \, dt.$$

When $n = 1$, the integral on the right is zero, and when $n = 2, 3, \cdots$, we readily obtain

$$a_n = \frac{\omega E}{2\pi} \left[-\frac{\cos(1+n)\omega t}{(1+n)\omega} - \frac{\cos(1-n)\omega t}{(1-n)\omega} \right]_0^{\pi/\omega}$$

$$= \frac{E}{2\pi} \left(\frac{-\cos(1+n)\pi + 1}{1+n} + \frac{-\cos(1-n)\pi + 1}{1-n} \right)$$

When n is odd, this is equal to zero, and for even n we have

$$a_n = \frac{E}{2\pi} \left(\frac{2}{1+n} + \frac{2}{1-n} \right) = -\frac{2E}{(n-1)(n+1)\pi} \qquad (n = 2, 4, \cdots).$$

In a similar fashion we find from (3*c*) that $b_1 = E/2$ and $b_n = 0$ for $n = 2, 3, \cdots$. Consequently,

$$u(t) = \frac{E}{\pi} + \frac{E}{2} \sin \omega t - \frac{2E}{\pi} \left(\frac{1}{1 \cdot 3} \cos 2\omega t + \frac{1}{3 \cdot 5} \cos 4\omega t + \cdots \right).$$

Problems

1. Show that every term in (4) has the period T.

2. Show that in (3) we may use any interval of length T as the interval of integration.

3. Obtain the Fourier series in Ex. 1 directly from that in Prob. 1, Sec. 9.2.

4. Find the Fourier series of the periodic function which is obtained by passing the voltage $v(t) = 2 \cos 100\pi t$ through a half-wave rectifier.

In each case find the Fourier series of the periodic function $f(t)$, of period T, and plot accurate graphs of $f(t)$ and the first three partial sums.

5. $f(t) = -1 \ (-1 < t < 0), \qquad f(t) = 1 \ (0 < t < 1), \qquad T = 2$

6. $f(t) = 1 \ (-1 < t < 1), \qquad f(t) = 0 \ (1 < t < 3), \qquad T = 4$

7. $f(\tau) = 0 \ (-2 < \tau < 0), \qquad f(\tau) = 1 \ (0 < \tau < 2), \qquad T = 4$

8. $f(t) = t^2 \ (-1 < t < 1), \qquad T = 2$

9. $f(t) = 0 \ (-1 < t < 0), \qquad f(t) = t \ (0 < t < 1), \qquad T = 2$

10. $f(t) = t \; (-1 < t < 1),$ $T = 2$

11. $f(t) = 1 - t^2 \; (-1 < t < 1),$ $T = 2$

12. $f(t) = \sin \pi t \; (0 < t < 1),$ $T = 1$

13. $f(t) = -1 \; (-1 < t < 0),$ $f(t) = 2t \; (0 < t < 1),$ $T = 2$

14. $f(t) = t \; (0 < t < 1),$ $f(t) = 1 - t \; (1 < t < 2),$ $T = 2$

15. Obtain the Fourier series in Prob. 5 directly from that in Ex. 1, Sec. 9.2.

16. Obtain the Fourier series in Prob. 8 directly from the result of Prob. 9, Sec. 9.2.

17. Obtain the Fourier series in Prob. 6 from that in Prob. 7 by setting $\tau = t + 1$.

18. Obtain the Fourier series in Prob. 11 from that in Prob. 8.

19. Obtain the Fourier series of $f(t) = (1 + t)^2 \; (-1 < t < 1)$, $T = 2$, from the results of Probs. 8 and 10.

20. Obtain the Fourier series in Prob. 13 from the results of Probs. 10 and 14.

9.4   *Even and Odd Functions*

Unnecessary work (and corresponding sources of errors) in determining Fourier coefficients of a function can be avoided if the function is odd or even.

We first remember that a function $y = g(x)$ is said to be **even** if

$$g(-x) = g(x) \qquad \text{for all } x,$$

The graph of such a function is symmetric with respect to the y-axis (Fig. 178). A function $h(x)$ is said to be **odd** if

$$h(-x) = -h(x) \qquad \text{for all } x,$$

(Cf. Fig. 179.) The function $\cos nx$ is even, while $\sin nx$ is odd.

If $g(x)$ is an even function, then

(1) $$\int_{-T/2}^{T/2} g(x) \, dx = 2 \int_{0}^{T/2} g(x) \, dx \qquad \text{(g even),}$$

If $h(x)$ is an odd function, then

(2) $$\int_{-T/2}^{T/2} h(x) \, dx = 0 \qquad \text{(h odd),}$$

The formulas (1) and (2) are obvious from the graphs of g and h, and we leave the formal proofs to the student.

Fig. 178. Even function   **Fig. 179.** Odd function

The product $q = gh$ of an even function g and an odd function h is odd, because

$$q(-x) = g(-x)h(-x) = g(x)[-h(x)] = -q(x).$$

Hence if $f(t)$ is even, then the integrand $f \sin (2n\pi t/T)$ in (3c) of the last section is odd, and $b_n = 0$. Similarly, if $f(t)$ is odd, then $f \cos (2n\pi t/T)$ in (3b), Sec. 9.3, is odd, and $a_n = 0$. From this and (1) we obtain

Theorem 1. *The Fourier series of an even function $f(t)$ having period T is a* **"Fourier cosine series"**

$$(3) \qquad\qquad f(t) = a_0 + \sum_{n=1}^{\infty} a_n \cos \frac{2n\pi}{T} t \qquad\qquad (f \text{ even})$$

with coefficients

$$(4) \qquad a_0 = \frac{2}{T} \int_0^{T/2} f(t) \, dt, \qquad a_n = \frac{4}{T} \int_0^{T/2} f(t) \cos \frac{2n\pi}{T} t \, dt, \qquad n = 1, 2, \cdots.$$

The Fourier series of an odd function $f(t)$ having period T is a **"Fourier sine series"**

$$(5) \qquad\qquad f(t) = \sum_{n=1}^{\infty} b_n \sin \frac{2n\pi}{T} t \qquad\qquad (f \text{ odd})$$

with coefficients

$$(6) \qquad\qquad b_n = \frac{4}{T} \int_0^{T/2} f(t) \sin \frac{2n\pi}{T} t \, dt.$$

In particular, the Fourier series of an even function $f(x)$ of period 2π is a Fourier cosine series

$$(3^*) \qquad f(x) = a_0 + a_1 \cos x + a_2 \cos 2x + a_3 \cos 3x + \cdots$$

with coefficients

$$(4^*) \qquad a_0 = \frac{1}{\pi} \int_0^\pi f(x) \, dx, \qquad a_n = \frac{2}{\pi} \int_0^\pi f(x) \cos nx \, dx, \qquad n = 1, 2, \cdots.$$

Similarly, the Fourier series of an odd function $f(x)$ of period 2π is a Fourier sine series

$$(5^*) \qquad f(x) = b_1 \sin x + b_2 \sin 2x + b_3 \sin 3x + \cdots$$

with coefficients

$$(6^*) \qquad\qquad b_n = \frac{2}{\pi} \int_0^\pi f(x) \sin nx \, dx, \qquad n = 1, 2, \cdots.$$

For instance, $f(x)$ in Ex. 1, Sec. 9.2, is odd and, therefore, is represented by a Fourier sine series.

Further simplifications result from

Theorem 2. *The Fourier coefficients of a sum $f_1 + f_2$ are the sums of corresponding Fourier coefficients of f_1 and f_2.*

Fig. 180. Example 1

Example 1 (*Rectangular pulse*). The function $f^*(x)$ in Fig. 180 is the sum of the function $f(x)$ in Ex. 1 of Sec. 9.2 and the constant k. Hence, from that example and Theorem 2 we conclude that

$$f^*(x) = k + \frac{4k}{\pi}\left(\sin x + \frac{1}{3}\sin 3x + \frac{1}{5}\sin 5x + \cdots\right).$$

Example 2 (*Saw-toothed wave*). Find the Fourier series of the function (Fig. 181)

$$f(x) = x + \pi \quad \text{when} \quad -\pi < x < \pi \quad \text{and} \quad f(x + 2\pi) = f(x).$$

We may write

$$f = f_1 + f_2 \quad \text{where} \quad f_1 = x \quad \text{and} \quad f_2 = \pi.$$

The Fourier coefficients of f_2 are zero, except for the first one (the constant term), which is π. Hence, by Theorem 2, the Fourier coefficients a_n, b_n are those of f_1, except for a_0 which is π.

(*a*) The function $f(x)$.

(*b*) Partial sums $S_n(x)$.

Fig. 181. Example 2

Since f_1 is odd, $a_n = 0$ for $n = 1, 2, \cdots$, and

$$b_n = \frac{2}{\pi} \int_0^\pi f_1(x) \sin nx \, dx = \frac{2}{\pi} \int_0^\pi x \sin nx \, dx.$$

Integrating by parts we obtain

$$b_n = \frac{2}{\pi} \left[\frac{-x \cos nx}{n} \Big|_0^\pi + \frac{1}{n} \int_0^\pi \cos nx \, dx \right] = -\frac{2}{n} \cos n\pi.$$

Hence $b_1 = 2$, $b_2 = -2/2$, $b_3 = 2/3$, $b_4 = -2/4, \cdots$, and the Fourier series representation of $f(x)$ is

$$f(x) = \pi + 2 \left(\sin x - \frac{1}{2} \sin 2x + \frac{1}{3} \sin 3x - + \cdots \right).$$

Problems

Are the following functions odd, even, or neither odd nor even?

1. e^x, e^{x^2}, $\sin nx$, $x \sin x$, $(\cos x)/x$, $\ln x$, $\sin x^2$, $\sin^2 x$
2. $|x|$, $x \cos nx$, $\sin x + \cos x$, c (a constant), $\ln (1 + e^x) - x/2$

Are the following functions $f(x)$ which are assumed to be periodic, of period 2π, even, odd or neither even nor odd?

3. $f(x) = x \quad (-\pi < x < \pi)$ **4.** $f(x) = x^2 \quad (0 < x < 2\pi)$

5. $f(x) = x|x| \, (-\pi < x < \pi)$ **6.** $f(x) = e^{|x|} \, (-\pi < x < \pi)$

7. $f(x) = \begin{cases} x & \text{if } 0 < x < \pi \\ 0 & \text{if } \pi < x < 2\pi \end{cases}$ **8.** $f(x) = \begin{cases} -x^3 & \text{if } -\pi < x < 0 \\ x^3 & \text{if } \quad 0 < x < \pi \end{cases}$

9. $f(x) = \begin{cases} x & \text{if } -\pi/2 < x < \pi/2 \\ 0 & \text{if } \quad \pi/2 < x < 3\pi/2 \end{cases}$ **10.** $f(x) = \begin{cases} x^2 & \text{if } -\pi/2 < x < \pi/2 \\ -x^2 & \text{if } \quad \pi/2 < x < 3\pi/2 \end{cases}$

11. Prove (1).
12. Find all functions which are both even and odd.

Prove:
13. Theorem 2.
14. The sum and the product of even functions are even functions.
15. The sum of odd functions is odd. The product of two odd functions is even.
16. If $f(x)$ is odd, then $|f(x)|$ and $f^2(x)$ are even functions.
17. If $f(x)$ is even, then $|f(x)|$, $f^2(x)$, and $f^3(x)$ are even functions.
18. If $g(x)$ is defined for all x, then the function $p(x) = [g(x) + g(-x)]/2$ is even and the function $q(x) = [g(x) - g(-x)]/2$ is odd.

Represent the following functions as the sum of an even and an odd function.

19. $1/(1 - x)$ **20.** $1/(1 - x)^2$ **21.** e^x **22.** $x/(x + 1)$

Find the Fourier series of the following functions which are assumed to have the period 2π.

23. $f(x) = \begin{cases} 1 & \text{if } -\pi/2 < x < \pi/2 \\ 0 & \text{if } \quad \pi/2 < x < 3\pi/2 \end{cases}$ **24.** $f(x) = \begin{cases} x & \text{if } 0 < x < \pi \\ \pi - x & \text{if } \pi < x < 2\pi \end{cases}$

25. $f(x) = \begin{cases} x & \text{if } -\pi/2 < x < \pi/2 \\ \pi - x & \text{if } \quad \pi/2 < x < 3\pi/2 \end{cases}$ **26.** $f(x) = \begin{cases} x^2 & \text{if } -\pi/2 < x < \pi/2 \\ \pi^2/4 & \text{if } \quad \pi/2 < x < 3\pi/2 \end{cases}$

27. $f(x) = \begin{cases} -x & \text{if } -\pi < x < 0 \\ x & \text{if } 0 < x < \pi \end{cases}$

28. $f(x) = \begin{cases} -x^2 & \text{if } -\pi < x < 0 \\ x^2 & \text{if } 0 < x < \pi \end{cases}$

29. $f(x) = x^2/4$ $\quad (-\pi < x < \pi)$

30. $f(x) = |\sin x|$ $\quad (-\pi < x < \pi)$

Show that

31. $1 - \frac{1}{3} + \frac{1}{5} - \frac{1}{7} + - \cdots = \dfrac{\pi}{4}$ $\quad$ (Use Prob. 23.)

32. $1 + \frac{1}{4} + \frac{1}{9} + \frac{1}{16} + \frac{1}{25} + \cdots = \dfrac{\pi^2}{6}$ $\quad$ (Use Prob. 29.)

33. $1 - \frac{1}{4} + \frac{1}{9} - \frac{1}{16} + - \cdots = \dfrac{\pi^2}{12}$ $\quad$ (Use Prob. 29.)

34. $1 + \dfrac{1}{3^2} + \dfrac{1}{5^2} + \dfrac{1}{7^2} + \cdots = \dfrac{\pi^2}{8}$ $\quad$ (Use Probs. 32, 33.)

35. Show that the familiar identity $\sin^3 x = \frac{3}{4} \sin x - \frac{1}{4} \sin 3x$ can be interpreted as a Fourier series expansion, and the same holds for $\cos^3 x = \frac{3}{4} \cos x + \frac{1}{4} \cos 3x$.

9.5 *Half-Range Expansions*

In various physical and engineering problems there is a practical need for applying Fourier series to functions $f(t)$ which are defined merely on some finite interval. Typical examples will arise in the next chapter (Secs. 10.3 and 10.5) in connection with partial differential equations. Then $f(t)$ will be defined on an interval $0 \leqq t \leqq l$, and on this interval we want to represent $f(t)$ by a Fourier series. For this purpose we may use Theorem 1 in the last section. We may let the interval $0 \leqq t \leqq l$ correspond to the interval of integration $0 \leqq t \leqq T/2$, that is, we may set $T/2 = l$ or $T = 2l$. Using (4) in Sec. 9.4, we then obtain a Fourier cosine series which represents an even periodic function $f_1(t)$ of period $T = 2l$. By construction, $f_1(t) = f(t)$ on $0 \leqq t \leqq l$. For this reason $f_1(t)$ is called the *even periodic extension* (or *continuation*) *of* $f(t)$ of period $T = 2l$. An illustration is given in Fig. 182*b*. From (3) and (4) in Sec. 9.4 with $T = 2l$ we see that

(1) $$f(t) = a_0 + \sum_{n=1}^{\infty} a_n \cos \frac{n\pi}{l} t \qquad (0 \leqq t \leqq l)$$

and the coefficients are

(2) $$a_0 = \frac{1}{l} \int_0^l f(t)\, dt, \qquad a_n = \frac{2}{l} \int_0^l f(t) \cos \frac{n\pi}{l} t\, dt, \qquad n = 1, 2, \cdots.$$

Instead of (4) in Sec. 9.4 we may equally well use (6) in Sec. 9.4 (with $T = 2l$, as before). Then we obtain a Fourier sine series, which represents an odd periodic function, say, $f_2(t)$, of period $T = 2l$. By construction, $f_2(t) = f(t)$ on $0 \leqq t \leqq l$, and $f_2(t)$ is called the *odd periodic extension of* $f(t)$ *of period* $T = 2l$. An illustration is given in Fig. 182*c*. From (5) and (6) in Sec. 9.4 we see that

(3) $$f(t) = \sum_{n=1}^{\infty} b_n \sin \frac{n\pi}{l} t \qquad (0 \leqq t \leqq l)$$

(a) The given function $f(t)$.

(b) $f(t)$ extended as an even periodic function of period $2l$.

(c) $f(t)$ extended as an odd periodic function of period $2l$.

Fig. 182. Periodic extensions

and the coefficients are

(4)
$$b_n = \frac{2}{l} \int_0^l f(t) \sin \frac{n\pi}{l} t \, dt, \qquad\qquad n = 1, 2, \cdots.$$

The series (1) and (3) with coefficients (2) and (4) are called **half-range expansions** of the given function $f(t)$.

Example 1. Find the half-range expansions of the function

$$f(t) = \begin{cases} \dfrac{2k}{l} t & \text{when} \quad 0 < t < \dfrac{l}{2} \\[2mm] \dfrac{2k}{l} (l - t) & \text{when} \quad \dfrac{l}{2} < t < l \end{cases}$$

shown in Fig. 183. From (2),

$$a_0 = \frac{1}{l} \left[\frac{2k}{l} \int_0^{l/2} t \, dt + \frac{2k}{l} \int_{l/2}^l (l - t) \, dt \right] = \frac{k}{2},$$

$$a_n = \frac{2}{l} \left[\frac{2k}{l} \int_0^{l/2} t \cos \frac{n\pi}{l} t \, dt + \frac{2k}{l} \int_{l/2}^l (l - t) \cos \frac{n\pi}{l} t \, dt \right].$$

Now by integration by parts,

$$\int_0^{l/2} t \cos \frac{n\pi}{l} t \, dt = \frac{lt}{n\pi} \sin \frac{n\pi}{l} t \Big|_0^{l/2} - \frac{l}{n\pi} \int_0^{l/2} \sin \frac{n\pi}{l} t \, dt$$

$$= \frac{l^2}{2n\pi} \sin \frac{n\pi}{2} + \frac{l^2}{n^2\pi^2} \left(\cos \frac{n\pi}{2} - 1 \right).$$

Similarly,

$$\int_{l/2}^l (l - t) \cos \frac{n\pi}{l} t \, dt = - \frac{l^2}{2n\pi} \sin \frac{n\pi}{2} - \frac{l^2}{n^2\pi^2} \left(\cos n\pi - \cos \frac{n\pi}{2} \right).$$

Fig. 183. The given function in Example 1

By inserting these two results we obtain

$$a_n = \frac{4k}{n^2\pi^2}\left(2\cos\frac{n\pi}{2} - \cos n\pi - 1\right).$$

Thus,

$$a_2 = -16k/2^2\pi^2, \qquad a_6 = -16k/6^2\pi^2, \qquad a_{10} = -16k/10^2\pi^2, \cdots$$

and $a_n = 0$ when $n \neq 2, 6, 10, 14, \cdots$. Hence the first half-range expansion of $f(t)$ is

$$f(t) = \frac{k}{2} - \frac{16k}{\pi^2}\left(\frac{1}{2^2}\cos\frac{2\pi}{l}t + \frac{1}{6^2}\cos\frac{6\pi}{l}t + \cdots\right).$$

This series represents the even periodic extension of $f(t)$ shown in Fig. 184a.
Similarly, from (4),

(5)
$$b_n = \frac{8k}{n^2\pi^2}\sin\frac{n\pi}{2}$$

and the other half-range expansion of $f(t)$ is

$$f(t) = \frac{8k}{\pi^2}\left(\frac{1}{1^2}\sin\frac{\pi}{l}t - \frac{1}{3^2}\sin\frac{3\pi}{l}t + \frac{1}{5^2}\sin\frac{5\pi}{l}t - + \cdots\right).$$

This series represents the odd periodic extension of $f(t)$ shown in Fig. 184b.

Problems

Represent the following functions $f(x)$ by a Fourier sine series and graph the corresponding
periodic extension of $f(x)$.

1. $f(x) = x \qquad (0 < x < \pi)$

2. $f(x) = 1 \qquad (0 < x < l)$

3. $f(x) = x \qquad (0 < x < 1)$

4. $f(x) = x^2 \qquad (0 < x < l)$

5. $f(x) = \begin{cases} x & \text{if } 0 < x < \pi/2 \\ \pi/2 & \text{if } \pi/2 < x < \pi \end{cases}$

6. $f(x) = \begin{cases} \pi/2 & \text{if } 0 < x < \pi/2 \\ \pi - x & \text{if } \pi/2 < x < \pi \end{cases}$

7. $f(x) = \begin{cases} x & \text{if } 0 < x < \pi/8 \\ \pi/4 - x & \text{if } \pi/8 < x < \pi/4 \end{cases}$

8. $f(x) = \begin{cases} \pi/2 - x & \text{if } 0 < x < \pi/2 \\ 0 & \text{if } \pi/2 < x < \pi \end{cases}$

(a) Even extension

(b) Odd extension

Fig. 184. Periodic extensions of $f(t)$ in Example 1

Represent the following functions $f(x)$ by a Fourier cosine series and graph the corresponding periodic extension of $f(x)$.

9. $f(x) = 1$ $(0 < x < l)$ **10.** $f(x) = x$ $(0 < x < l)$

11. $f(x) = x^2$ $(0 < x < l)$ **12.** $f(x) = e^x$ $(0 < x < l)$

13. $f(x) = \sin \dfrac{\pi}{l} x$ $(0 < x < l)$ **14.** $f(x) = \sin \dfrac{\pi x}{2l}$ $(0 < x < l)$

15. (**Complex form of the Fourier series, complex Fourier coefficients**) Using the formula $e^{i\theta} = \cos \theta + i \sin \theta$ (cf. Sec. 2.4), show that

$$\cos nx = \frac{1}{2}(e^{inx} + e^{-inx}), \qquad \sin nx = \frac{1}{2i}(e^{inx} - e^{-inx})$$

and the Fourier series

$$f(x) = a_0 + \sum_{n=1}^{\infty} (a_n \cos nx + b_n \sin nx)$$

may be written

(6)
$$f(x) = c_0 + \sum_{n=1}^{\infty} (c_n e^{inx} + k_n e^{-inx})$$

where $c_0 = a_0$, $c_n = (a_n - ib_n)/2$, $k_n = (a_n + ib_n)/2$, $n = 1, 2, \cdots$. Using (6), Sec. 9.2, show that

$$c_n = \frac{1}{2\pi} \int_{-\pi}^{\pi} f(x) e^{-inx}\, dx, \qquad k_n = \frac{1}{2\pi} \int_{-\pi}^{\pi} f(x) e^{inx}\, dx, \qquad n = 1, 2, \cdots.$$

Introducing the notation $k_n = c_{-n}$, show that (6) may be written

(7)
$$f(x) = \sum_{n=-\infty}^{\infty} c_n e^{inx}, \qquad c_n = \frac{1}{2\pi} \int_{-\pi}^{\pi} f(x) e^{-inx}\, dx, \qquad n = 0, \pm 1, \pm 2, \cdots.$$

This is the so-called *complex form* of the Fourier series, and the c_n are called the *complex Fourier coefficients* of $f(x)$.

16. Show that the complex Fourier coefficients of an odd function are pure imaginary and those of an even function are real.

Find the complex form of the Fourier series of the following periodic functions of period 2π. Obtain from it the corresponding real Fourier series and compare it with previous results.

17. $f(x) = x$ $(-\pi < x < \pi)$ (Cf. Prob. 1, this section.)

18. $f(x) = -1$ if $-\pi < x < 0$, $f(x) = 1$ if $0 < x < \pi$. (Cf. Ex. 1 in Sec. 9.2.)

19. Find the complex form of the Fourier series of $f(x) = e^x$ ˙ $(-\pi < x < \pi)$, $f(x + 2\pi) = f(x)$.

20. Obtain from the answer to Prob. 19 the corresponding real Fourier series.

9.6 Determination of Fourier Coefficients without Integration

In the previous sections relatively complicated and lengthy integrations led to relatively simple expressions for the Fourier coefficients a_n and b_n. This raises the question whether there might be a simpler way of obtaining Fourier coefficients. There is, and we want to show that the Fourier coefficients of a periodic function which is represented by polynomials can be obtained in terms of the jumps of the

Fig. 185. Jump of a function. (*a*) Positive jump. (*b*) Negative jump

Fig. 186. Example of a representation of the form (2) (with $m = 4$)

function and its derivatives. Of course, this is a big advantage, and the corresponding formulas are of great practical importance, because by applying them integrations are avoided (except for a_0, which must be determined as before).

By a *jump j* of a function $g(x)$ at a point x_0 we mean the difference between the right-hand and left-hand limits (Sec. 9.2) of $g(x)$ at x_0; that is (Fig. 185)

(1) $$j = g(x_0 + 0) - g(x_0 - 0).$$

It follows that an upward jump is positive and a downward jump is negative.

Let $f(x)$ be a function which has period 2π and is represented by polynomials $f_1, \cdots, p_m$ in the interval $-\pi < x < \pi$, say (Fig. 186)

(2) $$f(x) = \begin{cases} p_1(x) \text{ when } x_0 < x < x_1, \quad (x_0 = -\pi) \\ p_2(x) \text{ when } x_1 < x < x_2, \\ \quad \vdots \\ p_m(x) \text{ when } x_{m-1} < x < x_m \ (= \pi). \end{cases}$$

Then f may have jumps at $x_0, x_1, \cdots, x_m$, and the same is true for the derivatives $p', f'', \cdots$. We choose the following notation.

(3) $$\begin{aligned} j_s &= \text{jump of } f \text{ at } x_s, \\ j_s' &= \text{jump of } f' \text{ at } x_s, \qquad\qquad (s = 1, 2, \cdots, m) \\ j_s'' &= \text{jump of } f'' \text{ at } x_s, \qquad \text{etc.} \end{aligned}$$

Of course, if f is continuous at x_s, then $j_s = 0$, and for the derivatives the situation is similar, so that some of the numbers $j_s, j_s', \cdots$ in (3) may be zero.

Example 1. Let

$$f(x) = \begin{cases} 0 & \text{when} & -\pi < x < 0, \\ x^2 & \text{when} & 0 < x < \pi. \end{cases}$$

We plot graphs of f and its derivatives (Fig. 187),

$$f' = \begin{cases} 0 \\ 2x \end{cases} \qquad f'' = \begin{cases} 0 \\ 2 \end{cases} \qquad f''' = 0.$$

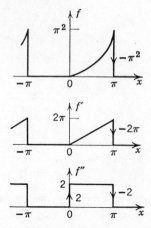

Fig. 187. $f(x)$ and derivatives in Example 1

We see that the jumps are

	Jump at $x_1 = 0$	Jump at $x_2 = \pi$
f	$j_1 = 0$	$j_2 = -\pi^2$
f'	$j_1' = 0$	$j_2' = -2\pi$
f''	$j_1'' = 2$	$j_2'' = -2$

Note that the jumps at $x = -\pi$ are not listed, because they are taken into account at $x = \pi$, the other end of the interval of periodicity.

To derive the desired formula for the Fourier coefficients $a_1, a_2, \cdots$ of f, given by (2), we start from the Euler formula (6*b*) in Sec. 9.2:

$$(4) \qquad \pi a_n = \int_{-\pi}^{\pi} f \cos nx \, dx.$$

Since f is represented by (2), we write the integral as the sum of m integrals:

$$(5) \qquad \pi a_n = \int_{x_0}^{x_1} + \int_{x_1}^{x_2} + \cdots + \int_{x_{m-1}}^{x_m} = \sum_{s=1}^{m} \int_{x_{s-1}}^{x_s} f \cos nx \, dx,$$

where $x_0 = -\pi$ and $x_m = \pi$. Integration by parts yields

$$(6) \qquad \int_{x_{s-1}}^{x_s} f \cos nx \, dx = \frac{f}{n} \sin nx \Big|_{x_{s-1}}^{x_s} - \frac{1}{n} \int_{x_{s-1}}^{x_s} f' \sin nx \, dx.$$

Now comes an important point: the evaluation of the first expression on the right. $f(x)$ may be discontinuous at x_s (Fig. 188), and we have to take the left-hand limit $f(x_s - 0)$ of f at x_s. Similarly, at x_{s-1} we have to take the right-hand limit $f(x_{s-1} + 0)$. Hence the first expression on the right-hand side of (6) equals

$$\frac{1}{n} [f(x_s - 0) \sin nx_s - f(x_{s-1} + 0) \sin nx_{s-1}].$$

Fig. 188. The formula (6)

Consequently, by inserting (6) into (5) and using the short notations $S_0 = \sin nx_0$, $S_1 = \sin nx_1$, etc., we obtain

(7)
$$\pi a_n = \frac{1}{n} [f(x_1 - 0)S_1 - f(x_0 + 0)S_0 + f(x_2 - 0)S_2 - f(x_1 + 0)S_1$$
$$+ \cdots + f(x_m - 0)S_m - f(x_{m-1} + 0)S_{m-1}] - \frac{1}{n} \sum_{s=1}^{m} \int_{x_{s-1}}^{x_s} f' \sin nx \, dx.$$

If we collect terms with the same S, the expression in the brackets becomes

(8)
$$-f(x_0 + 0)S_0 + [f(x_1 - 0) - f(x_1 + 0)]S_1$$
$$+ [f(x_2 - 0) - f(x_2 + 0)]S_2 + \cdots + f(x_m - 0)S_m.$$

The expressions in the brackets in (8) are the jumps of f, multiplied by -1. Furthermore, because of periodicity, $S_0 = S_m$ and $f(x_0) = f(x_m)$, so that we may combine the first and the last term in (8). Thus (8) is equal to

$$-j_1 S_1 - j_2 S_2 - \cdots - j_m S_m,$$

and from (7) we therefore have the intermediate result

(9)
$$\pi a_n = -\frac{1}{n} \sum_{s=1}^{m} j_s \sin nx_s - \frac{1}{n} \sum_{s=1}^{m} \int_{x_{s-1}}^{x_s} f' \sin nx \, dx.$$

By applying the same procedure to the integrals on the right-hand side of (9) we find

(10)
$$\sum_{s=1}^{m} \int_{x_{s-1}}^{x_s} f' \sin nx \, dx = \frac{1}{n} \sum_{s=1}^{m} j_s' \cos nx_s + \frac{1}{n} \sum_{s=1}^{m} \int_{x_{s-1}}^{x_s} f'' \cos nx \, dx.$$

Continuing in this fashion, we obtain integrals involving higher and higher derivatives of f. Since f is represented by polynomials and the $(r + 1)$th derivative of a polynomial of degree r is identically zero, we shall reach the point where no integrals are left, and this will happen after finitely many steps. By inserting (10) and the analogous formulas obtained in the further steps into (9) we obtain the desired formula

(11a)
$$a_n = \frac{1}{n\pi} \left[-\sum_{s=1}^{m} j_s \sin nx_s - \frac{1}{n} \sum_{s=1}^{m} j_s' \cos nx_s \right.$$
$$\left. + \frac{1}{n^2} \sum_{s=1}^{m} j_s'' \sin nx_s + \frac{1}{n^3} \sum_{s=1}^{m} j_s''' \cos nx_s - - + + \cdots \right],$$

where $n = 1, 2, \cdots$ (and a_0 must be obtained by integration as before). In precisely

Fig. 189. Example 2

the same fashion we obtain from (6c) in Sec. 9.2

(11b)
$$b_n = \frac{1}{n\pi} \left[\sum_{s=1}^{m} j_s \cos nx_s - \frac{1}{n} \sum_{s=1}^{m} j_s' \sin nx_s \right.$$

$$\left. - \frac{1}{n^2} \sum_{s=1}^{m} j_s'' \cos nx_s + \frac{1}{n^3} \sum_{s=1}^{m} j_s''' \sin nx_s + - - + + \cdots \right].$$

To avoid errors it is practical to graph $f(x)$ and its derivatives and list the jumps in a table as in Ex. 1.

Example 2.   Find the Fourier coefficients of the function (cf Fig. 189)

$$f(x) = \begin{cases} -k & \text{when} & -\pi < x < 0, \\ k & \text{when} & 0 < x < \pi. \end{cases}$$

We see that $f' \equiv 0$ and the jumps of f are

	Jump at $x_1 = 0$	Jump at $x_2 = \pi$
f	$j_1 = 2k$	$j_2 = -2k$

f is odd. Hence $a_n = 0$, and from (11b),

$$b_n = \frac{1}{n\pi} [j_1 \cos nx_1 + j_2 \cos nx_2] = \frac{1}{n\pi} [2k \cos 0 - 2k \cos n\pi]$$

$$= \frac{2k}{n\pi} (1 - \cos n\pi) = \begin{cases} 4k/n\pi & \text{for odd } n \\ 0 & \text{for even } n \end{cases} \quad \text{(cf. Ex. 1, Sec. 9.2).}$$

Example 3.   Find the Fourier series of the function in Ex. 1 of the current section. By integration,

$$a_0 = \frac{1}{2\pi} \int_0^{\pi} x^2 \, dx = \frac{\pi^2}{6}.$$

From (11a),

$$a_n = \frac{1}{n\pi} \left[\pi^2 \sin n\pi + \frac{2\pi}{n} \cos n\pi + \frac{1}{n^2} (2 \sin 0 - 2 \sin n\pi) \right] = \frac{2}{n^2} \cos n\pi.$$

Hence $a_1 = -2/1^2$, $a_2 = 2/2^2$, $a_3 = -2/3^2$, $\cdots$. From (11b) it follows that

$$b_n = \frac{1}{n\pi} \left[-\pi^2 \cos n\pi + \frac{2\pi}{n} \sin n\pi - \frac{1}{n^2} (2 \cos 0 - 2 \cos n\pi) \right]$$

$$= -\frac{\pi}{n} \cos n\pi + \frac{2}{n^3\pi} (\cos n\pi - 1).$$

Hence,

$$b_1 = \pi - \frac{4}{\pi}, \qquad b_2 = -\frac{\pi}{2}, \qquad b_3 = \frac{\pi}{3} - \frac{4}{3^3\pi}, \qquad b_4 = -\frac{\pi}{4}, \cdots,$$

and the Fourier series is

$$f(x) = \frac{\pi^2}{6} - 2\cos x + \left(\pi - \frac{4}{\pi}\right)\sin x + \frac{1}{2}\cos 2x - \frac{\pi}{2}\sin 2x + \cdots.$$

We mention that approximate values of the first twenty or thirty Fourier coefficients of a given function f may also be obtained from the graph of f by means of an instrument which is called a **harmonic analyzer** and is essentially a combination of a planimeter (or several planimeters) and some additional equipment (exchangeable gears, etc.). For details see Ref. [G15] in Appendix 1.

Problems

1. Obtain the Fourier series in Ex. 1 by the usual method and compare the amount of work with that in Ex. 3.
2. Derive (11b) from (6c) in Sec. 9.2.
3. Show that in the case of a function $f(t)$ having period T the formulas corresponding to (11) are

(12a)
$$a_n = \frac{1}{n\pi}\left[-\sum_{s=1}^{m} j_s \sin\frac{2n\pi}{T} t_s - \frac{T}{2n\pi}\sum_{s=1}^{m} j_s' \cos\frac{2n\pi}{T} t_s + \left(\frac{T}{2n\pi}\right)^2 \sum_{s=1}^{m} j_s'' \sin\frac{2n\pi}{T} t_s \right.$$
$$\left. + \left(\frac{T}{2n\pi}\right)^3 \sum_{s=1}^{m} j_s''' \cos\frac{2n\pi}{T} t_s - - + + \cdots \right],$$

(12b)
$$b_n = \frac{1}{n\pi}\left[\sum_{s=1}^{m} j_s \cos\frac{2n\pi}{T} t_s - \frac{T}{2n\pi}\sum_{s=1}^{m} j_s' \sin\frac{2n\pi}{T} t_s - \left(\frac{T}{2n\pi}\right)^2 \sum_{s=1}^{m} j_s'' \cos\frac{2n\pi}{T} t_s \right.$$
$$\left. + \left(\frac{T}{2n\pi}\right)^3 \sum_{s=1}^{m} j_s''' \sin\frac{2n\pi}{T} t_s + - - + + \cdots \right].$$

Using (11) or (12), find the Fourier series of the functions f in:
  4. Probs. 1–4, Sec. 9.2   5. Probs. 11–14, Sec. 9.2
  6. Probs. 5–7, Sec. 9.3   7. Probs. 8–11, Sec. 9.3

Using (11) or (12), find the Fourier sine series of
  8. f in Probs. 1 and 5 in Sec. 9.5   9. f in Probs. 7 and 8 in Sec. 9.5
  10. $f(x) = x^3 - x$ $(0 < x < \pi)$ 11. $f(x) = x(\pi^2 - x^2)$ $(0 < x < \pi)$

Using (12), find the Fourier cosine series of
  12. f in Probs. 10 and 11 in Sec. 9.5   13. $f(x) = x^3$ $(0 < x < l)$

14. Can the present method be applied to $f(x) = \cosh x$ $(-\pi < x < \pi)$, $f(x + 2\pi) = f(x)$?
15. Find the Fourier series of $f(x) = x^4$ $(-\pi < x < \pi)$, $f(x + 2\pi) = f(x)$ by the use of (6), Sec. 9.2, and (11), this section, and compare the amount of work.

Fig. 190. Vibrating system under consideration

9.7 Forced Oscillations

Fourier series have important applications in connection with differential equations. Let us consider an important practical problem involving an ordinary differential equation. (Partial differential equations will be considered in the next chapter.)

From Sec. 2.13 we know that forced oscillations of a body of mass m on a spring (cf. Fig. 190) are governed by the equation

$$(1) \qquad m\ddot{y} + c\dot{y} + ky = r(t),$$

where k is the spring modulus and c is the damping constant.

If the external force $r(t)$ is a sine or cosine function and the damping constant c is not zero, the steady-state solution represents a harmonic oscillation having the frequency of the external force.

We shall now see that if $r(t)$ is not a pure sine or cosine function but any other periodic function, then the steady-state solution will represent a superposition of harmonic oscillations having the frequency of $r(t)$ and multiples of this frequency. If the frequency of one of these oscillations is close to the resonant frequency of the vibrating system (cf. Sec. 2.13), then that oscillation may be the dominant part of the response of the system to the external force. Of course, this is quite surprising to an observer not familiar with the corresponding mathematical theory which is highly important in the study of vibrating systems and resonance. Let us illustrate the situation by an example.

Example 1. In (1), let $m = 1$ (gm), $c = 0.02$ (gm/sec), and $k = 25$ (gm/sec²), so that (1) becomes

$$(2) \qquad \ddot{y} + 0.02\dot{y} + 25y = r(t)$$

where $r(t)$ is measured in gm · cm/sec². Let (Fig. 191)

$$r(t) = \begin{cases} t + \dfrac{\pi}{2} & \text{when} \quad -\pi < t < 0, \\[2ex] -t + \dfrac{\pi}{2} & \text{when} \quad 0 < t < \pi, \end{cases} \qquad r(t + 2\pi) = r(t).$$

Find the steady-state solution $y(t)$.

Fig. 191. Force in Example 1

We represent $r(t)$ by a Fourier series, finding

$$(3) \qquad r(t) = \frac{4}{\pi}\left(\cos t + \frac{1}{3^2}\cos 3t + \frac{1}{5^2}\cos 5t + \cdots\right).$$

Then we consider the differential equation

$$(4) \qquad \ddot{y} + 0.02\dot{y} + 25y = \frac{4}{n^2\pi}\cos nt \qquad (n = 1, 3, \cdots)$$

whose right-hand side is a single term of the series (3). From Sec. 2.13 we know that the steady-state solution $y_n(t)$ of (4) is of the form

$$(5) \qquad y_n = A_n \cos nt + B_n \sin nt,$$

and by substituting this into (4) we find that

$$(6) \qquad A_n = \frac{4(25 - n^2)}{n^2\pi D}, \qquad B_n = \frac{0.08}{n\pi D}, \qquad \text{where} \qquad D = (25 - n^2)^2 + (0.02n)^2.$$

Since the differential equation (2) is linear, we may expect that the steady-state solution is

$$(7) \qquad y = y_1 + y_3 + y_5 + \cdots$$

where y_n is given by (5) and (6). In fact, this follows readily by substituting (7) into (2) and using the Fourier series of $r(t)$, provided that termwise differentiation of (7) is permissible. (Readers already familiar with the notion of uniform convergence [Sec. 15.4] may prove that (7) may be differentiated term by term.)

From (6) we find that the amplitude of (5) is

$$C_n = \sqrt{A_n{}^2 + B_n{}^2} = \frac{4}{n^2\pi\sqrt{D}}.$$

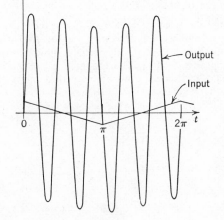

Fig. 192. Input and steady state output in Example 1

Numerical values are

$$C_1 = 0.0530$$
$$C_3 = 0.0088$$
$$C_5 = 0.5100$$
$$C_7 = 0.0011$$
$$C_9 = 0.0003$$

For $n = 5$ the quantity D is very small, the denominator of C_5 is small, and C_5 is so large that y_5 is the dominating term in (7). This implies that the steady-state motion is almost a harmonic oscillation whose frequency equals five times that of the exciting force (Fig. 192).

Problems

Find a general solution of the differential equation $\ddot{y} + \omega^2 y = r(t)$ where:

1. $r(t) = \sin t,$ $\omega = 0.5, 0.7, 0.9, 1.1, 1.5, 2.0, 10.0$

2. $r(t) = \cos \alpha t + \cos \beta t$ $(\omega^2 \neq \alpha^2, \beta^2)$

3. $r(t) = \sin t + \frac{1}{9} \sin 3t + \frac{1}{25} \sin 5t,$ $\omega = 0.5, 0.9, 1.1, 2, 2.9, 3.1, 4, 4.9, 5.1, 6, 8$

4. $r(t) = \sum_{n=1}^{N} a_n \cos nt,$ $|\omega| \neq 1, 2, \cdots, N$

5. $r(t) = \sum_{n=1}^{N} b_n \sin nt,$ $|\omega| \neq 1, 2, \cdots, N$

6. $r(t) = \begin{cases} t & \text{if } -\pi/2 < t < \pi/2 \\ \pi - t & \text{if } \pi/2 < t < 3\pi/2 \end{cases}$ and $r(t + 2\pi) = r(t),$ $|\omega| \neq 1, 3, 5, \cdots$

7. $r(t) = \dfrac{t^2}{4}$ when $-\pi < t < \pi$ and $r(t + 2\pi) = r(t),$ $|\omega| \neq 0, 1, 2, \cdots$

8. $r(t) = \dfrac{t}{12}(\pi^2 - t^2)$ when $-\pi < t < \pi$ and $r(t + 2\pi) = r(t),$ $|\omega| \neq 1, 2, \cdots$

Find the steady-state oscillation corresponding to $\ddot{y} + c\dot{y} + y = r(t)$ where $c > 0$ and

9. $r(t) = K \sin t$ **10.** $r(t) = \sin 3t$ **11.** $r(t) = a_n \cos nt$ **12.** $r(t) = \sum_{n=1}^{N} b_n \sin nt$

13. $r(t) = \begin{cases} \pi t/4 & \text{if } -\pi/2 < t < \pi/2 \\ \pi(\pi - t)/4 & \text{if } \pi/2 < t < 3\pi/2 \end{cases}$ and $r(t + 2\pi) = r(t)$

14. **(RLC-circuit)** Find the steady-state current $I(t)$ in the *RLC*-circuit in Fig. 193, where $R = 100$ ohms, $L = 10$ henrys, $C = 10^{-2}$ farads, $E(t) = 100t(\pi^2 - t^2)$ volts when $-\pi < t < \pi$ and $E(t + 2\pi) = E(t)$. Proceed as follows. Develop $E(t)$ in a Fourier series. $I(t)$ will appear in the form of a trigonometric series. Find the general formulas for the coefficients of this series. Compute numerical values of the first few coefficients. Graph the sum of the first few terms of that series.

Fig. 193. *RLC*-circuit

15. Same task as in Prob. 14 when R, L, C are as before, and

$$E(t) = \begin{cases} 100(\pi t + t^2) & \text{if } -\pi < t < 0 \\ 100(\pi t - t^2) & \text{if } 0 < t < \pi \end{cases} \quad \text{and} \quad E(t + 2\pi) = E(t).$$

9.8 Approximation by Trigonometric Polynomials. Square Error

Let $f(x)$ be a given function that has period 2π and can be represented by a Fourier series. Then the Nth partial sum of the Fourier series is an approximation of $f(x)$:

$$(1) \qquad f(x) \approx a_0 + \sum_{n=1}^{N} (a_n \cos nx + b_n \sin nx).$$

It is natural to ask whether (1) is the "best" approximation of f by a trigonometric polynomial

$$(2) \qquad F(x) = \alpha_0 + \sum_{n=1}^{N} (\alpha_n \cos nx + \beta_n \sin nx)$$

with the same fixed N, that is, an approximation for which the "error" is minimum.

Of course, we must first define what we mean by the error E of such an approximation. We want to choose a definition which measures the goodness of agreement between f and F *on the whole interval* $-\pi \leq x \leq \pi$. Obviously, the maximum of $|f - F|$ is not suitable for that purpose: in Fig. 194, the function F is a good approximation of f, but $|f - F|$ is large near x_0. We choose for our consideration

$$(3) \qquad E = \int_{-\pi}^{\pi} (f - F)^2 \, dx.$$

This is the so-called **total square error** of F relative to the function f over the interval $-\pi \leq x \leq \pi$. Clearly, $E \geq 0$.

N being fixed, we want to determine the coefficients in (2) such that E is minimum. We can write (3) in the form

$$(4) \qquad E = \int_{-\pi}^{\pi} f^2 \, dx - 2 \int_{-\pi}^{\pi} fF \, dx + \int_{-\pi}^{\pi} F^2 \, dx.$$

By inserting (2) into the last integral and evaluating the occurring integrals as in

Fig. 194. Error of approximation

Sec. 9.2 we readily obtain

$$\int_{-\pi}^{\pi} F^2 \, dx = \pi(2\alpha_0^2 + \alpha_1^2 + \cdots + \alpha_N^2 + \beta_1^2 + \cdots + \beta_N^2).$$

By inserting (2) into the second integral in (4) we see that the occurring integrals are those in the Euler formulas (6), Sec. 9.2, and

$$\int_{-\pi}^{\pi} fF \, dx = \pi(2\alpha_0 a_0 + \alpha_1 a_1 + \cdots + \alpha_N a_N + \beta_1 b_1 + \cdots + \beta_N b_N).$$

With these expressions (4) becomes

$$(5) \quad E = \int_{-\pi}^{\pi} f^2 \, dx - 2\pi\left[2\alpha_0 a_0 + \sum_{n=1}^{N}(\alpha_n a_n + \beta_n b_n)\right] + \pi\left[2\alpha_0^2 + \sum_{n=1}^{N}(\alpha_n^2 + \beta_n^2)\right].$$

If we take $\alpha_n = a_n$ and $\beta_n = b_n$ in (2), then from (5) we see that the square error corresponding to this particular choice of the coefficients of F is

$$(6) \qquad E^* = \int_{-\pi}^{\pi} f^2 \, dx - \pi\left[2a_0^2 + \sum_{n=1}^{N}(a_n^2 + b_n^2)\right].$$

By subtracting (6) from (5) we obtain

$$E - E^* = \pi\left\{2(\alpha_0 - a_0)^2 + \sum_{n=1}^{N}[(\alpha_n - a_n)^2 + (\beta_n - b_n)^2]\right\}.$$

The terms on the right are squares of real numbers and therefore nonnegative. Hence

$$E - E^* \geq 0 \qquad \text{or} \qquad E \geq E^*,$$

and $E = E^*$ if, and only if, $\alpha_0 = a_0, \cdots, \beta_N = b_N$. This proves

Theorem 1. *The total square error of F [cf. (2), N fixed] relative to f on the interval $-\pi \leq x \leq \pi$ is minimum if and only if the coefficients of F in (2) are the corresponding Fourier coefficients of f. This minimum value is given by (6).*

From (6) we see that E^* cannot increase as N increases, but may decrease. Hence, *with increasing N, the partial sums of the Fourier series of f yield better and better approximations for $f(x)$, considered from the viewpoint of the square error.*

Since $E^* \geq 0$ and (6) holds for any N, we obtain the important **Bessel inequality**[6]

$$(7) \qquad 2a_0^2 + \sum_{n=1}^{\infty}(a_n^2 + b_n^2) \leq \frac{1}{\pi}\int_{-\pi}^{\pi} f^2(x) \, dx$$

for the Fourier coefficients of any function f for which the integral on the right exists.

---

[6] It can be shown that for such a function f even the equality sign in (7) holds. Proof in Ref. [D4].

Problems

1. Let $f(x) = -1$ when $-\pi < x < 0$, $f(x) = 1$ when $0 < x < \pi$, and $f(x + 2\pi) = f(x)$. Find the function $F(x)$ of the form (2) for which the total square error (3) is minimum.
2. Compute the minimum square error in Prob. 1 for $N = 1, 3, 5, 7$. What is the smallest N such that $E^* \leq 0.2$?
3. Show that the minimum square error (6) is a monotone decreasing function of N.

In each case, find the function $F(x)$ of the form (2) for which the total square error E is minimum and compute this minimum value for $N = 1, 2, \cdots, 5$.

4. $f(x) = x \quad (-\pi < x < \pi)$ 5. $f(x) = x^2 \quad (-\pi < x < \pi)$

9.9   *The Fourier Integral*

Fourier series are powerful tools in treating various problems involving periodic functions. A first illustration of this fact was given in Sec. 9.7, and further important problems in connection with partial differential equations will be considered in Chap. 10. Since, of course, many practical problems do not involve periodic functions, it is desirable to generalize the method of Fourier series to include nonperiodic functions.

Roughly speaking, if we start with a periodic function $f_T(x)$ of period T and let T approach infinity, then the resulting function $f(x)$ is no longer periodic. Let us illustrate this with two examples.

Example 1.   Consider the function

$$f_T(x) = \begin{cases} 0 & \text{when} & -T/2 < x < -1 \\ 1 & \text{when} & -1 < x < 1 \\ 0 & \text{when} & 1 < x < T/2 \end{cases}$$

having period $T > 2$ (cf. Fig. 195 on p. 408). For $T \to \infty$ we obtain the function

$$f(x) = \lim_{T \to \infty} f_T(x) = \begin{cases} 1 & \text{when} & -1 < x < 1, \\ 0 & \text{otherwise.} \end{cases}$$

Example 2.   Let

$$f_T(x) = e^{-|x|} \quad \text{when} \quad -T/2 < x < T/2 \quad \text{and} \quad f_T(x + T) = f_T(x).$$

Then (cf. Fig. 196)

$$f(x) = \lim_{T \to \infty} f_T(x) = e^{-|x|}.$$

Now let us start from a periodic function $f_T(x)$ that has period T and can be represented by a Fourier series:

$$f_T(x) = a_0 + \sum_{n=1}^{\infty} \left(a_n \cos \frac{2n\pi}{T} x + b_n \sin \frac{2n\pi}{T} x \right).$$

Let us transform this representation as follows. If we use the short notation

$$w_n = \frac{2n\pi}{T}$$

Fig. 195. Example 1

and insert a_n and b_n according to the Euler formulas (3) in Sec. 9.3, denoting the variable of integration by v, we obtain

$$f_T(x) = \frac{1}{T} \int_{-T/2}^{T/2} f_T(v) \, dv + \frac{2}{T} \sum_{n=1}^{\infty} \left[\cos w_n x \int_{-T/2}^{T/2} f_T(v) \cos w_n v \, dv \right.$$

$$\left. + \sin w_n x \int_{-T/2}^{T/2} f_T(v) \sin w_n v \, dv \right].$$

Now,

$$w_{n+1} - w_n = \frac{2(n + 1)\pi}{T} - \frac{2n\pi}{T} = \frac{2\pi}{T},$$

and we set

$$\Delta w = w_{n+1} - w_n = \frac{2\pi}{T}.$$

Fig. 196. Example 2

Then $2/T = \Delta w/\pi$, and we may write that Fourier series in the form

$$f_T(x) = \frac{1}{T}\int_{-T/2}^{T/2} f_T(v)\,dv + \frac{1}{\pi}\sum_{n=1}^{\infty}\left[\cos(w_n x)\,\Delta w \int_{-T/2}^{T/2} f_T(v)\cos w_n v\,dv\right.$$
$$\left. + \sin(w_n x)\,\Delta w \int_{-T/2}^{T/2} f_T(v)\sin w_n v\,dv\right].$$

(1)

This representation is valid for any fixed T, arbitrarily large, but finite.

We now let T approach infinity and assume that the resulting nonperiodic function

$$f(x) = \lim_{T\to\infty} f_T(x)$$

is absolutely integrable on the x-axis, that is, the integral

$$(2) \qquad\qquad\qquad \int_{-\infty}^{\infty} |f(x)|\,dx$$

exists. Then $1/T \to 0$ and the value of the first term on the right side of (1) approaches zero. Furthermore, $\Delta w = 2\pi/T \to 0$ and it seems plausible that the infinite series in (1) becomes an integral from 0 to ∞, which represents $f(x)$, namely,

$$(3)\quad f(x) = \frac{1}{\pi}\int_0^{\infty}\left[\cos wx\int_{-\infty}^{\infty} f(v)\cos wv\,dv + \sin wx\int_{-\infty}^{\infty} f(v)\sin wv\,dv\right]dw.$$

If we introduce the short notations

$$(4) \qquad A(w) = \int_{-\infty}^{\infty} f(v)\cos wv\,dv, \qquad B(w) = \int_{-\infty}^{\infty} f(v)\sin wv\,dv,$$

this may be written in the form

$$(5) \qquad\qquad f(x) = \frac{1}{\pi}\int_0^{\infty}[A(w)\cos wx + B(w)\sin wx]\,dw.$$

This is a representation of $f(x)$ by a so-called **Fourier integral**.

It is clear that our naive approach merely *suggests* the representation (5), but by no means establishes it; in fact, the limit of the series in (1) as Δw approaches zero is not the definition of the integral (3). Sufficient conditions for the validity of (5) are as follows.

Theorem 1. *If $f(x)$ is piecewise continuous (cf. Sec. 4.1) in every finite interval and has a right- and left-hand derivative at every point (cf. Sec. 9.2) and the integral (2) exists, then $f(x)$ can be represented by a Fourier integral. At a point where $f(x)$ is discontinuous the value of the Fourier integral equals the average of the left- and right-hand limits of $f(x)$ at that point (cf. Sec. 9.2).* (Proof in Ref. [D3]; cf. Appendix 1.)

*Example 3 (**Single pulse, sine integral**).* Find the Fourier integral representation of the function (Fig. 197)

$$f(x) = \begin{cases} 1 & \text{when} & |x| < 1, \\ 0 & \text{when} & |x| > 1. \end{cases}$$

From (4) we obtain

$$A(w) = \int_{-\infty}^{\infty} f(v) \cos wv \, dv = \int_{-1}^{1} \cos wv \, dv = \frac{\sin wv}{w} \Big|_{-1}^{1} = \frac{2 \sin w}{w},$$

$$B(w) = \int_{-1}^{1} \sin wv \, dv = 0,$$

and (5) becomes

(6) $$f(x) = \frac{2}{\pi} \int_{0}^{\infty} \frac{\cos wx \sin w}{w} \, dw.$$

Fig. 197. Example 3

The average of the left- and right-hand limits of $f(x)$ at $x = 1$ is equal to $(1 + 0)/2$, that is, $1/2$. Hence, from (6) and Theorem 1 we obtain the desired answer

$$\int_{0}^{\infty} \frac{\cos wx \sin w}{w} \, dw = \begin{cases} \pi/2 & \text{when} & 0 \leqq x < 1, \\ \pi/4 & \text{when} & x = 1, \\ 0 & \text{when} & x > 1. \end{cases}$$

We mention that this integral is called **Dirichlet's**[7] **discontinuous factor.** Let us consider the case $x = 0$, which is of particular interest. When $x = 0$, then

(7) $$\int_{0}^{\infty} \frac{\sin w}{w} \, dw = \frac{\pi}{2}.$$

We see that this integral is the limit of the so-called **sine integral**

(8) $$\text{Si}(z) = \int_{0}^{z} \frac{\sin w}{w} \, dw$$

as $z \to \infty$ (z real). The graph of $\text{Si}(z)$ is shown in Fig. 198.

In the case of a Fourier series the graphs of the partial sums are approximation curves of the curve of the periodic function represented by the series. Similarly, in the case of the Fourier

Fig. 198. Sine integral

[7] PETER GUSTAV LEJEUNE DIRICHLET (1805–1859), German mathematician, known by his important research work on Fourier series and in number theory.

Fig. 199. The integral (9) for $a = 8$, 16, and 32

integral (5), approximations are obtained by replacing ∞ by numbers a. Hence the integral

$$(9) \qquad \int_0^a \frac{\cos wx \sin w}{w} \, dw$$

approximates the integral in (6) and therefore $f(x)$. Figure 199 shows oscillations near the points of discontinuity of $f(x)$.

We might expect that these oscillations disappear as a approaches infinity, but this is not true; with increasing a, they are shifted closer to the points $x = \pm 1$. This unexpected behavior, which also occurs in connection with Fourier series, is known as the **Gibbs**[8] **phenomenon**. It can be explained by representing (9) in terms of the sine integral as follows. Using (11) in Appendix 3, we first have

$$\frac{2}{\pi} \int_0^a \frac{\cos wx \sin w}{w} \, dw = \frac{1}{\pi} \int_0^a \frac{\sin (w + wx)}{w} \, dw + \frac{1}{\pi} \int_0^a \frac{\sin (w - wx)}{w} \, dw.$$

In the first integral on the right we set $w + wx = t$. Then $dw/w = dt/t$, and $0 \leqq w \leqq a$ corresponds to $0 \leqq t \leqq (x + 1)a$. In the last integral we set $w - wx = -t$. Then $dw/w = dt/t$, and the interval $0 \leqq w \leqq a$ corresponds to $0 \leqq t \leqq (x - 1)a$. Since $\sin (-t) = -\sin t$, we thus obtain

$$\frac{2}{\pi} \int_0^a \frac{\cos wx \sin w}{w} \, dw = \frac{1}{\pi} \int_0^{(x+1)a} \frac{\sin t}{t} \, dt - \frac{1}{\pi} \int_0^{(x-1)a} \frac{\sin t}{t} \, dt.$$

From this and (8) we see that our integral equals

$$\frac{1}{\pi} \text{Si}(a[x + 1]) - \frac{1}{\pi} \text{Si}(a[x - 1]),$$

[8] JOSIAH WILLARD GIBBS (1839–1903), American mathematician, whose work was of great importance to the development of vector analysis and mathematical physics.

and the oscillations in Fig. 199 result from those in Fig. 198. The increase of a amounts to a transformation of the scale on the axis and causes the shift of the oscillations. For further details see Ref. [D1] in Appendix 1.

If $f(x)$ is an even function, then $B(w) = 0$ in (4),

$$(10) \qquad\qquad A(w) = 2 \int_0^\infty f(v) \cos wv \, dv,$$

and (5) reduces to the simpler form

$$\textbf{(11)} \qquad\qquad f(x) = \frac{1}{\pi} \int_0^\infty A(w) \cos wx \, dw \qquad\qquad (f \text{ even}).$$

If $f(x)$ is odd, then $A(w) = 0$,

$$(12) \qquad\qquad B(w) = 2 \int_0^\infty f(v) \sin wv \, dv,$$

and (5) becomes

$$\textbf{(13)} \qquad\qquad f(x) = \frac{1}{\pi} \int_0^\infty B(w) \sin wx \, dw \qquad\qquad (f \text{ odd}).$$

These simplifications are quite similar to those in the case of a Fourier series discussed in Sec. 9.4.

Example 4 (Laplace integrals). Find the Fourier integral of

$$f(x) = e^{-kx} \quad \text{when} \quad x > 0 \quad \text{and} \quad f(-x) = f(x) \qquad (k > 0)$$

(cf. Fig. 196 in this section, where $k = 1$). Since f is even, we have from (10)

$$A(w) = 2 \int_0^\infty e^{-kv} \cos wv \, dv.$$

Now, by integration by parts,

$$\int e^{-kv} \cos wv \, dv = -\frac{k}{k^2 + w^2} e^{-kv} \left(-\frac{w}{k} \sin wv + \cos wv \right).$$

When $v = 0$, the expression on the right equals $-k/(k^2 + w^2)$; when v approaches infinity, it approaches zero because of the exponential factor. Thus

$$A(w) = \frac{2k}{k^2 + w^2},$$

and by substituting this in (11) we obtain the representation

$$f(x) = e^{-kx} = \frac{2k}{\pi} \int_0^\infty \frac{\cos wx}{k^2 + w^2} \, dw \qquad (x > 0, k > 0).$$

From this representation we see that

$$(14) \qquad\qquad \int_0^\infty \frac{\cos wx}{k^2 + w^2} \, dw = \frac{\pi}{2k} e^{-kx} \qquad (x > 0, k > 0).$$

Similarly, from the Fourier integral (13) of the odd function

$$f(x) = e^{-kx} \quad \text{when} \quad x > 0 \quad \text{and} \quad f(-x) = -f(x), \qquad (k > 0)$$

we obtain the result

(15)
$$\int_0^\infty \frac{w \sin wx}{k^2 + w^2} \, dw = \frac{\pi}{2} e^{-kx} \qquad\qquad (x > 0, k > 0).$$

The integrals (14) and (15) are the so-called *Laplace integrals*.

This example illustrates that the Fourier integral representation may be used for evaluating integrals.

Problems

Using the Fourier integral representation, show that

1. $\displaystyle\int_0^\infty \frac{\cos xw + w \sin xw}{1 + w^2} \, dw = \begin{cases} 0 & \text{when} \quad x < 0 \\ \pi/2 & \text{when} \quad x = 0 \\ \pi e^{-x} & \text{when} \quad x > 0 \end{cases}$ [Use (5).]

2. $\displaystyle\int_0^\infty \frac{w^3 \sin xw}{w^4 + 4} \, dw = \frac{\pi}{2} e^{-x} \cos x \qquad \text{when} \quad x > 0$ [Use (13).]

3. $\displaystyle\int_0^\infty \frac{\sin \pi w \sin xw}{1 - w^2} \, dw = \begin{cases} \dfrac{\pi}{2} \sin x & \text{when} \quad 0 \leqq x \leqq \pi \\ 0 & \text{when} \quad x > \pi \end{cases}$ [Use (13).]

4. $\displaystyle\int_0^\infty \frac{1 - \cos \pi w}{w} \sin xw \, dw = \begin{cases} \dfrac{\pi}{2} & \text{when} \quad 0 < x < \pi \\ 0 & \text{when} \quad x > \pi \end{cases}$ [Use (13).]

5. $\displaystyle\int_0^\infty \frac{\cos xw}{1 + w^2} \, dw = \frac{\pi}{2} e^{-x} \qquad (x > 0)$ [Use (11).]

6. $\displaystyle\int_0^\infty \frac{\sin w \cos xw}{w} \, dw = \begin{cases} \pi/2 & \text{when} \quad 0 \leqq x < 1 \\ \pi/4 & \text{when} \quad x = 1 \\ 0 & \text{when} \quad x > 1 \end{cases}$ [Use (11).]

7. $\displaystyle\int_0^\infty \frac{\cos (\pi w/2) \cos xw}{1 - w^2} \, dw = \begin{cases} \dfrac{\pi}{2} \cos x & \text{when} \quad |x| < \dfrac{\pi}{2} \\ 0 & \text{when} \quad |x| > \dfrac{\pi}{2} \end{cases}$ [Use (11).]

Represent the following functions $f(x)$ in the form (11).

8. $f(x) = \begin{cases} 1 & \text{when} \quad 0 < x < a \\ 0 & \text{when} \quad x > a \end{cases}$ **9.** $f(x) = \begin{cases} x^2 & \text{when} \quad 0 < x < a \\ 0 & \text{when} \quad x > a \end{cases}$

10. $f(x) = \begin{cases} x & \text{when} \quad 0 < x < 1 \\ 2 - x & \text{when} \quad 1 < x < 2 \\ 0 & \text{when} \quad x > 2 \end{cases}$ **11.** $f(x) = \begin{cases} x & \text{when} \quad 0 < x < a \\ 0 & \text{when} \quad x > a \end{cases}$

12. $f(x) = \dfrac{1}{1 + x^2}$ [cf. (14)]   **13.** $f(x) = e^{-x} + e^{-2x} \qquad (x > 0)$

If $f(x)$ has the representation (11), show that

14. $f(ax) = \dfrac{1}{\pi a} \displaystyle\int_0^\infty A\!\left(\dfrac{w}{a}\right) \cos xw \, dw \quad (a > 0)$

15. $x^2 f(x) = \dfrac{1}{\pi} \displaystyle\int_0^\infty A^*(w) \cos xw \, dw, \qquad A^* = -\dfrac{d^2 A}{dw^2}$

16. Solve Prob. 9 by applying the formula in Prob. 15 to the result of Prob. 8.

17. Show that $xf(x) = \dfrac{1}{\pi} \displaystyle\int_0^\infty B^*(w) \sin xw \, dw$ where $B^* = -\dfrac{dA}{dw}$ and A is given by (10).

18. Verify the formula in Prob. 17 for $f(x) = 1$ when $0 < x < a$ and $f(x) = 0$ when $x > a$.

19. Show that $f(x) = 1 \ (0 < x < \infty)$ cannot be represented by a Fourier integral.

20. **(Complex form of the Fourier integral, Fourier transform)**   Using (6) in Appendix 3, show that (5) can be written

(16) $$f(x) = \frac{1}{\pi} \int_0^\infty \left[\int_{-\infty}^\infty f(v) \cos (wx - wv) \, dv \right] dw.$$

Show that the integral from $-\infty$ to ∞ in (16) is an even function of w and (16) may be written

$$f(x) = \frac{1}{2\pi} \int_{-\infty}^\infty \left[\int_{-\infty}^\infty f(v) \cos (wx - wv) \, dv \right] dw.$$

Show that

(17) $$\frac{i}{2\pi} \int_{-\infty}^\infty \left[\int_{-\infty}^\infty f(v) \sin (wx - wv) \, dv \right] dw = 0,$$

so that by addition,

(18) $$f(x) = \frac{1}{2\pi} \int_{-\infty}^\infty \left[\int_{-\infty}^\infty f(v) e^{iw(x-v)} \, dv \right] dw.$$

This is the so-called *complex form* of the Fourier integral. Show that (18) can be written

(19) $$f(x) = \frac{1}{\sqrt{2\pi}} \int_{-\infty}^\infty C(w) e^{iwx} \, dw \quad \text{where} \quad C(w) = \frac{1}{\sqrt{2\pi}} \int_{-\infty}^\infty f(v) e^{-iwv} \, dv.$$

$f(x)$ is called the *Fourier transform* of $C(w)$, and $C(w)$ is called the *inverse Fourier transform* of $f(x)$. (Elaborate tables of Fourier transforms are listed in Part D of Appendix 1.)

CHAPTER 10

Partial Differential Equations

Partial differential equations arise in connection with various physical and geometrical problems when the functions involved depend on two or more independent variables. These variables may be the time and one or several coordinates in space. The present chapter will be devoted to some of the most important partial differential equations occurring in engineering applications. We shall derive these equations from physical principles and consider methods for solving initial and boundary value problems, that is, methods for obtaining solutions of those equations corresponding to the given physical situations.

In Sec. 10.1 we shall define the notion of a solution of a partial differential equation. Sections 10.2–10.4 will be devoted to the one-dimensional wave equation, governing the motion of a vibrating string. The heat equation will be considered in Secs. 10.5 and 10.6, the two-dimensional wave equation (vibrating membranes) in Secs. 10.7–10.10, and Laplace's equation in Secs. 10.11 and 10.12.

> *Prerequisites for this chapter:* ordinary linear differential equations (Chap. 2) and Fourier series (Chap. 9).
> *Sections which may be omitted in a shorter course:* 10.6, 10.9, 10.10.
> *References:* Appendix 1, Part E.
> *Answers to problems:* Appendix 2.

10.1 Basic Concepts

An equation involving one or more partial derivatives of an (unknown) function of two or more independent variables is called a **partial differential equation.** The order of the highest derivative is called the **order** of the equation.

Just as in the case of an ordinary differential equation, we say that a partial differential equation is **linear** if it is of the first degree in the dependent variable and its partial derivatives. If each term of such an equation contains either the dependent variable or one of its derivatives, the equation is said to be **homogeneous;** otherwise it is said to be **nonhomogeneous.**

Example 1. Some important linear partial differential equations of the second order are:

(1)
$$\frac{\partial^2 u}{\partial t^2} = c^2 \frac{\partial^2 u}{\partial x^2} \qquad\qquad (\textit{one-dimensional wave equation})$$

(2)
$$\frac{\partial u}{\partial t} = c^2 \frac{\partial^2 u}{\partial x^2} \qquad\qquad (\textit{one-dimensional heat equation})$$

(3)
$$\frac{\partial^2 u}{\partial x^2} + \frac{\partial^2 u}{\partial y^2} = 0 \qquad\qquad (\textit{two-dimensional Laplace equation})$$

(4)
$$\frac{\partial^2 u}{\partial x^2} + \frac{\partial^2 u}{\partial y^2} = f(x, y) \qquad\qquad (\textit{two-dimensional Poisson equation})$$

(5)
$$\frac{\partial^2 u}{\partial x^2} + \frac{\partial^2 u}{\partial y^2} + \frac{\partial^2 u}{\partial z^2} = 0 \qquad\qquad (\textit{three-dimensional Laplace equation});$$

here c is a constant, t is the time, and x, y, z are Cartesian coordinates. Equation (4) (with $f \not\equiv 0$) is nonhomogeneous, while the other equations are homogeneous.

A **solution** *of a partial differential equation in some region R of the space of the independent variables* is a function which has all the partial derivatives appearing in the equation in some domain containing R, and satisfies the equation everywhere in R. (Often one merely requires that that function is continuous at the boundary of R, has those derivatives in the interior of R, and satisfies the equation in the interior of R.)

In general, the totality of solutions of a partial differential equation is very large. For example, the functions

(6)
$$u = x^2 - y^2, \qquad u = e^x \cos y, \qquad u = \ln (x^2 + y^2),$$

which are entirely different from each other, are solutions of (3), as the student may verify. We shall see later that the unique solution of a partial differential equation corresponding to a given physical problem will be obtained by the use of additional information arising from the physical situation. For example, in some cases the values of the required solution of the problem on the boundary of some domain will be given ("**boundary conditions**"); in other cases when the time t is one of the variables, the values of the solution at $t = 0$ will be prescribed ("**initial conditions**").

We know that if an *ordinary* differential equation is linear and homogeneous, then from known solutions further solutions can be obtained by superposition. For a homogeneous linear *partial* differential equation the situation is quite similar. In fact, the following theorem holds.

Fundamental Theorem 1. *If u_1 and u_2 are any solutions of a linear homogeneous partial differential equation in some region, then*

$$u = c_1 u_1 + c_2 u_2,$$

where c_1 and c_2 are any constants, is also a solution of that equation in that region.

The proof of this important theorem is quite similar to that of Theorem 1 in Sec. 2.1 and is left to the student.

Problems

1. Prove Fundamental Theorem 1 for second-order differential equations in two and three independent variables.

2. Verify that the functions (6) are solutions of (3).

Verify that the following functions are solutions of Laplace's equation and sketch the surface in space represented by u.

3. $u = 2xy$
4. $u = e^x \sin y$
5. $u = \arctan (y/x)$
6. $u = x^3 - 3xy^2$
7. $u = \sin x \sinh y$
8. $u = x^4 - 6x^2y^2 + y^4$

9. Show that $u = 1/\sqrt{x^2 + y^2 + z^2}$ is a solution of Laplace's equation (5).

10. Verify that $u(x, y) = a \ln (x^2 + y^2) + b$ satisfies Laplace's equation (3) and determine a and b so that u satisfies the boundary conditions $u = 0$ on the circle $x^2 + y^2 = 1$ and $u = 5$ on the circle $x^2 + y^2 = 9$. Sketch the surface represented by u.

Verify that the following functions are solutions of the heat equation (2) for a suitable value of c.

11. $u = e^{-2t} \cos x$
12. $u = e^{-t} \sin 3x$
13. $u = e^{-4t} \cos \omega x$

Verify that the following functions are solutions of the wave equation (1) for a suitable value of c.

14. $u = x^2 + 4t^2$
15. $u = x^3 + 3xt^2$
16. $u = \sin \omega ct \sin \omega x$

17. Show that $u(x, t) = v(x + ct) + w(x - ct)$, where v and w are any twice differentiable functions, is a solution of the wave equation (1).

Relations to ordinary differential equations. If a partial differential equation involves derivatives with respect to one of the independent variables only, we may solve it like an ordinary differential equation, treating the other independent variables as parameters. Solve the following equations where $u = u(x, y)$.

18. $u_{xx} + 4u = 0$
19. $u_y + 2yu = 0$
20. $u_x = 2xyu$

Setting $u_x = p$, solve

21. $u_{xy} = 0$
22. $u_{xy} = u_x$
23. $u_{xy} + u_x = 0$

24. Solve the system of partial differential equations $u_{xx} = 0$, $u_{xy} = 0$, $u_{yy} = 0$, where $u = u(x, y)$.

25. Show that the solutions $z = z(x, y)$ of $yz_x - xz_y = 0$ represent surfaces of revolution. Give examples. *Hint.* Set $x = r \cos \theta$, $y = r \sin \theta$ and show that the equation becomes $z_\theta = 0$.

10.2 Vibrating String. One-Dimensional Wave Equation

As a first important partial differential equation, let us derive the equation governing small transverse vibrations of an elastic string, which is stretched to length l and then fixed at the end points. Suppose that the string is distorted and then at a certain instant, say, $t = 0$, it is released and allowed to vibrate. The problem is to determine the vibrations of the string, that is, to find its deflection $u(x, t)$ at any point x and at any time $t > 0$; cf. Fig. 200.

When deriving a differential equation corresponding to a given physical problem, we usually have to make simplifying assumptions in order that the resulting

Fig. 200. Vibrating string

equation does not become too complicated. We know this important fact from our considerations of ordinary differential equations, and for partial differential equations the situation is similar.

In our present case we make the following assumptions.

1. *The mass of the string per unit length is constant ("homogeneous string"). The string is perfectly elastic and does not offer any resistance to bending.*

2. *The tension caused by stretching the string before fixing it at the end points is so large that the action of the gravitational force on the string can be neglected.*

3. *The motion of the string is a small transverse vibration in a vertical plane, that is, each particle of the string moves strictly vertically, and the deflection and the slope at any point of the string are small in absolute value.*

These assumptions are such that we may expect that the solution $u(x, t)$ of the differential equation to be obtained will reasonably well describe small vibrations of the physical "nonidealized" string of small homogeneous mass under large tension.

To obtain the differential equation we consider the forces acting on a small portion of the string (Fig. 200). Since the string does not offer resistance to bending, the tension is tangential to the curve of the string at each point. Let T_1 and T_2 be the tensions at the end points P and Q of that portion. Since there is no motion in horizontal direction, the horizontal components of the tension must be constant. Using the notation shown in Fig. 200, we thus obtain

(1) $$T_1 \cos \alpha = T_2 \cos \beta = T = const.$$

In vertical direction we have two forces, namely the vertical components $-T_1 \sin \alpha$ and $T_2 \sin \beta$ of T_1 and T_2; here the minus sign appears because that component at P is directed downward. By Newton's second law the resultant of those two forces is equal to the mass $\rho \, \Delta x$ of the portion times the acceleration $\partial^2 u / \partial t^2$, evaluated at some point between x and $x + \Delta x$; here ρ is the mass of the undeflected string per unit length, and Δx is the length of the portion of the undeflected string. Hence

$$T_2 \sin \beta - T_1 \sin \alpha = \rho \, \Delta x \, \frac{\partial^2 u}{\partial t^2} \, .$$

By using (1) we obtain

(2)
$$\frac{T_2 \sin \beta}{T_2 \cos \beta} - \frac{T_1 \sin \alpha}{T_1 \cos \alpha} = \tan \beta - \tan \alpha = \frac{\rho \, \Delta x}{T} \frac{\partial^2 u}{\partial t^2} \, .$$

Now $\tan \alpha$ and $\tan \beta$ are the slopes of the curve of the string at x and $x + \Delta x$, that is,

$$\tan \alpha = \left(\frac{\partial u}{\partial x}\right)_x \quad \text{and} \quad \tan \beta = \left(\frac{\partial u}{\partial x}\right)_{x+\Delta x} \, .$$

Here we have to write *partial* derivatives because u depends also on t. Dividing (2) by Δx, we thus have

$$\frac{1}{\Delta x} \left[\left(\frac{\partial u}{\partial x}\right)_{x+\Delta x} - \left(\frac{\partial u}{\partial x}\right)_x \right] = \frac{\rho}{T} \frac{\partial^2 u}{\partial t^2} \, .$$

If we let Δx approach zero, we obtain the linear homogeneous partial differential equation

(3)
$$\frac{\partial^2 u}{\partial t^2} = c^2 \frac{\partial^2 u}{\partial x^2} \qquad\qquad c^2 = \frac{T}{\rho} \, .$$

This is the so-called **one-dimensional wave equation,** which governs our problem. The notation c^2 (instead of c) for the physical constant T/ρ has been chosen to indicate that this constant is positive.

Solutions of the equation will be obtained in the following section.

10.3 Separation of Variables (Product Method)

We have seen that the vibrations of an elastic string are governed by the one-dimensional wave equation

(1)
$$\frac{\partial^2 u}{\partial t^2} = c^2 \frac{\partial^2 u}{\partial x^2} \, ,$$

where $u(x, t)$ is the deflection of the string. Since the string is fixed at the ends $x = 0$ and $x = l$, we have the two **boundary conditions**

(2)
$$u(0, t) = 0, \qquad u(l, t) = 0 \qquad \text{for all } t.$$

The form of the motion of the string will depend on the initial deflection (deflection at $t = 0$) and on the initial velocity (velocity at $t = 0$). Denoting the initial deflection by $f(x)$ and the initial velocity by $g(x)$, we thus obtain the two **initial conditions**

(3)
$$u(x, 0) = f(x)$$

and

(4)
$$\frac{\partial u}{\partial t}\bigg|_{t=0} = g(x).$$

Our problem is now to find a solution of (1) satisfying the conditions (2)–(4). We shall proceed step by step, as follows.

First step. By applying the so-called *product method*, or *method of separating variables*, we shall obtain two ordinary differential equations.

Second step. We shall determine solutions of those equations that satisfy the boundary conditions.

Third step. Those solutions will be composed so that the result will be a solution of the wave equation (1), satisfying also the given initial conditions.

Let us start in detail with the first step.

First step. The product method yields solutions of (1) of the form

(5) $$u(x, t) = F(x)G(t)$$

which are a product of two functions, each depending only on one of the variables x and t. We shall see later that this method has various applications in engineering mathematics. By differentiating (5) we obtain

$$\frac{\partial^2 u}{\partial t^2} = F\ddot{G} \qquad \text{and} \qquad \frac{\partial^2 u}{\partial x^2} = F''G,$$

where dots denote derivatives with respect to t and primes derivatives with respect to x. By inserting this into (1) we have

$$F\ddot{G} = c^2 F''G.$$

Dividing by $c^2 FG$, we find

$$\frac{\ddot{G}}{c^2 G} = \frac{F''}{F}.$$

The expression on the left involves functions depending only on t while the expression on the right involves functions depending only on x. Hence both expressions must be equal to a constant, say, k, because if the expression on the left is not constant, then changing t will presumably change the value of this expression but certainly not that on the right, since the latter does not depend on t. Similarly, if the expression on the right is not constant, changing x will presumably change the value of this expression but certainly not that on the left. Thus

$$\frac{\ddot{G}}{c^2 G} = \frac{F''}{F} = k.$$

This yields immediately the two ordinary linear differential equations

(6) $$F'' - kF = 0$$

and

(7) $$\ddot{G} - c^2 kG = 0.$$

In these equations, k is still arbitrary.

Second step. We shall now determine solutions F and G of (6) and (7) so that $u = FG$ satisfies (2), that is,

$$u(0, t) = F(0)G(t) = 0, \qquad u(l, t) = F(l)G(t) = 0 \qquad \text{for all } t.$$

Clearly, if $G \equiv 0$, then $u \equiv 0$, which is of no interest. Thus $G \not\equiv 0$ and then

(8) $\qquad\qquad$ (a) $F(0) = 0$ $\qquad$ (b) $F(l) = 0$.

For $k = 0$ the general solution of (6) is $F = ax + b$, and from (8) we obtain $a = b = 0$. Hence $F \equiv 0$, which is of no interest because then $u \equiv 0$. For positive $k = \mu^2$ the general solution of (6) is

$$F = Ae^{\mu x} + Be^{-\mu x},$$

and from (8) we obtain $F \equiv 0$, as before. Hence we are left with the possibility of choosing k negative, say, $k = -p^2$. Then (6) takes the form

$$F'' + p^2 F = 0,$$

and the general solution is

$$F(x) = A \cos px + B \sin px.$$

From this and (8) we have

$$F(0) = A = 0 \qquad \text{and then} \qquad F(l) = B \sin pl = 0.$$

We must take $B \neq 0$ since otherwise $F \equiv 0$. Hence $\sin pl = 0$, that is,

(9) $\qquad\qquad\qquad pl = n\pi \qquad \text{or} \qquad p = \dfrac{n\pi}{l} \qquad\qquad$ (*n* integral).

Setting $B = 1$, we thus obtain infinitely many solutions $F(x) = F_n(x)$,

(10) $\qquad\qquad\qquad F_n(x) = \sin \dfrac{n\pi}{l} x \qquad\qquad n = 1, 2, \cdots,$

which satisfy (8). (For negative integral n we obtain essentially the same solutions, except for a minus sign, because $\sin(-\alpha) = -\sin \alpha$.)

k is now restricted to the values $k = -p^2 = -(n\pi/l)^2$, resulting from (9). For these k the equation (7) takes the form

$$\ddot{G} + \lambda_n^2 G = 0 \qquad \text{where} \qquad \lambda_n = \dfrac{cn\pi}{l}.$$

The general solution is

$$G_n(t) = B_n \cos \lambda_n t + B_n{}^* \sin \lambda_n t.$$

Hence the functions $u_n(x, t) = F_n(x)G_n(t)$, written out

(11) $\qquad u_n(x, t) = (B_n \cos \lambda_n t + B_n{}^* \sin \lambda_n t) \sin \dfrac{n\pi}{l} x \qquad (n = 1, 2, \cdots),$

are solutions of (1), satisfying the boundary conditions (2). These functions are called the **eigenfunctions,** or *characteristic functions*, and the values $\lambda_n = cn\pi/l$

Fig. 201. Normal modes of the vibrating string

are called the **eigenvalues,** or *characteristic values*, of the vibrating string. The set $\lambda_1, \lambda_2, \cdots$ is called the *spectrum*.

We see that each u_n represents a harmonic motion having the frequency $\lambda_n/2\pi = cn/2l$ cycles per unit time. The motion is called the nth **normal mode** of the string. The first normal mode is known as the *fundamental mode* ($n = 1$), and the others as *overtones;* musically they give the octave, octave plus fifth, etc. Since in (11)

$$\sin \frac{n\pi x}{l} = 0 \qquad \text{at} \qquad x = \frac{l}{n}, \frac{2l}{n}, \cdots, \frac{n-1}{n} l,$$

the nth normal mode has $n - 1$ so-called **nodes,** that is, points of the string which do not move (Fig. 201).

Figure 202 shows the second normal mode for various values of t. At any instant the string has the form of a sine wave. When the left part of the string is moving downward the other half is moving upward, and conversely. For the other modes the situation is similar.

Third Step. Clearly, a single solution $u_n(x, t)$ will, in general, not satisfy the initial conditions (3) and (4). Now, since the equation (1) is linear and homogeneous, it follows from Fundamental Theorem 1 in Sec. 10.1 that the sum of finitely many solutions u_n is a solution of (1). To obtain a solution that satisfies (3) and (4), we consider the infinite series

(12) $$u(x, t) = \sum_{n=1}^{\infty} u_n(x, t) = \sum_{n=1}^{\infty} (B_n \cos \lambda_n t + B_n{}^* \sin \lambda_n t) \sin \frac{n\pi}{l} x.$$

From this and (3) it follows that

(13) $$u(x, 0) = \sum_{n=1}^{\infty} B_n \sin \frac{n\pi}{l} x = f(x).$$

Hence, in order that (12) satisfy (3), the coefficients B_n must be chosen so that $u(x, 0)$ becomes a half-range expansion of $f(x)$, namely, the Fourier sine series of

Fig. 202. Second normal mode for various values of t

$f(x)$; that is [cf. (4) in Sec. 9.5]

(14) $$B_n = \frac{2}{l} \int_0^l f(x) \sin \frac{n\pi x}{l}\, dx, \qquad\qquad n = 1, 2, \cdots.$$

Similarly, by differentiating (12) with respect to t and using (4) we find

$$\left.\frac{\partial u}{\partial t}\right|_{t=0} = \left[\sum_{n=1}^{\infty} (-B_n \lambda_n \sin \lambda_n t + B_n{}^* \lambda_n \cos \lambda_n t) \sin \frac{n\pi x}{l}\right]_{t=0}$$

$$= \sum_{n=1}^{\infty} B_n{}^* \lambda_n \sin \frac{n\pi x}{l} = g(x).$$

Hence, in order that (12) satisfy (4), the coefficients $B_n{}^*$ must be chosen so that, for $t = 0$, $\partial u/\partial t$ becomes the Fourier sine series of $g(x)$; thus, by (4) in Sec. 9.5,

$$B_n{}^* \lambda_n = \frac{2}{l} \int_0^l g(x) \sin \frac{n\pi x}{l}\, dx$$

or, since $\lambda_n = cn\pi/l$,

(15) $$B_n{}^* = \frac{2}{cn\pi} \int_0^l g(x) \sin \frac{n\pi x}{l}\, dx, \qquad\qquad n = 1, 2, \cdots.$$

It follows that $u(x, t)$, given by (12) with coefficients (14) and (15), is a solution of (1) that satisfies the conditions (2)–(4), provided that the series (12) converges and also that the series obtained by differentiating (12) twice (termwise) with respect to x and t, converge and have the sums $\partial^2 u/\partial x^2$ and $\partial^2 u/\partial t^2$, respectively, which are continuous.

Hence the solution (12) is at first a purely formal expression, and we shall now establish it: For the sake of simplicity we consider only the case when the initial velocity $g(x)$ is identically zero. Then the $B_n{}^*$ are zero, and (12) reduces to the form

(16) $$u(x, t) = \sum_{n=1}^{\infty} B_n \cos \lambda_n t \sin \frac{n\pi x}{l}, \qquad \lambda_n = \frac{cn\pi}{l}.$$

It is possible to *sum* this series, that is, to write the result in a closed or finite form. We have [cf. (11) in Appendix 3]

$$\cos \frac{cn\pi}{l} t \sin \frac{n\pi}{l} x = \frac{1}{2}\left[\sin \left\{\frac{n\pi}{l}(x - ct)\right\} + \sin \left\{\frac{n\pi}{l}(x + ct)\right\}\right].$$

Consequently, we may write (16) in the form

$$u(x, t) = \frac{1}{2} \sum_{n=1}^{\infty} B_n \sin \left\{\frac{n\pi}{l}(x - ct)\right\} + \frac{1}{2} \sum_{n=1}^{\infty} B_n \sin \left\{\frac{n\pi}{l}(x + ct)\right\}.$$

These two series are those obtained by substituting $x - ct$ and $x + ct$, respectively, for the variable x in the Fourier sine series (13) for $f(x)$. Therefore,

(17) $$u(x, t) = \tfrac{1}{2}[f^*(x - ct) + f^*(x + ct)],$$

where f^* is the odd periodic extension of f with the period $2l$ (Fig. 203). Since the

Fig. 203. Odd periodic extension of $f(x)$

initial deflection $f(x)$ is continuous on the interval $0 \leq x \leq l$ and zero at the end points, it follows from (17) that $u(x, t)$ is a continuous function of both variables x and t for all values of the variables. By differentiating (17) we see that $u(x, t)$ is a solution of (1), provided $f(x)$ is twice differentiable on the interval $0 < x < l$, and has one-sided second derivatives at $x = 0$ and $x = l$, which are zero. Under these conditions $u(x, t)$ is established as a solution of (1), satisfying (2)–(4).

If $f'(x)$ and $f''(x)$ are merely piecewise continuous (cf. Sec. 4.1), or if those one-sided derivatives are not zero, then for each t there will be finitely many values of x at which the second derivatives of u appearing in (1) do not exist. Except at these points the wave equation will still be satisfied, and we may then regard $u(x, t)$ as a solution of our problem in a broader sense. For example, the case of a triangular initial deflection (Ex. 1, below) leads to a solution of this type.

Let us mention a very interesting physical interpretation of (17). The graph of $f^*(x - ct)$ is obtained from the graph of $f^*(x)$ by shifting the latter ct units to the right (Fig. 204). This means that $f^*(x - ct)$ $(c > 0)$ represents a wave which is traveling to the right as t increases. Similarly, $f^*(x + ct)$ represents a wave which is traveling to the left, and $u(x, t)$ is the superposition of these two waves.

Example 1.   Find the solution of the wave equation (1) corresponding to the triangular initial deflection

$$
f(x) =
\begin{cases}
\dfrac{2k}{l}\, x & \text{when} \quad 0 < x < \dfrac{l}{2} \\[2ex]
\dfrac{2k}{l}\, (l - x) & \text{when} \quad \dfrac{l}{2} < x < l
\end{cases}
$$

and initial velocity zero. Since $g(x) \equiv 0$, we have $B_n{}^* = 0$ in (12), and from Ex. 1 in Sec. 9.5 we see that the B_n are given by (5), Sec. 9.5. Thus (12) takes the form

$$
u(x, t) = \frac{8k}{\pi^2}\left[\frac{1}{1^2}\sin\frac{\pi}{l}x\cos\frac{\pi c}{l}t - \frac{1}{3^2}\sin\frac{3\pi}{l}x\cos\frac{3\pi c}{l}t + - \cdots\right].
$$

For plotting the graph of the solution we may use $u(x, 0) = f(x)$ and the above interpretation of the two functions in the representation (17). This leads to the graph shown in Fig. 205.

Fig. 204. Interpretation of (17) on p. 423

Fig. 205. Solution in Example 1 for various values of t

Problems

Find the deflection $u(x, t)$ of the vibrating string (length $l = \pi$, ends fixed, and $c^2 = T/\rho = 1$) corresponding to zero initial velocity and initial deflection:

1. $0.02 \sin x$

2. $k \sin 3x$

3. $k(\sin x - \sin 2x)$

4.

5.

6.

7. $0.01x(\pi - x)$

8. $k[(\tfrac{1}{2}\pi)^4 - (x - \tfrac{1}{2}\pi)^4]$

9. $0.01x(\pi^2 - x^2)$

10. In what manner does the frequency of the fundamental mode of the vibrating string depend on the length of the string, the tension, and the mass of the string per unit length?

11. What is the ratio of the amplitudes of the fundamental mode and the second overtone in Prob. 7? The ratio $a_1^2/(a_1^2 + a_2^2 + \cdots)$? *Hint.* Use (7) in Sec. 9.8, with the equality sign.

Find solutions $u(x, y)$ of the following equations by separating variables.

12. $u_x - u_y = 0$

13. $u_x + u_y = 0$

14. $xu_x - yu_y = 0$

15. $u_x - yu_y = 0$

16. $ayu_x - bxu_y = 0$

17. $u_x + u_y = 2(x + y)u$

18. $u_{xx} + u_{yy} = 0$

19. $u_{xy} - u = 0$

20. $x^2u_{xy} + 3y^2u = 0$

10.4 D'Alembert's Solution of the Wave Equation

It is interesting to note that the solution (17), Sec. 10.3, of the wave equation

$$(1) \qquad \frac{\partial^2 u}{\partial t^2} = c^2 \frac{\partial^2 u}{\partial x^2} \qquad\qquad c^2 = \frac{T}{\rho}$$

can be immediately obtained by transforming (1) in a suitable way, namely, by introducing the new independent variables[1]

$$(2) \qquad v = x + ct, \qquad z = x - ct.$$

u then becomes a function of v and z, and the derivatives in (1) can be expressed in terms of derivatives with respect to v and z by the use of the chain rule in Sec. 7.7. Denoting partial derivatives by subscripts, we see from (2) that $v_x = 1$ and $z_x = 1$. For simplicity let us denote $u(x, t)$, as a function of v and z, by the same letter u. Then

$$u_x = u_v v_x + u_z z_x = u_v + u_z.$$

By applying the chain rule to the right side and using $v_x = 1$ and $z_x = 1$ we find

$$u_{xx} = (u_v + u_z)_x = (u_v + u_z)_v v_x + (u_v + u_z)_z z_x = u_{vv} + 2u_{vz} + u_{zz}.$$

The other derivative in (1) is transformed by the same procedure, and the result is

$$u_{tt} = c^2(u_{vv} - 2u_{vz} + u_{zz}).$$

By inserting these two results in (1) we obtain

$$(3) \qquad u_{vz} \equiv \frac{\partial^2 u}{\partial v\, \partial z} = 0.$$

Obviously, the point of the present approach is that the resulting equation (3) can be readily solved by two successive integrations. In fact, integrating with respect to z we find,

$$\frac{\partial u}{\partial v} = h(v)$$

where $h(v)$ is an arbitrary function of v. Integrating this with respect to v, we have

$$u = \int h(v)\, dv + \psi(z)$$

where $\psi(z)$ is an arbitrary function of z. Since the integral is a function of v, say, $\phi(v)$, the solution u is of the form $u = \phi(v) + \psi(z)$. Because of (2) we may write

$$(4) \qquad u(x, t) = \phi(x + ct) + \psi(x - ct).$$

This is known as **d'Alembert's solution**[2] of the wave equation (1).

---

[1] We mention that the general theory of partial differential equations provides a systematic way for finding this transformation which will simplify the equation. Cf. Ref. [E14] in Appendix 1.

[2] JEAN-LE-ROND D'ALEMBERT (1717–1783), French mathematician, who is known for his important work in mechanics.

The functions ϕ and ψ can be determined from the initial conditions. Let us illustrate this in the case of zero initial velocity and given initial deflection $u(x, 0) = f(x)$.

By differentiating (4) we have

$$(5) \qquad\qquad \frac{\partial u}{\partial t} = c\phi'(x + ct) - c\psi'(x - ct)$$

where primes denote derivatives with respect to the *entire* arguments $x + ct$ and $x - ct$, respectively. From (4), (5), and the initial conditions we have

$$u(x, 0) = \phi(x) + \psi(x) = f(x)$$
$$u_t(x, 0) = c\phi'(x) - c\psi'(x) = 0.$$

From the last equation, $\psi' = \phi'$. Hence $\psi = \phi + k$, and from this and the first equation, $2\phi + k = f$ or $\phi = (f - k)/2$. With these functions ϕ and ψ the solution (4) becomes

$$(6) \qquad\qquad u(x, t) = \tfrac{1}{2}[f(x + ct) + f(x - ct)],$$

in agreement with (17) in Sec. 10.3. The student may show that because of the boundary conditions (2) in that section the function f must be odd and have period $2l$.

Our result shows that the two initial conditions and the boundary conditions determine the solution uniquely.

Problems

Using (6), sketch a figure (of the type of Fig. 205 in Sec. 10.3) of the deflection $u(x, t)$ of a vibrating string (length $l = 1$, ends fixed) starting with initial velocity zero and initial deflection $f(x)$, where

1. $f(x) = kx(1 - x)$ 2. $f(x) = k \sin 2\pi x$ 3. $f(x) = k(x - x^3)$
4. $f(x) = k(x^2 - x^4)$ 5. $f(x) = k \sin^2 \pi x$ 6. $f(x) = k(x^3 - x^5)$

7. **(Forced vibrations of an elastic string)** Show that the forced vibrations of an elastic string under an external force $P(x, t)$ per unit length acting normal to the string are governed by the equation

$$(7) \qquad\qquad u_{tt} = c^2 u_{xx} + \frac{P}{\rho}.$$

Consider and solve the equation in Prob. 7 for the sinusoidal force $P = A\rho \sin \omega t$, as follows.

8. Show that

$$P/\rho = A \sin \omega t = \sum_{n=1}^{\infty} k_n(t) \sin \frac{n\pi x}{l}$$

where $k_n(t) = (2A/n\pi)(1 - \cos n\pi) \sin \omega t$; consequently $k_n = 0$ (n even), and $k_n = (4A/n\pi) \sin \omega t$ (n odd). Show that substitution of

$$u(x, t) = \sum_{n=1}^{\infty} G_n(t) \sin \frac{n\pi x}{l}$$

into the wave equation (1) leads to the equation

$$\ddot{G}_n + \lambda_n{}^2 G = 0, \qquad \lambda_n = \frac{cn\pi}{l}.$$

9. Show that by substituting the expressions for u and P/ρ in Prob. 8 into (7) we obtain

$$\ddot{G}_n + \lambda_n{}^2 G_n = \frac{2A}{n\pi}(1 - \cos n\pi)\sin \omega t, \qquad\qquad \lambda_n = \frac{cn\pi}{l}.$$

Show that if $\lambda_n{}^2 \neq \omega^2$, the solution is

$$G_n(t) = B_n \cos \lambda_n t + B_n{}^* \sin \lambda_n t + \frac{2A(1 - \cos n\pi)}{n\pi(\lambda_n{}^2 - \omega^2)}\sin \omega t.$$

10. Determine B_n and $B_n{}^*$ in Prob. 9 so that u satisfies the initial conditions $u(x, 0) = f(x)$, $u_t(x, 0) = 0$.

11. Show that in the case of resonance $(\lambda_n = \omega)$,

$$G_n(t) = B_n \cos \omega t + B_n{}^* \sin \omega t - \frac{A}{n\pi\omega}(1 - \cos n\pi)t \cos \omega t.$$

12. (Vibrations of a beam) It can be shown that the small free vertical vibrations of a uniform cantilever beam are governed by the fourth-order equation

(8) $$\frac{\partial^2 u}{\partial t^2} + c^2 \frac{\partial^4 u}{\partial x^4} = 0 \qquad\qquad \text{(Ref. [E15])},$$

where $c^2 = EI/\rho A$ (E = Young's modulus of elasticity, I = moment of inertia of the cross section with respect to the y-axis in Fig. 206, ρ = density, A = cross-sectional area).

Fig. 206. Undeformed beam

Substituting $u = F(x)G(t)$ into (8) and separating variables, show that

$$F^{(4)}/F = -\ddot{G}/c^2 G = \beta^4 = const,$$
$$F(x) = A \cos \beta x + B \sin \beta x + C \cosh \beta x + D \sinh \beta x,$$
$$G(t) = a \cos c\beta^2 t + b \sin c\beta^2 t.$$

13. Find solutions $u_n = F_n(x)G_n(t)$ of (8) corresponding to zero initial velocity and satisfying the boundary conditions (Fig. 207)

$u(0, t) = 0, u(l, t) = 0$ (end fixed for all times t),
$u_{xx}(0, t) = 0, u_{xx}(l, t) = 0$ (zero moments and, hence, zero curvature at the ends).

Fig. 207. Beam in Problem 13

14. Find the solution of (8) that satisfies the conditions in Prob. 13 and the initial condition $u(x, 0) = f(x) = x(l - x)$.

15. What are the boundary conditions if the beam is clamped at both ends (Fig. 208)?

$x = 0$ $x = l$

Fig. 208. Beam clamped at both ends

16. Show that $F(x)$ in Prob. 12 satisfies the conditions in Prob. 15, if βl is a root of the equation

$$(9) \qquad\qquad \cosh \beta l \cos \beta l = 1.$$

17. Determine approximate solutions of (9).

18. If the beam is clamped at the left and free at the other end (Fig. 209), the boundary conditions are

$$u(0, t) = 0, \qquad u_x(0, t) = 0, \qquad u_{xx}(l, t) = 0, \qquad u_{xxx}(l, t) = 0.$$

Show that $F(x)$ in Prob. 12 satisfies these conditions, if βl is a root of the equation

$$(10) \qquad\qquad \cosh \beta l \cos \beta l = -1.$$

$x = 0$ $x = l$

Fig. 209. Beam clamped at one end and free at the other

19. Find approximate solutions of (10).

20. Compare the results of Prob. 14 and Prob. 7, Sec. 10.3. What is the basic difference between the frequencies of the normal modes of the vibrating string and the vibrating beam?

Using the indicated transformations, solve the following equations.

21. $x u_{xy} = y u_{yy} + u_y$ $(v = x, z = xy)$

22. $u_{xx} + 2u_{xy} + u_{yy} = 0$ $(v = x, z = x - y)$

23. **(Types of linear partial differential equations)** An equation of the form

$$(11) \qquad\qquad A u_{xx} + 2B u_{xy} + C u_{yy} = F(x, y, u, u_x, u_y)$$

is said to be **elliptic** if $AC - B^2 > 0$, **parabolic** if $AC - B^2 = 0$, and **hyperbolic** if $AC - B^2 < 0$. (Here A, B, C may be functions of x and y, and the type of (11) may be different in different parts of the xy-plane.) Show that

   Laplace's equation $u_{xx} + u_{yy} = 0$ is elliptic,

   the **heat equation** $u_t = c^2 u_{xx}$ is parabolic,

   the **wave equation** $u_{tt} = c^2 u_{xx}$ is hyperbolic,

   the **Tricomi equation** $y u_{xx} + u_{yy} = 0$ is of mixed type (elliptic in the upper half plane, parabolic on the x-axis, and hyperbolic in the lower half plane).

24. If (11) is *hyperbolic*, it can be transformed to the *normal form* $u_{vz} = F^*(v, z, u, u_v, u_z)$ by setting $v = \Phi(x, y)$, $z = \Psi(x, y)$ where $\Phi = const$ and $\Psi = const$ are the solutions $y = y(x)$ of $Ay'^2 - 2By' + C = 0$ (cf. Ref. [E14]). Show that in the case of the wave equation (1),

$$\Phi = x + ct, \qquad \Psi = x - ct.$$

25. If (11) is *parabolic*, the substitution $v = x$, $z = \Psi(x, y)$, with Ψ defined as in Prob. 24, reduces it to the *normal form* $u_{vv} = F^*(v, z, u, u_v, u_z)$. Verify this result for the equation in Prob. 22.

10.5   *One-Dimensional Heat Flow*

The heat flow in a body of homogeneous material is governed by the heat equation (cf. Sec. 8.9)

$$\frac{\partial u}{\partial t} = c^2 \nabla^2 u \qquad\qquad c^2 = \frac{K}{\sigma \rho}$$

where $u(x, y, z, t)$ is the temperature in the body, K is the thermal conductivity, σ is the specific heat, and ρ is the density of material of the body. $\nabla^2 u$ is the Laplacian of u, and with respect to Cartesian coordinates x, y, z,

$$\nabla^2 u = \frac{\partial^2 u}{\partial x^2} + \frac{\partial^2 u}{\partial y^2} + \frac{\partial^2 u}{\partial z^2}.$$

As an important application, let us consider the temperature in a long thin bar or wire of constant cross section and homogeneous material which is oriented along the x-axis (Fig. 210) and is perfectly insulated laterally, so that heat flows in the x-direction only. Then u depends only on x and time t, and the heat equation becomes the so-called **one-dimensional heat equation**

(1)
$$\frac{\partial u}{\partial t} = c^2 \frac{\partial^2 u}{\partial x^2}.$$

We shall solve (1) for some important types of boundary and initial conditions. The procedure of solving (1) will be similar to that in the case of the wave equation. The behavior of the solutions will be entirely different from that of the solutions of the wave equation, because (1) involves u_t whereas the wave equation involves u_{tt}. (Hence the classification in Prob. 23 of the last section is not merely a formal matter but has far-reaching consequences with respect to the general behavior of the solutions.)

Let us start with the case when the ends $x = 0$ and $x = l$ of the bar are kept at temperature zero. Then the *boundary conditions* are

(2) $u(0, t) = 0$, $u(l, t) = 0$ for all t.

Fig. 210. Bar under consideration

Note that this has the same form as (2) in Sec. 10.3. Let $f(x)$ be the initial temperature in the bar. Then the *initial condition* is

$$(3) \qquad u(x, 0) = f(x),$$

where $f(x)$ is a given function. We shall determine a solution $u(x, t)$ of (1), satisfying (2) and (3).

First step. By applying the method of separation of variables we first determine solutions of (1) that satisfy the boundary conditions (2). We start from

$$(4) \qquad u(x, t) = F(x)G(t).$$

Substituting this expression and its derivatives into (1), we obtain

$$F\dot{G} = c^2 F'' G$$

where dots denote derivatives with respect to t and primes denote derivatives with respect to x. We divide this equation by $c^2 FG$, finding

$$(5) \qquad \frac{\dot{G}}{c^2 G} = \frac{F''}{F}.$$

The expression on the left depends only on t, while the right side depends only on x. As in Sec. 10.3, we conclude that both expressions must be equal to a constant, say, k. The student may show that for $k \geq 0$ the only solution $u = FG$ that satisfies (2) is $u \equiv 0$. For negative $k = -p^2$ we have from (5)

$$\frac{\dot{G}}{c^2 G} = \frac{F''}{F} = -p^2.$$

We see that this yields the two ordinary linear differential equations

$$(6) \qquad F'' + p^2 F = 0$$

and

$$(7) \qquad \dot{G} + c^2 p^2 G = 0.$$

Second step. We consider (6). The general solution is

$$(8) \qquad F(x) = A \cos px + B \sin px.$$

From (2) it follows that

$$u(0, t) = F(0)G(t) = 0 \qquad \text{and} \qquad u(l, t) = F(l)G(t) = 0.$$

Since $G \equiv 0$ implies $u \equiv 0$, we require that $F(0) = 0$ and $F(l) = 0$. By (8), $F(0) = A$. Thus $A = 0$, and, therefore,

$$F(l) = B \sin pl.$$

We must have $B \neq 0$, since otherwise $F \equiv 0$. Hence the condition $F(l) = 0$ leads to

$$\sin pl = 0 \qquad \text{or} \qquad p = \frac{n\pi}{l}, \qquad n = 1, 2, \cdots.$$

Setting $B = 1$, we thus obtain the following solutions of (6) satisfying (2):

$$F_n(x) = \sin \frac{n\pi x}{l}, \qquad\qquad n = 1, 2, \cdots,$$

(As in Sec. 10.3, we need not consider negative integral values of n.)

We now consider the differential equation (7). For the values $p = n\pi/l$ it takes the form

$$\dot{G} + \lambda_n{}^2 G = 0 \qquad \text{where} \qquad \lambda_n = \frac{cn\pi}{l}.$$

The general solution is

$$G_n(t) = B_n e^{-\lambda_n{}^2 t}, \qquad\qquad n = 1, 2, \cdots,$$

where B_n is a constant. Hence the functions

(9) $$u_n(x, t) = F_n(x)G_n(t) = B_n \sin \frac{n\pi x}{l} e^{-\lambda_n{}^2 t} \qquad n = 1, 2, \cdots.$$

are solutions of the heat equation (1), satisfying (2).

Third step. To obtain a solution also satisfying (3), we consider the series

(10) $$u(x, t) = \sum_{n=1}^{\infty} u_n(x, t) = \sum_{n=1}^{\infty} B_n \sin \frac{n\pi x}{l} e^{-\lambda_n{}^2 t} \qquad \left(\lambda_n = \frac{cn\pi}{l}\right).$$

From this and (3),

$$u(x, 0) = \sum_{n=1}^{\infty} B_n \sin \frac{n\pi x}{l} = f(x).$$

Hence, for (10) to satisfy (3), the coefficients B_n must be chosen such that $u(x, 0)$ becomes a half-range expansion of $f(x)$, namely, the Fourier sine series of $f(x)$; that is [cf. (4) in Sec. 9.5],

(11) $$B_n = \frac{2}{l} \int_0^l f(x) \sin \frac{n\pi x}{l} \, dx \qquad n = 1, 2, \cdots.$$

The solution of our problem can be established, assuming that $f(x)$ is piecewise continuous on the interval $0 \leqq x \leqq l$ (cf. p. 150), and has one-sided derivatives[3] at all interior points of that interval; that is, under these assumptions the series (10) with coefficients (11) is the solution of our physical problem. The proof, which requires the knowledge of uniform convergence of series, will be given at a later occasion. (Probs. 19, 20 at the end of Sec. 15.4).

Because of the exponential factor all the terms in (11) approach zero as t approaches infinity. The rate of decay varies with n.

Example 1. If the initial temperature is

$$f(x) = \begin{cases} x & \text{when} & 0 < x < l/2, \\ l - x & \text{when} & l/2 < x < l \end{cases}$$

Cf. footnote 4 in Sec. 9.2.

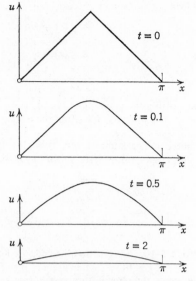

Fig. 211. Solution of Example 1 for various values of t

(cf. Fig. 211 where $l = \pi$ and $c = 1$), then we obtain from (11)

(12)
$$B_n = \frac{2}{l}\left(\int_0^{l/2} x \sin\frac{n\pi x}{l}\, dx + \int_{l/2}^l (l-x)\sin\frac{n\pi x}{l}\, dx\right).$$

Integration yields $B_n = 0$ when n is even and

$$B_n = \frac{4l}{n^2\pi^2} \qquad (n = 1, 5, 9, \cdots),$$

$$B_n = -\frac{4l}{n^2\pi^2} \qquad (n = 3, 7, 11, \cdots).$$

Hence the solution is

$$u(x, t) = \frac{4l}{\pi^2}\left[\sin\frac{\pi x}{l}\, e^{-(c\pi/l)^2 t} - \frac{1}{9}\sin\frac{3\pi x}{l}\, e^{-(3c\pi/l)^2 t} + \cdots\right].$$

The reader may compare Fig. 211 and Fig. 205 in Sec. 10.3.

Problems

1. Graph u_1, u_2, u_3 [cf. (9)] with $B_n = 1$, $c = 1$, $l = \pi$ as functions of x for $t = 0, 1, 2, 3$. Compare the behavior of these functions.
2. Represent the solutions in Prob. 1 graphically as surfaces over the xt-plane.
3. Show that the one-dimensional heat equation is parabolic, while the one-dimensional wave equation is hyperbolic (cf. Prob. 23 at the end of the last section).
4. How does the rate of decay of (9) for fixed n depend on the specific heat, the density, and the thermal conductivity of the material?

Find the temperature $u(x, t)$ in a bar of silver (length 10 cm, constant cross section of area 1 cm^2, density 10.6 gm/cm^3, thermal conductivity 1.04 cal/cm deg sec, specific heat 0.056 cal/gm deg) which is perfectly insulated laterally, whose ends are kept at temperature 0°C and whose initial temperature (in $°C$) is $f(x)$, where

5. $f(x) = \sin 0.1\pi x$

6. $f(x) = \sin 0.2\pi x$

7. $f(x) = \begin{cases} x & \text{if } 0 < x < 5 \\ 10 - x & \text{if } 5 < x < 10 \end{cases}$

8. $f(x) = \begin{cases} x & \text{if } 0 < x < 5 \\ 0 & \text{if } 5 < x < 10 \end{cases}$

9. $f(x) = x(10 - x)$

10. $f(x) = x(100 - x^2)$

11. The ends of a rod that satisfies the assumptions in the text are kept at different constant temperatures $u(0, t) = U_1$ and $u(l, t) = U_2$. Find the temperature $u_I(x)$ in the rod after a long time (theoretically: as $t \to \infty$).

12. In Prob. 11, let the initial temperature be $u(x, 0) = f(x)$. Show that the temperature for any time $t > 0$ is $u(x, t) = u_I(x) + u_{II}(x, t)$ with u_I as before and

$$u_{II} = \sum_{n=1}^{\infty} B_n \sin \frac{n\pi x}{l} e^{-(cn\pi/l)^2 t},$$

$$B_n = \frac{2}{l} \int_0^l [f(x) - u_I(x)] \sin \frac{n\pi x}{l} \, dx = \frac{2}{l} \int_0^l f(x) \sin \frac{n\pi x}{l} \, dx + \frac{2}{n\pi} [(-1)^n U_2 - U_1].$$

13. Consider the bar in Probs. 5–10. Assume that the ends are kept at $100°C$ for a long time. Then at some instant, say, at $t = 0$, the temperature at $x = l$ is suddenly changed to $0°C$ and kept at this value, while the temperature at $x = 0$ is kept at $100°C$. What are the temperatures in the middle of the bar at $t = 1, 2, 3, 10, 50$ sec?

14. Find the temperature $u(x, t)$ in a bar of length l which is perfectly insulated, also at the ends at $x = 0$ and $x = l$, assuming that $u(x, 0) = f(x)$. Physical information: the flux of heat through the faces at the ends is proportional to the values of $\partial u/\partial x$ there. Show that this situation corresponds to the conditions

$$u_x(0, t) = 0, \qquad u_x(l, t) = 0, \qquad u(x, 0) = f(x).$$

Show that the method of separating variables yields the solution

$$u(x, t) = A_0 + \sum_{n=1}^{\infty} A_n \cos \frac{n\pi x}{l} e^{-(cn\pi/l)^2 t}$$

where, by (2) in Sec. 9.5,

$$A_0 = \frac{1}{l} \int_0^l f(x) \, dx, \qquad A_n = \frac{2}{l} \int_0^l f(x) \cos \frac{n\pi x}{l} \, dx, \qquad n = 1, 2, \cdots.$$

15. In Prob. 14, $u \to A_0$ as $t \to \infty$. Does this agree with your physical intuition?

Find the temperature in the bar in Prob. 14, if $l = \pi$, $c = 1$, and

16. $f(x) = 1$

17. $f(x) = x^2$

18. $f(x) = \sin x$

19. $f(x) = \begin{cases} x & \text{when} \quad 0 < x < \pi/2 \\ 0 & \text{when} \quad \pi/2 < x < \pi \end{cases}$

20. $f(x) = \begin{cases} 1 & \text{when} \quad 0 < x < \pi/2 \\ 0 & \text{when} \quad \pi/2 < x < \pi \end{cases}$

10.6   Heat Flow in an Infinite Bar

We shall now consider solutions of the heat equation

(1)
$$\frac{\partial u}{\partial t} = c^2 \frac{\partial^2 u}{\partial x^2}$$

in the case of a bar which extends to infinity on both sides (and is laterally insulated, as before). In this case we do not have boundary conditions but only the initial condition

(2) $$u(x, 0) = f(x)$$ $(-\infty < x < \infty)$

where $f(x)$ is the given initial temperature of the bar.

To solve our present problem we start as in the last section, that is, we substitute $u(x, t) = F(x)G(t)$ into (1). This yields the two ordinary differential equations

(3) $$F'' + p^2 F = 0$$ [cf. (6), Sec. 10.5]

and

(4) $$\dot{G} + c^2 p^2 G = 0$$ [cf. (7), Sec. 10.5].

Solutions are

$$F(x) = A \cos px + B \sin px \quad \text{and} \quad G(t) = e^{-c^2 p^2 t}$$

respectively; here A and B are any constants. Hence a solution of (1) is

(5) $$u(x, t; p) = FG = (A \cos px + B \sin px) e^{-c^2 p^2 t}.$$

(As in the last section, we had to choose the constant of separation k negative, $k = -p^2$, because positive values of k lead to an increasing exponential function in (5), which has no physical meaning.)

Any series of functions (5), found in the usual manner by taking p as multiples of a fixed number, would lead to a function which is periodic in x when $t = 0$. However, since $f(x)$ in (2) is not assumed periodic, it is natural to use Fourier integrals in the present case instead of Fourier series.

Since A and B in (5) are arbitrary, we may consider these quantities as functions of p and write $A = A(p)$ and $B = B(p)$. Since the heat equation is linear and homogeneous, the function

(6) $$u(x, t) = \int_0^\infty u(x, t; p) \, dp = \int_0^\infty [A(p) \cos px + B(p) \sin px] e^{-c^2 p^2 t} \, dp$$

is then a solution of (1), provided this integral exists and can be differentiated twice with respect to x and once with respect to t.

From the representation (6) and the initial condition (2) it follows that

(7) $$u(x, 0) = \int_0^\infty [A(p) \cos px + B(p) \sin px] \, dp = f(x).$$

Using (4) and (5) in Sec. 9.9, we thus obtain

(8) $$A(p) = \frac{1}{\pi} \int_{-\infty}^\infty f(v) \cos pv \, dv, \qquad B(p) = \frac{1}{\pi} \int_{-\infty}^\infty f(v) \sin pv \, dv.$$

According to (16) in Prob. 20, Sec. 9.9, this Fourier integral may be written

$$u(x, 0) = \frac{1}{\pi} \int_0^\infty \left[\int_{-\infty}^\infty f(v) \cos (px - pv) \, dv \right] dp,$$

and (6), this section, thus becomes

$$u(x, t) = \frac{1}{\pi} \int_0^\infty \left[\int_{-\infty}^\infty f(v) \cos (px - pv) e^{-c^2 p^2 t} \, dv \right] dp.$$

Assuming that we may invert the order of integration, we obtain

(9) $$u(x, t) = \frac{1}{\pi} \int_{-\infty}^\infty f(v) \left[\int_0^\infty e^{-c^2 p^2 t} \cos (px - pv) \, dp \right] dv.$$

The inner integral can be evaluated by the use of the formula

(10) $$\int_0^\infty e^{-s^2} \cos 2bs \, ds = \frac{\sqrt{\pi}}{2} e^{-b^2}$$

which will be derived in Sec. 16.4 (Prob. 25). Introducing a new variable of integration p by setting $s = cp\sqrt{t}$ and choosing

$$b = \frac{x - v}{2c\sqrt{t}}$$

we find that (10) becomes

$$\int_0^\infty e^{-c^2 p^2 t} \cos (px - pv) \, dp = \frac{\sqrt{\pi}}{2c\sqrt{t}} e^{-(x-v)^2/4c^2 t}.$$

By inserting this result into (9) we obtain the representation

(11) $$u(x, t) = \frac{1}{2c\sqrt{\pi t}} \int_{-\infty}^\infty f(v) \exp \left\{ -\frac{(x - v)^2}{4c^2 t} \right\} dv.$$

We finally introduce the variable of integration $w = (v - x)/2c\sqrt{t}$. Then

(12) $$u(x, t) = \frac{1}{\sqrt{\pi}} \int_{-\infty}^\infty f(x + 2cw\sqrt{t}) e^{-w^2} \, dw.$$

If $f(x)$ is bounded for all values of x and integrable in every finite interval, it can be shown (cf. Ref. [D1]) that the function (11) or (12) satisfies (1) and (2). Hence this function is the required solution in the present case.

Example 1.   Find the temperature in the infinite bar if the initial temperature is (Fig. 212)

$$f(x) = \begin{cases} U_0 = const & when \quad |x| < 1, \\ 0 & when \quad |x| > 1. \end{cases}$$

From (11) we have

$$u(x, t) = \frac{U_0}{2c\sqrt{\pi t}} \int_{-1}^1 \exp \left\{ -\frac{(x - v)^2}{4c^2 t} \right\} dv.$$

Fig. 212. Initial temperature in Example 1

Fig. 213. Solution $u(x, t)$ of Example 1 for $U_0 = 100°C$, $c^2 = 1$ cm^2/sec, and several values of t

If we introduce the above variable of integration w, then the integration over v from -1 to 1 corresponds to the integration over w from $(-1 - x)/2c\sqrt{t}$ to $(1 - x)/2c\sqrt{t}$, and

$$(13) \qquad u(x, t) = \frac{U_0}{\sqrt{\pi}} \int_{-(1+x)/2c\sqrt{t}}^{(1-x)/2c\sqrt{t}} e^{-w^2} \, dw \qquad (t > 0).$$

We mention that this integral is not an elementary function, but can easily be expressed in terms of the error function, whose values have been tabulated. (Table A5 in Appendix 4 contains a few values; larger tables are listed in Ref. [J11] in Appendix 1. Cf. also Probs. 10–15, below.) Figure 213 shows $u(x, t)$ for $U_0 = 100°C$, $c^2 = 1$ cm^2/sec, and several values of t.

Problems

1. Graph the temperatures in Ex. 1 (with $U_0 = 100°C$ and $c^2 = 1$ cm^2/sec) at the points $x = 0.5$, 1, and 1.5 as functions of t. Do the results agree with your physical intuition?

2. If $f(x) = 1$ when $x > 0$ and $f(x) = 0$ when $x < 0$, show that (12) becomes

$$u(x, t) = \frac{1}{\sqrt{\pi}} \int_{-x/2c\sqrt{t}}^{\infty} e^{-w^2} \, dw \qquad (t > 0),$$

3. Show that for $x = 0$ the solution in Prob. 2 is independent of t. Can this result be expected for physical reasons?

4. If the bar is semi-infinite, extending from 0 to ∞, the end at $x = 0$ is held at temperature 0 and the initial temperature is $f(x)$, show that the temperature in the bar is

$$(14) \qquad u(x, t) = \frac{1}{\sqrt{\pi}} \left[\int_{-x/\tau}^{\infty} f(x + \tau w)e^{-w^2} dw - \int_{x/\tau}^{\infty} f(-x + \tau w)e^{-w^2} \, dw \right], \qquad \tau = 2c\sqrt{t}.$$

5. Obtain (14) from (11) by assuming that $f(v)$ in (11) is odd.

6. If in Prob. 4, $f(x) = 1$, show that

$$u(x, t) = \frac{2}{\sqrt{\pi}} \int_0^{x/\tau} e^{-w^2} dw \qquad (t > 0).$$

7. What form does (14) take if $f(x) = 1$ when $a < x < b$ (where $a > 0$) and $f(x) = 0$ otherwise?

8. Show that the result of Prob. 6 can be obtained from (11) or (12) by using $f(x) = 1$, when $x > 0$, and $f(x) = -1$ when $x < 0$. What is the reason?

9. Show that in Prob. 6 the times required for any two points to reach the same temperature are proportional to the squares of their distances from the boundary at $x = 0$.

Error function. The error function is defined by the integral

$$\operatorname{erf} x = \frac{2}{\sqrt{\pi}} \int_0^x e^{-w^2} \, dw.$$

It is important in engineering mathematics. To become familiar with the function, the student may solve the following problems. (A few formulas are included in Appendix 3, see (35)–(37). Cf. also Ex. 1 in Sec. 18.14.)

10. Show that $\operatorname{erf} x$ is odd.

11. Show that $\quad \displaystyle\int_a^b e^{-w^2} \, dw = \frac{\sqrt{\pi}}{2} \, (\operatorname{erf} b - \operatorname{erf} a), \qquad \int_{-b}^b e^{-w^2} \, dw = \sqrt{\pi} \operatorname{erf} b.$

12. Using a table of the exponential function, graph the integrand of $\operatorname{erf} x$ (the so-called *bell-shaped curve*).

13. Obtain a small table of $\operatorname{erf} x$ for $x = 0, 0.2, 0.4, \cdots, 1.0, 1.5, 2.0$ by counting squares under the curve in Prob. 12 or some other rough method of approximate integration, and compare the result with the actual values whose first two decimal places are 0.00, 0.22, 0.43, 0.60, 0.74, 0.84, 0.97, 1.00.

14. Obtain the Maclaurin series of $\operatorname{erf} x$ by term-by-term integration of that of the integrand.

15. Show that (13) may be written

$$u(x, t) = \frac{U_0}{2} \left[\operatorname{erf} \frac{1 - x}{2c\sqrt{t}} + \operatorname{erf} \frac{1 + x}{2c\sqrt{t}} \right] \qquad (t > 0).$$

10.7   Vibrating Membrane. Two-Dimensional Wave Equation

As another important problem from the field of vibrations, let us consider the motion of a stretched membrane, such as a drumhead. The reader will notice that our present considerations will be similar to those in the case of the vibrating string in Sec. 10.2.

We make the following assumptions:

1. *The mass of the membrane per unit area is constant ("homogeneous membrane"). The membrane is perfectly flexible and is so thin that it does not offer any resistance to bending.*

2. *The membrane is stretched and then fixed along its entire boundary in the xy-plane. The tension per unit length T caused by stretching the membrane is the same at all points and in all directions and does not change during the motion.*

3. *The deflection $u(x, y, t)$ of the membrane during the motion is small compared with the size of the membrane, and all angles of inclination are small.*

Fig. 214. Vibrating membrane

Although these assumptions cannot be realized in practice, small transverse vibrations of a thin physical membrane will satisfy these assumptions relatively accurately.

To derive the differential equation which governs the motion of the membrane, we consider the forces acting on a small portion of the membrane as shown in Fig. 214. Since the deflections of the membrane and the angles of inclination are small, the sides of the portion are approximately equal to Δx and Δy. The tension T is the force per unit length. Hence the forces acting on the edges of the portion are approximately $T \Delta x$ and $T \Delta y$. Since the membrane is perfectly flexible, these forces are tangent to the membrane.

We first consider the horizontal components of the forces. These components are obtained by multiplying the forces by the cosines of the angles of inclination. Since these angles are small, their cosines are close to 1. Hence the horizontal components of the forces at opposite edges are approximately equal. Therefore, the motion of the particles of the membrane in horizontal direction will be negligibly small. From this we conclude that we may regard the motion of the membrane as transversal, that is, each particle moves vertically.

The vertical components of the forces along the edges parallel to the yu-plane are[4] (Fig. 214)

$$T \Delta y \sin \beta \qquad \text{and} \qquad -T \Delta y \sin \alpha;$$

here the minus sign appears because the force on the left edge is directed downward. Since the angles are small, we may replace their sines by their tangents.

---

[4] Note that the angle of inclination varies along the edges, and α and β represent values of that angle at a suitable point of the edges under consideration.

Hence the resultant of those two vertical components is

(1)
$$T \, \Delta y (\sin \beta - \sin \alpha) \approx T \, \Delta y (\tan \beta - \tan \alpha)$$
$$= T \, \Delta y [u_x(x + \Delta x, y_1) - u_x(x, y_2)]$$

where subscripts x denote partial derivatives and y_1 and y_2 are values between y and $y + \Delta y$. Similarly, the resultant of the vertical components of the forces acting on the other two edges of the portion is

(2)
$$T \, \Delta x [u_y(x_1, y + \Delta y) - u_y(x_2, y)]$$

where x_1 and x_2 are values between x and $x + \Delta x$.

By Newton's second law (cf. Sec. 2.6), the sum of the forces given by (1) and (2) is equal to the mass $\rho \, \Delta A$ of the portion times the acceleration $\partial^2 u / \partial t^2$; here ρ is the mass of the undeflected membrane per unit area and $\Delta A = \Delta x \, \Delta y$ is the area of the portion when it is undeflected. Thus

$$\rho \, \Delta x \, \Delta y \frac{\partial^2 u}{\partial t^2} = T \, \Delta y [u_x(x + \Delta x, y_1) - u_x(x, y_2)]$$
$$+ T \, \Delta x [u_y(x_1, y + \Delta y) - u_y(x_2, y)]$$

where the derivative on the left is evaluated at some suitable point $(\tilde{x}, \tilde{y})$ corresponding to the portion. Division by $\rho \, \Delta x \, \Delta y$ yields

$$\frac{\partial^2 u}{\partial t^2} = \frac{T}{\rho} \left[\frac{u_x(x + \Delta x, y_1) - u_x(x, y_2)}{\Delta x} + \frac{u_y(x_1, y + \Delta y) - u_y(x_2, y)}{\Delta y} \right].$$

If we let Δx and Δy approach zero, then we obtain the partial differential equation

(3)
$$\frac{\partial^2 u}{\partial t^2} = c^2 \left(\frac{\partial^2 u}{\partial x^2} + \frac{\partial^2 u}{\partial y^2} \right) \qquad\qquad c^2 = \frac{T}{\rho}$$

This equation is called the **two-dimensional wave equation.** The expression in parentheses is the Laplacian $\nabla^2 u$ of u (cf. Sec. 7.8), and (3) can be written

(3')
$$\frac{\partial^2 u}{\partial t^2} = c^2 \nabla^2 u.$$

10.8   Rectangular Membrane

To solve the problem of a vibrating membrane, we have to determine a solution $u(x, y, t)$ of the two-dimensional wave equation

(1)
$$\frac{\partial^2 u}{\partial t^2} = c^2 \left(\frac{\partial^2 u}{\partial x^2} + \frac{\partial^2 u}{\partial y^2} \right)$$

that satisfies the boundary condition

(2) $u = 0$ on the boundary of the membrane for all $t \geq 0$

Fig. 215. Rectangular membrane

and the two initial conditions

(3) $u(x, y, 0) = f(x, y)$ [given initial displacement $f(x, y)$]

and

(4) $\left.\dfrac{\partial u}{\partial t}\right|_{t=0} = g(x, y)$ [given initial velocity $g(x, y)$].

These conditions are quite similar to those in the case of the vibrating string.

As a first important case, let us consider the rectangular membrane R shown in Fig. 215.

First step. By applying the method of separation of variables we first determine solutions of (1) that satisfy the condition (2). For this purpose we start from

(5) $u(x, y, t) = F(x, y)G(t).$

By substituting this into the wave equation (1) we have

$$F\ddot{G} = c^2(F_{xx}G + F_{yy}G)$$

where subscripts denote partial derivatives and dots denote derivatives with respect to t. Dividing both sides by c^2FG, we find

$$\frac{\ddot{G}}{c^2G} = \frac{1}{F}(F_{xx} + F_{yy}).$$

Since the functions on the left depend only on t while the functions on the right do not depend on t, the expressions on both sides must be equal to a constant. A little investigation shows that only negative values of that constant will lead to solutions which satisfy (2) without being identically zero; this is similar to the procedure in Sec. 10.3. Denoting this negative constant by $-\nu^2$, we thus have

$$\frac{\ddot{G}}{c^2G} = \frac{1}{F}(F_{xx} + F_{yy}) = -\nu^2.$$

From this we obtain the two ordinary linear differential equations

(6) $\ddot{G} + \lambda^2G = 0$ where $\lambda = c\nu,$

and

(7) $F_{xx} + F_{yy} + \nu^2F = 0.$

Now we consider (7) and apply the method of separating variables once more, that is, we determine solutions of (7) of the form

(8) $F(x, y) = H(x)Q(y)$

which are zero on the boundary of the membrane. Substitution of (8) into (7) yields

$$\frac{d^2H}{dx^2} Q = -\left(H \frac{d^2Q}{dy^2} + v^2 HQ\right).$$

By dividing both sides by HQ we find

$$\frac{1}{H} \frac{d^2H}{dx^2} = -\frac{1}{Q}\left(\frac{d^2Q}{dy^2} + v^2 Q\right).$$

The functions on the left depend only on x whereas the functions on the right depend only on y. Hence the expressions on both sides must be equal to a constant. This constant must be negative, say, $-k^2$, because only negative values will lead to solutions that satisfy (2) without being identically zero. Thus

$$\frac{1}{H} \frac{d^2H}{dx^2} = -\frac{1}{Q}\left(\frac{d^2Q}{dy^2} + v^2 Q\right) = -k^2.$$

This yields the ordinary differential equations

(9) $$\frac{d^2H}{dx^2} + k^2 H = 0$$

and

(10) $$\frac{d^2Q}{dy^2} + p^2 Q = 0 \qquad\qquad \text{where } p^2 = v^2 - k^2.$$

Second step. The general solutions of (9) and (10) are

$$H(x) = A \cos kx + B \sin kx \qquad \text{and} \qquad Q(y) = C \cos py + D \sin py$$

where A, B, C, and D are constants. From (5) and (2) it follows that $F = HQ$ must be zero on the boundary, which corresponds to $x = 0$, $x = a$, $y = 0$, and $y = b$; cf. Fig. 215. This yields the conditions

$$H(0) = 0, \qquad H(a) = 0, \qquad Q(0) = 0, \qquad Q(b) = 0.$$

Therefore, $H(0) = A = 0$, and then

$$H(a) = B \sin ka = 0.$$

We must take $B \neq 0$ since otherwise $H \equiv 0$ and $F \equiv 0$. Hence $\sin ka = 0$ or $ka = m\pi$, that is,

$$k = \frac{m\pi}{a} \qquad\qquad (m \text{ integral}).$$

In precisely the same fashion we conclude that $C = 0$ and p must be restricted to the values $p = n\pi/b$ where n is integral. In this way we obtain the solutions

$$H_m(x) = \sin \frac{m\pi x}{a} \qquad \text{and} \qquad Q_n(y) = \sin \frac{n\pi y}{b} \qquad \begin{array}{l} m = 1, 2, \cdots, \\ n = 1, 2, \cdots. \end{array}$$

(As in the case of the vibrating string, it is not necessary to consider m, $n = -1$,

$-2, \cdots$ because the corresponding solutions are essentially the same as for positive m and n, except for a factor -1.) It follows that the functions

(11) $$F_{mn}(x, y) = H_m(x)Q_n(y) = \sin \frac{m\pi x}{a} \sin \frac{n\pi y}{b} \qquad \begin{matrix} m = 1, 2, \cdots, \\ n = 1, 2, \cdots, \end{matrix}$$

are solutions of the equation (7) which are zero on the boundary of the rectangular membrane.

Since $p^2 = v^2 - k^2$ in (10) and $\lambda = cv$ in (6), we have

$$\lambda = c\sqrt{k^2 + p^2}.$$

Hence to $k = m\pi/a$ and $p = n\pi/b$ there corresponds the value

(12) $$\lambda = \lambda_{mn} = c\pi \sqrt{\frac{m^2}{a^2} + \frac{n^2}{b^2}} \qquad \begin{matrix} m = 1, 2, \cdots, \\ n = 1, 2, \cdots, \end{matrix}$$

in (6), and the corresponding general solution of (6) is

$$G_{mn}(t) = B_{mn} \cos \lambda_{mn}t + B_{mn}^* \sin \lambda_{mn}t.$$

It follows that the functions $u_{mn}(x, y, t) = F_{mn}(x, y)G_{mn}(t)$, written out

(13) $$u_{mn}(x, y, t) = (B_{mn} \cos \lambda_{mn}t + B_{mn}^* \sin \lambda_{mn}t) \sin \frac{m\pi x}{a} \sin \frac{n\pi y}{b}$$

with λ_{mn} according to (12) are solutions of the wave equation (1), which are zero on the boundary of the rectangular membrane in Fig. 215. These functions are called the **eigenfunctions** or *characteristic functions*, and the numbers λ_{mn} are called the **eigenvalues** or *characteristic values* of the vibrating membrane. The frequency of u_{mn} is $\lambda_{mn}/2\pi$.

It is interesting to note that, depending on a and b, several functions F_{mn} may correspond to the same eigenvalue. Physically this means that there may exist vibrations having the same frequency but entirely different **nodal lines** (curves of points on the membrane which do not move). This may be illustrated by the following example.

Example 1. Consider the square membrane for which $a = b = 1$. From (12) we see that the eigenvalues are

(14) $$\lambda_{mn} = c\pi \sqrt{m^2 + n^2}.$$

Hence

$$\lambda_{mn} = \lambda_{nm}$$

but for $m \neq n$ the corresponding functions

$$F_{mn} = \sin m\pi x \sin n\pi y \qquad \text{and} \qquad F_{nm} = \sin n\pi x \sin m\pi y$$

are certainly different. For example, to $\lambda_{12} = \lambda_{21} = c\pi\sqrt{5}$ there correspond the two functions

$$F_{12} = \sin \pi x \sin 2\pi y \qquad \text{and} \qquad F_{21} = \sin 2\pi x \sin \pi y.$$

Fig. 216. Nodal lines of the solutions u_{11}, u_{12}, u_{21}, u_{22}, u_{13}, and u_{31} in the case of the square membrane

Fig. 217. Nodal lines of the solution (15) for some values of B_{21}

Hence the corresponding solutions

$$u_{12} = (B_{12} \cos c\pi \sqrt{5}t + B_{12}^* \sin c\pi \sqrt{5}t)F_{12}$$

and

$$u_{21} = (B_{21} \cos c\pi \sqrt{5}t + B_{21}^* \sin c\pi \sqrt{5}t)F_{21}$$

have the nodal lines $y = \frac{1}{2}$ and $x = \frac{1}{2}$, respectively (cf. Fig. 216). Taking $B_{12} = 1$ and $B_{12}^* = B_{21}^* = 0$, we obtain

$$(15) \qquad u_{12} + u_{21} = \cos c\pi \sqrt{5}t \ (F_{12} + B_{21}F_{21}),$$

which represents another vibration corresponding to the eigenvalue $c\pi \sqrt{5}$. The nodal line of this function is the solution of the equation

$$F_{12} + B_{21}F_{21} = \sin \pi x \sin 2\pi y + B_{21} \sin 2\pi x \sin \pi y = 0$$

or, since $\sin 2\alpha = 2 \sin \alpha \cos \alpha$,

$$(16) \qquad \sin \pi x \sin \pi y \ (\cos \pi y + B_{21} \cos \pi x) = 0.$$

This solution depends on the value of B_{21} (Fig. 217).

From (14) we see that even more than two functions may correspond to the same numerical value of λ_{mn}. For example, the four functions F_{18}, F_{81}, F_{47}, and F_{74} correspond to $\lambda_{18} = \lambda_{81} = \lambda_{47} = \lambda_{74} = c\pi \sqrt{65}$, because

$$1^1 + 8^2 = 4^2 + 7^2 = 65.$$

This happens because 65 can be expressed as the sum of two squares of natural numbers in several ways. According to a theorem by Gauss, this is the case for every sum of two squares among whose prime factors there are at least two different ones of the form $4n + 1$ where n is a positive integer. In our case,

$$65 = 5 \cdot 13 = (4 + 1)(12 + 1).$$

Third step. To obtain the solution that also satisfies the initial conditions (3) and (4), we proceed in a similar fashion as in Sec. 10.3. We consider the double series[5]

$$(17) \qquad \begin{aligned} u(x, y, t) &= \sum_{m=1}^{\infty} \sum_{n=1}^{\infty} u_{mn}(x, y, t) \\ &= \sum_{m=1}^{\infty} \sum_{n=1}^{\infty} (B_{mn} \cos \lambda_{mn}t + B_{mn}^* \sin \lambda_{mn}t) \sin \frac{m\pi x}{a} \sin \frac{n\pi y}{b} . \end{aligned}$$

[5] We shall not consider the problems of convergence and uniqueness.

From this and (3) we obtain

(18) $$u(x, y, 0) = \sum_{m=1}^{\infty} \sum_{n=1}^{\infty} B_{mn} \sin \frac{m\pi x}{a} \sin \frac{n\pi y}{b} = f(x, y).$$

This series is called a **double Fourier series.** Suppose that $f(x, y)$ can be developed in such a series.[6] Then the Fourier coefficients B_{mn} of $f(x, y)$ in (18) can be determined as follows. Setting

(19) $$K_m(y) = \sum_{n=1}^{\infty} B_{mn} \sin \frac{n\pi y}{b}$$

we may write (18) in the form

$$f(x, y) = \sum_{m=1}^{\infty} K_m(y) \sin \frac{m\pi x}{a} .$$

For fixed y this is the Fourier sine series of $f(x, y)$, considered as a function of x, and from (4) in Sec. 9.5 it follows that the coefficients of this expansion are

(20) $$K_m(y) = \frac{2}{a} \int_0^a f(x, y) \sin \frac{m\pi x}{a} \, dx.$$

Furthermore, (19) is the Fourier sine series of $K_m(y)$, and from (4) in Sec. 9.5 it follows that the coefficients are

$$B_{mn} = \frac{2}{b} \int_0^b K_m(y) \sin \frac{n\pi y}{b} \, dy.$$

From this and (20) we obtain the *generalized Euler formula*

(21) $$B_{mn} = \frac{4}{ab} \int_0^b \int_0^a f(x, y) \sin \frac{m\pi x}{a} \sin \frac{n\pi y}{b} \, dx \, dy \qquad \begin{matrix} m = 1, 2, \cdots . \\ n = 1, 2, \cdots . \end{matrix}$$

for the Fourier coefficients of $f(x, y)$ in the double Fourier series (18).

The B_{mn} in (17) are now determined in terms of $f(x, y)$. To determine the B_{mn}^*, we differentiate (17) termwise with respect to t; using (4), we obtain

$$\frac{\partial u}{\partial t}\bigg|_{t=0} = \sum_{m=1}^{\infty} \sum_{n=1}^{\infty} B_{mn} \lambda_{mn}^* \sin \frac{m\pi x}{a} \sin \frac{n\pi y}{b} = g(x, y).$$

Suppose that $g(x, y)$ can be developed in this double Fourier series. Then, proceeding as before, we find

(22) $$B_{mn}^* = \frac{4}{ab\lambda_{mn}} \int_0^b \int_0^a g(x, y) \sin \frac{m\pi x}{a} \sin \frac{n\pi y}{b} \, dx \, dy \qquad \begin{matrix} m = 1, 2, \cdots . \\ n = 1, 2, \cdots , \end{matrix}$$

The result is that, for (17) to satisfy the initial conditions, the coefficients B_{mn} and B_{mn}^* must be chosen according to (21) and (22).

[6] Sufficient conditions: f, $\partial f/\partial x$, $\partial f/\partial y$, $\partial^2 f/\partial x \, \partial y$ continuous in the rectangle R under consideration.

Problems

1. Determine and graph the nodal lines of the solutions (13) with $m = 1, 2, 3, 4$ and $n = 1, 2, 3, 4$ in the case $a = b = 1$.
2. Same task as in Prob. 1 when $a = 2$ and $b = 1$.
3. How does the frequency of a solution (13) change, if the tension of the membrane is increased?
4. Show that, among all rectangular membranes of the same area $A = ab$ and the same c, the square membrane is that for which u_{11} [cf. (13)] has the lowest frequency.
5. Find a similar result as in Prob. 4 for the frequency of any solution (13) (m, n fixed).
6. Find further eigenvalues of the square membrane with side 1 such that four different eigenfunctions correspond to each such eigenvalue.
7. Find eigenvalues of the rectangular membrane of sides $a = 2$, $b = 1$ such that two or more different eigenfunctions correspond to each such eigenvalue.

Represent the following functions $f(x, y)$ $(0 < x < a, 0 < y < b)$ by a double Fourier series of the form (18).

    8. $f = 1$ **9.** $f = x + y$ **10.** $f = xy$
  11. $f = xy(a - x)(b - y)$ **12.** $f = (x + 1)(y + 1)$ **13.** $f = xy(a^2 - x^2)(b^2 - y^2)$

Find the deflection $u(x, y, t)$ of the square membrane with $a = b = 1$ and $c = 1$, if the initial velocity is zero and the initial deflection is $f(x, y)$, where

  14. $f = 0.01xy(1 - x)(1 - y)$ **15.** $f = k \sin \pi x \sin 2\pi y$ **16.** $f = k \sin^2 \pi x \sin^2 \pi y$

17. Show that

$$u_{tt} = c^2 \nabla^2 u + P/\rho$$

governs the forced vibrations of the membrane, where $P(x, y, t)$ is the external force per unit area acting normal to the xy-plane.

18. Find the eigenfunctions of the rectangular membrane in Fig. 218 which is fixed at the boundary.

Fig. 218. Rectangular membrane

Fig. 219. Square plate

19. The edges of a thin square plate (Fig. 219) are kept at temperature zero and the faces are perfectly insulated. The initial temperature is assumed to be $u(x, y, 0) = f(x, y)$. By applying the method of separating variables to the two-dimensional heat equation $u_t = c^2 \nabla^2 u$ show that the temperature in the plate is

$$u(x, y, t) = \sum_{m=1}^{\infty} \sum_{n=1}^{\infty} B_{mn} \sin mx \sin ny \, e^{-c^2(m^2 + n^2)t}$$

where

$$B_{mn} = \frac{4}{\pi^2} \int_0^\pi \int_0^\pi f(x, y) \sin mx \sin ny \, dx \, dy.$$

20. Find the temperature in the plate of Prob. 19, if $f(x, y) = x(\pi - x)y(\pi - y)$.

10.9　*Laplacian in Polar Coordinates*

In connection with boundary value problems for partial differential equations, it is a general principle to use coordinates with respect to which the boundary of the region under consideration has a simple representation. In the next section we shall consider circular membranes. Then polar coordinates r and θ, defined by

$$x = r \cos \theta, \qquad y = r \sin \theta,$$

will be appropriate, because the boundary of the membrane can then be represented by the simple equation $r = const$.

When using r and θ we have to transform the Laplacian

$$\nabla^2 u = \frac{\partial^2 u}{\partial x^2} + \frac{\partial^2 u}{\partial y^2}$$

in the wave equation into these new coordinates.

Transformations of differential expressions from one coordinate system into another are frequently required in applications. Therefore, the student should follow our present consideration with great attention.

As in Sec. 10.4, we shall use the chain rule. For the sake of simplicity we shall denote partial derivatives by subscripts and $u(x, y, t)$, as a function of r, θ, t, by the same letter u.

By applying the chain rule (4), Sec. 7.7, we obtain

$$u_x = u_r r_x + u_\theta \theta_x.$$

Differentiating this again with respect to x we first have

(1)
$$
\begin{aligned}
u_{xx} &= (u_r r_x)_x + (u_\theta \theta_x)_x \\
&= (u_r)_x r_x + u_r r_{xx} + (u_\theta)_x \theta_x + u_\theta \theta_{xx}.
\end{aligned}
$$

Now, by applying the chain rule again, we find

$$(u_r)_x = u_{rr} r_x + u_{r\theta} \theta_x \qquad \text{and} \qquad (u_\theta)_x = u_{\theta r} r_x + u_{\theta\theta} \theta_x.$$

To determine the partial derivatives r_x and θ_x, we have to differentiate

$$r = \sqrt{x^2 + y^2} \qquad \text{and} \qquad \theta = \text{arc tan}\,\frac{y}{x},$$

finding

$$r_x = \frac{x}{\sqrt{x^2 + y^2}} = \frac{x}{r}, \qquad \theta_x = \frac{1}{1 + (y/x)^2}\left(-\frac{y}{x^2}\right) = -\frac{y}{r^2}.$$

Differentiating these two formulas again, we obtain

$$r_{xx} = \frac{r - x r_x}{r^2} = \frac{1}{r} - \frac{x^2}{r^3} = \frac{y^2}{r^3}, \qquad \theta_{xx} = -y\left(-\frac{2}{r^3}\right) r_x = \frac{2xy}{r^4}.$$

We substitute all these expressions into (1). Assuming continuity of the first and second partial derivatives, we have $u_{r\theta} = u_{\theta r}$ and by simplifying we obtain

(2)
$$u_{xx} = \frac{x^2}{r^2} u_{rr} - 2\frac{xy}{r^3} u_{r\theta} + \frac{y^2}{r^4} u_{\theta\theta} + \frac{y^2}{r^3} u_r + 2\frac{xy}{r^4} u_\theta.$$

In a similar fashion it follows that

(3)
$$u_{yy} = \frac{y^2}{r^2} u_{rr} + 2\frac{xy}{r^3} u_{r\theta} + \frac{x^2}{r^4} u_{\theta\theta} + \frac{x^2}{r^3} u_r - 2\frac{xy}{r^4} u_\theta.$$

By adding (2) and (3) we see that the Laplacian in polar coordinates is

(4)
$$\nabla^2 u = \frac{\partial^2 u}{\partial r^2} + \frac{1}{r}\frac{\partial u}{\partial r} + \frac{1}{r^2}\frac{\partial^2 u}{\partial \theta^2}.$$

Problems

1. Transform (4) back into Cartesian coordinates.
2. Show that (4) may be written
$$\nabla^2 u = \frac{1}{r}\frac{\partial}{\partial r}\left(r\frac{\partial u}{\partial r}\right) + \frac{1}{r^2}\frac{\partial^2 u}{\partial \theta^2}.$$

3. If u is independent of θ, then (4) reduces to $\nabla^2 u = u_{rr} + u_r/r$. Derive this result directly from the Laplacian in Cartesian coordinates by assuming that u is independent of θ.
4. **(Laplacian in cylindrical coordinates)** Show that the Laplacian in cylindrical coordinates r, θ, z defined by $x = r\cos\theta$, $y = r\sin\theta$, $z = z$ is
$$\nabla^2 u = u_{rr} + \frac{1}{r}u_r + \frac{1}{r^2}u_{\theta\theta} + u_{zz}.$$

5. **(Laplacian in spherical coordinates)** Show that the Laplacian in spherical coordinates r, θ, ϕ defined by
$$x = r\cos\theta\sin\phi, \qquad y = r\sin\theta\sin\phi, \qquad z = r\cos\phi$$
is
$$\nabla^2 u = u_{rr} + \frac{2}{r}u_r + \frac{1}{r^2}u_{\phi\phi} + \frac{\cot\phi}{r^2}u_\phi + \frac{1}{r^2\sin^2\phi}u_{\theta\theta}.$$

6. Show that the Laplacian in spherical coordinates may be written
$$\nabla^2 u = \frac{1}{r^2}\left(\frac{\partial}{\partial r}(r^2 u_r) + \frac{1}{\sin\phi}\frac{\partial}{\partial \phi}(\sin\phi\, u_\phi) + \frac{1}{\sin^2\phi}u_{\theta\theta}\right).$$

7. Transform the Laplacian in Prob. 5 back into Cartesian coordinates.
8. If x, y are Cartesian coordinates, show that $x^* = x\cos\alpha - y\sin\alpha$, $y^* = x\sin\alpha + y\cos\alpha$ are Cartesian coordinates, and verify by calculation that $\nabla^2 u = u_{x^*x^*} + u_{y^*y^*}$.
9. If the surface of the homogeneous solid sphere $x^2 + y^2 + z^2 \leq R^2$ is kept at temperature zero and the initial temperature in the sphere is $f(r)$ where $r = \sqrt{x^2 + y^2 + z^2}$, show that the temperature $u(r, t)$ in the sphere is the solution of $u_t = c^2(u_{rr} + 2u_r/r)$, satisfying the conditions $u(R, t) = 0$, $u(r, 0) = f(r)$.
10. Show that by setting $v = ru$ the formulas in Prob. 9 take the form $v_t = c^2 v_{rr}$, $v(R, t) = 0$, $v(r, 0) = rf(r)$. Include the condition $v(0, t) = 0$ (which holds because u must be bounded at $r = 0$), and solve the resulting problem by separating variables.

10.10 Circular Membrane. Bessel's Equation

We shall now consider vibrations of the circular membrane of radius R (Fig. 220). Using polar coordinates defined by $x = r \cos \theta$, $y = r \sin \theta$, we see that the wave equation (3′) in Sec. 10.7 takes the form

$$\frac{\partial^2 u}{\partial t^2} = c^2 \left(\frac{\partial^2 u}{\partial r^2} + \frac{1}{r} \frac{\partial u}{\partial r} + \frac{1}{r^2} \frac{\partial^2 u}{\partial \theta^2} \right),$$

cf. (4) in Sec. 10.9. In the present section we shall consider solutions $u(r, t)$ of this equation which are radially symmetric, that is, do not depend on θ. Then the wave equation reduces to the simpler form

(1)
$$\frac{\partial^2 u}{\partial t^2} = c^2 \left(\frac{\partial^2 u}{\partial r^2} + \frac{1}{r} \frac{\partial u}{\partial r} \right).$$

Since the membrane is fixed along the boundary $r = R$, we have the boundary condition

(2) $u(R, t) = 0$ for all $t \geq 0$.

Solutions not depending on θ will occur if the initial conditions do not depend on θ, that is, if they are of the form

(3) $u(r, 0) = f(r)$ [initial deflection $f(r)$]

and

(4) $\left. \dfrac{\partial u}{\partial t} \right|_{t=0} = g(r)$ [initial velocity $g(r)$].

First step. Using the method of separating variables, we first determine solutions of (1) that satisfy the boundary condition (2). We start from

(5) $u(r, t) = W(r)G(t).$

By differentiating and inserting (5) into (1) and dividing the resulting equation by $c^2 WG$ we obtain

$$\frac{\ddot{G}}{c^2 G} = \frac{1}{W} \left(W'' + \frac{1}{r} W' \right)$$

Fig. 220. Circular membrane

where dots denote derivatives with respect to t and primes denote derivatives with respect to r. The expressions on both sides must be equal to a constant, and this constant must be negative, say, $-k^2$, in order to obtain solutions that satisfy the boundary condition without being identically zero. Thus,

$$\frac{\ddot{G}}{c^2 G} = \frac{1}{W}\left(W'' + \frac{1}{r}W'\right) = -k^2.$$

This yields the two ordinary linear differential equations

(6) $$\ddot{G} + \lambda^2 G = 0 \quad \text{where} \quad \lambda = ck$$

and

(7) $$W'' + \frac{1}{r}W' + k^2 W = 0.$$

Second step. We first consider (7). Introducing the new independent variable $s = kr$ we have $1/r = k/s$,

$$W' = \frac{dW}{dr} = \frac{dW}{ds}\frac{ds}{dr} = \frac{dW}{ds}k \quad \text{and} \quad W'' = \frac{d^2 W}{ds^2}k^2.$$

By substituting this into (7) and omitting the common factor k^2 we obtain

$$\frac{d^2 W}{ds^2} + \frac{1}{s}\frac{dW}{ds} + W = 0.$$

This is **Bessel's equation** (1), Sec. 3.5, with $\nu = 0$. The general solution is (cf. Sec. 3.6)

$$W = C_1 J_0(s) + C_2 Y_0(s)$$

where J_0 and Y_0 are the Bessel functions of the first and second kind of order zero. Since the deflection of the membrane is always finite while Y_0 becomes infinite as s approaches zero, we cannot use Y_0 and must choose $C_2 = 0$. Clearly $C_1 \neq 0$ since otherwise $W \equiv 0$. We may set $C_1 = 1$, and then

(8) $$W(r) = J_0(s) = J_0(kr).$$

On the boundary $r = R$ we must have $u(R, t) = W(R)G(t) = 0$. Since $G \equiv 0$ would imply $u \equiv 0$, we require that

$$W(R) = J_0(kR) = 0.$$

The Bessel function J_0 has (infinitely many) real zeros. Let us denote the positive zeros of $J_0(s)$ by $s = \alpha_1, \alpha_2, \cdots$ (cf. Fig. 221). We mention that numerical values (exact to 4 decimal places) are

$$\alpha_1 = 2.4048, \quad \alpha_2 = 5.5201, \quad \alpha_3 = 8.6537, \quad \alpha_4 = 11.7915, \quad \alpha_5 = 14.9309.$$

We see that these zeros are irregularly spaced, and the same is true for the further zeros $\alpha_6, \alpha_7, \cdots$. (More extended tables are included in Ref. [A8] in

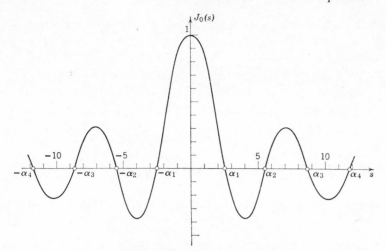

Fig. 221. Bessel function $J_0(s)$

Appendix 1.) Equation (8) now implies

(9) $$kR = \alpha_m \quad \text{or} \quad k = k_m = \frac{\alpha_m}{R}, \qquad m = 1, 2, \cdots.$$

Hence the functions

(10) $$W_m(r) = J_0(k_m r) = J_0\left(\frac{\alpha_m}{R} r\right) \qquad m = 1, 2, \cdots$$

are solutions of (7) which vanish at $r = R$.

The corresponding general solutions of (6) with $\lambda = \lambda_m = ck_m$ are

$$G_m(t) = a_m \cos \lambda_m t + b_m \sin \lambda_m t.$$

Hence the functions

(11) $$u_m(r, t) = W_m(r)\, G_m(t) = (a_m \cos \lambda_m t + b_m \sin \lambda_m t) J_0(k_m r)$$

where $m = 1, 2, \cdots$, are solutions of the wave equation (1), satisfying the boundary condition (2). These are the *eigenfunctions* of our problem, and the corresponding *eigenvalues* are λ_m.

The vibration of the membrane corresponding to u_m is called the mth **normal mode**; it has the frequency $\lambda_m/2\pi$ cycles per unit time. Since the zeros of J_0 are not regularly spaced on the axis (in contrast to the zeros of the sine functions appearing in the case of the vibrating string), the sound of a drum is entirely different from that of a violin. The forms of the normal modes can easily be obtained from Fig. 221 and are shown in Fig. 222. For $m = 1$, all the points of the membrane move upward (or downward) at the same time. For $m = 2$, the situation is as follows. The function

$$W_2(r) = J_0\left(\frac{\alpha_2}{R} r\right)$$

is zero for $\alpha_2 r/R = \alpha_1$ or $r = \alpha_1 R/\alpha_2$. The circle $r = \alpha_1 R/\alpha_2$ is, therefore, a nodal

$m = 1$ $m = 2$ $m = 3$

Fig. 222. Normal modes of the circular membrane in the case of vibrations independent of the angle

line, and when at some instant the central part of the membrane moves upward, the outer part $(r > \alpha_1 R/\alpha_2)$ moves downward, and conversely. The solution $u_m(r, t)$ has $m - 1$ nodal lines which are concentric circles (Fig. 222).

Third step. To obtain a solution that also satisfies the initial conditions (3) and (4), we may proceed as in the case of the vibrating string, that is, we consider the series[7]

$$(12) \qquad u(r, t) = \sum_{m=1}^{\infty} W_m(r) \, G_m(t) = \sum_{m=1}^{\infty} (a_m \cos \lambda_m t + b_m \sin \lambda_m t) J_0\left(\frac{\alpha_m}{R} r\right).$$

Setting $t = 0$ and using (3), we obtain

$$(13) \qquad u(r, 0) = \sum_{m=1}^{\infty} a_m J_0\left(\frac{\alpha_m}{R} r\right) = f(r).$$

Hence, for (12) to satisfy (3), the a_m must be the coefficients of the Fourier-Bessel series which represents $f(r)$ in terms of $J_0(\alpha_m r/R)$, that is [cf. (9) in Sec. 3.9],

$$a_m = \frac{2}{R^2 J_1^{\,2}(\alpha_m)} \int_0^R r f(r) \, J_0\left(\frac{\alpha_m}{R} r\right) dr \qquad m = 1, 2, \cdots.$$

Differentiability of $f(r)$ in the interval $0 \le r \le R$ is sufficient for the existence of the development (13), cf. Ref. [B14]. The coefficients b_m in (12) can be determined from (4) in a similar fashion. To obtain numerical values of a_m and b_m, we may apply one of the usual methods of approximate integration, using tables of J_0 and J_1.

Problems

1. What is the reason for using polar coordinates in the present consideration?
2. If the tension of the membrane is increased, how does the frequency of each normal mode (11) change?

[7] We shall not consider the problems of convergence and uniqueness.

3. Determine numerical values of the radii of the nodal lines of u_2 and u_3 [cf. (11)] when $R = 1$.
4. Same question as in Prob. 3, for u_4, u_5, and u_6.
5. Sketch a figure similar to Fig. 222, for u_4, u_5, u_6.
6. Is it possible that, for fixed c and R, two or more functions u_m [cf. (11)] that have different nodal lines correspond to the same eigenvalue?
7. Show that for (12) to satisfy (4),

$$b_m = \frac{2}{c\alpha_m R J_1{}^2(\alpha_m)} \int_0^R rg(r)J_0(\alpha_m r/R)\, dr, \qquad m = 1, 2, \cdots.$$

Find the deflection $u(r, t)$ of the circular membrane of radius $R = 1$, if $c = 1$, the initial velocity is zero, and the initial deflection is $f(r)$ where

8. $f = 0.1J_0(\alpha_2 r)$ 9. $f = k(1 - r^2)$ 10. $f = k(1 - r^4)$
   Hint. Remember Probs. 25–32 in Sec. 3.9.

11. **(Vibrations depending on r and θ)** Show that substitution of $u = F(r, \theta)G(t)$ into the wave equation

(14)
$$u_{tt} = c^2 \left(u_{rr} + \frac{1}{r} u_r + \frac{1}{r^2} u_{\theta\theta} \right)$$

leads to

(15)
$$\ddot{G} + \lambda^2 G = 0, \qquad \text{where } \lambda = ck,$$

(16)
$$F_{rr} + \frac{1}{r} F_r + \frac{1}{r^2} F_{\theta\theta} + k^2 F = 0.$$

12. Show that substitution of $F = W(r)Q(\theta)$ in (16) yields

(17)
$$Q'' + n^2 Q = 0,$$

(18)
$$r^2 W'' + rW' + (k^2 r^2 - n^2)W = 0.$$

13. Show that $Q(\theta)$ must be periodic with period 2π, and, therefore, $n = 0, 1, \cdots$ in (17) and (18). Show that this yields the solutions $Q_n = \cos n\theta$, $Q_n{}^* = \sin n\theta$, $W_n = J_n(kr)$, $n = 0, 1, \cdots$.
14. Show that the boundary condition

(19)
$$u(R, \theta, t) = 0$$

leads to the values $k = k_{mn} = \alpha_{mn}/R$ where $s = \alpha_{mn}$ is the mth positive zero of $J_n(s)$.
15. Show that solutions of (14) which satisfy (19) are

(20)
$$u_{mn} = (A_{mn} \cos ck_{mn}t + B_{mn} \sin ck_{mn}t)J_n(k_{mn}r)\cos n\theta,$$
$$u_{mn}^* = (A_{mn}^* \cos ck_{mn}t + B_{mn}^* \sin ck_{mn}t)J_n(k_{mn}r)\sin n\theta.$$

16. Show that $u_{m0}^* \equiv 0$ and u_{m0} is identical with (11) in the current section.
17. Show that u_{mn} has $m + n - 1$ nodal lines. (Cf. Fig. 223.)

u_{11} u_{12} u_{23}

Fig. 223. Nodal lines of some of the solutions (20)

18. Graph the nodal lines of u_{3m}, $m = 1, 2, 3$.

19. Graph the nodal lines of u_{mn}^*, $m, n = 1, 2, 3$.

20. Show that the initial condition $u_t(r, \theta, 0) = 0$ leads to $B_{mn} = 0$, $B_{mn}^* = 0$ in (20).

10.11 *Laplace's Equation. Potential*

One of the most important partial differential equations appearing in physics is Laplace's equation

$$\text{(1)} \qquad\qquad\qquad \nabla^2 u = 0.$$

Here $\nabla^2 u$ is the Laplacian of u. With respect to Cartesian coordinates x, y, z in space,

$$\text{(2)} \qquad\qquad \nabla^2 u = \frac{\partial^2 u}{\partial x^2} + \frac{\partial^2 u}{\partial y^2} + \frac{\partial^2 u}{\partial z^2}.$$

The theory of the solutions of Laplace's equation is called **potential theory.** Solutions of (1) that have *continuous* second-order partial derivatives are called **harmonic functions.**

The two-dimensional case, when u depends on two variables only, can most conveniently be treated by methods of complex analysis and will be considered in Sec. 11.3 and Chap. 17.

To illustrate the importance of Laplace's equation in engineering mathematics, let us mention some basic applications.

Laplace's equation occurs in connection with gravitational forces. In Ex. 4, Sec. 7.8, we have seen that if a particle A of mass M is fixed at a point (ξ, η, ζ) and another particle B of mass m is at a point (x, y, z), then A attracts B, the gravitational force being the gradient of the scalar function

$$u(x, y, z) = \frac{c}{r}, \qquad c = GMm = const,$$

$$r = \sqrt{(x - \xi)^2 + (y - \eta)^2 + (z - \zeta)^2} \qquad (> 0).$$

This function is called the potential of the gravitational field, and it satisfies Laplace's equation.

The extension to the potential and force due to a continuous distribution of mass is quite direct. If a mass of density $\rho(\xi, \eta, \zeta)$ is distributed throughout a region R in space, then the corresponding potential u at a point (x, y, z) not occupied by mass is defined to be

$$\text{(3)} \qquad\qquad u(x, y, z) = k \iiint\limits_R \frac{\rho}{r} \, d\xi \, d\eta \, d\zeta \qquad\qquad (k > 0).$$

Since $1/r$ $(r > 0)$ is a solution of (1), that is, $\nabla^2(1/r) = 0$, and ρ does not depend on x, y, z, we obtain

$$\nabla^2 u = k \iiint\limits_R \rho \nabla^2 \left(\frac{1}{r}\right) d\xi \, d\eta \, d\zeta = 0,$$

that is, the gravitational potential defined by (3) satisfies Laplace's equation at any point which is not occupied by matter.

In *electrostatics* the electrical force of attraction or repulsion between charged particles is governed by *Coulomb's law* (cf. Sec. 7.8), which is of the same mathematical form as Newton's law of gravitation. From this it follows that the field created by a distribution of electrical charges can be described mathematically by a potential function which satisfies Laplace's equation at any point not occupied by charges.

In Chap. 17 we shall see that Laplace's equation also appears in the theory of incompressible fluid flow.

Furthermore, the basic equation in problems of heat conduction is the heat equation (cf. Secs. 8.9 and 10.5)

$$u_t = c^2 \nabla^2 u,$$

and if the temperature u is independent of time t ("*steady state*"), this equation reduces to Laplace's equation.

In most applications leading to Laplace's equation, it is required to solve a **boundary value problem,** that is, to determine the solution of (1) satisfying given boundary conditions on certain surfaces. It is then necessary to introduce coordinates in space such that those surfaces can be represented in a simple manner. This requires the transformation of the Laplacian (2) into other coordinate systems. Of course, such a transformation is quite similar to that in the case of the Laplacian of a function of two variables (cf. Sec. 10.9).

From (4) in Sec. 10.9 it follows immediately that the Laplacian of a function u in **cylindrical coordinates**[8] (cf. Fig. 224)

(4) $$r = \sqrt{x^2 + y^2} \qquad \theta = \arctan \frac{y}{x}, \qquad z = z$$

is

(5) $$\nabla^2 u = \frac{\partial^2 u}{\partial r^2} + \frac{1}{r}\frac{\partial u}{\partial r} + \frac{1}{r^2}\frac{\partial^2 u}{\partial \theta^2} + \frac{\partial^2 u}{\partial z^2}.$$

Fig. 224. Cylindrical coordinates   **Fig. 225.** Spherical coordinates

---

[8] Observe that θ is not completely determined by the ratio y/x but we must also take the signs of x and y into account.

Other important coordinates are **spherical coordinates**[9] r, θ, ϕ which are related to Cartesian coordinates as follows (cf. Fig. 225):

(6) $x = r \cos \theta \sin \phi$, $y = r \sin \theta \sin \phi$, $z = r \cos \phi$.

The Laplacian of a function u in spherical coordinates is

(7) $$\nabla^2 u = \frac{\partial^2 u}{\partial r^2} + \frac{2}{r}\frac{\partial u}{\partial r} + \frac{1}{r^2}\frac{\partial^2 u}{\partial \phi^2} + \frac{\cot \phi}{r^2}\frac{\partial u}{\partial \phi} + \frac{1}{r^2 \sin^2 \phi}\frac{\partial^2 u}{\partial \theta^2}.$$

This may also be written

(7′) $$\nabla^2 u = \frac{1}{r^2}\left[\frac{\partial}{\partial r}\left(r^2 \frac{\partial u}{\partial r}\right) + \frac{1}{\sin \phi}\frac{\partial}{\partial \phi}\left(\sin \phi \frac{\partial u}{\partial \phi}\right) + \frac{1}{\sin^2 \phi}\frac{\partial^2 u}{\partial \theta^2}\right].$$

This formula can be derived in a manner similar to that in Sec. 10.9; the details are left as an exercise for the reader.

Problems

1. Verify that $u = c/r$ satisfies Laplace's equation in spherical coordinates.
2. Show that the only solution of $\nabla^2 u = 0$ depending only on $r = \sqrt{x^2 + y^2 + z^2}$ is $u = c/r + k$, where c and k are constants.
3. Determine c and k in Prob. 2 such that u represents the electrostatic potential between two concentric spheres of radii $r_1 = 10$ cm and $r_2 = 2$ cm kept at the potentials $U_1 = 110$ volts and $U_2 = 10$ volts, respectively.
4. Show that the only solution of the two-dimensional Laplace equation depending only on $r = \sqrt{x^2 + y^2}$ is $u = c \ln r + k$.
5. Find the electrostatic potential between two coaxial cylinders of radii $r_1 = 10$ cm and $r_2 = 2$ cm kept at the potentials $U_1 = 110$ volts and $U_2 = 10$ volts, respectively. Graph and compare the solutions of Probs. 3 and 5.
6. Express the spherical coordinates defined by (6) in terms of Cartesian coordinates.
7. Verify (5) by transforming $\nabla^2 u$ back into Cartesian coordinates.
8. Find the Laplacian in rectangular coordinates $x^* = ax$, $y^* = by$, $z^* = cz$ where x, y, z are Cartesian coordinates.
9. Show that substitution of $u = U(x, y, z)e^{-i\omega t}$ ($i = \sqrt{-1}$) into the three-dimensional wave equation $u_{tt} = c^2\nabla^2 u$ yields the so-called **Helmholtz equation**[10]

$$\nabla^2 U + k^2 U = 0 \qquad\qquad k = \omega/c.$$

Find the steady-state (time-independent) temperature distribution:

10. Between two parallel plates $x = x_0$ and $x = x_1$ kept at the temperatures u_0 and u_1, respectively.
11. Between two coaxial circular cylinders of radii r_0 and r_1 kept at the temperatures u_0 and u_1, respectively.
12. Between two concentric spheres of radii r_0 and r_1 kept at the temperatures u_0 and u_1, respectively.

Show that the following functions $u = f(x, y)$ satisfy the Laplace equation and plot some of the equipotential lines $u = const$.

13. $x^2 - y^2$ 14. $(x + 1)(y + 1)$ 15. $x^3 - 3xy^2$
16. $y/(x^2 + y^2)$ 17. $(x^2 - y^2)/(x^2 + y^2)^2$ 18. $x/(x^2 + y^2)$

[9] Sometimes also called *polar coordinates*.
[10] HERMANN VON HELMHOLTZ (1821–1894), German physicist.

19. If $u(r, \theta)$ satisfies $\nabla^2 u = 0$, show that $v(r, \theta) = u(r^{-1}, \theta)$ satisfies $\nabla^2 v = 0$. (r and θ are polar coordinates.)

20. Let r, θ, ϕ be spherical coordinates. If $u(r, \theta, \phi)$ satisfies $\nabla^2 u = 0$, show that $v(r, \theta, \phi) = r^{-1}u(r^{-1}, \theta, \phi)$ satisfies $\nabla^2 v = 0$.

10.12   Laplace's Equation in Spherical Coordinates. Legendre's Equation

Let us consider a typical boundary value problem that involves Laplace's equation in spherical coordinates. Suppose that a sphere S of radius R is kept at a fixed distribution of electric potential

$$(1) \qquad\qquad u(R, \theta, \phi) = f(\phi)$$

where r, θ, ϕ are the spherical coordinates defined in the last section, with the origin at the center of S, and $f(\phi)$ is a given function. We wish to find the potential u at all points in space which is assumed to be free of further charges. Since the potential on S is independent of θ, so is the potential in space. Thus $\partial^2 u/\partial\theta^2 = 0$, and from $(7')$ in Sec. 10.11 we see that Laplace's equation reduces to

$$(2) \qquad\qquad \frac{\partial}{\partial r}\left(r^2 \frac{\partial u}{\partial r}\right) + \frac{1}{\sin\phi}\frac{\partial}{\partial\phi}\left(\sin\phi \frac{\partial u}{\partial\phi}\right) = 0.$$

Furthermore, at infinity the potential will be zero:

$$(3) \qquad\qquad \lim_{r\to\infty} u(r, \phi) = 0.$$

We shall solve the boundary value problem consisting of the equation (2) and the boundary conditions (1) and (3) by the method of separating variables. Substituting a solution of the form

$$u(r, \phi) = G(r)H(\phi)$$

into (2), and dividing the resulting equation by GH, we obtain

$$\frac{1}{G}\frac{d}{dr}\left(r^2 \frac{dG}{dr}\right) = -\frac{1}{H\sin\phi}\frac{d}{d\phi}\left(\sin\phi \frac{dH}{d\phi}\right)$$

By the usual argument, the two sides of this equation must be equal to a constant, say, k, so that

$$(4) \qquad\qquad \frac{1}{\sin\phi}\frac{d}{d\phi}\left(\sin\phi \frac{dH}{d\phi}\right) + kH = 0$$

and

$$\frac{1}{G}\frac{d}{dr}\left(r^2 \frac{dG}{dr}\right) = k.$$

The last equation may be written

$$r^2 G'' + 2r G' - kG = 0.$$

This is Cauchy's equation, and from Sec. 2.7 we know that it has solutions of the form $G = r^\alpha$. These solutions will have a particularly simple form if we change our notation and write $n(n + 1)$ for k. In fact, then the equation becomes

(5) $$r^2 G'' + 2r G' - n(n + 1)G = 0$$

where n is still arbitrary. By substituting $G = r^\alpha$ into (5) we have

$$[\alpha(\alpha - 1) + 2\alpha - n(n + 1)]r^\alpha = 0.$$

The zeros of the expression in brackets are $\alpha = n$ and $\alpha = -n - 1$. Hence we obtain the solutions

(6) $$G_n(r) = r^n \quad \text{and} \quad G_n^*(r) = \frac{1}{r^{n+1}}.$$

Introducing $k = n(n + 1)$ in (4) and setting

$$\cos \phi = w,$$

we have $\sin^2 \phi = 1 - w^2$ and

$$\frac{d}{d\phi} = \frac{d}{dw}\frac{dw}{d\phi} = -\sin \phi \frac{d}{dw}.$$

Consequently, (4) takes the form

(7) $$\frac{d}{dw}\left[(1 - w^2)\frac{dH}{dw}\right] + n(n + 1)H = 0$$

or

(7') $$(1 - w^2)\frac{d^2 H}{dw^2} - 2w\frac{dH}{dw} + n(n + 1)H = 0.$$

This is *Legendre's equation* (cf. Sec. 3.3). For integral[11] $n = 0, 1, \cdots$, the Legendre polynomials

$$H = P_n(w) = P_n(\cos \phi), \qquad\qquad n = 0, 1, \cdots,$$

are solutions of (7). We thus obtain the following two sequences of solutions $u = GH$ of Laplace's equation (2):

$$u_n(r, \phi) = A_n r^n P_n(\cos \phi), \qquad u_n^*(r, \phi) = \frac{B_n}{r^{n+1}} P_n(\cos \phi)$$

where $n = 0, 1, \cdots$, and A_n and B_n are constants.

---

[11] So far, n was arbitrary since k was arbitrary. It can be shown that the restriction of n to real integral values is necessary to make the solution of (7) continuous, together with its derivative of the first order, in the interval $-1 \leqq w \leqq 1$ or $0 \leqq \phi \leqq \pi$.

To find a solution of (2), valid at points *inside* the sphere and satisfying (1), we consider the series[12]

$$(8) \qquad u(r, \phi) = \sum_{n=0}^{\infty} A_n r^n P_n(\cos \phi).$$

For (8) to satisfy (1) we must have

$$(9) \qquad u(R, \phi) = \sum_{n=0}^{\infty} A_n R^n P_n(\cos \phi) = f(\phi);$$

that is, (9) must be the generalized Fourier series of $f(\phi)$ in terms of Legendre polynomials. From (6) and (7) in Sec. 3.7 and (2) in Sec. 3.9 it follows that

$$A_n R^n = \frac{2n + 1}{2} \int_{-1}^{1} \tilde{f}(w) P_n(w) \, dw$$

where $\tilde{f}(w)$ denotes $f(\phi)$ as a function of $w = \cos \phi$. Since $dw = -\sin \phi \, d\phi$, and the limits of integration -1 and 1 correspond to $\phi = \pi$ and $\phi = 0$, respectively, we also have

$$(10) \qquad A_n = \frac{2n + 1}{2R^n} \int_{0}^{\pi} f(\phi) P_n(\cos \phi) \sin \phi \, d\phi, \qquad n = 0, 1, \cdots.$$

Thus the series (8) with coefficients (10) is the solution of our problem for points inside the sphere.

To find the solution *exterior* to the sphere, we cannot use the functions $u_n(r, \phi)$ because these functions do not satisfy (3), but we may use the functions $u_n{}^*(r, \phi)$, which satisfy (3), and proceed as before. This leads to the solution

$$(11) \qquad u(r, \phi) = \sum_{n=0}^{\infty} \frac{B_n}{r^{n+1}} P_n(\cos \phi) \qquad\qquad (r \geqq R)$$

with coefficients

$$(12) \qquad B_n = \frac{2n + 1}{2} R^{n+1} \int_{0}^{\pi} f(\phi) P_n(\cos \phi) \sin \phi \, d\phi.$$

Problems

1. Graph the functions $P_n(\cos \phi)$ for $n = 0, 1, 2, 3$ [cf. (11'), Sec. 3.3].
2. Find the surfaces on which the functions u_1, u_2, u_3 are zero.

Let r, θ, ϕ be the spherical coordinates used in the text. Find the potential in the interior of the sphere $R = 1$, assuming that there are no charges in the interior and the potential on the surface is $f(\phi)$ where

3. $f(\phi) = 1$	4. $f(\phi) = \cos \phi$	5. $f(\phi) = \cos^2 \phi$
6. $f(\phi) = \cos^3 \phi$	7. $f(\phi) = \cos 2\phi$	8. $f(\phi) = \cos 3\phi$

[12] Convergence will not be considered. It can be shown that if $f(\phi)$ and $f'(\phi)$ are piecewise continuous in the interval $0 \leqq \phi \leqq \pi$, the series (8) with coefficients (10) can be differentiated termwise twice with respect to r and with respect to ϕ and the resulting series converge and represent $\partial^2 u/\partial r^2$ and $\partial^2 u/\partial \phi^2$, respectively. Hence the series (8) with coefficients (10) is then the solution of our problem inside the sphere.

9. Show that in Prob. 3 the potential exterior to the sphere is the same as that of a point charge at the origin.

10. Find the potential exterior to the sphere in Probs. 4–8.

11. Verify by substitution that $u_n(r, \phi)$ and $u_n^*(r, \phi)$, $n = 0, 1, 2$ (p. 458) are solutions of (2).

12. **(Transmission line equations)**   Consider a long cable or telephone wire (Fig. 226) which is imperfectly insulated so that leaks occur along the entire length of the cable. The source S

Fig. 226. Transmission line

of the current $i(x, t)$ in the cable is at $x = 0$, the receiving end T at $x = l$. The current flows from S to T, through the load, and returns to the ground. Let the constants R, L, C, and G denote the resistance, inductance, capacitance to ground, and conductance to ground, respectively, of the cable per unit length. Show that

$$-\frac{\partial u}{\partial x} = Ri + L\frac{\partial i}{\partial t} \qquad \text{(First transmission line equation)}$$

where $u(x, t)$ is the potential in the cable. *Hint.* Apply Kirchhoff's second law to a small portion of the cable between x and $x + \Delta x$ (difference of the potentials at x and $x + \Delta x =$ resistive drop + inductive drop).

13. Show that for the cable in Prob. 12,

$$-\frac{\partial i}{\partial x} = Gu + C\frac{\partial u}{\partial t} \qquad \text{(Second transmission line equation)}.$$

Hint. Use Kirchhoff's first law (difference of the currents at x and $x + \Delta x =$ loss due to leakage to ground + capacitive loss).

14. Show that elimination of i from the transmission line equations leads to

$$u_{xx} = LCu_{tt} + (RC + GL)u_t + RGu.$$

15. Find a similar second-order differential equation for the current in the cable.

16. **(High frequency line equations)**   Show that in the case of alternating currents of high frequencies the equations in Probs. 14 and 15 can be approximated by the so-called *high frequency line equations*

$$u_{xx} = LCu_{tt}, \qquad i_{xx} = LCi_{tt}.$$

Solve the first of these equations, assuming that the initial potential is $U_0 \sin(\pi x/l)$, $\partial u/\partial t = 0$ when $t = 0$, and $u = 0$ at the ends $x = 0$ and $x = l$ for all t.

17. **(Telegraph equations)**   For a submarine cable G is negligible and the frequencies are low. Show that this leads to the so-called *submarine cable equations* or *telegraph equations*

$$u_{xx} = RCu_t, \qquad i_{xx} = RCi_t.$$

18. Find the potential in a submarine cable with ends ($x = 0$, $x = l$) grounded and initial voltage distribution $U_0 = const.$

19. What problem in heat conduction is the analogue of Prob. 18?

20. Under what conditions does the problem of a vibrating string lead to the same mathematical formulation as Prob. 16?

CHAPTER **11**

Complex Analytic Functions

Many engineering problems may be treated and solved by methods of complex analysis. Roughly speaking, these problems can be subdivided into two large classes. The first class consists of "elementary problems" for which the knowledge of complex numbers gained in college algebra and calculus is sufficient. For example, many applications in connection with electric circuits and mechanical vibrating systems are of this type. The second class of problems requires a detailed knowledge of the theory of complex analytic functions ("theory of functions," for short) and the powerful and elegant methods used in this branch of mathematics. Interesting problems in the theory of heat, in fluid dynamics, and in electrostatics belong to this category.

The following chapters will be devoted to the major parts of the theory of complex analytic functions and their applications. We shall see that the importance of these functions in engineering mathematics has the following three main roots.

1. The real and imaginary parts of an analytic function are solutions of Laplace's equation in two independent variables. Consequently, two-dimensional potential problems can be treated by methods developed in connection with analytic functions.

2. Many complicated real and complex integrals that occur in applications can be evaluated by methods of complex integration.

3. The majority of nonelementary functions appearing in engineering mathematics are analytic functions, and the consideration of these functions for complex values of the independent variable leads to a much deeper and more detailed knowledge of their properties.

In the present chapter we shall define analytic functions and consider their fundamental properties. The second half of the chapter will be devoted to the most important elementary complex functions.

Prerequisites for this chapter: elementary calculus.
References: Appendix 1, Part F.
Answers to problems: Appendix 2.

461

11.1 Complex Numbers

The fact that there are equations such as

$$x^2 + 3 = 0, \qquad x^2 - 10x + 40 = 0,$$

which are not satisfied by any real number, leads to the introduction of complex numbers.[1] These are ordered pairs of real numbers subject to rules of arithmetical operations defined below. They are usually written in the convenient form $x + iy$ (or $x + yi$), where x and y are real numbers. Of course, we may denote a complex number by a single letter, say, z; thus

$$z = x + iy.$$

The symbol i is called the **imaginary unit.** The number x is called the **real part** of the complex number z and y is called the **imaginary part** of z, and we write

$$\operatorname{Re} z = x, \qquad \operatorname{Im} z = y.$$

For example, the real part of $4 - 3i$ is 4, and the imaginary part is -3, and we may write $\operatorname{Re}(4 - 3i) = 4$, $\operatorname{Im}(4 - 3i) = -3$.

Complex numbers can be represented as points in the plane. For this purpose we choose two perpendicular coordinate axes, the horizontal x-axis and the vertical y-axis and on both axes the same unit of length (Fig. 227). The xy-coordinate system thus obtained is called a **Cartesian coordinate system.**[2] To the complex number $z = x + iy$ there corresponds the point P with Cartesian coordinates (x, y). The xy-plane in which the complex numbers are represented geometrically in this fashion is called the **complex plane** or *Argand diagram*.[3] The x-axis is called the **real axis,** and the y-axis the **imaginary axis.**

Fig. 227. The complex plane   Fig. 228. The number $4 - 3i$ in the
                                                        complex plane

---

[1] First to use complex numbers for this purpose was the Italian mathematician GIROLAMO CARDANO (1501–1576), who found the formula for solving cubic equations. The term "complex number" was introduced by the great German mathematician CARL FRIEDRICH GAUSS (cf. the footnote in Sec. 3.4), who also paved the way for a general and systematic use of complex numbers.

[2] Named after the French philosopher and mathematician RENATUS CARTESIUS (latinized for RENÉ DESCARTES (1596–1650)), who invented analytic geometry.

[3] JEAN ROBERT ARGAND (1768–1822), French mathematician. His paper on the complex plane appeared in 1806, nine years after a similar memoir by the Norwegian mathematician CASPAR WESSEL (1745–1818).

We shall now define equality of complex numbers and the rules of arithmetical operations.

Two complex numbers

$$z_1 = x_1 + iy_1 \quad \text{and} \quad z_2 = x_2 + iy_2$$

are defined to be **equal** if and only if their real parts are equal and their imaginary parts are equal, that is,

$$z_1 = z_2 \quad \textit{if and only if} \quad x_1 = x_2 \text{ and } y_1 = y_2.$$

If z_1 and z_2 are different (not equal), then we may write $z_1 \neq z_2$.

Inequalities between complex numbers, such as $z_1 < z_2$ or $z_1 \geqq z_2$, have no meaning. (Inequalities may hold between the *absolute values* of complex numbers, as we shall discuss later.)

Addition. The sum $z_1 + z_2$ of z_1 and z_2 is defined as the complex number obtained by adding the real parts and the imaginary parts of z_1 and z_2, that is,

(1) $$z_1 + z_2 = (x_1 + x_2) + i(y_1 + y_2).$$

We see that addition of complex numbers is in accordance with the *"parallelogram law"* by which forces are added in mechanics (Fig. 229).

Subtraction. This operation is defined as the inverse operation of addition; that is, the difference $z_1 - z_2$ is the complex number z for which $z_1 = z + z_2$. Obviously (cf. Fig. 230)

(2) $$z_1 - z_2 = (x_1 - x_2) + i(y_1 - y_2).$$

Multiplication. The product $z_1 z_2$ is defined as the complex number

(3) $$z_1 z_2 = (x_1 + iy_1)(x_2 + iy_2) = (x_1 x_2 - y_1 y_2) + i(x_1 y_2 + x_2 y_1),$$

which is obtained formally by applying the ordinary rules of arithmetic for real numbers, treating the symbol i as a number, and replacing $i^2 = ii$ by -1.

Division. This operation is defined as the inverse operation of multiplication; that is, the quotient $z = z_1/z_2$ is the complex number $z = x + iy$ for which

$$z_1 = zz_2 = (x + iy)(x_2 + iy_2).$$

Fig. 229. Addition of complex numbers

Fig. 230. Subtraction of complex numbers

By using (3) we see that this can be written

$$x_1 + iy_1 = (x_2 x - y_2 y) + i(y_2 x + x_2 y).$$

By definition of equality the real parts and the imaginary parts on both sides must be equal:

$$x_1 = x_2 x - y_2 y$$
$$y_1 = y_2 x + x_2 y.$$

This is a system of two linear equations in the unknowns x and y. Assuming that x_2 and y_2 are not both zero (briefly written $z_2 \neq 0$), we obtain the unique solution

$$x = \frac{x_1 x_2 + y_1 y_2}{x_2{}^2 + y_2{}^2}, \qquad y = \frac{x_2 y_1 - x_1 y_2}{x_2{}^2 + y_2{}^2} \qquad (z_2 \neq 0).$$

From this we see that the number $z = x + iy$ can be obtained formally by multiplying both the numerator and the denominator of the quotient z_1/z_2 by $x_2 - iy_2$; thus

(4) $$z = \frac{z_1}{z_2} = \frac{(x_1 + iy_1)(x_2 - iy_2)}{(x_2 + iy_2)(x_2 - iy_2)} = \frac{x_1 x_2 + y_1 y_2}{x_2{}^2 + y_2{}^2} + i \frac{x_2 y_1 - x_1 y_2}{x_2{}^2 + y_2{}^2}.$$

Example 1. Let $z_1 = 2 - 3i$ and $z_2 = -5 + i$. Then

$$z_1 + z_2 = -3 - 2i, \qquad z_1 - z_2 = 7 - 4i, \qquad z_1 z_2 = -7 + 17i,$$

$$\frac{z_1}{z_2} = \frac{(2 - 3i)(-5 - i)}{(-5 + i)(-5 - i)} = \frac{-10 - 2i + 15i - 3}{25 + 1} = -\frac{1}{2} + \frac{i}{2}.$$

For any complex numbers z_1, z_2, z_3 we have

$$z_1 + z_2 = z_2 + z_1$$
$$z_1 z_2 = z_2 z_1 \qquad \text{(\textit{Commutative laws})}$$

(5) $$(z_1 + z_2) + z_3 = z_1 + (z_2 + z_3)$$
$$(z_1 z_2) z_3 = z_1 (z_2 z_3) \qquad \text{(\textit{Associative laws})}$$
$$z_1 (z_2 + z_3) = z_1 z_2 + z_1 z_3 \qquad \text{(\textit{Distributive law})}.$$

These laws follow immediately from the corresponding laws for real numbers and the preceding definitions of the algebraic operations for complex numbers.

Let $z = x + iy$ be any complex number; then $x - iy$ is called the **conjugate** of z and is denoted by $\bar{z}$ (Fig. 231). Thus

$$z = x + iy, \qquad \bar{z} = x - iy.$$

For example, the conjugate of $4 + 7i$ is $4 - 7i$. Now we obtain

$$z + \bar{z} = 2x, \qquad z - \bar{z} = 2iy$$

and therefore

(6) $$\text{Re } z = x = \frac{1}{2}(z + \bar{z}), \qquad \text{Im } z = y = \frac{1}{2i}(z - \bar{z}).$$

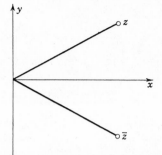

Fig. 231. Conjugate complex numbers

Obviously,

(7)
$$\overline{(z_1 + z_2)} = \bar{z}_1 + \bar{z}_2, \qquad \overline{(z_1 - z_2)} = \bar{z}_1 - \bar{z}_2,$$
$$\overline{(z_1 z_2)} = \bar{z}_1 \bar{z}_2, \qquad \overline{\left(\frac{z_1}{z_2}\right)} = \frac{\bar{z}_1}{\bar{z}_2}.$$

In the fundamental operations of arithmetic the complex number $x + i0$ behaves like the real number x. Indeed, equality, sum, and product as defined for two such numbers $x_1 + i0$ and $x_2 + i0$ carry over, respectively, into equality, sum, and product as defined for two real numbers x_1 and x_2. For this reason we may say that a complex number whose imaginary part is zero is a real number, and write simply x instead of $x + i0$. The complex number $0 + iy$ is written iy and is called a **pure imaginary number.** Defining i to mean $0 + i1$ we have

(8)
$$i^2 = (0 + i1)(0 + i1) = -1$$

and so i^2 may be replaced by -1 whenever it occurs. Consequently,

(9)
$$i^2 = -1, \qquad i^3 = -i, \qquad i^4 = 1, \qquad i^5 = i, \cdots$$
$$\frac{1}{i} = \frac{i}{i^2} = -i, \qquad \frac{1}{i^2} = -1, \cdots.$$

Polar form of complex numbers. If we introduce polar coordinates r, θ in the complex plane by setting

(10)
$$x = r \cos \theta, \qquad y = r \sin \theta,$$

then the complex number $z = x + iy$ may be written

(11)
$$z = r \cos \theta + ir \sin \theta = r(\cos \theta + i \sin \theta).$$

This is known as the **polar form** or *trigonometric form* of a complex number. r is called the **absolute value** or **modulus** of z and is denoted by $|z|$. Thus (Fig. 232)

(12)
$$|z| = r = \sqrt{x^2 + y^2} = \sqrt{z\bar{z}} \qquad (\geqq 0).$$

The directed angle measured from the positive x-axis to OP is called the **argument** of z and is denoted by arg z; *angles will be measured positive in the counterclockwise sense and in terms of radians.* As in trigonometry we have

(13)
$$\arg z = \theta = \arc \sin \frac{y}{r} = \arc \cos \frac{x}{r} = \arc \tan \frac{y}{x}.$$

Fig. 232. Complex plane, polar
form of a complex number

Fig. 233. Distance between
two points in the complex
plane

Geometrically, $|z|$ is the distance of the point z from the origin (Fig. 232). Hence the inequality

$$|z_1| > |z_2|$$

means that the point z_1 is farther from the origin than the point z_2, and $|z_1 - z_2|$ is the distance between the points z_1 and z_2 (Fig. 233).

For given z ($\neq 0$), the argument θ is determined only up to integral multiples of 2π. The value of θ which lies in the interval

$$-\pi < \theta \leq \pi$$

is called the **principal value** of the argument of z.

Example 2. Let $z = 1 + i$. Then

$$z = \sqrt{2}\left(\cos\frac{\pi}{4} + i\sin\frac{\pi}{4}\right), \qquad |z| = \sqrt{2}, \qquad \arg z = \frac{\pi}{4} \pm 2n\pi \qquad (n = 0, 1, \cdots).$$

The principal value of $\arg z$ is $\pi/4$ (Fig. 234).

Example 3. Let $z = 3 + 3\sqrt{3}\,i$. Then $z = 6\left(\cos\frac{\pi}{3} + i\sin\frac{\pi}{3}\right)$, $|z| = 6$, and the principal value of $\arg z$ is $\pi/3$.

The polar form of complex numbers is particularly useful in analyzing multiplication and division. Let

$$z_1 = r_1(\cos\theta_1 + i\sin\theta_1) \qquad \text{and} \qquad z_2 = r_2(\cos\theta_2 + i\sin\theta_2).$$

Then the product is

$$z_1 z_2 = r_1 r_2[(\cos\theta_1 \cos\theta_2 - \sin\theta_1 \sin\theta_2) + i(\sin\theta_1 \cos\theta_2 + \cos\theta_1 \sin\theta_2)].$$

Fig. 234. Example 2

If we apply the familiar addition theorems of the sine and the cosine, this assumes the simple form

(14) $$z_1 z_2 = r_1 r_2 [\cos (\theta_1 + \theta_2) + i \sin (\theta_1 + \theta_2)].$$

We thus obtain the important rules

(15) $$|z_1 z_2| = |z_1|\,|z_2|$$

and

(16) $$\arg (z_1 z_2) = \arg z_1 + \arg z_2 \qquad \text{(up to multiples of } 2\pi).$$

Similarly, from the definition of division it follows that

(17) $$\left| \frac{z_1}{z_2} \right| = \frac{|z_1|}{|z_2|}$$

and

(18) $$\arg \frac{z_1}{z_2} = \arg z_1 - \arg z_2 \qquad \text{(up to multiples of } 2\pi).$$

Example 4.   From (15) and (16) we obtain

$$z^n = r^n (\cos \theta + i \sin \theta)^n = r^n (\cos n\theta + i \sin n\theta)$$

and from this the so-called **formula of De Moivre**[4]

$$(\cos \theta + i \sin \theta)^n = \cos n\theta + i \sin n\theta.$$

Since for any complex number $z = x + iy$,

$$|z| = \sqrt{x^2 + y^2} \geq |x| \qquad \text{and similarly} \qquad |z| \geq |y|,$$

we obtain the useful inequalities

(19) $$|\operatorname{Re} z| \leq |z|, \qquad |\operatorname{Im} z| \leq |z|.$$

Let z_1 and z_2 be any complex numbers. Then the origin and the points z_1 and $z_1 + z_2$ are the vertices of a triangle[5] (Fig. 235) whose sides are $|z_1|$, $|z_2|$, and $|z_1 + z_2|$. We thus obtain the important **triangle inequality**

(20) $$|z_1 + z_2| \leq |z_1| + |z_2|,$$

which will be frequently used in our considerations. The formal proof is left to the student (Prob. 15).

Fig. 235. Triangle inequality

[4] ABRAHAM DE MOIVRE (1667–1754), French mathematician, who introduced imaginary quantities in trigonometry and contributed to the theory of mathematical probability.

[5] Which may degenerate if z_1 and z_2 lie on the same straight line through the origin.

By induction the triangle inequality can be extended to arbitrary sums:

(21) $$|z_1 + z_2 + \cdots + z_n| \leqq |z_1| + |z_2| + \cdots + |z_n|;$$

that is, *the absolute value of a sum is at most equal to the sum of the absolute values of the summands.*

Curves and regions in the complex plane will be frequently needed in our further discussion. To prepare for this, let us now consider some important types of curves and regions and their representations by equations and inequalities.

Since the distance between two points z and a is $|z - a|$, it follows that a circle C of radius ρ with center at a point a (Fig. 236) can be represented in the form

(22) $$|z - a| = \rho.$$

Consequently, the inequality

(23) $$|z - a| < \rho$$

holds for any point z inside C; that is, (23) represents the interior of C. Such a region is called a **circular disk,** or, more precisely, an *open* circular disk, in contrast to the *closed* circular disk

$$|z - a| \leqq \rho,$$

which consists of the interior of C and C itself. The open circular disk (23) is also called a **neighborhood** of the point a. Obviously, a has infinitely many such neighborhoods, each of which correponds to a certain value of ρ (>0).

Similarly, the inequality

$$|z - a| > \rho$$

represents the exterior of the circle C. Furthermore, the region between two concentric circles of radii ρ_1 and ρ_2 ($> \rho_1$) can be represented in the form

$$\rho_1 < |z - a| < \rho_2,$$

where a is the center of the circles. Such a region is called an *open circular ring* or *open* **annulus** (Fig. 237).

The equation

$$|z| = 1$$

Fig. 236. Circle in the complex plane

Fig. 237. Annulus in the complex plane

Fig. 238. Unit circle

represents the so-called **unit circle,** that is, the circle of radius 1 with center at the origin (Fig. 238). This circle will play an important role in various considerations.

It is important that the student becomes completely familiar with representations of curves and regions in the complex plane. Therefore, he should pay particular attention to the problems included in the present section. This will make his future work less difficult.

We finally define some notions which are of general importance and will be used in our further work.

The term **set of points** in the complex plane means any sort of collection of finitely or infinitely many points. For example, the solutions of a quadratic equation, the points on a line, and the points in the interior of a circle are sets.

A set S is called **open** if every point of S has a neighborhood every point of which belongs to S. For example, the points in the interior of a circle or a square form an open set, and so do the points of the "right half-plane" $\text{Re } z = x > 0$. The set consisting of the points in the interior of a circle C and on C ("closed circular disk") is not open, because no neighborhood of a point of C lies entirely in the set.

A set S in the complex plane is called **closed** if its complement is open. The **complement** of a set S in the complex plane is the set of all points in the complex plane which do not belong to S. For example, the points on and inside the unit circle form a closed set.

A set is called *bounded* if all of its points lie within a circle of sufficiently large radius. For example, the points in a rectangle form a bounded set, while the points on a straight line do not form a bounded set.

An open set S is said to be *connected* if any two of its points can be joined by a broken line of finitely many linear segments all of whose points belong to S. An open connected set is called a **domain.** Thus the interior of a circle is a domain.

A *boundary point* of a set S is a point every neighborhood of which contains both points which belong to S and points which do not belong to S. For example, the boundary points of an annulus are the points on the two bounding circles. Clearly, if a set S is open, then no boundary point belongs to S; if S is closed, then every boundary point belongs to S.

A **region** is a set consisting of a domian plus, perhaps, some or all of its boundary points. (The reader is warned that some authors use the term "region" for what we call a domain, and others make no distinction between the two terms.)

Problems

Find

1. $\text{Im } \dfrac{1}{1+i}$ **2.** $\text{Re } \dfrac{z}{\bar{z}}$ **3.** $\left|\dfrac{z}{\bar{z}}\right|$ **4.** $\text{Re } \dfrac{(1-i)^2}{4+2i}$

5. $\text{Im } z^3$ **6.** $\text{Re } z^4$ **7.** $|\cos\theta + i\sin\theta|$ **8.** $|z^6|$

9. $\left|\dfrac{z+1}{z-1}\right|$ **10.** $\left|\dfrac{(1+i)^6}{i^3(1+4i)^2}\right|$ **11.** $\left|\dfrac{1+3i}{3+i}\right|$ **12.** $\text{Re } \dfrac{1}{z^2}$

13. Verify (19) for $z = 3 - 4i$ and $z = (-2 + 5i)^3$.

14. Verify the triangle inequality for $z_1 = 3 - 2i$ and $z_2 = 4 + 2i$.

15. Prove the triangle inequality.

16. Using the triangle inequality, prove that $|z_1 + z_2| \geq |z_1| - |z_2|$.

17. Using the triangle inequality, prove that $|z_1 + z_2| \geq |z_2| - |z_1|$.

18. Under what conditions will $|z_1 + z_2| = |z_1| + |z_2|$?

19. Show that $(|x| + |y|)/\sqrt{2} \leq |z| \leq |x| + |y|$. Give examples.

20. Show that $|z_1 + z_2|^2 + |z_1 - z_2|^2 = 2|z_1|^2 + 2|z_2|^2$.

Represent the following complex numbers in polar form.

21. $1 + i, -1 + i, -1 - i$ 22. $3 + 4i, 4i, -i\sqrt{3}$

Determine and graph the loci represented by

23. $|z| \leq 1$ 24. $\mathrm{Re}\, z \geq -2$ 25. $\mathrm{Re}\,(z^2) \leq 1$ 26. $|\arg z| < \pi/2$

27. $\left|\dfrac{z+1}{z-1}\right| = 2$ 28. $\left|\dfrac{z+i}{z-i}\right| = 1$ 29. $\left|\dfrac{z+1}{z-1}\right| = 1$ 30. $\mathrm{Im}\,\dfrac{z+1}{z-1} \leq 2$

31. Show that multiplication of a complex number by i corresponds to a counterclockwise rotation of the corresponding vector through the angle $\pi/2$.

Determine the principal value of the arguments of

32. -7 33. $-4 - 4i$ 34. $1 + i\sqrt{3}$ 35. $-2i$

11.2   Limit. Derivative. Analytic Function

We shall now define some basic concepts of complex analysis. The reader will notice that these definitions are quite similar to those in real calculus.

If x and y are real variables, then $z = x + iy$ is said to be a *complex variable.*

Consider now two complex variables z and w, and suppose that a relation is given so that to each value in some region of the complex z-plane there is assigned a single value of w in a well-defined manner.[6] Then w is said to be a **function** of z, defined in that region, and we write

$$w = f(z)$$

or $w = g(z)$, etc., or sometimes simply $w(z)$. The set of complex numbers which $w = f(z)$ can assume as z varies in that region, is called the *range of values* of the function $w = f(z)$.

Let u and v be the real and imaginary parts of w. Then, since w depends on $z = x + iy$, it is clear that in general u depends on x and y, and so does v. We may, therefore, write

$$w = f(z) = u(x, y) + iv(x, y),$$

and this shows that a complex function $f(z)$ is equivalent to two real functions $u(x, y)$ and $v(x, y)$.

[6] In the literature on complex analysis, one sometimes uses relations such that to a value of z there may correspond more than one value of w, and it is customary to call such a relation a function (a "multivalued function"). We shall not adopt this convention but assume that all occurring functions are **single-valued** relations, that is, functions in the usual sense.

Example 1. Let

$$w = f(z) = z^2 + 3z.$$

Then

$$u(x, y) = \operatorname{Re} f(z) = x^2 - y^2 + 3x \quad \text{and} \quad v(x, y) = \operatorname{Im} f(z) = 2xy + 3y.$$

At $z = x + iy = 1 + 3i$ the function has the value

$$(1 + 3i)^2 + 3(1 + 3i) = -5 + 15i,$$

and we may write

$$f(1 + 3i) = -5 + 15i, \quad u(1, 3) = -5, \quad v(1, 3) = 15.$$

Similarly, $f(1 + i) = 3 + 5i$, etc. Obviously, our function is defined for all z.

Example 2. Determine the value of $f(z) = 3\bar{z} = 3x - 3iy$ at $z = 2 + 4i$. Since $x = 2$ and $y = 4$, we obtain $f(2 + 4i) = 6 - 12i$.

A function $f(z)$ is said to have the **limit** l as z approaches z_0 if $f(z)$ is defined in a neighborhood of z_0 (except perhaps at z_0) and if for every positive real number ϵ (no matter how small but not zero) we can find a positive real number δ such that for all values $z \neq z_0$ in the disk $|z - z_0| < \delta$,

$$(1) \qquad\qquad\qquad |f(z) - l| < \epsilon.$$

This means that the values of $f(z)$ are as close as desired to l for all z which are sufficiently close to z_0 (Fig. 239), and we write

$$(2) \qquad\qquad\qquad \lim_{z \to z_0} f(z) = l.$$

Note well that this definition of a limit implies that z may approach z_0 from *any* direction in the complex plane.

A function $f(z)$ is said to be **continuous** at $z = z_0$ if $f(z_0)$ is defined and

$$(3) \qquad\qquad\qquad \lim_{z \to z_0} f(z) = f(z_0).$$

Note that by the definition of a limit this implies that $f(z)$ is defined in some neighborhood of z_0. A *continuous function*, without further qualification, is one which is continuous at all points where it is defined.

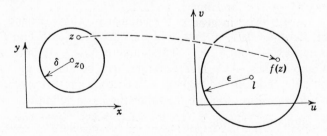

Fig. 239. Limit

A function $f(z)$ is said to be **differentiable** at a point $z = z_0$ if the limit

$$(4) \qquad f'(z_0) = \lim_{\Delta z \to 0} \frac{f(z_0 + \Delta z) - f(z_0)}{\Delta z}$$

exists. This limit is then called the **derivative** of $f(z)$ at the point $z = z_0$.
Setting $z_0 + \Delta z = z$, we have $\Delta z = z - z_0$ and may also write

$$(4') \qquad f'(z_0) = \lim_{z \to z_0} \frac{f(z) - f(z_0)}{z - z_0}.$$

Since the notion of a limit implies that z may approach z_0 from any direction, differentiability means that, along whatever path z approaches z_0, the quotient in (4') always approaches a certain value and all these values are equal. This fact will be quite important in our later considerations.

By the definition of a limit, (4') means that there is a complex number $f'(z_0)$ for which, an $\epsilon > 0$ being given, we can find a $\delta > 0$ such that

$$(5) \qquad \left| \frac{f(z) - f(z_0)}{z - z_0} - f'(z_0) \right| < \epsilon \qquad \text{when} \qquad |z - z_0| < \delta.$$

Example 3. The derivative of the function $f(z) = z^2$ is $2z$ at any point z, because

$$f'(z) = \lim_{\Delta z \to 0} \frac{(z + \Delta z)^2 - z^2}{\Delta z} = \lim_{\Delta z \to 0} (2z + \Delta z) = 2z.$$

All the familiar rules of real differential calculus, such as the rules for differentiating a constant, integer powers of z, sums, products, and quotients of differentiable functions, and the chain rule for differentiating a function of a function, continue to hold in complex.

In fact, the corresponding proofs are literally the same.

It is important to note that there are many simple functions which do not have a derivative at any point. For instance, $f(z) = \bar{z} = x - iy$ is such a function, because, setting $\Delta z = \Delta x + i\, \Delta y$, we see that, as Δz approaches zero, the quotient

$$(6) \qquad \frac{f(z + \Delta z) - f(z)}{\Delta z} = \frac{[(x + \Delta x) - i(y + \Delta y)] - (x - iy)}{\Delta x + i\, \Delta y} = \frac{\Delta x - i\, \Delta y}{\Delta x + i\, \Delta y}$$

approaches a value which depends on the manner in which Δx and Δy approach zero. In fact, if we let $\Delta y \to 0$ first, and then $\Delta x \to 0$ (path I in Fig. 240), the quotient approaches $+1$, but if $\Delta x \to 0$ first, and then $\Delta y \to 0$ (path II in Fig. 240), the quotient approaches -1. This example may be surprising, but it merely illustrates that differentiability of a complex function is a rather severe requirement.

We are now in a position to state the following basic definition.

Definition (Analyticity). *A function $f(z)$ is said to be analytic at a point $z = z_0$ if it is defined, and has a derivative, at every point in some neighborhood of z_0. It is said to be analytic in a domain D if it is analytic at every point in D.*

Fig. 240. Paths in (6)

We shall also use the term **analytic function,** without reference to a specific domain; by this term we shall mean a function which is analytic in *some* domain D.

Example 4. The integral powers $1, z, z^2, \cdots$ and, more generally, **polynomials,** that is, functions of the form

$$f(z) = c_0 + c_1 z + c_2 z^2 + \cdots + c_n z^n$$

where $c_0, \cdots, c_n$ are complex constants, are analytic in the entire plane. The function $1/(1 - z)$ is analytic everywhere except at $z = 1$.

Instead of *analytic in D* the terms **holomorphic** *in D* and *regular in D* are also used in the literature.

Complex analysis is concerned exclusively with analytic functions, and although many simple functions are not analytic, the large variety of remaining functions will yield a branch of mathematics which is most beautiful from the theoretical point of view and most useful for practical purposes.

Problems

Find $f(3 + i), f(-i), f(-4 + 2i)$ where $f(z)$ equals

   1. $z^2 - 2z$ **2.** $1/(1 - z)$ **3.** $1/z^3$

Find the real and imaginary parts of the following functions.

   4. $f(z) = z^2 + 4z - 1$ **5.** $f(z) = 2z^3 - 3z$ **6.** $f(z) = 1/(1 - z)$

In each case, find whether $f(z)$ is continuous at the origin, assuming that $f(0) = 0$ and, for $z \neq 0$, the function $f(z)$ equals

   7. Re $z/|z|$ **8.** Re $(z^2)/|z^2|$ **9.** $(\text{Re } z)^2/|z|$

   10. Using the definition of a limit, prove that $f(z) = z^2$ is continuous.

Differentiate

   11. $(z^2 - 4)^3$ **12.** $(z^2 - 4)/(z^2 + 1)$ **13.** $1/(1 - z)^2$

Find the value of the derivative of

   14. $(z^2 - i)^2$ at $3 - 2i$ **15.** $1/z^3$ at $3i$ **16.** $(z + i)/(z - i)$ at $-i$

   17. Show that $[af(z) + bg(z)]' = af'(z) + bg'(z)$, $[f(z)g(z)]' = f'(z)g(z) + f(z)g'(z)$.

   18. If $z_1, z_2, \cdots$ are complex numbers for which $\lim_{n \to \infty} z_n = a$ and $f(z)$ is continuous at $z = a$, show that $\lim_{n \to \infty} f(z_n) = f(a)$.

   19. Prove that (2) is equivalent to the pair of relations

$$\lim_{z \to z_0} \text{Re } f(z) = \text{Re } l, \qquad \lim_{z \to z_0} \text{Im } f(z) = \text{Im } l.$$

20. If $\lim\limits_{z \to z_0} f(z) = l$ and $\lim\limits_{z \to z_0} g(z) = p$, show that

$$\lim_{z \to z_0} [f(z) + g(z)] = \lim_{z \to z_0} f(z) + \lim_{z \to z_0} g(z) = l + p,$$

$$\lim_{z \to z_0} [f(z)g(z)] = \lim_{z \to z_0} f(z) \lim_{z \to z_0} g(z) = lp.$$

11.3 *Cauchy-Riemann Equations.* *Laplace's Equation*

We shall now derive a simple basic criterion for analyticity of a complex function

(1) $$w = f(z) = u(x, y) + iv(x, y).$$

We first suppose that $f(z)$ is analytic in a domain D of the z-plane. Then, by definition, $f(z)$ has a derivative

(2) $$f'(z) = \lim_{\Delta z \to 0} \frac{f(z + \Delta z) - f(z)}{\Delta z}$$

everywhere in D, where Δz may approach 0 along any path. We may set $\Delta z = \Delta x + i\,\Delta y$. Choosing path I in Fig. 241, we let $\Delta y \to 0$ first and then $\Delta x \to 0$. After Δy becomes zero, $\Delta z = \Delta x$, and by (1),

$$f'(z) = \lim_{\Delta x \to 0} \frac{u(x + \Delta x, y) + iv(x + \Delta x, y) - [u(x, y) + iv(x, y)]}{\Delta x}$$

$$= \lim_{\Delta x \to 0} \frac{u(x + \Delta x, y) - u(x, y)}{\Delta x} + i \lim_{\Delta x \to 0} \frac{v(x + \Delta x, y) - v(x, y)}{\Delta x}.$$

Since $f'(z)$ exists, the last two real limits exist. They are the partial derivatives of u and v with respect to x. Hence $f'(x)$ can be written

(3) $$f'(z) = \frac{\partial u}{\partial x} + i\frac{\partial v}{\partial x}.$$

Similarly, if we choose path II in Fig. 241, we let $\Delta x \to 0$ first and then $\Delta y \to 0$. After Δx becomes zero, $\Delta z = i\,\Delta y$, and

$$f'(z) = \lim_{\Delta y \to 0} \frac{u(x, y + \Delta y) - u(x, y)}{i\,\Delta y} + i \lim_{\Delta y \to 0} \frac{v(x, y + \Delta y) - v(x, y)}{i\,\Delta y};$$

Fig. 241. Paths in (2)

that is,

$$(4) \qquad f'(z) = -i\frac{\partial u}{\partial y} + \frac{\partial v}{\partial y}$$

because $1/i = -i$. The existence of $f'(z)$ thus implies the existence of the four partial derivatives in (3) and (4).

By equating the real and the imaginary parts of the right-hand sides of (3) and (4) we obtain

$$(5) \qquad \frac{\partial u}{\partial x} = \frac{\partial v}{\partial y} \quad \text{and} \quad \frac{\partial u}{\partial y} = -\frac{\partial v}{\partial x}.$$

These basic relations are called the **Cauchy-Riemann differential equations.**[7,8]

We may sum up our result as follows.

Theorem 1. *The real part and the imaginary part of an analytic function $f(z) = u(x, y) + iv(x, y)$ satisfy the Cauchy-Riemann equations (5) at each point where $f(z)$ is analytic.*

Example 1. The function $f(z) = z^2 = x^2 - y^2 + 2ixy$ is analytic for all z, and $f'(z) = 2z$. We have $u = x^2 - y^2$, $v = 2xy$,

$$\frac{\partial u}{\partial x} = 2x, \qquad \frac{\partial u}{\partial y} = -2y, \qquad \frac{\partial v}{\partial x} = 2y, \qquad \frac{\partial v}{\partial y} = 2x,$$

and the Cauchy-Riemann equations are satisfied for all x, y.

The Cauchy-Riemann equations are fundamental because they are not only necessary but also sufficient for a function to be analytic; more precisely:

Theorem 2. *If two real-valued functions $u(x, y)$ and $v(x, y)$ of two real variables x and y have continuous first partial derivatives that satisfy the Cauchy-Riemann equations in some domain D, then the complex function $f(z) = u(x, y) + iv(x, y)$ is analytic in D.*

Proof. Let $P: (x, y)$ be any fixed point in D. Since D is a domain, it contains a neighborhood of P. In this neighborhood we may choose a point $Q: (x + \Delta x, y + \Delta y)$ such that the segment PQ is in D. From the mean-value theorem in Sec. 7.7 it follows that

$$(6) \qquad \begin{aligned} u(x + \Delta x, y + \Delta y) - u(x, y) &= \Delta x \left(\frac{\partial u}{\partial x}\right)_{M_1} + \Delta y \left(\frac{\partial u}{\partial y}\right)_{M_1}, \\ v(x + \Delta x, y + \Delta y) - v(x, y) &= \Delta x \left(\frac{\partial v}{\partial x}\right)_{M_2} + \Delta y \left(\frac{\partial v}{\partial y}\right)_{M_2}, \end{aligned}$$

[7] Cf. the footnote on p. 69.

[8] BERNHARD RIEMANN (1826–1866), German mathematician, who developed what may be called the "geometrical approach" to complex analysis, based upon the Cauchy-Riemann equations and conformal mapping, in contrast to the German mathematician KARL WEIERSTRASS (1815–1897), who based complex analysis on power series (cf. Sec. 14.6, 15.1). Riemann developed also the so-called Riemannian geometry, which is the mathematical base of Einstein's theory of relativity. His important work gave the impetus to many ideas in modern mathematics (in particular, in topology and functional analysis). Cf. N. Bourbaki, *Elements of Mathematics, General Topology*, Part 1, pp. 162–166. Paris; Hermann, 1966.

the derivatives being evaluated at suitable points M_1 and M_2 of that segment. We set

$$f(z) = u(x, y) + iv(x, y), \qquad \Delta z = \Delta x + i \Delta y, \qquad \Delta f = f(z + \Delta z) - f(z).$$

Then we obtain from (6)

$$\Delta f = \Delta x \left(\frac{\partial u}{\partial x}\right)_{M_1} + \Delta y \left(\frac{\partial u}{\partial y}\right)_{M_1} + i\left[\Delta x \left(\frac{\partial v}{\partial x}\right)_{M_2} + \Delta y \left(\frac{\partial v}{\partial y}\right)_{M_2}\right].$$

Using the Cauchy-Riemann equations, we may replace $\partial u/\partial y$ by $-\partial v/\partial x$ and $\partial v/\partial y$ by $\partial u/\partial x$, finding

$$\Delta f = \Delta x \left(\frac{\partial u}{\partial x}\right)_{M_1} + i \Delta y \left(\frac{\partial u}{\partial x}\right)_{M_2} + i\left[\Delta x \left(\frac{\partial v}{\partial x}\right)_{M_2} + i \Delta y \left(\frac{\partial v}{\partial x}\right)_{M_1}\right].$$

Using $\Delta z = \Delta x + i\Delta y$, we see that this can be written

$$\Delta f = \Delta z \left(\frac{\partial u}{\partial x}\right)_{M_1} + i\Delta y\left\{\left(\frac{\partial u}{\partial x}\right)_{M_2} - \left(\frac{\partial u}{\partial x}\right)_{M_1}\right\}$$
$$+ i\left[\Delta z \left(\frac{\partial v}{\partial x}\right)_{M_1} + \Delta x\left\{\left(\frac{\partial v}{\partial x}\right)_{M_2} - \left(\frac{\partial v}{\partial x}\right)_{M_1}\right\}\right].$$

We divide by Δz on both sides and let Δz approach zero. Then, since the partial derivatives on the right are continuous, they approach $\partial u/\partial x$ and $\partial v/\partial x$, evaluated at (x, y). Furthermore, since $|\Delta x/\Delta z| \leqq 1$, $|\Delta y/\Delta z| \leqq 1$, the limit of the expression on the right exists and is independent of the path along which Δz approachs zero. We see that it equals the right-hand side of (3). This means that $f(z)$ is analytic in D, and the proof is complete.

These theorems are of great practical importance, and we may now easily decide whether or not a given complex function is analytic.

Example 2. Let
$$f(z) = \operatorname{Re} z = x.$$
Then
$$u = x \quad \text{and} \quad v = 0.$$

The equations (5) are not satisfied, that is, $\operatorname{Re} z$ is not analytic. Similarly, $\operatorname{Im} z$ is not analytic. Further simple functions which are not analytic are included in the problems.

We mention that if we use the polar form $z = r(\cos \theta + i \sin \theta)$ and set $f(z) = u(r, \theta) + iv(r, \theta)$, then the Cauchy-Riemann equations are

(7)
$$\frac{\partial u}{\partial r} = \frac{1}{r} \frac{\partial v}{\partial \theta}, \qquad \frac{\partial v}{\partial r} = -\frac{1}{r} \frac{\partial u}{\partial \theta}.$$

We shall prove later (in Sec. 13.6) that the derivative of an analytic function $f(z) = u(x, y) + iv(x, y)$ is itself analytic. By this important fact $u(x, y)$ and $v(x, y)$ will have continuous partial derivatives of all orders. In particular, the mixed second derivatives of these functions will be equal:

$$\frac{\partial^2 u}{\partial x\, \partial y} = \frac{\partial^2 u}{\partial y\, \partial x}, \qquad \frac{\partial^2 v}{\partial x\, \partial y} = \frac{\partial^2 v}{\partial y\, \partial x}.$$

Differentiating the Cauchy-Riemann equations, we thus obtain

$$\frac{\partial^2 u}{\partial x^2} = \frac{\partial^2 v}{\partial x\, \partial y}, \qquad \frac{\partial^2 u}{\partial y^2} = -\frac{\partial^2 v}{\partial x\, \partial y},$$

$$\frac{\partial^2 u}{\partial x\, \partial y} = \frac{\partial^2 v}{\partial y^2}, \qquad \frac{\partial^2 u}{\partial x\, \partial y} = -\frac{\partial^2 v}{\partial x^2}.$$

This yields the following important result.

Theorem 3. *The real part and the imaginary part of a complex function* $f(z) = u(x, y) + iv(x, y)$ *that is analytic in a domain D are solutions of Laplace's equation,*

$$\nabla^2 u = \frac{\partial^2 u}{\partial x^2} + \frac{\partial^2 u}{\partial y^2} = 0, \qquad \nabla^2 v = \frac{\partial^2 v}{\partial x^2} + \frac{\partial^2 v}{\partial y^2} = 0,$$

in D and have continuous second partial derivatives in D.

This is one of the main reasons for the great practical importance of complex analysis in engineering mathematics, as we shall see in Chaps. 12 and 17.

A solution of Laplace's equation having *continuous* second-order partial derivatives is called a **harmonic function** (cf. also Sec. 10.11). Hence the real and imaginary parts of an analytic function are harmonic functions.

If two harmonic functions $u(x, y)$ and $v(x, y)$ satisfy the Cauchy-Riemann equations in a domain D, that is, if u and v are the real and imaginary parts of an analytic function $f(z)$ in D, then $v(x, y)$ is said to be a **conjugate harmonic function** of $u(x, y)$ in D, and the pair u, v is called a *pair of conjugate harmonic functions* in D. (Of course this use of the word "conjugate" is different from that employed in defining $\bar{z}$, the conjugate of a complex number z.)

A conjugate of a given harmonic function can be obtained from the Cauchy-Riemann equations, as may be illustrated by the following example.

Example 3. The function $u = x^2 - y^2$ is harmonic, and we have $\partial u / \partial x = 2x$, $\partial u / \partial y = -2y$. Hence a conjugate of u must satisfy

$$\frac{\partial v}{\partial y} = 2x, \qquad \frac{\partial v}{\partial x} = 2y.$$

By integrating the first equation with respect to y we obtain

$$v = 2xy + h(x)$$

where $h(x)$ depends only on x. Substituting this in the last equation, we have $h'(x) = 0$ and, therefore, $h = c = const.$ Hence, the most general conjugate function of $x^2 - y^2$ is $2xy + c$, where c is a real constant, and the most general analytic function with the real part $x^2 - y^2$ is $x^2 - y^2 + i(2xy + c) = z^2 + ic$.

Problems

Using (3) or (4), find the derivative of the following functions.

1. $f(z) = az + b$ **2.** $f(z) = z^3 + 3z^2$ **3.** $f(z) = (z^2 + 3z)^2$

4. $f(z) = \dfrac{1}{1-z}$ **5.** $f(z) = z + \dfrac{1}{z}$ **6.** $f(z) = \dfrac{z+1}{z-2}$

7. Show that, in addition to (3) and (4),

$$\text{(8)}\qquad f'(z) = \frac{\partial u}{\partial x} - i\,\frac{\partial u}{\partial y}, \qquad f'(z) = \frac{\partial v}{\partial y} + i\,\frac{\partial v}{\partial x}.$$

Are the following functions analytic?

8. $f(z) = \operatorname{Im} z$ **9.** $f(z) = 1/(1+z)$ **10.** $f(z) = \operatorname{Re}(z^2)$
11. $f(z) = z + \bar{z}$ **12.** $f(z) = |z|^2$ **13.** $f(z) = 1/z$
14. $f(z) = \arg z$ **15.** $f(z) = e^x \cos y$ **16.** $f(z) = e^x(\cos y + i \sin y)$

Using (7), show that the following functions are analytic.

17. $f(z) = z^3$ **18.** $f(z) = \ln r + i\theta$ **19.** $f(z) = 1/z^4 \ (z \neq 0)$

Using (5), find the most general analytic function $f(z) = u(x, y) + iv(x, y)$ for which

20. $u = xy$ **21.** $v = xy$ **22.** $u = e^x \cos y$

Show that the following functions are harmonic and find a corresponding analytic function $f(z) = u(x, y) + iv(x, y)$.

23. $v = -\sin x \sinh y$ **24.** $u = \ln(x^2 + y^2)$ **25.** $v = \arg z$

26. Under what conditions is $u = ax^3 + bx^2y + cxy^2 + ky^3$ harmonic?
27. If $f(z)$ is analytic in a domain D and $|f(z)| = const$ in D, show that $f(z) = const$.
28. If $f(z)$ is analytic in a domain D and $\operatorname{Re} f(z) = const$ in D, show that $f = const$.
29. If $f(z)$ is analytic in a domain D and $f'(z) = 0$ everywhere in D, show that $f(z) = const$.
30. Derive (7) from (5).

11.4 Rational Functions. Root

The remaining sections of this chapter will be devoted to the most important elementary complex functions, such as powers, exponential function, logarithm, trigonometric functions, etc. We shall see that these functions can easily be defined in such a way that, for real values of the independent variable, the functions become identical with the familiar real functions. Some of the complex functions have interesting properties, which do not show when the independent variable is restricted to real values. The student should follow the consideration with great care, because these elementary functions will frequently be needed in applications. Furthermore, a detailed knowledge of these special functions will be helpful later for a better understanding of our more general considerations.

The powers

$$\text{(1)}\qquad\qquad w = z^n \qquad\qquad n = 0, 1, \cdots$$

are analytic in the entire plane. The same is true for a function of the form

$$\text{(2)}\qquad\qquad w = c_0 + c_1 z + c_2 z^2 + \cdots + c_n z^n \qquad\qquad (c_n \neq 0)$$

where $c_0, \cdots, c_n$ are (complex or real) constants. Such a function is called a **polynomial** or an *entire rational function*. The exponent n is called the *degree* of the polynomial. The study of polynomials is the principal subject of classical algebra.

A quotient of two polynomials $p(z)$ and $q(z)$ is called a (*fractional*) **rational function.** Such a function

$$(3) \qquad\qquad w = \frac{p(z)}{q(z)}$$

is analytic for every z for which $q(z)$ is not zero. A rational function of the particularly simple form

$$\frac{c}{(z - z_0)^m} \qquad\qquad (c \neq 0)$$

where both c and z_0 are complex numbers and m is a positive integer, is called a *partial fraction.* It is proved in algebra that every rational function can be represented as the sum of a polynomial and finitely many partial fractions.

If $z = w^n$ $(n = 1, 2, \cdots)$, then to each value of w there corresponds one value of z. We shall immediately see that to a given $z \neq 0$ there correspond precisely n distinct values of w. Each of these values is called an **nth root** of z, and we write

$$(4) \qquad\qquad w = \sqrt[n]{z}.$$

Hence this symbol is *multi-valued*, namely, *n-valued*, in contrast to the usual conventions made in real calculus. The n values of $\sqrt[n]{z}$ can easily be determined as follows. We set

$$w = R(\cos \phi + i \sin \phi) \qquad \text{and} \qquad z = r(\cos \theta + i \sin \theta).$$

Then from de Moivre's formula (cf. Sec. 11.1) it follows that

$$z = w^n = R^n(\cos n\phi + i \sin n\phi) = r(\cos \theta + i \sin \theta).$$

By equating the absolute values on both sides we have

$$R^n = r \qquad \text{or} \qquad R = \sqrt[n]{r}$$

where the root is real positive and thus uniquely determined. By equating the arguments we obtain

$$n\phi = \theta + 2k\pi \qquad \text{or} \qquad \phi = \frac{\theta}{n} + \frac{2k\pi}{n}$$

where k is an integer. Consequently, $\sqrt[n]{z}$, for $z \neq 0$, has the n distinct values

$$(5) \qquad \sqrt[n]{z} = \sqrt[n]{r}\left(\cos \frac{\theta + 2k\pi}{n} + i \sin \frac{\theta + 2k\pi}{n}\right), \qquad k = 0, 1, \cdots, n - 1.$$

These n values lie on a circle of radius $\sqrt[n]{r}$ with center at the origin and constitute the vertices of a regular polygon of n sides.

The value of $\sqrt[n]{z}$ obtained by taking the principal value of arg z (cf. Sec. 11.1) and $k = 0$ in (5) is called the **principal value** *of* $w = \sqrt[n]{z}$.

Example 1 (Square root). $w = \sqrt{z}$ has the values

$$z_1 = \sqrt{r}\left(\cos\frac{\theta}{2} + i\sin\frac{\theta}{2}\right)$$

and

$$z_2 = \sqrt{r}\left[\cos\left(\frac{\theta}{2} + \pi\right) + i\sin\left(\frac{\theta}{2} + \pi\right)\right] = -z_1$$

which lie symmetric with respect to the origin. For instance,

$$\sqrt{4i} = \pm 2\left(\cos\frac{\pi}{4} + i\sin\frac{\pi}{4}\right) = \pm(\sqrt{2} + i\sqrt{2}).$$

Example 2. If z is *positive real*, then $w = \sqrt[3]{z}$ has the real value $\sqrt[3]{r}$ and the conjugate complex values

$$\sqrt[3]{r}\left(\cos\frac{2\pi}{3} + i\sin\frac{2\pi}{3}\right) = \sqrt[3]{r}\left(-\frac{1}{2} + \frac{\sqrt{3}}{2}i\right)$$

and

$$\sqrt[3]{r}\left(\cos\frac{4\pi}{3} + i\sin\frac{4\pi}{3}\right) = \sqrt[3]{r}\left(-\frac{1}{2} - \frac{\sqrt{3}}{2}i\right).$$

For instance, $\sqrt[3]{1} = 1,\ -\frac{1}{2} \pm \frac{1}{2}\sqrt{3}\,i$ (Fig. 242). Clearly these are the roots of the equation $w^3 - 1 = 0$.

Example 3.   From (5) it follows that the **nth root of unity** is given by the formula

(6)
$$\sqrt[n]{1} = \cos\frac{2k\pi}{n} + i\sin\frac{2k\pi}{n}, \qquad k = 0, 1, \cdots n - 1.$$

If ω denotes the value corresponding to $k = 1$, then the n values of $\sqrt[n]{1}$ can be written as $1, \omega, \omega^2, \cdots, \omega^{n-1}$. These values are the vertices of a regular polygon of n sides inscribed in the unit circle, with one vertex at the point 1. For instance, $\sqrt[4]{1}$ has the values $1, i, -1$, and $-i$ (Fig. 243). Figure 244 shows $\sqrt[5]{1}$.

If w_1 is any nth root of an arbitrary complex number z, then

$$w_1, \qquad w_1\omega, \qquad w_1\omega^2, \qquad \cdots, \qquad w_1\omega^{n-1}$$

are the n values of $\sqrt[n]{z}$, because the multiplication of a number w_1 by ω^k corresponds to an increase of the argument of w_1 by $2k\pi/n$.

Fig. 242. $\sqrt[3]{1}$ Fig. 243. $\sqrt[4]{1}$ Fig. 244. $\sqrt[5]{1}$

Problems

Determine all values of the following roots and plot the corresponding points in the complex plane.

1. $\sqrt{\bar{i}}$ 2. $\sqrt{-i}$ 3. $\sqrt[3]{i}$ 4. $\sqrt[4]{1}$

5. $\sqrt{-25}$ 6. $\sqrt[8]{1}$ 7. $\sqrt[3]{1+i}$ 8. $\sqrt{\frac{1}{2}(1+i\sqrt{3})}$

9. $\sqrt[4]{-1}$ 10. $\sqrt[3]{-128}$ 11. $\sqrt[6]{-1}$ 12. $\sqrt{1-i\sqrt{3}}$

Find and plot all solutions of the following equations.

13. $z^3 = 64$ 14. $z^4 + 81 = 0$ 15. $z^4 + 5z^2 = 36$ 16. $z^6 + 7z^3 = 8$

17. Transform to a single equation for $x + iy$ and solve

$$x^4 - 6x^2y^2 + y^4 = 4, \qquad xy(x^2 - y^2) = 1 \qquad (x, y \text{ real}).$$

18. **(Square root)** Prove that

$$\sqrt{z} = \pm[\sqrt{(|z| + x)/2} + (\text{sign } y)i\sqrt{(|z| - x)/2}] \qquad (z = x + iy)$$

where sign $y = 1$ if $y \geq 0$, sign $y = -1$ if $y < 0$, and all square roots of positive numbers are taken with the positive sign. *Hint.* Set $\sqrt{z} = w = u + iv$, square, separate into two real equations, express u^2 and v^2 in terms of x and y.

Using the result of Prob. 18, find

19. $\sqrt{2i}$ 20. $\sqrt{-8i}$ 21. $\sqrt{5 + 12i}$ 22. $\sqrt{21 - 20i}$

Using the result of Prob. 18, solve the following equations.

23. $z^2 - (5 + i)z + 8 + i = 0$ 24. $z^4 - 3(1 + 2i)z^2 = 8 - 6i$

25. Represent $z^4 + 1$ as a product of quadratic factors with real coefficients.

11.5   Exponential Function

The real exponential function e^x (also written exp x) has the properties

(1) $(e^x)' = e^x$

(2) $e^{x_1 + x_2} = e^{x_1}e^{x_2}$

and the Maclaurin[9] series

(3) $e^x = 1 + x + \dfrac{x^2}{2!} + \dfrac{x^3}{3!} + \cdots.$

The exponential function for complex $z = x + iy$ is denoted by e^z and is defined in terms of the real functions e^x, cos y, and sin y, as follows:

(4) $e^z = e^x(\cos y + i \sin y)$ (also written exp z).

This definition is suggested by the following facts. When $z = x$ is real, then $e^z = e^x$. From the Cauchy-Riemann equations it follows that e^z is analytic for all z.

[9] COLIN MACLAURIN (1698–1746), Scots mathematician.

From (3) in Sec. 11.3 we see that

$$(e^z)' = \frac{\partial}{\partial x}(e^x \cos y) + i\frac{\partial}{\partial x}(e^x \sin y) = e^x \cos y + ie^x \sin y,$$

that is,

(5) $$(e^z)' = e^z.$$

Furthermore, setting $z_1 = x_1 + iy_1$ and $z_2 = x_2 + iy_2$ and applying the addition formulas for the sine and cosine [cf. (6) in Appendix 3], we readily find

(6) $$e^{z_1+z_2} = e^{z_1}e^{z_2},$$

which is the analogue of (2). In particular, when $z_1 = x$ and $z_2 = iy$, we have

(7) $$e^z = e^{x+iy} = e^x e^{iy}.$$

We shall see later (in Chap. 15) that complex analytic functions have Taylor series developments which are similar to the familiar Taylor series of real functions. It will then easily follow that if we replace x in (3) by z, we obtain the Maclaurin[10] series of e^z.

From (4) we obtain the **Euler formula**

(8) $$e^{iy} = \cos y + i \sin y.$$

This shows that the trigonometric representation of a complex number $x + iy$ (the polar form of z) may now be written

(9) $$z = r(\cos \theta + i \sin \theta) = re^{i\theta}.$$

Furthermore, from (8) we see that

(10) $$|e^{iy}| = \sqrt{\cos^2 y + \sin^2 y} = 1,$$

a result which the student should remember. Consequently, by (4),

(11) $$|e^z| = e^x, \qquad \arg e^z = y.$$

Since $\cos 2\pi = 1$ and $\sin 2\pi = 0$, we obtain from (8)

(12) $$e^{2\pi i} = 1.$$

Similarly,

(13) $$e^{\pi i} = e^{-\pi i} = -1, \qquad e^{\pi i/2} = i, \qquad e^{-\pi i/2} = -i.$$

From (12) and (6) we find

$$e^{z+2\pi i} = e^z e^{2\pi i} = e^z,$$

---

[10] This series may be used for *defining* e^z. For this purpose one would first have to consider Chap. 14 on series (which is independent of Chaps. 12 and 13 and does not refer to real series). The properties of e^z follow from the series, as shown in Ex. 2, Sec. 15.2. Our present approach has the advantage that integrals (Chap. 13) will appear earlier and we shall have a closer connection between the general consideration of series (Chap. 14) and the Taylor and Laurent series (Chap. 15), for which integrals will be needed. We also avoid irritating poorer students by series at such an early stage.

Fig. 245. Fundamental region of the exponential function e^z

which shows that e^z *is periodic with the imaginary period* $2\pi i$. Thus

(14) $e^{z \pm 2n\pi i} = e^z$ $(n = 0, 1, \cdots)$.

Because of the periodicity all the values which $w = e^z$ can assume are already assumed in the strip (Fig. 245)

(15) $-\pi < y \leqq \pi$.

This infinite strip is called a **fundamental region** of e^z.

From (6) we obtain $e^z e^{-z} = e^0 = 1$, which implies

(16) $e^z \neq 0$ for all z.

Problems

Find the value of e^z when z equals
   1. $-\pi i/4$ 2. $-\pi i/3$ 3. $1 + i$ 4. $2 + 5\pi i$

Find the real and imaginary parts of
   5. e^{-3z} 6. e^{z^2} 7. e^{z^3} 8. $e^{(e^z)}$

Write the following expressions in the form (9).
   9. $3 + 4i$ 10. $\sqrt{i}, \sqrt{-i}$ 11. $2 - 2i$ 12. $\sqrt{z}, \sqrt[3]{z}$

Find all solutions of the following equations and plot some of them in the complex plane.
   13. $e^z = 3$ 14. $e^z = 2$ 15. $e^z = -2$ 16. $e^{z^2} = 1$

Find all values of z such that
   17. $e^{\bar{z}} = \overline{e^z}$ 18. $e^{i\bar{z}} = \overline{e^{iz}}$ 19. $|e^{-2z}| < 1$ 20. e^z is real

   21. Using the Cauchy-Riemann equations, show that e^z is analytic for all z.
   22. Show that $f(z) = e^{\bar{z}}$ is nowhere analytic.
   23. Find all values of k such that $f(z) = e^x(\cos ky + i \sin ky)$ is analytic.
   24. Show that $f(z) = ce^z$ is the only analytic function for which $f'(z) = f(z)$.
   25. Consider the behavior of e^z as $|z| \to \infty$ along different rays, say, for arg $z = 0$, $\pi/2$, and π.

11.6 *Trigonometric and Hyperbolic Functions*

From Euler's formula (8) in the last section we obtain

$$\cos x = \frac{1}{2}(e^{ix} + e^{-ix}), \qquad \sin x = \frac{1}{2i}(e^{ix} - e^{-ix}) \qquad (x \text{ real}).$$

This suggests the following definitions for complex $z = x + iy$:

(1) $$\cos z = \frac{1}{2}(e^{iz} + e^{-iz}), \qquad \sin z = \frac{1}{2i}(e^{iz} - e^{-iz}).$$

Furthermore, in agreement with the familiar definitions from real calculus we define

(2) $$\tan z = \frac{\sin z}{\cos z}, \qquad \cot z = \frac{\cos z}{\sin z}$$

and

(3) $$\sec z = \frac{1}{\cos z}, \qquad \csc z = \frac{1}{\sin z}.$$

Since e^z is analytic for all z, the same is true for the functions $\cos z$ and $\sin z$. The functions $\tan z$ and $\sec z$ are analytic except at the points where $\cos z$ is zero, and $\cot z$ and $\csc z$ are analytic except where $\sin z$ is zero.

The functions $\cos z$ and $\sec z$ are even, and those other functions are odd:

(4)
$$\cos(-z) = \cos z \qquad \sin(-z) = -\sin z$$
$$\cot(-z) = -\cot z \qquad \tan(-z) = -\tan z$$

etc. Since the exponential function is periodic, the trigonometric functions are also periodic, and we have

(5)
$$\cos(z \pm 2n\pi) = \cos z \qquad \sin(z \pm 2n\pi) = \sin z$$
$$\tan(z \pm n\pi) = \tan z \qquad \cot(z \pm n\pi) = \cot z$$

where $n = 0, 1, \cdots$.

From the definitions it follows immediately that *all the familiar formulas for the real trigonometric functions continue to hold for complex values*. For example,

(6) $$\frac{d}{dz}\cos z = -\sin z, \qquad \frac{d}{dz}\sin z = \cos z, \qquad \frac{d}{dz}\tan z = \sec^2 z,$$

and

(7)
$$\cos(z_1 \pm z_2) = \cos z_1 \cos z_2 \mp \sin z_1 \sin z_2$$
$$\sin(z_1 \pm z_2) = \sin z_1 \cos z_2 \pm \sin z_2 \cos z_1$$
$$\sin^2 z + \cos^2 z = 1.$$

From (1) we see that *Euler's formula is valid for complex values*,

(8) $$e^{iz} = \cos z + i \sin z.$$

Using (7), we may readily represent $\cos z$ and $\sin z$ in terms of real functions. We first have

(9)
$$\cos(x + iy) = \cos x \cos iy - \sin x \sin iy$$
$$\sin(x + iy) = \sin x \cos iy + \cos x \sin iy.$$

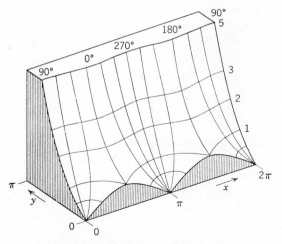

Fig. 246. Modular surface of sin z

Now from (1) and the definition of the hyperbolic cosine and sine it follows that

$$\cos iy = \frac{1}{2}(e^{-y} + e^{y}) = \cosh y, \qquad \sin iy = \frac{1}{2i}(e^{-y} - e^{y}) = i \sinh y.$$

This yields the desired representations

(10)
$$\cos (x + iy) = \cos x \cosh y - i \sin x \sinh y$$
$$\sin (x + iy) = \sin x \cosh y + i \cos x \sinh y$$

which are useful for numerical computation of $\cos z$ and $\sin z$.

The **hyperbolic cosine** and **sine** of a complex variable z are defined by the formulas

(11) $$\cosh z = \tfrac{1}{2}(e^{z} + e^{-z}), \qquad \sinh z = \tfrac{1}{2}(e^{z} - e^{-z}),$$

in agreement with the familiar definitions for a real variable [cf. (17) in Appendix 3]. These functions are analytic in the entire plane.

From (11) and (1) it follows that

(12) $$\cosh z = \cos (iz), \qquad \sinh z = -i \sin (iz).$$

Fig. 247. Curves $|\sin z| = const$ and arg $(\sin z) = const$ in the z-plane

As in real calculus we define

$$(13) \qquad \tanh z = \frac{\sinh z}{\cosh z}, \qquad \coth z = \frac{\cosh z}{\sinh z},$$

$$(14) \qquad \operatorname{sech} z = \frac{1}{\cosh z}, \qquad \operatorname{csch} z = \frac{1}{\sinh z}.$$

The absolute value $|f(z)|$ of an analytic function $f(z)$, $z = x + iy$, is a real function of the two real variables x and y and can, therefore, be represented by a surface in three-dimensional space; to each point (x, y) in the xy-plane there corresponds a point with Cartesian coordinates $(x, y, |f|)$ in space, and these points form that surface which is called the **modular surface** of $f(z)$. On this surface we may draw some of the curves $|f| = const$ and $\arg f = const$; this yields a useful graphical representation of an analytic function and a geometrical illustration of its behavior.

Figure 246 shows a portion of the modular surface of $f(z) = \sin z$ and some curves $|\sin z| = const$ and $\arg(\sin z) = const$ on the surface. Figure 247 shows the orthogonal projection of these curves into the xy-plane; this graphical representation of the complex sine function is well known and used in electrical engineering.

Modular surfaces have interesting geometrical properties which are considered in Ref. [C8] in Appendix 1.

Problems

1. Show that $\cos z$, $\sin z$, $\cosh z$, and $\sinh z$ are analytic for all z.
2. Verify by differentiation that Re $\cos z$, Im $\cos z$, Re $\sin z$, and Im $\sin z$ are harmonic.
3. Using (1), prove (4), (6), and (7).
4. Using (1) and (2), prove (5).
5. Show that $\overline{\cos \bar z} = \cos z$, $\overline{\sin \bar z} = \sin z$, $\overline{\tan \bar z} = \tan z$.

Find

6. $|\cos z|$ 7. $|\sin z|$ 8. $|\tan z|$
9. Re $\cot z$ 10. Re $\tan z$ 11. Im $\operatorname{csc} z$

Using Table A1 in Appendix 4, calculate

12. $\cos(1.7 + 1.5i)$ 13. $\cos 10i$ 14. $\cosh i$
15. $\sin(1.7 + 1.5i)$ 16. $\sin 10i$ 17. $\sinh(2 + i)$

Find all solutions of the following equations.

18. $\sin z = \cosh 3$ 19. $\cos z = 5$ 20. $\sin z = 1000$
21. $\sin z = i \sinh 1$ 22. $\cosh z = 0$ 23. $\cosh z = 0.5$

24. Find all values of z for which (a) $\cos z$, (b) $\sin z$ has real values.
25. Show that all the values for which (a) $\cos z = 0$, (b) $\sin z = 0$, are real.
26. Using (1), show that $\cos^4 \theta = \frac{1}{8}(\cos 4\theta + 4\cos 2\theta + 3)$.
27. Using (8), Sec. 11.5, and $1 + z + z^2 + \cdots = 1/(1 - z)$, $z = e^{i\theta}/2$, show that

$$1 + \frac{1}{2}\cos \theta + \frac{1}{4}\cos 2\theta + \frac{1}{8}\cos 3\theta + \cdots = \frac{4 - 2\cos \theta}{5 - 4\cos \theta}.$$

Show that

28. $\cosh z = \cosh x \cos y + i \sinh x \sin y$ **29.** $(\cosh z)' = \sinh z$
30. $\sinh z = \sinh x \cos y + i \cosh x \sin y$ **31.** $\tanh (z + i\pi) = \tanh z$
32. $\cot z \neq \pm i$ for all z **33.** $\cosh^2 z - \sinh^2 z = 1$
34. $\cos z = \cosh iz$ **35.** $\sin z = -i \sinh iz$

11.7   *Logarithm. General Power*

The *natural logarithm* of $z = x + iy$ is denoted by $\ln z$ (sometimes also by $\log z$) and is defined as the inverse of the exponential function; that is, $w = \ln z$ is defined by the relation

$$(1) \qquad\qquad e^w = z$$

for each $z \neq 0$.

Setting $w = u + iv$ and $z = |z| e^{i\theta} = re^{i\theta}$ in (1), we have

$$e^w = e^{u+iv} = e^u e^{iv} = re^{i\theta}.$$

It follows that

$$e^u = |z| = r \qquad \text{or} \qquad u = \ln |z|$$

where $\ln |z|$ is the elementary real natural logarithm of the positive number $|z|$, and

$$v = \theta = \arg z.$$

Therefore

$$(2) \qquad \ln z = \ln |z| + i \arg z = \ln \sqrt{x^2 + y^2} + i \arg (x + iy).$$

Since the argument of z is determined only up to multiples of 2π, *the complex natural logarithm is infinitely many-valued.*

The value of $\ln z$ corresponding to the principal value of $\arg z$, that is

$$-\pi < \arg z \leqq \pi \qquad\qquad \text{(Sec. 11.1)}$$

is called the **principal value** of $\ln z$ and is often denoted by Ln z.

Obviously the other values of $\ln z$ are then of the form

$$(3) \qquad\qquad \ln z = \text{Ln } z \pm 2n\pi i \qquad\qquad (n = 1, 2, \cdots);$$

they have the same real part, and their imaginary parts differ by multiples of 2π, in agreement with the fact that e^z is periodic with the imaginary period $2\pi i$.

Also, if z is real positive, the principal value of $\arg z$ is zero, and the principal value Ln z is identical with the real natural logarithm known from elementary calculus. If z is real negative, the principal value of $\arg z$ is π, and then

$$\text{Ln } z = \ln |z| + \pi i.$$

Example 1.

$$\ln (-1) = \pm \pi i, \pm 3\pi i, \pm 5\pi i, \cdots, \qquad \text{Ln } (-1) = \pi i$$

$$\ln i = \frac{\pi}{2} i, \quad -\frac{3\pi}{2} i, \quad \frac{5\pi}{2} i, \quad -\frac{7\pi}{2} i, \quad \frac{9\pi}{2} i, \cdots, \qquad \text{Ln } i = \frac{\pi}{2} i$$

$$\text{Ln } (-i) = -\frac{\pi}{2} i, \qquad \text{Ln } (-2 - 2i) = \ln \sqrt{8} - \frac{3}{4} \pi i.$$

The familiar relations for the natural logarithm continue to hold for complex values, that is,

(4)   (a) $\ln (z_1 z_2) = \ln z_1 + \ln z_2,$ (b) $\ln (z_1/z_2) = \ln z_1 - \ln z_2$

but these relations are to be understood in the sense that each value of one side is also contained among the values of the other side.

Example 2.   Let

$$z_1 = z_2 = e^{\pi i} = -1.$$

If we take

$$\ln z_1 = \ln z_2 = \pi i,$$

then (4a) holds provided we write $\ln (z_1 z_2) = \ln 1 = 2\pi i$; it is not true for the principal value, $\mathrm{Ln}\,(z_1 z_2) = \mathrm{Ln}\,1 = 0.$

By applying (3), Sec. 11.3, to (2), this section, we find

$$\frac{d}{dz} \ln z = \frac{\partial}{\partial x} \ln \sqrt{x^2 + y^2} + i \frac{\partial}{\partial x} (\arg z)$$

$$= \frac{x}{x^2 + y^2} + i \frac{1}{1 + (y/x)^2} \left(-\frac{y}{x^2}\right) = \frac{x - iy}{x^2 + y^2} = \frac{1}{z};$$

that is, the derivative of the natural logarithm is

(5) $\dfrac{d}{dz} \ln z = \dfrac{1}{z}$ $(z \neq 0).$

Hence the principal value $\mathrm{Ln}\,z$ $(z \neq 0)$, which is single-valued and thus a function in the usual sense, is analytic in the domain $-\pi < \arg z < \pi$ of the z-plane, that is, everywhere except at points of the negative real axis (where its imaginary part is not even continuous but has a jump of magnitude 2π).

General powers of a complex number $z = x + iy$ $(\neq 0)$ are defined by the formula

(6) $z^c = e^{c \ln z}$ $(c \text{ complex}, z \neq 0).$

Since $\ln z$ is infinitely many-valued, z^c will, in general, be multi-valued. The particular value

$$z^c = e^{c\,\mathrm{Ln}\,z}$$

is called the *principal value* of z^c.

If $c = n = 1, 2, \cdots$, then z^n is single-valued and identical with the usual nth power of z. If $c = -1, -2, \cdots$, the situation is similar.

If $c = 1/n$ where $n = 2, 3, \cdots$, then

$$z^c = \sqrt[n]{z} = e^{(1/n) \ln z}$$ $(z \neq 0),$

the exponent is determined up to multiples of $2\pi i/n$, and we obtain the n distinct values of the nth root, in agreement with the result in Sec. 11.4. If $c = p/q$, the quotient of two positive integers, the situation is similar, and z^c has only finitely many distinct values. However, if c is real irrational or genuinely complex, then z^c is infinitely many-valued.

Example 3.

$$i^i = e^{i \ln i} = e^{i[(\pi/2)i \pm 2n\pi i]} = e^{-(\pi/2) \mp 2n\pi}$$

All these values are real, and the principal value $(n = 0)$ is $e^{-\pi/2}$.

It is a *convention* that for real positive $z = x$ the expression z^c means $e^{c \ln x}$ where $\ln x$ is the elementary real natural logarithm (that is, the principal value of $\ln z$ $(z = x > 0)$ in the sense of our definition). Also, if $z = e$, the base of the natural logarithm, $z^c = e^c$ is *conventionally* regarded as the unique value obtained from (4) in Sec. 11.5.

From (6) we see that for any complex number a,

(7) $$a^z = e^{z \ln a}.$$

Problems

Find all values of the given expressions and plot some of them in the complex plane.

1. $\ln 1$ 2. $\ln e$ 3. $\ln (-e^2)$ 4. $\ln (e^{-2})$
5. $\ln (ie)$ 6. $\ln (-ie)$ 7. $\ln (e^i)$ 8. $\ln (-e^{-i})$

Using Table A1 in Appendix 4, calculate the principal value $\operatorname{Ln} z$ when z equals

9. $1 + i$ 10. $2 - 2i$ 11. -5 12. $-2 - i\sqrt{12}$

Solve the following equations for z.

13. $\ln z = -\pi i/2$ 14. $\ln z = \pi i/2$ 15. $\ln z = 1 + \pi i$ 16. $\ln z = (1 + i)\pi$

17. Verify (4) for $z_1 = i$, $z_2 = -1$.
18. Using (7) in Sec. 11.3, show that $\operatorname{Ln} z$ $(z \neq 0)$ is analytic in the region $-\pi < \theta < \pi$ where θ is the principal value of arg z.
19. Show that $\operatorname{Ln} z$ is not continuous on the negative real axis.
20. Show that $e^{\ln z} = z$, $\ln (e^z) = z \pm 2n\pi i$, $n = 0, 1, \cdots$.

Find the principal value of

21. $(1 + i)^{2-i}$ 22. $(2i)^{\frac{1}{2}}$ 23. $(1 + i)^i$ 24. $(2 - i)^{1+i}$
25. 3^{3-i} 26. 2^{2i} 27. $(1 - i)^{1+i}$ 28. 2^{3+2i}

The **inverse sine** $w = \sin^{-1} z$ is defined as the relation such that $\sin w = z$. The **inverse cosine** $w = \cos^{-1} z$ is defined as the relation such that $\cos w = z$. For the other trigonometric and hyperbolic functions, the inverse relations are defined and denoted in a similar fashion. Using representations in terms of exponential functions ($\sin w = (e^{iw} - e^{-iw})/2i$, etc.), show that

29. $\sin^{-1} z = -i \ln (iz + \sqrt{1 - z^2})$ 30. $\cos^{-1} z = -i \ln (z + \sqrt{z^2 - 1})$

31. $\cosh^{-1} z = \ln (z + \sqrt{z^2 - 1})$ 32. $\sinh^{-1} z = \ln (z + \sqrt{z^2 + 1})$

33. $\tan^{-1} z = \dfrac{i}{2} \ln \dfrac{i + z}{i - z}$ 34. $\tanh^{-1} z = \dfrac{1}{2} \ln \dfrac{1 + z}{1 - z}$

35. Show that $w = \sin^{-1} z$ is infinitely many-valued, and if w_1 is one of these values, the others are of the form $w_1 \pm 2n\pi$ and $\pi - w_1 \pm 2n\pi$, $n = 0, 1, \cdots$. (The *principal value* of $w = u + iv = \sin^{-1} z$ is defined to be the value for which $-\pi/2 \leq u \leq \pi/2$ when $v \geq 0$ and $-\pi/2 < u < \pi/2$ when $v < 0$.)

CHAPTER **12**

Conformal Mapping

If a complex function $w = f(z)$ is defined in a domain D of the z-plane, then to each point in D there corresponds a point in the w-plane. In this way we have a *mapping* of D onto the range of values of $f(z)$ in the w-plane. We shall see that if $f(z)$ is an analytic function, the mapping given by $f(z)$ is *conformal* (angle-preserving), except at points where the derivative $f'(z)$ is zero.

Conformal mapping is important in engineering mathematics, because it is a standard method for solving boundary value problems in two-dimensional potential theory by transforming a given complicated region into a simpler one.

We shall first define and explain the concept of mapping and then consider the mappings corresponding to elementary analytic functions. Applications will be included in this chapter as well as in Chap. 17.

Prerequisites for this chapter: Chap. 11.
References: Appendix 1, Part F.
Answers to problems: Appendix 2.

12.1 *Mapping*

A continuous *real* function $y = f(x)$ of a real variable x can be exhibited graphically by plotting a curve in the Cartesian xy-plane; this curve is called the *graph* of the function. In the case of a *complex* function

(1) $$w = f(z) = u(x, y) + iv(x, y) \qquad (z = x + iy)$$

the situation is more complicated, because each of the complex variables w and z is represented geometrically by the points in the complex plane. This suggests the use of two separate complex planes for the two variables: one the z-plane, in which the point $z = x + iy$ is to be plotted, and the other the w-plane, in which the corresponding point $w = u + iv$ is to be plotted. In this way the function f assigns to each z in its domain a value $w = f(z)$ in the w-plane. The relation thus defined is called a **mapping** (or *transformation*) of the domain of f **into** the w-plane or, more precisely, a mapping of the domain of f **onto**[1] the range of f in the w-plane.

---

[1] The general terminology is as follows. A mapping of a set A into a set B is called **surjective** or a mapping of A **onto** B if every element of B is the image of at least one element of A. The mapping is said to be **injective** or **one-to-one** if different elements of A have different images in B. The mapping is said to be **bijective** if it is both surjective and injective.

The point $w_0 = f(z_0)$ corresponding to a point z_0 is called the *image point* or **image** of the point z_0 with respect to the mapping defined by $f(z)$. If z moves along some curve C and $f(z)$ is continuous (not a constant), the corresponding point $w = f(z)$ will in general travel along a curve C^* in the w-plane. This curve is then called the *image* of the curve C, and the word "image" applies also to regions or other point sets.

We shall see that the properties of such mappings can be investigated by considering curves (and regions) in the z-plane and their images in the w-plane, and conversely. This will give more information about the functions than the consideration of individual points and their images.

Although two separate planes are used to represent w and z it is sometimes convenient to think of the mapping as effected in one plane and use such familiar terms as translation and rotation. For example, the mapping $w = z + 3$ may be interpreted as a translation which moves each point and configuration in the z-plane three units to the right.

To investigate the specific properties of a mapping defined by a given analytic function $w = u + iv = f(z)$, we may consider the images of the straight lines $x = const$ and $y = const$ in the w-plane. Another possibility is the study of the images of the circles $|z| = const$ and the straight lines through the origin. Conversely, we may consider the curves defined by $u(x, y) = const$ and $v(x, y) = const$ in the z-plane. These curves are called the **level curves** of u and v.

Example 1. Let us illustrate these various possibilities in the case of the mapping

(2) $$w = z^2.$$

In this case the simplest procedure is to set $z = re^{i\theta}$ and $w = Re^{i\phi}$. Then

$$R = r^2 \quad \text{and} \quad \phi = 2\theta.$$

We see that circles $r = r_0 = const$ are mapped onto circles $R = r_0{}^2 = const$, and rays $\theta = \theta_0 = const$ onto rays $\phi = 2\theta_0 = const$. In particular, the positive real axis ($\theta = 0$) is mapped onto the positive real axis in the w-plane, and the positive imaginary axis ($\theta = \pi/2$) in the z-plane is mapped onto the negative real axis in the w-plane. The angles at the origin are doubled under the mapping. The first quadrant $0 \leq \theta \leq \pi/2$ is mapped onto the entire upper half of the w-plane (Fig. 248).

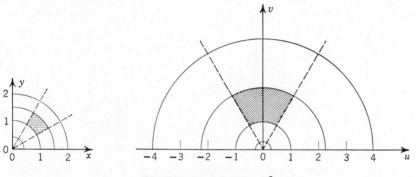

Fig. 248. Mapping $w = z^2$

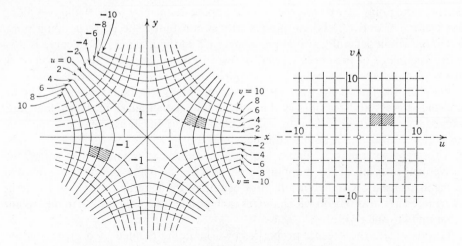

Fig. 249. Level curves of u and v in the case of the mapping $w = z^2$

In rectangular coordinates the transformation $w = z^2$ becomes

$$u + iv = x^2 - y^2 + 2xyi.$$

By separating the real and the imaginary parts we obtain

(3) $$u = x^2 - y^2, \qquad v = 2xy.$$

We see that the level curves of u and v are equilateral hyperbolas with the lines $y = \pm x$ and the coordinate axes for asymptotes. We observe that these curves are the orthogonal trajectories of each other (cf. Sec. 1.10). In Fig. 249, the two shaded domains in the z-plane are both mapped onto the shaded rectangle in the w-plane. Clearly, every point $w \neq 0$ is the image of precisely two points in the z-plane.

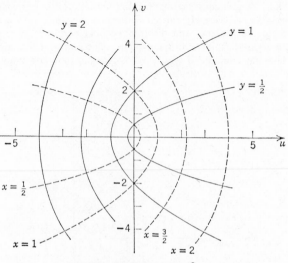

Fig. 250. Mapping $w = z^2$

Fig. 251. Mapping defined by $w = z^n$

Finally, we may use (3) for determining the images of the straight lines $x = const$ and $y = const$. The line $x = c = const$ has the image

$$u = c^2 - y^2, \qquad v = 2cy.$$

We can eliminate y from these equations, finding

$$v^2 = 4c^2(c^2 - u).$$

This is a parabola with focus at the origin and opening to the left. Similarly, the image of a line $y = k = const$ can be represented in the form

$$v^2 = 4k^2(k^2 + u).$$

This is a parabola with focus at the origin and opening to the right (Fig. 250).

The other powers

$$(4) \hspace{3cm} w = z^n, \qquad n = 3, 4, \cdots$$

may be considered in a similar fashion. Of course, the level curves, etc., are then represented by more complicated equations. The angular region $0 \leqq \arg z \leqq \pi/n$ is mapped onto the upper half of the w-plane (Fig. 251).

Problems

Find and plot the images of the following curves under the mapping $w = u + iv = z^2$.

1. $x = 1, 2, 3, 4$ **2.** $y = 1, 2, 3, 4$ **3.** $xy = 10$
4. $y = x + 1$ **5.** $y = x, y = -x$ **6.** $x^2 = 1 + y^2$

Plot the images of the following regions under the mapping $w = z^2$.

7. $|z| < \frac{1}{3}$ **8.** $|z| \geqq 5$ **9.** $|\arg z| < \pi/3$
10. $\operatorname{Re} z > 0$ **11.** $0 < y < 1$ **12.** $\pi/4 < \arg z < \pi/2$

Find the images of the curves $x = const$, $y = const$, $|z| = const$, $\arg z = const$, and the level curves of u and v in the case of the following functions.

13. $w = 2iz$ **14.** $w = (1 + i)z/\sqrt{2}$ **15.** $w = 1/z$

Give a geometrical description of the transformations defined by

16. $w = iz$ **17.** $w = az + b$ **18.** $w = 1/z$

19. Find and sketch the images of the angular region $0 \leqq \arg z \leqq \pi/8$ in the case of the mappings $w = iz$, $w = z^2$, $w = iz^2$, and $w = -z^2$.

20. Find and sketch the images of the regions in Prob. 19 under the transformations $w = 1/z$, $w = i/z$, $w = 1/z^2$, and $w = i/z^2$.

21. Find and sketch the images of the region $y > -1$ under the transformations $w = iz$, $w = (1 + i)z$, $w = (1 - i)z + 2i$, $w = 2iz - 4$, and $w = iz^2$.

22. Find and sketch the image of the region $x > 0$, $-\pi < y < \pi$ under the transformation $w = (1 + z/\pi)i$.
23. What set in the z-plane is mapped by $w = u + iv = z + 1/z$ onto the u-axis? Onto the v-axis?
24. Find an analytic function $w = u + iv = f(z)$ which maps the half plane $x \geqq 0$ onto the region $u \geqq 2$ such that $z = 0$ corresponds to $w = 2 + i$.
25. Find an analytic function $w = u + iv = f(z)$ which maps the region $0 < \arg z < \pi/3$ onto the region $u < 1$.

12.2 *Conformal Mapping*

We shall now consider the most important geometrical property of the mappings defined by analytic functions, namely, their conformality.

A mapping in the plane is said to be *angle-preserving*, or **conformal**, if it preserves angles between oriented curves in magnitude as well as in sense, that is, the images of any two intersecting oriented curves, taken with their corresponding orientation, make the same angle of intersection as the curves, both in magnitude and direction. Here the angle between two oriented curves is defined to be the angle α ($0 \leqq \alpha \leqq \pi$) between their oriented tangents at the point of intersection (Fig. 252).

A curve C in the complex z-plane can be represented in the form

(1) $$z(t) = x(t) + iy(t)$$

where t is a real parameter. For example, the function

$$z(t) = r \cos t + ir \sin t$$

represents the circle $|z| = r$, the function

$$z(t) = t + it^2$$

represents the parabola $y = x^2$, etc. The direction of increasing values of t in (1) is called the *positive direction* or *positive sense* on C. In this way (1) defines an *orientation* on C. We assume that $z(t)$ in (1) is differentiable and the derivative $\dot{z}(t)$ is continuous and nowhere zero. Then C has a unique tangent at each of its points and is called a **smooth curve.** The sense on each tangent corresponding to the positive sense on C is called the *positive sense* on that tangent, which is then oriented.

Fig. 252. Curves C_1 and C_2 and their respective images $C_1{}^*$ and $C_2{}^*$ under a conformal mapping

Fig. 253. Derivation of formula (2)

In fact, the tangent to C at a point $z_0 = z(t_0)$ is defined as the limiting position of the straight line through z_0 and another point $z_1 = z(t_0 + \Delta t)$ as z_1 approaches z_0 along C, that is, as $\Delta t \to 0$. (Cf. also Sec. 7.5.) Now the number $z_1 - z_0$ can be represented by the vector from z_0 to z_1 (Fig. 253), and the vector corresponding to $(z_1 - z_0)/\Delta t$, where $\Delta t > 0$, has the same direction as that vector. It follows that the vector corresponding to

$$(2) \qquad \dot{z}(t_0) = \frac{dz}{dt}\bigg|_{t_0} = \lim_{\Delta t \to 0} \frac{z_1 - z_0}{\Delta t} = \lim_{\Delta t \to 0} \frac{z(t_0 + \Delta t) - z(t_0)}{\Delta t}$$

is tangent to C at z_0, and the angle between this vector and the positive x-axis is $\arg \dot{z}(t_0)$.

Consider now the mapping given by a nonconstant analytic function $w = f(z) = u(x, y) + iv(x, y)$ defined in a domain containing C. Then the image of C under this mapping is a curve C^* in the w-plane represented by

$$w(t) = f[z(t)].$$

The point $z_0 = z(t_0)$ corresponds to the point $w(t_0)$ of C^*, and $\dot{w}(t_0)$ represents a tangent vector to C^* at this point. Now by the chain rule,

$$(3) \qquad \frac{dw}{dt} = \frac{df}{dz}\frac{dz}{dt}.$$

Hence, if $f'(z_0) \neq 0$, we see that $\dot{w}(t_0) \neq 0$ and C^* has a unique tangent at $w(t_0)$, the angle between the tangent vector $\dot{w}(t_0)$ and the positive u-axis being $\arg \dot{w}(t_0)$. Since the argument of a product equals the sum of the arguments of the factors, we have from (3)

$$\arg \dot{w}(t_0) = \arg f'(z_0) + \arg \dot{z}(t_0).$$

Thus under the mapping the directed tangent to C at z_0 is rotated through the angle

$$(4) \qquad \arg \dot{w}(t_0) - \arg \dot{z}(t_0) = \arg f'(z_0),$$

the angle between those two tangent vectors to C and C^*. Since the expression on the right is independent of the choice of C, we see that this angle is independent of C; that is, the transformation $w = f(z)$ rotates the tangents of all the curves through z_0 through the *same* angle $\arg f'(z_0)$. Hence, two curves through z_0 which form a certain angle at z_0 are mapped upon curves forming the same angle, in sense as well as in magnitude, at the image point w_0 of z_0. This proves the following basic result.

Theorem 1.   *The mapping defined by an analytic function $f(z)$ is conformal, except at points where the derivative $f'(z)$ is zero.*

A point at which $f'(z) = 0$ is sometimes called a **critical point.** At such a point the mapping is not conformal, as may be illustrated by the following example.

Example 1.   The mapping $w = z^2$ is conformal except at $z = 0$ where $w' = 2z = 0$. The conformality can be seen from the curves considered in the last section. At $z = 0$ the angles are doubled under the mapping, because each ray arg $z = c = $ *const* transforms into a ray arg $w = 2c$. Cf. Fig. 248 in the last section.

We further note that, by the definition of a derivative, we have

$$\lim_{z \to z_0} \left| \frac{j(z) - f(z_0)}{z - z_0} \right| = |f'(z_0)|.$$

Therefore, the mapping $w = f(z)$ magnifies the lengths of short lines by approximately the factor $|f'(z_0)|$. The image of a small figure *conforms* to the original figure in the sense that it has approximately the same shape. However, since $f'(z)$ varies from point to point, a *large* figure may have an image whose shape is quite different from that of the original figure.

We want to mention that, by (3) in Sec. 11.3 and the Cauchy-Riemann equations,

$$|f'(z)|^2 = \left| \frac{\partial u}{\partial x} + i \frac{\partial v}{\partial x} \right|^2 = \left(\frac{\partial u}{\partial x} \right)^2 + \left(\frac{\partial v}{\partial x} \right)^2 = \frac{\partial u}{\partial x} \frac{\partial v}{\partial y} - \frac{\partial u}{\partial y} \frac{\partial v}{\partial x},$$

that is,

(5)
$$|f'(z)|^2 = \begin{vmatrix} \dfrac{\partial u}{\partial x} & \dfrac{\partial u}{\partial y} \\ \dfrac{\partial v}{\partial x} & \dfrac{\partial v}{\partial y} \end{vmatrix} = \frac{\partial(u, v)}{\partial(x, y)}$$

where the determinant is the so-called *Jacobian* (cf. Sec. 8.3) of the transformation $w = f(z)$, written in real form

$$u = u(x, y), \qquad v = v(x, y).$$

Hence, the condition $f'(z_0) \neq 0$ implies that the Jacobian is not zero at z_0. This condition suffices that the restriction of the mapping $w = f(z)$ to a sufficiently small neighborhood N_0 of z_0 is **one-to-one** or injective, that is, different points in N_0 have different images; a proof is included in Ref. [A2] listed in Appendix 1. (Cf. also footnote 1 in Sec. 12.1.)

Example 2.   The mapping $w = z^2$ is one-to-one in a sufficiently small neighborhood of any point $z \neq 0$. In a neighborhood of $z = 0$ it is not one-to-one. The full z-plane is mapped onto the w-plane so that each point $w \neq 0$ is the image of two points in the z-plane. For instance, the points $z = 1$ and $z = -1$ are both mapped onto $w = 1$, and, more generally, z_1 and $-z_1$ have the same image point $w = z_1^2$.

The practical importance of conformal mapping results from the fact that harmonic functions of two real variables (cf. Sec. 11.3) remain harmonic under a change of variables arising from a conformal transformation (Theorem 2, below).

This has important consequences. Suppose that it is required to solve a **boundary value problem** in connection with a two-dimensional potential, that is, to determine a solution of Laplace's equation (in two independent variables) in a given region D which assumes given values on the boundary of D. It may be possible to find a conformal mapping which transforms D into some simpler region D^*, such as a circular disk or a half-plane. Then we may solve Laplace's equation subject to the transformed boundary conditions in D^*. The resulting solution when carried back to D by the inverse transformation will be the solution of the original problem. This powerful method is justified by the following theorem.

Theorem 2. *A harmonic function $h(x, y)$ remains harmonic under a change of the variables arising from a one-to-one conformal transformation $w = f(z)$.*

Proof. Let $h(x, y)$ be harmonic in a region D. Let $w = u + iv = f(z)$ be an analytic function which maps D onto a region D^* in the w-plane in a one-to-one and conformal fashion. Since the mapping is one-to-one and conformal, $f'(z) \neq 0$ in D, and the inverse function $z = F(w)$ which maps D^* onto D exists. This function is analytic, that is, it has a derivative in D^*, namely,

$$\frac{dF}{dw} = \frac{1}{df/dz} .$$

The proof of this formula is similar as in real calculus. Let $g(x, y)$ be a conjugate harmonic function of $h(x, y)$ (cf. Sec. 11.3). Then $h + ig$ is an analytic function of $z = x + iy$, say, $H(z)$. Since the function

$$z = F(w) = x(u, v) + iy(u, v)$$

is analytic, $H[F(w)]$ is an analytic function of $w = u + iv$. It has the real part $h[x(u, v), y(u, v)]$, which is thus a harmonic function of u and v in D^*, and the proof is complete.

When we use the method of conformal mapping in potential theory the difficulty is to find an analytic function which maps a given region onto a simpler one. For this purpose we need some experience and a detailed knowledge of the mapping properties of the elementary analytic functions. Therefore, we shall consider the most important elementary functions from this point of view.

Problems

Represent the following curves in the z-plane ($z = x + iy$) in the form (1).

1. $y = 2x^2$	**2.** $x^2 - y^2 = 1$	**3.** $y = ax + b$
4. $x^2 + y^2 = 9$	**5.** $x^2 + 4y^2 = 4$	**6.** $y = 1/x$

7. Does the mapping $w = \bar{z} = x - iy$ preserve angles in size as well as in sense?

8. Why do the images of the curves $|z| = $ *const* and arg $z = $ *const* under a mapping by an analytic function intersect at right angles?

9. What is the reason that the level curves $u = $ *const* and $v = $ *const* of an analytic function $w = u + iv = f(z)$ intersect at right angles at each point at which $f'(z) \neq 0$?

Find the points in the z-plane at which the following mappings fail to be conformal.

    10. $w = z^5$ **11.** $w = z^5 - 5z$ **12.** $w = e^{z^2}$

    13. $w = \sin \pi z$ **14.** $w = z + z^{-1}\ (z \neq 0)$ **15.** $w = \cosh z$

Verify (5) for the following functions $f(z) = u(x, y) + iv(x, y)$.

    16. $z^2 + z$ **17.** e^{-2z} **18.** $\sin z$

    19. Verify Theorem 2 for $h(x, y) = \ln (x^2 + y^2), f(z) = 1/z\ (z \neq 0)$.

    20. Verify Theorem 2 for $h(x, y) = x/(x^2 + y^2), f(z) = 2z + 1$.

12.3 *Linear Fractional Transformations*

We shall now consider an important class of conformal mappings, starting with the simplest types of transformations in this class.

The mapping

$$(1) \hspace{4cm} w = z + b$$

may be regarded as a *translation* in the direction defined by the argument of b through a distance equal to $|b|$. If $b = 0$, then we obtain the **identity transformation**

$$w = z.$$

A transformation of the form

$$(2) \hspace{4cm} w = az \hspace{4cm} (a \neq 0)$$

is a *rotation* through a fixed angle equal to arg a *combined with a uniform dilatation or contraction* such that $|w| = |a|\, |z|$. If a is real and positive, then (2) is a pure dilatation or contraction. If $|a| = 1$, then (2) is a pure rotation.

In the case of the transformation

$$(3) \hspace{4cm} w = \frac{1}{z}$$

it is useful to introduce polar coordinates by setting

$$z = re^{i\theta} \hspace{1cm} \text{and} \hspace{1cm} w = Re^{i\phi}.$$

Then our mapping may be written

$$(3') \hspace{4cm} R = \frac{1}{r}, \hspace{1cm} \phi = -\theta.$$

From this we see that a point $w = 1/z\ (z \neq 0)$ lies on the ray from the origin through $\bar{z}$, at the distance $1/|z|$ from the origin.

We mention that the image $w = 1/z$ of a given point z can be obtained by a simple geometric construction shown in Fig. 254. In this figure, P represents the given number z. The points T_1 and T_2 are those points of the unit circle C for which the tangents to C pass through P, and A is the point of intersection of the segments OP and T_1T_2. It can be shown by elementary methods that if OP has the length l, then OA has the length $1/l$. We say that A is obtained from P by **inversion** *with respect to the unit circle* (or "reflection" in the unit circle). Obviously, A represents the complex number $1/\bar{z}$. Hence, the point P^*, representing $w = 1/z$,

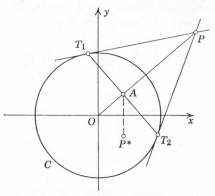

Fig. 254. Geometrical construction of $1/z$

is obtained by reflecting A in the real axis. This means that *the transformation $w = 1/z$ is an inversion with respect to the unit circle combined with a reflection in the real axis.*

A fixed point of a mapping $w = f(z)$ is a point whose image is the same complex number; that is, the fixed points are obtained from

$$w = f(z) = z.$$

The mapping $w = 1/z$ has the fixed points $z = \pm 1$, because these points are the solutions of the equation $z = 1/z$.

Figure 255 shows some of the straight lines $x = const$ and $y = const$ and their images under the mapping $w = 1/z$.

A transformation of the form

(4)
$$w = \frac{az + b}{cz + d} \qquad (ad - bc \neq 0)$$

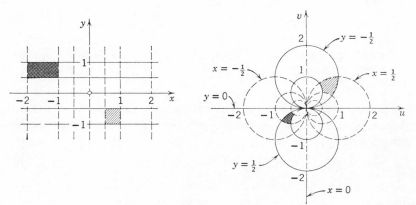

Fig. 255. Mapping $w = 1/z$

is called a **linear fractional transformation** or **Möbius transformation.** The constants a, b, c, and d in (4) are real or complex numbers.

The derivative is

$$w' = \frac{a(cz + d) - c(az + b)}{(cz + d)^2} = \frac{ad - bc}{(cz + d)^2},$$

and this shows that the condition $ad - bc \neq 0$ implies $w' \neq 0$, which entails conformality, whereas $ad - bc = 0$ leads to the uninteresting case $w' \equiv 0$ or $w = const$, which will be excluded in our further consideration.

From (4) we see that to each z for which $cz + d \neq 0$ there corresponds precisely one complex number w. Suppose that $c \neq 0$. Then to the value $z = -d/c$, for which $cz + d = 0$, there does not correspond a number w. This situation suggests that we attach an "improper point" to the w-plane. This point is called the **point at infinity** and is denoted by the symbol ∞ (*infinity*). The complex plane together with the point ∞ is called the **extended complex plane.** The complex plane without that improper point is called the **finite complex plane.** We now let $w = \infty$ be the image point of $z = -d/c$ ($c \neq 0$) under the mapping defined by (4). When $c = 0$ in (4), then $a \neq 0$ and $d \neq 0$ (why?), and we let $w = \infty$ be the image point of $z = \infty$, the improper point of the extended z-plane.

The inverse mapping of (4) is obtained by solving (4) for z; we find

(5)
$$z = \frac{-dw + b}{cw - a}.$$

When $c \neq 0$, then $cw - a = 0$ for $w = a/c$, and we let $w = a/c$ correspond to $z = \infty$. When $c = 0$, then $a \neq 0$ and $d \neq 0$, and we let $w = \infty$ correspond to $z = \infty$, as before. It follows that every mapping (4) is a one-to-one mapping of the extended z-plane onto the extended w-plane; we say, that *every linear fractional transformation* (4) *maps "the extended plane in a one-to-one manner onto itself."*

The fixed points of the mapping (4) are obtained from the equation

$$z = \frac{az + b}{cz + d}$$

or

(6)
$$cz^2 - (a - d)z - b = 0.$$

This is a quadratic equation in z whose coefficients all vanish if and only if the mapping is the identity (in this case, $a = d \neq 0$, $b = c = 0$). Hence we obtain the following result.

Theorem 1. *A linear fractional transformation, not the identity, has at most two fixed points. If a linear fractional transformation is known to have three or more fixed points, it must be the identity.*

Linear fractional transformations have the following interesting property.

Theorem 2. *Any linear fractional transformation (4) maps the totality of circles and straight lines in the z-plane onto the totality of circles and straight lines in the w-plane.*

Proof. Let $w = u + iv$. Any straight line or circle in the w-plane can be represented in the form

$$A(u^2 + v^2) + B_1 u + B_2 v + C = 0$$

where A, B_1, B_2, and C are real constants. Since

$$u^2 + v^2 = |w|^2 = w\bar{w}, \qquad u = \frac{1}{2}(w + \bar{w}), \qquad v = \frac{1}{2i}(w - \bar{w})$$

we may write this representation in the form

(7) $$A w\bar{w} + Bw + \bar{B}\bar{w} + C = 0$$

where $B = \frac{1}{2}(B_1 - iB_2)$. If we insert w and $\bar{w}$ according to (4) and multiply the resulting equation by $(cz + d)\overline{(cz + d)}$, we obtain

$$A(az + b)\overline{(az + b)} + B(az + b)\overline{(cz + d)}$$
$$+ \bar{B}\overline{(az + b)}(cz + d) + C(cz + d)\overline{(cz + d)} = 0.$$

If we perform the multiplication, this equation becomes

(8) $$\alpha z\bar{z} + \beta z + \bar{\beta}\bar{z} + \gamma = 0$$

where the coefficients are given by the formulas

$$\alpha = A\,|a|^2 + 2\,\mathrm{Re}\,(Ba\bar{c}) + C\,|c|^2$$
$$\beta = Aa\bar{b} + Ba\bar{d} + \bar{B}b\bar{c} + Cc\bar{d}$$
$$\gamma = A\,|b|^2 + 2\,\mathrm{Re}\,(Bb\bar{d}) + C\,|d|^2.$$

Since α and γ are real, (8) represents a circle or a straight line in the z-plane. Conversely, starting from (8) and using the inverse transformation (5), we obtain (7), and the proof is complete.

The calculations in the proof show that a straight line may very well be mapped onto a circle, but never onto any other curve. For example, $w = 1/z$ maps a straight line or a circle through the origin onto a straight line in the w-plane, while a straight line or circle not passing through $z = 0$ is mapped onto a circle in the w-plane.

Obviously, the mappings (1), (2), and (3) are special cases of (4). Conversely, *any mapping* (4) *is a composite of mappings of the forms* (1), (2), *and* (3). In fact, if $c = 0$, this is immediately clear. If $c \neq 0$, then we can write (4) in the form

(9) $$w = K\,\frac{1}{cz + d} + \frac{a}{c} \qquad \text{where} \qquad K = -\frac{ad - bc}{c}.$$

Setting

$$w_1 = cz, \qquad w_2 = w_1 + d, \qquad w_3 = \frac{1}{w_2}, \qquad w_4 = Kw_3,$$

we have $w = w_4 + a/c$, and the statement is proved.

Special linear fractional transformations of practical importance and further general properties of linear fractional transformations will be discussed in the following section.

Problems

1. Give a direct proof (without using Theorem 2) that $w = cz$ maps circles onto circles and straight lines onto straight lines.
2. Without using Theorem 2, prove that $w = 1/z$ maps circles onto circles (straight lines regarded as degenerate cases of circles).
3. How can Probs. 1 and 2 be used to prove Theorem 2?

Find the images of the following circles and straight lines under the mapping $w = 1/z$.

4. $|z| = 1$ 5. $|z - 1| = 1$ 6. $|z - 2| = 1$
7. $x = 1$ 8. $y = x + 1$ 9. $|z - 2i| = 1$

10. Find the image of the region $-2 < x < -1$, $-1 < y < 1$ under the mapping $w = 1/z$.
11. Show that under the mapping $w = 1/z$, the center of a given circle is not mapped onto the center of the image circle.
12. Find $1/(3 + 4i)$ from $3 + 4i$ by means of the geometrical construction explained in connection with the transformation $w = 1/z$.

Find the fixed points of the following mappings.

13. $w = z^2$ 14. $w = z + 2$ 15. $w = z^5$

16. $w = \dfrac{3z + 2}{z - 1}$ 17. $w = \dfrac{2iz - 1}{z + 2i}$ 18. $w = \dfrac{z - 1}{z + 1}$

Find a linear fractional transformation whose fixed points are

19. $-2, 2$ 20. $-i, i$ 21. 1

22. Find all linear fractional transformations whose fixed points are -1 and 1.
23. Find all linear fractional transformations whose fixed points are $-i$ and i.
24. Show that (4) with $c = 0$ can be represented as a composite of mappings of the types (1) and (2).
25. Represent $w = (z + i)/(iz + 4)$ as a composite of mappings of the types (1), (2), (3).

12.4 Special Linear Fractional Transformations

In applying conformal mapping to practical problems it is often required to map the interior of the unit circle or a half-plane onto itself, or the interior of the unit circle onto a half-plane. We want to show that this can be done by a suitable linear fractional transformation

(1)
$$w = \frac{az + b}{cz + d} \qquad (ad - bc \neq 0)$$

and in each specific case these transformations can be found in a systematic way.

General remark. If $z = \infty$, then the right-hand side of (1) becomes the meaningless expression $(a \cdot \infty + b)/(c \cdot \infty + d)$. We assign to it the value $w = a/c$ when $c \neq 0$ and $w = \infty$ when $c = 0$. This is suggested by our discussion of (4) and (5) in the last section.

The linear fractional transformation (1) depends on three essential constants, namely, the ratios of any three of the constants a, b, c, d to the fourth. The requirement that three distinct points in the z-plane have specified images in the w-plane leads to a unique linear fractional transformation, as follows.

Theorem 1. *Three given distinct points z_1, z_2, z_3 can always be mapped onto three prescribed distinct points w_1, w_2, w_3 by one, and only one, linear fractional transformation $w = f(z)$. This mapping is given implicitly by the equation*

$$(2) \qquad \frac{w - w_1}{w - w_3} \cdot \frac{w_2 - w_3}{w_2 - w_1} = \frac{z - z_1}{z - z_3} \cdot \frac{z_2 - z_3}{z_2 - z_1}.$$

(If one of these points is the point ∞, the quotient of those two differences which contain this point must be replaced by 1.)

Proof. Equation (2) is of the form $F(w) = G(z)$ where F and G denote fractional linear functions of the respective variables. From this we readily obtain $w = f(z) = F^{-1}[G(z)]$ where F^{-1} denotes the inverse function of F. Since the inverse of a linear fractional transformation and the composite of linear fractional transformations are linear fractional transformations (cf. Prob. 1 at the end of the section), $w = f(z)$ is a linear fractional transformation. Furthermore, from (2) we see that

$$F(w_1) = 0, \qquad F(w_2) = 1, \qquad F(w_3) = \infty,$$
$$G(z_1) = 0, \qquad G(z_2) = 1, \qquad G(z_3) = \infty.$$

Hence, $w_1 = f(z_1)$, $w_2 = f(z_2)$, $w_3 = f(z_3)$. This proves the existence of a linear fractional transformation $w = f(z)$ which maps z_1, z_2, z_3 onto w_1, w_2, w_3, respectively.

We prove that $w = f(z)$ is uniquely determined. Suppose that $w = g(z)$ is another linear fractional transformation which maps z_1, z_2, z_3 onto w_1, w_2, w_3, respectively. Then its inverse $g^{-1}(w)$ maps w_1 onto z_1, w_2 onto z_2, and w_3 onto z_3. Consequently, the composite mapping $H = g^{-1}[f(z)]$ maps each of the points z_1, z_2, and z_3 onto itself; that is, it has three distinct fixed points z_1, z_2, z_3. From Theorem 1 in the last section it follows that H is the identity mapping, and, therefore, $g(z) \equiv f(z)$.

The last statement of the theorem follows from the general remark at the beginning of this section. This completes the proof.

We shall now consider *linear fractional transformations which map the upper half-plane $y \geqq 0$ onto the unit disk $|w| \leqq 1$.* Clearly the x-axis, which is the boundary of that half-plane, must correspond to the unit circle $|w| = 1$ of that disk, and we have to make sure that the upper half-plane is mapped onto the interior but not onto the exterior of that unit circle.

Example 1. Find the linear fractional transformation (1) which maps $z_1 = -1$, $z_2 = 0$, $z_3 = 1$ onto $w_1 = -1$, $w_2 = -i$, $w_3 = 1$, respectively. From (2) we obtain

$$\frac{w - (-1)}{w - 1} \cdot \frac{-i - 1}{-i - (-1)} = \frac{z - (-1)}{z - 1} \cdot \frac{0 - 1}{0 - (-1)}.$$

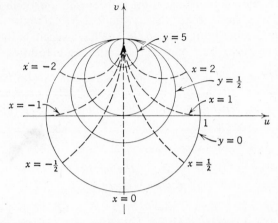

Fig. 256. Linear fractional transformation in Example 1

or

(3)
$$w = \frac{z - i}{-iz + 1}.$$

Let us show that we can determine the specific properties of such a mapping without difficult calculations. The images of the lines $x = const$ and $y = const$ are obtained as follows. The point $z = i$ corresponds to $w = 0$, and $z = \infty$ corresponds to $w = i$. If $z = iy$, then we have $w = i(y - 1)/(y + 1)$; that is, the positive imaginary axis is mapped onto the segment $u = 0$, $-1 \leqq v \leqq 1$. Since the mapping is conformal and straight lines are mapped onto circles or straight lines, the lines $y = const$ are mapped onto circles through the image of $z = \infty$, that is, onto circles through $w = i$ and with center on the v-axis. It follows that, for the same reasons, the lines $x = const$ are mapped onto circles which are orthogonal to the image circles of the lines $y = const$ (Fig. 256). The lower half-plane corresponds to the exterior of the unit circle $|w| = 1$.

Example 2. Determine the linear fractional transformation which maps $z_1 = 0$, $z_2 = 1$, $z_3 = \infty$ onto $w_1 = i$, $w_2 = -1$, $w_3 = -i$, respectively. From (2) we find that the desired mapping is

(4)
$$w = -i\frac{z - i}{z + i};$$

in this case we have to replace the quotient $(1 - \infty)/(z - \infty)$, which first occurs, by 1.

We shall now consider *linear fractional transformations which map the upper half-plane $y \geqq 0$ onto the upper half-plane $v \geqq 0$.* In this case the x-axis must be mapped onto the u-axis.

Example 3. Find the linear fractional transformation which maps the points $z_1 = -2$, $z_2 = 0$, $z_3 = 2$ onto the points $w_1 = \infty$, $w_2 = \frac{1}{4}$, $w_3 = \frac{3}{8}$, respectively. From (2) we obtain

(5)
$$w = \frac{z + 1}{2z + 4}.$$

We leave the discussion of the properties of the mapping to the reader (Probs. 19, 20).

As another important class let us consider the *linear fractional transformations which map the unit disk in the z-plane onto the unit disk in the w-plane.* We may

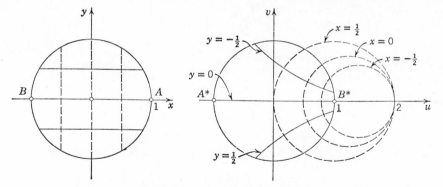

Fig. 257. Mapping in Example 4

readily verify that the function

(6) $$w = \frac{z - z_0}{cz - 1}, \qquad c = \bar{z}_0, \qquad |z_0| < 1$$

belongs to this class and maps the point z_0 onto $w = 0$ (cf. Prob. 22).

Example 4. Let $z_0 = \frac{1}{2}$. Then by (6),

$$w(z) = \frac{2z - 1}{z - 2}.$$

The real axes correspond to each other; in particular,

$$w(-1) = 1, \qquad w(0) = \tfrac{1}{2}, \qquad w(1) = -1.$$

Since the mapping is conformal and straight lines are mapped onto circles or straight lines and $w(\infty) = 2$, the images of the lines $x = const$ are circles through $w = 2$ with centers on the u-axis; the lines $y = const$ are mapped onto circles which are orthogonal to the aforementioned circles (Fig. 257).

By combining linear fractional transformations and transformations of the form $w = z^n$ ($n > 1$ and integral) we may obtain *mappings of angular regions onto the unit disk*.

Example 5. Map the angular region D: $-\pi/6 \leqq \arg z \leqq \pi/6$ onto the unit disk $|w| \leqq 1$. We may proceed as follows. The mapping $t = z^3$ maps D onto the right half of the t-plane. Then we may apply a linear fractional transformation which maps this half-plane onto the unit disk,

(z-plane)   (t-plane)   (w-plane)

Fig. 258. Mapping in Example 5

for example, the transformation

$$w = i \frac{t-1}{t+1}.$$

By inserting $t = z^3$ into this mapping we find

$$w = i \frac{z^3 - 1}{z^3 + 1};$$

this mapping has the required properties (Fig. 258).

Problems

1. Show that the composite of two linear fractional transformations is a linear fractional transformation.
2. In Prob. 1, does the composite depend on the order of composition?
3. Find the inverse of (3). Show that (3) maps the lines $x = c = const$ onto circles with centers on the line $v = 1$.

Find all linear fractional transformations $w(z)$ with the following property.

4. $0, 1, 2$ are mapped onto $2, 5, 8$, respectively.
5. $-1, 0, 1$ are mapped onto $-(3 - i)/5, -3/4, -1 - i/3$, respectively.
6. $0, 1, \infty$ are mapped onto $\infty, 1, 0$, respectively.
7. $\infty, 1, -1$ are mapped onto $1, (3 + 2i)/5, 3 - 2i$, respectively.
8. $-1, 0, 1$ are mapped onto $0, 1, -1$, respectively.
9. $\infty, 0, -1$ are mapped onto $1, 0, (1 + i)/2$, respectively.
10. $1, i, -1$ are mapped onto $i, -1, -i$, respectively.
11. $i, 0, 1$ are mapped onto $i/2, 0, (1 + i)/2$, respectively.
12. $0, 1, \infty$ are mapped onto $1, \infty, 0$, respectively.
13. z_1, z_2, z_3 are mapped onto $0, 1, \infty$, respectively.
14. $z_1 = 0$ is a fixed point.
15. $z_1 = 0$ and $z_2 = \infty$ are fixed points.
16. The x-axis is mapped onto the u-axis.
17. Derive (4) from (2). Graph the curves corresponding to the lines $u = const$ and $v = const$.
18. Find the inverse of (4) and graph the images of the lines $x = const$ and $y = const$.
19. Find and graph the images of the lines $x = -3, -2, \cdots, 2, 3$ and $y = -3, -2, \cdots$, $2, 3$ under the mapping (5).
20. Derive (5) from (2). Find the inverse of (5) and graph the curves corresponding to $u = const$ and $v = const$.
21. Find a linear fractional transformation which maps $|z| \leq 1$ onto $|w| \leq 1$ such that $z = i/4$ is mapped onto $w = 0$ and graph the images of the lines $x = const$ and $y = const$.
22. Prove the statement involving (6).
23. Find an analytic function which maps the second quadrant of the z-plane onto the interior of the unit circle in the w-plane.
24. Find an analytic function $w = f(z)$ which maps the region $2 \leq y \leq x + 1$ onto the unit disk $|w| \leq 1$.
25. Find an analytic function $w = f(z)$ which maps the region $0 \leq \arg z \leq \pi/4$ onto the unit disk $|w| \leq 1$.

12.5 Mapping by Other Elementary Functions

We shall now consider the mapping properties of some further important special functions.

The **exponential function** (Sec. 11.5)

$$(1) \hspace{4cm} w = e^z$$

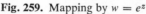

Fig. 259. Mapping by $w = e^z$

defines a mapping which is conformal everywhere, because its derivative is different from zero at any point. If we set $w = Re^{i\phi}$, then

$$Re^{i\phi} = e^{x+iy} = e^x e^{iy}$$

and (1) can be written in the form

(2) $$R = e^x, \qquad \phi = y.$$

From this we see that the lines $x = a = const$ are mapped onto the circles $R = e^a$, and the lines $y = c$ are mapped onto the rays $\phi = c$. Since $e^z \neq 0$ for all z, the point $w = 0$ is not an image of any point z. A rectangular region, say, $a \leq x \leq b$, $c \leq y \leq d$, is mapped onto the region

$$e^a \leq R \leq e^b, \qquad c \leq \phi \leq d$$

bounded by portions of rays and circles (Fig. 259).

The fundamental strip $-\pi < y \leq \pi$ is mapped onto the full w-plane (cut along the negative real axis). More generally, every horizontal strip bounded by two lines $y = c$ and $y = c + 2\pi$ is mapped onto the full w-plane. This illustrates the fact that e^z is periodic with the period $2\pi i$.

The horizontal strip $0 \leq y \leq \pi$ is mapped onto the upper half of the w-plane. The boundary $y = 0$ is mapped onto the positive half of the u-axis, and the line $y = \pi$ onto the negative half of the u-axis, as follows from (2). The segment from 0 to πi is mapped onto the semicircle $|w| = 1$, $v \geq 0$. The left half ($x \leq 0$) of our strip is mapped onto the region $|w| \leq 1$, $v \geq 0$, and the right half ($x \geq 0$) of the strip is mapped onto the exterior of that semi-circle $|w| = 1$ in the upper half of the w-plane (Fig. 260).

Since the **natural logarithm** $w = u + iv = \ln z$ is the inverse relation of the exponential function, the properties of the corresponding conformal mapping can be easily obtained from those of the exponential function by interchanging the

Fig. 260. Mapping by $w = e^z$

roles of the z and the w planes in the preceding considerations. The principal value $w = \text{Ln } z$ thus maps the z-plane (cut along the negative real axis) onto the horizontal strip $-\pi < v \leq \pi$ of the w-plane. Further details of the mapping will be discussed in Ex. 2 of the next section.

The sine function (Sec. 11.6)

(3) $$w = u + iv = \sin z = \sin x \cosh y + i \cos x \sinh y$$

where

(4) $$u = \sin x \cosh y, \qquad v = \cos x \sinh y,$$

is periodic. Hence, the mapping (4) is certainly not one-to-one if we consider it in the full xy-plane. We restrict z to the infinite strip D defined by $-\pi/2 \leq x \leq \pi/2$. Since $f'(z) = \cos z$ is zero at $z = \pm\pi/2$, the mapping is not conformal at these two points. From (4) we see that the boundary of D is mapped into the u-axis. The segment $-\pi/2 \leq x \leq \pi/2$ of the x-axis maps onto the segment $-1 \leq u \leq 1$ of the u-axis, the line $x = -\pi/2$ maps onto $u \leq -1$, $v = 0$, and the line $x = \pi/2$ maps onto $u \geq 1$, $v = 0$. The line segment $y = c > 0$, $-\pi/2 \leq x \leq \pi/2$ maps onto the portion of the ellipse

$$u = \cosh c \sin x, \qquad v = \sinh c \cos x$$

or

(5) $$\frac{u^2}{\cosh^2 c} + \frac{v^2}{\sinh^2 c} = 1$$

in the upper half of the w-plane. The line segment $y = -c$, $-\pi/2 \leq x \leq \pi/2$ ($c > 0$) maps onto the lower half of the ellipse (5). The foci of the ellipse are at $w = \pm 1$, and we see that they are independent of c. Consequently, if we let c vary, we obtain a family of confocal ellipses. The rectangular region D defined by $-\pi/2 < x < \pi/2$, $-c < y < c$ is thus mapped onto the interior of the ellipse (5); but note that the image of the boundary consists of the ellipse and the two segments of the x-axis, as shown in Fig. 261 (where $c = 1$). The image points of points on the vertical parts of the boundary coincide in pairs. In particular, $B^* = F^*$ and $C^* = E^*$.

The rectangle $-\pi < x < \pi$, $c < y < d$ ($c > 0$) maps onto an elliptic ring cut along the negative v-axis (Fig. 262). The straight lines $x = const$ $(-\pi/2 < x < \pi/2)$

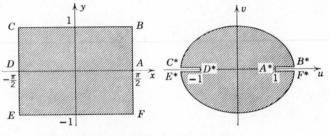

Fig. 261. Mapping by $w = \sin z$

Fig. 262. Mapping by $w = \sin z$

map onto confocal hyperbolas which intersect those ellipses at right angles, and the y-axis maps onto the v-axis.

The **cosine function**

$$(6) \qquad\qquad w = \cos z = \sin\left(z + \frac{\pi}{2}\right)$$

defines the same mapping as $\sin z$, preceded by a translation to the right through $\pi/2$ units.

The **hyperbolic function**

$$(7) \qquad\qquad w = \sinh z = -i \sin(iz)$$

defines a transformation which is a rotation $t = iz$ followed by the mapping $p = \sin t$ and another rotation $w = -ip$.

Similarly, the transformation

$$(8) \qquad\qquad w = \cosh z = \cos(iz)$$

is a rotation $t = iz$ followed by the mapping $w = \cos t$.

Example 1. Find the image of the semi-infinite strip $x \geq 0$, $0 \leq y \leq \pi$ (Fig. 263) under the mapping (8). We set $w = u + iv$. Since $\cosh 0 = 1$, the point $z = 0$ is mapped onto $w = 1$. For real $z = x \geq 0$, $\cosh z$ is real and increases monotone from 1 as x increases. Hence the positive x-axis is mapped onto the portion $u \geq 1$ of the u-axis. For pure imaginary $z = iy$ we have $\cosh iy = \cos y$. It follows that the left boundary of the strip is mapped onto the segment $1 \geq u \geq -1$ of the u-axis, the point $z = \pi i$ corresponding to

$$w = \cosh i\pi = \cos \pi = -1.$$

On the upper boundary of the strip, $y = \pi$, and since $\sin \pi = 0$, $\cos \pi = -1$, it follows that this part of the boundary is mapped onto the portion $u \leq -1$ of the u-axis. Hence the boundary

Fig. 263. Mapping in Example 1

of the strip is mapped onto the u-axis. It is not difficult to see that the interior of the strip is mapped onto the upper half of the w-plane, and the mapping is one-to-one.

Example 2.   Find the temperature $T(x, y)$ in the strip considered in Ex. 1, if the temperature on the boundary is

$$T = T_0 \text{ on the segment from 0 to } \pi i,$$

$$T = 0 \text{ on the upper and lower boundaries.}$$

Since T does not depend on time ("steady state temperature distribution"), the heat equation reduces to Laplace's equation (cf. Sec. 10.11)

$$\nabla^2 T = \frac{\partial^2 T}{\partial x^2} + \frac{\partial^2 T}{\partial y^2} = 0,$$

and we have to find a solution of this equation satisfying those boundary conditions.

To solve this problem we map the strip by means of (8) onto the upper half of the w-plane. Since the segment $0 \leq y \leq \pi$ of the y-axis is mapped onto the segment $-1 \leq u \leq 1$ of the u-axis, the boundary conditions in the w-plane are (Fig. 264):

$$T = T_0 \text{ on the segment from } -1 \text{ to } 1,$$

$$T = 0 \text{ on the other parts of the } u\text{-axis.}$$

The real and imaginary parts of analytic functions are solutions of Laplace's equation, and we have to find such a solution $T(u, v)$ in the upper half of the w-plane which satisfies those boundary conditions. For this purpose we consider the functions

(9*)

$$\text{Ln } (w + 1) = \ln |w + 1| + i\phi_1, \qquad \phi_1 = \arg (w + 1) = \arctan \frac{v}{u + 1},$$

$$\text{Ln } (w - 1) = \ln |w - 1| + i\phi_2, \qquad \phi_2 = \arg (w - 1) = \arctan \frac{v}{u - 1}.$$

Since $\phi_1(u, v)$ and $\phi_2(u, v)$ are harmonic functions, $\phi_2 - \phi_1$ is harmonic. If $w = u$ is real and smaller than -1, then $\phi_2 - \phi_1 = \pi - \pi = 0$; for real $w = u$ in the interval $-1 < u < 1$ we have $\phi_2 - \phi_1 = \pi - 0 = \pi$, and for real $w = u > 1$ we have $\phi_2 - \phi_1 = 0 - 0 = 0$. Hence the function

(9)

$$T(u, v) = \frac{T_0}{\pi} (\phi_2 - \phi_1)$$

is harmonic in the half-plane $v > 0$ and satisfies those boundary conditions in the w-plane. Since $\tan \phi_1 = v/(u + 1)$ and $\tan \phi_2 = v/(u - 1)$, it follows that

$$\tan (\phi_2 - \phi_1) = \frac{\tan \phi_2 - \tan \phi_1}{1 + \tan \phi_1 \tan \phi_2} = \frac{2v}{u^2 + v^2 - 1},$$

and (9) takes the form

(10)

$$T(u, v) = \frac{T_0}{\pi} \arctan \frac{2v}{u^2 + v^2 - 1}.$$

Fig. 264. Boundary conditions in Example 2

The function $w = \cosh z$ maps the strip under consideration onto the half-plane $v \geqq 0$, and we have

$$w = u + iv = \cosh(x + iy) = \cosh x \cos y + i \sinh x \sin y.$$

Separating the real and imaginary parts on both sides, we may write

$$u = \cosh x \cos y, \qquad v = \sinh x \sin y.$$

From these expressions for u and v it follows that in (10),

$$u^2 + v^2 - 1 = \cosh^2 x \cos^2 y + \sinh^2 x \sin^2 y - 1 = \sinh^2 x - \sin^2 y.$$

By inserting this and the expression for v into (10) and denoting $T(u(x, y), v(x, y))$ by $T^*(x, y)$ we have

$$T^*(x, y) = \frac{T_0}{\pi} \arctan \frac{2 \sinh x \sin y}{\sinh^2 x - \sin^2 y}.$$

Noting that the numerator and the denominator are the imaginary part and the real part of the function $(\sinh x + i \sin y)^2$, we may write

$$T^*(x, y) = \frac{T_0}{\pi} \arg [(\sinh x + i \sin y)^2] = \frac{2T_0}{\pi} \arg (\sinh x + i \sin y).$$

Hence the solution of our problem is

(11) $$T^*(x, y) = \frac{2T_0}{\pi} \arctan \frac{\sin y}{\sinh x}.$$

This function is harmonic in the interior of our strip (cf. Theorem 2 in Sec. 12.2) and it satisfies the boundary conditions. Indeed, $T^* = 0$ when $y = 0$ or $y = \pi$, and $T^* = T_0$ when $x = 0$. The isotherms (= curves of constant temperature) are the curves

$$\frac{\sin y}{\sinh x} = const.$$

In Ex. 2 we transformed the real potential $T(u, v)$ into the real potential $T^*(x, y)$ by using the function $w = u(x, y) + iv(x, y)$ which maps the half-plane onto the given region. In many cases problems of this type become simpler by working with a **complex potential,** that is, by taking a complex analytic function $F(w)$ such that the real potential $T(u, v)$ is the real or imaginary part of $F(w)$, and transforming F instead of T. It is clear that such a complex potential F may be readily obtained from T by determining a conjugate harmonic function of T (cf. at the end of Sec. 11.3). Let us illustrate this "method of complex potentials" in the case of our previous example. A detailed discussion of complex potentials will be presented later (in Chap. 17).

Example 3. From (9*) we see that the real potential

$$T(u, v) = \frac{T_0}{\pi} (\phi_2 - \phi_1)$$

in Ex. 2 [cf. (9)] is the imaginary part of the complex potential

(12) $$F(w) = \frac{T_0}{\pi} [\text{Ln}(w - 1) - \text{Ln}(w + 1)] = \frac{T_0}{\pi} \text{Ln} \frac{w - 1}{w + 1}.$$

The mapping function in Ex. 2 is

$$w = \cosh z = \tfrac{1}{2}(e^z + e^{-z}).$$

From this we see that in (12),

$$\frac{w-1}{w+1} = \frac{\cosh z - 1}{\cosh z + 1} = \frac{e^z + e^{-z} - 2}{e^z + e^{-z} + 2} = \frac{(e^{z/2} - e^{-z/2})^2}{(e^{z/2} + e^{-z/2})^2} = \tanh^2 \frac{z}{2}.$$

By inserting this into (12) and denoting $F(w(z))$ by $F^*(z)$ and $\tanh(z/2)$ by H, we have

(13) $$F^*(z) = \frac{T_0}{\pi} \operatorname{Ln} \tanh^2 \frac{z}{2} = \frac{2T_0}{\pi} \operatorname{Ln} \tanh \frac{z}{2} = \frac{2T_0}{\pi} \operatorname{Ln} H.$$

Because of (2) in Sec. 11.7 this can be written

(14) $$F^*(z) = \frac{2T_0}{\pi} \operatorname{Ln} H = \frac{2T_0}{\pi} \left(\ln |H| + i \arctan \frac{\operatorname{Im} H}{\operatorname{Re} H} \right).$$

$F^*(z)$ is the complex potential in the strip in Ex. 2, and its imaginary part is the solution of our problem. By using the definition of H in terms of exponential functions we find that

(15) $$H = \frac{\sinh x + i \sin y}{\cosh x + \cos y}.$$

From this and (14) we have

$$T^*(x, y) = \operatorname{Im} F^*(z) = \frac{2T_0}{\pi} \arctan \frac{\sin y}{\sinh x},$$

in agreement with our result in Ex. 2.

Furthermore, from (15) it follows that

$$|H|^2 = \frac{\sinh^2 x + \sin^2 y}{(\cosh x + \cos y)^2},$$

and from (14) we see that the real part of $F^*(z)$ becomes

$$S^*(x, y) = \operatorname{Re} F^*(z) = \frac{T_0}{\pi} \ln \frac{\sinh^2 x + \sin^2 y}{(\cosh x + \cos y)^2}.$$

The curves $S^* = const$ insersect the isotherms $T^* = const$ at right angles and, therefore, are the curves along which the heat flows. It is a great advantage that the *complex* potential yields *both* families of curves.

Problems

Find and graph the images of the following regions under the mapping $w = e^z$.

1. $0 \leq x \leq 1, 0 \leq y \leq \pi/2$ 2. $1 < x < 2, -\pi/2 < y < \pi/2$
3. $-2 < x < -1, 0 < y < \pi$ 4. $-1 < x < 1, \pi/4 < y < 3\pi/4$

5. Find an analytic function which maps the region R bounded by the positive x and y axes and the hyperbola $xy = c\ (>0)$ in the first quadrant onto the upper half-plane. *Hint.* First map R onto a horizontal strip.
6. Find and sketch the images of the lines $x = 0, \pm\pi/6, \pm\pi/3, \pm\pi/2$ under the mapping $w = \sin z$.
7. Determine all points at which the mapping $w = \sin z$ is not conformal.

Find and graph the images of the following regions under the mapping $w = \sin z$.

8. $-\pi/2 < x < \pi/2, 0 < y < 1$ 9. $-\pi/2 < x < \pi/2, 1 < y < 2$
10. $0 < x < \pi/2, 0 < y < 2$ 11. $0 < x < 2\pi, 1 < y < 2$

12. Describe the transformation $w = \cosh z$ in terms of the transformation $w = \sin z$ and rotations and translations.

13. Investigate the mapping $w = \sin^{-1} z$.

14. Find the temperature $T(x, y)$ in the strip in Ex. 1, assuming that $T = 0$ on the left boundary and $T = 1$ on the upper and lower boundaries of the strip.

15. Show that $w = \text{Ln}\,[(z - 1)/(z + 1)]$ maps the upper half-plane onto the horizontal strip $0 \leqq \text{Im}\,w \leqq \pi$.

12.6   Riemann Surfaces

We consider the mapping defined by

$$(1) \hspace{4cm} w = u + iv = z^2$$

(cf. Sec. 12.1). This mapping is conformal, except at $z = 0$ where $w' = 2z$ is zero. The angles at $z = 0$ are doubled under the mapping. The right half of the z-plane (including the positive y-axis) maps onto the full w-plane cut along the negative half of the u-axis; the mapping is one-to-one. Similarly, the left half of the z-plane (including the negative y-axis) is mapped onto the cut w-plane in a one-to-one manner.

Obviously, the mapping of the full z-plane is not one-to-one, because every point $w \neq 0$ corresponds to precisely two points z. In fact, if z_1 is one of these points, then the other is $-z_1$. For example, $z = i$ and $z = -i$ have the same image, namely, $w = -1$, etc. Hence, the w-plane is "covered twice" by the image of the z-plane. We say that the full z-plane is mapped onto the *doubly covered* w-plane. We can still give our imagination the necessary support as follows.

We imagine one of the two previously obtained copies of the cut w-plane to be placed upon the other so that the upper sheet is the image of the right half of the z-plane, and the lower sheet is the image of the left half of the z-plane; we denote these half-planes by R and L, respectively. When we pass from R to L, the corresponding image point should pass from the upper to the lower sheet. For this reason we join the two sheets crosswise along the cut, that is, along the negative real axis. (This construction can be carried out only in imagination, since the penetration of the two sheets of a material model can only be imperfectly realized.) The two origins are fastened together. The configuration thus obtained is called a **Riemann surface.** On it every point $w \neq 0$ appears twice, at superposed positions, and the origin appears precisely once. The function $w = z^2$ now maps the full z-plane onto this Riemann surface in a one-to-one manner, and the mapping is conformal, except for the "winding point" or **branch point** at $w = 0$ (Fig. 265).

Fig. 265. Example of a Riemann surface

Such a branch point connecting two sheets is said to be of the *first order*. (More generally, a branch point connecting n sheets is said to be of **order** $n - 1$.)

We now consider the double-valued relation

$$(2) \qquad\qquad w = \sqrt{z}.$$

To each $z \neq 0$ there correspond two values w, one of which is the principal value. If we replace the z-plane by the two-sheeted Riemann surface just considered, then each complex number $z \neq 0$ is represented by two points of the surface at superposed positions. We let one of these points correspond to the principal value—for example, the point in the upper sheet—and the other to the other value. Then (2) becomes single-valued, that is, (2) is a function of the points of the Riemann surface, and to any continuous motion of z on the surface there corresponds a continuous motion of the corresponding point in the w-plane. The function maps the sheet corresponding to the principal value onto the right half of the w-plane and the other sheet onto the left half of the w-plane.

Let us consider some further important examples.

Example 1. In the case of the relation

$$(3) \qquad\qquad w = \sqrt[n]{z} \qquad\qquad n = 3, 4, \cdots$$

we need a Riemann surface consisting of n sheets and having a branch point of order $n - 1$ at $z = 0$. One of the sheets corresponds to the principal value and the other $n - 1$ sheets to the other $n - 1$ values.

Example 2. For any $z \neq 0$ the relation

$$(4) \qquad\qquad w = \ln z = \operatorname{Ln} z + 2n\pi i \qquad (n = 0, \pm 1, \pm 2, \cdots, z \neq 0)$$

is infinitely many-valued. Hence (4) defines a function on a Riemann surface consisting of infinitely many sheets. The function $w = \operatorname{Ln} z$ corresponds to one of these sheets. On this sheet the argument θ of z ranges in the interval $-\pi < \theta \leqq \pi$ (cf. Sec. 11.7). The sheet is cut along the negative ray of the real axis, and the upper edge of the slit is joined to the lower edge of the next sheet, which corresponds to the interval $\pi < \theta \leqq 3\pi$, that is, to the function $w = \operatorname{Ln} z + 2\pi i$. In this way each value of n in (4) corresponds to precisely one of these infinitely many sheets. The function $w = \operatorname{Ln} z$ maps the corresponding sheet onto the horizontal strip $-\pi < v \leqq \pi$ in the w-plane. The next sheet is mapped onto the neighboring strip $\pi < v \leqq 3\pi$, etc. The function $w = \ln z$ thus maps all the sheets of the corresponding Riemann surface onto the entire w-plane, the correspondence between the points $z \neq 0$ of the Riemann surface and those of the w-plane being one-to-one.

Example 3. Let us consider the mapping defined by

$$(5) \qquad\qquad w = z + \frac{1}{z} \qquad\qquad (z \neq 0)$$

which is important in aerodynamics (see below). Since

$$w' = 1 - \frac{1}{z^2} = \frac{(z + 1)(z - 1)}{z^2}$$

the mapping is conformal except at the points $z = 1$ and $z = -1$; these points correspond to $w = 2$ and $w = -2$, respectively. From (5) we find

$$(6) \qquad\qquad z = \frac{w}{2} \pm \sqrt{\frac{w^2}{4} - 1} = \frac{w}{2} \pm \frac{1}{2}\sqrt{(w + 2)(w - 2)}.$$

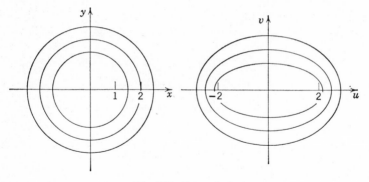

Fig. 266. Example 3

Hence, the points $w = 2$ and $w = -2$ are branch points of the first order of $z = z(w)$. To any value w ($\neq 2$, $\neq -2$) there correspond two values of z. Consequently, (5) maps the z-plane onto a two-sheeted Riemann surface, the two sheets being connected crosswise from $w = -2$ to $w = 2$ (Fig. 266), and this mapping is one-to-one. We set $z = re^{i\theta}$ and determine the images of the curves $r = const$ and $\theta = const$. From (5) we obtain

$$w = u + iv = re^{i\theta} + \frac{1}{r} e^{-i\theta} = \left(r + \frac{1}{r}\right) \cos \theta + i \left(r - \frac{1}{r}\right) \sin \theta.$$

By equating the real and imaginary parts on both sides we have

(7) $$u = \left(r + \frac{1}{r}\right) \cos \theta, \qquad v = \left(r - \frac{1}{r}\right) \sin \theta.$$

From this we find

$$\frac{u^2}{a^2} + \frac{v^2}{b^2} = 1, \qquad \text{where} \qquad a = r + \frac{1}{r}, \qquad b = \left| r - \frac{1}{r} \right|.$$

The circles $r = const$ are thus mapped onto ellipses whose principal axes lie in the u and v axes and have the lengths $2a$ and $2b$, respectively. Since $a^2 - b^2 = 4$, independent of r, these ellipses are confocal, with foci at $w = -2$ and $w = 2$. The unit circle $r = 1$ maps onto the line segment from $w = -2$ to $w = 2$. For every $r \neq 1$ the two circles with radii r and $1/r$ map onto the same ellipse in the w-plane, corresponding to the two sheets of the Riemann surface. Hence, the interior of the unit circle $|z| = 1$ corresponds to one sheet, and the exterior to the other.

Furthermore, from (7) we obtain

(8) $$\frac{u^2}{\cos^2 \theta} - \frac{v^2}{\sin^2 \theta} = -4.$$

The lines $\theta = const$ are thus mapped onto the hyperbolas which are the orthogonal trajectories of those ellipses. The real axis, that is, the rays $\theta = 0$ and $\theta = \pi$, are mapped onto the part of the real axis from $w = 2$ via ∞ to $w = -2$. The y-axis is mapped onto the v-axis. Any other pair of rays $\theta = \theta_0$ and $\theta = \theta_0 + \pi$ is mapped onto the two branches of the same hyperbola.

An exterior region of one of the above ellipses is free of branch points and corresponds to either the interior or the exterior of the corresponding circle in the z-plane, depending on the sheet of the Riemann surface to which the region belongs. In particular the full w-plane corresponds to the interior or the exterior of the unit circle $|z| = 1$, as was mentioned before.

The mapping (5) transforms suitable circles into airfoils with a sharp trailing edge whose interior angle is zero; these airfoils are known as **Joukowski**[2] **airfoils**. Since the airfoil to be obtained has a sharp edge, it is clear that the circle to be mapped must pass through one of the

[2] NIKOLAI JEGOROVICH JOUKOWSKI (1847–1921), Russian mathematician.

Fig. 267. Joukowski airfoil

points $z = \pm 1$ where the mapping is not conformal. Let us choose a circle C through $z = -1$ and such that $z = 1$ lies inside C. The simplest way of determining the image is the graphical vector addition of the vectors corresponding to z and $1/z$ (where the latter can be obtained from the former as shown in Sec. 12.3). This leads to the result shown in Fig. 267. Further details are included in Ref. [F8] in Appendix 1.

A more elaborate treatment of Riemann surfaces can be found in Ref. [F9].

Problems

1. Show that the Riemann surface of $w = \sqrt[3]{z}$ consists of three sheets and has a branch point of second order at $z = 0$. Find the path of the image point w of a point z which moves three times around the unit circle starting from the initial position $z = 1$.

2. Consider the Riemann surfaces of $w = \sqrt[4]{z}$ and $w = \sqrt[5]{z}$ in a fashion similar to that in Prob. 1.

3. Determine the path of the image of a point z under the mapping $w = \ln z$ as z moves several times around the unit circle.

4. Find the images of the annuli $\frac{1}{2} < |z| < 1, 1 < |z| < 2$, and $2 < |z| < 3$ under the mapping $w = z + 1/z$.

5. Show that the Riemann surface of $w = \sqrt{(z-1)(z-4)}$ has branch points at $z = 1$ and $z = 4$ and consists of two sheets which may be cut along the line segment from 1 to 4 and joined crosswise. *Hint.* Introduce polar coordinates $z - 1 = r_1 e^{i\theta_1}$ and $z - 4 = r_2 e^{i\theta_2}$.

6. Show that the Riemann surface of $w = \sqrt{(1-z^2)(4-z^2)}$ has four branch points and two sheets which may be joined crosswise along the segments $-2 \leqq x \leqq -1$ and $1 \leqq x \leqq 2$ of the x-axis.

Find the location of the branch points and the number of sheets of the Riemann surfaces of the following functions.

7. $\sqrt[3]{2z + 3i}$　　　　　　　8. $\sqrt{z^2 + 1}$　　　　　　　9. $\sqrt{z(z-1)(z+1)}$

10. $\sqrt{(z-a)(z-b)}, a \neq b \neq 0$　　11. $1 + z + \sqrt{z}$　　　　12. $\ln(z - a)$

13. $e^{\sqrt{z}}$　　　　　　　　　　14. $\sqrt{e^z}$　　　　　　　　15. $\sqrt{\sqrt[3]{z} - 1}$

CHAPTER 13

Complex Integrals

Integrals in the complex plane are important for two reasons. The practical reason is that in applications there occur real integrals which can be evaluated by complex integration, while the usual methods of real integral calculus are not successful. The other reason is of a theoretical nature and results from the fact that the method of complex integration yields proofs of some basic properties of analytic functions (in particular the existence of higher derivatives), which would be very difficult to prove without using integrals. This situation indicates a basic difference between real and complex calculus.

In this chapter we shall first define complex integrals. The most fundamental result will be Cauchy's integral theorem (Sec. 13.3) from which the important Cauchy integral formula (Sec. 13.5) will follow. In Sec. 13.6 we shall prove that if a function is analytic, it has derivatives of any order. This means that in this respect complex analytic functions behave much more simply than real functions of a real variable.

(Integration by means of residues and applications to real integrals will be considered in Chap. 16.)

Prerequisite for this chapter: Chap. 11.
References: Appendix 1, Part F.
Answers to problems: Appendix 2.

13.1 Line Integral in the Complex Plane

As in real calculus, we distinguish between definite integrals and indefinite integrals or antiderivatives. An **indefinite integral** is a function whose derivative equals a given analytic function in a region. By inverting known differentiation formulas we may find many types of indefinite integrals.

Let us now define *definite integrals*, or *line integrals*, of a complex function $f(z)$ where $z = x + iy$. We shall see that this definition will be a natural generalization of the familiar definition of a real definite integral, and the consideration will be similar to that in Sec. 8.1. In the case of a definite integral the path of integration

Fig. 268. Complex line integral

is an interval of the real axis. In the case of a complex definite integral we shall integrate along a curve[1] in the complex plane.

Let C be a smooth curve (cf. Sec. 12.2) in the complex z-plane. Then we may represent C in the form

(1) $$z(t) = x(t) + iy(t) \qquad\qquad (a \leq t \leq b)$$

where $z(t)$ has a continuous derivative $\dot{z}(t) \neq 0$ for all t, so that C is rectifiable (cf. Sec. 7.4) and has a unique tangent at each point. We remember that the positive direction along C corresponds to the sense of increasing values of the parameter t.

Let $f(z)$ be a continuous function which is defined (at least) at each point of C. We subdivide C by points $z_1, z_2, \cdots, z_{n-1}, z_n \, (= Z)$, cf. Fig. 268. On each portion of subdivision we choose an arbitrary point, say, a point ζ_1 between z_0 and z_1, a point ζ_2 between z_1 and z_2, etc., and form the sum

(2) $$S_n = \sum_{m=1}^{n} f(\zeta_m)\, \Delta z_m$$

where

$$\Delta z_m = z_m - z_{m-1}.$$

This we do for each $n = 2, 3, \cdots$ in a completely independent manner, but such that the greatest $|\Delta z_m|$ approaches zero as n approaches infinity. In this way we obtain a sequence of complex numbers $S_2, S_3, \cdots$. The limit of this sequence is called the **line integral** (or simply the integral) of $f(z)$ along the oriented curve C and is denoted by

(3) $$\int_C f(z)\, dz.$$

The curve C is called the **path of integration**.

Throughout the following considerations we shall assume that all paths of integration for complex line integrals are piecewise smooth, that is, consist of finitely many smooth curves.

From our assumptions the existence of the line integral (3) follows. In fact, let $f(z) = u(x, y) + iv(x, y)$ and set

$$\zeta_m = \xi_m + i\eta_m \qquad \text{and} \qquad \Delta z_m = \Delta x_m + i\, \Delta y_m.$$

---

[1] Actually along a portion, or arc, of a curve. For the sake of simplicity we shall use the single term "curve" to denote an entire curve as well as a portion of it.

Then (2) may be written

$$(4) \qquad S_n = \sum (u + iv)(\Delta x_m + i \, \Delta y_m)$$

where $u = u(\xi_m, \eta_m)$, $v = v(\xi_m, \eta_m)$ and we sum over m from 1 to n. We may now split up S_n into four sums:

$$S_n = \sum u \, \Delta x_m - \sum v \, \Delta y_m + i[\sum u \, \Delta y_m + \sum v \, \Delta x_m].$$

These sums are real. Since f is continuous, u and v are continuous. Hence, if we let n approach infinity in the aforementioned way, then the greatest Δx_m and Δy_m will approach zero and each sum on the right becomes a real line integral:

$$(5) \qquad \lim_{n \to \infty} S_n = \int_C f(z) \, dz = \int_C u \, dx - \int_C v \, dy + i\left[\int_C u \, dy + \int_C v \, dx\right].$$

This shows that the line integral (3) exists and its value is independent of the choice of subdivisions and intermediate points ζ_m.

Furthermore, as in Sec. 8.2 we may convert each of those real line integrals to a definite integral by the use of the representation (1) of the curve C:

$$(6) \qquad \int_C f(z) \, dz = \int_a^b u\dot{x} \, dt - \int_a^b v\dot{y} \, dt + i\left[\int_a^b u\dot{y} \, dt + \int_a^b v\dot{x} \, dt\right]$$

where $u = u[x(t), y(t)]$, $v = v[x(t), y(t)]$, and dots denote derivatives with respect to t.

Without causing misinterpretation, we may also write

$$\int_C f(z) \, dz = \int_a^b (u + iv)(\dot{x} + i\dot{y}) \, dt$$

or more briefly

$$(6^*) \qquad \int_C f(z) \, dz = \int_a^b f[z(t)] \, \dot{z}(t) \, dt.$$

Let us consider some basic examples.

Example 1. Integrate $f(z) = 1/z$ once around the unit circle C in the counterclockwise sense, starting from $z = 1$. We may represent C in the form

$$(7) \qquad z(t) = \cos t + i \sin t \qquad (0 \leq t \leq 2\pi).$$

Then we have

$$\dot{z}(t) = -\sin t + i \cos t,$$

and by (6*) the integral under consideration becomes

$$\int_C \frac{dz}{z} = \int_0^{2\pi} \frac{1}{\cos t + i \sin t} (-\sin t + i \cos t) \, dt = i \int_0^{2\pi} dt = 2\pi i,$$

a fundamental result which we shall use in our further consideration.

Clearly, instead of (7) we may write more simply

$$(7') \qquad z(t) = e^{it} \qquad (0 \leq t \leq 2\pi).$$

Then we obtain by differentiation

$$\dot{z}(t) = ie^{it}, \qquad dz = ie^{it}\, dt,$$

and from this the same result as before:

(8)
$$\int_C \frac{dz}{z} = \int_0^{2\pi} \frac{1}{e^{it}} ie^{it}\, dt = i \int_0^{2\pi} dt = 2\pi i.$$

Example 2. Integrate $f(z) = \operatorname{Re} z = x$ along the segment from $z_0 = 0$ to $Z = 1 + i$ (path C_1 in Fig. 269).

The segment may be represented in the form

$$z(t) = x(t) + iy(t) = (1 + i)t \qquad\qquad (0 \le t \le 1).$$

Then

$$f[z(t)] = \operatorname{Re} z(t) = x(t) = t, \qquad dz = (1 + i)\, dt.$$

Therefore, we obtain the result

$$\int_{C_1} \operatorname{Re} z\, dz = \int_0^1 t(1 + i)\, dt = (1 + i) \int_0^1 t\, dt = \tfrac{1}{2}(1 + i).$$

Let us now integrate $f(z) = \operatorname{Re} z = x$ along the real axis from 0 to 1, and then vertically to $1 + i$ (path C_2 in Fig. 269). We may represent the first part of C_2 by

$$z = z(t) = t \qquad\qquad (0 \le t \le 1)$$

and the last part in the form

$$z(t) = 1 + i(t - 1) \qquad\qquad (1 \le t \le 2).$$

Then the entire path corresponds to the interval $0 \le t \le 2$. On the first part, $\operatorname{Re} z = t$, $dz = dt$, and on the last part, $\operatorname{Re} z = 1$, $dz = i\, dt$. Therefore,

$$\int_{C_2} \operatorname{Re} z\, dz = \int_0^1 t\, dt + \int_1^2 i\, dt = \tfrac{1}{2} + i.$$

We note that the last part of C_2 could be represented equally well in the form

$$z(t) = 1 + it \qquad\qquad (0 \le t \le 1);$$

then the limits of the last integral are 0 and 1, and its value is the same as before.

By comparing our two results we see that in the case of the function $\operatorname{Re} z$ (which is not analytic) the value of the integral depends not only on the end points of the path, but also on its geometric shape.

Example 3. Let $f(z) = (z - z_0)^m$ where m is an integer and z_0 is a constant. Integrate in the counter-clockwise sense around the circle C of radius ρ with center at z_0 (cf. Fig. 270).

Fig. 269. Paths in Example 2   **Fig. 270.** Path in Example 3

We may represent C in the form

$$z(t) = z_0 + \rho(\cos t + i \sin t) = z_0 + \rho e^{it} \qquad (0 \leq t \leq 2\pi).$$

Then we have

$$(z - z_0)^m = \rho^m e^{imt}, \qquad dz = i\rho e^{it} \, dt,$$

and we obtain

$$\int_C (z - z_0)^m \, dz = \int_0^{2\pi} \rho^m e^{imt} i\rho e^{it} \, dt = i\rho^{m+1} \int_0^{2\pi} e^{i(m+1)t} \, dt.$$

The case $m = -1$ was considered in Ex. 2, and when $m \neq -1$, we obtain [cf. (14) in Sec. 11.5]

$$\int_0^{2\pi} e^{i(m+1)t} \, dt = \left[\frac{e^{i(m+1)t}}{i(m+1)} \right]_0^{2\pi} = 0 \qquad (m \neq -1).$$

The result is

(9)
$$\int_C (z - z_0)^m \, dz = \begin{cases} 2\pi i & (m = -1), \\ 0 & (m \neq -1 \text{ and integral}). \end{cases}$$

Example 4. Let $f(z) = k = const$, and C be any curve joining two points z_0 and Z. In this case we may use the definition of the line integral as the limit of the sum S_n given by (2). We have

$$S_n = \sum_{m=1}^n k \, \Delta z_m = k[(z_1 - z_0) + (z_2 - z_1) + \cdots + (Z - z_{n-1})] = k(Z - z_0).$$

From this we immediately obtain the result

$$\int_C k \, dz = \lim_{n \to \infty} S_n = k(Z - z_0).$$

We note that the value of this integral depends only on z_0 and Z, not on the choice of the path joining these points. If in particular C is closed, then $z_0 = Z$ and the integral is zero.

Example 5. Let $f(z) = z$ and C be any curve joining two points z_0 and Z. We again use (2). Taking $\zeta_m = z_m$ we obtain

$$S_n = \sum_{m=1}^n z_m \, \Delta z_m = z_1(z_1 - z_0) + z_2(z_2 - z_1) + \cdots + Z(Z - z_{n-1}).$$

Taking $\zeta_m = z_{m-1}$ we find

$$S_n{}^* = \sum_{m=1}^n z_{m-1} \, \Delta z_m = z_0(z_1 - z_0) + z_1(z_2 - z_1) + \cdots + z_{n-1}(Z - z_{n-1}).$$

Addition of these two sum yields $S_n + S_n{}^* = Z^2 - z_0{}^2$, as can easily be verified. Hence,

$$\lim_{n \to \infty} (S_n + S_n{}^*) = 2 \int_{z_0}^Z z \, dz = Z^2 - z_0{}^2.$$

It follows that

$$\int_{z_0}^Z z \, dz = \tfrac{1}{2}(Z^2 - z_0{}^2)$$

along every path joining z_0 and Z. If, in particular, C is a closed path, then $z_0 = Z$ and

(10)
$$\int_C z \, dz = 0.$$

Note that this result follows also from Green's theorem (cf. Sec. 8.4) by the use of formula (6) in this section.

Problems

Find a representation $z = z(t)$ of the line segments with endpoints
  1. $z = 0$ and $z = 3 - 12i$
  2. $z = 0$ and $z = 5 + 10i$
  3. $z = 1 - i$ and $z = 9 - 5i$
  4. $z = -3 + 2i$ and $z = -4 + 5i$
  5. $z = 4 + 2i$ and $z = 3 + 5i$
  6. $z = -4i$ and $z = -7 + 38i$

What curves are represented by the following functions?
  7. $3 - it, -1 \leq t \leq 1$
  8. $(1 + 2i)t, 0 \leq t \leq 3$
  9. $2 + i + 3e^{it}, 0 \leq t < 2\pi$
  10. $1 - i - 2e^{it}, 0 \leq t \leq \pi$
  11. $t + 2it^3, -2 \leq t \leq 2$
  12. $\cos t + 2i \sin t, -\pi < t < \pi$

Represent the following curves in the form $z = z(t)$.
  13. $|z - i| = 2$
  14. $|z - 3 + 4i| = 4$
  15. $y = x^2$ from $(0, 0)$ to $(2, 4)$
  16. $y = 1/x$ from $(1, 1)$ to $(3, \frac{1}{3})$
  17. $x^2 + 4y^2 = 4$
  18. $y = 1 + 2/x$ from $(1, 3)$ to $(3, \frac{5}{3})$

Integrate z^2 along the line segment:
  19. From 0 to $2 + i$
  20. From 0 to i
  21. From $-3 + 2i$ to $-4 + 5i$
  22. From 0 to $3 - 12i$

Integrate
  23. z^2 along the parabola in Prob. 15.
  24. $z + 1/z$ around the unit circle (clockwise).
  25. z from 1 vertically to $1 + i$ and then horizontally to $2 + i$.
  26. $az + b$ from 0 along the line segment to $-4 - 4i$.

Evaluate

  27. $\displaystyle\int_C (z - 3)^{-1} dz, C : |z - 3| = 2$, (a) counterclockwise, (b) clockwise.

  28. $\displaystyle\int_C \mathrm{Re}\, z\, dz$ around the circle $|z| = r$ (counterclockwise).

  29. $\displaystyle\int_C |z|\, dz$ from $-i$ to i along (a) the line segment, (b) the unit circle in the left half-plane, (c) the unit circle in the right half-plane.

  30. $\displaystyle\int_C (1/\sqrt{z})\, dz$ from 1 to -1 (a) along the upper semicircle $|z| = 1$, (b) along the lower semicircle $|z| = 1$, where $\sqrt{z}$ is the principal value of the square root.

13.2 Basic Properties of the Complex Line Integral

From the definition of a complex line integral as the limit of a sum we may immediately obtain the following properties.

If we decompose the path C into two portions C_1 and C_2 (Fig. 271), then

(1)
$$\int_C f(z)\, dz = \int_{C_1} f(z)\, dz + \int_{C_2} f(z)\, dz.$$

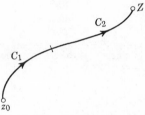

Fig. 271. Formula (1)

If we reverse the sense of integration, the sign of the value of the integral changes:

$$(2) \qquad \int_{z_0}^{Z} f(z)\, dz = -\int_{Z}^{z_0} f(z)\, dz;$$

here the path C with end points z_0 and Z is the same; on the left we integrate from z_0 to Z, and on the right from Z to z_0.

A sum of two (or more) functions may be integrated term by term, and constants may be taken out from under the integral sign, that is,

$$(3) \qquad \int_{C} [k_1 f_1(z) + k_2 f_2(z)]\, dz = k_1 \int_{C} f_1(z)\, dz + k_2 \int_{C} f_2(z)\, dz.$$

There will be a frequent necessity for estimating the absolute value of complex line integrals. The basic formula is

$$(4) \qquad \left| \int_{C} f(z)\, dz \right| \leq Ml$$

where l is the length of the path C and M is a real constant such that $|f(z)| \leq M$ everywhere on C.

To prove (4), we apply (21), Sec. 11.1, to the sum S_n defined by (2), Sec. 13.1, finding

$$|S_n| = \left| \sum_{m=1}^{n} f(\zeta_m)\, \Delta z_m \right| \leq \sum_{m=1}^{n} |f(\zeta_m)|\, |\Delta z_m| \leq M \sum_{m=1}^{n} |\Delta z_m|.$$

Now $|\Delta z_m|$ is the length of the chord whose end points are z_{m-1} and z_m, cf. Fig. 268 in Sec. 13.1. The sum on the right thus represents the length L of the broken line of chords whose end points are $z_0, z_1, \cdots, z_n \ (= Z)$. If n approaches infinity such that the greatest $|\Delta z_m|$ approaches zero, then L approaches the length l of the curve C, by the definition of the length of a curve (cf. Sec. 7.4). From this, formula (4) follows.

Example 1. Let $f(z) = 1/z$ and integrate once around the circle $|z| = \rho$. Then $l = 2\pi\rho$ and $|f(z)| = 1/\rho$ on the circle. Hence, by (4),

$$\left| \int_{C} \frac{dz}{z} \right| \leq \frac{1}{\rho} 2\pi\rho = 2\pi \qquad \text{(cf. Ex. 1., Sec. 13.1).}$$

Example 2. The path C_2 in Ex. 2, Sec. 13.1, has the length $l = 2$, and $|\text{Re } z| \leq 1$ on C_2. Therefore, by (4),

$$\left| \int_{C_2} \text{Re } z\, dz \right| \leq 2.$$

Problems

1. Verify (1) for $f(z) = 1/z$, C the unit circle, C_1 its upper half and C_2 its lower half.
2. Verify (2) for $f(z) = z^2 + 2z$ where C is the straight line segment from $1 + i$ to $3 + 3i$.
3. Verify (3) for $k_1 f_1 + k_2 f_2 = 2z^{-1} - 3z$ where C is the upper half of the unit circle from 1 to -1.

Using (3), evaluate $\displaystyle\int_C f(z)\, dz$, where

4. $f(z) = z^{-5} + z^3$, C the upper arc of the unit circle from 1 to -1.
5. $f(z) = z^{-5} + z^3$, C the lower arc of the unit circle from 1 to -1.
6. $f(z) = 2z^2 - z^{-1} + 2z^{-2}$, C the unit circle (clockwise).
7. $f(z) = z^4 - z^2 + z^{-1}$, C the unit circle (counterclockwise).
8. $f(z) = (z - 1)^{-1} + 2(z - 1)^{-2}$ where C is the circle $|z - 1| = 4$ (clockwise).

Evaluate $\displaystyle\int_C f(z)\, dz$ where

9. $f(z) = e^z$, C the line segment from 1 to $1 + \pi i/2$.
10. $f(z) = \sin z$, C the line segment from 0 to i.
11. $f(z) = e^z$, C the unit circle (clockwise).
12. $f(z) = \cosh z$, C as in Prob. 10.
13. $f(z) = \cos z$, C the line segment from i to $\pi + i$.

14. Verify (4) for $f(z) = 1/z$ and C as in Prob. 3.

Using (4), find upper bounds for the following integrals where C is the line segment from 0 to $3 + 4i$.

15. $\displaystyle\int_C z\, dz$ 16. $\displaystyle\int_C e^z\, dz$ 17. $\displaystyle\int_C e^{z^2}\, dz$

18. $\displaystyle\int_C \mathrm{Ln}\,(z + 1)\, dz$ 19. $\displaystyle\int_C (z + 1)^{-1}\, dz$

20. Find a better bound in Prob. 15 by decomposing C into two arcs.

13.3 *Cauchy's Integral Theorem*

Cauchy's integral theorem is very important in complex analysis and has various theoretical and practical consequences. To state this theorem, we shall need the following concept.

The interior of a simple closed curve (that is, a curve which is closed and has no self-intersections) is called a **simply connected domain**. A domain which is not simply connected is said to be *multiply connected*. For example, the interior of a circle ("circular disk"), ellipse, square, etc., is simply connected and a circular ring (cf. Sec. 11.1) is multiply connected (more precisely: doubly[2] connected).

---

[2] A domain is said to be *p-fold connected* if its boundary consists of p closed connected sets without common points. For the annulus, $p = 2$, because the boundary consists of two circles having no points in common.

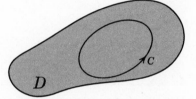

Fig. 272. Cauchy's integral theorem

Cauchy's integral theorem. *If $f(z)$ is analytic[3] in a simply connected bounded domain D, then for every simple closed path C in D,*

(1) $$\int_C f(z)\, dz = 0.$$

Cauchy's proof. From (5), Sec. 13.1, we have

$$\int_C f(z)\, dz = \int_C (u\, dx - v\, dy) + i\int_C (u\, dy + v\, dx).$$

$f(z)$ is analytic and, therefore $f'(z)$ exists. Cauchy made the *additional assumption* that $f'(z)$ is continuous. Then u and v have continuous first partial derivatives in D, as follows from (3) and (4), Sec. 11.3. Green's theorem (cf. Sec. 8.4) (with u and $-v$ instead of f and g) is applicable, and

$$\int_C (u\, dx - v\, dy) = \iint_R \left(-\frac{\partial v}{\partial x} - \frac{\partial u}{\partial y}\right) dx\, dy$$

where R is the region bounded by C. The second Cauchy-Riemann equation (Sec. 11.3) shows that the integrand on the right is identically zero. Hence, the integral on the left is zero. In the same fashion it follows by the use of the first Cauchy-Riemann equation that the last integral in the above formula is zero. This completes Cauchy's proof.

Goursat's[4] proof. Goursat proved Cauchy's theorem without assuming that $f'(z)$ is continuous. This progress is quite important. We start with the case when C is the boundary of a triangle. We orient it in the counterclockwise sense. By joining the midpoints of the sides of the triangle we subdivide it into four congruent triangles (Fig. 273). Then

$$\int_C f\, dz = \int_{C_{\mathrm{I}}} f\, dz + \int_{C_{\mathrm{II}}} f\, dz + \int_{C_{\mathrm{III}}} f\, dz + \int_{C_{\mathrm{IV}}} f\, dz$$

where $C_{\mathrm{I}}, \cdots, C_{\mathrm{IV}}$ are the boundaries of the triangles; in fact, on the right side we integrate along each of the three segments of subdivision in both possible directions, the corresponding integrals cancel out in pairs, and the sum of the integrals on the right equals the integral on the left. Among the four integrals on

---

[3] Remember that, by definition, a function is a single-valued relation. Cf. Sec. 11.2.
[4] EDOUARD GOURSAT (1858–1936), French mathematician.

Fig. 273. Proof of Cauchy's integral theorem

the right there must be one, call its boundary C_1, for which

$$\left| \int_C f \, dz \right| \leqq 4 \left| \int_{C_1} f \, dz \right|$$

because not each of the absolute values of the four integrals can be smaller than $1/4$ of the absolute value of the sum of those integrals. This follows readily from (21) in Sec. 11.1.

We subdivide the triangle bounded by C_1 as before and select a triangle of subdivision with boundary C_2 for which

$$\left| \int_{C_1} f \, dz \right| \leqq 4 \left| \int_{C_2} f \, dz \right|, \quad \text{and then} \quad \left| \int_C f \, dz \right| \leqq 4^2 \left| \int_{C_2} f \, dz \right|.$$

Continuing in this fashion, we obtain a sequence of triangles $T_1, T_2, \cdots$ with boundaries $C_1, C_2, \cdots$ which are similar and such that T_n lies in T_m when $n > m$, and

$$(2) \qquad\qquad \left| \int_C f \, dz \right| \leqq 4^n \left| \int_{C_n} f \, dz \right|, \qquad\qquad n = 1, 2, \cdots.$$

Let z_0 be the point which belongs to all these triangles. Since f is differentiable at $z = z_0$, the derivative $f'(z_0)$ exists, and we may write

$$(3) \qquad\qquad f(z) = f(z_0) + (z - z_0)f'(z_0) + h(z)(z - z_0).$$

Consequently, by integrating over the boundary C_n of the triangle T_n we have

$$\int_{C_n} f(z) \, dz = \int_{C_n} f(z_0) \, dz + \int_{C_n} (z - z_0)f'(z_0) \, dz + \int_{C_n} h(z)(z - z_0) \, dz.$$

Since $f(z_0)$ and $f'(z_0)$ are constants, it follows from the results in Exs. 4 and 5, Sec. 13.1, that the first two integrals on the right are zero. Hence,

$$\int_{C_n} f(z) \, dz = \int_{C_n} h(z)(z - z_0) \, dz.$$

By dividing (3) by $z - z_0$, transposing two terms to the left, and taking absolute values we obtain

$$\left| \frac{f(z) - f(z_0)}{z - z_0} - f'(z_0) \right| = |h(z)|.$$

From this and (5) in Sec. 11.2 we see that for a given positive number ϵ we can find a positive number δ such that

$$|h(z)| < \epsilon \qquad \text{when} \qquad |z - z_0| < \delta.$$

Fig. 274. Proof of Cauchy's integral theorem for a polygon

We may now take n so large that the triangle T_n lies in the disk $|z - z_0| < \delta$. Let l_n be the length of C_n. Then $|z - z_0| \leqq l_n/2$ for all z on C_n and z_0 in T_n. By applying (4) in Sec. 13.2 we thus obtain

$$(4) \qquad \left| \int_{C_n} f(z)\, dz \right| = \left| \int_{C_n} h(z)(z - z_0)\, dz \right| < \epsilon \frac{l_n}{2} l_n = \frac{\epsilon}{2} l_n{}^2.$$

Let l be the length of C. Then the path C_1 has the length $l_1 = l/2$, the path C_2 has the length $l_2 = l_1/2 = l/4$, etc., and C_n has the length

$$l_n = \frac{l}{2^n}.$$

From (2) and (4) we thus obtain

$$\left| \int_C f\, dz \right| \leqq 4^n \left| \int_{C_n} f\, dz \right| < 4^n \frac{\epsilon}{2} l_n{}^2 = 4^n \frac{\epsilon}{2} \frac{l^2}{4^n} = \frac{\epsilon}{2} l^2.$$

By choosing ϵ (>0) sufficiently small we can make the expression on the right as small as we please, while the expression on the left is the definite value of an integral. From this we conclude that this value must be zero, and the proof is complete.

The proof for *the case in which C is the boundary of a polygon* follows from the previous proof by subdividing the polygon into triangles (Fig. 274). The integral corresponding to each such triangle is zero. The sum of these integrals is equal to the integral over C, because we integrate along each segment of subdivision in both directions, the corresponding integrals cancel out in pairs, and we are left with the integral over C.

The case of a general simple closed path C can be reduced to the preceding one by inscribing in C a closed polygon P of chords, which approximates C "sufficiently accurately," and it can be shown that there is a polygon P such that the integral over P differs from that over C by less than any preassigned positive real number ϵ, no matter how small. The details of this proof are somewhat involved and can be found in Ref. [F6] in Appendix 1.

Example 1.

$$\int_C e^z\, dz = 0$$

for any closed path, because e^z is analytic for all z.

Example 2.

$$\int_C \frac{dz}{z^2} = 0,$$

where C is the unit circle (cf. Sec. 13.1). This result does not follow from Cauchy's theorem, because $f(z) = 1/z^2$ is not analytic at $z = 0$. Hence *the condition that f be analytic in D is sufficient rather than necessary for (1) to be true.*

Example 3.

$$\int_C \frac{dz}{z} = 2\pi i,$$

the integration being taken around the unit circle in the counterclockwise sense (cf. Sec. 13.1). C lies in the annulus $\frac{1}{2} < |z| < \frac{3}{2}$ where $1/z$ is analytic, but this domain is not simply connected, so that Cauchy's theorem cannot be applied. Hence *the condition that the domain D be simply connected is quite essential.*

If we subdivide the path C in Cauchy's theorem into two arcs C_1 and $C_2{}^*$ (Fig. 275), then (1) takes the form

$$\int_C f\,dz = \int_{C_1} f\,dz + \int_{C_2{}^*} f\,dz = 0.$$

Consequently,

(5′)
$$-\int_{C_2{}^*} f\,dz = \int_{C_1} f\,dz.$$

If we reverse the sense of integration along $C_2{}^*$, then the integral over $C_2{}^*$ is multiplied by -1, and we obtain (cf. Fig. 276)

(5)
$$\int_{C_2} f(z)\,dz = \int_{C_1} f(z)\,dz.$$

Hence, if f is analytic in D, and C_1 and C_2 are any paths in D joining two points in D and having no further points in common, then (5) holds.

Fig. 275. Formula (5′)

Fig. 276. Formula (5)

Fig. 277. Paths having finitely many intersections

Fig. 278. Continuous deformation of path

Fig. 279. Doubly connected domain

If those paths C_1 and C_2 have finitely many points in common (Fig. 277), then (5) continues to hold; this follows by applying the previous result to the portions of C_1 and C_2 between each pair of consecutive points of intersection.

It is even true that (5) *holds for any paths entirely in the domain D where $f(z)$ is analytic and joining any points z_1 and z_2 in D. Then we say that the integral of $f(z)$ from z_1 to z_2 is* **independent of path in D**. (Of course, the value of the integral depends on the choice of z_1 and z_2.)

The *proof* would require an additional consideration of the case in which C_1 and C_2 have infinitely many points of intersection, and will not be presented here.

We may imagine that the path C_2 in (5) was obtained from C_1 by a continuous deformation (Fig. 278). It follows that in a given integral we may impose a continuous deformation on the path of integration (keeping the end points fixed); as long as we do not pass through a point where $f(z)$ is not analytic, the value of the line integral will not change under such a deformation. This is often called the **principle of deformation of path**.

A *multiply connected domain* can be cut such that the resulting domain becomes simply connected. For a doubly connected domain D^* we need one cut $\tilde{C}$ (Fig. 279). If $f(z)$ is analytic in D^* and at each point of C_1 and C_2 then, since C_1, C_2, and $\tilde{C}$ bound a simply connected region, it follows from Cauchy's theorem that the integral of f taken over C_1, $\tilde{C}$, C_2 in the sense indicated by the arrows in Fig. 279 has the value zero. Since we integrate along $\tilde{C}$ in both directions, the corresponding integrals cancel out, and we obtain

(6) $$\int_{C_1} f(z)\, dz + \int_{C_2} f(z)\, dz = 0$$

where one of the curves is traversed in the counterclockwise sense and the other in the opposite sense.

Equation (6) may also be written

(7) $$\int_{C_1} f(z)\, dz = \int_{C_2} f(z)\, dz,$$

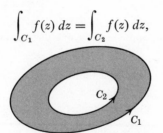

Fig. 280. Paths in (7)

Fig. 281. Triply connected domain

where C_1 and C_2 are now traversed in the same direction (Fig. 280). We remember that (7) holds under the assumption that $f(z)$ is analytic in the domain bounded by C_1 and C_2 and at each point of C_1 and C_2.

For more complicated domains we may need more than one cut, but the basic idea remains the same as before. For instance, for the domain in Fig. 281,

$$\int_{C_1} f(z)\,dz + \int_{C_2} f(z)\,dz + \int_{C_3} f(z)\,dz = 0$$

where C_2 and C_3 are traversed in the same sense and C_1 is traversed in the opposite sense.

We mention that a simple closed path is sometimes called a *contour*, and integrals along such a path are known as **contour integrals**.

Example 4. Let C_1 be the unit circle $|z| = 1$ and C_2 the circle $|z| = 1/2$. Then, by (7),

$$\int_{C_2} \frac{dz}{z} = \int_{C_1} \frac{dz}{z} = 2\pi i \qquad \text{(cf. Ex. 1 in Sec. 13.1),}$$

where both paths are traversed in the counterclockwise sense.

Example 5. From Ex. 3 in Sec. 13.1 and the principle of deformation of path it follows that

$$\int_C (z - z_0)^m \, dz = \begin{cases} 2\pi i & (m = -1) \\ 0 & (m \neq -1 \text{ and integral}) \end{cases}$$

where C is any contour containing the point z_0 in its interior and the integration is taken around C in the counterclockwise sense.

Problems

1. For what simple closed paths is the integral of $1/(1 - z)$ equal to zero?
2. Verify Cauchy's theorem for the integral of z^3 taken over the boundary of the rectangle with vertices $-1, 1, 1 + i, -1 + i$.

In each case find the value of the integral of the given function around the unit circle in the counterclockwise sense and indicate whether Cauchy's theorem may be applied.

3. $f(z) = 1/z^3$	**4.** $f(z) = e^{-z^2}$	**5.** $f(z) = |z|$
6. $f(z) = \operatorname{Re} z$	**7.** $f(z) = \operatorname{Im} z$	**8.** $f(z) = 1/(2z - 5)$
9. $f(z) = 1/|z|^3$	**10.** $f(z) = \bar{z}$	**11.** $f(z) = 1/\bar{z}$
12. $f(z) = z^2 \sec z$	**13.** $f(z) = \tanh z$	**14.** $f(z) = 1/(z^2 + 2)$

15. Obtain the answer to Prob. 6 from that to Prob. 7 and Cauchy's theorem applied to $f(z) = z$.

16. Using the principle of deformation of path and

$$\frac{2z-1}{z^2-z} = \frac{1}{z} + \frac{1}{z-1}, \qquad \text{show that} \qquad \int_C \frac{2z-1}{z^2-z}\,dz = \int_C \frac{dz}{z} + \int_C \frac{dz}{z-1} = 4\pi i$$

where C is the path in Fig. 282.

17. Integrate z^2 from 0 to $1 + i$ along C_1, C_2, and C_3 in Fig. 283.

Fig. 282. Path in Problem 16   **Fig. 283.** Problem 17

18. Integrate $f(z) = \bar{z}/|z|$ around the circle (*a*) $|z| = 2$, (*b*) $|z| = 4$, in the counterclockwise sense. Can the result in (*b*) be obtained from that in (*a*) by the principle of deformation of path?

Evaluate the following integrals. (*Hint.* If necessary, represent the integrand in terms of partial fractions.)

19. $\displaystyle\int_C \frac{2z+1}{z^2+z}\,dz,$ C: (*a*) $|z| = \frac{1}{4}$, (*b*) $|z - \frac{1}{2}| = \frac{1}{4}$, (*c*) $|z| = 2$ (clockwise).

20. $\displaystyle\int_C \frac{\sin z}{z + 3i}\,dz,$ $C: |z - 2 + 3i| = 1$ (counterclockwise).

21. $\displaystyle\int_C \frac{z^2 - \frac{1}{3}}{z^3 - z}\,dz,$ $C: |z - \frac{1}{2}| = 1$ (counterclockwise).

22. $\displaystyle\int_C \frac{dz}{z^2(z^2 + 4)},$ C consists of $|z| = 3$ (counterclockwise) and $|z| = 1$ (clockwise).

23. $\displaystyle\int_C \frac{dz}{1 + z^3},$ $C: |z + 1| = 1$ (counterclockwise).

24. $\displaystyle\int_C \frac{\cos z}{z}\,dz,$ C consists of $|z| = 1$ (counterclockwise) and $|z| = 3$ (clockwise).

25. $\displaystyle\int_C \frac{z^2 - z + 2}{z^3 - 2z^2}\,dz,$ C: the boundary of the rectangle with vertices $3 \pm i$, $-1 \pm i$ (clockwise).

13.4   *Evaluation of Line Integrals by Indefinite Integration*

Using Cauchy's integral theorem, we want to show that in many cases complex line integrals can be evaluated by a very simple method, namely, by indefinite integration.

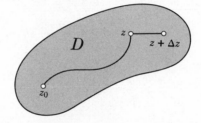

Fig. 284. Path of integration

Suppose that $f(z)$ is analytic in a simply connected domain D, and let z_0 be any fixed point in D. Then the integral

$$\int_{z_0}^{z} f(z^*) \, dz^*$$

is a function of z for all paths which lie in D and join z_0 and z, and we may write

(1)
$$F(z) = \int_{z_0}^{z} f(z^*) \, dz^*.$$

Let us prove that $F(z)$ *is an analytic function of z in D, and $F'(z) = f(z)$.*

We keep z fixed. Since D is a domain, a neighborhood N of z belongs to D. In N we choose a point $z + \Delta z$. Then the line segment with end points z and $z + \Delta z$ is in D, and from (1) we obtain

$$F(z + \Delta z) - F(z) = \int_{z_0}^{z+\Delta z} f(z^*) \, dz^* - \int_{z_0}^{z} f(z^*) \, dz^* = \int_{z}^{z+\Delta z} f(z^*) \, dz^*,$$

where we may integrate from z to $z + \Delta z$ along that segment (Fig. 284). Hence,

$$\frac{F(z + \Delta z) - F(z)}{\Delta z} - f(z) = \frac{1}{\Delta z} \int_{z}^{z+\Delta z} [f(z^*) - f(z)] \, dz^*$$

because z is kept fixed and, thus,

$$-\frac{1}{\Delta z} \int_{z}^{z+\Delta z} f(z) \, dz^* = -\frac{f(z)}{\Delta z} \int_{z}^{z+\Delta z} dz^* = -f(z).$$

Now $f(z)$ is continuous. Hence, an $\epsilon > 0$ being given, we can find a $\delta > 0$ such that

$$|f(z^*) - f(z)| < \epsilon \qquad \text{when} \qquad |z^* - z| < \delta.$$

Consequently, if $|\Delta z| < \delta$, then

$$\left| \frac{F(z + \Delta z) - F(z)}{\Delta z} - f(z) \right| = \frac{1}{|\Delta z|} \left| \int_{z}^{z+\Delta z} [f(z^*) - f(z)] \, dz^* \right|$$

$$< \frac{\epsilon}{|\Delta z|} \left| \int_{z}^{z+\Delta z} dz^* \right| = \epsilon;$$

that is,

(2)
$$F'(z) = \lim_{\Delta z \to 0} \frac{F(z + \Delta z) - F(z)}{\Delta z} = f(z).$$

From (1) it follows that if z_0 is replaced by another fixed point in D, the function $F(z)$ is changed by an additive constant. From (2) we see that $F(z)$ is an indefinite integral or antiderivative of $f(z)$, written

$$F(z) = \int f(z)\, dz;$$

that is, $F(z)$ is an analytic function in D whose derivative is $f(z)$.

If $F'(z) = f(z)$ and $G'(z) = f(z)$, then $F'(z) - G'(z) \equiv 0$ in D. Hence, the function $F(z) - G(z)$ is constant (cf. Prob. 29 at the end of Sec. 11.3). That is, the two indefinite integrals $F(z)$ and $G(z)$ differ only by a constant. In view of (1), we have for any points a and b in D and any path in D from a to b,

$$\int_a^b f(z)\, dz = \int_{z_0}^b f(z)\, dz - \int_{z_0}^a f(z)\, dz = F(b) - F(a),$$

as in the case of real definite integrals, but *it is important that the paths of integration lie in a simply connected domain D where $f(z)$ is analytic.*

We may sum up our result as follows.

Theorem 1. *If $f(z)$ is analytic in a simply connected domain D, then there exists an indefinite integral of $f(z)$ in the domain D, that is, an analytic function $F(z)$ such that $F'(z) = f(z)$ in D, and for all paths in D joining two points a and b in D,*

(3)
$$\int_a^b f(z)\, dz = F(b) - F(a).$$

This theorem enables us to evaluate complex line integrals by indefinite integration.

Example 1.

$$\int_i^{1+4i} z^2\, dz = \left[\frac{z^3}{3}\right]_i^{1+4i} = \frac{1}{3}[(1 + 4i)^3 - i^3] = -\frac{47}{3} - 17i.$$

Example 2.

$$\int_i^{\pi/2} \cos z\, dz = \sin z \Big|_i^{\pi/2} = \sin \frac{\pi}{2} - \sin i = 1 - i \sinh 1.$$

Problems

Evaluate the following integrals.

1. $\displaystyle\int_0^{1+i} z^2\, dz$ **2.** $\displaystyle\int_{2i}^{3i} (z + 4)^2\, dz$ **3.** $\displaystyle\int_i^{2i} (z^2 - 1)^3\, dz$ **4.** $\displaystyle\int_0^i (z^2 + z^3)\, dz$

5. $\displaystyle\int_0^{\pi i} e^z\, dz$ **6.** $\displaystyle\int_{-2\pi i}^{2\pi i} e^{-z/2}\, dz$ **7.** $\displaystyle\int_{-\pi i}^{3\pi i} e^{2z}\, dz$ **8.** $\displaystyle\int_1^i z e^{z^2}\, dz$

9. $\displaystyle\int_{-i}^{i} \sin z \, dz$ **10.** $\displaystyle\int_{-\pi i}^{\pi i} \cos z \, dz$ **11.** $\displaystyle\int_{\pi}^{\pi i} \sin 2z \, dz$ **12.** $\displaystyle\int_{0}^{i} \sinh \pi z \, dz$

13. $\displaystyle\int_{\pi i}^{0} \cosh 4z \, dz$ **14.** $\displaystyle\int_{-1}^{1} z \cosh z^2 \, dz$ **15.** $\displaystyle\int_{\pi i/6}^{0} \cosh 3z \, dz$ **16.** $\displaystyle\int_{\pi i}^{0} z \cos z \, dz$

17. $\displaystyle\int_{-\pi i}^{\pi i} \sin^2 z \, dz$ **18.** $\displaystyle\int_{-i}^{i} z \cosh z \, dz$ **19.** $\displaystyle\int_{0}^{\pi i} z \cos z^2 \, dz$ **20.** $\displaystyle\int_{1+i}^{1} z^3 e^{z^4} \, dz$

13.5 Cauchy's Integral Formula

The most important consequence of Cauchy's integral theorem is Cauchy's integral formula. This formula and its conditions of validity may be stated as follows.

Theorem 1. *Let $f(z)$ be analytic in a simply connected domain D. Then for any point z_0 in D and any simple closed path C in D which encloses z_0 (Fig. 285),*

(1) $$\int_{C} \frac{f(z)}{z - z_0} \, dz = 2\pi i f(z_0) \qquad \textbf{(Cauchy's integral formula)},$$

the integration being taken in the counterclockwise sense.

Proof. Writing $f(z) = f(z_0) + [f(z) - f(z_0)]$ and remembering that a constant may be taken out from under the integral sign, we have

(2) $$\int_{C} \frac{f(z)}{z - z_0} \, dz = f(z_0) \int_{C} \frac{dz}{z - z_0} + \int_{C} \frac{f(z) - f(z_0)}{z - z_0} \, dz.$$

From Ex. 5 in Sec. 13.3 it follows that the first term on the right is equal to $2\pi i f(z_0)$. Hence (1) holds, provided the last integral in (2) is zero. Now the integrand of this integral is analytic in D except at the point z_0. We may thus replace C by a small circle K with center at z_0, without changing the value of the integral (Fig. 286). Since $f(z)$ is analytic, it is continuous. Hence, an $\epsilon > 0$ being given, we can find a $\delta > 0$ such that

$$|f(z) - f(z_0)| < \epsilon \qquad \text{for all } z \text{ in the disk } |z - z_0| < \delta.$$

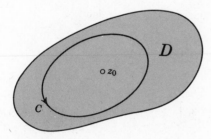

Fig. 285. Cauchy's integral formula

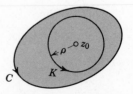

Fig. 286. Proof of Cauchy's integral formula

Choosing the radius ρ of K smaller than δ, we thus have the inequality

$$\left| \frac{f(z) - f(z_0)}{z - z_0} \right| < \frac{\epsilon}{\rho}$$

at each point of K. The length of K is $2\pi\rho$. Hence, by (4) in Sec. 13.2,

$$\left| \int_K \frac{f(z) - f(z_0)}{z - z_0} \, dz \right| < \frac{\epsilon}{\rho} 2\pi\rho = 2\pi\epsilon.$$

Since ϵ can be made arbitrarily small, it follows that the last integral on the right side of (2) has the value zero, and the theorem is proved.

Example 1. Integrate

$$\frac{z^2 + 1}{z^2 - 1}$$

in the counterclockwise sense around a circle of radius 1 with center at the point

$$(a) \ z = 1 \qquad (b) \ z = \tfrac{1}{2} \qquad (c) \ z = -1 \qquad (d) \ z = i.$$

(a) The integral under consideration may be written

$$\int_C \frac{z^2 + 1}{z^2 - 1} \, dz = \int_C \frac{z^2 + 1}{z + 1} \frac{dz}{z - 1} .$$

The expression on the right is of the form (1) where $z_0 = 1$ and

$$f(z) = \frac{z^2 + 1}{z + 1} .$$

The point $z_0 = 1$ lies inside the circle C under consideration, and $f(z)$ is analytic inside and on C. (The point $z = -1$, where $f(z)$ is not analytic, lies outside C.) Hence, by Cauchy's integral formula,

$$\int_C \frac{z^2 + 1}{z^2 - 1} \, dz = \int_C \frac{z^2 + 1}{z + 1} \frac{dz}{z - 1} = 2\pi i \left[\frac{z^2 + 1}{z + 1} \right]_{z=1} = 2\pi i.$$

(b) We obtain the same result as before because the given function is analytic except at the points $z = 1$ and $z = -1$, and we may obtain the circle (b) from that in the case (a) by a continuous deformation (even a translation) without passing over a point where the given function is not analytic.

(c) We may write

$$\int_C \frac{z^2 + 1}{z^2 - 1} \, dz = \int_C \frac{z^2 + 1}{z - 1} \frac{dz}{z + 1} .$$

The integral on the right is of the form (1) where $z_0 = -1$ and

$$f(z) = \frac{z^2 + 1}{z - 1} .$$

The point $z_0 = -1$ lies inside the circle C now under consideration, and $f(z)$ is analytic inside and on C. Therefore, by (1),

$$\int_C \frac{z^2 + 1}{z^2 - 1} \, dz = \int_C \frac{z^2 + 1}{z - 1} \frac{dz}{z + 1} = 2\pi i \left[\frac{z^2 + 1}{z - 1} \right]_{z=-1} = -2\pi i.$$

(d) The given function is analytic everywhere inside the circle now under consideration and on that circle. Hence, by Cauchy's integral theorem, the integral has the value zero.

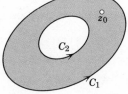

Fig. 287. Formula (3)

In the case of a *multiply connected domain* we may proceed in a manner similar to that in Sec. 13.3. For instance, if $f(z)$ is analytic on C_1 and C_2 and in the ring-shaped domain bounded by C_1 and C_2 (Fig. 287) and z_0 is any point in that domain, then

(3)
$$f(z_0) = \frac{1}{2\pi i} \int_{C_1} \frac{f(z)}{z - z_0} \, dz - \frac{1}{2\pi i} \int_{C_2} \frac{f(z)}{z - z_0} \, dz,$$

where both integrals are taken in the counterclockwise sense.

Problems

Integrate $(z^2 - 1)/(z^2 + 1)$ in the counterclockwise sense around the circle

1. $|z| = \frac{1}{2}$ **2.** $|z + i| = 1$ **3.** $|z - i| = 1$ **4.** $|z - 2i| = 2$

Integrate $1/(z^4 - 1)$ in the counterclockwise sense around the circle

5. $|z + 1| = 1$ **6.** $|z + 3| = 1$ **7.** $|z - i| = 1$ **8.** $|z - 1| = 1$

Integrate the following functions in the counterclockwise sense around the unit circle.

9. e^z/z **10.** $(e^z - 1)/z$ **11.** $(\cos z)/z$ **12.** $(z + 2)/(z - 2)$

13. $(\sin z)/2z$ **14.** $(z - 2)^{-1} \sin z$ **15.** $(z - \pi)^{-1} \cos z$ **16.** $z^3/(2z - i)$

17. $e^z \cos z$ **18.** $\tanh z$ **19.** $e^{3z}/(3z - i)$ **20.** $(\cosh z)/z$

13.6   The Derivatives of an Analytic Function

From the assumption that a *real* function of a real variable is once differentiable nothing follows about the existence of derivatives of higher order. We shall now see that from the assumption that a *complex* function has a first derivative in a domain D there follows the existence of derivatives of all orders in D. This means that in this respect complex analytic functions behave much more simply than real functions which are once differentiable.

Theorem 1.   *If $f(z)$ is analytic in a domain D, then it has derivatives of all orders in D which are then also analytic functions in D. The values of these derivatives at a point z_0 in D are given by the formulas*

(1')
$$f'(z_0) = \frac{1}{2\pi i} \int_C \frac{f(z)}{(z - z_0)^2} \, dz,$$

(1'')
$$f''(z_0) = \frac{2!}{2\pi i} \int_C \frac{f(z)}{(z - z_0)^3} \, dz,$$

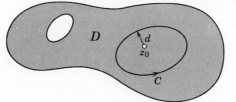

Fig. 288. Theorem 1

and in general

$$(1) \qquad f^{(n)}(z_0) = \frac{n!}{2\pi i} \int_C \frac{f(z)}{(z - z_0)^{n+1}} \, dz \qquad (n = 1, 2, \cdots);$$

here C is any simple closed path in D which encloses z_0 and whose full interior belongs to D; the curve C is traversed in the counterclockwise sense (Fig. 288).

Remark. For memorizing (1), it is useful to observe that these formulas are obtained formally by differentiating the Cauchy formula (1), Sec. 13.5, under the integral sign *with respect to z_0.*

Proof of Theorem 1. We prove (1′). By definition,

$$f'(z_0) = \lim_{\Delta z \to 0} \frac{f(z_0 + \Delta z) - f(z_0)}{\Delta z}.$$

From this and Cauchy's integral formula (1), Sec. 13.5, it follows that

$$(2) \qquad f'(z_0) = \lim_{\Delta z \to 0} \frac{1}{2\pi i \, \Delta z} \left[\int_C \frac{f(z)}{z - (z_0 + \Delta z)} \, dz - \int_C \frac{f(z)}{z - z_0} \, dz \right].$$

Straightforward calculation shows that

$$\frac{1}{\Delta z} \left[\frac{1}{z - (z_0 + \Delta z)} - \frac{1}{z - z_0} \right] = \frac{1}{(z - z_0)^2} + \frac{\Delta z}{(z - z_0 - \Delta z)(z - z_0)^2}.$$

Hence (2) can be written

$$f'(z_0) = \frac{1}{2\pi i} \int_C \frac{f(z)}{(z - z_0)^2} \, dz + \lim_{\Delta z \to 0} \frac{\Delta z}{2\pi i} \int_C \frac{f(z)}{(z - z_0 - \Delta z)(z - z_0)^2} \, dz.$$

Consequently, (1′) will be established if we show that the last expression on the right has the value zero. On C, $f(z)$ is continuous and, therefore, bounded in absolute value, say, $|f(z)| < M$. Let d be the distance of the point or points of C which are closest to z_0. Then for all z on C,

$$|z - z_0| \geqq d \qquad \text{and} \qquad \frac{1}{|z - z_0|} \leqq \frac{1}{d}.$$

Furthermore, if $|\Delta z| \leqq d/2$, then for all z on C we have the inequality

$$|z - z_0 - \Delta z| \geqq \frac{d}{2} \qquad \text{and} \qquad \frac{1}{|z - z_0 - \Delta z|} \leqq \frac{2}{d}.$$

Denoting the length of C by L and using (4) in Sec. 13.2, we thus obtain

$$\left| \frac{\Delta z}{2\pi i} \int_C \frac{f(z)}{(z - z_0 - \Delta z)(z - z_0)^2} \, dz \right| < \frac{|\Delta z|}{2\pi} \frac{M}{\frac{1}{2} d d^2} L \qquad \left(|\Delta z| \leq \frac{d}{2} \right).$$

If Δz approaches zero, the expression on the right approaches zero. This proves (1'). Note that we used Cauchy's integral formula (1), Sec. 13.5, but if all we had known about $f(z_0)$ is the fact that it can be represented by (1), Sec. 13.5, our argument would have established the existence of the derivative $f'(z_0)$ of $f(z)$. This implies that formula (1'') can be proved by a similar argument, and the general formula (1) follows by induction.

Using Theorem 1, let us prove the converse of Cauchy's theorem.

Morera's[5] theorem. *If $f(z)$ is continuous in a simply connected domain D and if*

$$\text{(3)} \qquad\qquad\qquad \int_C f(z) \, dz = 0$$

for every closed path in D, then $f(z)$ is analytic in D.

Proof. In Sec. 13.4 it was shown that if $f(z)$ is analytic in D, then

$$F(z) = \int_{z_0}^{z} f(z^*) \, dz^*$$

is analytic in D and $F'(z) = f(z)$. In the proof we used only the continuity of $f(z)$ and the property that its integral around every closed path in D is zero; from these assumptions we concluded that $F(z)$ is analytic. By Theorem 1 the derivative of $F(z)$ is analytic, that is, $f(z)$ is analytic in D, and Morera's theorem is proved.

We shall now derive an important inequality. In (1), let C be a circle of radius r with center at z_0, and let M be the maximum of $|f(z)|$ on C. Then, by applying (4), Sec. 13.2, to (1) we find

$$|f^{(n)}(z_0)| = \frac{n!}{2\pi} \left| \int_C \frac{f(z) \, dz}{(z - z_0)^{n+1}} \right| \leqq \frac{n!}{2\pi} M \frac{1}{r^{n+1}} 2\pi r.$$

This yields **Cauchy's inequality**

$$\text{(4)} \qquad\qquad\qquad |f^{(n)}(z_0)| \leqq \frac{n! \, M}{r^n}.$$

Using (4), let us prove the following interesting and basic result.

Liouville's theorem. *If $f(z)$ is analytic and bounded in absolute value for all (finite) z, then $f(z)$ is a constant.*

Proof. By assumption, $|f(z)|$ is bounded, say, $|f(z)| < K$ for all finite z. Using (4), we see that $|f'(z_0)| < K/r$. Since this is true for every r, we can take r as large as we please and conclude that $f'(z_0) = 0$. Since z_0 is arbitrary, $f'(z) = 0$ for all finite z, and $f(z)$ is a constant (cf. Prob. 29 at the end of Sec. 11.3). This completes the proof.

---

[5] GIACINTO MORERA (1856–1909), Italian mathematician.

Problems

Using Theorem 1, integrate the following functions in the counterclockwise sense around the circle $|z| = 2$. (n in Probs. 11–15 is a positive integer. a in Probs. 12 and 16 is any number.)

1. $z^2/(z - i)^2$ **2.** $z^4/(z - 3i)^2$ **3.** $e^{\pi z}/z^2$ **4.** $(\cos z)/z^2$

5. $(\cos z)/z^3$ **6.** $z^3/(z + 1)^3$ **7.** $(\sin \pi z)/z^3$ **8.** $(e^z \sin z)/z^2$

9. $e^{z^2}/(z - 1)^2$ **10.** e^{z^3}/z^3 **11.** e^z/z^n **12.** e^{az}/z^{n+1}

13. $(\sin z)/z^{2n}$ **14.** $(\cos z)/z^{2n}$ **15.** $z^n/(z + 1)^{n+1}$ **16.** $(\sinh az)/z^4$

17. If $f(z)$ is not a constant and is analytic for all (finite) z, and R and M are any positive real numbers (no matter how large), show that there exist values of z for which $|z| > R$ and $|f(z)| > M$. *Hint.* Use Liouville's theorem.

18. If $f(z)$ is a polynomial of degree $n > 0$ and M an arbitrary positive real number (no matter how large), show that there exists a positive real number R such that $|f(z)| > M$ for *all* $|z| > R$.

19. Show that $f(z) = e^z$ has the property characterized in Prob. 17, but does not have that characterized in Prob. 18.

20. Prove the **Fundamental theorem of algebra**: If $f(z)$ is a polynomial in z, not a constant, then $f(z) = 0$ for at least one value of z. *Hint.* Assume $f(z) \neq 0$ for all z and apply the result of Prob. 17 to $g = 1/f$.

CHAPTER 14

Sequences and Series

Infinite sequences and series play a basic role in complex analysis and its applications. We shall define sequences, series, and related notions, consider tests for convergence and divergence, and show that power series represent analytic functions. The fact that, conversely, analytic functions can always be represented by power series will be shown in the next chapter. The concepts in the present chapter are similar to those in real analysis, but the chapter is self-contained and independent.

Prerequisites for this chapter: Secs. 11.1–11.3.
References: Appendix 1, Part F.
Answers to problems: Appendix 2.

14.1 Sequences

We shall see that most of the definitions and theorems relating to complex sequences and series are quite similar to those in the case of real sequences and series.

If to each positive integer n there is assigned a number z_n, then these numbers

$$z_1, z_2, \cdots, z_n, \cdots$$

are said to form an *infinite sequence* or, briefly, a **sequence,** and the numbers z_n are called the **terms** of the sequence.

A sequence whose terms are real is called a *real sequence.*

At times it is convenient to number the terms of a sequence starting with 0, with 2, or with some other integer.

A sequence $z_1, z_2, \cdots$ is said to **converge** *to the number c,* or to *be convergent with the* **limit** *c,* if for each positive number ϵ (no matter how small but not zero) we can find a number N such that

(1) $$|z_n - c| < \epsilon \qquad \text{for each } n > N.$$

Then we write

$$\lim_{n \to \infty} z_n = c$$

Fig. 289. Convergent complex sequence

or

$$z_n \to c \qquad (n \to \infty).$$

A sequence which is not convergent is said to *be* **divergent** or to **diverge.**

The condition (1) has a simple geometrical interpretation. It means that each term z_n with $n > N$ lies in the open circular disk of radius ϵ with center at c (Fig. 289), and at most finitely many terms z_n do not lie in that disk, no matter how small the radius ϵ of the disk is chosen. Of course, N will depend on the choice of ϵ, in general.

In the case of a *real* sequence, the condition (1) means geometrically that each term z_n with $n > N$ lies between $c - \epsilon$ and $c + \epsilon$ (Fig. 290), and at most finitely many terms of the sequence do not lie in that interval.

Example 1. The sequence whose terms are $z_n = 1 + \dfrac{2}{n}$, $(n = 1, 2, \cdots)$, that is,

$$3, \ 2, \ \tfrac{5}{3}, \ \tfrac{6}{4}, \ \tfrac{7}{5}, \ \cdots$$

is convergent with the limit $c = 1$. In fact, we readily see that in (1),

$$z_n - c = 1 + \frac{2}{n} - 1 = \frac{2}{n} \quad \text{and} \quad \frac{2}{n} < \epsilon \quad \text{when} \quad \frac{n}{2} > \frac{1}{\epsilon} \quad \text{or} \quad n > \frac{2}{\epsilon}.$$

For example, choosing $\epsilon = 0.01$ we have $2/n < 0.01$ when $n > 200$.

Example 2. The sequences $1, 2, 3, \cdots$ and $\tfrac{1}{4}, \tfrac{3}{4}, \tfrac{1}{5}, \tfrac{4}{5}, \tfrac{1}{6}, \tfrac{5}{6}, \cdots$ are divergent.

Example 3. The sequence whose terms are

$$z_n = \frac{2n - 1}{n} + i\,\frac{n + 2}{n}$$

is

$$1 + 3i, \quad \tfrac{3}{2} + 2i, \quad \tfrac{5}{3} + \tfrac{5}{3}i, \quad \tfrac{7}{4} + \tfrac{3}{2}i, \cdots,$$

cf. Fig. 291. It is not difficult to see that the sequence is convergent with the limit $c = 2 + i$. In fact, in (1) we have

$$|z_n - c| = \left| \frac{2n - 1}{n} + i\,\frac{n + 2}{n} - (2 + i) \right| = \left| -\frac{1}{n} + \frac{2i}{n} \right| = \frac{\sqrt{5}}{n}$$

$$\text{and} \quad \frac{\sqrt{5}}{n} < \epsilon \quad \text{when} \quad \frac{n}{\sqrt{5}} > \frac{1}{\epsilon} \quad \text{or} \quad n > \frac{\sqrt{5}}{\epsilon} = \frac{2.236\cdots}{\epsilon}.$$

Fig. 290. Convergent real sequence

Fig. 291. Sequence in Example 3

For example, choosing $\epsilon = \frac{1}{100}$ we thus we obtain

$$|z_n - c| < \epsilon \quad \text{when} \quad n = 224, 225, \cdots.$$

A complex sequence $z_1, z_2, \cdots$ being given, we may consider the sequence of the real parts and the sequence of the imaginary parts. Setting

$$z_n = x_n + iy_n,$$

these sequences are

$$x_1, x_2, x_3, \cdots \quad \text{and} \quad y_1, y_2, y_3, \cdots.$$

For instance, in Ex. 3 the two sequences are

$$1, \tfrac{3}{2}, \tfrac{5}{3}, \tfrac{7}{4}, \cdots \quad \text{and} \quad 3, 2, \tfrac{5}{3}, \tfrac{3}{2}, \cdots,$$

and we observe that the sequence of the real parts converges to 2, the real part of c, while the sequence of the imaginary parts converges to 1, the imaginary part of c. This illustrates

Theorem 1. *A sequence $z_1, z_2, \cdots, z_n, \cdots$ of complex numbers $z_n = x_n + iy_n$ $(n = 1, 2, \cdots)$ converges to the limit $c = a + ib$ if and only if the sequence of the real parts $x_1, x_2, \cdots$ converges to a and the sequence of the imaginary parts $y_1, y_2, \cdots$ converges to b.*

Proof. If $|z_n - c| < \epsilon$, then $z_n = x_n + iy_n$ is within the circle of radius ϵ about $c = a + ib$ so that necessarily (Fig. 292a)

$$|x_n - a| < \epsilon, \qquad |y_n - b| < \epsilon.$$

Thus convergence $z_n \to c$ as $n \to \infty$ implies convergence $x_n \to a$ and $y_n \to b$.

Fig. 292. Proof of Theorem 1

Conversely, if $x_n \to a$ and $y_n \to b$ as $n \to \infty$, then for a given $\epsilon > 0$ we can choose N so large that, for every $n > N$,

$$|x_n - a| < \frac{\epsilon}{2}, \qquad |y_n - b| < \frac{\epsilon}{2}.$$

These two inequalities imply that $z_n = x_n + iy_n$ lies in a square with center c and side ϵ. Hence, z_n must lie within a circle of radius ϵ about c (Fig. 292*b*). This completes the proof.

This theorem shows that, by studying the real and imaginary parts, the convergence of complex sequences can be referred back to that of real sequences.

A point a is called a **limit point** of a sequence $z_1, z_2, \cdots$ if, given an $\epsilon > 0$ (no matter how small),

(2) $|z_n - a| < \epsilon$ for infinitely many n.

Geometrically speaking, this means that infinitely many terms of the sequence lie within the circle of radius ϵ about a, no matter how small the radius ϵ is chosen.

Note that if (2) holds, there may still be infinitely many terms which do not lie within that circle, and the sequence may not be convergent. In fact, if a sequence has more than one limit point, it diverges. The following example may illustrate the situation.

Example 4.

Sequence	Limit points at:	Convergent or divergent
$1, 2, 3, \cdots$	(none)	divergent
$\frac{1}{2}, \frac{2}{3}, \frac{3}{4}, \frac{4}{5}, \cdots$	1	convergent
$\frac{1}{2}, 2, \frac{1}{3}, 3, \frac{1}{4}, 4, \cdots$	0	divergent
$\frac{1}{4}, \frac{3}{4}, \frac{1}{5}, \frac{4}{5}, \frac{1}{6}, \frac{5}{6}, \cdots$	0 and 1	divergent

A number which appears infinitely often in a sequence is to be regarded as a limit point; this is a matter of convenience and convention.

A sequence $z_1, z_2, \cdots$ is said to be **bounded**, if there is a positive number K such that all the terms of the sequence lie in a disk of radius K about the origin, that is

$$|z_n| < K \qquad \text{for all } n.$$

For example, the second and the last sequences in Ex. 4 are bounded while the first and third are not. We observe that the two bounded sequences have limit points. This illustrates the following important theorem.

Fig. 293. Last sequence in Example 4

Theorem 2 (Bolzano[1] and Weierstrass[2]).   *A bounded infinite sequence has at least one limit point.*

Proof.   It is obvious that both conditions are necessary: a finite sequence cannot have a limit point, and the sequence $1, 2, 3, \cdots$, though infinite, has no limit point because it is not bounded. To prove the theorem, consider a bounded infinite sequence $z_1, z_2, \cdots$ and let K be such that $|z_n| < K$ for all n. If only finitely many values of the z_n are different, then, since the sequence is infinite, some number z must occur infinitely many times in the sequence, and, by definition, this number is a limit point of the sequence.

We may now turn to the case when the sequence contains infinitely many different terms. We draw the large square Q_0 in Fig. 294 which contains all z_n. We subdivide Q_0 into four congruent squares. Clearly, at least one of these squares (each taken with its complete boundary) must contain infinitely many terms of the sequence. The square of this type with the lowest number (1, 2, 3, or 4) will be denoted by Q_1. This is the first step. In the next step we subdivide Q_1 into four congruent squares and select a square Q_2 according to the same rule, and so on. This yields an infinite sequence of squares $Q_0, Q_1, Q_2, \cdots, Q_n, \cdots$ with the property that the side of Q_n approaches zero as n approaches infinity, and Q_m contains all Q_n with $n > m$. It is not difficult to see that the number which belongs to all these squares,[3] call it $z = a$, is a limit point of the sequence. In fact, given an $\epsilon > 0$, we can choose an N so large that the side of the square Q_N is less than ϵ and, since Q_N contains infinitely many z_n we have $|z_n - a| < \epsilon$ for infinitely many n. This completes the proof.

By the use of this theorem we shall now establish a general criterion for convergence which will be fundamental in our further consideration. The point is that this criterion can be applied without knowing in advance the value of the limit. This is a great advantage, compared with the direct application of the definition of convergence of a sequence.

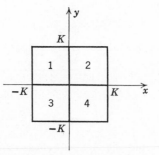

Fig. 294. Proof of Theorem 2

[1] BERNHARD BOLZANO (1781–1848), German mathematician, a pioneer in the study of point sets.
[2] Cf. footnote 8 in Sec. 11.3.
[3] The fact that such a unique number $z = a$ exists seems to be obvious, but it actually follows from an axiom of the real number system, the so-called *Cantor-Dedekind axiom.* Cf. the next footnote.

Theorem 3 (Cauchy's convergence principle). *A sequence $z_1, z_2, \cdots$ is convergent if and only if for every positive number ϵ we can find a number N (which may depend on ϵ) such that*

(3) $$|z_m - z_n| < \epsilon \qquad \text{when } m > N, n > N;$$

(that is, any two terms z_m, z_n with $m > N$, $n > N$ must have a distance of less than ϵ from each other).

Proof. (a) Let the sequence $z_1, z_2, \cdots$ be convergent with the limit c. Then, for a given $\epsilon > 0$, we can find an N such that

$$|z_n - c| < \frac{\epsilon}{2} \qquad \text{for every } n > N.$$

Hence, when $m > N$, $n > N$, then by the triangle inequality

$$|z_m - z_n| = |(z_m - c) - (z_n - c)| \leqq |z_m - c| + |z_n - c| < \frac{\epsilon}{2} + \frac{\epsilon}{2} = \epsilon,$$

which shows that, if the sequence converges, then (3) holds.

(b) Conversely, consider a sequence $z_1, z_2, \cdots$ for which (3) holds. We first show that the sequence is bounded. In fact, choose a fixed ϵ and a fixed $n = n_0 > N$ in (3). Then (3) implies that every z_m with $m > N$ lies in a disk of radius ϵ about z_{n_0}, and only finitely many terms of the sequence do not lie in that disk. It is clear that we can find a circle about the origin so large that the disk and those finitely many z_n lie within the circle. This shows that the sequence is bounded, and from the Bolzano-Weierstrass theorem we conclude that it has at least one limit point, call it L.

We show that the sequence is convergent with the limit L. The definition of a limit point implies that, an $\epsilon > 0$ being given, $|z_n - L| < \epsilon/2$ for infinitely many n. Since (3) holds for any $\epsilon > 0$, when an $\epsilon > 0$ is given we can find an N^* such that $|z_m - z_n| < \epsilon/2$ for $m > N^*$, $n > N^*$. Choose a fixed $n > N^*$ such that $|z_n - L| < \epsilon/2$, and let m be any integer greater than N^*. Then, by the triangle inequality,

$$|z_m - L| = |(z_m - z_n) + (z_n - L)| \leqq |z_m - z_n| + |z_n - L| < \frac{\epsilon}{2} + \frac{\epsilon}{2} = \epsilon,$$

that is, $|z_m - L| < \epsilon$ for all $m > N^*$; by definition, this means that the sequence is convergent with the limit L. This completes the proof.

This theorem will play an important role in our considerations of infinite series. The following property of a convergent sequence is almost obvious.

Theorem 4. *A convergent sequence is bounded. Consequently, if a sequence is not bounded, it diverges.*

Proof. Let the sequence $z_1, z_2, \cdots$ be convergent with the limit c. Then we may choose an $\epsilon > 0$ and find a corresponding N such that every z_n with $n > N$ lies in the disk of radius ϵ with center c, and at most finitely many terms may not lie in that disk. It is clear that we can find a circle of radius K about the origin so large that the disk as well as those finitely many terms lie within that circle. This proves that the sequence is bounded.

Note well that the condition of Theorem 4 is only necessary for convergence but not sufficient. This is illustrated by the last sequence in Ex. 4, which is bounded but has two limit points and, therefore, is divergent. The first and third sequences in Ex. 4 are not bounded, and from Theorem 4 it follows that they are divergent.

In the case of a *real sequence* we may find a simple condition sufficient for convergence. For this purpose we need the following concepts.

A **real** sequence $x_1, x_2, \cdots, x_n, \cdots$ is said to be **monotone increasing** if

$$x_1 \leqq x_2 \leqq x_3 \leqq \cdots.$$

Similarly, the sequence is said to be **monotone decreasing** if

$$x_1 \geqq x_2 \geqq x_3 \geqq \cdots.$$

A sequence which is either monotone increasing or monotone decreasing is called a **monotone sequence.**

For example, the first two sequences in Ex. 4 are monotone, while the last two are not monotone. We observe that the second sequence in Ex. 4 is bounded and monotone, and we shall now prove that these two properties together are sufficient for convergence.

Theorem 5. *If a real sequence is bounded and monotone, it converges.*

Proof. Let $x_1, x_2, \cdots$ be a bounded monotone increasing sequence. Then its terms are smaller than some number B and, since $x_1 \leqq x_n$ for all n, they lie in the interval $x_1 \leqq x_n \leqq B$, which will be denoted by I_0. We bisect I_0, that is, we subdivide it into two parts of equal length. If the right half (together with its end points) contains terms of the sequence, we denote it by I_1. If it does not contain terms of the sequence, then the left half of I_0 (together with its end points) is called I_1. This is the first step.

In the second step we bisect I_1, select one half by the same rule, and call it I_2, and so on (Fig. 295).

In this way we obtain shorter and shorter intervals $I_0, I_1, I_2, \cdots$ with the following properties. Each I_m contains all I_n for $n > m$. No term of the sequence lies to the right of I_m, and, since the sequence is monotone increasing, all x_n with n greater than some number N lie in I_m; of course, N will depend on m, in general.

Fig. 295. Proof of Theorem 5

The lengths of the I_m approach zero as m approaches infinity. Hence there is precisely one number, call it L, which lies in all those intervals,[4] and we may now easily prove that the sequence is convergent with the limit L.

In fact, given an $\epsilon > 0$, we choose an m such that the length of I_m is less than ϵ. Then L and all the x_n with $n > N(m)$ lie in I_m, and, therefore, $|x_n - L| < \epsilon$ for all those n. This completes the proof for an increasing sequence. For a decreasing sequence the proof is the same, except for a suitable interchange of "left" and "right" in the construction of those intervals.

Problems

Find and plot the first few terms of the sequence $z_1, z_2, \cdots, z_n, \cdots$ where z_n equals

1. $in/(n + 1)$
2. $i^n n^2/(n + i)$
3. $(-1)^n + 2n\pi i$
4. $(1 + i)^n/n^2$
5. $\sqrt[n]{n + 1}$
6. $e^{in\pi/4}$

7. Find the first few terms and the limit of the sequence $z_1 = 1$, $z_2 = i/2$, $z_n = iz_{n-2}z_{n-1}$ ($n = 3, 4, \cdots$).
8. Show that if a sequence converges, its limit is unique.

Are the following real sequences bounded? Convergent? Monotone?

9. $\frac{1}{2}, \frac{1}{4}, \frac{1}{8}, \frac{1}{16}, \cdots$
10. $1, -\frac{1}{5}, \frac{1}{25}, -\frac{1}{125}, \cdots$
11. $1, -2, 3, -4, 5, -6, \cdots$
12. $100^2/2!, 100^3/3!, 100^4/4!, \cdots$
13. $\sqrt{2}, \sqrt[3]{3}, \sqrt[4]{4}, \cdots$
14. $k, 2k^2, 3k^3, \cdots$ $(|k| < 1)$
15. $c, 2^2c^2, 3^2c^3, 4^2c^4, \cdots$ $(|c| < 1)$
16. $(\ln 2)/2, (\ln 3)/3, (\ln 4)/4, \cdots$

17. Show that a complex sequence $z_1, z_2, \cdots$ is bounded if and only if the two corresponding sequences of the real parts and the imaginary parts are bounded.
18. Does $\lim\limits_{n \to \infty} |z_n| = 1$ imply convergence of $z_1, z_2, \cdots$?

Are the following sequences $z_1, z_2, \cdots, z_n, \cdots$ bounded? Convergent? Find their limit points.

19. $z_n = (2i)^n$
20. $z_n = 1 + i^n$
21. $z_n = (-1)^n + \dfrac{i}{n}$
22. $z_n = e^{in\pi/2}$
23. $z_n = i^n \cos n\pi$
24. $z_n = i^n \cosh n\pi$
25. $z_n = (1 - i)^n$
26. $z_n = (1 + i)^{2n}/n$
27. $z_n = (3 + 4i)^n/n!$

28. If the sequence $z_1, z_2, \cdots$ is convergent with the limit 0, and the sequence $b_1, b_2, \cdots$ is such that $|b_n| \leqq K|z_n|$ for some fixed $K > 0$ and all n, show that the sequence $b_1, b_2, \cdots$ is convergent with the limit 0.
29. If $z_1, z_2, \cdots$ converges with the limit l and $z_1{}^*, z_2{}^*, \cdots$ converges with the limit l^*, show that $z_1 + z_1{}^*, z_2 + z_2{}^*, \cdots$ converges with the limit $l + l^*$.
30. Show that under the assumptions of Prob. 29 the sequence $z_1 z_1{}^*, z_2 z_2{}^*, \cdots$ converges with the limit ll^*.

---

[4] This statement seems to be obvious, but actually it is not; it may be regarded as an axiom of the real number system in the following form. Let $J_1, J_2, \cdots$ be closed intervals such that each J_m contains all J_n with $n > m$, and the lengths of the J_m approach zero as m approaches infinity. Then there is precisely one real number which is contained in all those intervals. This is the so-called **Cantor-Dedekind Axiom**, named after the German mathematicians, GEORG CANTOR (1845–1918), the creator of set theory, and RICHARD DEDEKIND (1831–1916). For further details see Ref. [A2] in Appendix 1. (An interval I is said to be **closed** if its two end points are regarded as points belonging to I. It is said to be **open** if the end points are not regarded as points of I.)

14.2 *Series*

Let $w_1, w_2, \cdots, w_m, \cdots$ be a sequence of numbers, complex or real. Then we may consider the **infinite series**

$$(1) \qquad \sum_{m=1}^{\infty} w_m = w_1 + w_2 + w_3 + \cdots.$$

The w_m are called the **terms** of the series. The sum of the first n terms is

$$(2) \qquad s_n = w_1 + w_2 + \cdots + w_n.$$

This expression is called the nth **partial sum** of the series (1). Clearly if we omit the terms of s_n from (1), the remaining expression is

$$(3) \qquad R_n = w_{n+1} + w_{n+2} + w_{n+3} + \cdots.$$

This is called the **remainder** *of the series* (1) *after the n-th term.*

In this way we have now associated with the series (1) the sequence of its partial sums $s_1, s_2, s_3, \cdots$. If this sequence is convergent, say,

$$\lim_{n \to \infty} s_n = s,$$

then the series (1) is said to **converge** or to *be convergent*, the number s is called its **value** or *sum*, and we write

$$s = \sum_{m=1}^{\infty} w_m = w_1 + w_2 + \cdots.$$

If the sequence of the partial sums diverges, the series (1) is said to **diverge** or to *be divergent.*

If the series (1) converges and has the value s, then

$$(4) \qquad s = s_n + R_n \qquad \text{or} \qquad R_n = s - s_n.$$

Example 1. The series

$$\sum_{m=1}^{\infty} \frac{1}{2^m} = \frac{1}{2} + \frac{1}{4} + \frac{1}{8} + \cdots$$

converges, because

$$s_n = \frac{1}{2} + \frac{1}{4} + \cdots + \frac{1}{2^n} = 1 - \frac{1}{2^n} \qquad \text{and} \qquad \lim_{n \to \infty} s_n = 1;$$

the series has the value 1.

Example 2. The series

$$\sum_{m=1}^{\infty} m = 1 + 2 + 3 + \cdots$$

diverges.

Example 3. The series

$$\sum_{m=0}^{\infty} (-1)^m = 1 - 1 + 1 - + \cdots$$

diverges, because

$$s_0 = 1, \qquad s_1 = 1 - 1 = 0, \qquad s_2 = 1 - 1 + 1 = 1, \qquad \text{etc.,}$$

and the sequence $1, 0, 1, 0, \cdots$ is divergent.

Fig. 296. Example 4

Example 4. The **harmonic series**

$$\sum_{m=1}^{\infty} \frac{1}{m} = 1 + \frac{1}{2} + \frac{1}{3} + \cdots$$

diverges. In fact,

$$s_n = 1 + \frac{1}{2} + \cdots + \frac{1}{n},$$

and s_n equals the sum of the areas of the n rectangles in Fig. 296. This area is greater than the area A_n under the corresponding portion of the curve $y = 1/x$, and

$$A_n = \int_1^{n+1} \frac{dx}{x} = \ln (n + 1) \to \infty \qquad \text{as} \quad n \to \infty.$$

Since $s_n > A_n$, it follows that $s_n \to \infty$ as $n \to \infty$, which means divergence.

Theorem 1. *If the series $w_1 + w_2 + \cdots$ converges, then*

$$(5) \qquad\qquad\qquad\qquad \lim_{m \to \infty} w_m = 0.$$

Consequently, a series which does not satisfy (5) diverges.

Proof. Let $w_1 + w_2 + \cdots$ be convergent with the sum s. Then

$$w_{n+1} = s_{n+1} - s_n,$$

and

$$\lim_{n \to \infty} (s_{n+1} - s_n) = \lim_{n \to \infty} s_{n+1} - \lim_{n \to \infty} s_n = s - s = 0.$$

Note well that the condition (5) is only necessary for convergence, but not sufficient. In fact, the harmonic series satisfies (5), but diverges. The series in Exs. 2 and 3 do not satisfy (5) and, therefore, are divergent.

If a given series $w_1 + w_2 + \cdots$ converges and has the value s, then

$$s = s_n + R_n \qquad \text{or} \qquad R_n = s - s_n \qquad\qquad \text{[cf. (4)]}$$

and from the definition of convergence of a sequence it follows that $|R_n|$ can be made as small as we please, by taking n large enough. In many cases it will be impossible to find the sum s of a convergent series. Then for computational purposes we must use a partial sum s_n as an approximation of s, and by estimating the remainder R_n we may obtain information about the degree of accuracy of the approximation.

In the case of a *real series* whose terms have alternating sign and decrease in absolute value, the following useful theorem may be applied for that purpose.

Theorem 2 (Leibniz test for real series).   *Let* $u_1, u_2, \cdots$ *be real and*

(6)   (a) $u_1 \geqq u_2 \geqq u_3 \geqq \cdots$,   (b) $\lim\limits_{m \to \infty} u_m = 0$.

Then the series

$$u_1 - u_2 + u_3 - u_4 + - \cdots$$

converges, and for the remainder R_n *after the nth term we have the estimate*

(7) $|R_n| \leqq u_{n+1}.$

Proof.   Let s_n be the nth partial sum of the series. Then, because of (6a),

$$s_1 = u_1, \qquad\qquad s_2 = u_1 - u_2 \leqq s_1,$$
$$s_3 = s_2 + u_3 \geqq s_2, \qquad s_3 = s_1 - (u_2 - u_3) \leqq s_1,$$

so that $s_2 \leqq s_3 \leqq s_1$. Proceeding in this fashion, we conclude that (Fig. 297)

(8) $s_1 \geqq s_3 \geqq s_5 \geqq \cdots \geqq s_6 \geqq s_4 \geqq s_2$

which shows that the odd partial sums form a bounded monotone sequence, and so do the even partial sums. Hence, by Theorem 5, Sec. 14.1, both sequences converge, say,

$$\lim_{n \to \infty} s_{2n+1} = s, \qquad \lim_{n \to \infty} s_{2n} = s^*.$$

Now, since $s_{2n+1} - s_{2n} = u_{2n+1}$, we readily see that (6b) implies

$$s - s^* = \lim_{n \to \infty} s_{2n+1} - \lim_{n \to \infty} s_{2n} = \lim_{n \to \infty} (s_{2n+1} - s_{2n}) = \lim_{n \to \infty} u_{2n+1} = 0.$$

Hence $s^* = s$, and the series converges with the sum s.

We shall now prove the estimate (7) for the remainder. Since $s_n \to s$, it follows from (8) that

$$s_{2n+1} \geqq s \geqq s_{2n} \qquad \text{and also} \qquad s_{2n-1} \geqq s \geqq s_{2n}.$$

By subtracting s_{2n} and s_{2n-1}, respectively, we obtain

$$s_{2n+1} - s_{2n} \geqq s - s_{2n} \geqq 0, \qquad 0 \geqq s - s_{2n-1} \geqq s_{2n} - s_{2n-1}.$$

In these inequalities, the first expression is equal to u_{2n+1}, the last is equal to $-u_{2n}$, and the expressions between the inequality signs are the remainders R_{2n} and

Fig. 297. Proof of Theorem 2 (Leibniz test)

R_{2-n1}. Thus the inequalities may be written

$$u_{2n+1} \geqq R_{2n} \geqq 0, \qquad 0 \geqq R_{2n-1} \geqq -u_{2n}$$

and we see that they imply (7). This completes the proof of Theorem 2.

Example 5. The series

$$1 - \tfrac{1}{2} + \tfrac{1}{3} - \tfrac{1}{4} + - \cdots$$

is convergent, as follows from Theorem 2.

From Theorem 1, Sec. 14.1, we immediately obtain the following relation between a complex series and the series of its real and imaginary parts.

Theorem 3. *Let* $w_m = u_m + iv_m$. *Then the series*

$$\sum_{m=1}^{\infty} w_m = w_1 + w_2 + w_3 + \cdots$$

converges and has the value $s = a + ib$ *if and only if the series of the real parts and the series of the imaginary parts, that is*

$$\sum_{m=1}^{\infty} u_m = u_1 + u_2 + \cdots \qquad and \qquad \sum_{m=1}^{\infty} v_m = v_1 + v_2 + \cdots,$$

converge and have the values a *and* b, *respectively.*

Furthermore, we may apply Theorem 3 in Sec. 14.1 to the sequence of the partial sums s_n of a given series $w_1 + w_2 + \cdots$. Then the inequality in that theorem becomes

$$|s_m - s_n| < \epsilon \qquad\qquad (m > N, n > N)$$

or, if we set $m = n + p$,

$$|s_{n+p} - s_n| < \epsilon \qquad\qquad (n > N, p = 1, 2, \cdots).$$

Now by the definition of a partial sum,

$$s_{n+p} - s_n = w_{n+1} + w_{n+2} + \cdots + w_{n+p}$$

and we immediately obtain the following basic result.

Theorem 4 (Cauchy's convergence principle for series). *A series* $w_1 + w_2 + \cdots$ *is convergent if and only if for any given* $\epsilon > 0$ *(no matter how small) we can find an* N *(which will depend on* ϵ, *in general) such that*

$$|w_{n+1} + w_{n+2} + \cdots + w_{n+p}| < \epsilon \qquad for\ every\ n > N\ and\ p = 1, 2, \cdots.$$

Theorem 3 establishes a relation between complex and real series. A far more important relation can be obtained by the use of the following basic notion.

A series $w_1 + w_2 + \cdots$ is said to be **absolutely convergent** if the corresponding series

$$(9) \qquad\qquad \sum_{m=1}^{\infty} |w_m| = |w_1| + |w_2| + \cdots$$

(whose terms are real and nonnegative) converges.

If the series $w_1 + w_2 + \cdots$ converges but (9) diverges, then the series is called, more precisely, *conditionally convergent*.

For instance, the series in Ex. 5 is conditionally convergent because the corresponding series (9) is the harmonic series (Ex. 4), which diverges. The series

$$\tfrac{1}{2} - \tfrac{1}{4} + \tfrac{1}{8} + \tfrac{1}{16} + - \cdots$$

is absolutely convergent, because the corresponding series (9) is convergent (cf. Ex. 1).

The following property of an absolutely convergent series is almost obvious.

Theorem 5. *If a series $w_1 + w_2 + \cdots$ is absolutely convergent, it is convergent.*

Proof. By the generalized triangle inequality (21) in Sec. 11.1 we have

$$(10) \qquad\qquad |w_{n+1} + \cdots + w_{n+p}| \leqq |w_{n+1}| + |w_{n+2}| + \cdots + |w_{n+p}|.$$

Since, by assumption, the series $|w_1| + |w_2| + \cdots$ converges, it follows from Theorem 4 that the right-hand side of (10) becomes less than any given $\epsilon > 0$ for each $n > N$ (N sufficiently large) and $p = 1, 2, \cdots$. Hence, the same is true for the left-hand side of (10), and by the same theorem, this proves convergence of the series $w_1 + w_2 + \cdots$.

14.3 *Tests for Convergence and Divergence of Series*

Before we use an infinite series for computational or other purposes we must know whether it converges. In most cases that arise in engineering mathematics, this question may be answered by applying one of the various tests for convergence and divergence. These tests, therefore, are of great practical interest.

A simple test for divergence (Theorem 1 in the last section) and the so-called Leibniz test for real series were already considered. The following theorem is a source of various criteria for convergence.

Theorem 1 (Comparison test). *If a series $w_1 + w_2 + \cdots$ is given and we can find a converging series $b_1 + b_2 + \cdots$ with nonnegative real terms such that*

$$(1) \qquad\qquad |w_n| \leqq b_n \qquad \text{for } n = 1, 2, \cdots,$$

then the given series is absolutely convergent.

Proof. Since the series $b_1 + b_2 + \cdots$ converges, it follows from Theorem 4 in the last section that for any given $\epsilon > 0$ we can find an N such that

$$b_{n+1} + \cdots + b_{n+p} < \epsilon \qquad \text{for every } n > N \text{ and } p = 1, 2, \cdots.$$

From this and (1) we conclude that

$$|w_{n+1}| + \cdots + |w_{n+p}| \leqq b_{n+1} + \cdots + b_{n+p} < \epsilon$$

for those n and p. Consequently, by Theorem 4 in the last section, the series $|w_1| + |w_2| + \cdots$ is convergent, and the given series converges absolutely.

For deriving two important tests from Theorem 1 let us prove

Theorem 2 (Geometric series).   *The geometric series*

$$\sum_{m=0}^{\infty} q^m = 1 + q + q^2 + \cdots$$

converves with the sum $1/(1 - q)$ *when* $|q| < 1$ *and diverges when* $|q| \geq 1$.

Proof.   When $|q| \geq 1$, then $|q^n| \geq 1$ and divergence follows from Theorem 1 in the last section.   Now let $|q| < 1$.   The nth partial sum is

$$s_n = 1 + q + \cdots + q^n.$$

From this, $$qs_n =   q + \cdots + q^n + q^{n+1}.$$

By subtraction,

$$s_n - qs_n = (1 - q)s_n = 1 - q^{n+1}$$

since the other terms cancel in pairs.   Now $q - 1 \neq 0$ because $q \neq 1$, and we may solve for s_n, finding

(2) $$s_n = \frac{1 - q^{n+1}}{1 - q} = \frac{1}{1 - q} - \frac{q^{n+1}}{1 - q}.$$

Since $|q| < 1$, the last term approaches zero as $n \to \infty$.   Hence the series is convergent and has the value $1/(1 - q)$.   This completes the proof.

From Theorems 1 and 2 we shall now derive two important tests, the ratio test and the root test.

Theorem 3 (Ratio test).   *Suppose that* $w_n \neq 0$ *for* $n = 1, 2, \cdots$ *and the sequence of the ratios*

$$\left| \frac{w_{n+1}}{w_n} \right| \qquad\qquad n = 1, 2, \cdots$$

converges with the limit L.   Then the series

$$w_1 + w_2 + w_3 + \cdots$$

is

absolutely convergent, if $L < 1$,
divergent, if $L > 1$.

(*If* $L = 1$, *the test fails*).

Proof.   By assumption

$$\lim_{n \to \infty} k_n = L \qquad \text{where} \qquad k_n = \left| \frac{w_{n+1}}{w_n} \right|.$$

Clearly, k_n and L are real.   By the definition of a limit, for a given $\epsilon > 0$ we can find an N such that every k_n with $n > N$ lies between $L - \epsilon$ and $L + \epsilon$, that is,

(3)   (a) $k_n < L + \epsilon$ (b) $k_n > L - \epsilon$ $(n > N)$.

We first consider the case $L < 1$. We set $L + \epsilon = q$ and choose $\epsilon = (1 - L)/2$. Then $\epsilon > 0$, and (3a) becomes

$$k_n < q = L + \frac{1 - L}{2} = \frac{1 + L}{2}.$$

Since $L < 1$, we have $q < 1$. Now we can write

(4)
$$|w_{N+1}| + |w_{N+2}| + |w_{N+3}| + \cdots$$
$$= |w_{N+1}|\left(1 + \left|\frac{w_{N+2}}{w_{N+1}}\right| + \left|\frac{w_{N+3}}{w_{N+2}}\right|\left|\frac{w_{N+2}}{w_{N+1}}\right| + \cdots\right)$$
$$= |w_{N+1}|(1 + k_{N+1} + k_{N+2}k_{N+1} + k_{N+3}k_{N+2}k_{N+1} + \cdots).$$

Since $k_n < q < 1$, each term of this series is less than the corresponding term of the geometric series

$$|w_{N+1}| (1 + q + q^2 + q^3 + \cdots).$$

By Theorem 2 this series converges because $q < 1$. From Theorem 1 it follows that the series in (4) is convergent. Hence the series $|w_1| + |w_2| + \cdots$ is convergent. This implies that the series $w_1 + w_2 + \cdots$ is absolutely convergent.

We now consider the case $L > 1$. We choose $\epsilon = (L - 1)/2$. Clearly, $\epsilon > 0$, and (3b) becomes

$$k_n > L - \epsilon = \frac{1 + L}{2} > 1 \qquad\qquad (n > N),$$

that is,

$$k_n = \left|\frac{w_{n+1}}{w_n}\right| > 1 \quad \text{or} \quad |w_{n+1}| > |w_n| \qquad\qquad (n > N).$$

The last inequality shows that the terms are increasing in absolute value. From this and Theorem 1, Sec. 14.2, it follows that the series is divergent.

For $L = 1$ the series may converge or diverge so that the test fails. This is illustrated by the harmonic series (Ex. 4 in Sec. 14.2)

$$\sum_{n=1}^{\infty} \frac{1}{n} = 1 + \frac{1}{2} + \frac{1}{3} + \cdots$$

which diverges and for which

$$\frac{w_{n+1}}{w_n} = \frac{n}{n + 1} \to L = 1 \qquad\qquad \text{as } n \to \infty,$$

and by the series

(5)
$$\sum_{n=1}^{\infty} \frac{1}{n^2} = 1 + \frac{1}{4} + \frac{1}{9} + \cdots$$

Fig. 298. Convergence of the series (5)

which converges and for which

$$\frac{w_{n+1}}{w_n} = \frac{n^2}{(n+1)^2} \to L = 1 \qquad \text{as } n \to \infty.$$

Convergence of (5) may be shown as follows. The nth partial sum is

$$s_n = 1 + \frac{1}{4} + \frac{1}{9} + \cdots + \frac{1}{n^2}.$$

Clearly, $s_n > 0$ and (Fig. 298)

$$s_n \leqq 1 + \int_1^n \frac{dx}{x^2} = 2 - \frac{1}{n},$$

which shows that the sequence of the partial sums is bounded. Since the terms of the series are positive, the sequence is monotone increasing, and convergence follows from Theorem 5 in Sec. 14.1.

The following test is more general than the ratio test, but its application is often more difficult.

Theorem 4 (Root test). *Suppose that the sequence*

$$\sqrt[n]{|w_n|} \qquad\qquad n = 1, 2, \cdots$$

is convergent with the limit L. Then the series

$$w_1 + w_2 + w_3 + \cdots$$

is

> *absolutely convergent, if $L < 1$,*
>
> *divergent, if $L > 1$.*

(If $L = 1$, the test fails).

Proof. If $L < 1$, then, as in the proof for the ratio test, we can choose $q < 1$ and find a corresponding N such that

$$k_n{}^* \equiv \sqrt[n]{|w_n|} < q < 1 \qquad\qquad \text{for every } n > N.$$

From this we obtain

$$|w_n| < q^n < 1 \qquad\qquad (n > N),$$

and the series $|w_N| + |w_{N+1}| + \cdots$ converges by comparison with the geometric series. Hence, the series $w_1 + w_2 + \cdots$ converges absolutely.

If $L > 1$, then $\sqrt[n]{|w_n|} > 1$ for all sufficiently large n. Hence $|w_n| > 1$ for those n, and divergence follows from Theorem 1 in the last section.

If $L = 1$, the test fails. This is illustrated by the harmonic series for which

$$\sqrt[n]{\frac{1}{n}} = \frac{1}{n^{1/n}} = \frac{1}{e^{(1/n)\ln n}} \to \frac{1}{e^0} = 1 \qquad (n \to \infty)$$

because $(1/n)\ln n \to 0$, and the series (5), for which

$$\sqrt[n]{\frac{1}{n^2}} = \frac{1}{n^{2/n}} = \frac{1}{e^{(2/n)\ln n}} \to \frac{1}{e^0} = 1 \qquad (n \to \infty)$$

because $(2/n)\ln n \to 0$ as $n \to \infty$.

Example 1.   Test the series

$$\sum_{n=1}^{\infty} \frac{n^2}{2^n} = \frac{1}{2} + 1 + \frac{9}{8} + 1 + \frac{25}{32} + \cdots.$$

We have

$$w_n = \frac{n^2}{2^n}, \qquad w_{n+1} = \frac{(n+1)^2}{2^{n+1}}, \qquad \frac{w_{n+1}}{w_n} = \frac{(n+1)^2}{2n^2} \to \frac{1}{2} \qquad (n \to \infty),$$

and convergence follows from the ratio test. We may also apply the root test:

$$\sqrt[n]{\frac{n^2}{2^n}} = \frac{n^{2/n}}{2} = \frac{e^{(2/n)\ln n}}{2} \to \frac{e^0}{2} = \frac{1}{2} \qquad (n \to \infty).$$

This confirms our result.

Example 2.   Is the series

$$\sum_{n=0}^{\infty} \frac{(3-4i)^n}{n!} = 1 + (3-4i) + \frac{1}{2!}(3-4i)^2 + \cdots$$

convergent or divergent? We have

$$|w_n| = \frac{|3-4i|^n}{n!} = \frac{5^n}{n!}, \qquad |w_{n+1}| = \frac{5^{n+1}}{(n+1)!}, \qquad \left|\frac{w_{n+1}}{w_n}\right| = \frac{5}{n+1} \to 0$$

as $n \to \infty$, and convergence follows from the ratio test.

We finally mention that both the ratio test and the root test can be generalized to the case when the sequences with the terms $|w_{n+1}/w_n|$ and $\sqrt[n]{|w_n|}$, respectively, are not convergent.

Theorem 5 (Ratio test). *If* $w_n \neq 0$ *for* $n = 1, 2, \cdots$, *and if for every n greater than some number N*

(6)
$$\left|\frac{w_{n+1}}{w_n}\right| \leqq q,$$

where q is some fixed number less than 1, then the series

$$w_1 + w_2 + w_3 + \cdots$$

converges absolutely. If

(7) $$\left| \frac{w_{n+1}}{w_n} \right| \geq 1 \qquad \text{for every } n > N,$$

the series diverges.

Proof. In the first part of the proof of Theorem 3, convergence followed from the existence of a number $q < 1$ such that (6) holds for all $n > N$. Hence, for the present theorem, convergence follows by the same argument. The last statement of the theorem follows from the fact that, by (7), $|w_{n+1}| \geq |w_n|$ and Theorem 1 in the last section.

Example 3. In the series

$$1 + \frac{1}{2} + \frac{1}{8} + \frac{1}{16} + \frac{1}{64} + \frac{1}{128} + \frac{1}{512} + \cdots$$

the odd terms and the even terms each form a geometric series with ratio $\frac{1}{8}$. Convergence follows from Theorem 5, because the ratios of successive terms are

$$\frac{1}{2}, \ \frac{1}{4}, \ \frac{1}{2}, \ \frac{1}{4}, \cdots .$$

We see that the sequence of these ratios does not converge. Hence Theorem 3 cannot be applied, and our example illustrates that Theorem 5 is more general than Theorem 3.

Theorem 6 (Root test).   *If for every n greater than some number N*

(8) $$\sqrt[n]{|w_n|} \leq q,$$

where q is some fixed number less than 1, then the series

$$w_1 + w_2 + w_3 + \cdots$$

converges absolutely. If for infinitely many n

(9) $$\sqrt[n]{|w_n|} \geq 1,$$

the series diverges.

Proof. If (8) holds, then

$$|w_n| \leq q^n < 1 \qquad\qquad (n > N)$$

and the series $|w_N| + |w_{N+1}| + \cdots$ converges by comparison with the geometric series. Hence, the series $w_1 + w_2 + \cdots$ converges absolutely. If (9) holds, then $|w_n| \geq 1$ for infinitely many n, and divergence follows from Theorem 1 in the last section.

In the last two theorems it is essential for convergence that $|w_{n+1}/w_n|$ and $\sqrt[n]{|w_n|}$, respectively, should be ultimately less than or equal to a definite number $q < 1$. It does not at all suffice for convergence that we should have

$$\left| \frac{w_{n+1}}{w_n} \right| < 1 \qquad \text{or} \qquad \sqrt[n]{|w_n|} < 1$$

for sufficiently large n. For example, in the case of the harmonic series (Ex. 4 in Sec. 14.2) we have

$$\frac{w_{n+1}}{w_n} = \frac{1/(n+1)}{1/n} = \frac{n}{n+1} < 1 \quad \text{and} \quad \sqrt[n]{\frac{1}{n}} < 1,$$

but the series diverges.

Problems

Show that the following series converge. How many terms are needed to compute s with an error less than 0.01?

1. $s = \frac{1}{2} - \frac{1}{4} + \frac{1}{8} - + \cdots$

2. $s = 1 - \frac{1}{3} + \frac{1}{6} - \frac{1}{12} + \frac{1}{24} - + \cdots$

3. $s = 1 - \frac{1}{3} + \frac{1}{9} - \frac{1}{27} + - \cdots$

4. $s = 1 - \frac{1}{2!} + \frac{1}{4!} - \frac{1}{6!} + - \cdots$

Are the following series convergent or divergent?

5. $1 + i + i^2 + i^3 + \cdots$

6. $1 - 3 + 5 - 7 + - \cdots$

7. $\frac{5i}{1!} - \frac{(5i)^3}{3!} + \frac{(5i)^5}{5!} - + \cdots$

8. $1 + \frac{8i}{1!} + \frac{(8i)^2}{2!} + \frac{(8i)^3}{3!} + \cdots$

9. $\frac{1}{1 \cdot 2} + \frac{1}{2 \cdot 3} + \frac{1}{3 \cdot 4} + \cdots$

10. $\frac{1}{\sqrt{1 \cdot 2}} + \frac{1}{\sqrt{2 \cdot 3}} + \frac{1}{\sqrt{3 \cdot 4}} + \cdots$

Are the following series convergent or divergent?

11. $\displaystyle\sum_{n=1}^{\infty} \frac{(-1)^{n-1}}{2n-1}$

12. $\displaystyle\sum_{n=0}^{\infty} \frac{100^{2n}}{n!}$

13. $\displaystyle\sum_{n=1}^{\infty} \frac{n^{2n}}{n!}$

14. $\displaystyle\sum_{n=1}^{\infty} \left(\frac{4}{5}\right)^n n^5$

15. $\displaystyle\sum_{n=1}^{\infty} \frac{n+1}{2^n n}$

16. $\displaystyle\sum_{n=0}^{\infty} \frac{(6+5i)^n}{8^n}$

17. $\displaystyle\sum_{n=1}^{\infty} \frac{(4-6i)^n}{7^n n}$

18. $\displaystyle\sum_{n=1}^{\infty} \frac{n-i}{n+i}$

19. $\displaystyle\sum_{n=0}^{\infty} \frac{(13i)^{2n+1}}{(2n+1)!}$

20. $\displaystyle\sum_{n=1}^{\infty} \frac{(n!)^2}{(2n)!}$

21. $\displaystyle\sum_{n=1}^{\infty} \frac{(2i)^n}{3^n n}$

22. $\displaystyle\sum_{n=1}^{\infty} \frac{2^n n!}{n^n}$

23. Determine how many terms are needed to compute the sum s of the geometric series $1 + q + q^2 + \cdots$ with an error less than 0.01, when $q = 0.25$, $q = 0.5$, $q = 0.9$.

24. If $|w_{n+1}/w_n| \leq q < 1$, so that the series $w_1 + w_2 + \cdots$ converges by the ratio test (Theorem 5), show that the remainder $R_n = w_{n+1} + w_{n+2} + \cdots$ satisfies $|R_n| \leq |w_{n+1}|/(1-q)$. *Hint.* Use the fact that the ratio test is a comparison of the series $w_1 + w_2 + \cdots$ with the geometric series.

25. Using Prob. 24, find how many terms suffice for computing the sum s of the series in Prob. 15 with an error not exceeding 0.05 and compute s to this accuracy.

14.4   *Operations on Series*

We shall now consider some simple operations which are frequently used in working with series.

Let us start with the addition of series and show that two convergent series may be added term by term.

Theorem 1. *If the series $w_1 + w_2 + \cdots$ and $z_1 + z_2 + \cdots$ are convergent with the respective sums s and s^*, then the series*

(1)
$$\sum_{n=1}^{\infty} (w_n + z_n), \qquad \sum_{n=1}^{\infty} (w_n - z_n), \qquad and \qquad \sum_{n=1}^{\infty} k w_n$$

where k is any constant, are convergent and have the sums $s + s^*$, $s - s^*$, and ks, respectively.

Proof. The partial sums of the two given series are

$$s_n = w_1 + \cdots + w_n, \qquad s_n{}^* = z_1 + \cdots + z_n,$$

and by the definition of convergence of a series we have

$$\lim_{n \to \infty} s_n = s, \qquad \lim_{n \to \infty} s_n{}^* = s^*.$$

Now the nth partial sum of the first series in (1) is

$$S_n = s_n + s_n{}^* = (w_1 + z_1) + \cdots + (w_n + z_n).$$

From this we obtain

$$\lim_{n \to \infty} S_n = \lim_{n \to \infty} (s_n + s_n{}^*) = \lim_{n \to \infty} s_n + \lim_{n \to \infty} s_n{}^* = s + s^*.$$

Hence this series converges and has the sum $s + s^*$. The other statements can be proved by a similar argument.

The next operation to be considered is that of inserting parentheses in a given series. This operation is called **grouping.**

For example, by grouping the terms of the series $w_1 + w_2 + w_3 + \cdots$ we may obtain the series $(w_1 + w_2) + (w_3 + w_4) + \cdots$ whose terms are $W_n = w_{2n-1} + w_{2n}$ where $n = 1, 2, \cdots$.

Of course, in the case of a *finite* series (that is, a series with only finitely many terms) we may insert parentheses without changing the sum of the series. However, this is not true in general for an infinite series. For instance, by grouping the divergent series

$$1 - 1 + 1 - 1 + - \cdots \qquad \text{(cf. Ex. 3 in Sec. 14.2)}$$

we may obtain the convergent series

$$(1 - 1) + (1 - 1) + \cdots = 0 + 0 + \cdots$$

whose sum is zero. However, we shall now prove that in the case of a convergent series, grouping is permissible.

Theorem 2. *If a series converges, then insertion of parentheses yields a new convergent series having the same sum as the original series.*

Proof. Obviously, the partial sums of the new series are obtained by skipping certain partial sums of the given series. For example, the partial sums of the series

$$(w_1 + w_2 + w_3) + (w_4 + w_5 + w_6) + \cdots$$

are the partial sums $s_3, s_6, s_9, \cdots$ of the series $w_1 + w_2 + \cdots$. Now if the partial sums s_n of the given series form a convergent sequence whose limit is s, it follows immediately from the definition of convergence of a sequence that the new sequence obtained by skipping must also converge to s. This proves the theorem.

Example 1. *By the Leibniz test* (cf. Sec. 14.2) the series

$$\sum_{n=1}^{\infty} \frac{(-1)^{n+1}}{n} = 1 - \frac{1}{2} + \frac{1}{3} - \frac{1}{4} + - \cdots$$

converges. Let s be its sum. Then by Theorem 2,

$$\frac{1}{1 \cdot 2} + \frac{1}{3 \cdot 4} + \frac{1}{5 \cdot 6} + \cdots = \left(1 - \frac{1}{2}\right) + \left(\frac{1}{3} - \frac{1}{4}\right) + \cdots$$

(2)

$$= \sum_{m=1}^{\infty} \left(\frac{1}{2m-1} - \frac{1}{2m}\right) = s$$

and, similarly,

$$\frac{1 \cdot 2 + 3 \cdot 4}{1 \cdot 2 \cdot 3 \cdot 4} + \frac{5 \cdot 6 + 7 \cdot 8}{5 \cdot 6 \cdot 7 \cdot 8} + \cdots = \left(1 - \frac{1}{2} + \frac{1}{3} - \frac{1}{4}\right) + \left(\frac{1}{5} - \frac{1}{6} + \frac{1}{7} - \frac{1}{8}\right) + \cdots$$

(3)

$$= \sum_{m=1}^{\infty} \left(\frac{1}{4m-3} - \frac{1}{4m-2} + \frac{1}{4m-1} - \frac{1}{4m}\right) = s.$$

The last operation to be considered in this section is that of changing the order of the terms in a series.

Clearly, if a series is finite, the order of its terms may be changed without altering the sum. Also, we may change the order of finitely many terms of a given infinite series: if the given series is divergent, so will be the new series, and if the given series converges, the new series will be convergent with the same sum. This follows immediately from the definitions of convergence and divergence.

We may now ask what happens if we "change the order of infinitely many terms" of a series. Of course we must first define what we mean by such an operation. This can be done as follows.

A series

$$\sum_{n=1}^{\infty} w_n{}^* = w_1{}^* + w_2{}^* + \cdots$$

is said to be a **rearrangement** of a series

$$\sum_{m=1}^{\infty} w_m = w_1 + w_2 + \cdots$$

if there is a one-to-one correspondence between the indices n and m such that $w_n{}^* = w_m$ for corresponding indices.

For example, the series

$$\frac{1}{2} + 1 + \frac{1}{4} + \frac{1}{3} + \frac{1}{6} + \frac{1}{5} + \cdots$$

is a rearrangement of the harmonic series

$$1 + \frac{1}{2} + \frac{1}{3} + \frac{1}{4} + \frac{1}{5} + \frac{1}{6} + \cdots .$$

The following example illustrates that by rearranging a convergent series we may obtain a convergent series whose sum is different from that of the original series.

Example 2. We rearrange the series

$$s = 1 - \frac{1}{2} + \frac{1}{3} - \frac{1}{4} + \frac{1}{5} - \frac{1}{6} + \frac{1}{7} - \frac{1}{8} + \frac{1}{9} - + \cdots$$

by writing first two positive terms, then one negative, then again two positive, etc., in the order of their occurrence. This yields the rearrangement

$$s^* = 1 + \frac{1}{3} - \frac{1}{2} + \frac{1}{5} + \frac{1}{7} - \frac{1}{4} + \frac{1}{9} + \frac{1}{11} - \frac{1}{6} + \frac{1}{13} + \cdots .$$

We want to show that the sums s and s^* are different. By inserting parentheses in the rearrangement we have

$$s^* = \left(1 + \frac{1}{3} - \frac{1}{2}\right) + \left(\frac{1}{5} + \frac{1}{7} - \frac{1}{4}\right) + \cdots = \sum_{m=1}^{\infty} \left(\frac{1}{4m-3} + \frac{1}{4m-1} - \frac{1}{2m}\right).$$

On the other hand, by multiplying the series (2) by $\frac{1}{2}$ and adding the resulting series and the series (3) term by term (cf. Theorem 1), we get

$$\frac{3s}{2} = \sum_{m=1}^{\infty} \left(\frac{1}{4m-3} - \frac{1}{4m-2} + \frac{1}{4m-1} - \frac{1}{4m} + \frac{1/2}{2m-1} - \frac{1/2}{2m}\right)$$

$$= \sum_{m=1}^{\infty} \left(\frac{1}{4m-3} + \frac{1}{4m-1} - \frac{1}{2m}\right) = s^*.$$

Hence $s^* = 3s/2$.

We note that the series in Ex. 2 is not absolutely convergent. Let us prove that an absolutely convergent series can always be rearranged without altering the sum.

Theorem 3. *If a series converges absolutely, then any of its rearrangements converges absolutely and has the same sum as the original series.*

Proof. Let $w_1^* + w_2^* + w_3^* + \cdots$ be any rearrangement of an absolutely convergent series $w_1 + w_2 + w_3 + \cdots$. Then, since every w_m^* equals a w_n for appropriate n and no two m's correspond to the same n, we clearly have

$$\sum_{m=1}^{n} |w_m^*| \leqq \sum_{k=1}^{\infty} |w_k| \qquad\qquad \text{for every } n.$$

The sum on the left is the nth partial sum of the series $|w_1^*| + |w_2^*| + \cdots$. Since these partial sums are nonnegative, the inequality shows that they form a bounded sequence. Since $|w_m^*| \geqq 0$, the sequence is monotone increasing and, therefore, convergent (cf. Theorem 5 in Sec. 14.1). Hence the rearrangement $w_1^* + w_2^* + \cdots$ converges absolutely. Let s^* be its sum, and let s be the sum of the original series. We show that $s^* = s$.

From the definition of convergence and from Theorem 4 in Sec. 14.2 when applied to the series $|w_1| + |w_2| + \cdots$ it follows that, when an $\epsilon > 0$ is given, we can find an N such that

(4)   (a) $|s_n - s| < \dfrac{\epsilon}{2}$ (b) $|w_{n+1}| + \cdots + |w_{n+p}| < \dfrac{\epsilon}{2}$

for every $n > N$ and $p = 1, 2, \cdots$; here s_n is the nth partial sum of the original series. Now for sufficiently large m, the mth partial sum $s_m{}^*$ of the rearrangement will contain all the terms $w_1, \cdots, w_n$ ($n > N$ and fixed) and perhaps some more terms w_r ($r > n$) of the original series. Hence $s_m{}^*$ will be of the form

(5) $$s_m{}^* = s_n + A_{mn}$$

where A_{mn} is the sum of those additional terms. Let $n + p$ be the greatest subscript of the terms in A_{mn}. Then by (4b), since $n > N$, we have

$$|A_{mn}| \leqq |w_{n+1}| + \cdots + |w_{n+p}| < \frac{\epsilon}{2}.$$

From this and (5) it follows that

$$|s_m{}^* - s_n| = |A_{mn}| < \frac{\epsilon}{2}.$$

Using (4a) and the triangle inequality, we thus obtain

$$|s_m{}^* - s| = |(s_m{}^* - s_n) + (s_n - s)| \leqq |s_m{}^* - s_n| + |s_n - s| < \frac{\epsilon}{2} + \frac{\epsilon}{2} = \epsilon$$

for every sufficiently large m. Hence, the sequence $s_1{}^*, s_2{}^*, \cdots$ converges to s and, therefore, $s^* = s$. This completes the proof.

14.5   *Power Series*

We have made the definitions for series of *constant* terms. If the terms of a series are *variable*, say, functions of a variable z, they assume definite values when z is given a fixed value and then all those definitions apply. It is obvious that in the case of such a series of functions of z the partial sums, the remainders, and the sum will be functions of z. Usually such a series will be convergent for some values of z, for example, for all z in some region, and divergent for the other values of z.

In complex analysis, the most important series with variable terms are power series. A **power series**[5] *in powers of* $z - a$ is an infinite series of the form

(1) $$\sum_{m=0}^{\infty} c_m(z - a)^m = c_0 + c_1(z - a) + c_2(z - a)^2 + \cdots$$

[5] It should be emphasized that the term "power series" alone usually refers to series of the form (1), including the particular case (2), but *does not include* series of negative powers of z such as $c_1 z^{-1} + c_2 z^{-2} + \cdots$ or series involving fractional powers of z.

where z is a variable, $c_0, c_1, \cdots$ are constants, called the **coefficients,** and a is a constant, called the **center** of the series.

If $a = 0$, we obtain as a particular case a *power series in powers of z*:

(2)
$$\sum_{m=0}^{\infty} c_m z^m = c_0 + c_1 z + c_2 z^2 + \cdots.$$

The convergence behavior of a power series can be characterized in a very simple way. Let us start with three examples, which will turn out to be typical.

Example 1.   The **geometric series**

$$\sum_{m=0}^{\infty} z^m = 1 + z + z^2 + \cdots$$

converges absolutely when $|z| < 1$ and diverges when $|z| \geqq 1$ (cf. Theorem 2 in Sec. 14.3).

Example 2.   The power series

$$\sum_{n=0}^{\infty} \frac{z^n}{n!} = 1 + z + \frac{z^2}{2!} + \frac{z^3}{3!} + \cdots$$

is absolutely convergent for each (finite) z, as follows from the ratio test. In fact, for any fixed z,

$$\left| \frac{z^{n+1}/(n + 1)!}{z^n/n!} \right| = \frac{|z|}{n + 1} \to 0 \qquad \text{as} \qquad n \to \infty.$$

Example 3.   The power series

$$\sum_{n=0}^{\infty} n!\, z^n = 1 + z + 2z^2 + 6z^3 + \cdots$$

converges only at the center $z = 0$, but diverges for any $z \neq 0$. In fact, this follows from the ratio test because

$$\left| \frac{(n + 1)!\, z^{n+1}}{n!\, z^n} \right| = (n + 1) |z| \to \infty \qquad \text{as} \qquad n \to \infty \qquad (z \text{ fixed and } \neq 0).$$

The power series (1) converges when $z = a$, because then $z - a = 0$ and the series reduces to the single term c_0. Example 3 illustrates that in some cases this may be the only value of z for which the series converges. However, if the series (1) converges for some $z_0 \neq a$, then it converges for every z whose distance from the center is less than that of z_0. In fact, the following theorem holds.

Theorem 1.   *If the power series* (1) *converges for $z = z_0$, it converges absolutely for every z for which $|z - a| < |z_0 - a|$, that is, for each z within the circle through z_0 about a.*

Proof.   Since the series (1) converges for z_0, Theorem 1 in Sec. 14.2 shows that

$$c_n (z_0 - a)^n \to 0 \qquad \text{as} \qquad n \to \infty.$$

This implies that for $z = z_0$ the terms of the series (1) are bounded, say,

$$|c_n (z_0 - a)^n| < M \qquad \text{for every } n = 0, 1, \cdots.$$

From this we obtain

$$|c_n(z - a)^n| = \left| c_n(z_0 - a)^n \left(\frac{z - a}{z_0 - a} \right)^n \right| < M \left| \frac{z - a}{z_0 - a} \right|^n$$

and, therefore,

$$(3) \qquad \sum_{n=0}^{\infty} |c_n(z - a)^n| < \sum_{n=0}^{\infty} M \left| \frac{z - a}{z_0 - a} \right|^n = M \sum_{n=0}^{\infty} \left| \frac{z - a}{z_0 - a} \right|^n.$$

From the assumption $|z - a| < |z_0 - a|$ we have the inequality

$$\left| \frac{z - a}{z_0 - a} \right| < 1,$$

and the series on the right-hand side of (3) is a geometric series which converges. Hence the series on the left-hand side of (3) converges, and the series (1) converges absolutely when $|z - a| < |z_0 - a|$. This completes the proof.

Examples 2 and 3 illustrate that a power series may converge for all z or only for $z = a$. Let us exclude these two cases for a moment. Then, if a power series (1) is given, we may consider all the points z in the complex plane for which the series converges. Let R be the smallest real number so that the distance of any of these points from the center a is at most equal to R. (For instance, in Ex. 1, $R = 1$.) Then from Theorem 1 it follows that the series converges for all z within the circle of radius R about the center a, that is, for all z for which

$$(4) \qquad\qquad\qquad |z - a| < R,$$

and diverges for all z for which

$$|z - a| > R.$$

The circle

$$|z - a| = R$$

is called the **circle of convergence**, and its radius R is called the **radius of convergence** of (1). Cf. Fig. 299.

At the points on the circle of convergence the series may converge or diverge. For instance, in Ex. 1 we have $R = 1$, and the series diverges at each point of the circle of convergence $|z| = 1$. The power series

$$\sum_{n=1}^{\infty} \frac{z^n}{n} = z + \frac{z^2}{2} + \frac{z^3}{3} + \cdots$$

Fig. 299. Circle of convergence

Fig. 300. Interval of convergence of a real power series

converges for $|z| < 1$ and diverges for $|z| > 1$, as follows by the ratio test. Thus $R = 1$. At $z = 1$ it becomes the harmonic series and diverges. At $z = -1$ it becomes the series in Ex. 5, Sec. 14.2, and converges. This illustrates that a series may converge at some points on the circle of convergence, while at others it may diverge.

Clearly, if we consider a *real* power series (1), that is, if the variable $z = x$, the center, and the coefficients are real, then (4) represents an interval of length $2R$ with midpoint at a on the x-axis, the so-called **interval of convergence** (Fig. 300).

If the series (1) converges for all z (as in Ex. 2), then we set

$$R = \infty \qquad \text{(and } 1/R = 0\text{);}$$

if it converges only at the center $z = a$ (as in Ex. 3), then we set

$$R = 0 \qquad \text{(and } 1/R = \infty\text{).}$$

Using these conventions, the radius of convergence R of the power series (1) may be determined from the coefficients of the series as follows.

Theorem 2 (Radius of convergence). *If the sequence $\sqrt[n]{|c_n|}, n = 1, 2, \cdots$, converges with the limit L, then the radius of convergence R of the power series (1) is*

(5a) $$R = \frac{1}{L} \qquad \text{(\textbf{Cauchy--Hadamard formula}[6]),}$$

including the case $L = 0$ where $R = \infty$ and the series (1) converges for all z.

If that sequence does not converge but is bounded, then

(5b) $$R = \frac{1}{l}$$

where l is the greatest of the limit points of the sequence.

If that sequence is not bounded, then $R = 0$, and the series converges only for $z = a$.

Proof. If

$$\lim_{n \to \infty} \sqrt[n]{|c_n|} = L \neq 0,$$

then

$$\lim_{n \to \infty} \sqrt[n]{|c_n(z - a)^n|} = |z - a| \lim_{n \to \infty} \sqrt[n]{|c_n|} = |z - a| \, L.$$

---

[6] Named after the French mathematicians, A. L. CAUCHY (cf. the footnote in Sec. 1.13) and JACQUES HADAMARD (1865–1964).

Since the terms of the series (1) are $w_n = c_n(z - a)^n$, the root test (Sec. 14.3) shows that the series converges absolutely when

$$|z - a| L < 1 \qquad \text{or} \qquad |z - a| < \frac{1}{L} = R$$

but diverges when

$$|z - a| L > 1 \qquad \text{or} \qquad |z - a| > \frac{1}{L} = R.$$

If

$$\lim_{n \to \infty} \sqrt[n]{|c_n|} = L = 0,$$

it follows from the definition of a limit that for any given $\epsilon > 0$, for example, for $\epsilon = 1/(2 |z_1 - a|)$ with arbitrary fixed z_1, we can find an N such that

$$\sqrt[n]{|c_n|} < \frac{1}{2 |z_1 - a|} \qquad \qquad \text{for every } n > N.$$

From this we have

$$|c_n| < \frac{1}{(2 |z_1 - a|)^n} \qquad \text{or} \qquad |c_n(z_1 - a)^n| < \frac{1}{2^n}.$$

Now, since $\Sigma \, 2^{-n}$ converges, the comparison test (cf. Sec. 14.3) shows that the series (1) converges absolutely for $z = z_1$. Since z_1 is arbitrary, this means absolute convergence for every finite z, and the proof of the statement involving (5a) is complete.

We next prove the statement involving (5b). The existence of l follows from the Bolzano-Weierstrass theorem (Sec. 14.1), and since $\sqrt[n]{|c_n|} \geqq 0$, we clearly have $l > 0$. From the definition of a limit point it follows that, an $\epsilon > 0$ being given,

$$l - \epsilon < \sqrt[n]{|c_n|} < l + \epsilon \qquad \qquad \text{for infinitely many } n.$$

Multiplying this by the positive quantity $|z - a|$, we obtain the inequalities

$$(6) \qquad \qquad |z - a| \, (l - \epsilon) < \sqrt[n]{|c_n(z - a)^n|}$$

and

$$(7) \qquad \qquad \sqrt[n]{|c_n(z - a)^n|} < |z - a| \, (l + \epsilon).$$

The inequality (7) holds even for all sufficiently large n, say, for $n > N$, because l is the greatest limit point and, therefore, at most finitely many terms can be greater than the expression on the right. We prove that for

$$(8) \qquad \qquad |z - a| < \frac{1}{l}$$

convergence of the power series (1) follows from (7). In fact, if we choose

$$\epsilon = \frac{1 - l \, |z - a|}{2 \, |z - a|},$$

then because of (8) we have $\epsilon > 0$, and the inequality (7) takes the form

$$\sqrt[n]{|c_n(z-a)^n|} < \frac{1 + l\,|z-a|}{2} \qquad (n > N).$$

From (8) we see that the expression on the right is less than 1, and convergence follows by means of the root test (Theorem 6, Sec. 14.3). On the other hand, when

$$|z - a| > \frac{1}{l},$$

then, choosing

$$\epsilon = \frac{l\,|z-a| - 1}{2\,|z-a|},$$

we have $\epsilon > 0$, and (6) becomes

$$\sqrt[n]{|c_n(z-a)^n|} > \frac{|z-a|\,l + 1}{2} > 1.$$

Hence, by Theorem 6 in Sec. 14.3 the series diverges for those z. This proves the statement involving (5b).

Finally if the sequence $\sqrt[n]{|c_n|}$ is not bounded, then, by definition, any K being given,

$$\sqrt[n]{|c_n|} > K \qquad \text{for infinitely many } n.$$

Choosing $K = 1/|z - a|$ where $z \neq a$, we see that this inequality becomes

$$\sqrt[n]{|c_n|} > \frac{1}{|z-a|} \qquad \text{or} \qquad \sqrt[n]{|c_n(z-a)^n|} > 1,$$

and divergence follows from Theorem 6 in Sec. 14.3. This completes the proof of Theorem 2.

We shall now consider the operations of addition and multiplication of power series.

From Theorem 1, Sec. 14.4, it is clear that *two power series may be added term by term for any z for which both series are convergent.*

Let us consider the term-by-term multiplication of two power series

$$(9) \qquad \sum_{k=0}^{\infty} a_k z^k = a_0 + a_1 z + \cdots \qquad \text{and} \qquad \sum_{m=0}^{\infty} c_m z^m = c_0 + c_1 z + \cdots.$$

If we multiply each term of the first series by each term of the second series and collect products of like powers of z, we obtain

$$a_0 c_0 + (a_0 c_1 + a_1 c_0) z + (a_0 c_2 + a_1 c_1 + a_2 c_0) z^2 + \cdots$$

(10)
$$= \sum_{n=0}^{\infty} (a_0 c_n + a_1 c_{n-1} + \cdots + a_n c_0) z^n.$$

This series is called the **Cauchy product** of the series (9).

Theorem 3.   *The Cauchy product of the two power series* (9) *is absolutely convergent for each z within the circle of convergence of each of the series* (9). *If the series have the sums g(z) and h(z), respectively, the Cauchy product has the sum*

$$(11) \qquad\qquad\qquad s(z) = g(z)h(z).$$

Proof.   The general term of the product series (10) is

$$p_n = (a_0c_n + a_1c_{n-1} + \cdots + a_nc_0)z^n.$$

Now, by the generalized triangle inequality (21) in Sec. 11.1, we obtain

$$|p_0| + |p_1| = |a_0c_0| + |(a_0c_1 + a_1c_0)z| \leqq (|a_0| + |a_1z|)(|c_0| + |c_1z|),$$

$$|p_0| + |p_1| + |p_2| \leqq (|a_0| + |a_1z| + |a_2z^2|)(|c_0| + |c_1z| + |c_2z^2|),$$

as can be verified by performing the multiplication on the right; in general

$$|p_0| + |p_1| + \cdots + |p_n|$$
$$\leqq (|a_0| + |a_1z| + \cdots + |a_nz^n|)(|c_0| + |c_1z| + \cdots + |c_nz^n|).$$

If z lies within the circle of convergence of each of the series (9), the sequence of the expressions on the right is bounded, and so is the sequence of the partial sums of the series $|p_0| + |p_1| + \cdots$. Since $|p_n| \geqq 0$, that sequence is monotone increasing, and from Theorem 5 in Sec. 14.1 it follows that it converges. Hence that series is convergent, and the product series (10) converges absolutely.

We prove (11). By Theorem 3 in Sec. 14.4 any rearrangement of (10) is absolutely convergent for precisely those z and has the same sum as (10). We consider the particular rearrangement $p_0{}^* + p_1{}^* + \cdots$ where $p_n{}^*$ equals (Fig. 301)

$$(a_nc_0 + a_0c_n)z^n + (a_nc_1 + a_1c_n)z^{n+1} + \cdots + (a_nc_{n-1} + a_{n-1}c_n)z^{2n-1} + a_nc_nz^{2n}.$$

Obviously,

$$a_0c_0 = p_0{}^*, \qquad (a_0 + a_1z)(c_0 + c_1z) = p_0{}^* + p_1{}^*$$

and, in general,

$$(a_0 + a_1z + \cdots + a_nz^n)(c_0 + c_1z + \cdots + c_nz^n) = p_0{}^* + p_1{}^* + \cdots + p_n{}^*.$$

By letting n approach infinity we obtain (11), and Theorem 3 is proved.

Fig. 301. Proof of Theorem 3

Example 4. The geometric series $1 + z + z^2 + \cdots$ has the sum $1/(1 - z)$ when $|z| < 1$ (cf. Theorem 2 in Sec. 14.3), and from Theorem 3 it follows that

$$\left(\frac{1}{1 - z}\right)^2 = \sum_{k=0}^{\infty} z^k \sum_{m=0}^{\infty} z^m = (1 + z + z^2 + \cdots)(1 + z + z^2 + \cdots)$$

$$= 1 + 2z + 3z^2 + \cdots = \sum_{n=0}^{\infty} (n + 1)z^n \qquad (|z| < 1).$$

Problems

1. Show that if the sequence $|c_{n+1}/c_n|$, $n = 1, 2, \cdots$, is convergent with the limit L, then the radius of convergence R of the power series (1) is $R = 1/L$ when $L > 0$ and is $R = \infty$ when $L = 0$.

Find the radius of convergence of the following series. (*a* in Prob. 9 is any number, and *m* in Prob. 13 is a positive integer.)

2. $\displaystyle\sum_{n=0}^{\infty} (z + i)^n$ 3. $\displaystyle\sum_{n=0}^{\infty} 3^n(z - i)^n$ 4. $\displaystyle\sum_{n=0}^{\infty} \frac{(z - i)^n}{5^n}$ 5. $\displaystyle\sum_{n=0}^{\infty} \left(\frac{\pi}{4}\right)^n z^{2n}$

6. $\displaystyle\sum_{n=0}^{\infty} \frac{(-1)^n}{n!} z^n$ 7. $\displaystyle\sum_{n=1}^{\infty} \frac{z^n}{n}$ 8. $\displaystyle\sum_{n=1}^{\infty} \frac{z^n}{n^2}$ 9. $\displaystyle\sum_{n=0}^{\infty} \frac{(z - a)^n}{n!}$

10. $\displaystyle\sum_{n=1}^{\infty} n^n z^n$ 11. $\displaystyle\sum_{n=1}^{\infty} n^{1000} z^n$ 12. $\displaystyle\sum_{n=0}^{\infty} \frac{z^{2n+1}}{(2n + 1)!}$ 13. $\displaystyle\sum_{n=0}^{\infty} \binom{n + m}{n} z^n$

14. $\displaystyle\sum_{n=0}^{\infty} \frac{(3n)!}{(n!)^3} z^n$ 15. $\displaystyle\sum_{n=1}^{\infty} \frac{(z - 2)^n}{n^n}$ 16. $\displaystyle\sum_{n=0}^{\infty} \frac{n^2}{3^n} z^n$ 17. $\displaystyle\sum_{n=0}^{\infty} \frac{(n + 1)^2}{2^{n+1}} z^{2n}$

18. At points on the circle of convergence a series may converge or diverge. Illustrate this for the geometric series and the series in Probs. 7 and 8 at $z = 1$ and $z = -1$.

19. If $\sum c_n z^n$ converges for all finite z, show that $\sqrt[n]{|c_n|} \to 0$ as $n \to \infty$. Give examples.

20. Show that if the radius of convergence of (2) is R, then the radius of convergence of $\sum c_m z^{2m}$ is $\sqrt{R}$.

14.6   Functions Represented by Power Series

Power series are basic in complex analysis, and the main objective of this section is to show the reasons for this fact.

Let $\sum_{n=0}^{\infty} c_n z^n$ be an arbitrary power series with nonzero radius of convergence R. Then the sum of this series is a function of z, say, $f(z)$, and we write

(1) $f(z) = \displaystyle\sum_{n=0}^{\infty} c_n z^n = c_0 + c_1 z + c_2 z^2 + \cdots$ $(|z| < R)$.

We say that $f(z)$ *is represented by the power series* or that *it is developed in the power series*. For instance, the geometric series in Ex. 1, Sec. 14.5, represents the function $f(z) = 1/(1 - z)$ in the interior of the unit circle $|z| = 1$. This follows from Theorem 2 in Sec. 14.3.

Theorem 1.   *The function $f(z)$ in (1) is continuous at $z = 0$.*

Proof.   We have to show that

$$\lim_{z \to 0} f(z) = f(0) = c_0. \tag{2}$$

We choose an arbitrary positive number $r < R$. Since the series in (1) is absolutely convergent in the disk $|z| < R$, it follows that the series

$$\sum_{n=1}^{\infty} |c_n| \, r^{n-1} = \frac{1}{r} \sum_{n=1}^{\infty} |c_n| \, r^n \qquad (0 < r < R)$$

is convergent. Let K denote its sum. Then we readily obtain

$$|f(z) - c_0| = \left| z \sum_{n=1}^{\infty} c_n z^{n-1} \right| \leqq |z| \sum_{n=1}^{\infty} |c_n| |z|^{n-1} \leqq |z| \, K \qquad (0 < |z| \leqq r).$$

An $\epsilon > 0$ being given, $|f(z) - c_0| < \epsilon$ for all z such that $|z| < \delta$ where δ is a positive real number less than both r and ϵ/K. By the definition of a limit, this means that (2) holds, and the proof is complete.

Next we consider the question of **uniqueness** and show that *the same function $f(z)$ cannot be represented by two different power series with the same center.* If $f(z)$ can be developed in a power series with center a, the development is unique. This important fact is frequently used in real and complex analysis. We may formulate it as follows (assuming that $a = 0$, without loss of generality).

Theorem 2 (Identity theorem for power series).   *Suppose that*

$$\sum_{n=0}^{\infty} a_n z^n \qquad and \qquad \sum_{n=0}^{\infty} b_n z^n$$

are power series which are convergent for $|z| < R$ where R is positive, and have the same sum for all these z. Then the series are identical, that is

$$a_n = b_n \qquad \qquad for\ all\ n = 0, 1, \cdots. \tag{3}$$

Proof.   We proceed by induction. By assumption,

$$a_0 + a_1 z + a_2 z^2 + \cdots = b_0 + b_1 z + b_2 z^2 + \cdots \qquad (|z| < R). \tag{4}$$

We let z approach zero. Then, by Theorem 1, we have $a_0 = b_0$. We assume that $a_n = b_n$ for $n = 0, 1, \cdots, m$. Then, by omitting the first $m + 1$ terms on both sides of (4) and dividing by z^{m+1} ($\neq 0$), we obtain

$$a_{m+1} + a_{m+2} z + a_{m+3} z^2 + \cdots = b_{m+1} + b_{m+2} z + b_{m+3} z^2 + \cdots.$$

By Theorem 1, each of these power series represents a function which is continuous at $z = 0$. Hence $a_{m+1} = b_{m+1}$. This completes the proof.

Let us now consider termwise differentiation and integration of power series. By differentiating the series $c_0 + c_1 z + c_2 z^2 + \cdots$ we obtain the series

$$(5) \qquad \sum_{n=1}^{\infty} n c_n z^{n-1} = c_1 + 2c_2 z + 3c_3 z^2 + \cdots.$$

This series is called the *derived series* of the given power series.

Theorem 3. *The derived series of a power series has the same radius of convergence as the original series.*

Proof. Let $n c_n = c_n{}^*$. Then $\sqrt[n]{|c_n{}^*|} = \sqrt[n]{n}\sqrt[n]{|c_n|}$. Since $\sqrt[n]{n} \to 1$ as $n \to \infty$, it follows that the sequences $\sqrt[n]{|c_n{}^*|}$ and $\sqrt[n]{|c_n|}$ either both converge with the same limit or both diverge. If they diverge, they are both unbounded or bounded, and in the latter case their greatest limit points are the same. From this and Theorem 2 in the last section the statement of our present theorem follows.

Example 1. The power series

$$\sum_{n=1}^{\infty} (n+1)z^n$$

has the radius of convergence $R = 1$. This follows by differentiating the geometric series and applying Theorem 3.

Theorem 4. *The series*

$$\sum_{n=0}^{\infty} \frac{c_n}{n+1} z^{n+1} = c_0 z + \frac{c_1}{2} z^2 + \frac{c_2}{3} z^3 + \cdots$$

obtained by integrating the series $c_0 + c_1 z + c_2 z^2 + \cdots$ *term by term has the same radius of convergence as the original series.*

The proof is similar to that of Theorem 3.

Power series represent analytic functions. More precisely:

Theorem 5. *A power series with a nonzero radius of convergence R represents an analytic function at every point interior to its circle of convergence. The derivatives of this function are obtained by differentiating the original series term by term; all the series thus obtained have the same radius of convergence as the original series.*

Proof. We first prove that for any integer $n \geq 2$,

$$(6a) \qquad \frac{b^n - a^n}{b - a} - na^{n-1} = (b - a)A_n$$

where

$$(6b) \qquad A_n = b^{n-2} + 2ab^{n-3} + 3a^2 b^{n-4} + \cdots + (n-1)a^{n-2}.$$

We proceed by induction. A simple calculation shows that (6) is true for $n = 2$. Assuming that (6) holds for $n = k$, we show that it holds for $n = k + 1$. We have

$$\frac{b^{k+1} - a^{k+1}}{b - a} = \frac{b^{k+1} - ba^k + ba^k - a^{k+1}}{b - a} = b\frac{b^k - a^k}{b - a} + a^k.$$

By the induction hypothesis, the right-hand side equals

$$b[(b - a)A_k + ka^{k-1}] + a^k.$$

Direct calculation shows that this is equal to

$$(b - a)\{bA_k + ka^{k-1}\} + ka^k + a^k.$$

From (6*b*) with $n = k$ we see that the expression in the braces equals

$$b^{k-1} + 2ab^{k-2} + \cdots + (k - 1)ba^{k-2} + ka^{k-1} = A_{k+1}.$$

Hence our result is

$$\frac{b^{k+1} - a^{k+1}}{b - a} = (b - a)A_{k+1} + (k + 1)a^k,$$

which is (6) with $n = k + 1$. This proves (6) for any integer $n \geq 2$.

To prove the statements in Theorem 5, we consider a representation

$$f(z) = \sum_{n=0}^{\infty} c_n z^n,$$

assuming that the radius of convergence R is not zero, and prove that, for any fixed z with $|z| < R$ and $\Delta z \to 0$ the difference quotient $[f(z + \Delta z) - f(z)]/\Delta z$ approaches the function represented by the derived series (5), which we denote by $f_1(z)$. By termwise addition we first have

$$\frac{f(z + \Delta z) - f(z)}{\Delta z} - f_1(z) = \sum_{n=2}^{\infty} c_n \left[\frac{(z + \Delta z)^n - z^n}{\Delta z} - nz^{n-1} \right].$$

From (6) with $b = z + \Delta z$, $a = z$, and $b - a = \Delta z$ we see that the series on the right can be written

$$\Delta z \sum_{n=2}^{\infty} c_n [(z + \Delta z)^{n-2} + 2z(z + \Delta z)^{n-3} + \cdots + (n - 1)z^{n-2}],$$

and for $|z| \leq R_0$ and $|z + \Delta z| \leq R_0$, $R_0 < R$, the absolute value of this cannot exceed

$$(7) \qquad\qquad |\Delta z| \sum_{n=2}^{\infty} |c_n| \, n(n - 1)R_0^{n-2},$$

where $n - 1$ is the largest of the coefficients $1, 2, \cdots, n - 1$, and n is the number of terms. The series in (7) is closely related to the second derived series of the series under consideration, evaluated at R_0; by Theorem 3 in this section and Theorem 1 in Sec. 14.5 it converges at R_0 ($< R$). Let $K(R_0)$ be its value. Then our result may be written

$$\left| \frac{f(z + \Delta z) - f(z)}{\Delta z} - f_1(z) \right| \leq |\Delta z| \, K(R_0).$$

Letting $\Delta z \to 0$ and noting that R_0 ($< R$) is arbitrary, we conclude that $f(z)$ is analytic at any point interior to the circle of convergence and its derivative is represented by the derived series. From this the statements about the higher derivatives follow by induction, and Theorem 5 is proved.

From Theorem 5 we see that the mth derivative $f^{(m)}(z)$ of the function $f(z)$ represented by (1) is

(8) $$f^{(m)}(z) = \sum_{n=m}^{\infty} n(n-1)\cdots(n-m+1)c_n z^{n-m} \qquad (|z| < R).$$

In the next section we shall see that *every analytic function can be represented by power series.*

Problems

1. If $f(z)$ in (1) is even, show that $c_n = 0$ for odd n. (Use Theorem 2.)
2. If $f(z)$ in (1) is odd, show that $c_n = 0$ for even n. Give examples.
3. Applying Theorem 2 to $(1 + z)^p(1 + z)^q = (1 + z)^{p+q}$ (p and q positive integers), show that

$$\sum_{n=0}^{r} \binom{p}{n}\binom{q}{r-n} = \binom{p+q}{r}.$$

4. Verify Theorems 3 and 4 for the geometric series and for the series in Ex. 2, Sec. 14.5.

Applying Theorems 3 and 4 to the geometric series, find the radius of convergence of the following series.

5. $\displaystyle\sum_{n=1}^{\infty} \frac{n}{5^n}(z - i)^n$ 6. $\displaystyle\sum_{n=1}^{\infty} \frac{(z+2)^n}{n}$ 7. $\displaystyle\sum_{n=1}^{\infty} \frac{4^n}{n}(z + 2i)^n$

8. $\displaystyle\sum_{n=k}^{\infty} \binom{n}{k}\left(\frac{z}{2}\right)^n$ 9. $\displaystyle\sum_{n=0}^{\infty}\left[\binom{n+k}{n}\right]^{-1} z^{n+k}$ 10. $\displaystyle\sum_{n=0}^{\infty}\left[\binom{n+k}{n}\right]^{-1}\left(\frac{z}{3}\right)^n$

CHAPTER **15**

Taylor and Laurent Series

In the last section we have seen that power series with nonzero radius of convergence represent analytic functions. We shall now see that *every* analytic function can be represented by power series. These representations are called Taylor series and are quite similar to the familiar Taylor series of real functions. In fact, by replacing the real variable in the latter series by a complex variable we may *"continue"* real functions analytically to the complex domain.

The last part of the chapter is devoted to the representation of analytic functions by Laurent series. These are series involving positive and negative integral powers of the independent variable. They are also useful for evaluating complex and real integrals, as we shall see in the next chapter.

> *Prerequisites for this chapter:* Chaps. 11, 13, 14.
> *Sections which may be omitted in a shorter course:* 15.4, 15.6.
> *References:* Appendix 1, Part F.
> *Answers to problems:* Appendix 2.

15.1 *Taylor Series*

The familiar Taylor series is an effective tool in real calculus and its applications. We shall now see that in complex analysis there is a Taylor expansion which is a generalization of that series and is even more important.

Let us consider a function $f(z)$ which is analytic in a neighborhood of a point $z = a$. Let C be a circle which lies in this neighborhood and has the center a. Then we may apply Cauchy's integral formula (1), Sec. 13.5; writing z and z^* in place of z_0 and z, we have

(1)
$$f(z) = \frac{1}{2\pi i} \int_C \frac{f(z^*)}{z^* - z} \, dz^*$$

where z is an arbitrary fixed point inside C and z^* is the complex variable of integration (cf. Fig. 302). We shall now develop $1/(z^* - z)$ in (1) in powers of

Fig. 302. Representation (1)

$z - a$. We first have

(2)
$$\frac{1}{z^* - z} = \frac{1}{z^* - a - (z - a)} = \frac{1}{(z^* - a)\left(1 - \dfrac{z - a}{z^* - a}\right)}.$$

For later use we note that since z^* is on C while z is inside C,

(3)
$$\left|\frac{z - a}{z^* - a}\right| < 1.$$

From the geometric progression

$$1 + q + q^2 + \cdots + q^n = \frac{1 - q^{n+1}}{1 - q} \qquad (q \neq 1)$$

we obtain the relation

$$\frac{1}{1 - q} = 1 + q + \cdots + q^n + \frac{q^{n+1}}{1 - q}.$$

By setting $q = (z - a)/(z^* - a)$ we find

$$\frac{1}{1 - [(z - a)/(z^* - a)]} = 1 + \frac{z - a}{z^* - a} + \left(\frac{z - a}{z^* - a}\right)^2 + \cdots + \left(\frac{z - a}{z^* - a}\right)^n$$
$$+ \frac{[(z - a)/(z^* - a)]^{n+1}}{(z^* - z)/(z^* - a)}.$$

We insert this into (2), and then (2) into (1). Since z and a are constant, we may take the powers of $z - a$ out from under the integral sign, and (1) takes the form

(4)
$$f(z) = \frac{1}{2\pi i} \int_C \frac{f(z^*)}{z^* - a}\, dz^* + \frac{z - a}{2\pi i} \int_C \frac{f(z^*)}{(z^* - a)^2}\, dz^* + \cdots$$
$$+ \frac{(z - a)^n}{2\pi i} \int_C \frac{f(z^*)}{(z^* - a)^{n+1}}\, dz^* + R_n(z)$$

where the last term is given by the formula

(5)
$$R_n(z) = \frac{(z - a)^{n+1}}{2\pi i} \int_C \frac{f(z^*)}{(z^* - a)^{n+1}(z^* - z)}\, dz^*.$$

Using (1), Sec. 13.6, we may write this expansion in the form

(6) $$f(z) = f(a) + \frac{z - a}{1!} f'(a) + \frac{(z - a)^2}{2!} f''(a) + \cdots + \frac{(z - a)^n}{n!} f^{(n)}(a) + R_n(z).$$

This representation is called **Taylor's[1] formula**. $R_n(z)$ is called the *remainder*. Since the analytic function $f(z)$ has derivatives of all orders, we may take n in (6) as large as we please. If we let n approach infinity, we obtain from (6) the power series

(7)
$$f(z) = \sum_{m=0}^{\infty} \frac{f^{(m)}(a)}{m!} (z - a)^m.$$

This series is called the **Taylor series** of $f(z)$ *with center at a*. The particular case where $a = 0$ is called the **Maclaurin[2] series** of $f(z)$.

Clearly, the series (7) will converge and represent $f(z)$ if and only if

(8)
$$\lim_{n \to \infty} R_n(z) = 0.$$

To prove (8), we consider (5). Since z^* is on C while z is inside C, we have $|z^* - z| > 0$. Since $f(z)$ is analytic inside C and on C, it follows that the absolute value of $f(z^*)/(z^* - z)$ is bounded, say,

$$\left| \frac{f(z^*)}{z^* - z} \right| < \tilde{M}$$

for all z^* on C. Let r be the radius of C. Then $|z^* - a| = r$ for all z^* on C, and C has the length $2\pi r$. Hence, by applying (4) in Sec. 13.2 to (5) in the current section we obtain

$$|R_n| = \frac{|z - a|^{n+1}}{2\pi} \left| \int_C \frac{f(z^*)}{(z^* - a)^{n+1}(z^* - z)} \, dz^* \right|$$

$$< \frac{|z - a|^{n+1}}{2\pi} \tilde{M} \frac{1}{r^{n+1}} 2\pi r = \tilde{M} r \left| \frac{z - a}{r} \right|^{n+1}.$$

If we let n approach infinity, it follows from (3) that the expression on the right approaches zero. This proves (8) for all z inside C. Since, by Theorem 2 in Sec. 14.6, the representation of $f(z)$ in the form (7) is unique in the sense that (7) is the only power series with center at a which represents the given function $f(z)$, we may sum up our result as follows.

Taylor's theorem. *Let $f(z)$ be analytic in a domain D and let $z = a$ be any point in D. Then there exists precisely one power series with center at a which represents $f(z)$; this series is of the form*

(9)
$$f(z) = \sum_{n=0}^{\infty} b_n(z - a)^n$$

where

$$b_n = \frac{1}{n!} f^{(n)}(a) \qquad\qquad n = 0, 1, \cdots;$$

this representation is valid in the largest open disk with center a contained in D. The

---

[1] BROOK TAYLOR (1685–1731), English mathematician, who introduced this formula for functions of a real variable.
[2] COLIN MACLAURIN (1698–1746), Scots mathematician.

remainders $R_n(z)$ of (9) can be represented in the form (5). The coefficients satisfy the inequality

(10)
$$|b_n| \leqq \frac{M}{r^n}$$

where M is the maximum of $|f(z)|$ on the circle $|z - a| = r$.

Relation (10) follows from Cauchy's inequality (4) in Sec. 13.6.

Practically speaking, (8) means that for all z for which (9) converges, the nth partial sum of (9) will approximate $f(z)$ to any assigned degree of accuracy; we just have to choose n large enough.

From Taylor's theorem we see that the radius of convergence of (9) is at least equal to the shortest distance from a to the boundary of D. It may be larger, but then the series may no longer represent $f(z)$ at all points of D which lie in the interior of the circle of convergence.

One surprising property of complex analytic functions is that they have derivatives of all orders, and now we have discovered the other surprising property that they can always be represented by power series of the form (9). This is not true in general for real functions; there are real functions which have derivatives of all orders but cannot be represented by a power series. (Example: $f(x) = \exp(-1/x^2)$ when $x \neq 0$ and $f(0) = 0$. For details, see Ref. [A2, p. 128] in Appendix 1.)

The relation between our present consideration and that in Sec. 14.6 about power series may be established by the following theorem.

Theorem 2. *Every power series with a nonzero radius of convergence is the Taylor series of the function represented by that power series.*

Proof. Let the power series
$$\sum_{n=0}^{\infty} b_n(z - a)^n$$

have a nonzero radius of convergence R. Then it represents some analytic function $f(z)$ in the disk $|z - a| < R$, that is,
$$f(z) = b_0 + b_1(z - a) + b_2(z - a)^2 + \cdots.$$

From Theorem 5 in Sec. 14.6 it follows that
$$f'(z) = b_1 + 2b_2(z - a) + \cdots$$

and more generally
$$f^{(n)}(z) = n! \, b_n + (n + 1)n \cdots 3 \cdot 2 b_{n+1}(z - a) + \cdots;$$

all these series converge in the disk $|z - a| < R$ and represent analytic functions. Hence these functions are continuous at $z = a$. By setting $z = a$ we thus obtain
$$f(a) = b_0, \qquad f'(a) = b_1, \qquad \cdots, \qquad f^{(n)}(a) = n! \, b_n, \cdots.$$

Since these formulas are identical with those in Taylor's theorem, the proof is complete.

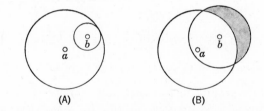

Fig. 303. Development around different centers and analytic continuation

A point at which a function $f(z)$ ceases to be analytic is called a **singular point** of $f(z)$; we also say that $f(z)$ has a **singularity** at such a point. More precisely: a point $z = z_0$ is called a *singular point* of $f(z)$, if $f(z)$ is not differentiable at z_0 but if every neighborhood of z_0 contains points at which $f(z)$ is differentiable.

Using this concept, we may say that there is at least one singular point of $f(z)$ on the circle of convergence[3] of the development (9).

Before we discuss examples and practical aspects of Taylor series, let us mention the concepts of **developments around different centers** and **analytic continuation**. Suppose that a power series of powers of $z - a$ with a nonzero radius of convergence R is given and denote its sum by $f(z)$; thus

$$(11) \qquad f(z) = \sum_{n=0}^{\infty} a_n(z - a)^n.$$

From Theorem 5 in Sec. 14.6 we know that $f(z)$ is analytic in the disk $|z - a| < R$. From Theorem 2 in this section we conclude that the series in (11) is the Taylor series of $f(z)$ with center a. We may now choose any point b in that disk and apply Taylor's theorem to obtain the development

$$(12) \qquad f(z) = \sum_{n=0}^{\infty} b_n(z - b)^n$$

with coefficients [cf. (11)]

$$b_n = \frac{1}{n!} f^{(n)}(b) = \sum_{k=n}^{\infty} \binom{n}{k} a_k(b - a)^k.$$

This new development is valid at least in the disk $|z - b| < R - |b|$ shown in Fig. 303A. However, in many cases the radius of convergence of (12) will be larger than $R - |b|$, as is illustrated in Fig. 303B, so that (12) yields an "*extension*" or "*continuation*" of $f(z)$ to points outside the disk $|z - a| < R$ (shaded in Fig. 303B). This process of extending an analytic function, at first given by a power series in a region of convergence, beyond that region is called **analytic continuation**.

---

[3] The radius of convergence of (9) will in general be equal to the distance from a to the nearest singular point of $f(z)$, but it may be larger; for example, Ln z is singular along the negative real axis, and the distance from $a = -1 + i$ to that axis is 1, but the Taylor series of Ln z with center $a = -1 + i$ has radius of convergence $\sqrt{2}$.

15.2　Taylor Series of Elementary Functions

Example 1 (Geometric series). Let $f(z) = 1/(1-z)$. Then we have $f^{(n)}(z) = n!/(1-z)^{n+1}$, $f^{(n)}(0) = n!$. Hence the Maclaurin expansion of $1/(1-z)$ is the geometric series

$$\text{(1)} \qquad \frac{1}{1-z} = \sum_{n=0}^{\infty} z^n = 1 + z + z^2 + \cdots \qquad (|z| < 1).$$

$f(z)$ is singular at $z = 1$; this point lies on the circle of convergence.

Example 2 (Exponential function). We know that the exponential function e^z (Sec. 11.5) is analytic for all z, and $(e^z)' = e^z$. Hence from (9), Sec. 15.1, with $a = 0$, we obtain the Maclaurin series

$$\text{(2)} \qquad e^z = \sum_{n=0}^{\infty} \frac{z^n}{n!} = 1 + z + \frac{z^2}{2!} + \cdots.$$

This series is also obtained if we replace x in the Maclaurin series of e^x by z.

Let us show how we can prove the multiplication formula

$$\text{(3)} \qquad e^{z_1} e^{z_2} = e^{z_1 + z_2}$$

by using (2). We have

$$e^{z_1} e^{z_2} = \sum_{k=0}^{\infty} \frac{z_1^{k}}{k!} \sum_{m=0}^{\infty} \frac{z_2^{m}}{m!}.$$

Since both series converge absolutely, we may multiply them term by term; the sum of the products for which $k + m = n$ is

$$\frac{z_1^{n}}{n!} + \frac{z_1^{n-1}}{(n-1)!} \frac{z_2}{1!} + \cdots + \frac{z_1}{1!} \frac{z_2^{n-1}}{(n-1)!} + \frac{z_2^{n}}{n!}$$

$$= \frac{1}{n!} \left[z_1^{n} + \binom{n}{1} z_1^{n-1} z_2 + \binom{n}{2} z_1^{n-2} z_2^2 + \cdots + z_2^{n} \right] = \frac{(z_1 + z_2)^n}{n!}.$$

Hence the product of the two series may be written

$$\sum_{n=0}^{\infty} \frac{(z_1 + z_2)^n}{n!} = e^{z_1 + z_2},$$

and (3) is proved.

Furthermore, by setting $z = iy$ in (2) and applying Theorem 1 in Sec. 14.4 we obtain

$$e^{iy} = \sum_{n=0}^{\infty} \frac{(iy)^n}{n!} = \sum_{k=0}^{\infty} (-1)^k \frac{y^{2k}}{(2k)!} + i \sum_{k=0}^{\infty} (-1)^k \frac{y^{2k+1}}{(2k+1)!}.$$

Since the series on the right are the familiar Maclaurin expansions of the real functions $\cos y$ and $\sin y$, this represents the Euler formula

$$\text{(4)} \qquad e^{iy} = \cos y + i \sin y;$$

cf. (8) in Sec. 11.5. Multiplying by e^x and using (3), we obtain the formula (4) in Sec. 11.5 which was used to define e^z. Our present consideration shows that one may use (2) for *defining* e^z and derive from (2) all the formulas in Sec. 11.5.

Example 3 (Trigonometric and hyperbolic functions). By substituting (2) into (1) of Sec. 11.6 we obtain

(5)
$$\cos z = \sum_{n=0}^{\infty} (-1)^n \frac{z^{2n}}{(2n)!} = 1 - \frac{z^2}{2!} + \frac{z^4}{4!} - + \cdots$$

$$\sin z = \sum_{n=0}^{\infty} (-1)^n \frac{z^{2n+1}}{(2n+1)!} = z - \frac{z^3}{3!} + \frac{z^5}{5!} - + \cdots.$$

When $z = x$ these are the familiar Maclaurin series of the real functions $\cos x$ and $\sin x$. Similarly, by substituting (2) into (11), Sec. 11.6, we obtain

(6)
$$\cosh z = \sum_{n=0}^{\infty} \frac{z^{2n}}{(2n)!} = 1 + \frac{z^2}{2!} + \frac{z^4}{4!} + \cdots$$

$$\sinh z = \sum_{n=0}^{\infty} \frac{z^{2n+1}}{(2n+1)!} = z + \frac{z^3}{3!} + \frac{z^5}{5!} + \cdots.$$

Example 4 (Logarithm). From (9) in Sec. 15.1 it follows that

(7)
$$\text{Ln}\,(1 + z) = z - \frac{z^2}{2} + \frac{z^3}{3} - + \cdots \qquad (|z| < 1).$$

Replacing z by $-z$ and multiplying both sides by -1, we get

(8)
$$-\text{Ln}\,(1 - z) = \text{Ln}\,\frac{1}{1 - z} = z + \frac{z^2}{2} + \frac{z^3}{3} + \cdots \qquad (|z| < 1).$$

By adding both series we obtain

(9)
$$\text{Ln}\,\frac{1 + z}{1 - z} = 2\left(z + \frac{z^3}{3} + \frac{z^5}{5} + \cdots\right) \qquad (|z| < 1).$$

Problems

1. Using (2), prove $(e^z)' = e^z$.
2. Derive (5) and (6) from (2). Obtain (7) from Taylor's theorem.
3. Using (5), show that $\cos z$ is even and $\sin z$ is odd.
4. Using (6), show that $\cosh z \neq 0$ for all real $z = x$.

Find the Taylor series of the following functions about the point $z = a$ and determine the radius of convergence.

5. e^{-z}, $a = 0$ 6. $1/z$, $a = 1$ 7. e^{2z}, $a = 2i$
8. $\cos z$, $a = -\pi/2$ 9. $\sin z$, $a = \pi/2$ 10. $1/(z + 2)$, $a = 1 + i$
11. $\sin \pi z$, $a = 0$ 12. $\cos z^2$, $a = 0$ 13. $\cos^2 z$, $a = 0$
14. $1/(1 - z)$, $a = -1$ 15. $1/(1 - z)$, $a = i$ 16. $\text{Ln}\,z$, $a = 1$

Find the first three terms of the Maclaurin series of the following functions.

17. $\tan z$ 18. $z \cot z$ 19. $e^z \sin z$

Find the Maclaurin series by integrating that of the integrand term by term. (erf z is called the **error function**, Si(z) the **sine integral**, and S(z) and C(z) the **Fresnel integrals**. Cf. also Sec. 18.14 and Appendix 3.)

20. $\displaystyle\int_0^z \frac{e^t - 1}{t}\, dt$ 21. $\displaystyle\int_0^z \frac{1 - \cos t}{t^2}\, dt$

22. $\text{erf}\,z = \dfrac{2}{\sqrt{\pi}} \displaystyle\int_0^z e^{-t^2}\, dt$ 23. $\text{Si}(z) = \displaystyle\int_0^z \frac{\sin t}{t}\, dt$

24. $S(z) = \displaystyle\int_0^z \sin t^2 \, dt$ **25.** $C(z) = \displaystyle\int_0^z \cos t^2 \, dt$

15.3 *Practical Methods for Obtaining Power Series*

In most practical cases the determination of the coefficients of a Taylor series by means of the formula in Taylor's theorem will be complicated or time-consuming. There are a number of simpler practical procedures for that purpose which may be illustrated by the following examples. The uniqueness of the representations thus obtained follows from Theorem 2 in Sec. 14.6.

Example 1 (Substitution). Find the Maclaurin series of $f(z) = 1/(1 + z^2)$. By substituting $-z^2$ for z in (1), Sec. 15.2, we obtain

(1)
$$\frac{1}{1 + z^2} = \frac{1}{1 - (-z^2)} = \sum_{n=0}^{\infty} (-z^2)^n = \sum_{n=0}^{\infty} (-1)^n z^{2n}$$

$$= 1 - z^2 + z^4 - z^6 + \cdots \qquad (|z| < 1).$$

Example 2 (Integration). Let $f(z) = \tan^{-1} z$. We have $f'(z) = 1/(1 + z^2)$. By integrating (1) term by term and noting that $f(0) = 0$ we find

$$\tan^{-1} z = \sum_{n=0}^{\infty} \frac{(-1)^n}{2n + 1} z^{2n+1} = z - \frac{z^3}{3} + \frac{z^5}{5} - + \cdots \qquad (|z| < 1);$$

this series represents the principal value of $w = u + iv = \tan^{-1} z$, defined as that value for which $|u| < \pi/2$.

Example 3 (Development by using the geometric series). Develop $1/(c - bz)$ in powers of $z - a$ where $c - ab \neq 0$ and $b \neq 0$. Obviously,

$$\frac{1}{c - bz} = \frac{1}{c - ab - b(z - a)} = \frac{1}{(c - ab)\left[1 - \dfrac{b(z - a)}{c - ab}\right]}$$

To the last expression we apply (1) in Sec. 15.2 with z replaced by $b(z - a)/(c - ab)$, finding

$$\frac{1}{c - bz} = \frac{1}{c - ab} \sum_{n=0}^{\infty} \left[\frac{b(z - a)}{c - ab}\right]^n = \sum_{n=0}^{\infty} \frac{b^n}{(c - ab)^{n+1}} (z - a)^n$$

$$= \frac{1}{c - ab} + \frac{b}{(c - ab)^2} (z - a) + \frac{b^2}{(c - ab)^3} (z - a)^2 + \cdots.$$

This series converges for

$$\left|\frac{b(z - a)}{c - ab}\right| < 1 \qquad \text{that is,} \qquad |z - a| < \left|\frac{c - ab}{b}\right| = \left|\frac{c}{b} - a\right|.$$

Example 4 (Binomial series, reduction by partial fractions). Find the Taylor series of the function

$$f(z) = \frac{2z^2 + 9z + 5}{z^3 + z^2 - 8z - 12}$$

with center at $z = 1$.

Given a rational function, we may first represent it as a sum of partial fractions and then apply the *binomial series*

$$\frac{1}{(1+z)^m} = (1+z)^{-m} = \sum_{n=0}^{\infty} \binom{-m}{n} z^n$$

(2)

$$= 1 - mz + \frac{m(m+1)}{2!} z^2 - \frac{m(m+1)(m+2)}{3!} z^3 + \cdots.$$

Since the function on the left is singular at $z = -1$, the series converges in the disk $|z| < 1$. In our case we obtain

$$f(z) = \frac{1}{(z+2)^2} + \frac{2}{z-3} = \frac{1}{[3+(z-1)]^2} - \frac{2}{2-(z-1)}.$$

This may be written in the form

$$f(z) = \frac{1}{9}\left(\frac{1}{[1+\frac{1}{3}(z-1)]^2}\right) - \frac{1}{1-\frac{1}{2}(z-1)}.$$

By using the binomial series we obtain

$$f(z) = \frac{1}{9} \sum_{n=0}^{\infty} \binom{-2}{n}\left(\frac{z-1}{3}\right)^n - \sum_{n=0}^{\infty} \left(\frac{z-1}{2}\right)^n.$$

We may add the two series on the right term by term. Since the binomial coefficient in the first series equals $(-1)^n(n+1)$, we find

$$f(z) = \sum_{n=0}^{\infty} \left[\frac{(-1)^n(n+1)}{3^{n+2}} - \frac{1}{2^n}\right] (z-1)^n = -\frac{8}{9} - \frac{31}{54}(z-1) - \frac{23}{108}(z-1)^2 - \cdots.$$

Since $z = 3$ is the singular point of $f(z)$ which is nearest to the center $z = 1$, the series converges in the disk $|z - 1| < 2$.

Example 5 (*Use of differential equations*). Find the Maclaurin series of $f(z) = \tan z$. We have $f'(z) = \sec^2 z$ and, therefore,

$$f'(z) = 1 + f^2(z), \qquad f'(0) = 1.$$

Observing that $f(0) = 0$, we obtain by successive differentiation

$$f'' = 2ff', \qquad\qquad f''(0) = 0,$$
$$f''' = 2f'^2 + 2ff'', \qquad f'''(0) = 2, \qquad f'''(0)/3! = 1/3,$$
$$f^{(4)} = 6f'f'' + 2ff''', \qquad f^{(4)}(0) = 0,$$
$$f^{(5)} = 6f''^2 + 8f'f''' + 2ff^{(4)}, \qquad f^{(5)}(0) = 16, \qquad f^{(5)}(0)/5! = 2/15, \qquad\qquad \text{etc.}$$

Hence the result is

(3)
$$\tan z = z + \frac{1}{3}z^3 + \frac{2}{15}z^5 + \frac{17}{315}z^7 + \cdots \qquad \left(|z| < \frac{\pi}{2}\right).$$

Example 6 (*Undetermined coefficients*). Find the Maclaurin series of $\tan z$ by using those of $\cos z$ and $\sin z$ (Sec. 15.2). Since $\tan z$ is odd, the desired expansion will be of the form

$$\tan z = b_1 z + b_3 z^3 + b_5 z^5 + \cdots.$$

Using $\sin z = \tan z \cos z$ and inserting those developments, we obtain

$$z - \frac{z^3}{3!} + \frac{z^5}{5!} - + \cdots = (b_1 z + b_3 z^3 + b_5 z^5 + \cdots)\left(1 - \frac{z^2}{2!} + \frac{z^4}{4!} - + \cdots\right).$$

Since $\tan z$ is analytic except at $z = \pm\pi/2, \pm 3\pi/2, \cdots$, its Maclaurin series converges in the disk $|z| < \pi/2$, and for these z we may multiply the two series on the right term by term and arrange the resulting series in powers of z (cf. Theorem 3 in Sec. 14.5). By Theorem 2 in Sec. 14.6 the coefficient of each power of z is the same on both sides. This yields

$$1 = b_1, \qquad -\frac{1}{3!} = -\frac{b_1}{2!} + b_3, \qquad \frac{1}{5!} = \frac{b_1}{4!} - \frac{b_3}{2!} + b_5, \qquad \text{etc.}$$

Therefore $b_1 = 1, b_3 = \frac{1}{3}, b_5 = \frac{2}{15}$, etc., as before.

Problems

Find the Maclaurin series of the following functions.

1. $\dfrac{1}{1 + z^4}$

2. $\dfrac{1}{1 - z^5}$

3. $\dfrac{z + 2}{1 - z^2}$

4. $\dfrac{z^2}{(1 - z^3)^2}$

5. $\dfrac{4 - 3z}{(1 - z)^2}$

6. $\dfrac{4z^2 + 30z + 68}{(z + 4)^2(z - 2)}$

7. $\sin 2z^2$

8. $e^{z^2} \displaystyle\int_0^z e^{-t^2}\, dt$

9. $e^{z^2 - z}$

Find the first few terms of the Maclaurin series of the following functions.

10. $\dfrac{\cos z}{1 - z^2}$

11. $e^{z^2}\sin z^2$

12. $e^{z^2}/\cos z$

13. $e^{1/(1-z)}$

14. $\cos\left(\dfrac{z}{1 - z}\right)$

15. $e^{(e^z)}$

16. Obtain the series in Ex. 2 from Prob. 33 in Sec. 11.7 and (9) in Sec. 15.2.

17. Developing $1/\sqrt{1 - z^2}$ and integrating, show that

$$\sin^{-1} z = z + \left(\frac{1}{2}\right)\frac{z^3}{3} + \left(\frac{1 \cdot 3}{2 \cdot 4}\right)\frac{z^5}{5} + \left(\frac{1 \cdot 3 \cdot 5}{2 \cdot 4 \cdot 6}\right)\frac{z^7}{7} + \cdots \qquad (|z| < 1).$$

18. Show that the series in Prob. 17 represents the principal value of $\sin^{-1} z$ (cf. Prob. 35 in Sec. 11.7).

19. Find a Maclaurin series for which the corresponding function has more than one singularity on the circle of convergence.

Find the Taylor series of the given function about $z = a$.

20. $\dfrac{1}{1 - z}$, $\quad a = 3i$

21. $\tan z$, $\quad a = \dfrac{\pi}{4}$

22. $\dfrac{2 - 3z}{2z^2 - 3z + 1}$, $\quad a = -1$

23. (Euler numbers) The Maclaurin series

$$(4) \qquad \sec z = E_0 - \frac{E_2}{2!}z^2 + \frac{E_4}{4!}z^4 - + \cdots$$

defines the *Euler numbers* E_{2n}. Show that $E_0 = 1, E_2 = -1, E_4 = 5, E_6 = -61$.

24. (Bernoulli numbers) The Maclaurin series

$$(5) \qquad \frac{z}{e^z - 1} = 1 + B_1 z + \frac{B_2}{2!}z^2 + \frac{B_3}{3!}z^3 + \cdots$$

defines the *Bernoulli numbers* B_n. Using the method of undetermined coefficients, show that[4]

(6) $$B_1 = -\frac{1}{2}, \quad B_2 = \frac{1}{6}, \quad B_3 = 0, \quad B_4 = -\frac{1}{30}, \quad B_5 = 0, \quad B_6 = \frac{1}{42}, \cdots.$$

25. Using (1), (2), Sec. 11.6, and (5), show that

(7) $$\tan z = \frac{2i}{e^{2iz} - 1} - \frac{4i}{e^{4iz} - 1} - i = \sum_{n=1}^{\infty} (-1)^{n-1} \frac{2^{2n}(2^{2n} - 1)}{(2n)!} B_{2n} z^{2n-1}.$$

15.4   Uniform Convergence

Suppose we know that a given series converges in a certain region R. Then the remaining question is whether the convergence is sufficiently rapid throughout the whole region or whether there are points near which the convergence becomes poor. The practical importance of this question in connection with computations is obvious, but we shall see that the theoretical aspect of the question is even more important. To illustrate the situation, let us start with some examples.

Example 1.   Suppose we want to compute a table of e^x for real x in the interval $0 \leq x \leq 1$, for example, for $x = 0, 0.1, 0.2, \cdots$ and the absolute value of the error of each value should be less than a given number ϵ, say, less than $\frac{1}{2}$ unit of the sixth decimal place. We may use a suitable partial sum

$$s_n = 1 + x + \cdots + \frac{x^n}{n!}$$

of the Maclaurin series. Then the absolute value of the error is equal to $|R_n| = |s - s_n|$ where $s = e^x$, the sum of the series, and we have to choose n such that

$$|s(x) - s_n(x)| < \epsilon \ (= 5 \cdot 10^{-7}).$$

From Prob. 24 in Sec. 14.3 it follows that when $x = 1$, then for $n = 10$, and thus for any $n > N = 9$, we obtain the required accuracy. Now the remainder decreases in absolute value as $x \ (\geq 0)$ decreases, and, therefore,

$$|s(x) - s_n(x)| < \epsilon \qquad \text{for} \qquad n > N(\epsilon) \ (= 9) \qquad \text{and all } x$$

under consideration. Note that, of course, N depends on ϵ, and if we want more accurate values so that ϵ is smaller, then N will be larger.

Example 2.   In the case of the geometric series $1 + z + z^2 + \cdots$ the remainder is

$$R_n(z) = s(z) - s_n(z) = \sum_{m=n+1}^{\infty} z^m = \frac{z^{n+1}}{1 - z}$$

and becomes arbitrarily large for real $z = x \ (<1)$ and sufficiently close to 1. Hence, a maximum error ϵ being prescribed, we cannot find an N *depending only on* ϵ such that we have $|R_n(x)| = |s(x) - s_n(x)| < \epsilon$ for $n > N$ and *all* x in the interval $0 \leq x < 1$. (Cf. also Prob. 23 at the end of Sec. 14.3.) This result is not quite unexpected because the series diverges at $z = 1$. A really surprising situation will be discussed in the next example.

Example 3.   Consider the series

$$x^2 + \frac{x^2}{1 + x^2} + \frac{x^2}{(1 + x^2)^2} + \frac{x^2}{(1 + x^2)^3} + \cdots.$$

[4] For tables, see J. W. L. Glaisher, *Quarterly Journal Math.* **29**, 1898, 1–168.

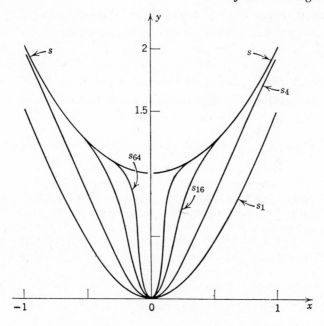

Fig. 304. Partial sums and sum in Example 3

Using the formula for the sum of a geometric progression, the student may readily verify that the nth partial sum is

$$s_n(x) = 1 + x^2 - \frac{1}{(1 + x^2)^n}.$$

(Some of these sums are shown in Fig. 304.) Hence, if $x \neq 0$, the series has the sum

$$s(x) = \lim_{n \to \infty} s_n(x) = 1 + x^2.$$

If $x = 0$, then $s_n = 0$ for all n and, therefore,

$$s(0) = \lim_{n \to \infty} s_n(0) = 0.$$

This shows that the series converges for all x (even absolutely), but we have the surprising result that the sum is discontinuous (at $x = 0$), although all the terms of the series are continuous functions. Furthermore, when $x \neq 0$ the absolute value of the remainder is

$$|R_n(x)| = |s(x) - s_n(x)| = \frac{1}{(1 + x^2)^n}$$

and we see that for a given ϵ (<1) we cannot find an N depending only on ϵ such that $|R_n| < \epsilon$ for all $n > N(\epsilon)$ and all x in the interval $0 \leq x \leq 1$.

The series in the examples are of the form

(1) $$\sum_{n=0}^{\infty} f_n(z) = f_0(z) + f_1(z) + f_2(z) + \cdots.$$

We assume that (1) converges for all z in a region G. Let $s(z)$ be the sum and

$s_n(z)$ the nth partial sum of (1). We know that convergence of (1) at a point z means that, an $\epsilon > 0$ being given, we can find an $N = N(\epsilon, z)$ such that

$$|s(z) - s_n(z)| < \epsilon \qquad \text{for all } n > N(\epsilon, z).$$

N depends on ϵ and will, in general, also depend on the point z which has been selected for consideration. Now, any $\epsilon > 0$ being given, it may happen that we can find a number $N(\epsilon)$, *independent of z*, such that

$$|s(z) - s_n(z)| < \epsilon \qquad \text{for all } n > N(\epsilon) \textbf{ and all } z \textbf{ in } G.$$

Then the series is said to be **uniformly convergent** *in G*.

Uniformity of convergence is thus a property that refers to an infinite set of values of z, whereas the convergence of a series may be considered for various particular values of z, without reference to other values.

The series in Ex. 1 is uniformly convergent in the interval $0 \leq x \leq 1$ (and, in fact, in any bounded interval). The series in Ex. 3 is not uniformly convergent in any interval containing the point 0. This shows that *an absolutely convergent series may not be uniformly convergent.* Conversely, *a uniformly convergent series may not be absolutely convergent.* This is illustrated by

Example 4. The series

$$\sum_{n=1}^{\infty} \frac{(-1)^{n-1}}{x^2+n} = \frac{1}{x^2+1} - \frac{1}{x^2+2} + \frac{1}{x^2+3} - + \cdots \qquad (x \text{ real}) \qquad \text{(x real)}$$

is uniformly convergent for all real x, but not absolutely convergent (cf. Prob. 16).

Example 2 is typical of power series, because for such a series the situation is very simple, as follows.

Theorem 1 (Power series).   *A power series*

(2)
$$\sum_{n=0}^{\infty} c_n (z - a)^n$$

with a nonzero radius of convergence R is uniformly convergent in every circular disk $|z - a| \leq r$ *of radius* $r < R$.

Proof.   For $|z - a| \leq r$ we have

(3) $|c_{n+1}(z - a)^{n+1} + \cdots + c_{n+p}(z - a)^{n+p}| \leq |c_{n+1}| r^{n+1} + \cdots + |c_{n+p}| r^{n+p}.$

Since (2) converges absolutely when $|z - a| = r < R$, it follows from Theorem 4 in Sec. 14.2 that, an $\epsilon > 0$ being given, we can find an $N(\epsilon)$ such that

$$|c_{n+1}| r^{n+1} + \cdots + |c_{n+p}| r^{n+p} < \epsilon \quad \text{for} \quad n > N(\epsilon) \quad \text{and} \quad p = 1, 2, \cdots.$$

From this and (3) we obtain

$$|c_{n+1}(z - a)^{n+1} + \cdots + c_{n+p}(z - a)^{n+p}| < \epsilon$$

for all z in the disk $|z - a| \leq r$, every $n > N(\epsilon)$, and every $p = 1, 2, \cdots$. Since $N(\epsilon)$ is independent of z, this shows uniform convergence, and the theorem is proved.

While, of course, the sum of finitely many continuous functions is continuous, Example 3 illustrates that the sum of an infinite series of continuous functions may be discontinuous, even if it converges absolutely. But if a series converges *uniformly*, this cannot happen. In fact the following important theorem holds.

Theorem 2 (Continuity). *Let the series*

$$\sum_{m=0}^{\infty} f_m(z) = f_0(z) + f_1(z) + \cdots$$

be uniformly convergent in a region G and let $F(z)$ be its sum. Then, if each term $f_m(z)$ is continuous at a point z_0 in G, the function $F(z)$ is continuous at z_0.

Proof. Let $s_n(z)$ be the nth partial sum of the series and $R_n(z)$ the corresponding remainder:

$$s_n = f_0 + f_1 + \cdots + f_n, \qquad R_n = f_{n+1} + f_{n+2} + \cdots.$$

Since the series converges uniformly, for a given $\epsilon > 0$ we can find an $n = N(\epsilon)$ such that

$$|R_N(z)| < \frac{\epsilon}{3} \qquad \qquad \text{for all } z \text{ in } G.$$

Since $s_N(z)$ is a sum of finitely many functions which are continuous at z_0, this sum is continuous at z_0. Therefore we can find a $\delta > 0$ such that

$$|s_N(z) - s_N(z_0)| < \frac{\epsilon}{3} \qquad \text{for all } z \text{ in } G \text{ for which } |z - z_0| < \delta.$$

By the triangle inequality (Sec. 11.1) for these z we thus obtain

$$|F(z) - F(z_0)| = |s_N(z) + R_N(z) - [s_N(z_0) + R_N(z_0)]|$$
$$\leq |s_N(z) - s_N(z_0)| + |R_N(z)| + |R_N(z_0)| < \frac{\epsilon}{3} + \frac{\epsilon}{3} + \frac{\epsilon}{3} = \epsilon.$$

This implies that $F(z)$ is continuous at z_0, and the theorem is proved.

We want to mention that in this theorem uniformity of convergence is a sufficient rather than a necessary condition. This may be illustrated by the following example.

Example 5. Let

$$u_m(x) = \frac{mx}{1 + m^2 x^2}$$

and consider the series

$$\sum_{m=1}^{\infty} f_m(x) \qquad \text{where} \qquad f_m(x) = u_m(x) - u_{m-1}(x).$$

The nth partial sum is

$$s_n = u_1 - u_0 + u_2 - u_1 + \cdots + u_n - u_{n-1} = u_n - u_0 = u_n.$$

Hence the series has the sum

$$F(x) = \lim_{n \to \infty} s_n(x) = \lim_{n \to \infty} u_n(x) = 0,$$

which is a continuous function. However, the series is not uniformly convergent in an interval $0 \leq x \leq a$, where $a > 0$. In fact, from

$$\left| F(x) - s_n(x) \right| = \frac{nx}{1 + n^2 x^2} < \epsilon$$

we obtain

$$\frac{nx}{\epsilon} < 1 + n^2 x^2 \quad \text{or} \quad n^2 x^2 - \frac{nx}{\epsilon} + 1 > 0$$

and from this

$$n > \frac{1}{2x\epsilon} (1 + \sqrt{1 - 4\epsilon^2}).$$

For fixed ϵ the right side approaches infinity as x appoaches zero, which shows that the series is not uniformly convergent in that interval.

Under what conditions may we integrate a series term by term?

Let us start our discussion with an example which illustrates the important fact that term by term integration of series is not always permissible.

Example 6. Let

$$u_m(x) = mxe^{-mx^2}$$

and consider the series

$$\sum_{m=1}^{\infty} f_m(x) \quad \text{where} \quad f_m(x) = u_m(x) - u_{m-1}(x)$$

in the interval $0 \leq x \leq 1$. The nth partial sum is

$$s_n = u_1 - u_0 + u_2 - u_1 + \cdots + u_n - u_{n-1} = u_n - u_0 = u_n.$$

Hence the series has the sum

$$F(x) = \lim_{n \to \infty} s_n(x) = \lim_{n \to \infty} u_n(x) = 0 \qquad (0 \leq x \leq 1).$$

From this we obtain

$$\int_0^1 F(x) \, dx = 0.$$

On the other hand, by integrating term by term,

$$\sum_{m=1}^{\infty} \int_0^1 f_m(x) \, dx = \lim_{n \to \infty} \sum_{m=1}^{n} \int_0^1 f_m(x) \, dx = \lim_{n \to \infty} \int_0^1 s_n(x) \, dx.$$

Now $s_n = u_n$ and the expression on the right becomes

$$\lim_{n \to \infty} \int_0^1 u_n(x) \, dx = \lim_{n \to \infty} \int_0^1 nxe^{-nx^2} \, dx = \lim_{n \to \infty} \tfrac{1}{2}(1 - e^{-n}) = \tfrac{1}{2},$$

but not 0. This shows that the series under consideration cannot be integrated term by term from $x = 0$ to $x = 1$.

The series in Ex. 6 is not uniformly convergent in that interval, and we shall now prove that in the case of a uniformly convergent series of continuous functions we may integrate term by term.

Theorem 3 (Termwise integration). *Let*

$$F(z) = \sum_{n=0}^{\infty} f_n(z) = f_0(z) + f_1(z) + \cdots$$

be a uniformly convergent series of continuous functions within a region G. Let C be any path in G. Then the series

(4)
$$\sum_{n=0}^{\infty} \int_C f_n(z)\, dz = \int_C f_0(z)\, dz + \int_C f_1(z)\, dz + \cdots$$

is convergent and has the sum $\int_C F(z)\, dz.$

Proof. From Theorem 2 it follows that $F(z)$ is continuous. Let $s_n(z)$ be the nth partial sum of the given series and $R_n(z)$ the corresponding remainder. Then $F = s_n + R_n$ and

$$\int_C F(z)\, dz = \int_C s_n(z)\, dz + \int_C R_n(z)\, dz.$$

Let l be the length of C. Since the given series converges uniformly, to every given $\epsilon > 0$ we can find a number N such that

$$|R_n(z)| < \frac{\epsilon}{l} \qquad \text{for all } n > N \text{ and all } z \text{ in } G.$$

By applying (4), Sec. 13.2, we thus obtain

$$\left| \int_C R_n(z)\, dz \right| < \frac{\epsilon}{l} l = \epsilon \qquad \text{for all } n > N.$$

Since $R_n = F - s_n$, this means that

$$\left| \int_C F(z)\, dz - \int_C s_n(z)\, dz \right| < \epsilon \qquad \text{for all } n > N.$$

Hence, the series (4) converges and has the sum indicated in the theorem. This completes the proof.

Theorem 2 and Theorem 3 characterize the two most important properties of uniformly convergent series.

Of course, since differentiation and integration are inverse processes, we readily conclude from Theorem 3 that a convergent series may be differentiated term by term, provided the terms of the given series have continuous derivatives and the resulting series is uniformly convergent; more precisely, the following theorem holds.

Theorem 4 (Termwise differentiation). *Let the series $f_0(z) + f_1(z) + f_2(z) + \cdots$ be convergent in a region G and let $F(z)$ be its sum. Suppose that the series $f_0'(z) + f_1'(z) + f_2'(z) + \cdots$ converges uniformly in G and its terms $f_0'(z)$, $f_1'(z), \cdots$ are continuous in G. Then*

$$F'(z) = f_0'(z) + f_1'(z) + f_2'(z) + \cdots \qquad \textit{for all } z \textit{ in } G.$$

The simple proof is left to the student (Prob. 2).

Usually uniform convergence is established by a comparison test, which is called the Weierstrass M-test.

Theorem 5 (Weierstrass M-test). *If, for all values of z within a region G, the absolute values of the terms of a given series of the form (1) are, respectively, less than or equal to the corresponding terms in a convergent series of constant terms,*

(5)
$$M_0 + M_1 + M_2 + \cdots,$$

then the series (1) converges uniformly in G.

The simple proof is left to the student (Prob. 3).

Example 7. The series

$$\sum_{m=1}^{\infty} \frac{\sin mx}{m^2} \qquad (x \text{ real})$$

converges uniformly in any interval. This follows from the Weierstrass test, because for real x,

$$\left| \frac{\sin mx}{m^2} \right| \leq \frac{1}{m^2}$$

and $\sum m^{-2}$ is convergent [cf. (5) in Sec. 14.3].

Problems

1. If the series (1) converges uniformly in a region G, show that it converges uniformly in any portion of G.
2. Derive Theorem 4 from Theorem 3.
3. Give a proof of the Weierstrass M-test (Theorem 5).

Prove that the following series converge uniformly in the given regions.

4. $\sum_{n=0}^{\infty} z^n, \quad |z| \leq 0.99$

5. $\sum_{n=0}^{\infty} \frac{z^n}{n!}, \quad |z| \leq 10^{30}$

6. $\sum_{n=1}^{\infty} \frac{(n!)^2}{(2n)!} z^n, \quad |z| \leq 3.9$

7. $\sum_{n=1}^{\infty} \frac{z^n}{n^2}, \quad |z| \leq 1$

8. $\sum_{n=1}^{\infty} \frac{\sin n|z|}{2^n}, \quad \text{all } z$

9. $\sum_{n=1}^{\infty} \frac{\tanh^n x}{n(n+1)}, \quad \text{all real } x$

10. Determine the smallest integer n such that $|R_n| < 0.01$ in Ex. 2, when $x = 0.5, 0.6, 0.7, 0.8, 0.9$. What does the result mean from the viewpoint of computing $1/(1 - x)$ with an absolute error less than 0.01 by means of the geometric series?
11. Find the series whose nth partial sum is $s_n(x) = nx/(nx + 1)$ and graph s_1, s_2, s_3, s_4, and $s = \lim_{n \to \infty} s_n$ for $x \geq 0$.
12. Prove that the series in Ex. 3 is not uniformly convergent in any interval containing the point $x = 0$.
13. Show that $x^2 \sum_{n=1}^{\infty} (1 + x^2)^{-n} = 1 \ (x \neq 0)$ and 0 when $x = 0$ and sketch a figure of the partial sums similar to Fig. 304.
14. Find the precise region of convergence of the series in Ex. 3 with x replaced by a complex variable z.

15. Show that $1 + \sum_{n=1}^{\infty} (x^n - x^{n-1})$ is not uniformly convergent in the interval $0 \leq x \leq 1$. Graph the partial sums s_1, s_2, s_3, s_4.

16. Prove the statement in Ex. 4.

17. Show that $s_n(x)$ in Ex. 5 has a maximum at $x = 1/n$ and $s_n(1/n) = 1/2$. Graph s_1, s_2, s_3, s_4 and $s = \lim\limits_{n \to \infty} s_n$.

18. Show that $\displaystyle\sum_{n=1}^{\infty} \left(\frac{x^{2n}}{1 + x^{2n}} - \frac{x^{2n-2}}{1 + x^{2n-2}} \right) = \begin{cases} -1 & \text{when } |x| < 1 \\ -\frac{1}{2} & \text{when } x = \pm 1 \\ 0 & \text{when } |x| > 1 \end{cases}$

Heat equation. Show that (10), Sec. 10.5, with coefficients (11) is a solution of the heat equation for $t > 0$, assuming that $f(x)$ is continuous on the interval $0 \leq x \leq l$ and has one-sided derivatives at all interior points of that interval. Proceed as follows.

19. Show that $|B_n|$ is bounded, say, $|B_n| < K$ for all n. Conclude that

$$|u_n| < K e^{-\lambda_n^2 t_0} \qquad \text{when} \qquad t \geq t_0 > 0$$

and, by the Weierstrass test, the series (10) converges uniformly with respect to x and t when $t \geq t_0$, $0 \leq x \leq l$. Using Theorem 2, show that $u(x, t)$ is continuous when $t \geq t_0$ and thus satisfies the boundary conditions (2) when $t \geq t_0$.

20. Show that $|\partial u_n/\partial t| < \lambda_n^2 K e^{-\lambda_n^2 t_0}$ when $t \geq t_0$ and the series of the expressions on the right converges, by the ratio test. Conclude from this, the Weierstrass test, and Theorem 4 that the series (10) can be differentiated term by term with respect to t and the resulting series has the sum $\partial u/\partial t$. Show that (10) can be differentiated twice with respect to x and the resulting series has the sum $\partial^2 u/\partial x^2$. Conclude from this and the result of Prob. 19 that (10) is a solution of the heat equation for all $t \geq t_0$. (The proof that (10) satisfies the given initial condition can be found in Ref. [D3].)

15.5 *Laurent Series*

In various applications it is necessary to expand a function $f(z)$ around points where $f(z)$ is singular. Taylor's theorem cannot be applied in such cases. A new type of series, known as *Laurent series*, is necessary. This will be a representation which is valid in an annulus bounded by two concentric circles C_1 and C_2 such that $f(z)$ is analytic in the annulus and at each point of C_1 and C_2 (Fig. 305). As in the case of the Taylor series, $f(z)$ may be singular at some points outside C_1 and, as the essentially new feature, it may also be singular at some points inside C_2.

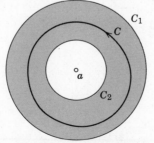

Fig. 305. Laurent's theorem

Laurent's theorem.[5] *If $f(z)$ is analytic on two concentric circles C_1 and C_2 with center a and in the annulus between them, then $f(z)$ can be represented by the Laurent series*

(1)
$$f(z) = \sum_{n=0}^{\infty} b_n(z-a)^n + \sum_{n=1}^{\infty} \frac{c_n}{(z-a)^n}$$

$$= b_0 + b_1(z-a) + b_2(z-a)^2 + \cdots + \frac{c_1}{z-a} + \frac{c_2}{(z-a)^2} + \cdots$$

where[6]

(2) $$b_n = \frac{1}{2\pi i} \int_C \frac{f(z^*)}{(z^*-a)^{n+1}} \, dz^*, \qquad c_n = \frac{1}{2\pi i} \int_C (z^*-a)^{n-1} f(z^*) \, dz^*,$$

each integral being taken in the counterclockwise sense around any simple closed path C which lies in the annulus and encircles the inner circle (Fig. 305).

This series converges and represents $f(z)$ in the open annulus obtained from the given annulus by continuously increasing the circle C_1 and decreasing C_2 until each of the two circles reaches a point where $f(z)$ is singular.

Remark.   Obviously, instead of (1) and (2) we may write simply

(1′)
$$f(z) = \sum_{n=-\infty}^{\infty} A_n(z-a)^n$$

where

(2′)
$$A_n = \frac{1}{2\pi i} \int_C \frac{f(z^*)}{(z^*-a)^{n+1}} \, dz^*.$$

Proof of Laurent's theorem.   Let z be any point in the given annulus. Then from Cauchy's integral formula [cf. (3) in Sec. 13.5] it follows that

(3)
$$f(z) = \frac{1}{2\pi i} \int_{C_1} \frac{f(z)^*}{z^*-z} \, dz^* - \frac{1}{2\pi i} \int_{C_2} \frac{f(z^*)}{z^*-z} \, dz^*,$$

where we integrate in the counterclockwise sense. We shall now transform these integrals in a fashion similar to that in Sec. 15.1. Since z lies inside C_1, the first of these integrals is precisely of the same type as the integral (1), Sec. 15.1. By expanding it and estimating the remainder as in Sec. 15.1 we obtain

(4)
$$\frac{1}{2\pi i} \int_{C_1} \frac{f(z^*)}{z^*-z} \, dz^* = \sum_{n=0}^{\infty} b_n(z-a)^n$$

where the coefficients are given by the formula

(5)
$$b_n = \frac{1}{2\pi i} \int_{C_1} \frac{f(z^*)}{(z^*-a)^{n+1}} \, dz^*$$

and we integrate in the counterclockwise sense. Since a is not a point of the annulus, the functions $f(z^*)/(z^*-a)^{n+1}$ are analytic in the annulus. Hence, we

---

[5] PIERRE ALPHONSE LAURENT (1813–1854), French mathematician.
[6] We denote the variable of integration by z^* because z is used in $f(z)$.

may integrate along the path C (cf. in the theorem) instead of C_1, without altering the value of b_n. This proves (2) for all $n \geq 0$.

In the case of the last integral in (3), the situation is different, since z lies outside C_2. Instead of (3), Sec. 15.1, we now have

(6)
$$\left| \frac{z^* - a}{z - a} \right| < 1,$$

that is, we now have to develop $1/(z^* - z)$ in powers of $(z^* - a)/(z - a)$ for the resulting series to be convergent. We find

$$\frac{1}{z^* - z} = \frac{1}{z^* - a - (z - a)} = \frac{-1}{(z - a)\left(1 - \dfrac{z^* - a}{z - a}\right)}.$$

By applying the formula for a finite geometric progression to the last expression we obtain

$$\frac{1}{z^* - z} = - \frac{1}{z - a} \left\{ 1 + \frac{z^* - a}{z - a} + \left(\frac{z^* - a}{z - a}\right)^2 + \cdots + \left(\frac{z^* - a}{z - a}\right)^n \right\}$$
$$- \frac{1}{z - z^*} \left(\frac{z^* - a}{z - a}\right)^{n+1}.$$

From this development we readily have

$$-\frac{1}{2\pi i} \int_{C_2} \frac{f(z^*)}{z^* - z} \, dz^*$$
$$= \frac{1}{2\pi i} \left\{ \frac{1}{z - a} \int_{C_2} f(z^*) \, dz^* + \frac{1}{(z - a)^2} \int_{C_2} (z^* - a) f(z^*) \, dz^* + \cdots \right.$$
$$\left. + \frac{1}{(z - a)^{n+1}} \int_{C_2} (z^* - a)^n f(z^*) \, dz^* \right\} + R_n^*(z);$$

in this representation the last term is of the form

(7)
$$R_n^*(z) = \frac{1}{2\pi i (z - a)^{n+1}} \int_{C_2} \frac{(z^* - a)^{n+1}}{z - z^*} f(z^*) \, dz^*.$$

In the integrals in the braces we may replace the circle C_2 by the aforementioned path C, without altering their values. This establishes Laurent's theorem provided that

(8)
$$\lim_{n \to \infty} R_n^*(z) = 0.$$

We prove (8). Since $z - z^* \neq 0$ and $f(z)$ is analytic in the annulus and on C_2, the absolute value of the expression $f(z^*)/(z - z^*)$ in (7) is bounded, say,

$$\left| \frac{f(z^*)}{z - z^*} \right| < \tilde{M} \qquad\qquad \text{for all } z^* \text{ on } C_2.$$

By applying (4) in Sec. 13.2 to (7) and denoting the length of C_2 by l we thus obtain

$$|R_n{}^*(z)| < \frac{1}{2\pi |z-a|^{n+1}} |z^* - a|^{n+1} \tilde{M}l = \frac{\tilde{M}l}{2\pi} \left| \frac{z^* - a}{z - a} \right|^{n+1}.$$

From (6) we see that the expression on the right approaches zero as n approaches infinity. This proves (8). The representation (1) with coefficients (2) is now established in the given annulus.

Finally let us prove convergence of (1) in the open annulus characterized at the end of the theorem.

We denote the sums of the two series in (1) by $g(z)$ and $h(z)$, and the radii of C_1 and C_2 by r_1 and r_2, respectively. Then $f = g + h$. The first series is a power series. Since it converges in the annulus, it must converge in the entire disk bounded by C_1, and g is analytic in that disk.

Setting $Z = 1/(z - a)$, the last series becomes a power series in Z. The annulus $r_2 < |z - a| < r_1$ then corresponds to the annulus $1/r_1 < |Z| < 1/r_2$, the new series converges in this annulus and, therefore, in the entire disk $|Z| < 1/r_2$. Since this disk corresponds to $|z - a| > r_2$, the exterior of C_2, the given series converges for all z outside C_2, and h is analytic for all these z.

Since $f = g + h$, it follows that g must be singular at all those points outside C_1 where f is singular, and h must be singular at all those points inside C_2 where f is singular. Consequently, the first series converges for all z inside the circle about a whose radius is equal to the distance of that singularity of f outside C_1 which is closest to a. Similarly, the second series converges for all z outside the circle about a whose radius is equal to the maximum distance of the singularities of f inside C_2. The domain common to both of those domains of convergence is the open annulus characterized at the end of the theorem, and the proof is complete.

It follows that if $f(z)$ is analytic inside C_2, the Laurent series reduces to the Taylor series of $f(z)$ with center a. In fact, by applying Cauchy's integral theorem to (2) we see that in this case all the coefficients of the negative powers in (1) are zero.

Furthermore, if $z = a$ is the only singular point of $f(z)$ in C_2, then the Laurent expansion (1) converges for all z in C_1 except at $z = a$. This case occurs frequently and, therefore, is of particular importance.

The Laurent series of a given analytic function $f(z)$ in its annulus of convergence is unique (cf. Prob. 7, this section). *However, $f(z)$ may have different Laurent series in two annuli with the same center* (cf. Ex. 2, below).

The uniqueness is important, because Laurent series usually are not obtained by using (2) for determining the coefficients, but by various other methods. Some of these methods are illustrated by the following examples. If a Laurent series is found by any such process, it must be *the* Laurent series of the given function in the given annulus.

Example 1. The Laurent series of $z^2 e^{1/z}$ with center 0 can be obtained from (2) in Sec. 15.2. Replacing z by $1/z$ in that series, we find

$$z^2 e^{1/z} = z^2 \left(1 + \frac{1}{1! \, z} + \frac{1}{2! \, z^2} + \cdots \right) = z^2 + z + \frac{1}{2} + \frac{1}{3! \, z} + \frac{1}{4! \, z^2} + \cdots \qquad (|z| > 0).$$

Example 2. Find all Laurent series of $f(z) = 1/(1 - z^2)$ with center at $z = 1$. We have $1 - z^2 = -(z - 1)(z + 1)$. Using the geometric series

$$\frac{1}{1 - q} = \sum_{n=0}^{\infty} q^n \qquad (|q| < 1),$$

we find

(a)
$$\frac{1}{z + 1} = \frac{1}{2 + (z - 1)} = \frac{1}{2} \frac{1}{\left[1 - \left(-\frac{z - 1}{2} \right) \right]}$$

$$= \frac{1}{2} \sum_{n=0}^{\infty} \left(-\frac{z - 1}{2} \right)^n = \sum_{n=0}^{\infty} \frac{(-1)^n}{2^{n+1}} (z - 1)^n ;$$

this series converges in the disk $|(z - 1)/2| < 1$, that is, $|z - 1| < 2$. Similarly,

(b)
$$\frac{1}{z + 1} = \frac{1}{(z - 1) + 2} = \frac{1}{(z - 1)} \frac{1}{\left(1 + \frac{2}{z - 1} \right)}$$

$$= \frac{1}{z - 1} \sum_{n=0}^{\infty} \left(-\frac{2}{z - 1} \right)^n = \sum_{n=0}^{\infty} \frac{(-2)^n}{(z - 1)^{n+1}} ;$$

this series converges for $|2/(z - 1)| < 1$, that is, $|z - 1| > 2$. Therefore, from (a) we obtain

$$f(z) = \frac{-1}{(z - 1)(z + 1)} = \sum_{n=0}^{\infty} \frac{(-1)^{n+1}}{2^{n+1}} (z - 1)^{n-1}$$

$$= \frac{-1/2}{z - 1} + \frac{1}{4} - \frac{1}{8} (z - 1) + \frac{1}{16} (z - 1)^2 - + \cdots ;$$

this series converges in the domain $0 < |z - 1| < 2$. Similarly, from (b) we obtain

$$f(z) = -\sum_{n=0}^{\infty} \frac{(-2)^n}{(z - 1)^{n+2}} = -\frac{1}{(z - 1)^2} + \frac{2}{(z - 1)^3} - \frac{4}{(z - 1)^4} + - \cdots \qquad (|z - 1| > 2).$$

Example 3. From Prob. 18 at the end of Sec. 15.2 we immediately have the Laurent series

$$\cot z = \frac{1}{z} - \frac{1}{3} z - \frac{1}{45} z^3 - \frac{2}{945} z^5 - \cdots \qquad (0 < |z| < \pi).$$

Problems

Expand the following functions in Laurent series which converge for $0 < |z| < R$ and determine the precise region of convergence.

1. $\dfrac{\sin 4z}{z^4}$

2. $\dfrac{1}{z^3(1 - z)}$

3. $\dfrac{e^{-1/z^2}}{z^6}$

4. $\dfrac{1}{z^2(1 + z)^2}$

5. $\dfrac{1}{z^5 + z}$

6. $\dfrac{1}{z^4 - 3z^3}$

7. Prove that the Laurent expansion of a given analytic function in a given annulus is unique.
8. Does tan $(1/z)$ have a Laurent series convergent in a region $0 < |z| < R$?

Find all Taylor series and Laurent series with center $z = a$ and determine the precise regions of convergence.

9. $\dfrac{1}{z^2 + 1}$, $a = i$ 10. $\dfrac{1}{z^2 + 1}$, $a = -i$ 11. $\dfrac{4z - 1}{z^4 - 1}$, $a = 0$

12. $\dfrac{1}{1 - z^4}$, $a = -1$ 13. $\dfrac{1}{1 - z^4}$, $a = 0$ 14. $\dfrac{1}{z(z - 1)^2}$, $a = 0$

15. $\dfrac{e^z}{(z - 1)^2}$, $a = 1$ 16. $\dfrac{\sin z}{(z - \frac{1}{4}\pi)^3}$, $a = \dfrac{\pi}{4}$ 17. $\dfrac{7z^2 + 9z - 18}{z^3 - 9z}$, $a = 0$

18. $\dfrac{1}{z^3}$, $a = i$ 19. $\dfrac{1}{z^4}$, $a = 1$ 20. $\dfrac{4z^2 + 2z - 4}{z^3 - 4z}$, $a = 2$

15.6 Behavior of Functions at Infinity

We shall see in the next section that the Laurent series can be used for classifying the singularities of analytic functions. In this connection we shall also consider the behavior of analytic functions $f(z)$ as $|z|$ approaches infinity.

For investigating a function $f(z)$ for large $|z|$, it is convenient to introduce a new variable w by setting $z = 1/w$ because large $|z|$ then corresponds to small $|w|$, and $|z| \to \infty$ corresponds to $|w| \to 0$.

The transformation $w = 1/z$ was considered in Sec. 12.3 in the **extended complex plane** which was obtained by attaching an improper point ("*point at infinity*") to the finite complex plane. We may then let $z = \infty$ correspond to $w = 0$ (and $w = \infty$ correspond to $z = 0$). The procedure of extending the complex plane may also be motivated by the following simple geometrical consideration.

The usual representation of complex numbers z in the complex plane is convenient as long as the absolute values of the numbers are not too large. For large $|z|$ the situation becomes inconvenient, and in this case we may prefer a representation of the complex numbers on a sphere, which was suggested by Riemann and is obtained as follows.

Let S be a sphere of diameter 1 which touches the complex z-plane at the origin (Fig. 306). Let N be the "North pole" of S (the point diametrically opposite to the point of contact between the sphere and the plane). Let P be any point in the complex plane. Then the straight segment with end points P and N intersects S at a point P^*. We let P and P^* correspond to each other. In this way we obtain a correspondence between the points in the complex plane and the points on S, and P^* is the image point of P with respect to this mapping. The complex numbers, first represented in the plane, are now represented by points on S. To each z there corresponds a point on S. Conversely, each point on S represents a complex number z, except for the point N which does not correspond to any point in the complex plane. But if we introduce the improper point $z = \infty$ and let this point correspond to N, the mapping becomes a one-to-one mapping of the extended

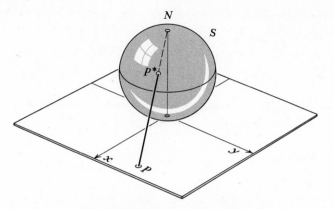

Fig. 306. Riemann number sphere

plane onto S. The sphere S is called the **Riemann number sphere.** The particular mapping which we have used is called a **stereographic projection.**

Obviously, the unit circle is mapped onto the "equator" of S. The interior of the unit circle corresponds to the "Southern hemisphere" and the exterior to the "Northern hemisphere." Numbers z whose absolute values are large lie close to the North pole N. The x and y axes (and, more generally, all the straight lines through the origin) are mapped onto "meridians," while circles with center at the origin are mapped onto "parallels." It can be shown that any circle or straight line in the z-plane is mapped onto a circle on S, and, furthermore, that stereographic projection is conformal.

Given a function $f(z)$ to be investigated for large $|z|$, we may now set $z = 1/w$ and investigate $f(z) = f(1/w) \equiv g(w)$ in a neighborhood of $w = 0$. We define

$$g(0) = \lim_{w \to 0} g(w),$$

and we say that $f(z)$ is **analytic or singular at infinity** according as $g(w)$ is analytic or singular, respectively, at $w = 0$.

Example 1. The function $f(z) = 1/z^2$ is analytic at infinity since $g(w) = f(1/w) = w^2$ is analytic at $w = 0$. The function $f(z) = z^3$ is singular at infinity because $g(w) = f(1/w) = 1/w^3$ is singular at $w = 0$. The exponential function e^z is singular at infinity since $e^{1/w}$ is singular at $w = 0$. Similarly, the trigonometric functions $\sin z$ and $\cos z$ are singular at infinity.

Let $f(z)$ be analytic outside a circle C, say, in the domain $|z - a| > R$, also at infinity. If we set

$$z = \frac{1}{w} + a, \qquad \text{then} \qquad z - a = \frac{1}{w},$$

and

$$f(z) = f\left(\frac{1}{w} + a\right) \equiv h(w)$$

is analytic in the disk $|w| < 1/R$. Let the Maclaurin series of $h(w)$ be

$$h(w) = \sum_{n=0}^{\infty} c_n w^n = c_0 + c_1 w + c_2 w^2 + \cdots \qquad \left(|w| < \frac{1}{R}\right).$$

Then by inserting $w = 1/(z - a)$ we readily obtain the Laurent series

(1) $$f(z) = \sum_{n=0}^{\infty} \frac{c_n}{(z-a)^n} = c_0 + \frac{c_1}{z-a} + \frac{c_2}{(z-a)^2} + \cdots \qquad (|z-a| > R).$$

Problems

Are the following functions analytic at infinity?

1. e^z, $e^{1/z}$ $e^{1/(z-1)}$ **2.** $\cos z$, $\sin z$, $\cosh z$ **3.** $ze^{1/z}$, $z^{-1} \sin z$

4. $z + z^{-1}$, $z^{-3} - z^{-4}$ **5.** $\tan z$, $\cot z$ **6.** $(z^2 + 1)/(z^2 - 1)$

Describe and sketch the images of the following regions on the Riemann number sphere.

7. $|z| > 7$ **8.** $|z| \leq 1$ **9.** $1 < |z| < 2$

10. $\operatorname{Re} z \geq 0$ **11.** $\operatorname{Im} z \leq 0$ **12.** $|\arg z| \leq \pi/2$

13. $\operatorname{Re} z > -1$ **14.** $\operatorname{Re} z > 0$, $\operatorname{Im} z > 0$ **15.** $|z| > 1$, $|\arg z| < \pi/4$

16. Using the method described at the end of this section, develop $1/z^4$ in a Laurent series of negative powers of $z - 1$.

17. Describe the relative positions of z, $-z$, $\bar{z}$, $-\bar{z}$ in the complex plane and on the Riemann sphere.

18. What is the length of the image of the unit circle on the Riemann sphere?

19. **(Chordal distance)**   Let z_1, z_2 be any points in the complex plane and Z_1, Z_2 their images on the Riemann sphere.   Then the distance in space of Z_1, Z_2 is called the *chordal distance* of z_1, z_2 and is denoted by $\chi(z_1, z_2)$.   It can be shown that (cf. Ref. [F5] in Appendix 1)

$$\chi(z_1, z_2) = \frac{|z_1 - z_2|}{\sqrt{(1 + |z_1|^2)(1 + |z_2|^2)}}.$$

Using this formula, show that $\chi(z, \bar{z}) = 2|y|/(1 + |z|^2)$.

20. Using the formula in Prob. 19, show that $\chi(z, -1/\bar{z}) = 1$.   What does this mean geometrically?

CHAPTER **16**

Integration by the Method of Residues

Since there are various methods for determining the coefficients of a Laurent series (1), Sec. 15.5, without using the integral formulas (2), Sec. 15.5, we may use the formula for c_1 for evaluating complex integrals in a very elegant and simple fashion. c_1 will be called the *residue* of $f(z)$ at $z = a$. This powerful method may also be applied for evaluating certain real integrals, as we shall see in Sec. 16.4.

> *Prerequisites for this chapter:* Chaps. 11, 13, 15.
> *References:* Appendix 1, Part F.
> *Answers to problems:* Appendix 2.

16.1 Zeros and Singularities

If a function $f(z)$ is analytic in a domain D and is zero at a point $z = a$ in D, then $f(z)$ is said to have a **zero** at that point $z = a$. If not only f but also the derivatives $f', \cdots, f^{(n-1)}$ are all zero at $z = a$ and $f^{(n)}(a) \neq 0$, then $f(z)$ is said to have a *zero of* **order** n at the point $z = a$.

For example, if $f(a) = 0$, $f'(a) \neq 0$, then f has a zero of first order or *simple zero* at $z = a$. If $f(a) = f'(a) = 0$, $f''(a) \neq 0$, then the zero of f at $z = a$ is of second order, etc. Furthermore, our consideration in the last section suggests the following definition.

An analytic function $f(z)$ is said to have a *zero of n-th order at infinity* if $f(1/z)$ has such a zero at $z = 0$.

Example 1. The function $\sin z$ has simple zeros at $z = 0, \pm\pi, \pm2\pi, \cdots$. The function $(z - a)^3$ has a zero of third order at $z = a$. The function $1 - \cos z$ has second-order zeros at $z = 0$, $\pm2\pi, \pm4\pi, \cdots$. The function $1/(1 - z)$ has a simple zero at infinity.

599

Clearly, if $f(z)$ is analytic in some neighborhood of a point $z = a$ and has a zero of nth order at a, it follows from Taylor's theorem in Sec. 15.1 that the coefficients $b_0, \cdots, b_{n-1}$ of its Taylor series with center $z = a$ are zero, and the series is of the form

(1)
$$f(z) = b_n(z - a)^n + b_{n+1}(z - a)^{n+1} + \cdots$$
$$= (z - a)^n[b_n + b_{n+1}(z - a) + b_{n+2}(z - a)^2 + \cdots] \qquad (b_n \neq 0).$$

A point of a point set S is called an **isolated point** of S, if it has a neighborhood which does not contain further points of S. A point b is called a **limit point** of S, if every neighborhood of b (no matter how small) contains at least one point ($\neq b$) of S (and hence infinitely many points of S).

Example 2. The set of the points $z = n$ ($n = 1, 2, \cdots$) consists wholly of isolated points and has no limit point in the finite plane. The set of the points $z = i/n$ ($n = 1, 2, \cdots$) on the imaginary axis consists wholly of isolated points and has one limit point, namely, $z = 0$; this point does not belong to the set. The set of all complex numbers z for which $|z| < 1$ has no isolated points. Every point of the set and also the points on the unit circle $|z| = 1$ (which do not belong to the set) are limit points of the set.

Theorem 1. *The zeros of an analytic function $f(z)$ ($\not\equiv 0$) are isolated.*

Proof. We consider (1). Let $g(z)$ be the analytic function represented by the series in the brackets [$\cdots$]. Since $b_n \neq 0$, we have $g(a) \neq 0$. Consequently, since $g(z)$ is continuous, it is not zero in some neighborhood of $z = a$. It follows that $f(z)$ is not zero in that neighborhood (except at $z = a$), so that $z = a$ is the only zero of $f(z)$ in that neighborhood and is, therefore, isolated. This completes the proof.

Analytic functions may have different types of singularities.[1] We first remember that a *singular point* of an analytic function $f(z)$ is a point where $f(z)$ ceases to be analytic (cf. Sec. 15.1). We also say that $f(z)$ *is singular*, or *has a singularity*, at that point. The function $f(z)$ is said to be *singular at infinity*, if $f(1/z)$ is singular at $z = 0$.

If $f(z)$ has an isolated singularity at a point $z = a$, then we can represent it by its **Laurent series** (cf. Sec. 15.5)

(2)
$$f(z) = \sum_{n=0}^{\infty} b_n(z - a)^n + \sum_{n=1}^{\infty} \frac{c_n}{(z - a)^n}$$

valid throughout some neighborhood of $z = a$ (except at $z = a$ itself). The last series in (2) is called the **principal part** of $f(z)$ near $z = a$.

It may happen that from some n on, all the coefficients c_n are zero, say, $c_m \neq 0$ and $c_n = 0$ for all $n > m$. Then (2) reduces to the form

(3)
$$f(z) = \sum_{n=0}^{\infty} b_n(z - a)^n + \frac{c_1}{z - a} + \cdots + \frac{c_m}{(z - a)^m} \qquad (c_m \neq 0).$$

In this case, where the principal part consists of finitely many terms, the singularity of f at $z = a$ is called a **pole,** and m is called the **order** of the pole. Poles of the first order are also known as *simple poles.*

---

[1] We recall, that, by definition, a function is a single-valued relation. (Cf. Sec. 11.2.)

If an analytic function f (single-valued in the complex plane) has a singularity other than a pole, this singularity is called an **essential singularity** of f.

Poles are, by definition, isolated singularities. All singularities which are not isolated (for instance, the singularity of $\tan (1/z)$ at $z = 0$) are thus essential singularities. An essential singularity may be isolated or not. If in (2) infinitely many c_n are different from zero, then the singularity of $f(z)$ at $z = a$ is not a pole but an isolated essential singularity.

Example 3. The function

$$f(z) = \frac{1}{z(z-2)^5} + \frac{3}{(z-2)^2}$$

has a simple pole at $z = 0$ and a pole of fifth order at $z = 2$. The functions

$$(4) \qquad e^{1/z} = \sum_{n=0}^{\infty} \frac{1}{n!\, z^n} = 1 + \frac{1}{z} + \frac{1}{2!\, z^2} + \cdots$$

and

$$(5) \qquad \sin \frac{1}{z} = \sum_{n=0}^{\infty} \frac{(-1)^n}{(2n+1)!\, z^{2n+1}} = \frac{1}{z} - \frac{1}{3!\, z^3} + \frac{1}{5!\, z^5} - + \cdots$$

have an isolated essential singularity at $z = 0$.

The function $\tan (1/z)$ has poles at

$$\frac{1}{z} = \pm \frac{\pi}{2}, \pm \frac{3\pi}{2}, \cdots \qquad \text{that is,} \qquad z = \pm \frac{2}{\pi}, \pm \frac{2}{3\pi}, \cdots.$$

The limit point $z = 0$ of these points is thus a nonisolated essential singularity of $\tan (1/z)$.

Example 4. The polynomial $f(z) = 2z + 6z^3$ has a pole of third order at infinity, because $f\left(\frac{1}{z}\right) = \frac{2}{z} + \frac{6}{z^3}$ has such a pole at $z = 0$. More generally, a polynomial of nth degree has a pole of nth order at infinity.

The functions e^z, $\sin z$, $\cos z$ have an isolated essential singularity at infinity, since $e^{1/z}$, $\sin (1/z)$, and $\cos (1/z)$ have an isolated essential singularity at $z = 0$.

A function $f(z)$ which is not analytic at a point $z = a$ but can be made analytic there by assigning some value to $f(z)$ at $z = a$, is said to have a *removable singularity* at $z = a$. Such singularities are not of interest, because they can be removed.

A function which is analytic everywhere in the finite plane is called an **entire function.**

If such a function is also analytic at infinity, it is bounded for all z, and from Liouville's theorem (cf. Sec. 13.6) it follows that it must be a constant. Hence, any entire function which is not a constant must be singular at infinity. For instance, polynomials (of at least first degree), e^z, $\sin z$, and $\cos z$ are entire functions, and they are singular at infinity.

An analytic function whose only singularities in the finite plane are poles is called a **meromorphic function.**

Example 5. Rational functions with nonconstant denominator, $\tan z$, $\cot z$, $\sec z$, and $\csc z$ are meromorphic functions.

The classification of singularities into poles and essential singularities is not merely a formal matter, because the behavior of an analytic function in a neighborhood of an essential singularity is entirely different from that in the neighborhood of a pole.

Example 6. The function $f(z) = 1/z^2$ has a pole at $z = 0$, and $|f(z)| \to \infty$ as $z \to 0$ in any manner.

This example illustrates

Theorem 2. *If $f(z)$ is analytic and has a pole at $z = a$, then $|f(z)| \to \infty$ as $z \to a$ in any manner.* (Cf. Prob. 12.)

Example 7. The function $f(z) = e^{1/z}$ has an essential singularity at $z = 0$. It has no limit for approach along the imaginary axis; it becomes infinite if $z \to 0$ through positive real values, but it approaches zero if $z \to 0$ through negative real values. It takes on any given value $c = c_0 e^{i\alpha} \neq 0$ in an arbitrarily small neighborhood of $z = 0$. In fact, setting $z = re^{i\theta}$, we must solve the equation

$$e^{1/z} = e^{(\cos\theta - i\sin\theta)/r} = c_0 e^{i\alpha}$$

for r and θ. Equating the absolute values and the arguments, we have $e^{(\cos\theta)/r} = c_0$ or

$$\cos\theta = r\ln c_0 \quad \text{and} \quad \sin\theta = -\alpha r.$$

From these two equations and $\cos^2\theta + \sin^2\theta = 1$ we obtain

$$r^2 = \frac{1}{(\ln c_0)^2 + \alpha^2} \quad \text{and} \quad \tan\theta = -\frac{\alpha}{\ln c_0}.$$

We see that r can be made arbitrarily small by adding multiples of 2π to α, leaving c unaltered.

This example illustrates the famous

Picard's theorem. *If $f(z)$ is analytic and has an isolated essential singularity at a point a, it takes on every value, with at most one exceptional value, in an arbitrarily small neighborhood of a.*

In Ex. 7, the exceptional value is $z = 0$. The proof of Picard's theorem is rather complicated; it can be found in Ref. [F10].

Problems

Find the location and order of the zeros of the following functions.

1. $z^4 - 1$ 2. $(z - 1)^4$ 3. $z^{-3}\cos^3 \pi z$

4. $z\sin\dfrac{1}{z}$ 5. $\cosh^2 2z$ 6. $\dfrac{(9 + z^2)^2}{(1 - z^2)^3}$

7. $\dfrac{z + i}{z^2 + 3}$ 8. $\dfrac{(z + 2i)^2}{z^5}$ 9. $\dfrac{(4 + z^2)^2}{(1 + z)^4}$

10. If $f(z)$ has a zero of order n at $z = a$, prove that $f^2(z)$ has a zero of order $2n$ at $z = a$. Give examples.

11. If $f(z)$ has a zero of order $n > 1$ at $z = a$, prove that $f'(z)$ has a zero of order $n - 1$ at $z = a$. Give examples.

12. Verify Theorem 2 for $f(z) = z^{-2} + z^{-1}$. Prove Theorem 2.

13. If $f(z)$ has a zero of order n at $z = a$, prove that $1/f(z)$ has a pole of order n at $z = a$. Give examples.

Find the location and type of singularities of the following functions.

14. $\sinh \pi z$
15. $z^2 + z^{-1}$
16. e^{-z^2}/z^2
17. $z^7/(1 + z^2)^3$
18. $z e^{1/(z-1)}$
19. $\cosh^2 [1/(z - \pi)]$
20. $\cos z - \sin z$
21. $\tan^2 (1/z)$
22. $z^4 e^{1/z}$
23. $z \cot z$
24. $\csc^2 \pi z$
25. $e^{1/(z-1)}/(e^z - 1)$

16.2   Residues

If $f(z)$ is analytic in a neighborhood of a point $z = a$, then, by Cauchy's integral theorem,

$$(1) \qquad\qquad \int_C f(z)\, dz = 0$$

for any contour in that neighborhood. If, however, $f(z)$ has a pole or an isolated essential singularity (cf. Sec. 16.1) at $z = a$ and a lies in the interior of C, then the integral in (1) will, in general, be different from zero. In this case we may represent $f(z)$ by a Laurent series

$$(2) \qquad\qquad f(z) = \sum_{n=0}^{\infty} b_n (z - a)^n + \sum_{n=1}^{\infty} \frac{c_n}{(z - a)^n}$$

which converges in the domain $0 < |z - a| < R$, where R is the distance from a to the nearest singular point of $f(z)$. From (2) in Sec. 15.5 we see that the coefficient c_1 of the power $1/(z - a)$ in this development is

$$c_1 = \frac{1}{2\pi i} \int_C f(z)\, dz,$$

and, therefore,

$$(3) \qquad\qquad \int_C f(z)\, dz = 2\pi i c_1,$$

the integration being taken in the counterclockwise sense around a simple closed path C which lies in the domain $0 < |z - a| < R$ and contains the point $z = a$ in its interior. The coefficient c_1 in the development (2) of $f(z)$ is called the **residue** of $f(z)$ at $z = a$, and we shall use the notation

$$(4) \qquad\qquad c_1 = \operatorname*{Res}_{z=a} f(z).$$

We have seen that Laurent expansions can be obtained by various methods, without using the integral formulas for the coefficients. *Hence, we may determine the residue by one of those methods and then use the formula* (3) *for evaluating contour integrals.*

Example 1. Integrate the function $f(z) = z^{-4} \sin z$ around the unit circle C in the counterclockwise sense.

From (5) in Sec. 15.2 we obtain the Laurent series

$$f(z) = \frac{\sin z}{z^4} = \frac{1}{z^3} - \frac{1}{3!\,z} + \frac{z}{5!} - \frac{z^3}{7!} + - \cdots.$$

We see that $f(z)$ has a pole of third order at $z = 0$, the corresponding residue is $c_1 = -1/3!$, and from (3) it follows that

$$\int_C \frac{\sin z}{z^4}\, dz = 2\pi i c_1 = -\frac{\pi i}{3}.$$

Before we proceed in evaluating integrals we shall develop a simple standard method for determining the residue in the case of a pole.

If $f(z)$ has a **simple pole** at a point $z = a$, the corresponding Laurent series is of the form [cf. (3) in the last section]

$$\cdot \ f(z) = \frac{c_1}{z - a} + b_0 + b_1(z - a) + b_2(z - a)^2 + \cdots \qquad (0 < |z - a| < R)$$

where $c_1 \neq 0$. Multiplying both sides by $z - a$, we have

(5) $$(z - a)f(z) = c_1 + (z - a)[b_0 + b_1(z - a) + \cdots].$$

If we let z approach a, the right-hand side approaches c_1 and we obtain

(6) $$\operatorname*{Res}_{z=a} f(z) = c_1 = \lim_{z \to a} (z - a)f(z).$$

In the case of a simple pole, another useful formula is obtained as follows. If $f(z)$ has a simple pole at $z = a$, we may set

$$f(z) = \frac{p(z)}{q(z)}$$

where $p(z)$ and $q(z)$ are analytic at $z = a$, $p(a) \neq 0$, and $q(z)$ has a simple zero at $z = a$. Consequently, $q(z)$ can be expanded in a Taylor series of the form

$$q(z) = (z - a)q'(a) + \frac{(z - a)^2}{2!} q''(a) + \cdots.$$

Hence, by (6),

$$\operatorname*{Res}_{z=a} f(z) = \lim_{z \to a} (z - a) \frac{p(z)}{q(z)} = \lim_{z \to a} \frac{(z - a)p(z)}{(z - a)[q'(a) + (z - a)q''(a)/2 + \cdots]} ;$$

that is, in the case of a simple pole we also have

(7) $$\operatorname*{Res}_{z=a} f(z) = \operatorname*{Res}_{z=a} \frac{p(z)}{q(z)} = \frac{p(a)}{q'(a)}.$$

Example 2. The function $f(z) = (4 - 3z)/(z^2 - z)$ has simple poles at $z = 0$ and $z = 1$. From (7) we obtain

$$\operatorname*{Res}_{z=0} f(z) = \left[\frac{4 - 3z}{2z - 1}\right]_{z=0} = -4, \qquad \operatorname*{Res}_{z=1} f(z) = \left[\frac{4 - 3z}{2z - 1}\right]_{z=1} = 1.$$

We shall now consider **poles of higher order.** If $f(z)$ has a pole of order $m > 1$ at a point $z = a$, the corresponding Laurent expansion is of the form

$$f(z) = \frac{c_m}{(z-a)^m} + \frac{c_{m-1}}{(z-a)^{m-1}} + \cdots + \frac{c_2}{(z-a)^2} + \frac{c_1}{z-a} + b_0 + b_1(z-a) + \cdots$$

where $c_m \neq 0$ and the series converges in some neighborhood of $z = a$, except at the point a itself. By multiplying both sides by $(z-a)^m$ we obtain

$$(z-a)^m f(z) = c_m + c_{m-1}(z-a) + \cdots + c_2(z-a)^{m-2} + c_1(z-a)^{m-1}$$
$$+ b_0(z-a)^m + b_1(z-a)^{m+1} + \cdots.$$

This shows that the residue c_1 of $f(z)$ at $z = a$ is now the coefficient of the power $(z-a)^{m-1}$ in the Taylor expansion of the function $g(z) = (z-a)^m f(z)$ with center at $z = a$. Therefore, from Taylor's theorem (Sec. 15.1),

$$c_1 = \frac{g^{(m-1)}(a)}{(m-1)!};$$

that is,

(8) $$\operatorname*{Res}_{z=a} f(z) = \frac{1}{(m-1)!} \lim_{z \to a} \left\{ \frac{d^{m-1}}{dz^{m-1}} [(z-a)^m f(z)] \right\}.$$

Example 3. The function

$$f(z) = \frac{2z}{(z+4)(z-1)^2}$$

has a pole of second order at $z = 1$, and from (8) we obtain the corresponding residue

$$\operatorname*{Res}_{z=1} f(z) = \lim_{z \to 1} \frac{d}{dz}[(z-1)^2 f(z)] = \lim_{z \to 1} \frac{d}{dz}\left(\frac{2z}{z+4}\right) = \frac{8}{25}.$$

Of course, in the case of a rational function $f(z)$ the residues can also be determined from the representation of $f(z)$ in terms of partial fractions.

Problems

Find the residues at the singular points of the following functions.

1. $\dfrac{3}{1-z}$ **2.** $\dfrac{z-1}{z+1}$ **3.** $\dfrac{3e^z}{z^4}$ **4.** $\dfrac{2}{1-e^z}$

5. $z^{-5} \cos z$ **6.** $\tan z$ **7.** $\cot z$ **8.** $\pi \csc \pi z$

9. $(z^2 - 1)^{-2}$ **10.** $\csc^2 z$ **11.** $\coth 2z$ **12.** $(z + \pi i)^{-6} e^z$

In each case, find the residues at those singular points which lie inside the circle $|z| = 1.5$.

13. $\dfrac{z^2}{1-z^4}$ **14.** $\dfrac{4z-1}{z^2 + 3z + 2}$ **15.** $\dfrac{3-2z}{z^3 + 3z^2}$

16. $\dfrac{3}{(z^4 - 1)^2}$ **17.** $\dfrac{z+2}{(z+1)(z^2 + 16)}$ **18.** $\dfrac{8-6z}{z^3 - 3z^2 + 2z}$

Evaluate the following integrals where C is the unit circle (counterclockwise).

19. $\displaystyle\int_C \frac{z}{2z-1}\, dz$ **20.** $\displaystyle\int_C \frac{3\cos z}{2i - 3z}\, dz$ **21.** $\displaystyle\int_C \frac{4z^3 + 8}{\pi + 4z}\, dz$

22. $\displaystyle\int_C \frac{z+1}{2z+i}\, dz$ **23.** $\displaystyle\int_C \frac{z^2+1}{z^2-2z}\, dz$ **24.** $\displaystyle\int_C \frac{e^{1/z}}{z}\, dz$

25. $\displaystyle\int_C \csc z\, dz$ **26.** $\displaystyle\int_C \cot z\, dz$ **27.** $\displaystyle\int_C \tan z\, dz$

28. $\displaystyle\int_C \coth 2z\, dz$ **29.** $\displaystyle\int_C \frac{dz}{1-e^z}$ **30.** $\displaystyle\int_C \frac{(z^2+1)}{e^z \sin z}\, dz$

16.3   *The Residue Theorem*

So far we are in a position to evaluate contour integrals whose integrands have only a single isolated singularity inside the contour of integration. We shall now see that our simple method may easily be extended to the case when the integrand has several isolated singularities inside the contour.

Residue theorem. *Let $f(z)$ be a function which is analytic inside a simple closed path C and on C, except for finitely many singular points $a_1, a_2, \cdots, a_m$ inside C. Then*

(1) $$\int_C f(z)\, dz = 2\pi i \sum_{j=1}^{m} \operatorname*{Res}_{z=a_j} f(z),$$

the integral being taken in the counterclockwise sense around C.

Proof. We enclose each of the singular points a_j in a circle C_j with radius small enough that those m circles and C are all separated (Fig. 307). Then $f(z)$ is analytic in the multiply connected domain D bounded by C and $C_1, \cdots, C_m$ and on the entire boundary of D. From Cauchy's integral theorem it follows that

(2) $$\int_C f(z)\, dz + \int_{C_1} f(z)\, dz + \int_{C_2} f(z)\, dz + \cdots + \int_{C_m} f(z)\, dz = 0,$$

the integral along C being taken in the counterclockwise sense and the other integrals in the clockwise sense (cf. Sec. 13.3). We now reverse the sense of integration along $C_1, \cdots, C_m$. Then the signs of the values of these integrals change, and we obtain from (2)

(3) $$\int_C f(z)\, dz = \int_{C_1} f(z)\, dz + \int_{C_2} f(z)\, dz + \cdots + \int_{C_m} f(z)\, dz,$$

all integrals now being taken in the counterclockwise sense. Since, by (3) in the

Fig. 307. Residue theorem

last section,

$$\int_{C_j} f(z)\, dz = 2\pi i \operatorname*{Res}_{z=a_j} f(z),$$

formula (3) yields (1), and the theorem is proved.

This important theorem has various applications in connection with complex and real integrals. We shall first consider some complex integrals.

Example 1. The function $(4 - 3z)/(z^2 - z)$ is analytic except at the points 0 and 1 where it has simple poles; the residues are -4 and 1, respectively (cf. Ex. 2 in the last section). Therefore,

$$\int_C \frac{4 - 3z}{z^2 - z}\, dz = 2\pi i(-4 + 1) = -6\pi i$$

for every simple closed path C which encloses the points 0 and 1, and

$$\int_C \frac{4 - 3z}{z^2 - z}\, dz = 2\pi i(-4) = -8\pi i$$

for any simple closed path C for which $z = 0$ lies inside C and $z = 1$ lies outside C, the integrations being taken in the counterclockwise sense.

Example 2. Integrate $1/(z^3 - 1)^2$ in the counterclockwise sense around the circle $|z - 1| = 1$. The function has poles of second order at $z = 1$, $z = e^{2\pi i/3}$, and $z = e^{-2\pi i/3}$. Only the pole at $z = 1$ lies inside C. Using (8), Sec. 16.2, we thus obtain

$$\int_C \frac{dz}{(z^3 - 1)^2} = 2\pi i \operatorname*{Res}_{z=1} \frac{1}{(z^3 - 1)^2} = 2\pi i\left(-\frac{2}{9}\right) = -\frac{4\pi i}{9}.$$

Example 3. Integrate $1/(z - a)^m$ (m a positive integer) in the counterclockwise sense around any simple closed path C enclosing the point $z = a$. We find

$$\operatorname*{Res}_{z=a} \frac{1}{z - a} = 1 \quad \text{and} \quad \operatorname*{Res}_{z=a} \frac{1}{(z - a)^m} = 0 \quad (m = 2, 3, \cdots).$$

Therefore, the result is

$$\int_C \frac{dz}{(z - a)^m} = \begin{cases} 2\pi i & (m = 1) \\ 0 & (m = 2, 3, \cdots), \end{cases}$$

in agreement with Ex. 3 in Sec. 13.1.

Problems

Evaluate the following integrals where C is the unit circle (counterclockwise).

1. $\displaystyle\int_C \frac{dz}{1 + 4z^2}$ **2.** $\displaystyle\int_C \frac{z}{1 + 9z^2}\, dz$ **3.** $\displaystyle\int_C \frac{dz}{z^2 + 6iz}$

4. $\displaystyle\int_C \frac{(2z + 1)^2}{4z^3 + z}\, dz$ **5.** $\displaystyle\int_C \frac{z^2 \sin z}{4z^2 - 1}\, dz$ **6.** $\displaystyle\int_C \frac{-z^5 + 3z^3 - 1}{(z^2 + 4)(z + \frac{1}{2})}\, dz$

7. $\displaystyle\int_C e^z \cot 4z\, dz$ **8.** $\displaystyle\int_C \frac{e^z}{\sin z}\, dz$ **9.** $\displaystyle\int_C \frac{1 - 4z + 6z^2}{(z^2 + \frac{1}{4})(2 - z)}\, dz$

10. $\displaystyle\int_C \frac{e^{z^2}}{\cos 2z}\, dz$ **11.** $\displaystyle\int_C \frac{e^z}{\cos \pi z}\, dz$ **12.** $\displaystyle\int_C \frac{(z^2 + 2)^3}{2z^2 - z}\, dz$

13. $\displaystyle\int_C \tan \pi z\, dz$
  14. $\displaystyle\int_C \tan 2\pi z\, dz$
  15. $\displaystyle\int_C \cot \frac{z}{4}\, dz$

16. $\displaystyle\int_C \frac{\sinh z}{4z^2+1}\, dz$
  17. $\displaystyle\int_C \frac{\cosh z}{z^2-3iz}\, dz$
  18. $\displaystyle\int_C \frac{(z+4)^3}{z^4+5z^3+6z^2}\, dz$

Integrate $\dfrac{3z^2+2z-4}{z^3-4z}$ in the counterclockwise sense around the path

19. $|z-i| = 3$
  20. $|z-3-i| = 2$
  21. $|z-1-i| = 2$

Integrate $\dfrac{z+1}{z(z-1)(z-2)}$ in the clockwise sense around the path

22. $|z-1| = \frac{1}{4}$
  23. $|z-2| = \frac{1}{2}$
  24. $|z| = \frac{3}{2}$
  25. $|z-\frac{1}{2}| = \frac{1}{4}$

16.4 *Evaluation of Real Integrals*

The residue theorem yields a very elegant and simple method for evaluating certain classes of complicated real integrals, and we shall see that in many cases this method of integration can be made a routine matter.

A. Integrals of rational functions of cos θ and sin θ. We first consider integrals of the type

(1)
$$I = \int_0^{2\pi} R(\cos\theta, \sin\theta)\, d\theta$$

where $R(\cos\theta, \sin\theta)$ is a real rational function of $\cos\theta$ and $\sin\theta$ finite on the interval $0 \leq \theta \leq 2\pi$. Setting $e^{i\theta} = z$, we obtain

$$\cos\theta = \tfrac{1}{2}(e^{i\theta} + e^{-i\theta}) = \frac{1}{2}\left(z + \frac{1}{z}\right),$$

$$\sin\theta = \frac{1}{2i}(e^{i\theta} - e^{-i\theta}) = \frac{1}{2i}\left(z - \frac{1}{z}\right),$$

and we see that the integrand becomes a rational function of z, say, $f(z)$. As θ ranges from 0 to 2π, the variable z ranges once around the unit circle $|z| = 1$ in the counterclockwise sense. Since $dz/d\theta = ie^{i\theta}$, we have $d\theta = dz/iz$, and the given integral takes the form

(2)
$$I = \int_C f(z)\frac{dz}{iz},$$

the integration being taken in the counterclockwise sense around the unit circle.

Example 1. Let p be a fixed real number in the interval $0 < p < 1$. We consider

$$\int_0^{2\pi} \frac{d\theta}{1 - 2p\cos\theta + p^2} = \int_C \frac{dz/iz}{1 - 2p\frac{1}{2}\left(z + \frac{1}{z}\right) + p^2} = \int_C \frac{dz}{i(1 - pz)(z - p)}.$$

The integrand has simple poles at $z = 1/p > 1$ and $z = p < 1$. Only the last pole lies inside the

unit circle C, and the residue is

$$\operatorname*{Res}_{z=p} \frac{1}{i(1 - pz)(z - p)} = \left[\frac{1}{i(1 - pz)} \right]_{z=p} = \frac{1}{i(1 - p^2)}.$$

The residue theorem yields

$$\int_0^{2\pi} \frac{d\theta}{1 - 2p \cos \theta + p^2} = 2\pi i \frac{1}{i(1 - p^2)} = \frac{2\pi}{1 - p^2} \qquad (0 < p < 1).$$

B. Improper integrals of rational functions. We consider real integrals of the type

(3)
$$\int_{-\infty}^{\infty} f(x)\, dx.$$

Such an integral, for which the interval of integration is not finite, is called an **improper integral,** and it has the meaning

(4')
$$\int_{-\infty}^{\infty} f(x)\, dx = \lim_{a \to -\infty} \int_a^0 f(x)\, dx + \lim_{b \to \infty} \int_0^b f(x)\, dx.$$

If both limits exist, we may couple the two independent passages to $-\infty$ and ∞, and write[2]

(4)
$$\int_{-\infty}^{\infty} f(x)\, dx = \lim_{r \to \infty} \int_{-r}^{r} f(x)\, dx.$$

We assume that the function $f(x)$ in (3) is a real rational function whose denominator is different from zero for all real x and is of degree at least two units higher than the degree of the numerator. Then the limits in (4') exist, and we may start from (4). We consider the corresponding contour integral

(4*)
$$\int_C f(z)\, dz$$

around a path C, as shown in Fig. 308. Since $f(x)$ is rational, $f(z)$ has finitely many poles in the upper half-plane, and if we choose r large enough, C encloses

Fig. 308. Path of the contour integral in (4*)

[2] The expression on the right side of (4) is called the *Cauchy principal value* of the integral; it may exist even if the limits in (4') do not exist. For instance,

$$\lim_{r \to \infty} \int_{-r}^{r} x\, dx = \lim_{r \to \infty} \left(\frac{r^2}{2} - \frac{r^2}{2} \right) = 0, \quad \text{but} \quad \lim_{b \to \infty} \int_0^b x\, dx = \infty.$$

all these poles. By the residue theorem we then obtain

$$\int_C f(z)\,dz = \int_S f(z)\,dz + \int_{-r}^r f(x)\,dx = 2\pi i \sum \text{Res}\, f(z)$$

where the sum consists of all the residues of $f(z)$ at the points in the upper half-plane where $f(z)$ has a pole. From this we have

(5) $$\int_{-r}^r f(x)\,dx = 2\pi i \sum \text{Res}\, f(z) - \int_S f(z)\,dz.$$

We prove that, if $r \to \infty$, the value of the integral over the semicircle S approaches zero. If we set $z = re^{i\theta}$, then S is represented by $r = const$, and as z ranges along S the variable θ ranges from 0 to π. Since the degree of the denominator of $f(z)$ is at least two units higher than the degree of the numerator, we have

$$|f(z)| < \frac{k}{|z|^2} \qquad\qquad (|z| = r > r_0)$$

for sufficiently large constants k and r_0. By applying (4) in Sec. 13.2 we thus obtain

$$\left| \int_S f(z)\,dz \right| < \frac{k}{r^2}\,\pi r = \frac{k\pi}{r} \qquad\qquad (r > r_0).$$

Hence, as r approaches infinity, the value of the integral over S approaches zero, and (4) and (5) yield the result

(6) $$\int_{-\infty}^\infty f(x)\,dx = 2\pi i \sum \text{Res}\, f(z),$$

the sum being extended over all the residues of $f(z)$ corresponding to the poles of $f(z)$ in the upper half-plane.

Example 2.   Using (6), let us show that

$$\int_0^\infty \frac{dx}{1 + x^4} = \frac{\pi}{2\sqrt{2}}.$$

Indeed, $f(z) = 1/(1 + z^4)$ has four simple poles at the points

$$z_1 = e^{\pi i/4}, \qquad z_2 = e^{3\pi i/4}, \qquad z_3 = e^{-3\pi i/4}, \qquad z_4 = e^{-\pi i/4}.$$

The first two of these poles lie in the upper half-plane (Fig. 309). From (7) in Sec. 16.2 we find

$$\operatorname*{Res}_{z=z_1} f(z) = \left[\frac{1}{(1 + z^4)'} \right]_{z=z_1} = \left[\frac{1}{4z^3} \right]_{z=z_1} = \frac{1}{4} e^{-3\pi i/4} = -\frac{1}{4} e^{\pi i/4},$$

$$\operatorname*{Res}_{z=z_2} f(z) = \left[\frac{1}{(1 + z^4)'} \right]_{z=z_2} = \left[\frac{1}{4z^3} \right]_{z=z_2} = \frac{1}{4} e^{-9\pi i/4} = \frac{1}{4} e^{-\pi i/4}.$$

By (1) in Sec. 11.6 and (6) in the current section,

(7) $$\int_{-\infty}^\infty \frac{dx}{1 + x^4} = \frac{2\pi i}{4} (-e^{\pi i/4} + e^{-\pi i/4}) = \pi \sin\frac{\pi}{4} = \frac{\pi}{\sqrt{2}}.$$

Fig. 309. Example 2

Since $1/(1 + x^4)$ is an even function,

$$\int_0^\infty \frac{dx}{1 + x^4} = \frac{1}{2}\int_{-\infty}^\infty \frac{dx}{1 + x^4}.$$

From this and (7) the result follows.

C. Fourier integrals. Real integrals of the form

(8) $$\int_{-\infty}^\infty f(x)\cos sx\, dx \qquad \text{and} \qquad \int_{-\infty}^\infty f(x)\sin sx\, dx \qquad (s \text{ real})$$

occur in connection with the Fourier integral (cf. Sec. 9.9).

If $f(x)$ is a rational function satisfying the assumptions stated in connection with (3), then the integrals (8) may be evaluated in a similar way as the integrals (3).

In fact, we may then consider the corresponding integral

$$\int_C f(z)\, e^{isz}\, dz \qquad (s \text{ real and positive})$$

over the contour C in Fig. 308, and instead of (6) we now obtain the formula

(9) $$\int_{-\infty}^\infty f(x)e^{isx}\, dx = 2\pi i \sum \text{Res}\,[f(z)e^{isz}] \qquad (s > 0)$$

where the sum consists of the residues of $f(z)e^{isz}$ at its poles in the upper half-plane. Equating the real and the imaginary parts on both sides of (9), we have

(10) $$\int_{-\infty}^\infty f(x)\cos sx\, dx = -2\pi \sum \text{Im Res}\,[f(z)e^{isz}],$$
$$\int_{-\infty}^\infty f(x)\sin sx\, dx = 2\pi \sum \text{Re Res}\,[f(z)e^{isz}]. \qquad (s > 0)$$

We remember that (6) was established by proving that the value of the integral over the semicircle S in Fig. 308 approaches zero as $r \to \infty$. To establish (9) we should now prove the same fact for our present contour integral. This can be done as follows. Since S lies in the upper half-plane $y \geq 0$ and $s > 0$, it follows that

$$|e^{isz}| = |e^{isx}|\,|e^{-sy}| = e^{-sy} \leq 1 \qquad (s > 0, y \geq 0).$$

From this we obtain the inequality

$$|f(z)e^{isz}| = |f(z)|\,|e^{isz}| \leq |f(z)| \qquad (s > 0, y \geq 0)$$

which reduces our present problem to that in part *B* of this section. Continuing as before, we see that the value of the integral under consideration approaches zero as *r* approaches infinity, and (9) is now established.

Example 3. Using (10), let us show that

$$\int_{-\infty}^{\infty} \frac{\cos sx}{k^2 + x^2} \, dx = \frac{\pi}{k} e^{-ks}, \qquad \int_{-\infty}^{\infty} \frac{\sin sx}{k^2 + x^2} \, dx = 0 \qquad (s > 0, k > 0).$$

In fact, $e^{isz}/(k^2 + z^2)$ has only one pole in the upper half-plane, namely, a simple pole at $z = ik$, and from (7) in Sec. 16.2 we obtain

$$\operatorname*{Res}_{z=ik} \frac{e^{isz}}{k^2 + z^2} = \left[\frac{e^{isz}}{2z} \right]_{z=ik} = \frac{e^{-ks}}{2ik}.$$

Therefore,

$$\int_{-\infty}^{\infty} \frac{e^{isx}}{k^2 + x^2} \, dx = 2\pi i \frac{e^{-ks}}{2ik} = \frac{\pi}{k} e^{-ks}$$

and this yields the above results [cf. also (14) in Sec. 9.9].

D. Other types of real improper integrals are definite integrals

$$(11) \qquad \qquad \int_{A}^{B} f(x) \, dx$$

whose integrand becomes infinite at a point *a* in the interval of integration, that is,

$$\lim_{x \to a} |f(x)| = \infty.$$

Then the integral (11) means

$$(12) \qquad \int_{A}^{B} f(x) \, dx = \lim_{\epsilon \to 0} \int_{A}^{a-\epsilon} f(x) \, dx + \lim_{\eta \to 0} \int_{a+\eta}^{B} f(x) \, dx$$

where both ϵ and η approach zero independently and through positive values. It may happen that neither of these limits exists when ϵ, $\eta \to 0$ independently, but

$$(13) \qquad \lim_{\epsilon \to 0} \left[\int_{A}^{a-\epsilon} f(x) \, dx + \int_{a+\epsilon}^{B} f(x) \, dx \right]$$

exists; this is called the **Cauchy principal value** of the integral and is often written

$$\operatorname{pr. v.} \int_{A}^{B} f(x) \, dx.$$

For example,

$$\operatorname{pr. v.} \int_{-1}^{1} \frac{dx}{x^3} = \lim_{\epsilon \to 0} \left[\int_{-1}^{-\epsilon} \frac{dx}{x^3} + \int_{\epsilon}^{1} \frac{dx}{x^3} \right] = 0;$$

the principal value exists although the integral itself has no meaning. The whole situation is quite similar to that discussed in part *B* of this section.

To evaluate improper integrals whose integrands have poles on the real axis, we use a path which avoids these singularities by following small semicircles with centers at the singular points; the procedure may be illustrated by the following example.

Example 4. Show that

$$\int_0^\infty \frac{\sin x}{x}\, dx = \frac{\pi}{2} \qquad\qquad \text{(cf. Sec. 9.9).}$$

We do not consider $(\sin z)/z$ because this function does not behave suitably at infinity. We consider e^{iz}/z, which has a simple pole at $z = 0$, and integrate around the contour shown in Fig. 310. Since e^{iz}/z is analytic inside and on C, it follows from Cauchy's integral theorem that

$$(14) \qquad\qquad \int_C \frac{e^{iz}}{z}\, dz = 0.$$

We show that the value of the integral over the large semicircle C_1 approaches zero as R approaches infinity. Setting $z = Re^{i\theta}$ we have $dz = iRe^{i\theta}\, d\theta$, $dz/z = i\, d\theta$ and therefore

$$\left| \int_{C_1} \frac{e^{iz}}{z}\, dz \right| = \left| \int_0^\pi e^{iz} i\, d\theta \right| \leqq \int_0^\pi |e^{iz}|\, d\theta \qquad\qquad (z = Re^{i\theta}).$$

In the integrand on the right,

$$|e^{iz}| = |e^{iR(\cos\theta + i\sin\theta)}| = |e^{iR\cos\theta}|\, |e^{-R\sin\theta}| = e^{-R\sin\theta}.$$

By inserting this and using $\sin(\pi - \theta) = \sin\theta$ we obtain

$$\int_0^\pi |e^{iz}|\, d\theta = \int_0^\pi e^{-R\sin\theta}\, d\theta = 2\int_0^{\pi/2} e^{-R\sin\theta}\, d\theta = 2\left[\int_0^\epsilon e^{-R\sin\theta}\, d\theta + \int_\epsilon^{\pi/2} e^{-R\sin\theta}\, d\theta \right]$$

where ϵ is any value between 0 and $\pi/2$. The absolute value of the integrand in the first and the last integral on the right is at most equal to 1 and $e^{-R\sin\epsilon}$, respectively, because the integrand is a monotone decreasing function of θ in the interval of integration. Consequently, the whole expression on the right is smaller than

$$2\left[\int_0^\epsilon d\theta + e^{-R\sin\epsilon} \int_\epsilon^{\pi/2} d\theta \right] = 2\left[\epsilon + e^{-R\sin\epsilon}\left(\frac{\pi}{2} - \epsilon \right) \right] < 2\epsilon + \pi e^{-R\sin\epsilon}.$$

Altogether,

$$\left| \int_{C_1} \frac{e^{iz}}{z}\, dz \right| < 2\epsilon + \pi e^{-R\sin\epsilon}.$$

We first take ϵ arbitrarily small. Then, having fixed ϵ, the last term can be made as small as we please by choosing R sufficiently large. Hence, the value of the integral along C_1 approaches zero as R approaches infinity.

For the integral over the small semicircle C_2 in Fig. 310 we have

$$\int_{C_2} \frac{e^{iz}}{z}\, dz = \int_{C_2} \frac{dz}{z} + \int_{C_2} \frac{e^{iz} - 1}{z}\, dz.$$

The first integral on the right has the value $-\pi i$. The integrand of the last integral is analytic at $z = 0$ and, therefore, bounded in absolute value as $\rho \to 0$. From this and (4) in Sec. 13.2 we

Fig. 310. Example 4

conclude that the value of that integral approaches zero as $\rho \to 0$. From (14) we thus obtain

$$\text{pr. v.} \int_{-\infty}^{\infty} \frac{e^{ix}}{x} \, dx = -\lim_{\rho \to 0} \int_{C_2} \frac{e^{iz}}{z} \, dz = +\pi i,$$

and by taking the imaginary parts on both sides

(15)
$$\text{pr. v.} \int_{-\infty}^{\infty} \frac{\sin x}{x} \, dx = \pi.$$

Now the integrand in (15) is not singular at $x = 0$. Furthermore, since for positive x the function $1/x$ decreases, the areas under the curve of the integrand between two consecutive positive zeros decrease in a monotone fashion, that is, the absolute values of the integrals

$$I_n = \int_{n\pi}^{n\pi+\pi} \frac{\sin x}{x} \, dx \qquad\qquad n = 0, 1, \cdots$$

form a monotone decreasing sequence $|I_1|, |I_2|, \cdots$, and $I_n \to 0$ as $n \to \infty$. Since these integrals have alternating sign, it follows from the Leibniz criterion in Sec. 14.2 that the infinite series $I_0 + I_1 + I_2 + \cdots$ converges. Clearly, the sum of the series is the integral

$$\int_0^\infty \frac{\sin x}{x} \, dx = \lim_{b \to \infty} \int_0^b \frac{\sin x}{x} \, dx,$$

which therefore exists. Similarly, the integral from 0 to $-\infty$ exists. Hence we need not take the principal value in (15), and

$$\int_{-\infty}^{\infty} \frac{\sin x}{x} \, dx = \pi.$$

Since the integrand is an even function, the desired result follows.

Problems

Evaluate the following real integrals.

1. $\int_0^{\pi} \dfrac{d\theta}{3 + 2\cos\theta}$

2. $\int_0^{2\pi} \dfrac{d\theta}{5 - 4\cos\theta}$

3. $\int_0^{2\pi} \dfrac{d\theta}{17 - 8\cos\theta}$

4. $\int_0^{2\pi} \dfrac{\cos\theta}{\cos\theta - 17/8} \, d\theta$

5. $\int_0^{2\pi} \dfrac{d\theta}{5 - 4\sin\theta}$

6. $\int_0^{2\pi} \dfrac{\sin^2\theta - 2\cos\theta}{2 + \cos\theta} \, d\theta$

7. $\int_0^{2\pi} \dfrac{\cos^2\theta}{13 - 5\cos\theta} \, d\theta$

8. $\int_0^{\pi} \dfrac{\cos^2 3\theta}{5 - 4\cos 2\theta} \, d\theta$

9. $\int_0^{2\pi} \dfrac{\cos\theta}{13 - 12\cos 2\theta} \, d\theta$

10. (*a*) Using direct calculation, (*b*) using Ex. 1, prove that

$$\int_0^{2\pi} \frac{d\theta}{1 + k^2 - 2k\sin\theta} = \frac{2\pi}{1 - k^2} \qquad\qquad (|k| < 1).$$

11. Derive (10) from (9).

Evaluate the following real integrals.

12. $\int_0^{\infty} \dfrac{dx}{x^2 + 1}$

13. $\int_{-\infty}^{\infty} \dfrac{dx}{(1 + x^2)^2}$

14. $\int_{-\infty}^{\infty} \dfrac{1 + x^2}{1 + x^4} \, dx$

15. $\displaystyle\int_{-\infty}^{\infty} \frac{dx}{(1+x^2)^3}$ **16.** $\displaystyle\int_{-\infty}^{\infty} \frac{x}{1+x^8}\, dx$ **17.** $\displaystyle\int_{-\infty}^{\infty} \frac{x}{(x^2-2x+2)^2}\, dx$

18. $\displaystyle\int_{-\infty}^{\infty} \frac{dx}{x^4+10x^2+9}$ **19.** $\displaystyle\int_{-\infty}^{\infty} \frac{dx}{(x^2+1)(x^2+4)(x^2+9)}$

20. $\displaystyle\int_{-\infty}^{\infty} \frac{\sin 2x}{x^2+x+1}\, dx$ **21.** $\displaystyle\int_{-\infty}^{\infty} \frac{\cos x}{x^4+1}\, dx$ **22.** $\displaystyle\int_{-\infty}^{\infty} \frac{\cos 4x}{(x^2+1)(x^2+4)}\, dx$

23. Solve Prob. 12 by elementary methods.

24. Find pr. v. $\displaystyle\int_{-\infty}^{\infty} \frac{dx}{(x+1)(x^2+2)}$ by integrating along the contour in Fig. 311.

Fig. 311. Problem 24

25. Integrating e^{-z^2} around the boundary of the rectangle with vertices $-a,\ a,\ a+ib,$ $-a+ib$, letting $a \to \infty$, and using

$$\int_0^\infty e^{-x^2}\, dx = \frac{\sqrt{\pi}}{2}, \qquad \text{show that} \qquad \int_0^\infty e^{-x^2}\cos 2bx\, dx = \frac{\sqrt{\pi}}{2}e^{-b^2}.$$

CHAPTER 17

Complex Analytic Functions and Potential Theory

Laplace's equation $\nabla^2 u = 0$ is one of the most important partial differential equations in engineering mathematics, because it occurs in connection with gravitational fields (Sec. 7.8), electrostatic fields (Sec. 10.11), steady-state heat conduction (Sec. 12.5), incompressible fluid flow, etc. The theory of the solutions of this equation is called *potential theory*.

In the "two-dimensional case" when u depends only on two Cartesian coordinates x and y, Laplace's equation becomes

$$\nabla^2 u = u_{xx} + u_{yy} = 0.$$

We know that then its solutions are closely related to complex analytic functions (cf. Sec. 11.3).[1] We shall now consider this connection and its consequences in more detail and illustrate it by practical examples taken from hydrodynamics and electrostatics. In Sec. 17.3 we shall see that results about analytic functions can be used for characterizing various general properties of harmonic functions. Finally, we shall derive an important general formula (Poisson's integral formula) for the solution of boundary value problems involving Laplace's equation in a circular disk.

> *Prerequisites for this chapter:* Chaps. 11–14.
> *References:* Appendix 1, Part F.
> *Answers to problems:* Appendix 2.

17.1 Electrostatic Fields

The electrical force of attraction or repulsion between charged particles is governed by Coulomb's law. This force is the gradient of a function u, called the *electrostatic potential*, and at any points free of charges u is a solution of Laplace's equation

$$\nabla^2 u = 0.$$

---

[1] No such close relation exists in the three-dimensional case.

Fig. 312. Potential in Example 1   **Fig. 313.** Potential in Example 2

Cf. Sec. 10.11. The surfaces $u = const$ are called **equipotential surfaces.** At each point P the gradient of u is perpendicular to the surface $u = const$ through P, that is, the electrical force has the direction perpendicular to the equipotential surface.

Example 1. Find the potential of the field between two parallel conducting plates extending to infinity (Fig. 312), which are kept at potentials U_1 and U_2, respectively. From the shape of the plates it follows that u depends only on x, and Laplace's equation becomes $u'' = 0$. By integrating twice we obtain $u = ax + b$ where the constants a and b are determined by the given boundary values of u on the plates. For example, if the plates correspond to $x = -1$ and $x = 1$, the solution is

$$u(x) = \tfrac{1}{2}(U_2 - U_1)x + \tfrac{1}{2}(U_2 + U_1).$$

The equipotential surfaces are parallel planes.

Example 2. Find the potential between two coaxial conducting cylinders which extend to infinity on both sides (Fig. 313) and are kept at potentials U_1 and U_2, respectively. Here u depends only on $r = \sqrt{x^2 + y^2}$, for reasons of symmetry, and Laplace's equation becomes

$$ru'' + u' = 0 \qquad\qquad \text{[cf. (4) in Sec. 10.9].}$$

By separating variables and integrating we obtain

$$\frac{u''}{u'} = -\frac{1}{r}, \qquad \ln u' = -\ln r + \tilde{a}, \qquad u' = \frac{a}{r}, \qquad u = a \ln r + b,$$

and a and b are determined by the given values of u on the cylinders. Although no infinitely extended conductors exist, the field in our idealized conductor will approximate the field in a long finite conductor in that part which is far away from the ends of the two cylinders.

If the potential u depends only on two Cartesian coordinates x and y, then Laplace's equation becomes

$$(1) \qquad\qquad \nabla^2 u = \frac{\partial^2 u}{\partial x^2} + \frac{\partial^2 u}{\partial y^2} = 0.$$

The equipotential surfaces $u = const$ appear as *equipotential lines* in the xy-plane.

We make the general assumption that $u(x, y)$ is *harmonic*, that is, its second partial derivatives are continuous. Then, if $v(x, y)$ is a conjugate harmonic function of $u(x, y)$ (cf. Sec. 11.3), the function

$$F(z) = u(x, y) + iv(x, y)$$

is an analytic function of $z = x + iy$. This function is called the **complex potential** corresponding to the real potential u. Remember that for a given u the conjugate is uniquely determined, except for an additive real constant.

Since the lines $v = const$ intersect the equipotential lines $u = const$ at right angles [except at points where $F'(z) = 0$], they have the direction of the electrical force and, therefore, are called **lines of force.**

Example 3. In Ex. 1, a conjugate is $v = ay$. It follows that the complex potential is

$$F(z) = az + b = ax + b + iay,$$

and the lines of force are straight lines parallel to the x-axis.

Example 4. In Ex. 2 we have

$$u = a \ln r + b = a \ln |z| + b.$$

A conjugate is $v = a \arg z$. The complex potential is $F(z) = a \ln z + b$, and the lines of force are straight lines through the origin. $F(z)$ may also be interpreted as the complex potential of a source line whose trace in the xy-plane is the origin.

Often more complicated potentials can be obtained by superposition. This may be illustrated by

Example 5. Determine the complex potential of a pair of oppositely charged source lines of the same strength at the points $z = x_1$ and $z = x_2$. From Exs. 2 and 4 it follows that the potential of each of the source lines is

$$u_1 = -c \ln |z - x_1| \quad \text{and} \quad u_2 = c \ln |z - x_2|,$$

respectively. These are the real parts of the complex potentials

$$F_1(z) = -c \ln (z - x_1) \quad \text{and} \quad F_2(z) = c \ln (z - x_2).$$

Hence, the complex potential of the combination of the two source lines is

(2) $$F(z) = F_1(z) + F_2(z) = c \ln \frac{z - x_2}{z - x_1}.$$

The equipotential lines are the circles

$$u = \operatorname{Re} F(z) = c \ln \frac{|z - x_2|}{|z - x_1|} = const,$$

and the lines of force are the curves

$$v = \operatorname{Im} F(z) = c \arg \frac{z - x_2}{z - x_1} = c[\arg (z - x_2) - \arg (z - x_1)] = const,$$

Fig. 314. Example 5

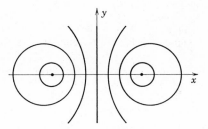

Fig. 315. Potential field of two oppositely charged source lines of the same strength

that is,

$$v = c(\theta_2 - \theta_1) = const$$

(cf. Fig. 314). Now, $|\theta_2 - \theta_1|$ is the angle between the line segments from z to x_1 and x_2. The lines of force, therefore, are curves along each of which the line segment $x_1 x_2$ appears under a constant angle; these curves are the totality of circular arcs over $x_1 x_2$, as is well known from elementary geometry. The function (2) may also be interpreted as the complex potential of a capacitor consisting of two circular cylinders whose axes are parallel but do not coincide (Fig. 315).

Problems

Find the potential u between two infinite coaxial cylinders of revolution of radii r_1 and r_2 ($>r_1$) which are kept on the potentials U_1 and U_2, respectively, where

1. $r_1 = 1$, $r_2 = 2$, $U_1 = 200$ V, $U_2 = 400$ V
2. $r_1 = 0.5$, $r_2 = 3$, $U_1 = 0$, $U_2 = 110$ V
3. $r_1 = 0.1$, $r_2 = 1$, $U_1 = 1000$ V, $U_2 = 100$ V

4. Show that $F(z) = \cos^{-1} z$ may be interpreted as the complex potential of the configurations in Figs. 316 to 318.
5. Show that $F(z) = \cosh^{-1} z$ may be interpreted as the complex potential between two confocal elliptic cylinders.
6. Find the potential of two oppositely charged source lines at the points $z = a$ and $z = -a$. Show the equipotential lines graphically.
7. Find the potential of two source lines at $z = a$ and $z = -a$ having the same charge.

Fig. 316. Problem 4   **Fig. 317.** Problem 4   **Fig. 318.** Problem 4

8. Find the potential between the cylinders $(x + \frac{5}{3})^2 + y^2 = 1$ and $(x - \frac{5}{3})^2 + y^2 = 1$, assuming that on these cylinders $u = -1$ and $u = 1$, respectively.

9. Find and sketch the equipotential lines of the complex potential $F(z) = z^2$.

10. Find and sketch the equipotential lines of the complex potential $F(z) = 1/z$.

17.2   *Two-Dimensional Fluid Flow*

Harmonic functions play an important role in hydrodynamics. To illustrate this, let us consider the two-dimensional steady motion of a nonviscous fluid. Here "two-dimensional" means that the motion of the fluid is the same in all planes parallel to the xy-plane, the velocity being parallel to that plane. It then suffices to consider the motion of the fluid in the xy-plane. "Steady" means that the velocity is independent of the time.

At any point (x, y) the flow has a certain velocity which is determined by its magnitude and its direction and is thus a vector. Since in the complex plane any number a represents a vector (the vector from the origin to the point corresponding to a), we may represent the velocity of the flow by a complex variable, say,

$$(1) \qquad\qquad V = V_1 + iV_2.$$

Then V_1 and V_2 are the components of the velocity in the x and y directions, and V is tangential to the paths of the moving particles of the fluid. Such a path is called a **streamline** of the motion. Cf. Fig. 319.

Consider now an arbitrary given smooth curve C. Let s denote the arc length of C, and let the real variable V_t be the component of the velocity V tangent to C (Fig. 320). Then the value of the line integral

$$(2) \qquad\qquad \int_C V_t \, ds$$

taken along C in the sense of increasing values of s is called the **circulation** of the fluid along C. Dividing the circulation by the length of C, we obtain the mean velocity[2] of the flow along the curve C. Now

$$V_t = |V| \cos \alpha \qquad\qquad \text{(cf. Fig. 320)}.$$

Consequently, V_t is the scalar product (Sec. 5.5) of V and the unit tangent vector (cf. Sec. 12.2)

$$\frac{dz}{ds} = \frac{dx}{ds} + i\frac{dy}{ds}$$

[2] *Definitions:* $\dfrac{1}{b-a} \displaystyle\int_a^b f(x)\, dx$ = mean value of f on the interval $a \leq x \leq b$.

$\qquad\qquad \dfrac{1}{l} \displaystyle\int_C f(s)\, ds$ = mean value of f on C. (l = length of C)

$\qquad\qquad \dfrac{1}{A} \displaystyle\iint_D f(x, y)\, dx\, dy$ = mean value of f on D. (A = area of D)

Fig. 319. Velocity

Fig. 320. Tangential component of the velocity with respect to a curve C

to C, where $z(s) = x(s) + iy(s)$ is a representation of C. Therefore, the product $V_t \, ds$ may be written

$$V_t \, ds = V \cdot dz = V_1 \, dx + V_2 \, dy \qquad (dz = dx + i \, dy).$$

Now let C be a *closed* curve, namely, the boundary curve of a simply connected domain D. Then, if V has continuous partial derivatives in a domain containing D and C it follows from Green's theorem (Sec. 8.4) that the circulation around C can be represented by a double integral,

$$(3) \qquad \int_C (V_1 \, dx + V_2 \, dy) = \iint_D \left(\frac{\partial V_2}{\partial x} - \frac{\partial V_1}{\partial y} \right) dx \, dy.$$

The function in the integral on the right permits a simple physical interpretation as follows. Let C be a circle of radius r. The circulation divided by $2\pi r$ then represents the mean velocity of the fluid along C, and the mean angular velocity ω_0 of the fluid about the axis of the circle is obtained by dividing the mean velocity by the radius r; thus

$$\omega_0 = \frac{1}{\pi r^2} \iint_D \frac{1}{2} \left(\frac{\partial V_2}{\partial x} - \frac{\partial V_1}{\partial y} \right) dx \, dy.$$

The expression on the right is the mean value[3] of the function

$$(4) \qquad \omega = \frac{1}{2} \left(\frac{\partial V_2}{\partial x} - \frac{\partial V_1}{\partial y} \right)$$

on the disk D bounded by the circle C. The function ω is called the **rotation**, and 2ω is called the **vorticity** of the motion. If $r \to 0$, the limit of that expression is the value of ω at the center of C. Hence, $\omega(x, y)$ is the limiting angular velocity of a circular element of the fluid as the circle shrinks to the point (x, y). Roughly speaking, if a spherical element of the fluid were suddenly solidified and the surrounding fluid simultaneously annihilated, the element would rotate with the angular velocity ω (cf. also Sec. 7.11).

[3] Cf. the previous footnote.

We consider only *irrotational flows*, that is, flows for which ω is zero everywhere in the region D of the flow, that is

(5)
$$\frac{\partial V_2}{\partial x} - \frac{\partial V_1}{\partial y} = 0,$$

the existence and continuity of the derivatives being assumed.

We further assume that the fluid is *incompressible*. Then

(6)
$$\frac{\partial V_1}{\partial x} + \frac{\partial V_2}{\partial y} = 0 \qquad \text{[cf. (9), Sec. 7.10]}$$

in every region which is free of **sources** or **sinks,** that is, points at which fluid is produced or disappears.

If D is a simply connected domain and the flow is irrotational, it follows from Theorem 3 in Sec. 8.12 that the line integral

(7)
$$\int_C (V_1 \, dx + V_2 \, dy)$$

is independent of path in D. If we integrate from a fixed point (a, b) in D to a variable point (x, y) in D, the integral thus becomes a function of the point (x, y), say, $\Phi(x, y)$:

(8)
$$\Phi(x, y) = \int_{(a,b)}^{(x,y)} (V_1 \, dx + V_2 \, dy).$$

The function $\Phi(x, y)$ is called the **velocity potential**[4] of the motion. Since the integral is independent of path, $V_1 \, dx + V_2 \, dy$ is an exact differential (Sec. 8.12), namely, the differential of the function $\Phi(x, y)$, that is,

(9)
$$V_1 \, dx + V_2 \, dy = \frac{\partial \Phi}{\partial x} \, dx + \frac{\partial \Phi}{\partial y} \, dy.$$

It follows that

(10)
$$V_1 = \frac{\partial \Phi}{\partial x}, \qquad V_2 = \frac{\partial \Phi}{\partial y},$$

and we see that the velocity vector is the gradient of $\Phi(x, y)$:

(11)
$$V = V_1 + iV_2 = \frac{\partial \Phi}{\partial x} + i \frac{\partial \Phi}{\partial y} \qquad \text{(cf. Sec. 7.8).}$$

The curves $\Phi(x, y) = const$ are called **equipotential lines.** Since V is the gradient of Φ, at each point V is perpendicular to the equipotential line through that point (provided $V \neq 0$).

Furthermore, by inserting (10) into (6) we see that Φ satisfies Laplace's equation

$$\nabla^2 \Phi = \frac{\partial^2 \Phi}{\partial x^2} + \frac{\partial^2 \Phi}{\partial y^2} = 0.$$

[4] Some authors use $-\Phi$ (instead of Φ) as the velocity potential.

Let $\Psi(x, y)$ be a conjugate harmonic function of $\Phi(x, y)$. Then at each point the curves

$$\Psi(x, y) = const$$

are perpendicular[5] to the equipotential lines $\Phi(x, y) = const$. Hence, their tangents have the direction of the velocity of the fluid. Consequently, *the curves* $\Psi(x, y) = const$ *are the streamlines of the flow*. The function $\Psi(x, y)$ is called the **stream function** of this flow.

We assume that both Φ and Ψ have continuous second partial derivatives. Then the complex function

(12) $$F(z) = \Phi(x, y) + i\Psi(x, y)$$

is analytic in the region of the flow. This function is called the **complex potential** of the flow. Working with the complex potential is simpler than working with Φ and Ψ separately.

The velocity of the flow can be obtained by differentiating (12) and using the Cauchy-Riemann equations; we find

$$F'(z) = \frac{\partial \Phi}{\partial x} + i\frac{\partial \Psi}{\partial x} = \frac{\partial \Phi}{\partial x} - i\frac{\partial \Phi}{\partial y} = V_1 - iV_2.$$

From this it follows that

(13) $$V = V_1 + iV_2 = \overline{F'(z)}.$$

In this way *two-dimensional irrotational steady flows of incompressible fluids may be described in terms of analytic functions*, and the methods of complex analysis, such as conformal mapping, may be employed.

In connection with boundary value problems the stream function Ψ is important, because a boundary across which fluid cannot flow is a streamline. Under conformal mapping, a streamline transforms into a streamline in the image plane. Another possibility for obtaining and investigating complicated flows is the *superposition of simple flows*. The sum $F = F_1 + F_2$ of the complex potentials F_1, F_2 of two flows is the complex potential of the flow which is obtained by vector addition of the velocity vectors of the two flows. Clearly, since Laplace's equation is linear and homogeneous, the sum of two harmonic functions is a harmonic function.

Note that while in electrostatics the given boundaries (conducting plates) are equipotential lines, in hydrodynamics the boundaries are streamlines and thus are orthogonal to the equipotential lines.

Problems

1. **(Parallel flow)** Show that $F(z) = Kz$ (K positive real) describes a uniform flow to the right, which can be interpreted as a uniform flow between two parallel lines (between two parallel planes in three-dimensional space). Cf. Fig. 321. Find the velocity vector, the streamlines, and the equipotential lines.

---

[5] Except for points where $F'(z) = 0$ [cf. (12)].

Fig. 321. Parallel flow   Fig. 322. Flow around a corner

2. **(Flow around a corner)** Show that $F(z) = z^2 = x^2 - y^2 + 2ixy$ describes a flow around a corner (Fig. 322). Find the streamlines, the equipotential lines, and the velocity vector V. Show that along any streamline S the speed has a minimum at the point P where the cross section of the channel is large.

3. Obtain the flow in Prob. 2 from that in Prob. 1 by a conformal mapping of the first quadrant onto the upper half-plane.

4. Sketch the streamlines and equipotential lines of the flow with complex potential $F(z) = z^3$. Find the velocity vector V and determine all points at which that vector is parallel to the x-axis.

5. **(Source and sink)** Consider the complex potential $F(z) = (c/2\pi) \ln z$ where c is positive real. Show that $V = (c/2\pi r^2)(x + iy)$ where $r = \sqrt{x^2 + y^2}$, and this implies that the flow is directed radially outward (Fig. 323), so that this potential corresponds to a **point source** at $z = 0$ (that is, a **source line** $x = 0, y = 0$ in space). (The constant c is called the **strength** or **discharge** of the source. If c is negative real, the flow is said to have a **sink** at $z = 0$, it is directed radially inward and fluid disappears at the singular point $z = 0$ of the complex potential.)

Fig. 323. Point source

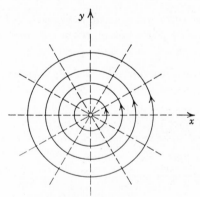

Fig. 324. Vortex flow

6. **(Vortex line)** Show that $F(z) = -(iK/2\pi) \ln z$ with real positive K describes a flow which circles around the origin in the counterclockwise sense (Fig. 324). (The point $z = 0$ is a **vortex**; the potential increases by K each time we travel around the vortex.)

7. Find the complex potential of a flow which has a point source of strength 1 at $z = -a$.

8. Find the complex potential of a flow which has a point sink of strength 1 at $z = a$.

9. Show that vector addition of the velocity vectors of two flows leads to a flow whose complex potential is obtained by adding those of the two flows.

10. Add the potentials in Probs. 7 and 8 and show the streamlines graphically.

11. Find the streamlines of the flow corresponding to $F(z) = 1/z$. Show that for small $|a|$ the streamlines in Prob. 10 look similar to those in the present problem.

12. Show that $F(z) = \cosh^{-1} z$ corresponds to a flow whose streamlines are confocal hyperbolas with foci at $z = \pm 1$, and the flow may be interpreted as a flow through an aperture (cf. Fig. 325).

 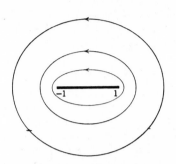

Fig. 325. Flow through an aperture   **Fig. 326.** Flow around a plate

13. Show that $F(z) = \cos^{-1} z$ can be interpreted as the complex potential of a flow circulating around an elliptic cylinder or a plate (the straight segment from $z = -1$ to $z = 1$). Show that the streamlines are confocal ellipses with foci at $z = \pm 1$. (Cf. Fig. 326.)

14. (Flow around a cylinder) Consider $F(z) = z + z^{-1}$. Setting $z = re^{i\theta}$, show that the streamlines are $(r - r^{-1}) \sin \theta = const$, the streamline $(r - r^{-1}) \sin \theta = 0$ consists of the x-axis and the unit circle, and for large $|z|$ the flow is nearly uniform and parallel, so that it may be interpreted as the flow around a long circular cylinder of unit radius (Fig. 327, left upper part). Find the **stagnation points** of the flow (points at which the velocity is zero).

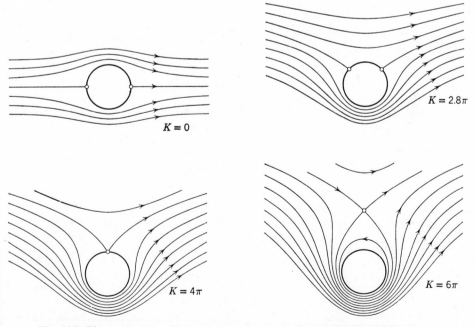

Fig. 327. Flow around a cylinder without circulation ($K = 0$) and with circulation

15. **(Flow with circulation around a cylinder)**   Show that addition of the potentials in Probs. 6 and 14 gives a flow such that the cylinder wall $|z| = 1$ is a streamline. Find the speed and show that the stagnation points are

$$z = \frac{iK}{4\pi} \pm \sqrt{\frac{-K^2}{16\pi^2} + 1};$$

when $K = 0$ they are at ± 1, as K increases they move up on the unit circle until they unite at $z = i$ ($K = 4\pi$, cf. Fig. 327), and when $K > 4\pi$ they lie on the imaginary axis (one lies in the field of flow and the other one lies inside the cylinder and has no physical meaning).

17.3 *General Properties of Harmonic Functions*

In this section we want to show how results about complex analytic functions can be used for deriving general properties of harmonic functions.

If $u(x, y)$ is any given harmonic function, we may determine a conjugate harmonic function $v(x, y)$ by the use of the Cauchy-Riemann equations, and $f(z) = u(x, y) + iv(x, y)$ is then an analytic function (cf. Sec. 11.3). This shows that any harmonic function can be regarded as the real part of a suitable analytic function. Since analytic functions have derivatives of all orders, we immediately obtain the following result.

Theorem 1.   *A function $u(x, y)$ which is harmonic in a domain D has partial derivatives of all orders in D.*

If $f(z)$ is analytic in a simply connected domain D, then, by Cauchy's integral formula (Sec. 13.5),

(1*)
$$f(z_0) = \frac{1}{2\pi i} \int_C \frac{f(z)}{z - z_0}\, dz$$

where C is a simple closed path in D and z_0 lies inside C. Choosing for C a circle

$$z = z_0 + re^{i\phi}$$

in D, we have

$$z - z_0 = re^{i\phi}, \qquad dz = ire^{i\phi}\, d\phi$$

and the integral formula becomes

(1)
$$f(z_0) = \frac{1}{2\pi} \int_0^{2\pi} f(z_0 + re^{i\phi})\, d\phi.$$

The right side is the mean value of f on the circle (= value of the integral divided by the length of the interval of integration). This proves

Theorem 2 (Mean value property).   *Let $f(z)$ be analytic in a simply connected domain D. Then the value of $f(z)$ at a point z_0 in D is equal to the mean value of $f(z)$ on any circle in D with center at z_0.*

Fig. 328. Proof of Theorem 3

Another important property of analytic functions is the following one.

Theorem 3 (Maximum modulus theorem). *Let D be a bounded region and suppose that $f(z)$ is analytic and nonconstant in D and on the boundary of D. Then the absolute value $|f(z)|$ cannot have a maximum at an interior point of D. Consequently, the maximum of $|f(z)|$ is taken on the boundary of D. If $f(z) \neq 0$ in D, the same is true with respect to the minimum of $|f(z)|$.*

Proof. We show that the assumption that $|f(z)|$ has a maximum at an interior point of D leads to a contradiction. Let z_0 be that point, and let $|f(z_0)| = M$ be that maximum. Since $f(z)$ is not constant, $|f(z)|$ is not constant. Consequently, we can find a circle C of radius r with center at z_0 such that the interior of C is in D and $|f(z)|$ is smaller than M at some point P of C. Since $|f(z)|$ is continuous, it will be smaller than M on an arc C_1 of C which contains P, say, $|f(z)| \leq M - \epsilon$ ($\epsilon > 0$) for all z on C_1 (Fig. 328). If C_1 has the length l_1, the complementary arc C_2 of C has the length $2\pi r - l_1$. By applying (4), Sec. 13.2, to (1*), this section, and noting that $|z - z_0| = r$, we thus obtain

$$M = |f(z_0)| \leq \frac{1}{2\pi} \left| \int_{C_1} \frac{f(z)}{z - z_0} \, dz \right| + \frac{1}{2\pi} \left| \int_{C_2} \frac{f(z)}{z - z_0} \, dz \right|$$

$$\leq \frac{1}{2\pi} (M - \epsilon) \frac{1}{r} l_1 + \frac{1}{2\pi} M \frac{1}{r} (2\pi r - l_1) = M - \frac{\epsilon l_1}{2\pi r} < M,$$

that is, $M < M$, which is impossible. Hence, our assumption is false and the first statement of the theorem is proved.

We prove the last statement. If $f(z) \neq 0$ in D, then $1/f(z)$ is analytic in D. From the statement already proved it follows that the maximum of $1/|f(z)|$ lies on the boundary of D. But this maximum corresponds to a minimum of $|f(z)|$. This completes the proof.

From these theorems we shall now derive corresponding results about harmonic functions.

Theorem 4. *Let D be a simply connected bounded domain and let C be its boundary curve. Then if $u(x, y)$ is harmonic in a domain containing D and C, it has the following properties.*

I. The value of $u(x, y)$ at a point (x_0, y_0) of D is equal to the mean value of $u(x, y)$ on any circle in D with center at (x_0, y_0).

II. The value of $u(x, y)$ at the point (x_0, y_0) is equal to the mean value of $u(x, y)$ in any circular disk in D with center at (x_0, y_0). [Cf. footnote 2 in Sec. 17.2.]

III. **(Maximum principle)**   *If $u(x, y)$ is nonconstant, it has neither a maximum nor a minimum in D. Consequently, the maximum and the minimum are taken on the boundary of D.*

IV. If $u(x, y)$ is constant on C, then $u(x, y)$ is a constant.

V. If $h(x, y)$ is harmonic in D and on C and if $h(x, y) = u(x, y)$ on C, then $h(x, y) = u(x, y)$ everywhere in D.

Proof.   Statement I follows from (1) by taking the real parts on both sides, that is,

$$u(x_0, y_0) = \operatorname{Re} f(x_0 + iy_0) = \frac{1}{2\pi} \int_0^{2\pi} u(x_0 + r \cos \phi, y_0 + r \sin \phi) \, d\phi.$$

If we multiply both sides by r and integrate over r from 0 to r_0 where r_0 is the radius of a circular disk in D with center at (x_0, y_0), then we obtain on the left-hand side $\frac{1}{2}r_0^2 u(x_0, y_0)$ and therefore

$$u(x_0, y_0) = \frac{1}{\pi r_0^2} \int_0^{r_0} \int_0^{2\pi} u(x_0 + r \cos \phi, y_0 + r \sin \phi) r \, d\phi \, dr.$$

This proves the second statement.

We prove statement III. Let $v(x, y)$ be a conjugate harmonic function of $u(x, y)$ in D. Then $f(z) = u(x, y) + iv(x, y)$ is analytic in D, and so is

$$F(z) = e^{f(z)}.$$

The absolute value is

$$|F(z)| = e^{\operatorname{Re} f(z)} = e^{u(x, y)}.$$

From Theorem 3 it follows that $|F(z)|$ cannot have a maximum at an interior point of D. Since e^u is a monotone increasing function of the real variable u, statement III about the maximum of u follows. From this, the statement about the minimum follows by replacing u by $-u$.

If u is constant on C, say, $u = k$, then by III the maximum and the minimum of u are equal. From this, statement IV follows.

If h and u are harmonic in D and on C, then $h - u$ is also harmonic in D and on C, and by assumption, $h - u = 0$ everywhere on C. Hence, by IV, we have $h - u = 0$ everywhere in D, and statement V is proved. This completes the proof of Theorem 4.

The last statement of Theorem 4 is very important. It means that *a harmonic function is uniquely determined in D by its values on the boundary of D.* Usually, $u(x, y)$ is required to be harmonic in D and continuous on the boundary[6] of D. Under these circumstances the maximum principle (Theorem 4, III) is still applicable. The problem of determining $u(x, y)$ when the boundary values are given is known as the **Dirichlet problem** for the Laplace equation in two variables. From Theorem 4, V we have

---

[6] That is, $\lim\limits_{\substack{x \to x_0 \\ y \to y_0}} u(x, y) = u(x_0, y_0)$ where (x_0, y_0) is on the boundary and (x, y) is in D.

Theorem 5. *If for a given region and given boundary values the Dirichlet problem for the Laplace equation in two variables has a solution, the solution is unique.*

Problems

1. Verify Theorem 2 for $f(z) = (z + 2)^2$, $z_0 = 1$ and a circle of radius 1 with center at z_0.
2. Verify Theorem 3 for $f(z) = z^2$ and the rectangle $-2 < x < 2$, $-1 < y < 1$.
3. Verify Theorem 3 for $f(z) = e^z$ and any bounded domain.
4. The function $f(x) = \cos x$ has a maximum at $x = 0$. Using Theorem 3, conclude that the modular surface of $f(z) = \cos z$ (cf. Sec. 11.6) cannot have a summit at $z = 0$.
5. If $f(z)$ is analytic (not a constant) in a simply connected domain D, and the curve given by $|f(z)| = c$ (c any fixed constant) lies in D and is closed, show that $f(z) = 0$ at a point in the interior of that curve. Give examples.

17.4 *Poisson's Integral Formula*

The Dirichlet problem for a circular disk can be solved by the use of the so-called Poisson[7] formula which represents a harmonic function in terms of its values given on the boundary circle of the disk. We shall derive this formula, starting from Cauchy's integral formula

$$(1) \qquad f(z) = \frac{1}{2\pi i} \int_C \frac{f(z^*)}{z^* - z} \, dz^*;$$

here C is the circle represented by

$$z^* = Re^{i\phi} \qquad\qquad (0 \leqq \phi \leqq 2\pi)$$

and the function

$$f(z) = u(r, \theta) + iv(r, \theta) \qquad\qquad (z = re^{i\theta})$$

is assumed to be analytic in a simply connected region containing C in its interior. Since $dz^* = iRe^{i\phi} \, d\phi = iz^* \, d\phi$, we obtain from (1)

$$(2) \qquad f(z) = \frac{1}{2\pi} \int_0^{2\pi} f(z^*) \frac{z^*}{z^* - z} \, d\phi \qquad (z^* = Re^{i\phi}, z = re^{i\theta}).$$

On the other hand, if we consider a point Z outside C, say, the point $Z = z^*\bar{z}^*/\bar{z}$ (whose absolute value is $R^2/r > R$), then the integrand in (1) is analytic in the disk $|z| \leqq R$ and the integral is zero:

$$0 = \frac{1}{2\pi i} \int_C \frac{f(z^*)}{z^* - Z} \, dz^* = \frac{1}{2\pi} \int_0^{2\pi} f(z^*) \frac{z^*}{z^* - Z} \, d\phi.$$

By inserting $Z = z^*\bar{z}^*/\bar{z}$ and simplifying the fraction we find

$$0 = \frac{1}{2\pi} \int_0^{2\pi} f(z^*) \frac{\bar{z}}{\bar{z} - \bar{z}^*} \, d\phi.$$

[7] SIMÉON DENIS POISSON (1781–1840), French mathematician and physicist.

Subtracting this from (2) and using

(3)
$$\frac{z^*}{z^* - z} - \frac{\bar{z}}{\bar{z} - \bar{z}^*} = \frac{z^*\bar{z}^* - z\bar{z}}{(z^* - z)(\bar{z}^* - \bar{z})},$$

we obtain

(4)
$$f(z) = \frac{1}{2\pi} \int_0^{2\pi} f(z^*) \frac{z^*\bar{z}^* - z\bar{z}}{(z^* - z)(\bar{z}^* - \bar{z})} \, d\phi.$$

From the polar representations of z and z^* we see that the quotient in the integrand is equal to

$$\frac{R^2 - r^2}{(Re^{i\phi} - re^{i\theta})(Re^{-i\phi} - re^{-i\theta})} = \frac{R^2 - r^2}{R^2 - 2Rr \cos(\theta - \phi) + r^2}.$$

Hence, by taking the real parts on both sides of (4) we obtain **Poisson's integral formula**

(5)
$$u(r, \theta) = \frac{1}{2\pi} \int_0^{2\pi} u(R, \phi) \frac{R^2 - r^2}{R^2 - 2Rr \cos(\theta - \phi) + r^2} \, d\phi,$$

which represents the harmonic function u in the disk $|z| \leq R$ in terms of its values $u(R, \phi)$ on the circle which bounds the disk.

It is of practical interest to note that in (5) we may use for $u(R, \phi)$ any function which is merely piecewise continuous on the interval of integration. Formula (5) then *defines* a function $u(r, \theta)$ which is harmonic in the open disk $|z| < R$ and continuous on the circle $|z| = R$. On this circle this function is equal to $u(R, \phi)$, except at points where $u(R, \phi)$ is discontinuous. The proof can be found in Ref. [F1].

From (5) we may obtain an important series development of u in terms of simple harmonic functions. We remember that the quotient in the integrand of (5) was derived from (3), and it is not difficult to see that the right side of (3) is the real part of $(z^* + z)/(z^* - z)$. By the use of the geometric series we obtain

(6)
$$\frac{z^* + z}{z^* - z} = \frac{1 + (z/z^*)}{1 - (z/z^*)} = \left(1 + \frac{z}{z^*}\right) \sum_{n=0}^{\infty} \left(\frac{z}{z^*}\right)^n = 1 + 2 \sum_{n=1}^{\infty} \left(\frac{z}{z^*}\right)^n.$$

Since $z = re^{i\theta}$ and $z^* = Re^{i\phi}$, we have

$$\text{Re} \left(\frac{z}{z^*}\right)^n = \text{Re} \left[\frac{r^n}{R^n} e^{in\theta} e^{-in\phi}\right] = \left(\frac{r}{R}\right)^n \cos(n\theta - n\phi)$$

(7)
$$= \left(\frac{r}{R}\right)^n (\cos n\theta \cos n\phi + \sin n\theta \sin n\phi).$$

From (6) and (7) we obtain

$$\text{Re} \frac{z^* + z}{z^* - z} = 1 + 2 \sum_{n=1}^{\infty} \frac{r^n}{R^n} (\cos n\theta \cos n\phi + \sin n\theta \sin n\phi).$$

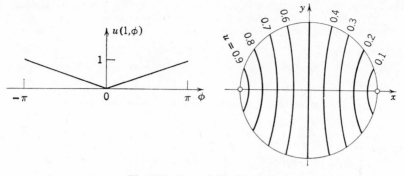

Fig. 329. Potential in Example 1

This expression is equal to the quotient in (5), as we have mentioned before, and by inserting the series into (5) and integrating term by term we find

(8)
$$u(r, \theta) = a_0 + \sum_{n=1}^{\infty} \left(\frac{r}{R}\right)^n (a_n \cos n\theta + b_n \sin n\theta)$$

where the coefficients are

(9)
$$a_0 = \frac{1}{2\pi} \int_0^{2\pi} u(R, \phi)\, d\phi, \qquad a_n = \frac{1}{\pi} \int_0^{2\pi} u(R, \phi) \cos n\phi\, d\phi,$$
$$b_n = \frac{1}{\pi} \int_0^{2\pi} u(R, \phi) \sin n\phi\, d\phi, \qquad\qquad n = 1, 2, \cdots,$$

the Fourier coefficients of $u(R, \phi)$. Note that for $r = R$ the series (8) becomes the Fourier series of $u(R, \phi)$, and therefore the representation (8) will be valid whenever $u(R, \phi)$ can be represented by a Fourier series.

Example 1. Find the potential $u(r, \theta)$ in the unit disk $r < 1$ having the boundary values (Fig. 329)

$$u(1, \phi) = \begin{cases} -\phi/\pi & \text{when } -\pi < \phi < 0 \\ \phi/\pi & \text{when } 0 < \phi < \pi. \end{cases}$$

Since $u(1, \phi)$ is even, $b_n = 0$, and from (9) we obtain $a_0 = \frac{1}{2}$ and

$$a_n = \frac{1}{\pi} \left[-\int_{-\pi}^0 \frac{\phi}{\pi} \cos n\phi\, d\phi + \int_0^\pi \frac{\phi}{\pi} \cos n\phi\, d\phi \right] = \frac{2}{n^2\pi^2} (\cos n\pi - 1).$$

Hence, $a_n = -4/n^2\pi^2$ when n is odd, $a_n = 0$ when $n = 2, 4, \cdots$, and the potential is

$$u(r, \theta) = \frac{1}{2} - \frac{4}{\pi^2} \left[r \cos \theta + \frac{r^3}{3^2} \cos 3\theta + \frac{r^5}{5^2} \cos 5\theta + \cdots \right].$$

Problems

1. Show that each term in (8) is a harmonic function in the disk $r^2 < R^2$.

2. Verify (3).

3. Derive Theorem 4I, Sec. 17.3, from (5).

Using (8), find the potential $u(r, \theta)$ in the unit disk $r < 1$ having the given boundary values $U(\theta) = u(1, \theta)$. Using the first few terms of the series, compute some values of u and sketch a figure of the equipotential lines.

4. $U(\theta) = 1 - \cos \theta$

5. $U(\theta) = 1 + \cos 2\theta$

6. $U(\theta) = \cos 2\theta - \cos 6\theta$

7. $U(\theta) = \theta/2 \; (-\pi < \theta < \pi)$

8. $U(\theta) = \sin^3 \theta$

9. $U(\theta) = \theta^2 \; (-\pi < \theta < \pi)$

10. $U(\theta) = \begin{cases} \theta & \text{if } -\pi/2 < \theta < \pi/2 \\ \pi - \theta & \text{if } \pi/2 < \theta < 3\pi/2 \end{cases}$

11. $U(\theta) = \begin{cases} 0 & \text{if } -\pi < \theta < 0 \\ 1 & \text{if } 0 < \theta < \pi \end{cases}$

12. $U(\theta) = \begin{cases} \theta & \text{if } -\pi/2 < \theta < \pi/2 \\ \pi/2 & \text{if } \pi/2 < \theta < \pi \\ -\pi/2 & \text{if } \pi < \theta < 3\pi/2 \end{cases}$

13. $U(\theta) = \begin{cases} 0 & \text{if } -\pi < \theta < 0 \\ 1 & \text{if } 0 < \theta < \pi/2 \\ -1 & \text{if } \pi/2 < \theta < \pi \end{cases}$

14. $U(\theta) = \begin{cases} -\theta^2 & \text{if } -\pi < \theta < 0 \\ \theta^2 & \text{if } 0 < \theta < \pi \end{cases}$

15. $U(\theta) = \begin{cases} -1 & \text{if } -\pi < \theta < 0 \\ 1 & \text{if } 0 < \theta < \pi \end{cases}$

16. Show that the potential in Prob. 7 may be written $u(r, \theta) = \operatorname{Im} \operatorname{Ln} (1 + z)$.

17. Using (9) in Sec. 15.2, show that the result of Prob. 13 may be written

$$u(r, \theta) = \frac{1}{\pi} \operatorname{Im} \operatorname{Ln} \frac{(1 + iz)(1 + z^2)}{(1 - iz)(1 - z^2)}.$$

18. Using (9), Sec. 15.2, show that in Prob. 15

$$u(r, \theta) = \frac{2}{\pi} \operatorname{Im} \operatorname{Ln} \frac{1 + z}{1 - z} = \frac{2}{\pi} [\arg (1 + z) - \arg (1 - z)]$$

and, applying a familiar theorem of elementary geometry to this result, show that the curves $u = const$ in Prob. 15 are circular arcs. Graph these curves.

19. Show that

$$H = 1 + \frac{2}{\pi} \operatorname{Im} \operatorname{Ln} \frac{w + 1}{w - 1} \qquad\qquad (w = u + iv)$$

is harmonic in the upper half-plane $v > 0$ and has the values -1 for $-1 < u < 1$ and $+1$ on the remaining part of the u-axis.

20. Show that the linear transformation which maps $w_1 = -1$, $w_2 = 0$, $w_3 = 1$ onto $z_1 = -1$, $z_2 = -i$, $z_3 = 1$, respectively, is

$$z = \frac{w - i}{-iw + 1}.$$

Find the inverse $w = w(z)$, insert it into H in Prob. 19, and show that the resulting harmonic function is that in Prob. 18.

CHAPTER **18**

Numerical Analysis

Engineering mathematics ultimately comes down to numerical results, and the engineering student should, therefore, include in his mathematical equipment a definite knowledge of some fundamental **numerical methods,** that is, methods for obtaining numerical results from given data. The branch of mathematics concerned with the development of such methods is called **numerical analysis.**

The need for such processes for obtaining practical answers to given problems is obvious, because in many cases the answers obtained from the theory may be almost useless for numerical purposes. Typical examples are the integral formula for the solution of a linear first-order differential equation (Sec. 1.7) and Cramer's rule for solving systems of linear algebraic equations in terms of determinants (Secs. 6.6, 6.11). In other cases the theory may merely guarantee the existence of a solution but give no indication of how to obtain it.

The development of numerical analysis is strongly influenced and determined by the advent and use of the automatic computer, which is now indispensable in the mathematical work of every engineer. This development includes the creation of new methods, modification of existing methods to make them more effective in automatic calculation, a theoretical and practical analyzation of algorithms for the standard computational processes and pointing out those algorithms which are satisfactory in various situations. Error analysis is part of that task for the numerical analyst, and he must try to characterize and eliminate arithmetic traps of all kinds which are ready for the unwary.

> *Prerequisite for this chapter:* elementary calculus; for Secs. 18.6 and 18.7 also the elements of differential equations, and for some of the topics in Secs. 18.8–18.13 the elements of matrix algebra.
> *References:* Appendix 1, Part G.
> *Answers to problems:* Appendix 2.

18.1 Errors and Mistakes. Automatic Computers

Since in computations we work with a finite number of digits and carry out finitely many steps, the methods of numerical analysis are **finite processes,** and a

numerical result is an **approximate value** of the (unknown) exact result, except
for the rare cases where the exact answer is a sufficiently simple rational number
and we can use a numerical method that gives the exact answer.

If a^* is an approximate value of a quantity whose exact value is a, then the
difference $\epsilon = a^* - a$ is called the *absolute error* of a^* or, briefly, the **error** of a^*.
Hence

$$a^* = a + \epsilon \qquad \text{Approximation} = \text{True value} + \text{error}.$$

The **relative error** ϵ_r of a^* is defined by

$$\epsilon_r = \frac{\epsilon}{a} = \frac{a^* - a}{a} \qquad\qquad (a \neq 0).$$

Clearly, $\epsilon_r \approx \epsilon/a^*$ if $|\epsilon|$ is much less than $|a^*|$. We may also introduce the quantity
$\gamma = a - a^* = -\epsilon$ and call it the *correction*, thus[1]

$$a = a^* + \gamma \qquad \text{True value} = \text{Approximation} + \text{correction}.$$

Finally, an **error bound** for a^* is a number β such that

$$|a^* - a| \leq \beta, \qquad \text{that is,} \qquad |\epsilon| \leq \beta.$$

Depending on the source of errors, we may distinguish between experimental
errors, truncation errors, and roundoff errors. **Experimental errors** are errors of
given data (probably arising from measurements). **Truncation errors** are errors
corresponding to the fact that a (finite or infinite) sequence of computational
steps necessary to produce an exact result is "truncated" prematurely after a
certain number of steps. These errors depend on the computational method and
have to be discussed individually with each method. **Rounding errors** are errors
arising from the process of rounding off during computation, and we shall now
discuss this type of error.

In the decimal notation, every real number is represented by a finite or infinite
sequence of decimal digits. For machine computation the number must be replaced
by a number of finitely many digits. In automatic digital computers there are two
ways in which numbers may be stored, as follows. In a **fixed point** system the
numbers are represented with a *fixed number of decimal places*, e.g., 62.358, 0.013,
1.000. In a **floating point** system the numbers are represented with a *fixed number
of significant digits*, e.g., 0.6236×10^2, -0.1714×10^{-3}, -0.2000×10^1.
Significant digit of a number c is any given digit of c, except possibly for zeros to
the left of the first nonzero digit that serve only to fix the position of the decimal
point. (Thus, any other zero is a significant digit of c.) For instance, each of the
numbers 1360, 1.360, 0.001 360 has 4 significant digits.[2]

---

[1] $\gamma = -\epsilon$ is sometimes called the *error;* the distinction is rather irrelevant, but one should be
consistent.

[2] In tables of functions showing k significant digits, it is conventionally assumed that any given
value a^* deviates from the corresponding exact value a by at most ± 0.5 unit of the last given
digit, unless otherwise stated; e.g. if $a = 1.1996$, then a table with 4 significant digits should
show $a^* = 1.200$. Correspondingly, if 12 000 is correct to three digits only, we should better
write 120×10^2, etc. "Decimal" is abbreviated by D and "significant digit" by S. For example,
5D means 5 decimals, and 8S means 8 significant digits.

We shall now state the **rule for rounding off** a number to k decimals. (The rule for rounding off to k significant figures is the same, with "decimal" replaced by "significant figure.")

Discard the $(k + 1)$th and all subsequent decimals. If the number thus discarded is less than half a unit in the kth place, leave the kth decimal unchanged ("*rounding down*"). If it is greater than half a unit in the kth place, add one to the kth decimal ("*rounding up*"). If it is exactly half a unit, round off to the nearest *even* decimal. (Example: Rounding off 3.45 and 3.55 to 1 decimal gives 3.4 and 3.6, respectively.)

The last part of the rule is supposed to ensure that in discarding exactly half a decimal, rounding up and rounding down happens about equally often, on the average.

If we round off 1.253 5 to 3, 2, 1 decimals, we get 1.254, 1.25, 1.3, but if 1.25 is rounded off to one decimal, without further information, we get 1.2.

Rounding errors may ruin a computation completely. In general, they are the more dangerous the more steps of computations we have to perform. It is therefore important to analyze computational programs for rounding errors to be expected and to find an arrangement of the computations such that the effect of rounding errors is as small as possible. **Guarding figures** are extra figures (one, two, or even more) carried in intermediate stages of a desk calculation with the intention to eliminate rounding errors to some extent. **Scaling** of variables may reduce the number of decimal places carried in a calculation. For instance, $0.002 : 900 = 0.000\ 002\ 222$ and we need 9 decimal places to have a result of 4 significant digits, but if we can scale the variables so that the calculation becomes $2 : 9 = 0.222\ 2$, we need only 4 decimals.

We want to emphasize that a careful selection of most effective methods is as important in automatic computation as it is in using a desk calculator, not merely in order to minimize time and cost but also errors. Even for simple problems there may be several ways which differ by their efficiency. Let us illustrate this with an example.

Example 1 (Quadratic equation). Find the roots of each of the equations

$$(1) \qquad (a)\ x^2 - 4x + 2 = 0 \quad \text{and} \quad (b)\ x^2 - 40x + 2 = 0,$$

using 4 significant figures in the calculation.

Formulas for the roots x_1, x_2 of a quadratic equation $ax^2 + bx + c = 0$ are

$$(2) \qquad x_1 = \frac{1}{2a}(-b + \sqrt{b^2 - 4ac}), \qquad x_2 = \frac{1}{2a}(-b - \sqrt{b^2 - 4ac})$$

and, since $x_1 x_2 = c/a$,

$$(3) \qquad\qquad x_1 \text{ as before}, \qquad x_2 = \frac{c}{ax_1}.$$

For (1a), formula (2) gives $x = 2 \pm \sqrt{2} = 2.000 \pm 1.414$, $x_1 = 3.414$, $x_2 = 0.586$, and (3) gives $x_1 = 3.414$, $x_2 = 2.000/3.414 = 0.5858$, in error by less than one unit in the last digit (cf. Prob.

10). For (1*b*), formula (2) gives $x = 20 \pm \sqrt{398} = 20.00 \pm 19.95$, $x_1 = 39.95$, $x_2 = 0.05$, which is poor, whereas (3) gives $x_1 = 39.95$, $x_2 = 2.000/39.95 = 0.05006$, in error by less than one unit in the last digit (cf. Prob. 11).

In hand computing we may vary the number of decimal places being carried in the light of intermediate results in order to obtain an answer of the required accuracy with a minimum of work. In automatic computing we can perform a calculation using the maximum possible number of digits (which depends on the design of the machine) and then determine the accuracy of the result obtained, rather than try to produce a result of prescribed accuracy.

In addition to errors a computation may contain **mistakes** because of the fallibility of the programmer, machine operator or user of the desk calculator, or the mechanical or electrical equipment used in the computation. Whereas errors cannot be avoided, mistakes are avoidable in principle. **Checking** of numerical results is quite important, and any numerical method should include checking procedures for final (and intermediate) results.

To safeguard against mistakes in **desk calculation,** any computation should be arranged in a systematic tabular form. We should not use scraps of paper but record all the intermediate results directly on the working sheet (squared paper of full notebook page size or paper ruled in rectangular boxes), and write the numbers in a neat and legible manner. We should not try to do too many steps at once. To correct a mistake we should draw a line through the wrong number and write the corrected value above the wrong number, thus keeping a record of the correction. Each sheet should have on it some explanations to keep the calculation understandable for months and years to come.

In **automatic computation** the situation can be roughly described as follows. An automatic computer can add, subtract, multiply, and divide. It has to be told in detail the steps necessary to solve a given problem. This we do by feeding into the computer a **program** punched on **punch cards** and consisting of all relevant given data and a sequence of **instructions** which the computer executes in a certain order, thereby performing the computational steps that lead to an answer which the computer prints on a sheet of paper. Hence the essential components of a computer are:

(I) an **input** for feeding punch cards and transferring the information on the cards into the memory;

(II) a **memory** or **store** for storing instructions and data, each of which, when stored, is allocated an *address* (position in the memory) and can be produced from the store at will;

(III) a **control unit** for organizing the calculations, that is, arranging for the input of information, the execution of arithmetic operations in the requested order specified by the program, and the output of information;

(IV) an **arithmetic unit** for performing arithmetical operations (addition, subtraction, multiplication, division); and

Fig. 330. Components of a computer and main machine functions. Solid arrows represent flow of data or instructions and dashed arrows represent flow of contol signals.

(V) an **output** for transferring information from the memory onto the output sheet.

Fig. 330 represents the main machine functions in the form of a block diagram. We first plan a computation in all details. Then we describe it formally in a **"problem-oriented language"** such as FORTRAN (FORmula TRANslation) or ALGOL (ALGOrithmic Language) and transcribe this description onto a deck of punch cards, which we feed into the computer. Using a **compiler,** the computer performs an **automatic translation** of our program into a **machine language program** and executes the latter, thereby carrying out the planned computation. (The machine language program may not appear outside the machine, but if we want to use it again, the computer can punch it on a deck of cards which we can keep.)

Experience shows that the occurrence of mistakes in programming cannot be discounted. The translation process usually includes searches for more or less clerical mistakes which are listed and corrected by the computer, and the corrected program is resubmitted to the computer. However, programs should be prepared with greatest care and checked as far as possible before they reach the stage of machine operation. Detection of mistakes by means of such early checking is more efficient than reliance on programmed checks. Furthermore, it must not be assumed that a program is correct merely because it can be translated and executed, but one should always use a new program to repeat the computations of **pilot programs** that have been thoroughly checked previously.

Problems

1. Compute a two-decimal table [3] of $f(x) = x/16$, $x = 0(1)20$ and find out how the rounding error is distributed.
2. Compute a three-decimal table of $f(x) = x/3$, $x = 0(1)100$ and find out how the rounding error is distributed.
3. Explain the difficulties in obtaining a four-place table by rounding off the values (*a*) in a five-place table, (*b*) in a six-place table.
4. Illustrate with an example that in computations with a fixed number of significant digits the result of adding numbers depends on the order in which they are added.
5. Give a simple example illustrating the use of guarding figures.

---

[3] $x = a(h)b$ means that function values are given for $x = a, a + h, a + 2h, \cdots, b$.

6. If β_1 and β_2 are error bounds for $a_1{}^*$ and $a_2{}^*$, show that $\beta = \beta_1 + \beta_2$ is an error bound for the sum $s^* = a_1{}^* + a_2{}^*$.
7. Show that β in Prob. 6 is an error bound for the difference $d^* = a_1{}^* - a_2{}^*$, and illustrate with an example that in general β cannot be replaced by a smaller number.
8. Show that for small relative errors the relative error of a product is approximately equal to the sum of the relative errors of the factors. Give a simple numerical example.
9. Show that for small relative errors the relative error of a quotient is approximately equal to the difference of the relative errors of the factors.
10. Using Prob. 9, show that the absolute value of the error of $x_2 = 2.000/3.414 = 0.5858$ in Ex. 1 is less than 0.0001.
11. Using Prob. 9, show that the absolute value of the error of $x_2 = 2.000/39.95 = 0.050\ 06$ in Ex. 1 is less than 0.000 01.

18.2 *Solution of Equations by Iteration*

In engineering mathematics we frequently have to find **solutions** of equations of the form

$$(1) \qquad\qquad\qquad f(x) = 0,$$

that is, numbers X_0 such that $f(X_0)$ equals zero; here f is a given function. Examples are $x^2 - 3x + 2 = 0$, $x^3 + x = 1$, $\sin x = 0.5x$, $\tan x = x$, $\cosh x = \sec x$, $\cosh x \cos x = -1$, which can all be written in the form (1). The first two are **algebraic equations** because the corresponding f is a polynomial, and in this case the solutions are also called **roots** of the equations. The other equations are **transcendental equations** because they involve transcendental functions. Formulas that give exact numerical values of the solutions will exist only in very simple cases. In most cases we have to use approximation methods, in particular, iteration methods.

A numerical **iteration method** is a method such that we choose an arbitrary x_0 and compute a sequence $x_0, x_1, x_2, \cdots$ recursively from a relation of the form

$$(2) \qquad\qquad\qquad x_{n+1} = g(x_n) \qquad\qquad (n = 0, 1, \cdots)$$

where g is defined in some interval containing x_0 and the range of g lies in that interval. Hence we compute successively $x_1 = g(x_0)$, $x_2 = g(x_1)$, $x_3 = g(x_2)$, $\cdots$.

In this section, both the domain and the range of $g(x)$ will be on the real line. In more general problems x or g or both may be vector variables.

We mention that iteration methods are very important for various types of problems in numerical analysis.

There are several ways of obtaining iteration methods for solving (1), and we shall discuss three particularly important ones.

Algebraic transformations. We may try to transform (1) *algebraically* into the form

$$(3) \qquad\qquad\qquad x = g(x).$$

Then the corresponding iteration procedure is given by (2) with the function g

Fig. 331. Example 1, iterations (4a) and (4b)

obtained in (3). A solution of (3) is called a **fixed point** of g. To a given equation (1) there may correspond several equations (3) and the behavior of iterative sequences x_0, x_1, x_2, $\cdots$ may differ accordingly (and depend on x_0). Let us illustrate this with a simple example.

Example 1. Set up an iteration process for the equation $f(x) = x^2 - 3x + 1 = 0$. Since we know the solutions

$$x = 1.5 \pm \sqrt{1.25}, \qquad \text{thus} \qquad 2.618\ 034 \qquad \text{and} \qquad 0.381\ 966,$$

we can watch the behavior of the error as the iteration proceeds. The equation may be written

(4a) $$x = g_1(x) = \tfrac{1}{3}(x^2 + 1), \qquad \text{thus} \qquad x_{n+1} = \tfrac{1}{3}(x_n{}^2 + 1),$$

and if we choose $x_0 = 1$, we obtain the sequence (cf. Fig. 331a)

$$x_0 = 1.000, \qquad x_1 = 0.667, \qquad x_2 = 0.481, \qquad x_3 = 0.411, \qquad x_4 = 0.390, \cdots$$

which seems to approach the smaller root. If we choose $x_0 = 2$ the situation is similar. If we choose $x_0 = 3$, we obtain the sequence (cf. Fig. 331a)

$$x_0 = 3.000, \qquad x_1 = 3.333, \qquad x_2 = 4.037, \qquad x_3 = 5.766, \qquad x_4 = 11.414, \cdots$$

which seems to diverge. Our equation may also be written

(4b) $$x = g_2(x) = 3 - \frac{1}{x}, \qquad \text{thus} \qquad x_{n+1} = 3 - \frac{1}{x_n},$$

and if we choose $x_0 = 1$, we obtain the sequence (cf. Fig. 331b)

$$x_0 = 1.000, \qquad x_1 = 2.000, \qquad x_2 = 2.500, \qquad x_3 = 2.600, \qquad x_4 = 2.615, \cdots$$

which seems to approach the larger solution. Similarly, if we choose $x_0 = 3$, we obtain the sequence (cf. Fig. 331b)

$$x_0 = 3.000, \qquad x_1 = 2.667, \qquad x_2 = 2.625, \qquad x_3 = 2.619, \qquad x_4 = 2.618, \cdots$$

Our figures indicate that convergence seems to depend on the fact that in a neighborhood of a solution the curve of $g(x)$ is less steep than the straight line $y = x$, and we shall now see that this condition $|g'(x)| < 1$ (slope of $y = x$) is sufficient for convergence.

An iteration process defined by (2) is said to be **convergent** for an x_0 if the corresponding sequence $x_0, x_1, \cdots$ is convergent.

Theorem 1 (Convergence).   *Let $x = s$ be a solution of $x = g(x)$ and suppose that g has a continuous derivative in some interval J containing s. Then if $|g'(x)| \leqq \alpha < 1$ in J, the iteration process defined by (2) converges for any x_0 in J.*

Proof.   By the mean value theorem of differential calculus there is a ξ between x and s such that

$$g(x) - g(s) = g'(\xi)(x - s) \qquad\qquad (x \text{ in } J).$$

Since $g(s) = s$ and $x_1 = g(x_0)$, $x_2 = g(x_1), \cdots$, we obtain

$$|x_n - s| = |g(x_{n-1}) - g(s)| = |g'(\xi)\|x_{n-1} - s| \leqq \alpha \, |x_{n-1} - s|$$
$$\leqq \alpha^2 \, |x_{n-2} - s| \leqq \cdots \leqq \alpha^n \, |x_0 - s|$$

and, since $\alpha < 1$, we have $\alpha^n \to 0$ and $|x_n - s| \to 0$ as $n \to \infty$. This completes the proof.

Example 2.   Find a solution of $f(x) = x^3 + x - 1 = 0$ by iteration. A rough sketch shows that a real solution lies near $x = 1$. We may write the equation in the form

$$x = g_1(x) = \frac{1}{1 + x^2}, \qquad \text{thus} \qquad x_{n+1} = \frac{1}{1 + x_n^2}.$$

Then $|g_1'(x)| = 2\,|x|/(1 + x^2)^2 < 1$ for any x, so that we have convergence for any x_0. Choosing $x_0 = 1$, we obtain (cf. Fig. 332)

$$x_1 = 0.500, \qquad x_2 = 0.800, \qquad x_3 = 0.610, \qquad x_4 = 0.729, \qquad x_5 = 0.653, \qquad x_6 = 0.701, \cdots.$$

The root exact to 6D is $s = 0.682\,328$. The equation may also be written

$$x = g_2(x) = 1 - x^3. \qquad\qquad \text{Then} \qquad\qquad |g_2'(x)| = 3x^2$$

and this is greater than 1 near the root, so that we cannot expect convergence. The reader may try $x_0 = 1$, $x_0 = 0.5$, $x_0 = 2$ and see what happens.

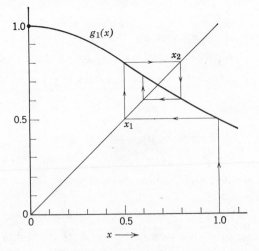

Fig. 332. Iteration in Example 2

Fig. 333. Newton's method

Newton's method (also called *Newton-Raphson method*) is another iteration method for solving equations $f(x) = 0$, where f is differentiable. The idea is that we approximate the graph of f by suitable tangents. Using a value x_0 obtained from the graph of f we let x_1 be the point of intersection of the x-axis and the tangent to the curve of f at x_0 (cf. Fig. 333). Then

$$\tan \beta = f'(x_0) = \frac{f(x_0)}{x_0 - x_1}, \qquad \text{hence} \qquad x_1 = x_0 - \frac{f(x_0)}{f'(x_0)}.$$

In the next step we compute (cf. Fig. 333)

$$x_2 = x_1 - \frac{f(x_1)}{f'(x_1)}$$

etc., and the general formula is

(5)
$$x_{n+1} = x_n - \frac{f(x_n)}{f'(x_n)} \qquad\qquad (n = 0, 1, \cdots).$$

Example 3 (Square root). Set up a Newton iteration for computing the square root x of a given positive number c and apply it to $c = 2$. We have $x = \sqrt{c}$, hence $f(x) = x^2 - c = 0$, $f'(x) = 2x$, and (5) takes the form

$$x_{n+1} = x_n - \frac{x_n^2 - c}{2x_n} = 0.5\left(x_n + \frac{c}{x_n}\right).$$

For $c = 2$, choosing $x_0 = 1$, we obtain

$$x_1 = 1.500\,000, \qquad x_2 = 1.416\,667, \qquad x_3 = 1.414\,216, \qquad x_4 = 1.414\,214, \cdots.$$

x_4 is exact to 6D.

Example 4. Find the positive solution of $2 \sin x = x$. Setting $f(x) = x - 2 \sin x$, we have $f'(x) = 1 - 2 \cos x$, and (5) gives

$$x_{n+1} = x_n - \frac{x_n - 2 \sin x_n}{1 - 2 \cos x_n} = \frac{2(\sin x_n - x_n \cos x_n)}{1 - 2 \cos x_n} = \frac{N_n}{D_n}.$$

From the graph of f we conclude that the solution is near $x_0 = 2$. Using tables of the sine and cosine functions, we compute:

n	x_n	N_n	D_n	x_{n+1}
0	2.000	3.483	1.832	1.901
1	1.901	3.125	1.648	1.896
2	1.896	3.107	1.639	1.896

(The solution to 4D is 1.8955.)

Example 5. Apply Newton's method to the equation $f(x) = x^3 + x - 1 = 0$. From (5) we have

$$x_{n+1} = x_n - \frac{x_n^3 + x_n - 1}{3x_n^2 + 1} = \frac{2x_n^3 + 1}{3x_n^2 + 1}.$$

Starting from $x_0 = 1$, we obtain

$$x_1 = 0.750\,000, \qquad x_2 = 0.686\,047, \qquad x_3 = 0.682\,340, \qquad x_4 = 0.682\,328, \cdots$$

where x_4 is exact to 6D. A comparison with Ex. 2 shows that the present convergence is much more rapid. This may motivate the concept of the order of an iteration process, which we shall now discuss.

Let $x_{n+1} = g(x_n)$ define an iteration method, and let x_n approximate a solution s of $x = g(x)$. Then $x_n = s + \epsilon_n$, where ϵ_n is the error of x_n. Suppose that g is differentiable a number of times, so that the Taylor formula gives

$$x_{n+1} = g(x_n) = g(s) + g'(s)(x_n - s) + \tfrac{1}{2}g''(s)(x_n - s)^2 + \cdots$$
$$= g(s) + g'(s)\epsilon_n + \tfrac{1}{2}g''(s)\epsilon_n^2 + \cdots.$$

The exponent of ϵ_n in the first nonvanishing term after $g(s)$ is called the **order** of the iteration process defined by g. Since $x_{n+1} - g(s) = x_{n+1} - s = \epsilon_{n+1}$, the error of x_{n+1}, and in the case of convergence ϵ_n will be small for large n, the order is a measure for the speed of convergence.

Newton's iteration is of second order.

In fact, we have

$$g(x) = x - \frac{f(x)}{f'(x)}, \qquad\qquad g'(x) = \frac{f(x)f''(x)}{f'(x)^2}$$

and $g'(s) = 0$ because $f(s) = 0$; hence the process is at least of second order. Another differentiation shows that $g''(s) = f''(s)/f'(s)$, which is not zero in general. In Ex. 2, the process given by $g_1(x)$ is only of first order because $g_1(x) = 1/(1 + x^2)$ and $g_1'(x) = -2x/(1 + x^2)^2$.

If $f'(x) = 0$ near a solution of $f(x) = 0$, then Newton's method may cause difficulties, but these can often be clarified by plotting the curve of $f(x)$ for values of x near the required solution and using the geometrical interpretation of Newton's method. If $f'(x)$ is small near a required solution of $f(x) = 0$, it may be necessary to compute $f(x_n)$ and $f'(x_n)$ to a high degree of accuracy in order to obtain an accurate approximation of x_{n+1}. This indicates that the equation is ill-conditioned. This concept is defined as follows.

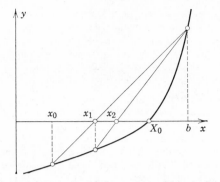

Fig. 334. Method of false position (*Regula falsi*)

An equation $f(x) = 0$ is said to be **ill-conditioned** if a small value of

$$R(s^*) = f(s^*)$$

can be produced by values s^* appreciably different from the exact solution s, or if small variations in the constants appearing in f can produce large variations in the solutions. $R(s^*) = f(s^*)$ is called the **residual** of the equation at s^*. Its smallness gives some indication of the accuracy of s^*, but it is important to realize that small $R(s^*)$ does not necessarily imply that the error of s^* is small.

A third iteration method for solving an equation $f(x) = 0$ is the **method of false position** (*regula falsi*). In this method we approximate the curve of $f(x)$ by a chord, as shown in Fig. 334. The chord intersects the x-axis at

(6a)
$$x_1 = \frac{x_0 f(b) - b f(x_0)}{f(b) - f(x_0)}$$

and this is an approximation of the solution X_0 of $f(x) = 0$. In the next step we compute the approximation

(6b)
$$x_2 = \frac{x_1 f(b) - b f(x_1)}{f(b) - f(x_1)}$$

etc. Convergence to X_0 can be accelerated if improvements in b can be made in the direction of bringing b closer to X_0. This can usually be done by guesswork.

Example 6. Determine an approximate value of the root of the equation $f(x) = x^3 + x - 1 = 0$ near $x = 1$ (cf. Ex. 2). Since $f(0.5) = -0.375$ and $f(1) = 1$, we may choose $x_0 = 0.5$ and $b = 1$. Formula (6a) then yields

$$x_1 = \frac{0.5 \cdot 1 - 1 \cdot (-0.375)}{1 - (-0.375)} = 0.64.$$

From (6b) we obtain the more accurate approximation $x_2 = 0.672$, and so on.

Problems

1. Show that if g is continuous in a closed interval I and its range lies in I, then the equation $x = g(x)$ has at least one solution in that interval. Illustrate that it may have more than one solution.

2. Show that in Ex. 2, $|g_1'(x)|$ is maximum at $\tilde{x} = \pm 1/\sqrt{3}$ and $|g'(\tilde{x})| = 3\sqrt{3}/8 = 0.65$.

3. Why do we obtain a monotone sequence in Ex. 1, but not in Ex. 2?

4. Perform the iterations indicated at the end of Ex. 2. Sketch a figure similar to Fig. 332.

5. Find the root of $x^5 = x + 0.2$ near $x = 0$ by transforming the equation algebraically into the form (2) and starting from $x_0 = 0$.

6. The equation in Prob. 5 has a root near $x = 1$. Find this root by writing the equation in the form $x = \sqrt[5]{x + 0.2}$ and iterating, starting with $x_0 = 1$.

7. What happens in Prob. 6 if you write the equation in the form $x = x^5 - 0.2$ and start from $x_0 = 1$?

8. Using iteration, show that the smallest positive root of the equation $x = \tan x$ is 4.49, approximately. *Hint.* Conclude from the graphs of x and $\tan x$ that a root lies close to $x_0 = 3\pi/2$; write the equation in the form $x = \arctan x$. (Why?)

9. Design a Newton iteration for computing the kth root of a positive number c.

Find a root of the following equations; carry out three steps of Newton's method, starting from the given x_0.

10. $x^3 - 1.2x^2 + 2x - 2.4 = 0$, $x_0 = 2$

11. $x^3 - 3.9x^2 + 4.79x - 1.881 = 0$, $x_0 = 1$

12. The roots of the equation in Prob. 11 are 0.9, 1.1, and 1.9. Although $x_0 = 1$ lies close to 0.9 and 1.1, Newton's method does not yield one of these roots. Why? Choose another x_0 such that the method yields approximations for the root 1.1.

Find all real roots of the following equations by Newton's method.

13. $x^4 - 0.1x^3 - 0.82x^2 - 0.1x - 1.82 = 0$

14. $\cos x = x$ **15.** $x + \ln x = 2$ **16.** $2x + \ln x = 1$

Find the real roots of the following equations by the method of false position.

17. $x^4 = 2$ **18.** $x^4 = 2x$ **19.** $3 \sin x = 2x$

20. In Prob. 17 the approximate values of the positive root will always be somewhat smaller than the exact value of the root. Why?

18.3 *Finite Differences*

Finite differences are basic in various branches of numerical analysis, such as interpolation, checking tables, approximation, differentiation, and solution of differential equations. We assume that we are given a **table** of numerical values $f_j = f(x_j)$ of a function f at equally spaced points

$$x_0, \qquad x_1 = x_0 + h, \qquad x_2 = x_0 + 2h, \qquad x_3 = x_0 + 3h, \cdots \qquad (h > 0, \text{ fixed})$$

where the $f(x_j)$ may result from a formula or may be obtained **empirically** by experiments. We may now form the **first differences** by subtracting each function value of $f(x)$ from that for the next greater value of x in the table. An example is shown in Table 3 where[3] $f(x) = x^3$, $x = -3(1)3$. Applying the same process of subtraction to the first differences, we obtain the **second differences** of f, and so on. In this **difference table**, each difference is entered, in the appropriate column,

---

[3] $x = a(h)b$ means that function values are given for $x = a, a + h, a + 2h, \cdots, b$.

Table 3. Difference Table of $f(x) = x^3$, $x = -3(1)3$

x	$f(x) = x^3$	First Diff.	Second Diff.	Third Diff.	Fourth Diff.
-3	-27				
		19			
-2	-8		-12		
		7		6	
-1	-1		-6		0
		1		6	
0	0		0		0
		1		6	
1	1		6		0
		7		6	
2	8		12		
		19			
3	27				

midway between the elements in the preceding column from which it is constructed. Decimal points and leading zeros of differences may be omitted (cf. Table 4).

There are three different **notations for the differences** occurring in a difference table. To avoid confusion, we should say at the beginning that in any concrete numerical case the same numbers occur in the same positions no matter which of the three notations we use. The first (and probably most important) notation is that for **central differences:**

$$
\begin{array}{llllll}
x_{-2} & f_{-2} & & & & \\
 & & \delta f_{-3/2} & & & \\
x_{-1} & f_{-1} & & \delta^2 f_{-1} & & \\
 & & \delta f_{-1/2} & & \delta^3 f_{-1/2} & \\
x_0 & f_0 & & \delta^2 f_0 & & \\
 & & \delta f_{1/2} & & \delta^3 f_{1/2} & \\
x_1 & f_1 & & \delta^2 f_1 & & \\
 & & \delta f_{3/2} & & & \\
x_2 & f_2 & & & &
\end{array}
$$

Thus $\delta f_{-3/2} = f_{-1} - f_{-2}$, $\delta f_{-1/2} = f_0 - f_{-1}$ and in general

(1) $$\delta f_{m+1/2} = f_{m+1} - f_m,$$

the subscript on the left is the mean of the subscripts on the right. Similarly,

$$\delta^2 f_m = \delta f_{m+1/2} - \delta f_{m-1/2},$$

and so on. Differences with the same subscript appear in a horizontal line. (Note that x_0 need not be the smallest x in the table. For instance, in Table 4 we may let $x_0 = 1.6$, then $f_0 = 0.6250$, $\delta f_{1/2} = -0.0694$, $\delta^2 f_0 = 0.0199$, etc.)

Table 4. Values and Differences of $f(x) = 1/x$,
$x = 1(0.2)2$, 4D

x	$f(x) = 1/x$	First Diff.	Second Diff.	Third Diff.
1.0	1.0000			
		-1667		
1.2	0.8333		477	
		-1190		-180
1.4	0.7143		297	
		-893		-98
1.6	0.6250		199	
		-694		-61
1.8	0.5556		138	
		-556		
2.0	0.5000			

The second notation is that for **forward differences:**

$$
\begin{array}{cccc}
x_{-2} & f_{-2} & & & \\
 & & \Delta f_{-2} & & \\
x_{-1} & f_{-1} & & \Delta^2 f_{-2} & \\
 & & \Delta f_{-1} & & \Delta^3 f_{-2} \\
x_0 & f_0 & & \Delta^2 f_{-1} & \\
 & & \Delta f_0 & & \Delta^3 f_{-1} \\
x_1 & f_1 & & \Delta^2 f_0 & \\
 & & \Delta f_1 & & \\
x_2 & f_2 & & &
\end{array}
$$

Thus $\Delta f_{-2} = f_{-1} - f_{-2}$, $\Delta f_{-1} = f_0 - f_{-1}$, $\Delta f_0 = f_1 - f_0$ and in general

(2) $\Delta f_m = f_{m+1} - f_m.$

Similarly,

$$\Delta^2 f_m = \Delta f_{m+1} - \Delta f_m,$$

and so on. For instance, in Table 4, if we let $x_0 = 1.6$, then $f_0 = 0.6250$, $\Delta f_0 = -0.0694$, $\Delta^2 f_0 = 0.0138$. Differences with the same subscript appear on lines sloping downward or *forward* into the table.

The third notation is that for **backward differences:**

$$
\begin{array}{cccc}
x_{-2} & f_{-2} & & & \\
 & & \nabla f_{-1} & & \\
x_{-1} & f_{-1} & & \nabla^2 f_0 & \\
 & & \nabla f_0 & & \nabla^3 f_1 \\
x_0 & f_0 & & \nabla^2 f_1 & \\
 & & \nabla f_1 & & \nabla^3 f_2 \\
x_1 & f_1 & & \nabla^2 f_2 & \\
 & & \nabla f_2 & & \\
x_2 & f_2 & & &
\end{array}
$$

Thus $\nabla f_{-1} = f_{-1} - f_{-2}$, $\nabla f_0 = f_0 - f_{-1}$, $\nabla f_1 = f_1 - f_0$ and in general

(3)
$$\nabla f_m = f_m - f_{m-1}.$$

Similarly,

$$\nabla^2 f_m = \nabla f_m - \nabla f_{m-1},$$

and so on. Differences with the same subscript lie on lines sloping upward or *backward* into the table. Backward difference notation is often convenient for calculations near the *end* of a tabulated range.

Any specified difference in a table can now be represented by three different symbols. For example, in Table 4, if we let $x_0 = 1.6$, then we obtain $-0.0893 = \delta f_{-1/2} = \Delta f_{-1} = \nabla f_0$. In general we have

$$\delta^n f_m = \Delta^n f_{m-n/2} = \nabla^n f_{m+n/2}.$$

Differences may be used for **detecting errors** in tables. An error ϵ in a function value spreads into the differences as shown in Table 5. It follows that large fluctuations in the differences indicate possible errors in the function values. Of course, small fluctuation may result from rounding.

Differences are also useful in the **approximation** *of functions by polynomials.* For a polynomial $p_n(x) = a_0 x^n + a_1 x^{n-1} + \cdots + a_n$ of degree n the nth differences in a table with step h are constant (equal to $n! h^n a_0$) and all higher differences are zero, because

$$p_n(x + h) - p_n(x) = a_0[(x + h)^n - x^n] + \cdots = a_0 n h x^{n-1} + \cdots$$

is a polynomial of degree $n - 1$; similarly, the second differences are represented by a polynomial of degree $n - 2$ with leading coefficient $a_0 n(n - 1)h^2$, etc. It follows that if the nth differences in a given table of a function f are approximately

Table 5. Spread of an Error of a Function Value in the Difference Table $f(x) = \sqrt{x}$, $x = 2.0(0.1)2.6$, 4D [Error in $f(2.3)$.]

x	$\sqrt{x}$	Differences		$\sqrt{x}$	Differences		Spread of error ϵ		
2.0	1.4142			1.4142					
		349			349				
2.1	1.4491		−8	1.4491		−8			
		341			341				ϵ
2.2	1.4832		−7	1.4832		3		ϵ	
		334	−1		344	−31	ϵ		−3ϵ
2.3	1.5166		−8	1.5176		−28	ϵ	−2ϵ	
		326	1		316	31	−ϵ		3ϵ
2.4	1.5492		−7	1.5492		3		ϵ	
		319	2		319	−8			−ϵ
2.5	1.5811		−5	1.5811		−5			
		314			314				
2.6	1.6125			1.6125					

constant over a given range, then the tabular values can be represented approximately in this range by an nth-degree polynomial p_n. Let us illustrate a method for finding p_n when f is given.

Example 1. In Table 5 the second differences are almost constant (equal to -7). Hence, we may try to find a quadratic approximating polynomial p_2. We first construct a difference table, assuming that the second differences are *exactly* equal to -7 and choosing a function value and a first difference in the middle of the range, say 1.5166 and 334. This gives Table 6. The leading coefficient a_0 of p_2 is obtained from $a_0 2! \, h^2 = a_0 \cdot 2 \cdot 0.1^2 = -0.0007$ ($=$ the second difference); thus $a_0 = -0.0007/0.02 = -0.035$. Hence $p_1(x) = p_2(x) + 0.035x^2$ is of first degree, and from Table 6 we calculate that its first differences are constant (0.04915), and we know that they must equal $a_1 h$. Thus $a_1 = 0.04915/0.1 = 0.4915$. Finally, $p_1(x) - 0.4915x = a_0 = 0.5713$, so that we have

$$p_2(x) = -0.0350x^2 + 0.4915x + 0.5713.$$

Table 6. Approximation of $f(x) = \sqrt{x}$ by a Quadratic Polynomial p_2

x	$p_2(x)$	Differences	
2.0	1.4143		
		348	
2.1	1.4491		-7
		341	
2.2	1.4832		-7
		334	
2.3	1.5166		-7
		327	
2.4	1.5493		-7
		320	
2.5	1.5813		-7
		313	
2.6	1.6126		

Problems

1. In Table 4, choose $x_0 = 1.2$ and write all the occurring numbers in terms of the notation (*a*) for central differences, (*b*) for forward differences, (*c*) for backward differences.

2. Show that $\delta^2 f_m = f_{m+1} - 2f_m + f_{m-1}$, $\delta^3 f_{m+1/2} = f_{m+2} - 3f_{m+1} + 3f_m - f_{m-1}$.

3. Calculate $f(x) = 1/(x + 1)$, $x = 0(0.2)1$ to (*a*) 2D, (*b*) 3D, (*c*) 4D, and compare the effect of the rounding errors in the corresponding difference tables.

4. Using differences, check the table:

x	4.0	4.1	4.2	4.3	4.4	4.5
$f(x)$	0.250	0.244	0.242	0.233	0.227	0.222

5. Perform the computations indicated in Ex. 1.

18.4 Interpolation

If a table of values of a function $f(x)$ is given, it is often necessary to obtain values of $f(x)$ for values of x between the x-values appearing in the table. This

Fig. 335. Linear interpolation

problem of obtaining such a value of f from the given tabular values is called **interpolation**. The tabular values of $f(x)$ used in the process are called **pivotal values**. The usual methods of interpolation are based on the assumption that in a neighborhood of the x-value in question, f can be approximated by a polynomial p, whose value at that x is then taken as an approximation of the value of f at that x.

The simplest method is **linear interpolation**, in which we approximate the curve of f by a chord at two adjacent tabular x-values x_0 and x_1 (cf. Fig. 335). Hence as the approximate value of f at $x = x_0 + rh$ we take

(1) $\quad f(x) \approx p_1(x) = f_0 + r(f_1 - f_0) = f_0 + r\Delta f_0 \quad \left(r = \dfrac{x - x_0}{h},\ 0 \leqq r \leqq 1 \right).$

This is known to us from elementary calculus in connection with tables of logarithms or trigonometric functions. For example, to obtain ln 9.2 from a table showing ln 9.0 = 2.197 and ln 9.5 = 2.251, we may compute $r = 0.2/0.5 = 0.4$ and

$$\ln 9.2 = \ln 9.0 + 0.4(\ln 9.5 - \ln 9.0) = 2.219.$$

Linear interpolation will be satisfactory as long as the x-values in a table are so close that chords deviate but little from the curve of $f(x)$, say, by less than $\frac{1}{2}$ unit of the last digit in the table for each x between x_0 and x_1.

In **quadratic interpolation** we approximate the curve of the function f between x_0 and $x_2 = x_0 + 2h$ by the quadratic parabola which passes through the points (x_0, f_0), (x_1, f_1), (x_2, f_2) and obtain the more accurate formula

(2) $\quad f(x) \approx p_2(x) = f_0 + r\Delta f_0 + \dfrac{r(r - 1)}{2}\Delta^2 f_0 \quad \left(r = \dfrac{x - x_0}{h}, \quad 0 \leqq r \leqq 2 \right)$

where $x = x_0 + rh$. For $x = x_0$ ($r = 0$) the right-hand side equals f_0; for $x = x_1$ ($r = 1$) it equals $f_0 + \Delta f_0 = f_1$, and for $x = x_2$ ($r = 2$) it equals

$$f_0 + 2(f_1 - f_0) + [(f_2 - f_1) - (f_1 - f_0)] = f_2.$$

Example 1. If ln 9.0 = 2.1972 and ln 9.5 = 2.2513 are given, then (1) yields ln 9.2 = 2.2188, exact to 3D only, whereas (2), with ln 10.0 = 2.3026, yields

$$\ln 9.2 = 2.1972 + 0.4 \cdot 0.0541 + \frac{0.4 \cdot (-0.6)}{2}(-0.0028) = 2.2192,$$

which is exact to 4D.

Still more accurate approximations are obtained if we use polynomials of higher degree. A polynomial $p_n(x)$ of degree n is uniquely determined by its values at

$n + 1$ distinct values of x. In the present case we need the polynomial p_n such that $p_n(x_0) = f_0, \cdots, p_n(x_n) = f_n$, where $f_0 = f(x_0), \cdots, f_n = f(x_n)$ are the values of f in the table. This polynomial is given in **Newton's** (or *Gregory-Newton's*) **forward-difference interpolation formula**

$$f(x) \approx p_n(x) = f_0 + r\,\Delta f_0 + \frac{r(r-1)}{2!}\,\Delta^2 f_0 + \cdots$$

(3)
$$+ \frac{r(r-1)\cdots(r-n+1)}{n!}\,\Delta^n f_0$$

$$\left(x = x_0 + rh,\ r = \frac{x - x_0}{h},\ 0 \leqq r \leqq n\right),$$

which includes (1) and (2) as special cases ($n = 1$, $n = 2$). We must prove that $p_n(x_k) = f_k$ ($k = 0, 1, \cdots, n$). The right-hand side of (3) shows that this follows from

(4)
$$f_k = \binom{k}{0} f_0 + \binom{k}{1} \Delta f_0 + \binom{k}{2} \Delta^2 f_0 + \cdots + \binom{k}{k} \Delta^k f_0$$

where the **binomial coefficients** are defined by

(5) $$\binom{k}{0} = 1, \quad \binom{k}{s} = \frac{k(k-1)(k-2)\cdots(k-s+1)}{s!} \qquad (s \geqq 0, \text{integral})$$

and $s! = 1 \cdot 2 \cdot 3 \cdots s$. In fact, if we set $r = k$ in (3), then the right-hand side of (3) becomes identical with that of (4). Formula (4) can be proved by induction. It is true for $k = 0$. Suppose that it holds for a $k = q$. Then, using (4) with $k = q$ and the formula obtained from it by applying Δ, we have

$$f_{q+1} = f_q + \Delta f_q$$

$$= \binom{q}{0} f_0 + \binom{q}{1} \Delta f_0 + \binom{q}{2} \Delta^2 f_0 + \cdots + \binom{q}{q} \Delta^q f_0$$

$$+ \binom{q}{0} \Delta f_0 + \binom{q}{1} \Delta^2 f_0 + \binom{q}{2} \Delta^3 f_0 + \cdots + \binom{q}{q} \Delta^{q+1} f_0.$$

In this formula, $\Delta^s f_0$ has the coefficient [cf. (5)]

$$\binom{q}{s} + \binom{q}{s-1} = \binom{q+1}{s},$$

which gives (4) with $k = q + 1$ and completes the proof by induction.

A formula similar to (3) but involving backward differences is **Newton's** (or *Gregory-Newton's*) **backward-difference interpolation formula**

$$f(x) \approx p_n(x) = f_0 + r\,\nabla f_0 + \frac{r(r+1)}{2!}\,\nabla^2 f_0 + \cdots$$

(6)
$$+ \frac{r(r+1)\cdots(r+n-1)}{n!}\,\nabla^n f_0$$

with $x = x_0 + rh$, $r = (x - x_0)/h$, $0 \leqq r \leqq n$, as in (3).

The literature on methods and formulas of interpolation using finite differences is enormous. For instance, there are formulas involving only even-order differences. The simplest **Everett formula**

$$(7) \quad f(x) \approx (1 - r)f_0 + rf_1 + \frac{(2 - r)(1 - r)(-r)}{3!} \delta^2 f_0 + \frac{(r + 1)r(r - 1)}{3!} \delta^2 f_1$$

where $r = (x - x_0)/h$, $0 \leq r \leq 1$, is of this type and is particularly useful. To facilitate its application, function tables often give second differences needed in (7).

Example 2. Find $e^{1.24}$ from (7) and the table

x	e^x	δ^2
1.2	3.3201	333
1.3	3.6693	367

We have $r = 0.04/0.1 = 0.4$, and (7) gives

$$e^{1.24} = 0.6 \cdot 3.3201 + 0.4 \cdot 3.6693 + \frac{1.6 \cdot 0 \cdot 0.6 \cdot (-0.4)}{6} \cdot 0.0333 + \frac{1.4 \cdot 0.4 \cdot (-0.6)}{6} \cdot 0.0367$$

$$= 3.4598 - 0.0021 - 0.0021 = 3.4556$$

exact to 4D. Note that linear interpolation gives 3.4598, which is exact to only 2D. (The reader may use $e^{1.1} = 3.0042$ and $e^{1.4} = 4.0552$ for checking the given second differences.)

We mention that the general **Everett formula** is

$$f(x) = qf_0 + rf_1 + \binom{q + 1}{3} \delta^2 f_0 + \binom{r + 1}{3} \delta^2 f_1$$

$$(8) \qquad\qquad\qquad + \binom{q + 2}{5} \delta^4 f_0 + \binom{r + 2}{5} \delta^4 f_1 + \cdots$$

where $r = (x - x_0)/h$, $0 \leq r \leq 1$, as before, and $q = 1 - r$. In this formula the ratio of the coefficients of $\delta^4 f_0$ and $\delta^2 f_0$ is

$$\binom{q + 2}{5} \Big/ \binom{q + 1}{3} = \frac{q^2 - 4}{20}$$

and, similarly, the ratio of the coefficients of $\delta^4 f_1$ and $\delta^2 f_1$ is $(r^2 - 4)/20$. Both ratios vary but little in the interval from 0 to 1. Hence if we choose a reasonable mean value μ of them, we can include the fourth-difference effect by using (7) with **modified second differences**

$$(9) \qquad\qquad \delta_m^2 f = \delta^2 f + \mu \delta^4 f, \qquad \mu = -0.18393,$$

where the given μ is the preferred value among various possible values discussed in the literature. This method of using modified differences is called **throwback**. The actual error incurred is less than half a unit if the fourth differences are less than 1000 units (of the last digit) and the fifth differences less than 70.

We mention without proof that if $x_0, x_1, \cdots, x_n$ are arbitrarily spaced, the polynomial of degree n through $(x_0, f_0), \cdots, (x_n, f_n)$, where $f_j = f(x_j)$, is given by the right-hand side of **Newton's divided difference interpolation formula**

(10a)
$$f(x) \approx f_0 + (x - x_0)f[x_0, x_1] + (x - x_0)(x - x_1)f[x_0, x_1, x_2] + \cdots$$
$$+ (x - x_0) \cdots (x - x_{n-1})f[x_0, \cdots, x_n]$$

involving **divided differences,** which are defined iteratively by the relations

(10b)
$$f[x_0, x_1] = \frac{f(x_1) - f(x_0)}{x_1 - x_0}, \qquad f[x_0, x_1, x_2] = \frac{f[x_1, x_2] - f[x_0, x_1]}{x_2 - x_0}, \cdots$$
$$f[x_0, \cdots, x_k] = \frac{f[x_1, \cdots, x_k] - f[x_0, \cdots, x_{k-1}]}{x_k - x_0}.$$

If $x_k = x_0 + kh$ (equal spacing), then $f[x_0, \cdots, x_k] = \Delta^k f_0/k! \, h^k$, and (10a) takes the form (3).

In interpolation by difference methods we have to calculate differences and may use them for checks if a table is new or doubtful. However, the question of what order of interpolation to use is left unanswered in most tables. **Lagrangian interpolation** is based on **Lagrange's interpolation formula**

(11a)
$$f(x) \approx L_n(x) = \sum_{k=0}^n \frac{l_k(x)}{l_k(x_k)} f_k$$

where $x_0, \cdots, x_n$ are not necessarily equally spaced and

(11b)
$$l_0(x) = (x - x_1)(x - x_2) \cdots (x - x_n)$$
$$l_k(x) = (x - x_0) \cdots (x - x_{k-1})(x - x_{k+1}) \cdots (x - x_n),$$
$$k = 1, \cdots, n - 1$$
$$l_n(x) = (x - x_0)(x - x_1) \cdots (x - x_{n-1}).$$

Formula (11a) is known as **Lagrange's $n + 1$ point formula.** It can easily be seen that $L_n(x_k) = f_k$ because inspection of (11b) shows that $l_k(x_j) = 0$ when $j \neq k$ and $(l_k(x)/l_k(x_k))f_k = f_k$ when $x = x_k$. The advantage is that we need not compute differences and we see the effect of the various f_k's more directly, but the calculations may be more laborious and the use of the method should be restricted to guaranteed tables. Tables of Lagrangian coefficients $l_k(x)/l_k(x_k)$ for equal spacing have been published; cf. Ref. [J12] in Appendix 1.

Example 3. Find ln 9.2, using (11) with $n = 3$ and the values

x	9.0	9.5	10.0	11.0
ln x	2.19722	2.25129	2.30259	2.39790

We have $l_0(x) = (x - 9.5)(x - 10)(x - 11)$, $l_1(x) = (x - 9)(x - 10)(x - 11)$, etc., and compute

$$\ln 9.2 = \frac{-0.43200}{-1.00000} \cdot 2.19722 + \frac{0.28800}{0.37500} \cdot 2.25129 + \frac{0.10800}{-0.50000} \cdot 2.30259 + \frac{0.04800}{3.00000} \cdot 2.39790$$

$$= 2.21920 \text{ (exact to 5D)}.$$

The problem of finding x for given $f(x)$ is known as **inverse interpolation.** If f is differentiable and df/dx is not zero near the point where the inverse interpolation is to be effected, the inverse $x = F(y)$ of $y = f(x)$ exists locally near the given value of f and it may happen that F can be approximated in that neighborhood by a polynomial of moderately low degree. Then we may effect the inverse interpolation by tabulating F as a function of y and apply methods of direct interpolation to F. If $df/dx = 0$ near or at the desired point, it may be useful to solve $p(x) = \tilde{f}$ by iteration; here $p(x)$ is a polynomial that approximates $f(x)$ and $\tilde{f}$ is the given value.

Problems

1. Verify that the parabola (2) passes through (x_0, f_0), (x_1, f_1), (x_2, f_2).
2. Verify the calculation in Ex. 2.
3. Find $f(3)$, using Lagrange's formula (11) and $f(1) = 2, f(2) = 11, f(4) = 77$.
4. Show that

$$\Delta^k f_n = \binom{k}{0} f_{n+k} - \binom{k}{1} f_{n+k-1} + - \cdots + (-1)^k \binom{k}{k} f_n.$$

5. Using ln $8.5 = 2.14007$ and ln 9, ln 10, ln 11 as in Ex. 3 and applying (11) with $n = 3$, compute ln 9.2 and compare the result with that in Ex. 3.
6. Using ln $8.5 = 2.14007$ and ln 9.0, ln 9.5, ln 10 as in Ex. 3, compute ln 9.2 (a) by means of (3) with $n = 3$, (b) by means of (6) with $n = 3$.
7. Using the same data as in Prob. 6, compute ln 9.2 (a) by means of (7), (b) by means of (11) with $n = 3$.
8. Assuming that $x_1 = x_0 + h$, $x_2 = x_0 + 2h$, $x_3 = x_0 + 3h$ and using $r = (x - x_0)/h$, show that (11) with $n = 3$ may be written

$$f(x) \approx -\binom{r-1}{3} f_0 + \frac{r(r-2)(r-3)}{2} f_1 - \frac{r(r-1)(r-3)}{2} f_2 + \binom{r}{3} f_3.$$

9. Applying the formula in Prob. 8, check the result in Prob. 7b.
10. **(Checking differences)** Show that the sum of the differences in a column is equal to the difference between the last and the first entries of the preceding column. Apply this partial check to Table 4 in Sec. 18.3.

18.5 *Numerical Integration and Differentiation*

The problem of **numerical integration** is the numerical evaluation of a definite integral

$$J = \int_a^b f(x)\, dx$$

where a and b are given and f is a function given analytically by a formula or empirically by a table of values.

We know that if f is such that we can find a differentiable function F whose derivative is f, then we can evaluate J by applying the familiar formula

$$J = \int_a^b f(x)\, dx = F(b) - F(a) \qquad\qquad [F'(x) = f(x)],$$

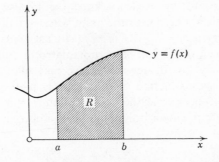

Fig. 336. Geometrical interpretation of a definite integral

and tables of integrals, such as Ref. [J3] listed in Appendix 1, may be helpful for that purpose.

However, in engineering applications there frequently occur integrals whose integrand is an empirical function given by a table, or is such that the integral cannot be represented in terms of finitely many elementary functions (examples in Appendix 3), or the explicit form of F is complicated and not very helpful. Then we may use a numerical method of approximate integration.

Since J equals the area of the region R under the curve of $f(x)$ between a and b (cf. Fig. 336), we may obtain rough approximate values of J by cutting R from cardboard, determine its weight, and divide the result by the weight of a square of this cardboard of area 1. Another simple method is to draw R on graph paper and count squares. More accurate results can be obtained by using a **planimeter,** whose theory and application are discussed in Ref. [G15] listed in Appendix 1.

Numerical integration methods may be obtained by approximating the integrand f by polynomials. To derive the simplest formula, we subdivide the interval of integration into n equal subintervals of length $h = (b - a)/n$ and approximate f in each such interval by a constant function $f(x_j^*)$, where x_j^* is the midpoint of the interval (cf. Fig. 337). Then f is approximated by a **step function** (piecewise constant function), the n rectangles in Fig. 337 have the areas $f(x_1^*)h, \cdots,$ $f(x_n^*)h$, and we obtain the **rectangular rule**

$$(1) \qquad J = \int_a^b f(x)\, dx \approx h[f(x_1^*) + f(x_2^*) + \cdots + f(x_n^*)] \qquad \left(h = \frac{b - a}{n} \right).$$

Fig. 337. Rectangular rule   **Fig. 338.** Trapezoidal rule

If we approximate f by a piecewise linear function (polygon of chords of the curve of f; cf. Fig. 338), we obtain the **trapezoidal rule**

$$(2) \quad J = \int_a^b f(x)\,dx \approx h[\tfrac{1}{2}f(a) + f(x_1) + f(x_2) + \cdots + f(x_{n-1}) + \tfrac{1}{2}f(b)],$$

where $h = (b - a)/n$, and $x_0 \,(= a)$, $x_1, x_2, \cdots, x_{n-1}, x_n \,(= b)$ are the end points of the intervals already used in (1); thus $x_j = x_0 + jh$. In fact, the n trapezoids in Fig. 338 have the areas

$$\tfrac{1}{2}[f(a) + f(x_1)]h, \quad \tfrac{1}{2}[f(x_1) + f(x_2)]h, \quad \cdots, \quad \tfrac{1}{2}[f(x_{n-1}) + f(b)]h,$$

and their sum J^* equals the right-hand side of (2).

The error ϵ of J^* is (cf. Sec. 18.1)

$$\epsilon = J^* - J.$$

If $f(x)$ is a linear function, then $\epsilon = 0$. In this case f' is constant, f'' is zero for all x, and it seems plausible that for a general function f (having a continuous second derivative) we may obtain **error bounds** (bounds for ϵ) involving f''. For this purpose we replace b by a variable t and apply (2) with $n = 1$. Then the corresponding error is

$$\epsilon(t) = \frac{t - a}{2}\,[f(a) + f(t)] - \int_a^t f(x)\,dx.$$

We note that $\epsilon(a) = 0$, which is trivial. Differentiation gives

$$\epsilon'(t) = \tfrac{1}{2}[f(a) + f(t)] + \frac{t - a}{2} f'(t) - f(t).$$

We see that $\epsilon'(a) = 0$. Differentiating once more, we have

$$\epsilon''(t) = \tfrac{1}{2}(t - a)f''(t)$$

and can obtain bounds for ϵ'' by replacing f'' by its smallest and largest value in the interval $a \leqq t \leqq b$. Denoting these values by M_2^* and M_2, respectively, and noting that $t - a \geqq 0$, we obtain

$$\tfrac{1}{2}(t - a)M_2^* \leqq \epsilon''(t) \leqq \tfrac{1}{2}(t - a)M_2$$

for all t in that interval. Integrating from a to t, we have

$$\tfrac{1}{4}(t - a)^2 M_2^* \leqq \epsilon'(t) - \epsilon'(a) \leqq \tfrac{1}{4}(t - a)^2 M_2.$$

Using $\epsilon'(a) = 0$, $\epsilon(a) = 0$, integrating again, and setting $t = a + h$, we see that

$$\tfrac{1}{12}h^3 M_2^* \leqq \epsilon(a + h) \leqq \tfrac{1}{12}h^3 M_2.$$

For the errors corresponding to the other $n - 1$ subintervals we obtain $n - 1$ similar inequalities, and addition of all the n inequalities gives an inequality for the error ϵ corresponding to the integration from a to b; since $h = (b - a)/n$, we obtain

(3) $\qquad\qquad KM_2{}^* \leqq \epsilon \leqq KM_2 \qquad$ where $\qquad K = \dfrac{(b - a)^3}{12n^2}$

and $M_2{}^*$ and M_2 are the smallest and largest values of the second derivative of f in the interval of integration.

Example 1. Evaluate $J = \int_0^1 e^{-x^2}\, dx$ by means of (2) with $n = 10$ and estimate the error.

From Table 7 we see that

$$J \approx 0.1(0.5 \cdot 1.367\ 879 + 6.778\ 167) = 0.746\ 211.$$

Since $f''(x) = 2(2x^2 - 1)e^{-x^2}$, we have $M_2{}^* = f''(0) = -2$, $M_2 = f''(1) = 0.735\ 759$. Furthermore $K = 1/1200$, and (3) gives

$$-0.001\ 667 \leqq \epsilon \leqq 0.000\ 614.$$

Hence the exact value of J must lie between

$$0.746\ 211 - 0.000\ 614 = 0.745\ 597 \quad \text{and} \quad 0.746\ 211 + 0.001\ 667 = 0.747\ 878.$$

(Actually, $J = 0.746\ 824$, exact to 6D.)

Table 7. Computation in Example 1

j	x_j	$x_j{}^2$		$e^{-x_j{}^2}$
0	0	0	1.000 000	
1	0.1	0.01		0.990 050
2	0.2	0.04		0.960 789
3	0.3	0.09		0.913 931
4	0.4	0.16		0.852 144
5	0.5	0.25		0.778 801
6	0.6	0.36		0.697 676
7	0.7	0.49		0.612 626
8	0.8	0.64		0.527 292
9	0.9	0.81		0.444 858
10	1.0	1.00	0.367 879	
		Sums	1.367 879	6.778 167

Piecewise constant approximation of f led to the rectangular rule (1), piecewise linear approximation to the trapezoidal rule (2), and piecewise quadratic approximation will give Simpson's rule, which is of great practical importance because it is sufficiently accurate for most problems, but still sufficiently simple. To derive this formula, we subdivide the interval of integration $a \leqq x \leqq b$ into an **even** number of equal subintervals, say, into $2n$ subintervals of length $h = (b - a)/2n$, with end points $x_0 (= a)$, $x_1, \cdots, x_{2n-1}, x_{2n} (= b)$. We consider the first two subintervals and approximate $f(x)$ in the interval $x_0 \leqq x \leqq x_2 = x_0 + 2h$ by the

Lagrange polynomial $p_2(x)$ through (x_0, f_0), (x_1, f_1), (x_2, f_2), where $f_j = f(x_j)$. From (11a) in Sec. 18.4 we obtain

$$(4^*) \quad p_2(x) = \frac{(x - x_1)(x - x_2)}{(x_0 - x_1)(x_0 - x_2)} f_0 + \frac{(x - x_0)(x - x_2)}{(x_1 - x_0)(x_1 - x_2)} f_1 + \frac{(x - x_0)(x - x_1)}{(x_2 - x_0)(x_2 - x_1)} f_2.$$

The denominators are $2h^2$, $-h^2$, and $2h^2$, respectively. Setting $s = (x - x_1)/h$, we have $x - x_0 = (s + 1)h$, $x - x_1 = sh$, $x - x_2 = (s - 1)h$, and

$$p_2(x) = \tfrac{1}{2}s(s - 1)f_0 - (s + 1)(s - 1)f_1 + \tfrac{1}{2}(s + 1)sf_2.$$

We now integrate with respect to x from x_0 to x_2. This corresponds to integrating with respect to s from -1 to 1. Since $dx = h\,ds$, the result of this simple integration is

$$\int_{x_0}^{x_2} f(x)\, dx \approx \int_{x_0}^{x_2} p_2(x)\, dx = h\left(\frac{1}{3}f_0 + \frac{4}{3}f_1 + \frac{1}{3}f_2\right).$$

A similar formula holds for the next two subintervals from x_2 and x_4, and so on. By summing all these n formulas we obtain **Simpson's rule**

$$(4) \quad \int_a^b f(x)\, dx \approx \frac{h}{3}(f_0 + 4f_1 + 2f_2 + 4f_3 + \cdots + 2f_{2n-2} + 4f_{2n-1} + f_{2n}),$$

where $h = (b - a)/2n$ and $f_j = f(x_j)$. Bounds for the error ϵ_S in (4) can be obtained by a method similar to that in the case of the trapezoidal rule (2), assuming that the fourth derivative of f exists and is continuous in the interval of integration. The result is

$$(5) \quad CM_4{}^* \leqq \epsilon_S \leqq CM_4, \quad \text{where} \quad C = \frac{(b - a)^5}{180(2n)^4}$$

and $M_4{}^*$ and M_4 are the smallest and the largest value of the fourth derivative of f in the interval of integration. It is trivial that $\epsilon_S = 0$ for a quadratic polynomial f, but (5) shows that $\epsilon_S = 0$ even for a cubic polynomial.

Example 2. Evaluate $J = \int_0^1 e^{-x^2}\, dx$ by Simpson's rule with $2n = 10$ and estimate the error. Since $h = 0.1$, Table 8 gives

$$J \approx \frac{0.1}{3}(1.367\,879 + 4 \cdot 3.740\,266 + 2 \cdot 3.037\,901) = 0.746\,825.$$

Estimate of error. Differentiation gives $f^{IV}(x) = 4(4x^4 - 12x^2 + 3)e^{-x^2}$. By considering the derivative of f^{IV} we find that the smallest value of f^{IV} in the interval of integration occurs at $x = x^* = 2.5 + 0.5\sqrt{10}$ and the largest value at $x = 0$. Computation gives $M_4{}^* = f^{IV}(x^*) = -7.359$ and $M_4 = f^{IV}(0) = 12$. Since $2n = 10$ and $b - a = 1$, we have $C = 1/1\,800\,000 = 0.000\,000\,56$. Therefore

$$-0.000\,004 \leqq \epsilon_S \leqq 0.000\,006, \quad \text{and} \quad 0.746\,818 \leqq J \leqq 0.746\,830,$$

which shows exactness of our approximation to at least 4D. Actually 0.746 825 is exact to 5D, because $J = 0.746\,824$ (exact to 6D). Note that our present result is much better than that in Ex. 1 obtained by the trapezoidal rule, whereas the work is nearly the same in both cases.

Table 8. Computations in Example 2

j	x_j	x_j^2	$e^{-x_j^2}$		
0	0	0	1.000 000		
1	0.1	0.01		0.990 050	
2	0.2	0.04			0.960 789
3	0.3	0.09		0.913 931	
4	0.4	0.16			0.852 144
5	0.5	0.25		0.778 801	
6	0.6	0.36			0.697 676
7	0.7	0.49		0.612 626	
8	0.8	0.64			0.527 292
9	0.9	0.81		0.444 858	
10	1.0	1.00	0.367 879		
Sums			1.367 879	3.740 266	3.037 901

Simpson's rule is sufficiently accurate for most engineering purposes, and in automatic computation it is preferred to more complicated formulas of higher precision, which are obtained by the use of approximating polynomials of higher degree. We want to mention two such formulas of higher precision, which are sometimes useful. A cubic through $(x_0, f_0), \cdots, (x_3, f_3)$ yields the **three-eights rule**

$$(6) \qquad \int_{x_0}^{x_3} f(x)\, dx \approx \frac{3h}{8}(f_0 + 3f_1 + 3f_2 + f_3),$$

which is exact for cubical polynomials (as is Simpson's rule) and has the advantage that it can be applied to an *odd* number of subintervals. The polynomial of sixth degree through $(x_0, f_0), \cdots, (x_6, f_6)$ yields a formula with rather complicated coefficients, but a much simpler formula of the same kind, and only slightly inferior in accuracy, is **Weddle's rule**

$$(7) \qquad \int_{x_0}^{x_6} f(x)\, dx \approx \frac{3h}{10}(f_0 + 5f_1 + f_2 + 6f_3 + f_4 + 5f_5 + f_6).$$

Our integration formulas discussed involve function values for equidistant x-values and give exact results for polynomials not exceeding a certain degree. More generally, we may set

$$(8) \qquad \int_{-1}^{1} f(x)\, dx \approx \sum_{j=1}^{n} A_j f_j \qquad\qquad [f_j = f(x_j)]$$

(choosing $a = -1$ and $b = 1$, which may be accomplished by a linear transformation of scale) and determine the $2n$ constants $A_1, \cdots, A_n, x_1, \cdots, x_n$ so that (8) gives exact results for polynomials of degree m as high as possible. Since $2n$ is the number of coefficients of a polynomial of degree $2n - 1$, it follows that $m \leq 2n - 1$. Gauss has shown that exactness for polynomials of degree $2n - 1$ can be attained if and only if $x_1, \cdots, x_n$ are the n zeros of the **Legendre polynomial** *of degree n* [cf. also Sec. 3.3]

$$P_n(x) = \frac{(2n)!}{2^n(n!)^2} x^n - \frac{(2n-2)!}{2^n 1!\,(n-1)!\,(n-2)!} x^{n-2}$$

$$+ \frac{(2n-4)!}{2^n 2!\,(n-2)!\,(n-4)!} x^{n-4} - + \cdots,$$

that is,

$$P_0 = 1, \quad P_1(x) = x, \quad P_2(x) = \tfrac{1}{2}(3x^2 - 1), \quad P_3(x) = \tfrac{1}{2}(5x^3 - 3x), \cdots,$$

and the coefficients A_j are suitably chosen. Then (8) is called a **Gaussian integration formula**. Its advantage is the high accuracy, its disadvantage the irregular spacing of $x_1, \cdots, x_n$. Hence such a formula may be practical when automatic computers are used and the computation of $f(x_j)$ does not depend on the number of decimals in the argument or when $f(x_j)$ is obtained from an experiment in which the x_j can be set once and for all. The formula is not likely to be practical when the integrand is tabulated at equal intervals, because then preliminary interpolations are necessary to evaluate the $f(x_j)$, and these may outweigh any gain due to the smaller error estimates.

Since the end points -1 and 1 of the interval of integration in (8) are not zeros of P_n, they do not occur among $x_1, \cdots, x_n$, and the Gauss formula (8) is called, therefore, an **open formula,** in contrast to a **closed formula** in which the end points of the interval of integration do occur. (For example, (2), (4), (6), and (7) are closed.)

We mention that, just as in the case of interpolation, there are numerical integration methods based on differences. A very efficient method uses the **central-difference formula by Gauss**

$$(9) \qquad \int_{x_0}^{x_1} f(x)\,dx \approx \frac{h}{2}\left(f_0 + f_1 - \frac{\delta^2 f_0 + \delta^2 f_1}{12} + \frac{11(\delta^4 f_0 + \delta^4 f_1)}{720} \right).$$

For more details see Ref. [G12]–[G14] in Appendix 1.

The problem of **numerical differentiation** is the determination of approximate values of the derivative of a function f which is given by a table. Numerical differentiation should be avoided whenever possible because approximate values of derivatives will in general be less accurate than the function values from which they are derived. In fact, the derivative is the limit of the difference quotient, and in the latter we normally subtract two large quantities and divide by a small one; furthermore if a given function f is approximated by a polynomial p, the difference in function values may be small but the derivatives may differ considerably. Hence it is plausible that numerical differentiation is delicate, in contrast to numerical integration, which is not much affected by inaccuracies of function values, because integration is essentially a smoothing process.

We use the notations $f_j' = f'(x_j)$, $f_j'' = f''(x_j)$, etc., and may obtain rough approximation formulas for derivatives by noting that

$$f'(x) = \lim_{h \to 0} \frac{f(x+h) - f(x)}{h}.$$

This suggests

(10)
$$f_{1/2}' \approx \frac{\delta f_{1/2}}{h} = \frac{f_1 - f_0}{h}.$$

Similarly, for the second derivative we obtain

(11)
$$f_1'' \approx \frac{\delta^2 f_1}{h^2} = \frac{f_2 - 2f_1 + f_0}{h^2}$$

and so on.

More accurate approximations are obtained by differentiating suitable Lagrange polynomials. Differentiating (4*) and remembering that the denominators in (4*) are $2h^2$, $-h^2$, $2h^2$, we have

$$f'(x) \approx p_2'(x) = \frac{2x - x_1 - x_2}{2h^2} f_0 - \frac{2x - x_0 - x_2}{h^2} f_1 + \frac{2x - x_0 - x_1}{2h^2} f_2.$$

Evaluating this at x_0, x_1, x_2, we obtain

$$f_0' \approx \frac{1}{2h}(-3f_0 + 4f_1 - f_2)$$

(12)
$$f_1' \approx \frac{1}{2h}(-f_0 + f_2)$$

$$f_2' \approx \frac{1}{2h}(f_0 - 4f_1 + 3f_2).$$

Applying the same idea to Lagrange's five-point formula, we obtain similar formulas, in particular

(13)
$$f_2' \approx \frac{1}{12h}(f_0 - 8f_1 + 8f_3 - f_4).$$

Further details and formulas are included in Ref. [G6] listed in Appendix 1.

Problems

Review some integration formulas and methods by integrating

1. $\displaystyle\int \frac{dx}{a^2 + x^2}$

2. $\displaystyle\int \frac{dx}{\sqrt{a^2 - x^2}}$

3. $\displaystyle\int e^{ax} \cos bx \, dx$

4. $\displaystyle\int e^{ax} \sin bx \, dx$

5. $\displaystyle\int \ln x \, dx$

6. $\displaystyle\int \cos^2 x \, dx$

7. $\displaystyle\int \sin^2 \omega x \, dx$

8. $\displaystyle\int \tan x \, dx$

9. $\displaystyle\int \frac{dx}{x^2(x^2 + 1)^2}$

10. Compute $\int_0^1 x^3 \, dx$ by the rectangular rule (1) with $n = 5$. What is the error?

11. Compute the integral in Prob. 10 by the trapezoidal rule (2) with $n = 5$. What error bounds are obtained from (3)? What is the actual error of the result? Why is this result larger than the exact value?

12. Compute the integral in Ex. 1 by using (1) with $n = 5$.

Using the values of $\sin x$ in Table A1, Appendix 4, evaluate $\displaystyle\int_0^1 \frac{\sin x}{x}\, dx$:

13. By the rectangular rule (1) with $n = 5$.
14. By the trapezoidal rule (2) with $n = 5$.
15. By (2) with $n = 10$.
16. By Simpson's rule with $2n = 2$.
17. By Simpson's rule with $2n = 10$.

18. Evaluate $\int_0^1 x^5\, dx$ by Simpson's rule with $2n = 10$. What error bounds are obtained from (5)? What is the actual error of the result?

19. Find an approximate value of $\ln 2 = \displaystyle\int_1^2 \frac{dx}{x}$ by Simpson's rule with $2n = 4$. Estimate the error by (5).

20. Derive (13) as explained in the text.

18.6 *Numerical Methods for First-Order Differential Equations*

This section is independent of the other sections in this chapter and may also be discussed immediately after Chap. 1.

From Chap. 1 we know that a *differential equation of the first order* is of the form $F(x, y, y') = 0$, and often it will be possible to write the equation in the *explicit form* $y' = f(x, y)$. An **initial value problem** consists of a differential equation and a condition which the solution must satisfy (or several conditions referring to the same value of x if the equation is of higher order). In this section we shall consider initial value problems of the form

(1) $$y' = f(x, y), \qquad y(x_0) = y_0,$$

assuming f to be such that the problem has a unique solution in some interval containing x_0, and we shall discuss methods for computing numerical values of the solution.

If we can obtain a formula for the solution, we may evaluate it numerically, either directly or by the use of tables. In this approach the book by Kamke (Ref. [B9] listed in Appendix 1) and an index of tables (cf. Ref. [A6] in Appendix 1) will be helpful. If that formula is too complicated or if no formula for the solution is available, we may apply one of the methods discussed in this section.

These methods are **step-by-step methods,** that is, we start from $y_0 = y(x_0)$ and proceed stepwise. In the first step we compute an approximate value y_1 of the solution y of (1) at $x = x_1 = x_0 + h$. In the second step we compute an approximate value y_2 of that solution at $x = x_2 = x_0 + 2h$, etc. Here h is a fixed number, for example, 0.2 or 0.1 or 0.01; principles for the choice of h will be discussed later in this section.

In each step the computations are done by the same formula. Such formulas are suggested by the Taylor series

$$y(x + h) = y(x) + hy'(x) + \frac{h^2}{2} y''(x) + \cdots.$$

From (1) we have $y' = f$. By differentiation, $y'' = f' = \partial f / \partial x + (\partial f / \partial y) y'$, etc., and the Taylor series becomes

$$(2) \qquad y(x + h) = y(x) + hf + \frac{h^2}{2} f' + \frac{h^3}{6} f'' + \cdots,$$

where $f, f', f'', \cdots$ are evaluated at $(x, y(x))$.

For small values of h, the higher powers $h^2, h^3, \cdots$ in (2) will be very small. This suggests the crude approximation

$$y(x + h) \approx y(x) + hf$$

and the following iteration process. In the first step we compute

$$y_1 = y_0 + hf(x_0, y_0),$$

which approximates $y(x_1) = y(x_0 + h)$. In the second step we compute

$$y_2 = y_1 + hf(x_1, y_1),$$

which approximates $y(x_2) = y(x_0 + 2h)$, etc., and in general

$$(3) \qquad y_{n+1} = y_n + hf(x_n, y_n) \qquad\qquad (n = 0, 1, \cdots).$$

This is called the **Euler method** or **Euler-Cauchy method**. Geometrically it is an approximation of the curve of $y(x)$ by a polygon whose first side is tangent to the curve at x_0 (cf. Fig. 339).

The method is called a **first-order method**, because in (2) we take only the constant term and the term containing the first power of h. The omission of the further terms in (2) causes an error, which is called the **truncation error** of the method. For small h, the third and higher powers of h will be small compared with h^2 in the first neglected term in (2), and we therefore say that the **truncation error per step** is of **order** h^2. In addition there are **rounding errors** in this and other methods, which may affect the accuracy of the values $y_1, y_2, \cdots$ more and more as n increases; we shall return to this point in the next section.

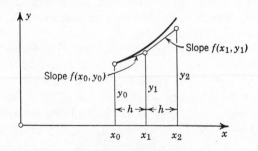

Fig. 339. Euler method

The practical value of the Euler method is limited, but since it is simple, it may be helpful for understanding the basic idea of the methods in this section.

Example 1. Apply the Euler method to the initial value problem

(4) $$y' = x + y, \qquad y(0) = 0,$$

choosing $h = 0.2$ and computing $y_1, \cdots, y_5$. Here $f(x, y) = x + y$, and (3) becomes

$$y_{n+1} = y_n + 0.2(x_n + y_n).$$

Table 9 shows the computations, the values of the exact solution

$$y(x) = e^x - x - 1$$

obtained from (4) in Sec. 1.7, and the error. In practice the exact solution is unknown, but an indication of the accuracy of the values can be obtained by applying the Euler method once more with step $2h = 0.4$ and comparing corresponding approximations. They differ by 0.040 ($x = 0.4$) and 0.110 ($x = 0.8$), as a simple calculation shows. Since the error is of order h^2, a switch from h to $2h$ corresponds to a multiplication by $2^2 = 4$, but since we need only half as many steps as before, the error will only be multiplied by $4/2 = 2$; hence those differences indicate the size of the error (cf. Table 9).

Table 9. Euler Method Applied to (4) and Error

n	x_n	y_n	$0.2(x_n + y_n)$	Exact Values	Absolute Value of Error
0	0.0	0.000	0.000	0.000	0.000
1	0.2	0.000	0.040	0.021	0.021
2	0.4	0.040	0.088	0.092	0.052
3	0.6	0.128	0.146	0.222	0.094
4	0.8	0.274	0.215	0.426	0.152
5	1.0	0.489		0.718	0.229

By taking more terms in (2) into account we obtain numerical methods of higher order and precision. The corresponding formulas can be represented in such a form that the complicated computation of derivatives of $f(x, y)$ is avoided and is replaced by computing f for one or several suitably chosen auxiliary values of (x, y). Let us discuss two such methods which are of practical importance.

The first method is the so-called **improved Euler method** or **improved Euler-Cauchy method**. In each step of this method we first compute the auxiliary value

(5a) $$y_{n+1}^* = y_n + hf(x_n, y_n)$$

and then the new value

(5b) $$y_{n+1} = y_n + \tfrac{1}{2}h[f(x_n, y_n) + f(x_{n+1}, y_{n+1}^*)].$$

This method has a simple geometric interpretation. In fact, we may say that in the interval from x_n to $x_n + \tfrac{1}{2}h$ we approximate the solution y by the straight line through (x_n, y_n) with slope $f(x_n, y_n)$, and then we continue along the straight line with slope $f(x_{n+1}, y_{n+1}^*)$ until x reaches x_{n+1} (cf. Fig. 340, p. 664, where $n = 0$).

Fig. 340. Improved Euler method

Example 2.   Apply the improved Euler method to the initial value problem (4) in Ex. 1, choosing $h = 0.2$, as before.   Here (5) becomes

$$y^*_{n+1} = y_n + 0.2(x_n + y_n)$$

$$y_{n+1} = y_n + 0.1[(x_n + y_n) + (x_{n+1} + y^*_{n+1})]$$

and is so simple that it pays to insert the first formula into the second.   Simplifying the result, we readily obtain

$$y_{n+1} = 0.12x_n + 0.1x_{n+1} + 1.22y_n.$$

From Table 10 we see that the present results are more accurate than those in Ex. 1; cf. also Table 12.

The improved Euler method is a **second-order method**, *because the truncation error per step is of order h^3.*

In fact, setting $\tilde{f}_n = f(x_n, y(x_n))$ and using (2), we have

(6a) $$y(x_n + h) - y(x_n) = h\tilde{f}_n + \tfrac{1}{2}h^2\tilde{f}_n' + \tfrac{1}{6}h^3\tilde{f}_n'' + \cdots.$$

Approximating the expression in the brackets in (5b) by $\tilde{f}_n + \tilde{f}_{n+1}$ and using again the Taylor expansion, we obtain from (5b)

(6b) $$y_{n+1} - y_n \approx \tfrac{1}{2}h(\tilde{f}_n + \tilde{f}_n + h\tilde{f}_n' + \tfrac{1}{2}h^2\tilde{f}_n'' + \cdots).$$

Subtraction of (6a) from (6b) gives the truncation error per step

$$\frac{h^3}{4}\tilde{f}_n'' - \frac{h^3}{6}\tilde{f}_n'' + \cdots = \frac{h^3}{12}\tilde{f}_n'' + \cdots.$$

We shall now discuss the **choice of the step** h, which is an important problem in applying a step-by-step method. h should not be too small, because otherwise the number of steps and the rounding errors become large.   On the other hand, h

Table 10. Improved Euler Method Applied to (4)

n	x_n	y_n	$0.12x_n$	$0.1x_{n+1}$	$1.22y_n$	y_{n+1}
0	0.0	0.0000	0.0000	0.0200	0.0000	0.0200
1	0.2	0.0200	0.0240	0.0400	0.0244	0.0884
2	0.4	0.0884	0.0480	0.0600	0.1078	0.2158
3	0.6	0.2158	0.0720	0.0800	0.2633	0.4153
4	0.8	0.4153	0.0960	0.1000	0.5067	0.7027
5	1.0	0.7027				

should not be too large, because a large h implies a large truncation error per step and, in addition to it, an error caused by the fact that f is evaluated at (x_n, y_n) instead of $(x_n, y(x_n))$. The latter would be zero if f were independent of y, and it will matter the more the faster f varies as y varies, that is, the larger the absolute value of the partial derivative $f_y = \partial f / \partial y$ is. More precisely, denoting that error by φ_n and applying the mean value theorem, we have

$$\varphi_n = f(x_n, y_n) - f(x_n, y(x_n)) = f_y(x_n, \tilde{y}) \eta_n$$

where $\eta_n = y_n - y(x_n)$ is the error of y_n and $\tilde{y}$ lies between y_n and $y(x_n)$. Hence the contribution of φ_n to the error of y_{n+1} is approximately $h\varphi_n = hf_y(x_n, \tilde{y})\eta_n$. This suggests to take a close upper bound K of $|f_y|$ in the region of interest and to choose h such that

$$\kappa = hK$$

is not too large. We see that if $|f_y|$ is large (strong dependence of f on y), then K is large and h must be small, which is understandable. (In Exs. 1 and 2, $f_y = 1$, $K = 1$, $hK = 0.2$.) If f_y varies very much, we may choose a close upper bound K_n of $|f_y(x_n, \tilde{y})|$ and choose two or even three different values of h in different regions, to keep

$$\kappa_n = hK_n$$

within a certain interval (e.g., $0.1 \leqq \kappa_n \leqq 0.2$), which depends on the desired accuracy; of course, because of the truncation error per step, we cannot let h increase beyond a certain value.

A still more accurate method is the **Runge-Kutta method**,[4] which is of great practical importance. In each step of this method, we first compute four auxiliary quantities

(7a)
$$A_n = hf(x_n, y_n), \qquad\qquad B_n = hf(x_n + \tfrac{1}{2}h, y_n + \tfrac{1}{2}A_n),$$
$$C_n = hf(x_n + \tfrac{1}{2}h, y_n + \tfrac{1}{2}B_n), \qquad D_n = hf(x_{n+1}, y_n + C_n),$$

and then the new value

(7b)
$$y_{n+1} = y_n + \tfrac{1}{6}(A_n + 2B_n + 2C_n + D_n).$$

It can be shown that the truncation error per step is of the order h^5 (cf. Ref. [G15], Appendix 1) and the method is, therefore, a fourth-order method. Note that if f depends only on x, this method reduces to Simpson's rule of integration (Sec. 18.5).

In desk calculations, the frequent calculation of $f(x, y)$ is laborious. In automatic computation this does not matter too much, and the method is well suited because it needs no special starting procedure, makes light demands on the store, requires no estimation, and uses the same straightforward computational procedure several times.

[4] CARL RUNGE (1856–1927) and WILHELM KUTTA (1867–1944), German mathematicians.

Table 11. Runge-Kutta Method Applied to (4); Computations by the Use of (8)

n	x_n	y_n	$x_n + y_n$	$0.2214(x_n + y_n)$	$0.2214(x_n + y_n)$ $+ 0.0214$
0	0.0	0	0	0	0.021 400
1	0.2	0.021 400	0.221 400	0.049 018	0.070 418
2	0.4	0.091 818	0.491 818	0.108 889	0.130 289
3	0.6	0.222 107	0.822 107	0.182 014	0.203 414
4	0.8	0.425 521	1.225 521	0.271 330	0.292 730
5	1.0	0.718 251			

Example 3. Apply the Runge-Kutta method to the initial value problem (4) in Ex. 1, choosing $h = 0.2$, as before and computing five steps. Here $f(x, y) = x + y$, and (7a) becomes

$$A_n = 0.2(x_n + y_n), \qquad\qquad B_n = 0.2(x_n + 0.1 + y_n + 0.5A_n),$$
$$C_n = 0.2(x_n + 0.1 + y_n + 0.5B_n), \qquad D_n = 0.2(x_n + 0.2 + y_n + C_n).$$

Since these expressions are so simple, we find it convenient to insert A_n into B_n, obtaining $B_n = 0.22\,(x_n + y_n) + 0.02$, insert this into C_n, finding $C_n = 0.222(x_n + y_n) + 0.022$, and finally insert this into D_n, finding $D_n = 0.2444(x_n + y_n) + 0.0444$. If we use these expressions, (7b) becomes

$$(8) \qquad\qquad y_{n+1} = y_n + 0.2214(x_n + y_n) + 0.0214.$$

Table 11 shows the corresponding computations, and from Table 12 we see that the values are much more accurate than those in Ex. 1 and 2.

The **step length** h should not be greater than a certain value H which depends on the accuracy and should otherwise be such that

$$\kappa = hK \qquad\qquad (K \text{ a small upper bound for } |\partial f/\partial y|)$$

lies about between 0.1 and 0.2; this is similar to the case of the improved Euler method discussed before. It is an advantage of the Runge-Kutta method that we may control h by means of A_n, B_n, C_n, because from the definition of f_y we have

$$\kappa = hK \approx h\,|f_y| \approx h\left|\frac{f(x, y^*) - f(x, y^{**})}{y^* - y^{**}}\right|,$$

Table 12. Comparison of the Accuracy of the Three Methods under Consideration in the Case of the Initial Value Problem (4), with $h = 0.2$

x	$y = e^x - x - 1$	Absolute Value of Error		
		Euler Method (Table 9)	Improved Euler Method (Table 10)	Runge-Kutta Method (Table 11)
0.2	0.021 403	0.021	0.0014	0.000 003
0.4	0.091 825	0.052	0.0034	0.000 007
0.6	0.222 119	0.094	0.0063	0.000 011
0.8	0.425 541	0.152	0.0102	0.000 020
1.0	0.718 282	0.229	0.0156	0.000 031

and if we choose $x = x_n + \frac{1}{2}h, y^* = y_n + \frac{1}{2}B_n, y^{**} = y_n + \frac{1}{2}A_n$, then we have $y^* - y^{**} = (B_n - A_n)/2$ and

$$(9) \qquad\qquad \kappa \approx \kappa_n = 2 \left| \frac{C_n - B_n}{B_n - A_n} \right|.$$

We may now make provision to leave h unchanged as long as, say, $0.05 \leqq \kappa_n \leqq 0.2$, to decrease h by 50% if $\kappa_n > 0.2$, and to double h if $\kappa_n < 0.05$ (if doubling is possible without increasing h beyond a suitably chosen number H, which depends on the desired accuracy).

Another control of h results from performing the computation simultaneously with step $2h$, which corresponds to increasing the truncation error per step by a factor $2^5 = 32$, but since the number of steps decreases, the actual increase is by a factor $2^5/2 = 16$. Hence the error of the first computation (with step h) equals about $\frac{1}{15}$ times the difference δ of corresponding values of y. We may now choose a number ϵ (for example, 1 unit of the last digit that is supposed to be significant) and leave h unchanged if $0.2\epsilon \leqq |\delta| \leqq 10\epsilon$, decrease h by 50% if $|\delta| > 10\epsilon$, and double h if $|\delta| < 0.2\epsilon$; of course, in doubling we must care that the step does not become larger than a suitable number H; this is as before.

Problems

1. Repeat the computations in Ex. 1, choosing $h = 0.1$ (instead of $h = 0.2$), and note that the errors of the values thus obtained are about 50% of those in Ex. 1 (which are listed in Table 12).
2. Same task as in Prob. 1, but with $h = 0.01$ (twenty steps). Compare the error of the value corresponding to $x = 0.2$ with that in Ex. 1.
3. Repeat the computations in Ex. 2, choosing $h = 0.1$, and note that the errors of the values thus obtained are about 25% of those in Ex. 2.
4. Same task as in Prob. 3, but with $h = 0.05$ (eight steps). Compare the error of the value corresponding to $x = 0.4$ with those in Ex. 2 and Prob. 3.

Apply the Euler method to the following differential equations. (Carry out ten steps.)

5. $y' = y, y(0) = 1, h = 0.1$
6. $y' = y, y(0) = 1, h = 0.01$
7. $y' = 2xy + 1, y(0) = 0, h = 0.1$
8. $y' = 2xy, y(0) = 1, h = 0.1$

9. Compute $y = e^x$ for $x = 0, 0.1, 0.2, \cdots, 1.0$ by applying the Runge-Kutta method to the initial value problem in Prob. 5. Show that the first five decimal places of the result are correct (cf. Table A1 in Appendix 4).
10. In Prob. 9 insert A_n into B_n, then B_n into C_n, etc., and show that the resulting formula becomes $y_{n+1} = 1.105\ 170\ 833y_n$. Repeat the computations in Prob. 9, using this formula.
11. Apply the Euler method to the initial value problem $y' = 1 + y^2$, $y(0) = 0$, choosing $h = 0.1$. Compute five steps and compare the results with the exact values (cf. Table A1 in Appendix 4).
12. Apply the improved Euler method to the initial value problem in Prob. 11, choosing $h = 0.1$. Compute two steps and compare the errors with those of the corresponding values in Prob. 11.
13. Apply the Runge-Kutta method to the initial value problem in Prob. 11, choosing $h = 0.1$. Compute two steps. Compare the results with the exact values $0.100\ 334\ 672$ and $0.202\ 710\ 036$.
14. Apply the improved Euler method to the initial value problem $y' = 2x, y(0) = 0$, choosing $h = 0.1$. Why are the errors zero?
15. Apply the Euler method with $h = 0.1$ to $y' = -100y$, $y(0) = 1$. Show that it gives $y = (-9)^{10x}$, compare this with the exact solution, and comment.

18.7 *Numerical Methods for Second-Order Differential Equations*

This section includes numerical methods similar to those in the preceding section, and is independent of the other sections in this chapter. Hence it may also be discussed immediately after Chap. 2.

An **initial value problem** for a second-order differential equation consists of that equation and two conditions (*initial conditions*) referring to the same point. In this section we shall consider two numerical methods for solving initial value problems of the form

(1) $y'' = f(x, y, y')$, $y(x_0) = y_0$, $y'(x_0) = y_0'$

assuming f to be such that the problem has a unique solution on some interval containing x_0. The first method is simple (but inaccurate) and serves to illustrate the principle, whereas the second method is of great precision and practical importance.

In both methods we shall obtain approximate values of the solution $y(x)$ of (1) at equidistant points $x_1 = x_0 + h$, $x_2 = x_0 + 2h$, $\cdots$; these values will be denoted by $y_1, y_2, \cdots$, respectively. Similarly, approximate values of the derivative $y'(x)$ at those points will be denoted by $y_1', y_2', \cdots$, respectively.

The methods in the last section were suggested by the Taylor expansion

(2) $y(x + h) = y(x) + hy'(x) + \dfrac{h^2}{2} y''(x) + \dfrac{h^3}{3!} y'''(x) + \cdots,$

which we shall now use for the same purpose, together with the expansion for the derivative

(3) $y'(x + h) = y'(x) + hy''(x) + \dfrac{h^2}{2} y'''(x) + \cdots.$

The roughest numerical method is obtained by neglecting the terms containing y''' and the further terms in (2) and (3); this yields the approximations

$$y(x + h) \approx y(x) + hy'(x) + \frac{h^2}{2} y''(x),$$

$$y'(x + h) \approx y'(x) + hy''(x).$$

In the first step of the method we compute

$$y_0'' = f(x_0, y_0, y_0')$$

from (1), then

$$y_1 = y_0 + hy_0' + \frac{h^2}{2} y_0''$$

which approximates $y(x_1) = y(x_0 + h)$, and furthermore

$$y_1' = y_0' + hy_0''$$

which will be needed in the next step. In the second step we compute

$$y_1'' = f(x_1, y_1, y_1')$$

from (1), then

$$y_2 = y_1 + hy_1' + \frac{h^2}{2} y_1''$$

which approximates $y(x_2) = y(x_0 + 2h)$, and furthermore

$$y_2' = y_1' + hy_1''.$$

In the $(n + 1)$th step we compute

$$y_n'' = f(x_n, y_n, y_n')$$

from (1), then the new value

(4a) $$y_{n+1} = y_n + hy_n' + \frac{h^2}{2} y_n''$$

which is an approximation for $y(x_{n+1})$, and furthermore

(4b) $$y_{n+1}' = y_n' + hy_n''$$

which is an approximation for $y'(x_{n+1})$ needed in the next step.

Note that, geometrically speaking, this method is an approximation of the curve of $y(x)$ by portions of parabolas.

Example 1. Apply (4) to the initial value problem

(5) $$y'' = \tfrac{1}{2}(x + y + y' + 2), \qquad y(0) = 0, \qquad y'(0) = 0,$$

choosing $h = 0.2$. Here (4) becomes

$$y_{n+1} = y_n + 0.2y_n' + 0.02y_n''$$
$$y_{n+1}' = y_n' + 0.2y_n''$$

where $y_n'' = \tfrac{1}{2}(x_n + y_n + y_n' + 2).$

The computations are shown in Table 13. The student may verify that the exact solution is

$$y = e^x - x - 1.$$

Table 15 shows that the errors of our approximate values are large. This is typical, because in most practical cases our present method will be too inaccurate.

Table 13. Computations in Example 1

n	x_n	y_n	y_n'	$0.2y_n'$	$x_n + y_n \\ + y_n' + 2$	$0.2y_n''$	$0.02y_n''$	$0.2y_n' \\ + 0.02y_n''$
0	0	0.0000	0.0000	0.0000	2.0000	0.0200	0.0200	0.0200
1	0.2	0.0200	0.2000	0.0400	2.4200	0.2420	0.0242	0.0642
2	0.4	0.0842	0.4420	0.0884	2.9262	0.2926	0.0293	0.1177
3	0.6	0.2019	0.7346	0.1469	3.5365	0.3537	0.0354	0.1823
4	0.8	0.3842	1.0883	0.2177	4.2725	0.4273	0.0427	0.2604
5	1.0	0.6446						

A much more accurate method is the **Runge-Kutta-Nyström method**[5] which is a generalization of the Runge-Kutta method in Sec. 18.6. We mention without proof that this is a *fourth-order method*, which means that in the Taylor formulas for y and y' the first terms up to and including the term that contains h^4 are given exactly.

In the general step [the $(n + 1)$th step] of the method, we first compute the auxiliary quantities

(6a)
$$A_n = \tfrac{1}{2}h f(x_n, y_n, y_n')$$
$$B_n = \tfrac{1}{2}h f(x_n + \tfrac{1}{2}h, y_n + \beta_n, y_n' + A_n) \qquad \text{where} \qquad \beta_n = \tfrac{1}{2}h(y_n' + \tfrac{1}{2}A_n)$$
$$C_n = \tfrac{1}{2}h f(x_n + \tfrac{1}{2}h, y_n + \beta_n, y_n' + B_n)$$
$$D_n = \tfrac{1}{2}h f(x_n + h, y_n + \delta_n, y_n' + 2C_n) \qquad \text{where} \qquad \delta_n = h(y_n' + C_n)$$

then the new value

(6b) $\qquad y_{n+1} = y_n + h(y_n' + K_n) \qquad$ where $\qquad K_n = \tfrac{1}{3}(A_n + B_n + C_n)$

which is an approximation for $y(x_{n+1})$, and furthermore,

(6c) $\qquad y'_{n+1} = y_n' + K_n{}^* \quad$ where $\quad K_n{}^* = \tfrac{1}{3}(A_n + 2B_n + 2C_n + D_n)$

which is an approximation for $y'(x_{n+1})$ needed in the next step.

h can be controlled in a fashion similar to that described at the end of the last section, where we now take for δ the larger of δ^* and δ^{**}; here δ^* is $\tfrac{1}{15}$ times the difference of corresponding values of y, and δ^{**} is $\tfrac{1}{15}$ times the difference of corresponding values of y'.

Example 2. Apply the Runge-Kutta-Nyström method to the initial value problem (5), choosing $h = 0.2$. Here $f = 0.5(x + y + y' + 2)$, and (6a) becomes

$$A_n = 0.05(x_n + y_n + y_n' + 2),$$
$$B_n = 0.05(x_n + 0.1 + y_n + \beta_n + y_n' + A_n + 2), \qquad \beta_n = 0.1(y_n' + \tfrac{1}{2}A_n),$$
$$C_n = 0.05(x_n + 0.1 + y_n + \beta_n + y_n' + B_n + 2),$$
$$D_n = 0.05(x_n + 0.2 + y_n + \delta_n + y_n' + 2C_n + 2), \qquad \delta_n = 0.2(y_n' + C_n).$$

In the present case the differential equation is simple, and so are the expressions for A_n, B_n, C_n, and D_n. Hence we may insert A_n into B_n, then B_n into C_n, and, finally, C_n into D_n. The result of this elementary calculation is

$$B_n = 0.05[1.0525(x_n + y_n) + 1.152\,5y_n' + 2.205],$$
$$C_n = 0.05[1.055\,125(x_n + y_n) + 1.160\,125y_n' + 2.215\,25],$$
$$D_n = 0.05[1.116\,063\,75(x_n + y_n) + 1.327\,613\,75y_n' + 2.443\,677\,5].$$

From this we may now determine K_n and $K_n{}^*$ and insert the resulting expressions into (6b) and (6c), finding

(7)
$$y_{n+1} = y_n + a(x_n + y_n) + by_n' + c$$
$$y'_{n+1} = y_n' + a^*(x_n + y_n) + b^*y_n' + c^*$$

[5] E. J. NYSTRÖM, Finnish mathematician. His paper on the method appeared in *Acta Soc. Sci. fennicae*, vol. 50 (1925).

Table 14. Runge-Kutta-Nyström Method (with $h = 0.2$) Applied to the Initial Value Problem (5); Five Steps Computed by the Use of (7)

n	x_n	y_n	$y_n{}'$	$a(x_n + y_n)$ $+ by_n{}' + c$	$a^*(x_n + y_n)$ $+ b^*y_n{}' + c^*$
0	0.0	0.000 0000	0.000 0000	0.021 4008	0.221 4030
1	0.2	0.021 4008	0.221 4030	0.070 4196	0.270 4220
2	0.4	0.091 8204	0.491 8250	0.130 2913	0.330 2940
3	0.6	0.222 1117	0.822 1190	0.203 4186	0.403 4219
4	0.8	0.425 5303	1.225 5409	0.292 7365	0.492 7403
5	1.0	0.718 2668	1.718 2812		

where

$$a = 0.010\ 3588 \qquad b = 0.211\ 0421 \qquad c = 0.021\ 4008$$
$$a^* = 0.105\ 5219 \qquad b^* = 0.115\ 8811 \qquad c^* = 0.221\ 4030$$

Table 14 shows the corresponding computations. The errors of the approximate values for $y(x)$ are much smaller than those in Ex. 1 (cf. Table 15).

In addition to the truncation error of a method, there are **rounding errors,** and we want to warn the reader that rounding errors may affect results to a large extent. For instance, the solution of the problem $y'' = y$, $y(0) = 1$, $y'(0) = -1$ is $y = e^{-x}$, but the rounding error will introduce a small multiple of the unwanted solution e^x, which may eventually (after sufficiently many steps) swamp the required solution. This is known as **building-up error.** In our simple example we may avoid it by starting from known values of e^{-x} and its derivative for a large x and compute in the reverse direction, but in more complicated cases considerable experience is needed for avoiding that phenomenon.

Table 15. Comparison of Accuracy of the Two Methods Under Consideration in the Case of the Initial Value Problem (5), with $h = 0.2$

x	$y = e^x - x - 1$	Absolute Value of Error	
		Ex. 1	Table 14
0.2	0.021 4028	0.0014	0.000 0020
0.4	0.091 8247	0.0076	0.000 0043
0.6	0.222 1188	0.0202	0.000 0071
0.8	0.425 5409	0.0413	0.000 0106
1.0	0.718 2818	0.0737	0.000 0150

Problems

1. Repeat the computation in Ex. 1, choosing $h = 0.1$, and compare the errors of the values thus obtained with those in Ex. 1 (listed in Table 15).

Apply (4) to the following initial value problems. (Carry out five steps.)

2. $y'' = -y$, $y(0) = 0$, $y'(0) = 1$, $h = 0.1$
3. $y'' = -y$, $y(0) = 1$, $y'(0) = 0$, $h = 0.1$

4. $y'' = -y$, $y(0) = 1$, $y'(0) = 0$, $h = 0.05$
5. $y'' = y$, $y(0) = 1$, $y'(0) = 1$, $h = 0.1$
6. $y'' = y$, $y(0) = 1$, $y'(0) = -1$, $h = 0.1$

7. Consider the initial value problem $(1 - x^2)y'' - 2xy' + 2y = 0$, $y(0) = 0$, $y'(0) = 1$. Show that (4) with $h = 0.1$ takes the form

$$y_{n+1} = y_n + 0.1y_n' + 0.01\frac{x_n y_n' - y_n}{1 - x_n^2}, \qquad y_{n+1}' = y_n' + 0.2\frac{x_n y_n' - y_n}{1 - x_n^2}.$$

Carry out five steps. Verify by substitution that the exact solution is $y = x$.

8. Apply (4) with $h = 0.1$ to $(1 - x^2)y'' - 2xy' + 6y = 0$, $y(0) = -0.5$, $y'(0) = 0$. Carry out five steps. Verify that the exact solution is $y = 0.5(3x^2 - 1)$.

9. Apply (4) with $h = 0.1$ to $y'' = xy' - 3y$, $y(0) = 0$, $y'(0) = -3$. Carry out five steps. Verify that the exact solution is $y = x^3 - 3x$.

10. Apply (4) with $h = 0.1$ to $y'' = xy' - 4y$, $y(0) = 3$, $y'(0) = 0$. Carry out five steps. Verify that the exact solution is $y = x^4 - 6x^2 + 3$.

11. Apply the Runge-Kutta-Nyström method to the initial value problem in Prob. 2. Choose $h = 0.2$, carry out five steps, and compare the results with the exact values (cf. Table A1 in Appendix 4).

12. In Prob. 11, replace $h = 0.2$ by $h = 0.1$, carry out four steps, and compare the results with the corresponding values in Prob. 11 and with the values (exact to 9D)

$$0.099\ 833\ 417, \qquad 0.198\ 669\ 331, \qquad 0.295\ 520\ 207, \qquad 0.389\ 418\ 342.$$

18.8 *Systems of Linear Equations. Gauss Elimination*

A **system of *m* linear equations** (or *set of m simultaneous linear equations*) *in n unknowns* $x_1, \cdots, x_n$ is a set of equations of the form

(1)
$$\begin{aligned}
a_{11}x_1 + \cdots + a_{1n}x_n &= b_1 \\
a_{21}x_1 + \cdots + a_{2n}x_n &= b_2 \\
&\cdots\cdots\cdots\cdots\cdots \\
a_{m1}x_1 + \cdots + a_{mn}x_n &= b_m
\end{aligned}$$

where the **coefficients** a_{jk} and the b_j are given numbers. The system is said to be **homogeneous** if all the b_j are zero; otherwise it is said to be **nonhomogeneous.** Readers familiar with matrix multiplication (Sec. 6.4) will see that (1) may be written as a single vector equation

(2) $$\mathbf{Ax} = \mathbf{b}$$

where the **coefficient matrix** $\mathbf{A} = (a_{ik})$ is the $(m \times n)$ matrix

$$\mathbf{A} = \begin{pmatrix} a_{11} & a_{12} & \cdots & a_{1n} \\ a_{21} & a_{22} & \cdots & a_{2n} \\ \cdot & \cdot & \cdots & \cdot \\ a_{m1} & a_{m2} & \cdots & a_{mn} \end{pmatrix}, \quad \text{and} \quad \mathbf{x} = \begin{pmatrix} x_1 \\ \cdot \\ \cdot \\ \cdot \\ x_n \end{pmatrix} \quad \text{and} \quad \mathbf{b} = \begin{pmatrix} b_1 \\ \cdot \\ \cdot \\ \cdot \\ b_m \end{pmatrix}$$

are column vectors. A **solution** of (1) is a set of numbers $x_1, \cdots, x_n$ which satisfy all the m equations, and a **solution vector** of (1) is a vector $\mathbf{x}$ whose components constitute a solution of (1).

The method of solving such a system by determinants (Cramer's rule in Sec. 6.11) is impracticable for large systems, even with efficient methods for evaluating the determinants.

A practical method for the solution of a system of linear equations is the **Gauss elimination.** It suffices to explain it in terms of an example.

Example 1. Solve the system

$$
\begin{array}{ll}
\begin{aligned}
2w + \ x + 2y + \ \ z &= \ \ 6 \\
6w - 6x + 6y + 12z &= 36 \\
4w + 3x + 3y - \ \ 3z &= -1 \\
2w + 2x - \ \ y + \ \ z &= 10
\end{aligned}
&
\begin{array}{c}
\textit{Check sum} \\
12 \\
54 \\
6 \\
14
\end{array}
\end{array}
$$

First step. We subtract multiples of the first equation from the other three so as to eliminate w in each of these, giving

$$
\begin{array}{ll}
\begin{aligned}
-9x \ \ \ \ \ \ + \ 9z &= \ \ 18 \\
x - \ y - \ 5z &= -13 \\
x - 3y \ \ \ \ \ \ &= \ \ \ 4
\end{aligned}
&
\begin{array}{c}
\textit{Check sum} \\
18 \\
-18 \\
2
\end{array}
\end{array}
$$

Second step. We subtract multiples of the first of these equations from the other two so as to eliminate x; we obtain

$$
\begin{array}{ll}
\begin{aligned}
-y - \ 4z &= -11 \\
-3y + \ \ z &= \ \ \ 6
\end{aligned}
&
\begin{array}{c}
\textit{Check sum} \\
-16 \\
4
\end{array}
\end{array}
$$

Third step. We subtract a multiple of the first of these equations from the other so as to eliminate y:

$$
\begin{array}{ll}
13z = 39
&
\begin{array}{c}
\textit{Check sum} \\
52
\end{array}
\end{array}
$$

Final step. By back-substituting we now obtain

$$
\begin{array}{ll}
\begin{aligned}
13z &= \ \ \ 39 \\
-y - 4 \cdot 3 &= -11 \\
-9x \ \ \ \ \ \ \ \ \ \ + 9 \cdot 3 &= \ \ \ 18 \\
2w + 1 + 2 \cdot (-1) + \ \ \ 3 &= \ \ \ \ 6
\end{aligned}
&
\begin{aligned}
z &= \ \ \ 3 \\
y &= -1 \\
x &= \ \ \ 1 \\
w &= \ \ \ 2
\end{aligned}
\end{array}
$$

An example in which the number of equations differs from the number of unknowns is included in Sec. 6.5.

A useful check is provided by carrying along, as indicated in the extra column on the right, the sum of the coefficients and terms on the right, operating on these numbers as on the equations, and then checking that the derived equations have the correct sum.

In practice, the full equations at each step are not recorded, but only the matrix of coefficients and the column of terms on the right.

In each step the coefficient of the first unknown in the first equation is called the **pivotal coefficient.** If a pivotal coefficient is small, we have to subtract large multiples of the corresponding equation from the others, and any uncertainty in coefficients will be amplified. This may affect the accuracy of the result. Consequently, in practice it is important that we choose pivotal coefficients which are not too small in absolute value. We can accomplish this by reordering the equations and the variables if necessary. This process is called **pivoting** or **positioning for size.**

Very small pivots can often be avoided by **scaling,** that is, by multiplying equations by suitable factors or changing some of the variables by setting $x_j{}^* = c_j x_j$ where c_j is a suitable number.

The **inverse** of a nonsingular square matrix A may now be determined by solving the n systems

$$(3) \qquad\qquad\qquad Ax = b_j \qquad\qquad (j = 1, \cdots, n),$$

where b_j is the jth column of the n-rowed unit matrix. (An iterative method for determining the inverse is included in the next section.)

There are various modifications of the Gauss method. We mention a method based on a result by **Cholesky,** who has shown that if a square matrix A and all its leading square submatrices are nonsingular, then

$$(4) \qquad\qquad\qquad A = LU$$

where L and U are lower and upper triangular matrices. L and U are essentially unique, and if we specify the diagonal elements of L (or of U), they are unique. The point is that L and U can be obtained without solving simultaneous equations (example below). To solve a system $Ax = b$ of n equations in n unknowns, we may now insert (4), having

$$LUx = b,$$

and premultiply by L^{-1}. This gives

$$(5) \qquad\qquad Ux = z \qquad \text{where} \qquad z = L^{-1}b$$

and we see that this is the triangular form of our system. We first determine z from [cf. (5)]

$$(6) \qquad\qquad\qquad Lz = b$$

and then we solve

$$(7) \qquad\qquad\qquad Ux = z$$

for x.

Example 2. The reader may verify that the system

$$x + 2y + 3z = 14$$
$$2x + 3y + 4z = 20$$
$$3x + 4y + z = 14$$

has the solution $x = 1, y = 2, z = 3$, and we shall see how to get this result by Cholesky's method. The coefficient matrix is symmetric. This implies $L = U^T$ and we may determine the elements of U from

$$\begin{pmatrix} 1 & 2 & 3 \\ 2 & 3 & 4 \\ 3 & 4 & 1 \end{pmatrix} = \begin{pmatrix} a_{11} & 0 & 0 \\ a_{12} & a_{22} & 0 \\ a_{13} & a_{23} & a_{33} \end{pmatrix} \begin{pmatrix} a_{11} & a_{12} & a_{13} \\ 0 & a_{22} & a_{23} \\ 0 & 0 & a_{33} \end{pmatrix}$$

by using the definition of matrix multiplication and equating corresponding elements. We obtain successively $a_{11}^2 = 1$, say, $a_{11} = 1$, from this $a_{11}a_{12} = a_{12} = 2$, $a_{11}a_{13} = a_{13} = 3$, $a_{12}^2 + a_{22}^2 = 4 + a_{22}^2 = 3$, say, $a_{22} = i \, (= \sqrt{-1})$, from this

$$a_{12}a_{13} + a_{22}a_{23} = 6 + ia_{23} = 4, \qquad a_{23} = 2i$$

and finally

$$a_{13}^2 + a_{23}^2 + a_{33}^2 = 9 - 4 + a_{33}^2 = 1, \qquad \text{say, } a_{33} = 2i.$$

Hence (6) has the form

$$\begin{pmatrix} 1 & 0 & 0 \\ 2 & i & 0 \\ 3 & 2i & 2i \end{pmatrix} \begin{pmatrix} z_1 \\ z_2 \\ z_3 \end{pmatrix} = \begin{pmatrix} 14 \\ 20 \\ 14 \end{pmatrix}, \qquad \text{and} \qquad \begin{pmatrix} z_1 \\ z_2 \\ z_3 \end{pmatrix} = \begin{pmatrix} 14 \\ 8i \\ 6i \end{pmatrix}.$$

We now solve

$$\begin{pmatrix} 1 & 2 & 3 \\ 0 & i & 2i \\ 0 & 0 & 2i \end{pmatrix} \begin{pmatrix} x_1 \\ x_2 \\ x_3 \end{pmatrix} = \begin{pmatrix} 14 \\ 8i \\ 6i \end{pmatrix}, \qquad \text{finding} \qquad \begin{pmatrix} x_1 \\ x_2 \\ x_3 \end{pmatrix} = \begin{pmatrix} 1 \\ 2 \\ 3 \end{pmatrix}.$$

Problems

Using the Gauss elimination, solve the following systems of linear equations.

1. $3x + y = 7$
   $x - 4y = -2$

2. $4x - 2y = 6$
   $6x + y = -3$

3. $2x - 7y = 5$
   $8x + y = -9$

4. $x - 3y + z = 3$
   $2x + y = 2$
   $3x + 7y - 4z = -5$

5. $4x - y + 2z = 15$
   $-x + 2y + 3z = 5$
   $5x - 7y + 9z = 8$

6. $y + 3z = 9$
   $2x + 2y - z = 8$
   $-x + 5z = 8$

7. $3x + 4y - 7z = -7$
   $x - 2y + z = 1$

8. $x - 2y + z = 0$
   $-x + 3y - 2z = 1$

9. $4x - 3y + 3z = 0$
   $8x + 7y - 7z = 0$

10. $2w - 4x + 3y - z = 3$
    $w - 2x + 5y - 3z = 0$
    $3w - 6x - y - z = 0$

11. $3w - x + 8y - 2z = -2$
    $-w + 2x - 13y + 3z = 3$
    $4w + 3x - 9y + z = 1$

12. Show that in the case of n equations in n unknowns the number of multiplications and divisions in the Gauss elimination is of the order of $n^3/3$. (The main labor in solving large systems of equations lies in the multiplications and divisions, and our result shows that the labor increases rapidly with the size of the system.)

18.9 *Systems of Linear Equations.* *Solution by Iteration*

The Gauss elimination in the previous section belongs to the **direct methods** for solving systems of linear equations; these are methods that yield solutions after an amount of computation that can be specified in advance. In contrast, an **indirect** or **iterative method** is one in which we start from an approximation to the true solution and, if successful, obtain better and better approximations from a computational cycle repeated as often as may be necessary for achieving a required accuracy, so that the amount of arithmetic depends upon the accuracy required.

Iterative methods are used mainly in those problems for which convergence is known to be rapid, so that the solution is obtained with much less work than that of a direct method, and for systems of large order but with many zero coefficients, the so-called "*sparse*" *systems*, for which elimination methods would be relatively laborious and need much storage. Such systems occur in connection with vibrational problems, partial differential equations, and other applications and have given rise to a rapidly growing literature.

An iteration process of practical importance is the **Gauss-Seidel iteration,** which we shall explain in terms of an example. Let us consider the system

$$
\begin{aligned}
w - 0.25x - 0.25y \qquad\qquad &= 0.50 \\
-0.25w + \quad x \qquad\quad - 0.25z &= 0.50 \\
-0.25w \qquad\quad + \quad y - 0.25z &= 0.25 \\
- 0.25x - 0.25y + \quad z &= 0.25
\end{aligned}
$$

(1)

(Equations of this form arise in the numerical solution of partial differential equations.) We write the system in the form

$$
\begin{aligned}
w &= \qquad\quad 0.25x + 0.25y \qquad\qquad + 0.50 \\
x &= 0.25w \qquad\qquad\qquad\quad + 0.25z + 0.50 \\
y &= 0.25w \qquad\qquad\qquad\quad + 0.25z + 0.25 \\
z &= \qquad\quad 0.25x + 0.25y \qquad\qquad + 0.25
\end{aligned}
$$

(2)

and use these equations for the iteration; that is, we start from an approximation to the solution, say, $w_0 = 1$, $x_0 = 1$, $y_0 = 1$, $z_0 = 1$, and compute a presumably better approximation

$$
\begin{aligned}
w_1 &= \qquad\quad 0.25x_0 + 0.25y_0 \qquad\qquad + 0.50 = 1.0000 \\
x_1 &= 0.25w_1 \qquad\qquad\qquad\quad + 0.25z_0 + 0.50 = 1.0000 \\
y_1 &= 0.25w_1 \qquad\qquad\qquad\quad + 0.25z_0 + 0.25 = 0.7500 \\
z_1 &= \qquad\quad 0.25x_1 + 0.25y_1 \qquad\qquad + 0.25 = 0.6875
\end{aligned}
$$

(3)

We see that these equations are obtained from (2) by substituting on the right the most recent approximations. In fact, corresponding elements replace previous ones as soon as they have been computed, so that in the second and third equations we use w_1 (not w_0), and in the last equation of (3) we use x_1 and y_1 (not x_0 and y_0). The next step yields

$$
\begin{aligned}
w_2 &= \qquad\quad 0.25x_1 + 0.25y_1 \qquad\qquad + 0.50 = 0.9375 \\
x_2 &= 0.25w_2 \qquad\qquad\qquad\quad + 0.25z_1 + 0.50 = 0.9063 \\
y_2 &= 0.25w_2 \qquad\qquad\qquad\quad + 0.25z_1 + 0.25 = 0.6563 \\
z_2 &= \qquad\quad 0.25x_2 + 0.25y_2 \qquad\qquad + 0.25 = 0.6407
\end{aligned}
$$

The reader may show that the exact solution is $w = x = 0.875$, $y = z = 0.625$.

We mention without proof that the Gauss-Seidel iteration converges for any choice of the initial approximation if and only if all the eigenvalues (Sec. 6.14) of the "iteration matrix" **C** [below in (5)] have absolute value less than 1, and the rate of convergence depends on the **spectral radius** ($=$ maximum of those absolute values). The matrix **C** is obtained as follows. Let

$$\mathbf{Ax} = \mathbf{b}$$

be the given system of n equations, where **x** is a column vector with components $x_1, \cdots, x_n$, the n unknowns. Let $\mathbf{x}_{(0)}, \mathbf{x}_{(1)}, \cdots$ be the iterative sequence of the successive Gauss-Seidel approximations corresponding to an initial approximation $\mathbf{x}_{(0)}$. The method is said to *converge* for an $\mathbf{x}_{(0)}$ if the corresponding iterative sequence converges to a solution of the given system. Without loss of generality we may assume $a_{jj} = 1$ for $j = 1, \cdots, n$, because this form of **A** can be achieved by rearranging the equations so that no diagonal coefficient vanishes, and then dividing each equation by the corresponding diagonal coefficient. Then we may write $\mathbf{A} = \mathbf{I} + \tilde{\mathbf{L}} + \tilde{\mathbf{U}}$, where **I** is the n-rowed unit matrix and $\tilde{\mathbf{L}}$ and $\tilde{\mathbf{U}}$ are respectively lower and upper triangular matrices with null principal diagonals. Substituting this form of **A** into $\mathbf{Ax} = \mathbf{b}$, we have $(\mathbf{I} + \tilde{\mathbf{L}} + \tilde{\mathbf{U}})\mathbf{x} = \mathbf{b}$. It is customary to set $\tilde{\mathbf{L}} = -\mathbf{L}$ and $\tilde{\mathbf{U}} = -\mathbf{U}$; then

$$(\mathbf{I} - \mathbf{L} - \mathbf{U})\mathbf{x} = \mathbf{b}, \qquad \text{thus} \qquad (\mathbf{I} - \mathbf{L})\mathbf{x} = \mathbf{b} + \mathbf{Ux}.$$

From this we obtain the Gauss-Seidel formula

$$(4) \qquad\qquad (\mathbf{I} - \mathbf{L})\mathbf{x}_{(m+1)} = \mathbf{b} + \mathbf{Ux}_{(m)} \qquad\qquad (m = 0, 1, \cdots).$$

In fact, **U** is upper triangular and its nonzero elements correspond to those positions at which we have still to use "old" approximations because corresponding "new" ones are not yet available. In contrast, **L** is lower triangular and its nonzero elements correspond to those positions at which elements of the "new" approximation $\mathbf{x}_{(m+1)}$ are already available. Solving (4) for $\mathbf{x}_{(m+1)}$, we have

$$(5) \qquad \mathbf{x}_{(m+1)} = (\mathbf{I} - \mathbf{L})^{-1}\mathbf{b} + \mathbf{Cx}_{(m)} \qquad \text{where} \qquad \mathbf{C} = (\mathbf{I} - \mathbf{L})^{-1}\mathbf{U},$$

the aforementioned matrix **C** whose eigenvalues matter in connection with convergence.

The Gauss-Seidel iteration is a method of **successive corrections** because we replace approximations by corresponding new ones as soon as the latter have been computed. A method is called a method of **simultaneous corrections** if no element of an approximation $\mathbf{x}_{(m+1)}$ is used until all the elements of $\mathbf{x}_{(m+1)}$ have been calculated. A method of this type is the **Jacobi iteration**, which is similar to the Gauss-Seidel iteration, but consists in *not* using improved values until a step has been completed and then replacing $\mathbf{x}_{(m)}$ by $\mathbf{x}_{(m+1)}$ entirely for the next cycle. Hence, if we write $\mathbf{Ax} = \mathbf{b}$ in the form $\mathbf{x} = \mathbf{b} + (\mathbf{I} - \mathbf{A})\mathbf{x}$, the Jacobi iteration in matrix notation is

$$(6) \qquad\qquad \mathbf{x}_{(m+1)} = \mathbf{b} + (\mathbf{I} - \mathbf{A})\mathbf{x}_{(m)}.$$

This method is largely of theoretical interest. It converges for every choice of $\mathbf{x}_{(0)}$ if and only if the spectral radius of $\mathbf{I} - \mathbf{A}$ is less than 1; here we again assume $a_{jj} = 1$ for $j = 1, \cdots, n$.

The **inverse** of a nonsingular square matrix **A** may also be determined by an iteration method suggested by the following idea. The reciprocal x of a given number a satisfies $xa = 1$, and if we want to determine x without division, we may apply Newton's method to the function $f(x) = x^{-1} - a$. Since $f'(x) = -1/x^2$, the Newton iteration is

$$x_{m+1} = x_m - (x_m^{-1} - a)(-x_m^2) = x_m(2 - ax_m).$$

This suggests an analogous iteration formula for determining the inverse $\mathbf{X} = \mathbf{A}^{-1}$ of **A**, namely,

$$(7) \qquad\qquad \mathbf{X}_{(m+1)} = \mathbf{X}_{(m)}(2\mathbf{I} - \mathbf{A}\mathbf{X}_{(m)}).$$

The process converges (produces $\mathbf{A}^{-1}$ as $m \to \infty$) if and only if an $\mathbf{X}_{(0)}$ is chosen so that each eigenvalue of $\mathbf{I} - \mathbf{A}\mathbf{X}_0$ is of absolute value less than 1 (cf. Ref. [G14] listed in Appendix 1). The method is suitable if the occurring multiplications are simple (e.g., many zeros in **A**). In practice a suitable choice of $\mathbf{A}_{(0)}$ is difficult, if not impossible, and the method is mostly used for improving an inaccurate inverse obtained by another method.

Problems

1. Verify the solution of (1) given in the text. Solve (1) by the Gauss elimination.
2. Compute two more steps of the Gauss-Seidel iteration in the text.
3. Find **C** [cf. (5)] for the system (1).
4. Apply the Jacobi iteration to the system (1); start from $w_0 = 1$, $x_0 = 1$, $y_0 = 1$, $z_0 = 1$, perform two steps, and compare the accuracy of the result with that in the text.

Apply the Gauss-Seidel iteration to the following systems. Perform three steps, starting from 1, 1, 1.

5.	$10x - y - z = 13$	**6.** $4x + y = -8$	**7.** $4x + 2y + z = 14$
	$x + 10y + z = 36$	$4y + z = 2$	$x + 5y - z = 10$
	$-x - y + 10z = 35$	$x + 2z = 2$	$x + y + 8z = 20$

8. Starting from 0, 0, 0, show that for the following system the Gauss-Seidel iteration converges, whereas the Jacobi iteration diverges.

$$\begin{aligned} 2x + y + z &= 4 \\ x + 2y + z &= 4 \\ x + y + 2z &= 4 \end{aligned}$$

9. It is plausible to think that the Gauss-Seidel iteration is better than the Jacobi iteration. Actually the methods are not comparable. Illustrate this surprising fact by showing that for the following system the Jacobi iteration converges whereas the Gauss-Seidel iteration diverges. (*Hint.* Use eigenvalues.)

$$\begin{aligned} x + z &= 2 \\ -x + y &= 0 \\ x + 2y - 3z &= 0 \end{aligned}$$

10. Apply (7) with

$$\mathbf{X}_{(0)} = \begin{pmatrix} 2.9 & -0.9 \\ -4.9 & 1.9 \end{pmatrix} \qquad \text{to the matrix} \qquad \mathbf{A} = \begin{pmatrix} 2 & 1 \\ 5 & 3 \end{pmatrix},$$

verify that the condition for convergence is satisfied, perform two steps, and compare the results with the exact inverse.

18.10 *Systems of Linear Equations. Ill-conditioning*

A system of linear equations is said to be **well-conditioned** if small errors in the coefficients or in the solving process have only a small effect on the solution. In this case the solution is relatively strongly indicated by the equations.

A system of linear equations is said to be **ill-conditioned** if small errors in the coefficients or in the solving process have a large effect on the solution. In this case the solution is relatively weakly indicated by the equations.

For instance, two linear equations in two unknowns represent two straight lines. Such a system is ill-conditioned if and only if the angle γ between the lines is small, that is, if and only if the lines are almost parallel. In fact, then a small change in a coefficient may cause a large displacement of the point of intersection of the lines (cf. Fig. 341). For larger systems of linear equations the situation is similar in principle, but no such simple geometrical interpretation is possible, and we would not be able to follow each detail of the situation.

Example 1. The reader may verify that the system

$$0.9999x - 1.0001y = 1$$
$$x - \quad y = 1$$

has the solution $x = 0.5$, $y = -0.5$, whereas the system

$$0.9999x - 1.0001y = 1$$
$$x - \quad y = 1 + \epsilon$$

has the solution $x = 0.5 + 5000.5\epsilon$, $y = -0.5 + 4999.5\epsilon$. This shows that the system is ill-conditioned because a change on the right of magnitude ϵ produces a change in the solution of magnitude 5000ϵ, approximately.

Ill-conditioning may be regarded as an approach toward singularity. It manifests itself by a loss of significant figures during computation, so that it is more difficult to get an accurate inverse or solution of the associated equations. Several measures for ill-conditioning have been proposed, but the common usage of the term is still qualitative.

In the case of ill-conditioning we must use a much larger number of decimal places in intermediate calculations necessary to determine the solution to a specific

(a)   (b)

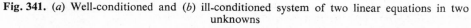

Fig. 341. (*a*) Well-conditioned and (*b*) ill-conditioned system of two linear equations in two unknowns

number of decimal places (when round-off error is present). If the coefficients of an ill-conditioned system as well as the numbers on the right-hand side can be computed from a formula, we can calculate them, at some small effort, to an arbitrary degree of accuracy, using single precision, double precision, triple precision, etc., so that the situation is not too serious. A much greater handicap arises in cases in which those coefficients and numbers are obtained experimentally with a necessarily limited accuracy. Then we have to accept the fact that inaccuracies in the given data permit the solution to be determined only within relatively wide error limits, and we should perhaps try to reformulate the original problem in terms of another system of equations which is reasonably well-conditioned.

Some symptoms for ill-conditioning are as follows. $|\det \mathbf{A}|$ is small compared with the size of the maximum of the $|a_{jk}|$'s and the terms on the right-hand side of the system. Poor approximations of the solution may produce merely very small residuals (see below). The elements of $\mathbf{A}^{-1}$ are large in absolute value compared with the absolute values of the elements of the solution.

If the principal diagonal elements are large in absolute value compared with the absolute values of the other elements, the system is well-conditioned. If a square matrix with largest elements between 0.1 and 10 has an inverse with largest elements also of order unity, then the associated system of linear equations is well-conditioned.

In the case of ill-conditioning we may want to obtain an improved solution of

$$(1) \qquad\qquad\qquad \mathbf{Ax} = \mathbf{b}$$

from an approximate solution $\mathbf{x}_{(1)}$. To $\mathbf{x}_{(1)}$ there corresponds the **residual**

$$\mathbf{r}_{(1)} = \mathbf{b} - \mathbf{Ax}_{(1)}.$$

We have

$$\mathbf{Ax}_{(1)} = \mathbf{b} - \mathbf{r}_{(1)}$$

and thus

$$(2) \qquad\qquad\qquad \mathbf{A}(\mathbf{x} - \mathbf{x}_{(1)}) = \mathbf{r}_{(1)}.$$

This shows that the correction to be applied to $\mathbf{x}_{(1)}$ to give the solution of (1) is the solution of (2). Unless the system is very ill-conditioned, the components of $\mathbf{r}_{(1)}$ will be smaller than those of $\mathbf{b}$. For details of corresponding automatic computation, see Ref. [G11] in Appendix 1.

Problems

1. In Ex. 1, compute the determinant of the system, divided by the coefficient which has the largest absolute value. Comment. Can the determinant of an ill-conditioned system have a large value?
2. Setting $\xi = x + y + 1$, $\eta = x - y - 1$, obtain another ill-conditioned system from that in Ex. 1.
3. Show that the angle γ between two lines represented by a system

$$a_{11}x_1 + a_{12}x_2 = b_1, \qquad a_{21}x_1 + a_{22}x_2 = b_2$$

is given by

$$\tan \gamma = (a_{11}a_{22} - a_{12}a_{21})/(a_{11}a_{21} + a_{12}a_{22})$$

and explain consequences of this formula in connection with ill-conditioning.

4. Show that the solution of the system

$$6x_1 + 7x_2 + 8x_3 = 21$$
$$7x_1 + 8x_2 + 9x_3 = 24$$
$$8x_1 + 9x_2 + 9x_3 = 26$$

is $x_1 = 1$, $x_2 = 1$, $x_3 = 1$. Compute the determinant of the system and the residual corresponding to $x_1 = -0.8$, $x_2 = 2.9$, $x_3 = 0.7$. Comment.

5. Let

$$A = \begin{pmatrix} 1.00 & 1.00 \\ 1.00 & 1.01 \end{pmatrix} \quad \text{and} \quad B = \begin{pmatrix} 111 & -100 \\ -110 & 100 \end{pmatrix}.$$

Show that **AB** is almost equal to the unit matrix, whereas **BA** is not. Comment.

6. Using the Gauss elimination, show that the system

$$x + \tfrac{1}{2}y + \tfrac{1}{3}z = 1$$
$$\tfrac{1}{2}x + \tfrac{1}{3}y + \tfrac{1}{4}z = 0$$
$$\tfrac{1}{3}x + \tfrac{1}{4}y + \tfrac{1}{5}z = 0$$

has the solution $x = 9$, $y = -36$, $z = 30$. Rework the problem, assuming that a computer capable of carrying only two significant digits at a time is to do the computations. Compare the results and comment. (The coefficient matrix is called the three-rowed *Hilbert matrix*.)

18.11 *Method of Least Squares*

In **curve fitting** we are given n points (pairs of numbers)

$$(x_1, y_1), \cdots, (x_n, y_n)$$

and we want to determine a function $f(x)$ such that $f(x_j) \approx y_j$, $j = 1, \cdots, n$. The type of function (for example, polynomials, exponential functions, sine and cosine functions) may be suggested by the nature of the problem (the underlying physical law, for instance), and in many cases a polynomial of a certain degree will be appropriate.

If we require strict equality $f(x_1) = y_1, \cdots, f(x_n) = y_n$ and use polynomials of sufficiently high degree, we may apply one of the methods discussed in Sec. 18.4 in connection with interpolation. However, there are situations where this would not be the appropriate solution of the actual problem. For instance, to the four points

(1) $(-1.0, 1.000)$, $(-0.1, 1.099)$, $(0.2, 0.808)$, $(1.0, 1.000)$

there corresponds the Lagrange polynomial $f(x) = x^3 - x + 1$ (Fig. 342), but if we graph the points, we see that they lie nearly on a straight line. Hence if these

Fig. 342. The approximate fitting of a straight line

values are obtained in an experiment and thus involve an experimental error, and if the nature of the experiment suggests a linear relation, we better fit a straight line through the points (Fig. 342). Such a line may be useful for predicting values to be expected for other values of x. In simple cases a straight line may be fitted by eye, but if the points are scattered, this becomes unreliable and we better use a mathematical principle. A widely used procedure of this type is the **method of least squares** by Gauss. In the present situation it may be formulated as follows.

The straight line

$$y = a + bx$$

should be fitted through the given points $(x_1, y_1), \cdots, (x_n, y_n)$ *so that the sum of the squares of the distances of those points from the straight line is minimum, where the distance is measured in the vertical direction (the y-direction).*

The point on the line with abscissa x_j has the ordinate $a + bx_j$. Hence its distance from (x_j, y_j) is $|y_j - a - bx_j|$ (cf. Fig. 343) and that sum of squares is

$$q = \sum_{j=1}^{n} (y_j - a - bx_j)^2.$$

q depends on a and b. A necessary condition for q to be minimum is

(2)
$$\frac{\partial q}{\partial a} = -2 \sum (y_j - a - bx_j) = 0$$

$$\frac{\partial q}{\partial b} = -2 \sum x_j(y_j - a - bx_j) = 0$$

Fig. 343. Vertical distance of a point (x_j, y_j) from a straight line $y = a + bx$

(where we sum over j from 1 to n). Thus

(3)
$$an + b \sum x_j = \sum y_j$$
$$a \sum x_j + b \sum x_j^2 = \sum x_j y_j.$$

These equations are called the **normal equations** of our problem.

Example 1. Using the method of least squares, fit a straight line to the four points in (1). We calculate

$$n = 4, \quad \sum x_j = 0.1, \quad \sum x_j^2 = 2.05, \quad \sum y_j = 3.907, \quad \sum x_j y_j = 0.0517$$

Hence the normal equations are
$$4a + 0.10b = 3.9070$$
$$0.1a + 2.05b = 0.0517$$

The solution is $a = 0.9773$, $b = -0.0224$, and we obtain the line (Fig. 342)

$$y = 0.9773 - 0.0224x.$$

Our approach of curve fitting can be generalized from a polynomial $y = a + bx$ to a polynomial degree m

$$p(x) = b_0 + b_1 x + \cdots + b_m x^m.$$

Then q takes the form

$$q = \sum_{j=1}^{n} (y_j - p(x_j))^2$$

and depends on $m + 1$ parameters $b_0, \cdots, b_m$. Instead of (2) we then have $m + 1$ conditions

$$\frac{\partial q}{\partial b_0} = 0, \cdots, \frac{\partial q}{\partial b_m} = 0,$$

which give a system of $m + 1$ normal equations. The reader may show that in the case of a quadratic polynomial

(4)
$$p(x) = b_0 + b_1 x + b_2 x^2$$

the normal equations are (summation from 1 to n)

(5)
$$b_0 n \quad + b_1 \sum x_j \; + b_2 \sum x_j^2 = \sum y_j$$
$$b_0 \sum x_j \; + b_1 \sum x_j^2 + b_2 \sum x_j^3 = \sum x_j y_j$$
$$b_0 \sum x_j^2 + b_1 \sum x_j^3 + b_2 \sum x_j^4 = \sum x_j^2 y_j$$

Problems

In each case plot the given points in the xy-plane and fit a straight line (a) by eye, (b) by the method of least squares.

1. $(0, 2.3), (2, 4.1), (4, 5.7), (6, 6.9)$
2. $(2, 5), (3, 9), (4, 15), (5, 21)$
3. $(4, 3), (15, 16), (30, 13), (100, 70), (200, 90)$
4. $(1, 1), (2, 2 + k), (3, 3)$, where k is a constant.

In each case determine the parabola (4) by the method of least squares.

5. $(-1, 2)$, $(0, 0)$, $(0, 1)$, $(1, 2)$
6. $(0, 3)$, $(1, 1)$, $(2, 0)$, $(4, 1)$, $(6, 4)$
7. $(1.4, 7400)$, $(1.8, 7500)$, $(2.3, 7600)$, $(3.0, 7500)$, $(4.0, 7200)$

8. Determine the normal equations in the case of a polynomial of the third degree.
9. In the least-squares principle involving a polynomial we seek to satisfy as well as possible

$$b_0 + b_1 x_j + b_2 x_j{}^2 + \cdots + b_m x_j{}^m = y_j \qquad\qquad (j = 1, \cdots, n).$$

Introduce a matrix **C** such that this can be written **Cb** = **y** and show that then the normal equations can be written $\mathbf{C^T C b} = \mathbf{C^T y}$.

10. In problems of growth it is often required to fit an exponential function $y = b_0 e^{bx}$ by the method of least squares. Show that by taking logarithms this task can be reduced to that of determining a straight line.

18.12   *Inclusion of Matrix Eigenvalues*

An **eigenvalue** or **characteristic value** (or *latent root*) of an n-rowed (real or complex) square matrix $\mathbf{A} = (a_{jk})$ is defined to be a number λ such that the vector equation

(1) $$\mathbf{Ax} = \lambda \mathbf{x}$$

has a nontrivial solution, that is, a solution $\mathbf{x} \neq \mathbf{0}$, which is then called an **eigenvector** or **characteristic vector** of **A** corresponding to that eigenvalue λ. The set of all eigenvalues of **A** is called the **spectrum** of **A**. The eigenvalues of **A** are the roots of the characteristic equation

(2) $$D(\lambda) = \det(\mathbf{A} - \lambda \mathbf{I}) = 0,$$

where **I** is the n-rowed unit matrix. $D(\lambda)$ is called the **characteristic determinant** and can be represented as a polynomial in λ of degree n, which is called the **characteristic polynomial** corresponding to **A**. It follows that **A** has at least one and at most n numerically different eigenvalues. For more details, see Sec. 6.14.

Given **A**, we may determine the coefficients of the characteristic polynomial and then determine the roots, for instance, by Newton's method. However, if n is large, both steps involve time-consuming computations, and it is desirable to proceed in a more efficient way. There are essentially two types of corresponding methods:

1. *Methods for obtaining bounds for eigenvalues,*
2. *Methods for computing approximate values for eigenvalues.*

We shall illustrate both types by some standard examples. For further methods see Ref. [C1] in Appendix 1.

The following interesting theorem by Gershgorin[6] yields a region consisting of closed circular disks and containing all the eigenvalues of a given matrix.

---

[6] *Bull. Acad. Sciences de l'URSS*, Classe mathém., 7-e série, Leningrad, 1931, p. 749.

Theorem 1.   *Let λ be an eigenvalue of an arbitrary n-rowed square matrix* $\mathbf{A} = (a_{jk})$. *Then for some integer* k $(1 \leqq k \leqq n)$,

$$(3) \qquad |a_{kk} - \lambda| \leqq |a_{k1}| + |a_{k2}| + \cdots + |a_{k,k-1}| + |a_{k,k+1}| + \cdots + |a_{kn}|.$$

Proof.   Let $\mathbf{x}$ be an eigenvector corresponding to that eigenvalue λ of $\mathbf{A}$. Then

$$(4) \qquad\qquad \mathbf{Ax} = \lambda\mathbf{x} \qquad \text{or} \qquad (\mathbf{A} - \lambda\mathbf{I})\mathbf{x} = \mathbf{0}.$$

Let x_k be a component of $\mathbf{x}$ which is the greatest in absolute value. Then we have $|x_m/x_k| \leqq 1$ $(m = 1, \cdots, n)$. The vector equation (2) is equivalent to a system of n equations for the n components of the vectors on both sides, and the kth of these n equations is

$$a_{k1}x_1 + \cdots + a_{k,k-1}x_{k-1} + (a_{kk} - \lambda)x_k + a_{k,k+1}x_{k+1} + \cdots + a_{kn}x_n = 0.$$

From this we have

$$a_{kk} - \lambda = -a_{k1}\frac{x_1}{x_k} - \cdots - a_{k,k-1}\frac{x_{k-1}}{x_k} - a_{k,k+1}\frac{x_{k+1}}{x_k} - \cdots - a_{kn}\frac{x_n}{x_k}.$$

By taking absolute values on both sides of this equation, applying the triangle inequality

$$|a + b| \leqq |a| + |b|$$

(where a and b are any complex numbers) and observing that

$$\left|\frac{x_1}{x_k}\right| \leqq 1, \cdots, \qquad \left|\frac{x_n}{x_k}\right| \leqq 1$$

we obtain (3), and the theorem is proved.

For each $k = 1, \cdots, n$ the inequality (3) determines a closed circular disk in the complex λ-plane whose center is at a_{kk} and whose radius is given by the expression on the right-hand side of (3). Theorem 1 states that each of the eigenvalues of $\mathbf{A}$ lies in one of these n disks.

Example 1.   From Theorem 1 it follows that the eigenvalues of the matrix

$$\mathbf{A} = \begin{pmatrix} 26 & -2 & 2 \\ 2 & 21 & 4 \\ 4 & 2 & 28 \end{pmatrix}$$

lie in the three disks (Fig. 344 on p. 686)

$$D_1: \quad \text{center at } 26, \quad \text{radius } |-2| + 2 = 4,$$
$$D_2: \quad \text{center at } 21, \quad \text{radius } \quad 2 + 4 = 6,$$
and $\qquad\qquad\quad D_3: \quad \text{center at } 28, \quad \text{radius } \quad 4 + 2 = 6.$

(The reader may verify that the eigenvalues of $\mathbf{A}$ are 30, 25, and 20.)

Bounds for the absolute value of the eigenvalues result from the following theorem by Schur[7] which we shall state without proof.

_____

[7] *Mathematische Annalen*, Vol. 66, 1909, p. 488.

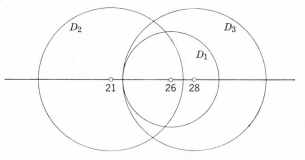

Fig. 344. Example 1

Theorem 2.   *Let* $\mathbf{A} = (a_{jk})$ *be an n-rowed square matrix, and let* $\lambda_1, \cdots, \lambda_n$ *be its eigenvalues. Then*

(5)
$$\sum_{i=1}^{n} |\lambda_i|^2 \leqq \sum_{j=1}^{n} \sum_{k=1}^{n} |a_{jk}|^2$$
   (Schur's inequality).

In (5) *the equality sign holds if and only if* $\mathbf{A}$ *is such that*

(6)
$$\bar{\mathbf{A}}^{\mathsf{T}}\mathbf{A} = \mathbf{A}\bar{\mathbf{A}}^{\mathsf{T}}.$$

Matrices which satisfy (6) are called **normal matrices.** It is not difficult to see that Hermitian, skew-Hermitian, and unitary matrices are normal, and so are real symmetric, skew-symmetric, and orthogonal matrices.

Let λ_m be any eigenvalue of the matrix $\mathbf{A}$ in Theorem 2. Then $|\lambda_m|^2$ is less than or equal to the sum on the left side of (5), and by taking square roots we obtain from (5)

(7)
$$|\lambda_m| \leqq \sqrt{\sum_{j=1}^{n} \sum_{k=1}^{n} |a_{jk}|^2}.$$

Example 2.   For the matrix $\mathbf{A}$ in Ex. 1 we obtain from (7)

$$|\lambda| \leqq \sqrt{1949} < 44.2.$$

(The eigenvalues of $\mathbf{A}$ are 30, 25, and 20; thus $30^2 + 25^2 + 20^2 = 1925 < 1949$; in fact, $\mathbf{A}$ is not normal.)

The preceding theorems are valid for any real or complex square matrix. There are other theorems which hold for special classes of matrices only. The following theorem by Frobenius,[8] which we state without proof, is of this type.

Theorem 3.   *Let* $\mathbf{A}$ *be a real square matrix whose elements are all positive. Then* $\mathbf{A}$ *has at least one real positive eigenvalue* λ, *and the corresponding eigenvector can be chosen real and such that all its components are positive.*

From this theorem we may derive the following useful result by Collatz.[9]

[8] *Sitzungsberichte Preuss. Akad. Wiss. Math.-phys. Klasse*, Berlin, 1908, p. 471.
[9] L. Collatz, *Eigenwertaufgaben mit technischen Anwendungen.* Leipzig: Akademische Verlagsgesellschaft, 1949, p. 291.

Theorem 4. *Let* $\mathbf{A} = (a_{jk})$ *be a real n-rowed square matrix whose elements are all positive. Let* $\mathbf{x}$ *be any real vector whose components* $x_1, \cdots, x_n$ *are all positive, and let* $y_1, \cdots, y_n$ *be the components of the vector* $\mathbf{y} = \mathbf{Ax}$. *Then the closed interval on the real axis bounded by the smallest and the largest of the n quotients* $q_j = y_j/x_j$ *contains at least one eigenvalue of* $\mathbf{A}$.

Proof. We have $\mathbf{Ax} = \mathbf{y}$ or

$$(8) \qquad\qquad \mathbf{y} - \mathbf{Ax} = \mathbf{0}.$$

The transpose $\mathbf{A}^\mathsf{T}$ satisfies the conditions of Theorem 3. Hence $\mathbf{A}^\mathsf{T}$ has a positive eigenvalue λ and, corresponding to this eigenvalue, an eigenvector $\mathbf{u}$ whose components u_j are all positive. Thus $\mathbf{A}^\mathsf{T}\mathbf{u} = \lambda\mathbf{u}$, and by taking the transpose we obtain $\mathbf{u}^\mathsf{T}\mathbf{A} = \lambda\mathbf{u}^\mathsf{T}$. From this and (8),

$$\mathbf{u}^\mathsf{T}(\mathbf{y} - \mathbf{Ax}) = \mathbf{u}^\mathsf{T}\mathbf{y} - \mathbf{u}^\mathsf{T}\mathbf{Ax} = \mathbf{u}^\mathsf{T}(\mathbf{y} - \lambda\mathbf{x}) = 0$$

or written out

$$\sum_{j=1}^{n} u_j(y_j - \lambda x_j) = 0.$$

Since all the components u_j are positive, it follows that

$$(9) \qquad \begin{array}{lll} y_j - \lambda x_j \geq 0, & \text{that is,} \quad q_j \geq \lambda & \text{for at least one } j, \\[6pt] y_j - \lambda x_j \leq 0, & \text{that is,} \quad q_j \leq \lambda & \text{for at least one } j. \end{array} \qquad \text{and}$$

Since $\mathbf{A}$ and $\mathbf{A}^\mathsf{T}$ have the same eigenvalues, λ is an eigenvalue of $\mathbf{A}$, and from (9) the statement of the theorem follows.

Example 3. Let

$$\mathbf{A} = \begin{pmatrix} 8 & 1 & 1 \\ 1 & 5 & 2 \\ 1 & 2 & 5 \end{pmatrix}. \quad \text{Choose} \quad \mathbf{x} = \begin{pmatrix} 1 \\ 1 \\ 1 \end{pmatrix}. \quad \text{Then} \quad \mathbf{y} = \begin{pmatrix} 10 \\ 8 \\ 8 \end{pmatrix}.$$

Hence $q_1 = 10$, $q_2 = 8$, $q_3 = 8$, and Theorem 4 implies that one of the eigenvalues of $\mathbf{A}$ must lie in the interval $8 \leq \lambda \leq 10$. Of course, the length of such an interval depends on the choice of $\mathbf{x}$. The student may show that $\lambda = 9$ is an eigenvalue of $\mathbf{A}$.

Problems

Using Theorem 1, determine and sketch disks which contain the eigenvalues of the following matrices.

1. $\begin{pmatrix} 1 & 2 & 3 \\ 2 & 4 & 6 \\ 3 & 6 & 1 \end{pmatrix}$ **2.** $\begin{pmatrix} 2 & -1 & 1 \\ -1 & 3 & 2 \\ 1 & 2 & 3 \end{pmatrix}$ **3.** $\begin{pmatrix} 0 & 0.5i & -i \\ 1-i & 1+i & 0 \\ 0.1i & 1 & -i \end{pmatrix}$

4. What conclusions about the eigenvalues of the matrix in Prob. 1 can be drawn by combining Theorem 1, this section, with Theorem 3 in Sec. 6.15?

Using (7), obtain an upper bound for the absolute value of the eigenvalues of the matrix:

5. In Prob. 1.   **6.** In Example 3.   **7.** In Prob. 3.

8. Show that the matrix in Ex. 1 is not normal.

9. Show that the matrix in Prob. 1 has the eigenvalues 10, 0, -4, and (5) holds with the equality sign.

10. Show that the matrix in Ex. 3 has the eigenvalues 9, 6, 3, and (5) holds with the equality sign.

11. Show that Hermitian, skew-Hermitian, and unitary matrices are normal.

12. Show that if **A** is normal and **U** is unitary and both matrices are n-rowed, then $\mathbf{B} = \mathbf{U}^{-1}\mathbf{A}\mathbf{U}$ is normal.

13. Illustrate the proposition in Prob. 12 with a simple example.

14. Show that the product of two n-rowed orthogonal matrices is orthogonal.

15. Show that if **A** is real symmetric and **B** is orthogonal, and both matrices are n-rowed, then $\mathbf{C} = \mathbf{B}^{-1}\mathbf{A}\mathbf{B}$ is real symmetric.

16. Find a two-rowed square matrix which is not normal.

17. Verify Theorem 3 for the matrix $\begin{pmatrix} 5 & 1 \\ 1 & 5 \end{pmatrix}$.

18. If in each row of a determinant the absolute value of the element in the principal diagonal is greater than the sum of the absolute values of the remaining elements in that row, show that the value of the determinant is different from zero.

Apply Theorem 4 to the following matrices, choosing the given vectors as vectors **x**.

19. $\begin{pmatrix} 17 & 8 & 1 \\ 8 & 18 & 8 \\ 1 & 8 & 17 \end{pmatrix}, \begin{pmatrix} 1 \\ 1 \\ 1 \end{pmatrix}, \begin{pmatrix} 1 \\ 2 \\ 1 \end{pmatrix}, \begin{pmatrix} 2 \\ 3 \\ 2 \end{pmatrix}$ **20.** $\begin{pmatrix} 3 & 1 & 1 \\ 1 & 3 & 1 \\ 1 & 1 & 3 \end{pmatrix}, \begin{pmatrix} 1 \\ 2 \\ 1 \end{pmatrix}, \begin{pmatrix} 1 \\ 1 \\ 1 \end{pmatrix}$

18.13   *Determination of Eigenvalues by Iteration*

A standard procedure for computing approximate values of the eigenvalues of an n-rowed square matrix $\mathbf{A} = (a_{jk})$ is the **iteration method for eigenvalues.** In this method we start from any vector $\mathbf{x}_0$ ($\neq \mathbf{0}$) with n components and compute successively

$$\mathbf{x}_1 = \mathbf{A}\mathbf{x}_0, \qquad \mathbf{x}_2 = \mathbf{A}\mathbf{x}_1, \cdots, \qquad \mathbf{x}_s = \mathbf{A}\mathbf{x}_{s-1}.$$

For simplifying notation, we denote $\mathbf{x}_{s-1}$ by $\mathbf{x}$ and $\mathbf{x}_s$ by $\mathbf{y}$, so that $\mathbf{y} = \mathbf{A}\mathbf{x}$. If **A** is real symmetric, the following theorem gives an approximation and error bounds.

Theorem 1.   *Let* **A** *be an n-rowed real symmetric matrix. Let* **x** *($\neq \mathbf{0}$) be any real vector with n components. Furthermore, let*

$$\mathbf{y} = \mathbf{A}\mathbf{x}, \qquad m_0 = \mathbf{x}^T\mathbf{x}, \qquad m_1 = \mathbf{x}^T\mathbf{y}, \qquad m_2 = \mathbf{y}^T\mathbf{y}.$$

Then the quotient

$$q = \frac{m_1}{m_0} \qquad\qquad \textbf{(Rayleigh quotient[10])}$$

---

[10] LORD RAYLEIGH (JOHN WILLIAM STRUTT) (1842–1919), English physicist and mathematician, who made important contributions to various branches of applied mathematics and theoretical physics, in particular the theory of waves, elasticity, and hydrodynamics.

is an approximation for an eigenvalue[11] λ of $\mathbf{A}$, and if we set $q = \lambda + \epsilon$, so that ϵ is the error of q, then

(1) $$|\epsilon| \leqq \sqrt{\frac{m_2}{m_0} - q^2}.$$

Proof. Let δ^2 denote the radicand in (1). Then, since $m_1 = qm_0$, we have

(2) $$(\mathbf{y} - q\mathbf{x})^{\mathsf{T}}(\mathbf{y} - q\mathbf{x}) = m_2 - 2qm_1 + q^2m_0 = m_2 - q^2m_0 = \delta^2m_0.$$

Since $\mathbf{A}$ is real symmetric, it has an orthogonal set of n real unit eigenvectors $\mathbf{z}_1, \cdots, \mathbf{z}_n$ corresponding to the eigenvalues $\lambda_1, \cdots, \lambda_n$, respectively (some of which may be equal). (Proof in Ref. [C14].) Then $\mathbf{x}$ has a representation of the form

$$\mathbf{x} = a_1\mathbf{z}_1 + \cdots + a_n\mathbf{z}_n.$$

Now $\mathbf{A}\mathbf{z}_1 = \lambda_1\mathbf{z}_1$, etc., and we obtain

$$\mathbf{y} = \mathbf{A}\mathbf{x} = a_1\lambda_1\mathbf{z}_1 + \cdots + a_n\lambda_n\mathbf{z}_n$$

and, since the $\mathbf{z}_j$ are orthogonal unit vectors,

(3) $$m_0 = \mathbf{x}^{\mathsf{T}}\mathbf{x} = a_1^2 + \cdots + a_n^2.$$

It follows that in (2),

$$\mathbf{y} - q\mathbf{x} = a_1(\lambda_1 - q)\mathbf{z}_1 + \cdots + a_n(\lambda_n - q)\mathbf{z}_n.$$

Since the $\mathbf{z}_j$ are orthogonal unit vectors, we thus obtain from (2)

$$\delta^2m_0 = a_1^2(\lambda_1 - q)^2 + \cdots + a_n^2(\lambda_n - q)^2.$$

Replacing each $(\lambda_j - q)^2$ by the smallest of these terms and using (3), we have

$$\delta^2m_0 \geqq (\lambda_c - q)^2(a_1^2 + \cdots + a_n^2) = (\lambda_c - q)^2m_0$$

where λ_c is the eigenvalue to which q is closest. From this, (1) follows, and the theorem is proved.

Example 1. As in Ex. 3 of the preceding section, let us consider

$$\mathbf{A} = \begin{pmatrix} 8 & 1 & 1 \\ 1 & 5 & 2 \\ 1 & 2 & 5 \end{pmatrix} \quad \text{and choose} \quad \mathbf{x}_0 = \begin{pmatrix} 1 \\ 1 \\ 1 \end{pmatrix}.$$

Then we obtain successively

$$\mathbf{x}_1 = \begin{pmatrix} 10 \\ 8 \\ 8 \end{pmatrix}, \quad \mathbf{x}_2 = \begin{pmatrix} 96 \\ 66 \\ 66 \end{pmatrix}, \quad \mathbf{x}_3 = \begin{pmatrix} 900 \\ 558 \\ 558 \end{pmatrix}, \quad \mathbf{x}_4 = \begin{pmatrix} 8316 \\ 4806 \\ 4806 \end{pmatrix}.$$

Taking $\mathbf{x} = \mathbf{x}_3$ and $\mathbf{y} = \mathbf{x}_4$, we have

$$m_0 = \mathbf{x}^{\mathsf{T}}\mathbf{x} = 1\ 432\ 728, \quad m_1 = \mathbf{x}^{\mathsf{T}}\mathbf{y} = 12\ 847\ 896, \quad m_2 = \mathbf{y}^{\mathsf{T}}\mathbf{y} = 115\ 351\ 128.$$

---

[11] Ordinarily that λ which is greatest in absolute value, but no general statements are possible.

From this we calculate

$$q = \frac{m_1}{m_0} = 8.967, \qquad |\epsilon| \leq \sqrt{\frac{m_2}{m_0} - q^2} = 0.311.$$

This shows that $q = 8.967$ is an approximation for an eigenvalue which must lie between 8.656 and 9.278. The student may show that an eigenvalue is $\lambda = 9$.

Problems

1. Choosing for x_0 the column vector with components 1, 1, 1, and using the matrix A in Prob. 2, Sec. 18.12, compute $x_1 = Ax_0$, $x_2 = Ax_1$, $x_3 = Ax_2$. Take x_2 for x and x_3 for y in Theorem 1 and calculate the Rayleigh quotient and an error bound (1).

2. Let A be the matrix in Prob. 2, Sec. 18.12. Choose $x_0{}^T = (1 \quad 0 \quad 0)$. Compute $x_1 = Ax_0$, the Rayleigh quotient and an error bound (1).

Choosing for x_0 the column vector with components 1, 1, 1, 1, compute x_1, x_2 and approximations $q = x_1{}^T x_0 / x_0{}^T x_0$, $q = x_2{}^T x_1 / x_1{}^T x_1$ and corresponding error bounds for an eigenvalue of each of the following symmetric matrices.

3. $$\begin{pmatrix} 3 & 2 & 0 & 1 \\ 2 & 0 & 5 & -1 \\ 0 & 5 & 2 & 1 \\ 1 & -1 & 1 & 4 \end{pmatrix}$$
4. $$\begin{pmatrix} 1 & 0 & 0 & 1 \\ 0 & 2 & -1 & 0 \\ 0 & -1 & 3 & 0 \\ 1 & 0 & 0 & -1 \end{pmatrix}$$
5. $$\begin{pmatrix} 2 & 0 & 1 & 0 \\ 0 & 0 & 3 & 1 \\ 1 & 3 & 4 & -2 \\ 0 & 1 & -2 & 0 \end{pmatrix}$$

6. Show that if x is an eigenvector, then $\epsilon = 0$ in (1).
7. To understand the importance of the error bound (1), consider the matrix

$$A = \begin{pmatrix} 3 & 4 \\ 4 & -3 \end{pmatrix}, \qquad \text{choose} \qquad x_0 = \begin{pmatrix} 3 \\ -1 \end{pmatrix}$$

   and show that $q = 0$ for all s. Find the eigenvalues and explain what happened. Start again, choosing another x_0.

8. To understand why the Rayleigh quotient q is generally an approximation of that eigenvalue, λ_1, which is largest in absolute value, let

$$x_0 = c_1 z_1 + \cdots + c_n z_n$$

   (with $z_1, \cdots, z_n$ as in the proof of Theorem 1) and show that

$$x = x_{s-1} = c_1 \lambda_1^{s-1} z_1 + \cdots + c_n \lambda_n^{s-1} z_n$$
$$y = x_s = c_1 \lambda_1^{s} z_1 + \cdots + c_n \lambda_n^{s} z_n$$

   and, consequently,

$$q = \frac{m_1}{m_0} = \frac{c_1{}^2 \lambda_1^{2s-1} + \cdots}{c_1{}^2 \lambda_1^{2s-2} + \cdots} \approx \lambda_1.$$

   Under what conditions will this be a good approximation?

9. Using $x_0{}^T = (3 \quad 1 \quad 2)$ and applying iteration to

$$A = \begin{pmatrix} -2 & -1 & 4 \\ 2 & 1 & -2 \\ -1 & -1 & 3 \end{pmatrix},$$

   show that $A^2 x_0 = x_0$ and draw conclusions from this result.
10. Can (1) be used in the case of the matrix in Prob. 9?

18.14   *Asymptotic Expansions*

Asymptotic expansions are (in general divergent) series which are of great practical importance for computing values of a function $f(x)$ for large x. It is clear that the Maclaurin series of $f(x)$, if it exists and converges for large x, is not suitable for that purpose, because the number of terms needed to obtain a certain number of significant digits increases rapidly as x increases. For a Taylor series with center a and large $|x - a|$, the situation is similar. We shall see that the larger x is, the fewer terms of an asymptotic expansion we need for obtaining a required accuracy. On the other hand, the accuracy is limited, and it decreases as x decreases so that asymptotic expansions can be used for large x only.

The variables and functions to be considered in this section are assumed to be real.

A series of the form

$$c_0 + \frac{c_1}{x} + \frac{c_2}{x^2} + \cdots \qquad (c_0, c_1, \cdots \text{constant})$$

(which need not converge for any value of x) *is called an* **asymptotic expansion,** *or* **asymptotic series,** *of a function $f(x)$ which is defined for every sufficiently large value of x if, for every fixed $n = 0, 1, 2, \cdots$,*

(1)
$$\left[f(x) - \left(c_0 + \frac{c_1}{x} + \frac{c_2}{x^2} + \cdots + \frac{c_n}{x^n} \right) \right] x^n \to 0 \qquad \text{as } x \to \infty,$$

and we shall then write

$$f(x) \sim c_0 + \frac{c_1}{x} + \frac{c_2}{x^2} + \cdots.$$

If a function $f(x)$ has an asymptotic expansion, then this expansion is unique, because its coefficients $c_0, c_1, \cdots$ are uniquely determined by (1). In fact, from (1) we obtain

$$f(x) - c_0 \to 0 \qquad \text{or} \qquad c_0 = \lim_{x \to \infty} f(x),$$

(1*)
$$\left[f(x) - c_0 - \frac{c_1}{x} \right] x \to 0 \qquad \text{or} \qquad c_1 = \lim_{x \to \infty} [f(x) - c_0]x, \quad \text{etc.}$$

On the other hand, different functions may have the same asymptotic expansion. In fact, let $f(x) = e^{-x}$. Then, since $e^{-x} \to 0$, $xe^{-x} \to 0$, etc., we see from (1*) that $c_0 = 0$, $c_1 = 0$, etc. Hence,

$$e^{-x} \sim 0 + \frac{0}{x} + \cdots.$$

Thus, if $g(x)$ has an asymptotic expansion, then $g(x) + e^{-x}$ has certainly the same asymptotic expansion.

For applications it is advantageous to *extend the definition* by writing

$$f(x) \sim g(x) + h(x)\left(c_0 + \frac{c_1}{x} + \frac{c_2}{x^2} + \cdots\right)$$

whenever

$$\frac{f(x) - g(x)}{h(x)} \sim c_0 + \frac{c_1}{x} + \frac{c_2}{x^2} + \cdots$$

in the sense of the preceding definition.

Only in rare cases may we determine the coefficients of an asymptotic expansion directly from (1*). In general, other methods will be more suitable, for example, successive integration by parts.

Example 1 (Asymptotic series of the error function).   The error function erf x is defined by the integral (cf. Fig. 390 in Appendix 3)

$$(2a) \qquad\qquad \text{erf } x = \frac{2}{\sqrt{\pi}} \int_0^x e^{-t^2}\, dt$$

and its complementary function erfc x by

$$(2b) \qquad \text{erfc } x = 1 - \text{erf } x = \frac{2}{\sqrt{\pi}} \int_x^\infty e^{-t^2}\, dt = \frac{1}{\sqrt{\pi}} \int_{x^2}^\infty e^{-\tau} \tau^{-1/2}\, d\tau,$$

where $t^2 = \tau$, and erf $\infty = 1$. Repeated integration by parts will lead to integrals of the form

$$(3) \qquad\qquad F_n(x) = \int_{x^2}^\infty e^{-\tau} \tau^{-(2n+1)/2}\, d\tau, \qquad\qquad n = 0, 1, \cdots.$$

Note that erfc $x = F_0(x)/\sqrt{\pi}$. By integration by parts we obtain

$$F_n(x) = -e^{-\tau} \tau^{-(2n+1)/2}\Big|_{x^2}^\infty - \frac{2n+1}{2} \int_{x^2}^\infty e^{-\tau} \tau^{-(2n+3)/2}\, d\tau.$$

The integral on the right is $F_{n+1}(x)$ and, therefore,

$$e^{x^2} F_n(x) = \frac{1}{x^{2n+1}} - \frac{2n+1}{2} e^{x^2} F_{n+1}(x), \qquad\qquad n = 0, 1, \cdots.$$

Repeated application of this formula yields

$$e^{x^2} F_0(x) = \frac{1}{x} - \frac{1}{2} e^{x^2} F_1(x)$$

$$\cdots \cdots \cdots \cdots \cdots$$

$$(4) \qquad = \left[\frac{1}{x} - \frac{1}{2x^3} + \frac{1\cdot 3}{2^2 x^5} - + \cdots + (-1)^{n-1} \frac{1\cdot 3 \cdots (2n-3)}{2^{n-1} x^{2n-1}}\right]$$

$$+ (-1)^n \frac{1\cdot 3 \cdots (2n-1)}{2^n} e^{x^2} F_n(x).$$

We show that the series obtained in this way is an asymptotic expansion,

$$(5) \qquad\qquad e^{x^2} F_0(x) \sim \frac{1}{x} - \frac{1}{2x^3} + \frac{1\cdot 3}{2^2 x^5} - + \cdots.$$

Let S_{2n-1} denote the expression in the brackets in (4). Then from (4) we obtain

(6) $$[e^{x^2}F_0(x) - S_{2n-1}]x^{2n-1} = K_n e^{x^2} x^{2n-1} F_n(x),$$

where $K_n = (-2)^{-n} 1 \cdot 3 \cdots (2n-1)$. We have to show that for each fixed $n = 1, 2, \cdots$ the expression on the right approaches zero as x approaches infinity. In (3) we have

$$\frac{1}{\tau^{(2n+1)/2}} \leqq \frac{1}{x^{2n+1}} \qquad \text{for all } \tau \geqq x^2.$$

Hence we obtain the inequality

(7) $$F_n(x) = \int_{x^2}^{\infty} \frac{e^{-\tau}}{\tau^{(2n+1)/2}} d\tau < \frac{1}{x^{2n+1}} \int_{x^2}^{\infty} e^{-\tau} d\tau = \frac{e^{-x^2}}{x^{2n+1}}.$$

From this we immediately see that

$$|K_n| e^{x^2} x^{2n-1} F_n(x) < \frac{|K_n|}{x^2} \to 0 \qquad\qquad (x \to \infty).$$

This proves that the series in (5) is an asymptotic expansion of the function on the left-hand side of (5). Since $\operatorname{erf} x = 1 - \operatorname{erfc} x = 1 - F_0(x)/\sqrt{\pi}$, it follows that the desired asymptotic series of the error function is

(8) $$\operatorname{erf} x \sim 1 - \frac{1}{\sqrt{\pi}} e^{-x^2} \left(\frac{1}{x} - \frac{1}{2x^3} + \frac{1 \cdot 3}{2^2 x^5} - \frac{1 \cdot 3 \cdot 5}{2^3 x^7} + - \cdots \right).$$

For large x we thus have the simple approximation

(8*) $$\operatorname{erf} x \approx 1 - \frac{1}{\sqrt{\pi} \, x} e^{-x^2}.$$

From (6) and (7) we see that

$$|e^{x^2}F_0(x) - S_{2n-1}| = \frac{1 \cdot 3 \cdots (2n-1)}{2^n} e^{x^2} F_n(x) < \frac{1 \cdot 3 \cdots (2n-1)}{2^n} \frac{1}{x^{2n+1}},$$

and for sufficiently large x the expression on the right is very small. This shows that for such x, S_{2n-1} is a very good approximation to $e^{x^2}F_0(x)$, and so, for large x, the error function can be computed very accurately by means of (7). It turns out that even for relatively small $|x|$ the results are surprisingly accurate. For instance, (8*) gives

$$\operatorname{erf} 5 \sim 1 - \frac{1}{\sqrt{\pi}} \frac{e^{-25}}{5} = 0.99999\ 99999\ 98433;$$

the first thirteen decimals are correct, the error is three units of the fourteenth decimal. However, we warn the reader that the present case may be misleading; for other asymptotic expansions a similarly favorable situation may arise not at $x = 5$, but only for much larger x, perhaps for $x = 20$ or $x = 100$.

For practical applications it is useful to know that asymptotic series may be added, multiplied, and under certain restrictions also integrated and differentiated term by term. Let us formulate these properties in a precise manner.

Theorem 1 (Addition and multiplication). *If*

$$f(x) \sim a_0 + \frac{a_1}{x} + \frac{a_2}{x^2} + \cdots \qquad \text{and} \qquad g(x) \sim b_0 + \frac{b_1}{x} + \frac{b_2}{x^2} + \cdots$$

then the function $Af + Bg$, where A and B are constants, has the asymptotic expansion

$$(9) \qquad Af(x) + Bg(x) \sim Aa_0 + Bb_0 + \frac{Aa_1 + Bb_1}{x} + \frac{Aa_2 + Bb_2}{x^2} + \cdots$$

and the function fg has the asymptotic expansion

$$(10) \qquad f(x)g(x) \sim c_0 + \frac{c_1}{x} + \frac{c_2}{x^2} + \cdots$$

whose coefficients are given by the formula

$$c_n = a_0 b_n + a_1 b_{n-1} + \cdots + a_n b_0.$$

Proof. The simple proof of (9) is left to the reader. We prove the last statement. We must show that for any fixed non-negative integer n,

$$(11) \qquad (fg - S_n)\, x^n \to 0 \qquad \text{as } x \to \infty,$$

where

$$S_n(x) = c_0 + \frac{c_1}{x} + \cdots + \frac{c_n}{x^n}$$

with the coefficients $c_0, \cdots, c_n$ given in the theorem.

We choose an arbitrary fixed n and write

$$f(x) = s_n(x) + \frac{h(x)}{x^n} \qquad \text{where} \qquad s_n(x) = a_0 + \frac{a_1}{x} + \cdots + \frac{a_n}{x^n}.$$

Then

$$[f(x) - s_n(x)]x^n = h(x).$$

From this and the definition of an asymptotic expansion it follows that $h(x)$ must approach zero as $x \to \infty$. Similarly, if we write

$$g(x) = s_n^*(x) + \frac{l(x)}{x^n} \qquad \text{where} \qquad s_n^*(x) = b_0 + \frac{b_1}{x} + \cdots + \frac{b_n}{x^n},$$

then $l(x) \to 0$ as $x \to \infty$. From these representations we obtain

$$fg = s_n s_n^* + \frac{h + l}{x^n} + \frac{hl}{x^{2n}}.$$

By multiplying term by term and collecting like powers of x we can readily verify that

$$s_n s_n^* = S_n + T_n$$

where T_n is the sum of terms involving the powers $1/x^{n+1}, \cdots, 1/x^{2n}$. Clearly, $x^n T_n \to 0$ as $x \to \infty$. Consequently, in (11),

$$fg - S_n = T_n + \frac{h + l}{x^n} + \frac{hl}{x^{2n}}.$$

We multiply by x^n on both sides. Then the resulting expression on the right approaches zero as $x \to \infty$, because $x^n T_n \to 0$, $l \to 0$, and $h \to 0$. Hence the expression on the left-hand side of (11) approaches zero, and the proof is complete.

Theorem 2 (Integration). *Let $f(x)$ be continuous for all sufficiently large x, and let*

$$f(x) \sim \frac{c_2}{x^2} + \frac{c_3}{x^3} + \cdots .$$

Then for those x,

(12) $$\int_x^\infty f(t)\, dt \sim \frac{c_2}{x} + \frac{c_3}{2x^2} + \frac{c_4}{3x^3} + \cdots .$$

Proof. Let $F(x)$ denote the integral in (12) and let $S_{n-1}(x)$ denote the integral of

$$s_n(x) = \frac{c_2}{x^2} + \cdots + \frac{c_n}{x^n},$$

that is,

$$S_{n-1}(x) = \int_x^\infty s_n(t)\, dt = \frac{c_2}{x} + \frac{c_3}{2x^2} + \cdots + \frac{c_n}{(n-1)x^{n-1}}.$$

By the definition of an asymptotic expansion, for every $n = 0, 1, \cdots$,

$$|f(x) - s_n(x)|\, x^n \to 0 \qquad \text{as } x \to \infty.$$

Because of the continuity of f this implies that, given any $\epsilon > 0$ we can find an x_0 such that for all $x > x_0$,

$$|f(x) - s_n(x)|\, x^n < \epsilon \qquad \text{or} \qquad |f(x) - s_n(x)| < \frac{\epsilon}{x^n}.$$

From this we obtain for all $x > x_0$

$$|F(x) - S_{n-1}(x)| = \left| \int_x^\infty f(t)\, dt - \int_x^\infty s_n(t)\, dt \right| = \left| \int_x^\infty [f(t) - s_n(t)]\, dt \right|$$

$$\leq \int_x^\infty |f(t) - s_n(t)|\, dt < \epsilon \int_x^\infty \frac{dt}{t^n} = \frac{\epsilon}{(n-1)x^{n-1}}.$$

By multiplying on both sides by the positive quantity x^{n-1} we have

$$|F(x) - S_{n-1}(x)|\, x^{n-1} < \frac{\epsilon}{n-1} \qquad \text{when} \quad x > x_0(\epsilon).$$

Since ϵ (> 0) may be chosen as small as we please, it follows that

$$|F(x) - S_{n-1}(x)|\, x^{n-1} \to 0 \qquad \text{as } x \to \infty,$$

and the proof is complete.

If $f(x)$ is continuous for all sufficiently large x and

$$f(x) \sim c_0 + \frac{c_1}{x} + \frac{c_2}{x^2} + \cdots ,$$

then we conclude from Theorem 2 that

(12*) $$\int_x^\infty \left[f(t) - c_0 - \frac{c_1}{t} \right] dt \sim \frac{c_2}{x} + \frac{c_3}{2x^2} + \frac{c_4}{3x^3} + \cdots .$$

If $f(x)$ has an asymptotic expansion, it does not follow that the derivative $f'(x)$ has an asymptotic expansion. For example, from (1*) we obtain

$$f(x) = e^{-x} \sin(e^x) \sim 0 + \frac{0}{x} + \frac{0}{x^2} + \cdots$$

but the derivative of $f(x)$, which is given by the formula

$$f'(x) = -e^{-x} \sin(e^x) + e^{-x} \cos(e^x) e^x = -f(x) + \cos(e^x),$$

does not have an asymptotic expansion. (Why?) However, if the derivative $f'(x)$ of a function $f(x)$ has an asymptotic expansion, it can be obtained by termwise differentiation of that of $f(x)$. In fact, the following theorem holds true.

Theorem 3 (Differentiation). *If*

(13) $$f(x) \sim c_0 + \frac{c_1}{x} + \frac{c_2}{x^2} + \cdots$$

and $f(x)$ has a continuous derivative $f'(x)$ which has an asymptotic expansion, then this expansion is

(14) $$f'(x) \sim -\frac{c_1}{x^2} - \frac{2c_2}{x^3} - \frac{3c_3}{x^4} - \cdots.$$

Proof. By assumption,

(15) $$f'(x) \sim a_0 + \frac{a_1}{x} + \frac{a_2}{x^2} + \cdots$$

and we have to show that the coefficients a_n are such that (14) and (15) are identical. We first show that $a_0 = 0$ and $a_1 = 0$. From (13) and the definition of an asymptotic expansion we have

(16) $\quad\quad$ (a) $\lim_{x \to \infty} f(x) = c_0,$ $\quad\quad$ (b) $\lim_{x \to \infty} [f(x) - c_0]x = c_1.$

The corresponding relations for the asymptotic series (15) are

(17) $\quad\quad$ (a) $\lim_{x \to \infty} f'(x) = a_0,$ $\quad\quad$ (b) $\lim_{x \to \infty} [f'(x) - a_0]x = a_1.$

We may relate f and f' by the formula

(18) $$f(x) = \int_{x_0}^{x} f'(t)\, dt + k \quad\quad\quad [x_0\ (> 0) \text{ and } k \text{ constant}].$$

From this and (16a) it follows that

$$\lim_{x \to \infty} \int_{x_0}^{x} f'(t)\, dt + k = c_0.$$

Now (17a) shows that if a_0 were not zero, the limit of the integral would not exist. Hence $a_0 = 0$. Then (17b) becomes

(17b*) $$\lim_{x \to \infty} x f'(x) = a_1.$$

By definition this means that for any preassigned $\epsilon > 0$ and all sufficiently large x,

(19) $a_1 - \epsilon < xf'(x) < a_1 + \epsilon$ or $\dfrac{a_1 - \epsilon}{x} < f'(x) < \dfrac{a_1 + \epsilon}{x}.$

From (16b) and (18) we obtain

$$\lim_{x \to \infty} \left(\int_{x_0}^{x} f'(t)\, dt + k - c_0 \right) x = c_1.$$

From (19) we see that for $a_1 \neq 0$ this limit does not exist. Hence $a_1 = 0$, and (15) becomes

$$f'(x) \sim \frac{a_2}{x^2} + \frac{a_3}{x^3} + \cdots.$$

Consequently, from (18) and Theorem 2 we now obtain

(20)
$$f(x) = \int_{x_0}^{\infty} f'(t)\, dt - \int_{x}^{\infty} f'(t)\, dt + k$$
$$\sim \int_{x_0}^{\infty} f'(t)\, dt + k - \frac{a_2}{x} - \frac{a_3}{2x^2} - \cdots,$$

the first integral on the right being a constant. If a given function has an asymptotic expansion, it is unique. We may thus compare corresponding terms in (13) and (20), finding $a_2 = -c_1$, $a_3 = -2c_2$, etc. With these coefficients the series (14) and (15) become identical, and the theorem is proved.

If we know that a function $f(x)$ satisfies the assumptions of Theorem 3 and is a solution of a first-order differential equation, then we may determine the coefficients of its asymptotic expansion by substituting (13) and (14) in the differential equation.

Example 2 (Exponential integral).   The exponential integral $\mathrm{Ei}(x)$ is defined by the formula

$$\mathrm{Ei}(x) = \int_{x}^{\infty} \frac{e^{-t}}{t}\, dt \qquad\qquad (x > 0).$$

To obtain the asymptotic expansion of $\mathrm{Ei}(x)$, we consider the function

$$y = f(x) = e^{x}\mathrm{Ei}(x) = e^{x}\int_{x}^{\infty} \frac{e^{-t}}{t}\, dt \qquad\qquad (x > 0).$$

By differentiation we see that $f(x)$ satisfies the linear differential equation

(21) $$y' - y + \frac{1}{x} = 0.$$

It may be proved directly that this equation has only one solution y such that y and y' exist for positive x and have an asymptotic expansion. By substituting (13) and (14) into (21) and equating the coefficient of each power of x to zero we obtain

$$-c_0 = 0, \qquad -c_1 + 1 = 0, \qquad -c_1 - c_2 = 0, \quad \cdots, \qquad -nc_n - c_{n+1} = 0, \quad \cdots,$$

that is,

$$c_0 = 0, \quad c_1 = 1, \quad c_2 = -1, \quad \cdots, \quad c_{n+1} = (-1)^n n!, \quad \cdots$$

and, therefore,

$$(22) \qquad \mathrm{Ei}(x) = e^{-x}f(x) \sim e^{-x}\left(\frac{1}{x} - \frac{1}{x^2} + \frac{2!}{x^3} - \frac{3!}{x^4} + - \cdots\right).$$

Problems

1. Show that the series $1 + \frac{1}{x} + \frac{1}{2!\,x^2} + \cdots$ [which converges for $|x| > 0$ and represents $e^{1/x}$] is an asymptotic expansion of $e^{1/x}$.

2. Show that $\sin\dfrac{1}{x} \sim \dfrac{1}{x} - \dfrac{1}{3!\,x^3} + \dfrac{1}{5!\,x^5} - + \cdots$.

3. Obtain the asymptotic expansion of erf x by showing that $y = \frac{1}{2}\sqrt{\pi}\,e^{x^2}$ erfc x satisfies $y' - 2xy + 1 = 0$.

Integrating by parts, find the asymptotic expansion of [12]

4. $\mathrm{si}(x) = \displaystyle\int_x^\infty \frac{\sin t}{t}\,dt$ (**Complementary sine integral**).

5. $\mathrm{c}(x) = \displaystyle\int_x^\infty \cos t^2\,dt$ (**Complementary Fresnel integral**).

6. $\mathrm{s}(x) = \displaystyle\int_x^\infty \sin t^2\,dt$ (**Complementary Fresnel integral**).

7. $Q(\alpha, x) = \displaystyle\int_x^\infty e^{-t}t^{\alpha-1}\,dt$ (**Incomplete gamma function**).

Using the results of Probs. 4, 5, and 7, find the asymptotic expansion of

8. $\mathrm{Si}(x) = \displaystyle\int_0^x \frac{\sin t}{t}\,dt$ (**Sine integral**). Use $\mathrm{si}(0) = \dfrac{\pi}{2}$.

9. $C(x) = \displaystyle\int_0^x \cos t^2\,dt$ (**Fresnel integral**). Use $\mathrm{c}(0) = \dfrac{1}{2}\sqrt{\dfrac{\pi}{2}}$.

10. $P(\alpha, x) = \displaystyle\int_0^x e^{-t}t^{\alpha-1}\,dt$ (**Incomplete gamma function**).

---

[12] Some formulas for these and related functions are included in Appendix 3. Further formulas can be found in Ref. [A5] (cf. Appendix 1) and references to tables of numerical values in Ref. [A6].

CHAPTER 19

Probability and Statistics

The importance of mathematical statistics in engineering is increasing, in particular in mass production and in the analysis of experimental data. In the present chapter we shall give an introduction to this field. We shall start with the tabular and graphical representation of data (Secs. 19.2, 19.3). Then we shall discuss some fundamental concepts and relations from the theory of mathematical probability (Secs. 19.4–19.11), because probability theory forms the basis of mathematical statistics. The remaining sections (19.12–19.20) will be devoted to some of the most important statistical methods.

> *Prerequisite for this chapter:* elementary differential and integral calculus.
> *Sections which may be omitted in a shorter course:* 19.17–19.20.
> *References:* Appendix 1, Part H.
> *Answers to problems:* Appendix 2.

19.1 Nature and Purpose of Mathematical Statistics

In engineering statistics we are concerned with methods for designing and evaluating experiments to obtain information about practical problems, for example, the inspection of quality of raw material or manufactured products, the comparison of machines and tools or methods used in production, the output of workers, the reaction of consumers to new products, the yield of a chemical process under various conditions, the relation between the iron content of iron ore and the density of the ore, the effectivity of air conditioning systems under various temperatures, the relation between the Rockwell hardness and the carbon content of steel, and so on.

For instance, in a process of mass production (of screws, bolts, light bulbs, typewriter keys, etc.) we ordinarily have *nondefective items*, that is, articles which satisfy the quality requirements, and *defective items*, those which do not satisfy the requirements. Such requirements include maximum and minimum diameters of axle shafts, minimum lifetimes of light bulbs, limiting values of the resistance of resistors in radio and TV production, maximum weights of airmail envelopes, minimum contents of automatically filled bottles, maximum reaction times of switches, and minimum values of strength of yarn.

The reason for the differences in the quality of products is *variation* due to numerous factors (in the raw material, function of automatic machinery, workmanship, etc.) whose influence cannot be predicted, so that the variation must be regarded as a *random variation*. In the case of evaluating the efficiency of production methods and the other preceding examples the situation is similar.

In most cases the inspection of each item of a production is prohibitively expensive and time-consuming. It may even be impossible if it leads to the destruction of the item. Hence instead of inspecting all the manufactured items just a few of them (a "*sample*") are inspected and from this inspection conclusions can be drawn about the totality (the "*population*"). If we draw 100 screws from a lot of 10,000 screws and find that 5 of the 100 screws are defective, we are inclined to conclude that *about* 5% of the lot are defective, provided the screws were drawn "*at random*", that is, so that each screw of the lot had the same "chance" of being drawn. It is clear that such a conclusion is not absolutely certain, that is, we cannot say that the lot contains *precisely* 5% defectives, but in most cases such a precise statement would not be of particular practical interest anyway. We also feel that the conclusion is the more dependable the more randomly selected screws we inspect. Using the theory of mathematical probability we shall see that those somewhat vague intuitive notions can be made precise. Furthermore, this theory will also yield measures for the reliability of conclusions about populations obtained from samples by statistical methods.

Similarly, to obtain information about the iron content μ of iron ore, we may pick a certain number n of specimens at random and measure the iron content. This yields a sample of n numbers $x_1, \cdots, x_n$ (the results of the n measurements) whose average $\bar{x} = (x_1 + x_2 + \cdots + x_n)/n$ is an approximate value for μ.

Problems of differing natures may require different methods and techniques, but the steps leading from the formulation to the solution of the problem are the same in almost all cases. They are as follows.

1. *Formulation of the problem.* It is important to formulate the problem in a precise fashion and to limit the investigation, so that we can expect a useful answer within a prescribed interval of time, taking into account the cost of the statistical investigation, the skill of the investigators, and the facilities available. This step must also include the creation of a mathematical model based on clear concepts. (For example, we must define what we mean by a defective item in a specific situation.)

2. *Design of experiment.* This step includes the choice of the statistical method to be used in the last step, the sample size n (number of items to be drawn and inspected or number of experiments to be made, etc.), and the physical methods and techniques to be used in the experiment. The goal is to obtain a maximum of information using a minimum of cost and time.

3. *Experimentation or collection of data.* This step should adhere to strict rules.

4. *Tabulation.* In this step we arrange the experimental data in a clear and simple tabular form, and we may represent them graphically by diagrams, bar charts, etc. We also compute numbers that characterize the average size and the spread of the sample values.

5. *Statistical inference.* In this step we use the sample and apply a suitable statistical method for drawing conclusions about the unknown properties of the population so that we obtain the answer to our problem.

19.2   Tabular and Graphical Representation of Samples

In the course of a statistical experiment we usually obtain a sequence of observations (numbers in most cases) which we write down in the order they occur. A typical example is shown in Table 16. These data were obtained by making standard test cylinders (diameter 6 in., length 12 in.) of concrete and splitting them 28 days later. We thus have a **sample** consisting of 100 **sample values,** so that the **size** of the sample is $n = 100$.

In the present section we want to learn how to represent samples in a suitable tabular and graphical form. It suffices to discuss the corresponding methods in terms of our sample in Table 16.

To see what information is contained in Table 16, we shall now order the data. We write 300 (the smallest value), 310, 320, $\cdots$, 440 (the largest value) in a column. Then we run through Table 16 line by line and make a tally mark for each occurring number. In this way we obtain a **tally chart** of our sample, consisting of the first two columns of Table 17. The number of tallies is then listed in the third column of Table 17. It indicates how often the corresponding value x occurs in the sample and is called the *absolute frequency* or, more briefly, the **frequency** of that value x in the sample. Dividing it by the size n of the sample, we obtain the **relative frequency** in column 4; in our case, $n = 100$. Thus $x = 330$ has the frequency 6 and the relative frequency 0.06 or 6%.

Table 16. Sample of 100 Values of the Splitting Tensile Strength (lb/in.²) of Concrete Cylinders

320	380	340	410	380	340	360	350	320	370
350	340	350	360	370	350	380	370	300	420
370	390	390	440	330	390	330	360	400	370
320	350	360	340	340	350	350	390	380	340
400	360	350	390	400	350	360	340	370	420
420	400	350	370	330	320	390	380	400	370
390	330	360	380	350	330	360	300	360	360
360	390	350	370	370	350	390	370	370	340
370	400	360	350	380	380	360	340	330	370
340	360	390	400	370	410	360	400	340	360

D. L. IVEY, Splitting tensile tests on structural lightweight aggregate concrete. Texas Transportation Institute, College Station, Texas, 1965

Table 17. Frequency Table of the Sample in Table 16

1 Tensile Strength x[lb/in.^{2}]	2 Absolute Frequency		4 Relative Frequency	5 Cumulative Frequency	6 Cumulative Relative Frequency
	Tallies				
300	||	2	0.02	2	0.02
310		0	0.00	2	0.02
320	||||	4	0.04	6	0.06
330	ЖҐ |	6	0.06	12	0.12
340	ЖҐ ЖҐ |	11	0.11	23	0.23
350	ЖҐ ЖҐ ||||	14	0.14	37	0.37
360	ЖҐ ЖҐ ЖҐ |	16	0.16	53	0.53
370	ЖҐ ЖҐ ЖҐ	15	0.15	68	0.68
380	ЖҐ |||	8	0.08	76	0.76
390	ЖҐ ЖҐ	10	0.10	86	0.86
400	ЖҐ |||	8	0.08	94	0.94
410	||	2	0.02	96	0.96
420	|||	3	0.03	99	0.99
430		0	0.00	99	0.99
440	|	1	0.01	100	1.00

If for a certain x we sum all the frequencies corresponding to the sample values which are smaller than or equal to that x, we obtain the **cumulative frequency** corresponding to that x. This yields column 5 in Table 17. For example, to $x = 350$ there corresponds the cumulative frequency 37, and this tells us that there are 37 sample values which are smaller than or equal to 350. Division by the size n of the sample yields the **cumulative relative frequency** in column 6. For example, from column 6 we see that 76% of the sample values are smaller than or equal to 380.

If a certain numerical value does not occur in a sample, its frequency is 0. If all the n values of the sample are numerically equal, then this number has the frequency n and, therefore, the relative frequency $n/n = 1$. Since these are the two extreme possible cases, we have

Theorem 1. *The relative frequency is at least equal to 0 and at most equal to 1.*

Suppose that a given sample of size n consists of m *numerically different* values

$$x_1, \qquad x_2, \qquad \cdots, \qquad x_m \qquad\qquad (m \leq n)$$

with corresponding relative frequencies

$$\tilde{f}_1, \qquad \tilde{f}_2, \qquad \cdots, \qquad \tilde{f}_m.$$

Then we may introduce the function[1]

(1) $$\tilde{f}(x) = \begin{cases} \tilde{f}_j & \text{when } x = x_j \qquad\qquad (j = 1, 2, \cdots, m) \\ 0 & \text{for any value } x \text{ not appearing in the sample.} \end{cases}$$

[1] We write $\tilde{f}$ because we want to reserve the simple symbol f for the theoretical counterpart of the frequency function. f will appear much more often in our text.

This function is called the **frequency function of the sample.** It shows how the values of the sample are distributed. We therefore say that it determines the **frequency distribution** of the sample.

For example, in Table 17 the values of the frequency function are shown in column 4, and we see that $\tilde{f}(300) = 0.2$, $\tilde{f}(310) = 0$, $\tilde{f}(320) = 0.04$, and so on.

The sum of all the frequencies in a sample of size n must equal n. (Why?) This yields

Theorem 2. *The sum of all the relative frequencies in a sample equals 1, that is,*

$$\sum_{j=1}^{m} \tilde{f}(x_j) = \tilde{f}(x_1) + \tilde{f}(x_2) + \cdots + \tilde{f}(x_m) = 1.$$

Graphical representations of samples are illustrated by Figs. 345 to 348. In Fig. 347 the area of each rectangle is equal to the corresponding relative frequency. Hence the ordinate should be labeled "*relative frequency per unit interval.*" Since in the present case the rectangles are equally wide, those values on the ordinate are proportional to $\tilde{f}(x)$ and we may label the ordinate in terms of $\tilde{f}(x)$. However, this would no longer be true if the rectangles were of different width. In Fig. 348 the situation is similar.

We now introduce the function

$$\tilde{F}(x) = sum \ of \ the \ relative \ frequencies \ of \ all \ the$$

$$values \ that \ are \ smaller \ than \ x \ or \ equal \ to \ x.$$

This function is called the **cumulative frequency function of the sample** or the **sample distribution function.** An example is shown in Fig. 349.

$\tilde{F}(x)$ is a *step function* (piecewise constant function) having jumps of magnitude $\tilde{f}(x)$ precisely at those x at which $\tilde{f}(x) \neq 0$. The first jump is at the smallest sample value and the last at the largest. Afterwards $\tilde{F}(x) = 1$.

The relation between $\tilde{f}(x)$ and $\tilde{F}(x)$ is

(2) $$\tilde{F}(x) = \sum_{t \leq x} \tilde{f}(t)$$

where $t \leq x$ means that for an x we have to form the sum of all those $\tilde{f}(t)$ for which t is less than x or equal to x.

If a sample consists of too many numerically different sample values, then its tabular and graphical representations are unnecessarily complicated, but may be simplified by the process of **grouping,** as follows.

A sample being given, we choose an interval I that contains all the sample values. We subdivide I into subintervals, which are called **class intervals.** The midpoints of these intervals are called **class midpoints** or **class marks.** The sample values in each such interval are said to form a **class.** Their number is called the

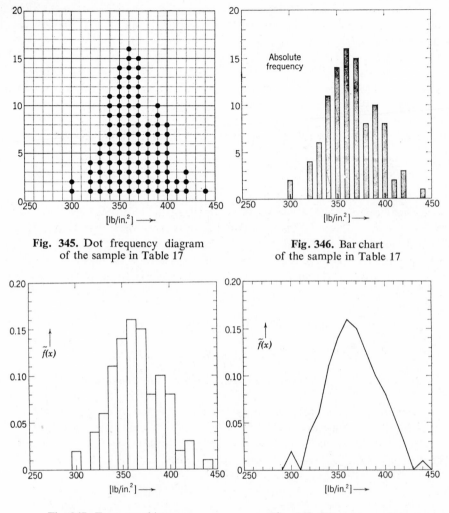

Fig. 345. Dot frequency diagram
of the sample in Table 17

Fig. 346. Bar chart
of the sample in Table 17

Fig. 347. Frequency histogram
of the sample in Table 17

Fig. 348. Frequency polygon
of the sample in Table 17

corresponding **class frequency.** Division by the sample size n gives the **relative class frequency.** This frequency, considered as a function $\tilde{f}(x)$ of the class marks is called the **frequency function of the grouped sample,** and the corresponding cumulative relative class frequency, considered as a function $\tilde{F}(x)$ of the class marks, is called the **distribution function of the grouped sample.** An example is shown in Tables 18 and 19.

The fewer classes we choose, the simpler the distribution of the grouped sample becomes but the more information we lose, because the original sample values do no longer appear explicitly. Grouping should be done so that only unessential

Fig. 349. Frequency function $\tilde{f}(x)$ and cumulative frequency function $\tilde{F}(x)$ of the sample in Table 17

details are eliminated. Unnecessary complications in the later use of a grouped sample are avoided by obeying the following rules.

1. All the class intervals should have the same length.

2. The class intervals should be chosen so that the class marks correspond to simple numbers (numbers with few nonzero digits).

3. If a sample value x_j coincides with the common endpoint of two class intervals, we take it into that class interval which extends from x_j to the right.

Table 18. Strength of 50 Lots of Cotton (lb required to break a skein)

114	118	86	107	87	94	82	81	98	84
120	126	98	89	114	83	94	106	96	111
123	110	83	118	83	96	96	74	91	81
102	107	103	80	109	71	96	91	86	129
130	104	86	121	96	96	127	94	102	87

Table 19. Frequency Table of the Sample in Table 18 (Grouped)

Class Interval	Class Mark x	Absolute Frequency		$\tilde{f}(x)$	$\tilde{F}(x)$
		Tallies			
65– 75	70	||	2	0.04	0.04
75– 85	80	ⅢⅢ Ⅲ	8	0.16	0.20
85– 95	90	ⅢⅢ ⅢⅢ |	11	0.22	0.42
95–105	100	ⅢⅢ ⅢⅢ ||	12	0.24	0.66
105–115	110	ⅢⅢ Ⅲ	8	0.16	0.82
115–125	120	ⅢⅢ	5	0.10	0.92
125–135	130	||||	4	0.08	1.00
		Sum	50	1.00	

Problems

In each case make a frequency table of the given sample and represent the sample by a dot frequency diagram, a bar chart, and a histogram.

1. Resistance [ohms] of resistors

$$99 \quad 100 \quad 102 \quad 101 \quad 98 \quad 103 \quad 100 \quad 102 \quad 99 \quad 101$$
$$100 \quad 100 \quad 99 \quad 101 \quad 100 \quad 102 \quad 99 \quad 101 \quad 98 \quad 100$$

2. Numbers that turned up on a die

$$6 \quad 2 \quad 4 \quad 1 \quad 2 \quad 4 \quad 3 \quad 3 \quad 2 \quad 1 \quad 6 \quad 5 \quad 6 \quad 3 \quad 4$$

3. Release time [sec] of a relay

$$1.3 \quad 1.4 \quad 1.1 \quad 1.5 \quad 1.4 \quad 1.3 \quad 1.2 \quad 1.4 \quad 1.5 \quad 1.3$$
$$1.2 \quad 1.3 \quad 1.5 \quad 1.4 \quad 1.4 \quad 1.6 \quad 1.3 \quad 1.5 \quad 1.1 \quad 1.4$$

4. Carbon content [%] of coal

$$87 \quad 86 \quad 85 \quad 87 \quad 86 \quad 87 \quad 86 \quad 81 \quad 77 \quad 85$$
$$86 \quad 84 \quad 83 \quad 83 \quad 82 \quad 84 \quad 83 \quad 79 \quad 82 \quad 73$$

5. Tensile strength [kg/mm^2] of sheet steel

$$44 \quad 43 \quad 41 \quad 41 \quad 44 \quad 44 \quad 43 \quad 44 \quad 42 \quad 45 \quad 43 \quad 43 \quad 44 \quad 45 \quad 46$$
$$42 \quad 45 \quad 41 \quad 44 \quad 44 \quad 43 \quad 44 \quad 46 \quad 41 \quad 43 \quad 45 \quad 45 \quad 42 \quad 44 \quad 44$$

6. Graph the cumulative frequency function of the sample in Prob. 3.

7. Graph the bar chart, the histogram, and the frequency polygon of the grouped sample in Table 19.

8. Graph the histogram of the following sample of lifetimes [hours] of light bulbs.

Lifetime	Absolute Frequency	Lifetime	Absolute Frequency	Lifetime	Absolute Frequency
950–1050	4	1350–1450	51	1750–1850	20
1050–1150	9	1450–1550	58	1850–1950	9
1150–1250	19	1550–1650	53	1950–2050	3
1250–1350	36	1650–1750	37	2050–2150	1

9. Group the sample in Table 16, using class intervals with midpoints 300, 320, 340, · · · ·. Make up the corresponding frequency table. Graph the histogram and compare it with Fig. 347. Graph the cumulative frequency function.

10. Group the sample in Table 18, using class intervals with midpoints 75, 85, 95, · · · ·. Make up the corresponding frequency table. Graph the histogram and compare it with that in Prob. 7.

19.3 *Sample Mean and Sample Variance*

The frequency function (or equally well the distribution function) characterizes a given sample in detail. From this function we may compute measures for certain properties of the sample, such as the average size of the sample values, the "spread," the "asymmetry," etc. In the present section we shall consider the two most important such quantities, which are called the sample mean and the sample variance.

The *mean value of a sample* $x_1, x_2, \cdots, x_n$ or, briefly, **sample mean,** is denoted by $\bar{x}$ and is defined by the formula

$$(1) \qquad \bar{x} = \frac{1}{n} \sum_{j=1}^{n} x_j = \frac{1}{n}(x_1 + x_2 + \cdots + x_n).$$

It is the sum of all the sample values, divided by the size n of the sample. Obviously, it measures the average size of the sample values, and sometimes the term *average* is used for $\bar{x}$.

The *variance of a sample* $x_1, x_2, \cdots, x_n$ or, briefly, **sample variance,** is denoted by s^2 and is defined by the formula

$$(2) \qquad \begin{aligned} s^2 &= \frac{1}{n-1} \sum_{j=1}^{n}(x_j - \bar{x})^2 \\ &= \frac{1}{n-1}[(x_1 - \bar{x})^2 + (x_2 - \bar{x})^2 + \cdots + (x_n - \bar{x})^2]. \end{aligned}$$

It is the sum of the squares of the deviations of the sample values from the mean $\bar{x}$, divided by $n - 1$. It measures the spread or dispersion of the sample values and is positive, except for the rare case when all the sample values are equal (and are then equal to $\bar{x}$). The positive square root of the sample variance s^2 is called the **standard deviation** of the sample and is denoted by s.

Example 1.   Ten randomly selected nails had the lengths (in.)

$$0.80 \quad 0.81 \quad 0.81 \quad 0.82 \quad 0.81 \quad 0.82 \quad 0.80 \quad 0.82 \quad 0.81 \quad 0.81.$$

From (1) we see that the sample mean is

$$\bar{x} = \tfrac{1}{10}(0.80 + 0.81 + 0.81 + 0.82 + \cdots + 0.81) = 0.811 \text{ [in.]}.$$

Applying (2), we thus obtain the sample variance

$$s^2 = \tfrac{1}{9}[(0.800 - 0.811)^2 + \cdots + (0.810 - 0.811)^2] = 0.000\ 054 \text{ [in.}^2].$$

This calculation becomes simpler if we take equal sample values together. Then

$$\bar{x} = \tfrac{1}{10}(2 \cdot 0.80 + 5 \cdot 0.81 + 3 \cdot 0.82) = 0.811.$$

In the parentheses we have the sum of the three *numerically different* sample values $x_1 = 0.80$, $x_2 = 0.81$, $x_3 = 0.82$, each multiplied by its frequency. Similarly,

$$s^2 = \tfrac{1}{9}[2(0.800 - 0.811)^2 + 5(0.810 - 0.811)^2 + 3(0.820 - 0.811)^2] = 0.000\,054.$$

The example illustrates how we may compute $\bar{x}$ and s^2 by the use of the frequency function $\tilde{f}(x)$ of the sample. If a sample of n values contains precisely m *numerically different* sample values

$$x_1, \qquad x_2, \qquad \cdots, \qquad x_m \qquad\qquad (m \leq n),$$

the corresponding relative frequencies are

$$\tilde{f}(x_1), \qquad \tilde{f}(x_2), \qquad \cdots, \qquad \tilde{f}(x_m).$$

Hence the corresponding frequencies needed in the computation are

$$n\tilde{f}(x_1), \qquad n\tilde{f}(x_2), \qquad \cdots, \qquad n\tilde{f}(x_m),$$

and (1) and (2) take the form

(3)
$$\bar{x} = \frac{1}{n} \sum_{j=1}^{m} x_j n\tilde{f}(x_j)$$

and

(4)
$$s^2 = \frac{1}{n-1} \sum_{j=1}^{m} (x_j - \bar{x})^2 n\tilde{f}(x_j).$$

Note that in (1) and (2) we sum over *all* the sample values, whereas now we sum over the *numerically different* sample values. The absolute frequencies $n\tilde{f}(x_j)$ are integers, while the relative frequencies $\tilde{f}(x_j)$ may be messy, for example, if $n = 23$ or $n = 84$, etc.

Our formulas for s^2 are impractical because $x_j - \bar{x}$ may be small in absolute value compared with the sample values, and this may cause a loss of significant digits (undetected in automatic calculation). We shall derive the formula for s^2 which is used in computations. Inserting

$$(x_j - \bar{x})^2 = x_j^2 - 2x_j\bar{x} + \bar{x}^2$$

into (2) and decomposing into three sums, we have

$$\sum (x_j - \bar{x})^2 = \sum x_j^2 - 2\bar{x} \sum x_j + \sum \bar{x}^2.$$

The last sum equals $n\bar{x}^2$. Replacing $\bar{x}$ according to (1), we obtain

$$-2\bar{x} \sum x_j = -\frac{2}{n} \left(\sum x_j \right)^2 \qquad \text{and} \qquad n\bar{x}^2 = \frac{1}{n} \left(\sum x_j \right)^2.$$

Taking these two expressions together, we finally have

(5)
$$s^2 = \frac{1}{n-1} \left[\sum_{j=1}^{n} x_j^2 - \frac{1}{n} \left(\sum_{j=1}^{n} x_j \right)^2 \right].$$

Similarly, (4) may be transformed to

(6)
$$s^2 = \frac{1}{n-1} \left[\sum_{j=1}^{m} x_j^2 n\tilde{f}(x_j) - \frac{1}{n} \left(\sum_{j=1}^{m} x_j n\tilde{f}(x_j) \right)^2 \right].$$

For instance, in Ex. 1 we obtain from (3) and (6) (cf. Table 20) the values $\bar{x} = 8.11/10 = 0.811$ and

$$s^2 = \frac{1}{9}\left(6.5777 - \frac{8.11^2}{10}\right) = \frac{0.00049}{9} = 0.000\ 054,$$

as before.

Table 20. Computation of Mean and Variance in Ex. 1 by the Use of (3) and (6)

x_j	$10\tilde{f}(x_j)$	$x_j \cdot 10\tilde{f}(x_j)$	$x_j{}^2$	$x_j{}^2 \cdot 10\tilde{f}(x_j)$
0.80	2	1.60	0.6400	1.2800
0.81	5	4.05	0.6561	3.2805
0.82	3	2.46	0.6724	2.0172
Sum		8.11		6.5777

Problems

1. Compute the mean and variance of the sample in Prob. 2, Sec. 19.2.
2. Compute the mean and variance of the sample in Prob. 4, Sec. 19.2.
3. Graph a histogram of the sample 2, 1, 4, 5 and guess $\bar{x}$ and s by inspecting the histogram. Then calculate $\bar{x}$, s^2, and s.
4. Show that $\bar{x}$ lies between the smallest and the largest sample values.
5. **(Range of a sample)** The difference between the largest and the smallest values in a sample is called the *range of the sample*. Find the range of the sample in Ex. 1.
6. **(Percentile, median)** The pth **percentile** of a sample is a number Q_p such that at least $p\%$ of the sample values are smaller than or equal to Q_p and also at least $(100 - p)\%$ of those values are larger than or equal to Q_p. If there is more than one such number (in which case there will be an interval of them), the pth percentile is defined as the average of the numbers (midpoint of the interval). In particular, Q_{50} is called the **middle quartile** or **median** and is denoted by $\bar{x}$. Find $\bar{x}$ for the sample in Table 17.
7. The percentiles Q_{25} and Q_{75} of a sample are called the **lower** and **upper quartiles** of the sample, and $Q_{75} - Q_{25}$, which is a measure for the spread, is called the **interquartile range**. Find Q_{25}, Q_{75}, and $Q_{75} - Q_{25}$ for the sample in Table 17.
8. Same tasks as in Probs. 6 and 7 for the sample in Table 18.
9. **(Mode)** A *mode* of a sample is a sample value that occurs most frequently in the sample. Find the mean, median, and mode of the following sample. Comment.

Value of Stocks Owned	100	1000	100,000
Frequency	100	90	20

10. **(Working origin)** If $x_j = x_j{}^* + c$, where $j = 1, \cdots, n$ and c is any constant, show that

$$\bar{x} = c + \bar{x}^*\qquad \left(\bar{x}^* = \frac{1}{n}\sum_{j=1}^{n} x_j{}^*\right),\qquad s^2 = s^{*2},$$

where s^{*2} is the variance of the $x_j{}^*$'s. (In practice, c is chosen so that the $x_j{}^*$'s are small in absolute value. Geometrically this is a shift of the origin and is called the *method of working origin*.)

11. Apply the method of working origin to the sample in Ex. 1.

12. **(Full coding)**   If $x_j = c_1 x_j^* + c_2$, where $j = 1, \cdots, n$ and c_1 and c_2 are constants, show that

$$\bar{x} = c_1 \bar{x}^* + c_2, \qquad s^2 = c_1^2 s^{*2},$$

where $\bar{x}^*$ and s^{*2} have the same meaning as in Prob. 10. (This is called the *method of full coding*. Obviously it is valuable in nonelectronic calculations, for instance for quick checks.)

13. Apply full coding to the sample in Ex. 1.
14. If a sample is being grouped, its mean will change, in general. Show that the change cannot exceed $l/2$ where l is the length of each class interval.
15. Compute the mean and the variance of the ungrouped sample in Table 18 and the grouped sample in Table 19 and compare the results.

19.4   *Random Experiments, Outcomes, Events*

Statistical experiments or observations yield samples from which we want to draw conclusions about the corresponding population. Before we can do so we must first develop mathematical models of populations, using the theory of mathematical probability. This theory, therefore, supplies an important basis in mathematical statistics, and we shall consider it to the extent needed. In this section we shall introduce some fundamental concepts.

A *random experiment* or *random observation*, briefly **experiment** or **observation**, is a process that has the following properties.

1. It is performed according to a set of rules that determines the performance completely.

2. It can be repeated arbitrarily often.

3. The result of each performance depends on "chance" (that is, on influences which we cannot control) and can therefore not be uniquely predicted.

The result of a single performance of the experiment is called the **outcome** of that trial.

Examples are games of chance such as rolling a die or flipping a coin, technical experiments such as the random selection and inspection of 10 screws from a box containing 100 screws or the determination of the yield of a chemical process under various conditions, and other experiments such as the random selection of 20 persons from a group and the determination of their blood pressures or their opinions about a certain movie.

The set of all possible outcomes of an experiment is called the **sample space** of the experiment and will be denoted by S. Each outcome is called an *element* or *point* of S. A sample space is said to be *finite* or *infinite*, depending on whether it consists of finitely many or infinitely many elements, respectively.

For instance, with the random experiment of rolling a die we may associate the sample space

$$S = \{1, 2, 3, 4, 5, 6\}$$

consisting of six elements corresponding to the six faces the die can turn up.

In connection with industrial production we may draw an item to find out whether it is defective or nondefective. Then S consists of the two elements D (defective) and N (nondefective), which may also be characterized by numbers, for example 0 (defective) and 1 (nondefective). If in this case we distinguish between several types of defects, we have a sample space consisting of more than two points.

The sample space of the experiment of measuring the strength of cotton (cf. Table 18 in Sec. 19.2) is infinite, because the outcome may be any positive number within a certain range.

In most practical problems we are not so much interested in the individual outcomes but in whether an outcome belongs (or does not belong) to a certain set of outcomes. Clearly, each such set A is a subset of the sample space S. It is called an **event.**

Since an outcome is a subset of S, it is an event, but a rather special one, sometimes called an *elementary event.* Similarly, the entire space S is another special event.

Example 1. If we draw 2 gaskets from a set of 5 gaskets (numbered from 1 to 5), the sample space consists of the 10 possible outcomes

$$1, 2 \quad 1, 3 \quad 1, 4 \quad 1, 5 \quad 2, 3 \quad 2, 4 \quad 2, 5 \quad 3, 4 \quad 3, 5 \quad 4, 5,$$

but we may be interested in the number of defective gaskets we get in the draws and thus distinguish among the 3 events

    A: No defective gasket,   *B: 1 defective gasket,*   *C: 2 defective gaskets.*

Assuming that 3 gaskets, say 1, 2, 3, are defective, we see that

    A occurs, if we draw   4, 5

    B occurs, if we draw   1, 4;   1, 5;   2, 4;   2, 5;   3, 4;   or 3, 5

    C occurs, if we draw   1, 2;   1, 3;   or 2, 3.

A sample space S and the events of an experiment can be represented graphically by a **Venn diagram,** as follows. Suppose that the set of points inside the rectangle in Fig. 350 represents S. Then the interior of a closed curve inside the rectangle represents an event which we denote by E. The set of all the elements (outcomes) not in E is called the **complement** *of E in S* and is denoted[2] by E^c.

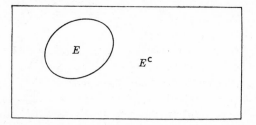

Fig. 350. Venn diagram representing a sample space S and events E and E^c

---

[2] Or by $\bar{E}$, but we shall not use this notation because it is used in set theory for another purpose (to denote the closure of a set).

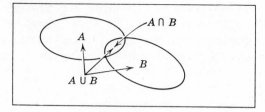

Fig. 351. Venn diagram representing two events A and B in a sample space S and their union $A \cup B$ and intersection $A \cap B$

For example, in the experiment of rolling a die the complement of the event

$$E: \text{ The die turns up an even number}$$

is the event

$$E^c: \text{ The die turns up an odd number.}$$

An event containing no element is called the **impossible event** or *empty event* and is denoted by ϕ.

Let A and B be any two events in an experiment. Then the event consisting of all the elements of the sample space S contained in A or B, or both, is called the **union** of A and B and is denoted by

$$A \cup B.$$

Figure 351 illustrates how to represent these two events by a Venn diagram. If A and B have no element in common, then $A \cap B = \phi$, and A and B are called **mutually exclusive events** or **disjoint events.**

For instance, in Ex. 1, $B \cap C = \phi$ and $B \cup C$: *1 or 2 defective gaskets.*

Example 2. In the random experiment of rolling a die, the events

$$A: \text{ The die turns up a number not smaller than 4}$$

$$B: \text{ The die turns up a number divisible by 3}$$

have the union $A \cup B = \{3, 4, 5, 6\}$ and the intersection $A \cap B = \{6\}$ (cf. Fig. 352).

If all the elements of an event A are also contained in an event B, then A is called a **subevent** of B, and we write

$$A \subset B \qquad \text{or} \qquad B \supset A.$$

Fig. 352. Venn diagram in Example 2

Clearly, if $A \subset B$, then if B occurs, A necessarily occurs. For instance, the event $D = \{4, 6\}$ is a subevent of A in Ex. 2.

Let $A_1, \cdots, A_m$ be several events in a sample space S. Then the event consisting of all elements contained in one or more of these m events is called the **union** of $A_1, \cdots, A_m$ and is denoted by

$$A_1 \cup A_2 \cup \cdots \cup A_m, \qquad \text{or more briefly} \qquad \bigcup_{j=1}^{m} A_j.$$

The event consisting of all elements contained in all those m events is called the **intersection** of $A_1, \cdots, A_m$ and is denoted by

$$A_1 \cap A_2 \cap \cdots \cap A_m, \qquad \text{or more briefly} \qquad \bigcap_{j=1}^{m} A_j.$$

More generally, let $A_1, A_2, \cdots, A_m, \cdots$ be infinitely many elements in S. Then the **union**

$$A_1 \cup A_2 \cup \cdots, \qquad \text{or more briefly} \qquad \bigcup_{j=1}^{\infty} A_j$$

is defined to be the event consisting of all the elements contained in at least one of those events, and the **intersection**

$$A_1 \cap A_2 \cap \cdots, \qquad \text{or more briefly} \qquad \bigcap_{j=1}^{\infty} A_j$$

is defined to be the event consisting of all the elements contained in all those events.

If events $A_1, A_2, \cdots, A_m \cdots$ are such that the occurrence of any of them makes the simultaneous occurrence of any other of them impossible, then we have $A_j \cap A_k = \phi$ for any j and $k \neq j$, and the events are called **mutually exclusive events** or **disjoint events.**

For instance, the events A, B, and C in Ex. 1 are mutually exclusive.

Suppose that we perform a random experiment n times and obtain a sample consisting of n values. Let A and B be events whose relative frequencies in those n trials are $\tilde{f}(A)$ and $\tilde{f}(B)$, respectively. Then the event $A \cup B$ has the relative frequency

(1) $$\tilde{f}(A \cup B) = \tilde{f}(A) + \tilde{f}(B) - \tilde{f}(A \cap B).$$

If A and B are mutually exclusive, then $\tilde{f}(A \cap B) = 0$, and

(2) $$\tilde{f}(A \cup B) = \tilde{f}(A) + \tilde{f}(B).$$

These formulas are rather obvious from the Venn diagram in Fig. 351; the formal proof is left to the reader (Prob. 1).

Problems

1. Prove (1).
2. Graph the sample space of the random experiment of tossing two coins.
3. Define the sample space of the experiment of recording the lifetime of each of three electronic components.

Fig. 353. Relative frequency of the event "birth of a boy". (Data from Graz, Austria, 1962)

4. A box of 20 fountain pens contains 10 nondefective pens, 8 with type A defects, 5 with type B defects, and 3 with both types of defects. Suppose that 1 pen is drawn at random. Graph a Venn diagram of the corresponding sample space S showing the events E_A of getting a type A defective, E_B of getting a type B defective, $E_A \cap E_B$, $E_A \cap E_B^c$, $E_A^c \cap E_B$, E_A^c, $\cap E_B^c$, $E_A \cup E_B$, $E_A^c \cup E_A$, $E_A \cup E_B^c$, and $E_A^c \cup E_B^c$. How many elements (outcomes) are in each event?

5. In the experiment of drilling a hole into a plate and measuring its diameter, find the complement of E: The diameter of the hole is at least 2.9 in. and at most 3.1 in.

19.5 *Probability*

Experience shows that most random experiments exhibit *statistical regularity* or *stability of relative frequencies;* that is, in several long sequences of such an experiment the corresponding relative frequencies of an event are almost equal. Typical examples are shown in Table 21 and Fig. 353. In Fig. 353 the percentage of boys seems to fluctuate less and less as the number of births increases. The percentage of defective items will show a similar behavior if the conditions of production are reasonably constant, and other examples can easily be found.

Since most random experiments exhibit statistical regularity, we may assert that for any event E in such an experiment there is a number $P(E)$ such that the relative frequency of E in a great number of performances of the experiment is approximately equal to $P(E)$.

For this reason we now postulate the existence of a number $P(E)$ which is called the **probability** *of the event E in that random experiment.* Note that this number is not an absolute property of E but refers to a certain sample space S, that is, to a certain random experiment.

Table 21. Coin Tossing

Experiments by	Number of Throws	Number of Heads	Relative Frequency of Heads
BUFFON	4,040	2,048	0.5069
K. PEARSON	12,000	6,019	0.5016
K. PEARSON	24,000	12,012	0.5005

The statement "E has the probability $P(E)$" then means that if we perform the experiment very often, it is practically certain that the relative frequency $\tilde{f}(E)$ is approximately equal to $P(E)$. (Here the term "approximately equal" must be made precise. This we have to postpone until Sec. 19.10.)

The probability thus introduced is the counterpart of the empirical relative frequency. It is therefore natural to require that it should have certain basic properties which the relative frequency has. These properties are suggested by Theorems 1 and 2, Sec. 19.2, and formula (2), Sec. 19.4, and may be formulated as so-called

Axioms of mathematical probability

1. *If E is any event in a sample space S, then*

(1) $$0 \leq P(E) \leq 1.$$

2. *To the entire sample space S there corresponds*

(2) $$P(S) = 1.$$

3. *If A and B are mutually exclusive events* (cf. Sec. 19.4), *then*

(3) $$P(A \cup B) = P(A) + P(B).$$

If the sample space is infinite, we must replace Axiom 3 by

3*. *If $E_1, E_2, \cdots$ are mutually exclusive events, then*

(3*) $$P(E_1 \cup E_2 \cup \cdots) = P(E_1) + P(E_2) + \cdots.$$

From Axiom 3 we obtain by induction the following

Theorem 1 (Addition rule for mutually exclusive events). *If $E_1, \cdots, E_m$ are mutually exclusive events, then*

(4) $$P(E_1 \cup E_2 \cup \cdots \cup E_m) = P(E_1) + P(E_2) + \cdots + P(E_m).$$

The reader may prove the following analogue of (1), Sec. 19.4.

Theorem 2 (Addition rule for arbitrary events). *If A and B are any events in a sample space S, then*

(5) $$P(A \cup B) = P(A) + P(B) - P(A \cap B).$$

Furthermore, an event E and its complement E^c (Sec. 19.4) are mutually exclusive, and $E \cup E^c = S$. Using Axioms 3 and 2, we thus have

$$P(E \cup E^c) = P(E) + P(E^c) = 1.$$

This yields

Theorem 3 (Complementation rule). *The probabilities of an event E and its complement E^c in a sample space S are related by the formula*

(6) $$P(E) = 1 - P(E^c).$$

This formula may be used if the computation of $P(E^c)$ is simpler than that of $P(E)$. An application will be shown in Ex. 2.

How can we assign probabilities to the various events in a sample space S?

If S is finite, consisting of k elements, and the nature of the experiment shows that these k outcomes are equally likely to occur, then we may assign the same probability to each outcome, and because of Axiom 2 this probability must equal $1/k$. In this case the computation of probabilities of events reduces to that of counting the elements that make up the events.

Example 1. We consider a **fair die**, that is, a die of homogeneous material and strictly cubical form, and roll it once. In this experiment, $S = \{1, 2, 3, 4, 5, 6\}$. We thus have $P(1) = 1/6$, $P(2) = 1/6$, $\cdots$, $P(6) = 1/6$. From this and Theorem 1 we see that, for instance, the event

$$A: \text{ An even number turns up}$$

has the probability $P(A) = P(2) + P(4) + P(6) = 1/2$, the event

$$B: \text{ A number greater than 4 turns up}$$

has the probability $P(B) = P(5) + P(6) = 1/3$, etc. Formulas for more complicated cases and further examples will be included in the next section.

Example 2. Five coins are tossed simultaneously. Find the probability of the event A: *At least 1 head turns up*. Since each coin can turn up heads or tails, the sample space consists of $2^5 = 32$ elements. Assuming that the coins are fair, we may assign the same probability $(1/32)$ to each outcome. Then the event A^c (*no heads turn up*) consists of only 1 outcome. Hence $P(A^c) = 1/32$, and the answer is $P(A) = 1 - P(A^c) = 31/32$.

If the nature of the experiment does not show that the finitely many outcomes are equally likely to occur, or if the sample space is not finite, we must assign probabilities by using relative frequencies observed in long sequences of trials in such a fashion that the axioms of probability are satisfied.

In this way we only obtain approximate values, but this does not matter. It resembles the situation in classical physics, where we postulate that each body has a certain mass but cannot determine that mass exactly. This is no handicap for developing the theory.

If we are in doubt whether we assigned the probabilities in the right fashion, we may apply a statistical test which we shall consider later (in Sec. 19.18).

Often it is required to find the probability of an event B if it is known that an event A has occurred. This probability is called the **conditional probability** *of B given A* and is denoted by $P(B \mid A)$. In this case A serves as a new (reduced) sample space, and that probability is the fraction of $P(A)$ which corresponds to $A \cap B$. Thus

$$(7) \qquad\qquad P(B \mid A) = \frac{P(A \cap B)}{P(A)} \qquad\qquad [P(A) \neq 0].$$

Similarly, the *conditional probability of A given B* is

$$(8) \qquad\qquad P(A \mid B) = \frac{P(A \cap B)}{P(B)} \qquad\qquad [P(B) \neq 0].$$

Solving (7) and (8) for $P(A \cap B)$, we obtain

Theorem 4 (Multiplication rule). *If A and B are events in a sample space S and* $P(A) \neq 0$, $P(B) \neq 0$, *then*

$$(9) \qquad P(A \cap B) = P(A)P(B \mid A) = P(B)P(A \mid B).$$

If the events A and B are such that

$$(10) \qquad P(A \cap B) = P(A)P(B),$$

they are called **independent events.** Assuming $P(A) \neq 0$, $P(B) \neq 0$, we see from (7)–(9) that in this case

$$P(A \mid B) = P(A), \qquad P(B \mid A) = P(B),$$

which means that the probability of A does not depend on the occurrence or nonoccurrence of B, and conversely. This justifies the terminology.

Similarly, m events $A_1, \cdots, A_m$ are said to be **independent** if for any k events $A_{j_1}, A_{j_2}, \cdots, A_{j_k}$ (where $1 \leqq j_1 < j_2 < \cdots < j_k \leqq m$ and $k = 2, 3, \cdots, m$)

$$(11) \qquad P(A_{j_1} \cap A_{j_2} \cap \cdots \cap A_{j_k}) = P(A_{j_1})P(A_{j_2}) \cdots P(A_{j_k}).$$

To prepare for an illustrative example, we first note that there are two ways of drawing objects for obtaining a sample from a given set of objects, briefly referred to as **sampling from a population,** at follows.

1. Sampling with replacement means that the object which was drawn at random is placed back to the given set and the set is mixed thoroughly. Then we draw the next object at random.

2. Sampling without replacement means that the object which was drawn is put aside.

Example 3 (Sampling with and without replacement). A box contains 10 screws, 3 of which are defective. Twò screws are drawn at random. Find the probability that none of the two screws is defective. We consider the events

> *A: First drawn screw nondefective*
>
> *B: Second drawn screw nondefective.*

Clearly, $P(A) = \frac{7}{10}$ because 7 of the 10 screws are nondefective and we sample at random, so that each screw has the same probability $(\frac{1}{10})$ of being picked. If we sample with replacement, the situation before the second drawing is the same as at the beginning, and $P(B) = \frac{7}{10}$. The events are independent, and the answer is

$$P(A \cap B) = P(A)P(B) = 0.7 \cdot 0.7 = 0.49 = 49\%.$$

If we sample without replacement, then $P(A) = \frac{7}{10}$, as before. If A has occurred, then there are 9 screws left in the box, 3 of which are defective. Thus $P(B \mid A) = \frac{6}{9} = \frac{2}{3}$, and Theorem 4 yields the answer

$$P(A \cap B) = \frac{7}{10} \cdot \frac{2}{3} \approx 47\%.$$

Problems

1. What is the probability of obtaining at least 1 head in tossing 5 fair coins?
2. In throwing three fair dice, find the probability of the event E that at least two of the three numbers obtained are different.

3. Three screws are drawn at random from a lot of 100 screws, 10 of which are defective. Find the probability of the event that all 3 screws drawn are nondefective, assuming that we draw (*a*) with replacement, (*b*) without replacement.

4. Three urns contain 5 chips each, numbered from 1 to 5, and 1 chip is drawn from each urn. Find the probability of the event *E* that the sum of the numbers on the drawn chips is greater than 3.

5. A box of 100 gaskets contains 10 gaskets with type *A* defects, 5 gaskets with type *B* defects, and 2 gaskets with both types of defects. Suppose that a gasket drawn is known to have a type *A* defect. What is the probability that it also has a type *B* defect?

6. Find the probability that in throwing 2 fair dice the sum of the faces exceeds 9, given that one of the faces is a 5.

7. Prove Theorem 1.

8. Prove Theorem 2.

9. Extending Theorem 4, show that

$$P(A \cap B \cap C) = P(A)P(B \mid A)P(C \mid A \cap B).$$

10. Show that if *B* is a subset of *A*, then $P(B) \leq P(A)$.

19.6 Permutations and Combinations

From the last section we know that in the case of a finite sample space *S* consisting of *k* equally likely outcomes, each outcome has probability $1/k$, and the probability of an event *A* is obtained by counting the number of outcomes of which *A* consists. Clearly, if that number is *m*, then $P(A) = m/k$. For the counting of outcomes the subsequent formulas will often be useful.

Suppose that there are given *n* things (*elements* or *objects*). We may lay them out in a row *in any order*. Each such arrangement is called a **permutation** of those things.

Theorem 1. *The number of permutations of n different things taken all at a time is*

(1) $n! = 1 \cdot 2 \cdot 3 \cdots n$ (read "*n factorial*").

In fact, there are *n* possibilities for filling the first place in the row; then $n - 1$ objects are still available for filling the second place, etc. In a similar fashion we obtain

Theorem 2. *If n given things can be divided into c classes such that things belonging to the same class are alike while things belonging to different classes are different, then the number of permutations of these things taken all at a time is*

(2) $$\frac{n!}{n_1! \, n_2! \cdots n_c!} \qquad (n_1 + n_2 + \cdots + n_c = n)$$

where n_j is the number of things in the jth class.

A **permutation of n things taken k at a time** is a permutation containing only *k* of the *n* given things. Two such permutations consisting of the same *k* elements, in a different order, are different, by definition. For example, there are 6 different permutations of the 3 letters *a*, *b*, *c*, taken 2 letters at a time, namely *ab*, *ac*, *bc*, *ba*, *ca*, *cb*.

A **permutation of *n* things taken *k* at a time with repetitions** is an arrangement obtained by putting any given thing in the first position, any given thing, including a repetition of the one just used, in the second, and continuing until *k* positions are filled. For example, there are $3^2 = 9$ different such permutations of *a*, *b*, *c*, taken 2 letters at a time, namely, the preceding 6 permutations and *aa*, *bb*, *cc*. The reader may prove the following theorem (cf. Prob. 7).

Theorem 3. *The number of different permutations of n different things taken k at a time without repetitions is*

(3a)
$$n(n-1)(n-2)\cdots(n-k+1) = \frac{n!}{(n-k)!}$$

and with repetitions is

(3b)
$$n^k.$$

In the case of a permutation not only are the things themselves important, but so also is their order. A **combination** of given things is any selection of one or more of the things *without regard to order*. There are two types of combinations, as follows.

The number of **combinations of *n* different things, *k* at a time, without repetitions** is the number of sets that can be made up from the *n* things, each set containing *k* different things and no two sets containing exactly the same *k* things.

The number of **combinations of *n* different things, *k* at a time, with repetitions** is the number of sets that can be made up of *k* things chosen from the given *n*, each being used as often as desired.

For example, there are 3 combinations of the 3 letters *a*, *b*, *c*, taken 2 letters at a time, without repetitions, namely, *ab*, *ac*, *bc*, and 6 such combinations with repetitions, namely, *ab*, *ac*, *bc*, *aa*, *bb*, *cc*.

Theorem 4. *The number of different combinations of n different things, k at a time, without repetitions, is*

(4a)
$$\binom{n}{k} = \frac{n!}{k!\,(n-k)!} = \frac{n(n-1)\cdots(n-k+1)}{1\cdot 2\cdots k},$$

and the number of those combinations with repetitions is

(4b)
$$\binom{n+k-1}{k}.$$

The statement involving (4a) follows from the first part of Theorem 3 by noting that there are *k*! permutations of *k* things from the given *n* things which differ by the order of the elements (cf. Theorem 1), but there is only a single combination of those *k* things of the type characterized in the first statement of Theorem 4. The last statement of Theorem 4 can be proved by induction (cf. Prob. 8).

Example 1. If there are 10 different screws in a box which are needed in a certain order for assembling a certain product, and these screws are drawn at random from the box, the probability *P* of picking them in the desired order is very small, namely (cf. Theorem 1),

$$P = 1/10! = 1/3628\,800 \approx 0.000\,03\%.$$

If the box contains 6 alike right-handed screws and 4 alike left-handed screws and the 6 right-handed ones are needed first and the 4 left-handed ones are needed afterwards, the probability P that random drawing yields the screws in the desired order is (cf. Theorem 2)

$$P = 6!\,4!/10! = 1/210 \approx 0.5\%.$$

Example 2. In a coded telegram the letters are arranged in groups of 5 letters, called *words*. From (3b) we see that there are

$$26^5 = 11\,881\,376$$

different such words. From (3a) it follows that there are

$$26!/(26 - 5)! = 26 \cdot 25 \cdot 24 \cdot 23 \cdot 22 = 7\,893\,600$$

different such words which contain each letter at most once.

Example 3. The number of samples of 5 light bulbs which can be selected from a lot of 500 bulbs is [cf. (4a)]

$$\binom{500}{5} = \frac{500!}{5!\,495!} = \frac{500 \cdot 499 \cdot 498 \cdot 497 \cdot 496}{1 \cdot 2 \cdot 3 \cdot 4 \cdot 5} = 255\,244\,687\,600.$$

Let us include some remarks about the **factorial function.** We define

$$(5) \qquad\qquad\qquad 0! = 1$$

and may compute further values by the relation

$$(6) \qquad\qquad\qquad (n + 1)! = (n + 1)n!.$$

For large n the function is very large (cf. Table A4 in Appendix 4). A convenient approximation for large n is the **Stirling formula**[3]

$$(7) \qquad\qquad n! \sim \sqrt{2\pi n}\left(\frac{n}{e}\right)^n \qquad\qquad (e = 2.718 \cdots),$$

where $\sim$ means that the ratio of the two sides of (7) approaches 1 as n approaches infinity.

The **binomial coefficients** are defined by the formula

$$(8) \qquad\qquad \binom{a}{k} = \frac{a(a - 1)(a - 2) \cdots (a - k + 1)}{k!} \qquad (k \geqq 0, \text{integral}).$$

The numerator has k factors. Furthermore, we define

$$(9) \qquad\qquad \binom{a}{0} = 1, \quad \text{in particular} \quad \binom{0}{0} = 1.$$

For integral $a = n$ we obtain from (8)

$$(10) \qquad\qquad \binom{n}{k} = \binom{n}{n - k} \qquad\qquad (n \geqq 0, 0 \leqq k \leqq n).$$

Binomial coefficients may be computed recursively, because

$$(11) \qquad\qquad \binom{a}{k} + \binom{a}{k + 1} = \binom{a + 1}{k + 1} \qquad\qquad (k \geqq 0, \text{integral}).$$

[3] JAMES STIRLING (1692–1770), English mathematician.

Formula (8) also yields

(12) $$\binom{-m}{k} = (-1)^k \binom{m + k - 1}{k} \qquad (k \geqq 0, \text{integral}, m > 0).$$

There are numerous further relations; we mention

(13) $$\sum_{s=0}^{n-1} \binom{k + s}{k} = \binom{n + k}{k + 1} \qquad (k \geqq 0, n \geqq 1, \text{both integral})$$

and

(14) $$\sum_{k=0}^{r} \binom{p}{k} \binom{q}{r - k} = \binom{p + q}{r}.$$

Problems

1. List all permutations of four digits 1, 2, 3, 4, taken all at a time.
2. List all permutations of five letters a, e, i, o, u, taken three at a time.
3. In how many ways can a committee of three be chosen from ten persons?
4. How many different license plates showing 5 symbols, namely, 2 letters followed by 3 digits, could be made?
5. Find the number of samples of 3 objects which can be drawn from a lot of 100 objects.
6. An urn contains 2 black, 3 white, and 4 red balls. We draw 1 ball at random and put it aside. Then we draw the next ball, and so on. Find the probability of drawing at first the 2 black balls, then the 3 white ones, and finally the red ones.
7. Prove Theorem 3.
8. Prove the last statement of Theorem 4. *Hint.* Use (13).
9. Using (7), compute approximate values of 4! and 8! and determine the absolute and relative errors.
10. Derive (11) from (8).

19.7   *Random Variables. Discrete and Continuous Distributions*

If we roll two dice, we know that the sum X of the two numbers which turn up must be an integer between 2 and 12, but we cannot predict which value of X will occur in the next trial, and we may say that X depends on "chance." Similarly, if we want to draw 5 bolts from a lot of bolts and measure their diameters, we cannot predict how many will be defective, that is, will not meet given requirements; hence X = *number of defectives* is again a function which depends on "chance." The lifetime X of a light bulb to be drawn at random from a lot of bulbs also depends on "chance," and so does the content X of a bottle of lemonade filled by a filling machine and selected at random from a given lot.

Roughly speaking, a **random variable** X (also called **stochastic variable** or **variate**) is a function whose values are real numbers and depend on "chance"; more precisely, it is a function X which has the following properties:

1. *X is defined on the sample space S of the experiment, and its values are real numbers.*

2. *Let a be any real number, and let I be any interval. Then the set of all outcomes in S for which X = a has a well-defined probability, and the same is true for the set of all outcomes in S for which the values of X are in I. These probabilities are in agreement with the axioms in Sec. 19.5.*

Although this definition is quite general and includes many functions, we shall see that in practice the number of important types of random variables and corresponding "probability distributions" is rather small.

If we perform a random experiment and the event corresponding to a number a occurs, then we say that in this trial the random variable X corresponding to that experiment has **assumed** the value a. We also say that we have **observed** the value $X = a$. Instead of "the event corresponding to a number a," we say, more briefly, "the event $X = a$." The corresponding probability is denoted by $P(X = a)$. Similarly, the probability of the event

$$X \text{ assumes any value in the interval } a < X < b$$

is denoted by $P(a < X < b)$. The probability of the event

$$X \leq c \ (X \text{ assumes any value smaller than } c \text{ or equal to } c)$$

is denoted by $P(X \leq c)$, and the probability of the event

$$X > c \ (X \text{ assumes any value greater than } c)$$

is denoted by $P(X > c)$.

The last two events are mutually exclusive. From Axiom 3 in Sec. 19.5 we thus obtain

$$P(X \leq c) + P(X > c) = P(-\infty < X < \infty).$$

From Axiom 2 we see that the right side equals 1, because $-\infty < X < \infty$ corresponds to the whole sample space. This yields the important formula

(1) $$P(X > c) = 1 - P(X \leq c)$$ (*c* arbitrary).

For example, if X is the number that turns up in rolling a fair die, $P(X = 1) = \frac{1}{6}$, $P(X = 2) = \frac{1}{6}$, etc., $P(1 < X < 2) = 0$, $P(1 \leq X \leq 2) = \frac{1}{3}$, $P(0 \leq X \leq 3.2) = \frac{1}{2}$, $P(X > 4) = \frac{1}{3}$, $P(X \leq 0.5) = 0$, etc.

In most practical cases the random variables are either *discrete* or *continuous*. These two important notions will now be considered one after the other.

A random variable X and the corresponding distribution are said to be **discrete,** if X has the following properties.

1. The number of values for which X has a probability different from 0 is finite or at most countably infinite.

2. Each finite interval on the real line contains at most finitely many of those values. If an interval $a < X \leq b$ does not contain such a value, then $P(a < X \leq b) = 0$.

Let

$$x_1, \qquad x_2, \qquad x_3, \qquad \cdots$$

be the values for which X has a positive probability, and let

$$p_1, \qquad p_2, \qquad p_3, \qquad \cdots$$

be the corresponding probabilities. Then $P(X = x_1) = p_1$, etc., We now introduce the function

$$
(2) \qquad\qquad f(x) =
\begin{cases}
p_j & \text{when } x = x_j \quad (j = 1, 2, \cdots) \\
0 & \text{otherwise.}
\end{cases}
$$

$f(x)$ is called the **probability function** of X.

Since $P(S) = 1$ (cf. Axiom 2 in Sec. 19.5), we must have

$$(3) \qquad\qquad \sum_{j=1}^{\infty} f(x_j) = 1.$$

If we know the probability function of a discrete random variable X, then we may readily compute the probability $P(a < X \leq b)$ corresponding to any interval $a < X \leq b$. In fact,

$$(4) \qquad\qquad P(a < X \leq b) = \sum_{a < x_j \leq b} f(x_j) = \sum_{a < x_j \leq b} p_j.$$

This is the sum of all the probabilities $f(x_j) = p_j$ for which the x_j lie in that interval. For a closed, open, or infinite interval the situation is quite similar. We express this by saying that the probability function $f(x)$ determines the **probability distribution** or, briefly, the **distribution** of the random variable X in a unique fashion.

If X is any random variable, not necessarily discrete, then for any real number x there exists the probability $P(X \leq x)$ corresponding to

$$X \leq x \text{ (X assumes any value smaller than } x \text{ or equal to } x).$$

Clearly, $P(X \leq x)$ depends on the choice of x; it is a function of x, which is called the **distribution function**[4] of X and is denoted by $F(x)$. Thus

$$(5) \qquad\qquad F(x) = P(X \leq x).$$

Since for any a and $b > a$ we have

$$P(a < X \leq b) = P(X \leq b) - P(X \leq a),$$

it follows that

$$(6) \qquad\qquad P(a < X \leq b) = F(b) - F(a).$$

This shows that the distribution function determines the distribution of X uniquely and that it can be used for computing probabilities.

---

[4] We note that some authors call $F(x)$ the *cumulative distribution function*, in particular those who use the term *distribution function* for the probability function $f(x)$.

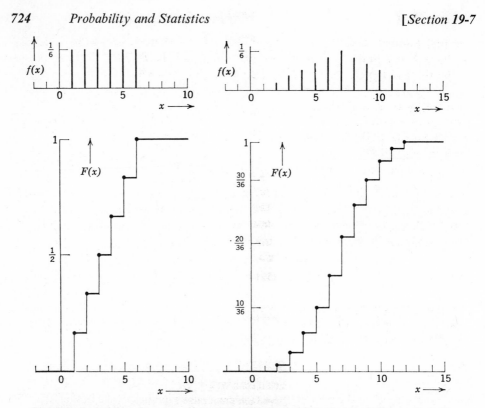

Fig. 354. Probability function $f(x)$ and distribution function $F(x)$ of the random variable $X = $ *number obtained in rolling a fair die once*

Fig. 355. Probability function $f(x)$ and distribution function $F(x)$ of the random variable $X = $ *sum of the two numbers obtained in rolling two fair dice once*

Suppose that X is a discrete variable. Then we may represent the distribution function $F(x)$ in terms of the probability function $f(x)$. In fact, by inserting (4) (with $a = -\infty$ and $b = x$) we have

$$(7) \qquad F(x) = \sum_{x_j \leq x} f(x_j)$$

where the right-hand side is the sum of all those $f(x_j)$ for which $x_j \leq x$. Simple examples are shown in Figs. 354 and 355. In both figures $f(x)$ is represented by a bar chart. In Fig. 354 we have $f(x) = \frac{1}{6}$ when $x = 1, 2, \cdots, 6$ and 0 otherwise. In Fig. 355 the function $f(x)$ has the values

x	2	3	4	5	6	7	8	9	10	11	12
$f(x)$	$\frac{1}{36}$	$\frac{2}{36}$	$\frac{3}{36}$	$\frac{4}{36}$	$\frac{5}{36}$	$\frac{6}{36}$	$\frac{5}{36}$	$\frac{4}{36}$	$\frac{3}{36}$	$\frac{2}{36}$	$\frac{1}{36}$

because there are $6 \cdot 6 = 36$ equally likely outcomes, each of which thus has probability $\frac{1}{36}$; we have $X = 2$ for just one outcome, namely, (1, 1) (where the first number refers to the first die and the second number to the other die); we have $X = 3$ in the case of the two outcomes $(1, 2)$ and $(2, 1)$; $X = 4$ in the case of the three outcomes $(1, 3)$, $(2, 2)$, and $(3, 1)$, etc.

The values $x_1, x_2, x_3, \cdots$ for which a discrete random variable X has positive probability (and only these values) are called the **possible values** of X. In each interval that contains no possible value the distribution function $F(x)$ is constant. Hence $F(x)$ is a **step function** (piecewise constant function) which has an upward jump of magnitude $p_j = P(X = x_j)$ at $x = x_j$ and is constant between two subsequent possible values. Figs. 354 and 355 are illustrative examples.

We shall now define and consider continuous random variables. A random variable X and the corresponding distribution are said to be *of continuous type* or, briefly, **continuous** if the corresponding distribution function $F(x) = P(X \leq x)$ can be represented by an integral in the form[5]

(8)
$$F(x) = \int_{-\infty}^{x} f(v) \, dv$$

where the integrand is continuous, possibly except for at most finitely many values of v. The integrand f is called the *probability density* or, briefly, the **density** of the distribution. Differentiating (8) we see that

$$F'(x) = f(x)$$

for every x at which $f(x)$ is continuous. In this sense *the density is the derivative of the distribution function.*

From (8) and Axiom 2, Sec. 19.5, we also have

(9)
$$\int_{-\infty}^{\infty} f(v) \, dv = 1.$$

Furthermore, from (6) and (8) we obtain the formula

(10)
$$P(a < X \leq b) = F(b) - F(a) = \int_{a}^{b} f(v) \, dv.$$

Hence this probability equals the area under the curve of the density $f(x)$ between $x = a$ and $x = b$, as shown in Fig. 356 on the next page.

Obviously, for any fixed a and b $(> a)$ the probabilities corresponding to the intervals $a < X \leq b$, $a < X < b$, $a \leq X < b$, and $a \leq X \leq b$ are all the same. This is different from the situation in the case of a discrete distribution.

Since probabilities are non-negative and (10) holds for every interval, we must have $f(x) \geq 0$ for all x.

Examples of continuous distributions will be considered later.

---

[5] $F(x)$ is continuous, but continuity of $F(x)$ does not imply the existence of a representation of the form (8). Since continuous distribution functions that cannot be represented in the form (8) are rare in practice, the widely accepted terms "continuous random variable" and "continuous distribution" should not be confusing.

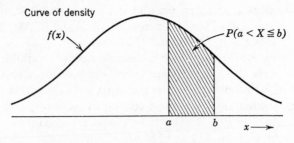

Fig. 356. Example illustrating formula (10)

Problems

1. Graph the probability function $f(x) = x^2/30$ $(x = 1, 2, 3, 4)$ and the distribution function.
2. Graph f and F when $f(0) = f(3) = \frac{1}{8}, f(1) = f(2) = \frac{3}{8}$.
3. Graph the density $f(x) = \frac{1}{2}$ $(1 < x < 3)$ and the distribution function.
4. Let $f(x) = ce^{-x}$ when $x > 0, f(x) = 0$ when $x < 0$. Find c. Graph f and F.
5. Let $f(x) = kx^2$ when $0 \leq x \leq 1$ and 0 otherwise. Find k. Find the number c such that $P(X \leq c) = 72.9\%$.
6. Let $f(x) = 6x(1 - x)$ when $0 \leq x \leq 1$ and 0 otherwise. Find $P(0 < X < 0.75)$.
7. Find and graph the probability function $f(x)$ of the random variable $X = $ *sum of the* 3 *numbers obtained in rollings* 3 *fair dice.*
8. A box contains 4 right-handed and 6 left-handed screws. Two screws are drawn at random without replacement. Let X be the number of left-handed screws drawn. Find the probabilities $P(X = 0)$, $P(X = 1)$, $P(X = 2)$, $P(1 < X < 2)$, $P(X \leq 1)$, $P(X \geq 1)$, $P(X > 1)$, and $P(0.5 < X < 10)$.
9. Show that the random variable $X = $ *number of times a fair coin is tossed until a head appears* has the probability function $f(x) = 2^{-x}$ $(x = 1, 2, \cdots)$. Show that $f(x)$ satisfies (3).
10. Show that $b < c$ implies $P(X \leq b) \leq P(X \leq c)$.

19.8 *Mean and Variance of a Distribution*

The *mean value* or **mean** of a distribution is denoted by μ and is defined by

$$(a) \quad \mu = \sum_{j} x_j f(x_j) \qquad \text{(discrete distribution)},$$

(1)

$$(b) \quad \mu = \int_{-\infty}^{\infty} x f(x) \, dx \qquad \text{(continuous distribution)}.$$

In (1a) the function $f(x)$ is the probability function of the random variable X considered, and we sum over all possible values (cf. Sec. 19.7) In (1b) the function $f(x)$ is the density of X. The mean is also known as the *mathematical expectation of X* and is sometimes denoted by $E(X)$. By definition it is assumed that the series in (1a) converges absolutely and the integral of $|x| f(x)$ from $-\infty$ to ∞ exists. If this does not hold, we say that the distribution does not have a mean; this case will rarely occur in engineering applications.

A distribution is said to be **symmetric** with respect to a number $x = c$ if for every real x,

$$(2) \qquad\qquad f(c + x) = f(c - x).$$

The reader may prove (cf. Prob. 1)

Theorem 1. *If a distribution is symmetric with respect to $x = c$ and has a mean μ, then $\mu = c$.*

The **variance** of a distribution is denoted by σ^2 and is defined by the formula

$$(a) \quad \sigma^2 = \sum_j (x_j - \mu)^2 f(x_j) \qquad \text{(discrete distribution)},$$

(3)

$$(b) \quad \sigma^2 = \int_{-\infty}^{\infty} (x - \mu)^2 f(x)\, dx \qquad \text{(continuous distribution)},$$

where, by definition, it is assumed that the series in (3a) converges absolutely and the integral in (3b) exists.

In the case of a discrete distribution with $f(x) = 1$ at a point and $f = 0$ otherwise, we have $\sigma^2 = 0$. This case is of no practical interest. In any other case

$$(4) \qquad\qquad\qquad \sigma^2 > 0.$$

The positive square root of the variance is called the **standard deviation** and is denoted by σ.

Roughly speaking, the variance is a measure of the spread or dispersion of the values which the corresponding random variable X can assume.

Example 1. The random variable

$$X = \textit{number of heads in a single toss of a fair coin}$$

has the possible values $X = 0$ and $X = 1$ with probabilities $P(X = 0) = 1/2$ and $P(X = 1) = 1/2$. From (1a) we thus obtain the mean value $\mu = 0 \cdot \frac{1}{2} + 1 \cdot \frac{1}{2} = \frac{1}{2}$, and (3$a$) yields

$$\sigma^2 = (0 - \tfrac{1}{2})^2 \cdot \tfrac{1}{2} + (1 - \tfrac{1}{2})^2 \cdot \tfrac{1}{2} = \tfrac{1}{4}.$$

Example 2 (Uniform distribution). The distribution with the density

$$f(x) = \frac{1}{b - a} \qquad \text{when} \qquad a < x < b$$

and $f = 0$ otherwise is called the *uniform* or *rectangular distribution* on the interval $a < x < b$. From Theorem 1 or from (1b) we find that $\mu = (a + b)/2$, and (3b) yields the variance

$$\sigma^2 = \int_a^b \left(x - \frac{a + b}{2} \right)^2 \frac{1}{b - a}\, dx = \frac{(b - a)^2}{12}.$$

Figure 357 on the next page shows special cases illustrating that σ^2 measures the spread.

Fig. 357. Uniform distributions having the same mean (0.5) but different variances σ^2

Theorem 2. *If a random variable X has mean μ and variance σ^2, the random variable $X^* = c_1 X + c_2$ $(c_1 \neq 0)$ has the mean*

$$(5) \qquad\qquad \mu^* = c_1 \mu + c_2$$

and the variance

$$(6) \qquad\qquad \sigma^{*2} = c_1^2 \sigma^2.$$

Proof. We prove (5) in the continuous case. For corresponding x and $x^* = c_1 x + c_2$ the densities $f(x)$ of X and $f^*(x^*)$ of X^* satisfy the relation $f^*(x^*) = f(x)/c_1$, because to a small interval of length Δx on the X-axis there corresponds the probability $f(x)\,\Delta x$ (approximately) and this must equal $f^*(x^*)\,\Delta x^*$ where $\Delta x^* = c_1\,\Delta x$ is the length of the corresponding interval on the X^*-axis. Since $dx^*/dx = c_1$, $dx^* = c_1\,dx$, we thus have $f^*(x^*)\,dx^* = f(x)\,dx$. Hence

$$\mu^* = \int_{-\infty}^{\infty} x^* f^*(x^*)\,dx^* = \int_{-\infty}^{\infty} (c_1 x + c_2) f(x)\,dx = c_1 \int_{-\infty}^{\infty} x f(x)\,dx + c_2 \int_{-\infty}^{\infty} f(x)\,dx.$$

The last integral equals 1, cf. (9), Sec. 19.7, and formula (5) is proved. Since

$$x^* - \mu^* = (c_1 x + c_2) - (c_1 \mu + c_2) = c_1 x - c_1 \mu,$$

the definition of the variance yields

$$\sigma^{*2} = \int_{-\infty}^{\infty} (x^* - \mu^*)^2 f^*(x^*)\,dx^* = \int_{-\infty}^{\infty} (c_1 x - c_1 \mu)^2 f(x)\,dx = c_1^2 \sigma^2.$$

The proof of Theorem 2 for a discrete distribution is quite similar.

From (5) and (6) we readily obtain

Theorem 3. *If a random variable X has mean μ and variance σ^2, then the corresponding variable $Z = (X - \mu)/\sigma$ has the mean 0 and the variance 1.*

Z is called the **standardized variable** corresponding to X.

If X is any random variable and $g(X)$ is any continuous function defined for all real X, then the number

(7)

$$(a) \quad E(g(X)) = \sum_j g(x_j) f(x_j) \qquad (X \text{ discrete})$$

$$(b) \quad E(g(X)) = \int_{-\infty}^{\infty} g(x) f(x)\, dx \qquad (X \text{ continuous})$$

is called the **mathematical expectation** of $g(X)$. Here f is the probability function or the density, respectively.

Taking $g(X) = X^k$ $(k = 1, 2, \cdots)$ in (7), we obtain

(8)
$$E(X^k) = \sum_j x_j^k f(x_j) \quad \text{and} \quad E(X^k) = \int_{-\infty}^{\infty} x^k f(x)\, dx,$$

respectively. $E(X^k)$ is called the **kth moment** of X. Taking $g(X) = (X - \mu)^k$ in (7), we have

(9)
$$E([X - \mu]^k) = \sum_j (x_j - \mu)^k f(x_j) \quad \text{and} \quad \int_{-\infty}^{\infty} (x - \mu)^k f(x)\, dx,$$

respectively. This expression is called the **kth central moment** of X. The reader may show that

(10)
$$E(1) = 1$$

(11)
$$\mu = E(X)$$

(12)
$$\sigma^2 = E([X - \mu]^2).$$

Problems

1. Prove Theorem 1.
2. Prove (5) and (6) in the discrete case.
3. Derive Theorem 3 from (5) and (6).
4. Prove (10) and (11).
5. Prove (12).
6. Show that $E(X - \mu) = 0$.
7. Show that $\sigma^2 = E(X^2) - \mu^2$.
8. Show that $\sigma^2 = E(X[X - 1]) + \mu - \mu^2$.

9. Show that $E(ag(X) + bh(X)) = aE(g(X)) + bE(h(X))$; [$a$ and b are constants].
10. $E([X - c]^2)$ is called the *second moment of X with respect to c*. If it exists, show that it is minimum when $c = \mu$.
11. Find the moments of the uniform distribution on the interval $0 \leqq x \leqq 1$.

12. **(Skewness)** The number

$$\gamma = \frac{1}{\sigma^3} E([X - \mu]^3)$$

   is called the *skewness* of X. Justify this term by showing that if X is symmetric with respect to μ and the third central moment exists, this moment is zero.
13. Find the skewness of the distribution with density $f(x) = xe^{-x}$ when $x > 0$ and $f(x) = 0$ otherwise. Graph $f(x)$.
14. Find the skewness of the distribution with density $f(x) = 2(1 - x)$ when $0 < x < 1$, $f(x) = 0$ otherwise.

15. (**Moment generating function**) The so-called *moment generating function* of a discrete or continuous random variable X is given by the formulas

$$G(t) = E(e^{tX}) = \sum_j e^{tx_j} f(x_j) \quad \text{and} \quad G(t) = \int_{-\infty}^{\infty} e^{tx} f(x) \, dx,$$

respectively. Assuming that differentiation under the summation sign and the integral sign is permissible, show that $E(X^k) = G^{(k)}(0)$, in particular $\mu = G'(0)$, where $G^{(k)}(t)$ is the kth derivative of G with respect to t.

19.9 *Binomial, Poisson, and Hypergeometric Distributions*

We shall now consider special discrete distributions which are particularly important in statistics. We start with the binomial distribution, which is obtained if we are interested in the number of times an event A occurs in n independent performances of an experiment, assuming that A has probability $P(A) = p$ in a single trial. Then $q = 1 - p$ is the probability that in a single trial the event A does not occur. We assume that the experiment is performed n times and consider the random variable

$$X = number \ of \ times \ A \ occurs.$$

Then X can assume the values $0, 1, \cdots, n$, and we want to determine the corresponding probabilities. For this purpose we consider any of these values, say, $X = x$, which means that in x of the n trials A occurs and in $n - x$ trials it does not occur. This may look as follows:

(1)
$$\underbrace{AA \cdots A}_{x \text{ times}} \underbrace{BB \cdots B}_{n-x \text{ times}}.$$

Here $B = A^c$, that is, A does not occur. We assume that the trials are *independent*, that is, do not influence each other. Then, since $P(A) = p$ and $P(B) = q$, we see that to (1) there corresponds the probability

$$\underbrace{pp \cdots p}_{x \text{ times}} \underbrace{qq \cdots q}_{n-x \text{ times}} = p^x q^{n-x}.$$

Clearly, (1) is merely *one* order of arranging x A's and $n - x$ B's, and the probability $P(X = x)$ thus equals $p^x q^{n-x}$ times the number of different arrangements of x A's and $n - x$ B's, as follows from Theorem 1 in Sec. 19.5. We may number the n trials from 1 to n and pick x of these numbers corresponding to those trials in which A happens. Since the order in which we pick the x numbers does not matter, we see from (4a) in Sec. 19.6 that we can pick those x numbers from the n numbers in $\binom{n}{x}$ different ways. Hence the probability $P(X = x)$ corresponding to $X = x$ equals

(2)
$$f(x) = \binom{n}{x} p^x q^{n-x} \qquad (x = 0, 1, \cdots, n)$$

$p = 0.1$　　　　$p = 0.2$　　　　$p = 0.5$　　　　$p = 0.8$　　　　$p = 0.9$

Fig. 358. Probability function (2) of the binomial distribution for $n = 5$ and various values of p

and $f(x) = 0$ for any other value of x. This is the probability that in n independent trials an event A occurs precisely x times where p is the probability of A in a single trial and $q = 1 - p$. The distribution determined by the probability function (2) is called the **binomial distribution** or *Bernoulli distribution*. The occurrence of A is called *success*, and the nonoccurrence is called *failure*. p is called the *probability of success in a single trial*. Figure 358 shows illustrative examples of $f(x)$.

The binomial distribution has the mean (cf. Prob. 11)

$$(3) \qquad\qquad \mu = np$$

and the variance (cf. Prob. 12)

$$(4) \qquad\qquad \sigma^2 = npq.$$

Note that when $p = 0.5$, the distribution is symmetric with respect to μ.

A table of the binomial distribution is included in Appendix 4. For more extended tables see Ref. [J10] in Appendix 1.

The distribution with the probability function

$$(5) \qquad\qquad f(x) = \frac{\mu^x}{x!} e^{-\mu} \qquad\qquad (x = 0, 1, \cdots)$$

is called the **Poisson distribution,** named after S. D. Poisson (cf. Sec. 17.4). Figure 359 shows (5) for some values of μ. It can be proved that this distribution may be obtained as a limiting case of the binomial distribution, if we let $p \to 0$ and $n \to \infty$ so that the mean $\mu = np$ approaches a finite value. The Poisson distribution has the mean μ and the variance (cf. Prob. 13)

$$(6) \qquad\qquad \sigma^2 = \mu.$$

This distribution provides probabilities, e.g., of given numbers of cars passing an intersection per unit interval of time, of given numbers of defects per unit length of wire or unit area of paper or textile, and so on.

The binomial distribution is important in **sampling with replacement** (cf. Ex. 3, Sec. 19.5), as follows. Suppose that a box contains N things, for example, screws,

Fig. 359. Probability function (5) of the Poisson distribution for various values of μ

M of which are defective. If we want to draw a screw at random, the probability of obtaining a defective screw is

$$p = \frac{M}{N}.$$

Hence in drawing a sample of n screws *with replacement*, the probability that precisely x of the n screws are defective is [cf. (2)]

$$(7) \qquad f(x) = \binom{n}{x}\left(\frac{M}{N}\right)^{x}\left(1 - \frac{M}{N}\right)^{n-x} \qquad (x = 0, 1, \cdots, n).$$

In **sampling without replacement** that probability is

$$(8) \qquad f(x) = \frac{\binom{M}{x}\binom{N - M}{n - x}}{\binom{N}{n}} \qquad (x = 0, 1, \cdots, n).$$

This distribution with the probability function (8) is called the **hypergeometric distribution.**[6]

To verify our statement we first note that, by (4a) in Sec. 19.6, there are

(a) $\binom{N}{n}$ different ways of picking n things from N,

(b) $\binom{M}{x}$ different ways of picking x defectives from M,

(c) $\binom{N - M}{n - x}$ different ways of picking $n - x$ nondefectives from $N - M$,

and each way in (b) combined with each way in (c) gives the total number of mutually exclusive ways of obtaining x defectives in n drawings without replacement. Since (a) is the total number of outcomes and we draw at random, each such way has the probability $1 \Big/ \binom{N}{n}$. From this, (8) follows.

---

[6] Because the moment generating function (cf. Prob. 15, Sec. 19.8) of this distribution can be represented in terms of the hypergeometric function.

The hypergeometric distribution has the mean (cf. Prob. 15)

(9)
$$\mu = n\frac{M}{N}$$

and the variance

(10)
$$\sigma^2 = \frac{nM(N - M)(N - n)}{N^2(N - 1)}.$$

Example 1. We want to draw random samples of 2 gaskets from a box containing 10 gaskets, 3 of which are defective. Find the probability function of the random variable

$$X = \textit{number of defectives in the sample.}$$

We have $N = 10$, $M = 3$, $N - M = 7$, $n = 2$. For sampling with replacement, formula (7) yields

$$f(x) = \binom{2}{x}\left(\frac{3}{10}\right)^x\left(\frac{7}{10}\right)^{2-x}, \quad f(0) = 0.49, \quad f(1) = 0.42, \quad f(2) = 0.09.$$

For sampling without replacement we have to use (8), finding

$$f(x) = \binom{3}{x}\binom{7}{2-x}\Big/\binom{10}{2}, \quad f(0) = f(1) = \frac{21}{45} \approx 0.47, \quad f(2) = \frac{3}{45} \approx 0.07.$$

If N, M, and N − M are large compared with n, then it does not matter too much whether we sample with or without replacement, and in this case the hypergeometric distribution may be approximated by the binomial distribution (with $p = M/N$), *which is somewhat simpler.*

Hence in sampling from an indefinitely large population ("**infinite population**") *we may use the binomial distribution, regardless of whether we sample with or without replacement.*

Problems

1. Find formulas for $f(x)$, μ, and σ^2 in the case of the binomial distribution with $p = 0.5$.
2. Five fair coins are tossed simultaneously. Find the probability function of the random variable $X = \textit{number of heads}$ and compute the probabilities of obtaining no heads, precisely 1 head, at least 1 head, and not more than 4 heads.
3. If, in an experiment, positive and negative values are equally likely, what is the probability of obtaining at most 1 negative value in 9 trials?
4. A carton contains 20 fuses, 5 of which are defective. Find the probability that, if a sample of 3 fuses is chosen at random from the carton, x will be defective.
5. If the probability of hitting a target is 10% and 10 shots are fired independently, what is the probability that the target will be hit at least once?
6. Let $p = 1\%$ be the probability that a certain type of light bulb will fail in a 24-hour test. Find the probability that a sign consisting of 10 such bulbs will burn 24 hours with no bulb failures.
7. Suppose that 3% of bolts made by a machine are defective, the defectives occurring at random during production. If the bolts are packaged 50 per box, what is the Poisson approximation of the probability that a given box will contain x defectives?
8. **(Multinomial distribution)** Suppose a trial can result in precisely one of k mutually exclusive events $A_1, \cdots, A_k$ with probabilities $p_1, \cdots, p_k$, respectively, where $p_1 + \cdots + p_k = 1$. Suppose that n independent trials are performed. Show that the

probability of getting x_1 A_1's, $\cdots$, x_k A_k's is

$$f(x_1, \cdots, x_k) = \frac{n!}{x_1! \cdots x_k!} p_1^{x_1} \cdots p_k^{x_k}$$

where $0 \leq x_j \leq n$, $j = 1, \cdots, k$, and $x_1 + \cdots + x_k = n$. The distribution with this probability function is called the *multinomial distribution*.

9. Suppose that in a production of electrical resistors the probability of producing a resistor of resistance $R < 198$ ohms or $R > 201$ ohms is 3% and 5%, respectively. Find the probability that a random sample of 20 resistors contains precisely x_1 resistors with $R < 198$ ohms and precisely x_2 resistors with $R > 201$ ohms.

10. Using the binomial theorem, show that the binomial distribution has the moment generating function (cf. Prob. 15, Sec. 19.8)

$$G(t) = \sum_{x=0}^{n} e^{tx} \binom{n}{x} p^x q^{n-x} = \sum_{x=0}^{n} \binom{n}{x} (pe^t)^x q^{n-x} = (pe^t + q)^n.$$

11. Using Prob. 10 and $p + q = 1$, prove (3).
12. Prove (4).
13. Show that the Poisson distribution has the moment generating function

$$G(t) = e^{-\mu} e^{\mu e^t}$$

and prove (6).

14. Show that $E([X - \mu]^3) = E(X^3) - 3\mu E(X^2) + 2\mu^3$. Using this and Prob. 13, show that the Poisson distribution has the skewness $\gamma = 1/\sqrt{\mu}$ and conclude that for large μ the distribution is almost symmetric (cf. Fig. 359).

15. Prove (9).

19.10   Normal Distribution

The continuous distribution having the density

$$(1) \qquad f(x) = \frac{1}{\sigma \sqrt{2\pi}} e^{-\frac{1}{2}\left(\frac{x-\mu}{\sigma}\right)^2} \qquad (\sigma > 0)$$

is called the **normal distribution** or *Gauss distribution*. A random variable having this distribution is said to be **normal** or *normally distributed*. This distribution is very important, because many random variables of practical interest are normal or approximately normal or can be transformed into normal random variables in a relatively simple fashion. Furthermore, the normal distribution is a useful approximation of more complicated distributions. It also appears in the mathematical proofs of various statistical tests.

In (1), μ is the mean and σ is the standard deviation of the distribution. The curve of $f(x)$ is called the *bell-shaped curve*. It is symmetric with respect to μ. Figure 360 shows $f(x)$ for $\mu = 0$. For $\mu > 0$ ($\mu < 0$) the curves have the same shape, but are shifted $|\mu|$ units to the right (to the left). The smaller σ^2 is, the higher is the peak at $x = 0$ in Fig. 360 and the steeper are the descends on both sides. This agrees with the meaning of the variance.

From (1) we see that the normal distribution has the distribution function

$$(2) \qquad F(x) = \frac{1}{\sigma \sqrt{2\pi}} \int_{-\infty}^{x} e^{-\frac{1}{2}\left(\frac{v-\mu}{\sigma}\right)^2} dv.$$

Fig. 360. Density (1) of the normal distribution with $\mu = 0$ for various values of σ

From this and (10) in Sec. 19.7 we obtain

(3) $$P(a < X \leq b) = F(b) - F(a) = \frac{1}{\sigma\sqrt{2\pi}} \int_a^b e^{-\frac{1}{2}\left(\frac{v-\mu}{\sigma}\right)^2} dv.$$

The integral in (2) cannot be evaluated by elementary methods, but can be represented in terms of the integral (Fig. 361)

(4) $$\Phi(z) = \frac{1}{\sqrt{2\pi}} \int_{-\infty}^z e^{-u^2/2} du,$$

which is the distribution function of the normal distribution with mean 0 and variance 1 and has been tabulated (Table A8 in Appendix 4). In fact, if we set $(v - \mu)/\sigma = u$, then $du/dv = 1/\sigma$, $dv = \sigma\, du$, and we now have to integrate from $-\infty$ to $z = (x - \mu)/\sigma$. From (2) we thus obtain

$$F(x) = \frac{1}{\sigma\sqrt{2\pi}} \int_{-\infty}^{(x-\mu)/\sigma} e^{-u^2/2} \sigma\, du.$$

σ drops out, and the expression on the right equals (4) where $z = (x - \mu)/\sigma$,

Fig. 361. Distribution function $\Phi(z)$ of the normal distribution with mean 0 and variance 1

Fig. 362. Illustration of formula (7)

that is,

(5)
$$F(x) = \Phi\left(\frac{x - \mu}{\sigma}\right).$$

From this important formula and (3) we obtain

(6)
$$P(a < X \le b) = F(b) - F(a) = \Phi\left(\frac{b - \mu}{\sigma}\right) - \Phi\left(\frac{a - \mu}{\sigma}\right).$$

In particular, when $a = \mu - \sigma$ and $b = \mu + \sigma$, the right hand side equals $\Phi(1) - \Phi(-1)$; to $a = \mu - 2\sigma$ and $b = \mu + 2\sigma$ there corresponds the value $\Phi(2) - \Phi(-2)$, etc. Using Table A8 in Appendix 4 we thus find (Fig. 362)

(7)	(a)	$P(\mu - \sigma < X \le \mu + \sigma) \approx 68\%,$
	(b)	$P(\mu - 2\sigma < X \le \mu + 2\sigma) \approx 95.5\%,$
	(c)	$P(\mu - 3\sigma < X \le \mu + 3\sigma) \approx 99.7\%.$

Hence we may expect that a large number of observed values of a normal random variable X will be distributed as follows:

(a) *About $\frac{2}{3}$ of the values will lie between $\mu - \sigma$ and $\mu + \sigma$.*
(b) *About 95% of the values will lie between $\mu - 2\sigma$ and $\mu + 2\sigma$.*
(c) *About $99\frac{3}{4}\%$ of the values will lie between $\mu - 3\sigma$ and $\mu + 3\sigma$.*

This may also be expressed as follows.

A value that deviates more than σ from μ will occur about once in 3 trials. A value that deviates more than 2σ or 3σ from μ will occur about once in 20 or 400 trials, respectively. Practically speaking, this means that all the values will lie between $\mu - 3\sigma$ and $\mu + 3\sigma$; these two numbers are called **three-sigma limits.**

In a similar fashion we obtain

(8)	(a)	$P(\mu - 1.96\sigma < X \le \mu + 1.96\sigma) = 95\%,$
	(b)	$P(\mu - 2.58\sigma < X \le \mu + 2.58\sigma) = 99\%,$
	(c)	$P(\mu - 3.29\sigma < X \le \mu + 3.29\sigma) = 99.9\%.$

The following typical examples should help the reader to understand the practical use of Tables A8 and A9 in Appendix 4.

Example 1. Determine the probabilities

(a) $P(X \leq 2.44)$, (b) $P(X \leq -1.16)$, (c) $P(X \geq 1)$, (d) $P(2 \leq X \leq 10)$

where X is assumed to be normal with mean 0 and variance 1.

Since $\mu = 0$ and $\sigma^2 = 1$, we may obtain the desired values directly from Table A8, finding

(a) 0.9927, (b) 0.1230, (c) $1 - P(X \leq 1) = 1 - 0.8413 = 0.1587$ (cf. (6), Sec. 19.5),

(d) $\Phi(10) = 1.0000$ (why?), $\Phi(2) = 0.9772$, $\Phi(10) - \Phi(2) = 0.0228$.

Example 2. Determine the probabilities in the previous example, assuming that X is normal with mean 0.8 and variance 4.

From (6) and Table A8 we obtain

$$(a) \ F(2.44) = \Phi\left(\frac{2.44 - 0.80}{2}\right) = \Phi(0.82) = 0.7939$$

(b) $F(-1.16) = \Phi(-0.98) = 0.1635$
(c) $1 - P(X \leq 1) = 1 - F(1) = 1 - \Phi(0.1) = 0.4602$
(d) $F(10) - F(2) = \Phi(4.6) - \Phi(0.6) = 1 - 0.7257 = 0.2743$.

Example 3. Let X be normal with mean 0 and variance 1. Determine the constant c such that

(a) $P(X \geq c) = 10\%$,   (b) $P(X \leq c) = 5\%$,
(c) $P(0 \leq X \leq c) = 45\%$,   (d) $P(-c \leq X \leq c) = 99\%$.

From Table A9 in Appendix 4 we obtain

(a) $1 - P(X \leq c) = 1 - \Phi(c) = 0.1$, $\Phi(c) = 0.9$, $c = 1.282$
(b) $c = -1.645$
(c) $\Phi(c) - \Phi(0) = \Phi(c) - 0.5 = 0.45$, $\Phi(c) = 0.95$, $c = 1.645$
(d) $c = 2.576$.

Example 4. Let X be normal with mean -2 and variance 0.25. Determine c such that

(a) $P(X \geq c) = 0.2$ (b) $P(-c \leq X \leq -1) = 0.5$
(c) $P(-2 - c \leq X \leq -2 + c) = 0.9$ (d) $P(-2 - c \leq X \leq -2 + c) = 99.6\%$.

Using Table A9 in Appendix 4, we obtain the following results.

$$(a) \ 1 - P(X \leq c) = 1 - \Phi\left(\frac{c + 2}{0.5}\right) = 0.2,$$

$$\Phi(2c + 4) = 0.8, \ 2c + 4 = 0.842, \ \ c = -1.579$$

$$(b) \ \Phi\left(\frac{-1 + 2}{0.5}\right) - \Phi\left(\frac{-c + 2}{0.5}\right) = 0.9772 - \Phi(4 - 2c) = 0.5,$$

$$\Phi(4 - 2c) = 0.4772, \ 4 - 2c = -0.057, \ c = 2.03$$

$$(c) \ \Phi\left(\frac{-2 + c + 2}{0.5}\right) - \Phi\left(\frac{-2 - c + 2}{0.5}\right)$$

$$= \Phi(2c) - \Phi(-2c) = 0.9, \ 2c = 1.645, \ c = 0.823$$

(d) $\Phi(2c) - \Phi(-2c) = 99.6\%$, $2c = 2.878$, $c = 1.439$.

Example 5. Suppose that in producing iron plates the plates are required to have a certain thickness, and the machine work is done by a shaper. In any case industrial products will differ slightly from each other because the properties of the material and the behavior of the machines and tools used show slight random variations caused by small disturbances which we cannot predict. We may therefore regard the thickness X [mm] of the plates as a random variable. We assume that for a certain setting the variable X is normal with mean $\mu = 10$ mm and standard deviation $\sigma = 0.02$ mm. We want to determine the percentage of defective plates to be expected, assuming that defective plates are (a) plates thinner than 9.97 mm, (b) plates thicker than 10.05 mm, (c) plates that deviate more than 0.03 mm from 10 mm. (d) How should we choose numbers $10 - c$ and $10 + c$ in order that the expected percentage of defectives be not greater than 5%? (e) How does the percentage of defectives in part (d) change if μ is shifted from 10 mm to 10.01 mm?

Using Table A8 in Appendix 4 and (6), we obtain the following solutions.

$$(a)\ P(X \leq 9.97) = \Phi\left(\frac{9.97 - 10.00}{0.02}\right) = \Phi(-1.5) = 0.0668 \approx 6.7\%$$

$$(b)\ P(X \geq 10.05) = 1 - P(X \leq 10.05) = 1 - \Phi\left(\frac{10.05 - 10.00}{0.02}\right)$$

$$= 1 - \Phi(2.5) = 1 - 0.9938 \approx 0.6\%$$

$$(c)\ P(9.97 \leq X \leq 10.03) = \Phi\left(\frac{10.03 - 10.00}{0.02}\right) - \Phi\left(\frac{9.97 - 10.00}{0.02}\right)$$

$$= \Phi(1.5) - \Phi(-1.5) = 0.8664.\quad \textit{Answer}\ 1 - 0.8664 \approx 13\%.$$

(d) From (8a) we obtain $c = 1.96\sigma = 0.039$.   *Answer* 9.961 mm and 10.039 mm.

$$(e)\ P(9.961 \leq X \leq 10.039) = \Phi\left(\frac{10.039 - 10.010}{0.02}\right) - \Phi\left(\frac{9.961 - 10.010}{0.02}\right)$$

$$= \Phi(1.45) - \Phi(-2.45) = 0.9265 - 0.0071 \approx 92\%.\quad \textit{Answer}\ 8\%.$$

We see that this slight change of the adjustment of the tool bit causes a considerable increase of the percentage of defectives.

Under a linear transformation, a normal random variable transforms into a normal random variable. In fact, using (5), the reader may prove the following theorem.

Theorem 1. *If X is normal with mean μ and variance σ^2, then $X^* = c_1 X + c_2$ $(c_1 \neq 0)$ is normal with mean $\mu^* = c_1\mu + c_2$ and variance $\sigma^{*2} = c_1^2\sigma^2$.*

The normal distribution may also be used for approximating the binomial distribution when n is large, so that the binomial coefficients and powers in the probability function

$$(9)\qquad\qquad f(x) = \binom{n}{x} p^x q^{n-x} \qquad\qquad (x = 0, 1, \cdots, n)$$

(cf. Sec. 19.9) of the binomial distribution become very inconvenient. In fact, the following important theorem holds true.

Theorem 2 (Limit theorem of De Moivre and Laplace). *For large n,*

$$f(x) \sim f^*(x) \qquad\qquad (x = 0, 1, \cdots, n),$$

where f is given by (9),

$$(10)\qquad\qquad f^*(x) = \frac{1}{\sqrt{2\pi}\sqrt{npq}}\, e^{-z^2/2}, \qquad z = \frac{x - np}{\sqrt{npq}}$$

is the density of the normal distribution with mean $\mu = np$ and variance $\sigma^2 = npq$ (the mean and variance of the binomial distribution), and the symbol $\sim$ (read

asymptotically equal) *means that the ratio of both sides approaches* 1 *as n approaches* ∞. *Furthermore*,

$$P(a \leq X \leq b) = \sum_{x=a}^{b} \binom{n}{x} p^x q^{n-x} \sim \Phi(\beta) - \Phi(\alpha),$$

(11)

$$\alpha = \frac{a - np - 0.5}{\sqrt{npq}}, \qquad \beta = \frac{b - np + 0.5}{\sqrt{npq}}.$$

A proof of this theorem can be found in Ref. [H4], cf. Appendix 1. The proof shows that the term 0.5 in α and β is a correction caused by the change from a discrete to a continuous distribution.

Problems

1. Show that the curve of (1) has two points of inflection corresponding to $x = \mu \pm \sigma$.
2. Considering $\Phi^2(\infty)$ and introducing polar coordinates in the double integral, prove

(12)
$$\Phi(\infty) = \frac{1}{\sqrt{2\pi}} \int_{-\infty}^{\infty} e^{-u^2/2} \, du = 1.$$

3. Show that $\Phi(-z) = 1 - \Phi(z)$.
4. Using (12) and integration by parts, show that σ in (1) is the standard deviation of the normal distribution.
5. Prove Theorem 1.
6. Let X be normal with mean 10 and variance 4. Find $P(X > 12)$, $P(X < 10)$, $P(X < 11)$, and $P(9 < X < 13)$.
7. Let X be normal with mean 50 and variance 9. Determine c such that $P(X < c) = 5\%$, $P(X > c) = 1\%$, and $P(50 - c < X < 50 + c) = 50\%$.
8. In Ex. 5, part (c), what value should σ have in order that the percentage of defective plates reduces to 1%?
9. What percentage of defective iron plates can we expect in question (c) of Ex. 5 if we use a better shaper so that $\sigma = 0.01$ mm?
10. A producer sells electric bulbs in cartons of 1000 bulbs. Using (11), find the probability that any given carton contains not more than 1% defective bulbs, assuming the production process to be a Bernoulli experiment with $p = 1\%$ (= probability that any given bulb will be defective).
11. A manufacturer produces airmail envelopes whose weight is normal with mean $\mu = 1.95$ grams and standard deviation $\sigma = 0.05$ grams. The envelopes are sold in lots of 1000. How many envelopes in a lot will be heavier than 2 grams?
12. In Prob. 11, how many envelopes weighing 2.05 grams or more can be expected in a given package of 100 envelopes?
13. What is the probability of obtaining at least 2048 heads if a coin is tossed 4040 times and heads and tails are equally likely? (Cf. Table 21, Sec. 19.5.)
14. If X is normal with mean μ and variance σ^2, what distribution does $-X$ have?
15. **(Bernoulli's law of large numbers)** In a random experiment let an event A have probability p $(0 < p < 1)$, and let X be the number of times A happens in n independent trials. Show that for any given $\epsilon > 0$,

$$P\left(\left| \frac{X}{n} - p \right| < \epsilon \right) \to 1 \qquad \qquad \text{as} \qquad n \to \infty.$$

19.11   *Distributions of Several Random Variables*

If in a random experiment we observe a single quantity, we have to associate with that experiment a single random variable, call it X. From Sec. 19.7 we remember that the corresponding distribution function $F(x) = P(X \leq x)$ determines the distribution completely, because for each interval $a < X \leq b$,

$$P(a < X \leq b) = F(b) - F(a).$$

If in a random experiment we observe two quantities, we have to associate with the experiment two random variables, say, X and Y. For example, X may correspond to the Rockwell hardness and Y to the carbon content of steel, if we measure these two quantities. Each performance of the experiment yields a pair of numbers $X = x$, $Y = y$, briefly (x, y), which may be plotted as a point in the XY-plane. We may now consider a rectangle $a_1 < X \leq b_1$, $a_2 < Y \leq b_2$ (Fig. 363). If for each such rectangle we know the corresponding probability

$$P(a_1 < X \leq b_1, a_2 < Y \leq b_2),$$

then we say that the **two-dimensional probability distribution** of the random variables X and Y or of the **two-dimensional random variable** (X, Y) is known. The function

(1) $$F(x, y) = P(X \leq x, Y \leq y)$$

is called the **distribution function** of that distribution or of (X, Y). It determines the distribution uniquely, because (cf. Prob. 1)

(2) $$P(a_1 < X \leq b_1, a_2 < Y \leq b_2) = F(b_1, b_2) - F(a_1, b_2) - F(b_1, a_2) + F(a_1, a_2).$$

The distribution and the variable (X, Y) are said to be **discrete** if (X, Y) has the following properties.

(X, Y) can assume only finitely many or at most countably infinitely many pairs of values (x, y), the corresponding probabilities being positive. Each bounded domain[7] contains at most finitely many such pairs, and to any domain containing no such pairs there corresponds the probability 0.

Let (x_i, y_j) be any such pair and let $P(X = x_i, Y = y_j) = p_{ij}$ (where we admit that p_{ij} may be 0 for certain pairs of subscripts i, j). The function

(3) $f(x, y) = p_{ij}$ when $x = x_i, y = y_j$ and $f(x, y) = 0$ otherwise

Fig. 363. Notion of a two-dimensional distribution

[7] That is, a domain which can be enclosed in a circle of sufficiently large radius.

is called the **probability function** of (X, Y); here $i = 1, 2, \cdots$ and $j = 1, 2, \cdots$ independently. The analogue of (7), Sec. 19.7, is

(4)
$$F(x, y) = \sum_{x_i \leq x} \sum_{y_j \leq y} f(x_i, y_j),$$

and instead of formula (3) in Sec. 19.7 we now have the condition

(5)
$$\sum_i \sum_j f(x_i, y_j) = 1.$$

For example, if we toss a dime and a nickel once and consider

$$X = \textit{number of heads the dime turns up},$$
$$Y = \textit{number of heads the nickel turns up},$$

hen X and Y can have the values 0 or 1, and the probability function is

$$f(0, 0) = f(1, 0) = f(0, 1) = f(1, 1) = \tfrac{1}{4}, \qquad f(x, y) = 0 \text{ otherwise.}$$

(X, Y) and its distribution are said to be **continuous** if the corresponding distribution function $F(x, y)$ can be represented by a double integral

(6)
$$F(x, y) = \int_{-\infty}^{y} \int_{-\infty}^{x} f(x^*, y^*) \, dx^* \, dy^*$$

where $f(x, y)$ is defined, nonnegative, and bounded in the entire plane, possibly except for finitely many continuously differentiable curves. $f(x, y)$ is called the **probability density** of the distribution. It follows that

(7)
$$P(a_1 < X \leq b_1, a_2 < Y \leq b_2) = \int_{a_2}^{b_2} \int_{a_1}^{b_1} f(x, y) \, dx \, dy.$$

For example (cf. Fig. 364)

(8) $f(x, y) = 1/k$ when (x, y) is in R, $f(x, y) = 0$ otherwise

defines the so-called *uniform distribution in the rectangle R;* here k is the area of R, that is, $k = (\beta_1 - \alpha_1)(\beta_2 - \alpha_2)$. The distribution function is shown in Fig. 365.

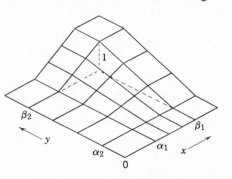

Fig. 364. Probability function (8) of the uniform distribution

Fig. 365. Distribution function of the uniform distribution defined by (8)

In the case of a *discrete* random variable (X, Y) with probability function $f(x, y)$ we may ask for the probability $P(X = x, Y \text{ arbitrary})$ that X assume a value x, while Y may assume any value, which we ignore. This probability is a function of x, say $f_1(x)$, and we have the formula

$$(9) \qquad f_1(x) = P(X = x, Y \text{ arbitrary}) = \sum_y f(x, y)$$

where we sum all the values of $f(x, y)$ which are not 0 for that x. Clearly, $f_1(x)$ is the probability function of the probability distribution of a single random variable. This distribution is called the **marginal distribution** *of X with respect to the given two-dimensional distribution.* It has the distribution function

$$(10) \qquad F_1(x) = P(X \leq x, Y \text{ arbitrary}) = \sum_{x^* \leq x} f_1(x^*).$$

Similarly, the probability function

$$(11) \qquad f_2(y) = P(X \text{ arbitrary}, Y = y) = \sum_x f(x, y)$$

determines the so-called **marginal distribution** *of Y with respect to the given two-dimensional distribution.* In (11) we sum all the values of $f(x, y)$ which are not 0 for the corresponding y. The distribution function of this distribution is

$$(12) \qquad F_2(y) = P(X \text{ arbitrary}, Y \leq y) = \sum_{y^* \leq y} f_2(y^*).$$

Obviously, both marginal distributions of a discrete random variable (X, Y) are discrete.

Table 22 shows an illustrative example. X is the number of queens and Y is the number of kings or aces obtained in drawing 3 cards with replacement from a bridge deck. Since the 52 cards of the deck contain 4 queens, 4 kings, and 4 aces, the probability of obtaining a queen if we draw a single card is $\frac{4}{52} = \frac{1}{13}$ and a king or ace is obtained with probability $\frac{8}{52} = \frac{2}{13}$. Hence to this random experiment there corresponds the probability function

$$f(x, y) = \frac{3!}{x! \, y! \, (3 - x - y)!} \left(\frac{1}{13}\right)^x \left(\frac{2}{13}\right)^y \left(\frac{10}{13}\right)^{3-x-y} \qquad (x + y \leq 3)$$

and $f(x, y) = 0$ otherwise. Table 22 shows the values of $f(x, y)$, $f_1(x)$, and $f_2(y)$.

Similarly, in the case of a *continuous* random variable (X, Y) with density $f(x, y)$ we may consider

$$(X \leq x, Y \text{ arbitrary}) \qquad \text{or} \qquad (X \leq x, -\infty < Y < \infty).$$

Obviously, the corresponding probability is

$$F_1(x) = P(X \leq x, -\infty < Y < \infty) = \int_{-\infty}^{x} \left(\int_{-\infty}^{\infty} f(x^*, y) \, dy \right) dx^*.$$

Setting

$$(13) \qquad f_1(x) = \int_{-\infty}^{\infty} f(x, y) \, dy$$

Table 22. Values of the Probability Functions $f(x, y)$, $f_1(x)$, and $f_2(y)$ in Drawing 3 Cards with Replacement from a Bridge Deck where X is the Number of Queens Drawn and Y is the Number of Kings or Aces Drawn

x \ y	0	1	2	3	$f_1(x)$
0	$\frac{1000}{2197}$	$\frac{600}{2197}$	$\frac{120}{2197}$	$\frac{8}{2197}$	$\frac{1728}{2197}$
1	$\frac{300}{2197}$	$\frac{120}{2197}$	$\frac{12}{2197}$	0	$\frac{432}{2197}$
2	$\frac{30}{2197}$	$\frac{6}{2197}$	0	0	$\frac{36}{2197}$
3	$\frac{1}{2197}$	0	0	0	$\frac{1}{2197}$
$f_2(y)$	$\frac{1331}{2197}$	$\frac{726}{2197}$	$\frac{132}{2197}$	$\frac{8}{2197}$	

we may write

$$(14) \qquad F_1(x) = \int_{-\infty}^{x} f_1(x^*) \, dx^*.$$

$f_1(x)$ is called the density and $F_1(x)$ the *distribution function* of the **marginal distribution** *of X with respect to the given continuous distribution.* The function

$$(15) \qquad f_2(y) = \int_{-\infty}^{\infty} f(x, y) \, dx$$

is called the *density* and

$$(16) \qquad F_2(y) = \int_{-\infty}^{y} f_2(y^*) \, dy^* = \int_{-\infty}^{y} \int_{-\infty}^{\infty} f(x, y^*) \, dx \, dy^*$$

is called the *distribution function* of the **marginal distribution** *of Y with respect to the given two-dimensional distribution.* We see that both marginal distributions of a continuous distribution are continuous.

The two random variables X and Y of a two-dimensional (X, Y)-distribution with distribution function $F(x, y)$ are said to be **independent** if

$$(17) \qquad F(x, y) = F_1(x)F_2(y)$$

holds for all (x, y). Otherwise these variables are said to be **dependent.**

Suppose that X and Y are either both discrete or both continuous. Then X and Y are independent if and only if the corresponding probability functions or densities $f_1(x)$ and $f_2(x)$ satisfy

$$(18) \qquad f(x, y) = f_1(x)f_2(y)$$

for all (x, y); cf. Prob. 2. For example, the variables in Table 22 are dependent. The variables $X = $ *number of heads on a dime*, $Y = $ *number of heads on a nickel* in tossing a dime and a nickel once may assume the values 0 or 1 and are independent.

The notions of independence and dependence may be extended to the n random variables of an n-dimensional $(X_1, \cdots, X_n)$-distribution with distribution function

$$F(x_1, \cdots, x_n) = P(X_1 \leqq x_1, \cdots, X_n \leqq x_n).$$

These random variables are said to be **independent** if for all $(x_1, \cdots, x_n)$,

(19) $$F(x_1, \cdots, x_n) = F_1(x_1)F_2(x_2) \cdots F_n(x_n),$$

where $F_j(x_j)$ is the distribution function of the marginal distribution of X_j, that is,

$$F_j(x_j) = P(X_j \leqq x_j, X_k \text{ arbitrary}, k \neq j).$$

Otherwise these random variables are said to be **dependent.**

Let (X, Y) be a random variable with probability function or density $f(x, y)$ and distribution function $F(x, y)$, and let $g(x, y)$ be any continuous function which is defined for all (x, y) and is not constant. Then $Z = g(X, Y)$ is a random variable, too. For example, if we roll two dice and X is the number that the first die turns up whereas Y is the number that the second die turns up, then $Z = X + Y$ is the sum of those two numbers (cf. Fig. 355 in Sec 19.7).

If $(X_1, \cdots, X_n)$ is an n-dimensional random variable and $g(x_1, \cdots, x_n)$ is a continuous function which is defined for all $(x_1, \cdots, x_n)$ and is not constant, then $Z = g(X_1, \cdots, X_n)$ is a random variable, too.

In the case of a *discrete* random variable (X, Y) we may obtain the probability function $f(z)$ of $Z = g(X, Y)$ by summing all $f(x, y)$ for which $g(x, y)$ equals the value of z considered; thus

(20) $$f(z) = P(Z = z) = \sum \sum_{g(x,y)=z} f(x, y).$$

The distribution function of Z is

(21) $$F(z) = P(Z \leqq z) = \sum \sum_{g(x,y) \leqq z} f(x, y),$$

where we sum all values of $f(x, y)$ for which $g(x, y) \leqq z$.

In the case of a *continuous* random variable (X, Y) we similarly have

(22) $$F(z) = P(Z \leqq z) = \iint_{g(x,y) \leqq z} f(x, y) \, dx \, dy,$$

where for each z we integrate over the region $g(x, y) \leqq z$ in the xy-plane.

The number

(23) $$E(g(X, Y)) = \begin{cases} \sum_x \sum_y g(x, y)f(x, y) & \text{(discrete distribution)} \\ \int_{-\infty}^{\infty} \int_{-\infty}^{\infty} g(x, y)f(x, y) \, dx \, dy & \text{(continuous distribution)} \end{cases}$$

is called the **mathematical expectation** or, briefly, the **expectation of** $g(X, Y)$. Here it is assumed that the double series converges absolutely and the integral of $|g(x, y)| f(x, y)$ over the xy-plane exists. The formula

(24) $$E(ag(X, Y) + bh(X, Y)) = aE(g(X, Y)) + bE(h(X, Y))$$

can be proved in a fashion similar to that in Prob. 9, Sec. 19.8. An important special case is $E(X + Y) = E(X) + E(Y)$, and by induction we have the following result.

Theorem 1 (Addition of means). *The mean (expectation) of a sum of random variables equals the sum of the means (expectations), that is,*

(25)　　　　$E(X_1 + X_2 + \cdots + X_n) = E(X_1) + E(X_2) + \cdots + E(X_n).$

Furthermore we readily obtain

Theorem 2 (Multiplication of means). *The mean (expectation) of the product of **independent** random variables equals the product of the means (expectations), that is,*

(26)　　　　$E(X_1 X_2 \cdots X_n) = E(X_1)E(X_2) \cdots E(X_n).$

Proof. If X and Y are independent random variables (both discrete or both continuous), then $E(XY) = E(X)E(Y)$. In fact, in the discrete case we have

$$E(XY) = \sum_x \sum_y xy f(x, y) = \sum_x x f_1(x) \sum_y y f_2(y) = E(X)E(Y),$$

and in the continuous case the proof of the relation is similar. Extension to n independent random variables gives (26), and Theorem 2 is proved.

We shall now discuss the addition of variances. Let $Z = X + Y$. We first have (cf. Prob. 7 in Sec. 19.8)

$$\sigma^2 = E([Z - \mu]^2) = E(Z^2) - [E(Z)]^2.$$

From (24) we see that the first term on the right equals

$$E(Z^2) = E(X^2 + 2XY + Y^2) = E(X^2) + 2E(XY) + E(Y^2),$$

and for the second term on the right, we obtain from Theorem 1

$$[E(Z)]^2 = [E(X) + E(Y)]^2 = [E(X)]^2 + 2E(X)E(Y) + [E(Y)]^2.$$

By substituting these expressions in the formula for σ^2 we have

$$\sigma^2 = E(X^2) - [E(X)]^2 + E(Y^2) - [E(Y)]^2$$
$$+ 2[E(XY) - E(X)E(Y)].$$

From Prob. 7, Sec. 19.8, we see that the expression in the first line on the right is the sum of the variances of X and Y, which we denote by σ_1^2 and σ_2^2, respectively. The quantity

(27)　　　　$\sigma_{XY} = E(XY) - E(X)E(Y)$

is called the **covariance** of X and Y. Consequently, our result is

(28)　　　　$\sigma^2 = \sigma_1^2 + \sigma_2^2 + 2\sigma_{XY}.$

If X and Y are independent, then $E(XY) = E(X)E(Y)$; hence $\sigma_{XY} = 0$, and

(29) $$\sigma^2 = \sigma_1{}^2 + \sigma_2{}^2.$$

Extension to more than two variables yields

Theorem 3 (Addition of variances). *The variance of the sum of* **independent** *random variables equals the sum of the variances of these variables.*

Problems

1. Prove (2).
2. Prove the statement involving (18).
3. Let $f(x, y) = 2$ when $x > 0$, $y > 0$, $x + y < 1$ and 0 otherwise. Graph f and the corresponding distribution function.
4. Find the marginal distributions of the distribution in Figs. 364 and 365 on p. 741.
5. Find $P(X > Y)$ when (X, Y) has the density $f(x, y) = e^{-(x+y)}$ if $x \geq 0$, $y \geq 0$ and 0 otherwise.
6. Find the densities of the marginal distributions in Prob. 5.
7. Give an example of two different continuous distributions that have the same marginal distributions.
8. Let (X, Y) have the probability function $f(0, 0) = f(1, 1) = 3/8, f(0, 1) = f(1, 0) = 1/8$. Are X and Y independent?
9. Using Theorem 1, obtain the formula for the mean μ of the binomial distribution.
10. Using Theorem 3, obtain the formula for the variance σ^2 of the binomial distribution.

19.12 *Random Sampling. Random Numbers*

The preceding sections (19.3–19.11) were devoted to probability theory and the remaining sections (19.12–19.20) deal with statistics. Probability theory helps to create mathematical models of populations, and the statistical methods to be discussed yield relations between the theory and the observable reality, conclusions about populations by means of samples (*statistical inference;* cf. Sec. 19.1).

So far it was sufficient to know that a sample from a population is a selection taken from a population (for examples, see Sec. 19.1), but from now on we have to define this notion in a precise fashion. In fact, in order to obtain meaningful information about populations from samples, a sample must be a **random selection;** that is, each element of the population must have a known probability of being taken into the sample. This condition must be satisfied (at least approximately), otherwise the methods to be discussed may yield completely meaningless and misleading results.

If the sample space is infinite, the sample values will be **independent,** that is, the results of the n performances of a random experiment made for obtaining n sample values will not influence each other. This certainly applies to samples from a normal population. If the sample space is finite, the sample values will still be independent if we sample with replacement; they will be *practically* independent if we sample without replacement and keep the size of the sample small compared with the size of the population (for instance, samples of 5 or 10 values from a population of 1000 values). However, if we sample without replacement and take large samples from a finite population, the dependence will matter considerably.

It is not so easy to satisfy the requirement that a sample be a random selection, because there are many and subtle factors of various types that can bias results of sampling. For example, if an interested purchaser wants to draw and inspect a sample of 10 items from a lot of 80 items before he decides whether to purchase the lot, how should he select physically those 10 items so that he can be reasonably sure that all possible $\binom{80}{10}$ samples of size 10 would be equally probable?

Methods to solve this problem have been developed, and we shall now describe such a procedure, which is frequently used.

We number the items of that lot from 1 to 80. Then we select 10 items using Table A10 in Appendix 4, which contains *random numbers* made up of sets of **random digits,** as follows. We first select a line number from 0 to 99 at random. This can be done by tossing a fair coin 7 times denoting heads by 1 and tails by 0, thus generating a 7-place binary number which will represent $0, 1, \cdots, 127$ with equal probabilities. We use this number if it is $0, 1, \cdots,$ or 99. Otherwise we ignore it and repeat this process. Then we select a column number from 0 to 9 by generating a 4-place binary number in a similar fashion. Suppose that we obtained 0011010 (= 26) and 0111 (= 7), respectively. In line 26 and column 7 of Table A10 we find 44973. We keep the first 2 digits, that is, 44. We move down the column, starting with 44973 and keeping only the first 2 digits. In this way we find

$$44 \qquad 44 \qquad 83 \qquad 91 \qquad 55 \qquad \text{etc.}$$

We omit numbers which are greater than 80 or occur for the second time and continue until 10 numbers are obtained. This yields

$$44 \quad 55 \quad 53 \quad 03 \quad 52 \quad 61 \quad 67 \quad 78 \quad 39 \quad 54.$$

The 10 items having these numbers represent the desired selection.

For a larger table of random digits see Ref. [J15], Appendix 1.

Problems

1. Suppose that in the example explained in the text we would start from line 83 and column 2 of Table A10 in Appendix 4, and move upward. What items would then be included in a sample of size 10?
2. Using Table A10 in Appendix 4, select a sample of size 20 from a lot of 250 given items.
3. How can fair dice be used in connection with random selection?

19.13 Estimation of Parameters

Quantities appearing in distributions, such as p in the binomial distribution and μ and σ in the normal distribution, are called **parameters.**

A **point estimate** of a parameter is a number (point on the real line), which is computed from a given sample and serves as an approximation of the unknown exact value of the parameter. An **interval estimate** is an interval ("*confidence interval*") obtained from a sample; such estimates will be considered in the next section. Estimation of parameters is an important practical problem.

As an approximation of the mean μ of a population we may take the mean $\bar{x}$ of a corresponding sample. This gives the estimate $\hat{\mu} = \bar{x}$ for μ, that is

$$(1) \qquad \hat{\mu} = \bar{x} = \frac{1}{n}(x_1 + \cdots + x_n) \qquad (n = \text{size of the sample}).$$

Similarly, an estimate $\hat{\sigma}^2$ for the variance of a population is the variance s^2 of a corresponding sample, that is,

$$(2) \qquad \hat{\sigma}^2 = s^2 = \frac{1}{n-1}\sum_{j=1}^{n}(x_j - \bar{x})^2.$$

Clearly, (1) and (2) are estimates of parameters for distributions in which μ or σ^2 appear explicitly as parameters, such as the normal and Poisson distributions. For the binomial distribution, $p = \mu/n$ [cf. (3) in Sec. 19.9]. In this case, $x_j = 1$ in (1) if the event A whose probability is p occurred in the jth trial, and $x_j = 0$ if A did not occur in that trial. From (1) we thus obtain for p the estimate

$$(3) \qquad \hat{p} = \frac{\bar{x}}{n}.$$

We mention that (1) is a very special case of the so-called **method of moments.** In this method the parameters to be estimated are expressed in terms of the moments of the distribution (cf. Sec. 19.8). In the resulting formulas those moments are replaced by the corresponding moments of the sample. This yields the desired estimates. Here the **kth moment of a sample** $x_1, \cdots, x_n$ is

$$m_k = \frac{1}{n}\sum_{j=1}^{n}x_j^{k}.$$

Another method for obtaining estimates is the so-called **maximum likelihood method** of R. A. Fisher (Messenger Math. **41**, 1912, 155–160). To explain it we consider a discrete (or continuous) random variable X whose probability function (or density) $f(x)$ depends on a single parameter θ and take a corresponding sample of n independent values $x_1, \cdots, x_n$. Then in the discrete case the probability that a sample of size n consists precisely of those n values is

$$(4) \qquad l = f(x_1)f(x_2)\cdots f(x_n).$$

In the continuous case the probability that the sample consists of values in the small intervals $x_i \leqq x \leqq x_i + \Delta x$ $(i = 1, 2, \cdots, n)$ is

$$(5) \qquad f(x_1)\,\Delta x f(x_2)\,\Delta x \cdots f(x_n)\,\Delta x = l(\Delta x)^n.$$

Since $f(x_i)$ depends on θ, the function l depends on $x_1, \cdots, x_n$ and θ. We imagine $x_1, \cdots, x_n$ to be given and fixed. Then l is a function of θ, which is called the

likelihood function. The basic idea of the maximum likelihood method is very simple, as follows. We choose that approximation for the unknown value of θ for which l is as large as possible. If l is a differentiable function of θ, a necessary condition for l to have a maximum (not at the boundary) is

(6) $$\frac{\partial l}{\partial \theta} = 0.$$

(We write a partial derivative, because l depends also on $x_1, \cdots, x_n$.) A solution of (6) depending on $x_1, \cdots, x_n$ is called a *maximum likelihood estimate* for θ. We may replace (6) by

(7) $$\frac{\partial \ln l}{\partial \theta} = 0,$$

because $f \geqq 0$, a maximum of f is in general positive, and $\ln l$ is a monotone increasing function of l. This often simplifies calculations.

If the distribution of X involves r parameters $\theta_1, \cdots, \theta_r$, then instead of (6) we have the r conditions $\partial l / \partial \theta_1 = 0, \cdots, \partial l / \partial \theta_r = 0$, and instead of (7) we have

(8) $$\frac{\partial \ln l}{\partial \theta_1} = 0, \cdots, \frac{\partial \ln l}{\partial \theta_r} = 0.$$

Example 1 (*Normal distribution*). Find maximum likelihood estimates for μ and σ in the case of the normal distribution. From (1), Sec. 19.10, and (4) we obtain

$$l = \left(\frac{1}{\sqrt{2\pi}}\right)^n \left(\frac{1}{\sigma}\right)^n e^{-h} \quad \text{where} \quad h = \frac{1}{2\sigma^2} \sum_{i=1}^{n} (x_i - \mu)^2.$$

Taking logarithms, we have

$$\ln l = -n \ln \sqrt{2\pi} - n \ln \sigma - h.$$

The first equation in (8) is $\partial \ln l / \partial \mu = 0$, written out

$$\frac{\partial \ln l}{\partial \mu} = -\frac{\partial h}{\partial \mu} = \frac{1}{\sigma^2} \sum_{i=1}^{n} (x_i - \mu) = 0, \quad \text{hence} \quad \sum_{i=1}^{n} x_i - n\mu = 0.$$

The solution is the desired estimate $\hat{\mu}$ for μ; we find

$$\hat{\mu} = \frac{1}{n} \sum_{i=1}^{n} x_i = \bar{x}.$$

The second equation in (8) is $\partial \ln l / \partial \sigma = 0$, written out

$$\frac{\partial \ln l}{\partial \sigma} = -\frac{n}{\sigma} - \frac{\partial h}{\partial \sigma} = -\frac{n}{\sigma} + \frac{1}{\sigma^3} \sum_{i=1}^{n} (x_i - \mu)^2 = 0.$$

Replacing μ by $\hat{\mu}$ and solving for σ^2, we obtain the estimate

$$\hat{\sigma}^2 = \frac{1}{n} \sum_{i=1}^{n} (x_i - \bar{x})^2,$$

which we shall use in Sec. 19.18. Note that this differs from (2). We cannot discuss criteria for the goodness of estimates, but we want to mention that for small n, formula (2) is preferable.

Fig. 366. Distribution function of the sample in Table 17 graphed on probability paper

There are **graphical methods** for obtaining estimates. As an important case, let us consider the normal distribution and obtain estimates for μ and σ from a sample $x_1, \cdots, x_n$. From (5) in Sec. 19.10 we know that the distribution function of the normal distribution is

$$F(x) = \Phi(z), \qquad z = \frac{x - \mu}{\sigma}.$$

Fig. 361 in Sec. 19.10 shows that the corresponding curve is s-shaped and intersects the 50%-line at $z = 0$, hence at $x - \mu = 0$, that is, $x = \mu$. Furthermore, from Table A9 in Appendix 4 we see that $\Phi = 84\%$ when $z = 1$ (approximately), hence $(x - \mu)/\sigma = 1$, that is, $x = \mu + \sigma$. This suggests the following method.

Graph the distribution function of that sample and fit an s-shaped curve C by eye; let x_0 and x_1 denote the abscissas of the points of intersection of C and the 50% and 84% lines, respectively. Then x_0 is an estimate for μ, and $x_1 - x_0$ is an estimate for σ.

This method can be improved. We may "stretch" the vertical scale so that afterwards the s-shaped curve of $F(x)$ appears as a straight line. This new paper is called **normal probability paper.** The advantage is obvious, because we now have to fit a straight line, which is easier than fitting an s-shaped curve. Figure 366 shows an example, where $\bar{x} \approx 365$ (computed 364.7) and $s \approx 28$ (computed $\sqrt{720.1} \approx 26.8$).

If the sample values are equally spaced (as in Table 17), work may be saved by plotting merely the values of the cumulative relative frequencies as points on the probability paper and fitting a straight line as best we can "by eye." From Fig. 366 it is obvious that if we would plot each value above the corresponding x-value, we would make a systematic error because the line would lie higher than that in Fig. 366. To compensate for this effect, we plot each value above the corresponding x-value augmented by $\frac{1}{2}$ the distance between consecutive x-values. (In Table 17 this distance is $\frac{10}{2} = 5$.) This yields Fig. 367.

Fig. 367. Cumulative relative frequencies of the sample in Table 17 graphed on probability paper

Similarly, if the sample is grouped, we graph the cumulative relative frequencies as points vertically above the right endpoints of the class intervals (not above the class marks). Then we fit a straight line to these points "by eye."

Normal probability paper can also be used for obtaining a rough check on whether a sample may have come from a normal population. A test for this problem will be considered in Sec. 19.18.

Problems

1. Find the maximum likelihood estimate of θ in the density $f(x) = \theta e^{-\theta x}$ when $x \geq 0$ and $f(x) = 0$ when $x < 0$.
2. In Prob. 1, find the mean μ, substitute it in $f(x)$, find the maximum likelihood estimate of μ, and show that it is identical with the estimate for μ which can be obtained from that for θ in Prob. 1.
3. Find the maximum likelihood estimate for the parameter μ of a normal distribution with known variance $\sigma^2 = \sigma_0^2$.
4. Apply the maximum likelihood method to the Poisson distribution.
5. Apply the maximum likelihood method to the normal distribution with $\mu = 0$.

Using probability paper, estimate μ and σ from the given sample, assuming that it comes from a normally distributed population.

6. The sample in Ex. 1, Sec. 19.3.
7. The (ungrouped) sample in Table 18, Sec. 19.2.
8. The grouped sample in Table 19, Sec. 19.2.
9. The sample in Prob. 1, Sec. 19.2.
10. The sample in Prob. 3, Sec. 19.2.

19.14 Confidence Intervals

The last section was devoted to point estimates of parameters, and we shall now discuss **interval estimates,** starting with a general motivation.

Whenever we use mathematical approximation formulas, we should try to find out how much the approximate value can at most deviate from the unknown true value. For example, in the case of numerical integration methods there exist "error formulas" from which we can compute the maximum possible error (that is, the difference between the approximate and the true value). Suppose that in a certain case we obtain 2.47 as an approximate value of a given integral and ± 0.02 as the maximum possible deviation from the unknown exact value. Then we are sure that the values $2.47 - 0.02 = 2.45$ and $2.47 + 0.02 = 2.49$ *"include"* the unknown exact value, that is, 2.45 is smaller than or equal to that value and 2.49 is larger than or equal to that value.

In estimating a parameter θ, the corresponding problem would be the determination of two numerical quantities that depend on the sample values and include the unknown value of the parameter with certainty. However, we already know that from a sample we cannot draw conclusions about the corresponding population that are 100% certain. So we have to be more modest and modify our problem, as follows.

Choose a probability γ close to 1 *(for example, $\gamma = 95\%$, 99%, or the like). Then determine two quantities Θ_1 and Θ_2 such that the probability that Θ_1 and Θ_2 include the exact unknown value of the parameter θ is equal to γ.*

Here the idea is that we replace the impossible requirement "with certainty" by the attainable requirement "with a preassigned probability close to 1."

Numerical values of those two quantities should be computed from a given sample $x_1, \cdots, x_n$. The n sample values may be regarded as observed values of n random variables $X_1, \cdots, X_n$. Then Θ_1 and Θ_2 are functions of these random variables and therefore random variables, too. Our above requirement may thus be written

$$P(\Theta_1 \leqq \theta \leqq \Theta_2) = \gamma.$$

If we know such functions Θ_1 and Θ_2 and a sample is given, we may compute a numerical value θ_1 of Θ_1 and a numerical value θ_2 of Θ_2. The interval with endpoints θ_1 and θ_2 is called a **confidence interval**[8] or *interval estimate* for the unknown parameter θ, and we shall denote it by

$$\text{CONF } \{\theta_1 \leqq \theta \leqq \theta_2\}.$$

The values θ_1 and θ_2 are called **lower** and **upper confidence limits** for θ. The number γ is called the **confidence level.** One chooses $\gamma = 95\%$, 99%, sometimes 99.9%.

Clearly, if we intend to obtain a sample and determine a corresponding confidence interval, then γ is the probability of getting an interval that will include the unknown exact value of the parameter.

For example, if we choose $\gamma = 95\%$, then we can expect that *about* 95% of the samples that we may obtain will yield confidence intervals that do include

---

[8] The modern theory and terminology of confidence intervals were developed by J. Neyman (*Annals Math. Stat.* **6,** 1935, 111–116).

In mathematics, $\theta_1 \leqq \theta \leqq \theta_2$ means that θ lies between θ_1 and θ_2, and, to avoid misunderstandings, it seems worthwhile to characterize a confidence interval by a special symbol, such as **CONF**.

Table 23. Determination of a Confidence Interval for the Mean μ of a Normal Distribution with Known Variance σ^2

1st step. Choose a confidence level γ (95%, 99% or the like).

2nd step. Determine the corresponding c:

γ	0.90	0.95	0.99	0.999
c	1.645	1.960	2.576	3.291

3rd step. Compute the mean $\bar{x}$ of the sample $x_1, \cdots, x_n$.

4th step. Compute $k = c\sigma/\sqrt{n}$. The confidence interval for μ is

(1)   CONF $\{\bar{x} - k \leq \mu \leq \bar{x} + k\}$.

---

the value θ, whereas the remaining 5% do not. Hence the statement "the confidence interval includes θ" will be correct in *about* 19 out of 20 cases, while in the remaining case it will be false.

Choosing $\gamma = 99\%$ instead of 95%, we may expect that statement to be correct even in *about* 99 out of 100 cases. But we shall see later that the intervals corresponding to $\gamma = 99\%$ are longer than those corresponding to $\gamma = 95\%$. This is the disadvantage of increasing γ.

What value of γ should we choose in a concrete case? This is not a mathematical question, but one that must be answered from the viewpoint of the application, taking into account what risk of making a false statement we can afford.

It is clear that the uncertainty involved in the present method as well as in the methods yet to be discussed comes from the sampling process, so that the statistican should be prepared for his share of mistakes. However, he is no worse off than a judge or a banker, who is subject to the laws of chance. Quite the contrary, he has the advantage that he can *measure* his chances of making a mistake.

We shall now consider methods for obtaining confidence intervals for the mean (Tables 23, 24) and the variance (Table 25) of the **normal distribution.** The corresponding theory will be presented in the last part of this section.

Example 1. Determine a 95%-confidence interval for the mean of a normal distribution with variance $\sigma^2 = 9$, using a sample of $n = 100$ values with mean $\bar{x} = 5$.

1st step. $\gamma = 0.95$ is required.

2nd step. The corresponding c equals 1.960.

3rd step. $\bar{x} = 5$ is given.

4th step. We need $k = 1.960 \cdot 3/\sqrt{100} = 0.588$. Hence $\bar{x} - k = 4.412$, $\bar{x} + k = 5.588$, and the confidence interval is

$$\text{CONF} \{4.412 \leq \mu \leq 5.588\}.$$

Sometimes this is written $\mu = 5 \pm 0.588$, but we shall not use this notation, which may be misleading.

Example 2 (*Determination of the sample size necessary for obtaining a confidence interval of prescribed length*). How large must n in the last example be, if we want to obtain a 95%-confidence interval of length $L = 0.4$?

Table 24. Determination of a Confidence Interval for the Mean μ of a Normal Distribution with Unknown Variance σ²

1st step. Choose a confidence level γ (95%, 99% or the like).

2nd step. Determine the solution c of the equation

(2) $F(c) = \frac{1}{2}(1 + \gamma)$

from the table of the *t*-distribution with $n - 1$ degrees of freedom (Table A11 in Appendix 4; n = sample size).

3rd step. Compute the mean $\bar{x}$ and the variance s^2 of the sample $x_1, \cdots, x_n$.

4th step. Compute $k = sc/\sqrt{n}$. The confidence interval is

(3)   CONF $\{\bar{x} - k \leq \mu \leq \bar{x} + k\}$.

The interval (1) has the length $L = 2k = 2c\sigma/\sqrt{n}$. Solving for n, we obtain

$$n = (2c\sigma/L)^2.$$

In the present case the answer is $n = (2 \cdot 1.960 \cdot 3/0.4)^2 \approx 870$.

Figure 368 shows that L decreases as n increases. The shorter the confidence interval should be, the larger the sample size n must be chosen.

Table 24 shows how to determine a confidence interval for the mean μ of a normal distribution with unknown variance σ^2. The steps are similar to those in Table 23, but k in Table 24 is different from that in Table 23. Moreover, c depends on n and must be determined from Table A11 in Appendix 4, which contains values z corresponding to given values of the distribution function

$$F(z) = K_m \int_{-\infty}^{z} \left(1 + \frac{u^2}{m}\right)^{-(m+1)/2} du$$

of the so-called **t-distribution** of Student (pseudonym for W. S. Gosset) where $K_m = \Gamma(\frac{1}{2}m + \frac{1}{2})/[\sqrt{m\pi}\, \Gamma(\frac{1}{2}m)]$, and $\Gamma(\alpha)$ is the gamma function (cf. (24) in Appendix 3). $m(=1, 2, \cdots)$ is a parameter which is called the *number of degrees of freedom* of the distribution.

Table 25. Determination of a Confidence Interval for the Variance σ² of the Normal Distribution, Whose Mean Need Not Be Known

---

1st step. Choose a confidence level γ (95%, 99% or the like).

2nd step. Determine solutions c_1 and c_2 of the equations

(4) $F(c_1) = \frac{1}{2}(1 - \gamma)$, $F(c_2) = \frac{1}{2}(1 + \gamma)$

from the table of the chi-square distribution with $n - 1$ degrees of freedom (Table A12 in Appendix 4; n = size of sample).

3rd step. Compute $(n - 1)s^2$ where s^2 is the variance of the sample $x_1, \cdots, x_n$.

4th step. Compute $k_1 = (n - 1)s^2/c_1$ and $k_2 = (n - 1)s^2/c_2$. The confidence interval is

(5)   CONF $\{k_2 \leq \sigma^2 \leq k_1\}$.

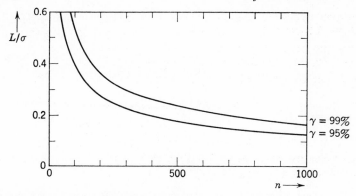

Fig. 368. Length of the confidence interval (1) (measured in multiples of σ) as a function of the sample size n

Example 3. Using the sample in Table 17, Sec. 19.2, determine a 99%-confidence interval for the mean μ of the corresponding population, assuming that the population is normal. (This assumption will be justified in Sec. 19.18.)

1st step. $\gamma = 0.99$ is required.

2nd step. Since $n = 100$, we obtain $c = 2.63$.

3rd step. Calculation gives $\bar{x} = 364.70$ and $s = \sqrt{720.1} = 26.83$.

4th step. We find $k = 26.83 \cdot 2.63/10 = 7.06$. Hence the confidence interval is

$$\text{CONF } \{357.64 \leqq \mu \leqq 371.76\}.$$

For comparison, if σ were known and equal to 26.83, Table 23 would give the value $k = 2.576 \cdot 26.83/\sqrt{100} = 6.91$, and CONF $\{357.79 \leqq \mu \leqq 371.61\}$. This differs but little from the preceding result because n is large. For smaller n the difference would be considerable, as Fig. 369 illustrates.

The steps in Table 25 for obtaining a confidence interval for the variance of a normal distribution are similar to those in Tables 23 and 24, but we now have to

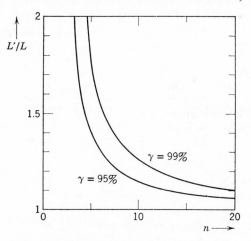

Fig. 369. Ratio of the lengths L' and L of the confidence intervals (3) and (1) as a function of the sample size n for equal s and σ

determine two numbers c_1 and c_2. Both numbers are obtained from Table A12 in Appendix 4, which contains values z corresponding to given values of the distribution function

$$F(z) = C_m \int_0^z e^{-u/2} u^{(m-2)/2}\, du \qquad \text{when} \qquad z \geqq 0,$$

$F(z) = 0$ when $z < 0$, of the so-called χ^2-**distribution** (*chi-square distribution*) where $C_m = 1/[2^{m/2}\Gamma(\tfrac{1}{2}m)]$ and $m\ (= 1, 2, \cdots)$ is a parameter which is called the *number of degrees of freedom* of the distribution.

Example 4. Using the sample in Table 17, Sec. 19.2, determine a 95%-confidence interval for the variance of the corresponding population.

1st step. $\gamma = 0.95$ is required.

2nd step. Since $n = 100$, we find $c_1 = 73.4$ and $c_2 = 128$.

3rd step. From Table 17 in Sec. 19.2 we compute $99s^2 = 71{,}291$.

4th step. The confidence interval is

$$\text{CONF } \{556 \leqq \sigma^2 \leqq 972\}.$$

Other distributions. Confidence intervals for the mean and variance of other distributions may be obtained by using the previous methods and *sufficiently large samples. Practically* speaking, if the sample indicates that the skewness of the unknown distribution is small, one should take samples of size $n = 20$ at least for obtaining confidence intervals for μ and samples of size $n = 50$ at least for obtaining confidence intervals for σ^2. The reason for this method will be explained at the end of this section.

Theoretical basis of the methods in Tables 23–25. We shall now discuss the theory which justifies our methods for obtaining confidence intervals, using the following simple but very important idea.

So far we have regarded the values $x_1, \cdots, x_n$ of a sample as n observed values of a single random variable X. We may equally well regard these n values as single observations of n random variables $X_1, \cdots, X_n$ which have the same distribution (the distribution of X) and are independent because the sample values are assumed to be independent.

For deriving (1) in Table 23 we need

Theorem 1 (Sum of independent normal random variables). *Suppose that X_1, $X_2, \cdots, X_n$ are independent normal random variables with means $\mu_1, \mu_2, \cdots, \mu_n$ and variances $\sigma_1{}^2, \sigma_2{}^2, \cdots, \sigma_n{}^2$, respectively. Then the random variable*

$$X = X_1 + X_2 + \cdots + X_n$$

is normal with the mean

$$\mu = \mu_1 + \mu_2 + \cdots + \mu_n$$

and the variance

$$\sigma^2 = \sigma_1{}^2 + \sigma_2{}^2 + \cdots + \sigma_n{}^2.$$

The statements about μ and σ follow directly from Theorems 1 and 3 in Sec. 19.11. The proof that X is normal can be found in Ref. [H4], cf. Appendix 1.

From this theorem, Theorem 1 in Sec. 19.10, and Theorem 3 in Sec. 19.8 we obtain

Theorem 2. *If $X_1, \cdots, X_n$ are independent normal random variables each of which has mean μ and variance σ^2, then the random variable*

$$\bar{X} = \frac{1}{n}(X_1 + \cdots + X_n) \tag{6}$$

is normal with the mean μ and the variance σ^2/n, and the random variable

$$Z = \sqrt{n}\,\frac{\bar{X} - \mu}{\sigma} \tag{7}$$

is normal with the mean 0 and the variance 1.

Let us derive (1). From the general motivation at the beginning of this section we remember that our goal is to find two random variables Θ_1 and Θ_2 such that

$$P(\Theta_1 \leqq \mu \leqq \Theta_2) = \gamma, \tag{8}$$

where γ is chosen, and the sample gives observed values θ_1 of Θ_1 and θ_2 of Θ_2, which then yield a confidence interval CONF $\{\theta_1 \leqq \mu \leqq \theta_2\}$. In the present case this can be done as follows. We choose a number γ between 0 and 1 and determine c from Table A9, Appendix 4, such that $P(-c \leqq Z \leqq c) = \gamma$. (When $\gamma = 0.90$, etc., we obtain the values c in Table 23.) The inequality $-c \leqq Z \leqq c$ with Z given by (7) is

$$-c \leqq \sqrt{n}\,(\bar{X} - \mu)/\sigma \leqq c$$

and can be transformed into an inequality for μ. In fact, multiplication by $\sigma/\sqrt{n}$ yields $-k \leqq \bar{X} - \mu \leqq k$ where $k = c\sigma/\sqrt{n}$. Multiplying by -1 and adding $\bar{X}$, we get

$$\bar{X} + k \geqq \mu \geqq \bar{X} - k. \tag{9}$$

Thus $P(-c \leqq Z \leqq c) = \gamma$ is equivalent to $P(\bar{X} - k \leqq \mu \leqq \bar{X} + k) = \gamma$. This is of the form (8) with $\Theta_1 = \bar{X} - k$ and $\Theta_2 = \bar{X} + k$. Under our assumptions it means that with probability γ the random variables $\bar{X} - k$ and $\bar{X} + k$ will assume values that include the unknown mean μ. Regarding the sample values $x_1, \cdots, x_n$ in Table 23 as observed values of n independent normal random variables $X_1, \cdots, X_n$, we see that the sample mean $\bar{x}$ is an observed value of (6), and by inserting this value into (9) we obtain (1).

For deriving (3) in Table 24 we need

Theorem 3. *Let $X_1, \cdots, X_n$ be independent normal random variables with the same mean μ and the same variance σ^2. Then the random variable*

$$T = \sqrt{n}\,\frac{\bar{X} - \mu}{S} \tag{10}$$

where $\bar{X}$ is given by (6) and

(11)
$$S^2 = \frac{1}{n-1} \sum_{j=1}^{n} (X_j - \bar{X})^2$$

has a t-distribution (cf. p. 754) with $n - 1$ degrees of freedom.

The proof can be found in Ref. [H4], cf. Appendix 1.

The derivation of (3) is similar to that of (1). We choose a number γ between 0 and 1 and determine a number c from Table A11, Appendix 4, with $n - 1$ degrees of freedom such that

(12) $$P(-c \leqq T \leqq c) = F(c) - F(-c) = \gamma.$$

Since the *t*-distribution is symmetric, we have $F(-c) = 1 - F(c)$, and (12) assumes the form (2). Transforming $-c \leqq T \leqq c$ in (12) as before, we get

(13) $$\bar{X} - K \leqq \mu \leqq \bar{X} + K \qquad \text{where} \qquad K = cS/\sqrt{n},$$

and (12) becomes $P(\bar{X} - K \leqq \mu \leqq \bar{X} + K) = \gamma$. By inserting the observed values $\bar{x}$ of $\bar{X}$ and s^2 of S^2 into (13) we obtain (3).

For deriving (5) in Table 25 we need

Theorem 4. *Under the assumptions in Theorem 3 the random variable*

(14)
$$Y = (n-1)\frac{S^2}{\sigma^2}$$

where S^2 is given by (11), *has a chi-square distribution (cf. p. 756) with $n - 1$ degrees of freedom.*

The proof can be found in Ref. [H4], cf. Appendix 1.

The derivation of (5) is similar to that of (1) and (3). We choose a number γ between 0 and 1 and determine c_1 and c_2 from Table A12, Appendix 4, such that [cf. (4)]

$$P(Y \leqq c_1) = F(c_1) = \tfrac{1}{2}(1 - \gamma), \qquad P(Y \leqq c_2) = F(c_2) = \tfrac{1}{2}(1 + \gamma).$$

Subtraction yields

$$P(c_1 \leqq Y \leqq c_2) = P(Y \leqq c_2) - P(Y \leqq c_1) = \gamma.$$

Transforming $c_1 \leqq Y \leqq c_2$ with Y given by (14) into an inequality for σ^2, we obtain

$$\frac{n-1}{c_2} S^2 \leqq \sigma^2 \leqq \frac{n-1}{c_1} S^2.$$

By inserting the observed value s^2 of S^2 we obtain (5).

Confidence intervals for the mean and variance of other distributions can be obtained by the methods in Tables 23 and 25, but we must use large samples. This follows from the basic

Theorem 5 (Central limit theorem). *Let* $X_1, \cdots, X_n, \cdots$ *be independent random variables that have the same distribution function and therefore the same mean* μ *and the same variance* σ^2. *Let* $Y_n = X_1 + \cdots + X_n$. *Then the random variable*

$$(15) \qquad\qquad Z_n = \frac{Y_n - n\mu}{\sigma\sqrt{n}}$$

is **asymptotically normal** *with the mean* 0 *and the variance* 1, *that is, the distribution function* $F_n(x)$ *of* Z_n *satisfies*

$$\lim_{n \to \infty} F_n(x) = \Phi(x) = \frac{1}{\sqrt{2\pi}} \int_{-\infty}^{x} e^{-u^2/2} \, du.$$

A proof can be found in Ref. [H4], cf. Appendix 1.

We know that if $X_1, \cdots, X_n$ are independent random variables with the same mean μ and the same variance σ^2, then their sum $X = X_1 + \cdots + X_n$ has the following properties.

(A) X has the mean $n\mu$ and the variance $n\sigma^2$ (cf. Theorems 1 and 3 in Sec. 19.11).
(B) If those variables are normal, then X is normal (cf. Theorem 1).

If those variables are not normal, then (B) fails to hold, but if n is large, then X is approximately normal (cf. Theorem 5) and this justifies the application of methods for the normal distribution to other distributions, but in such a case we have to use large samples.

Problems

1. Find a 99%-confidence interval for the mean μ of a normal population with variance $\sigma^2 = 1.69$, using a sample of size 36 with mean 18.4.
2. Obtain a 95%-confidence interval for the mean μ of a normal population with variance $\sigma^2 = 16$ from Fig. 368, using a sample of size 300 with mean 87.
3. Find a 95%-confidence interval for the mean μ of a normal population with standard deviation 2.5, using the sample 16, 12, 19, 10, 15.
4. What sample size would be needed to produce a 95%-confidence interval (1) of length (a) 2σ, (b) σ?

Assuming that the populations from which the following samples are taken are normal, determine a 95%-confidence interval for the mean μ of the population.

5. A sample of size 10 with mean 4.37 and standard deviation 0.157.
6. Density [grams per cm^3] of coke 1.40, 1.45, 1.39, 1.44, 1.38.
7. Nitrogenium content [%] of steel 0.74, 0.75, 0.73, 0.75, 0.74, 0.72.

8. What sample size should we use in Prob. 5 if we want to obtain a confidence interval of length 0.1?

Assuming that the populations from which the following samples are taken are normal, determine a 99%-confidence interval for the variance σ^2 of the population.

9. Rockwell hardness of tool bits 64.9, 64.1, 63.8, 64.0.
10. The sample in Prob. 7.
11. A sample of size $n = 128$ with variance $s^2 = 1.921$.

12. Why are interval estimates in most cases more useful than point estimates?
13. If X is normal with mean 12 and variance 25, what distributions do $-X$, $2X$, and $X4 - 1$ have?
14. Using Theorem 1, find out in what fraction of random matches of holes and bolts the bolts would fit, assuming that the diameter D_1 of the holes is normal with mean $\mu_1 = 1$ in. and standard deviation $\sigma_1 = 0.01$ in. and the diameter D_2 of the bolts is normal with mean $\mu_2 = 0.99$ in. and standard deviation $\sigma_2 = 0.01$ in.
15. If X_1 and X_2 are independent normal random variables with mean 6 and -2 and variance 4 and 6, respectively, what distribution does $4X_1 - X_2$ have?

19.15 *Testing of Hypotheses, Decisions*

A statistical **hypothesis** is an assumption about the distribution of a random variable, for example, that a certain distribution has mean 20.3, etc. A statistical **test** of a hypothesis is a procedure in which a sample is used to find out whether we may **"not reject" ("accept")** the hypothesis, that is, act as though it is true, or whether we should **"reject"** it, that is, act as though it is false.

These tests are applied quite frequently, and we may ask why they are important. Often we have to make decisions in situations where chance variation plays a role. If we have a choice between, say, two possibilities, the decision between them might be based on the result of some statistical test.

For example, if we want to use a certain lathe for producing bolts whose diameter should lie between given limits and we allow at most 2% defective bolts, we may take a sample of 100 bolts produced on that lathe and use it for testing the hypothesis $\sigma^2 = \sigma_0^2$ that the variance σ^2 of the corresponding population has a certain value σ_0^2 which we choose such that we can expect to obtain not more than 2% defectives. A meaningful *alternative* in this case is $\sigma^2 > \sigma_0^2$. Depending on the result of the test, we either do not reject the hypothesis $\sigma^2 = \sigma_0^2$ (and then use that lathe) or we reject it, assert that $\sigma^2 > \sigma_0^2$, and use another lathe. In the latter case we say that the test indicates a **significant deviation** of σ^2 from σ_0^2, that is, a deviation which is not merely caused by the unavoidable influence of chance factors but by the lack of precision of the lathe.

In other cases we may want to compare two things, for example, two different drugs, two methods of performing a certain work, the accuracy of two methods of measurement, the quality of products produced by two different tools, etc. Depending on the result of a suitable test we decide to use one of the two drugs, to introduce the better method of working, etc.

Typical sources for hypotheses are as follows.

1. The hypothesis may come from a quality requirement. (Experience about attainable quality may be gained by producing a large number of items with special care.)
2. The hypothesis is based on values known from previous experience.
3. The hypothesis results from a theory which one wants to verify.
4. The hypothesis is a pure guess caused by occasional observations.

Let us start with a simple introductory example.

Example 1. The birth of a single child may be regarded as a random experiment with two possible outcomes, namely, *B*: *Birth of a boy* and *G*: *Birth of a girl*. Intuitively we should feel that both outcomes are about equally likely. However, in the literature it is often claimed that births of boys are somewhat more frequent than births of girls. On the basis of this situation we want to test the hypothesis that the two outcomes *B* and *G* have the same probability. If we let *p* denote the probability of the outcome *B*, our hypothesis to be tested is $p = 50\% = 0.5$. Because of those claims we choose the alternative $p > 0.5$.

For the test we use a sample of $n = 3000$ babies born in 1962 in Graz, Austria; 1578 of these babies were boys.

If the hypothesis is true, we expect that in a sample of $n = 3000$ births *about* 1500 are boys. If the alternative holds, then we expect more than 1500 boys, on the average. Hence if the number of boys actually observed is much larger than 1500, we can use this as an indication that the hypothesis may be false, and we reject it.

To perform our test, we proceed as follows. We first determine a critical value *c*. Because of the alternative, *c* will be greater than 1500. (A method for determining *c* will be given below.) Then, if the observed number of boys is greater than *c*, we reject the hypothesis. If that number is not greater than *c*, we do not reject the hypothesis.

The basic question now is how we should choose *c*, that is, where we should draw the line between small random deviations and large significant deviations. Different people may have different opinions, and to answer the question we must use mathematical arguments. In the present case these are very simple, as will be seen in what follows.

We determine *c* such that if the hypothesis is true, the probability of observing more than *c* boys in a sample of 3000 single births is a very small number, call it α. It is customary to choose $\alpha = 1\%$ or 5%. Choosing $\alpha = 1\%$ (or 5%) we risk about once in 100 cases (in 20 cases, respectively) rejecting a hypothesis even though it is true. We shall return to this point later. Let us choose $\alpha = 1\%$ and consider the random variable

$$X = number\ of\ boys\ in\ 3000\ births.$$

Assuming that the hypothesis is true, we obtain the critical value *c* from the equation

$$(1) \qquad\qquad\qquad P(X > c)_{p=0.5} = \alpha = 0.01.$$

(That assumption is indicated by the subscript $p = 0.5$.) If the observed value 1578 is greater than *c*, we reject the hypothesis. If $1578 \leq c$, we do not reject the hypothesis.

To determine *c* from (1) we must know the distribution of *X*. For our purpose the binomial distribution is a sufficiently accurate model. Hence if the hypothesis is true, *X* has a binomial distribution with $p = 0.5$ and $n = 3000$. This distribution can be approximated by the normal distribution with mean $\mu = np = 1500$ and variance $\sigma^2 = npq = 750$; cf. Sec. 19.10. (For the sake of simplicity we shall disregard the term 0.5 in (11), Sec. 19.10.) The density is shown in Fig. 370 on the next page. Using (1), we thus obtain

$$P(X > c) = 1 - P(X \leq c) \approx 1 - \Phi\left(\frac{c - 1500}{\sqrt{750}}\right) = 0.01.$$

Table A9 in Appendix 4 yields $(c - 1500)/\sqrt{750} = 2.326$. Hence $c = 1564$. Since $1578 > c$, we reject the hypothesis and assert that $p > 0.5$. This completes the test.

For a sample of 300 values we would obtain from (1) with $X = $ *number of boys in* 300 *births* the critical value $c = 170$, and a sample in which 158 (the same percentage as in the large sample) are boys would give $158 < c$, so that the hypothesis would not be rejected. This is interesting, because it illustrates that the usefulness of the test increases with increasing sample size *n*. We should choose *n* so large that the test will yield information about the question which is of interest in a concrete case. On the other hand, *n* should not be larger than necessary, for reasons of time and money. In most cases an economical choice of *n* can be found from a small preliminary experiment.

Fig. 370. (Approximate) density of X in Ex. 1 if the hypothesis is true. Critical value $c = 1564$

The hypothesis to be tested is sometimes called the *null hypothesis*, and a counter-assumption (like $p > 0.5$ in Ex. 1) is called an *alternative hypothesis* or, briefly, an **alternative.** The number α (or $100\alpha\%$) is called the **significance level** of the test. c is called the *critical value*. The region containing the values for which we reject the hypothesis is called the **rejection region** or **critical region.** The region of values for which we do not reject the hypothesis is called the **acceptance region.** A frequent choice for α is 5%.

Let θ be an unknown parameter in a distribution, and suppose that we want to test the hypothesis $\theta = \theta_0$. There are three main types of alternatives, namely

(2) $\theta > \theta_0$

(3) $\theta < \theta_0$

(4) $\theta \neq \theta_0.$

(2) and (3) are called **one-sided alternatives,** and (4) is called a **two-sided alternative.** (2) is of the type considered in Ex. 1 (where $\theta_0 = p = 0.5$ and $\theta = p > 0.5$); c lies to the right of θ_0, and the rejection region extends from c to ∞ (Fig. 371, upper part). The test is called a **right-sided test.** In the case (3) the number c lies to the left of θ_0, the rejection region extends from c to $-\infty$ (Fig. 371, middle part), and the test is called a **left-sided test.** Both types of tests are called **one-sided tests.** In the case (4) we have two critical values c_1 and c_2 ($>c_1$), the rejection region extends from c_1 to $-\infty$ and from c_2 to ∞, and the test is called a **two-sided test.**

All three types of hypotheses are of practical importance. For example, (3) may appear in connection with testing strength of material. θ_0 may then be the required strength, and the alternative characterizes an undesirable weakness. The case that the material may be stronger than required is acceptable, of course, and does therefore not need special attention. (4) may be important, for example, in connection with the diameter of an axle shaft. θ_0 then is the required diameter, and slimmer axle-shafts are as bad as thicker ones, so that one has to watch for undesirable deviations from θ_0 in both directions.

We shall now consider the **risks of making false decisions** in a test of a hypothesis $\theta = \theta_0$ against an alternative which, for the sake of simplicity, is a single number

θ_1. Let $\theta_1 > \theta_0$ so that we have a right-sided test, the consideration for a left-sided or a two-sided test being similar. From a given sample $x_1, \cdots, x_n$ we compute a value $\hat{\theta} = g(x_1, \cdots, x_n)$. If $\hat{\theta} > c$, the hypothesis is rejected (as in Ex. 1). If $\hat{\theta} \leq c$, the hypothesis is not rejected. $\hat{\theta}$ can be regarded as an observed value of the random variable

$$\hat{\Theta} = g(X_1, \cdots, X_n)$$

because x_j may be regarded as an observed value of X_j, where $j = 1, \cdots, n$. In this test there are two possibilities of making an error, as follows.

Type I error (cf. Table 26, p. 764). The hypothesis is true, but is rejected because $\hat{\Theta}$ assumes a value $\hat{\theta} > c$. Obviously, the probability of making such an error equals

(5) $$P(\hat{\Theta} > c)_{\theta=\theta_0} = \alpha,$$

the significance level of the test.

Type II error (cf. Table 26). The hypothesis is false, but is not rejected because $\hat{\Theta}$ assumes a value $\hat{\theta} \leq c$. The probability of making such an error is denoted by β; thus

(6) $$P(\hat{\Theta} \leq c)_{\theta=\theta_1} = \beta.$$

$\eta = 1 - \beta$ is called the **power** of the test. Obviously, this is the probability of avoiding a Type II error.

Formulas (5) and (6) show that both α and β depend on c, and we would like to choose c so that these probabilities of making errors are as small as possible.

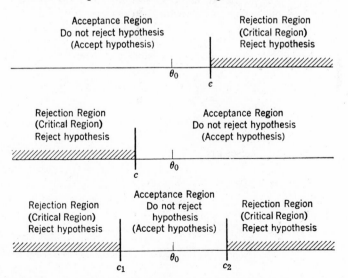

Fig. 371. Test in the case of alternative (2) (upper part of the figure), alternative (3) (middle part), and alternative (4)

Table 26. Type I and Type II Errors in Testing a Hypothesis
$\theta = \theta_0$ **Against an Alternative** $\theta = \theta_1$

		Unknown Truth	
		$\theta = \theta_0$	$\theta = \theta_1$
Not rejected (Accepted)	$\theta = \theta_0$	True decision $P = 1 - \alpha$	Type II error $P = \beta$
	$\theta = \theta_1$	Type I error $P = \alpha$	True decision $P = 1 - \beta$

But Fig. 372 indicates that these are conflicting requirements, since to let α decrease we must shift c to the right, but then β increases. In practice we first choose α (5%, sometimes 1%), then determine c, and finally compute β. If β is large so that the power $\eta = 1 - \beta$ is small, we should repeat the test, choosing a larger sample, for reasons that will appear shortly.

If the alternative is not a single number but is of the form (2)–(4), then β becomes a function of θ. This function $\beta(\theta)$ is called the **operating characteristic** (OC) of the test and its curve the **OC curve.** Clearly, in this case $\eta = 1 - \beta$ also depends on θ, and this function $\eta(\theta)$ is called the **power function** of the test.

Of course, from a test that leads to the acceptance of a certain hypothesis θ_0 it does *not* follow that this is the only possible hypothesis or the best possible hypothesis. Hence the terms **"not reject"** or **"fail to reject"** are perhaps better than the term **"accept."**

The following examples will explain tests of practically important hypotheses.

*Example 2 (**Mean of the normal distribution with known variance**).* Let X be a normal random variable with variance $\sigma^2 = 9$. Using a sample of size $n = 10$ with mean $\bar{x}$, test the hypothesis $\mu = \mu_0 = 24$ against the three types of alternatives, namely,

$$(a)\ \mu > \mu_0 \qquad (b)\ \mu < \mu_0 \qquad (c)\ \mu \neq \mu_0.$$

We choose the significance level $\alpha = 0.05$. An estimate of the mean will be obtained from

$$\bar{X} = \frac{1}{n}(X_1 + \cdots + X_n).$$

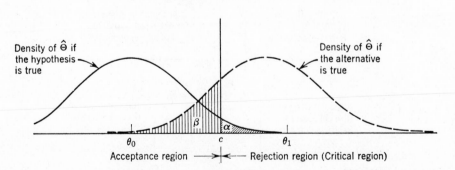

Fig. 372. Illustration of Type I and II errors in testing a hypothesis $\theta = \theta_0$ against an alternative $\theta = \theta_1 \, (> \theta_0)$

If the hypothesis is true, $\overline{X}$ is normal with mean $\mu = 24$ and variance $\sigma^2/n = 0.9$, cf. Theorem 2, Sec. 19.14. Hence we may obtain the critical value c from Table A9 in Appendix 4.

Case (a). We determine c from $P(\overline{X} > c)_{\mu=24} = \alpha = 0.05$, that is,

$$P(\overline{X} \le c)_{\mu=24} = \Phi\left(\frac{c-24}{\sqrt{0.9}}\right) = 1 - \alpha = 0.95.$$

Table A9, Appendix 4, yields $(c-24)/\sqrt{0.9} = 1.645$, and $c = 25.56$, which is greater than μ_0, as in the upper part of Fig. 371. If $\bar{x} \le 25.56$, the hypothesis is not rejected. If $\bar{x} > 25.56$, it is rejected. The power of the test is (cf. Fig. 373 on the next page)

$$\eta(\mu) = P(\overline{X} > 25.56)_\mu = 1 - P(\overline{X} \le 25.56)_\mu$$

(7)

$$= 1 - \Phi\left(\frac{25.56 - \mu}{\sqrt{0.9}}\right) = 1 - \Phi(26.94 - 1.05\mu).$$

Case (b). The critical value c is obtained from the equation

$$P(\overline{X} \le c)_{\mu=24} = \Phi\left(\frac{c-24}{\sqrt{0.9}}\right) = \alpha = 0.05.$$

Table A9, Appendix 4, yields $c = 24 - 1.56 = 22.44$. If $\bar{x} \ge 22.44$, we do not reject the hypothesis. If $\bar{x} < 22.44$, we reject it. The power of the test is

(8) $$\eta(\mu) = P(\overline{X} \le 22.44)_\mu = \Phi\left(\frac{22.44 - \mu}{\sqrt{0.9}}\right) = \Phi(23.65 - 1.05\mu).$$

Case (c). Since the normal distribution is symmetric, we choose c_1 and c_2 equidistant from $\mu = 24$, say, $c_1 = 24 - k$ and $c_2 = 24 + k$, and determine k from

$$P(24 - k \le \overline{X} \le 24 + k)_{\mu=24} = \Phi\left(\frac{k}{\sqrt{0.9}}\right) - \Phi\left(-\frac{k}{\sqrt{0.9}}\right) = 1 - \alpha = 0.95.$$

Using Table A9, Appendix 4, we obtain $k/\sqrt{0.9} = 1.960$, $k = 1.86$. Hence $c_1 = 24 - 1.86 = 22.14$ and $c_2 = 24 + 1.86 = 25.86$. If $\bar{x}$ is not smaller than c_1 and not greater than c_2, we do not reject the hypothesis. Otherwise we reject it. The power of the test is (cf. Fig. 373)

$$\eta(\mu) = P(\overline{X} < 22.14)_\mu + P(\overline{X} > 25.86)_\mu = P(\overline{X} < 22.14)_\mu + 1 - P(\overline{X} \le 25.86)_\mu$$

(9)

$$= 1 + \Phi\left(\frac{22.14 - \mu}{\sqrt{0.9}}\right) - \Phi\left(\frac{25.86 - \mu}{\sqrt{0.9}}\right)$$

$$= 1 + \Phi(23.34 - 1.05\mu) - \Phi(27.26 - 1.05\mu).$$

Consequently, the operating characteristic $\beta(\mu) = 1 - \eta(\mu)$ (see before) is (cf. Fig. 374)

$$\beta(\mu) = \Phi(27.26 - 1.05\mu) - \Phi(23.34 - 1.05\mu).$$

If we take a larger sample, say, of size $n = 100$ (instead of 10), then $\sigma^2/n = 0.09$ (instead of 0.9), and the critical values are $c_1 = 23.41$ and $c_2 = 24.59$, as can be verified by calculation. Then the operating characteristic of the test is

$$\beta(\mu) = \Phi\left(\frac{24.59 - \mu}{\sqrt{0.09}}\right) - \Phi\left(\frac{23.41 - \mu}{\sqrt{0.09}}\right) = \Phi(81.97 - 3.33\mu) - \Phi(78.03 - 3.33\mu).$$

Figure 374 shows that the corresponding OC curve is steeper than that for $n = 10$. This means that the increase of n has led to an improvement of the test. In any practical case, n is chosen as small as possible but so large that the test brings out deviations between μ and μ_0 that are of

Fig. 373. Power $\eta(\mu)$ in Ex. 2, case (*a*) (dashed) and case (*c*)

practical interest. For instance, if deviations of ± 2 units are of interest, we see from Fig. 374 that $n = 10$ is much too small because when $\mu = 24 - 2 = 22$ or $\mu = 24 + 2 = 26$, then β is almost 50%. On the other hand, we see that $n = 100$ will be sufficient for that purpose.

Example 3 (*Mean of the normal distribution with unknown variance*). The tensile strength of a sample of $n = 16$ manila ropes (diameter 3 in.) was measured. The sample mean was $\bar{x} = 4482$ kg, and the sample standard deviation was $s = 115$ kg (N. C. Wiley, 41st Annual Meeting of the Amer. Soc. for Test. Materials). Assuming that the tensile strength is a normal random variable, test the hypothesis $\mu_0 = 4500$ kg against the alternative $\mu_1 = 4400$ kg. Here μ_0 may be a value given by the manufacturer, while μ_1 may result from previous experience.

We choose the significance level $\alpha = 5\%$. If the hypothesis is true, it follows from Theorem 3 in Sec. 19.14, that the random variable

$$T = \sqrt{n}\,\frac{\bar{X} - \mu_0}{S} = 4\,\frac{\bar{X} - 4500}{S}$$

has a *t*-distribution with $n - 1 = 15$ degrees of freedom. The critical value c is obtained from the equation

$$P(T < c)_{\mu_0} = \alpha = 0.05.$$

Table A11 in Appendix 4 yields $c = -1.75$. As an observed value of T we obtain from the sample $t = 4(4482 - 4500)/115 = -0.626$. We see that $t > c$ and do not reject the hypothesis. For obtaining numerical values of the power of the test, we would need tables called noncentral Student *t*-tables; we shall not discuss this question here.

Example 4 (*Variance of the normal distribution*). Using a sample of size $n = 15$ and sample variance $s^2 = 13$ from a normal population, test the hypothesis $\sigma^2 = \sigma_0^2 = 10$ against the alternative $\sigma^2 = \sigma_1^2 = 20$.

Fig. 374. Curves of the operating characteristic (OC curves) in Ex. 2, case (*c*), for two different sample sizes *n*

We choose the significance level $\alpha = 5\%$. If the hypothesis is true, then

$$Y = (n - 1)\frac{S^2}{\sigma_0^2} = 14\frac{S^2}{10} = 1.4 S^2$$

has a chi-square distribution with $n - 1 = 14$ degrees of freedom, cf. Theorem 4, Sec. 19.14. From

$$P(Y > c) = \alpha = 0.05, \qquad \text{that is,} \qquad P(Y \leq c) = 0.95,$$

and Table A12, Appendix 4, with 14 degrees of freedom we obtain $c = 23.68$. This is the critical value of Y. Hence to $S^2 = \sigma_0^2 Y/(n - 1) = 0.714 Y$ there corresponds the critical value $c^* = 0.714 \cdot 23.68 = 16.91$. Since $s^2 < c^*$, we do not reject the hypothesis.

If the alternative is true, the variable

$$Y_1 = 14\frac{S^2}{\sigma_1^2} = 0.7 S^2$$

has a chi-square distribution with 14 degrees of freedom. Hence our test has the power

$$\eta = P(S^2 > c^*)_{\sigma^2 = 20} = P(Y_1 > 0.7c^*)_{\sigma^2 = 20} = 1 - P(Y_1 \leq 11.84)_{\sigma^2 = 20} \approx 62\%$$

and we see that the Type II risk is very large, namely, 38%. To make that risk smaller, we must increase the sample size.

Example 5 (Comparison of the means of two normal distributions). Using a sample $x_1, \cdots, x_{n_1}$ from a normal distribution with unknown mean μ_1 and a sample $y_1, \cdots, y_{n_2}$ from another normal distribution with unknown mean μ_2, we want to test the hypothesis that the means are equal, that is, $\mu_1 = \mu_2$, against an alternative, say, $\mu_1 > \mu_2$. The variances need not be known, but are assumed to be equal.[9] Two cases are of practical importance:

Case A. *The samples have the same size. Furthermore, each value of the first sample corresponds to precisely one value of the other*, because corresponding values result from the same person or thing **(paired comparison)**; for example, two measurements of the same thing by two different methods or two measurements from the two eyes of the same animal, and, more generally, they may result from pairs of *similar* individuals or things, for example, identical twins, pairs of used front tires from the same car, etc. Then we should form the differences of corresponding values and test the hypothesis that the population corresponding to the differences has mean 0, using the method in Ex. 3. If we have a choice, that method is better than the following one.

Case B. *The two samples are independent and not necessarily of the same size.* Then we may proceed, as follows. Suppose that the alternative is $\mu_1 > \mu_2$. We choose a significance level α. We compute the sample means $\bar{x}$ and $\bar{y}$ and $(n_1 - 1)s_1^2$, $(n_2 - 1)s_2^2$, where s_1^2 and s_2^2 are the sample variances. Using Table A11, Appendix 4, with $n_1 + n_2 - 2$ degrees of freedom, we determine c from

$$(10) \qquad\qquad P(T \leq c) = 1 - \alpha.$$

We finally compute

$$(11) \qquad\qquad t_0 = \sqrt{\frac{n_1 n_2 (n_1 + n_2 - 2)}{n_1 + n_2}} \frac{\bar{x} - \bar{y}}{\sqrt{(n_1 - 1)s_1^2 + (n_2 - 1)s_2^2}}.$$

It can be shown that this is an observed value of a random variable which has a t-distribution with $n_1 + n_2 - 2$ degrees of freedom, provided the hypothesis is true. If $t_0 \leq c$, the hypothesis is not rejected. If $t_0 > c$, it is rejected.

---

[9] If the test in the next example shows that the variances differ significantly, then choose two not too small samples of the same size $n_1 = n_2 = n$ (> 30, say), use the fact that (12) is an observed value of an approximately normal random variable with mean 0 and variance 1, and proceed as in Ex. 2.

If the alternative is $\mu_1 \neq \mu_2$, then (10) must be replaced by

(10*) $P(T \leq c_1) = 0.5\alpha, \qquad P(T \leq c_2) = 1 - 0.5\alpha.$

Note that for samples of equal size $n_1 = n_2 = n$, formula (11) reduces to

(12)
$$t_0 = \sqrt{n}\,\frac{\bar{x} - \bar{y}}{\sqrt{s_1{}^2 + s_2{}^2}}.$$

To illustrate the numerical work, let us consider the two samples

105	108	86	103	103	107	124	105

and

89	92	84	97	103	107	111	97

showing the relative output of tin plate workers under two different working conditions (J. J. B. WORTH, J. Indust. Eng. **9**, 1958, 249–253). Assuming that the corresponding populations are normal and have the same variance, let us test the hypothesis $\mu_1 = \mu_2$ against the alternative $\mu_1 \neq \mu_2$. (Equality of variances will be tested in the next example.) We find $\bar{x} = 105.125$, $\bar{y} = 97.500$, $s_1{}^2 = 106.125$, $s_2{}^2 = 84.000$. We choose the significance level $\alpha = 5\%$. From (10*) with $0.5\alpha = 2.5\%$, $1 - 0.5\alpha = 97.5\%$ and Table A11 in Appendix 4 with 14 degrees of freedom we obtain $c_1 = -2.15$ and $c_2 = 2.15$. Formula (12) with $n = 8$ gives the value $t_0 = \sqrt{8} \cdot 7.625 / \sqrt{190.125} = 1.56$. Since $c_1 \leq t_0 \leq c_2$, we do not reject the hypothesis $\mu_1 = \mu_2$ that under both conditions the mean output is the same.

Case A applies to the example, because the two first sample values correspond to a certain type of work, the next two were obtained in another kind of work, etc. So we may use the differences

16	16	2	6	0	0	13	8

of corresponding sample values and the method in Ex. 3 to test the hypothesis $\mu = 0$ where μ is the mean of the population corresponding to the differences. As a logical alternative we take $\mu \neq 0$. The sample mean is $d = 7.625$, and the sample variance is $s^2 = 45.696$. Hence $t = \sqrt{8}(7.625 - 0)/\sqrt{45.696} = 3.19$. From $P(T \leq c_1) = 2.5\%$, $P(T \leq c_2) = 97.5\%$ and Table A11 in Appendix 4 with $n - 1 = 7$ degrees of freedom we obtain $c_1 = -2.37$, $c_2 = 2.37$ and reject the hypothesis because $t = 3.19$ does not lie between c_1 and c_2. Hence our present test, in which we used more information (but the same samples), shows that the difference in output is significant.

Example 6 (Comparison of the variances of two normal distributions). Using the two samples in the last example, test the hypothesis $\sigma_1{}^2 = \sigma_2{}^2$; assume that the corresponding populations are normal and the nature of the experiment suggests the alternative $\sigma_1{}^2 > \sigma_2{}^2$. We find $s_1{}^2 = 106.125$, $s_2{}^2 = 84.000$. We choose the significance level $\alpha = 5\%$. Using $P(V \leq c) = 1 - \alpha = 95\%$ and Table A13 in Appendix 4, with $(n_1 - 1, n_2 - 1) = (7, 7)$ degrees of freedom, we determine $c = 3.79$. We finally compute $v_0 = s_1{}^2/s_2{}^2 = 1.26$. Since $v_0 \leq c$, we do not reject the hypothesis. If $v_0 > c$, we would reject it.

This test is justified by the fact that v_0 is an observed value of a random variable which has a so-called **F-distribution** with $(n_1 - 1, n_2 - 1)$ degrees of freedom, provided the hypothesis is true. (Proof in Ref. [H4], cf. Appendix 1.) The F-distribution with (m, n) degrees of freedom was introduced by R. A. Fisher and has the distribution function $F(z) = 0$ when $z < 0$ and

(13)
$$F(z) = K_{mn} \int_0^z t^{(m-2)/2}(mt + n)^{-(m+n)/2}\, dt \qquad (z \geq 0)$$

where $K_{mn} = m^{m/2} n^{n/2} \Gamma(\tfrac{1}{2}m + \tfrac{1}{2}n)/\Gamma(\tfrac{1}{2}m)\Gamma(\tfrac{1}{2}n)$.

Problems

1. Using the data of Buffon in Table 21, Sec. 19.5, test the hypothesis that the coin is fair, that is, that heads and tails have the same probability of occurrence, against the alternative that heads are more likely than tails.

2. Same test as in Prob. 1, using the data of Pearson in Table 21, Sec. 19.5.
3. Suppose that the sample in Ex. 1, Sec. 19.3, was taken from a normal population. Test the hypothesis $\mu = 0.80$ against the alternative $\mu > 0.80$.
4. Using the sample in Ex. 1, Sec. 19.3, test the hypothesis $\mu = 0.80$ in. (the length indicated on the box) against the alternative $\mu \neq 0.80$ in. (Assume normality.)
5. Test $\mu = 0$ against $\mu > 0$, assuming normality and using the sample $1, -1, 1, 3, -8, 6, 0$ (deviations of the azimuth (in multiples of 0.01 radians) in the 143rd revolution of the satellite Telstar).
6. Test the hypothesis that there is no significant difference between two methods of measuring the starch content of potatoes, assuming normality and using the sample (differences of corresponding values from 16 potatoes, measured in multiples of 0.1%)

$$2 \quad 0 \quad 0 \quad 1 \quad 2 \quad 2 \quad 3 \quad -3 \quad 1 \quad 2 \quad 3 \quad 0 \quad -1 \quad 1 \quad -2 \quad 1.$$

7. Graph the OC curves in Ex. 2, cases (a) and (b).
8. Show that for a normal distribution the two types of errors in a test of a hypothesis $H_0: \mu = \mu_0$ against an alternative $H_1: \mu = \mu_1$ can be made as small as one pleases (not zero) by taking the sample sufficiently large.
9. If a standard drug cures about 70% of patients with a certain disease and a new drug cured 148 of the first 200 patients on whom it was tried, can we conclude that the new drug is better?
10. Assuming normality and equal variance and using independent samples with $n_1 = 9$, $\bar{x} = 12$, $s_1 = 2$, $n_2 = 9$, $\bar{y} = 15$, $s_2 = 2$, test $H_0: \mu_1 = \mu_2$ against $\mu_1 \neq \mu_2$; choose $\alpha = 5\%$.
11. Three specimens of high quality concrete had compressive strength 357, 359, 413 (in kilogram/centimeter^2), and for three specimens of ordinary concrete the values were 346, 358, 302. Test for equality of the population means, $\mu_1 = \mu_2$, against the alternative $\mu_1 > \mu_2$. (Assume normality and equality of variances.)
12. For checking the reliability of a certain electrical method for measuring temperatures, two samples were obtained. Assuming normality and equality of variances of the corresponding populations, test the hypothesis that the population means are equal. The samples are [°C]

$$106.9 \quad 106.3 \quad 107.0 \quad 106.0 \quad 104.9$$
$$106.5 \quad 106.7 \quad 106.8 \quad 106.1 \quad 105.6$$

13. Construct two simple samples such that the hypothesis of equal population means is not rejected although the difference between the sample means is large.
14. Using samples of sizes 10 and 5 with variances $s_1^2 = 50$ and $s_2^2 = 20$ and assuming normality of the corresponding populations, test the hypothesis $H_0: \sigma_1^2 = \sigma_2^2$ against the alternative $\sigma_1^2 > \sigma_2^2$. Choose $\alpha = 5\%$.
15. Using two samples (weight of dust in certain tubes, measured in mg) obtained in an experiment by the British Coal Utilisation Research Association and assuming normality of the corresponding populations, find whether the variances of the populations differ significantly from each other. The samples are

$$75 \quad 20 \quad 70 \quad 70 \quad 85 \quad 90 \quad 100 \quad 40 \quad 35 \quad 65 \quad 90 \quad 35$$
$$20 \quad 35 \quad 55 \quad 50 \quad 65 \quad 40$$

19.16 *Quality Control*

No production process is so perfect that all the products are completely alike. There is always a small variation that is caused by a great number of small, uncontrollable factors and must therefore be regarded as a chance variation. It is important to make sure that the products have required values (for example, length, strength, or whatever property may be of importance in a particular case). For this purpose one makes a test of the hypothesis that the products have the required property, say $\mu = \mu_0$ where μ_0 is a required value. If this is done after an entire lot has been produced (for example, a lot of 100,000 screws), the test will tell us how good or how bad the products are, but it is obviously too late to alter undesirable results. It is much better to test during the production run. This is done at regular intervals of time (for example, every 30 minutes or every hour) and is called **quality control.** Each time a sample of the same size is taken, in practice, 3 to 10 items. If the hypothesis is rejected, we stop the process of production and look for the trouble that causes the deviation.

If we stop the production process even though it is progressing properly, we make a Type I error. If we do not stop the process even though something is not in order, we make a Type II error (cf. Sec. 19.15).

The result of each test is marked in graphical form on what is called a **control chart**. This was proposed by W. A. Shewhart in 1924 and makes quality control particularly effective.

Control chart for the mean. An illustration and example of a control chart is given in the upper part of Fig. 375. This control chart for the mean shows the **lower control limit** LCL and the **upper control limit** UCL. These two **control limits** correspond to the critical values c_1 and c_2 in Case (*c*) of Ex. 2 in Sec. 19.15. As soon as a sample mean falls outside the range between the control limits, we reject the hypothesis and assert that the process of production is "out of control," that is, we assert that there has been a shift in process level. Action is called for whenever a point exceeds the limits.

If we choose control limits that are too loose, we shall not detect process shifts. On the other hand, if we choose control limits that are too tight, we shall be unable to run the process because of a frequent search for nonexistent trouble. The usual significance level is $\alpha = 1\%$. From Theorem 2 in Sec. 19.14 and Table A9 in Appendix 4 we see that in the case of the normal distribution the corresponding control limits for the mean are

$$(1) \qquad \text{LCL} = \mu_0 - 2.58\,\frac{\sigma}{\sqrt{n}} \qquad \text{and} \qquad \text{UCL} = \mu_0 + 2.58\,\frac{\sigma}{\sqrt{n}}\,.$$

Here σ is assumed to be known. If σ is unknown, we may calculate the standard deviations of the first 20 or 30 samples and take their arithmetic mean as an approximation of σ. The broken line connecting the means in Fig. 375 is merely to display the results effectively.

Control chart for the variance. In addition to the mean, one often controls the variance, the standard deviation, or the range. To set up a control chart for the variance in the case of a normal distribution, we may employ the method in Ex. 4 of Sec. 19.15 for determining control limits. It is customary to use only one control limit, namely, an upper control limit. From Ex. 4 of Sec. 19.15 we see

Fig. 375. Control charts for the mean (upper part of figure) and the standard deviation in the case of the sample in Table 27

that this limit is

$$(2) \qquad\qquad UCL = \frac{\sigma^2 c}{n - 1}$$

where c is obtained from the equation

$$P(Y > c) = \alpha, \qquad \text{that is,} \qquad P(Y \le c) = 1 - \alpha$$

and the table of the chi-square distribution (Table A12 in Appendix 4) with $n - 1$ degrees of freedom; here α (5% or 1%, say) is the probability that an observed value s^2 of S^2 in a sample is greater than the upper control limit.

Table 27. Twelve Samples of 5 Values Each (Diameter of Small Cylinders, Measured in Millimeters)

Sample Number	Sample Values					$\bar{x}$	s	R
1	4.06	4.08	4.08	4.08	4.10	4.080	0.014	0.04
2	4.10	4.10	4.12	4.12	4.12	4.112	0.011	0.02
3	4.06	4.06	4.08	4.10	4.12	4.084	0.026	0.06
4	4.06	4.08	4.08	4.10	4.12	4.088	0.023	0.06
5	4.08	4.10	4.12	4.12	4.12	4.108	0.018	0.04
6	4.08	4.10	4.10	4.10	4.12	4.100	0.014	0.04
7	4.06	4.08	4.08	4.10	4.12	4.088	0.023	0.06
8	4.08	4.08	4.10	4.10	4.12	4.096	0.017	0.04
9	4.06	4.08	4.10	4.12	4.14	4.100	0.032	0.08
10	4.06	4.08	4.10	4.12	4.16	4.104	0.038	0.10
11	4.12	4.14	4.14	4.14	4.16	4.140	0.014	0.04
12	4.14	4.14	4.16	4.16	4.16	4.152	0.011	0.02

If we wanted a control chart for the variance with both an upper control limit UCL and a lower control limit LCL, these limits would be

$$(3) \qquad \text{LCL} = \frac{\sigma^2 c_1}{n-1} \quad \text{and} \quad \text{UCL} = \frac{\sigma^2 c_2}{n-1}$$

where c_1 and c_2 are obtained from the equations

$$(4) \qquad P(Y \leq c_1) = \frac{\alpha}{2} \quad \text{and} \quad P(Y \leq c_2) = 1 - \frac{\alpha}{2}$$

and Table A12 in Appendix 4 with $n-1$ degrees of freedom.

Control chart for the standard deviation. Similarly, to set up a control chart for the standard deviation, we need an upper control limit

$$(5) \qquad \text{UCL} = \frac{\sigma\sqrt{c}}{\sqrt{n-1}}$$

obtained from (2). For example, in Table 27 we have $n = 5$. Assuming that the corresponding population is normal with standard deviation $\sigma = 0.02$ and choosing $\alpha = 1\%$, we obtain from the equation

$$P(Y \leq c) = 1 - \alpha = 99\%$$

and Table A12 in Appendix 4 with 4 degrees of freedom the critical value $c = 13.28$ and from (5) the corresponding value

$$\text{UCL} = \frac{0.02\sqrt{13.28}}{\sqrt{4}} = 0.0365,$$

which is shown in the lower part of Fig. 375.

A control chart for the standard deviation with both an upper and a lower control limit is obtained from (3).

Control chart for the range. If we control σ^2 or σ, we must compute s^2 or s, respectively. Since this may be difficult for mathematically untrained persons, one likes to replace the control of the variance or standard deviation by that of the range R (= largest sample value minus smallest sample value). It can be shown that in the case of the normal distribution, the standard deviation σ is proportional to the expectation of the random variable R^* for which R is an observed value, say, $\sigma = \lambda_n E(R^*)$, where the factor of proportionality λ_n depends on the sample size n and has the values

n	2	3	4	5	6	7	8	9	10
$\lambda_n = \sigma/E(R^*)$	0.89	0.59	0.49	0.43	0.40	0.37	0.35	0.34	0.32

n	12	14	16	18	20	30	40	50
$\lambda_n = \sigma/E(R^*)$	0.31	0.29	0.28	0.28	0.27	0.25	0.23	0.22

Since R depends on two sample values only, it gives less information about a sample than s does. Clearly, the larger the sample size n is, the more information we shall lose in using R instead of s. A practical rule is to use s when n is larger than ten.

Note that the estimate of the standard deviation based on the range is an easy approximate check on the computation of s. It may be useful in cases when somebody else (a computing center, for example) did the numerical work for the statistical investigator. Then misunderstandings and mistakes are quite common, and quick checks are therefore of increasing importance.

Problems

1. Suppose a machine for filling cans with lubricating oil is set so that it will generate fillings which form a normal population with mean 1 gal and standard deviation 0.02 gal. Set up a control chart of the type shown in Fig. 375 for controlling the mean (that is, find LCL and UCL), assuming that the sample size is 4.
2. What LCL and UCL do we obtain in Prob. 1 if we require the significance level $\alpha = 0.3\%$? *Hint.* Cf. (7) in Sec. 19.10.
3. How does the meaning of the control limits (1) change if we apply a control chart with these limits in the case of a population which is not normal?
4. What LCL and UCL should we use instead of (1) if we use the sum $x_1 + \cdots + x_n$ of the sample values instead of $\bar{x}$?
5. Determine the LCL and UCL mentioned in Prob. 4 in the case of Fig. 375.
6. Since the presence of a point outside control limits for the mean indicates trouble, how often would we be making the mistake of looking for nonexistent trouble if we used (*a*) 1-sigma limits, (*b*) 2-sigma limits? (Assume normality.)
7. How would progressive tool wear in an automatic lathe operation be indicated by a control chart for the mean? Answer the same question for a sudden change in the position of the tool in that operation.

8. Ten samples of size 2 were taken from a production lot of bolts. The values (length in millimeters) are

Sample number	1	2	3	4	5	6	7	8	9	10
Length	27.4	27.4	27.5	27.3	27.9	27.6	27.6	27.8	27.5	27.3
	27.6	27.4	27.7	27.4	27.5	27.5	27.4	27.3	27.4	27.7

Assuming that the population is normal with mean 27.5 and variance 0.024 and using (1), set up a control chart for the mean and graph the sample means on the chart.

9. Graph the means of the following 10 samples (thickness of washers, coded values) on a control chart for means, assuming that the population is normal with mean 5 and standard deviation 1.55.

Time	8:00	8:30	9:00	9:30	10:00	10:30	11:00	11:30	12:00	12:30
	3	3	5	7	7	4	5	6	5	5
Sample	4	6	2	5	3	4	6	4	5	2
values	8	6	5	4	6	3	4	6	6	5
	4	8	6	4	5	6	6	4	4	3

10. Graph the ranges of the samples in Prob. 9 on a control chart for ranges.

19.17   *Acceptance Sampling*

Acceptance sampling is applied in mass production when a *producer* is supplying to a *consumer* lots of N items. In such a situation the decision to accept or reject an individual lot must be made. This decision is often based on the result of inspecting a sample of size n from the lot and determining the number of **defective items,** briefly called **defectives,** that is, items which do not meet the specifications (size, color, strength, or whatever may be important). If the number of defective items x in the sample is not greater than a specified number c $(< n)$, the lot is accepted. If $x > c$, the lot is rejected. c is called the *allowable number of defectives* or the **acceptance number.** It is clear that the producer and the consumer must agree on a certain **sampling plan,** that is, on a certain sample size n and an allowable number c. Such a plan is called a **single sampling plan** because it is based on a single sample. *Double sampling plans*, in which two samples are used, will be mentioned later.

Let A be the event that a lot is accepted. It is clear that the corresponding probability $P(A)$ depends not only on n and c but also on the number of defectives in the lot. Let M denote this number, let the random variable X be the number of defectives in a sample, and suppose that we sample without replacement. Then (cf. Sec. 19.9)

(1) $$P(A) = P(X \leqq c) = \sum_{x=0}^{c} \binom{M}{x} \binom{N - M}{n - x} \bigg/ \binom{N}{n}.$$

If $M = 0$ (no defectives in the lot), then X must assume the value 0, and

$$P(A) = \binom{0}{0} \binom{N}{n} \bigg/ \binom{N}{n} = 1.$$

For fixed n and c and increasing M the probability $P(A)$ decreases. If $M = N$ (all items of the lot defective), then X must assume the value n, and we have $P(A) = P(X \leq c) = 0$ because $c < n$.

The ratio $\theta = M/N$ is called the **fraction defective** in the lot. Note that $M = N\theta$, and (1) may be written

$$(2) \qquad P(A;\theta) = \sum_{x=0}^{c} \binom{N\theta}{x} \binom{N - N\theta}{n - x} \Big/ \binom{N}{n}.$$

Since θ can have one of the $N + 1$ values $0, 1/N, 2/N, \cdots, N/N$, the probability $P(A)$ is defined for these values only. For fixed n and c we may plot $P(A)$ as a function of θ. These are $N + 1$ points. Through these points we may then draw a smooth curve, which is called the **operating characteristic curve** (OC curve) of the sampling plan considered.

Example 1. Suppose that certain tool bits are packaged 20 to a box, and the following single sampling plan is used. A sample of 2 tool bits is drawn, and the corresponding box is accepted if and only if both bits in the sample are good. In this case, $N = 20, n = 2, c = 0$, and (2) takes the form

$$P(A;\theta) = \binom{20\theta}{0} \binom{20 - 20\theta}{2} \Big/ \binom{20}{2} = \frac{(20 - 20\theta)(19 - 20\theta)}{380}.$$

The numerical values are

θ	0.00	0.05	0.10	0.15	0.20	$\cdots$
$P(A;\theta)$	1.00	0.90	0.81	0.72	0.63	$\cdots$

The OC curve is shown in Fig. 376.

In most practical cases θ will be small (less than 10%). In many cases the lot size N will be very large (1000, 10,000, etc.), so that we may approximate the

Fig. 376. OC curve of the single sampling plan with $n = 2$ and $c = 0$ for lots of size $N = 20$

Fig. 377. OC curve in Example 2

hypergeometric distribution in (1) and (2) by the binomial distribution with $p = \theta$. Then if n is such that $n\theta$ is moderate (say, less than 20), we may approximate that distribution by the Poisson distribution with mean $\mu = n\theta$. From (2) we then have

(3) $$P(A; \theta) \sim e^{-\mu} \sum_{x=0}^{c} \frac{\mu^x}{x!} \qquad (\mu = n\theta).$$

Example 2. Suppose that for large lots the following single sampling plan is used. A sample of size $n = 20$ is taken. If it contains not more than 1 defective, the lot is accepted. If the sample contains 2 or more defectives, the lot is rejected. In this plan, we obtain from (3)

$$P(A; \theta) \sim e^{-20\theta}(1 + 20\theta).$$

The corresponding OC curve is shown in Fig. 377.

We shall now discuss the two types of possible errors in acceptance sampling and the related problem of choosing n and c. In acceptance sampling the producer and the consumer have different interests. The producer may require the probability of rejecting a "good" or "acceptable" lot to be a small number, call it α. The consumer (buyer) may demand the probability of accepting a "bad" or "unacceptable" lot to be a small number β. More precisely, suppose that the two parties agree that a lot for which θ does not exceed a certain number θ_0 is an *acceptable lot* while a lot for which θ is greater than or equal to a certain number θ_1 is an *unacceptable lot*. Then α is the probability of rejecting a lot with $\theta \leq \theta_0$ and is called **producer's risk.** This corresponds to a Type I error in testing a hypothesis (Sec. 19.15). β is the probability of accepting a lot with $\theta \geq \theta_1$ and is called **consumer's risk.** This corresponds to a Type II error in Sec. 19.15. Figure 378 shows an illustrative example. θ_0 is called the **acceptable quality level** (AQL), and θ_1 is called the **lot tolerance per cent defective** (LTPD) or the **rejectable quality level** (RQL). A lot with $\theta_0 < \theta < \theta_1$ may be called an *indifferent lot*.

From Fig. 378 we see that the points $(\theta_0, 1 - \alpha)$ and (θ_1, β) lie on the OC curve. It can be shown that for large lots we can choose θ_0, θ_1 $(>\theta_0)$, α, β and then determine n and c such that the OC curve runs very close to those prescribed points. Sampling plans for specified α, β, θ_0, and θ_1 have been published; cf. Ref. [J6], Appendix 1.

There is a close relationship between sampling inspection and testing a hypothesis, as follows:

Sampling Inspection	Hypothesis Testing
Acceptable quality level (AQL) $\theta = \theta_0$	Hypothesis $\theta = \theta_0$
Lot tolerance percent defectives (LTPD) $\theta = \theta_1$	Alternative $\theta = \theta_1$
Allowable number of defectives c	Critical value c
Producer's risk α of rejecting a lot with $\theta \leq \theta_0$	Probability α of making a Type I error (significance level)
Consumer's risk β of accepting a lot with $\theta \geq \theta_1$	Probability β of making a Type II error

Fig. 378. OC curve, producer's and consumer's risk

The sampling procedure by itself does not protect the consumer sufficiently well. In fact, if the producer is permitted to resubmit a rejected lot without telling that the lot has already been rejected, then even bad lots will eventually be accepted. To protect the consumer against this and other possibilities, the producer may agree with the consumer that a rejected lot is **rectified,** that is, is inspected 100%, item by item, and all defective items in the lot are removed and replaced by nondefective items.[10] Suppose that a plant produces $100\theta\%$ defective items and rejected lots are rectified. Then K lots of size N contain KN items, $KN\theta$ of which are defective. $KP(A;\ \theta)$ of the lots are accepted; these contain a total of $KPN\theta$ defective items. The rejected and rectified lots contain no defective items. Hence after the rectification the fraction defective in the K lots equals $KPN\theta/KN = \theta P(A;\ \theta)$. This function of θ is called the **average outgoing quality** (AOQ) and is denoted by $AOQ(\theta)$. Thus

(4) $$\mathrm{AOQ}(\theta) = \theta P(A;\ \theta).$$

A sampling plan being given, this function and its graph, the *average outgoing quality curve* (AOQ curve), can readily be obtained from $P(A;\ \theta)$ and the OC curve. An example is shown in Fig. 379.

[10] Of course, rectification is impossible if the inspection is destructive, or is not worthwhile if it is too expensive, compared with the value of the lot. The rejected lot may then be sold at a cut-rate price or scrapped.

Fig. 379. OC curve and AOQ curve for the sampling plan in Fig. 376

Clearly AOQ(0) = 0. Also, AOQ(1) = 0 because $P(A; 1) = 0$. From this and AOQ $\geqq 0$ we conclude that this function must have a maximum at some $\theta = \theta^*$. The corresponding value AOQ(θ^*) is called the **average outgoing quality limit** (AOQL). This is the worst average quality which may be expected to be accepted under the rectifying procedure.

It turns out that several single sampling plans may correspond to the same AOQL; see Ref. [J6] in Appendix 1. Hence if the AOQL is all the consumer cares about, the producer has some freedom in choosing a sampling plan and may select a plan which minimizes the amount of sampling, that is, the number of inspected items per lot. This number is

$$nP(A; \theta) + N(1 - P(A; \theta))$$

where the first term corresponds to the accepted lots and the last term to the rejected and rectified lots; in fact, rectification requires the inspection of all N items of the lot, and $1 - P(A; \theta)$ is the probability of rejecting a lot.

We mention that inspection work may be saved by using a **double sampling plan,** in which the sample of size n is broken up into two samples of sizes n_1 and n_2 (where $n_1 + n_2 = n$). If the lot is very good or very bad, it may then be possible to decide about acceptance or rejectance, using one sample only, so that the other sample will be necessary in the case of a lot of intermediate quality only. Ref. [J6] contains double sampling plans using rectifying inspection of the following type (where x_1 and x_2 are the numbers of defectives in those two samples).

(1) If $x_1 \leqq c_1$, accept the lot. If $x_1 > c_2$, reject the lot.
(2) If $c_1 < x_1 \leqq c_2$, use the second sample, too. If $x_1 + x_2 \leqq c_2$, accept the lot. If $x_1 + x_2 > c_2$, reject the lot.

Problems

1. Find the binomial approximation of the hypergeometric distribution in Ex. 1 and compare the approximate and the accurate values.
2. Using the Poisson approximation of the hypergeometric distribution, graph the OC curve corresponding to the following single sampling plan for large lots. A sample of $n = 30$ items is drawn from the lot and the 30 items are inspected. The lot is accepted if the sample contains not more than $c = 1$ defective. Otherwise the lot is rejected.
3. Graph the AOQ curve in Prob. 2. Determine the AOQL, assuming that rectification is possible.
4. Do the work required in Prob. 2 when $n = 50$ and $c = 0$.
5. Graph the OC curve and the AOQ curve for the single sampling plan for large lots with $n = 5$ and $c = 0$.
6. Same task as in Prob. 5 when $n = 4$ and $c = 1$.
7. Graph and compare sampling plans with $c = 0$ and increasing values of n, say, $n = 2, 3, 4$. (Use the binomial distribution.)
8. Find the risks in the single sampling plan with $n = 5$ and $c = 0$, assuming that the AQL is $\theta_0 = 1\%$ and the RQL is $\theta_1 = 15\%$.
9. Estimate θ_0 in Prob. 5 if the producer's risk is 5%. Estimate θ_1 in Prob. 5 if the consumer's risk is 10%.
10. Graph the OC curve and the AOQ curve for the single sampling plan for large lots with $n = 5$ and $c = 0$, and find the AOQL.

19.18 *Goodness of Fit.* χ^2-*Test*

Using a sample $x_1, \cdots, x_n$, we want to test the hypothesis that a certain function $F(x)$ is the distribution function of the population from which the sample was taken. Clearly, the sample distribution function $\tilde{F}(x)$ is an approximation of $F(x)$, and if it approximates $F(x)$ "sufficiently well," we shall not reject the hypothesis that $F(x)$ is the distribution function of that population. If $\tilde{F}(x)$ deviates "too much" from $F(x)$, we shall reject the hypothesis.

To decide in this fashion, we have to know how much $\tilde{F}(x)$ can differ from $F(x)$ if the hypothesis is true. Hence we first have to introduce a measure for the deviation of $\tilde{F}(x)$ from $F(x)$, and we have to know the probability distribution of this measure under the assumption that the hypothesis is true. Then we proceed as follows. We determine a number c such that if the hypothesis is true, a deviation greater than c has a small preassigned probability. If, nevertheless, a deviation larger than c occurs, we have reason to doubt that the hypothesis is true and we reject it. On the other hand, if the deviation does not exceed c, so that $\tilde{F}(x)$ approximates $F(x)$ sufficiently well, we do not reject the hypothesis. Of course, if we do not reject the hypothesis, this means that we have insufficient evidence to reject it, and does not exclude the possibility that there are other functions which would not be rejected in the test. In this respect the situation is quite similar to that in Sec. 19.15.

Table 28 shows a test of this type, which was introduced by R. A. Fisher. This test is justified by the fact that if the hypothesis is true, then χ_0^2 is an observed value of a random variable whose distribution function approaches that of the chi-square distribution with $K - 1$ degrees of freedom (or $K - r - 1$ degrees of freedom if r parameters are estimated) as n approaches infinity. The requirement that at least 5 sample values should lie in each interval in Table 28 results from the fact that for finite n that random variable has only approximately a chi-square distribution. A proof can be found in Ref. [H4], cf. Appendix 1. If the sample is so small that the requirement cannot be satisfied, one may continue with the test, but use the result with great care.

Table 28. Chi-square Test for the Hypothesis That $F(x)$ is the Distribution Function of a Population from Which a Sample $x_1, \cdots, x_n$ is Taken

---

1st step. Subdivide the x-axis into K intervals $I_1, I_2, \cdots, I_K$ such that each interval contains at least 5 values of the given sample $x_1, \cdots, x_n$. Determine the number b_j of sample values in the interval I_j ($j = 1, \cdots, K$). If a sample value lies at a common boundary point of two intervals, add 0.5 to each of the two corresponding b_j.

2nd step. Using $F(x)$, compute the probability p_j that the random variable X under consideration assumes any value in the interval I_j ($j = 1, \cdots, K$). Compute

$$e_j = np_j.$$

(This is the number of sample values theoretically expected in I_j if the hypothesis is true.)

3rd step. Compute the deviation

(1) $$\chi_0^2 = \sum_{j=1}^{K} \frac{(b_j - e_j)^2}{e_j}.$$

4th step. Choose a significance level (5%, 1%, or the like).

5th step. Determine the solution c of the equation

$$P(\chi^2 \leqq c) = 1 - \alpha$$

from the table of the chi-square distribution with $K - 1$ degrees of freedom (Table A12, Appendix 4). If r parameters of $F(x)$ are unknown and their maximum likelihood estimates (Sec. 19.13) are used, then use $K - r - 1$ degrees of freedom (instead of $K - 1$). If $\chi_0^2 \leqq c$, do not reject the hypothesis. If $\chi_0^2 > c$, reject the hypothesis.

Example 1. Test whether the population from which the sample in Table 17, Sec. 19.2, was taken is normal. The maximum likelihood estimates for μ and σ^2 are $\hat{\mu} = \bar{x} = 364.7$ and $\hat{\sigma}^2 = 712.9$. The calculation in Table 29 yields $\chi_0^2 = 2.790$. We choose $\alpha = 5\%$. Since $K = 10$ and we estimated $r = 2$ parameters, we have to use Table A12, Appendix 4, with $K - r - 1 = 7$ degrees of freedom, finding $c = 14.07$ as the solution of $P(\chi^2 \leqq c) = 95\%$. Since $\chi_0^2 < c$, we do not reject the hypothesis that the population is normal.

Table 29. Calculations in Example 1

x_j	$\dfrac{x_j - 364.7}{26.7}$	$\Phi\left(\dfrac{x_j - 364.7}{26.7}\right)$	$e_j = 100p_j$	b_j	Terms in (1)
$-\infty \cdots 325$	$-\infty \cdots -1.49$	$0.0000 \cdots 0.0681$	6.81	6	0.096
$325 \cdots 335$	$-1.49 \cdots -1.11$	$0.0681 \cdots 0.1335$	6.54	6	0.045
$335 \cdots 345$	$-1.11 \cdots -0.74$	$0.1335 \cdots 0.2296$	9.61	11	0.201
$345 \cdots 355$	$-0.74 \cdots -0.36$	$0.2296 \cdots 0.3594$	12.98	14	0.080
$355 \cdots 365$	$-0.36 \cdots \ 0.00$	$0.3594 \cdots 0.5000$	14.06	16	0.268
$365 \cdots 375$	$0.00 \cdots \ 0.39$	$0.5000 \cdots 0.6517$	15.17	15	0.002
$375 \cdots 385$	$0.39 \cdots \ 0.76$	$0.6517 \cdots 0.7764$	12.47	8	1.602
$385 \cdots 395$	$0.76 \cdots \ 1.13$	$0.7764 \cdots 0.8708$	9.44	10	0.033
$395 \cdots 405$	$1.13 \cdots \ 1.51$	$0.8708 \cdots 0.9345$	6.37	8	0.417
$405 \cdots \infty$	$1.51 \cdots \ \ \infty$	$0.9345 \cdots 1.0000$	6.55	6	0.046

$$\chi_0{}^2 = 2.790$$

Problems

1. In one of G. Mendel's classical experiments the result was 355 yellow peas and 123 green peas. Is this in agreement with Mendel's theory according to which in the present case the ratio yellow peas:green peas should be $3:1$?

2. A coin was tossed 50 times. Head was obtained 27 times. Using this result, test the hypothesis that heads and tails have the same probability.

3. In the sample in Prob. 2, what is the minimum number of heads (greater than 25) that would lead to the rejection of the hypothesis if $\alpha = 5\%$ is used?

4. The following sample consists of values of the mass of cotton material (measured in gram/m²). Test the hypothesis that the corresponding population is normal.

Class mark	96	98	100	102	104	106	108	110	112
Absolute class frequency	1	0	1	2	8	19	28	30	41

Class mark	114	116	118	120	122	124	126	128	130
Absolute class frequency	66	50	27	8	5	3	0	1	1

5. Is the sample in Table 19, Sec. 19.2, taken from a normal population?

19.19 Nonparametric Tests

The tests in Sec. 19.15 refer to normal populations. In many applications the populations will not be normal. Then we may apply a so-called **nonparametric test** or **distribution-free test,** which is based on order statistics and therefore is valid for any *continuous* distribution. These tests are simple. However, for a normal distribution the tests in Sec. 19.15 yield better results. To illustrate the basic idea of a nonparametric test, let us consider two typical examples.

Example 1 (Sign test for the median). A **median** is a solution $x = \tilde{\mu}$ of the equation $F(x) = 0.5$ where F is the distribution function. Using the sample of differences in Ex. 5, Sec. 19.15, that is,

$$16 \quad 16 \quad 2 \quad 6 \quad 0 \quad 0 \quad 13 \quad 8,$$

we shall test the hypothesis $\tilde{\mu} = 0$, that is, the outputs corresponding to the two different working conditions do not differ significantly. We choose the alternative $\tilde{\mu} > 0$ and the significance level $\iota = 5\%$. If the hypothesis is true, the probability p of a positive difference is the same as that of a negative difference. Hence in this case, $p = 0.5$, and the random variable

$$X = \textit{number of positive values among n values}$$

has a binomial distribution with $p = 0.5$. Our sample has 8 values. We omit the values 0 which do not contribute to the decision. Then 6 values are left, all of which are positive. Since

$$P(X = 6) = \binom{6}{6}(0.5)^6(0.5)^0 = 0.0156 = 1.56\% < \alpha,$$

we reject the hypothesis.

If 1 of those 6 values were negative, we would have

$$P(X \geqq 5) = \binom{6}{5}(0.5)^5 \cdot 0.5 + \binom{6}{6}(0.5)^6 = 10.9\%$$

and we would not reject the hypothesis.

Example 2 (*Test for arbitrary trend*). A certain machine is used for cutting lengths of wire. Five successive pieces had the lengths

$$29 \qquad 31 \qquad 28 \qquad 30 \qquad 32.$$

Using this sample, test the hypothesis that there is *no trend*, that is, the machine does not have the tendency of producing longer and longer pieces. Assume that the type of machine suggests the alternative that there is *positive trend*, that is, there is the tendency that successive pieces get longer.

We count the number of *transpositions* in the sample, that is, the number of times a larger value precedes a smaller value:

$$29 \text{ precedes } 28 \qquad\qquad (1 \text{ transposition}),$$
$$31 \text{ precedes } 28 \text{ and } 30 \qquad (2 \text{ transpositions}).$$

The remaining three sample values follow in ascending order. Hence in the sample there are $1 + 2 = 3$ transpositions.

We now consider the random variable

$$T = \textit{number of transpositions.}$$

If the hypothesis is true (no trend), then each of the $5! = 120$ permutations of 5 elements 1 2 3 4 5 has the same probability $(1/120)$. We arrange these permutations according to their number of transpositions:

T = 0	T = 1	T = 2	T = 3
1 2 3 4 5	1 2 3 5 4	1 2 4 5 3	1 2 5 4 3
	1 2 4 3 5	1 2 5 3 4	1 3 4 5 2
	1 3 2 4 5	1 3 2 5 4	1 3 5 2 4
	2 1 3 4 5	1 3 4 2 5	1 4 2 5 3
		1 4 2 3 5	1 4 3 2 5
		2 1 3 5 4	1 5 2 3 4
		2 1 4 3 5	2 1 4 5 3
		2 3 1 4 5	2 1 5 3 4 etc.
		3 1 2 4 5	2 3 1 5 4
			2 3 4 1 5
			2 4 1 3 5
			3 1 2 5 4
			3 1 4 2 5
			3 2 1 4 5
			4 1 2 3 5

From this we obtain

$$P(T \leqq 3) = \tfrac{1}{120} + \tfrac{4}{120} + \tfrac{9}{120} + \tfrac{15}{120} = \tfrac{29}{120} = 24\%.$$

Hence we do not reject the hypothesis.

Values of the distribution function of T in the case of no trend are shown in Table A14, Appendix 4. Our method and those values refer to *continuous* distributions. Theoretically we may then expect that all the values of a sample are different. Practically, some sample values may still be equal, because of rounding off. If m values are equal, add $m(m-1)/4$ ($=$ mean value of the transpositions in the case of the permutations of m elements), that is, $\tfrac{1}{2}$ for each pair of equal values, $\tfrac{3}{2}$ for each triple, etc.

Problems

1. Apply the sign test to Prob. 6, Sec. 19.15. (The necessary values of the binomial distribution are not contained in Table A6, Appendix 4, but must be computed.)
2. Each of 10 patients were given two different sedatives A and B. The following table shows the effect (increase of sleeping time, measured in hours).

A	1.9	0.8	1.1	0.1	−0.1	4.4	5.5	1.6	4.6	3.4
B	0.7	−1.6	−0.2	−1.2	−0.1	3.4	3.7	0.8	0.0	2.0
Difference	1.2	2.4	1.3	1.3	0.0	1.0	1.8	0.8	4.6	1.4

Using the sign test, find out whether the difference is significant.

3. Assuming that the populations corresponding to the samples in Prob. 2 are normal, apply the test explained in Ex. 3, Sec. 19.15.
4. Make a table similar to that in Ex. 2 when $n = 4$.
5. Apply the test in Ex. 2 to the following data ($x =$ disulphide content of a certain type of wool, measured in per cent of the content in unreduced fibres; $y =$ saturation water content of the wool, measured in per cent).

x	10	15	30	40	50	55	80	100
y	50	46	43	42	36	39	37	33

19.20 Pairs of Measurements. Fitting Straight Lines

We shall now discuss experiments in which we observe or measure two quantities simultaneously. In practice we may distinguish between two types of experiments, as follows.

1. In **correlation analysis** both quantities are random variables and we are interested in relations between them. (We shall not discuss this branch of statistics.)

2. In **regression analysis** one of the two variables, call it x, can be regarded as an ordinary variable, that is, can be measured without appreciable error. The other variable, Y, is a random variable. x is called the *independent* (sometimes the *controlled*) *variable*, and one is interested in the dependence of Y on x. Typical examples are the dependence of the blood pressure Y on the age x of a person or, as we shall now say, the regression of Y on x, the regression of the gain of weight

Y of certain animals on the daily ration of food x, the regression of the heat conductivity Y of cork on the specific weight x of the cork, etc.

In the experiment the experimenter first selects n values $x_1, \cdots, x_n$ of x and then observes Y at those values of x, so that he obtains a sample of the form (x_1, y_1), $(x_2, y_2), \cdots, (x_n, y_n)$. In regression analysis the mean μ of Y is assumed to depend on x, that is, is a function $\mu = \mu(x)$ in the ordinary sense. The curve of $\mu(x)$ is called the *regression curve of Y on x*. In the present section we shall discuss the simplest case, when $\mu(x)$ is a linear function, $\mu(x) = \alpha + \beta x$. Then we may want to plot the sample values as n points in the xY-plane, fit a straight line through them, and use this line for estimating $\mu(x)$ for given values of x, so that we know what values of Y we can expect if we choose certain values of x. If the points are scattered, fitting "by eye" becomes unreliable and we need a mathematical method for fitting lines that yields a unique result depending only on the points. A widely used procedure of this type is the **method of least squares** developed by Gauss. In our present situation it may be formulated as follows.

The straight line should be fitted through the given points so that the sum of the squares of the distances of those points from the straight line is minimum, where the distance is measured in the vertical direction (the y-direction).

General assumption (A1). *The x-values $x_1, \cdots, x_n$ of our sample $(x_1, y_1), \cdots,$ (x_n, y_n) are not all equal.*

Consider a sample $(x_1, y_1), \cdots, (x_n, y_n)$ of size n. The vertical distance (distance measured in the y-direction) of a sample value (x_j, y_j) from a straight line $y = a + bx$ is $|y_j - a - bx_j|$; cf. Fig. 380. Hence the sum of the squares of these distances is

(1)
$$q = \sum_{j=1}^{n}(y_j - a - bx_j)^2.$$

In the method of least squares we choose a and b such that q is minimum. q depends on a and b, and a necessary condition for q to be minimum is

(2)
$$\frac{\partial q}{\partial a} = 0 \quad \text{and} \quad \frac{\partial q}{\partial b} = 0.$$

We shall see that from this condition we obtain the formula

(3)
$$y - \bar{y} = b(x - \bar{x})$$

Fig. 380. Vertical distance of a point (x_j, y_j) from a line $y = a + bx$

where

(4) $$\bar{x} = \frac{1}{n}(x_1 + \cdots + x_n) \quad \text{and} \quad \bar{y} = \frac{1}{n}(y_1 + \cdots + y_n).$$

(3) is called the **regression line** *of the y-values of the sample on the x-values of the sample. Its slope b is called the* **regression coefficient** *of y on x,* and we shall see that

(5) $$b = \frac{s_{xy}}{s_1^{\,2}}.$$

Here,

(6) $$s_1^{\,2} = \frac{1}{n-1} \sum_{j=1}^{n} (x_j - \bar{x})^2 = \frac{1}{n-1}\left[\sum_{j=1}^{n} x_j^{\,2} - \frac{1}{n}\left(\sum_{j=1}^{n} x_j\right)^2\right]$$

and

(7) $$s_{xy} = \frac{1}{n-1} \sum_{j=1}^{n} (x_j - \bar{x})(y_j - \bar{y}) = \frac{1}{n-1}\left[\sum_{j=1}^{n} x_j y_j - \frac{1}{n}\left(\sum_{i=1}^{n} x_i\right)\left(\sum_{j=1}^{n} y_j\right)\right].$$

s_{xy} is called the **covariance** of the sample. Obviously, the regression line (3) passes through the point $(\bar{x}, \bar{y})$.

To derive (3), we use (1) and (2), finding

$$\frac{\partial q}{\partial a} = -2\sum (y_j - a - bx_j) = 0$$

$$\frac{\partial q}{\partial b} = -2\sum x_j (y_j - a - bx_j) = 0$$

(where we sum over j from 1 to n). Thus

$$na + b\sum x_j = \sum y_j$$

$$a\sum x_j + b\sum x_j^{\,2} = \sum x_j y_j.$$

Because of Assumption (A1), the determinant

$$n\sum x_j^{\,2} - (\sum x_j)^2 = n(n-1)s_1^{\,2}$$

[cf. (6)] of this system of linear equations is not zero, and the system has a unique solution [cf. (4), (6), (7)]

(8) $$a = \bar{y} - b\bar{x}, \qquad b = \frac{n\sum x_j y_j - \sum x_i \sum y_j}{n(n-1)s_1^{\,2}}.$$

This yields (3) with b given by (5)–(7).

Desk calculations can be simplified by *coding*, that is, by setting

(9) $$x_j = c_1 x_j^* + l_1, \qquad y_j = c_2 y_j^* + l_2$$

and choosing the constants c_1, c_2, l_1, l_2 such that the transformed values x_j^* and y_j^* are as simple as possible. We first compute the values $\bar{x}^*, \bar{y}^*, s_1^{*2}, s_{xy}^*$ corresponding to the transformed values and then

$$\bar{x} = c_1\bar{x}^* + l_1, \qquad \bar{y} = c_2\bar{y}^* + l_2$$

(10)

$$s_1^{\,2} = c_1^{\,2}s_1^{*2}, \qquad s_{xy} = c_1 c_2 s_{xy}^*.$$

Table 30. Regression of the Decrease of Volume y **[%] of
Leather on the Pressure** x **[Atmospheres]**

Given values		Auxiliary values	
x_j	y_j	x_j^2	$x_j y_j$
4,000	2.3	16,000,000	9,200
6,000	4.1	36,000,000	24,600
8,000	5.7	64,000,000	45,600
10,000	6.9	100,000,000	69,000
28,000	19.0	216,000,000	148,400

(C. E. WEIR, Journ. Res. Nat. Bureau of Standards **45**, 1950, 468)

Example 1. The decrease of volume y [%] of leather for certain fixed values of high pressure x [atmospheres] was measured. The results are shown in the first two columns of Table 30. Find the regression line of y on x. We see that $n = 4$ and obtain the values $\bar{x} = 28{,}000/4 = 7000$, $\bar{y} = 19.0/4 = 4.75$,

$$s_1^2 = \frac{1}{3}\left(216\,000\,000 - \frac{28\,000^2}{4}\right) = \frac{20\,000\,000}{3}$$

$$s_{xy} = \frac{1}{3}\left(148\,400 - \frac{28\,000 \cdot 19}{4}\right) = \frac{15\,400}{3}.$$

Hence $b = 15\,400/20\,000\,000 = 0.000\,77$, and the regression line is

$$y - 4.75 = 0.000\,77(x - 7000) \quad \text{or} \quad y = 0.000\,77\,x - 0.64.$$

We now make the following two assumptions.

Assumption (A2). *For each fixed x the random variable Y is normal with mean*

(11) $$\mu(x) = \alpha + \beta x$$

and variance σ^2 where the latter is independent of x.

Assumption (A3). *The n performances of the experiment by which we obtain a sample $(x_1, y_1), \cdots, (x_n, y_n)$ are independent (cf. p. 746).*

β in (11) is called the **regression coefficient** *of the population*, because it can be shown that under Assumptions (A1)–(A3) the maximum likelihood estimate of β is the sample regression coefficient b in (5).

Under Assumptions (A1)–(A3) we may obtain a confidence interval for β, as shown in Table 31.

Example 2. Using the sample in Table 30, determine a confidence interval for β by the method in Table 31.

1st step. We choose $\gamma = 0.95$.

2nd step. Equation (12) takes the form $F(c) = 0.975$, and Table A11 with $n - 2 = 2$ degrees of freedom yields $c = 4.30$.

Table 31. Determination of a Confidence Interval for β in (11) under Assumptions (A1)–(A3)

1st step. Choose a confidence level γ (95%, 99% or the like).
2nd step. Determine the solution c of the equation

(12)
$$F(c) = \tfrac{1}{2}(1 + \gamma)$$

from the table of the t-distribution with $n - 2$ degrees of freedom (Table A11 in Appendix 4; $n =$ sample size).
3rd step. Using a sample $(x_1, y_1), \cdots, (x_n, y_n)$, compute $(n - 1)s_1^2$ from (6), $(n - 1)s_{xy}$ from (7), b from (5),

(13)
$$(n - 1)s_2^2 = \sum_{j=1}^{n} y_j^2 - \frac{1}{n}\left(\sum_{j=1}^{n} y_j\right)^2$$

and

(14)
$$q_0 = (n - 1)(s_2^2 - b^2 s_1^2).$$

4th step. Compute $k = c\sqrt{q_0/(n - 2)(n - 1)s_1^2}$. The confidence interval is

(15)
$$\text{CONF } \{b - k \leq \beta \leq b + k\}.$$

3rd step. From Ex. 1 we have $3s_1^2 = 20\,000\,000$ and $b = 0.000\,77$. From Table 30 we compute

$$3s_2^2 = 102.2 - \frac{19^2}{4} = 11.95, \qquad q_0 = 11.95 - 20\,000\,000 \cdot 0.000\,77^2 = 0.092.$$

4th step. We thus obtain $k = 4.30\sqrt{0.092/2 \cdot 20\,000\,000} = 0.000\,206$ and

$$\text{CONF } \{0.000\,56 \leq \beta \leq 0.000\,98\}.$$

Problems

1. Obtain the results in Ex. 1 by coding; set $x_j = 2000x_j^* + 4000$, $y_j = 0.1y_j^* + 5$.
2. Fit a straight line by eye. Estimate the stopping distance of the car traveling at 35 miles per hour.

$x =$ Speed (miles/hour)	20	30	40	50
$y =$ Stopping distance (feet)	50	95	150	210

3. Construct a simple sample for which $b = 0$.

In each case find and graph the sample regression line of y on x.
4. The sample in Prob. 5, Sec. 19.19.
5. $(1, 1), (2, 1.7), (3, 3)$.
6. Mortality y [%] of termites (*Reticulitermes lucifugus Rossi*) as a function of the concentration x [%] of chloronaphthalene

x	0.04	0.15	0.30	1.00	2.00
y	3	16	13	70	90

7. Number of revolutions x (per minute) and power y (hp) of a Diesel engine

x	400	500	600	700	750
y	580	1030	1420	1880	2100

In each case find a 95%-confidence interval for the regression coefficient β, using the given sample and supposing that Assumptions (A2) and (A3) are satisfied.

8. $(1, 1)$, $(2, 2 + p)$, $(3, 3)$, where p is a constant.

9. The sample in Prob. 7.

10. $x =$ humidity of air (in percent), $y =$ expansion of gelatine (in percent)

x	10	20	30	40
y	0.8	1.6	2.3	2.8

APPENDIX 1

References

A General References

[A1] Abramowitz, M., and I. A. Stegun, *Handbook of Mathematical Functions.* New York: Dover, 1965.

[A2] Buck, R. C., *Advanced Calculus.* 2nd ed. New York: McGraw-Hill, 1965.

[A3] Courant, R., *Differential and Integral Calculus.* 2 vols. New York: Interscience, 1964, 1956.

[A4] Courant, R., and D. Hilbert, *Methods of Mathematical Physics.* 2 vols. New York: Interscience, 1953, 1962.

[A5] Erdélyi, A., W. Magnus, F. Oberhettinger, and F. G. Tricomi, *Higher Transcendental Functions*, 3 vols. New York: McGraw-Hill, 1953, 1955.

[A6] Fletcher, A., J. C. P. Miller, L. Rosenhead, and L. J. Comrie, *An Index of Mathematical Tables.* Oxford: Blackwell, 1962.

[A7] Fulks, W., *Advanced Calculus.* 2nd ed. New York: Wiley 1969.

[A8] Jahnke, E., F. Emde, and F. Lösch, *Tables of Higher Functions.* 7th ed. Stuttgart: Teubner, 1966.

[A9] Kaplan, W., *Advanced Calculus.* Reading, Mass.: Addison-Wesley, 1952.

[A10] Magnus, W., F. Oberhettinger, and R. P. Soni, *Formulas and Theorems for the Special Functions of Mathematical Physics.* 3rd ed. New York: Springer, 1966.

[A11] Protter, M. H., and C. B. Morrey, Jr., *Modern Mathematical Analysis.* Reading, Mass.: Addison-Wesley, 1964.

[A12] Rainville, E. D., *Special Functions.* New York: Macmillan, 1960.

[A13] Simmons, G. F., *Introduction to Topology and Modern Analysis.* New York: McGraw-Hill, 1963.

[A14] Thomas, G. B., *Calculus and Analytic Geometry.* 3rd ed. Reading, Mass.: Addison-Wesley, 1960.

[A15] Whittaker, E. T., and G. N. Watson, *A Course of Modern Analysis.* 4th ed. Cambridge: Harvard University Press, 1927 (reprinted 1965).

B Ordinary Differential Equations (Chaps. 1–4)

[A4], [A5], [A10], [A12], [A15]

[B1] Carslaw, H. S., and J. C. Jaeger, *Operational Methods in Applied Mathematics.* New York: Dover, 1963.

[B2] Churchill, R. V., *Operational Mathematics.* 2nd ed. New York: McGraw-Hill, 1958.

[B3] Collatz, L., *Numerical Treatment of Differential Equations.* 3rd ed. New York: Springer, 1960.

[B4] Coddington, E. A., and N. Levinson, *Theory of Ordinary Differential Equations.* New York: McGraw-Hill, 1955.

[B5] Erdélyi, A., W. Magnus, F. Oberhettinger, and F. Tricomi, *Tables of Integral Transforms.* 2 vols. New York: McGraw-Hill, 1954.

[B6] Fox, L., *Numerical Solution of Ordinary and Partial Differential Equations.* Reading, Mass.: Addison-Wesley, 1962.

[B7] Hartman, P., *Ordinary Differential Equations.* New York: Wiley, 1964.

[B8] Ince, E. L., *Ordinary Differential Equations.* New York: Dover, 1956.

[B9] Kamke, E., *Differentialgleichungen, Lösungsmethoden und Lösungen. I. Gewöhnliche Differentialgleichungen.* 8th ed. Leipzig: Akademische Verlagsgesellschaft, 1967. (This extremely useful book contains a systematic list of more than 1500 differential equations and their solutions.)

[B10] Kaplan, W., *Operational Methods for Linear Systems.* Reading, Mass.: Addison-Wesley, 1962.

[B11] Minorsky, N., *Nonlinear Oscillations.* New York: Van Nostrand, 1962.

[B12] Smith, M. G., *Laplace Transform Theory.* New York: Van Nostrand, 1966.

[B13] Stern, T. E., *Theory of Nonlinear Networks and Systems.* Reading, Mass.: Addison-Wesley, 1965.

[B14] Watson, G. N., *A Treatise on the Theory of Bessel Functions.* 2nd ed. Cambridge: University Press, 1944.

[B15] Widder, D. V., *The Laplace Transform.* Princeton, N.J.: Princeton University Press, 1941.

C   Linear Algebra: Vectors and Matrices.   Vector Calculus (Chaps. 5–8)

[C1] Bodewig, E., *Matrix Calculus.* 2nd ed. Amsterdam: North-Holland Pub. 1959.

[C2] Brand, L., *Vector Analysis.* New York: Wiley, 1957.

[C3] Frazer, R. A., W. J. Duncan, and A. R. Collar, *Elementary Matrices.* Cambridge: University Press, 1938 (reprinted 1963).

[C4] Gantmacher, F. R., *The Theory of Matrices.* 2 vols. New York: Chelsea, 1959.

[C5] Gantmacher, F. R., *Applications of the Theory of Matrices.* New York: Interscience, 1959.

[C6] Gel'fand, I. M., *Lectures on Linear Algebra.* New York: Interscience, 1961.

[C7] Hohn, F. E., *Elementary Matrix Algebra.* 2nd ed. New York: Macmillan, 1964.

[C8] Kreyszig, E., *Introduction to Differential Geometry and Riemannian Geometry.* Toronto: University of Toronto Press, 1968.

[C9] Lamb, H., *Hydrodynamics.* 6th ed. New York: Dover, 1945.

[C10] McDuffee, C. C., *The Theory of Matrices.* New York: Chelsea, 1946.

[C11] Milne, E. A., *Vectorial Mechanics.* London: Methuen, 1948.

[C12] Nering, E. D., *Linear Algebra and Matrix Theory.* 2nd ed. New York: Wiley, 1970.

[C13] Schreier, O., and E. Spencer, *Introduction to Modern Algebra and Matrix Theory.* 2nd ed. New York: Chelsea, 1959.

[C14] Schwerdtfeger, H., *Introduction to Linear Algebra and the Theory of Matrices.* Groningen: Noordhoff, 1950.

[C15] Wilkinson, J. H., *The Algebraic Eigenvalue Problem.* Oxford: Clarendon, 1965.

D   Fourier Series and Integrals (Chap. 9)

[D1] Carslaw, H. S., *Introduction to the Theory of Fourier's Series and Integrals.* 3rd ed. London: Macmillan, 1930.

[D2] Davis, H. F., *Fourier Series and Orthogonal Functions.* Boston: Allyn and Bacon, 1963.

[D3] Churchill, R. V., *Fourier Series and Boundary Value Problems.* 2nd ed. New York: McGraw-Hill, 1963.

[D4] Rogosinski, W., *Fourier Series.* New York: Chelsea, 1959.

[D5] Sneddon, I. N., *Fourier Transforms.* New York: McGraw-Hill, 1951.

[D6] Szegö, G., *Orthogonal Polynomials.* 3rd ed. New York: American Mathematical Society, 1967.

[D7] Titchmarsh, E. C., *Introduction to the Theory of Fourier Integrals.* 2nd ed. Oxford: Clarendon, 1948.

[D8] Tolstov, G. P., *Fourier Series.* Englewood Cliffs, N.J.: Prentice-Hall, 1962.

[D9] Zygmund, A., *Trigonometric Series.* 2 vols. 2nd ed. Cambridge: University Press, 1959.

[D10] Zygmund, A., *Trigonometrical Series.* New York: Dover, 1955.

E Partial Differential Equations (Chap. 10)

[A4], [A5], [B6], [D3]

[E1] Bateman, H., *Partial Differential Equations of Mathematical Physics.* Cambridge: University Press, 1944 (reprinted 1964).

[E2] Bergman, S., and M. Schiffer, *Kernel Functions and Elliptic Differential Equations in Mathematical Physics.* New York: Academic Press, 1953.

[E3] Duff, G. F. D., *Partial Differential Equations.* Toronto: Toronto University Press, 1956.

[E4] Duff, G. F. D., and D. Naylor, *Differential Equations of Applied Mathematics.* New York: Wiley, 1966.

[E5] Forsythe, G. E., and W. R. Wasow, *Finite-Difference Methods for Partial Differential Equations.* New York: Wiley, 1960.

[E6] Garabedian, P. R., *Partial Differential Equations.* New York: Wiley, 1964.

[E7] Hadamard, J., *Lectures on Cauchy's Problem.* New York: Dover, 1952.

[E8] Kellog, O. D., *Foundations of Potential Theory.* New York: Dover, 1953.

[E9] Mitchell, A. R., *Computational Methods in Partial Differential Equations.* New York: Wiley, 1969.

[E10] Petrovsky, I. G., *Lectures on Partial Differential Equations.* New York: Interscience, 1954.

[E11] Lord Rayleigh, *The Theory of Sound.* 2 vols. New York: Dover, 1945.

[E12] Sagan, H., *Boundary and Eigenvalue Problems in Mathematical Physics.* New York: Wiley, 1961.

[E13] Sneddon, I. N., *Elements of Partial Differential Equations.* New York: McGraw-Hill, 1957.

[E14] Sommerfeld, A., *Partial Differential Equations in Physics.* New York: Academic Press, 1949.

[E15] Webster, A. G., *Partial Differential Equations of Mathematical Physics.* 2nd ed. New York: Hafner, 1947.

F Complex Analysis (Chaps. 11-17)

[A15]

[F1] Ahlfors, L. V., *Complex Analysis.* 2nd ed. New York: McGraw-Hill, 1966.

[F2] Bieberbach, L., *Conformal Mapping.* New York: Chelsea, 1964.

[F3] Churchill, R. V., *Complex Variables and Applications.* 2nd ed. New York: McGraw-Hill, 1960.

[F4] Dettman, J. W., *Applied Complex Variables.* New York: Macmillan, 1965.

[F5] Hille, E., *Analytic Function Theory.* 2 vols. Boston: Ginn, 1959, 1962.

[F6] Knopp, K., *Theory of Functions.* 2 vols. New York: Dover, 1945.

[F7] Nehari, Z., *Conformal Mapping.* New York: McGraw-Hill, 1952.

[F8] Rothe, R., F. Ollendorf, and K. Pohlhausen, *Theory of Functions as Applied to Engineering Problems.* New York: Dover, 1961.

[F9] Springer, G., *Introduction to Riemann Surfaces.* Reading, Mass.: Addison-Wesley, 1957.

[F10] Titchmarsh, E. C., *The Theory of Functions.* 2nd ed. London: Oxford University Press, 1939.

G Numerical Analysis (Chap. 18)

[B3], [B6], [C15], [E5], [E9]

[G1] Alt, F. L., *Electronic Digital Computers.* New York: Academic Press, 1958.

[G2] Buckingham, R. A., *Numerical Methods.* London: Pitman, 1957.

[G3] Forsythe, G. E., and P. C. Rosenbloom, *Numerical Analysis and Partial Differential Equations.* New York: Wiley, 1958.

[G4] Henrici, P., *Elements of Numerical Analysis*. New York: Wiley, 1964.

[G5] Hartree, D. R., *Numerical Analysis*. 2nd ed. London: Oxford University Press, 1958.

[G6] Hildebrandt, F. B., *Introduction to Numerical Analysis*. New York: McGraw-Hill, 1956.

[G7] Isaacson, E., and H. B. Keller, *Analysis of Numerical Methods*. New York: Wiley, 1966.

[G8] Kopal, Z., *Numerical Analysis*. 2nd ed. London: Chapman and Hall, 1961.

[G9] McCracken, D. D., and W. S. Dorn, *Numerical Methods and FORTRAN Programming*. New York: Wiley, 1964.

[G10] Milne, W. E., *Numerical Calculus*. Princeton, N.J.: Princeton University Press, 1949.

[G11] National Physical Laboratory, *Modern Computing Methods*. 2nd ed. London: Her Majesty's Stationery Office, 1961.

[G12] Noble, B., *Numerical Methods*. 2 vols. New York: Interscience, 1964.

[G13] Todd, J., *Numerical Analysis*. In: Condon, E. U., and H. Odishaw, *Handbook of Physics*. 2nd ed. New York: McGraw-Hill, 1967.

[G14] Todd, J., *Survey of Numerical Analysis*. New York: McGraw-Hill, 1962.

[G15] Willers, F. A., *Practical Analysis*. New York: Dover, 1948.

H Probability and Statistics (Chap. 19)

[J2], [J6], [J8]–[J10], [J13]–[J15]

[H1] Cochran, W. G., *Sampling Techniques*. 2nd ed. New York: Wiley, 1963.

[H2] Cochran, W. G., and G. M. Cox, *Experimental Designs*. 2nd ed. New York: Wiley, 1957.

[H3] Cox, D. R., *Planning of Experiments*. New York: Wiley, 1958.

[H4] Cramér, H., *Mathematical Methods of Statistics*. Princeton: Princeton University Press, 1946.

[H5] Feller, W., *An Introduction to Probability and Its Applications*. Vol. I. 2nd ed. New York, Wiley, 1957.

[H6] Fisher, R. A., *Statistical Methods for Research Workers*. 13th ed. New York: Hafner, 1958.

[H7] Fraser, D. A. S., *Nonparametric Methods in Statistics*. New York: Wiley, 1957.

[H8] Grant, E. L., *Statistical Quality Control*. 2nd ed. New York: McGraw-Hill, 1952.

[H9] Huff, D., *How to Lie with Statistics*. New York: Norton, 1954.

[H10] Kreyszig, E., *Introductory Mathematical Statistics. Principles and Methods*. New York: Wiley, 1970.

[H11] Parzen, E., *Modern Probability Theory and Its Applications*. New York: Wiley, 1960.

[H12] Rice, W. B., *Control Charts in Factory Management*. New York: Wiley, 1947.

[H13] Scheffé, H., *The Analysis of Variance*. 2nd ed. New York: Wiley, 1963.

[H14] Wald, A., *Statistical Decision Functions*. New York: Wiley, 1950.

[H15] Wilks, S. S., *Mathematical Statistics*. New York: Wiley, 1962.

J Tables

[A6], [A8], [B5]

[J1] *Barlow's Tables of Squares, Cubes, Square Roots, Cube Roots, and Reciprocals*. 4th ed. London: Spon, 1941 (10th impression 1965).

[J2] British Association for the Advancement of Science, *Mathematical Tables*. Vol. VII. *The Probability Integral*. Cambridge: University Press, 1939 (reissued 1968).

[J3] *C.R.C. Standard Mathematical Tables*. 17th ed. Cleveland, Ohio: Chemical Rubber Co., 1969.

[J4] *C.R.C. Handbook of Tables for Mathematics*. 3rd ed. Cleveland, Ohio: Chemical Rubber Co., 1967.

[J5] Davis, H. T., *Tables of the Mathematical Functions*. 3 vols. Bloomington, Ind.: Principia Press, 1962–1963.

[J6] Dodge, H. F., and H. G. Romig, *Sampling Inspection Tables*. 2nd ed. New York: Wiley, 1959.

[J7] Dwight, H. B., *Mathematical Tables*. 3rd ed. New York: Dover 1961.

[J8] Fisher, R. A., and F. Yates, *Statistical Tables for Biological, Agricultural, and Medical Research.* 6th ed. London: Oliver and Boyd, 1963.

[J9] Hald, A., *Statistical Tables and Formulas.* New York: Wiley, 1952.

[J10] Harvard Computation Laboratory, *Tables of the Cumulative Binomial Probability Distribution.* Cambridge, Mass.: Harvard University Press, 1955.

[J11] National Bureau of Standards, *Tables of the Error Function and Its Derivative.* Washington, D.C.: US Gov. Printing Office, 1954.

[J12] National Bureau of Standards, *Tables of Lagrangian Interpolation Coefficients.* New York: Columbia University Press, 1944.

[J13] National Bureau of Standards, *Tables of Normal Probability Functions.* Washington, D.C.: US Gov. Printing Office, 1953.

[J14] Pearson, E. S., and H. O. Hartley, *Biometrika Tables for Statisticians.* Vol. I. 3rd ed. Cambridge: University Press, 1966.

[J15] Rand Corporation, *A Million Random Digits with 100,000 Normal Deviates.* Glencoe: Free Press, 1955.

APPENDIX 2

Answers to Odd-Numbered Problems

Remark. In the case of more complicated problems, we shall present not only the answer but also the solution. Odd-numbered problems not occurring in this Appendix are those in which obvious graphical work, verifications of statements and formulas in the text, or simple proofs are required, or the answer is already given in the enunciation of the problem.

SECTION 1.1

7. $y = e^x + c$

9. $y = -\frac{1}{4}\cos 2x + ax + b$

15. $y = \pi x/2$

17. $y = 0.5e^{-x^2}$

19. $y = x^3 - 2$

21. $xy' - 3y = 12$

23. $y = 1.5 + x^{-1}$

25. $s(t) = 0.5gt^2 + v_0t$

SECTION 1.2

1. $c = 1/\sqrt{e}$

17. Exact solution $y = e^{-x^2}$ (bell-shaped curve)

SECTION 1.3

1. Suppose you forgot $\tilde{c}$ in (4), wrote $\ln y = -x^2$, transformed this to $y = e^{-x^2}$ and then afterwards added a constant c, obtaining $y = e^{-x^2} + c$, which is not a solution of $y' = -2xy$.

3. $y = ce^{kx}$

5. $y = ce^{2x} + 0.5a$

7. $y = c(x - 1)^2 \exp(2x^3/3 + x^2 + 2x)$

9. $y = (\ln|x| + c)^2 + 1$

11. $y = c\sqrt{\sin 2x}$

13. $y = \tan(0.5x^2 + x + c)$

15. $y = c\cosh x$

17. $y = c\sqrt{\sin 2x}$

19. $y = c\ln x$

25. $y = 2/\sqrt{\cos 2x}$

27. $y = 4x^{3/2}$ (semicubical parabola)

29. $y^2 = 1/(8 - 4e^x)$

31. $r = r_0 e^{-t^2/2}$

33. $y = \arcsin x$

35. $y - 2\ln|y + 2| = \ln|x/8|$

37. $y' = y/x, \ y = cx$

39. A triangle with vertices $(x - 1, 0)$, $(x, 0)$, (x, y) shows that $y' = y/1$. Ans. $y = ce^x$.

Problem 43

41. $y' = x/y$, $x^2 - y^2 = c$

43. Let $y(x)$ be such a curve. Then $OM = 2x$, $ON = 2y$, and we see that the slope at (x, y) is $y' = -ON/OM = -y/x$. Thus $xy = c$. (Cf. the figure on p. 794.)

45. $k = (h^{3/2} - 2Rh^{1/2})h'$ where $k = 0.6r_0^2\sqrt{2g}$, $kt + c = 2h^{5/2}/5 - 4Rh^{3/2}/3$; $h(0) = R$ gives $c = -14R^{5/2}/15$; to $h = 0$ there corresponds $t = 1400/1.8\sqrt{g} \approx 25$ [sec].

47. $y' = ky$, $y = y_0e^{kt}$. At $t = 1590$, $y = y_0/2$. Consequently, $1590k = \ln \frac{1}{2}$, $k = -0.000\,436$, $y(1) = y_0e^{-0.000\,436} \approx y_0(1 - 0.000\,436)$. Ans. 0.04%.

49. $y = 0.01e^{-0.11t}$

51. $y' = -0.05y$, $y(0) = 10$, $y = 10e^{-0.05t}$, $y(120) = 0.025$

53. 25 min

55. $\Delta A = -kA\,\Delta x$ ($A =$ amount of incident light, $\Delta A =$ absorbed light, $\Delta x =$ thickness, $-k =$ constant of proportionality). Let $\Delta x \to 0$. Then $A' = -kA$, so $A(x) = A_0e^{-kx}$ is the amount of light in a thick layer at depth x from the surface of incidence.

SECTION 1.4

1. $y = cx^3 + 2x$

3. $y = cx^2 - 2x$

5. $y = x \arcsin(x + c)$

7. $y = x + x(c - x^2)^{-1/2}$

9. $\ln(x^2 + y^2) + 2\arctan\dfrac{y}{x} = c$

11. $(y - x)^2 + 10y - 2x = c$

13. $y = \arcsin ce^x - x$

15. $\sin(x^2y^2) = ce^x$

17. $r = l\exp(-\theta - \frac{1}{4}\pi)/\sqrt{2}$, etc., where l is the length of the table and r and θ are polar coordinates with center at the center of the table. Graph this logarithmic spiral.

SECTION 1.5

1. $x\,dx + y\,dy = 0$

3. $x^2\cosh(x^3y)(3y\,dx + x\,dy) = 0$

5. $(y - b)\,dx + (x - a)\,dy = 0$

7. $xy = c$

9. $\cosh x \cos y = c$

11. $x(x + e^y) = c$

13. $y = c/(1 + x^2)$

15. Not exact, $xye^y = c$

17. $y = c\sin 3x$

19. $x^2 + y^2 = c$

21. $ax^2 + 2kxy + ly^2 = c$

23. $\int dg/dy = \partial h/\partial x$

25. $y = 2e^{-x}$

27. $y = -\ln(x/2)/x$

29. $\cos \alpha x \sin \beta y = 1$

31. $e^{xy} + y^2 = 2$

33. $\exp(x^2 - y^2) - y^2 = 1$

35. $(x - 3)^2 + 9(y - 1)^2 = 9$

SECTION 1.6

1. $x^2y = c$

3. $y = cx$

5. $e^x\sin y = c$

7. $H(u)xy^2$ where $u = x^2y^3$ and H is arbitrary.

9. $F = x$, $y = c/x^2$

11. $F = 1/xy$, $x^3e^{4x} = cy^2$

13. $F = y$, $y^2\sin x = c$

15. $F = e^{2x}$, $e^{2x}\sin y = 1$

17. $F = \cos x$, $2\sin x + \sin y = 0$

19. $(FP)_y = (FQ)_x$, $P_y + mP/y = Q_x$

21. $P_y = Q + Q_x$

23. $xP + P_y = yQ + Q_x$

25. $(FP)_y = (FQ)_x$, $FP_y = F'Q + FQ_x$, $F'/F = (P_y - Q_x)/Q = f$, integrate.

SECTION 1.7

5. $y = ce^{-x^2}$

7. $y = ce^{-2x} + 2e^x$

9. $y = cx^{-1} + 0.25x^3 + 0.5x$

11. $y = cx^2 + x^2e^x$

13. $y = ce^{-x} + 0.5(\sin x - \cos x)$

15. $y = ce^{2x} - \frac{2}{13}\cos 3x + \frac{3}{13}\sin 3x$

17. $y = (c + 0.5e^{x^2})/x^2$

19. $y = -e^{-x} + x^2 + 1$

21. $y = (e^x + \sin x)x^3 + 2x^2$

SECTION 1.8

1. $y = ce^x - x - 1$

3. $y = 0.5x^4 + cx^2$

7. $x = ce^y + ye^y$

9. $x = 0.5y^4 + cy^2$

11. $y = 1/(1 + ce^x)$

13. $y^2 = 1 + ce^{-x^2}$

15. $y^{-4} = u$, $u' + \dfrac{2}{x}u = -20x^2$, $u = \dfrac{c}{x^2} - 4x^3$, $y = 1/\sqrt[4]{u}$

19. $\sin y = z$, $z = 2 + ce^{-x^2/2}$

SECTION 1.9

3. $(E_0/R)(1 - 0.5e^{-(R/L)t})$, E_0/R, $(E_0/R)(1 + e^{-(R/L)t})$

7. $9 \cdot 10^{-6}$ sec

9. $I(t) = E_0(R^2 + \omega^2 L^2)^{-1/2}[e^{-(R/L)t} \sin \delta + \sin(\omega t - \delta)]$

11. $I = 6$ when $0 \leqq t \leqq 10$, $I = 6e^{1-0.1t}$ when $t > 10$

13. $I(t) = 2 - 2 \cdot 10^{-4}(100 - t)^2$ when $0 \leqq t \leqq 100$, $I = 2$ when $t > 100$

17. $I(0) = \omega E_0 C/(1 + (\omega RC)^2)$

19. $I = 0.0035(\cos 314t + 314 \sin 314t - e^{-t}) \approx 1.1 \sin 314t$

SECTION 1.10

1. $c(x + 2) + y - 3 = 0$

3. $\dfrac{x^2}{c^2 + 1} + \dfrac{y^2}{c^2} - 1 = 0$

5. Parallel straight lines

7. Congruent parabolas

9. Circles that touch $y = \pm x$

11. $y = c/(1 + x^2)$

13. $y = ce^{2x} - \frac{2}{13}\cos 3x + \frac{3}{13}\sin 3x$

15. $xy' = y \ln y$

17. $y' = \sqrt{1 - y^2}$

19. $y' = 1 + y^2$

21. $x + 4yy' = 0$

23. $y = c^* - x$

25. $y = c^* x$

27. $x^2 + 2y^2 = c^*$

29. $y = c^* e^{-2x}$

31. $y = \sqrt{x + c^*}$

33. $\sin x \cosh y = c^*$

35. Circles $(x + c^*)^2 + y^2 = c^{*2} - 1$ (shown in Sec. 17.1, Fig. 315).

37. $dg = 0$ gives $dy/dx = -g_x/g_y$, etc.

39. $dv = 0$ gives $dy/dx = -v_x/v_y$ and this must equal u_y/u_x (cf. Prob. 37).

SECTION 1.11

1. $y_0 = 1$, $y_1 = 1 + \displaystyle\int_0^x dt = 1 + x$, $y_2 = 1 + \displaystyle\int_0^x (1 + t)\, dt = 1 + x + \dfrac{x^2}{2!}$, etc.

3. $y_n = \dfrac{x^2}{2!} + \dfrac{x^3}{3!} + \cdots + \dfrac{x^{n+1}}{(n + 1)!}$

5. $y_n = 1 + x + \displaystyle\int_0^x ty_{n-1}(t)\, dt$, $y_1 = 1 + x + \dfrac{x^2}{2}$, etc. $\left(y = e^{x^2/2}\left(\displaystyle\int_0^x e^{-t^2/2}\, dt + 1\right)\right)$.

7. $y_0 = 0$, $y_1 = -x$, $y_2 = -x - \frac{2}{3}x^3$, etc.

9. $y_0 = 0$, $y_1 = x^2 - \dfrac{x^4}{4}$, $y_2 = x^2 - \dfrac{x^6}{4 \cdot 6}$, $y_3 = x^2 - \dfrac{x^8}{4 \cdot 6 \cdot 8}$, $\cdots$

SECTION 1.12

1. $y = cx^2$, c arbitrary

3. $y = cx^2$, (a) $y(0) = k \neq 0$, (b) $y(0) = 0$, (c) $y(x_0) = y_0$, $x_0 \neq 0$

7. $y = (x - 1)^2$, $y \equiv 0$.

9. Yes, with $M = 1$. Not on the x-axis ($y = 0$).

SECTION 2.1

7. $y = c_1 e^x + c_2$

9. $y = c_1 e^{-9x} + c_2$

11. $y = c_1 \ln|x| + c_2$

13. $F(y, z, (dz/dy)z) = 0$

15. $y = \ln|c_1 x + c_2|$

17. $x = y - c_1 \ln|y + c_1| + c_2$

19. $(y - 1)e^y = c_1 x + c_2$

21. $y = 0.5e^{3t} + 0.5$, $t \approx 1$

23. $xy = 1$

25. $y = x^4$

SECTION 2.2

1. e^{2x}, e^{-x} **3.** $e^{-5x}, 1$ **5.** $e^{0.5ix}, e^{-0.5ix}$ **7.** e^{4ix}, e^{-4ix}
9. $e^{-(2+i)x}, e^{-(2-i)x}$ **11.** $e^{i\omega x}, e^{-i\omega x}$
13. (a) $e^{-3x}, 1$, (b) $c_1 e^{-3x} + c_2$
15. $y'' - y' - 6y = 0$ **17.** $y'' - 4y' + 5y = 0$
19. $y'' + 2y' = 0$

SECTION 2.3

1. $y = c_1 e^{3x} + c_2 e^x$ **3.** $y = c_1 e^{4ix} + c_2 e^{-4ix}$
5. $y = c_1 e^{-2x} + c_2$ **7.** $y = c_1 e^{-(2+i)x} + c_2 e^{-(2-i)x}$
9. $y = c_1 e^{i\omega x} + c_2 e^{-i\omega x}$ **11.** $y'' - 5y' + 4y = 0$
13. $y'' + 3y' + 2y = 0$ **15.** $y'' - 2y' + 5y = 0$
17. Linearly independent   **19.** Linearly independent   **21.** Linearly independent
23. Linearly dependent   **25.** Linearly independent   **27.** Linearly independent

SECTION 2.4

1. $y = A\cos x + B\sin x$ **3.** $y = e^x(A\cos x + B\sin x)$
5. $y = e^{-5x}(A\cos 2x + B\sin 2x)$ **7.** $y = A\cos \omega x + B\sin \omega x$
9. $y = A\cos 5x + B\sin 5x$ **11.** $y = A\cos \sqrt{3}x + B\sin \sqrt{3}x$
13. $y = c_1 e^{-2x} + c_2 e^x$ **15.** $y = e^{-x}(A\cos 3x + B\sin 3x)$
17. $y = e^{-2.5x}(A\cos 2.5x + B\sin 2.5x)$ **19.** $y = e^{-\alpha x}(A\cos x + B\sin x)$
21. $\lambda^2 + a\lambda + b = (\lambda - \lambda_1)(\lambda - \lambda_2) = (\lambda - p - iq)(\lambda - p + iq) = \lambda^2 - 2p\lambda + p^2 + q^2$, hence $a = -2p$ and $b = p^2 + q^2$.
23. $y'' - 2y' + 5y = 0$ **25.** $y'' + 2\alpha y' + (\alpha^2 + 9)y = 0$
27. $y'' - 2y' + 2y = 0, y = e^x(\cos x + 2\sin x)$ **29.** $y = e^{-x}\cos x$
31. $y = 2\cos \sqrt{3}x + 3\sin \sqrt{3}x$ **33.** $y = e^{2x} + e^{-2x} = 2\cosh 2x$

SECTION 2.5

1. xe^{-x} **5.** $y = (c_1 + c_2 x)e^{4x}$
7. $y = A\cos 3x + B\sin 3x$ **9.** $y = (c_1 + c_2 x)e^{-0.5x}$
11. $y = (c_1 + c_2 x)e^{2.5x}$ **13.** $y = e^{-x}(A\cos \sqrt{3}x + B\sin \sqrt{3}x)$
15. $y'' + 8y' + 16y = 0$ **17.** $y'' + 2\alpha y' + \alpha^2 y = 0$ **19.** $y'' + 0.25y = 0$
21. $y = -3xe^{2x}$ **23.** $y = \sqrt{2}\,e^{kx}$ **25.** $y = e^x\cos 2x$

SECTION 2.6

1. It decreases if m is increased. It is independent of the initial conditions.
3. $y(t) = \sqrt{y_0^2 + (v_0/\omega_0)^2}\cos(\omega_0 t - \arctan(v_0/\omega_0 y_0))$
5. $\cos t - \sin t, \cos t, \cos t + \sin t$
7. $\cos t, e^{-0.5t}(\cos 0.87t + 0.57\sin 0.87t), (t+1)e^{-t}, 1.01e^{-0.1t} - 0.01e^{-9.9t}$
9. From (9) and $y = 0$ we have $-c_1/c_2 = e^{-2\beta t}$. Since the exponential function is positive, the statement follows.
15. Consecutive maxima occur at t_0 and $t_1 = t_0 + (2\pi/\omega^*)$, cf. Prob. 14, and the ratio is $\exp(-\alpha t_0)/\exp(-\alpha t_1) = \exp(2\pi\alpha/\omega^*)$.
17. $\dot{y} = 0$ gives $\omega^* \tan \omega^* t = -\alpha = -c/2m$. Hence they move to smaller t as c increases.

SECTION 2.7

3. $y = c_1 x^2 + c_2 x^{-2}$ **5.** $y = (c_1 + c_2 \ln x)/x$ **7.** $y = c_1 x^3 + c_2 x^{-2}$
9. $y = c_1 x^k + c_2 x^{1/k}$ **11.** $y = c_1 + c_2 \ln x$ **15.** $y = 2x - x^3$
17. $y = x + 3x^4$ **19.** $y = (1 + \ln x)x^2$

SECTION 2.8

1. $y'' - y = 0, W = -2$ **3.** $y'' + 4y = 0, W = 2$
5. $x^2 y'' + xy' - 0.25y = 0, W = -1/x$ **7.** $y'' - 2y' + 2y = 0, W = e^{2x}$
11. x^{2m-1} **13.** $e^{2\lambda x}$ **15.** q **17.** qe^{2px}

SECTION 2.9

1. Linearly independent   **3.** Linearly independent   **5.** Linearly dependent
7. Linearly independent   **9.** Linearly dependent

SECTION 2.10

1. Substitute $y = uy_1$ to get $u''' = 0$ and $y = (c_1 + c_2 x + c_3 x^2)e^{-x}$.
3. $y''' - 6y'' + 11y' - 6y = 0$ **5.** $y''' = 0$
7. $c_1 + c_2 e^{-x} + c_3 e^x$ **9.** $(c_1 + c_2 x)e^x + c_3 e^{-x}$
11. $c_1 \cosh 2x + c_2 \sinh 2x + c_3 \cos 2x + c_4 \sin 2x$
19. $(c_1 + c_2 x + c_3 x^2)e^{-2x}$

SECTION 2.11

1. $y = c_1 e^x + c_2 e^{-x} - \cos x$ **3.** $y = A \cos 3x + B \sin 3x + 0.5e^{-5x}$
5. $y = c_1 e^x + c_2 e^{-4x} + \sin x + 0.6 \cos x$ **7.** $y = (c_1 + c_2 \ln x)\sqrt{x} + \cos 0.5x$
9. $y = (c_1 + c_2 x)e^{3x} + 2x^2 e^{3x}$ **15.** $y = c_1 e^x + c_2 e^{-4x} + \sin x + 0.6 \cos x + 2xe^x$

SECTION 2.12

1. $y = x^2 - 2$ **3.** $y = (3x - 2)/9$ **5.** $y = \sin 4x$ **7.** $y = e^x + x$
9. $y = e^x + 9e^{-x}$ **11.** $y = A \cos x + B \sin x + 0.5e^x$
13. $y = (c_1 + c_2 x)e^x + 0.5x^2 e^x$
15. $y = e^{2x}(A \cos \sqrt{5}\, x + B \sin \sqrt{5}\, x) + 2e^{2x} + \sin 3x$
17. $y = 5 \cos 5x - \sin 5x + 0.2x$ **19.** $y = e^{3x} - 2e^x + 2xe^{3x}$
21. $y = 2e^x + xe^x$ **23.** $y = e^x - 2x^2$

SECTION 2.13

1. $y = 4 \sin 2t - 0.5 \cos 2t$
3. $y = -\frac{12}{433} \cos 3t - \frac{17}{433} \sin 3t - \frac{12}{17} \cos t - \frac{3}{17} \sin t$
5. $y = -\frac{5}{87} \cos 3t + \frac{2}{87} \sin 3t + 2 \cos t - \sin t$
7. $y = e^{-t}(A \cos t + B \sin t) + 0.2 \cos t + 0.4 \sin t$
9. $y = e^{-t}(A \cos 2t + B \sin 2t) + 0.1 \cos t - 0.2 \sin t$
11. $y = \cos t - 6 \sin t + 3 \sin 2t$ **13.** $y = e^{-t} \sin t - 0.5 \sin 2t$
15. $y = e^{-2t} \cos 4t + \sin t - \cos t$
17. $y = A \cos \omega t + B \sin \omega t$ where $A = -b\omega c/D$, $B = bm(\omega_0^2 - \omega^2)/D$,
$D = m^2(\omega_0^2 - \omega^2)^2 + \omega^2 c^2$, $\omega_0 = \sqrt{k/m}$.

SECTION 2.14

1. $I = A \cos 5t + B \sin 5t + [275\omega/(25 - \omega^2)] \cos \omega t$
3. $I = A \cos 20t + B \sin 20t + 0.02t$ **5.** $I = 10(\cos t - \cos 2t)$
7. $I = 1 - \cos t$ **9.** $I = \cos t + 2 \sin t$
11. $I = e^{-2t}(A \cos 4t + B \sin 4t) + \cos 4t + 4 \sin 4t$
13. $I = (c_1 + c_2 t)e^{-t} + 4 \cos 2t + 3 \sin 2t$
15. $I = e^{-3t}(3 \sin 4t - 4 \cos 4t) + 4 \cos 5t$ **17.** $I = 5e^{-2t} \sin t$
19. $I = I_1 = 1 - \cos t$ when $0 < t < a$, and $I = I_2 = \cos (t - a) - a \sin (t - a) - \cos t$ when
$t > a$ are obtained from $I(0) = 0$, $Q(0) = 0$, $E(0) = 0$, $I_2(a) = I_1(a)$, $\dot{I}_2(a) = \dot{I}_1(a) - a$.
21. $I = \begin{cases} \frac{1}{2}(e^{-t} - \cos t + \sin t) & \text{when } 0 < t < \pi \\ -\frac{1}{2}(1 + e^{-\pi}) \cos t + \frac{1}{2}(3 - e^{-\pi}) \sin t & \text{when } t > \pi \end{cases}$
25. $I = 1 - e^{-t}$ when $0 < t < a$, $I = (e^a - 1)e^{-t}$ when $t > a$.

SECTION 2.15

1. $y_p = -(30/61) \cos 3t + (36/61) \sin 3t$ **3.** $y_p = (\sin 4t)/6$
5. $y_p = -0.1 \cos 2t + 0.2 \sin 2t$ **7.** $I_p = -0.9704 \cos 5t + 0.4621 \sin 5t$
9. $I_p = (50 \sin 4t - 110 \cos 4t)/73$

SECTION 2.16

1. $y = A \cos 2x + B \sin 2x + 0.25 \cos 2x \ln |\cos 2x| + 0.5x \sin 2x$
3. $y = (c_1 + c_2 x)e^{-x} + x^2 e^{-x}(2 \ln x - 3)$
5. $y = (c_1 + c_2 x)e^x + e^x/x$
7. $y = e^{-x}(A \cos x + B \sin x - 0.5 \cos 2x/\cos x)$
9. $y = (c_1 + c_2 x + \frac{4}{35}x^{7/2})e^x$
11. $y = (c_1 + c_2 x + x \ln x - x)e^{2x}$
13. $y = c_1 x^{-1} + c_2 x - 10$
15. $y = c_1 x + c_2 x^2 + x^{-1}$
17. $y = c_1 x^2 + c_2 x^3 + x^{-4}$
19. $y = c_1 x + c_2 x^2 - x \cos x$

SECTION 3.2

1. $y = -1 + c_3 x^3$
3. $y = c_0(1 + 3x + 4x^2 + \frac{10}{3}x^3 + 2x^4 + \frac{14}{15}x^5 + \cdots) = c_0(x + 1)e^{2x}$

5. $c_0 = 0, c_1 = 2, c_m = c_{m-1}/(m - 2), y = 2x + c_2 x^2\left(1 + x + \frac{x^2}{2!} + \cdots\right) = c_2 x^2 e^x + 2x$

7. $y = c_0 + c_1 x + (\frac{3}{2}c_1 - c_0)x^2 + (\frac{7}{6}c_1 - c_0)x^3 + \cdots$. If we set $c_0 = A + B$ and $c_1 = A + 2B$, this becomes $y = Ae^x + Be^{2x}$. This illustrates the fact that even if the solution of an equation is a known function, the power series method may not yield it immediately in the usual form.
9. $y = c_1 x + c_0(1 - x^2 - \frac{1}{3}x^4 - \frac{1}{5}x^6 - \frac{1}{7}x^8 - \cdots)$. (This is a particular case of Legendre's equation ($n = 1$) which will be considered in the next section.)

11. $(t + 1)\ddot{y} - y = t + 1, y = c_0(1 + t) + (1 + t)\left(t - \frac{t^2}{2} + - \cdots\right) = c_0 x + x \ln x$

13. $y = c_0\left(1 - \frac{t^2}{2!} + \frac{t^4}{4!} - + \cdots\right) + c_1\left(t - \frac{t^3}{3!} + - \cdots\right) = c_0 \cos (x - 1) + c_1 \sin (x - 1)$

15. $y = c_0\left(1 - \frac{x^2}{2^2(1!)^2} + \frac{x^4}{2^4(2!)^2} - \frac{x^6}{2^6(3!)^2} + - \cdots\right)$

17. 4   **19.** ∞ **21.** $\sqrt{|k|}$ **23.** $\frac{1}{3}$ **25.** ∞ **27.** ∞ **29.** 3

SECTION 3.4

1. $y_1 = c_0(1 + x + x^2 + \cdots) = \dfrac{c_0}{1 - x}, \qquad y_2 = \dfrac{c_0^*}{x}$

3. $y_1 = 1 + \dfrac{x^2}{2^2} + \dfrac{x^4}{(2 \cdot 4)^2} + \dfrac{x^6}{(2 \cdot 4 \cdot 6)^2} + \cdots, \qquad y_2 = y_1 \ln x - \dfrac{x^2}{4} - \dfrac{3x^4}{8 \cdot 16} - \cdots$

5. $y_1 = \dfrac{c_0}{x^2}\left[1 + \dfrac{x^2}{3!} + \dfrac{x^4}{5!} + \cdots\right] = \dfrac{c_0}{x^3} \sinh x, \qquad y_2 = \dfrac{c_0^*}{x^3} \cosh x$

9. $y_1 = e^x, y_2 = e^x \ln x$

11. $y_1 = \dfrac{c_0}{x^2}\left(1 - \dfrac{x^2}{2!} + \dfrac{x^4}{4!} - + \cdots\right) = c_0 \dfrac{\cos x}{x^2}, \qquad y_2 = c_0^* \dfrac{\sin x}{x^2}$

13. $y_1 = x^{-2} \sin x^2, y_2 = x^{-2} \cos x^2$
15. $y_1 = 1 + x, y_2 = (1 + x) \ln x$
17. $y_1 = \sqrt{x}\, e^x, y_2 = (x + 1)e^x$
19. $y_1 = \sqrt{x}\, e^x, y_2 = e^{2x}$
31. The quotient y_1/y_2 is not constant.
33. $y = AF(\frac{1}{2}, \frac{1}{2}, \frac{1}{2}; x) + B\sqrt{x}\, F(1, 1, \frac{3}{2}; x)$

SECTION 3.5

5. Approximate values $\frac{3}{4}\pi + k\pi = 2.356, 5.498, 8.639, 11.781, 14.923$.
7. $AJ_v(\sqrt{x}) + BJ_{-v}(\sqrt{x})$ **9.** $AxJ_1(x)$ **11.** $x^{-k}(AJ_k(x) + BJ_{-k}(x))$

21. $(x^{-\nu}J_\nu)' = \sum\limits_{m=1}^{\infty} \dfrac{(-1)^m x^{2m-1}}{2^{2m+\nu-1}(m-1)!\,\Gamma(\nu+m+1)} = \sum\limits_{s=0}^{\infty} \dfrac{(-1)^{s+1} x^{2s+1}}{2^{2s+\nu+1}s!\,\Gamma(\nu+s+2)}$

(where $m = s + 1$).

23. Perform the product differentiations indicated in (22) and (23). Multiply the first of the resulting equations by $x^{-\nu}$ and the second by x^ν, solve for $J_{\nu-1}$ and $-J_{\nu+1}$, respectively, and add.

29. $J_2(x) = 2x^{-1}J_1(x) - J_0(x)$; cf. (24) with $\nu = 1$.

37. $-2J_2(x) - J_0(x) + c$ **39.** $-x^{-3}J_3(x) + c$

SECTION 3.6

1. 1.1   **3.** $AJ_2(x) + BY_2(x)$ **5.** $AJ_0(\sqrt{x}) + BY_0(\sqrt{x})$

7. $AJ_2(2x) + BY_2(2x)$ **9.** $AJ_n(kx) + BY_n(kx)$ **11.** $x^n(AJ_n(x^n) + BY_n(x^n))$

13. Use (12) in Sec. 3.5.

SECTION 3.7

1. $1/\sqrt{2\pi}$, $(\cos x)/\sqrt{\pi}$, $(\cos 2x)/\sqrt{\pi}$, $\cdots$ **3.** $\sin \pi x$, $\sin 2\pi x$, $\cdots$

5. $1/\sqrt{T}$, $\sqrt{2/T}\cos(2n\pi x/T)$ $(n = 1, 2, \cdots)$ **7.** $P_0/\sqrt{2}$, $\sqrt{\tfrac{3}{2}}P_1(x)$, $\sqrt{\tfrac{5}{2}}P_2(x)$, $\sqrt{\tfrac{7}{2}}P_3(x)$

9. Set $x = ct + k$.

SECTION 3.8

3. $\sin((2n+1)\pi x/2l)$, $n = 0, 1, \cdots$ **5.** $y'' + \lambda y = 0$, $y'(0) = 0$, $y'(\pi) = 0$

SECTION 3.9

1. $4P_1 + 2P_3$ **3.** $x^2 = (P_0 + 2P_2)/3$, $x^3 = (3P_1 + 2P_3)/5$

5. $P_0/4 + P_1/2 + 5P_2/16 + \cdots$ **7.** $P_0/2 + 5P_2/8 + \cdots$

15. $\displaystyle\int_0^{\infty} e^{-x}x^k L_n(x)\,dx = \dfrac{1}{n!}\int_0^{\infty} x^k \dfrac{d^n}{dx^n}(x^n e^{-x})\,dx = -\dfrac{k}{n!}\int_0^{\infty} x^{k-1}\dfrac{d^{n-1}}{dx^{n-1}}(x^n e^{-x})\,dx$

$$= \cdots = (-1)^k \dfrac{k!}{n!}\int_0^{\infty} \dfrac{d^{n-k}}{dx^{n-k}}(x^n e^{-x})\,dx = 0 \text{ when } n > k.$$

19. $G_x = \sum a_n'(x)t^n = \sum H_n'(x)t^n/n! = tG = \sum H_{n-1}(x)t^n/(n-1)!$, etc.

25. By (22), Sec. 3.5, with $\nu = 1$,

$$c_m = \dfrac{2}{R^2 J_1^{\,2}(\alpha_{m0})}\int_0^R xJ_0\!\left(\dfrac{\alpha_{m0}}{R}x\right)dx = \dfrac{2}{\alpha_{m0}^{\,2}J_1^{\,2}(\alpha_{m0})}\int_0^{\alpha_{m0}} wJ_0(w)\,dw = \dfrac{2}{\alpha_{m0}J_1(\alpha_{m0})},$$

$$f = 2\left(\dfrac{J_0(\lambda_{10}x)}{\alpha_{10}J_1(\alpha_{10})} + \dfrac{J_0(\lambda_{20}x)}{\alpha_{20}J_1(\alpha_{20})} + \cdots\right)$$

27. $c_m = \dfrac{2akJ_1(\alpha_{m0}a/R)}{\alpha_{m0}RJ_1^{\,2}(\alpha_{m0})}$ **29.** $c_m = \dfrac{4J_2(\alpha_{m0})}{\alpha_{m0}^{\,2}J_1^{\,2}(\alpha_{m0})}$

31. $c_m = \dfrac{2R^2}{\alpha_{m0}J_1(\alpha_{m0})}\left[1 - \dfrac{2J_2(\alpha_{m0})}{\alpha_{m0}J_1(\alpha_{m0})}\right]$ **35.** $x^3 = 16\left[\dfrac{J_3(\alpha_{13}x/2)}{\alpha_{13}J_4(\alpha_{13})} + \dfrac{J_3(\alpha_{23}x/2)}{\alpha_{23}J_4(\alpha_{23})} + \cdots\right]$

SECTION 4.1

1. $3/s^2 + 4/s$ **3.** $2/s^3 + a/s^2 + b/s$ **5.** $2/(s-1)^2$ **7.** $e^b/(s-a)$

9. $2n\pi[T(s^2 + (2n\pi/T)^2)]^{-1}$ **11.** $(s\cos\theta - \omega\sin\theta)/(s^2 + \omega^2)$

13. $1/2s + s/(2s^2 + 8)$ **15.** $(\omega\cos\theta + (s+1)\sin\theta)/[(s+1)^2 + \omega^2]$

17. $k(1 - e^{-cs})/s$ **19.** $k/s - k(1 - e^{-cs})/cs^2$

37. $0.5\sin 4t$ **39.** $\cosh 2t - 2\sinh 2t$

41. $e^{-2t}\cos t$ **43.** $e^t(\cosh 2t + 0.5\sinh 2t)$

45. $(e^{at} - e^{bt})/(a - b)$

SECTION 4.2

3. $\dfrac{2\omega^2}{s(s^2 + 4\omega^2)}$

15. $f' = 1 \ (0 < t < 1), \ \mathscr{L}(f') = (1 - e^{-s})/s$ (by direct integration),
$\mathscr{L}(f') = s\mathscr{L}(f) - 0 - (0 - 1)e^{-s}.$ *Ans.* $(1 - e^{-s})/s^2 - e^{-s}/s.$

17. $(e^{-s} - e^{-2s})/s^2 - e^{-2s}/s$ **19.** $1 - e^{-t}$ **21.** $(1 - \cos 2t)/4$

23. $2e^{-at} - 1$ **25.** $\sinh 2t - 2t$ **27.** $(e^{2t} - 1 - 2t - 2t^2)/8$

29. $1 + t - \cos t - \sin t$

SECTION 4.3

1. $y = \cos t - \sin t$ **3.** $y = -3 \sinh 2t$ **5.** $y = A \cos \omega t + (B/\omega) \sin \omega t$

7. $y = e^{-t} \sin t$ **9.** $y = 3e^{-t} \sin 4t$ **11.** $y = 3e^{t/3}$

13. $y = e^{kt} + 1$ **15.** $y = 3e^{t/2} \cos 3t$ **17.** $y = e^{-t/2} \cos t$

19. $y = 2e^{kt}$

SECTION 4.5

1. $3 - 2e^{-4t}$ **3.** $2e^{-4t} + e^{2t}$

5. $1 - 2 \sinh 5t$ **7.** $3 \cos 3t - (\sin 3t)/3$

9. $e^t(\cos t + \sin t)$ **11.** $2e^{3t} + \cos t + \sin t$

13. $e^{2t}(2t - 4)$ **15.** $e^{-t}(\cos 2t - 2t \sin 2t)$

17. $te^{2t}(\cos t - \sin t)$ **27.** $y = \cos t - \cos 2t$

29. $y = e^{3t} - 2e^{-3t} + e^t$ **31.** $y = 2e^{3t} + e^t$

33. $y = e^t \sin 2t$ **35.** $y = e^{-t}(\cos t - 2 \sin t)$

37. $y = e^t - e^{-2t} + t$ **39.** $y = \cos 5t + t(2 \cos 5t + \sin 5t)$

41. $y = t(\sin t - \cos t)$ **43.** $y_1 = \cos t, \ y_2 = \sin t$

45. $y_1 = \sin t, \ y_2 = \sin 2t$ **47.** $y_1 = \sinh t, \ y_2 = \cosh t$

49. $y_1 = e^t, \ y_2 = e^{-t}, \ y_3 = e^t - e^{-t}$

SECTION 4.6

1. $n!/(s - a)^{n+1}$ **3.** $1/(s - 2)^2$ **5.** $2/(s - 1)^3$

7. $\dfrac{12s^2 + 16}{(s^2 - 4)^3}$ **9.** $\dfrac{2s(s^2 - 3\omega^2)}{(s^2 + \omega^2)^3}$ **11.** $y = \dfrac{t}{2} \sin t$

13. $t^2 e^{at}$ **15.** $(e^t - 1)/t$ **17.** $(e^{-bt} - e^{-at})/t$

19. $e^{-t}(\sin t)/t$

SECTION 4.7

1. $2(1 - e^{-\pi s})/s$ **3.** $(e^{-s} - 2e^{-2s} + e^{-3s})/s$

5. $k/s(1 + e^{-as})$ **7.** $\dfrac{3}{2s(1 - e^{-3s})} - \dfrac{1}{2s(1 - e^{-s})}$

9. $1/s(1 - e^{-s})$ **11.** $(e^{-s} - e^{-2s})/s^2$

13. $1/s - (e^{-s} - 2e^{-2s} + e^{-3s})/s^2$ **15.** $(e^{-s} - e^{-2s})/s^2 - e^{-2s}/s$

19. $f(t) = 2$ if $2 < t < 4$ and 0 otherwise

21. $f(t) = t - a$ when $t > a$ and 0 when $t < a$; cf. Theorem 3 in Sec. 4.2.

23. $f(t) = t - 1, 2t - 3, 6 - t$ when $1 < t < 2, 2 < t < 3, 3 < t < 6$, and 0 otherwise

25. $i(t) = V_0(1 - e^{-Rt/L})/R$

SECTION 4.8

1. $e^{-\pi s}/s^2$ **3.** $(1/s^2 + 2/s)e^{-2s}$

5. $-e^{-\pi s}/(s^2 + 1)$ **7.** $e^{-s-2}/(s + 2)$

9. $e^{-s}(2s^{-3} + 2s^{-2} + s^{-1})$ **11.** $2s^{-3} - e^{-s}(2s^{-3} + 2s^{-2} + s^{-1})$

13. $(1 - e^{1-s})/(s - 1)$ **15.** $Ks(1 + e^{-\pi s/\omega})/s^2 + \omega^2)$

17. $s^{-1} - (s + 1)^{-1} - e^{-\pi s}(s^{-1} - e^{-\pi}(s + 1)^{-1})$ **19.** $f(t) = \cos 2t$ when $t > \pi$ and 0 otherwise

21. $f(t) = e^{2t-4}$ when $t > 2$ and 0 otherwise   **23.** $f(t) = (t - 1)^3/6$ when $t > 1$ and 0 otherwise
25. $-e^{\pi-t} \sin t$ when $t > \pi$ and 0 otherwise

27. $\mathscr{L}[f(at)] = \displaystyle\int_0^\infty e^{-st} f(at)\, dt = \frac{1}{a}\int_0^\infty e^{-s\tau/a} f(\tau)\, d\tau = \frac{1}{a} F\left(\frac{s}{a}\right)$

29. $t^3 e^{2t}/96$ **31.** $e^{-3t/\pi} \cos t$
33. $1 - \cos t$ if $0 < t < 1$, $\cos (t - 1) - \sin (t - 1) - \cos t$ if $t > 1$
35. $(e^{-t} - \cos t + \sin t)/2$ if $0 < t < \pi$, $[-(1 + e^{-\pi}) \cos t + (3 - e^{-\pi}) \sin t]/2$ if $t > \pi$

SECTION 4.9

1. $\dfrac{\pi s - 1 + (\pi s + 1)e^{-2\pi s}}{s^2(1 - e^{-2\pi s})} = \dfrac{\pi}{s} \coth \pi s - \dfrac{1}{s^2}$

3. $\dfrac{2(e^{2\pi s} - 1 - 2\pi s - 2\pi^2 s^2)}{s^3(e^{2\pi s} - 1)}$ **5.** $\dfrac{e^{2(1-s)\pi} - 1}{(1 - s)(1 - e^{-2\pi s})}$

7. $\left[\dfrac{1}{s^2} (1 - e^{-\pi s}) - \dfrac{\pi}{s} e^{-\pi s}\right]\bigg/ (1 - e^{-2\pi s})$

9. $\left[\dfrac{\pi}{s} e^{-\pi s}(e^{-\pi s} - 1) + \dfrac{1}{s^2} (e^{-\pi s} - 1)^2\right]\bigg/ (1 - e^{-2\pi s})$

13. $\dfrac{\omega}{(s^2 + \omega^2)(e^{\pi s/\omega} - 1)}$ **15.** $\dfrac{1}{s^2 + \omega^2}\left(s + \dfrac{\omega}{\sinh \dfrac{\pi s}{2\omega}}\right)$

21. $i(t) = \dfrac{1}{R}\left(\dfrac{e^{-kt}}{1 - e^{-k}} + t - \dfrac{L}{R}\right)$ when $0 < t < 1$, and $i(t + 1) = i(t)$, $k = \dfrac{R}{L}$

23. $Ri + q/C = v(t)$, and since the charge q on the capacitor cannot change abruptly, i must have a jump.

25. $i(t) = \begin{cases} a \cos \omega t + b \sin \omega t - ae^{-t/RC} & \text{when } 0 < t < \pi/\omega \\ -a(1 + e^{\pi/\omega RC})e^{-t/RC} & \text{when } t > \pi/\omega \end{cases}$

where $a = \omega CK$, $b = \omega^2 RC^2 K$, $K = 1/[1 + (\omega RC)^2]$.
29. $y = P(1 - (1 + \alpha t)e^{-\alpha t})/K$
31. For $0 < t < 1$ the solution is as in Prob. 30. When $t > 1$,

$$v = \omega - \frac{P}{M_1 + M_2} + \frac{P}{k(M_1 + M_2)} [\sin kt - \sin (k(t - 1))], \qquad k^2 = c\left(\frac{1}{M_1} + \frac{1}{M_2}\right)$$

($c = $ stiffness of the shaft).

33. $i_{1} = V_0\left[e^{-\alpha t}\left(-\dfrac{1}{R_1} \cos \omega^* t + \dfrac{1}{\omega^*}\left(\dfrac{L_2}{A} - \dfrac{\alpha}{R_1}\right) \sin \omega^* t\right) + \dfrac{1}{R_1}\right]$

35. $L_2 i_2' + R_2 i_2 = M i_1'$ where $i_1(0) = V_0/R_1$, $i_1 = 0$ when $t > 0$ and thus

$$L_2 s I_2 + R_2 I_2 = M\mathscr{L}(i_1') = M[s\mathscr{L}(i_1) - i_1(0)] = -\frac{MV_0}{R_1}, i_2 = -\frac{MV_0}{R_1 L_2} e^{-R_2 t/L_2}$$

SECTION 5.2

1. $3, 2, 0, |\mathbf{v}| = \sqrt{13}$ **3.** $-3, 0, 3, |\mathbf{v}| = 3\sqrt{2}$
5. $0, b - a, 0, |\mathbf{v}| = |b - a|$ **7.** $-16, 0, 0, |\mathbf{v}| = 16$
9. $0, 0, 0, |\mathbf{v}| = 0$ **11.** $Q\colon (3, 0, 0), |\mathbf{v}| = \sqrt{2}$
13. $Q\colon (0, 0, 0), |\mathbf{v}| = \sqrt{41}$ **15.** $Q\colon (6, 2, 4), |\mathbf{v}| = \sqrt{14}$
17. $Q\colon (1, 2, 3), |\mathbf{v}| = \sqrt{14}$ **19.** $Q\colon (0, 2, 2), |\mathbf{v}| = \sqrt{7/2}$

SECTION 5.3

1. $3i + 2k$

3. $4i - 5j + 27k$

5. $3i + 3j - 27k$

7. $(2i - j + 3k)/\sqrt{14}$, $(-2i + j - 3k)/\sqrt{14}$

9. $\sqrt{13} + 4$, $\sqrt{14} + \sqrt{11}$

11. $3i + 6k$

13. 0

15. $-i - 2j + 4k$

17. $1 \leqq |p + q| \leqq 3$, nothing about the direction.

19. Diagonals and diagonals of the sides.

21. The vector r from 0 to P is of the form $r = k(a + b)$. Also, $r = a + l(b - a)$. Thus, $ka + kb = (1 - l)a + lb$. From this, $k = 1 - l$, $k = l$; hence $k = \frac{1}{2}$, $l = \frac{1}{2}$, and the proof is complete.

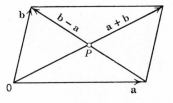

SECTION 5.4

1. $2, i, k$

3. $1, i + 2j - k$

5. $n, (1, 0, \cdots, 0), (0, 1, 0, \cdots, 0), \cdots, (0, \cdots, 0, 1)$

7. $2, i, j$

9. $2, \cos x, \sin x$

11. If not, $a = c_1 e_{(1)} + \cdots + c_n e_{(n)} = q_1 e_{(1)} + \cdots + q_n e_{(n)}$, $(c_1 - q_1)e_{(1)} + \cdots + (c_n - q_n)e_{(n)} = 0$, $c_j - q_j = 0$ for all j because the $e_{(j)}$ are linearly independent.

SECTION 5.5

1. -10

3. 6

5. 16

7. $\sqrt{11}, \sqrt{50}$

9. $\sqrt{37}$

11. $2, 0, -2$

13. $v = v_1 i - 0.5 v_1 j + v_3 k$

17. If and only if all four sides are equal.

19. $x + y - 2z = 0$

21. $v_1 i + (2 - 2v_1)j$

23. 0

25. $p_{(1)} \cdot d + p_{(2)} \cdot d = (p_{(1)} + p_{(2)}) \cdot d$

27. -5

29. 2

31. $2\sqrt{2}/3$

33. $-\sqrt{\frac{7}{15}}$

35. $\cos \alpha = 2/\sqrt{3}\sqrt{14} = 0.309$, $\alpha = 72°$

37. $|a - b|^2 = (a - b) \cdot (a - b) = |a|^2 + |b|^2 - 2|a||b| \cos \gamma$, etc.

39. $|a + b|^2 = (a + b) \cdot (a + b) \leqq |a|^2 + 2|a||b| + |b|^2 = (|a| + |b|)^2$

SECTION 5.8

1. $3k, -3k$

3. $i - j + k, \sqrt{3}, 5$

5. $3i - 6k$

7. $i - j + k$

9. $2i + 3j + k, 3$

13. $k, -k$

15. $i, -i$

17. $\pm(3j + 2k)/\sqrt{13}$

19. $(i + j + k)/\sqrt{3}$

21. $(15i - 13j + 6k)/\sqrt{430}$

23. $2i + 5j + 3k$

25. Orthogonal

27. No

29. Orthogonal

31. $\sqrt{109}$

33. $\sqrt{35}$

35. 1

37. $\sqrt{3}$

39. $\sqrt{8}$

41. 10

43. 9

45. $\sqrt{34}$

47. $\sqrt{234}/2$

49. Yes

53. $0, 0$

55. $13i - 10j + 14k, \sqrt{465}$

SECTION 5.9

1. 1   **3.** 0   **5.** 2   **7.** -3 **9.** 6   **11.** 40   **13.** 12   **15.** 7/3   **17.** 2/3
19. Linearly independent   **21.** Linearly dependent   **23.** Linearly dependent
25. Linearly independent   **27.** $\lambda = 5/3$ **29.** Yes
31. $\mathbf{a} \cdot [\mathbf{b} \times (\mathbf{c} \times \mathbf{d})]$ equals $(\mathbf{a} \quad \mathbf{b} \quad [\mathbf{c} \times \mathbf{d}]) = (\mathbf{a} \times \mathbf{b}) \cdot (\mathbf{c} \times \mathbf{d})$ as well as
$(\mathbf{a} \cdot \mathbf{c})(\mathbf{b} \cdot \mathbf{d}) - (\mathbf{a} \cdot \mathbf{d})(\mathbf{b} \cdot \mathbf{c})$.
33. $0, 0, 1, 1$ (right-handed coordinates)   **35.** $-396\mathbf{i} - 216\mathbf{j} - 12\mathbf{k}$
37. 53

SECTION 6.3

1. $\begin{pmatrix} -3 & -1 \\ 2 & 5 \end{pmatrix}$ **3.** $\begin{pmatrix} 7 & -5 \\ -2 & 3 \end{pmatrix}, \begin{pmatrix} -7 & 5 \\ 2 & -3 \end{pmatrix}$ **5.** $\begin{pmatrix} 24 & -14 \\ -8 & 4 \end{pmatrix}$

7. $\begin{pmatrix} 0 & 3 \\ 1 & 0 \\ -2 & 4 \end{pmatrix}, \begin{pmatrix} 1 & -1 \\ 2 & 5 \\ 4 & -3 \end{pmatrix}, \mathbf{D}$ **9.** $\begin{pmatrix} 6 & 0 \\ -9 & 12 \end{pmatrix}$ **11.** $\begin{pmatrix} 1 & 2 \\ 3 & 5 \\ 2 & 1 \end{pmatrix}$

13. No   **15.** **B** is symmetric.   **17.** Yes

19. $\begin{pmatrix} 7 & -3.5 \\ -3.5 & 3 \end{pmatrix} + \begin{pmatrix} 0 & -1.5 \\ 1.5 & 0 \end{pmatrix}$

21. $\begin{pmatrix} 3 & 3 & 5 \\ -1 & 5 & 2 \\ 4 & -3 & 6 \end{pmatrix}$ **23.** $\begin{pmatrix} -2 & 12 & 20 \\ 3 & 13 & 8 \\ -12 & 9 & 24 \end{pmatrix}$ **25.** $\begin{pmatrix} 3 & -1 & 4 \\ 3 & 5 & -3 \\ 5 & 2 & 6 \end{pmatrix}$

27. No, no, yes, no, yes.

SECTION 6.4

1. $\mathbf{AB} = \begin{pmatrix} 9 & 20 \\ 4 & 16 \\ -3 & 12 \end{pmatrix}, \mathbf{A}^{\mathsf{T}}\mathbf{B} = \begin{pmatrix} 7 & 21 \\ 14 & 20 \\ -7 & 8 \end{pmatrix}, \mathbf{B}^{\mathsf{T}}\mathbf{A} = \begin{pmatrix} 7 & 14 & -7 \\ 21 & 20 & 8 \end{pmatrix}$

3. $\mathbf{A}^2 = \begin{pmatrix} 33 & 26 & 3 \\ 14 & 14 & 7 \\ 3 & -4 & 20 \end{pmatrix}$ **5.** $\mathbf{BB}^{\mathsf{T}} = \begin{pmatrix} 20 & 4 & 6 \\ 4 & 1 & 2 \\ 6 & 2 & 5 \end{pmatrix}, \mathbf{B}^{\mathsf{T}}\mathbf{B} = \begin{pmatrix} 5 & 6 \\ 6 & 21 \end{pmatrix}$

7. $(\mathbf{AB})^{\mathsf{T}} = \begin{pmatrix} 9 & 4 & -3 \\ 20 & 16 & 12 \end{pmatrix} = \mathbf{B}^{\mathsf{T}}\mathbf{A}^{\mathsf{T}}$ **9.** $\begin{pmatrix} 43 & 44 & 0 \\ 23 & 12 & 13 \\ 6 & -10 & 33 \end{pmatrix}$

13. $\mathbf{AB} = \mathbf{BA} = \begin{pmatrix} -5 & 0 & 0 \\ 0 & -5 & 0 \\ 0 & 0 & -5 \end{pmatrix}$

19. $\mathbf{y} = \mathbf{ABz}, \mathbf{AB} = \begin{pmatrix} \cos & -2\sin\theta \\ \sin\theta & 2\cos\theta \end{pmatrix} \neq \mathbf{BA} = \begin{pmatrix} \cos\theta & -\sin\theta \\ 2\sin\theta & 2\cos\theta \end{pmatrix}$.

Geometrically this means that we obtain two different composite transformations, depending on the order of composition.

SECTION 6.5

1. $x = 1, y = 2$ **3.** $x = -2, y = 3$ **5.** $x = -4, y = 4$
7. $x = 2z, y = 3z$ **9.** $x = -2, y = 0, z = 4$ **11.** $x = 1, y = 2, z = 4$
13. $x = y + 1, z = 1$ **15.** $x = 3y + 2, z = 0$ **17.** $w = 2, y = x - 3z + 4$
19. $w = 0, x = 2y, z = 3y + 1$

SECTION 6.6

1. -5 **3.** 1   **5.** 0   **7.** 180   **9.** 0

SECTION 6.7

1. $4a^2b^2c^2$ **3.** 90

SECTION 6.8

5. 2   **7.** 2   **9.** 3

SECTION 6.10

1. Linearly independent   **3.** Linearly dependent
5. Linearly dependent   **7.** Linearly dependent
9. Linearly independent   **11.** $(0, 0, 0, 1)$
13. $k = 0$

SECTION 6.11

1. $x = 2, y = 1$ **3.** $x = 2, y = 2$ **5.** $x = -1, y = 0, z = 1$
7. $x = 1, y = 1, z = 1$ **9.** $x = 0, y = 0, z = 1$
11. $I_1 = (R_2 + R_3)E_0/D, I_2 = R_3E_0/D, I_3 = R_2E_0/D, D = (R_1 + R_4)(R_2 + R_3) + R_2R_3$

13. $I_1 = \dfrac{R_4E_1 + R_5(E_1 + E_2)}{(R_1 + R_2 + R_3)(R_4 + R_5) + R_4R_5}$, etc.

17. $3x + 7y = 44$ **19.** Use Prob. 16.

SECTION 6.12

3. $\begin{pmatrix} \cos\theta & -\sin\theta \\ \sin\theta & \cos\theta \end{pmatrix}$ **5.** $\begin{pmatrix} 3 & 0 & 1 \\ 0 & 5 & 0 \\ -1 & 1 & -1 \end{pmatrix}$ **7.** $\begin{pmatrix} 3 & -1 & 1 \\ -15 & 6 & -5 \\ 5 & -2 & 2 \end{pmatrix}$

9. $x = -2y^* + z^*$
$y = \frac{1}{2}(x^* + y^* - z^*)$
$z = -x^* + 2y^*$
11. Use (1), with $\mathbf{A}$ replaced by $\mathbf{C}$ and set $\mathbf{C} = \mathbf{A}^{-1}$.

13. $\begin{pmatrix} \cos 2\theta & -\sin 2\theta \\ \sin 2\theta & \cos 2\theta \end{pmatrix}$, $\begin{pmatrix} 29 & -11 & 10 \\ -160 & 61 & -55 \\ 55 & -21 & 19 \end{pmatrix}$

15. $\mathbf{I} = \mathbf{I}^\mathsf{T} = (\mathbf{A}^{-1}\mathbf{A})^\mathsf{T} = \mathbf{A}^\mathsf{T}(\mathbf{A}^{-1})^\mathsf{T}$

SECTION 6.13

1. $\begin{pmatrix} 6 & -2 \\ -2 & 2 \end{pmatrix}$ **3.** $\begin{pmatrix} 0 & 7 \\ 7 & -1 \end{pmatrix}$ **5.** $\begin{pmatrix} 1 & 1 & 1 \\ 1 & 1 & 1 \\ 1 & 1 & 1 \end{pmatrix}$

7. 7   **9.** $a|x_1|^2 + 2\operatorname{Re}[(b + ic)\bar{x}_1x_2] + k|x_2|^2$ **11.** 1
13. $i(|x_1|^2 + |x_2|^2)$ **15.** $32i$

17. $\mathbf{A} = \mathbf{H} + \mathbf{S}$, where $\mathbf{H} = (\mathbf{A} + \overline{\mathbf{A}}^{\mathsf{T}})/2$ is Hermitian and $\mathbf{S} = (\mathbf{A} - \overline{\mathbf{A}}^{\mathsf{T}})/2$ is skew-Hermitian.
19. $\mathbf{C}^{\mathsf{T}}$ must equal $-\overline{\mathbf{C}}$, which implies a, b real.
21. $S = \alpha + i\beta = \overline{\mathbf{x}}^{\mathsf{T}}\mathbf{A}\mathbf{x}$, $\overline{S} = \alpha - i\beta = \mathbf{x}^{\mathsf{T}}\overline{\mathbf{A}}\overline{\mathbf{x}} = -\mathbf{x}^{\mathsf{T}}\mathbf{A}^{\mathsf{T}}\overline{\mathbf{x}} = -(\overline{\mathbf{x}}^{\mathsf{T}}\mathbf{A}\mathbf{x})^{\mathsf{T}} = -S = -\alpha - i\beta$.
25. Not positive definite.

SECTION 6.14

1. $1, \begin{pmatrix} 1 \\ 0 \end{pmatrix}, -3, \begin{pmatrix} 0 \\ 1 \end{pmatrix}$

3. 0, any vector

5. $5, \begin{pmatrix} 2 \\ 1 \end{pmatrix}, -5, \begin{pmatrix} 1 \\ -2 \end{pmatrix}$

7. $3, \begin{pmatrix} 2 \\ 1 \end{pmatrix}, 2, \begin{pmatrix} 1 \\ 1 \end{pmatrix}$

9. 1, any vector

11. $3, \begin{pmatrix} 2 \\ -10 \\ -27 \end{pmatrix}, 4, \begin{pmatrix} 0 \\ 1 \\ 2 \end{pmatrix}, 1, \begin{pmatrix} 0 \\ 0 \\ 1 \end{pmatrix}$

13. $2, 1, -0.5$ **15.** 30, 25, 20
23. Use induction; premultiply $\mathbf{A}^k\mathbf{x}_j = \lambda_j{}^k\mathbf{x}_j$ by $\mathbf{A}$.
33. The characteristic equation is $\lambda^2 + (a + b + c)\lambda + ac = 0$ where $a = k_1/m_1$, $b = k_2/m_1$, $c = k_2/m_2$. The discriminant is

$$(a + b + c)^2 - 4ac = (a + b - c)^2 + 4bc > 0.$$

Thus the roots are real. Since their product equals ac (>0) and their sum equals $-(a + b + c)$, the statement follows.
35. $y_1 = \sin\sqrt{2}\,t$, $y_2 = 2\sin\sqrt{2}\,t$
37. $y_1 = 2e^{5t} - e^{-5t}$, $y_2 = e^{5t} + 2e^{-5t}$
39. $y_1 = 2\cosh 2t$, $y_2 = 4\sinh 2t$, $y_3 = 2\cosh 2t$

SECTION 6.15

1. Let $\mathbf{A}$ be unitary. Set $\mathbf{A}^{-1} = \mathbf{B}$. Then $\mathbf{B}^{\mathsf{T}} = (\mathbf{A}^{-1})^{\mathsf{T}} = (\mathbf{A}^{\mathsf{T}})^{-1} = (\overline{\mathbf{A}}^{-1})^{-1} = \overline{\mathbf{B}}^{-1}$.
3. $\mathbf{A}^{-1}\mathbf{A} = \mathbf{A}^{\mathsf{T}}\mathbf{A} = \mathbf{I}$. Thus $\det(\mathbf{A}^{-1}\mathbf{A}) = \det(\mathbf{A}^{\mathsf{T}}\mathbf{A}) = \det \mathbf{A}^{\mathsf{T}} \det \mathbf{A} = 1$.

5. $\dfrac{i}{2} + \dfrac{\sqrt{3}}{2}, \begin{pmatrix} 1 \\ 1 \end{pmatrix}, \dfrac{i}{2} - \dfrac{\sqrt{3}}{2}, \begin{pmatrix} 1 \\ -1 \end{pmatrix}$

7. $1, \begin{pmatrix} 1 \\ i - i\sqrt{2} \end{pmatrix}, -1, \begin{pmatrix} 1 \\ i + i\sqrt{2} \end{pmatrix}$

9. $1, -1$

11. $e^{i\theta} = \cos\theta + i\sin\theta$, $e^{-i\theta} = \cos\theta - i\sin\theta$
13. Let $\mathbf{A}\mathbf{x} = \lambda\mathbf{x}$ ($\mathbf{x} \neq 0$), $\mathbf{A}\mathbf{y} = \mu\mathbf{y}$ ($\mathbf{y} \neq 0$). Then $(\mathbf{A}\mathbf{x})^{\mathsf{T}} = \mathbf{x}^{\mathsf{T}}\mathbf{A}^{\mathsf{T}} = \mathbf{x}^{\mathsf{T}}\mathbf{A} = \lambda\mathbf{x}^{\mathsf{T}}$. Thus $\lambda\mathbf{x}^{\mathsf{T}}\mathbf{y} = \mathbf{x}^{\mathsf{T}}\mathbf{A}\mathbf{y} = \mathbf{x}^{\mathsf{T}}\mu\mathbf{y} = \mu\mathbf{x}^{\mathsf{T}}\mathbf{y}$. Hence, if $\lambda \neq \mu$, $\mathbf{x}^{\mathsf{T}}\mathbf{y} = 0$, which proves orthogonality.

17. $\lambda_1 = -k$, $\lambda_2 = -3k$, $\mathbf{x}_1 = \begin{pmatrix} 1 \\ 1 \end{pmatrix}$, $\mathbf{x}_2 = \begin{pmatrix} 1 \\ -1 \end{pmatrix}$, $\mathbf{y} = \mathbf{x}_1 \cos\sqrt{k}\,t + \mathbf{x}_2 \sin\sqrt{3k}\,t$
23. $\det(\tilde{\mathbf{A}} - \lambda\mathbf{I}) = \det(\mathbf{T}^{-1}\mathbf{A}\mathbf{T} - \mathbf{T}^{-1}\lambda\mathbf{I}\mathbf{T}) = \det[\mathbf{T}^{-1}(\mathbf{A} - \lambda\mathbf{I})\mathbf{T}]$
$\qquad\qquad\qquad = \det \mathbf{T}^{-1} \det(\mathbf{A} - \lambda\mathbf{I}) \det \mathbf{T} = \det(\mathbf{A} - \lambda\mathbf{I})$.
25. $\mathbf{T}^{-1}\mathbf{A}\mathbf{x} = \lambda\mathbf{T}^{-1}\mathbf{x} = \lambda\mathbf{y}$, $\mathbf{T}^{-1}\mathbf{A}\mathbf{x} = \mathbf{T}^{-1}\mathbf{A}\mathbf{T}\mathbf{y} = \tilde{\mathbf{A}}\mathbf{y}$, thus $\tilde{\mathbf{A}}\mathbf{y} = \lambda\mathbf{y}$.

SECTION 7.1

13. $4 - 3x^2$ **15.** Parallel planes   **17.** Ellipsoids
19. Paraboloids of revolution   **21.** Ellipsoids of revolution
23. Ellipses $3x^2 + y^2 = const$ **31.** $(x + 1/x)\mathbf{i} + (x - 1/x)\mathbf{j}$
33. Straight lines   **35.** Ellipsoids $4x^2 + z^2 + 9y^2 = const$

<h2 style="text-align:center">SECTION 7.2</h2>

1. $\mathbf{b}, 0, |\mathbf{b}|, 0$

3. $-\sin t\,\mathbf{i} + \cos t\,\mathbf{j}, -\cos t\,\mathbf{i} - \sin t\,\mathbf{j}, 1, 1$

5. $\mathbf{i} + 2t\mathbf{j} + 3t^2\mathbf{k}, 2\mathbf{j} + 6t\mathbf{k}, \sqrt{1 + 4t^2 + 9t^4}, \sqrt{4 + 36t^2}$

7. $e^t\mathbf{i} - e^{-t}\mathbf{j}, e^t\mathbf{i} + e^{-t}\mathbf{j}, \sqrt{e^{2t} + e^{-2t}}, \sqrt{e^{2t} + e^{-2t}}$

9. $3\cos 3t\,(\mathbf{i} + \mathbf{j}), -9\sin 3t\,(\mathbf{i} + \mathbf{j}), 3\sqrt{2}\,|\cos 3t|, 9\sqrt{2}\,|\sin 3t|$

11. $y\mathbf{i}, x\mathbf{i} + z\mathbf{j}, y\mathbf{j}$ **13.** $0, e^y\mathbf{i}, e^{-z}\mathbf{j}$

15. $2xy\mathbf{i} + z^2\mathbf{k}, x^2\mathbf{i} + 2yz\mathbf{j}, y^2\mathbf{j} + 2zx\mathbf{k}$

19. $(\mathbf{u} \cdot \mathbf{v})'' = \mathbf{u}'' \cdot \mathbf{v} + 2\mathbf{u}' \cdot \mathbf{v}' + \mathbf{u} \cdot \mathbf{v}'', (\mathbf{u} \times \mathbf{v})'' = \mathbf{u}'' \times \mathbf{v} + 2\mathbf{u}' \times \mathbf{v}' + \mathbf{u} \times \mathbf{v}''$

23. $-48t^3\mathbf{i} - (3t^2 + 2t)\mathbf{j} - 20t^4\mathbf{k}$ **25.** $-6t\mathbf{j}$

27. $(6t - 48t^3)\mathbf{i} - (2t + 30t^2 - 5t^4)\mathbf{j} + (3t^2 - 20t^4)\mathbf{k}$

<h2 style="text-align:center">SECTION 7.3</h2>

1. $\mathbf{r}(t) = \mathbf{i} + \mathbf{j} + t\mathbf{k}$ **3.** $\mathbf{r}(t) = (-3 + 3t)\mathbf{i} + \mathbf{j} - (2 + t)\mathbf{k}$

5. $\mathbf{r}(t) = (2 - t)\mathbf{i} + (4 + 2t)\mathbf{k}$ **7.** $\mathbf{r}(t) = (3 + 2t)\mathbf{i} - (1 + 4t)\mathbf{j} + 5\mathbf{k}$

9. $\mathbf{r}(t) = t(\mathbf{i} + \mathbf{j} + \mathbf{k})$ **11.** $\mathbf{r}(t) = t\mathbf{i} + (2t - 3)\mathbf{j} + (5 - t)\mathbf{k}$

13. $(x - 1)^2 + (y - 1)^2 = 1, z = 0$ **21.** $\mathbf{r}(t) = (-2 + 2\cos t)\mathbf{i} + (2 + 2\sin t)\mathbf{j}$

23. $\mathbf{r}(t) = t\mathbf{i} + t^2\mathbf{j} + t^3\mathbf{k}$ **25.** $2\cos t\,\mathbf{i} + 2\sin t\,\mathbf{j} + 4\cos^2 t\,\mathbf{k}$

<h2 style="text-align:center">SECTION 7.4</h2>

1. Start from $\mathbf{r}(t) = t\mathbf{i} + f(t)\mathbf{j}$.   **5.** $8a$

7. $\sinh 1$ **9.** $8(\sqrt{1000} - 1)/27$

<h2 style="text-align:center">SECTION 7.5</h2>

1. $\mathbf{q}(w) = (-1/\sqrt{2} + w)\mathbf{i} + (1/\sqrt{2} + w)\mathbf{j}$

3. $\mathbf{q}(w) = \sqrt{2}(1 - w)\mathbf{i} + (1/\sqrt{2})(1 + w)\mathbf{j}$

5. $\mathbf{q}(w) = (1 + w)\mathbf{i} + (2 + 8w)\mathbf{k}$

15. $\mathbf{p} = \mathbf{b} \times \mathbf{u}, \mathbf{p}' = \mathbf{b}' \times \mathbf{u} + \mathbf{b} \times \mathbf{u}' = -\tau\mathbf{p} \times \mathbf{u} + \mathbf{b} \times \kappa\mathbf{p} = -\tau(-\mathbf{b}) + \kappa(-\mathbf{u})$

<h2 style="text-align:center">SECTION 7.6</h2>

1. Positive x-axis, $3\mathbf{i}, 3, \mathbf{0}$ **3.** Positive z-axis, $8t\mathbf{k}, 8t, 8\mathbf{k}$

5. Segment $-1 \leqq y \leqq 1$ on the y-axis, $\cos t\,\mathbf{j}, |\cos t|, -\sin t\,\mathbf{j}$

7. $y = 2x, z = 0, (0 < x \leqq 10), -10e^{-t}\mathbf{i} - 5e^{-t}\mathbf{j}, e^{-t}\sqrt{125}, 10e^{-t}\mathbf{i} + 5e^{-t}\mathbf{j}$

9. Circle, $-2t\sin t^2\,\mathbf{i} + 2t\cos t^2\,\mathbf{j}, 2t, (-2\sin t^2 - 4t^2\cos t^2)\mathbf{i} + (2\cos t^2 - 4t^2\sin t^2)\mathbf{j}$

11. $\mathbf{v} = \mathbf{w} \times \mathbf{r}, \mathbf{w} = \omega\mathbf{k}, \dot{\mathbf{v}} = \mathbf{w} \times \dot{\mathbf{r}} = \mathbf{w} \times \mathbf{v}$. Since $\mathbf{v}$ is tangent to the path, $\mathbf{w} \times \mathbf{v}$ has the direction opposite to $\mathbf{r}$, and $|\mathbf{a}| = |\mathbf{w}|\,|\mathbf{v}| = \omega^2 R$, as before.

13. $|\mathbf{a}| = \omega^2 R = (2\pi/2.36 \cdot 10^6)^2 \cdot 3.85 \cdot 10^8 = 0.0027$ (meter/sec²), which is only $2.8 \cdot 10^{-4}g$ where g is the acceleration due to gravity.

15. $\mathbf{r}(t) = \mathbf{a}_0 t^2/2 + \mathbf{v}_0 t + \mathbf{r}_0$ ($\mathbf{a}_0, \mathbf{v}_0, \mathbf{r}_0$ constant vectors)

<h2 style="text-align:center">SECTION 7.7</h2>

3. $-4t/(\sin t^2 - \cos t^2)^2$

5. $(e^{2t} + \sin t \cos t)/\sqrt{e^{2t} + \sin^2 t}$

7. $(\cos t)^{\sin t - 1}[\cos^2 t \ln(\cos t) - \sin^2 t]$

9. $(1 - t - \sin t)\sin t + (1 + t + \cos t)\cos t$

11. $e^{4u}\exp(\tfrac{1}{4}e^{4u}\sin^2 2v)\sin^2 2v, \tfrac{1}{2}e^{4u}\exp(\tfrac{1}{4}e^{4u}\sin^2 2v)\sin 4v$

13. $r = \sqrt{x^2 + y^2}, \theta = \arctan(y/x), r_x = x/r, \theta_x = -y/r^2, r_{xx} = y^2/r^3$, etc.,

$w_{xx} = x^2 r^{-2}w_{rr} - 2xyr^{-3}w_{r\theta} + y^2 r^{-4}w_{\theta\theta} + y^2 r^{-3}w_r + 2xyr^{-4}w_\theta$, etc.

SECTION 7.8

1. $2y\mathbf{i} + 2x\mathbf{j}$
3. $2x\mathbf{i} + 2y\mathbf{j}$
5. $[(y^2 - x^2)\mathbf{i} - 2xy\mathbf{j}]/(x^2 + y^2)^2$
7. $e^x \sin y\,\mathbf{i} + e^x \cos y\,\mathbf{j}$
9. $4(x\mathbf{i} + y\mathbf{j} + z\mathbf{k})(x^2 + y^2 + z^2)$
11. $2(x\mathbf{i} + y\mathbf{j} + z\mathbf{k}) \cos (x^2 + y^2 + z^2)$
13. $\mathbf{n} = \nabla f = -4x\mathbf{i} + \mathbf{j}$ where $f = y - 2x^2$. At P, $\mathbf{n} = -8\mathbf{i} + \mathbf{j}$.
15. $-48\mathbf{i} - 16\mathbf{j}$
17. $12\mathbf{i} + 16\mathbf{j}$
19. $(\mathbf{i} + \mathbf{j} + \mathbf{k})/\sqrt{3}$
21. $(\mathbf{i} - 2\mathbf{j} + \mathbf{k})/\sqrt{6}$
23. $(2\mathbf{i} - 3\mathbf{j})/\sqrt{13}$
25. $(x^2 + y^2)/2$
27. $3x^2 - 2y^2 + z^2$
29. $xy + yz + zx$
33. $\pi/2$
41. $4, 4\sqrt{2}, 4, 0, -4, -4\sqrt{2}, -4, 0$
43. 0
45. $-\sqrt{2}/9$
47. $\frac{1}{3}$
49. $\frac{7}{3}$

SECTION 7.9

5. $a_{11} = a_{22} = a_{33} = 1$, $b_1 = 1$, $b_2 = 3$, $b_3 = -4$, all others zero.
7. $a_{12} = a_{21} = a_{33} = 1$, all others zero.
9. $a_{11} = 1$, $a_{22} = a_{33} = \cos \phi$, $a_{23} = -a_{32} = \sin \phi$, all others zero.

SECTION 7.10

3. $a + b + c$
5. 0
7. $6xyz$
9. 0

SECTION 7.11

1. $3y\mathbf{k}$
3. $-2z\mathbf{i} - 2x\mathbf{j} - 2y\mathbf{k}$
5. $-y\mathbf{i} + 2x\mathbf{j}$
15. $(2yz - zx + xy)\mathbf{i} + (2zx - xy + yz)\mathbf{j} + (2xy - yz + xz)\mathbf{k}$, $x^2 + y^2 + z^2 - yz - zx - xy$
17. Let c_1, c_2, c_3 and $c_1{}^*$, $c_2{}^*$, $c_3{}^*$ be the components of curl $\mathbf{v}$ with respect to right-handed Cartesian coordinates x_1, x_2, x_3 and $x_1{}^*$, $x_2{}^*$, $x_3{}^*$, respectively. Using (5'), Sec. 7.9, the chain rule for functions of several variables, and (9), Sec. 7.9, we find

$$c_1 = \frac{\partial v_3}{\partial x_2} - \frac{\partial v_2}{\partial x_3} = \sum_{m=1}^{3} \left(a_{m3} \frac{\partial v_m{}^*}{\partial x_2} - a_{m2} \frac{\partial v_m{}^*}{\partial x_3} \right)$$

$$= \sum_{m=1}^{3} \sum_{l=1}^{3} \left(a_{m3} \frac{\partial v_m{}^*}{\partial x_l{}^*} \frac{\partial x_l{}^*}{\partial x_2} - a_{m2} \frac{\partial v_m{}^*}{\partial x_l{}^*} \frac{\partial x_l{}^*}{\partial x_3} \right) = \sum_{m=1}^{3} \sum_{l=1}^{3} (a_{m3} a_{l2} - a_{m2} a_{l3}) \frac{\partial v_m{}^*}{\partial x_l{}^*}$$

$$= (a_{33} a_{22} - a_{32} a_{23}) \left(\frac{\partial v_3{}^*}{\partial x_2{}^*} - \frac{\partial v_2{}^*}{\partial x_3{}^*} \right) + \cdots$$

$$= (a_{33} a_{22} - a_{32} a_{23}) c_1{}^* + (a_{13} a_{32} - a_{12} a_{33}) c_2{}^* + (a_{23} a_{12} - a_{22} a_{13}) c_3{}^*.$$

Using (3) in Sec. 7.9, Lagrange's identity (Sec. 5.9), and $\mathbf{k}^* \times \mathbf{j}^* = -\mathbf{i}^*$, $\mathbf{k} \times \mathbf{j} = -\mathbf{i}$, we obtain

$$a_{33} a_{22} - a_{32} a_{23} = (\mathbf{k}^* \times \mathbf{j}^*) \cdot (\mathbf{k} \times \mathbf{j}) = \mathbf{i}^* \cdot \mathbf{i} = a_{11}, \text{ etc.}$$

Hence $c_1 = a_{11} c_1{}^* + a_{21} c_2{}^* + a_{31} c_3{}^*$, which is of the form (5'), Sec. 7.9. In the same way we may obtain corresponding formulas for c_2 and c_3. If the $x_1 x_2 x_3$-coordinates are left-handed, then $\mathbf{k} \times \mathbf{j} = +\mathbf{i}$, but then there is a minus sign in front of the determinant in (1). This proves the theorem.
19. curl $\mathbf{v} = 4\mathbf{k}$, div $\mathbf{v} = 0$, incompressible. If $\mathbf{r}(t) = x\mathbf{i} + y\mathbf{j} + z\mathbf{k}$ represents a path, then $\dot{\mathbf{r}} = \dot{x}\mathbf{i} + \dot{y}\mathbf{j} + \dot{z}\mathbf{k} = -2y\mathbf{i} + 2x\mathbf{j}$, $\dot{x} = -2y$, $\dot{y} = 2x$, $x\dot{x} + y\dot{y} = 0$, $x^2 + y^2 = const$, $z = const$. The paths are concentric circles.

SECTION 8.2

1. $\sqrt{8}$
3. $80\sqrt{10}$
5. 5
7. $3\pi/2$
9. $-\frac{4}{3}$
11. $\frac{2}{3}$
13. 72
15. -60
17. 0

SECTION 8.3

1. $\frac{4}{3}$ **3.** $\frac{1}{8}$ **5.** $3\pi/4$
7. $(1 - e^{-4})/2$ **9.** $\frac{1}{3}$ **11.** $\pi/2$
13. $\frac{16}{9}$ **15.** 1   **17.** 36
19. $\frac{4}{3}$, $\frac{8}{3}$ **21.** $8a/5\pi$, $8a/5\pi$ **23.** $I_x = bh^3/12$, $I_y = b^3h/4$
25. $I_x = (a + b)h^3/24$, $I_y = h(a^4 - b^4)/48(a - b)$

SECTION 8.4

1. 1   **3.** -1 **5.** $\frac{1}{4}$ **7.** $\frac{8}{5}$ **9.** -1 **11.** 0   **13.** 0   **15.** 4   **17.** 13.5
19. 2   **21.** $(e^2 - 1)/2$ **23.** 2π **25.** 1

SECTION 8.5

1. xy-plane, straight lines.
3. xy-plane, concentric circles, straight lines through the origin.
5. Cone of revolution, circles, straight lines.
7. Paraboloid of revolution, circles, parabolas of fourth order.
9. Torus, circles.
11. $\dfrac{x^2}{a^2} + \dfrac{y^2}{b^2} + \dfrac{z^2}{c^2} - 1 = 0$ **13.** $\dfrac{x^2}{a^2} - \dfrac{y^2}{b^2} - z = 0$
15. $\mathbf{r} = u\mathbf{j} + v\mathbf{k}$ **17.** $\mathbf{r} = u\mathbf{i} + v\mathbf{j} + (3 - u - v)\mathbf{k}$
19. $\mathbf{r} = (1 + 3\cos u)\mathbf{i} + (2 + 3\sin u)\mathbf{j} + v\mathbf{k}$ **21.** $\mathbf{r} = u\cos v\,\mathbf{i} + u\sin v\,\mathbf{j} + e^{-u^2}\mathbf{k}$
23. $(v\mathbf{i} + u\mathbf{j} - \mathbf{k})/\sqrt{1 + u^2 + v^2}$ **25.** $(g_x\mathbf{i} + g_y\mathbf{j} - \mathbf{k})/\sqrt{1 + g_x^2 + g_y^2}$

SECTION 8.6

3. $(\mathbf{r}^* - \mathbf{r}) \cdot \operatorname{grad} f = 0$, $\mathbf{n} = \operatorname{grad} f/|\operatorname{grad} f|$
5. $\mathbf{r}_{\tilde{u}} = -\mathbf{r}_u$, $\mathbf{r}_{\tilde{v}} = \mathbf{r}_v$ **7.** $z^* = 2\sqrt{2} + y^*$ **9.** $3x^* + 4y^* = 25$
11. $4x^* + 2y^* - z^* = 5$ **13.** $a^2\,du^2 + dv^2$ **15.** $(a + b\cos v)^2\,du^2 + b^2\,dv^2$
17. $2\,du^2 + u^2\,dv^2$ **21.** $\pi\sqrt{2}$ **23.** $\pi\sqrt{2}$
25. $\pi(\sqrt{125} - 1)/6$

SECTION 8.7

3. $(3^{5/2} - 2^{7/2} + 1)/15$ **5.** $\pi^2(17^{3/2} - 5^{3/2})/6$
7. $(5^{3/2} - 1)/2$ **13.** $I_{x=y} = \displaystyle\iint\limits_{S} \left[\frac{(x - y)^2}{2} + z^2\right] \sigma\,dA$
15. $ah\pi(a^2 + h^2/6)$ **17.** $2\pi^2ab(2a^2 + 3b^2)$
19. $ah\pi(a^2 + 2h^2/3)$

SECTION 8.9

1. $\frac{8}{27}$ **3.** $\frac{1}{120}$ **5.** $abc(b^2 + c^2)/12$ **7.** $\pi ha^4/2$
9. $8\pi a^5/15$ **11.** 3   **13.** 0   **15.** $3(e^2 - 1)$
17. Put $f = 1$ in (9).   **19.** Use Prob. 18.
23. Take $\mathbf{u} = x\mathbf{i}$, $y\mathbf{j}$, $z\mathbf{k}$, respectively, in (3), Sec. 8.8.

SECTION 8.11

1. ± 1 **3.** $\pm 3\pi/2$ **5.** $-18\pi\sqrt{2}$ **7.** 0   **11.** 21   **13.** 0   **15.** 0

SECTION 8.12

1. Exact   **3.** Exact   **5.** Not exact   **7.** $u = x + y + z + c$
9. $u = \sin x - 2\cos y$ **11.** $u = -1/(x^2 + y^2)$
13. $u = \sinh(y^2 + z^2)$ **15.** 5   **17.** -2 **19.** $-e^{-2} - 1$

SECTION 9.1

1. $2\pi, 2\pi, \pi, \pi, 2, 2, 1, 1$ 17. 0 (n even), $2/n$ (n odd)

19. 0 ($n = 0$), $2\pi/n$ ($n = 1, 3, \cdots$), $-2\pi/n$ ($n = 2, 4, \cdots$)

21. 0 23. $n[(-1)^n e^{-\pi} - 1]/(1 + n^2)$

25. $2\pi^3/3$ ($n = 0$), $(-1)^n 4\pi/n^2$ ($n = 1, 2, \cdots$)

SECTION 9.2

1. $\dfrac{1}{2} + \dfrac{2}{\pi}\left(\cos x - \dfrac{1}{3}\cos 3x + \dfrac{1}{5}\cos 5x - + \cdots\right)$

3. $\dfrac{4}{\pi}\left(\cos x - \dfrac{1}{3}\cos 3x + \dfrac{1}{5}\cos 5x - + \cdots\right)$

5. $\dfrac{2}{\pi}\left(\sin x + \dfrac{2}{2}\sin 2x + \dfrac{1}{3}\sin 3x + \dfrac{1}{5}\sin 5x + \dfrac{2}{6}\sin 6x + \cdots\right)$

7. $2\left(\sin x - \dfrac{1}{2}\sin 2x + \dfrac{1}{3}\sin 3x - \dfrac{1}{4}\sin 4x + - \cdots\right)$

9. $\dfrac{\pi^2}{3} - 4\left(\cos x - \dfrac{1}{4}\cos 2x + \dfrac{1}{9}\cos 3x - \dfrac{1}{16}\cos 4x + - \cdots\right)$

11. $\dfrac{\pi}{2} + \dfrac{4}{\pi}\left(\cos x + \dfrac{1}{9}\cos 3x + \dfrac{1}{25}\cos 5x + \cdots\right)$

13. $\dfrac{4}{\pi}\left(\sin x - \dfrac{1}{9}\sin 3x + \dfrac{1}{25}\sin 5x - + \cdots\right)$

SECTION 9.3

3. Use (1*b*).

5. $\dfrac{4}{\pi}\left(\sin \pi t + \dfrac{1}{3}\sin 3\pi t + \dfrac{1}{5}\sin 5\pi t + \cdots\right)$

7. $\dfrac{1}{2} + \dfrac{2}{\pi}\left(\sin \dfrac{\pi\tau}{2} + \dfrac{1}{3}\sin \dfrac{3\pi\tau}{2} + \dfrac{1}{5}\sin \dfrac{5\pi\tau}{2} + \cdots\right)$

9. $\dfrac{1}{4} - \dfrac{2}{\pi^2}\left(\cos \pi t + \dfrac{1}{9}\cos 3\pi t + \cdots\right) + \dfrac{1}{\pi}\left(\sin \pi t - \dfrac{1}{2}\sin 2\pi t + - \cdots\right)$

11. $\dfrac{2}{3} + \dfrac{4}{\pi^2}\left(\cos \pi t - \dfrac{1}{4}\cos 2\pi t + \dfrac{1}{9}\cos 3\pi t - + \cdots\right)$

13. $-\dfrac{4}{\pi^2}\left(\cos \pi t + \dfrac{1}{9}\cos 3\pi t + \cdots\right) + \dfrac{2}{\pi}\left(2\sin \pi t - \dfrac{1}{2}\sin 2\pi t + - \cdots\right)$

SECTION 9.4

1. Neither odd nor even, even, odd, even, odd, neither odd nor even, even, even.

3. Odd   5. Odd   7. Neither odd nor even

9. Odd   19. $1/(1 - x^2) + x/(1 - x^2)$ 21. $\cosh x + \sinh x$

23. $\dfrac{1}{2} + \dfrac{2}{\pi}\left(\cos x - \dfrac{1}{3}\cos 3x + \dfrac{1}{5}\cos 5x - + \cdots\right)$

25. $\dfrac{4}{\pi}\left(\sin x - \dfrac{1}{9}\sin 3x + \dfrac{1}{25}\sin 5x - + \cdots\right)$

27. $\dfrac{\pi}{2} - \dfrac{4}{\pi}\left(\cos x + \dfrac{1}{9}\cos 3x + \dfrac{1}{25}\cos 5x + \cdots\right)$

29. $\dfrac{\pi^2}{12} - \cos x + \dfrac{1}{4}\cos 2x - \dfrac{1}{9}\cos 3x + \dfrac{1}{16}\cos 4x - + \cdots$

SECTION 9.5

1. $2\left(\sin x - \dfrac{1}{2}\sin 2x + \dfrac{1}{3}\sin 3x - \dfrac{1}{4}\sin 4x + - \cdots\right)$

3. $\dfrac{2}{\pi}\left(\sin \pi x - \dfrac{1}{2}\sin 2\pi x + \dfrac{1}{3}\sin 3\pi x - + \cdots\right)$

5. $\left(1 + \dfrac{2}{\pi}\right)\sin x - \dfrac{1}{2}\sin 2x + \left(\dfrac{1}{3} - \dfrac{2}{9\pi}\right)\sin 3x - \dfrac{1}{4}\sin 4x + \cdots$

7. $\dfrac{1}{\pi}\left(\sin 4x - \dfrac{1}{9}\sin 12x + \dfrac{1}{25}\sin 20x - \dfrac{1}{49}\sin 28x + - \cdots\right)$

9. $f(x) = 1$

11. $\dfrac{l^2}{3} - \dfrac{4l^2}{\pi^2}\left(\cos\dfrac{\pi x}{l} - \dfrac{1}{4}\cos\dfrac{2\pi x}{l} + \dfrac{1}{9}\cos\dfrac{3\pi x}{l} - \dfrac{1}{16}\cos\dfrac{4\pi x}{l} + - \cdots\right)$

13. $\dfrac{2}{\pi} - \dfrac{4}{\pi}\left(\dfrac{1}{1\cdot 3}\cos\dfrac{2\pi x}{l} + \dfrac{1}{3\cdot 5}\cos\dfrac{4\pi x}{l} + \dfrac{1}{5\cdot 7}\cos\dfrac{6\pi x}{l} + \cdots\right)$

17. $i\displaystyle\sum_{n=-\infty}^{\infty}\dfrac{(-1)^n}{n}e^{inx}$

19. $\dfrac{\sinh\pi}{\pi}\displaystyle\sum_{n=-\infty}^{\infty}(-1)^n\dfrac{1+in}{1+n^2}e^{inx}$

SECTION 9.6

11. $f = 12\left(\sin x - \dfrac{1}{2^3}\sin 2x + \dfrac{1}{3^3}\sin 3x - + \cdots\right)$

13. $\dfrac{l^3}{4} + \dfrac{6l^3}{\pi^2}\left[\left(\dfrac{4}{\pi^2} - 1\right)\cos\dfrac{\pi x}{l} + \dfrac{1}{2^2}\cos\dfrac{2\pi x}{l} + \left(\dfrac{4}{3^4\pi^2} - \dfrac{1}{3^2}\right)\cos\dfrac{3\pi x}{l} + \cdots\right]$

15. $a_0 = \pi^4/5$, $a_n = (-1)^n 8(\pi^2 n^2 - 6)/n^4$, $b_n = 0$ $(n = 1, 2, \cdots)$

SECTION 9.7

1. $y = C_1\cos\omega t + C_2\sin\omega t + A(\omega)\sin t$, $A(\omega) = 1/(\omega^2 - 1)$, $A(0.5) = -1.33$, $A(0.7) = -0.20$, $A(0.9) = -5.3$, $A(1.1) = 4.8$, $A(1.5)=0.8$, $A(2) = 0.33$, $A(10) = 0.01$

3. $y = C_1\cos\omega t + C_2\sin\omega t + B_1\sin t + B_3\sin 3t + B_5\sin 5t$ where

ω	0.5	0.9	1.1	2.0	2.9	3.1	4.0	4.9	5.1	6.0	8.0
$B_1 = 1/(\omega^2 - 1)$	-1.33	-5.3	4.8	0.33	0.13	0.12	0.07	0.04	0.04	0.03	0.02
$B_3 = 1/9(\omega^2 - 9)$	-0.013	-0.014	-0.014	-0.02	-0.19	0.18	0.02	0.01	0.01	0.004	0.002
$B_5 = 1/25(\omega^2 - 25)$	-0.002	-0.002	-0.002	-0.002	-0.002	-0.003	-0.004	-0.04	0.04	0.004	0.001

5. $y = C_1\cos\omega t + C_2\sin\omega t + \displaystyle\sum_{n=1}^{N}\dfrac{b_n}{\omega^2 - n^2}\sin nt$

7. $y = C_1\cos\omega t + C_2\sin\omega t + \dfrac{\pi^2}{12\omega^2} - \dfrac{1}{\omega^2 - 1}\cos t + \dfrac{1}{4(\omega^2 - 4)}\cos 2t - + \cdots$

9. $y = -\dfrac{K}{c} \cos t$ **11.** $y = \dfrac{1 - n^2}{D} a_n \cos nt + \dfrac{nc}{D} a_n \sin nt, \; D = (1 - n^2)^2 + n^2c^2$

13. $y = A_1 \cos t + B_1 \sin t + A_3 \cos 3t + B_3 \sin 3t + \cdots$ where $A_n = -ncb_n/D$,
$B_n = (1 - n^2)b_n/D, \; D = (1 - n^2)^2 + n^2c^2, \; b_1 = 1, \; b_2 = 0, \; b_3 = -\frac{1}{9}, \; b_4 = 0, \; b_5 = \frac{1}{25}, \cdots$

15. $I = \sum\limits_{n=1}^{\infty} (A_n \cos nt + B_n \sin nt), \; A_n = \dfrac{80(10 - n^2)}{\pi n^2 D_n}, B_n = \dfrac{800}{n\pi D_n}$ $(n$ odd$)$,
   $A_n = 0, \; B_n = 0$ $(n$ even$)$, $D_n = (10 - n^2)^2 + 100n^2$,
   $I = 1.266 \cos t + 1.406 \sin t + 0.003 \cos 3t + 0.094 \sin 3t + \cdots$

SECTION 9.8

1. $F = \dfrac{4}{\pi}\left[\sin x + \dfrac{1}{3} \sin 3x + \cdots + \dfrac{1}{N} \sin Nx \right]$ $(N$ odd$)$

5. $F = \dfrac{\pi^2}{3} - 4\left(\cos x - \dfrac{1}{4} \cos 2x + \dfrac{1}{9} \cos 3x - \cdots + \dfrac{(-1)^{N+1}}{N^2} \cos Nx \right),$

   $E^* = \dfrac{2\pi^5}{5} - \pi\left(\dfrac{2\pi^4}{9} + 16 + 1 + \dfrac{16}{81} + \dfrac{1}{16} + \cdots \right)$

SECTION 9.9

9. $\dfrac{2}{\pi} \displaystyle\int_0^{\infty} \left[\left(a^2 - \dfrac{2}{w^2}\right) \sin aw + \dfrac{2a}{w} \cos aw \right] \dfrac{\cos wx}{w} \, dw$

11. $\dfrac{2}{\pi} \displaystyle\int_0^{\infty} \left[\dfrac{a \sin aw}{w} + \dfrac{\cos aw - 1}{w^2} \right] \cos xw \, dw$

13. $\dfrac{6}{\pi} \displaystyle\int_0^{\infty} \dfrac{2 + w^2}{4 + 5w^2 + w^4} \cos xw \, dw$

15. Differentiating (10) we have

$$\dfrac{d^2A}{dw^2} = -2\int_0^{\infty} f^*(v) \cos wv \, dv \quad \text{where} \quad f^*(v) = v^2 f(v),$$

and the result follows.

SECTION 10.1

19. $u(x, y) = c(x)e^{-y^2}$ **21.** $u(x, y) = v(x) + w(y)$ **23.** $u(x, y) = v(x)e^{-y} + w(y)$

SECTION 10.3

1. $u = 0.02 \cos t \sin x$ **3.** $u = k(\cos t \sin x - \cos 2t \sin 2x)$

5. $u = \dfrac{4}{5\pi}\left(\dfrac{1}{4} \cos 2t \sin 2x - \dfrac{1}{36} \cos 6t \sin 6x + \dfrac{1}{100} \cos 10t \sin 10x - + \cdots \right)$

7. $u = \dfrac{0.08}{\pi}\left(\cos t \sin x + \dfrac{1}{3^3} \cos 3t \sin 3x + \dfrac{1}{5^3} \cos 5t \sin 5x + \cdots \right)$

9. $u = 0.12\left(\cos t \sin x - \dfrac{1}{2^3} \cos 2t \sin 2x + \dfrac{1}{3^3} \cos 3t \sin 3x - + \cdots \right)$

11. $27,960/\pi^6 \approx 0.9986$ **13.** $u = ke^{c(x-y)}$ **15.** $u = ky^c e^{cx}$
17. $u = k \exp(x^2 + y^2 + c(x - y))$ **19.** $u = k \exp(cx + y/c)$

SECTION 10.4

13. $F_n = \sin(n\pi x/l)$, $G_n = a_n \cos(cn^2\pi^2t/l^2)$

15. $u(0, t) = 0$, $u(l, t) = 0$, $u_x(0, t) = 0$, $u_x(l, t) = 0$

17. $\beta l \approx (2n + 1)\pi/2$, $n = 1, 2, \cdots$ (more exactly 4.730, 7.853, 10.996, $\cdots$)

19. $\beta l \approx (2n + 1)\pi/2$, $n = 0, 1, \cdots$ (more exactly 1.875, 4.694, 7.855, $\cdots$)

21. $u = f_1(x) + f_2(xy)$

25. $y'^2 - 2y' + 1 = (y' - 1)^2 = 0$, $y = x + c$, $\Psi(x, y) = x - y$, $v = x$, $z = x - y$

SECTION 10.5

5. $u = \sin 0.1\pi x \, e^{-1.752\pi^2 t/100}$

7. $u = \dfrac{40}{\pi^2} \left(\sin 0.1\pi x \, e^{-0.01752\pi^2 t} - \dfrac{1}{9} \sin 0.3\pi x \, e^{-0.01752(3\pi)^2 t} + - \cdots \right)$

9. $u = \dfrac{800}{\pi^3} \left(\sin 0.1\pi x \, e^{-0.01752\pi^2 t} + \dfrac{1}{3^3} \sin 0.3\pi x \, e^{-0.01752(3\pi)^2 t} + \cdots \right)$

11. Since the temperatures at the ends are kept constant, the temperature will approach a steady-state (time-independent) distribution $u_I(x)$ as $t \to \infty$, and $u_I = U_1 + (U_2 - U_1)x/l$, the solution of (1) with $\partial u/\partial t = 0$, satisfying the boundary conditions.

13. $u(x, t) = 100 - 10x + \dfrac{200}{\pi} \sum\limits_{n=1}^{\infty} \dfrac{(-1)^{n+1}}{n} \sin \dfrac{n\pi x}{10} e^{-1.752(n\pi/10)^2 t}$

Obviously, as a good approximation we may take the sum of the first few terms and obtain the values $u(5, 1) = 99$, $u(5, 2) = 94$, $u(5, 3) = 88$, $u(5, 10) = 61$, $u(5, 50) = 50$.

17. $u = \dfrac{\pi^2}{3} - 4 \left(\cos x \, e^{-t} - \dfrac{1}{4} \cos 2x \, e^{-4t} + \dfrac{1}{9} \cos 3x \, e^{-9t} - + \cdots \right)$

19. $u = \dfrac{\pi}{8} + \left(1 - \dfrac{2}{\pi} \right) \cos x \, e^{-t} - \dfrac{1}{\pi} \cos 2x \, e^{-4t} - \left(\dfrac{1}{3} + \dfrac{2}{9\pi} \right) \cos 3x \, e^{-9t} + \cdots$

SECTION 10.6

7. $\dfrac{1}{\sqrt{\pi}} \displaystyle\int_{(a-x)/\tau}^{(b-x)/\tau} e^{-w^2} \, dw - \dfrac{1}{\sqrt{\pi}} \displaystyle\int_{(a+x)/\tau}^{(b+x)/\tau} e^{-w^2} \, dw$

SECTION 10.8

3. c increases and so does the frequency.

7. $c\pi\sqrt{260}$ (corresponding eigenfunctions $F_{4,16}$, $F_{16,14}$), etc.

9. $B_{mn} = \dfrac{8}{mn\pi^2}(k_n + l_m)$ where $k_n = \begin{cases} 0 & (n \text{ even}) \\ a & (n \text{ odd}) \end{cases}$ and $l_m = \begin{cases} 0 & (m \text{ even}) \\ b & (m \text{ odd}) \end{cases}$

11. $B_{mn} = 0$ (m or n even), $B_{mn} = \dfrac{64a^2b^2}{\pi^6 m^3 n^3}$ (m, n both odd)

13. $B_{mn} = (-1)^{m+n} \dfrac{144a^3b^3}{m^3 n^3 \pi^6}$

15. $u = k \cos \pi\sqrt{5}\, t \sin \pi x \sin 2\pi y$

SECTION 10.10

3. u_2: $r = \alpha_1/\alpha_2 = 0.43565$, u_3: $r = \alpha_1/\alpha_3 = 0.27789$, $r = \alpha_2/\alpha_3 = 0.63788$

9. $u = 4k \sum\limits_{m=1}^{\infty} \dfrac{J_2(\alpha_m)}{\alpha_m^2 J_1^2(\alpha_m)} \cos \alpha_m t \, J_0(\alpha_m r)$

SECTION 10.11

3. $u = 135 - \dfrac{250}{r}$ **5.** $u = \dfrac{100}{\ln 5} \ln \dfrac{r}{2} + 10$

11. $(u_1 - u_0) \ln r / \ln (r_1/r_0) + (u_0 \ln r_1 - u_1 \ln r_0)/\ln (r_1/r_0)$

SECTION 10.12

3. $u = 1$ **5.** $u = \frac{2}{3} r^2 P_2(\cos \phi) + \frac{1}{3} = r^2 (\cos^2 \phi - \frac{1}{3}) + \frac{1}{3}$
7. $\cos 2\phi = 2 \cos^2 \phi - 1,\ 2x^2 - 1 = \frac{4}{3} P_2(x) - \frac{1}{3},\ u = \frac{4}{3} r^2 P_2(\cos \phi) - \frac{1}{3}$
15. $i_{xx} = LCi_{tt} + (RC + GL)i_t + RGi$

SECTION 11.1

1. $-\frac{1}{2}$ **3.** 1 **5.** $3x^2 y - y^3$ **7.** 1
9. $\sqrt{(x + 1)^2 + y^2}/\sqrt{(x - 1)^2 + y^2}$ **11.** 1
15. Consider $c = a + ib = z_1/(z_1 + z_2)$, assuming $z_1 + z_2 \neq 0$. From (19), $|a| \leq |c|$, $|a - 1| \leq |c - 1|$. Thus $|a| + |a - 1| \leq |c| + |c - 1|$. Clearly $|a| + |a - 1| \geq 1$, and the last inequality becomes

$$1 \leq |c| + |c - 1| = \left| \frac{z_1}{z_1 + z_2} \right| + \left| \frac{z_2}{z_1 + z_2} \right|.$$

Multiplication by $|z_1 + z_2|$ yields (20).
17. By the triangle inequality, $|z_2| = |(z_1 + z_2) - z_1| \leq |z_1 + z_2| + |z_1|$. Subtract $|z_1|$ on both sides.
19. Square both sides of both inequalities.
21. $\sqrt{2}(\cos \pi/4 + i \sin \pi/4),\ \sqrt{2}(\cos 3\pi/4 + i \sin 3\pi/4),\ \sqrt{2}(\cos 5\pi/4 + i \sin 5\pi/4)$
25. The region between the two branches of the hyperbola $x^2 - y^2 = 1$.
27. $(x - 5/3)^2 + y^2 = \frac{16}{9}$ **29.** The y-axis
33. $-3\pi/4$ **35.** $-\pi/2$

SECTION 11.2

1. $2 + 4i, -1 + 2i, 20 - 20i$ **3.** $(9 - 13i)/500, -i, (-2 - 11i)/1000$
5. $2(x^3 - 3xy^2) - 3x, 2(3x^2y - y^3) - 3y$
7. No, because $f(z) \to 0\ [= f(0)]$ as $z \to 0$ along the y-axis, but $f(z) \to 1\ [\neq f(0)]$ as $z \to 0$ along the positive x-axis.
9. Yes, because for $z \neq 0$, $|f| = x^2/\sqrt{x^2 + y^2} \leq |x| \leq |z|$ and, therefore, $|f| \to 0\ [= f(0)]$ as $|z| \to 0$.
11. $6z(z^2 - 4)^2$ **13.** $2/(1 - z)^3$ **15.** $-\frac{1}{27}$
19. Use Re $f(z) = [f(z) + \overline{f(z)}]/2$, Im $f(z) = [f(z) - \overline{f(z)}]/2i$.

SECTION 11.3

9. Yes $(z \neq -1)$ **15.** No   **21.** $\frac{1}{2} z^2 + c$
23. $\cos x \cosh y - i \sin x \sinh y$ **25.** $\ln r + i\theta$
27. Let $f = u + iv$ and $|f| = c$. Then $u^2 + v^2 = c^2$. By differentiation, $uu_x + vv_x = 0$, $uu_y + vv_y = 0$. By (5), $uu_x - vu_y = 0$, $uu_y + vu_x = 0$. Hence

$$(uu_x - vu_y)^2 + (uu_y + vu_x)^2 = (u^2 + v^2)(u_x^2 + u_y^2) = c^2(u_x^2 + u_y^2) = 0.$$

If $c = 0$, then $u = v = 0$. If $c \neq 0$, then $u_x^2 + u_y^2 = 0$, i.e., $u_x = u_y = 0$ and, by (5), $v_x = 0$, $v_y = 0$. Hence, $u = const, v = const$.
29. Let $f = u + iv$. By (3), $f' = u_x + iv_x = 0$. Hence, $u_x = 0, v_x = 0$. By (4) or (5), $u_y = 0$, $v_y = 0$. Hence, $u = const, v = const, f = u + iv = const$.

SECTION 11.4

1. $\pm(1+i)/\sqrt{2}$ **3.** $(\pm\sqrt{3}+i)/2, -i$ **5.** $\pm 5i$

7. $\sqrt[6]{2}\left(\cos\dfrac{\pi}{12}+i\sin\dfrac{\pi}{12}\right)$, $\sqrt[6]{2}\left(\cos\dfrac{3\pi}{4}+i\sin\dfrac{3\pi}{4}\right)$, $\sqrt[6]{2}\left(\cos\dfrac{17\pi}{12}+i\sin\dfrac{17\pi}{12}\right)$

9. $\pm(1+i)/\sqrt{2}, \pm(-1+i)/\sqrt{2}$ **11.** $\pm(\sqrt{3}+i)/2, \pm i, \pm(\sqrt{3}-i)/2$

13. $4, -2\pm 2i\sqrt{3}$

15. Quadratic equation in $p=z^2$. Ans. $z=\sqrt{p}=\pm 2, \pm 3i$

17. $z^4=4+4i, \pm\sqrt[8]{32}(\cos\beta+i\sin\beta), \beta=\pi/16, 9\pi/16$

19. $\pm(2+i)$ **21.** $\pm(3+2i)$ **23.** $3+2i, 2-i$

25. $(z^2-z\sqrt{2}+1)(z^2+z\sqrt{2}+1)$

SECTION 11.5

1. $(1-i)/\sqrt{2}$ **3.** $e(\cos 1+i\sin 1)$ **5.** $e^{-3x}\cos 3y, -e^{-3x}\sin 3y$

7. $\exp(x^3-3xy^2)\cos(3x^2y-y^3), \exp(x^3-3xy^2)\sin(3x^2y-y^3)$

9. $5\exp(i\arctan\tfrac{4}{3})$ **11.** $\sqrt{8}\,e^{-\pi i/4}$

13. $\ln 3\pm 2n\pi i\,(n=0,1,\cdots)$ **15.** $z=\ln 2+(2n+1)\pi i, n=0, \pm 1, \pm 2, \cdots$

17. Any z **19.** $\operatorname{Re} z>0$ **23.** $k=1$

25. $e^z\to\infty$ when $\arg z=0$, e^{iy} has no limit, $e^z\to 0$ when $\arg z=\pi$.

SECTION 11.6

7. $\sqrt{\sin^2 x+\sinh^2 y}$ **9.** $\sin x\cos x/(\sin^2 x+\sinh^2 y)$

11. $-\cos x\sinh y/(\sin^2 x+\sinh^2 y)$ **13.** 11013

15. $2.333-0.274i$ **17.** $1.960+3.166i$

19. $\pm 2n\pi\pm 2.29i$ **21.** $\pm 2n\pi+i, \pm(2n+1)\pi-i$

23. $\pm(\pi/3)i\pm 2n\pi i, n=0,1,\cdots$

SECTION 11.7

1. $\pm 2n\pi i, n=0,1,\cdots$ **3.** $2\pm(2n+1)\pi i, n=0,1,\cdots$

5. $1+\left(\dfrac{\pi}{2}\pm 2n\pi\right)i, n=0,1,\cdots$ **7.** $(1\pm 2n\pi)i, n=0,1,\cdots$

9. $\dfrac{1}{2}\ln 2+\dfrac{\pi i}{4}\approx 0.347+0.785i$ **11.** $\ln 5+\pi i\approx 1.609+3.142i$

13. $-i$ **15.** $-e$

21. $\exp[(2-i)\operatorname{Ln}(1+i)]=\exp[(2-i)(\ln\sqrt{2}+\tfrac{1}{4}\pi i)]=\exp[\ln 2+\tfrac{1}{4}\pi+i(\tfrac{1}{2}\pi-\ln\sqrt{2})]$
$\quad = 2e^{\pi/4}[\sin(\tfrac{1}{2}\ln 2)+i\cos(\tfrac{1}{2}\ln 2)]\approx 1.49+4.13i$

23. $e^{-\pi/4}[\cos(\ln\sqrt{2})+i\sin(\ln\sqrt{2})]$

25. $27[\cos(\ln 3)-i\sin(\ln 3)]$

27. $\sqrt{2}\,e^{\pi/4}[\cos(\ln\sqrt{2}-\tfrac{1}{4}\pi)+i\sin(\ln\sqrt{2}-\tfrac{1}{4}\pi)]$

31. $\cosh w=\tfrac{1}{2}(e^w+e^{-w})=z, (e^w)^2-2ze^w+1=0, e^w=z+\sqrt{z^2-1}$

SECTION 12.1

3. $v=20$

5. The positive and the negative v-axis, respectively.

7. $|w|<\dfrac{1}{9}$ **9.** $|\arg w|<2\pi/3$

11. The region to the right of the parabola $v^2=4(1+u)$.

17. Translation combined with a rotation and a uniform dilatation or contraction.

19. $\pi/2\leqq\arg w\leqq 5\pi/8, 0\leqq\arg w\leqq\pi/4$, etc.

21. $u<1, v>u-2, v>-u, u<-2$, the entire w-plane.

23. The x-axis and the unit circle. The y-axis.

25. $w=iz^3+1$

SECTION 12.2

1. $z(t) = t + 2it^2$ **3.** $z(t) = t + i(at + b)$ **5.** $z(t) = 2\cos t + i\sin t$
7. No, the size is preserved, but the sense is reversed.
11. $z = \pm 1, \pm i$ **13.** $z = (2n + 1)/2, n = 0, \pm 1, \pm 2, \cdots$
15. $z = in\pi, n = 0, \pm 1, \pm 2, \cdots$

SECTION 12.3

3. By combining them with (9). **5.** $u = \frac{1}{2}$
7. $|w - 0.5| = 0.5$ **9.** $u^2 + (v + \frac{2}{3})^2 = \frac{1}{9}$
13. $0, 1$ **15.** $0, \pm 1, \pm i$
17. $\pm i$
19. For example, $w = 4/z$, $w = (z + 4)/(z + 1)$, etc.
21. $w = (2z - 1)/z$, $w = -1/(z - 2)$, etc. **23.** $w = (az + b)/(-bz + a)$
25. $w_1 = iz, w_2 = w_1 + 4, w_3 = 1/w_2, w_4 = 5iw_3, w = w_4 - i$

SECTION 12.4

3. $z = (w + i)/(iw + 1)$ **5.** $w = (iz + 3)/(z - 4)$
7. $w = (3z + 2i)/(3z + 2)$ **9.** $w = z/(z + i)$
11. $w = iz/(z + i)$ **13.** $w = (z - z_1)(z_2 - z_3)/(z - z_3)(z_2 - z_1)$
15. $w = az$ **19.** $z = (-4w + 1)/(2w - 1), x = \text{Re } z$, etc.
21. $w = (4z - i)/(-iz - 4)$ **23.** $w = -(z^2 + i)/(iz^2 + 1)$
25. $w = (z^4 - i)/(-iz^4 + 1)$

SECTION 12.5

1. $1 \leqq |w| \leqq e, 0 \leqq \arg w \leqq \pi/2$ **3.** $1/e^2 < |w| < 1/e, 0 < \arg w < \pi$
5. $w = \exp(\pi z^2/2c)$ **7.** $w' = \cos z = 0$ at $z = \pm(2n + 1)\pi/2,$
$$n = 0, 1, \cdots$$

9. The region in the upper half-plane bounded by the ellipses

$$\frac{u^2}{\cosh^2 1} + \frac{v^2}{\sinh^2 1} = 1 \quad \text{and} \quad \frac{u^2}{\cosh^2 2} + \frac{v^2}{\sinh^2 2} = 1$$

11. Elliptical annulus bounded by the ellipses in Prob. 9 and cut along the positive imaginary axis.
13. Interchange the w and z planes in the discussion of the sine function.

SECTION 12.6

7. $-3i/2$, 3 sheets **9.** $0, 1, -1$, 2 sheets **11.** 0, 2 sheets
13. 0, 2 sheets **15.** $0, 1$, 6 sheets

SECTION 13.1

1. $z = (1 - 4i)t, 0 \leqq t \leqq 3$ **3.** $z = 1 - i + (2 - i)t, 0 \leqq t \leqq 4$
5. $z = 4 + 2i + (-1 + 3i)t, 0 \leqq t \leqq 1$ **7.** Straight segment from $3 + i$ to $3 - i$
9. Circle (radius 3, center $2 + i$) **11.** Parabola $y = x^3$ from $(-2, -16)$ to $(2, 16)$.
13. $i + 2e^{it}, 0 \leqq t \leqq 2\pi$ **15.** $t + it^2, 0 \leqq t \leqq 2$
17. $2\cos t + i\sin t, 0 \leqq t \leqq 2\pi$ **19.** $(2 + 11i)/3$
21. $\frac{227}{3} + 23i$ **23.** $-(88 + 16i)/3$
25. $1 + 2i$ **27.** (a) $2\pi i$, (b) $-2\pi i$
29. (a) i, (b) $2i$, (c) $2i$

SECTION 13.2

5. 0 **7.** $2\pi i$ **9.** $e(i - 1)$ **11.** 0
13. $-2i\sinh 1$ **15.** 25 **17.** 5 **19.** 5

SECTION 13.3

3. 0, no **5.** 0, no **7.** $-\pi$, no **9.** 0, no
11. 0, no **13.** 0, yes **17.** $(-2 + 2i)/3$ **19.** $-2\pi i, 0, -4\pi i$
21. $4\pi i/3$ **23.** $2\pi i/3$ **25.** $-2\pi i$

SECTION 13.4

1. $(-2 + 2i)/3$ **3.** $-1566i/35$ **5.** -2 **7.** 0
9. 0 **11.** $(1 - \cosh 2\pi)/2$ **13.** 0 **15.** $-i/3$
17. $(\pi - \frac{1}{2}\sinh 2\pi)i$ **19.** $-\frac{1}{2}\sin \pi^2$

SECTION 13.5

1. 0 (by Cauchy's theorem).
3. $(z^2 - 1)/(z^2 + 1) = f(z)/(z - z_0)$, $z_0 = i$. Ans. -2π.
5. $-\pi i/2$ **7.** $-\pi/2$ **9.** $2\pi i$ **11.** $2\pi i$
13. 0 **15.** 0 **17.** 0 **19.** $2\pi i e^i/3$

SECTION 13.6

1. -4π **3.** $2\pi^2 i$ **5.** $-\pi i$
7. 0 **9.** $4e\pi i$ **11.** $2\pi i/(n - 1)!$
13. $(-1)^{n+1}2\pi i/(2n - 1)!$ **15.** $2\pi i$

SECTION 14.1

1. $i/2, 2i/3, 3i/4, \cdots$ **3.** $-1 + 2\pi i, 1 + 4\pi i, -1 + 6\pi i, \cdots$
5. $2, \sqrt{3}, \sqrt[3]{4}, \cdots$ **7.** $1, i/2, -\frac{1}{2}, \frac{1}{4}, -i/8, \cdots$
9. Yes, yes (limit 0), yes **11.** No, no, no
13. Yes, yes (limit 1), no **15.** Yes, yes (limit 0), monotone when
$$0 \leqq c \leqq 1/4$$
19. No, no, none
21. Yes, no, $-1, 1$ **23.** Yes, no, $-i, -1, i, 1$
25. No, no, none **27.** Yes, yes (limit 0)

SECTION 14.3

1. 6 terms **3.** 5 terms **5.** Divergent **7.** Convergent
9. Convergent [compare with (5)] **11.** Convergent **13.** Divergent
15. Convergent **17.** Divergent **19.** Convergent **21.** Convergent
23. 4, 8, 66
25. $\dfrac{(n + 2)n}{2(n + 1)^2} < \dfrac{1}{2}$, $|R_n| = \dfrac{|w_{n+1}|}{1 - q} = \dfrac{n + 2}{2^n(n + 1)} < 0.05$ when $n = 5$; $s \approx 1.657$.

SECTION 14.5

3. $\frac{1}{3}$ **5.** $2/\sqrt{\pi}$ **7.** 1 **9.** ∞ **11.** 1 **13.** 1 **15.** ∞ **17.** $\sqrt{2}$

SECTION 14.6

5. 5 **7.** $\frac{1}{4}$ **9.** 1

SECTION 15.2

5. $1 - z + \dfrac{z^2}{2!} - \dfrac{z^3}{3!} + - \cdots$, $R = \infty$

7. $e^{4i}\left[1 + 2(z - 2i) + \dfrac{2^2}{2!}(z - 2i)^2 + \cdots\right]$, $R = \infty$

9. $1 - \dfrac{1}{2!}\left(z - \dfrac{\pi}{2}\right)^2 + \dfrac{1}{4!}\left(z - \dfrac{\pi}{2}\right)^4 - + \cdots, R = \infty$

11. $\pi z - \dfrac{\pi^3 z^3}{3!} + \dfrac{\pi^5 z^5}{5!} - + \cdots, R = \infty$

13. $\dfrac{1}{2}(1 + \cos 2z) = \dfrac{1}{2}\left[2 - \dfrac{2^2}{2!}z^2 + \dfrac{2^4}{4!}z^4 - + \cdots\right], R = \infty$

15. $\dfrac{1+i}{2}\left[1 + \dfrac{1+i}{2}(z - i) + \left(\dfrac{1+i}{2}\right)^2(z - i)^2 + \cdots\right], R = \sqrt{2}$

17. $z + \dfrac{1}{3}z^3 + \dfrac{2}{15}z^5 + \cdots, R = \dfrac{\pi}{2}$ **19.** $z + z^2 + \dfrac{z^3}{3} + \cdots, R = \infty$

21. $\dfrac{z}{2!\,1} - \dfrac{z^3}{4!\,3} + \dfrac{z^5}{6!\,5} - + \cdots, R = \infty$

23. $z - \dfrac{z^3}{3!\,3} + \dfrac{z^5}{5!\,5} - \dfrac{z^7}{7!\,7} + - \cdots, R = \infty$

25. $z - \dfrac{z^5}{2!\,5} + \dfrac{z^9}{4!\,9} - \dfrac{z^{13}}{6!\,13} + - \cdots, R = \infty$

SECTION 15.3

1. $1 - z^4 + z^8 - z^{12} + - \cdots, |z| < 1$

3. $2 + z + 2z^2 + z^3 + 2z^4 + \cdots, |z| < 1$

5. $4 + 5z + 6z^2 + 7z^3 + \cdots, |z| < 1$

7. $2z^2 - \dfrac{2^3}{3!}z^6 + \dfrac{2^5}{5!}z^{10} - + \cdots, |z| < \infty$

9. $1 - z + \dfrac{3}{2}z^2 - \dfrac{7}{6}z^3 + \cdots, |z| < \infty$

11. $z^2 + z^4 + \dfrac{z^6}{3} + \cdots, |z| < \infty$

13. $e[1 + z + \dfrac{3}{2}z^2 + \dfrac{13}{6}z^3 + \cdots], |z| < 1$

15. $e[1 + z + z^2 + \dfrac{5}{6}z^3 + \cdots], |z| < \infty$

21. $1 + 2(z - \tfrac{1}{4}\pi) + 2(z - \tfrac{1}{4}\pi)^2 + \dfrac{8}{3}(z - \tfrac{1}{4}\pi)^3 + \cdots, |z - \tfrac{1}{4}\pi| < \tfrac{1}{4}\pi$

SECTION 15.4

3. Convergence follows from Theorem 1, Sec. 14.3. Let $R_n(z)$ and $R_n{}^*$ be the remainders of (1) and (5), respectively. Since (5) converges, for given $\epsilon > 0$ we can find an $N(\epsilon)$ such that $R_n{}^* < \epsilon$ for all $n > N(\epsilon)$. Since $|f_n(z)| \leqq M_n$ for all z in the region G, we have $|R_n(z)| \leqq R_n{}^*$ and therefore $|R_n(z)| < \epsilon$ for all $n > N(\epsilon)$ and all z in the region G. This proves that the convergence of (1) in the region G is uniform.

7. $|z^n| \leqq 1$ and $\sum 1/n^2$ converges.

9. $|\tanh^n x| \leqq 1$, $1/n(n + 1) < 1/n^2$, $\sum 1/n^2$ converges.

11. $f_n = s_n - s_{n-1} = x/(nx + 1)[(n - 1)x + 1]$

15. The sum $s = \lim s_n = \lim x^n$ is discontinuous at $x = 1$. Use Theorem 2.

SECTION 15.5

1. $\displaystyle\sum_{n=0}^{\infty} \dfrac{(-1)^n 4^{2n+1}}{(2n + 1)!} z^{2n-3}, \quad |z| > 0$ **3.** $\displaystyle\sum_{n=0}^{\infty} \dfrac{(-1)^n}{n!\, z^{2n+6}}, \quad |z| > 0$

5. $\displaystyle\sum_{n=0}^{\infty} (-1)^n z^{4n-1}, \quad 0 < |z| < 1$

7. Let $\sum\limits_{-\infty}^{\infty} A_n(z-a)^n$ and $\sum\limits_{-\infty}^{\infty} B_n(z-a)^n$ be two Laurent series of the same function $f(z)$ in the same annulus. We multiply both series by $(z-a)^{-k-1}$ and integrate along a circle with center at a in the interior of the annulus. Since the series converge uniformly, we may integrate term by term. This yields $2\pi i A_k = 2\pi i B_k$. Thus, $A_k = B_k$ for all $k = 0, \pm 1, \cdots$.

9. $-\sum\limits_{n=0}^{\infty} \left(\dfrac{i}{2}\right)^{n+1} (z-i)^{n-1}, \qquad 0 < |z-i| < 2, \qquad \sum\limits_{n=0}^{\infty} \dfrac{(-2i)^n}{(z-i)^{n+2}}, \qquad |z-i| > 2$

11. $(1-4z)\sum\limits_{n=0}^{\infty} z^{4n}, \qquad |z| < 1, \qquad \left(\dfrac{4}{z^3} - \dfrac{1}{z^4}\right) \sum\limits_{n=0}^{\infty} \dfrac{1}{z^{4n}}, \qquad |z| > 1$

13. $\sum\limits_{n=0}^{\infty} z^{4n}, \qquad |z| < 1, \qquad \sum\limits_{n=0}^{\infty} \dfrac{1}{z^{4n+4}}, \qquad |z| > 1$

15. $\sum\limits_{n=0}^{\infty} \dfrac{e}{n!}(z-1)^{n-2}, \qquad |z-1| > 0$

17. $\dfrac{2}{z} + \sum\limits_{n=0}^{\infty} \dfrac{(-1)^n - 4}{3^{n+1}} z^n, \qquad 0 < |z| < 3, \qquad \dfrac{2}{z} + \sum\limits_{n=0}^{\infty} \dfrac{3^n(4 + (-1)^n)}{z^{n+1}}, \qquad |z| > 3$

19. $\sum\limits_{n=0}^{\infty} \binom{-4}{n}(z-1)^n, \qquad |z-1| < 1, \qquad \sum\limits_{n=0}^{\infty} \binom{-4}{n} \dfrac{1}{(z-1)^{n+4}}, \qquad |z-1| > 1$

SECTION 15.6

1. No, yes, yes   **3.** No, no   **5.** No, no

SECTION 16.1

1. $\pm 1, \pm i$ (simple)
3. $(2n+1)/2$ (third order)
5. $(2n+1)\pi i/4, n = 0, \pm 1, \ldots$ (second order)
7. $-i, \infty$ (simple)
9. $\pm 2i$ (second order)
15. 0 (simple pole), ∞ (pole of second order)
17. $\pm i$ (pole of third order), ∞ (simple pole)
19. π (essential singularity)
21. $2/(2n+1)\pi$ ($n = 0, \pm 1, \cdots$, poles of second order), 0 (essential singularity)
23. $n\pi$ ($n = \pm 1, \pm 2, \cdots$, simple poles), ∞ (essential singularity)
25. $1, \infty$ (essential singularities), $\pm 2n\pi i$ ($n = 0, 1, \cdots$, simple poles)

SECTION 16.2

1. -3 (at $z = 1$)   **3.** $\frac{1}{2}$ (at $z = 0$)   **5.** $\frac{1}{24}$ (at $z = 0$)
7. 1 (at $z = 0, \pm\pi, \pm 2\pi, \cdots$)   **9.** $-\frac{1}{4}, \frac{1}{4}$ (at $z = 1, -1$)
11. $\frac{1}{2}$ (at $z = \pm in\pi/2, n = 0, 1, \cdots$)
13. $-\frac{1}{4}, i/4, \frac{1}{4}, -i/4$, (at $z = 1, i, -1, -i$)
15. -1 (at $z = 0$)   **17.** $\frac{1}{17}$ (at $z = -1$)   **19.** $\pi i/2$
21. $(4 - \pi^3/32)\pi i$ **23.** $-\pi i$ **25.** $2\pi i$
27. 0 **29.** $-2\pi i$

SECTION 16.3

1. 0 **3.** $\pi/3$ **5.** $(\pi i/4)\sin(\frac{1}{2})$ **7.** $(\frac{1}{2} + \cosh \pi/4)\pi i$
9. $-4\pi i$ **11.** $-4i \sinh(\frac{1}{2})$ **13.** $-4i$ **15.** $8\pi i$
17. $-2\pi/3$ **19.** $6\pi i$ **21.** $5\pi i$ **23.** $-3\pi i$
25. 0

SECTION 16.4

1. $\pi\sqrt{5}$ **3.** $2\pi/15$ **5.** $2\pi/3$

7. $\cos 2\theta = \dfrac{1}{2}\left(z^2 + \dfrac{1}{z^2}\right)$, $\displaystyle\int_0^{2\pi} \frac{\cos^2\theta}{13 - 5\cos 2\theta}\, d\theta = -\frac{1}{10i}\int_C \frac{(z^2+1)^2}{z(z^2-\frac{1}{5})(z^2-5)}\, dz = \frac{\pi}{10}$

9. 0   **13.** $\pi/2$ **15.** $3\pi/8$ **17.** $\pi/2$ **19.** $\pi/60$

21. $\pi e^{-1/\sqrt{2}}\,[\sin(1/\sqrt{2}) + \cos(1/\sqrt{2})]/\sqrt{2}$

SECTION 17.1

1. $u(r) = 200(\ln r/\ln 2) + 200 = 289\ln r + 200$
3. $u(r) = 100 - 900\ln r/\ln 10 = 100 - 391\ln r$
7. $u = c\,\mathrm{Re}\,\ln(z^2 - a^2) = c\ln|z^2 - a^2|$
9. $u = x^2 - y^2 = c$ (hyperbolas)

SECTION 17.2

1. $V = V_1 = K$, $Ky = \text{const}$, $Kx = \text{const}$
7. $F(z) = [\ln(z + a)]/2\pi$
11. The streamlines are circles $y/(x^2 + y^2) = c$ or $x^2 + (y - k)^2 = k^2$.
15. $|V| = \overline{|F'(z)|} = |F'(z)| = |(1 - z^{-2}) - iK/2\pi z|$

SECTION 17.3

3. Use that $|e^z| = e^x$ is monotone in x.

SECTION 17.4

3. Set $r = 0$ in (5).
5. $u = 1 + r^2\cos 2\theta = \mathrm{Re}\,(1 + z^2)$

7. $u = r\sin\theta - \dfrac{r^2}{2}\sin 2\theta + \dfrac{r^3}{3}\sin 3\theta - + \cdots$

9. $u = \dfrac{\pi^2}{3} - 4\left(r\cos\theta - \dfrac{r^2}{4}\cos 2\theta + \dfrac{r^3}{9}\cos 3\theta - + \cdots\right)$

11. $u = \dfrac{1}{2} + \dfrac{2}{\pi}\left(r\sin\theta + \dfrac{r^3}{3}\sin 3\theta + \dfrac{r^5}{5}\sin 5\theta + \cdots\right)$

13. $u = \dfrac{2}{\pi}\left(r\cos\theta - \dfrac{r^3}{3}\cos 3\theta + \dfrac{r^5}{5}\cos 5\theta - + \cdots + r^2\sin 2\theta + \dfrac{r^6}{3}\sin 6\theta + \dfrac{r^{10}}{5}\sin 10\theta + \cdots\right)$

15. $u = \dfrac{4}{\pi}\left(r\sin\theta + \dfrac{r^3}{3}\sin 3\theta + \dfrac{r^5}{5}\sin 5\theta + \cdots\right)$

SECTION 18.1

3. Rounding off a 5 in the last given place may introduce errors since no further digits are given. These errors will be more frequent if the five-place table is used.

7. $|d^* - d| = |\epsilon_1 - \epsilon_2| \leqq |\epsilon_1| + |\epsilon_2| \leqq \beta_1 + \beta_2 = \beta$, $a_1 = 1$, $a_1^* = 1.1$, $a_2 = 1$, $a_2^* = 0.9$ gives $|\epsilon| = |\epsilon_1| + |\epsilon_2|$.

9. $\dfrac{a_1^*}{a_2^*} = \dfrac{a_1 + \epsilon_1}{a_2 + \epsilon_2} = \dfrac{a_1 + \epsilon_1}{a_2}\left(1 - \dfrac{\epsilon_2}{a_2} + \dfrac{\epsilon_2^2}{a_2^2} - + \cdots\right) \approx \dfrac{a_1}{a_2} + \dfrac{\epsilon_1}{a_2} - \dfrac{\epsilon_2}{a_2}\cdot\dfrac{a_1}{a_2}$,

$\left(\dfrac{a_1^*}{a_2^*} - \dfrac{a_1}{a_2}\right)\bigg/\dfrac{a_1}{a_2} \approx \dfrac{\epsilon_1}{a_1} - \dfrac{\epsilon_2}{a_2}$

11. $|\epsilon(x_2)| = |\epsilon_r(x_2)\cdot x_2| = |\epsilon_r(x_1)\cdot x_2| \leqq (0.005/39.95)\cdot 0.050\,06 < 0.000\,01$

SECTION 18.2

1. This follows from the intermediate value theorem of calculus.
5. $x_1 = -0.2$, $x_2 = -0.200\,32$, $x_3 = -0.200\,323$, $\cdots$
7. Convergence to the root near -0.2.
9. $f(x) = x^k - c$, $x_{n+1} = (1 - k^{-1})x_n + k^{-1}cx_n^{1-k}$
11. $x_1 = 1.9$ (exact 1.9)   **13.** -1.3, 1.4 **15.** 1.557
17. ± 1.189 **19.** 0, ± 1.496

SECTION 18.4

3. 36   **5.** 2.21921, less exact (error 1 unit of the last digit).

7. (a) $\ln 9.2 = 0.6 \cdot 2.19722 + 0.4 \cdot 2.25129 + \dfrac{1.6 \cdot 0.6 \cdot (-0.4)}{6} \cdot (-0.00308)$
$+ \dfrac{1.4 \cdot 0.4 \cdot (-0.6)}{6} \cdot (-0.00277) = 2.21920$

(b) $\ln 9.2 = -0.064 \cdot 2.14007 + 0.672 \cdot 2.19722 + 0.448 \cdot 2.25129 - 0.056 \cdot 2.30259 = 2.21920$
(exact to 5D)

SECTION 18.5

11. 0.26, $0 \leqq \epsilon \leqq 0.02$ **13.** 0.9466 (0.9461 exact to 4D)
15. 0.9458   **17.** 0.9461
19. $\ln 2 \approx 0.693\,25$ (0.693 15 exact to 5D), $M_4{}^* = 0.75$, $M_4 = 24$, $0.000\,016 < \epsilon_S < 0.000\,53$
or $0.692\,72 < \ln 2 < 0.693\,24$.

SECTION 18.6

7. 0.1, 0.202, 0.31008, 0.42869 etc. Exact solution: $y = e^{x^2} \displaystyle\int_0^x e^{-t^2}\, dt$
11. $y = \tan x$

SECTION 18.8

1. $x = 2$, $y = 1$ **3.** $x = -1$, $y = -1$ **5.** $x = 4$, $y = 3$, $z = 1$
7. $x = y$, $z = y + 1$ **9.** $x = 0$, $z = y$ **11.** $w = 0$, $x = 2y$, $z = 3y + 1$

SECTION 18.9

3. $\begin{pmatrix} 0 & 0.25 & 0.25 & 0 \\ 0 & 0.0625 & 0.0625 & 0.25 \\ 0 & 0.0625 & 0.0625 & 0.25 \\ 0 & 0.03125 & 0.03125 & 0.125 \end{pmatrix}$ **5.** Exact 2, 3, 4   **7.** Exact 2, 2, 2

9. $\mathbf{C} = \begin{pmatrix} 0 & 0 & -1 \\ 0 & 0 & -1 \\ 0 & 0 & -1 \end{pmatrix}$, $\lambda_1 = -1$, $\lambda_2 = \lambda_3 = 0$, $\mathbf{I} - \mathbf{A} = \begin{pmatrix} 0 & 0 & -1 \\ 1 & 0 & 0 \\ \frac{1}{3} & \frac{2}{3} & 0 \end{pmatrix}$

$\lambda^3 + \lambda/3 + \frac{2}{3} = 0$, $\lambda_1 \approx -0.748$ (by Newton's method), hence $|\lambda_2| = |\lambda_3| = \sqrt{\frac{2}{3}/0.748} = 0.94$ because the constant term $\frac{2}{3}$ equals -1 times the product of all three eigenvalues, and λ_2 and λ_3 are conjugate.

SECTION 18.10

1. -0.0002 **5.** $\mathbf{AB} = \begin{pmatrix} 1 & 0 \\ -0.1 & 1 \end{pmatrix}$, $\mathbf{BA} = \begin{pmatrix} 11 & 10 \\ -10 & -9 \end{pmatrix}$

SECTION 18.11

1. $y = 2.44 + 0.77x$ **3.** $y = 6.497 + 0.457x$ **5.** $y = 0.5 + 1.5x^2$
7. $y = 6642 + 762.3x - 156.1x^2$
9. $\mathbf{C} = (c_{jk})$, $c_{jk} = x_j^{\ k}$, $\mathbf{b}^\mathsf{T} = (b_0 \cdots b_m)$

SECTION 18.12

1. Centers 1, 4, 1, radii 5, 8, 9   **3.** Centers 0, $1 + i$, $-i$, radii 1.5, $\sqrt{2}$, 1.1
5. $\sqrt{116} = 10.77$ **7.** $\sqrt{7.26} = 2.69$
15. $\mathbf{B}^{-1} = \mathbf{B}^\mathsf{T}$, $\mathbf{C}^\mathsf{T} = (\mathbf{B}^{-1}\mathbf{AB})^\mathsf{T} = \mathbf{B}^\mathsf{T}\mathbf{AB} = \mathbf{C}$
19. $26 \leqq \lambda \leqq 34$, $26 \leqq \lambda \leqq 34$, $28.66 \leqq \lambda \leqq 30$

SECTION 18.13

1. $m_0 = 1304$, $m_1 = 6412$, $m_2 = 31736$, $q = 4.9172$, $|\epsilon| \leqq \sqrt{24.3374 - 24.1788} \approx 0.398$
3. $q = 6.25, 6.46$, $|\epsilon| \leqq 1.09, 0.79$, respectively
5. $q = 3, 5.774$, $|\epsilon| \leqq 2.55, 1.337$

7. $\lambda = \pm 5$, $\mathbf{z}_1 = \begin{pmatrix} 2 \\ 1 \end{pmatrix}$, $\mathbf{z}_2 = \begin{pmatrix} 1 \\ -2 \end{pmatrix}$

9. A has the eigenvalue 1 or -1 (or both).

SECTION 18.14

1. Use (1).   **5.** $c(x) \sim -A(x) \sin x^2 + B(x) \cos x^2$ where

$$A(x) = \frac{1}{2}\left(\frac{1}{x} - \frac{1 \cdot 3}{4x^5} + \frac{1 \cdot 3 \cdot 5 \cdot 7}{16x^9} - + \cdots \right), B(x) = \frac{1}{2}\left(\frac{1}{2x^3} - \frac{1 \cdot 3 \cdot 5}{8x^7} + - \cdots \right).$$

7. $Q(\alpha, x) \sim x^\alpha e^{-x}\left[\dfrac{1}{x} - \dfrac{1 - \alpha}{x^2} + - \cdots + (-1)^{n-1}\dfrac{(1 - \alpha)(2 - \alpha) \cdots (n - 1 - \alpha)}{x^n} + \cdots \right]$

9. $C(x) \sim \pi^{1/2}2^{-3/2} + A(x) \sin x^2 - B(x) \cos x^2$ with A and B as in Prob. 5.

SECTION 19.3

1. $\bar{x} = 3.47$, $s^2 = 2.98$ **3.** $\bar{x} = 3$, $s^2 = 3.3$ **5.** 0.02
7. 350, 380, 30   **9.** 10 000, 1000, 100
15. $\bar{x} = 99.2$, $s^2 = 234.7$; grouped: $\bar{x} = 99.4$, $s^2 = 254.7$

SECTION 19.4

1. $A \cup B$ occurs if and only if $A \cap B$ or $A \cap B^c$ or $A^c \cap B$ occurs. These three events are mutually exclusive. Let n_1, n_2, n_3 be the corresponding absolute frequencies in the sample. Then $\tilde{f}(A) = (n_1 + n_2)/n$, $\tilde{f}(B) = (n_1 + n_3)/n$, $\tilde{f}(A \cap B) = n_1/n$, $\tilde{f}(A \cup B) = (n_1 + n_2 + n_3)/n$. From this, (1) follows.
3. The space of all ordered triples of nonnegative numbers.
5. E^c: That diameter is smaller than 2.9 in. or larger than 3.1 in.

SECTION 19.5

1. $\frac{31}{32}$ **3.** $(a)\, 0.9^3 = 72.9\%$. $(b)\, \dfrac{90}{100} \cdot \dfrac{89}{99} \cdot \dfrac{88}{98} = 72.65\%$
5. $P(E_B \mid E_A) = P(E_A \cap E_B)/P(E_A) = 0.02/0.10 = 20\%$

SECTION 19.6

3. $\dbinom{10}{3} = 120$ **5.** $\dbinom{100}{3} = 161\ 700$

7. The idea of proof is the same as that of Theorem 1, but instead of filling n places we now have to fill only k places. If repetitions are permitted, we have n elements for filling each of the k places.
9. 23.5, 0.5, 2%; 39 902, 400, 1%

SECTION 19.7

3. $F(x) = 0$, $(x - 1)/2$, 1 $(x < 1, 1 < x < 3, x > 3$, respectively)

5. $k = 3, c = 0.9$ **7.** $f(3) = \dfrac{1}{216}$, $f(4) = \dfrac{3}{216}$, etc.

SECTION 19.8

1. $\mu = \displaystyle\int_{-\infty}^{c} tf(t)\, dt + \int_{c}^{\infty} tf(t)\, dt = -\int_{\infty}^{0} (c - x)f(c - x)\, dx + \int_{0}^{\infty} (c + x)f(c + x)\, dx$

$\qquad = 2c\displaystyle\int_{0}^{\infty} f(c + x)\, dx = c$. For a discrete distribution the proof is similar.

3. Set $c_1 = 1/\sigma$ and $c_2 = -\mu/\sigma$ in (5) and (6).
11. $E(X^k) = 1/(k + 1)$
13. Integrate by parts, $\sigma^2 = 2$, $\gamma = 4/2\sqrt{2} = \sqrt{2}$

SECTION 19.9

1. $f(x) = \dbinom{n}{x}0.5^n$, $\mu = 0.5n$, $\sigma^2 = 0.25n$ **3.** $\frac{5}{256}$, that is, about 2%

5. $1 - 0.9^{10} \approx 65\%$ **7.** $f(x) = 1.5^x e^{-1.5}/x!$

9. $f(x_1, x_2) = \dfrac{20!}{x_1!\, x_2!\, (20 - x_1 - x_2)!} 0.03^{x_1} 0.05^{x_2} 0.92^{20 - x_1 - x_2}$

13. $G(0) = 1$, $\;G'(t) = e^{-\mu}\exp[\mu e^t] = \mu e^t G(t)$, $\;G''(t) = \mu e^t[G(t) + G'(t)]$, $\;E(X^2) = G''(0) =$
$\mu + \mu^2$, $\sigma^2 = E(X^2) - \mu^2 = \mu$

15. Use $x\dbinom{M}{x} = M\dbinom{M-1}{x-1}$ and (14) in Sec. 19.6 with $p = M - 1, k = x - 1, q = N - M$,
$r - k = n - x$.

SECTION 19.10

5. Set $x^* = c_1 x + c_2$, then $(x - \mu)/\sigma = (x^* - \mu^*)/\sigma^*$, $F(x^*) = P(X^* \leq x^*) = P(X \leq x) =$
$\Phi((x - \mu)/\sigma) = \Phi((x^* - \mu^*)/\sigma^*)$.
7. 45.065, 56.978, 2.022   **9.** 0.27%   **11.** About 160

13. $\Phi\left(\dfrac{4040 - 2020 + 0.5}{\sqrt{1010}}\right) - \Phi\left(\dfrac{2048 - 2020 - 0.5}{\sqrt{1010}}\right) = 19.3\%$

15. $P\left(\left|\dfrac{X}{n} - p\right| < \epsilon\right) = P([p - \epsilon]n < X < [p + \epsilon]n)$. Apply (11) with $a = (p - \epsilon)n$,

$b = (p + \epsilon)n$; then $\beta = (\epsilon n + 0.5)/\sqrt{npq}$, $\alpha = -\beta$ and $\alpha \to -\infty$, $\beta \to \infty$ as $n \to \infty$. Hence the above probability approaches $\Phi(\infty) - \Phi(-\infty) = 1 - 0 = 1$.

SECTION 19.11

1. (X, Y) takes a value in A, B, C, or D (cf. the figure on page 824) with probability $F(b_1, b_2)$, a value in A or C with probability $F(a_1, b_2)$, a value in C or D with probability $F(b_1, a_2)$, a value in C with probability $F(a_1, a_2)$, hence a value in B with probability given by the right-hand side of (2).
5. 50%

SECTION 19.12

1. The items 38, 69, 02, 49, 23, 52, 73, 29, 09, 05

SECTION 19.13

1. $\hat{\theta} = n/\Sigma\, x_j = 1/\bar{x}$

SECTION 19.14

1. CONF $\{17.84 \leqq \mu \leqq 18.96\}$ **3.** CONF $\{12.2 \leqq \mu \leqq 16.6\}$
5. CONF $\{4.25 \leqq \mu \leqq 4.49\}$ **7.** CONF $\{0.726 \leqq \mu \leqq 0.751\}$
9. CONF $\{0.05 \leqq \sigma^2 \leqq 10\}$
11. $c_1 = (\sqrt{253} - 2.58)^2/2 = 88.8$, $c_2 = 170.9$, CONF $\{1.42 \leqq \sigma^2 \leqq 2.75\}$
13. Normal distributions, means $-12, 24, 47$, variances $25, 100, 400$.
15. Normal distribution with mean 26 and variance 70.

SECTION 19.15

1. If the hypothesis $p = 0.5$ is true, $X = $ *number of heads in* 4040 *trials* is approximately normal with $\mu = 2020$, $\sigma^2 = 1010$ (Sec. 19.10). $P(X \leqq c) = \Phi([c - 2020]/\sqrt{1010}) = 0.95$, $c = 2072 > 2048$, do not reject the hypothesis.
3. $\alpha = 5\%$, $t = \sqrt{10}\,(0.811 - 0.800)/0.00735 \approx 4.7 > c = 1.83$. Hypothesis rejected.
5. $t = \sqrt{7}\,(0.286 - 0)/4.31 = 0.18 < c = 1.94$ ($\alpha = 5\%$), do not reject the hypothesis.
9. Hypothesis H_0: not better. Alternative H_1: better. Under H_0 the variable $X = $ *number of cases cured in* 200 *cases* is approximately normal, $\alpha = 5\%$ gives $(c - 140)/\sqrt{42} = 1.645$, $c = 150.65 > 148$, do not reject H_0.
11. $\bar{x} = 376.3$, $\bar{y} = 335.3$, $s_1{}^2 = 1009.3$, $s_2{}^2 = 869.3$, $t_0 = 1.64 < c = 2.13$ ($\alpha = 5\%$, 4 degrees of freedom), do not reject the hypothesis.
15. $v_0 = 679/254 = 2.67$. Choose $\alpha = 5\%$. Table A13 in Appendix 4 does not contain a value corresponding to $(11, 5)$ degrees of freedom, but neighboring values in the table show that it must be greater than v_0, hence do not reject the hypothesis.

SECTION 19.16

1. $LCL = 1 - 2.58 \cdot 0.02/2 = 0.974$, $UCL = 1.026$
3. LCL and UCL will correspond to a significance level α which, in general, is not equal to 1%.
5. $\mu_0 = 20.50$, $LCL = 4.05 \cdot 5 = 20.25$, $UCL = 20.75$
7. Trend of sample means to increase. Abrupt change of sample means.
9. $LCL = 3$, $UCL = 7$

SECTION 19.17

1. $(1 - \theta)^2$

5. $P(A; \theta) = \dbinom{N - N\theta}{5} \Big/ \dbinom{N}{5} \approx \theta^0(1 - \theta)^{5-0} = (1 - \theta)^5$, $AOQ(\theta) \approx \theta(1 - \theta)^5$

7. $P(A; \theta) = (1 - \theta)^n$ **9.** $\theta_0 \approx 0.01$, $\theta_1 \approx 0.37$

SECTION 19.18

1. $K = 2$, $n = 355 + 123 = 478$, $e_1 = 478 \cdot \frac{3}{4} = 358.5$, $e_2 = 478 \cdot \frac{1}{4} = 119.5$,

$$\chi_0^2 = \frac{(355.0 - 358.5)^2}{358.5} + \frac{(123.0 - 119.5)^2}{119.5} = 0.137.$$

We choose $\alpha = 5\%$. Then $c = 3.84$ (1 degree of freedom). Hence $\chi_0^2 < c$, that is, we assert that the deviation of the ratio in the sample from the theoretical ratio 3:1 is due to random fluctuation.

3. $a = 32$ because $\chi_0^2 = \frac{1}{25}[(a - 25)^2 + (50 - a - 25)^2] > c = 3.84$ when $a \geqq 32$ (or $a \leqq 18$).

5. $\bar{x} = 99.4$, $\tilde{\sigma} = 15.8$, $K = 5$ (endpoints $-\infty$, 85, 95, 105, 115, ∞), $\chi_0^2 = 0.7 < c = 5.99$ ($\alpha = 5\%$). Hypothesis not rejected.

SECTION 19.19

1. $\alpha = 5\%$, $P(X \geqq 10) = \sum_{x=10}^{13} \binom{13}{x} 0.5^{13} = 4.6\% < \alpha$. Hypothesis rejected.

3. Hypothesis $\tilde{\mu} = 0$. Alternative $\tilde{\mu} > 0$, $\bar{x} = 1.58$, $t = \sqrt{10} \cdot 1.58/1.23 = 4.06 > c = 1.83$ ($\alpha = 5\%$). Hypothesis rejected.

5. 2 transpositions, $n = 8$, $P(T \leqq 2) \approx 0.1\%$. We reject the hypothesis and assert that there is negative trend.

SECTION 19.20

1. $\bar{x}^* = 1.5$, $\bar{y}^* = -2.5$, $s_1^{*2} = 5/3$, $s_{xy}^* = 77/3$, etc.

5. $y = x - 0.1$ **7.** $y - 1402 = 4.32(x - 590)$

9. CONF $\{4.1 \leqq \beta \leqq 4.5\}$

APPENDIX **3**

Some Formulas for Special Functions

Tables of numerical values in Appendix 4.

Exponential function e^x (Fig. 381)

$$e = 2.71828\ 18284\ 59045\ 23536\ 02874\ 71353$$

(1) $\qquad e^x e^y = e^{x+y}, \qquad e^x/e^y = e^{x-y}, \qquad (e^x)^y = e^{xy}$

Fig. 381. Exponential function e^x **Fig. 382.** Natural logarithm $\ln x$

Natural logarithm (Fig. 382)

(2) $\ \ \ln (xy) = \ln x + \ln y, \qquad \ln (x/y) = \ln x - \ln y, \qquad \ln (x^a) = a \ln x$

$\ln x$ is the inverse of e^x, and $e^{\ln x} = x$, $e^{-\ln x} = e^{\ln (1/x)} = 1/x$.

Logarithm of base ten $\log_{10} x$ or simply $\log x$

(3) $\ \ \log x = M \ln x, \qquad M = \log e = 0.43429\ 44819\ 03251\ 82765\ 11289\ 18917$

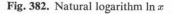

(4) $\ \ \ \ln x = \dfrac{1}{M} \log x, \qquad \dfrac{1}{M} = 2.30258\ 50929\ 94045\ 68401\ 79914\ 54684$

$\log x$ is the inverse of 10^x, and $10^{\log x} = x$, $10^{-\log x} = 1/x$.

826

Fig. 383. sin x **Fig. 384.** cos x

Sine and cosine functions (Figs. 383, 384). In calculus, angles are measured in radians, so that sin x and cos x have period 2π. sin x is odd, $\sin(-x) = -\sin x$, and cos x is even, $\cos(-x) = \cos x$.

(5) $$\sin^2 x + \cos^2 x = 1$$

(6)
$$\begin{cases} \sin(x+y) = \sin x \cos y + \cos x \sin y \\ \sin(x-y) = \sin x \cos y - \cos x \sin y \\ \cos(x+y) = \cos x \cos y - \sin x \sin y \\ \cos(x-y) = \cos x \cos y + \sin x \sin y \end{cases}$$

(7) $$\sin 2x = 2 \sin x \cos x, \qquad \cos 2x = \cos^2 x - \sin^2 x$$

(8)
$$\begin{cases} \sin x = \cos\left(x - \dfrac{\pi}{2}\right) = \cos\left(\dfrac{\pi}{2} - x\right) \\[2mm] \cos x = \sin\left(x + \dfrac{\pi}{2}\right) = \sin\left(\dfrac{\pi}{2} - x\right) \end{cases}$$

(9) $$\sin(\pi - x) = \sin x, \qquad \cos(\pi - x) = -\cos x$$

(10) $$\cos^2 x = \tfrac{1}{2}(1 + \cos 2x), \qquad \sin^2 x = \tfrac{1}{2}(1 - \cos 2x)$$

(11)
$$\begin{cases} \sin x \sin y = \tfrac{1}{2}[-\cos(x+y) + \cos(x-y)] \\ \cos x \cos y = \tfrac{1}{2}[\cos(x+y) + \cos(x-y)] \\ \sin x \cos y = \tfrac{1}{2}[\sin(x+y) + \sin(x-y)] \end{cases}$$

(12)
$$\begin{cases} \sin u + \sin v = 2 \sin\dfrac{u+v}{2} \cos\dfrac{u-v}{2} \\[3mm] \cos u + \cos v = 2 \cos\dfrac{u+v}{2} \cos\dfrac{u-v}{2} \\[3mm] \cos v - \cos u = 2 \sin\dfrac{u+v}{2} \sin\dfrac{u-v}{2} \end{cases}$$

(13) $$A \cos x + B \sin x = \sqrt{A^2 + B^2}\cos(x \pm \delta), \qquad \tan \delta = \frac{\sin \delta}{\cos \delta} = \mp\frac{B}{A}$$

(14) $$A \cos x + B \sin x = \sqrt{A^2 + B^2}\sin(x \pm \delta), \qquad \tan \delta = \frac{\sin \delta}{\cos \delta} = \pm\frac{A}{B}$$

Fig. 385. tan x

Fig. 386. cot x

Tangent, cotangent, secant, cosecant (Figs. 385, 386)

(15) $\tan x = \dfrac{\sin x}{\cos x}$, $\cot x = \dfrac{\cos x}{\sin x}$, $\sec x = \dfrac{1}{\cos x}$, $\csc x = \dfrac{1}{\sin x}$

(16) $\tan (x + y) = \dfrac{\tan x + \tan y}{1 - \tan x \tan y}$, $\tan (x - y) = \dfrac{\tan x - \tan y}{1 + \tan x \tan y}$

Hyperbolic functions (hyperbolic sine sinh x, etc.; Figs. 387, 388)

(17) $\sinh x = \frac{1}{2}(e^x - e^{-x})$, $\cosh x = \frac{1}{2}(e^x + e^{-x})$

(18) $\tanh x = \dfrac{\sinh x}{\cosh x}$, $\coth x = \dfrac{\cosh x}{\sinh x}$

(19) $\cosh x + \sinh x = e^x$, $\cosh x - \sinh x = e^{-x}$

(20) $\cosh^2 x - \sinh^2 x = 1$

(21) $\sinh^2 x = \frac{1}{2}(\cosh 2x - 1)$, $\cosh^2 x = \frac{1}{2}(\cosh 2x + 1)$

(22) $\begin{cases} \sinh (x \pm y) = \sinh x \cosh y \pm \cosh x \sinh y \\ \cosh (x \pm y) = \cosh x \cosh y \pm \sinh x \sinh y \end{cases}$

(23) $\tanh (x \pm y) = \dfrac{\tanh x \pm \tanh y}{1 \pm \tanh x \tanh y}$

Gamma function (cf. Fig. 389 and Table A3 in Appendix 4). The gamma function $\Gamma(\alpha)$ is defined by the integral

(24) $$\Gamma(\alpha) = \int_0^\infty e^{-t} t^{\alpha-1}\, dt \qquad\qquad (\alpha > 0)$$

which is meaningful only if $\alpha > 0$ (or, if we consider complex α, for those α whose real part is positive). Integration by parts gives the important *functional relation of the gamma function*,

(25) $$\Gamma(\alpha + 1) = \alpha \Gamma(\alpha).$$

Fig. 387. sinh x (dashed) and cosh x

Fig. 388. tanh x (dashed) and coth x

Fig. 389. Gamma function

From (24) we readily have $\Gamma(1) = 1$; hence if α is a positive integer, say k, then by repeated application of (25) we obtain

$$(26) \qquad\qquad \Gamma(k + 1) = k! \qquad\qquad (k = 0, 1, \cdots).$$

This shows that *the gamma function can be regarded as a generalization of the elementary factorial function.* (Sometimes the notation $(\alpha - 1)!$ for $\Gamma(\alpha)$ is used, even for noninteger values of α, and the gamma function is also known as the **factorial function**.)

By repeated application of (25) we obtain

$$\Gamma(\alpha) = \frac{\Gamma(\alpha + 1)}{\alpha} = \frac{\Gamma(\alpha + 2)}{\alpha(\alpha + 1)} = \cdots = \frac{\Gamma(\alpha + k + 1)}{\alpha(\alpha + 1)(\alpha + 2) \cdots (\alpha + k)},$$

and we may use this relation

$$(27) \qquad\qquad \Gamma(\alpha) = \frac{\Gamma(\alpha + k + 1)}{\alpha(\alpha + 1) \cdots (\alpha + k)} \qquad\qquad (\alpha \neq 0, -1, -2, \cdots)$$

for defining the gamma function for negative α ($\neq -1, -2, \cdots$), choosing for k the smallest integer such that $\alpha + k + 1 > 0$. *Together with* (1), *this then gives a definition of $\Gamma(\alpha)$ for all α not equal to zero or a negative integer* (Fig. 389).

It can be shown that the gamma function may also be represented as the limit of a product, namely, by the formula

$$(28) \qquad\qquad \Gamma(\alpha) = \lim_{n \to \infty} \frac{n! \, n^{\alpha}}{\alpha(\alpha + 1)(\alpha + 2) \cdots (\alpha + n)} \qquad\qquad (\alpha \neq 0, -1, \cdots).$$

From (27) or (28) we see that, for complex α, the gamma function $\Gamma(\alpha)$ is a meromorphic function which has simple poles at $\alpha = 0, -1, -2, \cdots$.

An approximation of the gamma function for large positive α is given by the **Stirling formula**

$$(29) \qquad\qquad \Gamma(\alpha + 1) \approx \sqrt{2\pi\alpha} \left(\frac{\alpha}{e}\right)^{\alpha}$$

where e is the base of the natural logarithm. We finally mention the special value

$$(30) \qquad\qquad \Gamma(\tfrac{1}{2}) = \sqrt{\pi}.$$

Incomplete gamma functions

$$(31) \qquad P(\alpha, x) = \int_0^x e^{-t} t^{\alpha - 1} \, dt, \qquad Q(\alpha, x) = \int_x^{\infty} e^{-t} t^{\alpha - 1} \, dt \qquad (\alpha > 0)$$

$$(32) \qquad\qquad \Gamma(\alpha) = P(\alpha, x) + Q(\alpha, x)$$

Beta function

$$(33) \qquad\qquad B(x, y) = \int_0^1 t^{x - 1}(1 - t)^{y - 1} \, dt \qquad\qquad (x > 0, y > 0)$$

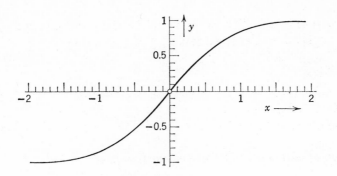

Fig. 390. Error function

Representation in terms of gamma functions:

$$(34) \qquad B(x, y) = \frac{\Gamma(x)\Gamma(y)}{\Gamma(x + y)}$$

Error function (cf. Fig. 390 and Table A5 in Appendix 4)

$$(35) \qquad \operatorname{erf} x = \frac{2}{\sqrt{\pi}} \int_0^x e^{-t^2}\, dt$$

$$(36) \qquad \operatorname{erf} x = \frac{2}{\sqrt{\pi}} \left(x - \frac{x^3}{1!\,3} + \frac{x^5}{2!\,5} - \frac{x^7}{3!\,7} + - \cdots \right)$$

$\operatorname{erf}(\infty) = 1$, *complementary error function*

$$(37) \qquad \operatorname{erfc} x = 1 - \operatorname{erf} x = \frac{2}{\sqrt{\pi}} \int_x^\infty e^{-t^2}\, dt$$

Fresnel integrals[1] (Fig. 391)

$$(38) \qquad C(x) = \int_0^x \cos(t^2)\, dt, \qquad S(x) = \int_0^x \sin(t^2)\, dt$$

Fig. 391. Fresnel integrals

---

[1] AUGUSTIN FRESNEL (1788–1827), French physicist and mathematician. For tables see Ref. [A6], [A7]; cf. also *Acta mathematica*, vol. 85 (1951), p. 180, and vol. 89 (1953), p. 130.

Fig. 392. Sine integral

$C(\infty) = \sqrt{\pi/8}$, $S(\infty) = \sqrt{\pi/8}$, *complementary functions*

(39)
$$c(x) = \sqrt{\frac{\pi}{8}} - C(x) = \int_x^\infty \cos(t^2)\, dt$$
$$s(x) = \sqrt{\frac{\pi}{8}} - S(x) = \int_x^\infty \sin(t^2)\, dt$$

Sine integral (cf. Fig. 392 and Table A5 in Appendix 4)

(40)
$$\mathrm{Si}(x) = \int_0^x \frac{\sin t}{t}\, dt$$

$\mathrm{Si}(\infty) = \pi/2$, *complementary function*

(41)
$$\mathrm{si}(x) = \frac{\pi}{2} - \mathrm{Si}(x) = \int_x^\infty \frac{\sin t}{t}\, dt$$

Cosine integral (cf. Table A5 in Appendix 4)

(42)
$$\mathrm{ci}(x) = \int_x^\infty \frac{\cos t}{t}\, dt$$

Tables

Tables of Laplace transforms in Sec. 4.10.

Table A1. Some Elementary Functions

x	$\sin x$	$\cos x$	$\tan x$	e^x	$\sinh x$	$\cosh x$
0.0	0.00000	1.00000	0.00000	1.00000	0.00000	1.00000
0.1	0.09983	0.99500	0.10033	1.10517	0.10017	1.00500
0.2	0.19867	0.98007	0.20271	1.22140	0.20134	1.02007
0.3	0.29552	0.95534	0.30934	1.34986	0.30452	1.04534
0.4	0.38942	0.92106	0.42279	1.49182	0.41075	1.08107
0.5	0.47943	0.87758	0.54630	1.64872	0.52110	1.12763
0.6	0.56464	0.82534	0.68414	1.82212	0.63665	1.18547
0.7	0.64422	0.76484	0.84229	2.01375	0.75858	1.25517
0.8	0.71736	0.69671	1.02964	2.22554	0.88811	1.33743
0.9	0.78333	0.62161	1.26016	2.45960	1.02652	1.43309
1.0	0.84147	0.54030	1.55741	2.71828	1.17520	1.54308
1.1	0.89121	0.45360	1.96476	3.00417	1.33565	1.66852
1.2	0.93204	0.36236	2.57215	3.32012	1.50946	1.81066
1.3	0.96356	0.26750	3.60210	3.66930	1.69838	1.97091
1.4	0.98545	0.16997	5.79788	4.05520	1.90430	2.15090
1.5	0.99750	0.07074	14.10142	4.48169	2.12928	2.35241
1.6	0.99957	−0.02920	−34.23253	4.95303	2.37557	2.57746
1.7	0.99166	−0.12884	−7.69660	5.47395	2.64563	2.82832
1.8	0.97385	−0.22720	−4.28626	6.04965	2.94217	3.10747
1.9	0.94630	−0.32329	−2.92710	6.68589	3.26816	3.41773
2.0	0.90930	−0.41615	−2.18504	7.38906	3.62686	3.76220

x	$\ln x$	x	$\ln x$	x	$\ln x$	x	$\ln x$
1.0	0.00000	2.0	0.69315	3.0	1.09861	5	1.60944
1.1	0.09531	2.1	0.74194	3.1	1.13140	7	1.94591
1.2	0.18232	2.2	0.78846	3.2	1.16315	11	2.39790
1.3	0.26236	2.3	0.83291	3.3	1.19392	13	2.56495
1.4	0.33647	2.4	0.87547	3.4	1.22378	17	2.83321
1.5	0.40547	2.5	0.91629	3.5	1.25276	19	2.94444
1.6	0.47000	2.6	0.95551	3.6	1.28093	23	3.13549
1.7	0.53063	2.7	0.99325	3.7	1.30833	29	3.36730
1.8	0.58779	2.8	1.02962	3.8	1.33500	31	3.43399
1.9	0.64185	2.9	1.06471	3.9	1.36098	37	3.61092

Table A1. (*continued*)

$\frac{y}{x}$	arc tan $\frac{y}{x}$	$\frac{y}{x}$	arc tan $\frac{y}{x}$	$\frac{y}{x}$	arc tan $\frac{y}{x}$	$\frac{y}{x}$	arc tan $\frac{y}{x}$
0.0	0.00000	1.0	0.78540	2.0	1.10715	4.0	1.32582
0.1	0.09967	1.1	0.83298	2.2	1.14417	4.5	1.35213
0.2	0.19740	1.2	0.87606	2.4	1.17601	5.0	1.37340
0.3	0.29146	1.3	0.91510	2.6	1.20362	5.5	1.39094
0.4	0.38051	1.4	0.95055	2.8	1.22777	6.0	1.40565
0.5	0.46365	1.5	0.98279	3.0	1.24905	7.0	1.42890
0.6	0.54042	1.6	1.01220	3.2	1.26791	8.0	1.44644
0.7	0.61073	1.7	1.03907	3.4	1.28474	9.0	1.46014
0.8	0.67474	1.8	1.06370	3.6	1.29985	10.0	1.47113
0.9	0.73282	1.9	1.08632	3.8	1.31347	11.0	1.48014

Table A2. Bessel Functions

For more extensive tables see Ref. [A8] in Appendix 1.

x	$J_0(x)$	$J_1(x)$	x	$J_0(x)$	$J_1(x)$	x	$J_0(x)$	$J_1(x)$
0.0	1.0000	0.0000	3.0	−0.2601	0.3391	6.0	0.1506	−0.2767
0.1	0.9975	0.0499	3.1	−0.2921	0.3009	6.1	0.1773	−0.2559
0.2	0.9900	0.0995	3.2	−0.3202	0.2613	6.2	0.2017	−0.2329
0.3	0.9776	0.1483	3.3	−0.3443	0.2207	6.3	0.2238	−0.2081
0.4	0.9604	0.1960	3.4	−0.3643	0.1792	6.4	0.2433	−0.1816
0.5	0.9385	0.2423	3.5	−0.3801	0.1374	6.5	0.2601	−0.1538
0.6	0.9120	0.2867	3.6	−0.3918	0.0955	6.6	0.2740	−0.1250
0.7	0.8812	0.3290	3.7	−0.3992	0.0538	6.7	0.2851	−0.0953
0.8	0.8463	0.3688	3.8	−0.4026	0.0128	6.8	0.2931	−0.0652
0.9	0.8075	0.4059	3.9	−0.4018	−0.0272	6.9	0.2981	−0.0349
1.0	0.7652	0.4401	4.0	−0.3971	−0.0660	7.0	0.3001	−0.0047
1.1	0.7196	0.4709	4.1	−0.3887	−0.1033	7.1	0.2991	0.0252
1.2	0.6711	0.4983	4.2	−0.3766	−0.1386	7.2	0.2951	0.0543
1.3	0.6201	0.5220	4.3	−0.3610	−0.1719	7.3	0.2882	0.0826
1.4	0.5669	0.5419	4.4	−0.3423	−0.2028	7.4	0.2786	0.1096
1.5	0.5118	0.5579	4.5	−0.3205	−0.2311	7.5	0.2663	0.1352
1.6	0.4554	0.5699	4.6	−0.2961	−0.2566	7.6	0.2516	0.1592
1.7	0.3980	0.5778	4.7	−0.2693	−0.2791	7.7	0.2346	0.1813
1.8	0.3400	0.5815	4.8	−0.2404	−0.2985	7.8	0.2154	0.2014
1.9	0.2818	0.5812	4.9	−0.2097	−0.3147	7.9	0.1944	0.2192
2.0	0.2239	0.5767	5.0	−0.1776	−0.3276	8.0	0.1717	0.2346
2.1	0.1666	0.5683	5.1	−0.1443	−0.3371	8.1	0.1475	0.2476
2.2	0.1104	0.5560	5.2	−0.1103	−0.3432	8.2	0.1222	0.2580
2.3	0.0555	0.5399	5.3	−0.0758	−0.3460	8.3	0.0960	0.2657
2.4	0.0025	0.5202	5.4	−0.0412	−0.3453	8.4	0.0692	0.2708
2.5	−0.0484	0.4971	5.5	−0.0068	−0.3414	8.5	0.0419	0.2731
2.6	−0.0968	0.4708	5.6	0.0270	−0.3343	8.6	0.0146	0.2728
2.7	−0.1424	0.4416	5.7	0.0599	−0.3241	8.7	−0.0125	0.2697
2.8	−0.1850	0.4097	5.8	0.0917	−0.3110	8.8	−0.0392	0.2641
2.9	−0.2243	0.3754	5.9	0.1220	−0.2951	8.9	−0.0653	0.2559

$J_0(x) = 0$ for $x = 2.405, 5.520, 8.654, 11.792, 14.931, \cdots J_1(x) = 0$ for $x = 0, 3.832, 7.016, 10.173, 13.324, \cdots$

Table A2. (*continued*)

x	$Y_0(x)$	$Y_1(x)$	x	$Y_0(x)$	$Y_1(x)$	x	$Y_0(x)$	$Y_1(x)$
0.0	$(-\infty)$	$(-\infty)$	2.5	0.498	0.146	5.0	−0.309	0.148
0.5	−0.445	−1.471	3.0	0.377	0.325	5.5	−0.339	−0.024
1.0	0.088	−0.781	3.5	0.189	0.410	6.0	−0.288	−0.175
1.5	0.382	−0.412	4.0	−0.017	0.398	6.5	−0.173	−0.274
2.0	0.510	−0.107	4.5	−0.195	0.301	7.0	−0.026	−0.303

Table A3. Gamma Function (cf. (24) in Appendix 3)

α	$\Gamma(\alpha)$	α	$\Gamma(\alpha)$	α	$\Gamma(\alpha)$	α	$\Gamma(\alpha)$	α	$\Gamma(\alpha)$
1.00	1.000 000	1.20	0.918 169	1.40	0.887 264	1.60	0.893 515	1.80	0.931 384
1.02	0.988 844	1.22	0.913 106	1.42	0.886 356	1.62	0.895 924	1.82	0.936 845
1.04	0.978 438	1.24	0.908 521	1.44	0.885 805	1.64	0.898 642	1.84	0.942 612
1.06	0.968 744	1.26	0.904 397	1.46	0.885 604	1.66	0.901 668	1.86	0.948 687
1.08	0.959 725	1.28	0.900 718	1.48	0.885 747	1.68	0.905 001	1.88	0.955 071
1.10	0.951 351	1.30	0.897 471	1.50	0.886 227	1.70	0.908 639	1.90	0.961 766
1.12	0.943 590	1.32	0.894 640	1.52	0.887 039	1.72	0.912 581	1.92	0.968 774
1.14	0.936 416	1.34	0.892 216	1.54	0.888 178	1.74	0.916 826	1.94	0.976 099
1.16	0.929 803	1.36	0.890 185	1.56	0.889 639	1.76	0.921 375	1.96	0.983 743
1.18	0.923 728	1.38	0.888 537	1.58	0.891 420	1.78	0.926 227	1.98	0.991 708
1.20	0.918 169	1.40	0.887 264	1.60	0.893 515	1.80	0.931 384	2.00	1.000 000

Table A4. Factorial Function

n	$n!$	$\log(n!)$	n	$n!$	$\log(n!)$	n	$n!$	$\log(n!)$
1	1	0.000 000	6	720	2.857 332	11	39 916 800	7.601 156
2	2	0.301 030	7	5 040	3.702 431	12	479 001 600	8.680 337
3	6	0.778 151	8	40 320	4.605 521	13	6 227 020 800	9.794 280
4	24	1.380 211	9	362 880	5.559 763	14	87 178 291 200	10.940 408
5	120	2.079 181	10	3 628 800	6.559 763	15	1 307 674 368 000	12.116 500

Table A5. Error Function, Sine and Cosine Integrals (cf. (35), (40), (42) in Appendix 3)

x	erf x	Si (x)	ci (x)	x	erf x	Si (x)	ci (x)
0.0	0.0000	0.0000	∞	2.0	0.9953	1.6054	−0.4230
0.2	0.2227	0.1996	1.0422	2.2	0.9981	1.6876	−0.3751
0.4	0.4284	0.3965	0.3788	2.4	0.9993	1.7525	−0.3173
0.6	0.6039	0.5881	0.0223	2.6	0.9998	1.8004	−0.2533
0.8	0.7421	0.7721	−0.1983	2.8	0.9999	1.8321	−0.1865
1.0	0.8427	0.9461	−0.3374	3.0	1.0000	1.8487	−0.1196
1.2	0.9103	1.1080	−0.4205	3.2	1.0000	1.8514	−0.0553
1.4	0.9523	1.2562	−0.4620	3.4	1.0000	1.8419	0.0045
1.6	0.9763	1.3892	−0.4717	3.6	1.0000	1.8219	0.0580
1.8	0.9891	1.5058	−0.4568	3.8	1.0000	1.7934	0.1038
2.0	0.9953	1.6054	−0.4230	4.0	1.0000	1.7582	0.1410

Table A6. Binomial Distribution

Probability function $f(x)$ (cf. (2), Sec. 19.9) and distribution function $F(x)$

n	x	$p = 0.1$ $f(x)$	$F(x)$	$p = 0.2$ $f(x)$	$F(x)$	$p = 0.3$ $f(x)$	$F(x)$	$p = 0.4$ $f(x)$	$F(x)$	$p = 0.5$ $f(x)$	$F(x)$
		0.		0.		0.		0.		0.	
1	0	9000	0.9000	8000	0.8000	7000	0.7000	6000	0.6000	5000	0.5000
	1	1000	1.0000	2000	1.0000	3000	1.0000	4000	1.0000	5000	1.0000
	0	8100	0.8100	6400	0.6400	4900	0.4900	3600	0.3600	2500	0.2500
2	1	1800	0.9900	3200	0.9600	4200	0.9100	4800	0.8400	5000	0.7500
	2	0100	1.0000	0400	1.0000	0900	1.0000	1600	1.0000	2500	1.0000
	0	7290	0.7290	5120	0.5120	3430	0.3430	2160	0.2160	1250	0.1250
	1	2430	0.9720	3840	0.8960	4410	0.7840	4320	0.6480	3750	0.5000
3	2	0270	0.9990	0960	0.9920	1890	0.9730	2880	0.9360	3750	0.8750
	3	0010	1.0000	0080	1.0000	0270	1.0000	0640	1.0000	1250	1.0000
	0	6561	0.6561	4096	0.4096	2401	0.2401	1296	0.1296	0625	0.0625
	1	2916	0.9477	4096	0.8192	4116	0.6517	3456	0.4752	2500	0.3125
4	2	0486	0.9963	1536	0.9728	2646	0.9163	3456	0.8208	3750	0.6875
	3	0036	0.9999	0256	0.9984	0756	0.9919	1536	0.9744	2500	0.9375
	4	0001	1.0000	0016	1.0000	0081	1.0000	0256	1.0000	0625	1.0000
	0	5905	0.5905	3277	0.3277	1681	0.1681	0778	0.0778	0313	0.0313
	1	3281	0.9185	4096	0.7373	3602	0.5282	2592	0.3370	1563	0.1875
	2	0729	0.9914	2048	0.9421	3087	0.8369	3456	0.6826	3125	0.5000
5	3	0081	0.9995	0512	0.9933	1323	0.9692	2304	0.9130	3125	0.8125
	4	0005	1.0000	0064	0.9997	0284	0.9976	0768	0.9898	1563	0.9688
	5	0000	1.0000	0003	1.0000	0024	1.0000	0102	1.0000	0313	1.0000
	0	5314	0.5314	2621	0.2621	1176	0.1176	0467	0.0467	0156	0.0156
	1	3543	0.8857	3932	0.6554	3025	0.4202	1866	0.2333	0938	0.1094
	2	0984	0.9841	2458	0.9011	3241	0.7443	3110	0.5443	2344	0.3438
6	3	0146	0.9987	0819	0.9830	1852	0.9295	2765	0.8208	3125	0.6563
	4	0012	0.9999	0154	0.9984	0595	0.9891	1382	0.9590	2344	0.8906
	5	0001	1.0000	0015	0.9999	0102	0.9993	0369	0.9959	0938	0.9844
	6	0000	1.0000	0001	1.0000	0007	1.0000	0041	1.0000	0156	1.0000
	0	4783	0.4783	2097	0.2097	0824	0.0824	0280	0.0280	0078	0.0078
	1	3720	0.8503	3670	0.5767	2471	0.3294	1306	0.1586	0547	0.0625
	2	1240	0.9743	2753	0.8520	3177	0.6471	2613	0.4199	1641	0.2266
	3	0230	0.9973	1147	0.9667	2269	0.8740	2903	0.7102	2734	0.5000
7	4	0026	0.9998	0287	0.9953	0972	0.9712	1935	0.9037	2734	0.7734
	5	0002	1.0000	0043	0.9996	0250	0.9962	0774	0.9812	1641	0.9375
	6	0000	1.0000	0004	1.0000	0036	0.9998	0172	0.9984	0547	0.9922
	7	0000	1.0000	0000	1.0000	0002	1.0000	0016	1.0000	0078	1.0000
	0	4305	0.4305	1678	0.1678	0576	0.0576	0168	0.0168	0039	0.0039
	1	3826	0.8131	3355	0.5033	1977	0.2553	0896	0.1064	0313	0.0352
	2	1488	0.9619	2936	0.7969	2965	0.5518	2090	0.3154	1094	0.1445
	3	0331	0.9950	1468	0.9437	2541	0.8059	2787	0.5941	2188	0.3633
8	4	0046	0.9996	0459	0.9896	1361	0.9420	2322	0.8263	2734	0.6367
	5	0004	1.0000	0092	0.9988	0467	0.9887	1239	0.9502	2188	0.8555
	6	0000	1.0000	0011	0.9999	0100	0.9987	0413	0.9915	1094	0.9648
	7	0000	1.0000	0001	1.0000	0012	0.9999	0079	0.9993	0313	0.9961
	8	0000	1.0000	0000	1.0000	0001	1.0000	0007	1.0000	0039	1.0000

Table A7. Poisson Distribution

Probability function $f(x)$ (cf. (5), Sec. 19.9) and distribution function $F(x)$

x	$\mu = 0.1$		$\mu = 0.2$		$\mu = 0.3$		$\mu = 0.4$		$\mu = 0.5$	
	$f(x)$	$F(x)$	$f(x)$	$F(x)$	$f(x)$	$F(x)$	$f(x)$	$F(x)$	$f(x)$	$F(x)$
0	0.9048	0.9048	0.8187	0.8187	0.7408	0.7408	0.6703	0.6703	0.6065	0.6065
1	0905	0.9953	1637	0.9825	2222	0.9631	2681	0.9384	3033	0.9098
2	0045	0.9998	0164	0.9989	0333	0.9964	0536	0.9921	0758	0.9856
3	0002	1.0000	0011	0.9999	0033	0.9997	0072	0.9992	0126	0.9982
4	0000	1.0000	0001	1.0000	0003	1.0000	0007	0.9999	0016	0.9998
5							0001	1.0000	0002	1.0000

x	$\mu = 0.6$		$\mu = 0.7$		$\mu = 0.8$		$\mu = 0.9$		$\mu = 1$	
	$f(x)$	$F(x)$	$f(x)$	$F(x)$	$f(x)$	$F(x)$	$f(x)$	$F(x)$	$f(x)$	$F(x)$
0	0.5488	0.5488	0.4966	0.4966	0.4493	0.4493	0.4066	0.4066	0.3679	0.3679
1	3293	0.8781	3476	0.8442	3595	0.8088	3659	0.7725	3679	0.7358
2	0988	0.9769	1217	0.9659	1438	0.9526	1647	0.9371	1839	0.9197
3	0198	0.9966	0284	0.9942	0383	0.9909	0494	0.9865	0613	0.9810
4	0030	0.9996	0050	0.9992	0077	0.9986	0111	0.9977	0153	0.9963
5	0004	1.0000	0007	0.9999	0012	0.9998	0020	0.9997	0031	0.9994
6			0001	1.0000	0002	1.0000	0003	1.0000	0005	0.9999
7									0001	1.0000

x	$\mu = 1.5$		$\mu = 2$		$\mu = 3$		$\mu = 4$		$\mu = 5$	
	$f(x)$	$F(x)$	$f(x)$	$F(x)$	$f(x)$	$F(x)$	$f(x)$	$F(x)$	$f(x)$	$F(x)$
0	0.2231	0.2231	0.1353	0.1353	0.0498	0.0498	0.0183	0.0183	0.0067	0.0067
1	3347	0.5578	2707	0.4060	1494	0.1991	0733	0.0916	0337	0.0404
2	2510	0.8088	2707	0.6767	2240	0.4232	1465	0.2381	0842	0.1247
3	1255	0.9344	1804	0.8571	2240	0.6472	1954	0.4335	1404	0.2650
4	0471	0.9814	0902	0.9473	1680	0.8153	1954	0.6288	1755	0.4405
5	0141	0.9955	0361	0.9834	1008	0.9161	1563	0.7851	1755	0.6160
6	0035	0.9991	0120	0.9955	0504	0.9665	1042	0.8893	1462	0.7622
7	0008	0.9998	0034	0.9989	0216	0.9881	0595	0.9489	1044	0.8666
8	0001	1.0000	0009	0.9998	0081	0.9962	0298	0.9786	0653	0.9319
9			0002	1.0000	0027	0.9989	0132	0.9919	0363	0.9682
10					0008	0.9997	0053	0.9972	0181	0.9863
11					0002	0.9999	0019	0.9991	0082	0.9945
12					0001	1.0000	0006	0.9997	0034	0.9980
13							0002	0.9999	0013	0.9993
14							0001	1.0000	0005	0.9998
15									0002	0.9999
16									0000	1.0000

Table A8. Normal Distribution

Values of the distribution function $\Phi(z)$ (cf. (4), Sec. 19.10)
$\Phi(-z) = 1 - \Phi(z)$, $\Phi(0) = 0.5000$

z	$\Phi(z)$	z	$\Phi(z)$	z	$\Phi(z)$	z	$\Phi(z)$	z	$\Phi(z)$	z	$\Phi(z)$
	0.		0.		0.		0.		0.		0.
0.01	5040	0.51	6950	1.01	8438	1.51	9345	2.01	9778	2.51	9940
0.02	5080	0.52	6985	1.02	8461	1.52	9357	2.02	9783	2.52	9941
0.03	5120	0.53	7019	1.03	8485	1.53	9370	2.03	9788	2.53	9943
0.04	5160	0.54	7054	1.04	8508	1.54	9382	2.04	9793	2.54	9945
0.05	5199	0.55	7088	1.05	8531	1.55	9394	2.05	9798	2.55	9946
0.06	5239	0.56	7123	1.06	8554	1.56	9406	2.06	9803	2.56	9948
0.07	5279	0.57	7157	1.07	8577	1.57	9418	2.07	9808	2.57	9949
0.08	5319	0.58	7190	1.08	8599	1.58	9429	2.08	9812	2.58	9951
0.09	5359	0.59	7224	1.09	8621	1.59	9441	2.09	9817	2.59	9952
0.10	5398	0.60	7257	1.10	8643	1.60	9452	2.10	9821	2.60	9953
0.11	5438	0.61	7291	1.11	8665	1.61	9463	2.11	9826	2.61	9955
0.12	5478	0.62	7324	1.12	8686	1.62	9474	2.12	9830	2.62	9956
0.13	5517	0.63	7357	1.13	8708	1.63	9484	2.13	9834	2.63	9957
0.14	5557	0.64	7389	1.14	8729	1.64	9495	2.14	9838	2.64	9959
0.15	5596	0.65	7422	1.15	8749	1.65	9505	2.15	9842	2.65	9960
0.16	5636	0.66	7454	1.16	8770	1.66	9515	2.16	9846	2.66	9961
0.17	5675	0.67	7486	1.17	8790	1.67	9525	2.17	9850	2.67	9962
0.18	5714	0.68	7517	1.18	8810	1.68	9535	2.18	9854	2.68	9963
0.19	5753	0.69	7549	1.19	8830	1.69	9545	2.19	9857	2.69	9964
0.20	5793	0.70	7580	1.20	8849	1.70	9554	2.20	9861	2.70	9965
0.21	5832	0.71	7611	1.21	8869	1.71	9564	2.21	9864	2.71	9966
0.22	5871	0.72	7642	1.22	8888	1.72	9573	2.22	9868	2.72	9967
0.23	5910	0.73	7673	1.23	8907	1.73	9582	2.23	9871	2.73	9968
0.24	5948	0.74	7704	1.24	8925	1.74	9591	2.24	9875	2.74	9969
0.25	5987	0.75	7734	1.25	8944	1.75	9599	2.25	9878	2.75	9970
0.26	6026	0.76	7764	1.26	8962	1.76	9608	2.26	9881	2.76	9971
0.27	6064	0.77	7794	1.27	8980	1.77	9616	2.27	9884	2.77	9972
0.28	6103	0.78	7823	1.28	8997	1.78	9625	2.28	9887	2.78	9973
0.29	6141	0.79	7852	1.29	9015	1.79	9633	2.29	9890	2.79	9974
0.30	6179	0.80	7881	1.30	9032	1.80	9641	2.30	9893	2.80	9974
0.31	6217	0.81	7910	1.31	9049	1.81	9649	2.31	9896	2.81	9975
0.32	6255	0.82	7939	1.32	9066	1.82	9656	2.32	9898	2.82	9976
0.33	6293	0.83	7967	1.33	9082	1.83	9664	2.33	9901	2.83	9977
0.34	6331	0.84	7995	1.34	9099	1.84	9671	2.34	9904	2.84	9977
0.35	6368	0.85	8023	1.35	9115	1.85	9678	2.35	9906	2.85	9978
0.36	6406	0.86	8051	1.36	9131	1.86	9686	2.36	9909	2.86	9979
0.37	6443	0.87	8078	1.37	9147	1.87	9693	2.37	9911	2.87	9979
0.38	6480	0.88	8106	1.38	9162	1.88	9699	2.38	9913	2.88	9980
0.39	6517	0.89	8133	1.39	9177	1.89	9706	2.39	9916	2.89	9981
0.40	6554	0.90	8159	1.40	9192	1.90	9713	2.40	9918	2.90	9981
0.41	6591	0.91	8186	1.41	9207	1.91	9719	2.41	9920	2.91	9982
0.42	6628	0.92	8212	1.42	9222	1.92	9726	2.42	9922	2.92	9982
0.43	6664	0.93	8238	1.43	9236	1.93	9732	2.43	9925	2.93	9983
0.44	6700	0.94	8264	1.44	9251	1.94	9738	2.44	9927	2.94	9984
0.45	6736	0.95	8289	1.45	9265	1.95	9744	2.45	9929	2.95	9984
0.46	6772	0.96	8315	1.46	9279	1.96	9750	2.46	9931	2.96	9985
0.47	6808	0.97	8340	1.47	9292	1.97	9756	2.47	9932	2.97	9985
0.48	6844	0.98	8365	1.48	9306	1.98	9761	2.48	9934	2.98	9986
0.49	6879	0.99	8389	1.49	9319	1.99	9767	2.49	9936	2.99	9986
0.50	6915	1.00	8413	1.50	9332	2.00	9772	2.50	9938	3.00	9987

Table A9. Normal Distribution

Values of z for given values of $\Phi(z)$ (cf. (4), Sec. 19.10) and $D(z) = \Phi(z) - \Phi(-z)$
Example: $z = 0.279$ when $\Phi(z) = 61\%$; $z = 0.860$ when $D(z) = 61\%$.

%	$z(\Phi)$	$z(D)$	%	$z(\Phi)$	$z(D)$	%	$z(\Phi)$	$z(D)$
1	−2.326	0.013	41	−0.228	0.539	81	0.878	1.311
2	−2.054	0.025	42	−0.202	0.553	82	0.915	1.341
3	−1.881	0.038	43	−0.176	0.568	83	0.954	1.372
4	−1.751	0.050	44	−0.151	0.583	84	0.994	1.405
5	−1.645	0.063	45	−0.126	0.598	85	1.036	1.440
6	−1.555	0.075	46	−0.100	0.613	86	1.080	1.476
7	−1.476	0.088	47	−0.075	0.628	87	1.126	1.514
8	−1.405	0.100	48	−0.050	0.643	88	1.175	1.555
9	−1.341	0.113	49	−0.025	0.659	89	1.227	1.598
10	−1.282	0.126	50	0.000	0.674	90	1.282	1.645
11	−1.227	0.138	51	0.025	0.690	91	1.341	1.695
12	−1.175	0.151	52	0.050	0.706	92	1.405	1.751
13	−1.126	0.164	53	0.075	0.722	93	1.476	1.812
14	−1.080	0.176	54	0.100	0.739	94	1.555	1.881
15	−1.036	0.189	55	0.126	0.755	95	1.645	1.960
16	−0.994	0.202	56	0.151	0.772	96	1.751	2.054
17	−0.954	0.215	57	0.176	0.789	97	1.881	2.170
18	−0.915	0.228	58	0.202	0.806	98	2.054	2.326
19	−0.878	0.240	59	0.228	0.824	99	2.326	2.576
20	−0.842	0.253	60	0.253	0.842			
21	−0.806	0.266	61	0.279	0.860	99.1	2.366	2.612
22	−0.772	0.279	62	0.305	0.878	99.2	2.409	2.652
23	−0.739	0.292	63	0.332	0.896	99.3	2.457	2.697
24	−0.706	0.305	64	0.358	0.915	99.4	2.512	2.748
25	−0.674	0.319	65	0.385	0.935	99.5	2.576	2.807
26	−0.643	0.332	66	0.412	0.954	99.6	2.652	2.878
27	−0.613	0.345	67	0.440	0.974	99.7	2.748	2.968
28	−0.583	0.358	68	0.468	0.994	99.8	2.878	3.090
29	−0.553	0.372	69	0.496	1.015	99.9	3.090	3.291
30	−0.524	0.385	70	0.524	1.036			
31	−0.496	0.399	71	0.553	1.058	99.91	3.121	3.320
32	−0.468	0.412	72	0.583	1.080	99.92	3.156	3.353
33	−0.440	0.426	73	0.613	1.103	99.93	3.195	3.390
34	−0.412	0.440	74	0.643	1.126	99.94	3.239	3.432
35	−0.385	0.454	75	0.674	1.150	99.95	3.291	3.481
36	−0.358	0.468	76	0.706	1.175	99.96	3.353	3.540
37	−0.332	0.482	77	0.739	1.200	99.97	3.432	3.615
38	−0.305	0.496	78	0.772	1.227	99.98	3.540	3.719
39	−0.279	0.510	79	0.806	1.254	99.99	3.719	3.891
40	−0.253	0.524	80	0.842	1.282			

Table A10. Random Digits

Cf. Sec. 19.12

Row No.	0	1	2	3	Column Number 4	5	6	7	8	9
0	87331	82442	28104	26432	83640	17323	68764	84728	·37995	96106
1	33628	17364	01409	87803	65641	33433	48944	64299	79066	31777
2	54680	13427	72496	16967	16195	96593	55040	53729	62035	66717
3	51199	49794	49407	10774	98140	83891	37195	24066	61140	65144
4	78702	98067	61313	91661	59861	54437	77739	19892	54817	88645
5	55672	16014	24892	13089	00410	81458	76156	28189	40595	21500
6	18880	58497	03862	32368	59320	24807	63392	79793	63043	09425
7	10242	62548	62330	05703	33535	49128	66298	16193	55301	01306
8	54993	17182	94618	23228	83895	73251	68199	64639	83178	70521
9	22686	50885	16006	04041	08077	33065	35237	02502	94755	72062
10	42349	03145	15770	70665	53291	32288	41568	66079	98705	31029
11	18093	09553	39428	75464	71329	86344	80729	40916	18860	51780
12	11535	03924	84252	74795	40193	84597	42497	21918	91384	84721
13	35066	73848	65351	53270	67341	70177	92373	17604	42204	60476
14	57477	22809	73558	96182	96779	01604	25748	59553	64876	94611
15	48647	33850	52956	454 0	88212	05120	99391	32276	55961	41775
16	86857	81154	22223	74950	53296	67767	55866	49061	66937	81818
17	20182	36907	94644	99122	09774	29189	27212	79000	50217	71077
18	83687	31231	01133	41432	54542	60204	81618	09586	34481	87683
19	81315	12390	46074	47810	90171	36313	95440	77583	28506	38808
20	87026	52826	58341	76549	04105	66191	12914	55348	07907	06978
21	34301	76733	07251	90524	21931	83695	41340	53581	64582	60210
22	70734	24337	32674	49508	49751	90489	63202	24380	77943	09942
23	94710	31527	73445	32839	68176	53580	85250	53243	03350	00128
24	76462	16987	07775	43162	11777	16810	75158	13894	88945	15539
25	14348	28403	79245	69023	34196	46398	05964	64715	11330	17515
26	74618	89317	30146	25606	94507	98104	04239	44973	37636	88866
27	99442	19200	85406	45358	86253	60638	38858	44964	54103	57287
28	26869	44399	89452	06652	31271	00647	46551	83050	92058	83814
29	80988	08149	50499	98584	28385	63680	44638	91864	96002	87802
30	07511	79047	89289	17774	67194	37362	85684	55505	97809	67056
31	49779	12138	05048	03535	27502	63308	10218	53296	48687	61340
32	47938	55945	24003	19635	17471	65997	85906	98694	56420	78357
33	15604	06626	14360	79542	13512	87595	08542	03800	35443	52823
34	12307	27726	21864	00045	16075	03770	86978	52718	02693	09096
35	02450	28053	66134	99445	91316	25727	89399	85272	67148	78358
36	57623	54382	35236	89244	27245	90500	75430	96762	71968	65838
37	91762	78849	93105	40481	99431	03304	21079	86459	21287	76566
38	87373	31137	31128	67050	34309	44914	80711	61738	61498	24288
39	67094	41485	54149	86088	10192	21174	39948	67268	29938	32476
40	94456	66747	76922	87627	71834	57688	04878	78348	68970	60048
41	68359	75292	27710	86889	81678	79798	58360	39175	75667	65782
42	52393	31404	32584	06837	79762	13168	76055	54833	22841	98889
43	59565	91254	11847	20672	37625	41454	86861	55824	79793	74575
44	48185	11066	20162	38230	16043	48409	47421	21195	98008	57305
45	19230	12187	86659	12971	52204	76546	63272	19312	81662	96557
46	84327	21942	81727	68735	89190	58491	55329	96875	19465	89687
47	77430	71210	00591	50124	12030	50280	12358	76174	48353	09682
48	12462	19108	70512	53926	25595	97085	03833	59806	12351	64253
49	11684	06644	57816	10078	45021	47751	38285	73520	08434	65627

Table A10. Random Digits (*continued*)

Row No.	0	1	2	3	4	5	6	7	8	9
					Column Number					
50	12896	36576	68686	08462	65652	76571	70891	09007	04581	01684
51	59090	05111	27587	90349	30789	50304	70650	06646	70126	15284
52	42486	67483	65282	19037	80588	73076	41820	46651	40442	40718
53	88662	03928	03249	85910	97533	88643	29829	21557	47328	36724
54	69403	03626	92678	53460	15465	83516	54012	80509	55976	46115
55	56434	70543	38696	98502	32092	95505	62091	39549	30117	98209
56	58227	62694	42837	29183	11393	68463	25150	86338	95620	39836
57	41272	94927	15413	40505	33123	63218	72940	98349	57249	40170
58	36819	01162	30425	15546	16065	68459	35776	64276	92868	07372
59	31700	66711	26115	55755	33584	18091	38709	57276	74660	90392
60	69855	63699	36839	90531	97125	87875	62824	03889	12538	24740
61	44322	17569	45439	41455	34324	90902	07978	26268	04279	76816
62	62226	36661	87011	66267	78777	78044	40819	49496	39814	73867
63	27284	19737	98741	72531	52741	26699	98755	19657	08665	16818
64	88341	21652	94743	77268	79525	44769	66583	30621	90534	62050
65	53266	18783	51903	56711	38060	69513	61963	˜80470	88018	86510
66	50527	49330	24839	42529	03944	95219	88724	37247	84166	23023
67	15655	07852	77206	35944	71446	30573	19405	57824	23576	23301
68	62057	22206	03314	83465	57466	10465	19891	32308	01900	67484
69	41769	56091	19892	96253	92808	45785	52774	49674	68103	65032
70	25993	72416	44473	41299	93095	17338	69802	98548	02429	85238
71	22842	57871	04470	37373	34516	04042	04078	35336	34393	97573
72	55704	31982	05234	22664	22181	40358	28089	15790	33340	18852
73	94258	18706	09437	96041	90052	80862	20420	24323	11635	91677
74	74145	20453	29657	98868	56695	53483	87449	35060	98942	62697
75	88881	12673	73961	89884	73247	97670	69570	88888	58560	72580
76	01508	56780	52223	35632	73347	71317	46541	88023	36656	76332
77	92069	43000	23233	06058	82527	25250	27555	20426	60361	63525
78	53366	35249	02117	68620	39388	69795	73215	01846	16983	78560
79	88057	54097	49511	74867	32192	90071	04147	46094	63519	07199
80	85492	82238	02668	91854	86149	28590	77853	81035	45561	16032
81	39453	62123	69611	53017	34964	09786	24614	49514	01056	18700
82	82627	98111	93870	56969	69566	62662	07353	84838	14570	14508
83	61142	51743	38209	31474	96095	15163	54380	77849	20465	03142
84	12031	32528	61311	53730	89032	16124	58844	35386	45521	59368
85	31313	59838	29147	76882	74328	09955	63673	96651	53264	29871
86	50767	41056	97409	44376	62219	35439	70102	99248	71179	26052
87	30522	95699	84966	26554	24768	72247	84993	85375	92518	16334
88	74176	19870	89874	64799	03792	57006	57225	36677	46825	14087
89	17114	93248	37065	91346	04657	93763	92210	43676	44944	75798
90	53005	11825	64608	87587	05742	31914	55044	41818	29667	77424
91	31985	81539	79942	49471	46200	27639	94099	42085	79231	03932
92	63499	60508	77522	15624	15088	78519	52279	79214	43623	69166
93	30506	42444	99047	66010	91657	37160	37408	85714	21420	80996
94	78248	16841	92357	10130	68990	38307	61022	56806	81016	38511
95	64996	84789	50185	32200	64382	29752	11876	00664	54547	62597
96	11963	13157	09136	01769	30117	71486	80111	09161	08371	71749
97	44335	91450	43456	90449	18338	19787	31339	60473	06606	89788
98	42277	11868	44520	01113	11341	11743	97949	49718	99176	42006
99	77562	18863	58515	90166	78508	14864	19111	57183	85808	59385

Table A11. *t*-Distribution

Values of z for given values of the distribution function $F(z)$ (cf. p. 754).
Example: For 9 degrees of freedom, $z = 1.83$ when $F(z) = 0.95$.

	Number of Degrees of Freedom									
$F(z)$	1	2	3	4	5	6	7	8	9	10
0.5	0.00	0.00	0.00	0.00	0.00	0.00	0.00	0.00	0.00	0.00
0.6	0.33	0.29	0.28	0.27	0.27	0.27	0.26	0.26	0.26	0.26
0.7	0.73	0.62	0.58	0.57	0.56	0.55	0.55	0.55	0.54	0.54
0.8	1.38	1.06	0.98	0.94	0.92	0.91	0.90	0.89	0.88	0.88
0.9	3.08	1.89	1.64	1.53	1.48	1.44	1.42	1.40	1.38	1.37
0.95	6.31	2.92	2.35	2.13	2.02	1.94	1.90	1.86	1.83	1.81
0.975	12.7	4.30	3.18	2.78	2.57	2.45	2.37	2.31	2.26	2.23
0.99	31.8	6.97	4.54	3.75	3.37	3.14	3.00	2.90	2.82	2.76
0.995	63.7	9.93	5.84	4.60	4.03	3.71	3.50	3.36	3.25	3.17
0.999	318.3	22.3	10.2	7.17	5.89	5.21	4.79	4.50	4.30	4.14

	Number of Degrees of Freedom									
$F(z)$	11	12	13	14	15	16	17	18	19	20
0.5	0.00	0.00	0.00	0.00	0.00	0.00	0.00	0.00	0.00	0.00
0.6	0.26	0.26	0.26	0.26	0.26	0.26	0.26	0.26	0.26	0.26
0.7	0.54	0.54	0.54	0.54	0.54	0.54	0.53	0.53	0.53	0.53
0.8	0.88	0.87	0.87	0.87	0.87	0.87	0.86	0.86	0.86	0.86
0.9	1.36	1.36	1.35	1.35	1.34	1.34	1.33	1.33	1.33	1.33
0.95	1.80	1.78	1.77	1.76	1.75	1.75	1.74	1.73	1.73	1.73
0.975	2.20	2.18	2.16	2.15	2.13	2.12	2.11	2.10	2.09	2.09
0.99	2.72	2.68	2.65	2.62	2.60	2.58	2.57	2.55	2.54	2.53
0.995	3.11	3.06	3.01	2.98	2.95	2.92	2.90	2.88	2.86	2.85
0.999	4.03	3.93	3.85	3.79	3.73	3.69	3.65	3.61	3.58	3.55

	Number of Degrees of Freedom									
$F(z)$	22	24	26	28	30	40	50	100	200	∞
0.5	0.00	0.00	0.00	0.00	0.00	0.00	0.00	0.00	0.00	0.00
0.6	0.26	0.26	0.26	0.26	0.26	0.26	0.26	0.25	0.25	0.25
0.7	0.53	0.53	0.53	0.53	0.53	0.53	0.53	0.53	0.53	0.52
0.8	0.86	0.86	0.86	0.86	0.85	0.85	0.85	0.85	0.84	0.84
0.9	1.32	1.32	1.32	1.31	1.31	1.30	1.30	1.29	1.29	1.28
0.95	1.72	1.71	1.71	1.70	1.70	1.68	1.68	1.66	1.65	1.65
0.975	2.07	2.06	2.06	2.05	2.04	2.02	2.01	1.98	1.97	1.96
0.99	2.51	2.49	2.48	2.47	2.46	2.42	2.40	2.37	2.35	2.33
0.995	2.82	2.80	2.78	2.76	2.75	2.70	2.68	2.63	2.60	2.58
0.999	3.51	3.47	3.44	3.41	3.39	3.31	3.26	3.17	3.13	3.09

Table A12. Chi-square Distribution

Values of z for given values of the distribution function $F(z)$ (cf. p. 756).
Example: For 3 degrees of freedom, $z = 11.34$ when $F(z) = 0.99$.

$F(z)$	Number of Degrees of Freedom									
	1	2	3	4	5	6	7	8	9	10
0.005	0.00	0.01	0.07	0.21	0.41	0.68	0.99	1.34	1.73	2.16
0.01	0.00	0.02	0.11	0.30	0.55	0.87	1.24	1.65	2.09	2.56
0.025	0.00	0.05	0.22	0.48	0.83	1.24	1.69	2.18	2.70	3.25
0.05	0.00	0.10	0.35	0.71	1.15	1.64	2.17	2.73	3.33	3.94
0.95	3.84	5.99	7.81	9.49	11.07	12.59	14.07	15.51	16.92	18.31
0.975	5.02	7.38	9.35	11.14	12.83	14.45	16.01	17.53	19.02	20.48
0.99	6.63	9.21	11.34	13.28	15.09	16.81	18.48	20.09	21.67	23.21
0.995	7.88	10.60	12.84	14.86	16.75	18.55	20.28	21.96	23.59	25.19

$F(z)$	Number of Degrees of Freedom									
	11	12	13	14	15	16	17	18	19	20
0.005	2.60	3.07	3.57	4.07	4.60	5.14	5.70	6.26	6.84	7.43
0.01	3.05	3.57	4.11	4.66	5.23	5.81	6.41	7.01	7.63	8.26
0.025	3.82	4.40	5.01	5.63	6.26	6.91	7.56	8.23	8.91	9.59
0.05	4.57	5.23	5.89	6.57	7.26	7.96	8.67	9.39	10.12	10.85
0.95	19.68	21.03	22.36	23.68	25.00	26.30	27.59	28.87	30.14	31.41
0.975	21.92	23.34	24.74	26.12	27.49	28.85	30.19	31.53	32.85	34.17
0.99	24.73	26.22	27.69	29.14	30.58	32.00	33.41	34.81	36.19	37.57
0.995	26.76	28.30	29.82	31.32	32.80	34.27	35.72	37.16	38.58	40.00

$F(z)$	Number of Degrees of Freedom									
	21	22	23	24	25	26	27	28	29	30
0.005	8.0	8.6	9.3	9.9	10.5	11.2	11.8	12.5	13.1	13.8
0.01	8.9	9.5	10.2	10.9	11.5	12.2	12.9	13.6	14.3	15.0
0.025	10.3	11.0	11.7	12.4	13.1	13.8	14.6	15.3	16.0	16.8
0.05	11.6	12.3	13.1	13.8	14.6	15.4	16.2	16.9	17.7	18.5
0.95	32.7	33.9	35.2	36.4	37.7	38.9	40.1	41.3	42.6	43.8
0.975	35.5	36.8	38.1	39.4	40.6	41.9	43.2	44.5	45.7	47.0
0.99	38.9	40.3	41.6	43.0	44.3	45.6	47.0	48.3	49.6	50.9
0.995	41.4	42.8	44.2	45.6	46.9	48.3	49.6	51.0	52.3	53.7

$F(z)$	Number of Degrees of Freedom								
	40	50	60	70	80	90	100	> 100 (Approximation)	
0.005	20.7	28.0	35.5	43.3	51.2	59.2	67.3	$\frac{1}{2}(h - 2.58)^2$	
0.01	22.2	29.7	37.5	45.4	53.5	61.8	70.1	$\frac{1}{2}(h - 2.33)^2$	
0.025	24.4	32.4	40.5	48.8	57.2	65.6	74.2	$\frac{1}{2}(h - 1.96)^2$	
0.05	26.5	34.8	43.2	51.7	60.4	69.1	77.9	$\frac{1}{2}(h - 1.64)^2$	
0.95	55.8	67.5	79.1	90.5	101.9	113.1	124.3	$\frac{1}{2}(h + 1.64)^2$	
0.975	59.3	71.4	83.3	95.0	106.6	118.1	129.6	$\frac{1}{2}(h + 1.96)^2$	
0.99	63.7	76.2	88.4	100.4	112.3	124.1	135.8	$\frac{1}{2}(h + 2.33)^2$	
0.995	66.8	79.5	92.0	104.2	116.3	128.3	140.2	$\frac{1}{2}(h + 2.58)^2$	

In the last column, $h = \sqrt{2m - 1}$ where m is the number of degrees of freedom.

Table A13. F-Distribution with (m, n) Degrees of Freedom

Values of z for which the distribution function $F(z)$ (cf. (13), Sec. 19.15) has the value **0.95**.
Example: For $(7, 4)$ degrees of freedom, $z = 6.09$ when $F(z) = 0.95$.

n	$m = 1$	$m = 2$	$m = 3$	$m = 4$	$m = 5$	$m = 6$	$m = 7$	$m = 8$	$m = 9$
1	161	200	216	225	230	234	237	239	241
2	18.5	19.0	19.2	19.2	19.3	19.3	19.4	19.4	19.4
3	10.1	9.55	9.28	9.12	9.01	8.94	8.89	8.85	8.81
4	7.71	6.94	6.59	6.39	6.26	6.16	6.09	6.04	6.00
5	6.61	5.79	5.41	5.19	5.05	4.95	4.88	4.82	4.77
6	5.99	5.14	4.76	4.53	4.39	4.28	4.21	4.15	4.10
7	5.59	4.74	4.35	4.12	3.97	3.87	3.79	3.73	3.68
8	5.32	4.46	4.07	3.84	3.69	3.58	3.50	3.44	3.39
9	5.12	4.26	3.86	3.63	3.48	3.37	3.29	3.23	3.18
10	4.96	4.10	3.71	3.48	3.33	3.22	3.14	3.07	3.02
11	4.84	3.98	3.59	3.36	3.20	3.09	3.01	2.95	2.90
12	4.75	3.89	3.49	3.26	3.11	3.00	2.91	2.85	2.80
13	4.67	3.81	3.41	3.18	3.03	2.92	2.83	2.77	2.71
14	4.60	3.74	3.34	3.11	2.96	2.85	2.76	2.70	2.65
15	4.54	3.68	3.29	3.06	2.90	2.79	2.71	2.64	2.59
16	4.49	3.63	3.24	3.01	2.85	2.74	2.66	2.59	2.54
17	4.45	3.59	3.20	2.96	2.81	2.70	2.61	2.55	2.49
18	4.41	3.55	3.16	2.93	2.77	2.66	2.58	2.51	2.46
19	4.38	3.52	3.13	2.90	2.74	2.63	2.54	2.48	2.42
20	4.35	3.49	3.10	2.87	2.71	2.60	2.51	2.45	2.39
22	4.30	3.44	3.05	2.82	2.66	2.55	2.46	2.40	2.34
24	4.26	3.40	3.01	2.78	2.62	2.51	2.42	2.36	2.30
26	4.23	3.37	2.98	2.74	2.59	2.47	2.39	2.32	2.27
28	4.20	3.34	2.95	2.71	2.56	2.45	2.36	2.29	2.24
30	4.17	3.32	2.92	2.69	2.53	2.42	2.33	2.27	2.21
32	4.15	3.30	2.90	2.67	2.51	2.40	2.31	2.24	2.19
34	4.13	3.28	2.88	2.65	2.49	2.38	2.29	2.23	2.17
36	4.11	3.26	2.87	2.63	2.48	2.36	2.28	2.21	2.15
38	4.10	3.24	2.85	2.62	2.46	2.35	2.26	2.19	2.14
40	4.08	3.23	2.84	2.61	2.45	2.34	2.25	2.18	2.12
50	4.03	3.18	2.79	2.56	2.40	2.29	2.20	2.13	2.07
60	4.00	3.15	2.76	2.53	2.37	2.25	2.17	2.10	2.04
70	3.98	3.13	2.74	2.50	2.35	2.23	2.14	2.07	2.02
80	3.96	3.11	2.72	2.49	2.33	2.21	2.13	2.06	2.00
90	3.95	3.10	2.71	2.47	2.32	2.20	2.11	2.04	1.99
100	3.94	3.09	2.70	2.46	2.31	2.19	2.10	2.03	1.97
150	3.90	3.06	2.66	2.43	2.27	2.16	2.07	2.00	1.94
200	3.89	3.04	2.65	2.42	2.26	2.14	2.06	1.98	1.93
1000	3.85	3.00	2.61	2.38	2.22	2.11	2.02	1.95	1.89
∞	3.84	3.00	2.60	2.37	2.21	2.10	2.01	1.94	1.88

Table A13. F-Distribution with (m, n) Degrees of Freedom (*continued*)

Values of z for which the distribution function $F(z)$ (cf. (13), Sec. 19.15) has the value **0.95**.

n	$m = 10$	$m = 15$	$m = 20$	$m = 30$	$m = 40$	$m = 50$	$m = 100$	∞
1	242	246	248	250	251	252	253	254
2	19.4	19.4	19.4	19.5	19.5	19.5	19.5	19.5
3	8.79	8.70	8.66	8.62	8.59	8.58	8.55	8.53
4	5.96	5.86	5.80	5.75	5.72	5.70	5.66	5.63
5	4.74	4.62	4.56	4.50	4.46	4.44	4.41	4.37
6	4.06	3.94	3.87	3.81	3.77	3.75	3.71	3.67
7	3.64	3.51	3.44	3.38	3.34	3.32	3.27	3.23
8	3.35	3.22	3.15	3.08	3.04	3.02	2.97	2.93
9	3.14	3.01	2.94	2.86	2.83	2.80	2.76	2.71
10	2.98	2.85	2.77	2.70	2.66	2.64	2.59	2.54
11	2.85	2.72	2.65	2.57	2.53	2.51	2.46	2.40
12	2.75	2.62	2.54	2.47	2.43	2.40	2.35	2.30
13	2.67	2.53	2.46	2.38	2.34	2.31	2.26	2.21
14	2.60	2.46	2.39	2.31	2.27	2.24	2.19	2.13
15	2.54	2.40	2.33	2.25	2.20	2.18	2.12	2.07
16	2.49	2.35	2.28	2.19	2.15	2.12	2.07	2.01
17	2.45	2.31	2.23	2.15	2.10	2.08	2.02	1.96
18	2.41	2.27	2.19	2.11	2.06	2.04	1.98	1.92
19	2.38	2.23	2.16	2.07	2.03	2.00	1.94	1.88
20	2.35	2.20	2.12	2.04	1.99	1.97	1.91	1.84
22	2.30	2.15	2.07	1.98	1.94	1.91	1.85	1.78
24	2.25	2.11	2.03	1.94	1.89	1.86	1.80	1.73
26	2.22	2.07	1.99	1.90	1.85	1.82	1.76	1.69
28	2.19	2.04	1.96	1.87	1.82	1.79	1.73	1.65
30	2.16	2.01	1.93	1.84	1.79	1.76	1.70	1.62
32	2.14	1.99	1.91	1.82	1.77	1.74	1.67	1.59
34	2.12	1.97	1.89	1.80	1.75	1.71	1.65	1.57
36	2.11	1.95	1.87	1.78	1.73	1.69	1.62	1.55
38	2.09	1.94	1.85	1.76	1.71	1.68	1.61	1.53
40	2.08	1.92	1.84	1.74	1.69	1.66	1.59	1.51
50	2.03	1.87	1.78	1.69	1.63	1.60	1.52	1.44
60	1.99	1.84	1.75	1.65	1.59	1.56	1.48	1.39
70	1.97	1.81	1.72	1.62	1.57	1.53	1.45	1.35
80	1.95	1.79	1.70	1.60	1.54	1.51	1.43	1.32
90	1.94	1.78	1.69	1.59	1.53	1.49	1.41	1.30
100	1.93	1.77	1.68	1.57	1.52	1.48	1.39	1.28
150	1.89	1.73	1.64	1.53	1.48	1.44	1.34	1.22
200	1.88	1.72	1.62	1.52	1.46	1.41	1.32	1.19
1000	1.84	1.68	1.58	1.47	1.41	1.36	1.26	1.08
∞	1.83	1.67	1.57	1.46	1.39	1.35	1.24	1.00

Table A13. F-Distribution with (m, n) Degrees of Freedom (*continued*)

Values of z for which the distribution function $F(z)$ (cf. (13), Sec. 19.15) has the value **0.99**.

n	$m = 1$	$m = 2$	$m = 3$	$m = 4$	$m = 5$	$m = 6$	$m = 7$	$m = 8$	$m = 9$
1	4052	4999	5403	5625	5764	5859	5928	5982	6022
2	98.5	99.0	99.2	99.3	99.3	99.3	99.4	99.4	99.4
3	34.1	30.8	29.5	28.7	28.2	27.9	27.7	27.5	27.3
4	21.2	18.0	16.7	16.0	15.5	15.2	15.0	14.8	14.7
5	16.3	13.3	12.1	11.4	11.0	10.7	10.5	10.3	10.2
6	13.7	10.9	9.78	9.15	8.75	8.47	8.26	8.10	7.98
7	12.2	9.55	8.45	7.85	7.46	7.19	6.99	6.84	6.72
8	11.3	8.65	7.59	7.01	6.63	6.37	6.18	6.03	5.91
9	10.6	8.02	6.99	6.42	6.06	5.80	5.61	5.47	5.35
10	10.0	7.56	6.55	5.99	5.64	5.39	5.20	5.06	4.94
11	9.65	7.21	6.22	5.67	5.32	5.07	4.89	4.74	4.63
12	9.33	6.93	5.95	5.41	5.06	4.82	4.64	4.50	4.39
13	9.07	6.70	5.74	5.21	4.86	4.62	4.44	4.30	4.19
14	8.86	6.51	5.56	5.04	4.70	4.46	4.28	4.14	4.03
15	8.68	6.36	5.42	4.89	4.56	4.32	4.14	4.00	3.89
16	8.53	6.23	5.29	4.77	4.44	4.20	4.03	3.89	3.78
17	8.40	6.11	5.18	4.67	4.34	4.10	3.93	3.79	3.68
18	8.29	6.01	5.09	4.58	4.25	4.01	3.84	3.71	3.60
19	8.18	5.93	5.01	4.50	4.17	3.94	3.77	3.63	3.52
20	8.10	5.85	4.94	4.43	4.10	3.87	3.70	3.56	3.46
22	7.95	5.72	4.82	4.31	3.99	3.76	3.59	3.45	3.35
24	7.82	5.61	4.72	4.22	3.90	3.67	3.50	3.36	3.26
26	7.72	5.53	4.64	4.14	3.82	3.59	3.42	3.29	3.18
28	7.64	5.45	4.57	4.07	3.75	3.53	3.36	3.23	3.12
30	7.56	5.39	4.51	4.02	3.70	3.47	3.30	3.17	3.07
32	7.50	5.34	4.46	3.97	3.65	3.43	3.26	3.13	3.02
34	7.44	5.29	4.42	3.93	3.61	3.39	3.22	3.09	2.98
36	7.40	5.25	4.38	3.89	3.57	3.35	3.18	3.05	2.95
38	7.35	5.21	4.34	3.86	3.54	3.32	3.15	3.02	2.92
40	7.31	5.18	4.31	3.83	3.51	3.29	3.12	2.99	2.89
50	7.17	5.06	4.20	3.72	3.41	3.19	3.02	2.89	2.79
60	7.08	4.98	4.13	3.65	3.34	3.12	2.95	2.82	2.72
70	7.01	4.92	4.08	3.60	3.29	3.07	2.91	2.78	2.67
80	6.96	4.88	4.04	3.56	3.26	3.04	2.87	2.74	2.64
90	6.93	4.85	4.01	3.54	3.23	3.01	2.84	2.72	2.61
100	6.90	4.82	3.98	3.51	3.21	2.99	2.82	2.69	2.59
150	6.81	4.75	3.92	3.45	3.14	2.92	2.76	2.63	2.53
200	6.76	4.71	3.88	3.41	3.11	2.89	2.73	2.60	2.50
1000	6.66	4.63	3.80	3.34	3.04	2.82	2.66	2.53	2.43
∞	6.63	4.61	3.78	3.32	3.02	2.80	2.64	2.51	2.41

Table A13. F-Distribution with (m, n) **Degrees of Freedom** (*continued*)

Values of z for which the distribution function $F(z)$ (cf. (13), Sec. 19.15) has the value **0.99**.

n	$m = 10$	$m = 15$	$m = 20$	$m = 30$	$m = 40$	$m = 50$	$m = 100$	∞
1	6056	6157	6209	6261	6287	6300	6330	6366
2	99.4	99.4	99.4	99.5	99.5	99.5	99.5	99.5
3	27.2	26.9	26.7	26.5	26.4	26.4	26.2	26.1
4	14.5	14.2	14.0	13.8	13.7	13.7	13.6	13.5
5	10.1	9.72	9.55	9.38	9.29	9.24	9.13	9.02
6	7.87	7.56	7.40	7.23	7.14	7.09	6.99	6.88
7	6.62	6.31	6.16	5.99	5.91	5.86	5.75	5.65
8	5.81	5.52	5.36	5.20	5.12	5.07	4.96	4.86
9	5.26	4.96	4.81	4.65	4.57	4.52	4.42	4.31
10	4.85	4.56	4.41	4.25	4.17	4.12	4.01	3.91
11	4.54	4.25	4.10	3.94	3.86	3.81	3.71	3.60
12	4.30	4.01	3.86	3.70	3.62	3.57	3.47	3.36
13	4.10	3.82	3.66	3.51	3.43	3.38	3.27	3.17
14	3.94	3.66	3.51	3.35	3.27	3.22	3.11	3.00
15	3.80	3.52	3.37	3.21	3.13	3.08	2.98	2.87
16	3.69	3.41	3.26	3.10	3.02	2.97	2.86	2.75
17	3.59	3.31	3.16	3.00	2.92	2.87	2.76	2.65
18	3.51	3.23	3.08	2.92	2.84	2.78	2.68	2.57
19	3.43	3.15	3.00	2.84	2.76	2.71	2.60	2.49
20	3.37	3.09	2.94	2.78	2.69	2.64	2.54	2.42
22	3.26	2.98	2.83	2.67	2.58	2.53	2.42	2.31
24	3.17	2.89	2.74	2.58	2.49	2.44	2.33	2.21
26	3.09	2.82	2.66	2.50	2.42	2.36	2.25	2.13
28	3.03	2.75	2.60	2.44	2.35	2.30	2.19	2.06
30	2.98	2.70	2.55	2.39	2.30	2.25	2.13	2.01
32	2.93	2.66	2.50	2.34	2.25	2.20	2.08	1.96
34	2.89	2.62	2.46	2.30	2.21	2.16	2.04	1.91
36	2.86	2.58	2.43	2.26	2.17	2.12	2.00	1.87
38	2.83	2.55	2.40	2.23	2.14	2.09	1.97	1.84
40	2.80	2.52	2.37	2.20	2.11	2.06	1.94	1.80
50	2.70	2.42	2.27	2.10	2.01	1.95	1.82	1.68
60	2.63	2.35	2.20	2.03	1.94	1.88	1.75	1.60
70	2.59	2.31	2.15	1.98	1.89	1.83	1.70	1.54
80	2.55	2.27	2.12	1.94	1.85	1.79	1.66	1.49
90	2.52	2.24	2.09	1.92	1.82	1.76	1.62	1.46
100	2.50	2.22	2.07	1.89	1.80	1.73	1.60	1.43
150	2.44	2.16	2.00	1.83	1.73	1.66	1.52	1.33
200	2.41	2.13	1.97	1.79	1.69	1.63	1.48	1.28
1000	2.34	2.06	1.90	1.72	1.61	1.54	1.38	1.11
∞	2.32	2.04	1.88	1.70	1.59	1.52	1.36	1.00

Table A14. Distribution Function $F(x) = P(T \leq x)$ of the Random Variable T in Section 19.19

When $n = 3$, $F(2) = 1 - 0.167 = 0.833$.

When $n = 4$, $F(3) = 1 - 0.375 = 0.625$, $F(4) = 1 - 0.167 = 0.833$, etc.

x	$n=3$ 0.
0	167
1	500

x	$n=4$ 0.
0	042
1	167
2	375

x	$n=5$ 0.
0	008
1	042
2	117
3	242
4	408

x	$n=6$ 0.
0	001
1	008
2	028
3	068
4	136
5	235
6	360
7	500

x	$n=7$ 0.
1	001
2	005
3	015
4	035
5	068
6	119
7	191
8	281
9	386
10	500

x	$n=8$ 0.
2	001
3	003
4	007
5	016
6	031
7	054
8	089
9	138
10	199
11	274
12	360
13	452

x	$n=9$ 0.
4	001
5	003
6	006
7	012
8	022
9	038
10	060
11	090
12	130
13	179
14	238
15	306
16	381
17	460

x	$n=10$ 0.
6	001
7	002
8	005
9	008
10	014
11	023
12	036
13	054
14	078
15	108
16	146
17	190
18	242
19	300
20	364
21	431
22	500

x	$n=11$ 0.
8	001
9	002
10	003
11	005
12	008
13	013
14	020
15	030
16	043
17	060
18	082
19	109
20	141
21	179
22	223
23	271
24	324
25	381
26	440
27	500

x	$n=20$ 0.
50	001
51	002
52	002
53	003
54	004
55	005
56	006
57	007
58	008
59	010
60	012
61	014
62	017
63	020
64	023
65	027
66	032
67	037
68	043
69	049
70	056
71	064
72	073
73	082
74	093
75	104
76	117
77	130
78	144
79	159
80	176
81	193
82	211
83	230
84	250
85	271
86	293
87	315
88	339
89	362
90	387
91	411
92	436
93	462
94	487

x	$n=19$ 0.
43	001
44	002
45	002
46	003
47	003
48	004
49	005
50	006
51	008
52	010
53	012
54	014
55	017
56	021
57	025
58	029
59	034
60	040
61	047
62	054
63	062
64	072
65	082
66	093
67	105
68	119
69	133
70	149
71	166
72	184
73	203
74	223
75	245
76	267
77	290
78	314
79	339
80	365
81	391
82	418
83	445
84	473
85	500

x	$n=18$ 0.
38	001
39	002
40	003
41	003
42	004
43	005
44	007
45	009
46	011
47	013
48	016
49	020
50	024
51	029
52	034
53	041
54	048
55	056
56	066
57	076
58	088
59	100
60	115
61	130
62	147
63	165
64	184
65	205
66	227
67	250
68	275
69	300
70	327
71	354
72	383
73	411
74	441
75	470
76	500

x	$n=17$ 0.
32	001
33	002
34	002
35	003
36	004
37	005
38	007
39	009
40	011
41	014
42	017
43	021
44	026
45	032
46	038
47	046
48	054
49	064
50	076
51	088
52	102
53	118
54	135
55	154
56	174
57	196
58	220
59	245
60	271
61	299
62	328
63	358
64	388
65	420
66	452
67	484

x	$n=16$ 0.
27	001
28	002
29	002
30	003
31	004
32	006
33	008
34	010
35	013
36	016
37	021
38	026
39	032
40	039
41	048
42	058
43	070
44	083
45	097
46	114
47	133
48	153
49	175
50	199
51	225
52	253
53	282
54	313
55	345
56	378
57	412
58	447
59	482

x	$n=15$ 0.
23	001
24	002
25	003
26	004
27	006
28	008
29	010
30	014
31	018
32	023
33	029
34	037
35	046
36	057
37	070
38	084
39	101
40	120
41	141
42	164
43	190
44	218
45	248
46	279
47	313
48	349
49	385
50	423
51	461
52	500

x	$n=14$ 0.
18	001
19	002
20	002
21	003
22	005
23	007
24	010
25	013
26	018
27	024
28	031
29	040
30	051
31	063
32	079
33	096
34	117
35	140
36	165
37	194
38	225
39	259
40	295
41	334
42	374
43	415
44	457
45	500

x	$n=13$ 0.
14	001
15	001
16	002
17	003
18	005
19	007
20	011
21	015
22	021
23	029
24	038
25	050
26	064
27	082
28	102
29	126
30	153
31	184
32	218
33	255
34	295
35	338
36	383
37	429
38	476

x	$n=12$ 0.
11	001
12	002
13	003
14	004
15	007
16	010
17	016
18	022
19	031
20	043
21	058
22	076
23	098
24	125
25	155
26	190
27	230
28	273
29	319
30	369
31	420
32	473

Index

Z